The Conceptual Framework of Quantum Field Theory

Anthony Duncan

OXFORD
UNIVERSITY PRESS

OXFORD
UNIVERSITY PRESS

Great Clarendon Street, Oxford, OX2 6DP,
United Kingdom

Oxford University Press is a department of the University of Oxford.
If furthers the University's objective of excellence in research, scholarship,
and education by publishing worldwide. Oxford is a registered trade mark of
Oxford University Press in the UK and in certain other countries

First Edition published in 2012

Reprinted with corrections 2013

Impression: 4

British Library Cataloguing in Publication Data

Data available

Library of Congress Cataloging in Publication Data

Data available

ISBN 978–0–19–957326–4

Printed and bound by
CPI Group (UK) Ltd, Croydon, CR0 4YY

Preface

In the roughly six decades since modern quantum field theory came of age with the introduction in the late 1940s of covariant field theory, supplemented by renormalization ideas, there has been a steady stream of expository texts aimed at introducing each new generation of physicists to the concepts and techniques of this central area of modern theoretical physics. Each decade has produced one or more "classics", attuned to the background, needs, and interests of students wishing to acquire a proficiency in the subject adequate for the beginning researcher at the time. In the 1950s the seminal text of Jauch and Rohrlich, *Theory of Photons and Electrons*, provided the first systematic textbook treatment of the Feynman diagram technique for quantum electrodynamics, while more or less simultaneously the first field-theoretic attacks on the strong interactions were presented in the two-volume *Mesons and Fields* of Bethe, de Hoffmann, and Schweber. The 1960s saw the appearance of the massive treatise by Schweber, *Introduction to Relativistic Quantum Field Theory*, which addressed in much greater detail formal aspects of the theory, including the LSZ asymptotic formalism and the Wightman axiomatic approach. The dominant text of the late 1960s was undoubtedly the two-volume text of Bjorken and Drell, *Relativistic Quantum Mechanics* and *Relativistic Quantum Fields*, which combined a thorough introduction to Feynman graph technology (in volume 1) with a more formal introduction to Lagrangian field theory (in volume 2). In the 1970s the emergence of non-abelian gauge theories as the overwhelmingly favored candidates for a successful field-theoretic description of weak and strong interactions coincided with the emergence of functional (path-integral) methods as the appropriate technical tool for quantization of gauge theories. In due course, these methods received full treatment with the appearance in 1980 of Itzykson and Zuber's encyclopedic *Quantum Field Theory*. In a similar way, the surge to prominence of supersymmetric field theories throughout the 1980s necessitated a full account of supersymmetry, which is the sole subject of the third volume of Weinberg's comprehensive three-volume *The Quantum Theory of Fields*, the first edition of which appeared in 1995.

With such a selection of classic expository treatises (not to mention many other fine texts not listed above—with apologies to authors of same!) one may well doubt the need for yet another introductory treatment of quantum field theory. Nevertheless, in the course of teaching the subject to graduate students (typically, second year) over the last 25 years, I have been struck by the number of occasions on which important conceptual issues are raised by questions in the classroom which require a careful explanation not to be found in any of the readily available textbooks on quantum field theory. To give just a small sample of the sort of questions one encounters in the classroom setting: "Of the plethora of quantum fields introduced to describe Nature at subatomic scales, why do so few (basically, only electromagnetism and gravity) have classical macroscopic correlates?"; "If there are many possible quantum fields available to 'represent' a given particle, can, or in what sense does, quantum field theory prescribe a unique all-time dynamics?"; "If the interaction picture does not exist, as implied by Haag's theorem, why (or in what sense) are the formulas derived in this picture for the S-matrix still valid?"; "Are there non-perturbative phenomena amenable to treatment using perturbative (i.e., graph-theoretical) methods?"; and so on. None of these questions require an answer if one's attitude in learning quantum field theory amounts to a purely pragmatic desire to "start with a Lagrangian and compute a process to two loops". However, if the aim is to arrive at a truly deep and satisfying comprehension of the most powerful, beautiful, and effective theoretical edifice ever constructed in the physical sciences, the pedagogical approach taken by the instructor has to be quite a bit different from that adopted in the "classics" enumerated above.

In the present work, an attempt is made to provide an introduction to quantum field theory emphasizing conceptual issues frequently neglected in more "utilitarian" treatments of the subject. The book is divided into four parts, entitled respectively, "Origins", "Dynamics", "Symmetries", and "Scales". Although the emphasis is conceptual—the aim is to build the theory up systematically from some clearly stated foundational concepts—and therefore to a large extent *anti*-historical, I have included two historical chapters in the "Origins" section which trace the evolution of the modern theory from the earliest "penumbra" of quantum-field-theoretical phenomena detected by Planck and Einstein in the early years of the twentieth century to the emergence, in the late 1940s, of the recognizable structure of modern quantum field theory, in the form of quantum electrodynamics. The reader anxious to proceed with the business of logically developing the framework of modern field theory is at liberty to skim, or even entirely omit, this historical introduction.

The three remaining sections of the book follow a step-by-step reconstruction of this framework beginning with just a few basic assumptions: relativistic invariance, the basic principles of quantum mechanics, and the prohibition of physical action at a distance embodied in the clustering principle. The way in which these physical ingredients combine to engender some of the most dramatic results of relativistic quantum field theory is outlined qualitatively in Chapter 3, which also contains a summary of the topics treated in later chapters. Subsequent chapters in the "Dynamics" section of the book lay out the basic structure of quantum field theory arising from the sequential insertion of quantum-mechanical, relativistic, and locality constraints. The rather extended treatment of free fields allows us to discuss important conceptual issues (e.g., the classical limit of field theory) in greater depth than usually found in the standard texts. Some applications of perturbation theory to some simple theories and processes are discussed in Chapter 7, after the construction of covariant fields for general spin has been explained. A deeper discussion of interacting field theories is initiated in Chapters 9, 10, and 11, where we treat first general features shared by all interacting theories (Chapter 9) and then aspects amenable to formal perturbation expansions (Chapter 10). The "Dynamics" section concludes with a discussion of "non-perturbative" aspects of field theory—a rather imprecise methodological term encompassing a wide variety of very different physical processes. In Chapter 11 we attempt to clarify the extent to which certain features of field theory are "intrinsically" non-perturbative, requiring methods complementary to the graphical expansions made famous by Feynman.

In the "Symmetries" section we explore the many important ways in which symmetry principles influence both our understanding and our use of quantum field theory. Of course, at the heart of relativistic quantum field theory lies an inescapable symmetry of critical importance: Lorentz-invariance, which, together with translational symmetry in space and time, makes up the larger symmetry of the Poincaré group. The centrality of this symmetry explains the dominance of Lagrangian methods in field theory, even though from a physical standpoint the Hamiltonian would appear (as is typically the case in non-relativistic quantum theory) to hold pride of place. The role played by Lorentz-invariance in restricting the dynamics of a field theory is the main topic of Chapter 12, which also includes an introduction to the extension of the Poincaré algebra to the graded superalgebra of supersymmetric field theory. Discrete spacetime symmetries, and the famous twin theorems of axiomatic field theory—the Spin-Statistics and TCP theorems—are the subject of Chapter 13. The discussion of global symmetries, exact and approximate, in Chapter 14 leads naturally into the very important topics of spontaneous symmetry-breaking and the Goldstone theorem. The "Symmetries" section of the book closes with a treatment of local gauge symmetries in Chapter 15, which imply remarkable new features not present in theories where the only symmetries are global (i.e., involve a finite-dimensional algebra of spacetime-independent transformations).

With the final section of the book, entitled "Scales", we come to perhaps the most characteristic conceptual feature of quantum field theory: the scale separation property exhibited by theories defined by an effective local Lagrangian. Given that essentially all of the information obtained from scattering experiments at accelerators concerns asymptotic transitions (i.e., the infinite time evolution of an appropriately prepared quantum state, terminated by a detection measurement) it is critically important for theoretical progress that the probabilities of such transitions not depend in a sensitive way on interaction details at much smaller distances than those presently accessible in accelerator experiments (roughly, the inverse of the center-of-mass energy of the collision process). The insensitivity of field theory amplitudes to our inescapable ignorance of the nature of the interactions at very short distances (or equivalently, high momentum) is therefore of central importance if we are to infer reliably an underlying microdynamics from the limited phenomenology available at any given time. Remarkably, in this respect quantum field theories are far kinder to us than their classical (particle or field) counterparts, where non-linearities almost always introduce chaotic behavior which effectively precludes the possibility of accurate predictions of state evolution over long time periods. The technical foundations needed for examining these issues are taken up in Chapter 16, which contains an account of regularization, power-counting, effective Lagrangians, and the renormalization group. Applications to the proof of perturbative renormalizability, and a discussion of the "triviality" phenomenon (the absence of a non-trivial continuum limit), follow in Chapter 17. Chapters 18 and 19 then explore important features of the behavior of quantum field theories at short distance (e.g., the operator product expansion and factorization) and long distance (in particular, the complications in defining the correct physical state space in unbroken abelian and non-abelian gauge field theories).

To the beginning student, quantum field theory all too often takes on the appearance of a multi-headed Hydra, with many intertwined parts, the understanding of any one of which seems to require a prior understanding of the rest of the frightening anatomy of the whole beast. The motivation for the present work was the author's desire to provide an introduction to modern quantum field theory in which this rich and complex structure is seen to arise naturally from a few basic conceptual inputs, in contrast to the more typical approach in which Lagrangian field theory is presented as a theoretical *fait accompli* and then subsequently shown to have the desired physical features.

Much (perhaps most) of the attitude towards quantum field theory expressed in this book is the result of innumerable conversations, over four decades, with colleagues and students. For laying the foundations of my knowledge of field theory I wish especially to thank my predoctoral and post-doctoral mentors (Steven Weinberg and Al Mueller, respectively). In the case of the present work I am extremely grateful to Estia Eichten, Michel Janssen, Adam Leibovich, Max Niedermaier, Sergio Pernice, and Ralph Roskies for reading extensive parts of the manuscript, and for many useful comments and suggestions. Any remaining solecisms of style or content are, of course, entirely the responsibility of the author.

Contents

1

Origins I: From the arrow of time to the first quantum field

1.1 Quantum prehistory: crises in classical physics

The first indications of serious inadequacies in the framework of classical physics—deficiencies which eventually could only be resolved by the introduction of quantum-theoretical concepts and methods—can already be found in Maxwell's discussion of the anomalously low specific heat of gases in the mid-1870s. Nevertheless, the birth of quantum theory as such, and in particular the clear identification of a new fundamental constant of Nature characteristic of quantum phenomena, is usually located in the year 1900 with Planck's invention (Planck, 1900b) ("derivation" would perhaps be too charitable a term) of a novel formula for the distribution of energy over frequencies in thermal radiation (namely, electromagnetic radiation in the interior of a sealed enclosure which has been allowed to come to thermal equilibrium with the walls of the enclosure, themselves maintained at a fixed temperature T). The problem of thermal cavity (or "blackbody") radiation would in modern terms be regarded as one of thermal quantum field theory, so we have the strange situation that the first historical impetus to the discovery of quantum principles actually lay in a problem of quantum field theory, which is generally supposed to be a much later invention arising from a fusion of quantum, relativistic, and locality principles. In fact, Planck's papers of 1900 make no reference to quantization of the electromagnetic field-energy (a concept which Planck would continue to resist strenuously until the mid-1920s): the energy quantization principle for Planck is strictly a statement about the distribution of energy among the idealized material oscillators constituting the walls of the enclosure, and general principles of thermodynamics are then brought to bear to fix the electromagnetic energy distribution which must necessarily obtain once equilibrium between these oscillators and the interior radiation has been achieved. Only five years later, in his remarkable paper entitled "A heuristic point of view concerning the creation and conversion of light" (Einstein, 1905a), Einstein was to extend boldly and explicitly the idea of energy quantization to the electromagnetic field itself, introducing the idea of "light quanta" (in modern language, photons). From a conceptual (if not technical) point of view, this paper therefore marks the true birth of quantum field theory.

In order to understand the origins of quantum theory in the problem of black-body radiation, which to physicists of Planck's generation must have appeared quintessentially classical (merging as it did well-established principles of Maxwellian electromagnetic theory and thermodynamics), we need to push our time horizon back a quarter century or so and survey the overall situation in classical physics at the start of the final quarter-century of the 1800s. The three great edifices of classical physics—Newtonian mechanics (amplified and deepened, of course, by the contributions of Laplace, Lagrange, Hamilton, and many others), electromagnetic theory, only recently completed by Maxwell (in *Philosophical Transactions*, vol. 155, 1865), and, somewhat later, put in a form recognizable to the modern student of the subject by Hertz, and thermodynamics, which reached essential conceptual completeness at the hands of Clausius, also around 1865—stood as precise descriptions of natural phenomena, each apparently unassailable in its natural domain of applicability. In a sense, further progress keeping strictly within the limits of each of these disciplines had become difficult or impossible. However, precisely at this time, natural phenomena requiring the simultaneous application of *more than one* of these formal structures began to demand the attention of physicists. It is possible to trace the origins of the core disciplines of twentieth-century physics—quantum theory, statistical mechanics, and relativity—to developments at the *interfaces* of the three basic classical frameworks. We can summarize these developments very briefly as follows:

1. **Developments at the interface of thermodynamics and electromagnetic theory.**

   The discovery and classification of solar spectral lines by Fraunhofer, and the development of spectroscopy as an analytical tool by Bunsen and Kirchhoff in the 1850s, led naturally to an investigation of the radiation emitted by hot bodies, and thus to the study of the relation between thermodynamics and electromagnetic phenomena. It was immediately recognized by Kirchhoff that the intensity and frequency of the radiation emitted by a perfectly absorbing body had a fundamental significance. Various arguments—some of a rigorous thermodynamic nature, others of an heuristic character—led, by 1896, to a widely accepted form for this "blackbody distribution" (the Wien Law). The experimental failure of this "law" was the final stimulus which led Planck to (reluctantly) advance the quantum hypothesis in his seminal papers (Planck, 1900*b*) in *Annalen der Physik*, 1900.

2. **Developments at the interface of mechanics and thermodynamics.**

   Attempts to reconcile classical mechanics, regarded as the underlying dynamical description of all phenomena, with the formal principles of thermodynamics led to the development of kinetic theory by Clausius, Maxwell, and Boltzmann. The more general framework developed by the last of these has since come to be called statistical mechanics. Boltzmann was the first to understand clearly the role of statistical and probabilistic considerations in reconciling mechanics with heat theory (as thermodynamics was called at the time). However, his views were highly controversial at the time, especially with regard to his claim of a statistical origin for the phenomenon of irreversibility in thermal physics. Although he would prove to be absolutely right on this point, Boltzmann's methods proved incapable of explaining the observed specific heats of gases: this difficulty, early (before 1875) recognized by Maxwell (Maxwell, 1875) as a serious anomaly in classical theory, would only finally be removed by the application of quantum ideas in the 1920s, fully fifty years after the problem was first recognized.

3. **Developments at the interface of electromagnetic theory and mechanics.**

The historical and conceptual preeminence of classical mechanics implied a special status which led naturally to an attempt to interpret all natural phenomena in mechanical terms. In particular, the attempt to weld electromagnetic theory with mechanics (begun by Maxwell himself) by main force led by the second half of the nineteenth century to the development of a profusion of aether theories of increasing complexity and artificiality. The failure to produce any direct evidence for the existence of an aether in optical experiments (by measuring relative motion of the aether and the optical apparatus of choice) grew from a mere annoyance into an outright crisis with the null result of the experiments of Michelson (1881) and Michelson and Morley (1887). The entire class of complicated and messy *dynamical* aether theories concocted to surmount this impasse were demolished with one stroke by Einstein (Einstein, 1905*b*) in 1905, by accepting the *kinematical* structure natural to electromagnetic phenomena as generally valid in the mechanical sphere also.

In our outline of the conceptual origins of quantum mechanics and quantum field theory, the first item above holds pride of place: firstly, because by common consent quantum theory begins with Planck's introduction of a new universal constant of Nature in his blackbody distribution formula of 1900, and secondly, because, as we shall see later, the first explicitly *quantum field-theoretic* calculation, Jordan's derivation of the mean-square energy fluctuations in a subvolume of a cavity containing (a one-dimensional version of) electromagnetic radiation in the final section of the *Drei-Männer-Arbeit* (1925) of Born, Heisenberg, and Jordan (Born *et al.*, 1926), came directly out of an attempt to reproduce a remarkable result of Einstein dating from 1909 (Einstein, 1909*b*,*a*) in which the apparent paradox of simultaneous wave and particle behavior of light was first exposed with full clarity. *The essence of quantum field theory is to provide a unified dynamical framework in which these apparently disparate behaviors can coexist in a conceptually consistent fashion.* The electromagnetic radiation contained in a cavity at thermal equilibrium therefore plays a central role in the conceptual origins both of quantum theory generally and quantum field theory in particular. For this reason we shall retell in this chapter the history of thermal radiation in some detail, paying particular attention to those aspects important for understanding the conceptual origins of quantum field theory. The story will lead us continuously from the early arguments surrounding the role of the Second Law of Thermodynamics (and the "arrow of time" it implies) in blackbody radiation, to the appearance of the first truly quantum-field-theoretical analysis of electromagnetic radiation. We begin in the next section by describing some important milestones on the way to the understanding of blackbody radiation as it stood in 1900 when Planck took the first steps along the road to modern quantum theory.

1.2 Early work on cavity radiation

The fact that heated bodies glow with a color and intensity varying with their temperature must surely have been apparent in prehistoric times (at least since the discovery of fire!). The precise nature of thermal radiation became the subject of intense study in the nineteenth century, and as we shall see, led directly to the discovery of the quantum principle. Wedgewood, the porcelain manufacturer, observed in the 1790s that heated bodies all became red at the same temperature. The Scottish physicist Balfour Stewart noted (in 1858) that a block of rock salt at 100° C strongly absorbs the radiation emitted by a similar block at the same temperature, and suggested the rule of equality of radiating and absorbing power of bodies for rays of any given type (i.e., wavelength). About a year later (independently of Stewart) Kirchhoff put all of this phenomenology into a comprehensible framework by the use of very general thermodynamic arguments (Kirchhoff, 1859, 1860).

The arguments given by Kirchhoff established the universal character of blackbody, or "cavity", radiation—the radiation filling the interior of a hollow material cavity, the walls of which are maintained at a fixed temperature T. It is important to understand at the outset the role of the cavity in these arguments. After heating the cavity to the desired temperature T, a fixed amount of radiant energy fills the interior. To examine the nature of this radiation we are at liberty to drill a very small hole in the walls, as the small amount of radiant energy emerging will not sensibly disturb the established equilibrium. Note that from the point of view of the external world, the punctured cavity is essentially "black", in the sense that any radiation entering through the pinhole will have to scatter around in the interior of the cavity for a very long time before having an opportunity to escape. Such a cavity is therefore (effectively) a perfect absorber, or "black body".[1] The essence of the problem is that radiation in the cavity is forced to interact with the walls (or contents, if any) of the cavity until thermal equilibrium is reached.

Kirchhoff showed that the energy (per unit volume, and per wavelength interval) of the cavity radiation was uniform and isotropic throughout the interior. The arguments he gave were subsequently simplified by Pringsheim. The latter observed that by inserting a reflecting plane surface at some point inside the cavity, the equality of radiation in opposite directions follows, as otherwise the unequal radiation pressure exerted on the two faces would allow the spontaneous conversion of heat to work, thereby violating the Second Law of Thermodynamics. The existence of a pressure exerted by radiation reflected from a surface is crucial to these arguments: the precise relation (to be discussed further below) between this pressure and the energy density of the radiation had been derived earlier by Maxwell (Maxwell, 1873). Similar arguments employing two mirrors can be used to prove the isotropy and homogeneity of the radiation in the cavity. The existence of filters selectively absorbing and transmitting radiation of different wavelengths means that these statements hold separately within each interval of wavelength. Thus, we can define a function $\phi(\lambda, T)$ such that $\phi(\lambda, T)\Delta\lambda$ is the radiant energy per unit volume between wavelengths $\lambda, \lambda + \Delta\lambda$ *anywhere* in the cavity.

---

[1] The use of the term "blackbody", though historically predominant, frequently confuses the beginning student, who wonders quite naturally how a truly *black* body can radiate! In the forthcoming discussion we prefer the use of the term "cavity (or thermal) radiation". The German terminology, "Normal Spektrum", is not particularly illuminating either.

Next, one easily sees that $\phi(\lambda, T)$ is the same (universal!) function for any cavity, irrespective of size, shape, or material constitution. This can be established by connecting two cavities at the same temperature by a thin tube allowing radiation to pass in either direction. Unequal radiation densities would result in unequal fluxes down the tube, and hence in a spontaneous flow of heat between two systems *at the same temperature*. Again, by placing filters which pass only a limited range of wavelengths in the tube, this equality is found to hold in each wavelength interval $\lambda, \lambda + \Delta\lambda$.

The universal function $\phi(\lambda, T)$ appears in the description of radiation emitted from any hot surface in the following way. At thermal equilibrium the radiation impinging on the surface must be balanced by the total radiation leaving, as otherwise the surface would grow progressively cooler or hotter. Suppose the surface (say, the interior wall of our cavity discussed above) absorbs a fraction A_λ of the incident radiation flux e_λ (at wavelength λ). It is clear that the radiation emitted (as opposed to *reflected*), or the "emissive power" of the surface E_λ must just equal $A_\lambda e_\lambda$, or equivalently, the ratio E_λ/A_λ of emissive power to absorption coefficient is a universal function (essentially our old friend ϕ) of wavelength and temperature. This result, of course, lies at the core of Stewart's observations mentioned above. It also explains why thermos flasks are silvered (making A_λ small) in order to get them to *radiate* less.

The total energy density of cavity radiation (at all wavelengths) is evidently just the integral of Kirchhoff's function ϕ:

$$\rho(T) = \int \phi(\lambda, T) d\lambda \tag{1.1}$$

In 1879 Josef Stefan proposed on the basis of some preliminary experiments the form

$$\rho(T) = aT^4 \tag{1.2}$$

with a a universal constant. In 1884 Boltzmann derived this formula thermodynamically, essentially by the following argument. Consider a spherical cavity of radius r. Maxwell had shown that radiation of energy density ρ exerts a radiation pressure $\frac{1}{3}\rho$. Thus the internal energy is $U = \frac{4}{3}\pi r^3 \rho$ and the heat absorbed in an infinitesimal reversible expansion is given by the First Law of Thermodynamics as[2]

$$dQ = dU + dW = d(\frac{4}{3}\pi r^3 \rho) + 4\pi r^2 \frac{1}{3}\rho dr \tag{1.3}$$

with the corresponding entropy change

$$dS = dQ/T = \frac{4}{3}\pi r^3 \frac{d\rho}{T} + \frac{16}{3}\pi r^2 \rho \frac{dr}{T} = \frac{4}{3}\pi r^3 \frac{d\rho}{dT}\frac{dT}{T} + \frac{16}{3}\pi r^2 \rho \frac{dr}{T} \tag{1.4}$$

---

[2] We remind the reader that U and S are state functions, unlike work W and heat Q: only the changes in the latter are meaningful, whence the difference in notation in the associated differentials dW, dQ.

Equating $\frac{\partial}{\partial T}\frac{\partial S}{\partial r}$ to $\frac{\partial}{\partial r}\frac{\partial S}{\partial T}$ we find

$$\frac{d\rho}{dT} = \frac{4}{T}\rho \tag{1.5}$$

which immediately implies the T^4 law stated above, and now known as the Stefan–Boltzmann Law.

Serious attempts to measure experimentally the intensity and spectral composition of blackbody radiation began with Langley, the American astronomer, who invented the bolometer, a device for measuring the intensity of radiation in the infrared, using the principle of temperature-dependent resistance of a thin filament placed in the path of the rays after they were refracted through a rock-salt prism. These measurements (1886) extended up to about wavelengths of 5μ. He found "a real though slight progression of the point of maximum heat towards the shorter wave-lengths as the temperature rises" and the asymmetric form of the maximum of $\phi(\lambda, T)$, steeper on the shorter wavelength side. By 1895 the greatly improved measurements of Paschen established the rule that the wavelength λ_m of maximum intensity was inversely proportional to the temperature T. Paschen's measurements led him to propose, in 1896, the form

$$\phi(\lambda, T) = B\lambda^{-C}\exp(-\frac{A}{\lambda T}) \tag{1.6}$$

with the constant C somewhere in the range of 5–6.

In 1893 Wien derived, using purely thermodynamic arguments, an important constraint on the Kirchhoff function $\phi(\lambda, T)$. Consider once again the spherical cavity used above in the derivation of the Stefan–Boltzmann Law. Imagine that the sphere undergoes slow, adiabatic compression, where the radius contracts steadily at speed v ($v << c$). At every reflection from this inwardly contracting sphere light of wavelength λ suffers a Doppler shift to wavelength $\lambda(1 - 2v/c)$. During a contraction by Δr, occurring in time $\Delta r/v$, the light undergoes $c\Delta r/2vr$ reflections across a diameter of the sphere (it can be shown that light not incident perpendicular to the walls suffers a smaller Doppler shift each time, but is reflected correspondingly more frequently: the net result is the same). The result is a total blue shift to wavelength

$$(1 - \frac{2v}{c})^{\frac{c\Delta r}{2vr}}\lambda \sim (1 - \frac{\Delta r}{r})\lambda$$

so the wavelength is shifted by $\frac{\Delta\lambda}{\lambda} = -\frac{\Delta r}{r}$ in this adiabatic compression. As there is no heat transfer $dS = 0$ and (see Eq. (1.4) above)

$$\frac{4}{3}\pi r^3\frac{4}{T}\rho\frac{dT}{T} = -\frac{16}{3}\pi r^2\rho\frac{dr}{T}$$

$$\Rightarrow \frac{dr}{r} = -\frac{dT}{T}$$

so $r \propto \frac{1}{T}$.

The essence of the thermodynamic argument given by Wien lies in the observation that the adiabatic process described here gives at every stage cavity radiation in equilibrium at the new temperature T inversely proportional to r. If the result were otherwise, producing more or less radiation in some wavelength band than appropriate for blackbody radiation at the new temperature, introduction of filters absorbing in this range immediately allows the generation of temperature differences and a consequent violation of the Second Law.

In the time interval Δt of the adiabatic compression, the radiant energy originally in the wavelength band $(\lambda, \lambda + \Delta\lambda)$, namely

$$\frac{4}{3}\pi r^3 \phi(\lambda, r)\Delta\lambda \tag{1.7}$$

is shifted down to the interval $(\lambda(1 - \frac{v\Delta t}{r}), (\lambda + \Delta\lambda)(1 - \frac{v\Delta t}{r}))$, and increased by the amount of adiabatic work done against the radiation pressure, which is

$$(4\pi r^2)(v\Delta t)(\frac{1}{3}\phi(\lambda, r)\Delta\lambda) \tag{1.8}$$

Thus

$$\frac{4}{3}\pi(r - v\Delta t)^3 \phi(\lambda(1 - \frac{v\Delta t}{r}), r - v\Delta t)\Delta\lambda(1 - \frac{v\Delta t}{r})$$
$$= \frac{4}{3}\pi r^3 \phi(\lambda, r) + (\frac{4}{3}\pi r^2)(v\Delta t)\phi(\lambda, r)\Delta\lambda \tag{1.9}$$

Expanding to first order in Δt, one finds the equation

$$\lambda\frac{\partial\phi}{\partial\lambda} + r\frac{\partial\phi}{\partial r} = -5\phi$$

or

$$(\lambda\frac{\partial}{\partial\lambda} + r\frac{\partial}{\partial r})(\lambda^5\phi) = 0$$

so that

$$\lambda^5\phi = f(\lambda/r) = f(\lambda T)$$

Wien phrased this result slightly differently. Since

$$\frac{1}{T^5}\phi = \frac{1}{(\lambda T)^5}f(\lambda T)$$

it follows that if $\lambda_1 T_1 = \lambda_2 T_2$, then

$$\frac{1}{T_1^5}\phi(\lambda_1, T_1) = \frac{1}{T_2^5}\phi(\lambda_2, T_2)$$

which Wien called the "T^5" law, but is now commonly called the Wien Displacement Law. If we know the radiation function $\phi(\lambda_1, T_1)$ at temperature T_1, the law allows us to "displace" it into the appropriate curve for any other temperature T_2, as

$$\phi(\lambda, T_2) = (\frac{T_2}{T_1})^5\phi(\lambda\frac{T_2}{T_1}, T_1)$$

It follows immediately from the Wien Displacement Law that the total energy density

$$\rho(T) = \int d\lambda \phi(\lambda, T)$$

$$= \int d\lambda \frac{1}{\lambda^5} f(\lambda T)$$

$$= T^4 \int_0^\infty dx \frac{1}{x^5} f(x) \equiv aT^4 \tag{1.10}$$

satisfies the Stefan–Boltzmann Law (provided, of course, that the integral converges: a condition by no means to be taken for granted, as we shall see). Another immediate corollary is the result later verified experimentally by Paschen (but suggested previously by several workers in this field, notably H. F. Weber) that the maximum in wavelength displaces inversely with the temperature. Finally, the Wien Displacement Law immediately fixes the value of the constant C in the form proposed by Paschen (see (1.6)) to be 5 exactly. Independently of Paschen's work, Wien in 1896 arrived at the form (1.6) on the basis of an *ad hoc* assumption concerning the emission of radiation by molecules distributed according to a Maxwellian velocity distribution. This form, which is now a complete specification of the Kirchhoff function ϕ, was called the Wien Distribution Law, and was to play, with its "corrected" version, the Planck Distribution Law, a critical role in the evolution of attempts to understand quantization of the electromagnetic field.

1.3 Planck's route to the quantization of energy

Max Planck, the father of quantum theory, was born in Kiel, Germany, in 1858, and attended the Gymnasium (high school) and University in Munich before going to Berlin for his doctoral degree, where he had classes from Helmholtz and Kirchhoff. His doctoral thesis concerned the application of thermodynamics (*à la* Clausius) to problems of "Evaporation, Melting, and Sublimation". Planck was fascinated by the extraordinary scope, power and (apparent) infallibility of the energy conservation principle—or First Law of Thermodynamics—and accorded to the Second Law of Thermodynamics (with its concomitant "arrow of time") an equal degree of validity. He was therefore convinced that the Second Law could not rest on the purely mechanical foundations of Boltzmannian gas theory, in which the reversibility objections of Loschmidt and Zermelo (the latter a Planck assistant in Berlin) would necessarily lead to spontaneous processes (albeit rare) where the entropy *decreased*.[3] The consideration of the paradox of Maxwell's demon also led Planck to the rather peculiar conclusion that the Second Law could never be valid in a system comprised of discrete particles.

Instead, Planck began to investigate (in 1897) the possibility that irreversible thermal phenomena could somehow be traced back to irreversible processes in a *continuous* medium—in particular, the electromagnetic field. The archetypal process considered by Planck was the apparently irreversible conversion of plane radiation

---

[3] For a beautiful retelling of this remarkable period in the development of statistical heat theory, see the biography by Martin Klein of Paul Ehrenfest (Klein, 1970).

incident on a charged oscillator into outgoing spherical waves. This subject was explored with great thoroughness in a series of five papers in the *Berliner Berichte* (1897–99) entitled "Über irreversible Strahlungsvorgänge" ("On irreversible radiation processes") (Planck, 1900*a*). The subject of absorption and re-emission of electromagnetic radiation from an oscillator led Planck naturally into the subject of thermal cavity radiation. Here the oscillators constitute the material of the walls of the cavity, absorbing and re-emitting the radiation in the interior. The universal character of the thermal radiation discussed above allowed Planck the freedom of making a very simple model of the constituent particles of the walls (essentially charged simple harmonic oscillators), as the spectral distribution of the cavity radiation would have to be independent of the specific material constitution of the cavity once equilibrium is reached.

The irreversibility that Planck relies upon in his radiation studies can be seen clearly in the damped oscillator equation that he derived as a prelude to his studies of the coupled field-oscillator problem:

$$m\frac{d^2x}{dt^2} + kx - \frac{2e^2}{3c^3}\frac{d^3x}{dt^3} = eE\cos(2\pi\nu t) \tag{1.11}$$

The third ("radiation damping") term has three time-derivatives and evidently changes sign under time-reversal. It arises because the damping force times the velocity must give the power lost to radiation, which is proportional to the acceleration of the charged particle squared. The average of the third term above times the velocity $\frac{dx}{dt}$ over a cycle of the periodic system is easily seen to be the same as the average power radiated, by a single integration by parts. Planck was particularly impressed by the fact that the irreversibility in this system arises without any recourse to non-conservative processes, in which ordered energy is lost (as in friction or air resistance) to disordered heat. Instead, the energy appears to flow irreversibly from an ordered source (an incoming plane wave incident on the oscillator) to an equally ordered form: outgoing spherical radiation.

In 1898 Boltzmann succeeded in convincing Planck (in a paper entitled "On the *supposedly*(!) irreversible radiation processes" (Boltzmann, 1898)) that the hope of deriving irreversible phenomena from electromagnetic theory without additional statistical assumptions was bound to fail, as Maxwell's equations are just as invariant under time-reversal as those of classical mechanics. In fact (Boltzmann claimed), in the course of a careful derivation of radiation damping one is forced to apply boundary conditions to the fields which amount to a field analog of the assumption of molecular disorder implicit in the Boltzmann approach to gas theory.[4] Planck admitted this promptly and abandoned the attempt at a "microscopic" explanation of irreversibility based on electrodynamics.

In the fifth of his papers on irreversible radiation processes (Planck, 1899), Planck derived a crucial formula relating the distribution function for cavity radiation to the average energy of his fictional oscillators (at equilibrium). Before stating this formula,

---

[4] See (Klein, 1970), (Kuhn, 1978) for masterful expositions of the remarkable developments at the interface of mechanics and heat theory summarized all too briefly above.

a slight change in notation will be convenient. Let $\rho(\nu, T)d\nu$ be the energy/unit volume of cavity radiation in the *frequency* interval $(\nu, \nu + d\nu)$, where $\nu\lambda = c$, $|d\nu| = \frac{c}{\lambda^2}d\lambda$, so that

$$\rho(\nu, T)d\nu = \rho(\nu, T)\frac{c}{\lambda^2}d\lambda = \phi(\lambda, T)d\lambda$$

$$\rho(\nu, T) = \frac{\lambda^2}{c}\phi(\lambda, T) = \frac{c}{\nu^2}\phi(\frac{c}{\nu}, T) \tag{1.12}$$

In terms of $\rho(\nu, T)$, the T^5 law takes the form

$$\rho(\nu, T) = \nu^3 f(\frac{\nu}{T}) \tag{1.13}$$

and the Wien Distribution Law is

$$\rho(\nu, T) = \alpha\nu^3 \exp(-\beta\nu/T) \tag{1.14}$$

where the new constants α, β are related to those appearing in the Paschen result (1.6) by $\alpha = \frac{B}{c^4}, \beta = \frac{A}{c}$ (recall that the constant C was fixed previously by purely thermodynamic reasoning to be 5). The equation derived by Planck, obtained by equating at equilibrium the energy absorbed and emitted by the oscillator, stated simply

$$E(\nu_o, T) = \frac{c^3}{8\pi\nu^2}\rho(\nu_o, T) \tag{1.15}$$

It relates the average energy of an oscillator of natural frequency $\nu_o = \frac{1}{2\pi}\sqrt{k/m}$ to the blackbody distribution function. That such a relation must exist is physically clear. Planck showed that if the left-hand side exceeded the right, energy would flow from the oscillators to the electromagnetic field, while if the intensity of radiation at ν_0 became large enough that the right-hand side exceeded the left the oscillators would tend to absorb energy from the field. The importance of this equation (a full derivation of which we must unfortunately forego, in the interests of brevity) in Planck's intellectual journey *can scarcely be overemphasized*: it allowed him to restrict the application of energy quantization to the material oscillators alone (left-hand side of (1.15)), while relying on the equilibrium condition to transfer the resultant average distribution of energy by main force, as it were, to the *continuous* electromagnetic radiation (right-hand side) in the interior of the cavity. Planck would continue to insist on the continous, purely classical character of electromagnetic radiation for the next 25 years.

In his final paper on irreversible radiation processes (see (Planck, 1900a)), Planck gave a "derivation" of the Wien Law based on purely thermodynamic arguments together with the crucial formula (1.15) above. This was done by making a plausible assumption for the form of the entropy S of the oscillator as a function of energy E, using for the inverse temperature $T^{-1} = \frac{\partial S}{\partial E}$, and solving for E as a function of T. Planck showed that his assumption for $S(E)$ implied that the entropy of the whole system (oscillators plus radiation) would necessarily increase in time, in agreement with the Second Law of Thermodynamics. He was also under the (as it later turned out, erroneous) impression that this was the only possible choice for $S(E)$ consistent with the Second Law. Consequently, at this point Planck was quite convinced that he had finally managed a complete derivation of the blackbody spectrum from pure thermodynamics (even if he had now to agree with Boltzmann that the Second Law had a statistical rather than absolute significance, even in radiation phenomena).

On the afternoon of Sunday, 7 October 1900, Planck was visited at home by an experimental colleague from the Physikalische-Technische Reichsanstalt (the Physical-Technical Imperial Institute, or PTR), H. Rubens. He learnt from Rubens that recent experiments at the PTR had established incontrovertible deviations from the Wien Distribution Law on the infrared (low-frequency) side. In particular, the intensity was roughly proportional to temperature in this regime, instead of the saturation at high temperatures implied by the Wien Law (1.14). Planck realized that a more general form for the oscillator entropy $S(E)$ would in turn allow the derivation of a modified distribution law

$$\rho(\nu, T) = \alpha\nu^3 \frac{1}{\exp(\beta\nu/T) - 1} \tag{1.16}$$

which clearly reproduces the Wien Law at higher frequencies, but behaves like

$$\rho(\nu, T) \sim \alpha\nu^3 \frac{1}{(\beta\nu/T)} \sim \frac{\alpha}{\beta}T\nu^2 \tag{1.17}$$

in the infrared (small ν), showing the desired linear behavior with T. This interpolating formula, which Planck appears to have constructed in the few hours following the visit of Rubens, was checked within the next week and a half and found to match exactly the experimental data.

Planck was perfectly aware that his interpolating formula was nothing more than an enlightened guess at this stage, and he began right away to search for a proper understanding of the formula (1.16). His strategy was precisely the inverse of the one he had followed heretofore. He used (1.15) to obtain the average oscillator energy, *assuming* the validity of the Planck distribution (1.16):

$$E(\nu, T) = \frac{h\nu}{\exp(\beta\nu/T) - 1} \tag{1.18}$$

where $h \equiv \frac{\alpha c^3}{8\pi}$. He then reconstructed the corresponding expression for oscillator entropy as a function of energy, using the thermodynamic relation $TdS = dE$ valid for a reversible transformation involving transfer of heat but no external work. Here, oscillators of a fixed natural frequency ν (called ν_o above) are considered. Solving (1.18) for $1/T$ as a function of E:

$$\frac{1}{T} = \frac{1}{\beta\nu} \ln(1 + \frac{h\nu}{E}) \tag{1.19}$$

and integrating, one obtains,

$$S = \int \frac{1}{T(E)} dE$$

$$= \frac{1}{\beta \nu} \int (\ln(E + h\nu) - \ln(E)) dE$$

$$= \frac{h}{\beta} \{(1 + \frac{E}{h\nu}) \ln(1 + \frac{E}{h\nu}) - \frac{E}{h\nu} \ln(\frac{E}{h\nu})\} \tag{1.20}$$

apart from an irrelevant integration constant. The problem now shifted to finding a "fundamental" explanation for this last expression.

By this point in late 1900, Planck had been converted to Boltzmann's statistical approach, and he now adopted the techniques used by the latter for gas theory in an attempt to establish (1.20) by microstatistical reasoning. Thus, the entropy was to be determined by taking the logarithm of the number of available microscopic states consistent with the stated macroscopic parameters, $S = k \ln(W)$ (here k is Boltzmann's constant). Like Boltzmann, Planck introduced a finite-energy unit ϵ to facilitate the counting. The total energy E_N shared by N oscillators was a (large!) integer P number of these units, $E_N = P\epsilon$. Planck then "counted" W by simply computing the number of ways in which P units of energy could be distributed among the N oscillators. The combinatorial formula needed for this can be derived rapidly using a characteristically elegant trick due to Ehrenfest. Write out a string of P energy units ϵ, with dividers to indicate how many units belong to the first, second, etc., oscillator:

$$\epsilon\epsilon|\epsilon\epsilon\epsilon|\epsilon|\epsilon\epsilon...|\epsilon\epsilon$$

There are P of the ϵ symbols and $N - 1$ dividers. First assume that all these symbols are distinguishable. There are then $(P + N - 1)!$ ways of ordering them. As the dividers and energy units are (separately) indistinguishable, we have overcounted by a factor $(N - 1)!P!$. Thus the desired result (using Stirling's approximation to evaluate the factorials of large numbers) is

$$S = k \ln(\frac{(P + N - 1)!}{P!(N - 1)!}) \sim k((N + P) \ln(N + P) - P \ln(P) - N \ln(N))$$

The average entropy of each oscillator $S = \frac{1}{N} S_N$ while the average energy of a single oscillator is $E = \frac{1}{N} E_N = \frac{P}{N} \epsilon$. Consequently

$$S = k\{(1 + \frac{E}{\epsilon}) \ln(1 + \frac{E}{\epsilon}) - \frac{E}{\epsilon} \ln(\frac{E}{\epsilon})\} \tag{1.21}$$

This is exactly the relation (1.20), provided we identify $\epsilon = h\nu$, $\frac{h}{\beta} = k$. In other words, the derivation of the new distribution formula *forced Planck to keep the energy units ϵ finite, even at the end of the calculation*. Setting ϵ to zero here, as Boltzmann had done at the end of his gas theory calculations, would be equivalent to setting the constant h to zero, which would lead to an incorrect distribution law (the Rayleigh–Jeans Law, to be discussed further below). Apparently, the oscillators in the walls of a cavity were only allowed to have energies in integer multiples of the basic energy "quantum" $\epsilon = h\nu$! The arguments outlined above were presented in Planck's paper in *Annalen der Physik* 4 (1901),p. 553, "Über das Gesetz der Energieverteilung im Normalspectrum" ("On the law of energy distribution for the normal [i.e., blackbody] spectrum"). The famous Planck's constant h appears here for the first time. In modern notation, the blackbody distribution thus takes the form

$$\rho(\nu, T) = \frac{8\pi\nu^2}{c^3} \frac{h\nu}{\exp(h\nu/kT) - 1} \tag{1.22}$$

From the experimental fits, Planck determined $h = 6.55 \times 10^{-27}$ erg/sec, and k (Boltzmann's constant)= 1.346×10^{-16} ergs/degree. The latter value allowed Planck to obtain the first decently accurate value for Avogadro's number $N = R/k$ (where R is the gas constant).

It is a strange historical irony that Planck's modification of the Wien's Law, motivated by the pressure of the Kurlbaum–Rubens experimental results, was actually a move towards a *"more classical"* result: as Einstein was to emphasize in his epochal 1905 paper (Einstein, 1905*a*), in which the revolutionary idea of field quantization was introduced, the Wien Law is in a sense an extreme manifestation of the quantal properties of light. The deviations observed from this law in the infrared by Kurlbaum and Rubens are harbingers of the reappearance of the classical wave-like aspects of electromagnetic phenomena. To understand this we must realize that despite Planck's heroic efforts to obtain a rigorous and unique *classical* result for the distribution function of cavity radiation throughout the 1890s, leading up to the *quantum-theoretically correct* Planck distribution (1.22), the first derivation of the blackbody distribution based on a consistent and full application of classical principles is actually due to Lord Rayleigh. In a short (two-page) paper published in 1900 (Rayleigh, 1900) Rayleigh derived the correct classical form of the distribution function from the classical equipartition theorem applied directly to the electromagnetic modes in the cavity. Consider a cubical *LxLxL* box containing electromagnetic radiation in the form of standing waves. A typical standing wave mode takes the form

$$\sin(\frac{n_1\pi x}{L}) \sin(\frac{n_2\pi y}{L}) \sin(\frac{n_3\pi z}{L})$$

where the associated frequency is $\nu = \frac{c}{2L}|\vec{n}|$ and $\vec{n}$ is the vector with (positive) integer Cartesian components (n_1, n_2, n_3). The number of such modes in the shell $(|\vec{n}|, |\vec{n}| + d|\vec{n}|)$ (octant of positive components only!—an error of Rayleigh's later corrected by Jeans, see below) is evidently

$$\frac{1}{8}4\pi|n|^2 d|n| = 4\pi\frac{L^3}{c^3}\nu^2 d\nu$$

and each of these modes receives a total of $2kT$ at equilibrium by the equipartition principle (namely, $\frac{1}{2}kT$ each into electric and magnetic field energy, and each of two polarization modes). Thus the energy per unit volume in the field in the frequency interval $(\nu, \nu + d\nu)$ is

$$\rho(\nu, T)d\nu = \frac{1}{L^3}2kT(4\pi\frac{L^3}{c^3}\nu^2)d\nu = 8\pi\frac{\nu^2}{c^3}kTd\nu \qquad (1.23)$$

—a result which has since become known as the Rayleigh–Jeans Law. (The error mentioned above of an overall factor of eight made by Rayleigh in his original paper was subsequently corrected by Jeans. As Pais points out in his biography of Einstein (Pais, 1982), the correction was made also in Einstein's 1905 paper on the light quantum, so the result should perhaps more properly be called the Rayleigh–Jeans–Einstein Law.) Rayleigh was perfectly aware that this result could not be correct: the total energy contained in the cavity radiation, when integrated over all frequencies, would then be infinite! Instead, he assumed that it was correct only for the "graver modes" (i.e., lower frequencies) and that the distribution was modified for some as yet unknown reason at higher frequencies (Rayleigh simply inserted an exponential suppression factor at high frequencies, and the resultant formula was in fact his final result). In any event the simple linear dependence on temperature in the Rayleigh–Jeans Law flies in the face of experience: a bar of steel at room temperature (300 K, say) *does not* emit radiation at one-tenth the blinding intensity of a bar at 3000 K ! The infinite amount of energy present in the classical radiation field under equipartition would later (1911) be referred to by Ehrenfest (Ehrenfest, 1911) as the "ultraviolet catastrophe".

Of course, if Planck had finished his Boltzmannian calculation of the average oscillator energy by taking the energy units ϵ to zero, as Boltzmann had done previously in his discussion of gas theory, he would have arrived precisely at Rayleigh's result (though he does not seem to have been aware of Rayleigh's work during the critical period leading up to the 1901 paper), as the Rayleigh–Jeans Law is simply the $h \to 0$ limit of the Planck distribution. That he did not do so is probably due to a combination of reasons:

1. He does not seem to have regarded equipartition as a fundamental guiding principle to the same extent as other physicists of a more "mechanist" bent.
2. Planck attacked the problem from the point of view of the behavior of the oscillators at thermal equilibrium, rather than by directly considering the modes of the electromagnetic field itself, which would have led much more quickly to the (*wrong!*) classical result.
3. The result obtained by setting the energy units to zero would not have agreed with the Wien Law, with which Planck had started and which he knew to be empirically correct at higher frequencies.

1.4   First inklings of field quantization: Einstein and energy fluctuations

Although Planck succeeded in obtaining an absolutely correct expression for the equilibrium thermal frequency distribution of electromagnetic radiation in a cavity, there is absolutely no indication that he supposed any sort of energy quantization to hold for the electromagnetic field itself. Instead, the (at this point frankly magical) *effect* of the energy quantization imposed on the material oscillators receiving from and transferring energy to the radiation in the interior was forcibly transferred to the electromagnetic field via the equilibrium formula (1.15). The field itself, Planck was to insist for almost another full quarter century, was a continuous, fully classical entity regulated by Maxwell's equations. The situation was to change dramatically with Einstein's remarkable 1905 paper, "On a heuristic point of view concerning the creation and conversion of light" (Einstein, 1905*a*). Although this paper is now commonly referred to as the "photoelectric paper", Einstein spends much more time in it on an analysis of the volume-dependence of blackbody radiation (pp. 92–102) than on the brief discussion (pp. 104–105) of the photoelectric effect.

After pointing out that a strictly classical analysis must necessarily lead to the Rayleigh–Jeans result (1.23), with its inescapable concomitant ultraviolet catastrophe, Einstein goes on to analyse cavity radiation in the high-frequency domain, drawing some extraordinarily non-classical conclusions from the quintessentially "classical" (at least from an historical point of view) Wien Law. Einstein's approach in this paper is radically different from Planck's. He focusses first and foremost on the thermodynamic and statistical properties of the electromagnetic radiation in the interior of the cavity. Taking a cavity of volume V_0 and considering only the electromagnetic radiation in the frequency interval $(\nu, \nu + d\nu)$, the energy E of such radiation in the high-frequency domain where Wien's Law (1.14) holds is given by

$$E = \frac{8\pi h \nu^3}{c^3} V_0 e^{-\frac{h\nu}{kT}} d\nu \tag{1.24}$$

Solving this equation for $\frac{1}{T}$ and repeating the integration procedure of (1.20) to obtain an expression for the entropy S *of the electromagnetic radiation in this frequency interval*, one finds (the 0 subscript indicates that the radiation in the entire cavity of volume V_0 is being considered—we shall shortly consider radiation in a subcavity)

$$S_0 = -\frac{kE}{h\nu}\left\{\ln\left(\frac{E}{V_0 \frac{8\pi h\nu^3 d\nu}{c^3}}\right) - 1\right\} \tag{1.25}$$

The same amount of radiation confined to a smaller volume V would lead to an entropy S with exactly the same form as (1.25) but with V_0 replaced with V. Accordingly, the difference in entropy for the two situations is

$$S - S_0 = \frac{kE}{h\nu}\ln\left(\frac{V}{V_0}\right) = k\ln\left(\frac{V}{V_0}\right)^{E/h\nu} \tag{1.26}$$

The fundamental Boltzmannian association of entropy with the probability W of the associated microstates of the system, $S = k\ln W$, then leads to the conclusion that

$$W = \left(\frac{V}{V_0}\right)^{E/h\nu} \tag{1.27}$$

i.e., that the probability of an energy fluctuation leading to a concentration of all the electromagnetic radiation in the frequency interval $(\nu, \nu + d\nu)$ in the subvolume V of the full cavity V_0 takes exactly the form which we would expect if that radiation consisted of $\frac{E}{h\nu}$ "mutually independent energy quanta" (each of energy $h\nu$) moving freely throughout the cavity, in complete analogy to the behavior of molecules in a gas. This is as far from the classical picture of electromagnetic radiation as extended waves subject to mutual (destructive and constructive) interference as it is possible to get. The result (1.26)—extraordinarily simple, but profoundly baffling, from a classical point of view—clearly had a deep impact on Einstein's thinking. He was to hold firmly to the concept of energy (and later momentum) quantization of the electromagnetic field over the next 20 years—a period of time in which the majority of physicists were firmly on Planck's side and resistant to any notion of quantization of the sacred classical Maxwellian fields.

The centrality of blackbody radiation to Einstein's thinking about the nature of the electromagnetic field is clear once one reflects on the number of occasions on which he would return to the subject: to take the most prominent cases, in 1909 in two papers (Einstein, 1909b,a) (one entitled "On the present status of the radiation problem", the other "On the development of our conceptions on the nature and constitution of radiation") in which energy fluctuations were once more used as a diagnostic for exposing the underlying properties of radiation, and in 1917, in the famous "A-B coefficients" paper (Einstein, 1916, 1917), of critical importance in the later development of dispersion theory by Kramers, and thereafter in the 1925 development of matrix mechanics at the hands of Heisenberg, Born, and Jordan.[5] Here we briefly review Einstein's results of 1909, which proved to be a critical inspiration for Jordan's introduction in 1925, in the last section of the "Three-Man" paper of Born, Heisenberg, and Jordan (Born *et al.*, 1926), of the first true quantum field.

In returning to the problem of energy fluctuations in cavity radiation, Einstein decided to relax the simplifying assumption of high-frequency (or low-density) radiation described by the Wien Law, and to enquire into the implications of the full Planck distribution (1.16), valid at all densities and frequencies, for the fluctuation properties of thermal radiation. In this case, instead of considering the highly non-Gaussian process whereby a fluctuation would concentrate 100% of the radiation energy in a given interval $(\nu, \nu + d\nu)$ in a subvolume V (giving the result (1.27), later to be called "Einstein's first fluctuation theorem" by Jordan) Einstein decided to calculate the mean-square energy fluctuation of the energy in this interval in the subvolume V. The formula for such mean-square fluctuations is a standard result of statistical mechanics:

$$\langle (\Delta E)^2 \rangle = kT^2 \frac{d\langle E \rangle}{dT} \tag{1.28}$$

where T is the temperature and $\langle E \rangle$ the mean energy, which in this case is clearly just $V\rho(\nu, T)d\nu$. We can distinguish three interesting choices for the energy distribution $\rho(\nu, T)$ and corresponding mean-square energy fluctuation. We shall distinguish the results obtained for the mean-square energy fluctuation in each case by a subscript indicating the assumed form for the universal Kirchhoff distribution function $\rho(\nu, T)$: "RJ" for the completely classical Rayleigh–Jeans form, "W" for the Wien Law, and "P" for the final result of Planck. In the case of the Rayleigh–Jeans Law valid at low frequencies,

---

[5] For a thorough study of the role played by dispersion theory in the birth of modern quantum mechanics, see the two-part paper by M. Janssen and the present author (Duncan and Janssen, 2007a,b).

$$\rho_{RJ} = \frac{8\pi}{c^3}\nu^2 kT \tag{1.29}$$

$$\Rightarrow \quad \langle(\Delta E)^2\rangle_{RJ} = \frac{c^3}{8\pi\nu^2}\frac{\langle E\rangle_{RJ}^2}{V\,d\nu} \tag{1.30}$$

while the Wien case, valid at high frequencies, yields

$$\rho_W = \frac{8\pi h}{c^3}\nu^3 e^{-h\nu/kT} \tag{1.31}$$

$$\Rightarrow \quad \langle(\Delta E)^2\rangle_W = h\nu\langle E\rangle_W \tag{1.32}$$

and, using the Planck distribution formula valid at all frequencies, one obtains instead

$$\rho_P = \frac{8\pi h}{c^3}\frac{\nu^3}{e^{h\nu/kT}-1} \tag{1.33}$$

$$\Rightarrow \quad \langle(\Delta E)^2\rangle_P = \frac{c^3}{8\pi\nu^2}\frac{\langle E\rangle_P^2}{V\,d\nu} + h\nu\langle E\rangle_P \tag{1.34}$$

The energy fluctuation result (1.30) for the classical Rayleigh–Jeans regime would be reproduced by Lorentz (Lorentz, 1916) a few years later independently of thermo-dynamic considerations by considering the fluctuations of energy of electromagnetic radiation due to constructive and destructive interference in a subvolume (with an assumption of random phases of the component waves). The fact that the mean-square energy fluctuation comes out to be proportional to the square of the mean energy can be considered to be the characteristic feature of classical waves in this context.

The *linear* energy dependence of the Wien result for the squared energy fluctuation is on the other hand immediately suggestive of quantization, as we would expect a fluctuation of $\sqrt{N}$ from N particles of energy $h\nu$ to lead to a mean-square energy fluctuation $(\sqrt{N}h\nu)^2 = h\nu \cdot Nh\nu = h\nu\langle E\rangle$. The full Planck result (1.34), however, leads to a mean-square fluctuation which appears to be the *purely additive* result of wave and particle contributions.[6] In 1909 Einstein was to interpret this remarkable result (later dubbed by Jordan "Einstein's second fluctuation theorem", to distinguish it from (1.27), the "first fluctuation theorem") as evidence for two statistically—*and therefore structurally*—independent causes for energy fluctuation, and insisted in a lecture at the Salzburg *Naturforscherversammlung* (1909) that "the next phase of the development of theoretical physics will bring us a theory of light that can be interpreted as a kind of fusion of the wave and emission (i.e., particle) theories". In the next section we shall see that Pascual Jordan, in his introduction of the first quantum field, was to establish conclusively that two separate physical mechanisms for energy fluctuation are *not necessary*: rather, once the kinematic demands of the new quantum theory are properly implemented in the description of the modes of the electromagnetic field, the result (1.34) emerges precisely and naturally from a unified dynamical framework.

---

[6] Note, however, that in the Wien regime of large ν, the first term on the right-hand side of (1.34) is exponentially smaller than the second, agreeing with Einstein's 1905 assertion of purely particle behavior in this limit.

1.5 The first true quantum field: Jordan and energy fluctuations

The evolution of understanding of quantum physics in the twenty years between Einstein's introduction of light quanta and the development of the modern formal structure of quantum mechanics initiated by Heisenberg's famous *Umdeutung* (="Reinterpretation") paper of 1925 (Heisenberg, 1925) is a fascinating and complex story of frequent frustration punctuated by occasional leaps of understanding leading to yet further frustration. The final stages of this development, a fusion of correspondence principle arguments with Einstein's radiation theory (the "A and B coefficients") of 1917 leading to the quantum dispersion theory of Kramers, and thence to Heisenberg's reinterpretation of the kinematics of electrons in terms of non-commuting quantities, have been described many times (and in great detail in some recent work of M. Janssen and the present author (Duncan and Janssen, 2007*a,b*)). Here, we shall assume that the reader is familiar with the basic principles of quantum mechanics and focus our attention on the developments directly related to the quantization of fields—specifically, of the electromagnetic field, as this was the classical field of immediate phenomenological importance at the time, given the need to understand the interactions of atomic systems with light quanta.

The immediate successor to Heisenberg's *Umdeutung* paper—the work in which Heisenberg, without any reliance on the mathematics of matrix algebra, introduced the basic ideas of matrix mechanics—is the "Two-Man" paper of Born and Jordan "On Quantum Mechanics" (Born and Jordan, 1925), written in the late summer of 1925 as a formal amplification of the Heisenberg approach. In this paper, Born and Jordan derive the commutation relations of coordinates and momenta that we now regard as the fundamental defining characteristic of quantum phenomena. The first part of the paper clarifies, using explicit matrix methods, many of the "magical" results obtained by Heisenberg, but it is in the fourth chapter (entitled "Observations on electrodynamics") that the subject of quantization of fields is raised for the first time in the context of the new mechanics. The way in which this idea is introduced is of some relevance to the explicit calculations to follow in the "Three-Man" paper of Born, Heisenberg, and Jordan, and deserves an extensive quote:

A cavity with electromagnetic oscillations constitutes a system of infinitely many degrees of freedom. Nevertheless, the basic principles developed in the preceding sections, which admittedly only concern systems of a single degree of freedom, are sufficient to handle this case as well, given that it goes over to a system of *uncoupled* oscillators once analyzed in terms of eigenmodes. *There is hardly any possible doubt, how such a system is to be treated* (our emphasis). In particular, the circumstance that the basic equations of electromagnetism are linear is of importance, for it then follows that the virtual oscillators (eigenmodes) are *harmonic*, and it is precisely for harmonic oscillators, in contradistinction to other systems, that the validity of energy conservation is independent of the quantum condition.

In other words, after analysing (essentially by Fourier transformation) the electromagnetic field in terms of eigenmodes of specific frequencies, one is led to an infinite set of uncoupled harmonic oscillators, each of which can then be subjected to quantization (as a system of a single degree of freedom) by the methods outlined earlier. Moreover, for such harmonic systems, it turns out that energy conservation can be established *without* explicit use of the quantum condition $[q, p] = i\hbar$: although non-commuting quantities appear in the expression for the energy H, they do not have to be interchanged (requiring the use of the commutation relation of p and q) in the proof of energy conservation $\dot{H} = 0$. The rest of chapter 4 of the Born–Jordan paper[7] is then dedicated to a demonstration of energy and momentum conservation in linear electrodynamics along these lines. As the results do not at any point involve the quantum condition (or Planck's constant!) one is left with the impression that if this is indeed a calculation in quantum field theory (which in some sense it surely is), it is a peculiarly stillborn product, lacking any really characteristically *quantum* features. This defect would be dramatically, and spectacularly, remedied in the final section of the follow-up paper (the famous *Drei-Männer-Arbeit* of Born, Heisenberg, and Jordan (Born *et al.*, 1926)), where a truly quantum-field-theoretic calculation is employed to resolve the conundrum of the wave–particle duality of light first raised by Einstein's results of 1909 discussed in the previous section.

The *Drei-Männer-Arbeit* of Born, Heisenberg, and Jordan (entitled "On Quantum Mechanics: II") contains a thorough discussion of the fundamental principles of matrix mechanics: the kinematical quantities of classical mechanics are subjected to a reinterpretation in which they are replaced by non-commuting matrices (necessarily infinite) subject to the quantum condition specifying the commutator of canonically conjugate coordinates and momenta. The final chapter 4 of this work, devoted to physical applications, consists of three sections. In the first section the commutation relations for angular momentum are derived and selection rules for angular momentum discussed. The second section contains a short (and given the lack of understanding of electron spin at this time, not very productive) discussion of the Zeeman effect. In the final section of this long paper, entitled "Coupled harmonic resonators: Statistics of wavefields", the authors come to grips with the problem of energy fluctuations in the electromagnetic field. In particular, the aim is to show that Einstein's baffling result (1.34) giving the mean-square energy fluctuation in a subvolume of cavity radiation as a sum of wave and particle terms is a natural and inescapable consequence of subjecting the harmonic eigenmodes of the electromagnetic field to the same kinematic shift employed in the quantization of a single harmonic oscillator: i.e., the corresponding momentum p and coordinate q variables are replaced by matrices satisfying $[q, p] = i\hbar$. In contrast to the procedure of Planck, in which only the material oscillators in the wall were quantized systems, *all* physical entities, matter and radiation in equal measure, were now to be subjected to a quantization treatment.

It is commonly agreed that the material in chapter 4, section 3 of the *Drei-Männer-Arbeit* (henceforth referred to as the "3M paper") is entirely the work of Jordan: indeed, his two co-authors were later, for differing reasons, to disavow their involvement and even to criticize the validity of the result (see (Duncan and Janssen, 2008) for a detailed discussion of the historical background, as well as a careful mathematical reconstruction of the argument). With one exception, to be pointed out below, the criticisms are unwarranted: Jordan's calculation is technically correct and provides a

---

[7] It is best to draw the veil of charity over the attempts made at the very end to derive Heisenberg's connection between matrix elements of the electron coordinate and the transition amplitude for atomic transitions. This connection would first be properly elucidated in Dirac's seminal work of 1927, to be discussed in Chapter 2.

true insight into the physics of wave–particle duality. We will present a brief summary of Jordan's argument in the following (for more details, see the work cited above).

The problem of interpretation of "Einstein's second fluctuation theorem" (1.34) (as Jordan termed Einstein's 1909 result) involving wave and particle terms had been addressed by Ehrenfest just prior to the 3M paper in a paper (Ehrenfest, 1925) which is of interest to us only in one aspect: Ehrenfest introduces a one-dimensional model of cavity radiation which leads to a considerable technical simplification in the calculation of energy fluctuations. Unfortunately (for Ehrenfest), the paper precedes the *Umdeutung* paper of Heisenberg, so the non-commutativity of the eigenmode variables is unrecognized, and incorrect results necessarily follow in the quantum case. Still, the model of Ehrenfest was exactly the technical tool Jordan used to carry through a correct, post-*Umdeutung* calculation of the energy fluctuations in cavity radiation. The model of Ehrenfest which Jordan uses imagines a string of length l, fixed at both ends, and of constant elasticity and constant mass density. This is simply a one-dimensional analog of an electromagnetic field where the fixing of the string at the ends corresponds to an electric field component forced to vanish at the conducting sides of a box. The displacement of the string at location x (with $0 \leq x \leq l$) and time t is denoted $u(x, t)$. The wave equation for the string (the analog of the free Maxwell equations for this simple model) is then

$$\frac{\partial^2 u}{\partial t^2} - \frac{\partial^2 u}{\partial x^2} = 0 \tag{1.35}$$

Note that the velocity of propagation is set to unity here. The boundary conditions $u(0, t) = u(l, t) = 0$ for all times t express that the string is fixed at both ends. The general solution of this problem can be written as a Fourier series

$$u(x, t) = \sum_{k=1}^{\infty} q_k(t) \sin(\omega_k x), \quad \omega_k \equiv \frac{k\pi}{l} \tag{1.36}$$

$$q_k(t) = a_k \cos(\omega_k t + \varphi_k). \tag{1.37}$$

The classical Hamiltonian for a one-dimensional string is the well-known expression

$$H = \frac{1}{2} \int_0^l dx \, (\dot{u}^2 + u_x^2) \tag{1.38}$$

where the dot indicates a time-derivative and the subscript x a partial derivative with respect to x. The terms $\dot{u}^2$ and u_x^2 are the analogs of the densities of the electric and the magnetic field energy, respectively, in this simple model of blackbody radiation. Inserting (1.36) for $u(x, t)$ in (1.38), one finds

$$H = \frac{1}{2} \int_0^l dx \sum_{j,k=1}^{\infty} (\dot{q}_j(t) \dot{q}_k(t) \sin(\omega_j x) \sin(\omega_k x)$$

$$+ \omega_j \omega_k q_j(t) q_k(t) \cos(\omega_j x) \cos(\omega_k x)) \tag{1.39}$$

The functions $\{\sin(\omega_k x)\}_k$ in (1.36) are orthogonal on the interval $(0, l)$, i.e.,

$$\int_0^l dx \sin(\omega_j x) \sin(\omega_k x) = \frac{l}{2} \delta_{jk} \qquad (1.40)$$

The same is true for the functions $\{\cos(\omega_k x)\}_k$. It follows that the integral in (1.39) only gives contributions for $j = k$. The double sum thus turns into the single sum:

$$H = \sum_{j=1}^{\infty} \frac{l}{4} \left(\dot{q}_j^2(t) + \omega_j^2 q_j^2(t) \right) = \sum_{j=1}^{\infty} H_j \qquad (1.41)$$

This expression shows that the vibrating string can be replaced by an infinite number of uncoupled oscillators, one for every mode of the string, just as described in the extended quote from chapter 4 of the Born–Jordan paper given previously. Moreover, the distribution of the energy over the frequencies of these oscillators is constant in time. Since there is no coupling between the oscillators, there is no mechanism for transferring energy from one mode to another. The spatial distribution of the energy in a given frequency range over the length of the string, however, varies in time. In analogy to Einstein's considerations of 1909, Jordan now sets out to study the fluctuations of the energy in a narrow frequency interval $(\omega, \omega + \Delta\omega)$ in a small segment of the string, namely the region $0 \leq x \leq a, a << l$. The total energy in that frequency range will be constant but the fraction located in that small segment will fluctuate. Jordan derived an expression for the mean-square energy fluctuation of this energy, first in classical theory, then in matrix mechanics. Here we shall abbreviate the discussion by going directly to the quantum mechanical case. Accordingly, quantities like $q(t), \dot{q}(t)$ in the forthcoming equations must be considered to be non-commuting matrices. However, the *time-development* of these matrices involves exactly the usual periodic functions as in the classical case.

Changing the upper boundary of the integral in (1.39) from l to a ($a \ll l$) and restricting the sums over j to correspond to a narrow angular frequency range $(\omega, \omega + \Delta\omega)$ (i.e., $\omega < j(\pi/l) < \omega + \Delta\omega$ and $\omega < k(\pi/l) < \omega + \Delta\omega$), we find the instantaneous energy in that frequency range in a small segment $(0, a) \subset (0, l)$ of the string, here denoted $E_{(a,\omega)}$:

$$E_{(a,\omega)}(t) = \frac{1}{2} \int_0^a dx \sum_{j,k} (\dot{q}_j(t)\dot{q}_k(t) \sin(\omega_j x) \sin(\omega_k x)$$

$$+ \ \omega_j \omega_k q_j(t) q_k(t) \cos(\omega_j x) \cos(\omega_k x)) \qquad (1.42)$$

The functions $\{\sin(\omega_k x)\}_k$ and the functions $\{\cos(\omega_k x)\}_k$ are not orthogonal on the interval $(0, a)$, so both terms with $j = k$ and terms with $j \neq k$ will contribute to the instantaneous energy $E_{(a,\omega)}(t)$ in (1.42). First consider the $(j = k)$ terms. On the assumption that a is large enough for the integrals over $\sin^2(\omega_j x)$ and $\cos^2(\omega_j x)$ to range over many periods corresponding to ω_j, these terms are given by

$$E_{(a,\omega)}^{(j=k)}(t) \approx \frac{a}{4} \sum_j \left(\dot{q}_j^2(t) + \omega_j^2 q_j^2(t) \right) = \frac{a}{l} \sum_j H_j(t). \tag{1.43}$$

Since we are dealing with a system of uncoupled oscillators, the energy of the individual oscillators is constant, even at the quantum level (recall the emphasis on this point in the Born–Jordan paper). Since all terms $H_j(t)$ are constant, $E_{(a,\omega)}^{(j=k)}(t)$ is constant too and equal to its time average:

$$E_{(a,\omega)}^{(j=k)}(t) = \overline{E_{(a,\omega)}^{(j=k)}(t)}. \tag{1.44}$$

Since the time averages $\overline{\dot{q}_j(t)\dot{q}_k(t)}$ and $\overline{q_j(t)q_k(t)}$ vanish for $j \neq k$, the $(j \neq k)$ terms in (1.42) do not contribute to its time average:

$$\overline{E_{(a,\omega)}^{(j\neq k)}(t)} = 0. \tag{1.45}$$

The time average of (1.42) is thus given by the $(j = k)$ terms:

$$E_{(a,\omega)}^{(j=k)}(t) = \overline{E_{(a,\omega)}(t)}. \tag{1.46}$$

From (1.46) it follows that the $(j \neq k)$ terms in (1.42) give the instantaneous deviation $\Delta E_{(a,\omega)}(t)$ of the energy in this frequency range in the segment $(0, a)$ of the string from its mean (time average) value:

$$\Delta E_{(a,\omega)}(t) \equiv E_{(a,\omega)}(t) - \overline{E_{(a,\omega)}(t)} = E_{(a,\omega)}^{(j\neq k)}(t). \tag{1.47}$$

We now integrate the $(j \neq k)$ terms in (1.42) to find $\Delta E_{(a,\omega)}$. From now on, we suppress the explicit display of the time-dependence of $\Delta E_{(a,\omega)}$, q_j and $\dot{q}_j$.

$$\Delta E_{(a,\omega)} = \frac{1}{4} \int_0^a dx \sum_{j \neq k} \left(\dot{q}_j \dot{q}_k \left[\cos\left((\omega_j - \omega_k)x \right) - \cos\left((\omega_j + \omega_k)x \right) \right] \right.$$

$$+ \left. \omega_j \omega_k q_j q_k \left[\cos\left((\omega_j - \omega_k)x \right) + \cos\left((\omega_j + \omega_k)x \right) \right] \right)$$

$$\tag{1.48}$$

$$= \frac{1}{4} \sum_{j \neq k} \left(\dot{q}_j \dot{q}_k \left[\frac{\sin\left((\omega_j - \omega_k)a \right)}{\omega_j - \omega_k} - \frac{\sin\left((\omega_j + \omega_k)a \right)}{\omega_j + \omega_k} \right] \right.$$

$$+ \left. \omega_j \omega_k q_j q_k \left[\frac{\sin\left((\omega_j - \omega_k)a \right)}{\omega_j - \omega_k} + \frac{\sin\left((\omega_j + \omega_k)a \right)}{\omega_j + \omega_k} \right] \right).$$

Defining the expressions within square brackets as (cf. 3M paper, ch. 4, Eq. (45′))

$$K_{jk} \equiv \frac{\sin\left((\omega_j - \omega_k)a \right)}{\omega_j - \omega_k} - \frac{\sin\left((\omega_j + \omega_k)a \right)}{\omega_j + \omega_k},$$

$$K'_{jk} \equiv \frac{\sin\left((\omega_j - \omega_k)a \right)}{\omega_j - \omega_k} + \frac{\sin\left((\omega_j + \omega_k)a \right)}{\omega_j + \omega_k},$$

$$\tag{1.49}$$

we can write this as (cf. ch. 4, Eq. (45)):

$$\Delta E_{(a,\omega)} = \frac{1}{4} \sum_{j \neq k} \left(\dot{q}_j \dot{q}_k K_{jk} + \omega_j \omega_k q_j q_k K'_{jk} \right). \tag{1.50}$$

Recalling that we are working in a small frequency interval, $\omega_j - \omega_k << \omega_j + \omega_k$, it turns out that it is permissible to neglect the second term in the expressions for K_{jk} and K'_{jk} above, so we set $K'_{jk} = K_{jk}$ and obtain

$$\Delta E_{(a,\omega)} = \frac{1}{4} \sum_{j \neq k} K_{jk} \left(\dot{q}_j \dot{q}_k + \omega_j \omega_k q_j q_k \right) \tag{1.51}$$

where the coefficients K_{jk} now mean

$$K_{jk} \equiv \frac{\sin \left((\omega_j - \omega_k)a \right)}{\omega_j - \omega_k} \tag{1.52}$$

From (1.41) above it is apparent that the individual oscillators are formally identical to point particles of mass $m = l/2$. The subsequent calculations can be considerably simplified by introducing the now familiar[8] raising and lowering operators $a_j^\dagger(t), a_j(t)$

$$a_j(t) = \sqrt{\frac{l\omega}{4\hbar}} q_j(t) + i\sqrt{\frac{1}{l\hbar\omega}} p_j(t) = a_j(0)e^{-i\omega_j t}$$

$$a_j^\dagger(t) = \sqrt{\frac{l\omega}{4\hbar}} q_j(t) - i\sqrt{\frac{1}{l\hbar\omega}} p_j(t) = a_j^\dagger(0)e^{i\omega_j t} \tag{1.53}$$

satisfying

$$[a_j(t), a_k^\dagger(t)] = \delta_{jk} \tag{1.54}$$

In fact, the calculations of the 3M paper involve only the p_j and q_j matrices, and their commutation relation, and are mathematically perfectly equivalent to results obtained with the linear combinations defined in (1.53). The introduction of operators which raise or lower the excitation level of the individual eigenmodes will become central in our later development of the modern formalism of quantum field theory. To the extent that the excitation levels $\{n_j\}$ are identified (as they clearly are in the 3M paper) with the number of light quanta (i.e., photons, in modern terminology) with frequency ω_j, operators raising and lowering these levels are clearly identifiable as the particle creation and destruction operators of modern field theory. Later, in our systematic development of field theory, they will turn out to be the technical tool ideally suited to the introduction of physically sensible local interactions as well as dealing effortlessly with the statistics of properly symmetrized multi-particle states. Here they are introduced simply in order to allow us to write the expression for the energy fluctuation in a maximally compact fashion. We remind the reader (Baym,

---

[8] See (Baym, 1990) for a discussion of this now standard method for solving the quantized harmonic oscillator.

1990) that the effect of these operators for a single mode (the jth, say) on an eigenstate $|n_j\rangle$ of H_j is (at time 0)

$$a_j(0)|n_j\rangle = \sqrt{n_j}|n_j - 1\rangle \tag{1.55}$$

$$a_j^\dagger(0)|n_j\rangle = \sqrt{n_j + 1}|n_j + 1\rangle \tag{1.56}$$

$$a_j^\dagger(0)a_j(0)|n_j\rangle = n_j|n_j\rangle \tag{1.57}$$

In terms of the a_j and $a_j^\dagger$ operators the instantaneous energy fluctuation takes the very simple form:

$$\Delta E_{(a,\omega)} = \frac{\hbar}{l} \sum_{j \neq k} K_{jk} \sqrt{\omega_j \omega_k} a_j^\dagger(t) a_k(t) \tag{1.58}$$

In other words, the operator (or "matrix", in the language of the 3M paper) representing energy fluctuations is simply a sum of terms each of which takes a single photon in a given energy level and transfers it to a *different* energy level. What could be more natural?

The question now arises concerning in which state to evaluate the squared energy fluctuation $\Delta E^2_{(a,\omega)}$. The 1909 calculations of Einstein refer to cavity radiation in thermal equilibrium at a specified temperature T—i.e., to the evaluation of the mean-square fluctuation in a canonical thermal ensemble of states—but a careful perusal of the Jordan calculations of the 3M paper shows that the temperature never enters! Instead, Jordan calculates the quantum dispersion of the energy in a *single, pure state of the field* $|\{n_l\}\rangle$, characterized by specifying all excitation levels $n_l, l = 1, 2, 3, \ldots$ of the field (recall (1.41)):

$$H|\{n_l\}\rangle = \sum_j (n_j + \frac{1}{2})\hbar\omega_j |\{n_l\}\rangle \tag{1.59}$$

The expectation value of $\Delta E^2_{(a,\omega)}$ in this eigenstate of the full energy operator H is necessarily time-independent, so the time-averaging is moot. One has simply

$$\langle\{n_l\}|\Delta E^2_{(a,\omega)}|\{n_l\}\rangle = \frac{\hbar^2}{l^2} \sum_{j \neq k, j' \neq k'} K_{jk} K_{j'k'} \sqrt{\omega_j \omega_k \omega_{j'} \omega_{k'}} \langle\{n_l\}|a_j^\dagger a_k a_{j'}^\dagger a_{k'}|\{n_l\}\rangle \tag{1.60}$$

The diagonal matrix element in (1.60) only receives non-vanishing contributions when the indices j, k, j', k' satisfy $j = k' \neq k = j'$, i.e., when the photon destroyed in mode k' (by $a_{k'}$) and the photon created into the different mode j' (by $a_{j'}^\dagger$) are replaced in mode $j = k'$ (by $a_j^\dagger$) and removed in mode $k = j'$ (by a_k). Thus we obtain

$$\langle\{n_l\}|\Delta E^2_{(a,\omega)}|\{n_l\}\rangle = \frac{\hbar^2}{l^2}\sum_{j\neq k} K^2_{jk}\omega_j\omega_k\langle\{n_l\}|a^\dagger_j a_k a^\dagger_k a_j|\{n_l\}\rangle \tag{1.61}$$

$$= \frac{\hbar^2}{l^2}\sum_{j\neq k} K^2_{jk}\omega_j\omega_k\langle\{n_l\}|a^\dagger_j a_j(a^\dagger_k a_k + 1)|\{n_l\}\rangle \tag{1.62}$$

$$= \frac{\hbar^2}{l^2}\sum_{j\neq k} K^2_{jk} n_j(n_k + 1)\omega_j\omega_k \tag{1.63}$$

where in going from (1.61) to (1.62) we have used the commutation relation (1.54), and used the fact that the operator $a^\dagger_j a_j$ has eigenvalue n_j (i.e., the excitation level) for the jth mode. At this point a modern derivation of the thermal fluctuation would immediately perform a canonical ensemble average (by multiplying by the Boltzmann weight $e^{-\beta\sum_j n_j\hbar\omega_j}$ and summing over excitation levels to obtain the weighted average of the quantity in (1.63)). As the double sum is over non-identical indices $j\neq k$, the thermal averages factorize and the result is to replace the fixed occupation number n_j (resp. n_k) by the Planck mean occupation number of that mode $\bar{n}_j = \frac{1}{e^{\beta\hbar\omega_j}-1}$ (resp. $\bar{n}_k$). This is manifestly a smooth function of the discrete index j, a condition required by the next step Jordan takes in simplifying (1.63): namely, the (implicit) assumption that the dependence of the summand is sufficiently smooth that we can, with negligible error, replace the double sum with a double frequency integral. It should be emphasized once again that Jordan does *not* perform a thermal average in the 3M paper: rather, his derivation must be considered as valid for the pure state quantum dispersion, *assuming that the given pure state has a photon occupation number dependence which is adequately smooth with respect to the variation of the discrete mode index over the finite interval under consideration to allow the replacement of sums by integrals.* The further simplification of (1.63) therefore proceeds by the replacements

$$\sum_j \rightarrow \frac{l}{\pi}\int d\omega \tag{1.64}$$

$$\sum_k \rightarrow \frac{l}{\pi}\int d\omega' \tag{1.65}$$

where $\omega = \frac{j\pi}{l}, \omega' = \frac{k\pi}{l}$.

Resuming our calculation, we first note that if a is very large compared to the wavelengths associated with the frequencies in the narrow range $(\omega, \omega + \Delta\omega)$, we can set[9]

$$\int d\omega' f(\omega')\frac{\sin^2((\omega-\omega')a)}{(\omega-\omega')^2} = \int d\omega' f(\omega')\pi a\delta(\omega-\omega') = \pi a f(\omega) \tag{1.66}$$

---

[9] Here we use the fact that the integrand becomes highly peaked for $a \rightarrow \infty$, with total weight determined by the definite integral $\int_{-\infty}^{+\infty}\frac{\sin^2(x)}{x^2}dx = \pi$.

where $\delta(x)$ is the Dirac δ-function and $f(x)$ is an arbitrary function. As pointed out previously, the sine function in (1.66) is the dominant part of the K_{jk}^2 factor in (1.63) in a narrow frequency interval, so we finally obtain, with the indicated translations from sums to integrals,

$$
\begin{aligned}
(\{n_l\}|\Delta E_{(a,\omega)}^2|\{n_l\}) &= \frac{\hbar^2}{\pi^2} \int d\omega \int d\omega' \pi a \delta(\omega - \omega') n(\omega)(n(\omega') + 1)\omega\omega' \\
&= \frac{a}{\pi} \int_\omega^{\omega+\Delta\omega} (n(\omega)^2 + n(\omega))\hbar^2\omega^2 d\omega \\
&\simeq \frac{a}{\pi} (n(\omega)^2 + n(\omega))\hbar^2\omega^2 \Delta\omega
\end{aligned}
\tag{1.67}
$$

We once again (and for the last time!) re-emphasize that the smooth variation of the occupation numbers n_j is a precondition for the validity of the result (1.67). This smoothness is, of course, guaranteed once the thermal average is performed to replace fixed occupation numbers n_j by their thermal averages $\bar{n}_j$, as the latter are just the smooth Planck function $1/(e^{\beta\hbar\omega_j} - 1)$.

The result (1.67) is just Einstein's second fluctuation theorem (*à la* Jordan) in slightly disguised form. Writing all quantities as functions of the cyclic frequency ν rather than angular frequency ω, and using an overbar as a shorthand for the diagonal expectation values in Eqs. (1.60)–(1.67), we find:

$$
\overline{\Delta E_{(a,\nu)}^2} = 2a\Delta\nu \left((n(\nu)h\nu)^2 + (n(\nu)h\nu)h\nu\right)
\tag{1.68}
$$

We now introduce the excitation energy—the difference between the total energy and the zero-point energy. Jordan and his co-authors call this the "thermal energy" (p. 377, p. 384). Although the intuition behind it is clear, this terminology is misleading. The term "thermal energy" suggests that the authors consider a thermal ensemble of energy eigenstates—what we would call a mixed state—while as has been made clear in the preceding, in fact they are dealing with individual energy eigenstates, i.e., pure states. The term "excitation energy" is therefore preferable in this context. The excitation energy $E(\nu)$ in the narrow frequency range $(\nu, \nu + \Delta\nu)$ in the entire string in the state $\{n_\nu\}$ is

$$
E(\nu) = N(\nu)(n(\nu)h\nu) = 2l\Delta\nu(n(\nu)h\nu)
\tag{1.69}
$$

where we used that $N(\nu) = 2l\Delta\nu$ is the number of modes between ν and $\nu + \Delta\nu$ for our one-dimensional string. On average there will be a fraction a/l of this energy in the small segment $(0, a)$ of the string:

$$
\overline{E_{(a,\nu)}} = \frac{a}{l} E(\nu) = 2a\Delta\nu(n(\nu)h\nu)
\tag{1.70}
$$

Substituting $\overline{E_{(a,\nu)}}/2a\Delta\nu$ for $n(\nu)h\nu$ in (1.68), we arrive at the final result of this section of the *Dreimännerarbeit* (ch. 4, Eq. (55)):

$$\overline{\Delta E^2_{(a,\nu)}} = \frac{\overline{E_{(a,\nu)}}^2}{2a\Delta\nu} + h\nu\overline{E_{(a,\nu)}} \tag{1.71}$$

—precisely the analog, for a one-dimensional system of waves (with unit wave speed), of the Einstein result (1.34) (recall that in three dimensions, the number of electromagnetic modes in volume V in a narrow frequency interval is just $8\pi\nu^2 V d\nu/c^3$).

Jordan's derivation of Einstein's peculiar "hybrid" formula for the energy fluctuations in cavity radiation is used in the final paragraph of the 3M paper as further, as it were independent (of the dispersion considerations that had originally motivated Heisenberg) evidence for the validity of the matrix mechanical procedure of maintaining the form of the classical dynamical equations while reinterpreting as non-commuting matrix quantities the kinematical ingredients of these equations. But Jordan himself regarded the result as having far greater significance, insofar as it pointed the way to a general procedure for extending the principles of quantum mechanics to field systems with infinitely many degrees of freedom, including, as Jordan was to show a few years later, systems of particles satisfying Fermi statistics. Jordan was later (in 1962, in a comment to van der Waerden) to refer to his fluctuation calculation in the 3M paper as "almost the most important contribution I ever made to quantum mechanics."

The negative reaction of Jordan's contemporaries—including his co-authors on the 3M paper!—to the fluctuation calculation is of interest in its own right. Heisenberg seems to have been worried about potential divergences, and later (in 1930) was to publish a paper (Heisenberg, 1931) showing that the mean-square fluctuation in a subvolume of the cavity is in fact infinite *if one considers the electromagnetic energy integrated over all frequencies*. Although mathematically correct, this calculation is irrelevant as a criticism of Jordan's result in the 3M paper, which explicitly considers the energy fluctuations in a finite frequency interval, which are perfectly finite, as we have seen. It is also irrelevant from a physical point of view: the isolation of energy in a subvolume requires the introduction of enclosing physical filters of small but necessarily finite thickness, and the structure of these filters will eventually be resolved if we consider photons of arbitrarily high frequency and small wavelength. In fact, Heisenberg found that if the walls of the enclosing subvolume are smeared out (for example, if we replace the θ-function $\theta(a-x)$ implementing the restriction of the range of the integral (1.42) by a smooth function interpolating between 0 and 1), the integral over frequencies can be extended to infinity with a finite result for the mean-square energy fluctuation. Born was initially less vocal in opposition to Jordan's result, but later (in 1939), in exile in Edinburgh, published in collaboration with his assistant Klaus Fuchs (later famous after being discovered as a Soviet spy!) a paper (Born and Fuchs, 1939*a*,*b*) essentially retracting the entire calculation. The retraction had itself in short order to be retracted when serious technical errors were discovered by Pauli's assistant Markus Fierz.[10]

---

[10] For a detailed discussion of the reception of the Jordan fluctuation calculation, see the previously cited work of Duncan and Janssen (Duncan and Janssen, 2008).

In a letter to Jordan (1926), Einstein complained that Jordan's methods, while perfectly adequate in reproducing the second fluctuation theorem of 1909 (no mean feat in itself), must somehow fail in explaining the (at first sight simpler) result of 1905, for the quasi-molecular behavior of radiation in the Wien regime. It is difficult to infer at this remove what specific difficulty Einstein is referring to here: perhaps he was somehow under the impression that the zero-point energy should be included in the energy of an individual photon (making it $\frac{3}{2}h\nu$ instead of $h\nu$), which would then make the result (1.27) incomprehensible. It is certainly true that from a technical standpoint, using the methods of the 3M paper, the calculation of large non-Gaussian fluctuations of the kind considered in the first fluctuation theorem, where all the radiation energy is concentrated in the subvolume, is highly non-trivial. Such a calculation would require the evaluation of all (not just the quadratic) moments of the energy fluctuation. In principle, one could (with considerable effort!) explicitly calculate all higher moments and conclude that the behavior of light in the Wien regime does indeed mimic perfectly the statistical behavior of a gas of point-like particles—in particular in the extreme case of a fluctuation localizing all the energy in a subvolume—but in fact the technically superior route to this conclusion is simply to employ the modern methods of thermal statistical quantum field theory to calculate the expression for the entropy of the field, and then to proceed exactly as Einstein did in 1905!

To summarize the situation as the epochal year 1925 drew to a close, we have a clear example at last of a true quantum-field-theoretical calculation giving a meaningful resolution to the long-standing paradox of the wave–particle duality of light. This resolution can be achieved at the level of the non-interacting electromagnetic field: although true thermal equilibrium requires interactions of the light with the material container (the "heat bath" at temperature T), the quantum dispersion of a pure state of free photons in fact displays the same dual form for the mean-square energy fluctuation as the thermal case (with the caveats emphasized previously on smoothness of the occupation number distribution). This is the quantum analog of the fact that the classical fluctuation result (1.30) was also found by Lorentz (Lorentz, 1916) to hold rather generally (i.e., with no appeal to thermal concepts) provided a "reasonable" random phase assumption was made with regard to the classical modes of the electromagnetic field. Later, when we return to the issue of the classical limit of quantum fields, we shall see that this analogy is not accidental, but based on a deep principle of number–phase complementarity for fields.

While the fluctuation calculations of the final section of the *Drei-Männer-Arbeit* bring to a perfectly satisfactory conclusion the twenty-year controversy over wave–particle duality in electromagnetic phenomena initiated by Einstein's 1905 paper on light, it marks only the very beginning of the development of quantum field theory as we understand it today. In particular, two very important tasks required, and were soon to receive, the attention of physicists. First, there was the matter of interactions of light with matter. Indeed, the role of dispersion theory as the midwife to the new matrix mechanics of Heisenberg, Born, and Jordan indicates the primary role that the interaction between atomic systems (i.e., bound electrons in atoms) and light played in the thinking of physicists in the late 1920s. The 1927 paper of Dirac on the interactions between electrons and the quantized electromagnetic field was to begin the tortuous journey to a physically consistent quantum electrodynamics which would finally emerge by mid-century. The second task, also beginning in 1927, was taken up with great intensity and focus by Jordan and collaborators: the extension of the notion of field quantization to the treatment of matter fields, in particular fields with elementary excitations of fermionic character, which in consequence could never possess a classical counterpart analogous to the electromagnetic field of Maxwell. Our review of the historical evolution of quantum field theoretical concepts continues in the next chapter with a discussion of these developments.

2

Origins II: Gestation and birth of interacting field theory: from Dirac to Shelter Island

The convoluted evolution of quantum field theory in the period from the emergence of modern quantum mechanics in the late 1920s until the completion of the fully covariant and renormalizable quantum electrodynamics of the early 1950s seems at first sight a reprise of the extended birth pangs of quantum mechanics itself in the period from Planck's quantization of the distribution of energy among thermally equilibrated oscillators in 1900 to the development of matrix mechanics by Heisenberg in 1925. In both cases, and in contrast to the development of special and general relativity by Einstein, progress (often slow and halting) was due to the efforts of many physicists working along several lines of enquiry, with each new insight often opening up new questions and new difficulties.

However, at least in hindsight, it is apparent that there is a considerable difference in the intellectual background of the two efforts. The physicists struggling with the development of quantum mechanics in the first quarter of the twentieth century were faced with the need to construct an entirely novel, and in many respects completely counter-intuitive, *type of physical theory*, in which many of the basic concepts of classical physics seemed no longer applicable. Indeed, adherence to these concepts frequently obstructed rather than assisted the understanding of the complex of microscopic phenomena steadily being uncovered on the experimental front. As with relativity, and even more radically, the new theory seemed to demand the demolition of some of the most deeply held presuppositions of classical physics, and it was totally unclear, almost to the very end (i.e., the Heisenberg–Schrödinger revolution of 1925–26), what nexus of consistently interrelated concepts could replace them.

By contrast, at least from 1930 on, the physical requirements and conceptual structure needed for an adequate quantum theory of fields were fairly clear: the quantum mechanical substructure needed to follow a clear set of by now well-established rules, and the resultant theory should obviously respect the precepts of special relativity, yielding transition probabilities indifferent to one's choice of inertial frame. In a way, the fact that the quantum mechanical and relativistic foundations needed for an adequate physical theory were completely clear by the early 1930s made the apparent inconsistencies and frequently infinite results obtained in early calculations in quantum electrodynamics even more frustrating for the leading theorists in the field, who (as in the case of Heisenberg's willingness to introduce a discretization of space and a universal length unit) often felt the need for radical modifications at a fundamental level—modifications which we now understand to have been quite unnecessary.

A comprehensive (and comprehensible) account of the history of quantum field theory from the *Dreimännerarbeit* of 1925 to Dyson's renormalization analysis of 1949, effectively completing the formal framework of perturbative quantum electrodynamics (QED), would require an entire treatise.[1] In this chapter, constraints of space will limit us to a highly selective account of some of the major breakthroughs along the way. Many important and interesting contributions made to the early development of QED will be passed over in silence. The papers discussed are primarily those which had a definitive impact in (a) the development of the formal structure of quantum field theory, and (b) uncovering and (partially) resolving the conceptual difficulties occasioned by the lack of explicit relativistic covariance and by the appearance of ultraviolet divergencies in early calculations of quantum electrodynamic processes.

2.1 Introducing interactions: Dirac and the beginnings of quantum electrodynamics

In early 1927 Dirac took the first steps (Dirac, 1927*b*) towards a formulation of the theory of interaction of charged particles (specifically, electrons) with a quantized electromagnetic field which would eventually lead to the fully covariant, and renormalizable, quantum electrodynamics of the late 1940s: a formalism whose extraordinary quantitative successes established once and for all that local relativistic field theories are an essential ingredient in any precise description of the microworld, at least as far as electrodynamic effects are concerned. The motivation for Dirac's work was clear: the basis of the quantum dispersion theory which led to Heisenberg's founding paper on matrix mechanics, and indeed the interpretation of the matrices appearing in the latter as connected to the transition amplitudes for electrons jumping between distinct bound states in the atom, with the concomitant absorption or emission of photons, still rested on a completely *ad hoc* set of assumptions concerning the nature of the interactions of electrons in atoms with the electromagnetic field. The latter had indeed been subjected, in Jordan's calculations of energy fluctuations described in the preceding chapter, to the quantization procedure, but only in the approximation in which photons remain free particles, oblivious to the presence of matter. The establishment of the relation between matrix elements of the electron's coordinate operator in various atomic states, and the probabilities for transitions between these states (a relation simply posited by fiat by the founders of matrix mechanics), required an explicit theory of the interaction of an atomic system with a *quantized* electromagnetic field, and it was this theory which Dirac set out to develop in his seminal work. As a byproduct, the famous Einstein formulas for the A and B coefficients describing spontaneous and induced radiation—derived originally with the help of thermodynamic arguments—appear almost effortlessly once the appropriate formalism is in place.

---

[1] Indeed, the subject has been tackled already, in the excellent book by Schweber (Schweber, 1994), which the reader is encouraged to consult for more extensive details on many of the developments discussed below. See also the book by Miller (Miller, 1994), containing a number of the original papers in translation.

The problem addressed by Dirac in (Dirac, 1927b) was that of determining the "perturbation of an assembly satisfying the Einstein–Bose statistics" due to the interaction of the assembly with an atomic system (in fact, with an electron in such a system). The terminology here seems somewhat strange from a modern point of view: Dirac refers frequently to the set of independent modes of the non-interacting electromagnetic field (say, quantized in a box of finite volume so that the allowed modes form a discrete set, as in the Jordan calculation in the 3M paper) as an "assembly of independent systems". He then sets about to write a Hamiltonian for the electromagnetic field in terms of these mode variables—one sufficiently general to accommodate both the free electromagnetic field and the possible perturbation which might be induced by inserting an atomic system (described with the usual *non-relativistic* quantum formalism) into the field.

At this point Dirac introduces two technical devices which would become central features of quantum field theory: firstly, the use of a canonical transformation from the (p_r, q_r) type variables (such as those employed by Jordan in describing the individual harmonic oscillator systems for the rth field mode; cf. Section 1.5) to amplitude ($\sqrt{N_r}$) and phase (θ_r) variables, and thence to destruction b_r and creation $b_r^\dagger$ operators which lower and raise, respectively, by one the number of photons associated with the rth mode. The use of such variables actually goes back to a paper by London (London, 1926),[2] where the harmonic oscillator spectrum (and eigenfunctions) are derived using this technique.

Secondly, he employs for the first time the *interaction picture* of time development wherein the operators of the theory evolve via time-dependent unitary transformations due only to the free part of the Hamiltonian. The Hamiltonian introduced by Dirac for his Einstein–Bose "assembly" (i.e., the electromagnetic field) takes the form

$$H = \sum_{r,s} b_r^\dagger H_{rs} b_s, \quad H_{rs} = W_r \delta_{rs} + v_{rs} \tag{2.1}$$

where W_r is the energy of a free photon in the rth mode (namely, $W_r = h\nu_r$), and the v_{rs} form a matrix representing the effect of a perturbation (for example, due to the presence of an atomic electron in some atomic state) on the electromagnetic field. The operators b_r are defined in terms of the aforesaid conjugate amplitude and angle variables[3]

$$b_r = e^{-i\theta_r/\hbar} \sqrt{N_r}, \quad [N_r, \theta_s] = -i\hbar \delta_{rs} \tag{2.2}$$

whence one finds, using the representation $\theta_r = i\hbar \frac{\partial}{\partial N_r}$, and consequently $e^{-i\theta_r/\hbar} = e^{\frac{\partial}{\partial N_r}}$. Thus, $e^{\frac{\partial}{\partial N_r}} f(N_r) = f(N_r + 1) e^{\frac{\partial}{\partial N_r}}$, and we find

---

[2] See (Duncan and Janssen, 2009, p. 358) for a discussion of this important but not well known paper.

[3] The proper definition of a well-defined phase operator involves some subtle mathematical problems which at that time were not appreciated by Dirac, and which will be addressed in detail in Section 8.2. Here, we note simply that Dirac's results can be obtained without relying on the ill-defined phase operators θ_r, but purely in terms of the algebra of creation $b_r^\dagger$ and annihilation b_r operators, which are perfectly well defined.

$$b_r = e^{-i\theta_r/\hbar}\sqrt{N_r} = \sqrt{N_r + 1}\,e^{-i\theta_r/\hbar} \tag{2.3}$$

$$b_r^\dagger = \sqrt{N_r}\,e^{i\theta_r/\hbar} = e^{i\theta_r/\hbar}\sqrt{N_r + 1} \tag{2.4}$$

$$N_r = b_r^\dagger b_r \tag{2.5}$$

with the positive semidefinite operators N_r taking eigenvalues equal to zero or a positive integer value, clearly to be interpreted as the number of photons in mode r. In terms of the creation and destruction operators, the Hamiltonian (2.1) takes the form

$$H = \sum_{rs} H_{rs}\sqrt{N_r}\,e^{i\theta_r/\hbar}\sqrt{N_s + 1}\,e^{-i\theta_s/\hbar} = \sum_{rs} H_{rs}\sqrt{N_r}\sqrt{N_s + 1 - \delta_{rs}}\,e^{i(\theta_r - \theta_s)/\hbar} \tag{2.6}$$

A particular pure state $|\psi; t\rangle$ at time t (we shall use Dirac notation here, even though it did not yet exist in the modern form in the paper under discussion!) of the electromagnetic field may be identified by giving the "wave function" of the field in occupation number representation, i.e., by specifying the set of amplitudes

$$\psi(n_1, n_2, .., n_r, ...; t) \equiv \langle n_1, n_2, ... | \psi; t\rangle \tag{2.7}$$

(where n_r is the integer eigenvalue of N_r), and satisfies the time-dependent Schrödinger equation

$$i\hbar\frac{\partial}{\partial t}\psi(n_1, n_2, .., n_r, ..; t) = \langle n_1, n_2, ... | H | \psi; t\rangle$$

$$= \sum_{r,s} H_{rs}\sqrt{n_r}\sqrt{n_s + 1 - \delta_{rs}}\,\psi(n_1, ..., n_r - 1, ..., n_s + 1, ...; t) \tag{2.8}$$

The bilinear (in creation and destruction operators) form (2.1) chosen by Dirac for the Hamiltonian of the electromagnetic field means, of course, that the number of photons is necessarily *conserved*: the perhaps most characteristic feature of quantum field theories—particle creation and/or annihilation—is completely missing in this first attempt at an interacting theory of photons! Indeed, each destruction operator b_s is accompanied by a creation operator $b_r^\dagger$, so the time evolution of the system under the Hamiltonian (2.1) amounts simply to the continual transitioning of photons from one mode to another. This proves to be something of an embarrassment when Dirac addresses the basic object of the paper: the calculation of probabilities for the emission *or* absorption of photons by an electron (in a bound state of an atom). This difficulty is finessed by assuming that the "disappearance" of a photon in a pure absorption process really corresponds to the transition of the photon from a finite-energy (and hence detectable) state to a *zero* energy (and hence undetectable) state, which is possible, of course, for a massless particle with zero momentum. Indeed, Dirac assumes the existence of an omnipresent "sea" of infinitely many zero-energy photons,[4] with the addition or subtraction of a single (or any finite number) of zero-energy photons a physically unobservable event.

In hindsight, this assumption is completely unnecessary: indeed, one could easily have added terms linear in the creation and destruction operators directly to the right-hand side of (2.1) (symmetrically, in order to preserve hermiticity of the Hamiltonian), in which case the desired emission and absorption events would have appeared directly, with no need to resort to the device of an invisible sea of photons. The speculation is irresistible that the notion of an invisible sea of particles in their ground (lowest available energy) state which later re-emerges in Dirac's hole theory of positrons can be traced back to exactly this aspect of his first stab at quantum electrodynamics. Unfortunately, the hole theory in the context of charged massive particles (electrons and positrons) turned out to be a red herring of sinister vitality in the 1930s (and even into the 1940s), leading to a huge amount of fruitless (and complicated) formalism, as we shall see below, before the transition was made to a formulation in terms of a Fock space approach in which particles and antiparticles appear in a completely symmetric way.

Dirac circumvents the absence of pure emission or absorption terms in the proposed Hamiltonian (2.1) by a formal device which corresponds in modern terms to assuming that the vacuum is in a coherent state (cf. Section 8.2) of zero-energy photons, so that (for the destruction operator b_0 of a zero-energy photon) the photon states of the system satisfy[5]

$$b_0|\psi\rangle = \alpha|\psi\rangle, \quad \langle\psi|b_0^\dagger = \alpha^*\langle\psi| \tag{2.9}$$

with α a complex number with modulus going to infinity if (as Dirac assumes) there are infinitely many zero-energy photons present. Dirac assumes that the terms in H in (2.1) with $r \neq 0, s = 0$ (pure emission events) then lead (via vanishingly small v_{r0} coefficients) to finite matrix elements by setting the limit (for infinitely many zero-energy photons)

$$v_{r0}\alpha \to v_r = \text{finite} \tag{2.10}$$

which then implies a correspondingly finite amplitude $v_{0s}\alpha^* = v_s^*$ for a pure absorption event, in which a finite-energy photon in mode s is transferred to the zero-energy reservoir. The net effect of these shenanigans is the appearance of a term (necessarily hermitian, given the hermiticity of the original interaction matrix v_{rs}) linear in the

---

[4] In modern terminology, Dirac was imagining a Bose–Einstein condensate of zero-energy photons. While this idea is physically incorrect in the present context, Dirac's intuition of the physical imperceptibility of very-low-energy photons is a critical component in a proper understanding of the problem of infrared divergences in scattering amplitudes in quantum electrodynamics, as we shall see in Section 19.2.

[5] Strictly speaking, this is *not* what Dirac does. He assumes a definite number N_0 of zero-energy photons, in which case the matrix elements of the Hamiltonian between initial and final states would necessarily vanish! We are "fixing up" his argument here to yield the desired result, which he could, of course, have obtained directly by assuming from the outset an hermitian term linear in creation and destruction operators. Dirac was perfectly aware that the desired interaction Hamiltonian was necessarily linear in the electromagnetic vector potential and hence in the amplitude variable $\sqrt{N_r}$, thence also in the creation–destruction variables, as he shows in the final Section 7 of his paper (see below).

creation and destruction operators, and capable of initiating the desired pure emission and absorption events:

$$H = H_0 + H_{\text{lin}} + H_{\text{scat}} \tag{2.11}$$

$$H_0 = \sum_r W_r b_r^\dagger b_r = \sum_r W_r N_r \tag{2.12}$$

$$H_{\text{lin}} = \sum_{r \neq 0} (v_r b_r^\dagger + v_r^* b_r) \tag{2.13}$$

$$H_{\text{scat}} = \sum_{r,s \neq 0} v_{rs} b_r^\dagger b_s \tag{2.14}$$

For an interaction term of linear type, the Einstein results for the A and B coefficients describing spontaneous emission and absorption (Einstein, 1916, 1917) follow immediately once we re-express H_{lin} in terms of occupation number/phase variables:

$$H_{\text{lin}} = \sum_r (v_r e^{i\theta_r/\hbar} \sqrt{N_r + 1} + v_r^* e^{-i\theta_r/\hbar} \sqrt{N_r}) \tag{2.15}$$

Recalling (cf. (2.3, 2.4)) that the operators $e^{i\theta_r/\hbar}$ (resp. $e^{-i\theta_r/\hbar}$) increase (resp. decrease) the associated mode number eigenvalue n_r by one and hence effect the emission (resp. absorption) of a photon in the rth mode: the corresponding probabilities are therefore proportional to the initial-state occupation numbers $n_r + 1$ (n_r resp.). Note that the same factor $|v_r|^2$ appears in both probabilities (in Einstein's language, this is the equality of the coefficients for absorption and induced emission): it involves an atomic state matrix element which Dirac will identify once the interaction Hamiltonian is fully specified in terms of electron and electromagnetic field quantities. In any event, with the usual mode counting of classical plane wave modes in a box of volume V, giving $\frac{V}{c^3} \nu^2 d\nu d\Omega$ modes of a given polarization into solid angle $d\Omega$ in the frequency interval $d\nu$, one finds for the total energy of the radiation in this interval (assuming the occupation numbers n_r smoothly varying),

$$n_r \cdot h\nu_r \cdot \frac{V}{c^3} \nu_r^2 d\nu_r d\Omega \equiv \frac{V}{c} I(\nu_r) d\nu_r d\Omega \Rightarrow n_r = \frac{c^2}{h\nu_r^3} I(\nu_r) \tag{2.16}$$

where the specific intensity of the ambient radiation (in the initial state) $I(\nu_r)$ corresponds to the radiative flux in the particular frequency interval and solid angle (with flux defined as usual as the energy density times the speed of light). Thus, if the absorption rate is proportional to n_r and hence to $I(\nu_r)$, the emission rate is correspondingly proportional to $I(\nu_r) + \frac{h\nu_r^3}{c^2}$, with the first and second terms corresponding respectively to induced and spontaneous emission (the latter present even in the absence of ambient radiation). These are Einstein's laws for the emission and absorption of radiation, in the form presented by Dirac.

The remaining task facing Dirac was to go beyond the results of Einstein by providing a complete route to a calculation of the absolute (rather than relative) absorption and emission (spontaneous and induced) rates for specified transitions of an electron between distinct atomic stationary states, thereby putting on a firm basis (at last!) Heisenberg's basic, but as yet purely hypothetical, intuition in his seminal *Umdeutung* paper connecting the transition amplitudes for radiative processes in atoms to the matrix elements of the electron's coordinate operator. Dirac accomplishes this, in the final section of (Dirac, 1927*b*), by a correspondence principle argument identical in spirit to those used by Kramers, Born, and Heisenberg in the dispersion theory precursors to matrix mechanics. We shall briefly reprise the argument here, in modern notation.

We begin with the Fourier expansion for the classical vector potential in radiation (i.e., Coulomb) gauge, $\vec{\nabla} \cdot \vec{A} = 0$, in a finite box of volume V,

$$\vec{A}(\vec{r}, t) = \frac{1}{\sqrt{V}} \sum_{\vec{k}, \lambda} \frac{1}{\sqrt{2\omega_{\vec{k}}}} (\vec{\epsilon}_{\vec{k},\lambda} a_{\vec{k},\lambda} e^{i\vec{k}\cdot\vec{r} - i\omega_{\vec{k}} t} + c.c.) \tag{2.17}$$

where we are using natural units (so $\hbar = c = 1$), the polarization index λ takes two values, with a transverse polarization vector $\vec{k} \cdot \vec{\epsilon}_{\vec{k},\lambda} = 0$, and the energy associated with the discrete wavevector $\vec{k}$ is, for massless photons, $\omega_{\vec{k}} = |\vec{k}|$, and "c.c" means simply the complex conjugate, as $\vec{A}$ is a real c-number field. In Dirac's notation, the combination of the discrete momentum $\vec{k}$ and polarization index λ are denoted by a single mode index such as r, as previously. With the scalar potential zero, as appropriate for a free Coulomb gauge field, the electric field is $\vec{E} = -\frac{\partial \vec{A}}{\partial t}$ and the magnetic field is $\vec{B} = \vec{\nabla} \times \vec{A}$, whence the total energy of the free electromagnetic field in the box is, after some simple algebra,

$$H_0 = \frac{1}{2} \int_V (\vec{E}^2 + \vec{B}^2) d^3r = \sum_{\vec{k}, \lambda} \omega_{\vec{k}} a^*_{\vec{k},\lambda} a_{\vec{k},\lambda} \tag{2.18}$$

The interpretation of the amplitudes $a_{\vec{k},\lambda}$ is now clear: the absolute square is just the number of photons in the mode $\vec{k}, \lambda$, i.e., in Dirac's notation, after quantization, just the N_r operator. Dirac now completes the quantization of the electromagnetic field by *replacing* the classical coefficient $a_{\vec{k},\lambda}$ by the destruction operator $b_r = b_{\vec{k},\lambda}$, and its complex conjugate $a^*_{\vec{k},\lambda}$ by the hermitian conjugate operator $b^\dagger_{\vec{k},\lambda}$. The classical interaction energy between a particle of charge e and mass m in the presence of the vector potential is then (Dirac ignores the term quadratic in the vector field arising from the expansion of the gauged kinetic term $\frac{1}{2m}(\vec{p} - e\vec{A})^2$, keeping only the terms linear in the vector potential, which give the leading contribution in perturbation theory in e)

$$H_{\text{int}} = -\frac{e}{m} \vec{p} \cdot \vec{A}(\vec{r}, t) \tag{2.19}$$

where now $\vec{p}$ and $\vec{r}$ refer to the momentum and coordinate of the electron interacting with the electromagnetic field. Inserting (2.17) into (2.19) gives an operator precisely of the form H_{lin} in (2.15), but with the coefficient v^*_r of the destruction operator b_r now identified, up to unimportant constants, as $\vec{\epsilon}_{\vec{k},\lambda} \cdot \vec{p} e^{-i\vec{k}\cdot\vec{r}}$ for the mode r corresponding to wavevector $\vec{k}$ and polarization λ. In the electric dipole approximation the wavelength

k^{-1} of the photon is assumed much larger than atomic dimensions, and the exponential can be replaced by unity. The matrix element of this interaction operator between distinct initial and final atomic states $|i\rangle, |f\rangle$ thus involves matrix elements of the component of the electron momentum in the direction of the photon polarization, $\langle f|\vec{\epsilon}_{\vec{k},\lambda} \cdot \vec{p}|i\rangle$, multiplied by the dependence on the photon occupation numbers of the electromagnetic field found earlier by considering matrix elements of the b_r and $b_r^\dagger$ photon destruction and creation operators.[6] Dirac shows that these results agree in detail with the matrix-mechanical expressions for the Einstein A and B coefficients.

Dirac was perfectly aware that his fully quantum-mechanical discussion of the interaction of electrons with the electromagnetic field suffered from a serious shortcoming: the continuing treatment of the electrons as non-relativistic particles, which in particular made a proper derivation of the relativistic fine-structure effects of electrons in atomic bound states impossible. Initial attempts to produce a fully relativistic wave equation proceeded by replacing the non-relativistic (free-particle) Schrödinger equation

$$i\hbar\frac{\partial}{\partial t}\psi(\vec{r},t) = -\frac{\hbar^2}{2m}\vec{\nabla}^2\psi(\vec{r},t) \tag{2.20}$$

implementing (via the standard associations $H \to i\hbar\frac{\partial}{\partial t}, \vec{p} \to \frac{\hbar}{i}\vec{\nabla}$) the non-relativistic energy-momentum relation $H = \vec{p}^2/2m$, with the fully relativistic *Klein–Gordon* equation (we take $c=1$)[7]

$$\{\hbar^2(\frac{\partial^2}{\partial t^2} - \vec{\nabla}^2) + m^2\}\psi(\vec{r},t) = 0 \tag{2.21}$$

incorporating the correct relativistic relation $H^2 = \vec{p}^2 + m^2$. These attempts having failed in the archetypal test case of the hydrogen atom spectrum, Dirac tackled the problem of the coupling of a relativistic electron to the electromagnetic radiation in a seminal paper (Dirac, 1928) which settled once and for all the *kinematical* aspects of the relativistic treatment of spin-$\frac{1}{2}$ particles. The incorporation of an electromagnetic coupling in (2.21), by the usual minimal coupling replacement $\vec{p} \to \vec{p} - e\vec{A}, H \to H - eA_0$ (with $\vec{A}, A_0$ the vector and scalar potentials respectively) led to a wave-equation for the electron with two immediate, and serious, drawbacks:

1. The presence of a second-order derivative in time ran counter to the fundamental quantum-mechanical principle that the state of a quantum system at any time is determined solely by knowledge of its state at any given earlier time (unlike the situation in the configuration space formulation of classical mechanics, say, where both coordinates and velocities needed to be specified at an initial time to allow the subsequent time evolution to be computed). All of the highly successful quantum transformation theory developed up to this point, primarily by Dirac

---

[6] The matrix elements of the electron momentum operator can be converted to matrix elements of the coordinate operator by a simple commutator trick; see (Baym, 1990), chapter 13.

[7] We note for historical accuracy, that Schrödinger had actually tried the relativistic Klein–Gordon equation *first* in his treatment of the hydrogen atom, only to discover to his dismay that it led to incorrect fine-structure predictions.

(Dirac, 1927a) and Jordan (Jordan, 1927a,b), was predicated on this assumption, which clearly required the dynamical evolution equation of a quantum system to be *first order* in time.

2. Secondly, the second-order time-derivative had the highly unpleasant ancillary effect of introducing solutions with both positive and negative energy (with the latter, in the minimally coupled case, corresponding to particles of opposite charge to those of positive energy). This is due simply to the fact that the equation (2.21) determines only the square of the energy, but not its sign.

Dirac showed that it was possible to solve the first of these problems, and obtain a fully relativistic equation *first order in the time-derivative*, with the highly desirable byproduct of providing a natural explanation of the hitherto mysterious double-valuedness (rather quaintly termed "duplexity" by Dirac) of electron states, due to the intrinsic spin. However, as Dirac frankly admitted, the resultant equation, involving a four-component wavefunction (in other words, twice the desired "duplexity" accounting for spin), was still plagued with the unwanted appearance of negative-energy, opposite-charge solutions. The desired relativistic equation for the free particle case, if linear in the time-derivative (hence in the energy operator $p_0 \equiv H = i\hbar\frac{\partial}{\partial t}$), must necessarily be linear also in the spatial momentum operators $\vec{p} = -i\hbar\vec{\nabla}$, and Dirac showed that the simplest possibility, unique up to obvious similarity transformations, was obtained by setting

$$\{\gamma_\mu p^\mu - m\}\psi(\vec{r}, t) = 0 \tag{2.22}$$

where the $\gamma_\mu, \mu = 0, 1, 2, 3$ were algebraic objects satisfying the anticommutation algebra[8]

$$\{\gamma_\mu, \gamma_\nu\} = 2g_{\mu\nu} \tag{2.23}$$

where $g_{\mu\nu}$ is the flat-space metric tensor, with $g_{00} = -g_{ii} = 1, g_{\mu\nu} = 0, \mu \neq \nu$. The solutions to (2.22) necessarily also satisfied the Klein–Gordon equation (2.21), as

$$0 = (\gamma_\nu p^\nu + m)(\gamma_\mu p^\mu - m)\psi = (\gamma_\nu \gamma_\mu p^\nu p^\mu - m^2)\psi$$

$$= (\frac{1}{2}\{\gamma_\mu, \gamma_\nu\}p^\mu p^\nu - m^2)\psi$$

$$= (g_{\mu\nu}p^\mu p^\nu - m^2)\psi = (p_0^2 - \vec{p}^2 - m^2)\psi \tag{2.24}$$

which is identical with (2.21). A straightforward algebraic analysis led Dirac to the conclusion that the simplest realization of the Clifford algebra (2.23) was in terms of a set of four 4x4 matrices γ_μ, so that the wavefunction ψ necessarily had to carry an internal discrete index taking *four* values. This certainly gave room for the two-fold "duplexity" due to electron spin, and Dirac soon realized that the additional surplus "duplexity" corresponded exactly to the unwelcome reappearance in the theory of negative-energy solutions—basically because the γ_0 matrix associated

---

[8] Our notation *vis-à-vis* the Dirac equation and algebra differs slightly from Dirac's, in accord with modern usage.

with the energy term necessarily produced two $+1$ and two -1 eigenvalues on diagonalization.

The successful application of the Dirac equation (with the coupling of the electron to the electromagnetic field accomplished via the usual *minimal coupling* procedure of replacing $p_\mu \to p_\mu - \frac{e}{c} A_\mu$ in the free particle equation (2.22)) to the relativistic fine-structure of the hydrogen atom, and the fact that the new theory automatically yielded the correct, and previously utterly mysterious, gyromagnetic ratio of 2 for the electron, led to the immediate acceptance of the Dirac equation as the correct foundation for a fully relativistic treatment of the interaction of electrons and photons. However, the conceptual wave-mechanical framework in which the equation was conceived and born was to prove a stubborn hindrance to the early acceptance of a fully second-quantized formalism for matter fields such as the electron.

The irony is that Dirac, having pioneered the application of a second-quantized formalism (with creation and annihilation operators at the center) for dealing with the electromagnetic field, continued to insist on the use of first-quantization ideas for the electron. In particular, multi-electron states would be described relativistically in the same way as one was now accustomed to treat them in non-relativistic contexts, such as in multi-electron atoms, using wavefunctions defined on a $3N$-dimensional coordinate space (for N electrons) and with interactions handled approximately via Hartree–Fock mean field techniques.

The absence of disastrous instabilities incurred by the unavoidable transitions between positive- and negative-energy states once the Dirac electron was coupled to the electromagnetic field was legislated by fiat, by the assertion that the physical "vacuum" (i.e., no-particle) state actually consisted in having all negative-energy electron states filled, so that the normal positive electron states were simply (by the Pauli exclusion principle) denied the opportunity of dropping down to any of the infinitely many otherwise available negative-energy states (with the release of electromagnetic gamma-radiation of energy $\geq 2mc^2$). The *absence* of a negative-energy electron was then interpreted as the *presence* of a positive-energy particle of positive charge, interpreted first by Weyl and Dirac as the proton, but as difficulties arose with this proposal, as a (so far) unseen particle of equal mass and opposite charge to the electron.

This approach was soon confirmed by Anderson's experimental discovery of the positron in 1931. However, the idea of an unobservable "filled sea" of negative-energy electrons—in some sense the conceptual progeny of Dirac's earlier idea, discussed above, of a sea of zero-energy photons—would prove to be an extremely persistent distraction during the 1930s, leading to a vast amount of ultimately unprofitable, and extremely complicated, formalism, before the whole rickety framework[9] could be thrown overboard and replaced by a conceptually clean and technically efficient Fock

---

[9] Among many other artificialities, the negative-sea idea required the introduction of mathematically murky subtractions in the operators representing the total energy and charge of the system in order to implement the absence of energy and charge in the vacuum state—the dubious character of the necessary subtractions being considerably amplified by the assumption that the infinitely many occupied negative-energy electron states were *interacting* and the corresponding multi-particle wavefunction therefore had to be treated by extremely unconvincing Hartree–Fock approximation techniques incorporating, at least roughly, the interactions of the occupied negative-energy electrons.

space (i.e., occupation number) formalism in which positive-energy electron states and "negative-energy hole" states (i.e., positive-energy positron states) could be treated in a completely symmetric way. This required that the ideas of second-quantization, fully accepted for radiation after Dirac's 1927 paper, should be applied with equal force to matter (i.e., fermionic) fields. The unquestionable leader in this point of view, as we shall now see, was Pascual Jordan.

2.2   Completing the formalism for free fields: Jordan, Klein, Wigner, Pauli, and Heisenberg

In contrast to Dirac, Jordan was an early champion (by which we mean already in the period 1926–27 immediately following the breakthroughs of Heisenberg and Schrödinger) of the notion that wave–particle duality extended to a coherence in the mathematical formalisms used to describe radiation (specifically, the electromagnetic field) and matter (which seems to denote for Jordan and co-workers the aggregate behavior of massive particles of either bosonic or fermionic type). As we saw in Chapter 1, Jordan was the first to realize that the introduction of a quantum field for electromagnetism, in which the classical fields as c-number functions of space and time were replaced by q-number spacetime functions built from independent modes quantized as quantum harmonic oscillators, led to an explanation of the mysterious Einstein formula for the energy fluctuations in blackbody radiation, where the result involved a sum of wave and particle terms.

Jordan adopted very early the point of view that a similar approach, in which a single q-number function of space and time would incorporate all the physics of a system of arbitrarily many (massive) particles, should be applied universally. The mathematical advantage of such an approach was very clear (at least to Jordan!): the treatment of a many (N) particle system in terms of Schrödinger wavefunctions involving the abstract multi-dimensional coordinate space of N coordinate three-vectors (as well as the time t, of course) would be replaced by a dynamics specified by a single q-number field $\phi(\vec{r}, t)$ defined on the physical spacetime. The promise of such an approach, from a physical point of view, was twofold: (i) the treatment of interactions between matter and electromagnetism should be possible in a more natural way if both systems were subjected to a similar quantization technique, and (ii) the appearance of functions of spacetime $(\vec{r}, t)$, in lieu of the abstract $3N+1$-dimensional configuration space of many-body Schrödinger theory, suggested that the requirements of relativistic invariance could be more readily imposed on the theory. In the words of Jordan and Klein, "... this point of view seems especially suited to an attack on the relativistic many-body problem, because it describes matter and radiation in a mathematically equivalent way, namely through partial differential equations" (Jordan and Klein, 1927, p. 752).

In fact, in the paper by Jordan and Klein where a unified field-theoretic philosophy is first spelled out in detail, the matter particles are still treated non-relativistically (and are bosonic in nature). The problem addressed by Jordan and Klein amounts to a transcription into field-theoretic terms of the basic problem of atomic physics: the quantum mechanics of a system of non-relativistic charged particles (of charge e and mass μ) interacting via mutual electrostatic forces, and perhaps also subject to applied external fields. At this point, however, the incorporation of Fermi statistics in a second-quantized formalism was not yet fully understood (a deficit soon to be erased in the paper of Jordan and Wigner discussed below), so the particles are assumed to obey bosonic statistics. By writing the Hamiltonian operator of such a system (we assume that no external fields are present to simplify slightly the formalism) in the form

$$H = \int \frac{\hbar^2}{2\mu} \vec{\nabla}\phi^\dagger(\vec{r},t) \cdot \vec{\nabla}\phi(\vec{r},t) d^3r + \frac{e^2}{2} \int \frac{: \phi^\dagger(\vec{r},t)\phi(\vec{r},t)\phi^\dagger(\vec{r}',t)\phi(\vec{r}',t) :}{|\vec{r}-\vec{r}'|} d^3r d^3r'$$

$$(2.25)$$

with the field operator expanded in discrete plane wave modes (say, by quantizing the system in a box of volume V)

$$\phi(\vec{r},t) = \frac{1}{\sqrt{V}} \sum_{\vec{k}} e^{i\vec{k}\cdot\vec{r}} b_{\vec{k}}(t) \tag{2.26}$$

and where the $: \ldots :$ notation appearing in (2.25) embodies the instruction[10] to move all conjugate operators $b_{\vec{k}}^\dagger$ to the left of the $b_{\vec{k}}$ when (2.26) is inserted into (2.25), giving

$$H = \sum_{\vec{k}} E(k) b_{\vec{k}}^\dagger b_{\vec{k}} + \frac{e^2}{2} \sum_{\vec{k}\vec{k}'\vec{q}\vec{q}'} A(\vec{k}'\vec{q}'|\vec{k}\vec{q}) b_{\vec{k}}^\dagger b_{\vec{q}}^\dagger b_{\vec{k}'} b_{\vec{q}'} \tag{2.27}$$

$$A(\vec{k}'\vec{q}'|\vec{k}\vec{q}) \equiv \frac{1}{V^2} \int \frac{e^{-i(\vec{k}-\vec{k}')\cdot\vec{r} - i(\vec{q}-\vec{q}')\cdot\vec{r}'}}{|\vec{r}-\vec{r}'|} d^3r d^3r' \tag{2.28}$$

In analogy to Dirac's treatment of the electromagnetic field, the Hamiltonian equations for the classical Fourier amplitudes $b_{\vec{k}}(t), b_{\vec{k}}^*(t)$ are shown by Jordan and Klein to imply that these quantities form classically canonical pairs, whence the quantization of the system requires the commutator Ansatz for their quantum analogs:

$$[b_{\vec{k}}(t), b_{\vec{k}'}^\dagger(t)] = \delta_{\vec{k}\vec{k}'} \tag{2.29}$$

$$[b_{\vec{k}}(t), b_{\vec{k}'}(t)] = 0 = [b_{\vec{k}}^\dagger(t), b_{\vec{k}'}^\dagger(t)] \tag{2.30}$$

The transition to the occupation number basis, and the interpretation of the bs and $b^\dagger$s as destruction and creation operators, is then made exactly as in Dirac's work, via the transcription (see (2.2, 2.3, 2.4, 2.5))

$$b_{\vec{k}} \to e^{-\frac{i}{\hbar}\theta_{\vec{k}}} \sqrt{N_{\vec{k}}} \ , \quad b_{\vec{k}}^\dagger \to \sqrt{N_{\vec{k}}} e^{+\frac{i}{\hbar}\theta_{\vec{k}}} \tag{2.31}$$

---

[10] This instruction, in modern terminology the "normal-ordering" of the Hamiltonian operator, is essential to remove infinite contributions to the energy arising from the Coulomb self-energy of the individual particles, leaving only the electrostatic interaction energy of charged particle pairs (see (Jordan and Klein, 1927), Section 3). In the tangled history of 1930s field theory, the normal-ordering instruction was frequently referred to as the "Klein–Jordan trick".

The algebra (2.29, 2.30) implies that the Hamiltonian equation of motion of the field operator $\phi(\vec{r}, t)$, expressed in coordinate space, amounts to the field equations

$$i\hbar \frac{\partial}{\partial t} \phi(\vec{r}, t) = -\frac{\hbar^2}{2\mu} \Delta\phi(\vec{r}, t) + eV(\vec{r}, t)\phi(\vec{r}, t) \tag{2.32}$$

$$\Delta V(\vec{r}, t) = -4\pi e \phi^\dagger \phi(\vec{r}, t) \tag{2.33}$$

One sees immediately that (2.32) is formally identical with the non-relativistic (time-dependent) Schrödinger equation for a *single* particle of mass μ and charge e interacting with a Coulomb potential V, if we interpret $\phi(\vec{r}, t)$ as a c-number wavefunction. Eq. (2.33) is then simply the Poisson equation giving the electrostatic potential in terms of the particle probability density $|\phi(\vec{r}, t)|^2$. Jordan and Klein refer to the process of replacing the c-number coordinate space Schrödinger wavefunction by a q-number field operator as a "quantization of de Broglie waves". This terminology would soon be replaced by the denotation *second quantization*,[11] expressing the fact that the original c-number (or "first-quantized") equations of wave mechanics would lead to a correct treatment of multi-particle systems, including the proper treatment of the statistics (i.e., bosonic or fermionic), by the simple expedient of quantizing (i.e., replacing by operators satisfying suitable commutation relations) the expansion amplitudes of the single-particle wavefunctions.

The form of the Hamiltonian (2.27), when expressed in terms of creation and destruction operators, makes it clear that the theory (like Dirac's first version of the Hamiltonian for quantum electrodynamics) conserves particle number: each term contains an equal number of bs and $b^\dagger$s. Thus the time-dependent Schrödinger equation acts independently in each sector of the state space corresponding to a given fixed number of particles. Introducing a Schrödinger wavefunction for a given total number of particles N (in analogy to (2.7)), Jordan and Klein show after a simple calculation that the Hamiltonian (2.27) generates precisely the appropriate non-relativistic time-dependent Schrödinger equation (analogous to (2.8)) for the given N-particle system. Moreover, the Bose–Einstein symmetry of the multi-particle wavefunction is manifest throughout.

The extension of the second-quantization procedure to systems of particles obeying fermionic statistics followed within a few months of the Jordan–Klein paper, in an article of Jordan and E. Wigner entitled "On the Pauli Exclusion Principle" (Jordan and Wigner, 1928). The non-relativistic Schrödinger wave mechanical formalism for a system of N fermions, wherein the non-relativistic Schrödinger Hamiltonian acts on fully antisymmetric wavefunctions in the N-particle coordinate space, was shown to be equivalent to a second-quantized version in which the Hamiltonian appears in the form (2.27)[12] but with creation and destruction operators $a_\kappa^\dagger, a_\kappa$ subject to *anticommutation relations*[13]

---

[11] In an interview with Thomas Kuhn in June 1963 for the Archive for the History of Quantum Physics (session 3, p. 9), Jordan claims to have been the first to employ this terminology. See (Duncan and Janssen, 2008, p. 642).

[12] See (Jordan and Wigner, 1928), Eqs. (66a), (66b).

[13] The anticommutator of two operators A and B is defined as $\{A, B\} \equiv AB + BA$¿

$$\{a_\kappa, a_\lambda^\dagger\} = \delta_{\kappa\lambda} \tag{2.34}$$

$$\{a_\kappa, a_\lambda\} = 0 = \{a_\kappa^\dagger, a_\lambda^\dagger\} \tag{2.35}$$

with the indices κ, λ labeling a complete set of single-particle states (which may require both a continuous as well as discrete specification, with a corresponding interpretation of the Kronecker δ as a Dirac δ-function). The vanishing of the square of the creation operator $a_\kappa^\dagger$ (implied by (2.35), setting $\kappa = \lambda$) immediately incorporates the Pauli exclusion principle denying the possibility of multiply occupied fermionic states. However, the treatment of Jordan and Wigner is still entirely within the framework of non-relativistic physics: their paper appeared almost simultaneously with Dirac's relativistic equation for the electron, the correct second-quantized formulation of which, as we shall see, was as yet still several years in the future.

The formal developments outlined so far, while of critical importance in establishing the role of second quantization in enforcing the wave–particle connection while maintaining the correct symmetry of multi-particle states, left the issue of relativistic invariance essentially untouched. This shortcoming was remedied by the seminal paper of Heisenberg and Pauli (Heisenberg and Pauli, 1929), which put in place the formalism of Lagrangian field theory, still (eighty years later) at the core of modern field theory.

Heisenberg and Pauli begin with the canonical formalism of classical field theory, where an action functional defined as the spacetime integral of the Lagrangian gives rise to field equations via a variational principle:

$$\delta \int \mathcal{L}(\phi_\alpha, \vec{\nabla}\phi_\alpha, \dot{\phi}_\alpha)d^4x = 0 \quad \Rightarrow \quad \frac{\partial \mathcal{L}}{\partial \phi_\alpha} = \frac{\partial}{\partial x_\mu}\frac{\partial \mathcal{L}}{\partial(\frac{\partial \phi_\alpha}{\partial x_\mu})} \tag{2.36}$$

Here $\phi_\alpha(x^\mu)$ are a set of spacetime fields labeled by a discrete index α, and the covariant form of the Euler–Lagrange equation on the right-hand side of (2.36) suggests that, at least classically, relativistic invariance of the field equations will follow once the Lagrangian $\mathcal{L}$ is chosen to be a Lorentz scalar functional. The transition to a Hamiltonian framework is made by the standard procedure: one introduces canonically conjugate "momentum" fields $\pi_\alpha = \frac{\partial \mathcal{L}}{\partial \dot{\phi}_\alpha}$[14] and sets the Hamiltonian density equal to the Legendre transform of the Lagrangian density,

$$\mathcal{H}(\phi_\alpha, \vec{\nabla}\phi_\alpha, \pi_\alpha) \equiv \sum_\alpha \pi_\alpha \dot{\phi}_\alpha - \mathcal{L} \tag{2.37}$$

Exactly as in classical point mechanics, one is then easily able to verify that, in the absence of explicit time-dependence, the spatial integral of the density $\mathcal{H}$ is temporally constant, and hence deserves the appellation "total energy of the system". The fields ϕ_α and π_α at all spatial points but on the same time-slice are subjected to quantization by the usual replacement of the classical Poisson bracket by commutators

---

[14] Some notational warnings: Heisenberg and Pauli use Q_α (resp. P_α) for ϕ_α (resp. π_α) to emphasize the analogy to Q, P variables in classical mechanics: the analogy is made even more complete by a lattice formulation wherein the spatial continuum is discretized so that the $\vec{x}$ dependence also becomes discrete.

$$[\pi_\alpha(\vec{x}, t), \phi_\beta(\vec{y}, t)] = \frac{\hbar}{i} \delta_{\alpha\beta} \delta^3(\vec{x} - \vec{y}) \tag{2.38}$$

with the desirable consequence that the Hamiltonian operator H defined as the (conserved) spatial integral of the Hamiltonian density, $H \equiv \int \mathcal{H} d^3x$ generates the time evolution of any dynamical variable $F(\pi_\alpha, \phi_\alpha)$ constructed from the fields and their conjugates on a time-slice:

$$\frac{\partial F}{\partial t} = \frac{i}{\hbar}[H, F] \tag{2.39}$$

The article of Heisenberg and Pauli runs for 61 pages in the *Zeitschrift für Physik*, and we can do no more here than to summarize a few of the critical contributions it makes to the conceptual development of quantum field theory:

1. The relativistic invariance of the action (i.e., choice of a scalar Lagrange function) is shown to lead to relativistically invariant commutation relations, in the sense that the vanishing of the equal-time commutators of fields and conjugate (momentum) fields in one inertial frame implies the vanishing at equal time in another inertial frame (*vis-à-vis* the time coordinate of the new frame). From this follows the vanishing of local observables built from the fields and their conjugates at pairs of points with strictly space-like separation (as such point pairs can always be brought to equal time by a suitable Lorentz transformation). This is the first clear statement of the microcausality principle which lies at the heart of relativistic field theory.[15]

2. The canonical formalism was applied to the electromagnetic field, starting from the classically well-known Lagrangian function

$$\mathcal{L} = -\frac{1}{4}F_{\mu\nu}(x)F^{\mu\nu}(x), \quad F_{\mu\nu} \equiv \partial_\mu A_\nu(x) - \partial_\nu A_\mu(x) \tag{2.40}$$

where we use the convenient abbreviation $\partial_\mu \equiv \frac{\partial}{\partial x^\mu}$ for a spacetime-derivative. As Heisenberg and Pauli point out, the canonical program runs into immediate difficulties as the scalar potential A_0 (time component of the four-vector potential) has a vanishing conjugate momentum, $\pi_0 \equiv \partial \mathcal{L}/\partial \dot{A}_0 = F_{00} = 0$, which immediately derails the imposition of canonical commutation relations of the form (2.38). The canonical treatment of systems with a local gauge symmetry, leading to a degeneracy in the Legendre transform connecting the Lagrangian and Hamiltonian, is now well understood (cf. Chapter 15), but in 1929 the best that Heisenberg and Pauli could do was to suggest the addition of a term $\frac{1}{2}\epsilon(\partial_\mu A^\mu)^2$ to the Lagrangian, breaking the gauge symmetry $A_\mu \to A_\mu + \partial_\mu \Lambda$ (where Λ is an arbitrary scalar function), but allowing the definition of a sensible conjugate

---

[15] The earlier derivation of field commutation relations for the electromagnetic field by Jordan and Pauli (Jordan and Pauli, 1928) had considered only the free electromagnetic field, and the commutators were evaluated at arbitrary spacetime separations, so the specific aspect of space-like commutativity was not emphasized.

momentum field $\pi_0 = \epsilon \dot{A}_0$ for the scalar potential A_0.[16] Heisenberg and Pauli then suggest that the parameter ϵ be set to zero at the end of all calculations to recover the correct gauge-invariant results.

3. The Lagrangian formalism is developed for the Dirac relativistic electron equation. In modern notation, one writes an action (spacetime integral of the Lagrangian),

$$S_{\text{Dirac}} = \int \mathcal{L}_{\text{Dirac}} d^4 x = \int \bar{\psi}(x)(i\gamma^\mu \partial_\mu - m)\psi(x) d^4 x, \quad \bar{\psi} \equiv \psi^\dagger \gamma_0 \qquad (2.41)$$

which, by varying with respect to field $\bar{\psi}$ (assumed independent of ψ) yields the desired Dirac equation $(i\gamma^\mu \partial_\mu - m)\psi(x) = 0$. The requirement of Fermi–Dirac statistics then lead Heisenberg and Pauli to the choice of equal-time *anticommutation* relations to replace the commutators (2.38),

$$\{\psi_\alpha(\vec{x}, t), \psi_\beta^\dagger(\vec{y}, t)\} = \delta_{\alpha\beta}\delta^3(\vec{x} - \vec{y}) \qquad (2.42)$$

which follows formally by identifying the conjugate field to ψ as $\pi \equiv \partial\mathcal{L}_{\text{Dirac}}/\partial\dot{\psi} = i\bar{\psi}\gamma_0$ and replacing commutators by anticommutators. The identification of the electromagnetic charge current of the electron as $j^\mu(x) = e\bar{\psi}(x)\gamma^\mu\psi(x)$ is also made in this paper, as well as the assertion that the coupling between the electron and electromagnetic field is accomplished by including the term $j_\mu A^\mu$ in the Lagrangian. The *necessity* for the choice of fermionic statistics (specifically, anticommutators) for the electron is regarded by Heisenberg and Pauli as fundamentally mysterious: "a satisfying explanation for Nature's preference for the second possibility [i.e., anticommutators] can not be given". The Spin-Statistics theorem, establishing this necessity on the basis of the general tenets of relativistic invariance, quantum theory, and microcausality, is still a decade in the future (Pauli, 1940).

4. General formulas are adduced for the energy-momentum tensor $T_{\mu\nu}$ giving via spatial integrals the total four-momentum operators $P_0 = H$ (total energy), $\vec{P}$ (total spatial momentum) of the system, namely $P_\mu = \int T_{\mu 0} d^3 x$, the conservation of which are assured by the divergenceless property $\partial_\nu T^{\mu\nu} = 0$. Specific forms are given for $T_{\mu\nu}$ for both electromagnetism and the Dirac theory.

To these unquestionable triumphs we must also add the difficulties and failures which emerge in the course of applying the Heisenberg–Pauli formalism for interacting electrons and photons to some fundamental questions. Of particular historical interest is their second-order calculation of the self energy of the electron (i.e., the correction to the rest mass of an electron of second order in the charge e when the

---

[16] In two extremely influential papers in 1929 and 1930, Enrico Fermi (Fermi, 1929, 1930) was able to develop a consistent Hamiltonian quantum electrodynamics in the covariant Lorentz gauge $\partial_\mu A^\mu = 0$, and to establish its equivalence to the transverse gauge $\vec{\nabla} \cdot \vec{A} = 0$ used by Dirac. In a second paper on the quantization of wave fields (Heisenberg and Pauli, 1930), Heisenberg and Pauli showed that Fermi's covariant gauge results corresponded to a choice of unity for their parameter ϵ, and that the gauge condition $\partial_\mu A^\mu = 0$ should be interpreted as a constraint on the allowed states, not as an operator identity (a "q-Zahlrelation").

interaction with the electromagnetic field is turned on). The calculation yields (Eq. (115) in (Heisenberg and Pauli, 1929)) a linear divergence identical with the classical Coulomb self-energy result: namely, $\frac{e^2}{|\vec{x}-\vec{x}|} = \infty$. This divergent energy (together with the ubiquitous vacuum energy arising from the sum of $\frac{1}{2}h\nu$ zero-point energies from each mode of the electromagnetic field) would plague attempts to arrive at a mathematically consistent quantum electrodynamics throughout the 1930s and early 1940s. As only electron states are considered (the interpretation of the negative-energy solutions in terms of positrons, with a correct treatment of the latter, is yet to come), the Heisenberg–Pauli quantum electrodynamics conserves electron number, and the additive infinity in the electron energy can therefore be dropped "as one is only interested in energy differences". The development of positron theory, and its proper application by Weisskopf to the electron self-energy problem in 1939, would lead to a softening of the divergence from linear to logarithmic, but the extreme discomfort felt by physicists working on early quantum electrodynamics in the presence of numerous divergent expressions would only be assuaged in the late 1940s with the development of a covariant renormalization procedure.

2.3   Problems with interacting fields: infinite seas, divergent integrals, and renormalization

The history of quantum electrodynamics throughout the 1930s and into the mid-1940s is replete with false starts, conceptual confusions, and the frequent appearance of increasingly radical suggestions for the abandonment of "sacred" principles in a desperate attempt to stay afloat in a rising tide of physical and mathematical inconsistencies. The main problems were of two distinct, but related, types:

1. The persistence of the "negative sea" hole theory of Dirac, wherein positron states were interpreted as holes in an infinite background of negative-energy Dirac electrons. Conceptually, this theory was the unfortunate consequence of a stubborn persistence of the many-body configuration-space thinking, fully appropriate in the case of non-relativistic multi-electron systems such as atoms, in the treatment of relativistic fermionic matter on the part of Dirac and his followers, despite the fact that Dirac had pioneered the treatment of the electromagnetic field in terms of a second-quantized formalism involving photon creation and destruction operators. Despite the resistance of prominent theorists (especially Heisenberg and Pauli), who properly objected to the highly *ad hoc* way in which the physical effects of an infinite background of charged particles were disposed of, many of the detailed calculations of quantum electrodynamic processes in the 1930s were carried out on the basis of (or at the least, paying lip service to) the Dirac hole theory.[17]

2. The appearance of ultraviolet divergences (i.e., at the high-momentum or short-distance end of integrals appearing in the calculation of various quantum electrodynamic quantities) which seemed unavoidable without introducing fundamental distortions of underlying physical principles (such as the introduction of

---

[17] For a convenient compendium of many of the important papers, see (Miller, 1994).

a "universal length" cutoff (Heisenberg, 1938)) led to increasing dissatisfaction with the theoretical frameworks available (whether hole theory or fully second-quantized). The zero-point (vacuum) energy in the electromagnetic field already visible in Jordan's fluctuation calculations in the *Dreimännerarbeit* were soon joined by divergences in the electron self energy (i.e., the corrections to the energy of an isolated electron due to its interactions with the electromagnetic field), and by a divergent screening correction to the electric charge on an isolated electron due to the unavoidable presence in the vacuum of virtual electron–positron pairs, essentially corresponding to an infinite dielectric constant of the vacuum.

The way out of the impasse created by the fundamentally untenable notion of a negative-energy sea of electrons was shown very clearly in a paper by V. A. Fock in 1933 (Fock, 1933), and somewhat less clearly (as the basic idea is submerged in a large quantity of speculation on other unrelated topics) in an almost simultaneous article by Furry and Oppenheimer (Furry and Oppenheimer, 1934), although it must be admitted that the basic lessons of both papers seem to have been pretty much ignored by the theoretical community, which for the most part went right on calculating in terms of electrons and holes.

We shall describe briefly the ideas of Fock here, as this paper can be regarded as the seminal work responsible for the term "Fock space", which provides the basic kinematical scaffolding for modern (operator) formulations of relativistic field theory. Fock proposed that instead of treating electrons as the primary objects of the theory and positrons as derived concepts (i.e., holes in a negative-energy sea of electrons), the latter should appear in the theory along with electrons in a completely symmetrical way. By this time, the experimentally well-established phenomenological symmetry between the two types of particle (identical mass, opposite charge) certainly made this proposal a plausible one. Thus, the Hamiltonian of the *free* theory (i.e., with electromagnetic interactions switched off) was assumed to take the form[18]

$$H_0 = \int d^3p \sum_{\sigma} E(p)(b^{\dagger}(\vec{p}, \sigma)b(\vec{p}, \sigma) + d^{\dagger}(\vec{p}, \sigma)d(\vec{p}, \sigma)) + \text{infinite constant} \quad (2.43)$$

where the creation and destruction operators for electrons $(b^{\dagger}, b)$ and positrons $(d^{\dagger}, d)$, as already clear from the work of Jordan and Wigner, must obey anticommutation relations to enforce Fermi–Dirac statistics:

$$\{b(\vec{p}, \sigma), b^{\dagger}(\vec{p}', \sigma')\} = \delta_{\sigma\sigma'}\delta^3(\vec{p} - \vec{p}') \quad (2.44)$$

$$\{d(\vec{p}, \sigma), d^{\dagger}(\vec{p}', \sigma')\} = \delta_{\sigma\sigma'}\delta^3(\vec{p} - \vec{p}') \quad (2.45)$$

$$\{b(\vec{p}, \sigma), b(\vec{p}', \sigma')\} = \{d(\vec{p}, \sigma), d(\vec{p}', \sigma')\} = 0 \quad (2.46)$$

---

[18] We have taken the liberty of introducing modern notation here: in Fock's paper, the momentum-spin pair $\vec{p}, \sigma$ is denoted by the single variable q, the electron (resp. positron) destruction operators $b(\vec{p}, \sigma)$ (resp. $d(\vec{p}, \sigma)$) are denoted $\phi(q, 1)$ (resp. $\phi(q, 2)$), and the relativistic energy $E(p) = \sqrt{\vec{p}^{\,2} + m^2}$ is given as a matrix element of the single-particle Dirac Hamiltonian.

The infinite constant appearing in the Hamiltonian in (2.43) is a consequence of Fock having started with a Hamiltonian built from an electron field containing both positive and negative electron states, and then having defined the positron destruction operator $d(\vec{p}, \sigma)$ as the creation operator for a negative-energy electron state. Once the Hamiltonian is obtained in the symmetrical form (2.43), this infinite-energy term (basically the energy of the filled sea of negative-energy electrons in the Dirac hole theory) is immediately discarded by Fock as physically irrelevant.

The form of the remaining operator is completely transparent: the energy of a system of a finite number of free electrons and positrons is given simply by multiplying the possible energies by the number operators $b^\dagger b, d^\dagger d$ for electrons and positrons, with the energy of the vacuum state (no electrons or positrons) automatically zero. In particular, *no primacy of place* is given to either electrons or positrons in this approach, both types of particle appearing on an absolutely equal footing in the formalism. With the benefit of hindsight we now know, of course, that this was exactly the right attitude to adopt: the apparent preference of Nature for electrons over positrons in the world around us being a purely historical accident occasioned by the presence of a tiny CP violation (namely, a breaking of the particle–antiparticle symmetry) in the early Universe.

The free Hamiltonian H_0 in (2.43) preserves separately the number of electrons and positrons in any state. However, as Fock points out, the addition of interaction terms involving *odd* operators (i.e., those in which electron and positron creation or destruction operators appear multiplied, corresponding in the Dirac theory to transitions between positive- and negative-energy electron states) such as

$$H_u = \int d^3p\, d^3p' \sum_{\sigma\sigma'} (U(\vec{p}, \sigma; \vec{p}', \sigma') b^\dagger(\vec{p}, \sigma) d^\dagger(\vec{p}', \sigma') + h.c.) \tag{2.47}$$

(here *h.c.* denotes "hermitian conjugate") while conserving electric charge (as electrons and positrons are created or destroyed in tandem by the operator H_u), necessarily results in a theory in which the number of particles (either electrons or positrons) is *not* conserved. The importance of these prescient remarks would soon become clear when it became apparent that the gauge-invariant treatment of the coupling to electromagnetism involved a four-vector charge current $j^\mu(x)$ containing exactly odd terms of this sort (in addition to even terms conserving separately electron and positron number).

In retrospect, the advantages of a charge symmetric quantum electrodynamics should certainly have become completely manifest after the appearance of the paper of Pauli and Weisskopf in 1934 (Pauli and Weisskopf, 1934),   in which a fully gauge-invariant theory of massive charged scalar (i.e., spinless) particles coupled to electromagnetism was written down classically and then subjected to canonical quantization *à la* Heisenberg and Pauli. Thus, temporarily switching off the coupling to the electromagnetic field (and setting $\hbar = c = 1$), one starts with a free Hamiltonian for the scalar field ψ (where for c-number fields the † means simply complex conjugation)

$$H_0 = \int \{ \frac{\partial \psi^\dagger}{\partial t} \frac{\partial \psi}{\partial t} + \vec{\nabla}\psi^\dagger \cdot \vec{\nabla}\psi + m^2 \psi^\dagger \psi \} d^3x \tag{2.48}$$

and then (via the usual identification of conjugate momentum fields) introduces quantization via the equal-time commutation relations

$$[\pi(\vec{x},t),\psi(\vec{x}',t)] = -i\delta^3(\vec{x}-\vec{x}'), \quad \pi \equiv \frac{\partial\psi^\dagger}{\partial t} \tag{2.49}$$

This theory, with fields satisfying a Klein–Gordon equation (which follows from (2.48) and (2.49)) with classical solutions of both negative and positive energy, provided Pauli and Weisskopf with a clear analog of the problems in electron theory which led Dirac to the desperate expedient of a negative-energy sea: with the crucial difference that the absence of an exclusion principle made the notion of viewing the physical vacuum as a state with all negative-energy states filled (each one infinitely many times, as we are dealing with bosons here!) even more manifestly absurd than in the fermionic case.

Fortunately, Pauli and Weisskopf were able to show that the quantized version of the theory possessed a perfectly sensible interpretation, *provided the particles and antiparticles of the theory were put on a completely equal footing* (as in the work of Fock, which, however, is not referenced in (Pauli and Weisskopf, 1934)). A Fourier expansion of the scalar field $\psi(\vec{r},0)$ (at time zero), incorporating the commutation relations (2.49), and working in a box of volume V so that the allowed momenta are discrete, gives (with some slight modifications of notation to accommodate modern taste)

$$\psi(\vec{r},0) = \frac{i}{\sqrt{V}} \sum_{\vec{k}} \frac{1}{\sqrt{2E(k)}}(a(\vec{k})e^{i\vec{k}\cdot\vec{r}} - b^\dagger(-\vec{k})e^{-i\vec{k}\cdot\vec{r}}) \tag{2.50}$$

where $a(\vec{k})$ (resp. $b(\vec{k})$) are now interpreted as destruction operators for a particle of spatial momentum $\vec{k}$ (resp. an antiparticle of momentum $-\vec{k}$) and the hermitian conjugate operators are the corresponding creation operators. The Hamiltonian (2.48) then becomes (cf. Fock's (2.43))

$$H_0 = \sum_{\vec{k}} E(k)(a^\dagger(\vec{k})a(\vec{k}) + b^\dagger(\vec{k})b(\vec{k}) + 1) \tag{2.51}$$

with the infinite term $\sum_{\vec{k}} E(k)\cdot 1$ interpreted as a vacuum zero-point energy which "can be deleted in all applications". Once this is done, the vacuum, with zero energy, is simply the state $|0\rangle$ annihilated by H_0 via $a(\vec{k})|0\rangle = b(\vec{k})|0\rangle = 0$ for all $\vec{k}$: the noisome negative-energy sea is simply banished from the theory. The divergenceless four-vector $J_\mu(x) \equiv i(\frac{\partial\psi^\dagger}{\partial x^\mu}\psi - \psi^\dagger\frac{\partial\psi}{\partial x^\mu})$ (with $\partial^\mu J_\mu = 0$ following from the Klein–Gordon equations of the free theory) then leads in the usual way to a conserved charge operator

$$Q \equiv \int J_0(\vec{r},t)d^3r = \sum_{\vec{k}}(a^\dagger(\vec{k})a(\vec{k}) - b^\dagger(\vec{k})b(\vec{k})) \tag{2.52}$$

making clear the identification of the a and b operators as destruction operators for particles of opposite charge (but identical mass: the energy function $E(k) = \sqrt{\vec{k}^2 + m^2}$ in (2.51) involves the same mass m throughout). The coupling to the electromagnetic field (via the usual mimimal coupling replacement $\partial_\mu \psi \to (\partial_\mu - ieA_\mu)\psi$ in the Lagrangian, where A_μ is the electromagnetic four-vector potential) can be carried through in a straightforward way. One arrives at an interacting theory in which (a) charge conservation (and the vanishing of the four-divergence $\partial^\mu J_\mu$) is still exact, and (b) new photon-mediated pair-creation and annihilation processes appear in the theory, exactly of the sort expected in the Dirac hole theory from transitions between positive- and negative-energy electron states (but, finally, without the need for an invisible infinite background of charged particles!).

In hindsight, the advantages of a second-quantized formalism in which electrons and positrons are treated symmetrically seem so compelling that it is difficult to understand the persistence of the hole-theory perspective years after the works of Fock and Pauli–Weisskopf discussed above. Nevertheless, the hole-theory point of view remained prominent even up to the late 1940s, and the troublesome charge and mass divergences which would undermine the confidence of many of the early practitioners of quantum electrodynamics first made their appearance in the context of calculations performed on the basis of a vacuum consisting of an invisible filled sea of negative-energy electrons. By 1930, divergent field-theoretic quantities had already made their appearance in the form of the zero-point energy of the electromagnetic field and the infinite sea of negative-energy electrons, as well as in the linear divergence in the self-energy of the electron encountered by Heisenberg and Pauli. In his presentation at the 1933 Solvay conference (Dirac, 1933), Dirac pointed out that the alteration in the charge density of the background sea of filled electron states induced by the insertion of a test charge could be interpreted as a polarizability of the vacuum, leading to an effective screening of the bare test charge by a factor $(1 - \frac{2\alpha}{3\pi} \ln \frac{\Lambda}{mc})$,[19] where α is the fine-structure constant and Λ is a momentum cutoff which Dirac assumed should correspond to the inverse electron Compton wavelength, above which the theory was presumably unreliable. A perturbative calculation involving an intermediate state in which a negative electron changes state—necessarily to a positive-energy state, as all other negative-energy states are filled—corresponds in the Fock point of view to the appearance of a virtual electron–positron pair, so that the screening can alternatively be viewed as due to the preferential orientation of these virtual dipoles with respect to the applied field, much as in the classical theory of polarization. Dirac made clear that the "observed" charges measured on electrically charged particles necessarily differed, as a result of this polarization of the vacuum, from the "true" charges carried by these particles. This observation clearly contains the germ of the idea of charge renormalization, and more generally the realization that physically observed properties may—indeed must—contain built-in modifications as a consequence of radiative interaction effects, necessarily complicating the interpretation of the "true" (or, in modern terminology, "bare") parameters appearing in the fundamental Hamiltonian of the theory.

Further calculations of vacuum polarization in the mid-1930s, by Furry and Oppenheimer (Furry and Oppenheimer, 1934), Peierls (Peierls, 1934), and Weisskopf (Weisskopf, 1936), confirmed the presence of a logarithmic ultraviolet divergence in the

---

[19] The order α correction appears to differ from the correct value by a factor of 2, but the reason for this is unclear.

charge screening factor. However, it was generally accepted (perhaps "hoped" would be more accurate here) that the screening of the "true" charges would operate in a universal and field-independent way, and could therefore be consistently absorbed once and for all into a uniform redefinition of electric charge. Of course, this maneuver had the inevitable consequence of making the "true" charges appearing in the Hamiltonian cutoff dependent (a situation which persists to the present in local quantum field theories), and the unconscious presupposition that these "true" charges were somehow physically meaningful could only be satisfied by the expedient, considered desperate at the time, but now understood (cf. Chapter 16) to be an ineluctable feature of any realistic field theory, of assuming an actual breakdown of the theory at some high momentum, which would then cut off the divergent integrals and allow these underlying charges to take finite values.

Another classic example of the dominance of the hole-theory *language*, even when the results were equivalent to those obtainable via a second-quantized formalism with only positive-energy electrons and positrons, is Weisskopf's own calculation of the divergent self-energy of the electron in quantum electrodynamics in 1939 (Weisskopf, 1939), which is phrased throughout in hole-theory language, despite the fact that the subtractions performed to remove the unpleasant—and clearly unobserved—attributes of the negative-energy sea precisely correspond to the rewriting in terms of electron and positron operators suggested by Fock, suggesting that the lessons of second quantization have, at least subliminally, been absorbed. The second-order (in the electron charge e) correction to the energy of an electron at rest (with momentum $\vec{0}$ and spin σ) arises from a Coulomb self-energy term, corresponding to the diagonal matrix element of the Coulomb energy in first order,

$$\Delta E_{\text{Coul}} = \langle \vec{0}\sigma | H_{\text{coul}} | \vec{0}\sigma \rangle \tag{2.53}$$

$$H_{\text{coul}} = \frac{1}{2} \int \frac{\rho(\vec{r})\rho(\vec{r}')}{4\pi|\vec{r} - \vec{r}'|} d^3r\, d^3r' \tag{2.54}$$

where $\rho(\vec{r}) \equiv J_0(\vec{r})$ is the charge density operator (zeroth component of the four-vector current J_μ), and a transverse part ΔE_{tr} coming from the appearance (to second order) of the interaction Hamiltonian U_{tr} for the coupling of physical transverse photons to the electron charge current $\vec{J}$ (cf. Section 15.3),

$$H_{\text{tr}} = \int \vec{J} \cdot \vec{A}\, d^3r \tag{2.55}$$

From (2.53, 2.54) follows directly

$$\Delta E_{\text{Coul}} = \frac{1}{2} \int \frac{\tilde{G}(\vec{\xi})}{4\pi|\vec{\xi}|} d^3\xi \tag{2.56}$$

$$\tilde{G}(\vec{\xi}) \equiv \int \langle \vec{0}\sigma | \rho(\vec{r} - \frac{1}{2}\vec{\xi}) \rho(\vec{r} + \frac{1}{2}\vec{\xi}) | \vec{0}\sigma \rangle d^3r \tag{2.57}$$

The charge density operator is given in terms of the quantized field for the electron $\psi(\vec{r})$ (for the present calculation of the electrostatic self energy, only the field at time zero is needed) by $\rho(\vec{r}) = e\psi^\dagger(\vec{r})\psi(\vec{r})$. For practitioners of Dirac hole theory, this field was written as a single sum (over discrete momentum modes, with box normalization, where the single index q contains a spatial momentum $\vec{q}$, a spin index σ and a discrete energy sign index to distinguish between positive-energy and negative-energy modes)

$$\psi(\vec{r}) = \sum_q \phi_q(\vec{r})a_q \tag{2.58}$$

involving only destruction operators for either positive or negative-energy electrons. Here, the c-number coefficient functions $\phi_q(\vec{r})$ are single-particle solutions to the free-particle Dirac equation. In second-quantized language, introducing destruction operators for electrons $a_q \to b_{\vec{q}\sigma}$ (for the positive-energy modes in (2.58)) and positrons $a_q \to d^\dagger_{\vec{q}\sigma}$ (for the negative-energy modes, destruction of one of which in the filled Dirac sea corresponding to positron—i.e., hole—creation), the field is written instead

$$\psi(\vec{r}) = \sum_{\vec{q}\sigma}(u_{\vec{q}\sigma}(\vec{r})b_{\vec{q}\sigma} + v_{\vec{q}\sigma}(\vec{r})d^\dagger_{\vec{q}\sigma}) \tag{2.59}$$

with the wavefunctions ϕ_q relabeled as u (resp. v) for positive (resp. negative) energy solutions. The expression $\rho(\vec{r}) = e\psi^\dagger(\vec{r})\psi(\vec{r})$, taken literally, of course, contains an infinite background charge in the vacuum due to the negative-energy sea. If we insert (2.59) in this formula, and reorder the charge density via the "Klein–Jordan trick" of normal-ordering, whereby all destruction operators are moved to the right of all creation operators (with a change of sign for each interchange required, as we are dealing with fermions), one finds

$$\rho(\vec{r}) =: \rho(\vec{r}) : + \sum_{\vec{q}\sigma} ev^\dagger_{\vec{q}\sigma}v_{\vec{q}\sigma} =: \rho(\vec{r}) : + \sum_{\vec{q}\sigma} e \tag{2.60}$$

with the divergent second term on the right-hand side the sum of the (negative) charges for each electron in a filled negative-energy state. This term arises from reordering terms of the form $d_{\vec{q}\sigma}d^\dagger_{\vec{q}'\sigma'} = \delta_{\vec{q}\vec{q}'}\delta_{\sigma\sigma'} - d^\dagger_{\vec{q}'\sigma'}d_{\vec{q}\sigma}$ appearing in $\rho(\vec{r})$, using the anticommutation relations (2.45). By contrast, the normal-ordered charge density $: \rho(\vec{r}) :$ vanishes as physically required in the vacuum state, as the destruction (resp. creation) operators are deployed on the right (resp. left) side of the expression, and therefore encounter immediately the vacuum state $|0\rangle$ (resp. $\langle 0|$), giving zero. The subtractions performed by Weisskopf amount to the replacement of the charge-density operator $\rho(\vec{r})$ in (2.54) by its normal-ordered version $: \rho(\vec{r}) :$. When this normal-ordered expression is used in the evaluation of the charge–charge correlation function $\tilde{G}(\vec{\xi})$ defined in (2.57), one finds, returning to infinite volume and continuous momenta,[20]

$$\tilde{G}(\vec{\xi}) = e^2 \int \frac{d^3q}{(2\pi)^3} \frac{m}{E(q)} e^{i\vec{q}\cdot\vec{\xi}} \tag{2.61}$$

---

[20] The calculation is considerably simplified by using Wick expansion techniques described in Chapter 10; one also needs the appropriate normalization properties of the Dirac spinor functions $u_{\vec{q}\sigma}, v_{\vec{q}\sigma}$, defined and discussed in Chapter 7. See Chapter 10, Problem 5.

On the other hand, if the positron contributions are ignored, one finds a contribution to $\tilde{G}(\vec{\xi})$ proportional to $\delta^3(\vec{\xi})$ (due to the appearance of an extra factor of $\frac{E(q)}{m}$ in the integral), which leads, when substituted into (2.56), to the classic linear divergence in the Coulomb self-energy, as found previously by Heisenberg and Pauli. Inserting (2.61) into (2.56), one finds instead the logarithmically divergent integral[21]

$$\Delta E_{\text{Coul}} = \frac{e^2}{2} \int \frac{d^3q}{(2\pi)^3} \frac{m}{E(q)} \int \frac{e^{i\vec{q}\cdot\vec{\xi}}}{4\pi|\vec{\xi}|} d^3\xi = \frac{e^2}{2} \int \frac{d^3q}{(2\pi)^3} \frac{m}{E(q)q^2} \qquad (2.62)$$

Weisskopf was also able to show in his 1930 paper (correcting an earlier error pointed out by Furry in which he had found a quadratic divergence) that the other, transverse contribution to the electron self-energy was likewise given in terms of a logarithmically divergent integral, and that logarithmic divergences of this kind persisted to all higher orders of perturbation theory (in the electron charge).

The lack of manifest covariance in Weisskopf's calculation,[22] performed only for an electron at rest, with the electromagnetic field in radiation gauge, concealed the crucial fact that the lowest-order correction to the self-energy of an electron in motion, with momentum $\vec{p} \neq 0$, would take the form $\delta E(p) \sim \frac{1}{2E(p)}\delta m^2$, with δm^2 a divergent shift in the squared rest-mass, corresponding to the change $E(p) = \sqrt{\vec{p}^{\,2} + m^2} \rightarrow \sqrt{\vec{p}^{\,2} + m^2 + \delta m^2}$. In other words, the disturbing ultraviolet divergences appearing in the electron self-*energy* were really divergences in the (Lorentz-invariant) rest-*mass*, and could therefore be removed by a (admittedly divergent) redefinition—or *renormalization*—of the "bare" mass m appearing in the defining Hamiltonian of the theory. As we shall now see, this crucial realization, essential for a consistent formulation of quantum electrodynamics, would come only after another decade had passed, with the appearance of a fully covariant formulation, and more importantly, a *transparent calculational scheme* vastly simplifying the otherwise onerous higher-order calculations needed for a full understanding of the theory.

The wartime years 1939–45 brought an almost complete halt to research in fundamental issues in physics—such as the issues of consistency and calculability in quantum electrodynamics—as the discovery of nuclear fission in 1939 redirected the attention of the leading practitioners of subatomic physics to the urgent question of the military applicability of the potentially vast (and perhaps accessible) stores of energy in the nuclei of atoms. One important development in this period was Heisenberg's introduction (Heisenberg, 1943a,b, 1944) of the concept of the S-matrix, which attempted to replace a detailed microscopic prescription of the Hamiltonian dynamics of a quantum system (with the concomitant appearance of apparently intractable divergences) with a specification of only the phenomenologically "observable" aspects—in particular, the unitary scattering (or S-) matrix encoding the amplitudes with which particular

---

[21] Inserting a cutoff at $|\vec{q}| = \Lambda$ in the integral, with $\Lambda >> m$, one finds $\int \frac{d^3q}{E(q)q^2} \sim 4\pi \int_m^\Lambda \frac{dq}{q} \sim 4\pi \ln(\Lambda/m)$.

[22] In a footnote, Weisskopf admits that the direct calculation of the energy shift for electrons in motion is complicated by ambiguities in the subtraction of quadratic divergences appearing at various stages of his calculation. These concerns could, and would only, be put to rest with the development of a manifestly Lorentz-covariant formulation of QED in the late 1940s.

incoming states resolve to particular outgoing states in a scattering event. This project would receive an extended (but nonetheless finite) rebirth in the late 1950s and 1960s in the S-matrix theory approach to strong interactions, as frustration with the inability of field-theoretic methods to yield useful quantitative descriptions of strong-interaction processes mounted. But the resolution of the divergence difficulties of quantum electrodynamics in the late 1940s and early 1950s, together with the new beautiful and powerful perturbation-theoretic apparatus which (owing to the small value of the fine-structure constant, $\alpha \sim 1/137$), allowed ever more accurate calculation (in agreement with ever more accurate experiments!) of measured quantities such as the hydrogen fine-structure and electron magnetic moment, meant that the S-matrix would remain a useful auxiliary, if not the central, quantity in quantum electrodynamics.

In June 1947, the National Academy of Sciences of the US sponsored a three-day conference (25 participants, with an emphasis given to the younger generation of theorists, who dominated the conference numerically, although, as we shall see, it was the experimental results reported there by Rabi and Lamb that had a really dramatic effect in stimulating theoretical progress) on "The Foundations of Quantum Mechanics", to be held on a small island, Shelter Island, at the tip of Long Island in New York State. The three rapporteurs chosen to lead the discussions were V. Weisskopf, J. R. Oppenheimer, and H. A. Kramers. For these physicists (and in contrast to present usage) the term "foundations of quantum mechanics" meant primarily, not quantum measurement theory, but rather the accumulated difficulties and confusions of the preceding two decades in developing consistent quantum field theories to describe electrodynamics (and to a lesser extent, the strong and weak interactions). Weisskopf, in his abstract prepared and distributed in advance of the meeting, was quite explicit about these failures: "Certain well known attempts have been made in the last fifteen years to overcome a series of fundamental problems. All these attempts seem to have failed at an early stage."[23] Weisskopf explicitly mentions the need for obtaining finite results in a "reliable" way in the presence of divergent contributions to the electron self-energy and to the vacuum polarization. Kramers, in his abstract, also emphasized the need for a consistent treatment of divergences in "hole theory", and mentions in passing that the meson theory of nuclear forces offered no respite from similar difficulties, but rather "brought new divergence sorrows".

The first day of the Shelter Island conference was primarily given over to the new experimental results of Lamb and Rabi on hydrogen spectroscopy. In particular, the discovery of an unequivocal deviation from the hydrogen fine structure given by the Dirac theory, in which states of equal j (electron total, i.e., spin plus orbital, angular momentum) and neighboring l (orbital angular momentum) were exactly degenerate, was presented by Lamb in his measurement of the 2S–2P splitting (for $j = \frac{1}{2}$), which corresponded to an energy of order α^3 Rydbergs, in contrast to the Dirac formula for the hydrogen relativistic fine structure, in which only even powers of the fine-structure constant appear:

---

[23] See Schweber, op. cit., for a detailed account of the run-up to Shelter Island and the discussions in the conference itself.

$$E_{nj} = mc^2\{(1 + (\frac{\alpha}{n - j - \frac{1}{2} + \sqrt{(j + \frac{1}{2})^2 - \alpha^2}})^2)^{-1/2} - 1\}, \ j + \frac{1}{2} \leq n, \ n = 1, 2, 3, \ldots$$

$$(2.63)$$

It was clear to all the participants that a reliable calculation of this new "Lamb shift", requiring the subtraction of the divergent self-energy corrections for the electron in two distinct atomic bound states, would be an ideal test of the adequacy of any proposed quantum electrodynamic theory, inasmuch as the desired finite-energy shifts would have to be very carefully disentangled from the divergent electron self-energy contributions which were sure to appear in any higher-order calculation.

On the second day of the conference, Kramers gave a very important talk in which the essential conceptual content of mass renormalization was very clearly laid out, albeit in the context of a purely classical theory of a non-relativistic electron interacting with the electromagnetic field. Kramers emphasized—and Bethe's calculation of the Lamb shift just two days later, on the train home, showed that his arguments fell on fertile ground—that the measured mass of the electron should be regarded as already containing the divergent self-energy contributions, and that calculations should be reorganized to express the desired physical observables in terms of this physical mass, rather than the "intrinsic" or "bare" mass appearing in the Hamiltonian. During the conference it was realized that the weak logarithmic divergence in the electron self-energy would in fact cancel in the calculation of the energy difference ΔE between the $2S_{1/2}$ and $2P_{1/2}$ states (as the electron in both states receives the same self-energy correction)—a point emphasized by Weisskopf in his report on the divergence difficulties of hole theory. On the final day of the conference, Feynman presented his spacetime (in modern language, "path-integral") approach to quantum mechanics, which would lead within a year to his reformulation of quantum electrodynamics in terms of Feynman diagrams and a set of explicitly relativistically covariant calculational rules.

The calculations by Bethe of the Lamb shift (immediately following the Shelter Island conference) were performed for a non-relativistic electron, for which the self-energy corrections are *linearly* rather than logarithmically divergent (as the momentum integral in (2.62) becomes linearly divergent in the non-relativistic limit when we replace $E(q) \to mc^2$), with the result that the energy shift ΔE calculated by Bethe still contained a logarithmic divergence. Given that the correct relativistic treatment converts the linearly divergent behavior of the integral in (2.62) to logarithmic once $q > mc$, Bethe simply introduced a cutoff in his logarthmically divergent integral at $q \sim mc$, obtaining a finite result which agreed very well with Lamb's measurements (1040 MHz for the associated frequency, as compared to the observed 1000 MHz).

But the need for a relativistically correct calculation was urgently felt by all the theorists now engaged in the hunt for a fully consistent quantum electrodynamics. Bethe's conversations with Feynman at Cornell in the next few months provided a strong impetus for the latter's development (Feynman, 1949*a,b*) of a manifestly relativistic classical *Lagrangian* formalism, extended to quantum electrodynamics by the sum-over-histories (path-integral) approach that Feynman had already developed for ordinary quantum mechanics.

By the time of the Pocono conference in April 1948, relativistically invariant formulations of the theory (operator rather than path-integral based) had been independently developed by Julian Schwinger (who applied his methods to the calculation of the order α correction to the magnetic moment of the electron (Schwinger, 1948a,b)) and by Sin-itiro Tomonaga in Japan, who had already in 1943 produced a relativistically invariant formulation of field theory, applicable to QED, and only belatedly published in the West in 1946 (Tomonaga, 1946). Schwinger's presentation of his interaction-picture calculations at Pocono were widely regarded as a *tour de force* of computational power and elegance, while Feynman's much more intuitive spacetime diagrams met with considerable suspicion. Within a few years, however (and especially after the contributions of Dyson (Dyson, 1949), involving a graph-theoretic analysis of the divergence structure of the theory to all orders), the theoretical community had wholeheartedly embraced the Feynman approach, which came to be regarded (much in the same way as Schrödinger's wave mechanics superseded by sheer intuitive transparency, as well as computational efficacy, the matrix mechanical approach) as a far more usable technology in practical calculations than the highly formal Schwinger approach.[24]

The full story of the final transition to a relativistically invariant quantum electrodynamics, in which all divergences are absorbed into redefinitions of the physical constants defining the theory, has received an excellent and comprehensive treatment in the previously cited work of Schweber (Schweber, 1994), so we will leave our historical account at this point. In the constructive (and ahistorical!) rebuilding of quantum field theory which follows in the rest of the book, the need to maintain Lorentz-invariance will be a central part of our development of the theory. The other critical physical input—locality, or the absence of action at a distance (in the relativistic sense, as transmission of physical effects over space-like separations)—will also be inserted at a very early stage, and with it the particle–antiparticle symmetry which, given the persistence of hole theory up to the late 1940s, required almost two decades to be fully appreciated in the early history of quantum electrodynamics.

---

[24] For excellent treatments of the genesis and later spread of the use of the Feynman diagrammatic approach, see (Wüthrich, 2010) and (Kaiser, 2005).

3

Dynamics I: The physical ingredients of quantum field theory: dynamics, symmetries, scales

In the preceding chapters we have presented an all too brief review of some of the critical episodes in the historical evolution of modern quantum field theory, up to the point where renormalized covariant quantum field theory, epitomized by the astonishing quantitative successes of quantum electrodynamics beginning in the late 1940s and continuing to the present day, reached a state of technical (if not conceptual) completion. While this historical account is remarkably fascinating in its own right, it runs somewhat at cross-purposes to the account of field theory which is the major motivation of this book: namely, to present local quantum field theory as the *natural, and in a certain sense, almost inevitable* framework arising from the application of a few basic principles which lie at the very core of modern physical science. These principles fall into three basic categories: those involved in the specification of the *dynamics* of the sought-for microphysical theory, those concerned with the specification of the *symmetries* of the theory, and finally, those principles having to do with the behavior of the theory at different distance (or energy/momentum) *scales.* The rest of the book is therefore organized with a view to exploring how different conceptual strands in each of these three areas are woven together to produce the fabric of modern quantum field theory. In contrast to the procedure followed in the first two chapters, our approach for the rest of the book will be resolutely *antihistorical*: we shall introduce the basic principles from which relativistic quantum field theory can be constructed with little or no attention to the role played (explicitly or implicitly) by the invocation of such principles in the actual historical record. In particular, the order in which topics are discussed will have in general no connection to the actual historical sequence of events discussed in the "Origins" section of the book. In this chapter the main themes will be introduced, as far as possible, in a non-technical and qualitative (but, we hope, illuminating) manner. The technicalities will appear in proper course in the ensuing chapters!

Local relativistic quantum field theory is based on three basic principles which in combination lead to a powerful and elegant formalism which appears to allow a remarkably accurate description (the so-called "Standard Model") of at least three of the four fundamental forces in Nature: the strong, weak, and electromagnetic interactions.

1. Quantum mechanics: in a nutshell, the notion of linear superposition of amplitudes, the probability interpretation of these amplitudes (squared), and unitary evolution of the quantum state to implement the dynamics of the theory.

2. Special relativity: the symmetry of Lorentz-invariance. While many other symmetries play a crucial role in the quantum field theories of present importance, it is the fundamental symmetry of invariance under the homogeneous Lorentz group (more completely, under its inhomogeneous extension, the Poincaré group) which gives quantum field theory many of its most characteristic features.

3. Clustering: insensitivity of local processes to the distant environment. Here, the issue of the behavior of the theory at different distance scales (specifically, at long distances) becomes the crucial constraining factor, leading, in combination with the first two principles, to the characteristic features of relativistic quantum field theory. The clustering property as such is not specific to relativistic theories: we shall see that in the restricted context of such theories it is intimately linked to (but not synonymous with!) a special property of "locality" (or "microcausality"), which will ensure, among other things, the "Einstein causality" of the theory: namely, the absence of faster-than-light propagation of physical signals or effects.

Items 1 and 2 above are, of course, the basic ingredients of modern physics. One often encounters the assertion that quantum field theory arises from the marriage of these two. In fact, the addition of special relativity to quantum mechanics leads to no remarkably novel physics. Later we shall see that it is quite easy to write down scattering amplitudes which fulfill both the requirements of unitarity and Lorentz invariance. In a sense, such theories are just as unconstrained as non-relativistic quantum theory prior to the addition of the principle of special relativity: e.g., Hamiltonians can be written in terms of essentially arbitrary covariant functions of momenta, much as we are allowed to invent potential energy functions with abandon in elementary non-relativistic quantum theory.

The characteristic phenomena of relativistic field theory only appear once we insist on the third principle: clustering, i.e., the factorization of the S-matrix[1] containing the scattering amplitude for an arbitrary process as the product of two independent amplitudes in the event of two spatially far separated scattering subprocesses. This principle, which seems intuitively obvious, is surely a precondition for the success of experimental science. It relieves us of the obligation to specify completely the state of the entire world outside the laboratory prior to a correct interpretation of the results of an experiment. However, the inclusion of item 3 greatly increases the complexity of the resultant formalism, and means that it is no longer possible to write exactly S-matrices satisfying all the desired properties in spacetimes of more than 1 space-1 time dimension. Rather, we must resort to various approximative schemes. This is the bad part. On the other hand, the inclusion of the clustering requirement means, as we shall see, that the construction of an appropriate Hamiltonian (dynamics) is

---

[1] A precise definition of this object will be provided later, in Chapter 4.

now far more constrained. Arbitrary interaction potentials are no longer allowed: the potential between far-separated electric charges is *forced* to be $1/r$ and not $r^{-3.5}$, etc. Moreover, we are led ineluctably to the formalism of local quantum field theories, with two immediate and unavoidable consequences:[2]

(a) an explanation of the existence of antimatter, with each particle having an antiparticle of exactly equal mass and opposite additive quantum numbers,[3] and

(b) the Spin-Statistics theorem, which clarifies one of the great mysteries of non-relativistic quantum theory: the contrasting symmetry properties of the wavefunctions of particles of integer (bosonic) versus half-integer (fermionic) spin.

A simple and intuitive picture of the emergence of antimatter as a natural consequence of the basic physical ingredients of local field theory goes back to the work of Feynman (Feynman, 1949b) on quantum electrodynamics in the late 1940s. The results cited in item (a) above are special cases of the more general TCP theorem valid in any local relativistic quantum field theory: the invariance of scattering amplitudes under simultaneous interchange of particles with antiparticles (the "C" operation), spatial reflection (or parity, the "P" operation), and time reversal (the "T" operation). A beautifully simple argument to illustrate property (a) has been given by Weinberg (Weinberg, 1972), although, as pointed out above, the underlying ideas were first elucidated by Feynman. Consider a process such as that illustrated in Fig. 3.1, where a positive pion (π^+) emitted by a proton (P) at spacetime point x travels to a neutron (N) and is absorbed at spacetime point y. The idea of locality here amounts to the statement that the neutron and proton interact via *local* emission and absorption events of a third intermediary particle. On the one hand, the mutual indeterminacy

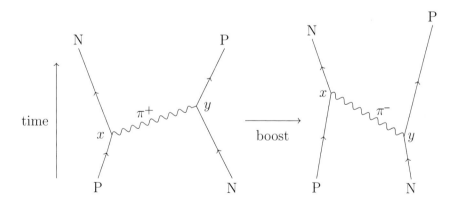

Fig. 3.1 Frame-dependence of a simple exchange process.

[2] The primary character of these results is emphasized, and rigorous proofs given, in the seminal work of Streater and Wightman (Streater and Wightman, 1978).

[3] Recently, the charge-to-mass ratio of the antiproton and proton was measured by Penning trap techniques to be equal to within about 1 part in 10^{12}!

of position and velocity in quantum mechanics allows for the possibility that the spacetime points x and y are actually space-like separated ("tunneling outside the classical light-cone", as it were).[4]

But if the interval between the emission and absorption is space-like, relativity tells us that it is possible to find an inertial frame in which $y^0 < x^0$, i.e., spacetime point y *precedes* spacetime point x, so the *same* event appears in this new frame as in the figure on the right. An observer in the new frame will naturally interpret this as the *emission* of a particle (at y) from the neutron, turning it into a proton. Such a particle must be *negatively* charged, if we are to maintain charge conservation, but with the same mass (as its kinematics is identical to that of the original π^+, with a spatially and temporally reversed path). This particle is just the π^-, the antiparticle for the original positively charged pion. This example contains in a nutshell the intimate association between spatiotemporal reflection and particle–antiparticle interchange characteristic of local theories and exemplified in the TCP theorem.

Insofar as the characteristic features of relativistic field theory require at a minimum the implementation of unitary quantum dynamics, Lorentz symmetry, and locality, our exploration of the conceptual framework of field theory must begin with a detailed examination of these physical ingredients. This will allow us to build up the technical framework appropriate to the task of weaving together the desired physical properties into a unified and consistent dynamical theory. This will be our object in this second section of the book (entitled "Dynamics", Chapters 3–11), where we shall concentrate on the most general features shared by essentially all relativistic local quantum field theories (which we henceforth denote "LQFTs").

The necessary input from quantum theory will be reviewed in Chapter 4, which will also contain a brief review of those results from quantum scattering theory needed for later development of the theory. Chapter 5 describes the kinematics of relativistic quantum mechanics, which incorporates the requirements of Lorentz symmetry (but not yet the clustering principle), leading to an enormous class of interacting theories almost all of which display bizarre and completely unphysical long-distance behavior. A natural way to restrict the form of the interactions—by introducing local fields— is introduced here and shown to incorporate the requirements of Lorentz-invariance (though, as yet, with no proof of the desired clustering properties). The restriction to physically sensible theories compatible with the clustering principle is effected in Chapter 6, which shows how the huge class of quantum theories incorporating special relativity can be systematically "pruned" to yield theories which display

---

[4] The reader may be momentarily disturbed by the apparent superluminal transmission of influence by the exchanged pion, which would seem to run counter to the requirement that physical signals/effects can only be transmitted at most at the speed of light ("Einstein causality"). Here, as in the EPR paradox, it is important to keep in mind that quantum theory is fundamentally a theory of the statistics of microscopic processes, and that the formalism can (and does!) contain apparently non-local features on an event by event basis, provided only that these features do not result in a measurable transmission of statistically measurable properties at faster than light speed. In the quantum information community, this is referred to as the "no-signalling" property of quantum mechanics. We shall see later, in Chapter 9, that measurements performed in two space-like separated domains of spacetime are guaranteed to yield statistically independent results, as a rigorous consequence of microcausality: i.e., the property of space-like commutativity of local field operators used to construct the hermitian operators embodying the said measurements.

sensible long-distance behavior, purged of bizarre action-at-a-distance effects.[5] This chapter also explores the connection between the concepts of clustering and locality (or microcausality), which are intimately related but not synonymous, the connection between locality and the smoothness (i.e., analyticity) properties of amplitudes in field theory, and the limitations of the localization concept in regards to particle states (as opposed to fields).

The construction of local covariant fields which incorporate in a natural way the requirements of Lorentz-invariance is the topic in Chapter 7, where we see that the mysterious plethora of *ad hoc* field equations (Klein–Gordon, Dirac, Maxwell–Proca, etc.) encountered in many texts arise inescapably from a straightforward analysis of the unitary representations of the Poincaré group. Here we shall see that such fields provide a convenient set of ingredients for the construction of local Hamiltonian energy densities, describing particles of arbitrary mass and spin, with the important special cases of massive particles with spin 0, $\frac{1}{2}$, and 1 worked out in detail (the peculiarities of the massless case are also discussed).

In Chapter 8 the question of the classical limit of quantum fields is examined: we discuss the mutual measurability of field observables, and the types of states of the field for which quasi-classical behavior is recovered. Here also, we discuss the energetic stability of quantum field theories, and encounter the related phenomenon of spontaneous symmetry-breaking for the first (but not the last!) time.

In Chapter 9 we come to grips for the first time with the intricacies of interacting field theories: our emphasis again will be on very general aspects common to all LQFTs. The basic concept here is that of the interpolating Heisenberg field in terms of which the dynamics of the theory is specified, but which may be connected in a variety of ways[6] to the actual physical particle states. At this point a characteristic (and for beginning students, frequently baffling) feature of LQFTs becomes apparent: namely, the absence of any preferred, one-to-one connection between particles and fields. The discussion of field theory in the Heisenberg picture is first carried out in an "heuristic" fashion, ignoring some important mathematical fine points, and then from a rigorous axiomatic point of view, starting with the Wightman axioms (spectral and field) (Wightman, 1956), and proceeding, via the Haag–Ruelle formulation of scattering theory, to the asymptotic formalism of Lehmann, Symanzik, and Zimmermann (Lehmann *et al.*, 1955). The latter is treated in some detail, as it is central to subsequent discussions in this chapter of the nature of the state space of field theory (as it depends on the presence of stable or unstable, elementary or composite particles in the theory). In Chapter 9 we also discuss the spectral properties of field theory, and the connection between the internal dynamics as specified by the interpolating fields and the phenomenological content of the theory as encapsulated in the asymptotic particle states and the S-matrix.

---

[5] By "action-at-a-distance" effects, we do *not* refer here to the psychologically unsettling effects involving non-local transitions in entangled wavefunctions, commonly referred to as the "EPR paradox", but to *physically observable* non-local phenomena: namely, those leading to superluminal transmission of physical signals. See the preceding footnote.

[6] For example, in the case of confinement, discussed in Chapter 19, the theory contains fields which do not correspond to finite-energy particle states at all!

Chapter 10 provides an introduction to perturbative aspects of interacting field theory: namely, the techniques appropriate for studying those aspects of LQFTs which emerge from a formal asymptotic expansion in some parameter (typically, a coupling constant) of the theory, both from an operatorial as well as a path-integral point of view. Topological aspects of the graphical expansion of field theory amplitudes are discussed in some detail, as well as the psychologically disturbing (but ultimately irrelevant) Haag's theorem. Some of the material here is, of course, to be found in essentially all introductory texts on quantum field theory, but is reviewed in order to lay the basis for important conceptual discussions later in the book. In particular, the technology of perturbative expansions of field theory becomes essential in the discussion in Chapter 11 of "non-perturbative" aspects of interacting field theory—those phenomena which are *not* appropriately described by a finite number of terms in the asymptotic expansion of a field theory amplitude in powers of coupling (or interaction) strength. Such expansions are *always* asymptotic only, corresponding to divergent series lacking a well-defined sum to all orders. One can therefore make a useful distinction between phenomena which (a) require the summation of an infinite number of terms extracted from a perturbative expansion, but (b) in which the extracted terms form a convergent series, with a well-defined summand, which represents in some precise sense the leading contribution to the desired amplitude (the "perturbatively non-perturbative" processes discussed in Section 11.2), and processes which require information beyond that available in even an infinite number of terms in the perturbative expansion (the "essentially non-perturbative" processes of Section 11.3). Non-relativistic threshhold bound states in gauge theories provide an example of the former, while quark confinement in four-dimensional quantum chromodynamics is an example of the latter.

The central role of symmetry and invariance principles in quantum theory generally, and in quantum field theory in particular, is now considered self-evident. One tends to forget nowadays that the use of symmetry ideas, fully exploiting the natural mathematical framework of group theory in order to express these ideas, was in the early days of quantum mechanics (the late 1920s and 1930s) quite controversial. Many atomic theorists of this period viewed the introduction of group-theoretical ideas as a "Group-plague" ("Gruppenpest" in German): an excessively abstract formalism quite irrelevant to the extraction of useful information about atomic spectra—an attitude which pervades, for example, (Condon and Shortley, 1935). And indeed, to the extent that a quantum dynamical system can be fully specified and then solved (by either analytic or numerical methods), say by a complete diagonalization of the underlying Hamiltonian and direct computation of all needed matrix elements of relevant physical observables, symmetry arguments and group-theoretical reasoning are, strictly speaking, unnecessary. Needless to say, this is rarely the case in non-relativistic quantum theory, and essentially never the case for relativistic field theory.

Eugene Wigner, one of the pioneers in developing group theoretical methods in quantum theory, has emphasized (Wigner, 1979*a*) the special efficacy of such methods in quantum (as opposed to classical) physics. The increased complexity of the quantum mechanical state space for even a point particle (an infinite-dimensional complex Hilbert space) over the classical phase-space (a six-dimensional real space) together with the linear structure of the quantum theory (allowing linear superposition of quantum states) means that symmetry arguments can play a much more significant role in the resolution of the dynamics in a quantum mechanical problem than in a comparable classical problem.[7] For example, the invariance of the Hamiltonian under a symmetry operation means that it must commute with the operator generating the transformation. Such a commutation property already implies a partial diagonalization of the Hamiltonian, as matrix elements connecting states with different eigenvalues of the symmetry generator must vanish. In some cases (e.g., the $O(4)$ symmetry of the hydrogen atom, or in completely integrable quantum systems) the symmetry is sufficiently large to allow a complete resolution of the spectrum, or the dynamics. The third section of this book (entitled "Symmetries") will therefore examine some of the important ways in which symmetry considerations are woven into the fabric of modern quantum field theory.

There is an important distinction between *spacetime symmetries*, involving symmetry transformations affecting exclusively the universal underlying spatiotemporal framework of field theories, and *internal symmetries* in which the symmetry transformations act in specific non-geometrical ways on the assorted fields in the theory.[8] By far the most important example of the first type of symmetry is Poincaré invariance— item 2 in the discussion at the beginning of this Chapter—whereby the physical content of special relativity is injected into relativistic field theory. The extension of this symmetry to supersymmetry (SUSY), wherein the Poincaré group is enlarged to a graded extension and spacetime to an enlarged "superspace" containing conventional space and time as well as a Grassmannian component, should probably be included in this category, purely on the basis of the extremely powerful formal analogy between operations carried out in normal spacetime and the extended superspace of SUSY. In Chapter 12, devoted to continuous spacetime symmetry, we develop the canonical formalism of Lagrangian field theory as the natural solution to the problem of generating, in as painless a process as possible, Hamiltonian energy densities that lead to a quantum field theory with fully Lorentz-invariant dynamics. The general connection between symmetries and conservation laws, expressed in the form most natural to field theory (Noether's theorem) is also given here, together with its application to the case of Poincaré symmetry, conformal symmetry, and global internal symmetries. Chapter 12 concludes with an introduction to the extension of Poincaré symmetry to the super-Poincaré algebra of supersymmetry.

Discrete spacetime symmetries (reflection or parity symmetry P, and time-reversal symmetry T) are treated in Chapter 13, together with charge-conjugation invariance symmetry C (a symmetry under interchange of particles and antiparticles): despite the "internal" appearance of the latter, the fact that we are dealing throughout with *local* theories immediately introduces an intimate and unbreakable connection with the P and T symmetries, making it natural to treat the C symmetry on the same

---

[7] For example, the application of a symmetry transformation to a possible classical phase-space trajectory will, of course, yield another possible classical trajectory, but does not directly assist in the explicit solution of either: on the other hand, the fact that symmetries imply conservation laws and hence invariants of the motion is clearly of great utility in resolving the dynamics in many important classical problems.

[8] The terminology here has evolved over time: Wigner (Wigner, 1979b) speaks in the first case of "classical" or "geometric" symmetries, and in the second, of "dynamical" or "non-geometric" symmetries.

footing with these. Our treatment of discrete spacetime symmetries concludes with proofs of the TCP and Spin-Statistics theorems, using techniques of axiomatic field theory introduced in Chapter 9.

Although discrete internal symmetries have played some role in constructing models of elementary particle interactions beyond the Standard Model, by far the most important internal symmetries have turned out to be the continuous ones corresponding to transformations which form compact, finite-dimensional Lie groups. Internal symmetries may either be "global", where the dynamics is invariant under an application of the same symmetry transformation to the field quantities at all spacetime points, or "local", in which the invariance persists even for spacetime dependent transformations. Evidently, every local (or "gauge") symmetry contains *ipso facto* a global subsymmetry. However, the presence of a local symmetry has extraordinarily deep ramifications for the dynamics of the theory displaying such a symmetry, far beyond the comparatively simple implications of global symmetry.

In Chapter 14 the role of global symmetries in LQFT is examined. We shall see that *exact* global symmetries are rare, indeed, if one takes gravitational effects into account, probably non-existent! Nevertheless, approximate global symmetries play an enormously important role in modern field theory. The appearance of massless Goldstone particles once an exact global symmetry is spontaneously broken is of enormous importance in modern field theory, and a proof of the Goldstone theorem embodying this phenomenon is given in Section 14.2. Dynamical aspects of spontaneous symmetry-breaking (SSB) are examined in Section 14.3, where we see that the essence of SSB resides in the energetics of the theory in the infrared (i.e., at long distances).

The additional rich structure introduced when a LQFT displays a *local* gauge symmetry is studied in Chapter 15, where we show how such symmetries require a generalization of the canonical Lagrangian/Hamiltonian formalism discussed in Section 12.3 in order to handle the presence of constraints entailed by the presence of local symmetries. The concept of a local symmetry is introduced in Section 15.1 with a simple example from classical mechanics, the lessons of which are extrapolated to a wide class of constrained Hamiltonian systems in Section 15.2, where we introduce the Dirac constrained Hamiltonian theory, and the Faddeev–deWitt functional quantization method for such systems. The quantization of gauge theories using this functional (path-integral) method is then explained, first using abelian gauge theory in Section 15.3, where the technical complications are minimal. The extension to non-abelian gauge theories is performed, again using path-integral methods (which in this case are vastly more efficient than the canonical operator approach) applied to the constrained Hamiltonian in Section 15.4, leading to the Feynman rules for general (unbroken) non-abelian gauge theories. The existence of quantum anomalies in the chiral currents of internal global symmetries is explored in Section 15.5, where we see that the classical current conservation implied by Noether's theorem may be violated by quantum effects, yielding a non-vanishing divergence of the Noether current explicitly proportional to Planck's constant. The peculiar features of spontaneous symmetry breaking in the presence of local (as opposed to global) gauge symmetry are the subject of Section 15.6, where we explain the famous "Higgs phenomenon" in the context of the electroweak sector of the Standard Model, and outline the derivation of the Feynman rules for a general spontaneously broken local gauge theory.

The class of theories obtained from the three basic requirements discussed at the beginning of this chapter turn out to display a very important feature: that of scale separation, which will here be vaguely defined as the weak coupling of physics at widely varying distance scales. This property, and its consequences, will be the central theme of the fourth, and final, section of this book, entitled "Scales". Much of the confusion over troublesome "infinities" which plagued the development of interacting field theory in the 1930s and 1940s, as described in Chapter 2, derived from a failure to appreciate this characteristic property of LQFTs.

Unlike the situation in classically chaotic systems, where small perturbations at very short distance scales can propagate rapidly up to much longer scales (the famous "butterfly in China leading to a hurricane in the Atlantic" effect), LQFTs can be "tailored" to accurately reflect the physics in some given range of length scales even if we are completely ignorant of the "true microphysics" which obtains at much shorter distances. This property is as indispensable to the theoretical success of field theory as the cluster decomposition property is for the practicability of experimental science. Our unavoidable ignorance—in a direct empirical sense—of the behavior of matter at distance scales much smaller than the reciprocal of the highest experimentally attainable particle momenta would be disastrous if we were dealing with theories in which complicated (and unknown!) details of the interactions at very short distances propagated up to the much longer scales presently accessible.

For example, there is no doubt that quantum gravity effects will drastically alter the structure of spacetime on distance scales corresponding to the inverse of the Planck mass, i.e., at distances below about 10^{-34} cm.[9] Nevertheless, quantum electrodynamics correctly predicts the anomalous magnetic moment of the electron to an astonishing nine significant figures, all in terms of integrals extending in principle up to infinite energy (or, in coordinate space, down to zero distance). Evidently, this remarkably accurate result means that the long-distance behavior of quantum electrodynamic systems must be *insensitive* to the detailed structure of the interactions at such very short distances.

Obviously, theories in which unknown short-distance structure infects the behavior of amplitudes at much longer scales would be as intractable from the point of view of theoretical predictability in quantum physics as chaotic systems are in the classical arena. So scale separation in the sense of the isolation of very short-distance physics (from the behavior at accessible scales) is as crucial to the formulation of successful theories as the isolation of long-distance effects entailed by clustering (item 3 above) was for the correct interpretation of experimental results. In Chapter 16 we discuss various aspects of scale separation: the critical role it plays in leading to quantitative predictions at accessible energy scales, the introduction of regularization techniques to quantify and simplify the study of scale sensitivity, the relevance of power counting methods in LQFTs, the extremely important concept of effective Lagrangians, and the classification of operators into relevant, marginal, and irrelevant on the basis of their scaling behavior. At this stage, the point of view first introduced by Wilson

---

[9] In theories with extra dimensions, the effective distance scale at which quantum gravity effects become significant can in fact be much larger than this value.

in the 1970s, whereby the physics of a system is described in terms of an "effective Lagrangian" which incorporates, via a set of "renormalization group" equations, the behavior over a strictly limited range of distance (or energy) scales, moves to the center of our discussion of quantum field theory.

Although, as just indicated, a general LQFT is only "designed" to represent microphysics in a limited range of length scales (typically, only down to a lower limit in distance, or up to a finite cutoff energy), there is a small subclass of local field theories in which the insensitivity to short-distance structure can be pushed up to very high energy scales indeed—in some cases, to infinity! Such theories are usually called "renormalizable" quantum field theories. This subclass can further be subdivided into "weakly" and "strongly" renormalizable theories (this is my language, not to be found in standard texts!). In weakly renormalizable theories, the insensitivity to short-distance structure of the interacting theory at arbitrarily small distances is valid only within the context of perturbation theory (an asymptotic expansion in the interaction strength in the theory), but fails when the full, non-perturbatively defined theory is considered. A famous example is the self-coupled $\lambda\phi^4$ interacting scalar field theory, to which we shall later return many times. In strongly renormalizable theories, the insensitivity to short-distance structure at arbitrarily short scales is valid even non-perturbatively: the effective field theory in such cases *could* in principle (ignoring inevitable quantum gravitational effects!) be regarded as a correct microphysics down to arbitrarily short distances, without any inconsistencies appearing in the quantum amplitudes. There appears to be only a single known example of a strongly renormalizable theory in this sense (in 3+1 spacetime dimensions): non-abelian gauge theory. This should not particularly worry us: as mentioned above, quantum gravity effects *necessarily* obliterate the Minkowski structure of spacetime assumed (item 2) in the whole construction of LQFTs anyway, once we reach distances on the order of the Planck length, and a completely new type of theory must emerge at that point. It is best to think of renormalizable theories (both kinds) just as LQFTs with a particularly weak (logarithmic) coupling between low and high energies (or distances). In fact, it can be shown (cf. Section 17.4) that the low-energy "residue" of an arbitrary LQFT is in fact *necessarily* a renormalizable (possibly free) field theory. More than anything else, this accounts for the historically central role played by renormalizable theories since the development of quantum electrodynamics in the late 1940s.

The proper technical instrument for the understanding of the scale separation features of LQFTs is called the *renormalization group*. This concept has found wide applications not only in elementary particle theory, but in the modern theory of critical phenomena[10] in condensed-matter physics, where the importance of scale separation can be seen directly from the existence of universal scaling laws for the long-distance behavior of correlation functions *independent* of fine details of the microscopic interactions in the system. In Section 16.4 we introduce the renormalization group in its most general form, appropriate for discussing its implications for LQFTs viewed as Wilsonian effective theories. In Chapter 17 the technical tools needed for the analysis of the perturbative renormalizability of a specific LQFT are introduced, and a proof

---

[10] The central role of the renormalization group in the understanding of second-order phase transitions was first set forth in the seminal work of K. Wilson (Wilson, 1971; Wilson and Kogut, 1974).

of cutoff-insensitivity is given, both using traditional graphical methods (i.e., the subtraction formalism of BPHZ (Zimmermann, 1969)), and from the point of view of effective Lagrangian theory.

The final two chapters of the book introduce the reader to some important aspects of the short-distance (Chapter 18) and long-distance (Chapter 19) structure of quantum field theory. In Chapter 18 we explain one of the most fertile (from the point of view of phenomenological impact) manifestations of scale separation in field theory: the Wilson operator product expansion (OPE) which provides a precise characterization of the short-distance asymptotics of field theory amplitudes in terms of factorized products of "short-" and "long-"distance terms. The useful application of the OPE in particular processes depends on the presence and structure of mass singularities in the relevant amplitudes—a topic which we address in Section 18.2. The role of the renormalization group in studying high-energy (or short-distance) behavior is outlined in Section 18.3.

Aspects of the long-distance behavior of field theory are studied in Chapter 19. In a theory with only massive fields, this behavior is essentially trivial: clustering is exponentially rapid, with the inverse of the smallest mass providing a length scale over which spatially separated processes decouple. The situation for theories with massless fields is radically different. Here, we need to distinguish between two important cases: when the massless field interpolates for a physical particle, and alternatively, when massless fields are present in the underlying Lagrangian dynamics but do not interpolate for physical particles. The former case corresponds to quantum electrodynamics, where we have, as far as we know, an exactly massless photon (and photon field). Indeed, the photon is the *only* massless particle for which we have any empirical evidence. The specific problematic issues arising with respect to introducing massless fields, charged particle states, and a well-defined S-matrix in this situation are explored in Sections 19.1 and 19.2. The second case mentioned above, where massless fields exist in the theory, but do not interpolate for physical particles, corresponds to quantum chromodynamics (QCD) and the physical phenomenon of color confinement. The massless gluon fields of this theory, as well as the massive quark fields, specify the Lagrangian dynamics of the theory but do not interpolate for finite-energy asymptotic states. This extraordinary behavior—apart from superconductivity, perhaps the most amazing and counter-intuitive phenomenon to emerge in twentieth-century physics—is explored in general terms in Section 19.3, where we introduce the basic concepts and techniques of lattice gauge theory, and in more detail in a toy model where the physical mechanism of confinement can be clearly exhibited using semiclassical arguments: namely, three-dimensional gauge theory.

We close this chapter with a comment on the role of LQFTs in the context of the focus in recent years on superstring theories as providing a possible framework for an ultimate microphysics (or "Theory of Everything"). In the last twenty years the attempt to develop a consistent quantum theory of gravity has led to the introduction of string theories which incorporate two additional physical principles which have recently attained central importance: supersymmetry (a symmetry between bosonic and fermionic particles), and duality (a symmetry connecting the weak and strong coupling sectors of the theory). The dynamics of such theories is fundamentally different from that of local quantum field theories, but to the extent to which the quantum amplitudes in a string theory display unitarity, Poincaré invariance, and cluster decomposition (as everyone certainly expects), we can rest assured that these amplitudes correspond *at long distance scales* to those derivable from some effective LQFT. At the present time the LQFT believed to describe microphysics up to energies of about one hundred GeV is called the "Standard Model", and appears to describe almost all (neutrino masses and the existence of a massive, stable dark-matter particle are intriguing exceptions!) the known features of strong, weak, and electromagnetic interactions in the experimentally accessible range. The dream of a TOE ("Theory of Everything")—a consistent microtheory including quantum gravity, and therefore capable of accurately describing physics at (and beyond!) the Planck scale, and also yielding the observed Standard Model at low energies—is the motivating force for the study of superstrings and their descendants (M-theory, p-branes, etc.). However, our ability to build up local field theory from just a few basic principles, which seem likely to be conserved[11] in any future theory, suggests that local quantum field theories will continue to provide an indispensable conceptual framework for understanding the vast majority of accessible microphysical processes.

---

[11] From this point of view LQFTs may be the analog in physics of the "conserved core processes" in Kirschner and Gerhart's theory of facilitated biological evolution (Kirschner and Gerhart, 2005).

4

Dynamics II: Quantum mechanical preliminaries

At one level, quantum field theories can be regarded as a very special subclass of all quantum theories: theories based on a kinematical structure consisting of a state space which is typically an infinite-dimensional complex Hilbert space, and a dynamical structure in which time-evolution is effected by a deterministic unitary transformation of the state vectors determined by the linear operator representing the energy of the system (the "Hamiltonian").[1]

In addition, this theoretical scaffolding needs to be supplemented with the stochastic postulate of quantum mechanical measurement theory: "detection" of a state $|\Psi\rangle$ (via interaction with a suitable macroscopic measurement apparatus) given a previously prepared state $|\Phi\rangle$ occurs with probability given by the absolute square of the inner product of the (suitably normalized) state vectors: $|\langle\Psi|\Phi\rangle|^2$. From the vast variety of possible quantum theories (distinguished by the structure of the Hilbert space representing the particular system under study, as well as by the variety of possible physically sensible Hamiltonians, measurable physical quantities, etc.) our object in this book is to select the minuscule subset of theories in which the relativistic invariance of special relativity is implemented exactly, and in which physical processes localized in space-like separated regions are strictly independent (i.e., no faster-than-light transmission of *physically measurable effects*). Our task in this chapter is to review and assemble just those parts of the basic underlying quantum-mechanical structure which will be critical in realizing these relativity and locality constraints in the following chapters. This will also serve as a convenient opportunity to introduce the reader to the particular notational idiosyncrasies of the author. We will begin with a review of the basic operator formalism underlying standard quantum mechanics, paying particular attention to dynamics (time evolution) and symmetries. Then we turn to the reformulation of quantum dynamics as a sum over histories (the "path-integral" approach) due to Feynman and Dirac, which has turned out to be of enormous conceptual and technical utility in quantum field theory. Finally, we review those aspects of quantum scattering theory which will be central in teasing out the intricate physical content of field theory.

---

[1] This text assumes that the reader is familiar with the basic formalism and technical apparatus of non-relativistic quantum mechanics, at the level of an advanced undergraduate or beginning graduate level. Conventions and notation used throughout generally coincide with those of Gordon Baym's excellent text "Lectures in Quantum Mechanics" (Baym, 1990).

4.1 The canonical (operator) framework

The modern formulation of quantum mechanics evolved over a period of roughly four years as a conceptual clarification and completion of the seminal papers of Heisenberg (June 1925) and Schrödinger (January 1926). In Heisenberg's formulation, which after formal amplification by Born and Jordan came to be called "matrix mechanics", the physical observables of classical mechanics (position, momentum, angular momentum, energy, and so on) were replaced by time-dependent matrices, and the normal algebraic operations whereby these quantities (real valued numerical functions of time in classical physics) are manipulated in classical theory were replaced by the corresponding matrix operations. The concept of a physical state as a vector in a Hilbert space is at most highly implicit in the founding papers of matrix mechanics, but gradually emerged in the subsequent years as the transformation theory of Dirac and Jordan took hold and was put on a rigorous mathematical basis by von Neumann. In particular, the wave-mechanical approach of Schrödinger (which was the original inspiration for the Dirac–Jordan transformation theory) made it natural to associate the dynamical development of a quantum system with the evolution of its state vector (or equivalently, the associated wavefunction) rather than with the physical observables, as in the Heisenberg–Born–Jordan approach. The dual character of quantum theory (involving both states and observables, which play different but complementary roles) makes a certain fluidity in the representation of the dynamics inevitable, as we shall now see.

4.1.1 Quantum dynamics: the Heisenberg, Schrödinger, and Dirac (interaction) pictures

In this section we are concerned only with describing the deterministic evolution of quantum systems isolated from the macroworld—in particular the stochastic modifications of the state arising from "measurements", i.e., interactions of the microsystem with a macroscopic apparatus capable of registering macroscopically distinguishable effects depending on the interaction with the microsystem, are not included in the description of the time-evolution of the system. As in Heisenberg's original matrix mechanics, we may assign the entire time-development of the quantum system to the operators corresponding to the physical observables, with the quantum state of the system fixed once and for all, say by specifying the state at time $t = 0$ (since the evolution is deterministic, the specification of the state at any time suffices to determine it at all other times, a situation with which Laplace would have been very happy!). The time-evolution of an operator O associated with an arbitrary physical observable is then given by

$$O_H(t) = e^{iHt/\hbar}O_H(0)e^{-iHt/\hbar} \tag{4.1}$$

where H is the self-adjoint operator representing the energy of the system: the "Hamiltonian". The Planck constant $\hbar = \frac{h}{2\pi}$ will henceforth be set to unity (for the rest of this book, with a few exceptions, natural units will hold sway: $\hbar = c = 1$). The subscript H appearing in (4.1) reminds us that we are in the "Heisenberg picture" of time development. In this picture, the quantum state of a particular system is a fixed vector $|\alpha\rangle$ in the Hilbert space appropriate for the system in question.[2] Taking the time-derivative of the finite time-evolution (4.1) yields the commutation property

$$\frac{\partial O_H(t)}{\partial t} = i[H, O_H(t)] \tag{4.2}$$

To put some meat on these rather abstract bones, consider a spinless point particle of mass m, described by non-relativistic kinematics, and moving on a one-dimensional line (say, the x-axis). In this case the Hilbert space of states is just the linear space of complex, Lebesgue square-integrable functions,

$$|\alpha\rangle \rightarrow \psi_\alpha(x), \quad \int_{-\infty}^{+\infty} |\psi_\alpha(x)|^2 dx < \infty \tag{4.3}$$

which is commonly denoted $L^2(R)$ (the R refers to the functions ψ_α being defined on the entire real axis: if our particle were constrained to move in the interval $a < x < b$, we would denote the corresponding Hilbert space $L^2(a,b)$). Of course, our linear space needs an inner product to be a Hilbert space, so if $|\beta\rangle$ is another state vector, representing the square-integrable function $\psi_\beta(x)$,

$$\langle \beta | \alpha \rangle = \int_{-\infty}^{+\infty} \psi_\beta(x)^* \psi_\alpha(x) dx \tag{4.4}$$

The Hilbert space $L^2(R)$ is separable (i.e., is spanned by a countable basis of orthonormal square-integrable functions), so if $|n\rangle \rightarrow \psi_n(x), n = 1, 2, 3, \ldots$ is such a basis, each physical observable, represented in the theory by a self-adjoint operator $O_H(t)$ can be completely specified[3] at time t by giving its *matrix*, i.e., the set of numbers

$$O_{H\,nm}(t) = \langle n | O_H(t) | m \rangle, \quad n, m = 1, 2, 3, \ldots \tag{4.5}$$

For the special case of systems such as the harmonic (or anharmonic) oscillators studied in Heisenberg's original work, the energy eigenstates of the Hamiltonian

$$H|n\rangle = E_n|n\rangle \tag{4.6}$$

form such a complete orthonormal basis (in other words, the spectrum of the Hamiltonian is completely discrete), and the matrix elements of any physical observable O in this basis have a purely oscillatory time-dependence determined by the energy differences between the states:

$$\langle n | O_H(t) | m \rangle = \langle n | e^{iHt} O_H(0) e^{-iHt} | m \rangle = e^{i(E_n - E_m)t} \langle n | O_H(0) | m \rangle \tag{4.7}$$

---

[2] We will use the Dirac bra-ket notation throughout this book: see Baym, op. cit.

[3] Contrary to assertions in many texts, this is true even for operators with a partially or fully continuous spectrum: matrix mechanics is *not* restricted to situations where the spectrum is fully discrete! Of course, in many, indeed most, cases the coordinate space representation of wave mechanics is technically more convenient.

The discovery of matrix mechanics by Heisenberg was occasioned by his recognition of the appearance in (4.7) of the appropriate time-dependence for emitted radiation in atomic transitions: an electron in an atom transitioning from a state $|n\rangle$ to a state $|m\rangle$ emits electromagnetic radiation with frequency given by the Bohr condition $\nu_{n \to m} = (E_n - E_m)/h$ (rather than at the frequencies given by Fourier analyzing the classical motion in Bohr orbits, as predicted by the old quantum theory).

In the Schrödinger approach to quantum mechanics, the dynamical evolution of a quantum system is incorporated in the state vector, which now evolves according to

$$|\alpha; t\rangle_S = e^{-iHt}|\alpha; 0\rangle_S \tag{4.8}$$

where the subscript S indicates that we are in the Schrödinger picture. In this picture, physical observables are represented by time-independent self-adjoint operators O_S. By convention, states and observables in the Heisenberg and Schrödinger picture coincide at time $t = 0$:

$$|\alpha; 0\rangle_S = |\alpha\rangle \tag{4.9}$$

$$O_S = O_H(0) \tag{4.10}$$

The expectation value of a Heisenberg observable in a Heisenberg state coincides with the corresponding expectation value in the Schrödinger picture:

$$_S\langle\alpha; t|O_S|\alpha; t\rangle_S = \langle\alpha|e^{iHt}O_S e^{-iHt}|\alpha\rangle = \langle\alpha|O_H(t)|\alpha\rangle \tag{4.11}$$

In both the Heisenberg and Schrödinger pictures, the time-evolution of the system is treated exactly, i.e., with the full energy operator H of the system. However, it is frequently the case—for quantum field theories, almost always the case—that the exact dynamics is too complicated for an analytic solution to be available. A standard tactic is then to split the full Hamiltonian H into "free" (H_0) and "interaction" (V) parts

$$H = H_0 + V \tag{4.12}$$

There are obviously an infinite number of ways in which such a split can be done, but the split is only useful if (a) H_0 generates an analytically simple dynamics, and (b) the effects of V represent a quantitatively small "perturbation" on the evolution induced by H_0. Then one can hope to obtain useful results by expanding the desired physical quantities in powers of the "small" interaction V. In order to facilitate such an expansion, Dirac introduced a third version of quantum-mechanical time-development, which is now universally referred to as the "interaction picture". In the interaction picture, states and observables share the burden of carrying the time development of the system. In particular, operators retain the time-development characteristic of the Heisenberg picture, but only the free part H_0 of the Hamiltonian is used:

$$O_{\mathrm{ip}}(t) = e^{iH_0 t}O_S e^{-iH_0 t} \tag{4.13}$$

while the states evolve unitarily according to

$$|\alpha; t\rangle_{\mathrm{ip}} = e^{iH_0 t}e^{-iHt}|\alpha\rangle \tag{4.14}$$

Once again, these choices ensure that expectation values of an observable are the same as those computed in (say) the Heisenberg picture:

$$
\begin{aligned}
{}_{\text{ip}}\langle \alpha; t | O_{\text{ip}}(t) | \alpha; t \rangle_{\text{ip}} &= \langle \alpha | e^{iHt} e^{-iH_0 t} e^{iH_0 t} O_S e^{-iH_0 t} e^{iH_0 t} e^{-iHt} | \alpha \rangle \\
&= \langle \alpha | e^{iHt} O_S e^{-iHt} | \alpha \rangle \\
&= \langle \alpha | O_H(t) | \alpha \rangle
\end{aligned}
\tag{4.15}
$$

From (4.14) it follows that time evolution within the interaction picture (say from time t_0 to a later time t) is accomplished by the unitary operators

$$
U(t, t_0) \equiv e^{iH_0 t} e^{-iH(t - t_0)} e^{-iH_0 t_0}
\tag{4.16}
$$

$$
|\alpha; t\rangle_{\text{ip}} = U(t, t_0) |\alpha; t_0\rangle_{\text{ip}}
\tag{4.17}
$$

The unitary operator (4.16) also gives directly the transformation of the operators from interaction to Heisenberg picture:

$$
O_H(t) = U^\dagger(t, 0) O_{\text{ip}}(t) U(t, 0)
\tag{4.18}
$$

We will now derive a *formal* expression for $U(t, t_0)$ as an expansion in powers of the interaction part of the Hamiltonian V. Here is as good a place as any to remark (and we will return to this issue on several occasions later in the book) that this expansion is in general (and in fact, in any interesting physical case) *not* a convergent Taylor expansion, but *at best* an asymptotic expansion.[4] Such an expansion is useful only in those situations in which the contribution of the initial few terms in the series to the physical quantity of interest decrease sufficiently rapidly to give a sufficiently accurate estimate of the exact answer. For the time being we will ignore this issue and show how to develop a formal expansion for the interaction-picture time-development operator $U(t, t_0)$. First, observe that

$$
\frac{d}{dt} U(t, t_0) = e^{iH_0 t} (iH_0 - iH) e^{-iH(t - t_0)} e^{-iH_0 t_0}
\tag{4.19}
$$

$$
= -i V_{\text{ip}}(t) U(t, t_0)
\tag{4.20}
$$

where

$$
V_{\text{ip}}(t) \equiv e^{iH_0 t} V e^{-iH_0 t}
\tag{4.21}
$$

is simply the interaction part of the Hamiltonian transformed to the interaction picture, and U obviously satisfies the initial condition $U(t_0, t_0) = 1$. A solution of (4.20) satisfying the boundary condition $U(t_0, t_0) = 1$ is clearly

---

[4] For an introduction to asymptotic expansions, the short treatise of Erdelyi, "Asymptotic Expansions" (Dover, 1956) is very useful.

$$U(t, t_0) = 1 - i \int_{t_0}^{t} V_{\text{ip}}(t_1) U(t_1, t_0) dt_1 \tag{4.22}$$

which can be straightforwardly iterated (by reinserting the right-hand side in the integral) to yield the *formal* expansion

$$U(t, t_0) = \sum_{n=0}^{\infty} (-i)^n \int_{t_0}^{t} dt_1 \int_{t_0}^{t_1} dt_2 \dots \int_{t_0}^{t_{n-1}} dt_n V_{\text{ip}}(t_1) V_{\text{ip}}(t_2) \dots V_{\text{ip}}(t_n) \tag{4.23}$$

In fact, if we just differentiate (4.23) once with respect to t, the t_1 integral is removed in each term, t_1 is replaced by t in the rest, and an extra factor of $-iV_{\text{ip}}(t)$ appears in front of the original series. So the desired differential equation is satisfied. The first $(n = 0)$ term in the series is interpreted as 1, and all subsequent terms vanish (the integrals collapse) for $t = t_0$, so the initial condition is also satisfied.

The formula (4.23) is so important that we pause here to comment on its structure. The operator $V_{\text{ip}}(t)$ depends in general on time in a highly non-trivial way (although V, the original interaction part of the Hamiltonian, does not). This would not be so if H_0 and V, the free and interacting parts of the Hamiltonian commuted, but this is (alas!) never the case in any interesting situation. Consequently, the *order* in which the V_{ip} operators appear in (4.23) is crucial, since V_{ip} at one time will not in general commute with V_{ip} at another time. Note that the operators are arranged in order of *time*, with the earliest operator at the right end and the latest at the left. The temptation is irresistible to interpret the n-th term in (4.23) as corresponding to n sequential interactions induced by V_{ip}, with free propagation of the system in the intervening time intervals.

Consider the $n = 2$ term in the series (4.23). The integration region is $t_0 \le t_2 < t_1 \le t$ (see Fig. 4.1). If we introduce a time-ordering symbol T which reorders a product of interaction-picture operators in a decreasing time sequence (left to right)

$$T(V_{\text{ip}}(t_1) V_{\text{ip}}(t_2)) \equiv V_{\text{ip}}(t_1) V_{\text{ip}}(t_2) , \ t_1 > t_2 \tag{4.24}$$

$$\equiv V_{\text{ip}}(t_2) V_{\text{ip}}(t_1) , \ t_2 > t_1 \tag{4.25}$$

we can expand the region of integration so that both t_1 and t_2 run from t_0 to t. In fact,

$$\int_{t_0}^{t} dt_1 \int_{t_0}^{t_1} dt_2 \, V_{\text{ip}}(t_1) V_{\text{ip}}(t_2) = \frac{1}{2} \int_{t_0}^{t} dt_1 \int_{t_0}^{t} dt_2 \, T(V_{\text{ip}}(t_1) V_{\text{ip}}(t_2)) \tag{4.26}$$

The factor of $\frac{1}{2}$ just compensates for the inclusion of the upper triangular region in the figure, which contributes equally (by the reordering action of the T symbol) to the lower. This can obviously be generalized to the nth term in the series. We simply allow all the integrations to go from t_0 to t, and compensate with a factor $\frac{1}{n!}$. A T symbol must be included to ensure that the operators are always in the proper time-sequence, no matter what sector of the multi-dimensional integration region we happen to be in. In other words,

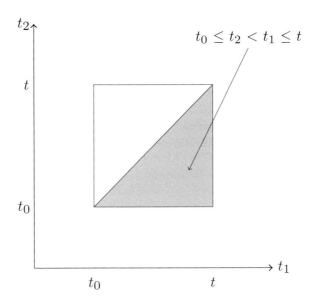

Fig. 4.1 Integration region for second-order term.

$$U(t, t_0) = \sum_{n=0}^{\infty} \frac{(-i)^n}{n!} \int_{t_0}^{t} dt_1 dt_2 .. dt_n T\{V_{\text{ip}}(t_1) .... V_{\text{ip}}(t_n)\} \qquad (4.27)$$

This formula will play a central role in our discussion of scattering theory, both for non-relativistic quantum systems later in this chapter, and for relativistic quantum field theories, where it will afford us a simple criterion for understanding how Lorentz-invariance can be guaranteed in the simplest field theories (cf. Chapter 5, Section 5). The resemblance of (4.27) to an exponential series suggests the convenient notation:

$$U(t, t_0) = T\{\exp\left(-i \int_{t_0}^{t} V_{\text{ip}}(\tau) d\tau\right)\} \qquad (4.28)$$

4.1.2   Propagators and kernels in quantum mechanics

As we saw in the preceding section, in the Heisenberg picture of quantum-mechanical time evolution, all of the dynamics takes place at the operator level, with state vectors fixed in time, and conventionally chosen to agree with the time-dependent state vectors of the Schrödinger picture at time $t = 0$: thus, the variable t appears in the Dirac ket $|\alpha; t\rangle_S$ for a Schrödinger state, but not in the corresponding Heisenberg state $|\alpha\rangle \equiv |\alpha; 0\rangle_S$. However, it is frequently more convenient to specify the Heisenberg state of the system as a condition on the system at a general time. For example, we may consider a non-relativistic particle moving in one dimension which is known to be exactly localized at position q_i at time t_i, and denote the corresponding *Heisenberg* state by $|q_i, t_i\rangle$, with

$$\mathbf{q}_H(t_i)|q_i, t_i\rangle = q_i|q_i, t_i\rangle \qquad (4.29)$$

where $\mathbf{q}_H(t)$ is the position operator for the particle in Heisenberg representation. The connection of such states to the conventionally defined Heisenberg states (specified at $t = 0$) follows immediately from (4.1):

$$|q_i, t_i\rangle = e^{iHt_i}|q_i\rangle \tag{4.30}$$

We distinguish between the implicit time-dependence of states $|q_i, t_i\rangle$ due to an initial condition and the explicit time-dependence of the states in Schrödinger picture $|q_i; t\rangle_S$ by using a comma in the former and a semicolon in the latter case to separate the time variable. The states (4.30) allow a simple and direct definition for the *propagator of a particle* $K(q_f, t_f; q_i, t_i)$ as the amplitude for detecting the particle at position q_f at time t_f given that the particle was previously localized at q_i at time t_i:

$$K(q_f, t_f; q_i, t_i) = \langle q_f, t_f | q_i, t_i \rangle = \langle q_f | e^{-iH(t_f - t_i)} | q_i \rangle \tag{4.31}$$

This result is clearly translation-invariant: the amplitude depends only on the elapsed time $T = t_f - t_i$, so we may as well consider simply $K(q_f, T; q_i, 0)$, with no loss of generality. The simple harmonic oscillator provides a concrete example: the Hamiltonian is[5]

$$H_{\text{sho}} = \frac{\mathbf{p}^2}{2m} + \frac{1}{2}m\omega^2\mathbf{q}^2 \tag{4.32}$$

so the propagator satisfies the differential equation

$$i\frac{\partial}{\partial T}K(q_f, T; q_i, 0) = -\frac{1}{2m}\frac{\partial^2 K(q_f, T; q_i, 0)}{\partial q_f^2} + \frac{1}{2}m\omega^2 q_f^2 K(q_f, T; q_i, 0) \tag{4.33}$$

with the initial condition

$$K(q_f, T; q_i, 0) \to \delta(q_f - q_i), \quad T \to 0 \tag{4.34}$$

The solution in this case is well known:

$$K(q_f, T; q_i, 0) = \left(\frac{m\omega}{2\pi i \sin \omega T}\right)^{1/2} \exp\left(\frac{im\omega}{2\sin \omega T}((q_i^2 + q_f^2)\cos \omega T - 2q_i q_f)\right) \tag{4.35}$$

The verification of (4.33) is a matter of some straightforward, if tedious, algebra. The zero time limit, yielding the δ-function normalization of the position eigenstates, is a more subtle matter, given the oscillatory behavior of the exponential in (4.35). This is the first indication of technical difficulties which will persist (in highly amplified degree!) in quantum field theory. Circumventing these problems leads to the imaginary-time formulation, which we now discuss briefly.

The propagator $K(q_f, T; q_i, 0)$ in (4.35) is evidently an analytic function of the time variable T (with essential singularities at $T = n\pi/\omega$). We may therefore analytically continue it by the replacement $T \to -iT$, leading[6] to the *Euclidean (or imaginary*

---

[5] We employ the standard device of bold-face notation to distinguish operators from c-numbers.

[6] This rotation by 90 degrees in the complex plane of the time variable is called a "Wick rotation".

time) propagator $K_E(q_f, T; q_i, 0)$ defined as

$$K_E(q_f, T; q_i, 0) = \langle q_f | e^{-H_{\text{sho}} T} | q_i \rangle$$

$$= (\frac{m\omega}{2\pi \sinh \omega T})^{1/2} \exp\left(-\frac{m\omega}{2 \sinh \omega T}((q_i^2 + q_f^2) \cosh \omega T - 2q_i q_f)\right) \quad (4.36)$$

The desired zero time limit is now easily demonstrated: as the time variable T only appears in the combination ωT, in this limit K_E for the harmonic oscillator coincides with the Euclidean propagator for a free particle ($\omega = 0$):

$$K_E(q_f, T; q_i, 0) \to K_E^{(0)} \equiv (\frac{m}{2\pi T})^{1/2} \exp\left(-\frac{m}{2T}(q_f - q_i)^2\right), \quad \omega \to 0 \quad (4.37)$$

$$\to \delta(q_f - q_i), \quad T \to 0 \quad (4.38)$$

where the δ-function limit is now apparent in the increasingly peaked Gaussians (normalized to unity) appearing on the right-hand side of (4.37).

The Euclidean propagators K_E are sometimes referred to as "heat kernels". Indeed, the free-particle propagator $K_E^{(0)}$ satisfies the one-dimensional diffusion equation

$$\frac{\partial^2 K_E(q_f, T; q_i, 0)}{\partial q_f^2} = 2m \frac{\partial K_E(q_f, T; q_i, 0)}{\partial T} \quad (4.39)$$

corresponding to diffusion in one dimension with diffusion constant $\kappa = \frac{1}{2m}$. Later in this chapter we will show how mathematically well-defined integral representations of such heat kernels (referred to generically as "Feynman–Kac" formulae) lead to the path integral formulation of quantum theory.

4.1.3 Quantum symmetries

The probability that a measurement of some observable O performed on a quantum system prepared in a specified (pure) quantum state $|\Psi\rangle$ yields a specified value (or range of values) is given by the absolute square of a probability amplitude which can typically be expressed as the Hilbert space inner product of the state $|\Psi\rangle$ and the eigenstate $|\Phi\rangle$ of O corresponding to the specified value:

$$P = |\langle \Phi | \Psi \rangle|^2 = |(|\Phi\rangle, |\Psi\rangle)|^2 \quad (4.40)$$

where we have introduced an alternative notation $(|\Phi\rangle, |\Psi\rangle)$ for the complex inner product of two Hilbert space vectors which will be useful in the following. Translational invariance of the laws of physics is a fundamental symmetry which has survived from the times of Galileo and Newton, and we are certainly entitled to expect that if the entire apparatus that prepared the system in the state $|\Psi\rangle$ and the detection apparatus which on interaction with the system will project it onto the eigenstate $|\Phi\rangle$ are both translated by the same fixed spatial vector $\vec{a}$, the measurement probability P in (4.40) should be unchanged. The translation of a physical system by displacement $\vec{a}$ is of course effected by the unitary operator

$$U_{\text{trans}}(\vec{a}) = e^{-i\vec{a} \cdot \vec{\mathbf{p}}} \quad (4.41)$$

where $\vec{p}$ is the total 3-momentum vector for the system. Similarly, a rotation of a physical system around the direction of the vector $\vec{\alpha}$ by an angle given by the magnitude $|\vec{\alpha}|$ is implemented on the Hilbert space of states by the unitary operator $U_{\text{rot}}(\vec{\alpha}) = e^{-i\vec{\alpha}\cdot\vec{\mathbf{J}}}$, where $\vec{\mathbf{J}}$ is the total angular momentum of the system.[7] Returning to the translation case, it is no surprise that if we replace

$$|\Phi\rangle \to U_{\text{trans}}(\vec{a})|\Phi\rangle, \quad |\Psi\rangle \to U_{\text{trans}}(\vec{a})|\Psi\rangle \tag{4.42}$$

then (by the unitarity of U_{trans})

$$P \to |(U_{\text{trans}}(\vec{a})|\Phi\rangle, U_{\text{trans}}(\vec{a})|\Psi\rangle)|^2 = |(\langle\Phi|, U^\dagger_{\text{trans}}(\vec{a})U_{\text{trans}}(\vec{a})|\Psi\rangle)|^2 = |\langle\Phi|\Psi\rangle|^2 = P \tag{4.43}$$

In general, the symmetries of physics can be expressed mathematically as groups of transformations (e.g., in the case just above, the succession of two translations $\vec{a}$, $\vec{b}$ is equivalent to the combined translation $\vec{a} + \vec{b}$, the translation $-\vec{a}$ is the inverse of the translation $\vec{a}$, etc.). A particular element g of such a symmetry transformation group will be associated with some Hilbert space operator $S(g)$ (just as translation by $\vec{a}$ above was associated with the unitary operator $e^{-i\vec{p}\cdot\vec{a}}$). And the statement that physics is invariant under such a group of transformations amounts to the requirement

$$|(S(g)|\Phi\rangle, S(g)|\Psi\rangle)|^2 = |(\langle\Phi|, |\Psi\rangle)|^2 \tag{4.44}$$

for *arbitrary states* $|\Phi\rangle, |\Psi\rangle$ and *arbitrary group elements* g. This requirement clearly holds if the symmetry group is implemented by unitary operators $S(g)$, such as the operator $U_{\text{trans}}(\vec{a})$ discussed above.

In fact, there is another option, as Wigner was the first to demonstrate, in his famous *unitarity–antiunitarity theorem* (Wigner, 1959). The other option—indeed the only other possibility compatible with (4.44)—is that $S(g)$ be an *antiunitary* operator. An operator $\mathcal{T}$ is antiunitary if, for some complete orthonormal basis $\{|n\rangle\}$ of the state space (which we shall assume here to be separable, i.e., to allow a denumerable basis),

$$(\mathcal{T}|n\rangle, \mathcal{T}|m\rangle) = \delta_{nm} \tag{4.45}$$

$$\mathcal{T}\sum_n a_n|n\rangle = \sum_n a_n^*\mathcal{T}|n\rangle \tag{4.46}$$

The property (4.46) indicates that $\mathcal{T}$ is an antilinear operator. The symmetry requirement (4.44) follows immediately:

$$|\Phi\rangle = \sum_n a_n|n\rangle, \quad |\Psi\rangle = \sum_m b_m|m\rangle \tag{4.47}$$

$$|(\mathcal{T}|\Phi\rangle, \mathcal{T}|\Psi\rangle)| = |(\sum_n a_n^*\mathcal{T}|n\rangle, \sum_m b_m^*\mathcal{T}|m\rangle)|$$

$$= |\sum_{n,m} a_n b_m^* (\mathcal{T}|n\rangle, \mathcal{T}|m\rangle)| \tag{4.48}$$

---

[7] See Baym, op. cit., Chapter 17.

Using (4.45), this becomes

$$|(\mathcal{T}|\Phi\rangle, \mathcal{T}|\Psi\rangle)| = |\sum_n a_n b_n^*|$$

$$= |\sum_n a_n^* b_n|$$

$$= |(|\Phi\rangle, |\Psi\rangle)| \qquad (4.49)$$

A symmetry group cannot consist purely of antiunitary operators, for the simple reason that the product of two antilinear operators must be linear. Indeed, the only case of physical interest in which the antiunitary option is required is for the discrete group consisting of (i) the identity and (ii) the time-reversal operation $t \to -t$. That time reversal should entail a complex conjugation is plausible once we consider that the time-dependence of quantum states in the energy basis involves the factor e^{-iEt} with the energy eigenvalue E real. For a classical particle the time-reversal operation is easily described in phase-space as the mapping taking $\vec{q}(t), \vec{p}(t)$ to

$$\vec{q}_{\mathrm{tr}}(t) = \vec{q}(-t) \qquad (4.50)$$

$$\vec{p}_{\mathrm{tr}}(t) = -\vec{p}(-t) \qquad (4.51)$$

where the subscript "tr" denotes the time-reversed trajectory. In quantum mechanics, the corresponding mapping is realized by an antiunitary operator $\mathcal{T}$ (the need for the "anti" will be shortly apparent) with the Heisenberg operators (we omit the "H" subscript to avoid clogging the notation) transforming like

$$\vec{q}_{\mathrm{tr}}(t) = \mathcal{T}\vec{q}(t)\mathcal{T}^{-1} \qquad (4.52)$$

$$= \mathcal{T}e^{iHt}\vec{q}(0)e^{-iHt}\mathcal{T}^{-1} \qquad (4.53)$$

$$= \mathcal{T}e^{iHt}\mathcal{T}^{-1}\vec{q}_{\mathrm{tr}}(0)\mathcal{T}e^{-iHt}\mathcal{T}^{-1} \qquad (4.54)$$

$$= \vec{q}(-t) = e^{-iHt}\vec{q}_{\mathrm{tr}}(0)e^{iHt} \qquad (4.55)$$

from which we conclude that the time-reversal operator $\mathcal{T}$ must satisfy

$$\mathcal{T}(iHt)\mathcal{T}^{-1} = -iHt \Rightarrow \mathcal{T}iH = -iH\mathcal{T} \qquad (4.56)$$

Since $\mathcal{T}$ has to commute with H (for example, if the Hamiltonian is quadratic in momenta), it must anticommute with i: in other words, $\mathcal{T}$ contains a complex conjugation and must be antilinear. Let us define a complex conjugation operator K as an antilinear operator performing complex conjugation on the components of a state vector *in a preferred basis* $|n\rangle$: thus,

$$K|n\rangle = |n\rangle \Rightarrow K\left(\sum_n a_n|n\rangle\right) = \sum_n a_n^* K|n\rangle = \sum_n a_n^*|n\rangle \qquad (4.57)$$

It can be readily shown (see, for example, (Messiah, 1966), Chapter XV) that the most general antilinear operator satisfying (4.49) takes the form

$$\mathcal{T} = UK \tag{4.58}$$

where U is a conventional (linear) unitary operator, and K is the complex conjugation operator (in a specified basis) described above. Typically, the basis chosen is the coordinate space basis in which the $\vec{q}$ operators are diagonal, and for the spin degrees of freedom (if any), the basis in which S_3 is diagonal and real (and S_1 and S_2 are purely real and purely imaginary matrices, respectively). Note that the orbital angular momentum $\vec{q} \times \vec{p}$ reverses sign under time-reversal, so we must require the same of spin angular momentum:

$$\mathcal{T}\vec{S}\mathcal{T}^{-1} = -\vec{S} \tag{4.59}$$

which can be achieved, for a particle of spin j, by choosing

$$\mathcal{T} = e^{-i\pi S_2}K = Y^{(j)}K \tag{4.60}$$

clearly of the form (4.58). The complex conjugation operator K takes $S_1 \to S_1, S_2 \to -S_2, S_3 \to S_3$, and then the subsequent rotation by π around the y-axis reverses the sign of S_1 and S_3, yielding the desired result (4.59). Explicitly, the matrix of $Y^{(j)}$ in the standard spin representation is given by

$$Y^{(j)}_{mm'} = (-1)^{j+m}\delta_{m,-m'} \quad (= -i\sigma_2 \text{ for spin} -\frac{1}{2}) \tag{4.61}$$

The discussion of symmetries so far has been essentially at a kinematic level: unitary operators $U(g)$ representing a particular group element g of a symmetry group $\mathcal{G}$ can be defined on the Hilbert space of states of a quantum mechanical particle quite independently of whether the dynamics (i.e., the time evolution) of the system respects the symmetry. We shall say that the "dynamics respects the symmetry"[8] if for any initial and final states $|\Psi\rangle$ and $|\Phi\rangle$, any time lapse T, and any group element g

$$(U(g)|\Phi\rangle, e^{-iHT}U(g)|\Psi\rangle) = (|\Phi\rangle, e^{-iHT}|\Psi\rangle) \tag{4.62}$$

In other words, the amplitude that the initial state $|\Psi\rangle$ will be found to have evolved to the state $|\Phi\rangle$ after time T is equal to the corresponding amplitude for the symmetry rotated states $U(g)|\Psi\rangle$ and $U(g)|\Phi\rangle$. Now suppose that $\mathcal{G}$ is a finite-dimensional linear Lie group: namely, a group of matrices parameterizable by a finite set of group parameters ω^α and finite-dimensional generator matrices T_α,

$$g = \exp\left(-i\omega^\alpha T_\alpha\right) \tag{4.63}$$

If the Lie group in question in unitary or orthogonal (e.g., the rotation group) then the group parameters ω^α can be chosen to be real and the generators T_α to be hermitian

---

[8] This definition of a "dynamical symmetry" differs from the usage introduced by Wigner (cf. Chapter 3), where the term was reserved for symmetries of a non-geometrical character.

matrices. For non-unitary groups, such as the homogeneous Lorentz group, discussed in greater detail in the following chapter, some of the generators must be non-hermitian (if we follow usual convention and continue to parameterize the group in terms of real parameters ω^α). In either case, the discussion of Wigner's theorem above makes clear that individual group operations g must be represented on the Hilbert space of quantum states by *unitary* operators (putting aside the special case of time reversal for the moment) $U(g)$, with

$$U(g) = \exp\left(-i\omega^\alpha \mathbf{J}_\alpha\right) \tag{4.64}$$

where the $\mathbf{J}_\alpha$ are self-adjoint operators in the quantum state space.

As $|\Phi\rangle$ and $|\Psi\rangle$ are arbitrary states in (4.62) we conclude that at the operator level

$$U^\dagger(g)e^{-iHT}U(g) = e^{-iHT} \tag{4.65}$$

or taking the derivative with respect to T at $T=0$

$$U^\dagger(g)HU(g) = H \Rightarrow [U(g), H] = 0 \tag{4.66}$$

The commutativity of the Hamiltonian with arbitrary group operations $U(g)$ then implies, for symmetry operations infinitesimally close to the identity, $g = 1 - i\omega^\alpha T_\alpha$, $U(g) = 1 - i\omega^\alpha \mathbf{J}_\alpha$, that

$$[\mathbf{J}_\alpha, H] = 0 \tag{4.67}$$

By a standard quantum mechanical argument, the generators $\mathbf{J}_\alpha$ of any dynamical symmetry therefore represent conserved observables of the theory. The dual character of (4.67)—simultaneously expressing the invariance of the Hamiltonian under the infinitesimal group transformations generated by the $\mathbf{J}_\alpha$ (a symmetry requirement) and the time-independence of physical observables associated with the $\mathbf{J}_\alpha$ (a conservation principle)—will receive an elegant transcription in Noether's theorem (cf. Chapter 12) for the conserved currents associated with continuous symmetries in quantum field theory.

At this point it will be convenient to give some concrete examples of the implementation of classical symmetries in a quantum theory. As we shall see, in some cases subtleties arise which obstruct a straightforward transferral of a classical symmetry to a unitary symmetry of the corresponding quantum system, a situation which will become even more prevalent in quantum field theory, with important physical consequences.

Example 1: Rotational symmetry

We begin with an elementary case, a (non-relativistic) spinless point particle in three dimensions subject to a rotationally invariant potential $V(r)$, $r = |\vec{r}|$. The Hamiltonian is

$$H = \frac{1}{2m}\vec{\mathbf{p}}^{\,2} + V(\mathbf{r}) \tag{4.68}$$

and clearly satisfies

$$[\vec{\mathbf{J}}, H] = 0 \tag{4.69}$$

where $\vec{\mathbf{J}}$ is the (orbital) angular momentum three-vector. Corresponding to a three-dimensional rotation $R(\vec{\alpha})$ by the angle $|\vec{\alpha}|$ around the direction of $\vec{\alpha}$ (which we can realize as a 3x3 real orthogonal matrix) is the unitary representative $U_{\mathrm{rot}}(\vec{\alpha}) = e^{-i\vec{\alpha}\cdot\vec{\mathbf{J}}}$, acting on states in Hilbert space. The general requirement of dynamical rotational invariance (4.62) implies in particular that the propagator $K(\vec{r}_{\mathrm{f}}, T; \vec{r}_{\mathrm{i}}, 0) = \langle \vec{r}_{\mathrm{f}} | e^{-iHT} | \vec{r}_{\mathrm{i}} \rangle$ for detecting the particle at $\vec{r}_{\mathrm{f}}$ at time T if localized initially at time zero at $\vec{r}_{\mathrm{i}}$ should satisfy, for any fixed rotation $R(\vec{\alpha})$

$$\langle R(\vec{\alpha})\vec{r}_{\mathrm{f}} | e^{-iHT} | R(\vec{\alpha})\vec{r}_{\mathrm{i}} \rangle = \langle \vec{r}_{\mathrm{f}} | e^{-iHT} | \vec{r}_{\mathrm{i}} \rangle \tag{4.70}$$

the physical significance of which is obvious.

Example 2: Canonical symmetry.

The canonical symmetry of classical mechanics[9] asserts the possibility of equivalent representations of the dynamical Hamiltonian evolution of a classical system in phase-space in terms of alternative choices of coordinate and momentum variables. Here we consider a particle moving in one space dimension, with a Hamiltonian dynamics determined by a function $H(p, q)$ of momentum (coordinate) variables p (q). An equivalent description is obtained by choosing a different canonical pair P, Q related to p, q by a *generating function* $F(q, P)$[10] as follows:

$$Q = \frac{\partial F}{\partial P} \tag{4.71}$$

$$p = \frac{\partial F}{\partial q} \tag{4.72}$$

In principle, solution of the equation pair (4.71, 4.72) allows us to express the initial canonical pair p, q in terms of the new pair P, Q, or *vice versa*. The problem of identifying unitary representatives of the classical canonical transformation $(p, q) \to (P, Q)$ in the Hilbert space of quantum states for our particle was first solved by Jordan (Jordan, 1926), in a paper which played a crucial role in the development

---

[9] See Goldstein (Goldstein, 2002), Chapter 9 for a review of the essential properties of canonical transformations.

[10] This is a generating function of the second type, F_2, in the notation of Goldstein, op. cit. We only consider time-independent generating functions here: the new Hamiltonian is then equal to the old one, re-expressed in terms of the new canonical variables.

of quantum transformation theory in the late 1920s. Jordan showed that the operator U_{can} implementing this transformation for the quantum kinematic variables, for the special case of classical generating functions of the form

$$F(q, P) = \sum_n f_n(q) g_n(P) \tag{4.73}$$

takes the form (again temporarily reintroducing Planck's constant)

$$U_{\text{can}}(P, Q) = C \exp\left(\frac{i}{\hbar}\{-(Q, P) + \sum_n (f_n(Q), g_n(P))\}\right) \tag{4.74}$$

with C an arbitrary constant. With this form Jordan could show (cf. Problem 2 at the end of this chapter)

$$q = U_{\text{can}}^{-1} Q U_{\text{can}} \tag{4.75}$$

$$p = U_{\text{can}}^{-1} P U_{\text{can}} \tag{4.76}$$

In the formula (4.74) the round bracket expressions (Q, P) and $(f_n(Q), g_n(P))$ imply a specific ordering of the non-commuting operators Q and P: one is instructed to order all Qs to the left of all Ps in the formally expanded exponential in (4.74). Although a formal demonstration of (4.74, 4.75, 4.76) is straightforward, the actual existence of U_{can} is not guaranteed, and in general the operator obtained in this fashion is not even unitary! We will shortly provide an example of the problems that can arise in this connection—but first, a "nice" canonical transformation where all the desired properties of a quantum symmetry obtain in an unproblematic way. We consider the generating function

$$F(q, P) = \frac{1}{2d}(bP^2 + 2qP - cq^2) \tag{4.77}$$

where a, b, c, d are real constants satisfying $ad - bc = 1$. The result is a *linear canonical transformation*

$$Q = aq + bp \tag{4.78}$$

$$P = cq + dp \tag{4.79}$$

In this case the general formula for the symmetry representative U_{can} (4.74) gives

$$U_{\text{can}} = \exp\left(\frac{i}{\hbar}\{\frac{b}{2d}P^2 - \frac{c}{2d}Q^2 + (\frac{1}{d} - 1)(Q, P)\}\right) \tag{4.80}$$

We now derive an explicit formula for this operator as an integral kernel $\mathcal{U}(Q, Q')$ acting on coordinate wavefunctions $\psi(Q)$ (so that the operator P in (4.80) becomes $\frac{\hbar}{i}\frac{\partial}{\partial Q}$):

$$U_{\text{can}}\psi(Q) = \int \mathcal{U}(Q, Q')\psi(Q') dQ' \tag{4.81}$$

A short exercise in Fourier transformation (cf. Problem 2) then shows

$$U(Q,Q') = \sqrt{\frac{1}{2\pi\hbar b}} \exp\{-\frac{i}{2\hbar b}(aQ^2 - 2QQ' + dQ'^2)\} \tag{4.82}$$

where the arbitrary constant C in (4.74) is chosen to ensure unitarity of the resultant kernel:

$$\int U(Q,Q')U^*(Q'',Q')dQ' = \delta(Q - Q'') \tag{4.83}$$

The special case $a = d = 0$, $b = -c = 1$, corresponding to a canonical transformation interchanging the roles of coordinate and momentum variables leads, of course, to the Fourier kernel $U = \sqrt{\frac{1}{2\pi\hbar}}e^{\frac{i}{\hbar}QQ'}$, which is of great importance in the development of transformation theory.

As mentioned above, it is important to realize that the transferral of a non-linear classical canonical transformation to the quantum arena as a unitary symmetry group is far from automatic: indeed, it is the exception rather than the rule (Anderson, 1994). Consider, for example, another class of canonical transformations of great importance classically: the *point transformations*

$$q \rightarrow Q = f(q) \tag{4.84}$$

$$p \rightarrow P = \frac{1}{f'(q)}p \tag{4.85}$$

corresponding to the generating function

$$F(q,P) = f(q)P \tag{4.86}$$

and the Jordan operator

$$U_{\text{can}} = C \exp\left(\frac{i}{\hbar}(f(Q) - Q, P)\right) \tag{4.87}$$

We assume that $f(q)$ is (a) monotone increasing, and (b) invertible. Unfortunately, irrespective of the choice of C, U_{can} fails to be norm-preserving for general non-linear choices of the reparameterization function $f(q)$, as

$$U_{\text{can}}\psi(Q) = C\psi(f(Q)) \tag{4.88}$$

and in general we certainly do not have $\int |\psi(Q)|^2 dQ = |C|^2 \int \psi(f(Q))|^2 dQ$! As Jordan showed, the non-unitarity of the symmetry representative in this case can be traced back to the non-hermiticity of the new momentum variable P in (4.85). The problem can be fixed by adding a *quantum correction* to the generating function (4.86)

$$F(q,P) = f(q)P + \frac{\hbar}{2i}\ln|f'(q)| \tag{4.89}$$

The relation between the new and old momentum variables is now

$$p = \frac{1}{2}(f'(q)P + Pf'(q)) \tag{4.90}$$

consistent with hermiticity of both p and P; the symmetry representative becomes

$$U_{\text{can}} = |f'(Q)|^{1/2} \exp\left(\frac{i}{\hbar}(f(Q) - Q, P)\right) \tag{4.91}$$

generating the norm-preserving action

$$U_{\text{can}}\psi(Q) = |f'(Q)|^{1/2}\psi(f(Q)) \tag{4.92}$$

The presence of a quantum obstruction (typically signalled by the appearance of a term proportional to Planck's constant) in a classical symmetry is generally referred to as an *anomaly*. Some further examples of quantum-mechanical anomalies in the implementation of classical canonical transformations are discussed in (Swanson, 1993). In Chapter 15 we shall see that quantum anomalies appear quite commonly in quantum field theories when we attempt to implement certain classical symmetries (such as dilatation and axial symmetry).

Before leaving the subject of the role of symmetries in quantum theory we should briefly discuss an important restriction (originally pointed out by Wick, Wightman, and Wigner (Wick *et al.*, 1952)) on the fundamental superposition principle of quantum mechanics, arising in connection with certain important physical symmetries. An unrestricted application of the superposition principle would imply that a linear combination of any two physical states $\alpha|\psi_1\rangle + \beta|\psi_2\rangle$, $\alpha, \beta \neq 0$ must necessarily also correspond to a physically realizable state. In particular, the *relative* phase of the two components of the state (i.e., of the complex numbers α and β) must be physically meaningful, measurable for example by some suitable suitable interference experiment, in contrast to the overall phase (common phase of α and β) which disappears from all expectation values. Wick *et al.*, in the aforesaid reference, point out that certain linear combinations are in fact not permitted, as the resultant state displays unphysical phase correlations. For example, a state represented by the linear combination $\alpha|B\rangle + \beta|F\rangle$ of a state $|B\rangle$ with integral total angular momentum and one $|F\rangle$ with half-integral total angular momentum (say, a state with an odd number of spin-$\frac{1}{2}$ particles) will be converted to the state $\alpha|B\rangle - \beta|F\rangle$ under a spatial rotation by 360 degrees around any axis, which clearly corresponds to no physical change to the state. Accordingly, the relative phase of the "bosonic" and "fermionic" components of the state is unobservable. Combinations of this kind are said to be excluded as a consequence of a *superselection principle*: the physical Hilbert space of the theory is constructed as a direct sum of distinct *superselection sectors*. In the example just given there are two such sectors: the Hilbert space $\mathcal{H}_B$ of all states with integral angular momentum, and the space $\mathcal{H}_F$ of all states with half-integral total angular momentum. Linear combinations of states in distinct superselection sectors are forbidden. The exact conservation of angular momentum in the theory ensures that such a prohibition, asserted at any initial time, will be respected by the dynamics of the theory, as states evolve independently within their superselection sectors, with transitions between sectors forbidden by a conservation principle. Exact electric charge conservation is similarly associated with a superselection rule: superposition of states with different total electric charge is proscribed, as the relative phase will be altered by an (unobservable) gauge transformation. Finally, we observe that in situations where superselection rules are operative, the hermitian operators corresponding to physically realizable measurements must have vanishing matrix elements between states in different superselection sectors.

4.2 The functional (path-integral) framework

The operator formulation of quantum mechanics employed in the preceding sections took shape in the late 1920s, within a few years of Heisenberg's 1925 breakthrough, and was put on a mathematically firm foundation by von Neumann (Von Neumann, 1996), who developed in the course of his formalization of quantum mechanics much of the essential machinery of modern functional analysis (e.g., the spectral theory of unbounded self-adjoint operators in Hilbert space). It therefore came as quite a surprise to many physicists when a completely different formulation of quantum theory in terms of integrals over infinite-dimensional function spaces emerged from work of Feynman (Feynman, 1948) and Dirac (Dirac, 1945). This "path-integral" approach was initially viewed with great suspicion by more mathematically inclined physicists, despite the clear intuitive power and formal elegance of this approach. Nevertheless, just as the initially disturbing odor of the "improper" Dirac δ-function was dispelled by the development of a mathematically rigorous theory of distributions, the Euclidean (imaginary-time) version of the Feynman path integral has been completely absorbed into a rigorous theory of conditional Wiener measures, which has become an absolutely indispensable technical tool in the arsenal of the constructive quantum field theorist.[11] We will begin our discussion of this approach by returning to our favorite toy in this chapter, the simple harmonic oscillator, where the availability of transparent analytic expressions will allow us to confront and resolve the troubling convergence issues which are unavoidable in any treatment of the path-integral method.

4.2.1 Path-integral formulation for the simple harmonic oscillator

We return to the Euclidean kernel for the simple harmonic oscillator,

$$K_E(q_f, t_f; q_i, t_i) = \langle q_f | e^{-H_{\text{sho}}(t_f - t_i)} | q_i \rangle \tag{4.93}$$

defined in (4.36), but with the initial and final (imaginary) times shifted to general values, so that the elapsed time $T = t_f - t_i$. We divide the finite interval (t_i, t_f) into N subintervals of size $\tau = T/N$ by introducing $N - 1$ intermediate times

$$t_n = t_i + n\tau, \quad n = 1, 2, 3, \ldots N - 1 \tag{4.94}$$

with $t_0 \equiv t_i, t_N \equiv t_f$. Likewise, $N - 1$ new coordinate variables $q_1, q_2, \ldots q_{N-1}$ (with $q_0 \equiv q_i, q_N \equiv q_f$) are introduced, corresponding to the position of our particle at

---

[11] For a rigorous introduction to conditional Wiener measures, including appropriate convergence proofs for potential theory, see Chapter 3 of (Glimm and Jaffe, 1987).

these intermediate times. Inserting a complete set of intermediate states, the finite-time heat kernel (4.93), can be written as a multiple convolution of the N kernels which accomplish the time-evolution of the system over the temporal subintervals $(t_n, t_{n+1}), n = 0, 1, 2, \ldots N - 1$:

$$K_E(q_f, t_f; q_i, t_i) = \int \prod_{n=1}^{N-1} dq_n \langle q_f | e^{-\tau H_{\text{sho}}} | q_{N-1} \rangle \langle q_{N-1} | e^{-\tau H_{\text{sho}}} | q_{N-2} \rangle$$

$$\cdots \langle q_2 | e^{-\tau H_{\text{sho}}} | q_1 \rangle \langle q_1 | e^{-\tau H_{\text{sho}}} | q_i \rangle \tag{4.95}$$

One easily sees, from the exact result (4.36), that in the limit of large N (T fixed, $\tau \to 0$), the individual kernel factors in (4.95) become

$$\langle q_{n+1} | e^{-\tau H_{\text{sho}}} | q_n \rangle$$

$$= (\frac{m}{2\pi\tau})^{1/2} \exp\left\{-\tau \frac{m}{2} \frac{(q_{n+1} - q_n)^2}{\tau^2} - \tau \frac{m\omega^2}{2} \frac{q_{n+1}^2 + q_n^2}{2}\right)(1 + O(\omega^2 \tau^2)\} \tag{4.96}$$

Inserting this result in the expression (4.95) for the finite-time evolution kernel, we find

$$K_E(q_f, t_f; q_i, t_i) = \lim_{N \to \infty} \int \prod_{n=1}^{N-1} dq_n e^{-S_E} \tag{4.97}$$

$$S_E \equiv \tau \sum_{n=0}^{N-1} (\frac{m}{2} (\frac{q_{n+1} - q_n}{\tau})^2 + \frac{1}{2} m\omega^2 \frac{q_{n+1}^2 + q_n^2}{2}) \tag{4.98}$$

The limit on the right-hand side of (4.97), of course, just yields our Euclidean propagator (4.36). Formally, identifying $q_n = q(t_n)$, the corresponding limit of the exponent in (4.98) yields the Euclidean action

$$S_E = \int_{t_i}^{t_f} \{\frac{1}{2} m\dot{q}^2(t) + \frac{1}{2} m\omega^2 q^2(t)\} dt \tag{4.99}$$

and the above-mentioned limit can therefore by interpreted as an integral of e^{-S_E} over all "paths" $q(t)$ subject to the boundary conditions $q(t_i) = q_i, q(t_f) = q_f$. In fact, it can be shown that this limit defines a countably additive measure (called a conditional Wiener measure to indicate the boundary conditions) over the space of continuous functions $q(t)$ defined on the interval (t_i, t_f) (see (Glimm and Jaffe, 1987)), whence the full weight of Lebesgue integration theory can be brought to bear to give a rigorous meaning to this "functional integral" over the space of continuous functions. We shall use the notation $\mathbf{D}q(t)$ to indicate the measure defining this integral, as follows:

$$K_E(q_f, t_f; q_i, t_i) = \int_{q(t_i) = q_i, q(t_f) = q_f} e^{-S_E} \mathbf{D}q(t) \tag{4.100}$$

The connection of S_E to the conventional action S, defined as the real time integral of the Lagrangian

$$S = \int_{t_i}^{t_f} \{\frac{1}{2}m\dot{q}^2(t) - \frac{1}{2}m\omega^2 q^2(t)\}dt \qquad (4.101)$$

can be seen if we make the reverse analytic continuation back to real time

$$t \rightarrow e^{i(\frac{\pi}{2}-\epsilon)}t = (i+\epsilon)t \qquad (4.102)$$

$$\tau = \frac{t}{N} \rightarrow (i+\epsilon)\tau \qquad (4.103)$$

where the rotation is by an angle $\frac{\pi}{2} - \epsilon$ in the complex plane, with ϵ a positive infinitesimal quantity, to be set to zero *after* the integrations in (4.97) are done. The need for this maneuver is apparent if we examine the discretized Euclidean action (4.98) after the continuation (4.103):

$$-S_E \rightarrow i\tau \sum_{n=0}^{N-1}(\frac{m}{2}(\frac{q_{n+1}-q_n}{\tau})^2 - \frac{1}{2}m\omega^2 \frac{q_{n+1}^2 + q_n^2}{2})$$

$$- \epsilon\tau \sum_{n=0}^{N-1}(\frac{m}{2}(\frac{q_{n+1}-q_n}{\tau})^2 + \frac{1}{2}m\omega^2 \frac{q_{n+1}^2 + q_n^2}{2}) \qquad (4.104)$$

As long as ϵ is retained as a small positive quantity, the integrals in (4.97) remain absolutely convergent in virtue of the negative real contribution of the second line of (4.104). If we set ϵ to zero prematurely, the integrals become undamped oscillatory ones and therefore ill-defined. Nevertheless, the real-time path-integral representation of the propagator is often written ignoring the converging factor involving ϵ, in which case we recognize the first line of (4.104) as giving (in the continuum limit $N \rightarrow \infty$) just $i/\hbar$ times the real-time action (time integral of the Lagrangian) $S = \int \mathcal{L}dt$:

$$K((q_f, t_f; q_i, t_i) = \int e^{\frac{i}{\hbar}\int \mathcal{L}dt}\mathbf{D}q(t) \qquad (4.105)$$

$$\mathcal{L} = \frac{1}{2}m\dot{q}^2(t) - \frac{1}{2}m\omega^2 q^2(t) \qquad (4.106)$$

We again emphasize that (4.105) must be interpreted as containing a hidden regularizing "ϵ" term in the exponent in order to be meaningful. Note that we have temporarily reintroduced Planck's constant $\hbar$ in (4.105), abandoning the choice of natural units ($\hbar = 1$) for a moment. The classical limit $\hbar \rightarrow 0$ clearly results in strong damping except where the phase of the exponent is stationary, which is the extremal action principle of classical mechanics selecting the classical path $q_{cl}(t)$:

$$\frac{\delta}{\delta q(t)} \int \mathcal{L}dt\bigg|_{q=q_{cl}} = 0 \qquad (4.107)$$

The variational principles of classical mechanics (specifically, Hamilton's principle) can therefore be regarded as simply the result of applying the stationary-phase

approximation to the path-integral representation of the quantum mechanical amplitudes of the system.

An important byproduct of the functional integral representation (4.100) of the
kernel (4.93) is obtained by taking the trace, as follows:

$$\text{Tr}(e^{-H_{\text{sho}}(t_f-t_i)}) = \int dQ \langle Q | e^{-H_{\text{sho}}(t_f-t_i)} | Q \rangle \tag{4.108}$$

$$= \int dQ \int_{q(t_i)=q(t_f)=Q} e^{-S_E} \mathbf{D}q(t) \tag{4.109}$$

$$= \int_{q(t_i)=q(t_f)} e^{-S_E} \mathbf{D}q(t) \tag{4.110}$$

In the last line the integration over all paths which begin and end at a given coordinate
Q, followed by an integration over all Q, has been replaced by a functional integral
over *all periodic paths* satisfying $q(t_i) = q(t_f)$. Choosing $t_i = -\beta/2$, $t_f = +\beta/2$, we see
that this functional integral actually provides a simple reformulation of the finite-
temperature partition function $Z \equiv \text{Tr}(e^{-\beta H_{\text{sho}}})$ of the simple harmonic oscillator.
Thus the path integral also provides, in its Euclidean version, an alternative tool
for computations in quantum statistical mechanics as well as for real-time quantum
mechanics.

4.2.2 Path-integral formulation of quantum mechanics: Hamiltonian version

We shall now abandon the simple harmonic oscillator in favor of a more general
discussion, where we consider an integral representation for the propagation of a
particle of mass m in one dimension under the influence of a general potential $V(q)$.
Thus, the Hamiltonian is now

$$H = \frac{1}{2m}\mathbf{p}^2 + V(\mathbf{q}) = H_0 + V \tag{4.111}$$

Under rather loose conditions on the potential $V(q)$ (e.g., it is sufficient that V be a
polynomial in q, bounded below so as to guarantee a "bottom" to the energy spectrum)
it is possible to show (Glimm and Jaffe, 1987) that for infinitesimal time intervals
$\tau = (t_f - t_i)/N$, $e^{-H\tau} \simeq e^{-H_0\tau}e^{-V\tau}$, in the sense that

$$\langle q_f | e^{-H(t_f-t_i)} | q_i \rangle = \langle q_f | (e^{-H\tau})^N | q_i \rangle = \lim_{N\to\infty} \langle q_f | (e^{-H_0\tau}e^{-V\tau})^N | q_i \rangle \tag{4.112}$$

As before, we introduce $N-1$ intermediate completeness sums for the kernel in
(4.112):

$$K_E(q_f, t_f; q_i, t_i) = \lim_{N \to \infty} \int \prod_{n=1}^{N-1} dq_n \langle q_f | e^{-H_0 \tau} e^{-V\tau} | q_{N-1} \rangle$$

$$\cdot \langle q_{N-1} | e^{-H_0 \tau} e^{-V\tau} | q_{N-2} \rangle \cdots \langle q_{n+1} | e^{-H_0 \tau} e^{-V\tau} | q_n \rangle \cdots \langle q_1 | e^{-H_0 \tau} e^{-V\tau} | q_i \rangle \ (4.113)$$

Once again we focus attention on the individual matrix elements representing the propagation amplitude over the infinitesimal intervals τ:

$$\langle q_{n+1} | e^{-H_0 \tau} e^{-V(\mathbf{q})\tau} | q_n \rangle = e^{-V(q_n)\tau} \langle q_{n+1} | e^{-H_0 \tau} | q_n \rangle$$

$$= e^{-V(q_n)\tau} \langle q_{n+1} | e^{-\frac{1}{2m} \mathbf{p}^2 \tau} | q_n \rangle$$

$$= e^{-V(q_n)\tau} \int \frac{dp_n}{2\pi} \langle q_{n+1} | p_n \rangle \langle p_n | e^{-\frac{1}{2m} \mathbf{p}^2 \tau} | q_n \rangle$$

$$= e^{-V(q_n)\tau} \int \frac{dp_n}{2\pi} e^{ip_n(q_{n+1} - q_n) - \frac{1}{2m} p_n^2 \tau} \qquad (4.114)$$

Notice that each time interval is now associated with an auxiliary *momentum* variable $p_n, n = 0, 1, \ldots N-1$. Inserting the result (4.114) into the integral representation (4.113), we find

$$K_E(q_f, t_f; q_i, t_i) = \lim_{N \to \infty} \int \prod_{n=1}^{N-1} dq_n \prod_{n=0}^{N-1} \frac{dp_n}{2\pi} e^{i \sum_{n=0}^{N-1} p_n(q_{n+1} - q_n) - \sum_{n=0}^{N-1} H(p_n, q_n)\tau}$$

$$(4.115)$$

where $H(p_n, q_n) = \frac{1}{2m} p_n^2 + V(q_n)$ is the c-number valued classical energy associated with momentum p_n and coordinate q_n. Although the integrals in (4.115) involve an oscillatory factor, the second part of the exponent is negative and real, resulting in absolute convergence of the integrals (provided, as mentioned above, that the potential $V(q)$ is bounded below!). As in the special case of the simple harmonic oscillator, the limit defines a conditional Wiener measure over continuous *phase-space paths* ($\{q(t), p(t)\}, t_i < t < t_f$) with the boundary condition $q(t_i) = q_i, q(t_f) = q_f$ and $p(t_i), p(t_f)$ unrestricted (all momentum integrals in (4.115) range from $-\infty$ to $+\infty$).

Of course, the Gaussian integral over the intermediate momentum p_n in (4.114) can easily be evaluated to leave us with

$$\int \frac{dp_n}{2\pi} e^{ip_n(q_{n+1} - q_n) - \frac{1}{2m} p_n^2 \tau} = (\frac{m}{2\pi\tau})^{1/2} e^{-\frac{1}{2} m \frac{(q_{n+1} - q_n)^2}{\tau}} \qquad (4.116)$$

which we immediately recognize as the first factor in (4.96), the infinitesimal time propagator for the simple harmonic oscillator. The product over all time intervals of the remaining factor in (4.96) gives (with $V(q) = \frac{1}{2} m\omega^2 q^2$ for the simple harmonic oscillator)

$$\prod_{n=0}^{N-1} e^{-\tau \frac{m\omega^2}{2} \frac{q_{n+1}^2 + q_n^2}{2}} = e^{-\tau \sum_{n=0}^{N-1} V(q_n) - \frac{m\omega^2}{4}(q_f^2 - q_i^2)\tau} \rightarrow e^{-\tau \sum_{n=0}^{N-1} V(q_n)}, \quad \tau \to 0$$

$$(4.117)$$

corresponding to the product of the potential terms: i.e., the first factor (outside the integral) in (4.114). This establishes the equivalence of our new integral representation (4.115), involving the Hamiltonian of the system and an integration over phase-space paths in coordinate and momentum, with the results obtained previously (for the harmonic oscillator) in which the exponent involved the Lagrangian (a function of coordinates and velocities) and an integration over paths in coordinate space solely.

The Wiener measure defined by the limit in (4.115) will be indicated by a notation analogous to that used previously in the Lagrangian formulation (4.100): namely,

$$K_E(q_f, t_f; q_i, t_i) = \int \mathbf{D}p\mathbf{D}q e^{\int_{t_i}^{t_f} (ip(t)\dot{q}(t) - H(p(t), q(t))) dt} \tag{4.118}$$

Implicit in this expression are (i) the boundary conditions $q(t_i) = q_i$, $q(t_f) = q_f$, and (ii) the $\frac{1}{2\pi}$ factors in the measure for the momentum integrations, visible in (4.115).

A Hamiltonian path integral representation for the real-time propagation amplitude $K(q_f, t_f; q_i, t_i)$ can be recovered from the Euclidean version (4.115) by the analytic continuation discussed above in the Lagrangian case: namely, we rotate $\tau \rightarrow (i + \epsilon)\tau$, obtaining

$$K(q_f, t_f; q_i, t_i) = \lim_{\backslash N \rightarrow \infty} \int \prod_{n=1}^{n=N-1} dq_n \prod_{n=0}^{N-1} \frac{dp_n}{2\pi} e^{i \sum_{n=0}^{N-1} (p_n(q_{n+1} - q_n) - (1 - i\epsilon)H(p_n, q_n)\tau)} \tag{4.119}$$

The integrals here are oscillatory except for the real factors $e^{-\epsilon H(p_n, q_n)\tau}$, which ensure absolute convergence provided H is bounded below (and grows for large q_n, p_n). Again, one typically uses an abbreviated notation for this real-time version of the Hamitonian path integral:

$$K(q_f, t_f; q_i, t_i) = \int \mathbf{D}p\mathbf{D}q e^{i \int_{t_i}^{t_f} (p(t)\dot{q}(t) - H(p(t), q(t))) dt} \tag{4.120}$$

but it must be understood that this integral is given meaning by a hidden $i\epsilon$ factor, as in (4.119). If Planck's constant is once more made explicit (as in (4.105)) and the stationary phase approximation applied to the classical limit $\hbar \rightarrow 0$, we find the *modified Hamilton's principle* (Leech, 1965) as the condition selecting the classical path in phase space:

$$\delta \int_{t_i}^{t_f} (p(t)\dot{q}(t) - H(p(t), q(t))) dt = 0 \tag{4.121}$$

4.2.3 Time-ordered products and operator ordering in the path-integral method

The path-integral formulations introduced in the preceding two sections provided integral representations for a particular quantum-mechanical quantity: the propagation amplitude for a particle localized at a specified position at some initial time (i.e., the state $|q_i, t_i\rangle$) to be detected at another position at some later time (corresponding to the state $|q_f, t_f\rangle$). This overlap amplitude can be generalized to allow the consideration of matrix elements between these initial and final states of products of Heisenberg operators. It turns out that the path-integral method is ideally suited for the representation of such matrix elements, *but only if the corresponding operator product is time-ordered.* As a simple example, consider the propagation of the system from initial time t_i to final time t_f with two intermediate times t_1, t_2 specified, and $t_1 > t_2$. The product $\mathbf{q}_H(t_1)\mathbf{q}_H(t_2)$ is then time-ordered (later operator to the left), and

$$\langle q_f, t_f | \mathbf{q}_H(t_1)\mathbf{q}_H(t_2) | q_i, t_i \rangle = \langle q_f | e^{-iHt_f} e^{iHt_1} \mathbf{q} e^{-iHt_1} e^{iHt_2} \mathbf{q} e^{-iHt_2} e^{iHt_i} | q_i \rangle$$

$$= \langle q_f | e^{-iH(t_f - t_1)} \mathbf{q} e^{-iH(t_1 - t_2)} \mathbf{q} e^{-iH(t_2 - t_i)} | q_i \rangle$$

$$(4.122)$$

We now repeat the steps leading from (4.112) to (4.113), dividing the time interval (t_i, t_f) into N subintervals. For very large N, we may assume that the times t_1, t_2 are arbitrarily close to the discrete times t_{n_1}, t_{n_2}. The only modifications to the previous calculation are therefore the appearance of the position operator $\mathbf{q}$ before the states $|q_{n_1}\rangle$ and $|q_{n_2}\rangle$ in the matrix elements (4.114), leading to the additional c-number factor $q_{n_1} q_{n_2}$ in the full functional integral. In the continuum limit, this additional factor is just $q(t_1)q(t_2)$, so we obtain, in analogy to (4.120)

$$\langle q_f, t_f | \mathbf{q}_H(t_1)\mathbf{q}_H(t_2) | q_i, t_i \rangle = \int \mathbf{D}p\mathbf{D}q \; q(t_1)q(t_2) e^{i \int_{t_i}^{t_f} (p(t)\dot{q}(t) - H(p(t), q(t))) dt} \quad (4.123)$$

It is important to realize that the order of the products $q(t_1)q(t_2)$ inside the path integral (4.123) is irrelevant, as at this stage we are dealing with c-number real-valued functions which multiply commutatively. However, the above argument shows that the path integral *automatically* computes the matrix element of the time-ordered product of the corresponding Heisenberg operators. In other words, *irrespective of the time order of t_1 and t_2*, we have

$$\langle q_f, t_f | T(\mathbf{q}_H(t_1)\mathbf{q}_H(t_2)) | q_i, t_i \rangle = \int \mathbf{D}p\mathbf{D}q \; q(t_1)q(t_2) e^{i \int_{t_i}^{t_f} (p(t)\dot{q}(t) - H(p(t), q(t))) dt}$$

$$(4.124)$$

The arguments of the preceding section leading to the result (4.118) for the imaginary time kernel may be repeated with the insertion of imaginary time Heisenberg operators (which evolve according to $\mathbf{q}_H(t) = e^{Ht}\mathbf{q}_H(0)e^{-Ht}$): unsurprisingly, the corresponding matrix element (which we distinguish with the subscript E) has the path-integral representation

$$\langle q_f, t_f | T(\mathbf{q}_H(t_1)\mathbf{q}_H(t_2)) | q_i, t_i \rangle_E = \int \mathbf{D}p\mathbf{D}q \; q(t_1)q(t_2) e^{\int_{t_i}^{t_f} (ip(t)\dot{q}(t) - H(p(t), q(t))) dt}$$

$$(4.125)$$

It is, of course, straightforward to repeat the above argument to establish that the insertion of arbitrary multi-nomials in both the coordinate $q(t)$ and momentum $p(t)$ values at distinct times in the integral (4.124) results in an integral representation for the matrix element of the corresponding time-ordered products of the Heisenberg coordinate $\mathbf{q}_H(t)$ and momentum $\mathbf{p}_H(t)$ operators. The same will be true were terms in the exponent of (4.124) involving products of $p(t)$ and $q(t)$ at different times to be present. Such terms do not appear in (4.124) as it stands, and one may wonder why they ever would! However, the fact that the path-integral formulation involves only commuting c-number functions—either on coordinate or on phase-space—leads to the following puzzle, which perhaps has already occurred to the reader. If the Hamiltonian $H(\mathbf{p}, \mathbf{q})$ contains terms with both coordinate and momentum operators, with some specified ordering (chosen, of course, to maintain the self-adjoint property of H), how can this be reflected in the path integral where different orders of multiplication of the c-number valued $p(t)$ and $q(t)$ seem manifestly equivalent? For example, if at the operator level the Hamiltonian contained a term (λ a real constant)

$$V_1 \equiv \lambda \mathbf{p}\mathbf{q}^2\mathbf{p} \tag{4.126}$$

the corresponding c-number term in the exponent of (4.124) would be indistinguishable from the path integral corresponding to a term

$$V_2 \equiv \frac{\lambda}{2}(\mathbf{p}^2\mathbf{q}^2 + \mathbf{q}^2\mathbf{p}^2) \tag{4.127}$$

despite the fact that (temporarily abandoning natural units to restore Planck's constant)

$$V_1 = V_2 + \lambda\hbar^2 \tag{4.128}$$

so that the two Hamiltonians definitely lead to different *quantum* dynamics. The solution to this quandary is to realize that in situations like this the apparently innocent continuum limit (4.119) develops ambiguities which must be resolved by temporarily separating the times of the coordinates and momenta to indicate the desired ordering. Thus propagators for a Hamiltonian involving the term V_1 above would be generated by a path integral in which the c-number Hamiltonian $H(p(t), q(t))$ in (4.120) contains a term

$$\lambda p(t + \delta)q(t)^2 p(t - \delta) \tag{4.129}$$

where δ is a small time-interval which is only set to zero *after* the $N \to \infty$ limit in (4.119) is carried out. Likewise, if we desire the propagator in a theory containing the term V_2 above, we need to regularize the path integral by adding a term

$$\frac{\lambda}{2}(p(t + \delta)^2 q(t)^2 + p(t)^2 q(t + \delta)^2) \tag{4.130}$$

to $H(p(t), q(t))$ in (4.120), with the limit $\delta \to 0$ performed only after the discrete time-limit defining the path integral is performed (i.e., $\tau \to 0$).

4.2.4 Ground-state expectation values from path integrals

One of the most important applications of the path-integral method lies in the evaluation of the ground-state expectation value of observables. Essentially all non-perturbative formulations of quantum field theory begin with the study of such expectation values of (products of) the field operators. This technique also lies at the heart of modern approaches to the numerical computation of the spectrum of theories such as quantum chromodynamics (the field theory describing the strong interactions of quarks and gluons), where perturbative methods fail. We begin with the result (4.124) for the real-time expectation value of the product of two Heisenberg coordinate operators:

$$\langle q_f, +t | T(\mathbf{q}_H(t_1)\mathbf{q}_H(t_2)) | q_i, -t \rangle = \int \mathbf{D}p\mathbf{D}q \; q(t_1)q(t_2) e^{i\int_{-t}^{+t}(p(t)\dot{q}(t) - H(p(t),q(t)))dt}$$

$$= \langle q_f, 0 | e^{-iHt(1-i\epsilon)} T(\mathbf{q}_H(t_1)\mathbf{q}_H(t_2)) e^{-iHt(1-i\epsilon)} | q_i, 0 \rangle$$

$$(4.131)$$

Note that we have chosen to evolve the system over the symmetric time-interval from $-t$ to $+t$, and that the $1 - i\epsilon$ factor needed to make the path integral well-defined (see (4.119)) is explicitly displayed in the corresponding matrix element. We now assume that our quantum system has a unique ground state $|0\rangle$ of energy E_0, separated in energy from the first excited state (or states) by a finite gap $E_1 - E_0$. If we insert a complete set of energy eigenstates $1 = \sum|n\rangle\langle n|$ (where $H|n\rangle = E_n|n\rangle$), we find

$$e^{-iHt(1-i\epsilon)} |q_i, 0\rangle = \sum_n e^{-iE_n t(1-i\epsilon)} \langle n|q_i, 0\rangle |n\rangle$$

$$\to e^{-iE_0 t(1-i\epsilon)} \langle 0|q_i, 0\rangle |0\rangle + O(e^{-\epsilon(E_1 - E_0)t}), \; t \to \infty$$

$$\langle q_f, 0 | e^{-iHt(1-i\epsilon)} \to \langle 0|\langle q_f, 0|0\rangle e^{-iE_0 t(1-i\epsilon)} + O(e^{-\epsilon(E_1 - E_0)t}), \; t \to \infty \quad (4.132)$$

In other words, the infinite time limit of the path integral acts as a "low-pass filter", effectively selecting out the ground-state component of the initial and final states. If we divide the matrix element in (4.131) by the same quantity without the T-product (i.e., by $\langle q_f, t | q_i, -t \rangle$) and take the large time limit, the exponential time factors and overlap factors $\langle 0|q_i, 0\rangle$ and $\langle q_f, 0|0\rangle$ cancel, leaving

$$\lim_{t \to \infty} \frac{\langle q_f, +t | T(\mathbf{q}_H(t_1)\mathbf{q}_H(t_2)) | q_i, -t \rangle}{\langle q_f, t | q_i, -t \rangle} = \frac{\langle 0 | T(\mathbf{q}_H(t_1)\mathbf{q}_H(t_2)) | 0 \rangle}{\langle 0|0 \rangle}$$

$$= \frac{\int \mathbf{D}p\mathbf{D}q \; q(t_1)q(t_2) e^{i\int_{-\infty}^{+\infty}(p(t)\dot{q}(t) - H(p(t),q(t)))dt}}{\int \mathbf{D}p\mathbf{D}q \; e^{i\int_{-\infty}^{+\infty}(p(t)\dot{q}(t) - H(p(t),q(t)))dt}} \quad (4.133)$$

This result generalizes in an obvious way to time-ordered products of n $\mathbf{q}_H(t)$ operators, which suggests the introduction of the following *generating functional*:

$$Z[j] \equiv \int \mathbf{D}p\mathbf{D}q \; e^{i\int_{-\infty}^{+\infty}(p(t)\dot{q}(t) - H(p(t),q(t)) - j(t)q(t))dt} \quad (4.134)$$

allowing us to generate arbitrary ground-state expectations of time-ordered coordinate operators by functional differentiation[12] with respect to the c-number source function $j(t)$:

$$\frac{\langle 0|T(\mathbf{q}_H(t_1)\dots\mathbf{q}_H(t_n))|0\rangle}{\langle 0|0\rangle} = \frac{1}{Z[j]}\frac{i^n\delta^n Z[j]}{\delta j(t_1)\dots\delta j(t_n)}\bigg|_{j=0} \qquad (4.135)$$

4.2.5 Quantum symmetries: path-integral aspects

The reformulation of operator quantum theory in the quasiclassical c-number framework of the path integral suggests that the transferral of classical symmetries to their quantum analogs should be particularly straightforward using path integral techniques. And in many cases of interest, this is exactly what we find. For the case of three-dimensional rotation symmetry discussed in Section 4.1.3, for example, where a particle moves in a rotationally invariant potential, subject to the Lagrangian

$$\mathcal{L}(\dot{\vec{r}},\vec{r}) = \frac{1}{2}m\frac{d\vec{r}}{dt}\cdot\frac{d\vec{r}}{dt} - V(\vec{r}(t)) \qquad (4.136)$$

the Lagrangian path-integral representation of the propagator (4.70) takes the form

$$\langle\vec{r}_{\mathrm{f}}|e^{-iHT}|\vec{r}_{\mathrm{i}}\rangle = \int \exp\{\frac{i}{\hbar}\int_0^T \mathcal{L}(\dot{\vec{r}},\vec{r})dt\}\,\mathbf{D}\vec{r},\ \ \vec{r}(0) = \vec{r}_{\mathrm{i}}, \vec{r}(T) = \vec{r}_{\mathrm{f}} \qquad (4.137)$$

with $V(\vec{r})$ a rotationally invariant potential, $V(R(\vec{\alpha})\vec{r}) = V(\vec{r})$. The dynamical rotational invariance expressed by (4.70) follows immediately by making a change of integration variable in the functional integral $\vec{r}(t) \to R(\vec{\alpha})\vec{r}(t)$ (rendered precise by discretization, as in (4.104)). All that is required is (a) the invariance property of the Lagrangian, $\mathcal{L}(R(\vec{\alpha})\dot{\vec{r}}, R(\vec{\alpha})\vec{r}) = \mathcal{L}(\dot{\vec{r}},\vec{r})$, and (b) the invariance of the functional measure, which follows from the unimodular property of the rotation matrices, $\det(R(\vec{\alpha})) = 1$.

For the canonical symmetries discussed previously, in which new coordinates and momenta are introduced which are in general non-linear functions of the old ones (cf. Section 4.1.3), the realization of the canonical symmetry at the quantum level can involve some subtle issues. Classically, for generating functions lacking explicit time-dependence, the new Hamiltonian (expressed in the new variables) is algebraically equal to the old Hamiltonian. In the quantum case there may be additional "anomalous" terms of order $\hbar$ (or higher powers of $\hbar$). In the path-integral formalism these terms appear as (a) a consequence of a non-trivial Jacobian in the change of variables q_n, p_n in the (discretized) Hamiltonian version of the path integral (4.119), and (b) time reordering of coordinate and momentum variables when a discretized version of the continuous classical contact transformation is implemented (recall our discussion above of operator ordering in the path-integral context). We shall return later in the book to the important issue of quantum anomalies in classical symmetries in quantum

---

[12] For a review of the basic elements of functional calculus, see Appendix A.

field theory: the reader interested in a more thorough discussion of these issues in non-relativistic quantum mechanics is referred to the work of Swanson (Swanson, 1993).

4.3   Scattering theory

The importance of scattering theory in the development of quantum field theory is to some extent a technological accident, stemming from the development of particle accelerators in this century as the primary experimental tool of subatomic physics. The primary source of phenomenological information concerning the behavior of matter at very short distance scales remains the observation of collision processes of elementary particles in high-energy accelerators. A typical such process may be described qualitatively in the following terms. A short time (in macroscopic units, but effectively at time $t = -\infty$ in time units appropriate for elementary particle interactions) before the collision, some number (typically two) of particles are travelling freely towards each other. A short time after the collision (effectively, at $t = +\infty$) the state of the system may be resolved into a complicated linear combination of states, each of which corresponds to some definite number of particles with various momenta and quantum numbers. The coefficient of each such state is a complex amplitude, the square of which is just the probability that this state was the end result of the collision. Our ultimate objective is to gain an understanding of the underlying dynamics giving rise to these experimentally measurable amplitudes.

Let us be a little more specific. The notation $|\alpha\rangle_{in}$ will refer to the state described above, where α is a shorthand for the complete specification of the momenta, spins, etc., of the incoming particles. For example, if we were colliding two spinless particles, α would simply specify (k_1, k_2), the two momenta of the incoming particles.

Strictly speaking, the state in which the incoming particles are initially localized, far apart, and moving towards one another must be obtained by constructing wave-packets. This is done in the usual way: one must fold the momentum eigenstates (which are infinitely extended plane waves) with a smooth (e.g., Gaussian) function of momentum to produce a wavefunction of finite extent in coordinate space. In other words, the physical "in-state" which is really prepared in an accelerator is actually

$$\int g(\alpha)|\alpha\rangle_{in} d\alpha \tag{4.138}$$

where $g(\alpha)$ is the folding function. One may also define "out-states" correspondingly as physical states in which the system goes over to a definite number of free outgoing particles after the collision. These are *not* the states prepared in any conceivable accelerator, but they *are* the states measured by the detectors after the collision has taken place. The amplitude that a given incoming state $|\alpha\rangle_{in}$ will then be found to be in the state $|\beta\rangle_{out}$ by a detector measurement after the collision is just the overlap

$$_{out}\langle\beta|\alpha\rangle_{in} \equiv S_{\beta\alpha} \tag{4.139}$$

The quantity $S_{\beta\alpha}$ is called the S-matrix and is of fundamental importance. It is convenient to think of it as a matrix, although in general the indices α, β may contain continuous (e.g., momenta of the incoming and outgoing particles) as well as discrete (e.g., spin and isospin quantum numbers) variables. A column of this matrix lists all the amplitudes for possible final states β arising from a given initial state α. The probability interpretation of quantum mechanics requires that the sum of the absolute squares of these amplitudes must be unity (*something* has to happen!). This is just the property of a unitary matrix. The unitarity of the S-matrix, which we shall see below follows from the hermiticity of the Hamiltonian, is one of the fundamental constraints which we will have to keep in mind when building quantum field theories.

4.3.1 Convergence notions for states in Hilbert space

It will be useful to present a more detailed account of these admittedly rather vague concepts in a familiar context: non-relativistic potential scattering of a particle in three dimensions. Thus, the dynamics of our particle of mass m is determined by a Hamiltonian

$$H = H_0 + V = \frac{\vec{\mathbf{p}}^2}{2m} + V(\vec{\mathbf{r}}) \tag{4.140}$$

In the qualitative discussion of scattering theory given above, the concepts of free particles localized far apart and moving towards or away from each other play a prominent role, as does the notion of the infinite time limits (both at $t \to -\infty$ and at $t \to +\infty$) of the quantum state of the scattered particle(s). The latter notion is particularly subtle, and we must pause here to remind the reader[13] of the differing notions of convergence which can apply to vectors in a Hilbert space. Recall that a Hilbert space with a complex inner product is *ipso facto* a *normed* space, where a "norm" (or "length") of a state vector $|\Psi\rangle$ can be defined as

$$|||\Psi\rangle|| \equiv \sqrt{\langle\Psi|\Psi\rangle} \tag{4.141}$$

In order to avoid the unsightly concatenation of vertical bars in (4.141), we shall temporarily abandon Dirac notation and denote state vectors by capital Greek letters, inner products by round brackets, and the norm by a double bar as above:

$$|\Psi\rangle \to \Psi$$
$$\langle\Phi|\Psi\rangle \to (\Phi, \Psi)$$
$$|||\Psi\rangle|| \to ||\Psi|| \tag{4.142}$$

In the Schrödinger picture we have time-dependent states $\Psi(t)$, and the question arises whether it makes sense to consider infinite time limits ("far past" or "far future") of such states. In general, a sequence of states Ψ_n is said to *converge weakly* to Υ if

$$\lim_{n\to\infty} (\Phi, \Psi_n) = (\Phi, \Upsilon) \tag{4.143}$$

for any fixed Hilbert space vector Φ. In particular, a sequence of states has a weak limit if each component of these states in a complete orthonormal basis converges

---

[13] For a more detailed introduction to the relevant concepts from functional analysis, see (Newton, 1966), Chapter 6.

separately to a finite limit. In this case we write simply

$$\Psi_n \to \Upsilon \tag{4.144}$$

The reason for the appelation "weak" becomes apparent when we consider that the sequence of states $(1, 0, 0, 0, \ldots), (0, 1, 0, 0, \ldots), (0, 0, 1, 0, \ldots)$, etc., specified by listing their components in a denumerable orthonormal basis, converges weakly to zero,[14] even though the norm of each state in the sequence is unity! Another equally off-putting case is given by the wavefunction for a localized wave-packet, given in coordinate space by

$$\Psi(\vec{r}, t) = \langle \vec{r} | \Psi, t \rangle = \int g(\vec{p}) e^{i\vec{p} \cdot \vec{r} - i \frac{p^2}{2m} t} d^3 p \tag{4.145}$$

which also converges weakly to zero at large (negative or positive) times, due to the famous spreading of the wave-packet, which implies that the overlap of $\Psi(t)$ with any (normalizable, and hence essentially localized) state must vanish at large times. Our intuitive feeling of "convergence" conforms more closely to a stronger requirement than that implied by (4.143). We say that the sequence of states Ψ_n converges *strongly* to the state vector Υ if the norm of the difference vectors converges to zero:

$$\lim_{n \to \infty} ||\Psi_n - \Upsilon|| = 0 \tag{4.146}$$

written concisely with a double-arrow as

$$\lim_{n \to \infty} \Psi_n \Rightarrow \Upsilon \tag{4.147}$$

With this definition, neither of the examples of weak convergence given in the preceding paragraph survive (as the individual states in the sequence have fixed norm and can clearly not converge in the strong sense to zero!). Correspondingly, weak (resp. strong) convergence of a sequence of *operators* O_n to a limit operator O can be defined by requiring that for every fixed state Φ, $O_n\Phi \to O\Phi$ (resp. $O_n\Phi \Rightarrow O\Phi$).

4.3.2   In- and out-states in potential scattering

Keeping the preceding mathematical niceties in mind, we now return to the somewhat slippery problem of specifying precisely the initial conditions of a scattering event. The limit $t \to -\infty$ implied so far in the specification of an "in-state" should not be taken literally: in that limit any particle described by a wave-packet of finite width at finite time will be spread over all of space with vanishing probability density at any given point, which hardly corresponds to the origin of electrons or protons used to form the beams in present-day high-energy accelerators, which are extracted from perfectly sensible localized states in originally neutral atoms. Nevertheless, we may retain the useful mathematical fiction of "infinite past (or future)" by noting that free and interacting solutions of the time-dependent Schrödinger equation can be

---

[14] Indeed, the overlap of these vectors with any finite norm vector with components c_n gives just $c_n \to 0, n \to \infty$, as the c_n must go to zero if $\sum_n |c_n|^2 < \infty$.

t→-∞
H and H₀
converge strongly

map to interacting region (provided V short range)

tint.

Scattering theory   **99**

found which approach each other arbitrarily closely in the "strong" sense in the limit $t \to -\infty$, and that this unique association of free with interacting states allows us to define a mapping from an arbitrary (finite norm) free state $\Psi(t)$ evolving according to H_0 to an interacting state $\Psi_{\text{in}}(t)$ evolving according to the full Hamiltonian H, *provided that the interaction potential V is sufficiently short-ranged.* To see how to do this, we work for the moment in coordinate representation and imagine that we are provided with a normalizable solution $\Psi(\vec{r}, t)$ of the free time-dependent Schrödinger equation

$$-\frac{1}{2m}\Delta\Psi(\vec{r}, t) = i\frac{\partial\Psi(\vec{r}, t)}{\partial t} \tag{4.148}$$

For definiteness, such a solution is given by (4.145), with $g(\vec{p})$ a smearing function peaked at some 3-momentum $\vec{p}_0$ and some width Δp, such that the resultant wave-packet is localized in the neighborhood of the spatial origin at time $t = 0$. For any given large negative time T, we now define an interacting solution $\Psi^{(T)}(\vec{r}, t)$ associated with this freely evolving state as the solution of the interacting Schrödinger equation

$$-\frac{1}{2m}\Delta\Psi^{(T)}(\vec{r}, t) + V(\vec{r})\Psi^{(T)}(\vec{r}, t) = i\frac{\partial\Psi^{(T)}(\vec{r}, t)}{\partial t} \tag{4.149}$$

subject to the boundary condition

$$\Psi^{(T)}(\vec{r}, T) = \Psi(\vec{r}, T) \tag{4.150}$$

It is intuitively plausible that for potentials which fall off sufficiently rapidly at large distances from the origin (where we imagine the interaction potential to be concentrated) the effect of the potential should be increasingly negligible as T goes to $-\infty$, as the center of the wave-packet recedes from the center of the potential. The rigorous proof of this hypothesis clearly requires a careful consideration of the spreading of the wave-packet in relation to the falloff of the potential. As shown by Brenig and Haag (Brenig and Haag, 1963), it suffices that the potential $V(\vec{r})$ falls off faster than $r^{-(1+\epsilon)}, \epsilon > 0$. Assuming this to be the case, it can then be shown that the interacting and free solutions approach one another strongly in the sense that the norm of the difference of free and interacting solutions is uniformly bounded in the far past. Thus, defining the shift $\chi^{(T)}(t)$ in the state due to the potential

$$\chi^{(T)}(t) \equiv \Psi^{(T)}(t) - \Psi(t), \quad \chi^{(T)}(T) = 0 \tag{4.151}$$

we expect that in the far past the influence of a localized potential must vanish for wave-packets then localized far from the center of the potential, i.e.,

$$\|\chi^{(T)}(t)\| < F(T), \quad t < T, \quad \text{with } F(T) \to 0, \ T \to -\infty \tag{4.152}$$

and that in this limit $\Psi^{(T)}(0)$ (i.e., the interacting solution matched in the far past to a specified free wave-packet, run forward to time zero) converges to a well-defined Heisenberg state (recall that the various representations are defined to coincide at time zero):

$$\Psi^{(T)}(0) \Rightarrow \Psi_{\text{in}}, \quad T \to -\infty \tag{4.153}$$

Of course, as emphasized in the preceding section, the limit $T \to -\infty$ corresponds physically to a very short time on any macroscopic time-scale.[15]

Note that since $\Psi^{(T)}(t) = e^{-iHt}\Psi^{(T)}(0)$, and $\Psi(t) = e^{-iH_0 t}\Psi(0)$, the boundary condition (4.150) implies

$$\Psi^{(T)}(0) = e^{iHT}e^{-iH_0 T}\Psi(0) = U(0,T)\Psi(0) \tag{4.154}$$

where $U(t, t_0)$ is the time-development operator in the interaction picture (see (4.16)). It then follows from (4.153) that $U(0,T)$ has a strong limit when T is taken to the infinite past:

$$U(0,T) \Rightarrow \Omega^-, \quad T \to -\infty \tag{4.155}$$

where the Møller wave operator Ω^- maps the free state Ψ onto the in-state Ψ_{in} associated with it by the procedure described above:

$$\Psi_{in} = \Omega^- \Psi \tag{4.156}$$

An exactly analogous procedure, this time taking the limit $T \to +\infty$, can be used to associate any freely evolving wave-packet solution Ψ with an interacting state converging strongly to it in the far future: the associated Heisenberg state is then called Ψ_{out}:

$$\Psi_{out} = \Omega^+ \Psi \tag{4.157}$$

$$U(0,T) \Rightarrow \Omega^+, \quad T \to +\infty \tag{4.158}$$

In the case that the potential V admits bound states (a not infrequent situation!), the above discussion conceals some subtleties which we shall mention briefly here. In this situation the scattering eigenstates of the full Hamiltonian H are not complete, as the bound state(s) are missing. Indeed, the Møller wave operators are in this case norm-preserving maps from the full Hilbert space $\mathcal{H}(=L^2(R^3))$ spanned by the free solutions Ψ to the subspace $\mathcal{H}_{scat}$ spanned by the interacting scattering states Ψ_{in} (or Ψ_{out}). Indeed, for a bound state of energy E_b, normalized eigenstate Ψ_b, consider the overlap matrix element

$$(U(0,T)\Psi, \Psi_b) = e^{-iE_b T}(e^{-iH_0 T}\Psi, \Psi_b) \tag{4.159}$$

Recall that Ψ here represents a localized wave-packet (e.g., the state given in (4.145)) so that $e^{-iH_0 T}\Psi$ will spread in the limit $T \to -\infty$ to a state with pointwise vanishing probability density, and hence vanishing overlap with any stationary, localized bound-state wavefunction Ψ_b. Thus, the right-hand side of (4.159) vanishes in the limit $T \to -\infty$. As Ω^- is the strong limit of $U(0,T)$ as $T \to -\infty$, it follows that this Møller operator maps an arbitrary free state onto the proper subspace of states orthogonal

---

[15] We also note here that the Coulomb potential fails to satisfy the falloff condition posited above, and indeed, non-relativistic Coulomb scattering exhibits a number of subtleties, which, however, will not concern us further here. Related field-theoretic subtleties in defining the scattering matrix for theories with massless particles will be considered explicitly in Section 19.2.

to the discrete bound states of V. As $U(0, T)$ is norm-preserving at any finite T, its strong limit Ω^- is also norm-preserving, but not onto, and hence not unitary.[16]

Returning to the Møller operators defined by the limits (4.155, 4.158), it is clear that the strong limit $T \to \pm\infty$ is unaffected by any fixed finite time-shift t; accordingly

$$\Omega^\pm = e^{iHt}\Omega^\pm e^{-iH_0 t} \tag{4.160}$$

for all t, whence, taking t infinitesimal

$$H\Omega^\pm = \Omega^\pm H_0 \tag{4.161}$$

We are finally equipped with the necessary tools to introduce the central concept of scattering theory: the S-matrix, which we discussed qualitatively in the preceding section. We recall that the Hilbert space $L^2(R^3)$ is separable, so the free solutions Ψ may be expanded in a discrete orthonormal basis of finite-norm states Ψ_α (α a discrete index)

$$\Psi = \sum_\alpha (\Psi_\alpha, \Psi)\Psi_\alpha \tag{4.162}$$

In particular, any localized wave-packet describing either the state prepared by an accelerator or the state detected after a collision can be so expanded. Each free basis state is associated with an in- or out-state ($\Psi_{\alpha,\text{in}}$ or $\Psi_{\alpha,\text{out}}$) via the Møller operators introduced above. The quantum-mechanical amplitude that a state prepared in the far past to match the behavior of the free state Ψ_α will evolve into the detected state Ψ_β in the far future is then given by the overlap

$$(\Psi_{\beta,\text{out}}, \Psi_{\alpha,\text{in}}) \equiv S_{\beta\alpha} = (\Omega^+ \Psi_\beta, \Omega^- \Psi_\alpha) = (\Psi_\beta, \Omega^{+\dagger}\Omega^- \Psi_\alpha) \tag{4.163}$$

Using (4.155, 4.158), this becomes

$$S_{\beta\alpha} = \lim_{T' \to \infty, T \to -\infty} (\Psi_\beta, U(0, T')^\dagger U(0, T)\Psi_\alpha)$$

$$= \lim_{T' \to \infty, T \to -\infty} (\Psi_\beta, U(T', T)\Psi_\alpha) \equiv (\Psi_\beta, U(+\infty, -\infty)\Psi_\alpha) \tag{4.164}$$

where the limits in the last two lines are strong, which together with the unitarity of $U(t, t_0)$ at finite t, t_0 and the completeness of the free states Ψ_α ensures that the infinite discrete matrix S is unitary in the standard way (*even in the presence of bound states*).[17] In practice, it is more convenient for obvious reasons to use continuum-normalized states: we go over to wave-packets of arbitrarily well-defined momentum, for example, in which limit the S-matrix amplitudes remain well defined (again, with

---

[16] As a simple example of a norm-preserving but non-unitary operator, consider the operator represented by the infinite discrete matrix $O_{nm} = \delta_{n,m+1}$, which maps an arbitrary vector in l^2, the Hilbert space of square-summable infinite complex sequences, into a shifted vector of equal norm but one with no first component.

[17] The unitarity of the S-matrix defined as an overlap of in- and out-states even in the presence of unitary defects in various interaction-picture operators will (fortunately) persist in quantum field theory, where the defect will become total, given Haag's theorem for the non-existence of the interaction picture, discussed below in Chapter 10.

the proviso of suitably localized interaction potentials). Then the discrete index α above is replaced by a specification of the momentum (and if present, spins) of the incoming particle. It should be emphasized that the matrix S defined above is to be thought of as the matrix of an operator S acting in the full Hilbert space spanned by freely-evolving wave-packets. Energy conservation requires commutation of S with H_0, the free Hamiltonian. Indeed, Ψ_α and Ψ_β can be chosen to be free wave-packets of arbitrarily well-defined H_0 eigenvalue, in which case we certainly require $S_{\beta\alpha}$ to vanish if $E_\beta \neq E_\alpha$, and this is ensured by the intertwining property (4.161):

$$H_0 S = H_0 \Omega^{+\dagger}\Omega^- = \Omega^{+\dagger}H\Omega^- = \Omega^{+\dagger}\Omega^- H_0 = SH_0 \tag{4.165}$$

4.3.3 Time-independent scattering theory

The derivation of many important theorems of scattering theory—in particular, the generalized optical theorem, which we prove below—is facilitated by another important result, the Lipmann–Schwinger equation, encapsulating the relation between free states and the interacting states matched to them in the far past or future. We derive this equation here in the case of in-states (the analogous result for out-states following from a completely parallel argument). From the integral equation (4.22) it follows that

$$U(t, -\infty) = U(t,0)\Omega^- = e^{iH_0 t}e^{-iHt}\Omega^- = e^{iH_0 t}\Omega^- e^{-iH_0 t} \tag{4.166}$$

satisfies

$$U(t, -\infty) = 1 - i \int_{-\infty}^{t} V_{\mathrm{ip}}(t_1) U(t_1, -\infty) dt_1 \tag{4.167}$$

whence, inserting (4.166),

$$e^{iH_0 t}\Omega^- e^{-iH_0 t} = 1 - \int_{-\infty}^{t} V_{\mathrm{ip}}(t_1) e^{iH_0 t_1}\Omega^- e^{-iH_0 t_1} dt_1 \tag{4.168}$$

or equivalently

$$\Omega^- = 1 - i \int_{-\infty}^{t} e^{-iH_0 t} V_{\mathrm{ip}}(t_1) e^{iH_0 t_1}\Omega^- e^{-iH_0(t_1-t)} dt_1 \tag{4.169}$$

$$= 1 - i \int_{-\infty}^{t} e^{iH_0(t_1-t)} V\Omega^- e^{-iH_0(t_1-t)} dt_1 \tag{4.170}$$

$$= 1 - i \int_{-\infty}^{0} e^{iH_0 t_1} V\Omega^- e^{-iH_0 t_1} dt_1 \tag{4.171}$$

We now take the matrix element of this result between two free wave-packet solutions Ψ_β and Ψ_α which are arbitrarily close to energy eigenstates with energy E_β, E_α. The free time-development factors $e^{\pm iH_0 t_1}$ mean that the centroids of these packets are moved very far from our presumably localized potential V at large negative times t_1, so we may at no cost insert an adiabatic switching factor $e^{\epsilon t_1}$ ($\epsilon > 0$) multiplying the potential: for very small ϵ (ϵ will be taken to zero at the very end), switching off the interaction potential at very large negative times will have no effect if the wave-packets are still very far from the potential center. We then obtain

$$(\Psi_\beta, \Omega^- \Psi_\alpha) = (\Psi_\beta, \Psi_\alpha) - i \int_{-\infty}^{0} e^{i(E_\beta - E_\alpha - i\epsilon)t_1}(\Psi_\beta, V\Omega^- \Psi_\alpha) \qquad (4.172)$$

$$= (\Psi_\beta, \Psi_\alpha) - \frac{1}{E_\beta - E_\alpha - i\epsilon}(\Psi_\beta, V\Omega^- \Psi_\alpha) \qquad (4.173)$$

$$= (\Psi_\beta, \Psi_\alpha) + (\Psi_\beta, \frac{1}{E_\alpha - H_0 + i\epsilon}V\Omega^- \Psi_\alpha) \qquad (4.174)$$

As the Ψ_β can be chosen to run over a complete basis of the Hilbert space, we may remove it, obtaining the desired *Lipmann–Schwinger* equation, relating free to interacting scattering states:

$$\Psi_{\alpha,\text{in}} = \Psi_\alpha + \frac{1}{E_\alpha - H_0 + i\epsilon}V\Psi_{\alpha,\text{in}} \qquad (4.175)$$

In a similar fashion one may derive the Lipmann–Schwinger equation for out-states:

$$\Psi_{\alpha,\text{out}} = \Psi_\alpha + \frac{1}{E_\alpha - H_0 - i\epsilon}V\Psi_{\alpha,\text{out}} \qquad (4.176)$$

Note that by multiplying both sides of (4.175) by $E_\alpha - H_0$ (at which point the $i\epsilon$ becomes irrelevant) we find

$$(E_\alpha - H_0)\Psi_{\alpha,\text{in}} = V\Psi_{\alpha,\text{in}} \Rightarrow H\Psi_{\alpha,\text{in}} = E_\alpha \Psi_{\alpha,\text{in}} \qquad (4.177)$$

so that the interacting scattering state has the same energy relative to the full Hamiltonian H as the free state Ψ_α which matches it in the far past has relative to H_0.

By carrying the time evolution in (4.166) all the way forward to $t = +\infty$ we of course obtain the S-matrix element $S_{\beta\alpha}$:

$$S_{\beta\alpha} = \lim_{t \to +\infty} (\Psi_\beta, U(t, -\infty)\Psi_\alpha) \qquad (4.178)$$

$$= (\Psi_\beta, \Psi_\alpha) - i \int_{-\infty}^{+\infty} (\Psi_\beta, e^{iH_0 t_1}V\Omega^- e^{-iH_0 t_1}\Psi_\alpha)dt_1 \qquad (4.179)$$

$$= (\Psi_\beta, \Psi_\alpha) - 2\pi i \delta(E_\beta - E_\alpha)(\Psi_\beta, V\Omega^- \Psi_\alpha) \qquad (4.180)$$

Defining the *T-matrix element* $T_{\beta\alpha}$

$$T_{\beta\alpha} \equiv (\Psi_\beta, V\Omega^- \Psi_\alpha) = (\Psi_\beta, V\Psi_{\alpha,\text{in}}) \qquad (4.181)$$

this becomes, if we choose the Ψ_α from an orthonormal set,

$$S_{\beta\alpha} = \delta_{\beta\alpha} - 2\pi i \delta(E_\beta - E_\alpha)T_{\beta\alpha} \qquad (4.182)$$

At this point it will be convenient to return to Dirac notation for states and matrix elements, as the fine points of convergence that infest time-dependent scattering theory have already been discussed adequately for our purposes. The Lipmann–Schwinger equations (4.175, 4.176) will henceforth be written

$$|\alpha\rangle_{\text{in}} = |\alpha\rangle + \frac{1}{E_\alpha - H_0 + i\epsilon} V|\alpha\rangle_{\text{in}}, \quad |\alpha\rangle_{\text{out}} = |\alpha\rangle + \frac{1}{E_\alpha - H_0 - i\epsilon} V|\alpha\rangle_{\text{out}} \quad (4.183)$$

We will also abandon our previous insistence on normalizable (wave-packet) states and allow the label α to denote continuum orthonormalized states of well-defined momentum and energy (possibly containing a discrete spin index as well). The usual completeness relations will then involve integrals (as well as spin sums), denoted formally $\int d\alpha \ldots$ Similarly, the notation $\delta_{\alpha\beta}$ will denote a product of continuous δ-functions (in the momentum variables) and discrete Kronecker δs (for any discrete spin indices).

We now turn to the derivation of some formal scattering theorems of great importance. By taking the adjoint of the Lippmann–Schwinger (LS) equation for an in-state, we find

$$_{\text{in}}\langle\beta| = \langle\beta| + {}_{\text{in}}\langle\beta|V\frac{1}{E_\beta - H_0 - i\epsilon}$$

$$\Rightarrow \quad {}_{\text{in}}\langle\beta|V|\alpha\rangle_{\text{in}} = {}_{\text{in}}\langle\beta|V|\alpha\rangle + {}_{\text{in}}\langle\beta|V\frac{1}{E_\alpha - H_0 + i\epsilon}V|\alpha\rangle_{\text{in}}$$

$$= T^*_{\alpha\beta} + {}_{\text{in}}\langle\beta|V\frac{1}{E_\alpha - H_0 + i\epsilon}V|\alpha\rangle_{\text{in}} \quad (4.184)$$

$$= \langle\beta|V|\alpha\rangle_{\text{in}} + {}_{\text{in}}\langle\beta|V\frac{1}{E_\beta - H_0 - i\epsilon}V|\alpha\rangle_{\text{in}}$$

$$= T_{\beta\alpha} + {}_{\text{in}}\langle\beta|V\frac{1}{E_\beta - H_0 - i\epsilon}V|\alpha\rangle_{\text{in}} \quad (4.185)$$

In (4.184) we have used the LS equation on the right, and the definition (4.181) of the T-matrix; in (4.185) one uses the LS equation on the left. Subtract the right-hand sides of (4.184) and (4.185) to obtain

$$T_{\beta\alpha} - T^*_{\alpha\beta} = {}_{\text{in}}\langle\beta|V\{\frac{1}{E_\alpha - H_0 + i\epsilon} - \frac{1}{E_\beta - H_0 - i\epsilon}\}V|\alpha\rangle_{\text{in}} \quad (4.186)$$

If we insert a complete set of free eigenstates of H_0, with $\int d\gamma |\gamma\rangle\langle\gamma| = 1$, on the right of the second V in (4.186), we find

$$T_{\beta\alpha} - T^*_{\alpha\beta} = \int d\gamma \ {}_{\text{in}}\langle\beta|V|\gamma\rangle\langle\gamma|V|\alpha\rangle_{\text{in}}$$

$$\cdot\{\frac{1}{E_\alpha - E_\gamma + i\epsilon} - \frac{1}{E_\beta - E_\gamma - i\epsilon}\} \quad (4.187)$$

At this point we remind the reader of a famous identity: observe that

$$\frac{1}{x + i\epsilon} = \frac{x}{x^2 + \epsilon^2} - i\frac{\epsilon}{x^2 + \epsilon^2} \quad (4.188)$$

The function $\frac{\epsilon}{x^2+\epsilon^2}$ is a highly peaked function of x (for ϵ small) which integrates to π: in other words, it is just $\pi\delta(x)$. The odd function $\frac{x}{x^2+\epsilon^2}$ is a regularized form of $1/x$ which goes by the name of "principal-part of $\frac{1}{x}$", or simply $P(\frac{1}{x})$. Now by energy conservation, $E_\beta = E_\alpha$ and (4.187) becomes the so-called "Generalized Optical theorem":

$$T_{\beta\alpha} - T^*_{\alpha\beta} = -2i\pi \int d\gamma T^*_{\gamma\beta}\delta(E_\gamma - E_\alpha)T_{\gamma\alpha} \qquad (4.189)$$

A particularly important special case of the above result occurs for *forward scattering*: for example, when α, β refer to two-particle states with identical momenta, i.e., the elastic scattering amplitude in the limit of zero exchanged momentum. Then (4.189) becomes

$$\text{Im}(T_{\alpha\alpha}) = -\pi \int d\gamma \delta(E_\gamma - E_\alpha)|T_{\gamma\alpha}|^2 \qquad (4.190)$$

Evidently, the right-hand side of this relation describes the integrated probability (or total cross-section) for the given initial state $|\alpha\rangle_{\text{in}}$ to evolve into an arbitrary final state $|\gamma\rangle_{\text{out}}$ (naturally, of the same energy, hence the δ-function). The optical theorem relates this to the imaginary part of the the forward scattering amplitude $T_{\alpha\alpha}$. The term "optical" derives from the special case where the scattering is that of photons off neutral atoms in a medium, in which case the forward scattering amplitude is related to the index of refraction and the integrated scattering cross-section to the absorption.

The Generalized Optical theorem (G.O.T.) is really nothing but the previously advertised unitarity of the S-matrix, somewhat disguised, as the following brief computation shows:

$$(S^\dagger S)_{\beta\alpha} = \int d\gamma S^\dagger_{\beta\gamma} S_{\gamma\alpha} = \int d\gamma S^*_{\gamma\beta} S_{\gamma\alpha} \qquad (4.191)$$

Inserting (4.182), this becomes

$$(S^\dagger S)_{\beta\alpha} = \int d\gamma (\delta_{\gamma\beta} + 2i\pi\delta(E_\gamma - E_\beta)T^*_{\gamma\beta})(\delta_{\gamma\alpha} - 2i\pi\delta(E_\gamma - E_\alpha)T_{\gamma\alpha})$$

$$= \delta_{\beta\alpha} + 2i\pi\delta(E_\alpha - E_\beta)T^*_{\alpha\beta}$$

$$-2i\pi\delta(E_\beta - E_\alpha)T_{\beta\alpha} - 2i\pi\delta(E_\beta - E_\alpha).2i\pi \int d\gamma T^*_{\gamma\beta}\delta(E_\gamma - E_\alpha)T_{\gamma\alpha}$$

$$= \delta_{\beta\alpha} \qquad (4.192)$$

where all terms save the final Kronecker δ cancel in the penultimate line, courtesy of the Generalized Optical theorem (4.189).

Of course, the above results still leave us a fair way from an actual scattering experiment. In particular, we need to be able to convert information about S-matrix elements into a precise statement of how many particles of given momentum and type will emerge per unit time when a given target is placed in a particle beam of a given intensity. Also of interest are the cases in which the $\Psi_{\alpha,\text{in}}$ state in (4.164) is a *one*-particle state, while the $\Psi_{\beta,\text{out}}$ state may contain two, three, or more particles—corresponding to the decay of an unstable particle. The relevant formulas connecting the S-matrix elements to the desired phenomenological cross-sections and rates are derived in Appendix B.

4.4 Problems

1. For the evolution operator $U(t, t_0)$ (4.16), verify the semigroup property

$$U(t_2, t_0) = U(t_2, t_1)U(t_1, t_0)$$

2. (a) Show that the Jordan operator U_{can} given in (4.74) effects the appropriate similarity transformation between old and new canonical coordinates (4.75, 4.76).
   (b) Show that the operator (4.80) implementing linear canonical transformations is (up to a multiplicative constant) equivalent to the unitary kernel (4.81). (Hint: apply the operator (4.80) to the coordinate space wavefunction written as a Fourier transform of the momentum-space wavefunction.)

3. The object of the following exercise is to build intuition for the very important concepts of in/out states and the S-matrix, in a simple example where everything can be worked out explicitly. The model being considered is that of a one-dimensional repulsive δ-function potential, $V(x) = g\delta(x)$, $g > 0$ (see Baym (Baym, 1990), p.113). Scattering experiments are performed by firing particles of mass m and momentum k ($\hbar$ is set to unity throughout!) in from the left or the right. By energy conservation, this results in outgoing particles (either to the left or right) with the same magnitude of momentum k. Use a normalization where the free particle plane wave moving left to right is given by a wavefunction $\langle x|k \rangle = e^{ikx}$.

   (a) Show that in coordinate space the Lippmann–Schwinger equation for right-moving in-states takes the form ($k > 0$)

   $$\Psi_{\text{in}}^{(+)}(x; k) \equiv \langle x|k\rangle_{\text{in}} = e^{ikx} + \int dy\, G_{\text{in}}(x, y) V(y) \Psi_{\text{in}}^{(+)}(y; k) \qquad (4.193)$$

   where the Lippmann–Schwinger in-state kernel is

   $$G_{\text{in}}(x, y) \equiv \langle x| \frac{1}{\frac{k^2}{2m} - H_0 + i\epsilon} |y\rangle$$

   $$H_0 \equiv \frac{p^2}{2m}$$

   (b) Show that the kernel G_{in} takes the explicit form

   $$G_{\text{in}}(x, y) = -\frac{im}{k} e^{ik|x-y|}$$

   (c) Use the results of (a) and (b) to obtain a completely explicit expression for $\Psi_{\text{in}}^{(+)}(x; k)$. Interpret the various terms in your result (for $x < 0$, $x > 0$

separately) in terms of the conventional transmission and reflection coefficients for one-dimensional scattering.

(d) Repeat (a–c) for left-moving in-states: i.e., compute $\Psi_{\text{in}}^{(-)}(x; k)$.

(e) Repeat (a–d) for left and right-moving out-states: i.e., compute $\Psi_{\text{out}}^{(\pm)}(x; k)$.

(f) Suppressing the momentum (magnitude) k, assumed fixed throughout, we may label the in/out states as simply $|\pm\rangle_{\text{in,out}}$, leading to a 2×2 S-matrix with elements $S_{++}, S_{+-}, \ldots$ etc. Calculate this 2×2 matrix (hint: it is convenient to use $\Psi_{\text{in}}^{(+)} = S_{++}\Psi_{\text{out}}^{(+)} + S_{-+}\Psi_{\text{out}}^{(-)}$, etc.) and verify that it is *unitary*.

4. Use the Lipmann–Schwinger equation to calculate $\Psi_{\text{in}}(x)$ for an incoming particle of momentum $\hbar k$ $(k > 0)$ scattering in one dimension off the double δ-potential

$$V(x) = -g(\delta(x - a) + \delta(x + a))$$

5. (a) For the case of a *repulsive* δ-function potential $(g > 0)$:

$$V(x) = g\delta(x)$$

show that the scattering states by themselves form a complete set (there is no bound-state in this case!).

(b) Now suppose that the δ-function potential in part (a) is attractive $(g < 0)$. There is now a single normalizable bound-state energy eigenstate $\Psi_{\text{b}}(x)$. Show that the completeness relation now requires inclusion of the bound-state: i.e.,

$$\int_0^\infty \{\Psi_{\text{in}}^{(+)}(x; k)\Psi_{\text{in}}^{(+)*}(y; k) + \Psi_{\text{in}}^{(-)}(x; k)\Psi_{\text{in}}^{(-)*}(y; k)\}\frac{dk}{2\pi}$$

$$+\Psi_{\text{b}}(x)\Psi_{\text{b}}^*(y) = \delta(x - y) \tag{4.194}$$

5

Dynamics III: Relativistic quantum mechanics

In this chapter we shall begin to explore the implications of the basic underlying symmetry of relativistic quantum field theory: the Poincaré group incorporating the symmetry of such theories under Lorentz transformations and translations in spacetime. This is, of course, the second of the three fundamental physical ingredients of quantum field theory discussed at length in Chapter 3 (the other two being quantum theory and the locality principle). Minkowski's introduction in 1908 of a four-dimensional space, in which the Lorentz transformations of special relativity could be interpreted as rotations preserving an indefinite metric, does not expand the physical content of relativity, but it vastly improves our ability to visualize the physical structure of the theory, and just as importantly, greatly simplifies the search for theories satisfying the constraints of relativistic invariance.

Although we assume the reader to be familiar with the basic tenets of special relativity, as formulated in spacetime concepts, we begin this chapter with a brief review of the Lorentz and Poincaré groups, which provide the kinematic underpinnings for the description of particle states in field theory. This will also provide a convenient opportunity for introducing the reader to the notational conventions that will prevail in the rest of this book.

5.1 The Lorentz and Poincaré groups

In special relativity we characterize *events* as points in a four-dimensional spacetime specified by four coordinates which identify the event by readings of clocks synchronized in and meter sticks attached to a given inertial frame $\mathcal{S}$. Typically, we use Cartesian coordinates x^1, x^2, x^3 for the spatial location (x, y, z are more familiar, but are unsuited to the general spacetime notation we shall use), and $x^0 = t$ for the time. The event as a whole is associated with the spacetime *contravariant four-vector*[1] x^μ, μ=0,1,2,3. The same physical event as viewed by meter sticks and clocks in another inertial frame $\mathcal{S}'$ will be identified by coordinates x'^μ. The linear operations relating the old and new spacetime four-vectors are called *homogeneous Lorentz transformations* (together constituting the *homogeneous Lorentz group*, or HLG):

$$x'^\mu = \Lambda^\mu{}_\nu x^\nu \tag{5.1}$$

---

[1] Note that the physical energy and three spatial momentum components are conventionally taken as the 0,1,2,3 components respectively of a contravariant vector k^μ.

The scalar product left invariant by a Lorentz transformation is most compactly expressed by introducing a covariant four-vector x_μ. Indices are lowered by $x_0 = x^0, x_1 = -x^1, x_2 = -x^2, x_3 = -x^3$ or equivalently, by using the Minkowski metric tensor $g_{\mu\nu}$,

$$x_\mu = g_{\mu\nu} x^\nu$$

The components of $g_{\mu\nu}$ are $g_{00} = 1, g_{11} = g_{22} = g_{33} = -1$, with all off-diagonal elements zero. The Lorentz transformation (5.1) preserves the relativistic scalar product of any two four-vectors:

$$x' \cdot y' \equiv x'^\mu y'_\mu = \Lambda^\mu{}_\nu \Lambda_{\mu\rho} x^\nu y^\rho = x_\rho y^\rho \tag{5.2}$$

for all x_ρ, y_ρ. Differentiating first with respect to y^ρ and then with respect to x^ν, one finds

$$\Lambda^\mu{}_\nu \Lambda_{\mu\rho} = g_{\nu\rho} \tag{5.3}$$

4x4 matrices satisfying (5.3) are said to lie in the fundamental representation of the Lorentz group. By raising the index ρ we may rewrite this as

$$\Lambda^\mu{}_\nu \Lambda_\mu{}^\rho = \delta^\rho_\nu \tag{5.4}$$

which implies

$$(\Lambda^{-1})_\nu{}^\mu = \Lambda_\mu{}^\nu \tag{5.5}$$

Because of the signs involved in raising and lowering indices, it should be noted that the Λ matrices are not orthogonal in general (this will hold only for the subgroup of purely spatial rotations).

Condition (5.3) can be written as the matrix equation

$$\Lambda^T g \Lambda = g \tag{5.6}$$

from which we find, by taking the determinant of both sides, that $\det(\Lambda)^2 = 1$, $\det(\Lambda) = \pm 1$. The set of Lorentz transformations with $\det(\Lambda) = +1$ form the "proper Lorentz transformations" (obviously a subgroup). The class of transformations considered can be refined further by considering only those transformations corresponding to physically realizable changes of inertial frame. Such "orthochronous" transformations leave the sign of the time component unchanged and are characterized by $\Lambda^0{}_0 > 0$. The combined application of the proper and orthochronous requirements lead us to the *restricted* Lorentz group, which contains all (and only) physically accessible[2] Lorentz transformations: namely, those corresponding to physically accessible changes of inertial frame. The unimodularity of the proper Lorentz transformations implies the invariance of four-dimensional spacetime integrals under change of variable corresponding to a Lorentz transform (as the Jacobian $\det(\Lambda)$ is unity):

---

[2] We exclude "looking in the mirror" (parity transformations with $\Lambda^0{}_0 = 1, \det(\Lambda) = -1$) from the set of physically accessible changes of frame.

$$\int d^4x \ldots = \int d^4x' \ldots \tag{5.7}$$

for $x' = \Lambda x$.

The kinematic discussion above focussed on descriptions of coordinates of events: in particle physics, the description of scattering processes is almost always exclusively in terms of energies and momenta of the scattering particles.[3] Using natural units, $\hbar = c = 1$, an isolated stable particle of mass m and well-defined spatial momentum $\vec{k}$ has energy $E(\vec{k}) = \sqrt{\vec{k}^2 + m^2}$. This "mass-shell condition" relating the relativistic energy and momentum is more simply written $k^2 \equiv k \cdot k = m^2$. We shall frequently encounter integrals over the possible four-momenta of a stable particle subject to (i) the mass-shell condition, and (ii) positivity of the energy. Under proper orthochronous transformations, the invariance of the relativistic product (and the sign of the energy) therefore implies the (not at all obvious!) invariance of the measure $\frac{d^3k}{E(\vec{k})}$:

$$\int d^4k\,\theta(k_0)\delta(k^2 - m^2)f(k_0, \vec{k}) = \int \frac{d^3k}{2E(\vec{k})}\, f(E(\vec{k}), \vec{k}) \tag{5.8}$$

—a result which we shall employ on numerous occasions.

We list here some special Lorentz transformations of particular interest:

1. Rotation by θ around the z-axis

$$\Lambda^\mu{}_\nu = R(\theta)^\mu{}_\nu = \begin{pmatrix} 1 & 0 & 0 & 0 \\ 0 & \cos(\theta) & \sin(\theta) & 0 \\ 0 & -\sin(\theta) & \cos(\theta) & 0 \\ 0 & 0 & 0 & 1 \end{pmatrix} \tag{5.9}$$

2. Boost with rapidity ω along z-axis

$$\Lambda^\mu{}_\nu = B(\omega)^\mu{}_\nu = \begin{pmatrix} \cosh(\omega) & 0 & 0 & -\sinh(\omega) \\ 0 & 1 & 0 & 0 \\ 0 & 0 & 1 & 0 \\ -\sinh(\omega) & 0 & 0 & \cosh(\omega) \end{pmatrix} \tag{5.10}$$

where the rapidity is related to the velocity of the boost by

$$\cosh(\omega) = \frac{1}{\sqrt{1 - v^2}} = \gamma$$

$$\sinh(\omega) = \frac{v}{\sqrt{1 - v^2}} = v\gamma \tag{5.11}$$

Note that the boost matrix $B(\omega)$ (in contrast to the rotation matrix $R(\theta)$) is *not* orthogonal.

The *restricted Poincaré group* (sometimes referred to as the *inhomogeneous (restricted) Lorentz group*) is the set of transformations consisting of the combined

---

[3] Oscillation experiments, such as in the neutral kaon system, are a notable exception.

effect of restricted Lorentz transformations (i.e., those Λ satisfying $\det(\Lambda) = +1, \Lambda^0{}_0 > 0$) and spacetime displacements. Unless otherwise explicitly stated, we shall assume henceforth that the restriction to proper, orthochronous Lorentz transformations applies in our further discussions of both the homogeneous Lorentz group and its inhomogeneous extension, the Poincaré group. In defining the latter, by convention, the Lorentz transformation is performed first, followed by the displacement. Thus an element of the Poincaré group is specified by a pair (Λ, a), with action

$$x^\mu \to x'^\mu = \Lambda^\mu{}_\nu x^\nu + a^\mu \qquad \Lambda_1 a_1 (\Lambda_2 x + a_2) \tag{5.12}$$

The composition law for the Poincaré group follows immediately: $\quad L_3 \ \Lambda_1 \Lambda_2 x + \Lambda_1 a_2$

$$(\Lambda_1, a_1) \cdot (\Lambda_2, a_2) = (\Lambda_1 \Lambda_2, a_1 + \Lambda_1 a_2) \qquad + a_1 \tag{5.13}$$

In accordance with the Wigner unitarity–antiunitarity theorem discussed in Chapter 3, the Poincaré symmetry of a relativistic quantum theory is realized by a unitary representation (recall that for a continuous group the antiunitary option is disallowed) associating the unitary operator $U(\Lambda, a)$ with the Poincaré element (Λ, a), with the algebra (following from (5.13))

$$U(\Lambda_1, a_1)U(\Lambda_2, a_2) = U(\Lambda_1 \Lambda_2, a_1 + \Lambda_1 a_2) \tag{5.14}$$

For the special case of the subgroup of proper Lorentz transformations ($a = 0$), the element Λ is represented on the state space by the unitary operator $U(\Lambda)$. We now turn to the properties of the $U(\Lambda)$, beginning with their action on states with spinless particles.

5.2 Relativistic multi-particle states (without spin)

Consider a state of a single stable massive particle (for simplicity, spinless, and ignoring all internal quantum numbers) with four-momentum k^μ. There is always a frame (unique up to rotations) in which the spatial components of the four-momentum vanish. Define $k'^0 \equiv m$ in this frame as the mass. In any other frame obtained by a proper orthochronous transformation from this one, the zeroth component will be given by $E(\vec{k}) = \sqrt{\vec{k}^2 + m^2}$. Thus the state of our *spinless* particle is completely characterized by giving just the spatial momentum, and may be written simply $|\vec{k}\rangle$.

There are two normalization conventions commonly used for states in relativistic quantum theory. The *covariant normalization* convention defines

$$\langle \vec{k}'|\vec{k}\rangle = 2E(\vec{k})\delta^3(\vec{k} - \vec{k}') \tag{5.15}$$

The "covariant" appellation is justified by the following invariance property:[4]

$$2E(\vec{k})\delta^3(\vec{k} - \vec{k}') = 2E(\Lambda\vec{k})\delta^3(\Lambda\vec{k} - \Lambda\vec{k}') \tag{5.16}$$

---

[4] The spatial vector $\Lambda\vec{k}$ is to be interpreted as the spatial part of the four-vector Λk, where the zeroth component k^0 is set equal to the on-mass-shell value $E(\vec{k})$.

To prove this, integrate both sides of (5.16) with a smooth test function $f(\vec{k})/2E(\vec{k})$:

$$\int d^3k\, 2E(\vec{k})\delta^3(\vec{k}-\vec{k}')\frac{f(\vec{k})}{2E(\vec{k})} = f(\vec{k}') \tag{5.17}$$

is obtained on the left side, while the result on the right is

$$\int d^3k\, 2E(\Lambda\vec{k})\delta^3(\Lambda\vec{k}-\Lambda\vec{k}')\frac{f(\vec{k})}{2E(\vec{k})} = \int d^4k\,\delta(k^2-m^2)2E(\Lambda\vec{k})\delta^3(\Lambda\vec{k}-\Lambda\vec{k}')f(k)$$

$$= \int d^4k\,\delta(k^2-m^2)2E(k)\delta^3(\vec{k}-\Lambda\vec{k}')f(\Lambda^{-1}\vec{k})$$

$$= \int d^3k\,\delta^3(\vec{k}-\Lambda\vec{k}')f(\Lambda^{-1}\vec{k}) = f(\vec{k}') \tag{5.18}$$

where on the second line we have made a change of variable $k \to \Lambda^{-1}k$. This ensures the invariance of the Hilbert-space inner product under Lorentz transformations:

$$\langle \vec{k}|\vec{k}'\rangle = \langle \Lambda\vec{k}|\Lambda\vec{k}'\rangle \tag{5.19}$$

With *non-covariant normalization*, we simply drop the factor of $2E(\vec{k})$ on the right-hand side of (5.15).

A state $|\vec{k}\rangle$ will look like $|\Lambda\vec{k}\rangle$ to a boosted or rotated observer. The Lorentz group is realized on the space of states by operators $U(\Lambda)$ defined to effect precisely this change:

$$U(\Lambda)|\vec{k}\rangle \equiv |\Lambda\vec{k}\rangle \tag{5.20}$$

In particle physics one is accustomed to the phenomena of particle creation (and annihilation): processes occur in which the number of particles of a given type changes, making it essential that we formulate the theory in a multi-particle state space, or *Fock space*, constructed mathematically as the infinite direct sum of Hilbert space sectors containing 0 (the "vacuum state", $|0\rangle$), 1 (the single particle states $|\vec{k}\rangle$ discussed above), 2 particles ($|\vec{k}_1, \vec{k}_2\rangle$), and so on. The inner product is defined naturally on the basis of multi-particle states of well defined momentum by asserting (i) orthogonality of sectors with differing number of particles, and (ii) for particles of the same type, Bose or Fermi symmetry depending on the spin.[5] For the N-particle sector, the latter requirement (using non-covariant normalization) amounts to the condition

$$\langle \vec{k}'_1, ..\vec{k}'_N|\vec{k}_1, ..\vec{k}_N\rangle = \sum_P (\pm)^P \delta^3(\vec{k}_1 - \vec{k}'_{P(1)}) ... \delta^3(\vec{k}_N - \vec{k}'_{P(N)}) \tag{5.21}$$

Here $\sum_P$ denotes a sum over all permutations P of the integers $1, 2, \ldots N$, with $(\pm)^P$ inserting a minus sign for fermions if the permutation P is odd. Of course,

---

[5] The *Spin-Statistics theorem* asserting the necessity of Bose statistics for particles of integer spin, and Fermi statistics for particles of half-integral spin, is one of the seminal results of local field theory: we shall discuss it in Chapters 7 and 13.

for the spinless bosonic particles under consideration in this section, the positive sign is to be taken in the symmetrization. Note that the completeness sums acquire extra combinatoric factors for identical particles: for example, the decomposition of the identity in the multi-particle Fock space of a single boson takes the form

$$\mathbf{1} = \sum_N \frac{1}{N!} \int d^3k_1 d^3k_2 \dots d^3k_N |\vec{k}_1, \vec{k}_2, .., \vec{k}_N\rangle\langle \vec{k}_1, \vec{k}_2, .., \vec{k}_N| \tag{5.22}$$

as the reader may readily verify by squaring the above expression. Rather surprisingly, the very "big" Fock space obtained by the construction outlined here is still nevertheless, like L^2, a separable Hilbert space: it is spanned by a *denumerable* orthonormal basis.[6]

The operator implementing a Lorentz transformation Λ on a multi-particle state takes the unsurprising form:

$$U(\Lambda)|\vec{k}_1, \vec{k}_2, ..\rangle \equiv |\Lambda\vec{k}_1, \Lambda\vec{k}_2, ..\rangle \tag{5.23}$$

With the covariant normalization of states defined above these operators are in fact unitary. For example, on single-particle states,

$$
\begin{aligned}
(U(\Lambda)|\vec{k}'\rangle, U(\Lambda)|\vec{k}\rangle) &= \langle \vec{k}'|U^\dagger(\Lambda)U(\Lambda)|\vec{k}\rangle \\
&= \langle \Lambda\vec{k}'|\Lambda\vec{k}\rangle \\
&= 2E(\Lambda\vec{k})\delta^3(\Lambda\vec{k}' - \Lambda\vec{k}) \\
&= 2E(\vec{k})\delta^3(\vec{k} - \vec{k}') \\
&= \langle \vec{k}'|\vec{k}\rangle
\end{aligned} \tag{5.24}
$$

and similarly for multi-particle states.

In practice it is generally more convenient to use non-covariantly normalized states, which are simply $\frac{1}{\sqrt{2E(k)}}$ (for each particle in the state) times the old covariantly normalized states. Thus, the non-covariant normalization convention is simply

$$\langle \vec{k}'|\vec{k}\rangle = \delta^3(\vec{k} - \vec{k}') \tag{5.25}$$

The modification in the Lorentz transformation rule is trivial:

$$U(\Lambda)|\vec{k}\rangle = \sqrt{\frac{E(\Lambda\vec{k})}{E(\vec{k})}}|\Lambda\vec{k}\rangle \tag{5.26}$$

and correspondingly, for a multi-particle state:

$$U(\Lambda)|\vec{k}_1, \vec{k}_2, .., \vec{k}_N\rangle = \prod_{i=1}^N \sqrt{\frac{E(\Lambda\vec{k}_i)}{E(\vec{k}_i)}}|\Lambda\vec{k}_1, \Lambda\vec{k}_2, .., \Lambda\vec{k}_N\rangle \tag{5.27}$$

---

[6] For a clear discussion of this frequently misunderstood feature, see (Streater and Wightman, 1978).

In either case, we shall often abbreviate the Lorentz action on a multi-particle state as $U(\Lambda)|\alpha\rangle = |\Lambda\alpha\rangle$.

We can view the multi-particle Fock space described above as a basis of eigenstates of a *free Hamiltonian* H_0 corresponding to having all self-interactions of our putative stable, spinless particle switched off. The energy of a multi-particle state is then simply

$$H_0|\vec{k}_1, \vec{k}_2, .., \vec{k}_N\rangle = \sum_{i=1}^{N} E(\vec{k}_i)|\vec{k}_1, \vec{k}_2, .., \vec{k}_N\rangle \tag{5.28}$$

and the momentum of such a state is clearly

$$\vec{P}^{(0)}|\vec{k}_1, \vec{k}_2, .., \vec{k}_N\rangle = \sum_{i=1}^{N} \vec{k}_i|\vec{k}_1, \vec{k}_2, .., \vec{k}_N\rangle \tag{5.29}$$

We can consider H_0 and $\vec{P}^{(0)}$ as the time and spatial components respectively of an energy-momentum four-vector operator $P^{(0)\mu}$ for free (hence the superscript $^{(0)}$ notation) particles. As we know from general quantum theory, these operators generate infinitesimal translations in time and space. Taking the matrix element of the Lorentz transformed energy-momentum operator between covariantly normalized single particle states:

$$\langle \vec{k}'|U^{\dagger}(\Lambda)P^{(0)\mu}U(\Lambda)|\vec{k}\rangle = (\Lambda k)^{\mu} \cdot 2E(\Lambda\vec{k})\delta^3(\Lambda\vec{k}' - \Lambda\vec{k})$$

$$= \Lambda^{\mu}{}_{\nu}k^{\nu} \cdot 2E(\vec{k})\delta^3(\vec{k}' - \vec{k})$$

$$= \Lambda^{\mu}{}_{\nu}\langle \vec{k}'|P^{(0)\nu}|\vec{k}\rangle \tag{5.30}$$

with a similar result for the matrix element in a general multi-particle state. Thus the four operators $(P^{(0)0}, P^{(0)i}, i = 1, 2, 3)$ transform as expected under Lorentz transformations

$$U^{\dagger}(\Lambda)P^{(0)\mu}U(\Lambda) = \Lambda^{\mu}{}_{\nu}P^{(0)\nu} \tag{5.31}$$

Finally, recalling that a general element (Λ, a) of the Poincaré group is defined as a Lorentz transformation Λ followed by a translation by the four-vector a^{μ}, we see that the unitary representation of (Λ, a) in the free Fock space is given by

$$U(\Lambda, a) = e^{iP^{(0)} \cdot a}U(\Lambda) \tag{5.32}$$

A few lines of algebra, employing the transformation property (5.31), confirms that these operators do indeed furnish a unitary representation of the full Poincaré group (cf. (5.13))

$$U(\Lambda_1, a_1)U(\Lambda_2, a_2) = U(\Lambda_1\Lambda_2, a_1 + \Lambda_1 a_2) \tag{5.33}$$

5.3 Relativistic multi-particle states (general spin)

The analysis of the transformation behavior of single particle states in relativistic quantum mechanics for general mass and spin was pioneered by Wigner in a classic

paper (Wigner, 1939). Mathematically, this amounts to a classification of the irreducible unitary representations of the Poincaré group, which may even be viewed as
furnishing a *definition* of a particle. Note that there is no claim to elementarity (absence
of internal structure) for such an object: any quantum state which is an eigenvalue of
the squared mass operator $P_\mu P^\mu$ and (for massive particles) of the angular momentum
operators $\vec{J}^2, J_z$ in the inertial frame where it is at rest, corresponds to a single particle
state associated with a specific irreducible representation of the Poincaré group. Thus,
from this point of view, the proton (a complicated bound state of quarks and gluons)
is just as much a particle as the (as far as we know) structureless electron.[7] Wigner's
technique for exposing the structure of these representations can be summarized as
follows. One first removes the "continuous" part of the specification of the state (i.e.,
the spatial momentum) by performing a Lorentz boost which (for massive particles)
reduces the particle to rest. The remaining discrete ("spin") degrees of freedom are
then exposed by analysing the properties of the "little group": i.e., the subgroup of the
full Poincaré group which leaves the particle in the standard state—i.e., for massive
particles, at rest, which means that the little group is simply the three-dimensional
rotation group. The analysis for massless particles is somewhat more subtle, as the
choice of standard state (which cannot be at rest) leads to a little group which is
not semisimple.[8] We shall concentrate on massive particles for most of this section,
returning briefly to the massless case at the end. There is some phenomenological
justification for this: we know of many massive but presumably only two exactly
massless particles (the photon, and hypothetically, the graviton). The interactions of
photons are subject to an abelian gauge symmetry (cf. Chapter 15) which renders
the massless limit smooth, so that in fact it is operationally impossible to distinguish
between the interactions of a photon of exactly zero mass and one with mass equal to
10^{-100} eV, for example. Accordingly, for this one case of a (possibly) massless particle,
we might as well provisionally take the photon to have non-zero mass and pass to the
zero mass limit at the end of the calculations.

Any massive particle state can be viewed from a frame at which it is at rest, so
there is no loss of generality in considering first only states with $\vec{p} = 0$, any other
one-particle state being obtainable by applying $U(L(\vec{k}))$ where $L(\vec{k})$ boosts from
the frame where the particle is at rest to the one where the momentum is $\vec{k}$. Of
course, $L(\vec{k})$ is not unique, as any Lorentz transformation from rest to a non-zero
momentum $\vec{k}$ can be preceded by an arbitrary rotation, or followed by an arbitrary
rotation around the direction of $\vec{k}$ without altering the final momentum. This will
lead to the freedom to define alternative basis sets for single (or multi-) particle states
with spin.

The subgroup of the homogeneous Lorentz group (HLG) leaving a state with zero
spatial momentum at zero momentum is the three-dimensional rotation group, with
generators J_x, J_y, J_z. The irreducible representations of this group are characterized
by the eigenvalues of $\vec{J}^2(= j(j + 1))$, with the states within a given representation

---

[7] Of course, in string theory even supposedly point-like quarks and leptons acquire a one-dimensional
string structure in ten dimensions, or a membrane structure in eleven-dimensional M-theory.

[8] The reader may recall that a semisimple group is one with no invariant abelian subgroups.

distinguished by the eigenvalue of $J_z (\equiv \sigma)$. We shall customarily omit the specification of j, and write a one-particle state of a particle at rest as simply $|\vec{0}, \sigma\rangle$, with

$$J_z |\vec{0}, \sigma\rangle = \sigma |\vec{0}, \sigma\rangle$$

$$(J_x \pm iJ_y)|\vec{0}, \sigma\rangle = \sqrt{(j \mp \sigma)(j \pm \sigma + 1)}|\vec{0}, \sigma \pm 1\rangle$$

$$\vec{J}^2 |\vec{0}, \sigma\rangle = j(j+1)|\vec{0}, \sigma\rangle \tag{5.34}$$

Of course, j is conventionally referred to as the *spin* of the particle in question, and must take either integer or half-integer values by the general representation theory of the rotation group.[9]
Let $R(\hat{k})$ be the rotation operator taking the z-axis into the $\hat{k}$ direction via a rotation around the axis perpendicular to both. Further, define $B(|\vec{k}|)$ (cf. (5.10)) as the boost from rest to the state with momentum $|\vec{k}|$ *in the z-direction*. It is immediately obvious that we arrive at a state with momentum $\vec{k}$ by applying *either* of the following Lorentz transformations

$$L(\vec{k}) \equiv R(\hat{k})B(|\vec{k}|)R^{-1}(\hat{k}) \tag{5.35}$$

$$\mathcal{L}(\vec{k}) \equiv R(\hat{k})B(|\vec{k}|) \tag{5.36}$$

to the state at rest $|\vec{0}, \sigma\rangle$. These two transformations give two possible definitions for general one-particle spin-j states: including the $\sqrt{m/E(\vec{k})}$ factor for non-covariant normalization (cf. (5.26)), we may either use states defined as

$$|\vec{k}, \sigma\rangle \equiv \sqrt{\frac{m}{E(\vec{k})}} U(L(\vec{k}))|\vec{0}, \sigma\rangle \tag{5.37}$$

or

$$|\vec{k}, \lambda\rangle \equiv \sqrt{\frac{m}{E(\vec{k})}} U(\mathcal{L}(\vec{k}))|\vec{0}, \lambda\rangle \tag{5.38}$$

The states defined by (5.37) are conventionally referred to as "spin" states: as in ordinary non-relativistic quantum theory, the discrete spin index σ refers to the component of angular momentum (of the particle in its rest frame) along a standard axis (typically, the z-axis). States defined as in (5.38) are called "helicity" states, as the λ quantum number actually refers to the projection of the angular momentum in the direction of motion of the particle:

---

[9] Throughout this section we shall make heavy use of the machinery of angular momentum and the rotation group: for a review, see Chapters 15 and 17 of (Baym, 1990).

$$\vec{J} \cdot \hat{k} |\vec{k}, \lambda\rangle = \sqrt{\frac{m}{E(\vec{k})}} U(R(\hat{k})) U^\dagger(R(\hat{k})) \vec{J} \cdot \hat{k} U(R(\hat{k})) U(B(|k|)|\vec{0}, \lambda\rangle$$

$$= \sqrt{\frac{m}{E(\vec{k})}} U(R(\hat{k})) J_z U(B(|k|))|\vec{0}, \lambda\rangle$$

$$= \sqrt{\frac{m}{E(\vec{k})}} U(R(\hat{k})) U(B(|k|)) J_z |\vec{0}, \lambda\rangle$$

$$= \lambda |\vec{k}, \lambda\rangle \tag{5.39}$$

where in the third line we have used the fact that rotations around the z axis (affecting only the transverse x and y coordinates) commute with boosts along the z axis.

Of course, both the $|\vec{k}, \sigma\rangle, \sigma = -j, -j+1, .., +j$ and the $|\vec{k}, \lambda\rangle, \lambda = -j, .. + j$ states form complete sets and can be expressed as linear combinations of one another (specifically: $|\vec{k}, \lambda\rangle = \sum_\sigma D^j_{\sigma\lambda}(R(\hat{k}))|\vec{k}, \sigma\rangle$). The helicity states $|\vec{k}, \lambda\rangle$ are of particular utility in dealing with massless or highly energetic particles, as we shall see below when we address the massless case explicitly. To summarize our options, we may either label the discrete spin state of our particle by the J_z eigenvalue of a comoving observer (as in (5.37)), or by the eigenvalue of the component of angular momentum $\vec{J} \cdot \hat{k}$ along the direction of motion of the particle (as in (5.38)).

Now we turn to the Lorentz transformation properties of these states. Evidently

$$U(\Lambda)|\vec{k}, \sigma\rangle = \sqrt{\frac{m}{E(\vec{k})}} U(\Lambda) U(L(\vec{k}))|\vec{0}, \sigma\rangle = \sqrt{\frac{m}{E(\vec{k})}} U(\Lambda L(\vec{k}))|\vec{0}, \sigma\rangle$$

$$= \sqrt{\frac{m}{E(\vec{k})}} U(L(\Lambda\vec{k})) U(L^{-1}(\Lambda\vec{k})\Lambda L(\vec{k}))|\vec{0}, \sigma\rangle$$

$$\equiv \sqrt{\frac{m}{E(\vec{k})}} U(L(\Lambda\vec{k})) U(W(\Lambda, \vec{k}))|\vec{0}, \sigma\rangle \tag{5.40}$$

The Lorentz transformation $W(\Lambda, \vec{k}) \equiv U(L^{-1}(\Lambda\vec{k})\Lambda L(\vec{k}))$ is actually a rotation, as it takes a particle at rest first to momentum $\vec{k}$, then to momentum $\Lambda\vec{k}$, and then finally back to rest, and therefore must be a pure rotation: it is called the *Wigner rotation* (the complete unitary representation theory of the Poincaré group was first developed by Eugene Wigner in the previously mentioned seminal paper in 1939 (Wigner, 1939)). The action of a pure rotation on a state of a spin-j particle at rest is given[10] by the rotation matrices $D^j_{\sigma'\sigma}$. So we have

---

[10] See Baym (Baym, 1990), Chapter 17. Our notation *vis-à-vis* rotation group matters agrees with this reference.

$$U(\Lambda)|\vec{k},\sigma\rangle = \sqrt{\frac{m}{E(\vec{k})}} U(L(\Lambda\vec{k})) \sum_{\sigma'} D^j_{\sigma'\sigma}(W(\Lambda,\vec{k}))|\vec{0},\sigma'\rangle$$

$$= \sqrt{\frac{E(\Lambda\vec{k})}{E(\vec{k})}} \sum_{\sigma'} D^j_{\sigma'\sigma}(W(\Lambda,\vec{k}))|\Lambda\vec{k},\sigma'\rangle \tag{5.41}$$

For $j = 0$ we have trivially $D^j_{\sigma'\sigma} = \delta_{\sigma'0}\delta_{\sigma0}$, and we regain the transformation law (5.26) for spinless particle states. A straightforward calculation, employing the unitarity of the D^j matrices and $U(\Lambda)$ operators, confirms the non-covariant normalization of these states:

$$\langle \vec{k}',\sigma'|\vec{k},\sigma\rangle = \delta_{\sigma'\sigma}\delta(\vec{k}'-\vec{k}) \tag{5.42}$$

Finally, we note that the extension of the unitary representation $U(\Lambda)$ for the homogeneous Lorentz group to the full Poincaré group including translations follows exactly the same lines as in the preceding section for spinless particles: in particular, the spin indices are unaffected by the application of the energy-momentum operator $P^{(0)\mu}$.

The above discussion for massive particles undergoes significant modifications for massless particles. As far as we know, the only exactly massless particles in Nature are the photon (spin $j = 1$) and the (hypothetical) graviton (spin $j = 2$). In either case, the one particle states are labeled by a discrete helicity index which takes only two possible values, $+j, -j$: the intermediate values $-j+1,\ldots, j-1$ which would be present for a massive particle are missing. The reason for this, as we shall now see, is that the little group for massless particles is quite different from that for massive particles (the rotation group O(3), as discussed above).

First, let us revert to covariant normalization of states, to avoid inconvenient factors of $\sqrt{E(k)}$ in the transformation formulas. We cannot choose the state of the particle at rest as our standard state for massless particles; instead, our standard state will be defined as the particle with a standard momentum (and energy) $+\mu$ in the z-direction, i.e., with four-momentum $k_0 = (\mu, 0, 0, \mu)$. The boost operator $B(k)$ will now be defined as the boost in the z-direction with rapidity $\omega = \ln\frac{k}{\mu}$, i.e., it takes the standard state with four-momentum $k_0 = (\mu, 0, 0, \mu)$ to four-momentum $(k, 0, 0, k)$. The rotation $R(\hat{k})$ then takes this to a general light-like state with four-momentum $(|\vec{k}|, \vec{k})$. Thus, if we label the (as yet unspecified) spin quantum number(s) needed to specify fully the one-particle state τ, a general massless particle state can be defined as

$$|\vec{k},\lambda\rangle \equiv U(\mathcal{L}(\vec{k}))|\vec{k}_0,\tau\rangle \tag{5.43}$$

$$\mathcal{L}(\vec{k}) \equiv R(\hat{k})B(|\vec{k}|) \tag{5.44}$$

Note that we must use the helicity type transformation of (5.36) here, rather than the spin type transformation (5.35), as our standard state has non-zero momentum. Following exactly the steps performed leading to (5.40), this time in covariant normalization (so the square-root factors are absent), we find the general Lorentz transformation law for such states to be

$$U(\Lambda)|\vec{k}, \tau\rangle = U(\mathcal{L}(\Lambda\vec{k}))U(\mathcal{W}(\Lambda, \vec{k}))|\vec{k}_0, \tau\rangle \qquad (5.45)$$

where the Lorentz transformation

$$\mathcal{W}(\Lambda, \vec{k}) \equiv B^{-1}(|\Lambda\vec{k}|)R^{-1}(\hat{\Lambda k})\Lambda R(\hat{k})B(|\vec{k}|)$$

must now be an element of the little group for massless particle states: namely, it must be a Lorentz transformation leaving the standard vector $k_0 = (\mu, 0, 0, \mu)$ invariant. As in the massive case, this condition specifies a three-parameter subgroup of the homogeneous Lorentz group. However, this group, as we shall now see, is *not* the conventional O(3) rotation group (obviously, as rotations around the x and y directions manifestly do not leave the standard vector unchanged), with its attendant fully understood finite-dimensional representations labeled by j, σ quantum numbers. Instead, it turns out to be a non-compact, non-semisimple group: namely, the Euclidean group of rotations and translations in two dimensions, with a completely different representation structure to the O(3) rotation group.

The little group of Lorentz transformations leaving our standard light-like four-vector $k_0 = (\mu, 0, 0, \mu)$ invariant clearly still retains a remnant of the full rotation group: namely, rotations $R(\theta)$ by an angle θ around the z-axis, as given in (5.9). Defining a two-dimensional vector $\vec{\xi} \equiv (\xi_1, \xi_2)$, the reader may easily verify that the two-parameter set of Lorentz transformations defined by

$$T(\vec{\xi})^\mu{}_\nu = \begin{pmatrix} 1 + \frac{1}{2}\vec{\xi}^2 & \xi_1 & \xi_2 & -\frac{1}{2}\vec{\xi}^2 \\ \xi_1 & 1 & 0 & -\xi_1 \\ \xi_2 & 0 & 1 & -\xi_2 \\ \frac{1}{2}\vec{\xi}^2 & \xi_1 & \xi_2 & 1 - \frac{1}{2}\vec{\xi}^2 \end{pmatrix} \qquad (5.46)$$

also leaves k_0 invariant. The transformations $T(\vec{\xi})$ form an abelian subgroup, as one easily sees that

$$T(\vec{\xi})T(\vec{\chi}) = T(\vec{\xi} + \vec{\chi}) \qquad (5.47)$$

and is therefore structurally identical to the group of translations in the x-y plane. Indeed, the two-vector $\vec{\xi}$ transforms as expected under a rotation around the z-axis:

$$R(\theta)T(\vec{\xi})R^{-1}(\theta) = T(R_\theta\vec{\xi}) \qquad (5.48)$$

where $R_\theta\vec{\xi} \equiv (\cos(\theta)\xi_1 + \sin(\theta)\xi_2, -\sin(\theta)\xi_1 + \cos(\theta)\xi_2)$. Moreover, (5.48) implies that these two-dimensional translations form an invariant abelian subgroup of the full three-parameter *Euclidean group* of two-dimensional translations and rotations $W(\vec{\xi}, \theta) \equiv T(\vec{\xi})R(\theta)$ which comprise the little group for a massless particle:

$$W(\vec{\xi}, \theta)T(\vec{\chi})W^{-1}(\vec{\xi}, \theta) = T(R_\theta\vec{\chi}) \qquad (5.49)$$

so, as advertised previously, our little group is definitely not semisimple, nor compact (as the translations are unbounded).

What is the representation structure of this Euclidean group? The unitary representative of the rotation part $R(\theta)$ is just $U(R(\theta)) = e^{i\theta J_z}$ so if we label the eigenvalue of the hermitian generator J_z by λ, we must include this "helicity" value (the component of angular momentum along the direction of motion) in the specification of quantum numbers τ which give a full labeling of our standard state $|k_0, \tau\rangle$. In principle, the hermitian generators of the translation part of the little group could also take on non-vanishing "momentum" eigenvalues $\vec{\pi} = (\pi_1, \pi_2)$, with $U(T(\vec{\xi}))|k_0, \tau\rangle = e^{i\vec{\pi}\cdot\vec{\xi}}|k_0, \tau\rangle$, in which case the full set of quantum numbers τ would be labeled by the three parameter set $(\vec{\pi}, \lambda)$. The one massless particle with which we have (extensive!) experience, the photon, definitely has non-zero helicity states, but there is no empirical evidence for any further degrees of freedom corresponding to the $\vec{\pi}$ variables, which we must therefore set to zero. Accordingly, we shall assume that one-particle massless states in general are fully specified by a momentum and *single* helicity variable, with a standard state $|k_0, \lambda\rangle$ satisfying

$$J_z|k_0, \lambda\rangle = \lambda|k_0, \lambda\rangle \tag{5.50}$$

The only remnant of the rotation group here is the one-dimensional abelian subgroup of rotations around the z-axis, for the standard state, or more generally, around the direction of motion of the massless particle. In particular, *we are missing* the raising and lowering operators constructed from the other two generators (J_x, J_y) which would normally allow us to (a) move stepwise from helicity λ to $\lambda \pm 1$, thereby filling out a full $2j + 1$-dimensional O(3) representation of spin j states, and (b) establish the integral or half-integral quantization of the maximal helicity. In fact, for massless particles the helicity is actually a Lorentz-invariant: it is exactly the same number for the state $|k, \lambda\rangle = U(\mathcal{L}(\vec{k}))|k_0, \lambda\rangle$ obtained by boosting the standard state $|k_0, \lambda\rangle$ for a particle moving in the z-direction with the same helicity, as we see by the calculation leading to (5.39). Physically, massless particles of different helicity are, from the standpoint of the *proper* Lorentz group (i.e., absent improper parity transformations), completely distinct and unrelated objects. If we include parity transformations, of course, which reverse momentum but leave angular momentum unchanged, the helicity changes sign, so if a massless particle participates in parity conserving interactions (as the photon certainly does), then if the particle exists with helicity $+\lambda$ it must also exist with helicity $-\lambda$. But the representation theory of the little group clearly says nothing about intermediate helicity values (such as 0 for the photon, or +1,0, and –1 for the spin-2 graviton).

What about the quantization condition for maximal helicity ($\equiv$ spin of the particle), given the absence of the full O(3) rotation group? At first, one might suppose that as a rotation by angle θ around the momentum vector of our helicity λ state $|\vec{k}, \lambda\rangle$ generates a phase factor $e^{i\lambda\theta}$, a rotation by 2π must return us to the original state, ensuring integer quantization of helicity. Of course, the doubly connected structure of the rotation group is the critical property saving us from this conclusion, and reinstating the possibility of the spinor type half-integral representations. Recall that whereas any continuous one-dimensional path $R(\alpha), 0 \leq \alpha \leq 1$ through the group manifold of O(3) in which the starting element $R(0)$ is the identity but the final rotation $R(1)$ is by an odd multiple of 2π (around the z-axis, say) *cannot* be continuously shrunk to a point (the identity) in the group manifold (i.e., to the path $R(\alpha) = R(0) = 1, \alpha \leq 1$): the rotation group in three dimensions is not simply connected. On the other hand, paths starting at the identity and ending at rotations involving a multiple of 4π can be so shrunk.[11] The corresponding unitary representatives $U(R(\alpha))$ must generate continuously varying phases applied to our massless one-particle state, which then implies that

$$e^{i(4\pi n)\lambda} = 1 \Rightarrow \lambda = 0, \pm\frac{1}{2}, \pm 1, \pm\frac{3}{2}, \dots \qquad (5.51)$$

In other words, as we expect, only integer or half-integer values for the helicity of a massless particle are allowed. This is an example of a topological, rather than an algebraic, quantization of a quantum number.

5.4 How not to construct a relativistic quantum theory

Having specified the kinematic structure of the multi-particle Hilbert space for relativistic particles, one may set about the task of constructing a sensible "relativistic quantum mechanics": at the very least we should expect to be able to construct, at least in principle, theories in which a unitary S-matrix satisfies the requirements of special relativity. Our general discussion of quantum symmetries in Section 4.1.3 (and in particular, the Wigner unitarity–antiunitarity theorem discussed there) require equality of the S-matrix amplitudes (and not just their absolute squares, giving the probabilities of the corresponding scattering) for a process in which a prepared state $|\alpha\rangle_{\text{in}}$ evolves to a detected state $|\beta\rangle_{\text{out}}$ in two inertial frames related by a Lorentz transformation Λ. Specifically, we must have

$$_{\text{out}}\langle\beta|\alpha\rangle_{\text{in}} = S_{\beta\alpha} = S_{\Lambda\beta,\Lambda\alpha} \qquad (5.52)$$

which ensures the inability of an observer to detect absolute inertial motion by performing scattering experiments. Writing out the matrix elements above more explicitly

$$S_{\beta\alpha} = \langle\beta|S|\alpha\rangle = \langle\Lambda\beta|S|\Lambda\alpha\rangle$$
$$= \langle\beta|U^\dagger(\Lambda)SU(\Lambda)|\alpha\rangle \qquad (5.53)$$

for all states $|\beta\rangle, |\alpha\rangle$. Consequently, we must have

$$S = U^\dagger(\Lambda)SU(\Lambda) \qquad (5.54)$$

for an arbitrary Lorentz transformation Λ. Equivalently, S *commutes* with the representatives $U(\Lambda)$ of Lorentz transformations on the quantum-state space:

$$[U(\Lambda), S] = 0 \qquad (5.55)$$

---

[11] The reader may verify this immediately by the magical "twisted belt" experiment: a belt held flat at both ends, but with a single twist by 360 degrees in the middle cannot be flattened by moving only the right end around (while held flat), whereas if subjected to a double twist of two full rotations, the belt "unwinds" easily if we move its right end appropriately, keeping the left end fixed.

We remind the reader (cf. Section 4.3) that the S-matrix is defined as a unitary operator in the basis of *free particle states*: the behavior of these states under the action of $U(\Lambda)$ is precisely the content of the preceding sections of this chapter. The requirement of Poincaré invariance adds invariance under spacetime translations to the above (i.e., we require commutativity of S with the larger class of unitary symmetry operators $U(\Lambda, a)$). This, of course, implies commutativity of S with the infinitesimal generators of spacetime translations, namely the (free) energy-momentum operator $P^{(0)\mu}$:

$$[P^{(0)\mu}, S] = 0 \tag{5.56}$$

There is, in fact, a trivial way to ensure both unitarity of S and the invariance requirements (5.55, 5.56). Let us write the full Hamiltonian of the theory as

$$H = H_0 + V \tag{5.57}$$

where the free part H_0 is defined by (5.28). Recall that the S-matrix is obtained as the infinite time limit of the interaction-picture time-evolution operator: $S = \lim_{T \to \infty} U(T, -T)$. Moreover, for general times t, t_0, this operator is constructed (cf. (4.27)) as a sum of products of the interaction operator $V_{\text{ip}}(t)$ (i.e., the interaction part of the Hamiltonian, in the interaction picture):

$$V_{\text{ip}}(t) \equiv e^{iH_0 t} V e^{-iH_0 t} \tag{5.58}$$

Thus, the S-matrix involves a sum of products of the operators H_0 and V. Energy-momentum conservation (5.56) is therefore ensured if[12]

$$[\vec{P}^{(0)}, V] = 0 \tag{5.59}$$

What about Lorentz-invariance? It would certainly follow if we could arrange

$$[U(\Lambda), V_{\text{ip}}(t)] = 0 \tag{5.60}$$

for all t. In fact, a very simple Ansatz would seem to do the trick. Consider the operator (for the rest of this section we use covariant normalization throughout, and drop vector symbols for three-momenta in the states, assumed to be multi-particle states of a single spinless particle):

$$V = \int \prod_{i=1}^{2} \frac{d^3 k_i}{2E(k_i)} \frac{d^3 k_i'}{2E(k_i')} \delta^4(k_1' + k_2' - k_1 - k_2) h(k_1', k_2', k_1, k_2) |k_1' k_2'\rangle \langle k_1 k_2| \tag{5.61}$$

The physical meaning of this interaction operator is clear: an incoming two-particle state with momenta k_1, k_2 is converted to an outgoing state k_1', k_2' with amplitude $h(k_1', k_2', k_1, k_2)$ subject to the constraint of energy-momentum conservation, implemented by the four-dimensional δ-function.

---

[12] Energy conservation is automatic given the time-invariance of H; cf. (4.182).

If $h(k_1', k_2', k_1, k_2)^* = h(k_1, k_2, k_1', k_2')$, it is easy to see that V is hermitian, so unitarity is guaranteed. The δ-function in (5.61) guarantees that energy is conserved by the interaction, so V commutes with H_0 and $V_{int}(t) = V$. Lorentz-invariance is established by checking

$$U(\Lambda)VU^\dagger(\Lambda) = V \tag{5.62}$$

With covariant normalization

$$U(\Lambda)|k_1 k_2\rangle = |\Lambda k_1, \Lambda k_2\rangle \tag{5.63}$$

$$\langle k_1', k_2'|U^\dagger(\Lambda) = \langle \Lambda k_1', \Lambda k_2'| \tag{5.64}$$

so

$$U(\Lambda)VU^\dagger(\Lambda) = \int \prod_i \frac{d^3 k_i}{2E(k_i)} \frac{d^3 k_i'}{2E(k_i')} \delta^4(k_1' + k_2' - k_1 - k_2) h(k_1', ..)|\Lambda k_1', \Lambda k_2'\rangle\langle \Lambda k_1, \Lambda k_2| \tag{5.65}$$

Changing variables from p_i to Λp_i and using the invariance of $\frac{d^3 p}{2E(p)}$ and the four-dimensional δ-function, we find

$$U(\Lambda)VU^\dagger(\Lambda) = \int \prod_i \frac{d^3 k_i}{2E(k_i)} \frac{d^3 k_i'}{2E(k_i')} \delta^4(k_1' + k_2' - k_1 - k_2)$$

$$\cdot h(\Lambda^{-1}k_1', \Lambda^{-1}k_2', \Lambda^{-1}k_1, \Lambda^{-1}k_2)|k_1', k_2'\rangle\langle k_1, k_2| \tag{5.66}$$

so invariance of V under arbitrary Lorentz transformations (or equivalently, commutativity of V with $U(\Lambda)$) follows provided

$$h(\Lambda^{-1}k_1', \Lambda^{-1}k_2', \Lambda^{-1}k_1, \Lambda^{-1}k_2) = h(k_1', k_2', k_1, k_2) \tag{5.67}$$

for general Lorentz transformations Λ. This is trivial to arrange: we simply make h an *arbitrary* function of the Lorentz-invariant scalar products $k_1' \cdot k_2'$, etc. Obviously there is a great deal of freedom here, rather as in non-relativistic quantum mechanics, where one is at liberty to construct potential functions with only very loose constraints. The above procedure for generating Lorentz-invariant 2-2 particle scattering can clearly be imitated for N-N particle scattering, so we can generalize the above Ansatz to

$$V = \sum_{N=2}^\infty \int \prod_{i=1}^N \frac{d^3 k_i}{2E(k_i)} \frac{d^3 k_i'}{2E(k_i')} \delta^4(\sum_i k_i' - \sum_i k_i) h^{(N)}(k_i', k_i)|k_1' k_2'..k_N'\rangle\langle k_1 k_2..k_N| \tag{5.68}$$

Alas, this seemingly trivial method of generating a profusion of theories with Lorentz-invariant interactions (and number-conserving Lorentz-invariant scattering for any number of particles) possesses some fatal flaws:

1. The "convenient" property that the free and interaction parts of the Hamiltonian commute, $[H_0, V] = 0$, induced by the need to complete the 3-momentum conservation δ-function in (5.68) to a four-dimensional energy-momentum δ-function

invariant under general Lorentz transformations, is in fact a disaster. The resultant time-independence of the interaction in interaction picture means that the infinite time limit giving the S-matrix (4.164) is divergent. One could obtain a finite result by adiabatically switching off the interaction $V \to e^{-\epsilon|t|}V$, but the resultant theory would violate time-translation invariance, and the results would in any case be sensitive to the switch-off rate ϵ.[13]

2. Even if we ignore the problem of defining a sensible asymptotic limit for the scattering, the theory defined by (5.68) conceals another potentially fatal disease: the presence of "spooky action-at-a-distance effects".[14] The interaction operator V commutes with Lorentz transformations for any choice of the functions $h^{(N)}$, provided that these functions are constructed from Lorentz-invariant dot-products of the four-momenta. However, the physical requirement that a localized particle (or particles) very far separated from a set of scattering particles should not affect the interactions of the latter means that there must in fact be intricate relations between the $h^{(N)}$ for different values of N. This requirement is generally referred to as the *clustering principle*, and we shall discuss it in great detail in the following Chapter. Here, it suffices to note that the scattering of two particles in a universe emptied of all other matter would be determined by the $h^{(2)}$ functions, but that the introduction of a third particle localized very far away (and propagating therefore freely) would mean that the scattering amplitude for the first two is now determined by the $h^{(3)}$ function, which must therefore be related to $h^{(2)}$. If we had set $h^{(3)}$ to zero, for example, the introduction of a third particle localized arbitrarily far from two other localized interacting particles would have instantly caused these two to cease interacting—very clearly a "spooky action-at-a-distance" effect! In fact, the cluster decomposition principle vetoing such effects implies an infinite set of recursive relations between the coefficient functions $h^{(N)}$, as well as specifying certain smoothness properties for them, as we shall see in Chapter 6. The dyadic notation used to specify the interaction V in (5.68) turns out to be extremely inefficient in expressing these contraints: instead, we shall soon see that the introduction of creation and annihilation operators is just the technical tool needed to facilitate the construction of clustering interaction theories.

The two pathologies outlined above are really different symptoms of the same underlying disease: with an interaction Hamiltonian of the form (5.68), our particles obstinately *refuse to stop interacting*, either after separating (as naturally must occur in quantum theory due to wave-packet spreading—cf. also Section 9.3, where the

---

[13] We shall later encounter "persistent interactions" in local quantum field theory which also result in divergent contributions to the infinite-time propagation of the system encapsulated by the S-matrix, although these effects can be removed by appropriate choice of the interaction operator V. However, for interaction operators V of the general class specified by (5.68), it is impossible to avoid a divergent, and hence physically meaningless, S-matrix.

[14] Einstein's original use of the term "spukhafte Fernwirkung" referred to the peculiar (from a classical standpoint) statistical correlations of entangled quantum states, as in the famous EPR effect. These correlations, while perhaps psychologically disturbing, do not lead to the physically unacceptable action-at-a-distance effect of the kind discussed here (which we might perhaps call an "entsetzliche Fernwirkung"!).

rigorous Haag–Ruelle theory of scattering asymptotics is introduced) over a long time period, or when separated spatially by a large distance from one another (the unacceptable "action-at-a-distance" effects). We have evidently failed to construct an appropriately *local*, either in space or in time, theory of interactions.

5.5 A simple condition for Lorentz-invariant scattering

The preceding section exposed some of the difficulties one faces in attempting to construct, *directly from a particle point of view*, a physically sensible relativistic quantum mechanics—a theory, in other words, leading to a unitary S-matrix subject to the constraints of special relativity. The concept of a field—a dynamical entity allowing us to associate physical energy and momentum to domains in spacetime rather than directly to multi-particle states—was nowhere in sight. But from the time of Maxwell's great work on electromagnetism, the central role of the field concept in implementing the idea of local transmission of physical effects (eschewing "action-at-a-distance" effects) has been clear. In this section we shall show that the introduction of the field idea, and specifically, the notion of a *local* field, commuting with itself at space-like separations, allows a rather simple implementation of the requirements of Lorentz-invariance. The extent to which this approach also provides a solution to the requirements of the clustering principle (and if so, a unique solution) will be discussed in the next chapter.

Let us suppose that the interaction energy operator V in (5.57) can be written as the spatial integral of an *interaction energy density operator* $\mathcal{H}_{\text{int}}(\vec{x})$:

$$V = \int d^3x \, \mathcal{H}_{\text{int}}(\vec{x}) \tag{5.69}$$

Correspondingly, the interaction operator in interaction picture $V_{\text{ip}}(t)$ will be given by a spatial integral of a spacetime field $\mathcal{H}_{\text{int}}(\vec{x}, t)$ ($\equiv \mathcal{H}_{\text{int}}(x)$: as usual, coordinate vectors without three-vector symbols are to be taken to be spacetime four-vectors):

$$V_{\text{ip}}(t) = \int d^3x \, \mathcal{H}_{\text{int}}(\vec{x}, t) \tag{5.70}$$

$$\mathcal{H}_{\text{int}}(\vec{x}, t) = e^{iH_0 t} \mathcal{H}_{\text{int}}(\vec{x}) e^{-iH_0 t} \tag{5.71}$$

Accordingly, the formal expansion (4.27) for the S-matrix $S = U(+\infty, -\infty)$ becomes

$$S = \sum_{n=0}^{\infty} \frac{(-i)^n}{n!} \int_{-\infty}^{+\infty} dt_1 dt_2 .. dt_n T\{V_{\text{ip}}(t_1) \dots V_{\text{ip}}(t_n)\} \tag{5.72}$$

$$= \sum_{n=0}^{\infty} \frac{(-i)^n}{n!} \int d^4x_1 d^4x_2 .. d^4x_n T\{\mathcal{H}_{\text{int}}(x_1) .. \mathcal{H}_{\text{int}}(x_n)\} \tag{5.73}$$

Ignore for the time being the presence of a time-ordering symbol in (5.73). Then the desired Lorentz-invariance of the S-matrix

$$S = U(\Lambda) S U^\dagger(\Lambda) \tag{5.74}$$

would follow directly if we could ensure that

$$U(\Lambda)\mathcal{H}_{\text{int}}(x)U^{\dagger}(\Lambda) = \mathcal{H}_{\text{int}}(\Lambda x) \tag{5.75}$$

since the change $x_i \to \Lambda x_i$ could be erased by the change of variable (with unit Jacobian) $x_i \to \Lambda^{-1}x_i$.[15]

Unfortunately, such a change of variable can in general change the ordering in time of the various interaction operators, which will not in general commute at different times (this would require $[H_0, V] = 0$ and we have already seen in the preceding section why this is untenable). When can this happen? Only if two of the spacetime arguments, x_i and x_j say, differ by a *space-like* interval, $(x_i - x_j)^2 < 0$. There would then exist Lorentz transformations Λ for which the time-ordering symbol will reverse the order of the corresponding interaction Hamiltonians. Unless we insist that these operators *commute* in this situation, the argument leading to (5.74) will break down. We may clearly avoid this outcome by insisting on *commutativity of interaction energy densities at space-like separations*. This result is, of course, intuitively plausible, as it asserts the non-interference of measurements of (interaction) energy performed at space-like separations. The hand-waving argument presented above suggests that the two conditions

$$U(\Lambda)\mathcal{H}_{\text{int}}(x)U^{\dagger}(\Lambda) = \mathcal{H}_{\text{int}}(\Lambda x) \tag{5.76}$$

$$[\mathcal{H}_{\text{int}}(x), \mathcal{H}_{\text{int}}(y)] = 0, \quad (x - y)^2 < 0 \tag{5.77}$$

might be sufficient to ensure the Lorentz-invariance of the S-matrix (5.72). We shall refer to a field satisfying (5.76) as a *Lorentz scalar field*, while fields satisfying the space-like commutativity requirement (5.77) will be termed *local fields*.

The admittedly heuristic argument given above turns out to require some careful fine-tuning due to subtleties in the behavior of the operator products in (5.73) at point coincidences in the multi-dimensional integral: i.e., when $x_i = x_j$ for $i \neq j$. Let us focus our attention on the $n = 2$ term in the expansion of the S-matrix (5.73), involving the time-ordered product of two interaction energy density operators. Recall the definition of the time-ordered product *in a particular inertial frame* (and hence, with a definite choice of time variable)

$$T\{\mathcal{H}_{\text{int}}(\vec{x}_1, t_1)\mathcal{H}_{\text{int}}(\vec{x}_2, t_2)\} = \theta(t_1 - t_2)\mathcal{H}_{\text{int}}(\vec{x}_1, t_1)\mathcal{H}_{\text{int}}(\vec{x}_2, t_2)$$
$$+ \theta(t_2 - t_1)\mathcal{H}_{\text{int}}(\vec{x}_2, t_2)\mathcal{H}_{\text{int}}(\vec{x}_1, t_1) \tag{5.78}$$

or, introducing the time-like unit vector $n^{\mu} = g_0^{\mu} = (1, 0, 0, 0)$,

$$\tau(x_1, x_2; n) \equiv T\{\mathcal{H}_{\text{int}}(x_1)\mathcal{H}_{\text{int}}(x_2)\} = \theta(n \cdot (x_1 - x_2))\mathcal{H}_{\text{int}}(x_1)\mathcal{H}_{\text{int}}(x_2)$$
$$+ \theta(n \cdot (x_2 - x_1))\mathcal{H}_{\text{int}}(x_2)\mathcal{H}_{\text{int}}(x_1) \tag{5.79}$$

---

[15] The order in which the similarity transformation is performed, with the $U(\Lambda)$ operator on the left and the $U^{\dagger}(\Lambda)$ on the right, is dictated by the need for two successive transformations to follow the group composition law: $U(\Lambda_2)U(\Lambda_1) = U(\Lambda_2\Lambda_1)$.

The failure of the above product to be covariant under Lorentz transformations, i.e., to satisfy $U(\Lambda)\tau(x_1, x_2; n)U^\dagger(\Lambda) = \tau(\Lambda x_1, \Lambda x_2; n)$, is clearly due to the explicit presence of the time-like frame-dependent vector n^μ in the θ-functions. An infinitesimal variation of the time-like vector n^μ lies in the space-like hypersurface orthogonal to n, on to which the operator $\Pi^\mu{}_\nu \equiv g^\mu{}_\nu - n^\mu n_\nu$ projects (as $\Pi^\mu{}_\nu n^\nu = 0$). The frame-dependence of the T-product $\tau(x_1, x_2)$ therefore involves the projected derivative

$$\Pi^{\mu\nu}\frac{\partial}{\partial n^\nu}\tau(x_1, x_2; n) = \Pi^{\mu\nu}(x_1 - x_2)_\nu \delta(n \cdot (x_1 - x_2))[\mathcal{H}_{\text{int}}(x_1), \mathcal{H}_{\text{int}}(x_2] \qquad (5.80)$$

In our original frame, $n^\mu = (1, 0, 0, 0)$, the δ-function sets the times of spacetime points x_1, x_2 equal, so the commutator of the interactions energy densities in (5.80) is an *equal-time commutator* (ETC). The points x_1 and x_2 must therefore either be coincident or space-like separated, $(x_1 - x_2)^2 < 0$. Locality (5.77) implies the vanishing of the commutator in the latter case, but we still have the possibility of δ-function-type singularities in the coincidence limit at equal time $x_1^0 = x_2^0 = t$, including possible terms with spatial derivatives of the coincidence δ-function $\delta^3(\vec{x}_1 - \vec{x}_2)$,

$$[\mathcal{H}_{\text{int}}(\vec{x}_1, t), \mathcal{H}_{\text{int}}(\vec{x}_2, t)] = C(\vec{x}_1, t)\delta^3(\vec{x}_1 - \vec{x}_2) + D_\rho(\vec{x}_1, t)\Pi^{\rho\sigma}\frac{\partial}{\partial x_1^\sigma}\delta^3(\vec{x}_1 - \vec{x}_2) + \dots$$
$$(5.81)$$

where the ellipsis indicates the possible presence of higher (spatial) derivatives. The fact that the commutator in (5.81) involves the same field operator $\mathcal{H}_{\text{int}}$ twice actually eliminates the first term on the right-hand side as the commutator must be antisymmetric under the exchange of $\vec{x}_1$ and $\vec{x}_2$ (in other words, in this case, $C(\vec{x}, t) = 0$). The second (and if present, higher) terms on the right-hand side of (5.81), involving derivatives of δ-functions in an equal-time commutator, are called *Schwinger terms*. If no such terms are present, the commutator is called "ultralocal". Inserting (5.81) in the result (5.80) we find

$$\Pi^{\mu\nu}\frac{\partial}{\partial n^\nu}\tau(x_1, x_2; n) = \Pi^{\mu\nu}(x_1 - x_2)_\nu \Pi^{\rho\sigma}D_\rho(x_1)\frac{\partial}{\partial x_1^\sigma}\delta^4(x_1 - x_2) + \dots$$
$$= -\Pi^{\mu\rho}D_\rho(x_1)\delta^4(x_1 - x_2) + \dots \qquad (5.82)$$

(where we have used $(x_1 - x_2)_\nu \delta^4(x_1 - x_2) = 0$). Note that the preceding equations are to be interpreted in the usual distributional sense: as equalities holding after integration with appropriately smooth, rapidly decreasing c-number functions. This allows us to remove the derivative from the δ-function in the Schwinger term and apply it to $D_\rho(x_1)(x_1 - x_2)_\nu$, obtaining the final result (5.82). We see that if $D_\rho(x) \neq 0$ the time-ordered product $\tau(x_1, x_2; n)$ contributing to the second-order term in the S-matrix is indeed frame-dependent—thereby ruining the Lorentz-invariance of the S-matrix—in the presence of Schwinger terms in the equal-time commutator of $\mathcal{H}_{\text{int}}$. This is not an empty possibility: we shall see in Chapter 12 that this is exactly the situation in derivatively-coupled field theories, for example.[16] For the time being we shall satisfy ourselves with the requirement of "ultralocality" in the equal-time commutators of $\mathcal{H}_{\text{int}}$: namely, the absence of any derivative terms in the general form of the ETC (5.81). With this proviso, it can then be shown that the qualitative argument given previously for the Lorentz-invariance of the S-matrix (5.73) is indeed correct.

Having satisfied the requirements of invariance under the homogeneous portion of the Poincaré group (i.e., the homogeneous Lorentz group) by choosing the interaction energy density to be an ultralocal Lorentz scalar field, what can we say about the inhomogeneous part: namely, invariance under translations in space and time? The corresponding conservation laws require the S-matrix to commute with the free state energy $(P_0^{(0)} = H_0)$ and momentum $(P_i^{(0)}, i = 1, 2, 3)$ operators. The fact that our free and interaction operators H_0 and V are time-independent ensures energy conservation (cf. (4.182)); but what about spatial momentum conservation? Recall that this is assured once we have $[P_i^{(0)}, V] = 0$ (5.59). It turns out that we get this for free once $\mathcal{H}_{\text{int}}$ is chosen to be a Lorentz scalar field satisfying (5.76). First, we note that by the usual property of interaction-picture operators,

$$i[H_0, \mathcal{H}_{\text{int}}(\vec{x}, t)] = i[P_0^{(0)}, \mathcal{H}_{\text{int}}(\vec{x}, t)] = \frac{\partial}{\partial t}\mathcal{H}_{\text{int}}(\vec{x}, t) \tag{5.83}$$

We saw previously (cf. 5.31) that the energy-momentum four-vector operator $P_\mu^{(0)}$ transforms in the expected way under Lorentz transformations

$$U^\dagger(\Lambda)P_\mu^{(0)}U(\Lambda) = \Lambda_\mu^{\ \nu}P_\nu^{(0)} \tag{5.84}$$

For example, for a boost along the z-axis, choosing Λ to be the Lorentz transformation given by (5.10), and writing $U(\Lambda) = U(\omega)$,

$$U^\dagger(\omega)H_0U(\omega) = U^\dagger(\omega)P_0^{(0)}U(\omega)$$
$$= \Lambda_0^{\ \nu}P_\nu^{(0)} = \cosh(\omega)H_0 + \sinh(\omega)P_3^{(0)} \tag{5.85}$$

Thus

$$iU^\dagger(\omega)[H_0, \mathcal{H}_{\text{int}}(\vec{x}, t)]U(\omega) = i[U^\dagger(\omega)H_0U(\omega), U^\dagger(\omega)\mathcal{H}_{\text{int}}(\vec{x}, t)U(\omega)]$$
$$= i[\cosh(\omega)H_0 + \sinh(\omega)P_3^{(0)}, \mathcal{H}_{\text{int}}(\Lambda^{-1}x)] \tag{5.86}$$

---

[16] For such theories it is possible (cf. Section 12.1) to concoct an additional non-covariant term in the interaction density which cancels everywhere the effect of the Schwinger term—a so-called "covariantizing seagull". The Lagrangian formalism developed in Chapter 12 provides an automatic and foolproof procedure for generating the correct form of such covariantizing terms, if needed.

On the other hand,

$$iU^\dagger(\omega)[H_0, \mathcal{H}_{\text{int}}(\vec{x},t)]U(\omega) = \frac{\partial}{\partial t}\mathcal{H}_{\text{int}}(\Lambda^{-1}x)$$

$$= \frac{\partial}{\partial t}\mathcal{H}_{\text{int}}(x^1, x^2, \cosh(\omega)x^3 + \sinh(\omega)t, \cosh(\omega)t + \sinh(\omega)x^3)$$

$$= (\sinh(\omega)\frac{\partial}{\partial x'^3} + \cosh(\omega)\frac{\partial}{\partial t'})\mathcal{H}_{\text{int}}(\vec{x}',t'), \quad x' \equiv \Lambda^{-1}x \qquad (5.87)$$

Comparing coefficients of $\sinh(\omega)$ in (5.86) and (5.87), we find

$$i[P_3^{(0)}, \mathcal{H}_{\text{int}}(\Lambda^{-1}x)] = i[P_3^{(0)}, \mathcal{H}_{\text{int}}(x')] = \frac{\partial}{\partial x'^3}\mathcal{H}_{\text{int}}(x') \qquad (5.88)$$

Evidently, for any of the spatial components

$$i[P_i^{(0)}, \mathcal{H}_{\text{int}}(x)] = \frac{\partial}{\partial x^i}\mathcal{H}_{\text{int}}(x) \qquad (5.89)$$

which implies

$$i[P_i^{(0)}, V] = i[P_i^{(0)}, \int d^3x \mathcal{H}_{\text{int}}(x)] = \int d^3x \frac{\partial}{\partial x^i}\mathcal{H}_{\text{int}}(x) = 0 \qquad (5.90)$$

exactly the condition for momentum conservation. We shall henceforth consider the definition of a general *Lorentz scalar field* $A(x)$ to include the transformation property (5.76) under the HLG *as well as* the commutation relations with the energy-momentum four-vector

$$i[P_\mu^{(0)}, A(x)] = \frac{\partial}{\partial x^\mu}A(x) \qquad (5.91)$$

From (5.91), a standard application of the Baker–Campbell–Hausdorff formula

$$e^P Q e^{-P} = Q + [P,Q] + \frac{1}{2!}[P,[P,Q]] + \frac{1}{3!}[P,[P,[P,Q]]] + \dots \qquad (5.92)$$

leads to the very important translation property for scalar fields in the interaction picture

$$e^{iP_\mu^{(0)}a^\mu}A(x)e^{-iP_\mu^{(0)}a^\mu} = A(x+a) \qquad (5.93)$$

for any fixed displacement four-vector a^μ. We can combine the transformation requirements under the HLG (5.76) with the translation property (5.93) to obtain the transformation of our scalar field under a general element (Λ, a) of the Poincaré group, with unitary representative $U(\Lambda, a) = e^{iP^{(0)}\cdot a}U(\Lambda)$:

$$U(\Lambda, a)A(x)U^\dagger(\Lambda, a) = A(\Lambda x + a) \qquad (5.94)$$

5.6 Problems

1. Calculate explicitly (as a 4x4 matrix) the commutator $[\Lambda_1, \Lambda_2]$ of a boost Λ_1 with rapidity ω_1 along the x axis with a boost Λ_2 with rapidity ω_2 along the y axis. If the rapidities ω_1, ω_2 are infinitesimal, what type of transformation does $[\Lambda_1, \Lambda_2]$ induce?

2. (a) Show the following connection between helicity and spin states:

$$|\vec{p}, \lambda\rangle = \sum_\sigma D^j_{\sigma\lambda}(R(\hat{p}))|\vec{p}, \sigma\rangle$$

(b) Prove that helicity states transform as follows under Lorentz transformations:

$$U(\Lambda)|\vec{p}, \lambda\rangle = \sqrt{\frac{E(\Lambda p)}{E(p)}} \sum_{\lambda'} D^j_{\lambda'\lambda}(\mathcal{W}(\Lambda, \vec{p}))|\Lambda\vec{p}, \lambda'\rangle$$

where

$$\mathcal{W}(\Lambda, \vec{p}) \equiv B^{-1}(|\Lambda\vec{p}|)R^{-1}(\widehat{\Lambda p})\Lambda R(\hat{p})B(|\vec{p}|)$$

(c) From (b), show that

$$U(\Lambda)a^\dagger(\vec{p}, \lambda)U^\dagger(\Lambda) = \sqrt{\frac{E(\Lambda p)}{E(p)}} \sum_{\lambda'} D^j_{\lambda'\lambda}(\mathcal{W}(\Lambda, \vec{p}))a^\dagger(\Lambda\vec{p}, \lambda')$$

3. Verify that the massless particle little group displacement transformations $T(\vec{\xi})$ in (5.46) are indeed Lorentz transformations. Also, verify the group composition rule (5.47) and the transformation law (5.48).

4. The transformation property of scalar fields under spatial rotations follows from the infinitesimal version of the transformation property:

$$U(\Lambda)A(x)U^\dagger(\Lambda) = A(\Lambda x)$$

Choosing Λ to be the infinitesimal rotation $(R_i\vec{x})^j = x^j + \epsilon_{ijk}x^k \delta\theta$, show that the scalar field $A(x)$ has the following commutation relation with the angular momentum components $J_i, i = 1, 2, 3$:

$$[J_i, A(x)] = i\epsilon_{ijk}x^k\partial_j A(x) \tag{5.95}$$

5. Show that a product $H(x) = A(x)B(x)C(x)..$ of local, Lorentz scalar fields $A(x)$, $B(x)$, $C(x)$,.. is itself a local, Lorentz scalar field: i.e., verify

$$U(\Lambda)H(x)U^\dagger(\Lambda) = H(\Lambda x)$$

$$[H(x), H(y)] = 0, \quad (x - y)^2 < 0$$

$$i[P^{(0)}_\mu, H(x)] = \frac{\partial}{\partial x^\mu}H(x)$$

6. A Lorentz vector field $V^\mu(x)$ is defined as a field transforming as follows:

$$U(\Lambda)V^\mu(x)U^\dagger(\Lambda) = (\Lambda^{-1})^\mu{}_\nu V^\nu(\Lambda x)$$

(a) Show that if $A(x)$ is a Lorentz scalar field, then $V^\nu(x) \equiv \partial^\nu A(x)$ is a Lorentz vector field.

(b) If $V^\nu(x), W^\nu(x)$ are Lorentz vector fields, show that $A(x) \equiv V_\nu W^\nu(x)$ is a Lorentz scalar field.

(c) Verify that the group composition law works out properly for the product of two successive Lorentz transformations Λ_1, Λ_2 of the vector field $V^\nu(x)$.

6

Dynamics IV: Aspects of locality: clustering, microcausality, and analyticity

---

In the preceding chapter we discussed two possible approaches to constructing a relativistically invariant theory of particle scattering. Our first attempt—a frontal assault in which one directly writes down for each scattering sector (i.e., with a specified number of incoming and outgoing particles) a manifestly Lorentz-invariant interaction operator containing momentum-dependent Lorentz scalar amplitudes—led to disaster. The resultant theory led to particle interactions which could not be confined to finite regions of spacetime, as all our experience with subatomic phenomena has led to us to conclude that they are. In particular, the constraints of *cluster decomposition*—which we now define as the necessity for scattering amplitudes of groups of far separated particles to factorize in such a way that the probability of the overall process is a product of the probability of scattering in the separate groups—do not appear to be satisfied in any natural way in such a framework. Our second attempt, in which the interaction Hamiltonian is written as a spatial integral of a local, Lorentz (ultra-)scalar field, accomplishes the primary goal of producing a Lorentz-invariant set of scattering amplitudes, but we are as yet unsure as to its compliance with the clustering principle. Our goal in this chapter is to put this latter requirement into a precise mathematical framework, called *second quantization*, so that the process of identifying clustering relativistic scattering theories can be simplified and even to some degree automated.

Before introducing the technical tools needed to accomplish this task, it may be best to outline in a completely qualitative way the essential conceptual elements that will be woven together by writing our theories in the language of second quantization. When one considers the nature of interactions in the subatomic world, the following two intuitions seem so completely natural that we can hardly imagine them to be violated in a physically sensible theory:

1. *Intuition A* (the clustering principle):[1] the scattering amplitude for two groups of particles contained (and localized, in the wave-packet sense) in distinct finite spatial regions, with the two regions separated by a distance R, should, in the

---

[1] The cluster decomposition principle for the S-matrix seems to have been first articulated by Wichmann and Crichton (Wichmann and Crichton, 1963): the proof that factorization of scattering probabilities extends to the scattering amplitudes was given by Taylor (Taylor, 1966).

limit $R \to \infty$, approach the product of the independent scattering amplitudes for the particles in each region to scatter among themselves. As we shall see shortly, this intuition can be formulated as a completely precise condition which a physically sensible multi-particle S-matrix must necessarily obey.

2. *Intuition B*: the quantum-mechanical amplitude for particle scattering can be constructed as a sum of terms in which particular particles can scatter not at all, once, twice, or indeed arbitrarily many times. If no interaction occurs between two particles, their initial and final spatial momenta must be *precisely unchanged*. Likewise, if no interaction occurs between two *groups* of particles, we expect the total momentum of each group to be exactly preserved. On the other hand, if even a single interaction, however weak, occurs between two particles, we expect the probability for their individual spatial momenta to be *exactly* preserved to be zero: the interaction will presumably end up transferring some non-zero, albeit very small, momentum between the two particles (all this with an obvious generalization to two groups of particles).

The amazing thing, as we shall see in the very next section, is that these two intuitions are in fact *one and the same*! By the magic of Fourier transformation, the spatial considerations of intuition A turn out to be precisely equivalent to the momentum considerations of intuition B. This realization will allow us to rephrase the requirement of clustering in a mathematically transparent and readily implementable way.

6.1   Clustering and the smoothness of scattering amplitudes

We now come to the constraints placed on the S-matrix by the requirement of cluster decomposition. To express the idea of factorization of far-separated processes in a simple but precise way, it is convenient to introduce the concept of *connected parts* of the S-matrix. This allows us to separate out the parts of the S-matrix corresponding to a non-trivial interaction among some subset of particles from those parts corresponding to the trivial situation where the particles pass through without interaction. The connected part S^c of S may be defined by induction on the number of particles in the initial and final states as follows. If both initial and final states contain only a single *stable* particle, no interactions occur and in this subspace the S-matrix is the identity, or (with non-covariant normalization of the states)

$$S_{\vec{k}',\vec{k}} = \delta^3(\vec{k}' - \vec{k}) \equiv S^c_{\vec{k}',\vec{k}} \tag{6.1}$$

which we may express pictorially as in Fig. 6.1.

For 2-2 scattering, the connected part is defined by the equation

$$S_{k'_1 k'_2, k_1 k_2} = S^c_{k'_1, k_1} S^c_{k'_2, k_2} + S^c_{k'_1, k_2} S^c_{k'_2, k_1} + S^c_{k'_1 k'_2, k_1 k_2} \tag{6.2}$$

which can be expressed graphically as indicated in Fig. 6.2. Note that in the present discussion we assume that the incoming particles are identical (hence, indistinguishable) Bose particles.

The three-particle connected part is similarly defined by subtracting off those parts of the full 3-3 scattering amplitude in which some proper subset of the particles pass through unaffected by interactions. Again, this can be pictorially represented as shown

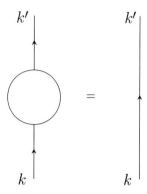

Fig. 6.1 S-matrix in the one-particle sector.

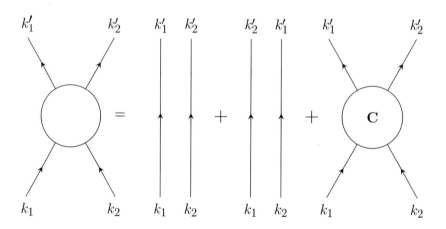

Fig. 6.2 Decomposition of 2-2 S-matrix amplitude into connected parts.

in Fig. 6.3 (here, "perms" refers to the appropriate set of terms with initial and final particle momenta permuted to ensure Bose symmetry of the amplitude). As the connected 2-2 amplitude has already been defined in (6.2), the fully connected 3-3 amplitude $S^c_{k'_1 k'_2 k'_3, k_1 k_2 k_3}$ is defined inductively as the full 3-3 amplitude *minus* the contributions from situations in which some proper subset of particles interacts separately from the others.

Since $S^c_{k'_1 k'_2, k_1 k_2}$, $S^c_{k'_1 k'_2 k'_3, k_1 k_2 k_3}$, etc., all come from the $\delta(E_\alpha - E_\beta) T_{\beta\alpha}$ (i.e., interaction) part of S (recall (4.182)), and T is assumed to conserve total spatial momentum, we expect these connected parts all to have a full four-dimensional δ-function of four-momentum conservation: $\delta^4(P' - P)$, with P, P' the total initial and final four-momenta for the connected subprocess. For clustering to hold, as we shall now see, it is *crucial* that this be the *only* δ-function in the connected parts of the S-matrix. Of course, the disconnected parts may have many more, as energy-momentum must

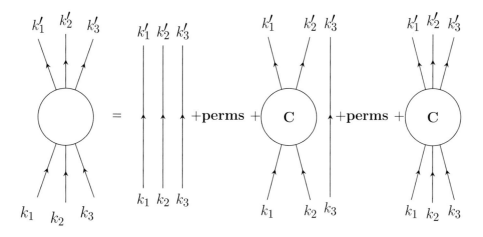

Fig. 6.3 Decomposition of 3-3 S-matrix amplitude into connected parts.

be conserved separately in each disconnected subprocess. To see the necessity for the above assertion, consider a N→N process (N particles in, N out) where N=$n_1 + n_2$, with n_1 particles scattering far from the other n_2. Cluster decomposition requires that the S-matrix factor into a product of S-matrices for $n_1 \to n_1$ and $n_2 \to n_2$ scattering separately. (We are assuming here for simplicity of notation only that the scatterings conserve the number of particles, which is certainly not the case in general in relativistic quantum theory.) The general expansion of the S-matrix in terms of connected parts means that we can write the graphical representation for the full N→N process as in Fig. 6.4. For the S-matrix to cluster—in other words, for the overall process to factorize into a product of independent $n_1 \to n_1$ and $n_2 \to n_2$ scatterings when the two sets of particles are spatially far separated—the extra terms containing connected parts with both qs and ks must vanish when we form wave-packets for the incoming and outgoing particles and then move all q-type particles far from all k-type ones.

The separation can be achieved by introducing a large three-vector $\vec{\Delta}$ and positioning one subset of particles around $-\vec{\Delta}$ and the other around $\vec{\Delta}$. The wave-packets are constructed in the usual way: replace[2] the plane wave $e^{i\vec{k}\cdot\vec{x}}$ by $\int d^3\tilde{k}g(\tilde{k};k)e^{i\vec{k}\cdot(x-[\Delta+\xi])}$, i.e., a wave-packet of momentum centered around k (if g is strongly peaked there) and peaked in coordinate space at $\Delta + \xi$. A typical connected transition amplitude for such a set of wave-packets will thus take the form:

$$\int d^3\tilde{k}_1 d^3\tilde{k}_2..d^3\tilde{q}_{n_2} d^3\tilde{k}'_1..d^3\tilde{q}'_{n_2} g(\tilde{k}'_1; k'_1)..g(\tilde{q}_{n_2}; q_{n_2})$$

$$e^{-i\sum_j \tilde{k}_j\cdot(\Delta+\xi_j)-i\sum_j \tilde{q}_j\cdot(-\Delta+\eta_j)+i\sum_j \tilde{k}'_j\cdot(\Delta+\xi'_j)+i\sum_j \tilde{q}'_j\cdot(-\Delta+\eta'_j)} S^c_{\tilde{k}'_1..\tilde{q}'_{n_2},\tilde{k}_1..\tilde{q}_{n_2}} \quad (6.3)$$

---

[2] For the rest of this section we drop three-vector notation to avoid overcrowding the formulas—but re-emphasize to the reader that the requirements of clustering are exclusively *spatial* ones!

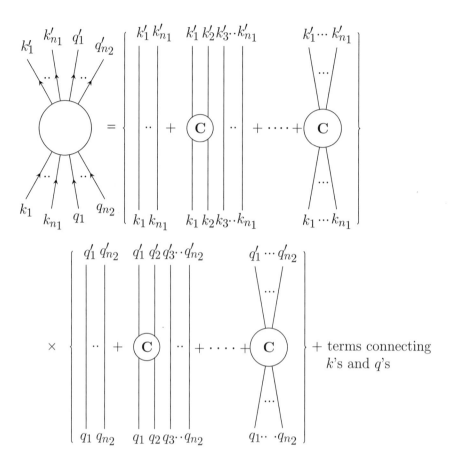

Fig. 6.4 Decomposition of a general S-matrix amplitude into separated subprocesses.

The Δ-dependence in the exponent of (6.3) is just

$$e^{i\Delta\cdot(\sum \tilde{k}'_j - \sum \tilde{q}'_j - \sum \tilde{k}_j + \sum \tilde{q}_j)} \tag{6.4}$$

On the other hand, momentum conservation says

$$\sum \tilde{k}_j + \sum \tilde{q}_j = \sum \tilde{k}'_j + \sum \tilde{q}'_j$$

$$\Rightarrow \sum \tilde{k}'_j - \sum \tilde{k}_j = \sum \tilde{q}_j - \sum \tilde{q}'_j \tag{6.5}$$

so the Δ-dependence may also be written $e^{2i\Delta\cdot(\sum \tilde{q}_j - \sum \tilde{q}'_j)}$. If the connected part S^c in (6.3) contained a partial δ-function such as $\delta(\sum \tilde{q}_j - \sum \tilde{q}'_j)$, the amplitude would lose Δ-dependence and S^c could not vanish in the limit $\Delta \to \infty$. In fact, we must demand that S^c contain only *a single overall* δ-function of energy-momentum conservation, times a sufficiently smooth function of momenta (where "sufficiently smooth" ensures that the Fourier transform falls sufficiently rapidly in coordinate space).

To summarize our main result from this section: the requirement of cluster factorization of the S-matrix for far-separated processes amounts to a momentum smoothness condition on the *connected* parts of the multi-particle scattering amplitudes. In particular, these connected parts must contain at most a single (three-dimensional) δ-function ensuring spatial momentum conservation, multiplied by a reasonably smooth momentum-dependent amplitude. The behavior of the Fourier transform of the latter in coordinate space determines the detailed spatial falloff (power versus exponential, for example) of the corrections to factorization for far-separated processes.

Before proceeding to a derivation of the constraints imposed on the dynamics of the theory (specifically, on the form of the Hamiltonian) by clustering, it will be convenient to introduce a technical tool which will greatly simply the combinatoric aspects of the argument, and which will in any event be needed later in our systematic study of the clustering properties of local field theory in the axiomatic Wightman framework in Chapter 9, and of perturbative quantum field theory in Chapter 10. As above, we will reduce inessential algebraic complications to a minimum by dealing with the scattering amplitudes of a single spinless boson, with Bose-symmetric multi-particle free states $|k_1, k_2, \ldots k_N\rangle$ specified entirely by listing the momenta of the particles (in any order). We can then associate any operator S acting in the Fock space of these states with a generating functional $\mathcal{S}(j^*, j)$, where $j(k)$ is a complex source function[3] of momentum, and

$$\mathcal{S}(j^*,j) \equiv \sum_{M,N} \int \frac{d^3k'_1...d^3k'_M}{M!} \frac{d^3k_1...d^3k_N}{N!} j^*(k'_1)...j^*(k'_M) S_{k'_1..k'_M,k_1..k_N} j(k_1)...j(k_N) \tag{6.6}$$

where $S_{k'_1..k'_M,k_1..k_N} = \langle k'_1..k'_M|S|k_1..k_N\rangle$. The association is unique, as we may extract any multi-particle amplitude by an appropriate multiple functional derivative:[4]

$$S_{k'_1..k'_M,k_1..k_N} = \frac{\delta^{M+N}}{\delta j^*(k'_1)..\delta j^*(k'_M)\delta j(k_1)..\delta j(k_N)} \mathcal{S}(j^*,j)\Big|_{j=j^*=0} \tag{6.7}$$

The utility of the generating functional defined in this way follows from the ease with which it allows us to extract the connected amplitudes. Indeed, if we define a *connected functional* $\mathcal{S}^c(j^*, j)$ as the logarithm of the full generating functional $\mathcal{S}$ defined above, so that

$$\mathcal{S}(j^*,j) \equiv \exp\left(\mathcal{S}^c(j^*,j)\right) \tag{6.8}$$

one easily sees that (for the case where S is the scattering operator) the amplitudes encoded in $\mathcal{S}^c$ are exactly the connected amplitudes defined above. For example,

---

[3] For fermions, a similar procedure can be used, but with source functions which take values in a anticommuting Grassmann field. We shall return to this later in the book when we discuss the path-integral formulation of quantum field theory.

[4] For a review of functional calculus, see Appendix A.

(6.2) corresponds to the result obtained if we assume only particle-number-conserving interactions, in which case only the $M = N = 1$ and $M = N = 2$ terms of (6.6) need be included in the exponent, and the functional derivative $\frac{\delta^4}{\delta j^*(k_1')\delta j^*(k_2')\delta j(k_1)\delta j(k_2)}$ applied to $\mathcal{S}(j^*, j)$ (with j and j^* then set to zero) then yields (6.2) after a short calculation. The exponential relation here between full and connected amplitudes is the precise analog of the corresponding relation in statistical mechanics between the partition function and the extensive (i.e., linear in the system volume) free energy of the system.

6.2   Hamiltonians leading to clustering theories

Matrix elements of the Hamiltonian also display a connectedness structure. We shall now see that the desired cluster decomposition property of the S-matrix is inherited from a corresponding characteristic of the underlying (interaction) Hamiltonian. To begin the discussion, imagine first a non-relativistic quantum-mechanical system where the particles interact by a two-body potential. The 3-3 matrix element of the potential operator will be

$$\langle k_1' k_2' k_3' | V | k_1 k_2 k_3 \rangle = \langle k_1' k_2' k_3' | V_{12} + V_{23} + V_{31} | k_1 k_2 k_3 \rangle \tag{6.9}$$

For example, the system being considered might be the three electrons of a lithium atom, and V above the total electrostatic interaction energy of the electrons. Explicitly,

$$\langle k_1' k_2' k_3' | V_{12} | k_1 k_2 k_3 \rangle$$

$$= \int d^3x_1 d^3x_2 d^3x_3 V_{12}(x_1 - x_2) e^{i[(k_1 - k_1')\cdot x_1 + (k_2 - k_2')\cdot x_2 + (k_3 - k_3')\cdot x_3]}$$

$$= (2\pi)^6 \delta^3(k_3 - k_3') \delta^3(k_1 + k_2 - k_1' - k_2') \int d^3x_- V_{12}(x_-) e^{i(k_1 - k_1')\cdot x_-} \tag{6.10}$$

which may be expressed graphically as shown in Fig. 6.5.

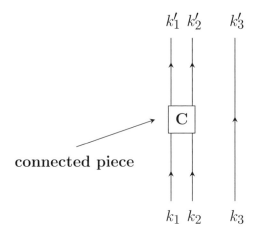

Fig. 6.5 Decomposition of a 3-3 matrix element of V.

Note that the remaining integral in (6.10) defines a smooth function of $k_1 - k_1'$ provided V_{12} goes to zero sufficiently rapidly for $x_- \to \infty$. In general, the multi-particle matrix elements of H decompose as follows:

$$H_{\vec{k}',\vec{k}} = < \vec{k}'|H|\vec{k}> \equiv E(k)\delta^3(\vec{k}' - \vec{k}) \equiv H^c_{\vec{k}',\vec{k}} \tag{6.11}$$

In the two-body sector,

$$H_{\vec{k}_1'\vec{k}_2',\vec{k}_1\vec{k}_2} \equiv H^c_{\vec{k}_1',\vec{k}_1}\delta^3(\vec{k}_2' - \vec{k}_2) + H^c_{\vec{k}_1',\vec{k}_2}\delta^3(\vec{k}_2' - \vec{k}_1) + H^c_{\vec{k}_2',\vec{k}_1}\delta^3(\vec{k}_1' - \vec{k}_2)$$

$$+ H^c_{\vec{k}_2',\vec{k}_2}\delta^3(\vec{k}_1' - \vec{k}_1) + H^c_{\vec{k}_1'\vec{k}_2',\vec{k}_1\vec{k}_2} \tag{6.12}$$

which defines the 2-2 connected piece $H^c_{k_1'k_2',k_1k_2}$. Graphically, (6.12) may be represented as shown in Fig. 6.6. Similarly, in the three-body sector we have a decomposition

$$H_{\vec{k}_1'\vec{k}_2'\vec{k}_3',\vec{k}_1\vec{k}_2\vec{k}_3} \equiv H^c_{\vec{k}_1',\vec{k}_1}\delta^3(\vec{k}_2' - \vec{k}_2)\delta^3(\vec{k}_3' - \vec{k}_3) + \text{perms.}$$

$$+ H^c_{\vec{k}_1'\vec{k}_2',\vec{k}_1\vec{k}_2}\delta^3(\vec{k}_3' - \vec{k}_3) + \text{perms.}$$

$$+ H^c_{\vec{k}_1'\vec{k}_2'\vec{k}_3',\vec{k}_1\vec{k}_2\vec{k}_3} \tag{6.13}$$

Note that in the simple non-relativistic example mentioned above, with only two-body forces present, the connected part of the Hamiltonian in the one-particle sector is given just by the matrix elements of the *free* Hamiltonian H_0 (see (6.11)), while the connected 3-3 part $H^c_{k_1'k_2'k_3',k_1k_2k_3}$ in (6.13) is actually zero. In more general situations in many-body theory, there are intrinsically three (or higher) body forces. Intuitively, this corresponds to a situation where a single interaction alters the momenta of all

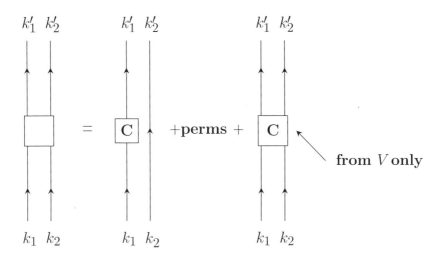

Fig. 6.6 Decomposition of 2-2 matrix element of H.

three (or more) participating particles. In relativistic quantum field theory, particle number is no longer conserved in interactions, and the existence of persistent interactions (cf. Chapter 10) can lead to interaction contributions even in the one-particle sector (depending on exactly how the full Hamiltonian is split into free and interacting parts).

To summarize the foregoing discussion for non-relativistic scattering, we expect that the connected N-N part of the Hamiltonian should not contain any partial δ-functions conserving momentum of any subset $M < N$ of the N interacting particles. This intuitive expectation turns out to be *precisely* the requirement that the resultant S-matrix possess the desired cluster-decomposition properties. More formally, the connected part of the Hamiltonian has matrix elements obtained by Fourier-transforming some multi-particle potential-energy function

$$H^c_{k'_1 k'_2 .., k_1 k_2} = \int d^3x'_1 d^3x'_2 .. d^3x_1 d^3x_2 .. V(x_1, x_2, .., x'_1 ..)$$

$$e^{-ik'_1 \cdot x'_1} e^{-ik'_2 \cdot x'_2} .. e^{ik_1 \cdot x_1} e^{ik_2 \cdot x_2} .. \qquad (6.14)$$

where the potential energy function V should only depend on differences of coordinates (translation invariance). The invariance of V under an equal shift of all coordinates leads directly to momentum conservation: H^c must contain an overall conservation δ-function, namely $\delta^3(\vec{k}'_1 + \vec{k}'_2 + .. - \vec{k}_1 - \vec{k}_2 - ..)$. The only way to have additional δ-functions conserving subsets of momenta is to have V constant when some subset of coordinates is moved en-bloc far away from some other subset. We can prevent this by insisting that V is a smooth[5] function of differences of coordinates falling to zero when *any two* coordinates separate to infinity with the others fixed.

The basic theorem specifying the smoothness properties needed in the connected Hamiltonian amplitudes to guarantee the clustering property for the S-matrix was proven by Weinberg (Weinberg, 1964b), in a classic study of multi-particle scattering,:

Theorem 6.1 *If* $H^c_{k'_1 k'_2 .., k_1 k_2} = \delta^3(\sum k' - \sum k)$ *times a smooth function of the momenta* k, k', *then* $S^c_{k'_1 k'_2 .., k_1 k_2} = \delta^3(\sum k' - \sum k)$ *times a smooth function.*

We shall derive this result for the same system used in our discussions in this and the preceding section: namely, for multi-particle scattering of a single species of spinless boson. However, there will be no restriction to non-relativistic scattering. In particular, the scattering amplitudes need not conserve particle number. To facilitate the otherwise rather tricky combinatorics, we shall use the generating functional technique described in the preceding section. As in the case of S-matrix amplitudes, the introduction of associated functionals

---

[5] Here "smooth" is being used in a somewhat loose sense: we are certainly *not* requiring analyticity in the momentum variables, for example. Rather, here and henceforth in our discussion of clustering, the reader should interpret the property "smooth" as simply implying the absence of singularities of δ-function strength.

$$\mathcal{H}(j^*,j) \equiv \sum_{M,N} \int \frac{d^3k_1'...d^3k_M'}{M!} \frac{d^3k_1...d^3k_N}{N!} j^*(k_1')...j^*(k_M') H_{k_1'..k_M',k_1..k_N} j(k_1)...j(k_N)$$

(6.15)

$$\mathcal{H}^c(j^*,j) \equiv \sum_{M,N} \int \frac{d^3k_1'...d^3k_M'}{M!} \frac{d^3k_1...d^3k_N}{N!} j^*(k_1')...j^*(k_M') H^c_{k_1'..k_M',k_1..k_N} j(k_1)...j(k_N)$$

(6.16)

allows the inductive sequence defining connected parts of the Hamiltonian to be solved very simply. Namely, defining

$$\mathcal{F}(j^*,j) \equiv \int d^3k j^*(k) j(k)$$

(6.17)

one finds simply

$$\mathcal{H}(j^*,j) = \mathcal{H}^c(j^*,j) \exp\left(\mathcal{F}(j^*,j)\right)$$

(6.18)

For example, the reader may easily check (take again number conserving theories with $M = N$ for simplicity) that applying the functional derivative $\frac{\delta^4}{\delta j^*(k_1')\delta j^*(k_2')\delta j(k_1)\delta j(k_2)}$ to (6.18) (and then setting $j = j^* = 0$) leads directly to the connectedness structure (6.12) for the two-body sector (as illustrated in Fig. 6.6).

We are interested in the clustering properties of the S-matrix, which the reader will recall from Section 4.3 is defined as the set of multi-particle matrix elements of the infinite time limit $U(+\infty, -\infty)$ of the finite time evolution operator $U(t,t_0)$ in the interaction picture, satisfying (4.20):

$$i\frac{\partial}{\partial t}U(t,t_0) = V_{\text{ip}}(t)U(t,t_0)$$

(6.19)

It will suffice to establish the desired smoothness property for the matrix elements of $U(t,t_0)$ at finite times t, t_0 as it will then carry over trivially to the large time limit. We need just one further technical tool to carry through the argument. It is easy to see that the functional associated with the product VU of two linear operators V and U is given by

$$(\mathcal{VU})(j^*,j) = \mathcal{LV}(j^*,J)\mathcal{U}(J^*,j)|_{J=J^*=0}$$

(6.20)

where $\mathcal{V}$ (resp. $\mathcal{U}$) are the functionals associated with the Fock space operators V (resp. U), and $\mathcal{L}$ is the *linking operator* defined as

$$\mathcal{L} \equiv \exp\left(\int d^3k \frac{\delta^2}{\delta J(k)\delta J^*(k)}\right)$$

(6.21)

The role of the exponential in (6.21) is to produce (on expansion) a series of terms, each of which ties together the momentum of the final-state particles in the matrix element of U with the initial-state particles in the matrix element of V (for some specific intermediate state in the operator product VU). There is no substitute here for the reader writing out a few simple examples to verify this result (see Problem 1 at the end of this chapter).

Suppressing the passive initial time t_0, which plays no role in the following, we shall denote the generating functional for the matrix elements of $U(t, t_0)$ by $\mathcal{U}(j^*, j; t)$, the functional for the matrix elements of the interaction operator $V_{\text{ip}}(t)$ by $\mathcal{V}(j^*, j; t)$, and the corresponding connected quantities as usual by attaching a superscript "c". Then the equation of motion (6.19) translates to a corresponding time-development equation for the associated functionals:

$$
\begin{aligned}
i\frac{\partial}{\partial t}\mathcal{U}(j^*, j; t) &= i\frac{\partial}{\partial t}\exp\left(\mathcal{U}^c(j^*, j; t)\right) \\
&= i\frac{\partial \mathcal{U}^c(j^*, j; t)}{\partial t}\exp\left(\mathcal{U}^c(j^*, j; t)\right) \\
&= (\mathcal{V}\mathcal{U})(j^*, j; t) \\
&= \mathcal{L}\mathcal{V}(j^*, J; t)\mathcal{U}(J^*, j; t)|_{J=J^*=0} \\
&= \mathcal{L}\{e^{\mathcal{F}(j^*, J)}\mathcal{V}^c(j^*, J; t)e^{\mathcal{U}^c(J^*, j; t)}\}\Big|_{J=J^*=0}
\end{aligned}
\tag{6.22}
$$

Note that the operators

$$
A \equiv \int d^3k\,\frac{\delta^2}{\delta J(k)\delta J^*(k)}, \qquad B \equiv \mathcal{F}(j^*, J) = \int d^3k\, j^*(k)J(k)
\tag{6.23}
$$

have the commutator

$$
[A, B] = \int d^3k\, j^*(k)\frac{\delta}{\delta J^*(k)}
\tag{6.24}
$$

which (a) commutes with A and B (remember that the functional derivatives with respect to J and J^* act independently), and (b) acts as the generator of translations on functionals of J^*:

$$
\exp\left(\int d^3k\, j^*(k)\frac{\delta}{\delta J^*(k)}\right)\mathcal{G}(J^*) = \mathcal{G}(J^* + j^*)
\tag{6.25}
$$

From (a), we have the Baker–Campbell formula $e^A e^B = e^B e^A e^{[A,B]}$, so

$$
\mathcal{L}e^{\mathcal{F}(j^*, J)} = e^{\mathcal{F}(j^*, J)}\mathcal{L}\exp\left(\int d^3k\, j^*(k)\frac{\delta}{\delta J^*(k)}\right)
\tag{6.26}
$$

Once the factor $e^{\mathcal{F}(j^*, J)}$ is to the left of the linking operator (containing derivatives with respect to J) the source J can be set to zero, leaving only the second two terms on the right-hand side of (6.26). Inserting this result on the final line of (6.22),

$$i\frac{\partial \mathcal{U}^c(j^*,j;t)}{\partial t} \, \exp\left(\mathcal{U}^c(j^*,j;t)\right)$$

$$= \mathcal{L}\exp\left(\int d^3k j^*(k)\frac{\delta}{\delta J^*(k)}\right)\{\mathcal{V}^c(j^*,J;t)e^{\mathcal{U}^c(J^*,j;t)}\}\bigg|_{J=J^*=0}$$

$$= \mathcal{L}\{\mathcal{V}^c(j^*,J;t)e^{\mathcal{U}^c(J^*+j^*,j;t)}\}\bigg|_{J=J^*=0} \qquad (6.27)$$

whence

$$i\frac{\partial \mathcal{U}^c(j^*,j;t)}{\partial t} = \mathcal{L}\{\mathcal{V}^c(j^*,J;t)e^{\mathcal{U}^c(J^*+j^*,j;t)-\mathcal{U}^c(j^*,j;t)}\}\bigg|_{J=J^*=0} \qquad (6.28)$$

The essential result we desire is already contained, in disguised form, in Eq. (6.28). On the right-hand side, we find an exponential which if expanded gives a sum of terms corresponding to a product of r (say) terms of the form $\mathcal{U}^c(J^*+j^*,j;t) - \mathcal{U}^c(j^*,j;t)$, each of which would vanish when we evaluate the final expression at $J^* = 0$, as indicated in the formula. The only way to avoid this is if each such term receives at least one derivative $\frac{\delta}{\delta J^*}$ from the linking operator on the left, which will then attach the corresponding connected $\mathcal{U}^c$ factor to the connected Hamiltonian term $\mathcal{V}^c(j^*,J;t)$. In other words, the only terms which survive involve r connected factors of the evolution amplitude U^c connected to a single factor of the connected interaction Hamiltonian V^c, as indicated schematically in Fig. 6.7, for the special case of $r =3$.

Now suppose that at a given time U^c has only a single overall δ-function of momentum conservation. By assumption, so does V^c. It is apparent that each term on the right of Fig. 6.7 is completely connected, so that after integration over internal momenta it can have only a single overall δ-function. Thus, $\frac{\partial}{\partial t}U^c$ has only a single δ-function, which implies that U^c retains this property for all time. At time $t = t_0$ however, $U(t_0,t_0) = 1$, so initially the only non-vanishing connected piece of U is the one-particle matrix element $\langle k'|U^c|k\rangle = \delta^3(k'-k)$, which indeed has but a single δ-function. Apart from this single overall δ-function then, we have (by assumption) only smooth functions of momenta, which remain smooth when combined and integrated via the linking operator. This establishes the desired result, as stated in theorem 6.1.

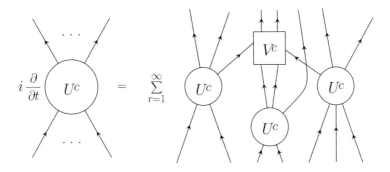

Fig. 6.7 Time development of connected U^c amplitude.

The importance of the above result is clear: we now have a precise criterion for choosing interaction Hamiltonians which lead to properly clustering S-matrices. In particular, the connected part of the matrix elements of V_{ip} should contain no dangerous δ-functions (in the terminology of (Weinberg, 1964b)). It turns out that the extraction of the connected part of the Hamiltonian by either the inductive scheme described above, or by functional methods, is rather inconvenient in general. Instead, a technical device known as "second quantization"[6] allows us to display the connected parts of H with great ease. This device, involving the introduction of the famous "creation" and "annihilation" operators, will also have an important added bonus: it will be trivial to ensure the proper symmetry (bosonic or fermionic) of the states. But the real advantage of the creation–annihilation technology is the ease with which clustering can be incorporated in the theory.

Much of the preceding discussion of clustering applies quite generally to scattering processes in non-relativistic quantum mechanics, as well as to scattering in relativistic quantum field theory. Certainly, we do not expect locality, in the sense of space-like commutativity of field operators, to play any role in the non-relativistic case. However, we shall soon see that this much more special property, sometimes termed *microcausality*, dovetails effortlessly with the clustering requirement once we add the condition of Lorentz-invariance of scattering amplitudes. The natural structure of local quantum field theory emerges inexorably once these basic concepts are fused.

6.3   Constructing clustering Hamiltonians: second quantization

We have arrived at a point where the mathematical preconditions for a clustering theory have been brought into a direct association with the structure of the interaction Hamiltonian of the theory: we must ensure that all *connected* matrix elements of this Hamiltonian are appropriately smooth, apart from the inevitable δ-function of spatial momentum conservation implied by the translation invariance of the theory. In order to do this, one more indispensable technical tool—second quantization—will be required. We shall introduce a class of operators, called "creation" and "annihilation" operators, which will vastly simplify the search for clustering theories. Using these operators will also allow us to incorporate trivially the correct (i.e., fermionic or bosonic) symmetry in multi-particle states. In the forthcoming discussion, the *upper* sign, e.g., as in $\pm$ or $\mp$, will always be taken to refer to bosons, the *lower* sign to fermions.

On a general multi-particle state $|k_1, k_2, , , k_N\rangle$,[7] define operators $a(k), a^\dagger(k)$ by the following action:

---

[6] The terminology "second quantization" arose in the late 1920s to describe the process, pioneered by Jordan and Dirac, of replacing the c-number wavefunctions for multi-particle systems by a single spacetime-dependent q-number operator field: it was soon realized that the creation and annihilation algebra provided an extremely convenient operator basis for constructing such fields. See Section 2.2.

[7] Again, in the interests of avoiding notational overload, the three-vector arrow indications on spatial momenta are omitted in what follows. Whether a momentum label refers to a spatial, or four momentum, should (we hope) be obvious from context.

$$a(k)|k_1, k_2, ..k_N\rangle \equiv \sum_{r=1}^{N} (\pm)^{r-1} \delta^3(k - k_r)|k_1, ..k_{r-1}, k_{r+1}, ..k_N\rangle \qquad (6.29)$$

$$a^\dagger(k)|k_1, k_2, ..k_N\rangle \equiv |k, k_1, k_2, ..k_N\rangle \qquad (6.30)$$

Intuitively, $a(k)$ attempts to remove a particle of momentum k (for simplicity of notation, three-vector symbols are omitted) while $a^\dagger(k)$ adds a particle of momentum k. We shall use non-covariant normalization throughout. It will also be assumed that all particles in the state are identical: non-identical particles simply act independently. However, for identical particles we have to be careful to build in the right Bose or Fermi symmetry. This appears in the normalization formula as follows:

$$\langle k'_1, ..k'_N|k_1, ..k_N\rangle = \sum_{P} (\pm)^P \delta^3(k_1 - k'_{P(1)})...\delta^3(k_N - k'_{P(N)}) \qquad (6.31)$$

Here $\sum_P$ denotes a sum over all permutations P of the integers $1, 2, ...N$, with $(\pm)^P$ inserting a minus sign for fermions if the permutation P is odd.

As suggested by the notation, $a^\dagger$ is really the hermitian conjugate of a, as the following computation shows:

$$\langle q_2, ..q_N|a(q_1)|k_1, k_2, ..k_N\rangle$$

$$= \sum_{r=1}^{N} (\pm)^{r-1} \delta^3(k_r - q_1) \sum_{P} (\pm)^P \delta^3(k_1 - q_{P(2)})..\delta^3(k_{r-1} - q_{P(r)}) \delta^3(k_{r+1} - q_{P(r+1)})..$$

$$= \sum_{P} (\pm)^P \delta^3(k_1 - q_{P(1)}) \delta^3(k_2 - q_{P(2)})..\delta^3(k_N - q_{P(N)})$$

$$= \langle q_1, q_2, ..q_N|k_1, k_2, ..k_N\rangle$$

$$= \langle k_1, k_2, ..k_N|q_1, q_2, ..q_N\rangle^*$$

$$= \langle k_1, k_2, ..k_N|a^\dagger(q_1)|q_2, ..q_N\rangle^* \qquad (6.32)$$

This implies that if $|0\rangle$ is the state with no particles (the "vacuum") and $\langle\psi|$ an arbitrary bra state

$$\langle\psi|a(k)|0\rangle = (a^\dagger(k)|\psi\rangle, |0\rangle) = \langle k, \psi|0\rangle = 0 \qquad (6.33)$$

as the bra state $\langle k, \psi|$ must contain at least one particle. Since $\langle\psi|$ is arbitrary, it follows that the annihilation operators $a(k)$ all annihilate the vacuum, $a(k)|0\rangle = 0$. (Note that we shall always normalize the vacuum to unity, $\langle 0|0\rangle = 1$.)

The commutator properties of the creation and annihilation operators will be crucial. Thus, observe

$$a^\dagger(k')a(k)|k_1..k_N\rangle$$

$$= a^\dagger(k')\sum_{r=1}^{N}(\pm)^{r-1}\delta^3(k-k_r)|k_1k_2..k_{r-1}k_{r+1}..k_N\rangle$$

$$= \sum_{r=1}^{N}\delta^3(k-k_r)|k_1..k_{r-1}k'k_{r+1}..k_N\rangle \tag{6.34}$$

whereas

$$a(k)a^\dagger(k')\ |k_1..k_N\rangle = a(k)|k'k_1..k_N\rangle$$

$$= \delta^3(k-k')|k_1..k_N\rangle + \sum_{r=1}^{N}(\pm)^r\delta^3(k-k_r)|k'k_1..k_{r-1}k_{r+1}...\rangle$$

$$= \delta^3(k-k')|k_1..k_N\rangle \pm \sum_{r=1}^{N}\delta^3(k-k_r)|k_1..k_{r-1}k'k_{r+1}..\rangle \tag{6.35}$$

Hence

$$(a(k)a^\dagger(k') \mp a^\dagger(k')a(k))|k_1..k_N\rangle = \delta^3(k-k')|k_1..k_N\rangle \tag{6.36}$$

More concisely, we have derived, for bosons (resp. fermions), the fundamental commutation (resp. anticommutation) relations

$$[a(k),a^\dagger(k')]_\mp = \delta^3(k-k') \tag{6.37}$$

By similar computations, it is trivial to show that creation and annihilation operators commute (resp. anticommute) among themselves:

$$[a(k),a(k')]_\mp = [a^\dagger(k),a^\dagger(k')]_\mp = 0 \tag{6.38}$$

For the case of fermions, this implies $a^\dagger(k)a^\dagger(k) + a^\dagger(k)a^\dagger(k) = 2a^\dagger(k)a^\dagger(k) = 0$, i.e., the Pauli exclusion principle forbidding the addition of two identical fermionic particles to any state.

Next we derive the behavior of the creation and annihilation operators under Lorentz transformations. Recall that with non-covariant normalization of the states,

$$U(\Lambda)|k,k_1,...\rangle = \sqrt{\frac{E(\Lambda k)}{E(k)}}\prod_i\sqrt{\frac{E(\Lambda k_i)}{E(k_i)}}|\Lambda k,\Lambda k_1,...\rangle \tag{6.39}$$

This may alternatively be written

$$U(\Lambda)a^\dagger(k)|k_1..\rangle = \sqrt{\frac{E(\Lambda k)}{E(k)}}a^\dagger(\Lambda k)\prod_i\sqrt{\frac{E(\Lambda k_i)}{E(k_i)}}|\Lambda k_1,..\rangle$$

$$= \sqrt{\frac{E(\Lambda k)}{E(k)}}a^\dagger(\Lambda k)U(\Lambda)|k_1,..\rangle \tag{6.40}$$

As the ket in (6.40) is arbitrary, we have the operator relation

$$U(\Lambda)a^\dagger(k) = \sqrt{\frac{E(\Lambda k)}{E(k)}}a^\dagger(\Lambda k)U(\Lambda) \tag{6.41}$$

Multiplying (6.41) on the right by $U^\dagger(\Lambda)$ and using unitarity of the $U(\Lambda)$

$$U(\Lambda)a^\dagger(k)U^\dagger(\Lambda) = \sqrt{\frac{E(\Lambda k)}{E(k)}}a^\dagger(\Lambda k) \tag{6.42}$$

with an exactly similar equation (by hermitian conjugation) for $a(k)$.

Remembering that the states $|k_1, ..\rangle$ are eigenstates of the free Hamiltonian $H_0 = P_0^{(0)}$ and the free momentum operator $P_i^{(0)}, i = 1, 2, 3$,

$$P_\mu^{(0)}|k, k_1, ..\rangle = (k_\mu + k_{1\mu} + ..)|k, k_1, ..\rangle \tag{6.43}$$

we have

$$P_\mu^{(0)}a^\dagger(k)|k_1, ..\rangle = k_\mu a^\dagger(k)|k_1, ..\rangle + a^\dagger(k)P_\mu^{(0)}|k_1, ..\rangle \tag{6.44}$$

Rearranging (6.44), and removing the arbitrary ket,

$$[P_\mu^{(0)}, a^\dagger(k)] = k_\mu a^\dagger(k) \tag{6.45}$$

By taking the hermitian conjugate of (6.45),

$$[P_\mu^{(0)}, a(k)] = -k_\mu a(k) \tag{6.46}$$

We emphasized above that the creation–annihilation formalism provides a convenient, as it were automated, mechanism for ensuring that the states in the theory are properly symmetrized (or, for fermions, antisymmetrized). As long as the fundamental commutation relations (6.37, 6.38) hold, the multi-particle states obtained by applying any number of creation operators to the vacuum will have the right symmetry under particle exchange. Even more important, however, is the fact that the creation–annihilation operator formalism affords a compact notation for writing multi-particle matrix elements in a way which exposes the connectedness structure with great clarity. To see this, first observe that an *arbitrary operator H* acting in the Fock space of multi-particle states may be expanded as a series in multi-nomials of $a, a^\dagger$:

$$H = \sum_{M,M'} \frac{1}{M!M'!} \int d^3 k_1' .. d^3 k_{M'}' d^3 k_1 .. d^3 k_M$$

$$h_{M'M}(k_1', .. k_{M'}', k_1, .. k_M) a^\dagger(k_1') .. a^\dagger(k_{M'}') a(k_1) .. a(k_M) \tag{6.47}$$

The coefficient functions $h_{M'M}$ encode the complete set of multi-particle matrix elements of the Fock space operator H. If $h_{M'M} = 0$ for all $M, M' \neq 0$, we call H a "c-number": in other words, it commutes with all operators in the theory. Otherwise, H is a "q-number". The entire operatorial structure of the theory can thereby be encoded by particle creation and annihilation operators.

The proof of the above assertion proceeds by induction. First, determine h_{11} by computing the matrix elements of H between two single-particle states:

$$\langle k'|H|k \rangle = \int d^3 k_1' d^3 k_1 h_{11}(k_1', k_1) \langle k'|a^\dagger(k_1') a(k_1)|k \rangle$$

$$= \int d^3 k_1' d^3 k_1 h_{11}(k_1', k_1) \delta^3(k_1' - k') \delta^3(k - k_1)$$

$$= h_{11}(k', k) \tag{6.48}$$

Similarly, $\langle k_1' k_2'|H|k_1 k_2 \rangle = h_{22}(k_1', k_2', k_1, k_2) +$ terms involving $h_{11} \delta^3(..)$. Having determined h_{11} in the preceding step, this fixes h_{22}, and so on. We may now state the critical result which validates the importance of an expansion in terms of creation and annihilation operators:

Theorem 6.2

$$\langle k_1', k_2', .. k_{N'}'|H^c|k_1, k_2, .. k_N \rangle = h_{N'N}(k_1', .. k_{N'}', k_1, .. k_N)$$

This remarkable theorem shows that the expansion (6.47) *directly* yields the connected matrix elements of the Hamiltonian in terms of the expansion functions $h_{N'N}$. The proof is simple. Consider a general (N, N') matrix element of H, expressed graphically as shown in Fig. 6.8. The first terms on the right-hand side indicate possible disconnected contributions: in particular, each term here must contain at least two δ-functions (of course, some of the terms, for example the first term on the right-hand side, only appear if $N = N'$). The last term on the right is the fully connected piece, with only a single overall δ-function of momentum conservation. The terms in the expansion $H = \sum_{MM'} \cdots$ with $M < N$ or $M' < N'$ do not contain enough creation or annihilation operators to affect all the particles in the final and initial states: thus such terms contribute only to the disconnected part. The terms with $M > N$ or $M' > N'$ give a vanishing contribution, by attempting to destroy more particles than are present in the initial state or create more than are present in the final state. So the only part of H to contribute to the fully connected matrix element is $h_{N'N}$! Q.E.D.

The theorem of the preceding section requiring H^c to contain at most a single delta-function may now be applied directly to the coefficient functions $h_{N'N}$ to obtain the following immediate corollary of theorem 6.2:

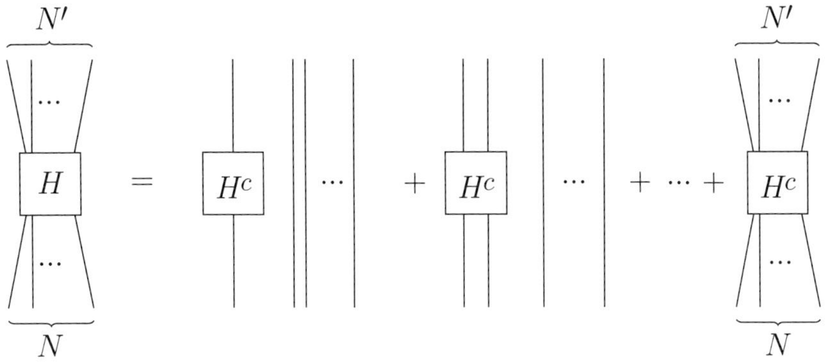

Fig. 6.8 Connectedness structure of Hamiltonian matrix elements.

Corollary 6.3 $h_{N'N} (k_1', k_2', .. \ k_1, k_2..) = \delta^3(k_1' + k_2' + .. - k_1 - k_2 - ..) \cdot f(k_1', k_2', ..)$
$\Longrightarrow$ *clustering property of the S-matrix, provided f is a smooth function of momenta.*

6.4   Constructing a relativistic, clustering theory

Although the general expansion (6.47) of a Hamiltonian operator in terms of creation and annihilation operators simplifies the task of enforcing proper clustering properties of our theory, the requirements of Lorentz-invariance of the resultant interactions are clearly far from obviously satisfied in such a framework: the ubiquitous spatial momentum integrals make the behavior of the theory under general Lorentz transformations quite obscure. On the other hand, in Section 5.5 we saw that relativistic invariance follows immediately if the interaction Hamiltonian is constructed as the spatial integral of an ultralocal scalar field. But do such theories satisfy the desired cluster decomposition properties? In this section we shall see that interaction Hamiltonians built as polynomials out of a basic set of *canonical* local fields linear in creation and annihilation operators do indeed have the right smoothness in momentum space to guarantee clustering of the resultant S-matrix. We shall assume throughout that we are dealing with spinless bosons, as the complications of spin are an unnecessary distraction from our immediate task, which is to understand the intimate way in which relativistic and locality considerations intertwine in the construction of quantum field theory.

Recall the lesson of Section 5.5: we wish to construct an interaction energy density $\mathcal{H}_{int}(x)$ which is a local[8] scalar field, i.e., with the properties

$$i[P_\mu^{(0)}, \mathcal{H}_{int}(x)] = \frac{\partial}{\partial x^\mu} \mathcal{H}_{int}(x) \tag{6.49}$$

---

[8] Strictly speaking, *ultralocal*, with no spacetime-derivatives appearing in the commutator contact terms (cf. the discussion in Section 5.5).

$$U(\Lambda)\mathcal{H}_{int}(x)U^\dagger(\Lambda) = \mathcal{H}_{int}(\Lambda x) \tag{6.50}$$

$$[\mathcal{H}_{int}(x_1), \mathcal{H}_{int}(x_2)] = 0, \quad (x_1 - x_2)^2 < 0 \tag{6.51}$$

We also know that $\mathcal{H}_{int}$ may be written as an expansion in destruction and creation operators (6.47), $a, a^\dagger$s. Let us therefore try to construct an object from a *single a* (or $a^\dagger$) which satisfies the above conditions. Then we can rely on the fact that local scalar fields form an algebraic "ring": the product of a set of local scalar fields is again a local scalar field. To prove this, suppose $A(x), B(x), C(x), \ldots$ comprise a set of local scalar fields. We want to show that $A(x)B(x)C(x)\ldots$ satisfies (6.49, 6.50, 6.51).

1. To check (6.49), observe that

$$i\left[P_\mu^{(0)}, A(x)B(x)C(x)..\right]$$

$$= i\left[P_\mu^{(0)}, A(x)\right]B(x)C(x).. + A(x)\left[iP_\mu^{(0)}, B(x)\right]C(x).. + ..$$

$$= \frac{\partial A(x)}{\partial x^\mu}B(x)C(x).. + A(x)\frac{\partial B}{\partial x^\mu}C(x).. + ..$$

$$= \frac{\partial}{\partial x^\mu}(A(x)B(x)C(x)..)$$

We remind the reader that all operators considered here are *in the interaction picture*: taking $\mu = 0$ we see that $P_0^{(0)} = H_0$, the free Hamiltonian, generates the time development for all fields in the theory.

2. From unitarity of $U(\Lambda)$, $U^\dagger(\Lambda)U(\Lambda) = 1$:

$$U(\Lambda)A(x)B(x)C(x)..U^\dagger(\Lambda) = U(\Lambda)A(x)U^\dagger(\Lambda)U(\Lambda)B(x)U^\dagger(\Lambda)...$$

$$= A(\Lambda x)B(\Lambda x)C(\Lambda x)...$$

3. If the points x_1, x_2 are space-like separated, then each of the operators $A(x_1)$, $B(x_1), C(x_1),\ldots$ commutes with each of $A(x_2), B(x_2), C(x_2), \ldots$, by (6.51), whence

$$[A(x_1)B(x_1)C(x_1).., A(x_2)B(x_2)C(x_2)..] = 0$$

For the time being imagine that we are dealing with a single spinless boson, with $a(\vec{k})$ the destruction operator for a particle of spatial momentum $\vec{k}$.[9] The most general operator linear in the destruction operator must take the form:[10]

$$\phi^{(+)}(x) = \int d^3k\, f(x; \vec{k})a(\vec{k}) \tag{6.52}$$

---

[9] In this section we restore the arrows to distinguish spatial from four-momenta.

[10] The "plus" superscript anticipates the fact, shortly to emerge, that the time-dependence of this operator involves positive frequencies, which by convention means that it is a linear combination of plane waves of the form $e^{+i(\vec{k}\cdot\vec{r}-\omega t)}$, $\omega > 0$.

First, impose the translation condition (6.49):

$$i\left[P_\mu^{(0)}, \phi^{(+)}(x)\right] = i\int d^3k f(x; \vec{k})\left[P_\mu^{(0)}, a(\vec{k})\right]$$

$$= -i\int d^3k\, k_\mu f(x; \vec{k})a(\vec{k})$$

$$= \int d^3k\, \partial_\mu f(x; \vec{k})a(\vec{k}) \tag{6.53}$$

which implies

$$\partial_\mu f(x; \vec{k}) = -ik_\mu f(x; \vec{k}) \tag{6.54}$$

Solving this differential equation, we find

$$f(x; \vec{k}) = f(\vec{k})e^{-ik\cdot x} \tag{6.55}$$

Note that the k in the exponential factor in (6.55) is the *four-vector* momentum, with $k_0 \equiv E(k) \equiv \sqrt{\vec{k}^2 + m^2}$: such a four-vector momentum is said to be "on-mass-shell". The result is that $f(x; k)$, and therefore $\phi^{(+)}(x)$, necessarily satisfy the *Klein–Gordon equation* (Klein, 1926; Gordon, 1926)

$$(\Box + m^2)\phi^{(+)}(x) = 0 \tag{6.56}$$

It is apparent from the preceding argument that the appearance of a relativistically invariant wave equation for our field is a direct consequence of the imposition of the spacetime translation property.

Next we must impose the Lorentz transformation property (6.50) which defines a scalar field:

$$U(\Lambda)\phi^{(+)}(x)U^\dagger(\Lambda) = \int d^3k f(\vec{k})e^{-ik\cdot x}\sqrt{\frac{E(\Lambda k)}{E(k)}}a(\Lambda\vec{k})$$

$$= \int \frac{d^3k}{E(k)} f(\vec{k})e^{-ik\cdot x}\sqrt{E(\Lambda k)E(k)}a(\Lambda\vec{k})$$

$$= \int \frac{d^3k}{E(k)} f(\vec{k})e^{-i(\Lambda k\cdot\Lambda x)}\sqrt{E(\Lambda k)E(k)}a(\Lambda\vec{k}) \tag{6.57}$$

On the other hand,

$$U(\Lambda)\phi^{(+)}(x)U^\dagger(\Lambda) = \phi^{(+)}(\Lambda x)$$

$$= \int \frac{d^3k}{E(k)} E(k)f(\vec{k})e^{-ik\cdot\Lambda x}a(\vec{k})$$

$$= \int \frac{d^3k}{E(k)} E(\Lambda k)f(\Lambda\vec{k})e^{-i(\Lambda k\cdot\Lambda x)}a(\Lambda\vec{k}) \tag{6.58}$$

Comparing the right-hand sides of (6.57) and (6.58), we conclude that

$$f(\Lambda \vec{k})E(\Lambda k) = f(\vec{k})\sqrt{E(\Lambda k)E(k)} \tag{6.59}$$

which implies that $f(\vec{k}) \propto \frac{1}{\sqrt{E(k)}}$. The conventional normalization is to take

$$f(\vec{k}) = \frac{1}{(2\pi)^{3/2}} \frac{1}{\sqrt{2E(k)}} \tag{6.60}$$

so that we obtain finally the desired result (unique, up to an overall constant!):

$$\phi^{(+)}(x) = \frac{1}{(2\pi)^{3/2}} \int \frac{d^3k}{\sqrt{2E(k)}} e^{-ik\cdot x} a(\vec{k}) \tag{6.61}$$

$\phi^{(+)}(x)$ and its hermitian conjugate $\phi^{(+)\dagger}$ (given by an analogous formula involving the creation operator $a^\dagger$) will be the basic ingredients out of which we shall construct our (necessarily hermitian) interaction Hamiltonians.

We must still address the question of locality, (6.51). For bosons we have the commutation relations $[a(\vec{k}), a(\vec{k}')] = [a^\dagger(\vec{k}), a^\dagger(\vec{k}')] = 0$ among creation and destruction operators separately, so automatically

$$\left[\phi^{(+)}(x), \phi^{(+)}(y)\right] = 0 = \left[\phi^{(+)\dagger}(x), \phi^{(+)\dagger}(y)\right] \tag{6.62}$$

On the other hand, the interaction Hamiltonian must be hermitian, and must therefore be built out of *both* $\phi^{(+)}$ and $\phi^{(+)\dagger}$. Thus, we must also check the commutation relation

$$\left[\phi^{(+)}(x), \phi^{(+)\dagger}(y)\right] = \frac{1}{(2\pi)^3} \int \frac{d^3k}{\sqrt{2E(k)}} \frac{d^3k'}{\sqrt{2E(k')}} e^{-i(k\cdot x - k'\cdot y)} \delta^3(k-k')$$

$$= \frac{1}{(2\pi)^3} \int \frac{d^3k}{2E(k)} e^{-ik\cdot(x-y)}$$

$$\equiv \Delta_+(x-y;m) \tag{6.63}$$

Here Δ_+ is a c-number *Lorentz-invariant* function of $x-y$ (and the mass m, through $E(k) = \sqrt{k^2 + m^2}$) as

$$\Delta_+(\Lambda x - \Lambda y; m) = \left[\phi^{(+)}(\Lambda x), \phi^{(+)\dagger}(\Lambda y)\right]$$

$$= U(\Lambda)\left[\phi^{(+)}(x), \phi^{(+)\dagger}(y)\right]U^\dagger(\Lambda)$$

$$= U(\Lambda)\Delta_+(x-y;m)U^\dagger(\Lambda)$$

$$= \Delta_+(x-y;m)$$

where the last line follows because Δ_+ is a c-number (no creation or annihilation operators), and therefore commutes with all operators in the theory, in particular with the $U(\Lambda)$. The frame-independence of this invariant function means that for $z \equiv x - y$

space-like we can evaluate it by choosing a frame in which the time component of z vanishes, with $|\vec{z}| = \sqrt{-z_\mu z^\mu}$. Then

$$\Delta_+(z;m) = \frac{1}{(2\pi)^3} \int \frac{4\pi k^2 dk}{2E(k)} \frac{\sin(k|\vec{z}|)}{k|\vec{z}|} \tag{6.64}$$

where we have used the result that the average over directions of $\vec{k}$ of $e^{i\vec{k}\cdot\vec{z}}$ is $\frac{\sin(k|\vec{z}|)}{k|\vec{z}|}$. Changing variables in the radial momentum integral to $\beta \equiv k/m$,

$$\Delta_+(z;m) = \frac{m^2}{4\pi^2} \int_0^\infty \frac{\beta d\beta}{\sqrt{\beta^2+1}} \frac{\sin(\beta m|\vec{z}|)}{m|\vec{z}|}$$

$$= \frac{m^2}{4\pi^2} \frac{1}{m\sqrt{-z^2}} K_1(m\sqrt{-z^2}) \tag{6.65}$$

where K_1 is a modified Bessel function. *In particular, the commutator (6.63) does not vanish when the separation $z = x - y$ is space-like.* Instead, the commutator falls off exponentially with a scale determined by the Compton wavelength m^{-1} of the associated particle (see Fig. 6.9):[11]

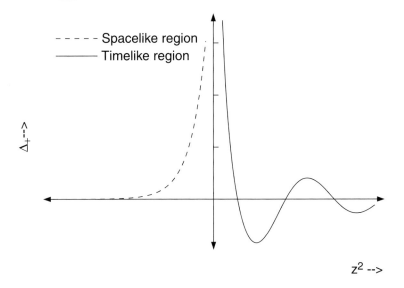

Fig. 6.9 Behavior of the invariant function Δ_+ (real part shown for $z^2 > 0$).

[11] In the time-like region with $z^2 > 0$, Δ_+ can be shown to take the form

$$\Delta_+(z;m) = \frac{m}{8\pi\sqrt{z^2}} (N_1(m\sqrt{z^2}) \pm iJ_1(m\sqrt{z^2})), \quad z_0 = \pm$$

which displays oscillatory behavior, in contrast to the exponential decrease of K_1 in the space-like region (Fig. 6.9).

$$\Delta_+(z; m) \sim \frac{m^2}{4\sqrt{2}} (\pi m \sqrt{-z^2})^{-3/2} e^{-m\sqrt{-z^2}}, \quad \sqrt{-z^2} \to \infty \tag{6.66}$$

Unfortunately for the Lorentz-invariance of the theory, exponential falloff is not good enough here: a non-zero commutator will result in non-Lorentz-invariant contributions to the S-matrix, as the argument following (5.75) in Section 5.5 makes clear. An important example of such a failure is provided by scalar *number-conserving* interaction Hamiltonians. Consider a theory in which the interaction Hamiltonian has an expansion of the form (6.47) with the only non-zero term having $M = M' = 2$. The perturbative expansion of the S-matrix then yields amplitudes for processes which can arise from a succession of 2-2 particle scatterings. Writing the interaction Hamiltonian as a spatial integral of an hermitian scalar density, the interaction energy density in the simplest such theory must then take the form

$$\mathcal{H}_{int}(x) = \lambda(\phi^{(+)\dagger}(x))^2(\phi^{(+)}(x))^2 \tag{6.67}$$

where λ is a real "coupling" constant parameterizing the strength of the interaction. In this theory the number of particles is strictly conserved, as the interaction operator (6.67) exactly commutes with the number operator

$$\mathcal{N} \equiv \int a^\dagger(\vec{k}) a(\vec{k}) d^3 k \tag{6.68}$$

Unfortunately, a number conserving interaction of this type (although itself a Lorentz scalar field) violates locality, and hence the Lorentz-invariance of the S-matrix. A straightforward calculation of the commutator of $\mathcal{H}_{int}$ at two space-like separated points gives

$$[\mathcal{H}_{int}(x), \mathcal{H}_{int}(y)] = 2\lambda^2 \Delta_+(x - y; m)$$
$$\times \{(\phi^{(+)\dagger}(x))^2(\phi^{(+)}(x)\phi^{(+)\dagger}(y) + \phi^{(+)\dagger}(y)\phi^{(+)}(x))(\phi^{(+)}(y))^2\} - (x \leftrightarrow y) \tag{6.69}$$

Neither the c-number function $\Delta_+(x - y; m)$ nor the associated operator (cubic in a's and $a^\dagger$'s) vanish for $x - y$ space-like, so the terms of second and higher order in the S-matrix expansion (5.73), which contain time-ordering instructions sensitive to frame for space-like separated points, will necessarily contain Lorentz-invariance violating terms.

The above example shows that we are likely to encounter violations of Lorentz-invariance if we construct the interaction Hamiltonian carelessly out of arbitrary combinations of $\phi^{(+)}$ and $\phi^{(+)\dagger}$. Physically, Lorentz-invariance really requires the inclusion of *both destruction and creation* parts in the *same* local field, as we saw in the description of pion exchange in Chapter 3 (recall the discussion of the process pictured in Fig. 3.1). There the existence of a process of π^+ exchange between nucleons initiated and terminated by local interaction events forced us to include a particle of opposite quantum numbers and equal mass: the π^-, in the example shown. In the general situation the particle and antiparticle are *distinct* (e.g., have opposite values for any conserved additive quantum number, such as electric charge), so we really have separate and independent destruction $a^c(\vec{k})$ and creation $a^{c\dagger}(\vec{k})$ operators for the antiparticle, satisfying (for bosons)

$$\left[a^c(\vec{k}), a^{c\dagger}(\vec{k}')\right] = \delta^3(\vec{k} - \vec{k}') \tag{6.70}$$

$$\left[a^c(\vec{k}), a(\vec{k}')\right] = \left[a^c(\vec{k}), a^{\dagger}(\vec{k}')\right] = 0 \tag{6.71}$$

$$\left[a^c(\vec{k}), a^c(\vec{k}')\right] = 0 \tag{6.72}$$

etc. Although the heuristic argument given in Chapter 3 implies equality of particle and antiparticle masses, we temporarily allow the antiparticle to have an independent mass m^c. Now destruction of a particle and creation of the antiparticle at a spacetime point x must be treated on the same footing (as they both describe the same physical event in different frames), which suggests that we write a single field (which we shall call a *canonical scalar field*) containing both terms:

$$\phi(x) = \frac{1}{(2\pi)^{3/2}} \int \frac{d^3k}{\sqrt{2E(k)}} (a(\vec{k})e^{-ik\cdot x} + a^{c\dagger}(\vec{k})e^{ik\cdot x}) \tag{6.73}$$

$$\equiv \phi^{(+)}(x) + \phi^{(-)}(x) \tag{6.74}$$

We note here in passing that $\phi(x)$, like its positive frequency part $\phi^{(+)}(x)$, automatically satisfies the Klein–Gordon equation (6.56): $(\Box + m^2)\phi(x) = 0$, as only on-mass-shell four-momenta occur in the plane-wave exponentials $e^{\pm ik\cdot x}$.

The basic commutation relations (6.70–6.72) immediately imply

$$[\phi(x), \phi(y)] = [\phi(x)^{\dagger}, \phi(y)^{\dagger}] = 0 \tag{6.75}$$

for *all* x, y (i.e., not necessarily space-like separated). The necessity to include both ϕ and $\phi^{\dagger}$ in an hermitian Hamiltonian requires that we also check for locality in the commutator

$$[\phi(x), \phi^{\dagger}(y)] = \frac{1}{(2\pi)^3} \int \frac{d^3k}{2E(k)} (e^{-ik\cdot(x-y)} - e^{ik\cdot(x-y)})$$

$$= \Delta_+(x - y; m) - \Delta_+(y - x; m^c) \tag{6.76}$$

If the particle and antiparticle masses m, m^c differ, this commutator will still fail to vanish even for $x - y$ space-like. On the other hand, *exact equality of particle and antiparticle masses* $m = m^c$ implies (since $\Delta_+(z; m)$ in (6.65) is *even* in z for z space-like) that the full invariant function

$$\Delta(z; m) \equiv \Delta_+(z; m) - \Delta_+(-z; m) \tag{6.77}$$

vanishes for space-like separation z ($\Delta_+(z; m)$ is even in z for z space-like), and we have the desired locality

$$[\phi(x), \phi^\dagger(y)] = 0 \quad \text{for} \quad (x - y)^2 < 0 \tag{6.78}$$

Of course, as a consequence of (6.62), we already have $[\phi(x), \phi(y)] = [\phi^\dagger(x), \phi^\dagger(y)] = 0$ for *all* x, y. The fact that strict locality of our fields implies an exact equality of particle and antiparticle masses is a consequence of the famous "TCP theorem" of local quantum field theory, which we shall discuss further in Section 13.4. For the proton and antiproton, the equality has been established experimentally to nine significant figures.

With the local fields $\phi(x)$ and $\phi^\dagger(x)$ in hand we may multiply them freely to obtain an hermitian scalar interaction Hamiltonian density guaranteed to have the right locality and Lorentz transformation properties, thereby ensuring Lorentz-invariance of the scattering amplitudes of the theory.[12] A famous example is "phi-4" theory, with interaction

$$\mathcal{H}_{\text{int}}(x) = \lambda \phi^\dagger(x)^2 \phi(x)^2 \tag{6.79}$$

Note the difference with our previous number-conserving interaction Hamiltonian (6.67): if we break up each of the fields in (6.79) into creation and annihilation parts, we find terms with unequal numbers of creation and annihilation operators, so the interaction induces processes in which the total number of particles in the state changes, which is surely one of the most characteristic features of relativistic quantum field theories. However, one easily sees that any additive quantum number corresponding to an operator of the following form, counting the *difference* between the number of particles and antiparticles,

$$Q = e \int (a^\dagger(\vec{k})a(\vec{k}) - a^{c\dagger}(\vec{k})a^c(\vec{k}))d^3k \tag{6.80}$$

commutes with $\mathcal{H}_{\text{int}}(x)$, and is thus exactly conserved by the dynamics of this theory (see also Problem 3).

In certain cases a particle may have identical quantum numbers to the antiparticle (e.g., photons, π^0, ρ^0,..): in other words, $a = a^c$ and the canonical local field ϕ representing this particle is hermitian, $\phi(x) = \phi^\dagger(x)$. Such particles (and their associated canonical field) are called "self-conjugate". A "real" version of the phi-4 theory (6.79) describing the self-interactions of a self-conjugate neutral boson would therefore take the form

$$\mathcal{H}_{\text{int}}(x) = \lambda \phi(x)^4 \tag{6.81}$$

With this lengthy digression into the requirements of locality (as a way to ensure Lorentz-invariance of the S-matrix) concluded, we may finally return to examine the constraints due to cluster decomposition. Recall that clustering required the interaction part of the Hamiltonian to take the form

---

[12] Of course, we must still ensure *ultralocality* of the equal-time commutators, as discussed in Section 5.5. In particular, derivatively-coupled theories will in general produce Schwinger terms in the commutator, requiring additional non-covariant "seagull" terms in the interaction to correct the resultant defects in Lorentz-invariance. We shall see in Chapter 12 how the Lagrangian formalism automatically solves the problem of generating the appropriate seagull terms.

$$V = \sum_{M,M'} \frac{1}{M!M'!} \int h_{M,M'}(k_1', ..k_{M'}', k_1, ..k_M)$$

$$\cdot a^\dagger(k_1')..a^\dagger(k_{M'}')a(k_1)..a(k_M) \frac{d^3 k_i'}{\sqrt{2E(k_i')}} \frac{d^3 k_i}{\sqrt{2E(k_i)}}$$

$$= \sum \frac{1}{M!M'!} \int \delta^3(k_1' + k_2' + ..k_{M'}' - k_1 - ..k_M)$$

$$\cdot f_{MM'}(k_1'..k_M) a^\dagger(k_1')....a(k_M) \frac{d^3 k_i'}{\sqrt{2E(k_i')}} \frac{d^3 k_i}{\sqrt{2E(k_i)}} \qquad (6.82)$$

with $f_{MM'}$ a smooth function of momenta. From (6.82) it is apparent that our interaction Hamiltonian is automatically the spatial integral of an energy density (introduced as a pure hypothesis in Section 5.5): $V = \int d^3 x \mathcal{H}_{int}(\vec{x})$, where

$$\mathcal{H}_{int}(\vec{x}) = \sum_{MM'} \frac{1}{M!M'!} \int \frac{d^3 k_1'}{\sqrt{2E(k_1')}} .. \frac{d^3 k_M}{\sqrt{2E(k_M)}}$$

$$\cdot f_{MM'}(k_1', .., k_M) e^{-i\vec{k}_1' \cdot \vec{x}} a^\dagger(k_1')..e^{i\vec{k}_M \cdot \vec{x}} a(k_M) \qquad (6.83)$$

From the fact that $a(k)$ $(a^\dagger(k))$ removes (resp. adds) a particle of energy $E(k)$ from any state on which it acts, it follows that

$$e^{iH_0 t} a^\dagger(k) e^{-iH_0 t} = e^{iE(k)t} a^\dagger(k)$$

$$e^{iH_0 t} a(k) e^{-iH_0 t} = e^{-iE(k)t} a(k) \qquad (6.84)$$

Thus, transforming (6.83) to the interaction picture, one obtains

$$\mathcal{H}_{int}(\vec{x}, t) = \sum_{MM'} \frac{1}{M!M'!} \int \frac{d^3 k_1'}{\sqrt{2E(k_1')}} .. \frac{d^3 k_M}{\sqrt{2E(k_M)}}$$

$$f_{MM'}(k_1', .., k_M) e^{ik_1' \cdot x} a^\dagger(k_1')..e^{-ik_M \cdot x} a(k_M) \qquad (6.85)$$

where now the exponential factors contain four-vector dot-products. It is now straightforward to show that the scalar-field transformation property $U(\Lambda) \mathcal{H}_{int}(x) U^\dagger(\Lambda) = \mathcal{H}_{int}(\Lambda x)$ requires that

$$f_{MM'}(\Lambda k_1', ..\Lambda k_M) = f_{MM'}(k_1', ..k_M) \qquad (6.86)$$

Thus the function $f_{MM'}$ should be cooked up from invariant dot-products of the various four-vector momenta $k_1, ..k_{M'}'$ available.

Clustering requires that the functions $f_{MM'}$ should be smooth functions of momenta. If we make the very strong assumption that $f_{MM'}$ are *analytic*, and therefore expandable in a series around zero momentum, then $\mathcal{H}_{int}$ must be a series in the scalar fields $\phi^{(+)}, \phi^{(-)}, \phi^{(+)\dagger}, \phi^{(-)\dagger}$ with four-derivatives, if any, coupled to an overall scalar (all four-vector indices contracted). *Building $\mathcal{H}_{int}$ out of the combinations $\phi = \phi^{(+)} + \phi^{(-)}$ and $\phi^\dagger = \phi^{(+)\dagger} + \phi^{(-)\dagger}$ ensures that the locality prop-*

erty $[\mathcal{H}_{int}(x), \mathcal{H}_{int}(y)] = 0$, $(x - y)$ space-like, will hold, giving us a Lorentz-invariant theory.

There is one tricky point here which has perhaps already occurred to the reader: namely, the form (6.83) in which we wrote the interaction Hamiltonian. In this expression, all creation operators appear to the left of all destruction operators. This is called the *normal-ordered* product. Given any product of fields, $ABC..$, in which each field $A, B, C..$ is a linear combination of creation and annihilation parts, we form the normal-ordered product $: ABC.. :$ by multiplying out the field product in the normal way and then moving all $a^\dagger$s to the left and all as to the right in each of the resultant terms, *ignoring any resultant commutator terms!* For example (for a self-conjugate field):

$$: \phi(x)\phi(y) : = : (\phi^{(+)}(x) + \phi^{(-)}(x))(\phi^{(+)}(y) + \phi^{(-)}(y)) :$$
$$\equiv \phi^{(+)}(x)\phi^{(+)}(y) + \phi^{(-)}(y)\phi^{(+)}(x) + \phi^{(-)}(x)\phi^{(+)}(y) + \phi^{(-)}(x)\phi^{(-)}(y)$$

This differs from the ordinary product $\phi(x)\phi(y)$ by a commutator

$$[\phi^{(+)}(x), \phi^{(-)}(y)] = \Delta_+(x - y; m) \tag{6.87}$$

which is, of course, a c-number. For field reorderings at the same spacetime point we encounter a *divergent* c-number $\Delta_+(0; m)$ (the momentum integral in (6.64) is quadratically divergent for $z = 0$)—our first encounter with the ubiquitous ultraviolet delicacies of local field theory. We shall see how to deal with the sensitivity of the theory to high-momentum contributions in the fourth section of this book, but for the time being we may simply imagine inserting a cutoff on divergent momentum integrals at some very high value, reflecting our inevitable ignorance of very-short-distance physics. Thus, the rearrangement implied by normal ordering within a single interaction term does not affect the locality properties of the interaction Hamiltonian density, as a normal-ordered product can be rearranged into a linear combination of ordinary powers of the local field. For example (see Problem 4):

$$: \phi(x)^4 := \phi(x)^4 + A\phi(x)^2 + B \tag{6.88}$$

with A, B c-number constants related to $\Delta_+(0; m)$. Accordingly, we are free to take

$$\mathcal{H}_{int}(x) = \frac{\lambda}{4!} : \phi(x)^4 :$$
$$\equiv \frac{\lambda}{4!}(\phi^{(+)4} + 4\phi^{(-)}\phi^{(+)3} + 6\phi^{(-)2}\phi^{(+)2} + 4\phi^{(-)3}\phi^{(+)} + \phi^{(-)4})$$

which is, despite the reorderings, still a local field in virtue of (6.88). The free Hamiltonian, $H_0 = \int d^3k E(k)a^\dagger(k)a(k)$, is also given (see Problem 5) by the integral of a spatial energy density, itself a sum of normal-ordered products

$$H_0 = \int d^3x \mathcal{H}_0(\vec{x}, t) = \int d^3x : \frac{\dot{\phi}^2}{2} + \frac{1}{2}|\vec{\nabla}\phi|^2 + \frac{1}{2}m^2\phi^2 : \tag{6.89}$$

where the divergent c-number commutator term by which the normal-ordered expression differs from the corresponding expression without normal ordering is just the infamous "zero-point" energy of a free scalar field theory, which we have already encountered in Chapter 1 in Jordan's seminal calculation of field energy fluctuations. The dependence on the time t at which the field operators in (6.89) are taken is spurious—not surprisingly, as H_0 is clearly time-independent in the interaction picture.

In summary, we have a theory with complete Hamiltonian

$$ H = \int d^3x : \frac{1}{2}\dot{\phi}^2 + \frac{1}{2}|\vec{\nabla}\phi|^2 + \frac{1}{2}m^2\phi^2 + \frac{\lambda}{4!}\phi^4 : \tag{6.90} $$

where all the field operators are taken at $t = 0$, say.

Let us return briefly to the very strong analyticity constraint assumed above for $f_{MM'}$. The Taylor series for $f_{MM'}$ in powers of momenta translates to higher derivatives of the fields in coordinate space. The immediate problem induced by such derivatives—failure of ultralocality—is not, in fact, fatal: as we shall see in Chapter 12, the Lagrangian formalism enables the construction of Lorentz invariant theories even for interaction Hamiltonians with arbitrarily many spacetime-derivatives. The justification for this assumption really lies in the basic feature needed to insulate us from our ignorance of physics at very short distance scales—the scale separation property—discussed in Chapter 3 (and in much greater detail in Chapter 16), and in the further property of *renormalizability* possessed by a small subclass of local quantum field theories, in which the low-momentum physics can, at least perturbatively (and in a few special cases exactly, i.e., even non-perturbatively), be completely isolated from the behavior of the theory at arbitrarily high momentum. In fact, the requirement of renormalizability will constrain the suitable range of theories in any more than one spacetime dimension to *polynomial* interactions in the fields: the sum over M, M' must terminate! In addition, for theories of spinless particles in four spacetime dimensions, the coefficient function $f_{MM'}$ must actually be momentum-independent. The necessity for this will be examined in great detail in the fourth section of this book when we consider the physical origin and role of renormalizability. We shall also see then that more general *effective field theories*, with arbitrarily high derivatives, and powers of the field, which are the natural end results of the imposition only of Lorentz-invariance and clustering requirements, lie inevitably at the microscopic "core" of any local quantum field theory—even the perturbatively renormalizable ones.

6.5 Local fields, non-localizable particles!

The commutativity of local fields at space-like separated points, commonly referred to as the property of *microcausality*, is a natural implementation at the quantum level of our macroscopic intuition that propagation of physical effects at superluminal speeds should be prohibited (which we may term *macrocausality*). We have seen that it is quite straightforward to construct fields which exactly satisfy microcausality, and commute even at arbitrarily small space-like separations. All that is required is that the positive frequency (or destruction) part of the field be balanced symmetrically by a negative frequency (creation) part involving a particle of equal mass and opposite additive quantum numbers. More generally, if v_1, v_2 are two compact regions of

spacetime such that for all $x \in v_1, y \in v_2$ the separation $x - y$ is space-like, and $f_1(x), (f_2(x))$ are c-number functions with support in v_1, (resp v_2), then the smeared fields $\phi_{f_i} \equiv \int d^4x f_i(x)\phi(x)$, $i = 1, 2$ satisfy ($\phi(x)$ a self-conjugate, hence hermitian, scalar field)

$$[\phi_{f_1}, \phi_{f_2}] = 0 \qquad (6.91)$$

and are thus, in the general sense of quantum measurement theory, mutually compatible hermitian (hence, measurable) observables of the theory: states can be constructed in which ϕ_{f_1} and ϕ_{f_2} take simultaneously sharp values.[13]

The exact localizability of relativistic quantum fields does not, however, extend to the quantal manifestations of these fields: the particles whose interactions have motivated the introduction of the field concept in the first place! Recall the situation in non-relativistic quantum theory, where a massive particle like an electron can be assigned a wavefunction $\psi(\vec{x}, t)$ which at some time t_0 exactly vanishes outside an arbitrarily small bounded spatial region v_1, thereby localizing the particle exactly inside the given region, at least at some instant of time.[14] More generally, for a non-relativistic many particle quantum system, a number density operator $N(\vec{x}, t)$ can be defined such that the operators $N_v \equiv \int_v d^3x N(\vec{x}, t_0)$ defined for arbitrary spatial regions at some instant t_0 exactly commute with each other for non-overlapping spatial regions. Thus, it makes perfect sense in such a theory to speak of a definite number of particles in a precisely well defined spatial volume at a given time.

All of this falls apart in relativistic field theory. Let us illustrate the basic issues in the simplest case, that of a massive spinless boson described by a self-conjugate scalar field $\phi(x)$ (as given by (6.73) with $a^c = a$). We first note that the failure of strict localizability for relativistic particles is a completely kinematic issue: interactions of the field are not relevant here. The number operator counting the total number of particles in a state can be written as a spatial integral of a number density $N(\vec{x}, t)$ (see Problem 6)

$$N \equiv \int d^3k\, a^\dagger(k)a(k) \qquad (6.92)$$

$$= \int d^3x\, N(\vec{x}, t), \qquad (6.93)$$

$$N(\vec{x}, t) = i\phi^{(-)}(\vec{x}, t) \overset{\leftrightarrow}{\frac{\partial}{\partial t}} \phi^{(+)}(\vec{x}, t) \qquad (6.94)$$

where the antisymmetric time-derivative symbol in (6.94) is defined as

$$A(t) \overset{\leftrightarrow}{\frac{\partial}{\partial t}} B(t) \equiv A(t)\frac{\partial B(t)}{\partial t} - \frac{\partial A(t)}{\partial t}B(t) = A(t)\dot{B}(t) - \dot{A}(t)B(t) \qquad (6.95)$$

---

[13] We shall see how to do this explicitly in Chapter 8 when we discuss coherent states of a quantum field.

[14] Of course, the instantaneous spreading of the wave-packet allowed in non-relativistic theory will produce a non-vanishing wavefunction outside of v_1 for $t > t_0$.

In this relativistic theory, the equal-time commutator of the number density operator at spatially distinct points $\vec{x} \neq \vec{y}$ does not vanish. We shall need the basic commutators (Problem 7)

$$[\phi^{(+)}(\vec{x}, t), \phi^{(-)}(\vec{y}, t] = \Delta_+(\vec{x} - \vec{y}, 0; m) \tag{6.96}$$

$$[\dot{\phi}^{(+)}(\vec{x}, t), \dot{\phi}^{(-)}(\vec{y}, t] \equiv \tilde{\Delta}_+(\vec{x} - \vec{y}, 0; m) \tag{6.97}$$

$$[\phi^{(+)}(\vec{x}, t), \dot{\phi}^{(-)}(\vec{y}, t] = \frac{i}{2}\delta^3(\vec{x} - \vec{y}) = 0 \quad (\vec{x} \neq \vec{y}) \tag{6.98}$$

$$[\phi^{(-)}(\vec{x}, t), \dot{\phi}^{(+)}(\vec{y}, t] = \frac{i}{2}\delta^3(\vec{x} - \vec{y}) = 0 \quad (\vec{x} \neq \vec{y}) \tag{6.99}$$

where, reintroducing the speed of light c in the energy-momentum dispersion formula $E(k) = \sqrt{m^2c^4 + k^2c^2}$ to facilitate a non-relativistic limit,

$$\Delta_+(\vec{x}, 0; m) = \int \frac{d^3k}{(2\pi)^3} \frac{1}{2\sqrt{m^2c^4 + k^2c^2}} e^{i\vec{k}\cdot\vec{x}} \tag{6.100}$$

and

$$\tilde{\Delta}_+(\vec{x}, 0; m) = \int \frac{d^3k}{(2\pi)^3} \frac{1}{2}\sqrt{m^2c^4 + k^2c^2}\, e^{i\vec{k}\cdot\vec{x}} \tag{6.101}$$

$$= (m^2c^4 - c^2\vec{\nabla}^2)\Delta_+ \tag{6.102}$$

Note that although the relativistic function Δ_+ is not zero, but rather falls exponentially (recall (6.66)), implying the same behavior for the function $\tilde{\Delta}_+$, by (6.102), in the formal non-relativistic limit $c \to \infty$, we may expand

$$\Delta_+(\vec{x}, 0; m) \sim \frac{1}{mc^2}(\delta^3(\vec{x}) + \frac{1}{2m^2c^2}\vec{\nabla}^2\delta^3(\vec{x}) + ..) \tag{6.103}$$

thereby recovering local behavior for all relevant commutators. However, in the relativistic case, both $\Delta_+(\vec{x}, 0; m)$ and $\tilde{\Delta}_+(\vec{x}, 0; m)$ have a dominant asymptotic exponential falloff $\sim e^{-m|\vec{x}|}$.

With these ingredients, a short calculation yields

$$[\mathcal{N}(\vec{x}, t), \mathcal{N}(\vec{y}, t)] = \Delta_+(\vec{x} - \vec{y}, 0; m)(\dot{\phi}^{(-)}(\vec{x}, t)\dot{\phi}^{(+)}(\vec{y}, t) - \dot{\phi}^{(-)}(\vec{y}, t)\dot{\phi}^{(+)}(\vec{x}, t))$$
$$+ \tilde{\Delta}_+(\vec{x} - \vec{y}, 0; m)(\phi^{(-)}(\vec{x}, t)\phi^{(+)}(\vec{y}, t) - \phi^{(-)}(\vec{y}, t)\phi^{(+)}(\vec{x}, t)) \tag{6.104}$$

Accordingly, measurements of the number of particles $N_{v_1} \equiv \int_{v_1} d^3x \mathcal{N}(\vec{x}, t)$, $N_{v_2} \equiv \int_{v_2} d^3x \mathcal{N}(\vec{x}, t)$ in two spatially non-overlapping volumes v_1, v_2 will mutually interfere, with the level of interference falling exponentially as the separation (smallest distance between points in v_1 and v_2) is increased, on the scale of the Compton wavelength m^{-1} of the particle. In the non-relativistic limit, after appropriately absorbing the leading $\frac{1}{mc^2}$ factor in (6.103) into the normalization of the operators, the commutator (6.104) vanishes as expected for all $\vec{x} \in v_1 \neq \vec{y} \in v_2$, ensuring $[N_{v_1}, N_{v_2}] = 0$ for non-overlapping volumes v_1, v_2.

The difficulty in defining localized states for relativistic particles can be seen with even greater clarity if we examine the behavior of various field observables for a one-particle state $|\psi\rangle$, defined by specifying a momentum wavefunction $\psi(\vec{k})$:

$$|\psi\rangle \equiv \int d^3k \, \psi(\vec{k}) a^\dagger(\vec{k})|0\rangle = \int d^3k \, \psi(\vec{k})|\vec{k}\rangle, \quad \langle\psi|\psi\rangle = \int d^3k |\psi(\vec{k})|^2 = 1 \quad (6.105)$$

We now consider the expectation value of various field observables in this one-particle state at some fixed time, say $t = 0$. For example, the expectation value of the number density is (suppressing the time-variable in the fields, as we are at $t = 0$)

$$\langle\psi|\mathcal{N}(\vec{x},0)|\psi\rangle = i \int d^3k \, d^3k' \psi^*(\vec{k}')\psi(\vec{k})\langle\vec{k}'|\phi^{(-)}(\vec{x})\dot\phi^{(+)}(\vec{x}) - \dot\phi^{(-)}(\vec{x})\phi^{(+)}(\vec{x})|\vec{k}\rangle$$

$$= \frac{1}{2(2\pi)^3} \int d^3k \, d^3k' \psi^*(\vec{k}')\psi(\vec{k}) e^{i(\vec{k}-\vec{k}')\cdot\vec{x}} \{ \sqrt{\frac{E(k)}{E(k')}} + \sqrt{\frac{E(k')}{E(k)}} \}$$

$$= \mathrm{Re}(\chi^*(\vec{x})\tilde\chi(\vec{x})) \quad (6.106)$$

involving two distinct coordinate space wavefunctions

$$\chi(\vec{x}) \equiv \frac{1}{(2\pi)^{3/2}} \int d^3k \frac{1}{\sqrt{E(k)}} \psi(\vec{k}) e^{i\vec{k}\cdot\vec{x}} \quad (6.107)$$

$$\tilde\chi(\vec{x}) \equiv \frac{1}{(2\pi)^{3/2}} \int d^3k \sqrt{E(k)} \psi(\vec{k}) e^{i\vec{k}\cdot\vec{x}} \quad (6.108)$$

In the non-relativistic limit (formally, $c \to \infty$) these wavefunctions revert, up to normalization, to the usual non-relativistic coordinate space wavefunction of a massive particle:

$$\chi(\vec{x}) \to \frac{1}{\sqrt{mc}}\psi(\vec{x}), \quad \psi(\vec{x}) = \frac{1}{(2\pi)^{3/2}} \int d^3k \psi(\vec{k}) e^{i\vec{k}\cdot\vec{x}} \quad (6.109)$$

$$\tilde\chi(\vec{x}) \to \sqrt{mc}\psi(\vec{x}) \quad (6.110)$$

and the expectation value of the number density (6.106) reduces to the conventional probability density of non-relativistic quantum mechanics $|\psi(\vec{x})|^2$, as first proposed by Max Born in 1926.

In the relativistic case, the non-commutativity of number operators for non-overlapping regions means, however, that even if we choose $\psi(\vec{k})$ in (6.105) so that the expectation value of $\mathcal{N}$ vanishes exactly outside some compact spatial volume v_1, this does not mean that the state is an eigenstate, with zero eigenvalue, of the number operator for another non-overlapping spatial volume v_2. In fact, such a state will even have a non-zero-energy distribution outside the volume v_1. This is easily seen by examining the energy density, given by the operator (6.89) (recall that we are interested only in a free system here, and that we are reinstating explicit factors of the velocity of light to facilitate the non-relativistic limit, while keeping natural units

$\hbar = 1$ for Planck's constant)

$$\mathcal{H}_0(\vec{x}, 0) = \frac{1}{2} : (\dot{\phi}(\vec{x}, 0)^2 + c^2 |\vec{\nabla}\phi(\vec{x}, 0)|^2 + m^2 c^4 \phi(\vec{x}, 0)^2) : \qquad (6.111)$$

A short calculation (see Problem 8) gives for the expectation value of the energy density in the state (6.105)

$$\langle \psi | \mathcal{H}_0(\vec{x}, 0) | \psi \rangle = \frac{1}{2} (|\tilde{\chi}(\vec{x})|^2 + c^2 \vec{\nabla}\chi^*(\vec{x}) \cdot \vec{\nabla}\chi(\vec{x}) + m^2 c^4 |\chi(\vec{x})|^2) \qquad (6.112)$$

We can localize our one-particle state exactly with respect to the number density operator either by choosing $\psi(\vec{k}) \propto \sqrt{E(k)}$ (in which case $\chi(\vec{x}) \propto \delta^3(\vec{x})$ but $\tilde{\chi}$ is not a point distribution, but rather our old friend $\tilde{\Delta}_+$ from (6.101)) or by choosing $\psi(\vec{k}) \propto \frac{1}{\sqrt{E(k)}}$ (in which case $\tilde{\chi}(\vec{x}) \propto \delta^3(\vec{x})$, and χ reduces to Δ_+), but *in either case* the energy density involves terms which do not vanish for $\vec{x} \neq 0$, but rather fall off exponentially away from the origin at a rate determined once again by the Compton wavelength of the particle, specifically (apart from power prefactors) like $e^{-2m|\vec{x}|}$. The non-relativistic limit, using (6.109, 6.110), is just as expected:

$$\langle \psi | \mathcal{H}_0(\vec{x}, 0) | \psi \rangle \to mc^2 |\psi(\vec{x})|^2 + \frac{1}{2m} |\vec{\nabla}\psi(\vec{x})|^2 \qquad (6.113)$$

which, on integration over space, gives the rest energy mc^2 plus the expectation value of the non-relativistic kinetic energy operator $-\frac{1}{2m}|\vec{\nabla}|^2$.

The peculiar resistance of relativistic particles described by local fields[15] to localization of their physical attributes is really a manifestation of a deep complementarity principle at play between particle and field aspects in relativistic quantum field theories. Indeed, the fluctuations of the energy of blackbody radiation in a subvolume of a cavity even for states in which the number of photons of each mode (in the full system) is completely definite was the critical piece of information used by Jordan to carry through the first real calculation in quantum field theory, as we saw in Chapter 1. We also recall from Heisenberg's original "gamma-ray microscope" argument for the uncertainty principle that we can often gain some physical understanding of complementary quantities in quantum physics by thought experiments in which physical processes are invoked to effect a "measurement" of a given quantity. From the field point of view, attempts to localize physical attributes of a relativistic particle (energy, number of quanta) in a finite volume of order the Compton wavelength necessarily entail interaction with other fields with momentum (and hence, relativistically, energy) components on the order of the inverse Compton wavelength. Such interactions can produce additional "virtual" particle–antiparticle

---

[15] One of the most remarkable examples of the "fuzzy" character of localization in quantum field theories is the Reeh–Schlieder theorem (see (Reeh and Schlieder, 1961); also (Streater and Wightman, 1978), theorem 4-2) of axiomatic quantum field theory, which asserts that an arbitrary physical state can be approximated arbitrarily well by applying polynomial functions of field operators localized in *any* finite open region of spacetime to the vacuum: even a region arbitrarily far separated from the "location" of the particles in the given state!

pairs, vitiating the desired localization of the original particle.[16] Perhaps the confusing disparity between exactly localizable fields and our stubbornly "fuzzy" particle states is best understood in the following terms. The locality principle operating at the field level is really an implementation of the point-like nature of the *particle interactions*: we construct interaction Hamiltonians by multiplying the relevant fields at exactly the same spacetime point. The structureless character of an elementary particle is, from this point of view, a statement about the way it interacts with other elementary particles (or, in the case of purely self-coupled theories, with itself), and *not* a statement about our ability to localize the physical characteristics (energy, momentum, charge, etc.) of the associated particle at a dimensionless spatial point, which, as the preceding discussion shows, is intrinsically impossible in a relativistic theory.

6.6   From microcausality to analyticity

As we have seen, the construction of relativistic interactions using local fields can be motivated by invoking the spatial clustering principle for relativistic scattering amplitudes, which in turn leads to a smoothness requirement on matrix elements of the interaction Hamiltonian in momentum space. In this section we shall, in a sense, complete the circle by exhibiting a deep connection between the microcausality of the underlying local fields and analyticity properties of amplitudes constructed from such fields.

The analyticity of S-matrix amplitudes is so stringent a constraint that it was at one time[17] believed (in conjunction with Lorentz-invariance, unitarity, and the crossing symmetry property of amplitudes, which we shall discuss later in Chapter 7) to provide an adequate basis for a complete theory of strong interactions, with no need for the explicit introduction of local fields. While this point of view has essentially been abandoned, the dispersion relations for amplitudes derivable on the basis of their analyticity properties still form a very important part of the armory of the field theorist.

The connection between causality and analyticity was first emphasized by Kramers in the Como conference of 1927 (Kramers, 1927).[18] Perhaps the simplest physical example, which nevertheless contains all essential features, is one familiar to electrical engineers: a causal linear device is one in which an output signal $O(t)$ (t is the time variable) is related linearly to an input signal $I(t)$ by a causal transfer function T,

$$O(t) = \int_{-\infty}^{+\infty} T(t - t')I(t')dt' \tag{6.114}$$

---

[16] We shall later return to the underlying particle–field complementarity (in the form of the number-phase mutual uncertainty principle) at work here in Chapter 8, when we examine the classical limit of quantum field theory.

[17] The S-matrix approach to strong interactions, pioneered in the late 1950s and 1960s by Chew, Mandelstam, Regge, and many others, while leading to many important and lasting results, faded once the efficacy of quantum chromodynamics in addressing a much wider variety of strong dynamics processes became apparent in the 1970s and 1980s.

[18] See the paper of Toll (Toll, 1956), for a review and careful discussion of the logical foundations.

where causality requires $T(t - t') = 0$, $t < t'$: the output at any time t can only depend on prior values of the input signal at time t'. The Fourier transform of the transfer function is given by

$$\tilde{T}(\omega) = \int_{-\infty}^{+\infty} e^{i\omega t} T(t) dt = \int_{0}^{+\infty} e^{i\omega t} T(t) dt \qquad (6.115)$$

The restriction of the time variable t to positive values in the above integral means that $\tilde{T}(\omega)$ may be analytically continued from real values of ω to the upper-half-plane of the complex frequency plane, $\text{Im}(\omega) > 0$, as the integral acquires an additional convergence factor $e^{-\text{Im}(\omega)t}$ as we move off the real ω axis into the upper half plane of ω. In fact, for square-integrable transfer functions, the connection goes both ways, as has been shown by Titchmarsh (Titchmarsh, 1948): upper-half-plane analytic functions square integrable along any line parallel to the real axis are inevitably the Fourier transforms of causal transfer functions (i.e., functions vanishing for negative time). This upper-half-plane analyticity of $\tilde{T}(\omega)$ allows the derivation of important *dispersion relations*, relating the real and imaginary parts of $\tilde{T}(\omega)$ for real values of ω. Such dispersion relations are of enormous phenomenological importance in high-energy physics: they played a central role in the S-matrix approach to strong interaction physics which dominated particle physics in the late 1950s and through the 1960s.

The connection between analyticity and causality is also familiar in ordinary quantum scattering theory. Form a wave-packet of incoming waves by smearing plane waves in the usual way:

$$\psi_{\text{in}}(z, t) = \int d\omega g(\omega) e^{i\omega(\frac{z}{c} - t)}$$

The scattering amplitude $f(\omega)$ then gives the outgoing spherical scattered wave as (see Fig. 6.10)

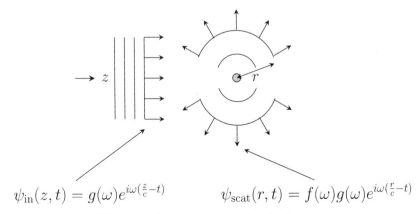

$$\psi_{\text{in}}(z, t) = g(\omega) e^{i\omega(\frac{z}{c} - t)} \qquad \psi_{\text{scat}}(r, t) = f(\omega) g(\omega) e^{i\omega(\frac{r}{c} - t)}$$

Fig. 6.10 Scattering from a localized target.

$$\psi_{\text{scat}}(r, t) = \int d\omega f(\omega) g(\omega) e^{i\omega(\frac{r}{c} - t)}$$

Now choose $g(\omega)$ so the incoming packet arrives at the scattering center at $t = 0$:

$$\psi_{\text{in}}(0, t) = 0, \quad t < 0, \quad \text{with } g(\omega) = \int_0^{+\infty} \frac{dt}{2\pi} \psi_{\text{in}}(0, t) e^{i\omega t}$$

$g(\omega)$ certainly exists for real ω, *a fortiori* for $\text{Im}(\omega) > 0$, where the integral has an additional real exponential convergence. In fact, this implies (by reasoning exactly analogous to that presented previously for causal signals) analyticity of $g(\omega)$ in the entire upper-half-plane. But causality requires that there be no scattered wave ahead of the incident one! In other words,

$$\psi_{\text{scat}}(r, t) = 0, \quad t - \frac{r}{c} < 0$$

so by the same argument, $f(\omega) g(\omega)$ is upper-half-plane analytic. This implies that the scattering amplitude $f(\omega)$ cannot have singularities in the upper-half-plane.

In field theory, a simple, and physically important, illustration of the connection between microcausality and analyticity can be found in the process of forward scattering, in which a massless particle (e.g., a photon) scatters with zero momentum exchange off a target (e.g., a neutral atom, or a proton). The complications of spin are completely irrelevant to the argument we shall present, so we shall discuss this process for a massless *spinless* boson described by a field $\phi(x)$. The target particle is assumed to be stable with respect to emission of the ϕ-particle. We also assume that the interaction between our scalar "photon" and the target can be treated perturbatively to second order in an interaction Hamiltonian of form

$$\mathcal{H}_{\text{int}}(x) = eJ(x)\phi(x) \tag{6.116}$$

where $\phi(x)$ is the canonical scalar field for the "photon"—in the interaction picture, and hence given explicitly by (6.73) (with $a = a^c$, as we assume a self-conjugate boson).

The fields describing the internal dynamics of the target are all contained in the "current"[19] $J(x)$ which evolves dynamically according to a Hamiltonian H_0 which, despite the suggestive subscript, is "free" only in the sense that the field ϕ is absent, while all other aspects of the internal dynamics of the target are treated exactly. For example, if the target is a proton, $J(x)$ would be constructed from the appropriate quark fields and would evolve dynamically with the full Hamiltonian of quantum chromodynamics, the gauge theory assumed to describe strong dynamics. We need only assume that $J(x)$ is a local scalar field with the usual translation property (cf (5.93))

$$J(x) = e^{iP_\mu^{(0)} \cdot x^\mu} J(0) e^{-iP_\mu^{(0)} \cdot x^\mu} \tag{6.117}$$

---

[19] The terminology is justified by the form of the interaction in the physical case of electron–proton scattering, where $\mathcal{H}_{\text{int}}(x) = eJ_\mu(x)A^\mu(x)$, with $J_\mu(x)$ the electromagnetic current for the strongly interacting fields, and $A^\mu(x)$ is the vector potential field mediating the photon. Here, for simplicity, we assume that the current is a Lorentz scalar field.

and microcausality

$$[J(x), J(y)] = 0, \quad (x - y)^2 < 0 \tag{6.118}$$

We note again that the energy-momentum operators $P_\mu^{(0)}$ in (6.117), despite the superscript, contain the *full* dynamics of the target system, but no interactions for the "photon".

The scattering S-matrix amplitude of a scalar photon of momentum q from a target particle of momentum p to second order in the interaction (6.116) is, by (5.72),

$$S^{(2)}(k', q'; k, q) = \frac{(-ie)^2}{2} \int d^4x \, d^4y \langle k', q' | T\{J(x)\phi(x)J(y)\phi(y)\}|k, q\rangle \tag{6.119}$$

or, writing out the T-product explicitly,

$$S^{(2)}(k', q'; k, q) = -\frac{e^2}{2} \int d^4x \, d^4y \{\theta(x^0 - y^0) \langle k' | J(x)J(y)|k\rangle \langle q' | \phi(x)\phi(y)|q\rangle \; + x \leftrightarrow y\} \tag{6.120}$$

Here the field ϕ is free and commutes with the current J, as the dynamics of the latter by definition lacks all reference to ϕ. This allows the factorization of the matrix element indicated in (6.120). From the explicit expression for ϕ in terms of creation and annihilation operators, we find

$$\langle q' | \phi(x)\phi(y)|q\rangle = \frac{1}{(2\pi)^3} \frac{1}{2E(q)} (e^{iq' \cdot x - iq \cdot y} + x \leftrightarrow y) \tag{6.121}$$

whence

$$S^{(2)}(k', q'; k, q) = \frac{-e^2}{2(2\pi)^3} \frac{1}{2E(q)} \int d^4x \, d^4y$$

$$\cdot (e^{iq' \cdot x - iq \cdot y} + x \leftrightarrow y) \langle k' | T\{J(x)J(y)\}|k\rangle \tag{6.122}$$

Changing spacetime variables from x, y to $Z \equiv \frac{x+y}{2}$, $z \equiv x - y$, and using the translation property

$$T\{J(Z + \frac{z}{2})J(Z - \frac{z}{2})\} = e^{iP^{(0)} \cdot Z} T\{J(\frac{z}{2})J(-\frac{z}{2})\} e^{-iP^{(0)} \cdot Z} \tag{6.123}$$

this becomes

$$S^{(2)}(k', q'; k, q) = \frac{-e^2}{2(2\pi)^3} \frac{1}{2E(q)} \int d^4z \, d^4Z$$

$$\cdot (e^{i(q'+q) \cdot z/2} + e^{-i(q'+q) \cdot z/2}) e^{i(q'+k'-q-k) \cdot Z} \langle k' | T\{J(\frac{z}{2})J(-\frac{z}{2})\}|k\rangle$$

$$= \frac{-e^2}{(2\pi)^3} \frac{1}{2E(q)} (2\pi)^4 \delta^4(k' + q' - k - q) T(k', q'; k, q) \tag{6.124}$$

where the invariant amplitude $T(k', q'; k, q)$ is defined as

$$T(k', q'; k, q) \equiv \int d^4 z \frac{1}{2} (e^{i(q'+q)\cdot z/2} + z \to -z) \langle k' | T\{J(\frac{z}{2}) J(-\frac{z}{2})\} | k \rangle$$

$$= \int d^4 z e^{i(q+q')\cdot z/2} \langle k' | T\{J(\frac{z}{2}) J(-\frac{z}{2})\} | k \rangle \tag{6.125}$$

with the last line following from the symmetry of the T-product: $T\{J(\frac{z}{2})J(-\frac{z}{2})\} = T\{J(-\frac{z}{2})J(\frac{z}{2})\}$. In the forward scattering limit $q' \to q, k' \to k$, we find

$$T(k, q; k, q) = \int d^4 z e^{iq \cdot z} \langle k | T\{J(\frac{z}{2}) J(-\frac{z}{2})\} | k \rangle = \int d^4 z e^{iq \cdot z} \langle k | T\{J(z) J(0)\} | k \rangle \tag{6.126}$$

where the translation of the T-product by $\frac{z}{2}$ (with no phase factor) is allowed by equality of the initial- and final-state momenta. Next, we note that

$$T\{J(z)J(0)\} = \theta(z^0) J(z) J(0) + \theta(-z^0) J(0) J(z)$$

$$= \theta(z^0)[J(z), J(0)] + (\theta(z^0) + \theta(-z^0)) J(0) J(z)$$

$$= \theta(z^0)[J(z), J(0)] + J(0) J(z) \tag{6.127}$$

and that the second term on the right-hand side of (6.127) makes no contribution to the forward scattering amplitude. This follows from the following brief computation:

$$\int d^4 q e^{iq \cdot z} \langle k | J(0) J(z) | k \rangle = \sum_n \int d^4 z e^{iq \cdot z} \langle k | J(0) | n \rangle \langle n | J(z) | k \rangle$$

$$= \sum_n \int d^4 z e^{i(q + P_n - k) \cdot z} \langle k | J(0) | n \rangle \langle n | J(0) | k \rangle$$

$$= \sum_n (2\pi)^4 \delta^4(q + P_n - k) \langle k | J(0) | n \rangle \langle n | J(0) | k \rangle \tag{6.128}$$

where we have inserted a complete set $|n\rangle$ of eigenstates of the target system energy-momentum $P^{(0)}$, $P^{(0)\mu} | n \rangle = P_n^\mu | n \rangle$, and used the translation property (6.117). By assumption, the stability of the target system $|p\rangle$ to emission of a "photon" implies that there are no states $|n\rangle$ with momentum $k - q$ and non-vanishing matrix element $\langle n | J(0) | k \rangle$.[20] Thus, the forward scattering amplitude $T(k, q; k, q)$ can be written as the Fourier transform of a *retarded commutator*:

$$T(k, q; k, q) = \int d^4 z e^{iq \cdot z} \theta(z^0) \langle k | [J(z), J(0)] | k \rangle \tag{6.129}$$

The commutator appearing in (6.129) is the necessary ingredient for invoking microcausality: together with the θ-function, the vanishing of the commutator for

---

[20] For example, our target system might be an atom in its ground state, or the proton.

space-like z ensures that the spacetime integral over z is restricted to the forward light-cone, i.e., coordinates z with $z^0 > 0$, $z^2 \geq 0$. Our massless "photon" has energy-momentum $q^\mu = (\omega, \omega\hat{q})$, with $\hat{q}$ a unit spatial vector, so (6.129) can be written (suppressing the dependence on the target momentum k and photon direction $\hat{q}$)

$$T(\omega) = \int d^4z\, e^{i\omega(z^0 - \hat{q}\cdot\vec{z})}\theta(z^0)\langle k|[J(z), J(0)]|k\rangle \tag{6.130}$$

In the forward light-cone for z we necessarily have that $z^0 - \hat{q}\cdot\vec{z} \geq 0$, so the Fourier transform in (6.130) has exactly the same property as $\tilde{T}(\omega)$ in (6.115): it may be smoothly analytically continued to the upper-half-plane of the frequency variable ω, as $\mathrm{Im}(\omega) > 0$ implies an exponential suppression factor $e^{-\mathrm{Im}(\omega)(z^0 - \hat{q}\cdot\vec{z})}$.

That the connection between microcausality (space-like commutativity) and analyticity runs very deep in quantum field theory has been demonstrated by rigorous proofs of very powerful analyticity theorems for the vacuum-expectation-values of local fields, based only on very general assumptions on the spectral and field properties of the theory. These assumptions, which have come to be called the *Wightman axioms*, and their connection to the analyticity properties of field-theory amplitudes, will be explained in our discussion of the axiomatic framework of field theory in the Heisenberg picture in Section 9.2. At that point we shall be in a position to introduce, and prove, the *Ruelle Clustering theorem*, which exhibits the clustering principle of local quantum field theory in its most general and powerful form.

6.7 Problems

1. Verify that the linking operator $\mathcal{L}$ defined in (6.21) correctly constructs the contribution to the 2-2 matrix elements of the product VU of two Fock space operators V and U arising from two-(bosonic)particle intermediate states. You will only need the $M = N = 2$ terms in the expansions of the $\mathcal{V}(j^*, j)$ and $\mathcal{U}(j^*, j)$ functionals. Also, recall the form of the completeness relation for the Fock space of a single boson, (5.22).

2. Consider a bosonic theory with Hamiltonian

$$H = \int d^3k_1' d^3k_1 h_{11}(k_1', k_1) a^\dagger(\vec{k}_1') a(\vec{k}_1)$$

$$+ \frac{1}{4}\int d^3k_1' d^3k_2' d^3k_1 d^3k_2 h_{22}(k_1', k_2', k_1, k_2) a^\dagger(\vec{k}_1') a^\dagger(\vec{k}_2') a(\vec{k}_1) a(\vec{k}_2)$$

Calculate the 3-3 matrix element $\langle q_1' q_2' q_3' | H | q_1 q_2 q_3 \rangle$ explicitly in terms of the functions h_{11}, h_{22}. Express your result graphically to display the connectedness structure of this matrix element.

3. A conserved four-vector field (called a "current") $J^\mu(x)$ can be defined as follows for a free complex scalar field $\phi(x)$:

$$J^\mu(x) = ie : (\phi^\dagger(x)\partial^\mu\phi(x) - (\partial^\mu\phi^\dagger(x))\phi(x)) : \tag{6.131}$$

If we attribute an electric charge e (resp. $-e$) to the particles (resp. antiparticles) of the theory, show that J^μ may be regarded as the electromagnetic current for

this field by demonstrating
(a) current conservation (use the Klein–Gordon equation for ϕ):

$$\partial_\mu J^\mu(x) = 0$$

(b) that the charge Q defined as

$$Q \equiv \int d^3x J^0(\vec{x}, t)$$

counts electric charge (defined as e times the difference in the number of particles and antiparticles), and is time-independent.

4. Show that the normal-ordered product of four self-conjugate scalar fields at the same spacetime point can be written in terms of even powers of the field (and is therefore itself local); i.e., show

$$: \phi(x)^4 := \phi(x)^4 + A\phi(x)^2 + B$$

where A, B are c-numbers (independent of x).

5. The following exercise shows that the *free* Hamiltonian H_0 can be written as an integral of a density constructed from the local self-conjugate field ϕ.
(a) Show that

$$H_0 = \int d^3k E(k) a^\dagger(\vec{k}) a(\vec{k})$$

i.e., show that an arbitrary free state $|k_1, k_2, ...k_n\rangle$ has the appropriate eigenvalue relative to this operator.
(b) Next, show that

$$\frac{1}{2} \int d^3x \{(\frac{\partial \phi}{\partial t})^2 + |\vec{\nabla}\phi|^2 + m^2\phi^2\} = \int d^3k E(k)(a^\dagger(\vec{k})a(\vec{k}) + a(\vec{k})a^\dagger(\vec{k}))/2$$

$$= H_0 + \frac{1}{2}\delta^3(0) \int d^3k E(k)$$

Note that the singular zero-point energy is removed by normal-ordering H_0, as in (6.89).

6. Verify the expressions for the number operator (6.92, 6.93, 6.94).
7. Verify the commutator results listed in Eqs. (6.96–6.99).
8. Show that the expectation value of the free scalar energy density in the state (6.105) is as given in (6.112).
9. Let $g_k(x) = \frac{1}{\sqrt{(2\pi)^3 2E(k)}} e^{-ik\cdot x}$ (k and x are four-vectors, with $k_0 = E(k)$). Show that the destruction operator $a(\vec{k})$ may be reconstructed from a self-conjugate scalar field by

$$a(\vec{k}) = i \int d^3x (g_k^*(\vec{x}, t) \partial_0 \phi(\vec{x}, t) - \phi(\vec{x}, t) \partial_0 g_k^*(\vec{x}, t)) \tag{6.132}$$

7
Dynamics V: Construction of local covariant fields

In many of the standard texts on quantum field theory, the introduction of fields representing particles of low spin (zero, $\frac{1}{2}$, or one—there is no direct phenomenological evidence for *elementary* particles of any higher spin) is a fairly *ad hoc* matter. Relativistic wave equations are introduced and shown to have "nice" covariance properties. A Lagrangian formalism is then constructed for which these equations are just the Euler–Lagrange equations of the theory, corresponding to the extremal condition on the classical action. Finally, a canonical quantization procedure is carried out: conjugate momentum fields are introduced, and the resultant Hamiltonian is shown to be the appropriate energy operator for particles of the desired mass and spin.

In this chapter we shall eschew this *ad hoc* methodology in favor of a more direct, constructive approach. The relativistic wave equations satisfied by the covariant fields representing particles of low spin are shown to be automatic consequences of the representation theory of the Poincaré group, which can be used to write a completely general expression for the fields transforming according to an arbitrary finite-dimensional representation of the Lorentz group *and* representing particles of arbitrary mass and spin.

The advantage of this approach is twofold. In the first place, it will become apparent in a completely natural way that there is an inevitable, but completely classifiable, fluidity in the association of fields to particles: many different covariant fields may be used with equal validity to represent a particle of a given spin, although there is usually a "best" (i.e., most convenient) choice. Secondly, the formalism allows us to solve the problem of constructing covariant fields in one fell swoop: the special cases of spin zero, $\frac{1}{2}$, or 1 then follow from the general result simply by inserting $j = 0, \frac{1}{2}, 1$ into the master formula. In particular, the Spin-Statistics theorem associating particles with integral (resp. half-integral) spin with fields of bosonic (resp. fermionic) type (at the free field level) will emerge naturally in the framework of this formalism.

7.1 Constructing local, Lorentz-invariant Hamiltonians

We saw previously in Chapter 5 that one simple way to achieve unitarity, Lorentz invariance and locality of a theory of interacting particles was to express the interaction Hamiltonian as the spatial integral of a Lorentz scalar local field. In Chapter 6 we saw that theories of interacting spinless particles are readily constructed by taking the Hamiltonian density as a polynomial in *canonical* scalar fields which are linear in creation and annihilation operators of the particles in question. Unfortunately, such simple scalar canonical fields can never produce particles of *non-zero* spin acting on the vacuum. From the defining transformation law for a scalar field

$$U(\Lambda)\phi(x)U^\dagger(\Lambda) = \phi(\Lambda x) \tag{7.1}$$

one finds, choosing Λ to be the infinitesimal rotation by $\delta\theta$ around the ith axis, $(R_i\vec{x})^j = x^j + \epsilon_{ijk}x^k\delta\theta$, and $U(\Lambda) = e^{-i\delta\theta J_i}$,

$$[J_i, \phi(x)] = i\epsilon_{ijk}x^k\partial_j\phi(x) \tag{7.2}$$

Recall that the spin of a particle is simply the residual angular momentum it possesses when at rest: i.e., at zero linear momentum. Any state, formed by ϕ acting on the vacuum *with zero spatial momentum* (which we achieve by integrating over 3-space to project out the creation operator at zero momentum), must then have zero *angular* momentum as well, as we discover by a simple integration by parts:

$$J_i \int d^3x\,\phi(x)|0\rangle = i \int d^3x\,\epsilon_{ijk}x^k\partial_j\phi(x)|0\rangle$$

identically zero

$$= -i \int d^3x\,\phi(x)\epsilon_{ijk}\partial_j x^k|0\rangle$$

$$= 0 \tag{7.3}$$

It is clear that more general fields are needed to describe particles of non-vanishing spin. Nevertheless, it will be important to be able to combine such fields to again produce an interaction Hamiltonian density which is itself a *Lorentz scalar field*, satisfying (7.1) above. The solution is to construct fields that transform according to definite finite-dimensional representations of the homogeneous Lorentz group. For example, two vector fields $A^\mu(x), B^\mu(x)$ transforming like

$$U(\Lambda)(A, B)^\mu(x)U^\dagger(\Lambda) = (\Lambda^{-1})^\mu_{\ \nu}(A, B)^\nu(\Lambda x) \tag{7.4}$$

can be coupled together to make a scalar field $C(x) \equiv A^\mu(x)B_\mu(x)$, which is easily seen to satisfy (7.1). Moreover, $C(x)$ is local (commutes with itself at space-like separation) if A, B are. The original four-fermion theory of the weak interactions, dating from the 1930s, involved a weak interaction Hamiltonian of precisely this form.

A general covariant field will be a set of field operators transforming according to a general finite-dimensional representation of the homogeneous Lorentz group realized by the finite-dimensional matrices $M_{nm}(\Lambda)$:

$$U(\Lambda)\phi_n(x)U^\dagger(\Lambda) = M_{nm}(\Lambda^{-1})\phi_m(\Lambda x) \tag{7.5}$$

Lorentz scalar fields can be constructed from such covariant fields by coupling them together with invariant tensors $t_{n_1 n_2\ldots}$ of the HLG; namely

$$\mathcal{H}(x) = t_{n_1 n_2\ldots}\phi_{n_1}\phi_{n_2}\cdots \tag{7.6}$$

is a Lorentz scalar field if

$$t_{n_1 n_2\ldots}M_{n_1 m_1}(\Lambda^{-1})M_{n_2 m_2}(\Lambda^{-1})\ldots = t_{m_1 m_2\ldots}, \quad \text{all } \Lambda \tag{7.7}$$

Of course, the simplest example of such a tensor is the two index Minkowski-space metric tensor $g_{\mu\nu}$, employed above to construct a scalar field C as the invariant dot-product of two vector fields A^μ, B^ν.

In addition we shall require translation invariance just as for scalar fields (cf(5.91)):

$$[P^{(0)}_\mu, \phi_n(x)] = -i\partial_\mu \phi_n(x) \tag{7.8}$$

and locality

$$[\phi_n(x), \phi_m(y)] = [\phi_n(x), \phi^\dagger_m(y)] = 0, \quad (x-y)^2 < 0 \tag{7.9}$$

7.2 Finite-dimensional representations of the homogeneous Lorentz group

In this section we shall develop the basic theory of finite-dimensional representations of the homogeneous Lorentz group (HLG), which will lead us to a complete classification of local covariant fields. Consider an infinitesimal Lorentz transformation

$$\Lambda^\mu_{\;\nu} = g^\mu_{\;\nu} + \Omega^\mu_{\;\nu} + O(\Omega^2) \tag{7.10}$$

where $\Omega^\mu_{\;\nu}$ is a matrix of infinitesimally small parameters ("rotation angles"). In fact, this matrix is forced to be antisymmetric:

$$\Lambda^\mu_{\;\nu}\Lambda_{\mu\rho} = (g^\mu_{\;\nu} + \Omega^\mu_{\;\nu})(g_{\mu\rho} + \Omega_{\mu\rho}) + O(\Omega^2) = g_{\nu\rho} \tag{7.11}$$

$$\Longrightarrow \Omega_{\rho\nu} + \Omega_{\nu\rho} = 0 \tag{7.12}$$

which implies that there are six independent parameters in the specification of an arbitrary infinitesimal Lorentz transformation (three angles and three boosts). In a general finite-dimensional representation, Λ is represented by the matrix

$$M_{nm}(\Lambda) = \delta_{nm} + \frac{i}{2}\Omega_{\mu\nu}(J^{\mu\nu})_{nm} + O(\Omega^2) \tag{7.13}$$

where the six independent matrices $J^{12}, J^{23}, J^{31}, J^{01}, J^{02}, J^{03}$ are the generators of rotations (around the z, x, and y axes, respectively) and boosts (along the x, y, and z axes, respectively). Now consider some definite fixed $\bar\Lambda = 1 + \bar\Omega$ which we subject to a similarity transformation with a general $M(\Lambda)$:

$$M(\Lambda)M(\bar\Lambda)M(\Lambda^{-1}) = M(\Lambda\bar\Lambda\Lambda^{-1}) \tag{7.14}$$

$$= M(1 + \Lambda\bar\Omega\Lambda^{-1}) \tag{7.15}$$

$$= 1 + \frac{i}{2}(\Lambda\bar\Omega\Lambda^{-1})_{\rho\sigma}J^{\rho\sigma} \tag{7.16}$$

$$= 1 + \frac{i}{2}\bar\Omega_{\mu\nu}\Lambda^\mu_{\;\rho}\Lambda^\nu_{\;\sigma}J^{\rho\sigma} \tag{7.17}$$

On the other hand,

$$M(\Lambda)M(\bar\Lambda)M^{-1}(\Lambda) = 1 + M(\Lambda)\frac{i}{2}\bar\Omega_{\mu\nu}J^{\mu\nu}M^{-1}(\Lambda) \tag{7.18}$$

Comparing (7.17) and (7.18), we see that the generators $J^{\mu\nu}$ transform as second-rank contravariant tensors under the Lorentz group:

$$M(\Lambda)J^{\mu\nu}M^{-1}(\Lambda) = \Lambda^\mu_\rho\Lambda^\nu_\sigma J^{\rho\sigma} \qquad (7.19)$$

Next, starting from (7.19), choose Λ itself infinitesimal, $\Lambda^\mu_\nu = g^\mu_\nu + \omega^\mu_\nu$, so that keeping terms of $O(\omega)$, we obtain

$$\frac{i}{2}[\omega_{\rho\sigma}J^{\rho\sigma}, J^{\mu\nu}] = (\omega^\mu_\rho g^\nu_\sigma + g^\mu_\rho \omega^\nu_\sigma)J^{\rho\sigma}$$

$$= \omega_{\rho\sigma}(g^{\sigma\mu}J^{\rho\nu} - g^{\rho\nu}J^{\mu\sigma})$$

$$= \frac{1}{2}\omega_{\rho\sigma}(g^{\sigma\mu}J^{\rho\nu} + g^{\sigma\nu}J^{\mu\rho} - g^{\rho\nu}J^{\mu\sigma} - g^{\rho\mu}J^{\sigma\nu}) \qquad (7.20)$$

from which follows immediately the full Lie algebra of the HLG:

$$[J^{\mu\nu}, J^{\rho\sigma}] = i(g^{\mu\sigma}J^{\rho\nu} + g^{\nu\sigma}J^{\mu\rho} - g^{\rho\mu}J^{\sigma\nu} - g^{\rho\nu}J^{\mu\sigma}) \qquad (7.21)$$

If μ, ν are both spatial indices, we are dealing with the rotation subgroup of the HLG, and the corresponding generators are therefore just the familiar angular momentum operators:

$$J_1 \equiv J^{32} = J_{32} \qquad (7.22)$$

$$J_2 \equiv J^{13} = J_{13} \qquad (7.23)$$

$$J_3 \equiv J^{21} = J_{21} \qquad (7.24)$$

which satisfy, as an immediate consequence of (7.21), the usual SU(2) algebra

$$[J_i, J_j] = i\epsilon_{ijk}J_k \qquad (7.25)$$

The physical content of the algebra is made more transparent by defining boost generators

$$K_1 \equiv J_{10} \qquad (7.26)$$

$$K_2 \equiv J_{20} \qquad (7.27)$$

$$K_3 \equiv J_{30} \qquad (7.28)$$

which transform as three-vectors under the rotation group

$$[J_i, K_j] = i\epsilon_{ijk}K_k \qquad (7.29)$$

while satisfying the following commutation relation among themselves:

$$[K_i, K_j] = -i\epsilon_{ijk}J_k \qquad (7.30)$$

The structure of the HLG is further clarified by considering the linear combinations

$$A_i \equiv \frac{1}{2}(J_i - iK_i). \quad B_i \equiv \frac{1}{2}(J_i + iK_i) \qquad (7.31)$$

whereupon one finds

$$[A_i, A_j] = i\epsilon_{ijk}A_k, \quad [B_i, B_j] = i\epsilon_{ijk}B_k \tag{7.32}$$

together with a complete decoupling of the A and B parts of the algebra:

$$[A_i, B_j] = 0 \tag{7.33}$$

Evidently, the HLG can be regarded as the direct product of two "angular momentum groups"! A note of caution is necessary here: we parenthesize "angular momentum groups" because the generators A_i, B_i are *not* hermitian, so that the groups they generate are not, strictly speaking, the usual unitary SU(2) group, but rather a complexified version. Nevertheless, the resolution of the full HLG algebra into two commuting subgroups tremendously simplifies (in fact, effectively solves) the problem of classifying all the finite-dimensional representations of HLG.

General representations of the HLG can be labeled by a spin pair (A, B), where A, B are integers or half-integers. Within a representation (A, B), a state is labeled by (a, b), where as usual $a = -A, -A + 1, ..., +A$, $b = -B, -B + 1, ..., +B$. Since the usual angular momentum $\vec{J} = \vec{A} + \vec{B}$, we can only describe particles of spin j by fields (A, B) where A and B can be coupled together to make angular momentum j, i.e,

$$|A - B| \leq j \leq |A + B| \tag{7.34}$$

On the other hand, *any* field satisfying this constraint can be used to represent a particle of the given spin! This is a central feature of field theory which appears at this point clearly for the first time: there is no unique correspondence between particles and fields. A given particle can be represented by a variety of covariant fields, and (as we shall see later) a given field can also represent many particle states.

The construction of scalar quantities under the HLG is analogous to the problem of coupling non-zero angular momenta to net zero spin in the theory of the rotation group—except that we have here to worry about two "rotation groups"! We shall soon see that the behavior of these representations under (i) complex conjugation, and (ii) spatial inversion (parity) play a particularly important role in understanding how to accomplish this. First, we make some comments about the effect of complex conjugation. This becomes an issue for spinorial (half-integral) representations: in the fundamental (spin-$\frac{1}{2}$) representation, for example, which is complex, but *pseudoreal*. In other words, the 2-spinor representation of SU(2) is isomorphic to its complex conjugate. In the case of the HLG, with its doubled SU(2) structure, conjugation has the additional effect of interchanging the A and B quantum numbers. To see this, consider the $M(\Lambda)$ representation matrix for the $(\frac{1}{2},0)$ representation, with Λ a *finite* HLG element with finite parameters $\Omega_{\mu\nu}$ (cf. Eq(7.13) for the infinitesimal case):

$$M^{(\frac{1}{2},0)}(\Lambda) = e^{\frac{i}{2}\Omega_{\mu\nu}J^{\mu\nu}} \tag{7.35}$$

Defining angle and boost vectors by $\vec{\phi} = (\Omega_{23}, \Omega_{31}, \Omega_{12}), \vec{\xi} = (\Omega_{10}, \Omega_{20}, \Omega_{30})$, and recalling that $J^{i0} = -K_i = -\frac{i}{2}\vec{\sigma}$ $(i=1,2,3)$, $J^{ij} = -\epsilon_{ijk}J_k = -\frac{1}{2}\epsilon_{ijk}\sigma_k$ for the $(\frac{1}{2},0)$

representation, with σ_i the conventional Pauli σ matrices, this becomes

$$M^{(\frac{1}{2},0)}(\Lambda) = e^{\frac{1}{2}\vec{\xi}\cdot\vec{\sigma} - \frac{i}{2}\vec{\phi}\cdot\vec{\sigma}} \qquad (7.36)$$

Likewise, the corresponding finite transformation matrix for the $(0,\frac{1}{2})$ representation is

$$M^{(0,\frac{1}{2})}(\Lambda) = e^{-\frac{1}{2}\vec{\xi}\cdot\vec{\sigma} - \frac{i}{2}\vec{\phi}\cdot\vec{\sigma}} \qquad (7.37)$$

Thus if a $(\frac{1}{2},0)$ spinor $\chi_\alpha, \alpha = 1, 2$ transforms under a Lorentz transformation as

$$\chi_\alpha \rightarrow (e^{\frac{1}{2}\vec{\xi}\cdot\vec{\sigma} - \frac{i}{2}\vec{\phi}\cdot\vec{\sigma}})_{\alpha\beta}\chi_\beta \qquad (7.38)$$

the conjugate spinor will transform as

$$\chi_\alpha^* \rightarrow (e^{\frac{1}{2}\vec{\xi}\cdot\vec{\sigma}^* + \frac{i}{2}\vec{\phi}\cdot\vec{\sigma}^*})_{\alpha\beta}\chi_\beta^* \qquad (7.39)$$

The pseudoreality property of the spinor representation alluded to above amounts to the following conjugation property of the Pauli matrices: defining the spinor conjugation matrix

$$C_s \equiv i\sigma_2 = \begin{pmatrix} 0 & 1 \\ -1 & 0 \end{pmatrix} \qquad (7.40)$$

we have

$$C_s\vec{\sigma}C_s^{-1} = -\vec{\sigma}^*, \quad C_s\vec{\sigma}^*C_s^{-1} = -\vec{\sigma} \qquad (7.41)$$

whence the conjugation transformation (7.39) becomes

$$(C_s\chi^*)_\alpha \rightarrow (C_s e^{\frac{1}{2}\vec{\xi}\cdot\vec{\sigma}^* + \frac{i}{2}\vec{\phi}\cdot\vec{\sigma}^*} C_s^{-1})_{\alpha\beta}(C_s\chi^*)_\beta$$
$$= (e^{-\frac{1}{2}\vec{\xi}\cdot\vec{\sigma} - \frac{i}{2}\vec{\phi}\cdot\vec{\sigma}})_{\alpha\beta}(C_s\chi^*)_\beta \qquad (7.42)$$

Comparing this result with (7.37), we see that the conjugate spinor $C_s\chi^*$ transforms appropriately for a $(0, \frac{1}{2})$ representation: conjugation has reversed the roles of the A and B quantum numbers. This is hardly surprising, given Eqs. (7.31), as the $\vec{J}$ (resp. $\vec{K}$) generators are represented by hermitian (resp. antihermitian) matrices.

Next, we consider the effects of a *parity transformation*: namely, an improper Lorentz transformation (i.e., with determinant equal to negative unity) reversing the sign of the three spatial coordinates while leaving the time coordinate unchanged. Evidently, the angular momentum generators $\vec{J}$, with two spatial indices (see (7.22)) are even under a parity transformation, while the boost generators $\vec{K}$, with a single spatial index (see (7.26)) are odd. In common parlance, angular momentum $\vec{J}$ is an *axial* vector, the boost vector $\vec{K}$ a *polar* vector. A glance at the definitions (7.31) shows that the parity transformation has the effect of *interchanging* the A and B labels of a given irreducible representation (A, B) of the HLG. If the interactions of our particles exactly conserve parity (as in the strong and electromagnetic interactions), then the Hamiltonian cannot change under a parity transformation, so that if we employ fields transforming under a representation (A, B), with $A \neq B$ (note: such fields are termed *chiral*), then the Hamiltonian must also contain, symmetrically, fields transforming according to the representation (B, A). Indeed, for fermions, the Spin-Statistics theorem (discussed below) implies j half-integral, whence, by (7.34), we necessarily have $A \neq B$. Parity may be preserved in this case either by employing conjugate representations as discussed above (leading to *Majorana* fermions, cf. Section 7.4.1), or by the use of *reducible* representations of the HLG, such as $(A, B) \oplus (B, A)$ (*Dirac* fermions, Section 7.4.2).

7.3   Local covariant fields for massive particles of any spin: the Spin-Statistics theorem

We now turn to the task of explicitly constructing local canonical covariant fields, linear in creation and annihilation operators, for (massive) particles of *any spin*. This section will correspondingly be algebraically somewhat more dense than most, but the end results will more than merit the effort expended: just a few pages will suffice to establish the general form of covariant fields of any spin, from which with very little further effort flows the whole panoply of relativistic wave equations (Klein–Gordon, Dirac, Maxwell–Proca, etc.) typically introduced in a more or less *ad hoc* fashion in earlier generations of field theory texts.[1] Moreover, a profound consequence of relativistic field theory—the Spin-Statistics connection—will emerge naturally as part of the construction.

The most general canonical field linear in creation and annihilation operators can be written

$$\phi_n(x) = \sum_\sigma \int \frac{d^3k}{(2\pi)^{3/2}\sqrt{2E(k)}}(u_n(\vec{k},\sigma)a(\vec{k},\sigma)e^{-ik\cdot x} + v_n(\vec{k},\sigma)a^{c\dagger}(\vec{k},\sigma)e^{ik\cdot x}) \quad (7.43)$$

Our task is to determine the coefficient functions $u_n(\vec{k},\sigma), v_n(\vec{k},\sigma)$ in order to satisfy the requirements of Poincaré covariance and locality, namely (7.5, 7.8, 7.9). First, note

$$U(\Lambda)\phi_n(x)U^\dagger(\Lambda) = \sum_{\sigma\sigma'} \int \frac{d^3k}{(2\pi)^{3/2}\sqrt{2E(k)}}\sqrt{\frac{E(\Lambda k)}{E(k)}}$$

$$\cdot \{u_n(\vec{k},\sigma)e^{-ik\cdot x}D^j_{\sigma\sigma'}(W^{-1}(\Lambda,\vec{k}))a(\Lambda\vec{k},\sigma')$$

$$+ v_n(\vec{k},\sigma)e^{ik\cdot x}D^j_{\sigma'\sigma}(W(\Lambda,\vec{k}))a^{c\dagger}(\Lambda\vec{k},\sigma')\} \quad (7.44)$$

On the other hand, this must equal, by (7.5),

$$\sum_m M_{nm}(\Lambda^{-1}) \int \frac{d^3k}{(2\pi)^{3/2}\sqrt{2E(k)}}\sqrt{\frac{E(\Lambda k)}{E(k)}} \{ u_m(\Lambda\vec{k},\sigma')e^{-ik\cdot x}a(\Lambda\vec{k},\sigma')$$

$$+ v_m(\Lambda\vec{k},\sigma')e^{ik\cdot x}a^{c\dagger}(\Lambda\vec{k},\sigma')\}$$

---

[1] The construction of covariant fields for any spin in a unified way employing the representation theory of the Poincaré group was first carried out in a seminal paper by Weinberg (Weinberg, 1964a).

where we have made a change of variable $\vec{k} \to \Lambda\vec{k}$ in the momentum integration. Comparing coefficents of $a(\Lambda\vec{k}, \sigma'), a^{c\dagger}(\Lambda\vec{k}, \sigma')$ we find

$$\sum_\sigma D^j_{\sigma\sigma'}(W^{-1}(\Lambda, \vec{k}))u_n(\vec{k}, \sigma) = \sum_m M_{nm}(\Lambda^{-1})u_m(\Lambda\vec{k}, \sigma') \tag{7.45}$$

$$\sum_\sigma D^j_{\sigma'\sigma}(W(\Lambda, \vec{k}))v_n(\vec{k}, \sigma) = \sum_m M_{nm}(\Lambda^{-1})v_m(\Lambda\vec{k}, \sigma') \tag{7.46}$$

The $u_n(\vec{k}, \sigma), v_n(\vec{k}, \sigma)$ should be regarded as connection coefficients between the finite-dimensional (n index) field representations of the HLG and the infinite-dimensional unitary Fock-space representation of the single-particle states (labeled by $\vec{k}$ and σ). The constraints (7.45) and (7.46) will turn out to uniquely determine these coefficients and hence the desired covariant field operators for any spin, up to an obvious normalization and phase freedom. They also imply, as we shall see, that the covariant field operators necessarily satisfy certain partial differential equations in coordinate space (commonly called "relativistic wave equations"): in conventional presentations of field theory these equations arise (somewhat magically) as Euler–Lagrange equations of relativistically invariant actions. Here the constraints appear naturally as a consequence of connecting the covariance of the field operator to the underlying unitary Fock-space structure. We shall now show how these constraints may be explicitly solved for arbitrary spin j.

Recall the definition of the Wigner rotation: for general Lorentz transformation Λ,

$$W(\Lambda, \vec{k}) = L^{-1}(\Lambda\vec{k})\Lambda L(\vec{k}) \tag{7.47}$$

Suppose $\vec{k} = 0$. Then, if $\Lambda = R$, a rotation, $W(R, 0) = R$ and we obtain from (7.45)

$$\sum_\sigma D^j_{\sigma\sigma'}(R^{-1})u_n(0, \sigma) = \sum_m M_{nm}(R^{-1})u_m(0, \sigma') \tag{7.48}$$

so that a unitary rotation on the field (m) index can be transferred to a unitary spin-j rotation on the particle-spin index (σ). In an irreducible representation, this constraint will determine $u_n(0, \sigma)$ up to overall normalization.

Similarly, choosing $\vec{k} \neq 0$, $\Lambda = L^{-1}(\vec{k})$,

$$W(\Lambda, \vec{k}) = L^{-1}(\Lambda\vec{k}) = L^{-1}(\vec{0}) = 1 \tag{7.49}$$

so

$$u_n(\vec{k}, \sigma) = \sum_m M_{nm}(L(\vec{k}))u_m(\vec{0}, \sigma) \tag{7.50}$$

so the full coefficient function $u_n(\vec{k}, \sigma)$ is determined by a boost once we have used (7.48) to fix the u_ns at zero momentum.

For the v_n coefficient functions, using (7.46), and the property of rotation matrices

$$D^j_{\sigma'\sigma}(W(\Lambda, \vec{k})) = (-)^{\sigma'-\sigma}D^j_{-\sigma,-\sigma'}(W^{-1}(\Lambda, \vec{k})) \tag{7.51}$$

one finds

$$\sum_\sigma D^j_{\sigma\sigma'}(R)(-)^{j+\sigma}v_n(0,-\sigma) = \sum_m M_{nm}(R)(-)^{j+\sigma'}v_m(0,-\sigma') \qquad (7.52)$$

which implies $(-)^{j+\sigma}v_n(0,-\sigma) = \xi u_n(0,\sigma)$ (ξ a so far arbitrary constant) as both satisfy (7.48), while the v_ns at non-zero momentum are again obtained by a boost:

$$v_n(\vec{k},\sigma) = \sum_m M_{nm}(L(\vec{k}))v_m(0,\sigma) \qquad (7.53)$$

so that finally

$$v_n(\vec{k},\sigma) = \xi(-)^{j-\sigma}u_n(\vec{k},-\sigma) \qquad (7.54)$$

We now turn to the task of explicitly solving the constraints (7.48, 7.50). Here we shall need the results of the preceding section, in which the finite-dimensional representations of the HLG (i.e., the structure of the representation matrices $M_{nm}(\Lambda)$) were classified. We shall use the $(AB; ab)$ notation described above to identify specific representations of the HLG and components within a representation. Thus

$$\phi_n \to \phi^{(AB)}_{ab}, \quad u_n \to u^{(AB)}_{ab} \qquad (7.55)$$

With this notation, the constraint (7.48) now reads

$$\sum_{\sigma'} D^j_{\sigma'\sigma}(R)u_{ab}(0,\sigma') = \sum_{a'b'}(e^{-i\vec{A}\cdot\vec{\alpha}})_{aa'}(e^{-i\vec{B}\cdot\vec{\alpha}})_{bb'}u_{a'b'}(0,\sigma) \qquad (7.56)$$

$$= \sum_{a'b'} D^A_{aa'}(R)D^B_{bb'}(R)u_{a'b'}(0,\sigma) \qquad (7.57)$$

where R is the rotation $e^{-i\vec{J}\cdot\vec{\alpha}}$. Now suppose that $|Aa, Bb\rangle$ denotes a set of eigenstates of $\vec{A}^2, A_3, \vec{B}^2, B_3$ and one constructs the state

$$|\sigma\rangle \equiv \sum_{ab} u_{ab}(0,\sigma)|Aa, Bb\rangle \qquad (7.58)$$

Then (7.57) just says that the $|\sigma\rangle$ states transform as an irreducible representation of the rotations induced by the angular momentum operator $\vec{J} = \vec{A} + \vec{B}$, corresponding to eigenvalue $\vec{J}^2 = j(j+1)$. Proof: under a simultaneous rotation $e^{-i\vec{A}\cdot\vec{\alpha}}e^{-i\vec{B}\cdot\vec{\alpha}}$

$$|Aa, Bb\rangle \to \sum_{a'b'} D^A_{a'a}(\vec{\alpha})D^B_{b'b}(\vec{\alpha})|Aa', Bb'\rangle \qquad (7.59)$$

so the above constraint amounts to

$$e^{-i\vec{J}\cdot\vec{\alpha}}|\sigma\rangle = \sum_{ab} u_{ab}(0,\sigma)e^{-i\vec{A}\cdot\vec{\alpha}}e^{-i\vec{B}\cdot\vec{\alpha}}|Aa, Bb\rangle \qquad (7.60)$$

$$= \sum_{ab,a'b'} u_{ab}(0,\sigma)D^A_{a'a}(\vec{\alpha})D^B_{b'b}(\vec{\alpha})|Aa', Bb'\rangle \qquad (7.61)$$

$$= \sum_{\sigma',a'b'} D^j_{\sigma'\sigma}(\tilde{\alpha}) u_{a'b'}(0,\sigma')|Aa',Bb'\rangle \tag{7.62}$$

$$= \sum_{\sigma'} D^j_{\sigma'\sigma}(\tilde{\alpha})|\sigma'\rangle \tag{7.63}$$

i.e., the $|\sigma\rangle$ span a spin j irreducible representation of the rotation group. The coefficients which perform the desired coupling (7.58) are just the familiar Clebsch–Gordon coefficients (unique up to a phase), so we now know the coefficient functions at zero momentum:

$$u^{AB}_{ab}(0,\sigma) = \langle A \, B \, a \, b \, | j \, \sigma \rangle \tag{7.64}$$

which (by (7.54)) implies

$$v^{AB}_{ab}(0,\sigma) = \xi^{AB}(-1)^{j-\sigma} \langle A \, B \, a \, b | j \, -\sigma \rangle \tag{7.65}$$

From (7.50), once the coefficient functions are known at zero momentum, they can be boosted to any non-zero momentum. For this, we need the general expression for the boost $L_{nm}(L(\vec{k}))$ in the (AB) representation. An infinitesimal boost in the ith direction is realized by the $J_{i0} = K_i$ generator. For example, if $\vec{k}$ is in the z-direction, and θ is the rapidity angle of the boost,

$$L^\rho_\sigma(\vec{k}) \equiv B^\rho_\sigma(\theta) = \begin{pmatrix} \cosh\theta & 0 & 0 & \sinh\theta \\ 0 & 1 & 0 & 0 \\ 0 & 0 & 1 & 0 \\ \sinh\theta & 0 & 0 & \cosh\theta \end{pmatrix}$$

One easily sees that $B(\theta) = e^{\theta M}$, where M is the matrix

$$M^\rho_\sigma = \begin{pmatrix} 0 & 0 & 0 & 1 \\ 0 & 0 & 0 & 0 \\ 0 & 0 & 0 & 0 \\ 1 & 0 & 0 & 0 \end{pmatrix}$$

which is equal to (in the four-vector representation) $i(J_{30})^\rho_\sigma = iK_3$. (In the four-vector representation, the explicit formula for the generators is $(J_{\mu\nu})^\rho_\sigma = i(g^\rho_\nu g_{\mu\sigma} - g^\rho_\mu g_{\nu\sigma})$; some boring algebra establishes that these matrices satisfy the Lie algebra (7.21)). It follows that z-boosts in a general (AB) representation are realized by

$$L_{nm}(B(\theta)) = (e^{i\theta K_3})_{nm} \tag{7.66}$$

$$= (e^{i\theta(i)(A_3 - B_3)})_{nm} \tag{7.67}$$

$$= (e^{-\theta A_3} e^{+\theta B_3})_{nm} \tag{7.68}$$

with $\cosh\theta = \frac{E(k)}{m}$. For a boost in a general direction we just use $\hat{k} \cdot \vec{A}, \hat{k} \cdot \vec{B}$ instead of A_3, B_3, obtaining finally the general solution for the u coefficient function in an arbitrary representation of the HLG and for arbitrary spin particles:

$$u_{ab}^{AB}(\vec{k}, \sigma) = \sum_{a'b'} (e^{-\theta\hat{k}\cdot\vec{A}})_{aa'} (e^{+\theta\hat{k}\cdot\vec{B}})_{bb'} \langle A\,B\,a'\,b'|j\,\sigma\rangle \tag{7.69}$$

with (using (7.54)) an obvious corresponding equation for $v_{ab}^{AB}(\vec{k}, \sigma)$.

Now that we know how to build covariant fields, transforming according to definite representations of the HLG, it remains to be seen whether we can also arrange for microcausality. For interaction Hamiltonians built purely from scalar fields, we saw that Lorentz-invariance of the S-matrix hinged on the local commutativity property:

$$[\mathcal{H}_{int}(x), \mathcal{H}_{int}(y)]_- = 0 \ , \ (x-y)^2 < 0 \tag{7.70}$$

which could be arranged if $\mathcal{H}_{int}$ was built as a polynomial of local fields:

$$[\phi(x), \phi^\dagger(y)]_- = 0 \ , \ (x-y)^2 < 0 \tag{7.71}$$

For *fermions*, $[\phi(x), \phi^\dagger(y)]_-$ is not even a c-number, but something quadratic in creation and annihilation operators, so it certainly cannot vanish identically in the space-like region. However $[\phi(x), \phi^\dagger(y)]_+ \equiv \{\phi(x), \phi^\dagger(y)\}$ is a c-number, so it is at least possible for the anticommutator of fermionic fields to vanish in the space-like region. Therefore, recalling that the commutator of products of an even number of fields can always be rewritten as a sum of terms involving only anticommutators (for example, $[AB, CD]_- = A\{B, C\}D - AC\{B, D\} + \{A, C\}DB - C\{A, D\}B$), it follows that local commutativity of $\mathcal{H}_{int}$ can be assured simply by building it out of an even number of fermionic fields, and insisting on space-like *anti*commutativity for the elementary fermionic fields.

The general structure of a commutator (or anticommutator—as usual, upper signs refer to bosons, lower to fermions) of two covariant fields is

$$[\phi_{a_1b_1}^{A_1B_1}(x), \phi_{a_2b_2}^{A_2B_2\dagger}(y)]_\mp = \int \frac{d^3k_1 d^3k_2}{(2\pi)^3\sqrt{2E(k_1)2E(k_2)}} \sum_{\sigma_1\sigma_2} [u_{a_1b_1}^{A_1B_1}(\vec{k_1}, \sigma_1)e^{-ik_1\cdot x}a(\vec{k_1}, \sigma_1)$$

$$+ v_{a_1b_1}^{A_1B_1}(\vec{k_1}, \sigma_1)e^{ik_1\cdot x}a^{c\dagger}(\vec{k_1}, \sigma_1), u_{a_2b_2}^{A_2B_2}(\vec{k_2}, \sigma_2)^*e^{ik_2\cdot y}a^\dagger(\vec{k_2}, \sigma_2)$$

$$+ v_{a_2b_2}^{A_2B_2}(\vec{k_2}, \sigma_2)^*e^{-ik_2\cdot y}a^c(\vec{k_2}, \sigma_2)]_\mp$$

$$= \int \frac{d^3k}{(2\pi)^3 2E(k)} \{\sum_\sigma u_{a_1b_1}^{A_1B_1}(\vec{k}, \sigma)u_{a_2b_2}^{A_2B_2}(\vec{k}, \sigma)^*e^{-ik\cdot(x-y)}$$

$$\mp \sum_\sigma v_{a_1b_1}^{A_1B_1}(\vec{k}, \sigma)v_{a_2b_2}^{A_2B_2}(\vec{k}, \sigma)^*e^{ik\cdot(x-y)}\} \tag{7.72}$$

Define

$$N^{A_1 B_1 A_2 B_2}_{a_1 b_1 a_2 b_2}(k) \equiv \sum_\sigma u^{A_1 B_1}_{a_1 b_1}(\vec{k}, \sigma) u^{A_2 B_2}_{a_2 b_2}(\vec{k}, \sigma)^*$$

$$= \sum_{a_1' b_1' a_2' b_2'} (e^{-\theta \hat{k} \cdot \vec{A}})_{a_1 a_1'} (e^{\theta \hat{k} \cdot \vec{B}})_{b_1 b_1'} (e^{-\theta \hat{k} \cdot \vec{A}})^*_{a_2 a_2'} (e^{\theta \hat{k} \cdot \vec{B}})^*_{b_2 b_2'}$$

$$\cdot \sum_\sigma \langle A_1 B_1 a_1' b_1' | j\sigma \rangle \langle A_2 B_2 a_2' b_2' | j\sigma \rangle \qquad (7.73)$$

where $\cosh(\theta) = E(k)/m, \sinh(\theta) = |\vec{k}|/m, e^\theta = \frac{E+|\vec{k}|}{m} = \frac{k^0+|\vec{k}|}{m}$. As a function of k^μ, $N(k)$ has a definite parity under $k^\mu \to -k^\mu$. This is easiest to see by choosing $\hat{k}$ along the z axis. Then $\hat{k} \cdot \vec{A} = A_3$, and recalling that the small a (resp. b) indices refer to eigenvalues of A_3 (resp. B_3) we easily find in this case

$$N^{A_1 B_1 A_2 B_2}_{a_1 b_1 a_2 b_2}(k) = e^{\theta(b_1+b_2-a_1-a_2)} \sum_\sigma \langle A_1 B_1 a_1 b_1 | j\sigma \rangle \langle A_2 B_2 a_2 b_2 | j\sigma \rangle \qquad (7.74)$$

while

$$N^{A_1 B_1 A_2 B_2}_{a_1 b_1 a_2 b_2}(-k) = e^{\theta'(a_1+a_2-b_1-b_2)} \sum_\sigma \langle A_1 B_1 a_1 b_1 | j\sigma \rangle \langle A_2 B_2 a_2 b_2 | j\sigma \rangle \qquad (7.75)$$

where $e^{\theta'} = \frac{-k^0+|\vec{k}|}{m} = -e^{-\theta}$, and the interchange of as and bs results from the change $\hat{k} \to -\hat{k}$ in (7.73). We thus obtain the reflection property

$$N^{A_1 B_1 A_2 B_2}_{a_1 b_1 a_2 b_2}(-k) = (-1)^{(a_1+a_2-b_1-b_2)} N^{A_1 B_1 A_2 B_2}_{a_1 b_1 a_2 b_2}(k)$$

$$= (-1)^{(2\sigma-2b_1-2b_2)} N^{A_1 B_1 A_2 B_2}_{a_1 b_1 a_2 b_2}(k)$$

$$= (-1)^{(2j-2B_1-2B_2)} N^{A_1 B_1 A_2 B_2}_{a_1 b_1 a_2 b_2}(k) \qquad (7.76)$$

Since $v^{A_1 B_1}_{a_1 b_1}(\vec{k}, \sigma) = \xi^{A_1 B_1}(-1)^{j-\sigma} u^{A_1 B_1}_{a_1 b_1}(\vec{k}, -\sigma)$, we have

$$\sum_\sigma v^{A_1 B_1}_{a_1 b_1}(\vec{k}, \sigma) v^{A_2 B_2}_{a_2 b_2}(\vec{k}, \sigma)^* = \xi^{A_1 B_1} \xi^{A_2 B_2 *} \sum_\sigma u^{A_1 B_1}_{a_1 b_1}(\vec{k}, -\sigma) u^{A_2 B_2}_{a_2 b_2}(\vec{k}, -\sigma)^*$$

$$= \xi^{A_1 B_1} \xi^{A_2 B_2 *} N^{A_1 B_1 A_2 B_2}_{a_1 b_1 a_2 b_2}(k) \qquad (7.77)$$

Using (7.73) and (7.77) in (7.72), we obtain

$$[\phi^{A_1 B_1}_{a_1 b_1}(x), \phi^{A_2 B_2 \dagger}_{a_2 b_2}(y)]_\mp = \int \frac{d^3 k}{(2\pi)^3 2E(k)} N^{A_1 B_1 A_2 B_2}_{a_1 b_1 a_2 b_2}(k)$$

$$(e^{-ik \cdot (x-y)} \mp \xi^{A_1 B_1} \xi^{A_2 B_2 *} e^{+ik \cdot (x-y)}) \qquad (7.78)$$

We can now use the parity property (7.76) to write this result as

$$[\phi^{A_1 B_1}_{a_1 b_1}(x), \phi^{A_2 B_2 \dagger}_{a_2 b_2}(y)]_{\mp} = \int \frac{d^3 k}{(2\pi)^3 2E(k)} (N^{A_1 B_1 A_2 B_2}_{a_1 b_1 a_2 b_2}(k) e^{-ik\cdot(x-y)}$$

$$\mp (-1)^{2B_1 + 2B_2 + 2j} N^{A_1 B_1 A_2 B_2}_{a_1 b_1 a_2 b_2}(-k) \xi^{A_1 B_1} \xi^{A_2 B_2 *} e^{+ik\cdot(x-y)})$$

$$= N^{A_1 B_1 A_2 B_2}_{a_1 b_1 a_2 b_2}(i\frac{\partial}{\partial x}) \mathcal{F}(x-y) \tag{7.79}$$

where

$$\mathcal{F}(x-y) \equiv \int \frac{d^3 k}{(2\pi)^3 2E(k)} (e^{-ik\cdot(x-y)} \mp (-1)^{2B_1} \xi^{A_1 B_1} (-1)^{2B_2} \xi^{A_2 B_2 *} (-1)^{2j} e^{+ik\cdot(x-y)}) \tag{7.80}$$

If we compare this result with (6.76) we see immediately that space-like commutativity (resp. anticommutativity) is assured if and only if

$$\pm(-1)^{2B_1} \xi^{A_1 B_1} (-1)^{2B_2} \xi^{A_2 B_2 *} (-1)^{2j} = 1 \tag{7.81}$$

Take first $(A_1, B_1) = (A_2, B_2)$ and recall that ξ^{AB} is a pure phase. Then we conclude

$$(-1)^{2j} = \pm 1 \tag{7.82}$$

This is the celebrated Pauli–Lüders *Spin-Statistics theorem*. Bosonic commutation relations, at least for free fields, require integral spin, fermionic anticommutation relations half-integral spin—a result completely incomprehensible within the framework of non-relativistic quantum theory. Now inserting $(-1)^{2j} = \pm 1$ into (7.80), we find that for any pair of covariant fields $(A_1, B_1), (A_2, B_2)$ appearing in the theory we must have

$$(-1)^{2B_1} \xi^{A_1 B_1} (-1)^{2B_2} \xi^{A_2 B_2 *} = 1 \tag{7.83}$$

which is solved by setting all $\xi^{AB} = (-1)^{2B}\xi$, with $\xi = e^{i\theta}$ a universal phase factor. In fact, an overall universal phase in all destruction parts is unobservable, so with no loss of generality we can set $\xi = 1$.[2] Referring back to (7.54), we thus obtain a completely general expression for a local covariant field of *any spin and Lorentz representation*:

$$\phi^{AB}_{ab}(x) = \int \frac{d^3 k}{(2\pi)^{3/2}\sqrt{2E(k)}} \sum_\sigma (u^{AB}_{ab}(\vec{k}, \sigma) e^{-ik\cdot x} a(\vec{k}, \sigma)$$

$$+ (-)^{2B}(-)^{j-\sigma} u^{AB}_{ab}(\vec{k}, -\sigma) e^{ik\cdot x} a^{c\dagger}(\vec{k}, \sigma)) \tag{7.84}$$

with $u^{AB}_{ab}(\vec{k}, \sigma)$ given by (7.69).

For a spinless particle, $j = 0$, the simplest covariant field satisfying (7.34) is clearly obtained by taking $A = B = 0$, whence the u coefficient function in (7.84) becomes unity (by (7.64) and (7.69)), and we recover the canonical scalar field (6.73) of Chapter 6, satisfying (automatically) the *Klein–Gordon* equation $(\Box + m^2)\phi(x) = 0$. The only other cases of real importance in the Standard Model of elementary particle physics

---

[2] In certain cases, e.g., the Majorana field discussed below, the choice $\xi = -1$ will be more convenient.

are for $j = \frac{1}{2}, 1$. These two important special cases are therefore given separate and detailed attention in the two following sections.

7.4 Local covariant fields for spin-$\frac{1}{2}$ (spinor fields)

From the constraint (7.34) we see immediately that the simplest covariant fields capable of describing spin-$\frac{1}{2}$ particles evidently transform according to the $(\frac{1}{2},0)$ and $(0,\frac{1}{2})$ representations of the HLG. Our simplest option is clearly to use a single two-component field $\chi_a(x), \alpha = 1, 2$ transforming, say, according to the $(\frac{1}{2},0)$ representation of the HLG (we shall see below that we could just as well use the $(0,\frac{1}{2})$ representation). In addition, we assume that our particle is massive: the special issues that arise for particles with zero mass and non-zero spin are deferred to Section 7.6. This approach leads to the Majorana field (Majorana, 1937), and as we shall see shortly, involves *self-conjugate* fermions, with the particle indistinguishable from the antiparticle. Consequently, such fields cannot be used to describe particles endowed with a non-vanishing conserved additive quantum number (such as electric charge). The only known electrically neutral elementary spin-$\frac{1}{2}$ particles are neutrinos, and it is indeed possible that neutrinos exist that are described by such a Majorana field. All other (charged) spin-$\frac{1}{2}$ particles are described by four component *Dirac* fields $\psi_a(x), \alpha = 1, 2, 3, 4$ transforming according to reducible representations $(\frac{1}{2},0) \oplus (0,\frac{1}{2})$ of the HLG. We will first discuss the construction and properties of the Majorana field,[3] and then proceed to the Dirac case.

7.4.1 The Majorana field

Choosing $A = \frac{1}{2}$, $B=0$, our general covariant field ϕ_{ab}^{AB} becomes a two-component field χ_a, with $\chi_1 = \phi_{\frac{1}{2}0}^{\frac{1}{2}0}, \chi_2 = \phi_{-\frac{1}{2}0}^{\frac{1}{2}0}$. The zero-momentum u-spinors are trivial Clebsch–Gordon coefficients

$$u(0, \frac{1}{2}) = \begin{pmatrix} 1 \\ 0 \end{pmatrix}$$

$$u(0, -\frac{1}{2}) = \begin{pmatrix} 0 \\ 1 \end{pmatrix}$$

With choice of phase $\xi^{\frac{1}{2}0} = -1$, the zero momentum v-spinor given by (7.65) is easily seen to be

$$v(0, \sigma) = i\sigma_2 u(0, \sigma) = C_s u(0, \sigma) \tag{7.85}$$

The HLG generators are $\vec{A} = \frac{1}{2}\vec{\sigma}, \vec{B} = 0, \vec{J} = \frac{1}{2}\vec{\sigma}, \vec{K} = \frac{i}{2}\vec{\sigma}$, with $\vec{\sigma}$ the conventional Pauli matrices. Thus the finite-momentum spinors are given (cf. (7.69)) by

$$u(\vec{k}, \sigma) = e^{-\theta \hat{k} \cdot \vec{\sigma}/2} u(0, \sigma)$$

$$v(\vec{k}, \sigma) = e^{-\theta \hat{k} \cdot \vec{\sigma}/2} C_s u(0, \sigma) \tag{7.86}$$

[3] Section 7.4.1 may be omitted on a first reading of the book.

where the rapidity of the boost is given by $\cosh \theta = \frac{E(k)}{m}$. Up to normalization, the unique covariant field of type $(\frac{1}{2},0)$ is given by (7.84). It is conventional for spinor fields of non-zero mass to include an additional $\sqrt{m}$ in the definition of the field. We thus obtain

$$\chi(x) = \int \frac{d^3k}{(2\pi)^{3/2}} \sqrt{\frac{m}{2E(k)}} \sum_\sigma (e^{-\theta \hat{k}\cdot\vec{\sigma}/2} u(0,\sigma) b(\vec{k},\sigma) e^{-ik\cdot x}$$

$$+ e^{-\theta \hat{k}\cdot\vec{\sigma}/2} C_s u(0,\sigma) b^\dagger(\vec{k},\sigma) e^{ik\cdot x}) \tag{7.87}$$

Note that the antiparticle creation operator $b^\dagger$ is just the hermitian adjoint of the particle annihilation operator: our field is self-conjugate. The spinor index (on χ and $u(0,\sigma)$) has been suppressed: the reader is invited to visualize the left- and right-hand sides of (7.87) as a column two-vector of field operators. Next we introduce the augmented Pauli matrices $\sigma_\mu, \mu = 0,1,2,3$ where $\sigma_0 \equiv 1$ and the $\sigma_i, i = 1,2,3$ are the usual Pauli matrices. Thus, $\sigma^\mu \partial_\mu = \partial_0 - \vec{\sigma}\cdot\vec{\nabla}$, and a straightforward calculation, using the conjugation properties (7.41), leads to the *Majorana equation*

$$i\sigma^\mu \partial_\mu \chi(x) = -mC_s \chi^* \tag{7.88}$$

Here, and throughout this section, we use the asterisk of complex conjugation to indicate both normal complex conjugation (of numbers) and hermitian conjugation of operators, while the $\dagger$ symbol is reserved for the combination of conjugation (both types) and transposition of 2-spinors or 2x2 matrices.

The free Hamiltonian H_0 for our particle is given by the usual expression in terms of creation and annihilation operators:

$$H_0 = \int d^3k \sum_\sigma E(k) b^\dagger(\vec{k},\sigma) b(\vec{k},\sigma) \tag{7.89}$$

It is not difficult to show that this operator can be written as a spatial integral of a free energy density

$$H_0 = \int d^3x : \chi^\dagger i\partial_0 \chi + \frac{m}{2}(\chi^\dagger C_s \chi^* + \chi^T C_s \chi) : \tag{7.90}$$

Note that the normal-ordering symbol :....: is defined for fermions by moving all creation operators to the left of all annihilation operators (as for bosons), but *with an extra minus sign for each transposition*. Use of the Majorana equation (7.88) allows us to rewrite this in the more usual form (see Problem 1):

$$H_0 = \int d^3x : \chi^\dagger i\vec{\sigma}\cdot\vec{\nabla}\chi + \frac{m}{2}(\chi^T C_s \chi - \chi^\dagger C_s \chi^*) : \tag{7.91}$$

Note that both the kinetic (spatial derivative) term and the mass term in this expression are hermitian operators (in particular, $\chi^\dagger C_s \chi^* = -(\chi^T C_s \chi)^\dagger$).

The mass term in (7.91) is actually the integral of a Lorentz scalar field. Recall that a general covariant field transforms under the HLG as

$$U(\Lambda)\phi_\alpha(x)U^\dagger(\Lambda) = M_{\alpha\beta}(\Lambda^{-1})\phi_\beta(\Lambda x) \tag{7.92}$$

In this case the representation matrix is (recall (7.36)) $M(\Lambda^{-1}) = e^{-\frac{1}{2}\vec{\xi}\cdot\vec{\sigma}+\frac{i}{2}\vec{\phi}\cdot\vec{\sigma}}$, which satisfies the quasi-orthogonality property

$$M^T(\Lambda^{-1})C_s M(\Lambda^{-1}) = C_s \tag{7.93}$$

Accordingly

$$
\begin{aligned}
U(\Lambda)\chi_\alpha(x)(C_s)_{\alpha\beta}\chi_\beta(x)U^\dagger(\Lambda) &= \chi_\alpha(\Lambda x)(M^T(\Lambda^{-1})C_s M(\Lambda^{-1}))_{\alpha\beta}\chi_\beta(\Lambda x)\\
&= \chi^T(\Lambda x)C_s\chi(\Lambda x) \tag{7.94}
\end{aligned}
$$

so the bilinear $S(x) = \chi^T(x)C_s\chi(x)$ is indeed a Lorentz scalar field. Such a bilinear can therefore be used to construct Lorentz scalar interaction densities of Yukawa form, coupling the Majorana fermion to a self-conjugate spinless scalar ϕ, for example:

$$\mathcal{H}_{\text{int}}(x) = \lambda_{\text{Yuk}}\phi(x)(\chi^T(x)C_s\chi(x) + \text{h.c.}) = \lambda_{\text{Yuk}}\phi(\chi^T C_s\chi - \chi^\dagger C_s\chi^*) \tag{7.95}$$

What if our $(\frac{1}{2},0)$ field is not self-conjugate? Then it is easy to see that the mass term in the Hamiltonian must be constructed as a bilinear in χ and χ^* (to reproduce the required number operators $b^\dagger b, b^{c\dagger}b^c$ in H_0), but then the resultant field is not a Lorentz scalar (specifically, the scalar property fails under boosts: see Problem 2). This still leaves the option of a massless spin-$\frac{1}{2}$ non-self-conjugate fermion transforming according to $(\frac{1}{2},0)$. Such fermions are called *left-handed Weyl fermions*:[4] we return to them in subsection 7.4.5, when we address the massless case and introduce the two-component Weyl field (Weyl, 1929). On the other hand, massive non-self-conjugate spin-$\frac{1}{2}$ particles are easily treated by using a reducible representation of the HLG containing both $(\frac{1}{2},0)$ and $(0,\frac{1}{2})$ components, as first introduced by Dirac (Dirac, 1928) in the late 1920s. We now turn to a study of such reducible fields.

7.4.2   The Dirac field

Another option for massive spin-$\frac{1}{2}$ particles—indeed, by far the most common in the Standard Model describing known elementary particle interactions up to the several hundred GeV scale—is to build a field transforming according to a reducible representation containing both $(\frac{1}{2},0)$ and $(0,\frac{1}{2})$ representations. The resultant reducible representation may be denoted $(\frac{1}{2},0) \oplus (0,\frac{1}{2})$ and is evidently four-dimensional, and

---

[4] Correspondingly, two-component massless fields transforming according to the $(0,\frac{1}{2})$ representation of HLG are *right-handed* Weyl fields.

we may conveniently display the corresponding field as a column 4-spinor ψ_{ab}^{AB} (generally called a "Dirac 4-spinor", or "bispinor"), as follows:

$$\psi = \begin{pmatrix} \psi_{\frac{1}{2}\ 0}^{\frac{1}{2}\ 0} \\ \psi_{-\frac{1}{2}0}^{\frac{1}{2}\ 0} \\ \psi_{0\ \ \frac{1}{2}}^{0\ \ \frac{1}{2}} \\ \psi_{0-\frac{1}{2}}^{0\ \ \frac{1}{2}} \end{pmatrix}$$

where the "A" (resp. "B") generators of HLG act only on the top (resp. bottom) two components of the 4-spinor:

$$\vec{A} = \begin{pmatrix} \frac{1}{2}\vec{\sigma} & 0 \\ 0 & 0 \end{pmatrix}, \quad \vec{B} = \begin{pmatrix} 0 & 0 \\ 0 & \frac{1}{2}\vec{\sigma} \end{pmatrix}$$

which then gives (recall (7.31)) the following 4x4 matrices for the generators of rotations and boosts:

$$\vec{J} = \begin{pmatrix} \frac{1}{2}\vec{\sigma} & 0 \\ 0 & \frac{1}{2}\vec{\sigma} \end{pmatrix}, \tag{7.96}$$

$$\vec{K} = \begin{pmatrix} \frac{i}{2}\vec{\sigma} & 0 \\ 0 & -\frac{i}{2}\vec{\sigma} \end{pmatrix} \tag{7.97}$$

Our next task is to construct the connection coefficients $u(\vec{k},\sigma), v(\vec{k},\sigma)$ which determine the canonical field operator for this representation. As usual, we start at zero momentum: recalling the trivial Clebsch–Gordon values $\langle \frac{1}{2}0\frac{1}{2}0|\frac{1}{2}\frac{1}{2}\rangle = 1$ etc., one finds

$$u(0,\frac{1}{2}) = \begin{pmatrix} \frac{1}{\sqrt{2}} \\ 0 \\ \frac{1}{\sqrt{2}} \\ 0 \end{pmatrix}$$

$$u(0,-\frac{1}{2}) = \begin{pmatrix} 0 \\ \frac{1}{\sqrt{2}} \\ 0 \\ \frac{1}{\sqrt{2}} \end{pmatrix}$$

where the additional $\sqrt{2}$ factors are conventional to normalize the 4-spinors. Recall (cf. (7.83)) that in general we have $v_{ab}(0,\sigma) = \xi(-1)^{2B}(-1)^{j-\sigma}u_{ab}(0,-\sigma)$. For Dirac spinors it is conventional to choose the arbitrary phase $\xi = -1$, giving

$$v(0, \tfrac{1}{2}) = \begin{pmatrix} 0 \\ -\tfrac{1}{\sqrt{2}} \\ 0 \\ \tfrac{1}{\sqrt{2}} \end{pmatrix}$$

$$v(0, -\tfrac{1}{2}) = \begin{pmatrix} \tfrac{1}{\sqrt{2}} \\ 0 \\ -\tfrac{1}{\sqrt{2}} \\ 0 \end{pmatrix}$$

The spinor coefficient functions at non-zero momentum are obtained from these by a boost:

$$u(\vec{k}, \sigma) = e^{-\theta \hat{k} \cdot \vec{A}} e^{\theta \hat{k} \cdot \vec{B}} u(0, \sigma)$$

$$= \begin{pmatrix} e^{-\theta \hat{k} \cdot \vec{\sigma}/2} & 0 \\ 0 & e^{\theta \hat{k} \cdot \vec{\sigma}/2} \end{pmatrix} u(0, \sigma) \tag{7.98}$$

Defining the matrix (4x4, but written in 2×2 blocks!)

$$\beta \equiv \begin{pmatrix} 0 & 1 \\ 1 & 0 \end{pmatrix}$$

one easily verifies that $\beta u(0, \sigma) = u(0, \sigma)$, whence $B(\vec{k}) u(\vec{k}, \sigma) = u(\vec{k}, \sigma)$ where

$$B(\vec{k}) \equiv \begin{pmatrix} 0 & e^{-\theta \hat{k} \cdot \vec{\sigma}} \\ e^{\theta \hat{k} \cdot \vec{\sigma}} & 0 \end{pmatrix}$$

Expanding the exponential, one finds

$$e^{\theta \hat{k} \cdot \vec{\sigma}} = \frac{E(k)}{m} + \frac{\vec{k} \cdot \vec{\sigma}}{m} \tag{7.99}$$

We may now establish contact with the conventional Dirac formalism by introducing the *Dirac matrices*:

$$\gamma_0 = \beta = \begin{pmatrix} 0 & 1 \\ 1 & 0 \end{pmatrix}$$

$$\gamma_i = \begin{pmatrix} 0 & -\sigma_i \\ \sigma_i & 0 \end{pmatrix} \tag{7.100}$$

in terms of which $B(\vec{k}) = k^\mu \gamma_\mu / m$ and the u-spinor then satisfies

$$(k^\mu \gamma_\mu - m) u(\vec{k}, \sigma) = (\slashed{k} - m) u(\vec{k}, \sigma) = 0 \tag{7.101}$$

(Note: the abbreviation $\slashed{k} \equiv k^\mu \gamma_\mu$ is ubiquitous in Dirac theory.) Using $\beta v(0, \sigma) = -v(0, \sigma)$ an exactly analogous argument reveals that the v-spinors satisfy

$$(\not{k} + m)v(\vec{k}, \sigma) = 0 \tag{7.102}$$

If we now recall the general expression for a covariant field, we find that the *Dirac field* $\psi(x)$, given by

$$\psi(x) = \int \frac{d^3 k}{(2\pi)^{3/2}} \sqrt{\frac{m}{E(k)}} \sum_\sigma (u(\vec{k}, \sigma)e^{-ik\cdot x}b(\vec{k}, \sigma) + v(\vec{k}, \sigma)e^{ik\cdot x}d^\dagger(\vec{k}, \sigma)) \tag{7.103}$$

satisfies the famous Dirac equation

$$(i\gamma^\mu \frac{\partial}{\partial x^\mu} - m)\psi(x) = 0 \tag{7.104}$$

in virtue of (7.101–7.102). The normalization factor of $\sqrt{\frac{m}{E(k)}}$ instead of $\sqrt{\frac{1}{2E(k)}}$ is conventional for *massive* Dirac fields (for massless fields, one customarily returns to the previous normalization, for obvious reasons: see subsection 7.4.5 below), as is the notation b, d (instead of a, a^c) for the particle and antiparticle destruction operators.

7.4.3 Diracology

The Dirac matrices introduced in (7.100) are readily seen to *anticommute* with each other; together with the fact that $\gamma_0^2 = 1, \gamma_i^2 = -1$, we conclude that

$$\{\gamma_\mu, \gamma_\nu\} = 2g_{\mu\nu} \tag{7.105}$$

where the 4x4 identity matrix is understood on the right-hand side of (7.105), multiplying the metric element $2g_{\mu\nu}$. The antiparticle coefficient functions $v(\vec{k}, \sigma)$ are conventionally defined as

$$v(\vec{k}, \sigma) = \gamma_5(-1)^{\frac{1}{2}+\sigma}u(\vec{k}, -\sigma) \tag{7.106}$$

where the matrix

$$\gamma_5 = \begin{pmatrix} 1 & 0 \\ 0 & -1 \end{pmatrix} \tag{7.107}$$

incorporates the $(-)^{2B}$ factor in (7.84). In fact, $\gamma_5 = -i\gamma^0\gamma^1\gamma^2\gamma^3$ and anticommutes with each of the γ^μ:

$$\{\gamma_5, \gamma^\mu\} = 0 \tag{7.108}$$

(NB: spacetime indices are raised and lowered on the γ-matrices in the usual way: the spatial ones get a minus sign).

In almost any calculation involving Dirac fields, we encounter *spin sums* analogous to (7.73):

$$N(\vec{k})_{nm} \equiv \sum_\sigma u_n(\vec{k}, \sigma)u_m^*(\vec{k}, \sigma) \tag{7.109}$$

At zero momentum, from the explicit result for the zero-momentum spinors derived above we have

$$N(0)_{nm} \equiv \sum_{\sigma} u_n(0, \sigma) u_m^*(0, \sigma) = \frac{1}{2}(1 + \gamma_0)_{nm} \tag{7.110}$$

Applying the boost matrix in (7.98) on the left and the right of $N(0)$, we find

$$N(\vec{k}) = \frac{1}{2} B(\vec{k}) \gamma_0 + \frac{1}{2} \gamma_0 = \frac{1}{2}(\frac{\not{k}}{m} + 1) \gamma_0 \tag{7.111}$$

Another very useful concept in Dirac theory is the "Dirac adjoint" of a 4-spinor. Namely:

$$\bar{u}_n \equiv u_m^* \gamma_{mn}^0$$
$$\bar{\psi}_n \equiv \psi_m^\dagger \gamma_{mn}^0 \tag{7.112}$$

etc. The spin sum (7.109) usually is needed in the equivalent form

$$\sum_{\sigma} u_n(\vec{k}, \sigma) \bar{u}_m(\vec{k}, \sigma) = \frac{1}{2}(\frac{\not{k}}{m} + 1)_{nm} \tag{7.113}$$

which follows from (7.111) as $\gamma_0^2 = 1$. For the antiparticle spinors, the result is similar, with an obvious change of sign:

$$\sum_{\sigma} v_n(\vec{k}, \sigma) \bar{v}_m(\vec{k}, \sigma) = \frac{1}{2}(\frac{\not{k}}{m} - 1)_{nm} \tag{7.114}$$

Finally, there are the overlap normalization properties such as

$$\bar{v}(\vec{k}, \sigma) \gamma_0 u(-\vec{k}, \sigma') = v^\dagger(\vec{k}, \sigma) u(-\vec{k}, \sigma') \tag{7.115}$$
$$= v^\dagger(0, \sigma) u(0, \sigma') = 0 \tag{7.116}$$

where the second line follows from the boost equation (7.98):

$$u(-\vec{k}, \sigma') = \begin{pmatrix} e^{\theta \hat{k} \cdot \vec{\sigma}/2} & 0 \\ 0 & e^{-\theta \hat{k} \cdot \vec{\sigma}/2} \end{pmatrix} u(0, \sigma') \tag{7.117}$$

$$v(\vec{k}, \sigma) = \begin{pmatrix} e^{-\theta \hat{k} \cdot \vec{\sigma}/2} & 0 \\ 0 & e^{\theta \hat{k} \cdot \vec{\sigma}/2} \end{pmatrix} v(0, \sigma) \tag{7.118}$$

Similarly, one easily verifies

$$\bar{u}(\vec{k}, \sigma) \gamma_0 u(\vec{k}, \sigma') = \frac{E(k)}{m} \delta_{\sigma \sigma'} = \bar{v}(\vec{k}, \sigma) \gamma_0 v(\vec{k}, \sigma') \tag{7.119}$$

With the help of (7.115–7.119) one may easily establish the field-theoretic formula for the free Dirac Hamiltonian:

$$H_0 = \int d^3k E(k) \sum_\sigma (b^\dagger(\vec{k},\sigma)b(\vec{k},\sigma) + d^\dagger(\vec{k},\sigma)d(\vec{k},\sigma)) \tag{7.120}$$

or, in terms of the Dirac field,

$$H_0 = \int d^3x : \bar{\psi}(x)(i\vec{\gamma}\cdot\vec{\nabla} + m)\psi(x) : \tag{7.121}$$

Similarly, one finds the following expression for the charge operator (for a Dirac particle of charge e; see Problem 4):

$$Q = \int d^3k \sum_\sigma (eb^\dagger(\vec{k},\sigma)b(\vec{k},\sigma) - ed^\dagger(\vec{k},\sigma)d(\vec{k},\sigma)) \tag{7.122}$$

$$= \int d^3x e : \bar{\psi}(x)\gamma^0\psi(x) : \tag{7.123}$$

The four-component bispinor notation has achieved such predominance in dealing with spin-$\frac{1}{2}$ fields that it is even commonly used for Majorana fields, which, as we saw in the preceding section, are strictly speaking two-component objects. However, we can continue to use the extremely convenient (and familiar!) apparatus of the Dirac algebra even in the Majorana case by defining a four-component field[5]

$$\psi_M = \begin{pmatrix} \chi \\ -C_s\chi^* \end{pmatrix}$$

with $\chi(x)$ the two-component field of Section 7.4.1, satisfying the Majorana equation (7.88) and its conjugate:[6]

$$i(\partial_0 - \vec{\sigma}\cdot\vec{\nabla})\chi = -mC_s\chi^* \tag{7.124}$$

$$-i(\partial_0 - \vec{\sigma}^*\cdot\vec{\nabla})\chi^* = -mC_s\chi \tag{7.125}$$

whence one easily verifies that the four-component field ψ_M satisfies the *Dirac* equation (7.104):

$$(i\gamma^\mu \frac{\partial}{\partial x^\mu} - m)\psi_M(x) = 0 \tag{7.126}$$

---

[5] Alternatively, one can take the $(0,\frac{1}{2})$ field as the starting point and write

$$\psi_M = \begin{pmatrix} C_s\chi^* \\ \chi \end{pmatrix}$$

[6] We temporarily return to the convention of Section 7.4.1, where an asterisk is used to represent both complex conjugation and hermitian conjugation.

Likewise, the free Hamiltonian (7.91) for the Majorana field takes exactly the same form as the Dirac Hamiltonian (7.121), with an extra factor of $\frac{1}{2}$ to compensate for the doubling of the Majorana field in the four-component notation:

$$H_0 = \frac{1}{2} \int d^3x : \bar{\psi}_M(x)(i\vec{\gamma} \cdot \vec{\nabla} + m)\psi_M(x) : \tag{7.127}$$

Indeed, one can think of the Dirac field as a 4-spinor composed of two *independent* Majorana fields

$$\psi_M = \begin{pmatrix} \chi_1 \\ -C_s\chi_2^* \end{pmatrix}$$

—a point of view which becomes extremely useful in supersymmetric theories (cf. Section 12.6), where the Majorana field serves as a basic "unit" for constructing spin-$\frac{1}{2}$ fields of all types.

7.4.4   Lorentz transformation properties

The Lorentz group is a non-compact group, with three hermitian generators (the $\vec{J}$ rotation generators) and three *anti-hermitian* generators (the $\vec{K}$ boost generators). From the explicit expressions for $\vec{J}, \vec{K}$ in the Dirac case (see Eqs. (7.96,7.97)), we see that the γ_0 matrix can be used to effect a conjugation:

$$\vec{J}^\dagger = \vec{J}$$
$$= \gamma_0 \vec{J} \gamma_0 \tag{7.128}$$
$$\vec{K}^\dagger = -\vec{K}$$
$$= \gamma_0 \vec{K} \gamma_0 \tag{7.129}$$

Recall that the matrix representing an element of HLG can be expressed as an exponential:

$$M(\Lambda) = e^{-i(\Omega_{10}K_1 + \Omega_{20}K_2 + \Omega_{30}K_3 + \Omega_{12}J_3 + \Omega_{23}J_1 + \Omega_{31}J_2)} \tag{7.130}$$

Using (7.128–7.129), it therefore follows that

$$M^\dagger(\Lambda) = \gamma_0 M^{-1}(\Lambda)\gamma_0 \tag{7.131}$$

Next, note that the matrix element $N(\vec{k})_{nm} = \sum_\sigma u_n(\vec{k},\sigma)u_m^*(\vec{k},\sigma)$ of the spin-sum matrix introduced above is a spinor dot-product of u_n and u_m, and therefore invariant under simultaneous rotations with the rotation matrix $D_{\sigma\sigma'}^j$. From (7.45), one finds

$$\sum_{\sigma,m_1 m_2} M_{n_1 m_1}(\Lambda^{-1}) M^*_{n_2 m_2}(\Lambda^{-1}) u_{m_1}(\Lambda \vec{k}, \sigma) u^*_{m_2}(\Lambda \vec{k}, \sigma)$$

$$= (M(\Lambda^{-1}) N(\Lambda \vec{k}) M^{\dagger}(\Lambda^{-1}))_{n_1 n_2}$$

$$= \sum_{\sigma} u_{n_1}(\vec{k}, \sigma) u^*_{n_2}(\vec{k}, \sigma) = N(\vec{k})_{n_1 n_2} \tag{7.132}$$

Using the conjugation property (7.131) in (7.132), we obtain

$$M(\Lambda^{-1}) N(\Lambda \vec{k}) \gamma_0 M(\Lambda) \gamma_0 = N(\vec{k}) \tag{7.133}$$

Inserting the explicit expression (7.111),

$$M(\Lambda^{-1})(\Lambda^{\mu}{}_{\nu} k^{\nu} \gamma_{\mu} + m) \gamma_0 \gamma_0 M(\Lambda) \gamma_0 = (\slashed{k} + m) \gamma_0 \tag{7.134}$$

from which

$$\Lambda^{\mu}{}_{\nu} k^{\nu} \gamma_{\mu} = M(\Lambda) \slashed{k} M^{-1}(\Lambda) \tag{7.135}$$

Differentiating (7.135) with respect to k^{ν}, we finally obtain the desired transformation property of the Dirac γ matrices under finite HLG transformations:

$$M(\Lambda) \gamma_{\nu} M^{-1}(\Lambda) = \Lambda^{\mu}{}_{\nu} \gamma_{\mu} \tag{7.136}$$

$$M(\Lambda) \gamma^{\nu} M^{-1}(\Lambda) = \Lambda_{\mu}{}^{\nu} \gamma^{\mu} \tag{7.137}$$

As expected, the four γ matrices transform as a four-vector under the HLG. The matrix γ_5 introduced in (7.107) transforms as a *pseudoscalar*:

$$\gamma_5 \equiv -i\gamma^0 \gamma^1 \gamma^2 \gamma^3 = -\frac{i}{4!} \epsilon_{\mu\nu\rho\sigma} \gamma^{\mu} \gamma^{\nu} \gamma^{\rho} \gamma^{\sigma}$$

$$M(\Lambda) \gamma_5 M^{-1}(\Lambda) = -\frac{i}{4!} \epsilon_{\mu\nu\rho\sigma} \Lambda_{\mu'}{}^{\mu} \Lambda_{\nu'}{}^{\nu} \Lambda_{\rho'}{}^{\rho} \Lambda_{\sigma'}{}^{\sigma} \gamma^{\mu'} \gamma^{\nu'} \gamma^{\rho'} \gamma^{\sigma'}$$

$$= -\frac{i}{4!} \epsilon_{\mu'\nu'\rho'\sigma'} \det(\Lambda) \gamma^{\mu'} \gamma^{\nu'} \gamma^{\rho'} \gamma^{\sigma'}$$

$$= \det(\Lambda) \gamma_5 \tag{7.138}$$

From the general transformation property of a covariant field (7.5)

$$U(\Lambda) \psi_n(x) U^{\dagger}(\Lambda) = M_{nm}(\Lambda^{-1}) \psi_m(\Lambda x) \tag{7.139}$$

follows the corresponding transformation for the adjoint field

$$U(\Lambda) \psi^{\dagger}_n(x) U^{\dagger}(\Lambda) = M^*_{nm}(\Lambda^{-1}) \psi^{\dagger}_m(\Lambda x) \tag{7.140}$$

Adding a γ_0 to convert the regular adjoint to the Dirac adjoint:

$$
\begin{aligned}
U(\Lambda)\bar{\psi}_n(x)U^\dagger(\Lambda) &= M^*_{mm'}(\Lambda^{-1})\psi^\dagger_{m'}(\Lambda x)(\gamma_0)_{mn} \\
&= (M^\dagger(\Lambda^{-1})\gamma_0)_{m'n}\psi^\dagger_{m'}(\Lambda x) \\
&= (\gamma_0 M(\Lambda))_{m'n}\psi^\dagger_{m'}(\Lambda x) \\
&= M_{mn}(\Lambda)\bar{\psi}_m(\Lambda x)
\end{aligned}
\tag{7.141}
$$

where we have used the conjugation property (7.131) between the second and third lines. From (7.139) and (7.141) follow directly

$$
U(\Lambda)\bar{\psi}_n(x)\psi_n(x)U^\dagger(\Lambda) = \bar{\psi}_n(\Lambda x)\psi_n(\Lambda x) \equiv \bar{\psi}(\Lambda x)\psi(\Lambda x)
\tag{7.142}
$$

Accordingly, the field $S(x) \equiv \bar{\psi}(x)\psi(x)$ is a *scalar field*. On the other hand, using (7.138),

$$
U(\Lambda)\bar{\psi}(x)\gamma_5\psi(x)U^\dagger(\Lambda) = \det(\Lambda)\bar{\psi}(\Lambda x)\gamma_5\psi(\Lambda x)
\tag{7.143}
$$

we see that there is an extra minus sign for transformations including a spatial reflection, so that $P(x) \equiv \bar{\psi}\gamma_5\psi$ is a *pseudoscalar field*. Not surprisingly, we can construct a vector field employing the Dirac matrices in the obvious way:

$$
\begin{aligned}
U(\Lambda)\bar{\psi}_{n'}(x)(\gamma_\mu)_{n'n}\psi_n(x)U^\dagger(\Lambda) &= M_{m'n'}(\Lambda)(\gamma_\mu)_{n'n}M_{nm}(\Lambda^{-1})\bar{\psi}_{m'}(\Lambda x)\psi_m(\Lambda x) \\
&= \Lambda^\rho{}_\mu\bar{\psi}_{m'}(\Lambda x)(\gamma_\rho)_{m'm}\psi_m(\Lambda x)
\end{aligned}
\tag{7.144}
$$

Omitting the matrix indices for clarity, this is

$$
U(\Lambda)\bar{\psi}(x)\gamma_\mu\psi(x)U^\dagger(\Lambda) = \Lambda^\rho{}_\mu\bar{\psi}(\Lambda x)\gamma_\rho\psi(\Lambda x)
\tag{7.145}
$$

Thus, $V_\mu(x) \equiv \bar{\psi}\gamma_\mu\psi(x)$ is indeed a vector field. The corresponding result for the four-vector field $A_\mu(x) \equiv \bar{\psi}\gamma_5\gamma_\mu\psi$ contains an extra $\det(\Lambda)$ factor (which would be -1 if the transformation Λ is improper, containing a spatial reflection), so we conclude that A_μ is an *axial vector field*.

As the Dirac field has four components, it is apparent that there must be sixteen independent bilinears constructible as linear combinations of $\bar{\psi}_n(x)\psi_m(x)$. So far, the S, P, V_μ, and A_μ fields provide us with $1+1+4+4=10$ independent operators. The six remaining independent operators form a second-rank antisymmetric Lorentz tensor

$$
T_{\mu\nu}(x) \equiv \bar{\psi}(x)\frac{i}{4}[\gamma_\mu, \gamma_\nu]\psi(x)
\tag{7.146}
$$

Of course, fields with non-trivial Lorentz transformation properties, such as V_μ, A_μ, and $T_{\mu\nu}$ can still be contracted in the standard way to produce Lorentz scalar fields suitable for use in an interaction Hamiltonian: e.g., $V_\mu V^\mu, T_{\mu\nu}T^{\mu\nu}$ etc. Combinations like $V_\mu A^\mu$ are *pseudoscalar* fields and will lead to parity violation if included in the interaction Hamiltonian density. Precisely this occurs in the effective Fermi theory of the weak interactions, as we shall see. Another important type of interaction is the *Yukawa* interaction between a scalar field $\phi(x)$(or pseudoscalar $\pi(x)$) and a Dirac field $\psi(x)$:

$$\mathcal{H}_{\text{int}} = \lambda \bar{\psi}(x)\psi(x)\phi(x) \tag{7.147}$$

$$\mathcal{H}_{\text{int}} = \lambda \bar{\psi}(x)\gamma_5\psi(x)\pi(x) \tag{7.148}$$

Note that both (7.147) and (7.148) are *parity conserving* interactions (as $\mathcal{H}_{\text{int}}$ is even under parity in both cases)!

7.4.5   Massless spin-$\frac{1}{2}$ fields: the Weyl field

For many years the neutrinos associated with the weak interactions were believed to be massless spin-$\frac{1}{2}$ fermions. The situation changed with the discovery of neutrino oscillations, beginning with the solar neutrino deficit experiment of Davis in the 1970s and 1980s. Although oscillation experiments typically measure mass differences between neutrino species, and the possibility still exists that one of the neutrino species may be exactly massless, there is no natural reason for this to be so, and the working assumption in the field is that all three generations of neutrino are in fact massive, albeit with very small (in some cases, considerably smaller than an electron-volt) masses. Nevertheless, for much of weak interaction phenomenology, the effects of the neutrino masses are completely negligible, so it is quite useful to consider the massless limit for a non-self-conjugate Dirac field (we know that neutrinos and antineutrinos are distinct particles). If we return to the Dirac four-component field of (7.103) we see right away that we had better transfer a factor of $\sqrt{2m}$ into the normalization of the u and v spinors before taking the massless limit. Accordingly, instead of (7.113) and (7.114), our new spinors are normalized by

$$\sum_\sigma u_n(\vec{k},\sigma)\bar{u}_m(\vec{k},\sigma) = (\not{k}+m)_{nm} \rightarrow \not{k}_{nm}, \; m=0$$

$$\sum_\sigma v_n(\vec{k},\sigma)\bar{v}_m(\vec{k},\sigma) = (\not{k}-m)_{nm} \rightarrow \not{k}_{nm}, \; m=0 \tag{7.149}$$

while the Dirac field now takes the form (with $E(k) = |\vec{k}|$)

$$\psi(x) = \int \frac{d^3k}{(2\pi)^{3/2}\sqrt{2E(k)}} \sum_\sigma (u(\vec{k},\sigma)e^{-ik\cdot x}b(\vec{k},\sigma) + v(\vec{k},\sigma)e^{ik\cdot x}d^\dagger(\vec{k},\sigma)) \tag{7.150}$$

There is one further subtlety of which we must be cognizant in taking the massless limit of such a field. The discussion of massless particle states in Section 5.3 made it clear that such states always appear with a definite value of helicity—i.e., of their angular momentum resolved along the direction of motion of the particle—rather than along an arbitrary z-axis, as in the spin states employed in the discussion of the massive Dirac field above. The relation between spin states $|\vec{k},\sigma\rangle$ and helicity states $|\vec{k},\lambda\rangle$ is very simple (recall Problem 2(a) of Chapter 5):

$$|\vec{k}, \lambda\rangle = \sum_\sigma D^{\frac{1}{2}}_{\sigma\lambda}(R(\hat{k}))|\vec{k}, \sigma\rangle$$

where $R(\hat{k})$ is the rotation from the z-axis into the direction of momentum $\hat{k}$ (with a similar relation for the antiparticle states). There is a corresponding relation between the creation operators

$$b^\dagger(\vec{k}, \lambda) = \sum_\sigma D^{\frac{1}{2}}_{\sigma\lambda}(R(\hat{k}))b^\dagger(\vec{k}, \sigma)$$

$$d^\dagger(\vec{k}, \lambda) = \sum_\sigma D^{\frac{1}{2}}_{\sigma\lambda}(R(\hat{k}))d^\dagger(\vec{k}, \sigma)$$

So, if we define helicity spinors

$$u(\vec{k}, \lambda) = \sum_\sigma D^{\frac{1}{2}}_{\sigma\lambda}(R(\hat{k}))u(\vec{k}, \sigma) \tag{7.151}$$

$$v(\vec{k}, \lambda) = \sum_\sigma D^{\frac{1}{2}*}_{\sigma\lambda}(R(\hat{k}))v(\vec{k}, \sigma) \tag{7.152}$$

the Dirac field (7.150) can be rewritten entirely in terms of helicity spinors and creation–annihilation operators:

$$\psi(x) = \int \frac{d^3k}{(2\pi)^{3/2}} \sqrt{\frac{1}{2E(k)}} \sum_\lambda (u(\vec{k}, \lambda)e^{-ik\cdot x}b(\vec{k}, \lambda) + v(\vec{k}, \lambda)e^{ik\cdot x}d^\dagger(\vec{k}, \lambda)) \tag{7.153}$$

Recall that the Dirac field transforms according to the reducible representation $(\frac{1}{2},0)\oplus(0,\frac{1}{2})$, with the upper (resp. lower) two components transforming as $(\frac{1}{2},0)$ (resp. $(0,\frac{1}{2})$). If we label the upper two components ψ_L and the lower two ψ_R, then it is easy to see that in the limit $m \to 0$ the Dirac equation (7.104) decouples into separate equations for these two *chiral* fields:[7]

$$\sigma^\mu \partial_\mu \psi_L(x) = 0 \tag{7.154}$$

$$\bar{\sigma}^\mu \partial_\mu \psi_R(x) = 0 \tag{7.155}$$

where $\sigma_\mu = (1, \sigma_i), \bar{\sigma}_\mu = (1, -\sigma_i)$. These are called the "left-handed" and "right-handed" *Weyl equations*,[8] respectively (Weyl, 1929). The first equation (for the $(\frac{1}{2},0)$ field) coincides in form, not surprisingly, with the previously discussed Majorana equation (7.88). Here, however, our field is not assumed to be self-conjugate: our particle is allowed to carry a non-zero additive conserved quantum number, for example, opposite in sign to that of the antiparticle (e.g., lepton number in the

---

[7] Recall from the discussion in Section 7.2 that half-integral spin fields must be described by an (A, B) representation with $A \neq B$. In the spin-$\frac{1}{2}$ case, the chiral character (left- or right-handed) of the Weyl field is directly correlated with the eigenvalue of γ_5; cf. (7.107).

[8] These equations, although evidently Lorentz-covariant, were rejected by Pauli in his famous Handbuch article, (Pauli, 1933), p. 226, for violating invariance under spatial reflections (parity)—a shortcoming which was later transformed into a virtue when parity non-conservation was discovered in the 1950s.

original form of weak interaction theory, with massless neutrinos). The change in sign of the spin matrices between the equations for ψ_L and ψ_R suggests that they describe (massless) particles of opposite helicity. We shall now demonstrate this explicitly.

Return for a moment to the massive case, in the original spin representation. We shall concentrate on the upper two components, so there will be a "L" subscript everywhere. From (7.98), the finite momentum spinor is obtained by a boost from the zero-momentum state:

$$u_L(\vec{k}, \sigma) = \sqrt{2m}\, e^{-\theta \hat{k} \cdot \vec{\sigma}/2} u_L(0, \sigma) \tag{7.156}$$

with $\cosh \theta = E(k)/m$, whence

$$e^{-\theta \hat{k} \cdot \vec{\sigma}/2} = \cosh\frac{\theta}{2} - \sinh\frac{\theta}{2}\hat{k} \cdot \vec{\sigma} = \frac{1}{\sqrt{2}}(\sqrt{\frac{E(k)}{m} + 1} - \sqrt{\frac{E(k)}{m} - 1}\,\hat{k} \cdot \vec{\sigma}) \tag{7.157}$$

Inserting this result in (7.156) we obtain

$$u_L(\vec{k}, \sigma) = (\sqrt{(E(k) + m)} - \sqrt{(E(k) - m)}\hat{k} \cdot \vec{\sigma})u_L(0, \sigma) \tag{7.158}$$

where the zero-momentum spinor in spin representation is given as discussed previously in terms of simple Clebsch–Gordon coefficients, namely $u_L(0, \sigma)_n = \frac{1}{\sqrt{2}}\delta_{n\sigma}$, if we interpret the Kronecker δ as giving one if $n = 1, \sigma = +\frac{1}{2}$ and $n = 2, \sigma = -\frac{1}{2}$. The corresponding result for the helicity representation spinor follows from (7.151):

$$u_L(\vec{k}, \lambda)_n = \sum_\sigma D^{\frac{1}{2}}_{\sigma\lambda}(R(\hat{k}))u_L(\vec{k}, \sigma)_n$$

$$= \frac{1}{\sqrt{2}}\sum_\sigma D^{\frac{1}{2}}_{\sigma\lambda}(R(\hat{k}))(\sqrt{(E(k) + m)} - \sqrt{(E(k) - m)}\hat{k} \cdot \vec{\sigma})_{n\sigma}$$

$$= \frac{1}{\sqrt{2}}\{(\sqrt{(E(k) + m)} - \sqrt{(E(k) - m)}\hat{k} \cdot \vec{\sigma})D^{\frac{1}{2}}_{\sigma\lambda}(R(\hat{k}))\}_{n\lambda}$$

whence, by commuting the rotation matrix through the helicity matrix $\hat{k} \cdot \vec{\sigma}$,

$$u_L(\vec{k}, \lambda)_n = \frac{1}{\sqrt{2}}\{D^{\frac{1}{2}}_{\sigma\lambda}(R(\hat{k}))(\sqrt{(E(k) + m)} - \sqrt{(E(k) - m)}\sigma_3)\}_{n\lambda} \tag{7.159}$$

again with the understanding that spin or helicity values $+\frac{1}{2}$ (resp. $-\frac{1}{2}$) need to be translated appropriately into row or column indices 1 (resp. 2) when evaluating matrix elements of the rotation or spin matrices. Here we have used the definition of $R(\hat{k})$ as the rotation taking the z-axis into the $\hat{k}$ direction, whence

$$D^{\frac{1}{2}}(R(\hat{k}))^{-1}\hat{k} \cdot \vec{\sigma}D^{\frac{1}{2}}(R(\hat{k})) = \sigma_3 \tag{7.160}$$

In particular, the positive helicity $\lambda = +\frac{1}{2}$ spinor is given by

$$u_L(\vec{k}, +\frac{1}{2})_n = \frac{1}{\sqrt{2}}D^{\frac{1}{2}}_{n1}(R(\hat{k}))(\sqrt{E(k) + m} - \sqrt{E(k) - m}) \tag{7.161}$$

and clearly *vanishes* in the zero-mass (or high-energy) limit. On the other hand, the negative helicity (or "left-handed") spinor survives in this limit (whence our choice of the subscript "L" in labeling the upper two components of the Dirac 4-spinor):

$$u_L(\vec{k}, -\tfrac{1}{2})_n = \frac{1}{\sqrt{2}} D_{n2}^{\frac{1}{2}}(R(\hat{k}))(\sqrt{E(k) + m} + \sqrt{E(k) - m})$$

$$\rightarrow \sqrt{2E(k)} D_{n2}^{\frac{1}{2}}(R(\hat{k})), \quad m \to 0 \tag{7.162}$$

In other words, the *chirality* of the field (defined as the eigenvalue of γ_5) and the *helicity* of the particle (the eigenvalue of $\hat{k} \cdot \vec{\sigma}$) have become linked in the high-energy, or zero-mass limit.

The helicity interpretation of these spinors is again a consequence of (7.160), which implies that the first column of the 2x2 rotation matrix $D^{\frac{1}{2}}(R(\hat{k}))$ is an eigenvector of the helicity operator $\hat{k} \cdot \vec{\sigma}$ with eigenvalue $+1$, while the second column appearing in (7.162) is an eigenvector with eigenvalue -1. A similar examination of the properties of the v spinors associated with the antiparticle states shows that in the massless limit only the *positive* helicity state survives. Thus our left-handed Weyl field ψ_L describes a massless particle with negative helicity, but whose antiparticle is necessarily positive helicity. The right-handed Weyl field ψ_R correspondingly describes a positive helicity particle, paired with a negative helicity antiparticle. As we shall see later, only the ψ_L-type neutrino fields participate in weak interactions in the Standard Model of elementary particle interactions, even though the particles themselves now appear to have non-zero mass—suggesting that we are dealing with Dirac fermions after all!

7.5   Local covariant fields for spin-1 (vector fields)

Next, let us construct a suitable covariant field for a particle of spin $j = 1$. As for spin-$\frac{1}{2}$, we begin with the massive case, and return later to the special features that apply for an exactly massless spin-1 particle. The simplest *symmetric* choice of HLG representation (A, B), with $A = B$, compatible with $j = 1$, is to take $A = B = \frac{1}{2}$. This four-dimensional representation turns out to be none other (see Problem 6) than the defining fundamental representation of the Lorentz group, with representation vectors labeled by a spacetime index μ. We shall use the notation $W^\mu(x)$ for a massive vector field. The condition (7.48) then amounts to

$$\sum_\sigma D_{\sigma\sigma'}^{(1)}(R) u^\mu(\vec{0}, \sigma) = R^\mu_{\ \nu} u^\nu(\vec{0}, \sigma') \tag{7.163}$$

where $M(\Lambda)^\mu_{\ \nu} = \Lambda^\mu_{\ \nu}$ is simply $R^\mu_{\ \nu}$ for a pure rotation R. For $\mu = 0$ this reduces to

$$\sum_\sigma D_{\sigma\sigma'}^{(1)}(R) u_0(0, \sigma) = u_0(0, \sigma') \tag{7.164}$$

which implies that $u_0(0, \sigma) = 0$. As (7.50) implies that the coefficient functions $u^\mu(\vec{k}, \sigma)$ at non-zero momentum are obtained from those at zero momentum by a boost

$$u^\mu(\vec{k}, \sigma) = L^\mu_{\ \nu}(\vec{k}) u^\nu(\vec{0}, \sigma) \tag{7.165}$$

we must have the transversality condition

$$k_\mu u^\mu(\vec{k}, \sigma) = 0 \tag{7.166}$$

as the covariant dot-product vanishes in the rest frame. From (7.54) it follows that the v^μ coefficients are also transverse. Recalling that $k_0 \equiv E(k) \equiv \sqrt{\vec{k}^2 + m^2}$ in the expression for the vector field

$$W^\mu(x) = \sum_\sigma \int \frac{d^3k}{(2\pi)^{3/2}\sqrt{2E}} (u^\mu(\vec{k}, \sigma)a(\vec{k}, \sigma)e^{-ik\cdot x} + v^\mu(\vec{k}, \sigma)a^{c\dagger}(\vec{k}, \sigma)e^{ik\cdot x}) \tag{7.167}$$

we have

$$(\partial^\rho \partial_\rho + m^2)W^\mu(x) = 0 \tag{7.168}$$
$$\partial_\mu W^\mu(x) = 0 \tag{7.169}$$

where the vanishing divergence of W^μ follows directly from the transversality property (7.166). The field equations (7.168), (7.169) can be summarized succintly by defining the tensor field

$$F^{\mu\nu}(x) \equiv \partial^\mu W^\nu(x) - \partial^\nu W^\mu(x) \tag{7.170}$$

The single field equation

$$\partial_\mu F^{\mu\nu} + m^2 W^\nu = 0 \tag{7.171}$$

referred to as the Maxwell–Proca equation (Proca, 1936), is easily seen to yield both the constraints (7.168), (7.169) which build in the correct transformation properties of the covariant field W^μ under Lorentz transformations. In the special case of a self-conjugate vector field, W^μ is hermitian and the antiparticle piece is just the hermitian adjoint of the particle piece, so (7.167) simplifies to

$$Z^\mu(x) = \sum_\sigma \int \frac{d^3k}{(2\pi)^{3/2}\sqrt{2E}} (\epsilon^\mu(\vec{k}, \sigma)a(\vec{k}, \sigma)e^{-ik\cdot x} + \epsilon^\mu(\vec{k}, \sigma)^* a^\dagger(\vec{k}, \sigma)e^{ik\cdot x}) \tag{7.172}$$

where we have (borrowing from electroweak theory) used the letter Z for a neutral (i.e., self-conjugate) vector boson, and (from quantum electrodynamics) the more conventional notation ϵ^μ for the corresponding polarization vector.

Once again, as in the case of spin-$\frac{1}{2}$ particles, the massless limit exhibits special features. We can proceed as before by examining the behavior of the massive field discussed above in the limit $m \to 0$. The discussion of massless particle representations in Section 5.3 again indicates that the helicity representation provides the appropriate description in this limit. From Eqs. (7.64, 7.65) the zero-momentum polarization vectors for a massive $j = 1$ particle (with $\xi^{AB} = (-1)^{2B}$) are easily seen to be

$$\epsilon^\mu(0, \sigma = 0) = \begin{pmatrix} 0 \\ 0 \\ 0 \\ -1 \end{pmatrix}$$

$$\epsilon^\mu(0, \sigma = \pm 1) = \begin{pmatrix} 0 \\ \pm\frac{1}{\sqrt{2}} \\ \frac{i}{\sqrt{2}} \\ 0 \end{pmatrix}$$

with the polarization vectors $v^\mu(0, \sigma)$ for the antiparticle term given by the conjugates $\epsilon^\mu{}^*(0, \sigma)$ as in (7.172).

The corresponding vectors for a particle moving in the positive z-direction (so that the spin σ and helicity λ specifications are equivalent) with momentum $\vec{k} = k\hat{z}$ are

$$\epsilon^\mu(0, \lambda = 0) = \begin{pmatrix} -\frac{k}{m} \\ 0 \\ 0 \\ -\frac{E(k)}{m} \end{pmatrix}$$

$$\epsilon^\mu(0, \lambda = \pm 1) = \begin{pmatrix} 0 \\ \pm\frac{1}{\sqrt{2}} \\ \frac{i}{\sqrt{2}} \\ 0 \end{pmatrix}$$

Note that the maximal helicity states $\lambda = \pm 1$ are unaffected by the boost. On the other hand, the zero-helicity mode is *singular* in the massless limit (or, for fixed mass, in the high-energy limit). This seems quite at variance with the situation for massless spin-$\frac{1}{2}$ Weyl fields, where, for example, the $(\frac{1}{2}, 0)$ left-handed field gave a well-defined left-handed spinor in the massless limit, while the right-handed spinor vanished as the mass was taken to zero (cf. (7.161)). Of course, if we start with an exactly massless $j = 1$ particle from the outset, the considerations of Section 5.3 assure us that helicity $\lambda = +1$ and $\lambda = -1$ states transform separately and irreducibly under the proper HLG, and in particular do not mix with a zero-helicity mode, which can be eliminated completely from the theory. On the other hand, if we wish to have parity conserving interactions, as is certainly the case for the photon, then both $\lambda = +1$ and $\lambda = -1$ states must appear.

Still, the singularity of the massless limit for the zero-helicity mode is somewhat unsettling: it would seem to imply that we could distinguish between an *exactly* massless photon and one with a mass of 10^{-80} eV, for example. In fact, the interactions of a photon with charged particles enjoy a *gauge symmetry* which ensures the decoupling of the zero-helicity mode and restores the smoothness of the massless limit, as well as softening the high-energy behavior of the theory, rendering it renormalizable. We shall return to these issues (of gauge symmetry and renormalizability) in much greater detail in the "Symmetries and Scales" sections of the book, but it may be useful here to provide a brief explanation of the role of gauge symmetry in taming the massless limit for the unwanted $\lambda = 0$ mode.

Suppose the interactions of our spin-1 field Z^μ (we use the Z notation as our field is still massive) are insensitive to altering Z^μ by longitudinal (i.e., gradient) fields of the form $\partial^\mu \Lambda(x)$, for *arbitrary* $\Lambda(x)$. Choose $\Lambda(x)$ to be the hermitian field

$$\Lambda(x) = i\frac{1}{m}\sum_\lambda \int \frac{d^3k}{(2\pi)^{3/2}\sqrt{2E}}(a(\vec{k},\lambda)e^{-ik\cdot x} - a^\dagger(\vec{k},\lambda)e^{ik\cdot x}) \tag{7.173}$$

The replacement $Z^\mu \to Z^\mu + \partial^\mu \Lambda$ (referred to as a *gauge transformation* of the vector field Z^μ) then clearly amounts to a shift of polarization vector $\epsilon^\mu(\vec{k},\lambda) \to \epsilon^\mu(\vec{k},\lambda) + \frac{k^\mu}{m}$. This corresponds to an alteration of the zero-helicity polarization vector (choosing the direction of $\vec{k}$ along the z-axis) to

$$\epsilon^\mu(0,\lambda=0) = \begin{pmatrix} \frac{E(k)-k}{m} \\ 0 \\ 0 \\ -\frac{E(k)-k}{m} \end{pmatrix}$$

But, in the massless limit $m \to 0$, $k \equiv |\vec{k}|$ fixed, $\frac{E(k)-k}{m} \to \frac{m}{2k} \to 0$, so we see that the interactions of the zero-helicity mode do indeed disappear as promised in the massless limit. We are therefore at liberty to define a massless spin-1 field, for which we shall now adopt the standard notation $A^\mu(x)$, entirely in terms of the two transverse polarization vectors $\epsilon(\vec{k},\lambda=\pm 1)$. For a general direction of motion $\hat{k}$ of our massless particle, these vectors are given by choosing real unit vectors $\vec{\epsilon}_1, \vec{\epsilon}_2$ with $\vec{\epsilon}_1 \times \vec{\epsilon}_2 = \hat{k}$, whence the spatial parts of the positive (resp. negative) helicity polarization vectors become $\vec{\epsilon}(\vec{k},+1) = \frac{1}{\sqrt{2}}(\vec{\epsilon}_1 + i\vec{\epsilon}_2)$ (resp. $\vec{\epsilon}(\vec{k},-1) = -\vec{\epsilon}(\vec{k},+1)^*$). The zeroth (time) component of these polarization vectors is itself zero, so this construction leads to a purely spatial field

$$\vec{A}(x) = \sum_{\lambda=\pm 1} \int \frac{d^3k}{(2\pi)^{3/2}\sqrt{2E}}(\vec{\epsilon}(\vec{k},\lambda)a(\vec{k},\lambda)e^{-ik\cdot x} + \vec{\epsilon}(\vec{k},\lambda)^* a^\dagger(\vec{k},\lambda)e^{ik\cdot x}) \tag{7.174}$$

satisfying the *radiation* (or *Coulomb*) gauge condition

$$\vec{\nabla}\cdot\vec{A}(x) = 0 \tag{7.175}$$

The appearance of a purely spatial field in a (hopefully!) Lorentz-invariant theory may seem puzzling at first sight, but recall from classical electromagnetic theory that an arbitrary four-vector potential $A^\mu(x)$ satisfying the source-free Maxwell equations can be put into radiation gauge $A^0 = 0$, $\vec{\nabla}\cdot\vec{A} = 0$ by precisely a gauge transformation of the sort $A^\mu \to A^\mu + \partial^\mu \Lambda$.

From (7.174), the recovery of the usual form for the free Hamiltonian as a sum of electric and magnetic field energies is straightforward (see Problem 9):

$$H_0 = \sum_{\lambda=\pm 1} \int d^3k \, E(k) a^\dagger(\vec{k}, \lambda) a(\vec{k}, \lambda) = \frac{1}{2} \int d^3x : (\vec{E}^2 + \vec{B}^2) : \qquad (7.176)$$

where $\vec{E} \equiv -\frac{\partial \vec{A}}{\partial t}, \vec{B} \equiv \vec{\nabla} \times \vec{A}$.

Instead of developing the concept of covariant fields starting from the somewhat *ad hoc* assumption of relativistic wave equations such as (7.168, 7.169, 7.171), the systematic construction followed in the preceding sections shows clearly that the form of the field equations, and the corresponding free Hamiltonian, satisfied by local covariant fields is actually specified uniquely by the underlying particle state transformation properties, once we insist on choosing the simplest representation of the HLG (for each spin) that will do the job.

7.6   Some simple theories and processes

At this stage, to give the reader a more concrete idea of how the covariant fields constructed in the preceding sections can be put to use, we describe some simple theories of interacting spinless and spin-$\frac{1}{2}$ particles, where the calculation of elementary scattering and decay processes can be carried out to low orders in perturbation theory without invoking the full technology of covariant perturbation theory, Wick's theorem, etc. (which we shall return to later in the book). Why not spin-1 particles as well? We shall see later, in Section 12.1, that our strategy so far for generating Lorentz-invariant theories, by writing the interaction Hamiltonian as an integral over an ultralocal scalar density, fails for theories involving particles of spin-1 or higher, or for theories involving interactions with spacetime-derivatives: the resultant interaction Hamiltonians typically contain non-covariant Schwinger terms in their space-like commutators (cf. discussion in Section 5.5). For such theories, the Lagrangian formalism developed later in Chapter 12 is ideally suited to ensuring that our theory indeed yields Lorentz-invariant scattering amplitudes: this will follow automatically simply by constructing the Lagrangian in a Lorentz-invariant way.

Restricting ourselves, therefore, to theories with only scalar or Dirac particles, we shall consider some simple processes in theories defined by the following three simple interaction Hamiltonians:

1. (Theory A) For coupling constant λ real, and $\phi(x) = \phi^\dagger(x)$ (a self-conjugate scalar field of mass M), assume an interaction Hamiltonian density

$$\mathcal{H}_{\text{int}}^{(A)}(x) = \frac{\lambda}{4!} \phi(x)^4 \qquad (7.177)$$

2. (Theory B) With λ, $\phi(x)$ as in theory A above, and with $\psi(x)$ a non-self-conjugate scalar field of mass m

$$\mathcal{H}_{\text{int}}^{(B)}(x) = \lambda \psi^\dagger(x) \psi(x) \phi(x) \qquad (7.178)$$

3. (Theory C) With $\lambda, \phi(x)$ as in theory B, but with $\psi(x)$ a spin-$\frac{1}{2}$ Dirac field of mass m, subject to a *Yukawa interaction*

$$\mathcal{H}_{\text{int}}^{(C)}(x) = \lambda \bar{\psi}(x)\psi(x)\phi(x) \tag{7.179}$$

We shall be calculating some simple decay and scattering processes in these theories, so we will need the results on rates and cross-sections from Appendix B. To employ these results we must first extract the non-singular part $\mathcal{T}_{\beta\alpha}$ of the T-matrix element of the process via

$$S_{\beta\alpha} = \delta_{\beta\alpha} - 2\pi i \delta^4(P_\alpha - P_\beta)\mathcal{T}_{\beta\alpha} \tag{7.180}$$

The desired phenomenological quantities are then given by (see Appendix B for a derivation of these essential results, specifically (B.17) and (B.27)):

1. The differential decay rate of a one-particle state α to final states β:

$$d\Gamma(\alpha \rightarrow \beta) = 2\pi \delta^4(P_\alpha - P_\beta)|\mathcal{T}_{\beta\alpha}|^2 d\beta \tag{7.181}$$

where the final-state phase-space factor is just

$$d\beta = \sum_{\text{spins}} \prod_i d^3 k_i' \tag{7.182}$$

with k_i' the final state momenta. Of course, when considering spinless particles, the sum over final-state spins will be irrelevant.

2. The differential cross-section for a two-particle state α to scatter into final states β:

$$d\sigma(\alpha \rightarrow \beta) = \frac{(2\pi)^4}{v_\alpha} \delta^4(P_\alpha - P_\beta)|\mathcal{T}_{\beta\alpha}|^2 d\beta \tag{7.183}$$

where the relative velocity in the initial state v_α is given by

$$v_\alpha = \frac{\sqrt{(k_1 \cdot k_2)^2 - m_1^2 m_2^2}}{E_1 E_2} \tag{7.184}$$

and the final-state phase-space is as above.

The perturbative expansion of the S-matrix

$$S = \sum_{n=0}^{\infty} \frac{(-i)^n}{n!} \int d^4 x_1 .. d^4 x_n T\{\mathcal{H}_{\text{int}}(x_1)....\mathcal{H}_{\text{int}}(x_n)\} \tag{7.185}$$

will be used to extract the T-matrix element for some simple processes in the theories specified above. We will evaluate the lowest-order non-trivial contribution to S for each of the following processes:

1. $\phi - \phi$ scattering in theory A (first order in λ).
2. ϕ decay in theory B (first order in λ).

3. $\psi - \psi^c$ scattering in theory B (second order in λ).
4. $\psi - \psi^c$ scattering in theory C (second order in λ).

7.6.1 2-2 scattering in $\lambda\phi^4$ theory

Non-trivial 2-2 scattering in theory A already occurs at first order in λ. We have deliberately chosen not to normal-order the interaction Hamiltonian (cf. the discussion in Section 6.4) in order to display certain features of the scattering amplitude which are characteristic of relativistic field theory, in contradistinction to non-relativistic scattering. For the elastic scattering process $\vec{k}_1 + \vec{k}_2 \rightarrow \vec{k}'_1 + \vec{k}'_2$ we have

$$S_{\vec{k}'_1\vec{k}'_2,\vec{k}_1\vec{k}_2} = \frac{-i\lambda}{4!} \int d^4x \langle k'_1 k'_2 | \phi(x)^4 | k_1 k_2 \rangle + O(\lambda^2) \tag{7.186}$$

with

$$\phi(x) = \int \mathcal{D}p (a(p)e^{-ip\cdot x} + a^\dagger(p)e^{ip\cdot x}) \tag{7.187}$$

$$\mathcal{D}p \equiv \frac{1}{(2\pi)^{3/2}\sqrt{2E(p)}} d^3p \tag{7.188}$$

In order to change the incoming state $|k_1 k_2\rangle$ to the outgoing one $|k'_1 k'_2\rangle$ (with *different* momenta), we must choose the annihilation part of two of the $\phi(x)$ fields to get rid of the incoming particles and the creation part of the remaining two to produce the outgoing ones. A typical matrix element appearing in (7.186) might, for example, be

$$\langle k'_1 k'_2 | a^\dagger(p_1) a(p_2) a^\dagger(p_3) a(p_4) | k_1 k_2 \rangle \tag{7.189}$$

If we rearrange the product of creation and annihilation operators to produce the normal-ordered product, we find an additional term:

$$a^\dagger(p_1)a(p_2)a^\dagger(p_3)a(p_4) = a^\dagger(p_1)a^\dagger(p_3)a(p_2)a(p_4) + \delta^3(p_2 - p_3)a^\dagger(p_1)a(p_4) \tag{7.190}$$

which clearly leads to a disconnected term, as

$$\langle k'_1 k'_2 | a^\dagger(p_1)a(p_4) | k_1 k_2 \rangle = \delta^3(k'_1 - p_1)\delta^3(k_1 - p_4)\delta^3(k'_2 - k_2) + \text{permutations} \tag{7.191}$$

corresponding to disconnected terms of the structure displayed in Fig. 6.6, in which one of the particles passes through the process completely unaffected by the interaction, while the other particle suffers a *self-interaction* induced by the interaction. Such persistent self-interactions, present even for an isolated particle, are typically absent in non-relativistic scattering, but are an intrinsic feature of relativistic field theory. As we shall see in our more detailed discussion of perturbation theory in Chapter 10, and in even more detail in Part 4 of the book (on "Scales"), they result in renormalizations of the attributes of single-particle states: specifically, of the mass and normalization of the one particle states of the theory. For now, we ignore such effects, concentrating on the fully connected contributions to the scattering exemplified by the right-most diagram in Fig. 6.6.

Fully connected contributions to the scattering arise from keeping the normal-ordered piece from each of the six possible ways in which we can pick two creation and two annihilation terms from the product of four fields in (7.186). Relabeling the momentum integration variables, we find six equivalent contributions. For example, we can take the term

$$
\int d^4x \prod_{i=1}^{4} \mathcal{D}p_i \langle k_1' k_2' | a^\dagger(p_1) e^{ip_1 \cdot x} a^\dagger(p_2) e^{ip_2 \cdot x} a(p_3) e^{-ip_3 \cdot x} a(p_4) e^{-ip_4 \cdot x} | k_1 k_2 \rangle \quad (7.192)
$$

and then multiply by 6. In (7.192) the annihilation operator $a(p_4)$ can remove either the particle k_1, in which case $a(p_3)$ must remove k_2, or the other way around. Likewise, there are two possibilities for creating the final state. Recalling that $a(p_4)|k_1....\rangle = \delta^3(p_4 - k_1)|...\rangle$, $\langle k_1'...|a^\dagger(p_1) = \delta^3(p_1 - k_1')\langle...|$, we see that the only interesting parts of (7.192) evaluate to

$$
\int d^4x \prod_{i=1}^{4} \mathcal{D}p_i (\delta^3(p_1 - k_1')\delta^3(p_2 - k_2') + p_1 \leftrightarrow p_2)
$$

$$
\times (\delta^3(p_3 - k_1)\delta^3(p_4 - k_2) + p_3 \leftrightarrow p_4) e^{i(p_1 + p_2 - p_3 - p_4)\cdot x} \quad (7.193)
$$

$$
= (2\pi)^4 \prod_{i=1}^{4} \mathcal{D}p_i \delta^4(p_1 + p_2 - p_3 - p_4)(\delta^3(p_1 - k_1')\delta^3(p_2 - k_2') + p_1 \leftrightarrow p_2)
$$

$$
\times (\delta^3(p_3 - k_1)\delta^3(p_4 - k_2) + p_3 \leftrightarrow p_4) \quad (7.194)
$$

$$
= \frac{4}{(2\pi)^6 \sqrt{2E_1 \cdot 2E_2 \cdot 2E_1' \cdot 2E_2'}} (2\pi)^4 \delta^4(k_1 + k_2 - k_1' - k_2') \quad (7.195)
$$

To summarize, the connected S-matrix scattering amplitude (putting in the factor of 6 discussed above) becomes

$$
S^c_{k_1'k_2',k_1k_2} = \frac{-i\lambda}{(2\pi)^6 \sqrt{2E_1 \cdot 2E_2 \cdot 2E_1' \cdot 2E_2'}} (2\pi)^4 \delta^4(k_1 + k_2 - k_1' - k_2') + O(\lambda^2)
$$

$$
\quad (7.196)
$$

Extracting the T-matrix element with (7.180):

$$
T^c_{k_1'k_2',k_1k_2} = \frac{\lambda}{(2\pi)^3 \sqrt{2E_1 \cdot 2E_2 \cdot 2E_1' \cdot 2E_2'}} + O(\lambda^2) \quad (7.197)
$$

so the differential cross-section is (from (7.183))

$$
d\sigma = \frac{\lambda^2}{64\pi^2} \frac{1}{\sqrt{(k_1 \cdot k_2)^2 - M^4}} \delta^4(k_1 + k_2 - k_1' - k_2') \frac{d^3k_1'}{E_1'} \frac{d^3k_2'}{E_2'} \quad (7.198)
$$

At this point we have to be a little more specific about the kinematic conditions for the scattering. Let us then assume we are performing the experiment in a collider, in the center-of-mass frame of the two incoming particles. Thus the four-vector momenta take the form $k_1 = (E, \vec{k}), k_2 = (E, -\vec{k}), k_1' = (E_1', \vec{k}_1'), k_2' = (E_2', \vec{k}_2')$. One finds

$(k_1 \cdot k_2)^2 - M^4 = 4|\vec{k}|^2 E^2$, while the energy-momentum conservation δ-function is $\delta^4(..) = \delta(E_1' + E_2' - 2E)\delta^3(\vec{k}_1' + \vec{k}_2')$. Inserting these into (7.198), and doing the $\vec{k}_2'$ integration using the δ-function, we find

$$do = \frac{\lambda^2}{64\pi^2} \frac{1}{2E|\vec{k}|} \int \frac{|\vec{k}_1'|^2}{(E_1')^2} \delta(2E_1' - 2E)d|\vec{k}_1'|d\Omega_{\hat{k}_1'} \tag{7.199}$$

If we have detectors deployed with angular resolution, we should leave the integral over solid angles undone; on the other hand, the detector will "blip" once the particle enters the opening angle of the detector, irrespective of the magnitude of momentum $|k_1'|$, so we should integrate over this variable. This gives the final result—a differential cross-section

$$\frac{d\sigma}{d\Omega_{\hat{k}_1'}} = \frac{\lambda^2}{256\pi^2 E^2} + O(\lambda^3) = \frac{\lambda^2}{256\pi^2(\vec{k}^2 + M^2)} + O(\lambda^3) \tag{7.200}$$

Note that the scattering (to lowest order) is *isotropic* ("s-wave"): independent of the angle between $\vec{k}_1$ and $\vec{k}_1'$.

7.6.2 ϕ decay in theory B

In theory B we have two distinct spinless particles, the ϕ particle with mass M, and a non-self-conjugate ψ particle with mass m. We now assume that $M > 2m$. We shall see that the interaction (7.178) induces a decay process $\phi \to \psi + \psi^c$. Denote the momentum of the initial ϕ particle by k, with k_1' (resp. k_2') the momenta of the outcoming ψ (resp. ψ^c). We need to calculate

$$S_{k_1'k_2',k} = -i\lambda \int d^4x \langle k_1'k_2'|\psi^\dagger(x)\psi(x)\phi(x)|k\rangle + O(\lambda^2) \tag{7.201}$$

The relevant pieces of the various field operators are clearly

$$\phi(x) \to \int \mathcal{D}p\, a(p)e^{-ip\cdot x}$$

$$\psi(x) \to \int \mathcal{D}p_2'\, a^{c\dagger}(p_2')e^{ip_2'\cdot x}$$

$$\psi^\dagger(x) \to \int \mathcal{D}p_1'\, a^\dagger(p_1')e^{ip_1'\cdot x}$$

Inserting these forms in (7.201) and performing the spacetime and momentum integrals, we obtain

$$S_{k_1'k_2',k} = \frac{-i\lambda}{(2\pi)^{9/2}\sqrt{2E\cdot 2E_1'\cdot 2E_2'}}(2\pi)^4\delta^4(k_1' + k_2' - k) + O(\lambda^2) \tag{7.202}$$

Stripping off the δ-function, this gives the T-matrix element

$$T_{k_1'k_2',k} = \frac{\lambda}{(2\pi)^{3/2}\sqrt{2E\cdot 2E_1'\cdot 2E_2'}} + O(\lambda^2) \tag{7.203}$$

The general decay formula (7.181) can now be applied to get the differential decay rate:

$$d\Gamma = 2\pi\delta^4(k_1' + k_2' - k)\frac{\lambda^2}{(2\pi)^3}\frac{1}{2E \cdot 2E_1' \cdot 2E_2'}d^3k_1'd^3k_2' + O(\lambda^3) \tag{7.204}$$

Once again we have reached the point where a choice of frame needs to be made. Clearly, it is easiest to pick the rest frame for the decaying particle (simple time-dilation arguments give the decay rate in other frames). So let us take $k = (M, 0)$, $k_1' = (E_1', \vec{k}_1'), k_2' = (E_2', \vec{k}_2')$. One then finds (integrating out $\vec{k}_2'$, and neglecting higher orders)

$$d\Gamma = \frac{\lambda^2}{(2\pi)^2 2M}\int\frac{1}{(2E_1')^2}\delta(2E_1' - M)|k_1'|^2 d|k_1'|d\Omega_{\hat{k}_1'} \tag{7.205}$$

As in the calculation of the $\phi - \phi$ scattering, we leave the angular integral undone to get a differential decay rate per unit solid angle, and perform the integration over the radial momentum component; a short calculation gives

$$\frac{d\Gamma}{d\Omega_{\hat{k}_1'}} = \frac{\lambda^2}{64\pi^2}\frac{\sqrt{M^2 - 4m^2}}{M^2} \tag{7.206}$$

which, of course, makes sense only if $M > 2m$, as assumed initially (otherwise the energy δ-function constraint in the k_1' integral is never satisfied, and we simply get zero: the ϕ particle is stable). Note that the above calculation gives the rate for detecting ψ particles at any angle (in this theory, the particle and antiparticle are in principle distinguishable; of course, the ψ^c simply comes out in the opposite direction to the ψ in the rest frame of the ϕ). The decay is isotropic, so the total decay rate (#decays/sec in all directions) is just, to second order in λ,

$$\Gamma_{\text{tot}} = \frac{\lambda^2}{16\pi}\frac{\sqrt{M^2 - 4m^2}}{M^2} \tag{7.207}$$

7.6.3 $\psi - \psi^c$ scattering in theory B

Referring to the form of the interaction Hamiltonian (7.178) in theory B, we see that the destruction of two incoming ψ particles and the creation of two outgoing ones requires at least four ψ-type fields, i.e., the interaction Hamiltonian must appear to at least *second* order. Moreover, as there are no ϕ particles in either the initial or final state, only terms of *even* order in $\mathcal{H}_{\text{int}}$ are relevant to this process:

$$S_{\vec{k}_1'\vec{k}_2',\vec{k}_1\vec{k}_2} = -\frac{\lambda^2}{2}\int d^4x_1 d^4x_2\langle k_1'k_2'|T\{\psi^\dagger(x_1)\psi(x_1)\phi(x_1)\psi^\dagger(x_2)\psi(x_2)\phi(x_2)\}|k_1k_2\rangle$$
$$+ O(\lambda^4) \tag{7.208}$$

The matrix element in the joint Fock space of multi-particle ψ and ϕ states can be factorized in an obvious way:

$$\langle k_1' k_2'|T\{ \ \ldots \ \}|k_1 k_2\rangle$$
$$= \{\theta(t_1 - t_2)\langle 0|\phi(x_1)\phi(x_2)|0\rangle \langle k_1' k_2'|\psi^\dagger(x_1)\psi(x_1)\psi^\dagger(x_2)\psi(x_2)|k_1 k_2\rangle$$
$$+ \theta(t_2 - t_1)\langle 0|\phi(x_2)\phi(x_1)|0\rangle \langle k_1' k_2'|\psi^\dagger(x_2)\psi(x_2)\psi^\dagger(x_1)\psi(x_1)|k_1 k_2\rangle\}$$

$$(7.209)$$

As in the calculation of $\phi - \phi$ scattering, we are only interested in the connected part of the ψ matrix element, in which every field operator associates with one of the incoming or outgoing particles. Only ψ fields contain annihilation parts for ψ particles (or creation parts for ψ^c), so these fields associate with the particles with momenta k_1 (resp. k_2'). Likewise, the $\psi^\dagger$ fields take care of the particles with momenta k_2, k_1'. The result clearly contains four terms:

$$\langle k_1' k_2'|\psi^\dagger(x_1)\psi(x_1)\psi^\dagger(x_2)\psi(x_2)|k_1 k_2\rangle = \text{PSF} \cdot (e^{-i(k_1+k_2)\cdot x_2 + i(k_1'+k_2')\cdot x_1}$$
$$+ e^{i(k_1'-k_1)\cdot x_2 + i(k_2'-k_2)\cdot x_1} + e^{i(k_2'-k_2)\cdot x_2 + i(k_1'-k_1)\cdot x_1} + e^{i(k_1'+k_2')\cdot x_2 - i(k_1+k_2)\cdot x_1})$$

$$(7.210)$$

where "PSF" is an elegant notation for the phase-space factor

$$\text{PSF} \equiv \frac{1}{(2\pi)^6 \sqrt{2E_1 \cdot 2E_2 \cdot 2E_1' \cdot 2E_2'}} \qquad (7.211)$$

Note that the result (7.210) is symmetric under the interchange $x_1 \leftrightarrow x_2$, so that the two matrix elements involving the ψ fields in (7.209) are in fact equal, and can be taken out as a common factor. The matrix elements involving the ϕ field can then be recombined into a single time-ordered product, as follows:

$$S_{\vec{k}_1' \vec{k}_2', \vec{k}_1 \vec{k}_2} = -\frac{\lambda^2}{2}(\text{PSF}) \int d^4x_1 d^4x_2 \langle 0|T(\phi(x_1)\phi(x_2))|0\rangle$$
$$\times \{e^{-i(k_1+k_2)\cdot x_2 + i(k_1'+k_2')\cdot x_1} + \ldots\} + O(\lambda^4) \qquad (7.212)$$

The vacuum expectation value (vev) of the time-ordered product of two fields appears in almost every perturbative calculation in field theory: it has been dubbed the *Feynman propagator* (see Fig. 7.1). Conventionally, a factor of i is included in the definition, as follows:

$$i\Delta_F(x_1, x_2) \equiv \langle 0|T(\phi(x_1)\phi(x_2))|0\rangle \qquad (7.213)$$
$$= \theta(t_1 - t_2)\int \frac{d^3k}{(2\pi)^3 2E(k)} e^{+ik\cdot(x_2 - x_1)} + \theta(t_2 - t_1)\int \frac{d^3k}{(2\pi)^3 2E(k)} e^{-ik\cdot(x_2 - x_1)}$$
$$\equiv i\Delta_F(x_1 - x_2) \qquad (7.214)$$

(Strictly speaking, this is the propagator for the ϕ field which should be distinguished from a similar object for the ψ field, not needed in this calculation.) Note that the

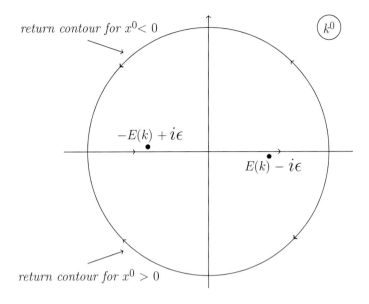

Fig. 7.1 Contours for the Feynman propagator.

Feynman propagator contains *both* time orderings $t_1 \leq t_2$ and $t_1 \geq t_2$; as we should expect from the discussion in Chapter 3 (see Fig. 3.1), a frame-independent description of particle exchange must necessarily include a symmetrical treatment of the emission and absorption events for the exchanged particle.

The coordinate space version of the propagator given above is less useful in practice than the Fourier transform, as we are typically more interested in the behavior of amplitudes in momentum space. In fact, the Fourier transform is remarkably simple:

$$\Delta_F(x) = \int \frac{d^4 k}{(2\pi)^4} \frac{e^{-ik \cdot x}}{k_\mu k^\mu - M^2 + i\epsilon} \tag{7.215}$$

where ϵ is a positive infinitesimal quantity. To see this, first note that

$$k^2 - M^2 + i\epsilon = (k^0)^2 - E(k)^2 + i\epsilon = (k^0 - E(k) + i\epsilon)(k^0 + E(k) - i\epsilon) \tag{7.216}$$

Accordingly, if $x^0 > 0$ (resp. $x^0 < 0$), the integrand of (7.215) vanishes exponentially fast in the negative (resp. positive) imaginary direction in the complex plane of the k^0 integration variable, and we can use Cauchy's theorem to close the contour in the lower (resp. upper) half-plane, in which case we pick up the residue of the pole at $k^0 = E(k) - i\epsilon$ (resp. $k^0 = -E(k) + i\epsilon$). The result of the k^0 integration is then

$$\Delta_F(x) = -i \int \frac{d^3 k}{(2\pi)^3 2E(k)} e^{-ik \cdot x}, \quad x^0 > 0 \tag{7.217}$$

$$\Delta_F(x) = -i \int \frac{d^3 k}{(2\pi)^3 2E(k)} e^{ik \cdot x}, \quad x^0 < 0 \tag{7.218}$$

agreeing with (7.214).

The Feynman propagator is the Green function for an extremely important differential operator, the *Klein–Gordon* operator $\Box + M^2$ (cf. (6.56)):

$$(\Box + M^2)\Delta_F(x) = -\delta^4(x) \tag{7.219}$$

It is apparent from the Fourier representation (7.215) that the Feynman propagator is a Lorentz-invariant function of its spacetime argument x, despite its definition (7.214) involving θ-functions of a frame-dependent time. The resolution of this apparent paradox becomes clear if we recall the locality of the $\phi(x)$ field: under a Lorentz transformation Λ altering the time-ordering of $\phi(x_1)$ and $\phi(x_2)$, the space-like separation of x_1 and x_2 ensures that the order of field operators is irrelevant, as the field is local. This is just a special case of the argument given in Section 5.5 for the Lorentz-invariance of an S-matrix constructed perturbatively in terms of the time-ordered products of local scalar interaction Hamiltonians.

Returning to the calculation of the S-matrix element for elastic $\psi - \psi^c$ scattering in (7.212), we can now write (at second order in λ)

$$S_{\vec{k}_1' \vec{k}_2', \vec{k}_1 \vec{k}_2} = -i\frac{\lambda^2}{2}(\text{PSF})\int d^4x_1 d^4x_2 \frac{d^4k}{(2\pi)^4}\frac{1}{k^2 - M^2 + i\epsilon}$$

$$\times\, e^{ik\cdot(x_2-x_1)}\left(e^{-i(k_1+k_2)\cdot x_2 + i(k_1'+k_2')\cdot x_1} + e^{-i(k_1-k_1')\cdot x_2 + i(k_2'-k_2)\cdot x_1} + x_1 \leftrightarrow x_2\right)$$

$$= -i\lambda^2(\text{PSF})\int \frac{d^4k}{(2\pi)^4}\frac{1}{k^2 - M^2 + i\epsilon}(2\pi)^8\{\delta^4(k - k_1 - k_2)\delta^4(k - k_1' - k_2')$$

$$+ \delta^4(k - k_1 + k_1')\delta^4(k + k_2 - k_2')\}$$

$$= -i\lambda^2(\text{PSF})(2\pi)^4\delta^4(k_1 + k_2 - k_1' - k_2')$$

$$\times\left\{\frac{1}{(k_1 + k_2)^2 - M^2 + i\epsilon} + \frac{1}{(k_1 - k_1')^2 - M^2 + i\epsilon}\right\} \tag{7.220}$$

We see that the final amplitude is the sum of two distinct pieces, depending respectively on the square of the total incoming four-momentum, $s \equiv (k_1 + k_2)^2$ (equal to four times the square of the particle energy in the CM frame) and on the square of the four-momentum transferred from the ψ to ψ^c, $t \equiv (k_1 - k_1')^2$. In the CM frame, the second term will, of course, lead to an angular dependence of the differential cross-section. The two terms have a simple graphical interpretation (see Fig. 7.2: note that the direction of the arrows for the ψ particles indicates "charge", rather than momentum, flow, if we associate positive charge with the ψ and negative charge with the antiparticle ψ^c). The result for the S-matrix amplitude can be read off immediately from the following simple *Feynman rules*:

1. $-i\lambda$ at each vertex.
2. $i\Delta_F(p) \equiv \frac{i}{p^2 - M^2 + i\epsilon}$ for each internal ϕ line carrying momentum p.
3. Apply four-momentum conservation at each vertex.

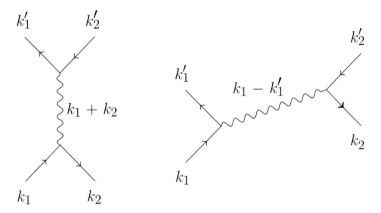

Fig. 7.2 Feynman Graphs for $\psi - \psi^c$ scattering.

4. Each external line has an associated factor of $\dfrac{1}{(2\pi)^{3/2}\sqrt{2E(k)}}$: the product for all external lines yields the phase-space factor called PSF above (see discussion of Lorentz-invariance below!).

5. There is an overall energy-momentum conservation factor $(2\pi)^4\delta^4(P_\alpha - P_\beta)$.

A real derivation of these rules, valid to all orders of perturbation theory, will be given later in Chapter 10, once we have Wick's theorem at our disposal.

One final comment on the Lorentz-invariance of our results. With the exception of the phase-space factor associated with the external lines (see (7.211)), our final result for the S-matrix element (7.220) is manifestly a Lorentz-invariant function of the momenta. The presence of the non-Lorentz-invariant phase-space factor is due to our use of non-covariantly normalized particle states: if we return to covariantly normalized states, the "PSF" factor disappears (as there is now an extra factor of $(2\pi)^{3/2}\sqrt{2E(k)}$ for each particle in the initial or final state, exactly cancelling the PSF), and we once again have the critical invariance property which we set out to ensure many pages ago: namely, for 2-2 scattering

$$\text{cov}\langle k_1' k_2'|S|k_1 k_2\rangle_{\text{cov}} = \text{cov}\langle k_1' k_2'|U^\dagger(\Lambda)SU(\Lambda)|k_1 k_2\rangle_{\text{cov}}$$
$$= \text{cov}\langle \Lambda k_1' \Lambda k_2'|S|\Lambda k_1 \Lambda k_2\rangle_{\text{cov}}$$

7.6.4 $\psi - \psi^c$ scattering in theory C

We now turn to a theory with massive Dirac particles interacting via a Yukawa interaction term, (7.179). The Dirac fields appearing in the interaction are

$$\psi(x) = \int Dp \sum_\sigma (u(p,\sigma)b(p,\sigma)e^{-ip\cdot x} + v(p,\sigma)d^\dagger(p,\sigma)e^{ip\cdot x}) \qquad (7.221)$$

$$\bar{\psi}(x) = \int Dp \sum_\sigma (\bar{u}(p,\sigma)b(p,\sigma)^\dagger e^{ip\cdot x} + \bar{v}(p,\sigma)d(p,\sigma)e^{-ip\cdot x}) \qquad (7.222)$$

with

$$\mathcal{D}p \equiv \frac{1}{(2\pi)^{3/2}} \sqrt{\frac{m}{E(p)}} d^3p \tag{7.223}$$

We shall consider the scattering of an incoming ψ particle (with momentum and spin p_1, σ_1) on an anti-ψ particle (momentum and spin p_2, σ_2). The outgoing particles are similarly labeled, with primes added. As before, we work to second order in the interaction (7.179). Following steps exactly analagous to those leading to the term (7.210) in theory B, we encounter a matrix element of the form

$$\langle p'_1\sigma'_1, p'_2\sigma'_2 | \bar{\psi}(x_1)\psi(x_1)\bar{\psi}(x_2)\psi(x_2) | p_1\sigma_1, p_2\sigma_2\rangle$$

$$= \mathrm{PSF}\{\bar{u}(p'_1\sigma'_1)v(p'_2\sigma'_2)\bar{v}(p_2\sigma_2)u(p_1\sigma_1)(e^{i(p'_1+p'_2)\cdot x_1 - i(p_1+p_2)\cdot x_2} + x_1 \leftrightarrow x_2)$$

$$- \bar{u}(p'_1\sigma'_1)u(p_1\sigma_1)\bar{v}(p_2\sigma_2)v(p'_2\sigma'_2)(e^{i(p'_1-p_1)\cdot x_1 + i(p'_2-p_2)\cdot x_2} + x_1 \leftrightarrow x_2)\} \tag{7.224}$$

where the phase-space factor now takes the form

$$\mathrm{PSF} = \frac{m^2}{(2\pi)^6 \sqrt{E(p_1)E(p_2)E(p'_1)E(p'_2)}} \tag{7.225}$$

Note the appearance of a minus sign in two of the four terms, due to the need for an odd number of transpositions of fermionic creation and destruction operators in the fields in order to destroy the incoming particles and create the final-state ones. As we shall see shortly when we discuss the crossing symmetry of these amplitudes, the minus sign is an indication of a generalized notion of Fermi antisymmetrization of amplitudes, applicable in local relativistic theories to exchange of particles *between the initial and final states* (and not only, as in non-relativistic quantum mechanics, to exchange of identical particles in either the initial or final state). Nevertheless, the entire expression is symmetric under $x_1 \leftrightarrow x_2$, just as in the bosonic case of Theory B. This allows us, as there, to factor the fermionic matrix element entirely from the vacuum expectation of time-ordered ϕ fields, so we recover, as before, a Feynman propagator $i\Delta_F(x_1 - x_2)$ for the ϕ field. After substituting the Fourier transform expression (7.215) and performing the spacetime integrals over x_1 and x_2, we obtain the final result for the second-order scattering amplitude:

$$S_{p'_1\sigma'_1 p'_2\sigma'_2, p_1\sigma_1 p_2\sigma_2} = i\lambda^2(\mathrm{PSF})(2\pi)^4\delta^4(p'_1 + p'_2 - p_1 - p_2)$$

$$\times \{\frac{\bar{u}(p'_1\sigma'_1)u(p_1\sigma_1)\bar{v}(p_2\sigma_2)v(p'_2\sigma'_2)}{(p_1 - p'_1)^2 - M^2 + i\epsilon}$$

$$- \frac{\bar{u}(p'_1\sigma'_1)v(p'_2\sigma'_2)\bar{v}(p_2\sigma_2)u(p_1\sigma_1)}{(p_1 + p_2)^2 - M^2 + i\epsilon}\} \tag{7.226}$$

Apart from the minus sign, the result (7.226) is very similar to the result (7.220), with a contribution from scalar exchange interfering with one from particle–antiparticle annihilation, as indicated in Fig. 7.3. Our previous list of Feynman rules needs only the following obvious additions for spin-$\frac{1}{2}$ Dirac particles:

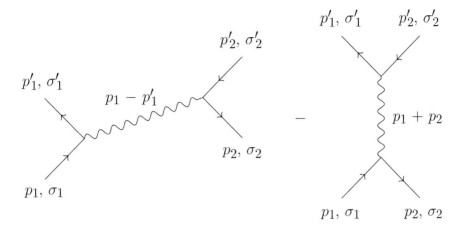

Fig. 7.3 Feynman graphs for $\psi - \psi^c$ scattering in Theory C.

1. Drawing particle lines with the arrow pointing upward (i.e., from the initial to the final state), each final-state particle is associated with a factor $\frac{1}{(2\pi)^{3/2}}\sqrt{\frac{m}{E(p')}}\bar{u}(p'\sigma')$, and each initial-state particle with a factor $\frac{1}{(2\pi)^{3/2}}\sqrt{\frac{m}{E(p)}}u(p\sigma)$.

2. Adopting the convention that antiparticle lines should be drawn with the arrow pointing downward (i.e., from final to initial state), each initial-state antiparticle is associated with a factor $\frac{1}{(2\pi)^{3/2}}\sqrt{\frac{m}{E(p)}}\bar{v}(p\sigma)$, and each final-state antiparticle with a factor $\frac{1}{(2\pi)^{3/2}}\sqrt{\frac{m}{E(p')}}v(p'\sigma')$.

In other processes in Theory C, such as ϕ-ψ scattering (see Problem 12), the Feynman–Dirac propagator $iS_F(x_1 - x_2)$, involving the time-ordered product of a ψ with a $\bar{\psi}$ field (see Problem 6), will make its appearance. After all Fourier transforms are performed, this leads, as expected, to a factor of $iS_F(p)$ for every internal fermion line carrying four-momentum p.

7.6.5   Crossing symmetry

The fact that the multi-particle scattering amplitudes of identical bosons (or fermions) are symmetric (resp. antisymmetric) under exchange of any two momenta in *either* the initial *or* the final state is a basic and irreducible postulate of non-relativistic quantum mechanics: it cannot be derived from any more fundamental concepts in the non-relativistic arena, as both relativistic invariance and the locality requirement are absent. In particular, particles and antiparticles are distinct and *totally unrelated* objects from a non-relativistic point of view. In local field theory, on the other hand, we have seen that particles and antiparticles are intimately related by their symmetrical appearance in the local fields used to construct the fundamental interaction processes of the theory. This leads to a generalization of the Bose–Fermi exchange symmetry corresponding to exchange of initial-state particles with final-state antiparticles which very strongly constrains the structure of a relativistic S-matrix. This generalization, for reasons that will shortly become apparent, is called *crossing symmetry*. We are not yet in a position to give a general proof of crossing symmetry, valid beyond perturbation theory (i.e., for the exact amplitudes of the theory): this will come later, in Chapter 9, when we have in hand the master ("LSZ") formula derived by Lehmann, Symanzik, and Zimmermann (Lehmann *et al.*, 1955) for arbitrary scattering amplitudes in a relativistic field theory. At that point, the relation between the crossing symmetry of the scattering amplitudes and the Bose (or Fermi) character of the underlying local fields will become immediate and manifest. Here, we shall content ourselves with a few simple examples based on the toy theories of the preceding sections.

A simple example of crossing symmetry can be seen by examining the second-order S-matrix amplitude for elastic $\psi - \psi$ scattering in theory B. Consider the process

$$\psi(p_1) + \psi(p_2) \rightarrow \psi(p_1') + \psi(p_2') \tag{7.227}$$

(we are using momenta labeled by "p"s rather than "k"s to avoid confusion in the crossing rules given below). A calculation along similar lines to that carried out above for $\psi - \psi^c$ scattering yields an amplitude for this process proportional to

$$\frac{1}{(p_1 - p_1')^2 - M^2 + i\epsilon} + \frac{1}{(p_1 - p_2')^2 - M^2 + i\epsilon} \tag{7.228}$$

corresponding to the Feynman graphs shown in Fig. 7.4. One easily sees that this amplitude transforms to the corresponding result for $\psi - \psi^c$ scattering in (7.220) with the simple replacements:

$$p_1 \rightarrow k_1$$
$$p_1' \rightarrow k_1'$$
$$p_2 \rightarrow -k_2'$$
$$p_2' \rightarrow -k_2$$

i.e., by twisting the antiparticle lines around and changing the sign of their momenta.

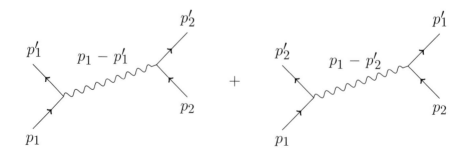

Fig. 7.4 Feynman graphs for $\psi - \psi$ scattering in Theory B.

Similar crossing rules apply in multi-fermion scattering amplitudes, with the expected supplemental rules for dealing with the external line spinors: u spinors are replaced by v spinors when an initial-state particle is crossed to a final-state antiparticle, etc. The appearance of the negative sign in (7.226) for ψ-anti-ψ scattering in Theory C is a necessary consequence of the Fermi antisymmetry of the ψ-ψ scattering amplitude under exchange of initial- (or final-)state particle momenta, once we subject that amplitude to the crossing transformation which converts it to a ψ-ψ^c scattering amplitude.

7.7 Problems

1. Verify the expression given in the text for the free Hamitonian of a massive Majorana field (given in (7.87)):

$$H_0 = \int d^3x : (\chi^\dagger i \vec{\sigma} \cdot \vec{\nabla}\chi + \frac{m}{2}(\chi^T C_s \chi - \chi^\dagger C_s \chi^*) :$$

Namely, show that the above expression reduces to the desired energy summation formula in terms of creation and annihilation operators:

$$H_0 = \int d^3k \sum_\sigma E(k) b^\dagger(\vec{k}, \sigma) b(\vec{k}, \sigma)$$

Remember to include an extra minus sign whenever a pair of creation or annihilation operators must be transposed in order to effect the normal ordering (i.e., when moving $b^\dagger$s to the left of bs). Hint: it is easiest to begin with the form (7.90); one also needs the result $e^{-\theta \hat{k} \cdot \vec{\sigma}} = \frac{1}{m}(E(k) - \vec{k} \cdot \vec{\sigma})$ for $\cosh\theta = \frac{E(k)}{m}$, and the conjugation property $C_s \vec{\sigma} C_s = \vec{\sigma}^*$.

2. Show that for a non-self-conjugate $(\frac{1}{2},0)$ field χ_α, the bilinear $\chi_\alpha^* \chi_\alpha$ does not transform as a scalar under general HLG transformations (see (7.38)).

3. Show that the free Hamiltonian for a Dirac particle of mass m is given by

$$H_0 = \int d^3k E(k) \sum_\sigma (b^\dagger(\vec{k}, \sigma) b(\vec{k}, \sigma) + d^\dagger(\vec{k}, \sigma) d(\vec{k}, \sigma))$$

$$= \int d^3x : \bar{\psi}(x)(i\vec{\gamma} \cdot \vec{\nabla} + m)\psi(x) :$$

As for the Majorana case, remember that for fermions the normal ordering includes an extra minus sign for each transposition of fermion operators, by definition.

4. Convince yourself that $J^\mu(x) \equiv e : \bar{\psi}(x)\gamma^\mu \psi(x) :$ is the quantized version of the conventional electric current four-vector for a charged Dirac particle described by a free Dirac field $\psi(x)$, by showing
(a) current conservation (use the Dirac equation for ψ):

$$\partial_\mu J^\mu(x) = 0$$

(b) that the charge Q defined as

$$Q \equiv \int d^3x \, J_0(\vec{x}, t)$$

counts electric charge (defined as e times the difference in the number of particles and antiparticles), and is time-independent.

5. Prove the following equal-time anticommutator relation for the Dirac field:

$$\{\psi_n(\vec{x}, t), \psi_m^\dagger(\vec{y}, t)\} = \delta^3(\vec{x} - \vec{y})\delta_{nm}$$

6. Use the result (7.113) for the spin sums for a Dirac particle

$$\sum_\sigma u(p, \sigma)\bar{u}(p, \sigma) = (\not{p} + m)/2m$$

(and the corresponding result (7.114) for the v spin functions) to derive the momentum-space formula for the Dirac propagator, defined as follows (note the minus sign!):

$$S_F(x - y)_{mn} \equiv \langle 0|T(\psi_m(x)\bar{\psi}_n(y))|0\rangle$$
$$\equiv \theta(x^0 - y^0)\langle 0|\psi_m(x)\bar{\psi}_n(y)|0\rangle - \theta(y^0 - x^0)\langle 0|\bar{\psi}_n(y)\psi_m(x)|0\rangle$$

Show that the Fourier transform of S_F is

$$S_F(p) = i\frac{\not{p} + m}{p^2 - m^2 + i\epsilon}$$

7. The relation between the four-vector notation and the (A=1/2,B=1/2) notation for the fundamental representation of the Lorentz group is given by

$$v_{\frac{1}{2},\frac{1}{2}}^{\frac{1}{2}\frac{1}{2}} = \frac{1}{\sqrt{2}}(v^1 - iv^2)$$

$$v_{\frac{1}{2},-\frac{1}{2}}^{\frac{1}{2}\frac{1}{2}} = \frac{1}{\sqrt{2}}(v^0 - v^3)$$

$$v_{-\frac{1}{2},\frac{1}{2}}^{\frac{1}{2}\frac{1}{2}} = -\frac{1}{\sqrt{2}}(v^0 + v^3)$$

$$v_{-\frac{1}{2},-\frac{1}{2}}^{\frac{1}{2}\frac{1}{2}} = -\frac{1}{\sqrt{2}}(v^1 + iv^2)$$

Recalling that $J_3 = A_3 + B_3$, $K_3 = i(A_3 - B_3)$, where both $\vec{A}$ and $\vec{B}$ are represented by $\frac{1}{2}\vec{\sigma}$ (one-half the Pauli matrices) for the (1/2,1/2) representation, show that the action of $A_3 + B_3$ on the left-hand sides is equivalent to the action of J_3 on the right-hand sides above, and likewise for the boost generator in the z direction.

8. Construct the spin functions $u^\mu(\vec{k}, \sigma), v^\mu(\vec{k}, \sigma)$, in four-vector notation, for a $j = 1$ massive boson, starting from the spin functions $u^{(\frac{1}{2}\frac{1}{2})}_{\pm\frac{1}{2},\pm\frac{1}{2}}$ etc., in (AB) notation. Do this in the following steps:

(a) First, construct the four-vectors $u^\mu(\vec{0}, \sigma), v^\mu(\vec{0}, \sigma)$ for the particle at rest, using (7.64, 7.54), with $\xi^{AB} = (-1)^{2B}$), and the translation dictionary supplied in Problem 7. Verify that $v^\mu(\vec{0}, \sigma) = (u^\mu(\vec{0}, \sigma))^*$.

(b) Show that the polarization vectors derived in (a) satisfy

$$\sum_\sigma u^\mu(\vec{0}, \sigma)u^{\nu*}(\vec{0}, \sigma) = -g^{\mu\nu} + g_0^\mu g_0^\nu$$

(Consider the cases of μ, ν both spatial, then one time, one spatial, etc.)

(c) Now, show that for non-zero momentum,

$$\sum_\sigma u^\mu(\vec{k}, \sigma)u^{\nu*}(\vec{k}, \sigma) = -(g^{\mu\nu} - \frac{k^\mu k^\nu}{m^2})$$

Use the boost operator $L^\mu_\nu(k)$, recalling that $L^\mu_0 = k^\mu/m$, and that $L^\mu_\nu(k)$ is a Lorentz transformation.

(d) The polarization vectors in helicity representation are

$$u^\mu(\vec{k}, \lambda) = \sum_\sigma D^j_{\sigma\lambda}(R(\hat{k}))u^\mu(\vec{k}, \sigma)$$

Show that they satisfy an identical equation to part (c).

9. Verify the expression (7.176) for the free-photon Hamiltonian, starting from (7.174) for the massless spin-1 field in radiation gauge. (Note the following properties for the radiation gauge polarization vectors: $\sum_\lambda \epsilon_i(\vec{k}, \lambda)\epsilon_j(\vec{k}, \lambda)^* = \delta_{ij} - \hat{k}_i\hat{k}_j$, $\sum_{i=1}^3 \epsilon_i(\vec{k}, \lambda)\epsilon_i(\vec{k}, \lambda')^* = \delta_{\lambda\lambda'}$.)

10. In Theory B of Section 7.6, with interaction Hamiltonian $\mathcal{H}_{\text{int}} = \lambda\psi^\dagger\psi\phi$, where ϕ is a self-conjugate field of mass M and ψ a non-self-conjugate field of mass m, calculate the *connected* S-matrix element for elastic $\psi - \phi$ scattering to lowest (i.e., second) order in λ:

$$\psi(p) + \phi(k) \rightarrow \psi(p') + \phi(k')$$

Note that in this case the vacuum expectation value of the time-ordered product of the ψ field occurs, in the form

$$\langle 0|T\{\psi(x_1)\psi^\dagger(x_2)\}|0\rangle \equiv i\Delta^{(\psi)}_F(x_1 - x_2)$$
$$= i\int \frac{d^4q}{(2\pi)^4} \frac{e^{-iq\cdot(x_1-x_2)}}{q^2 - m^2 + i\epsilon}$$

Interpret your result graphically.

11. Again, in Theory B of Section 7.6, calculate the lowest-order connected S-matrix element for the annihilation process

$$\psi(p_1) + \psi^c(p_2) \to \phi(k_1) + \phi(k_2)$$

Interpret your result graphically.

12. Calculate the second-order scattering amplitude for $\phi - \psi$ scattering in Theory C: i.e.,

$$\psi(p, \sigma) + \phi(k) \to \psi(p', \sigma') + \phi(k')$$

You should find again that the amplitude is the sum of two terms: discuss its crossing symmetry under the transposition of the bosonic particle momenta.

13. The Higgs particle (which recent results from the Large Hadron Collider at CERN suggest may finally have been discovered at a mass of around 125 GeV) couples to leptons (and quarks) via our Theory C; i.e., a standard Yukawa interaction term $\mathcal{H}_{\text{int}} = \lambda\bar{\psi}\psi\phi$, where the ϕ Higgs field represents a spinless self-conjugate particle of mass M and the lepton Dirac field ψ has mass $m < M/2$. Calculate the lowest-order total decay rate of the Higgs to a lepton-antilepton pair, to lowest order in λ. Spin sums such as (7.113) and (7.114) will be useful.

8

Dynamics VI: The classical limit of quantum fields

The precise way in which underlying microphysical processes, governed by the laws of quantum mechanics, merge into a phenomenal realm describable by classical laws has been a source of intense discussion and controversy from the very earliest days of quantum theory. At its core, this subject leads inexorably to quantum measurement theory, a subject long regarded by physicists of a more practical bent as an intellectual black hole, quite capable of permanently absorbing any physicist careless enough to stray within its event horizon. Nevertheless, much can be understood of the way in which quantum phenomena merge into and "mimic" classical physics without making a definite commitment to the ultimate role or character of measurement processes in quantum theory. Typically, one approaches this topic by identifying a set of "complementarity" relations between quantities which have a precise meaning in classical physics but cannot be simultaneously "sharp" once the theory is quantized. The archetype of such relations is, of course, the Heisenberg uncertainty principle relating the dispersion (or roughly, the "uncertainty") in the position and momentum observables for a non-relativistic point particle. Such relations follow directly, by a straightforward exercise in linear algebra, from the non-commuting character of the associated quantum-mechanical operators. From the complementarity point of view, the classical limit amounts to a regime in which the dispersion of the relevant observables (in the states of interest) becomes much smaller than their mean values, although, of course, still restricted by the appropriate quantum-mechanical uncertainty principle. Our task in this chapter will be to identify and examine those states in a relativistic quantum field theory for which the field observables of interest take on an essentially classical character.

8.1 Complementarity issues for quantum fields

As already seen in Chapter 1 in Jordan's treatment of a one-dimensional quantized massless field (his toy model for quantized electrodynamics), a quantum field theory is structurally a system of infinitely many quantized degrees of freedom. If we restrict the system to a finite-sized box, outside of which the fields vanish, then each classical eigenmode on quantization becomes (for the free field) an independent quantum-mechanical degree of freedom, describable either in terms of the familiar p and q operators, with $[p, q] = -i\hbar$, or equivalently in terms of their non-hermitian linear combinations, the destruction and creation operators $a, a^\dagger$, with commutation relation $[a, a^\dagger] = 1$ (independently for each mode). As these box eigenmodes are essentially momentum eigenstates, this description obscures the spacetime aspects of the field. From a spatiotemporal point of view, the conjugate variables analogous to the "q"s and "p"s of non-relativistic quantum theory are, for a spinless field ϕ, the field variables $\phi(\vec{x}, t)$ and their time-derivatives $\pi(\vec{x}, t) \equiv \frac{\partial \phi(\vec{x}, t)}{\partial t}$ at *each spatial point* $\vec{x}$ at some fixed time t. Locality ensures that the $\phi(\vec{x}, t)$ commute among themselves (i.e., for any $\vec{x}, \vec{y}, [\phi(\vec{x}, t), \phi(\vec{y}, t)] = 0$), as do the $\pi(\vec{x}, t)$, while (with $x^0 = y^0 = t$, and the measure $\mathcal{D}k \equiv \frac{d^3k}{\sqrt{(2\pi)^3 2E(k)}}$)

$$[\pi(\vec{x}, t), \phi(\vec{y}, t)] = \int \mathcal{D}k\mathcal{D}q[-iE(k)a(\vec{k})e^{-ik\cdot x} + iE(k)a^\dagger(\vec{k})e^{ik\cdot x}, a(\vec{q})e^{-iq\cdot y} + a^\dagger(\vec{q})e^{iq\cdot y}]$$

$$= \int \frac{d^3k}{(2\pi)^3 2E(k)}(-iE(k)e^{i\vec{k}\cdot(\vec{x}-\vec{y})} - iE(k)e^{-i\vec{k}\cdot(\vec{x}-\vec{y})})$$

$$= -i \int \frac{d^3k}{(2\pi)^3}e^{i\vec{k}\cdot(\vec{x}-\vec{y})} = -i\delta^3(\vec{x} - \vec{y}) \tag{8.1}$$

Note that the analogy to the fundamental $[p, q]$ commutation relation is somewhat obscured here by the use of natural units in which $\hbar$ is set to unity: restoring the $\hbar$, it appears on the right-hand side of (8.1) as expected.

The next step, in analogy to the well-known procedure in non-relativistic quantum mechanics, would naturally be a derivation of an inequality for the product of the dispersion of the two non-commuting field operators, analogous to $\Delta p \cdot \Delta x \geq \hbar/2$ in particle quantum mechanics. At this point we encounter an embarassment: the dispersions of the field operators $\phi(x)$, or $\pi(x)$, defined at a single spacetime point, are typically infinite! In the simplest possible Fock-space state, for example—the vacuum—one finds that $\langle 0|\phi(x)^2|0\rangle = \infty$. Mathematically, the reason for this is that the field operator $\phi(x)$ is in fact not well-defined on the Hilbert space of normalizable Fock-space states: the unit norm state $|0\rangle$ is taken into an infinite-norm state $\phi(x)|0\rangle$ by the action of the local field $\phi(x)$, whence the divergence of the vacuum expectation value $\langle 0|\phi(x)^2|0\rangle = (\phi(x)|0\rangle, \phi(x)|0\rangle)$. Local fields such as $\phi(x)$ and $\pi(x)$ should instead be regarded as operator-valued distributions: they yield well-defined[1] operators only after smearing with sufficiently smooth c-number test functions (cf. the discussion of localization in Section 6.5). At the very least, we must construct spatially smeared operators: e.g., we might replace the field at the origin by

$$\bar{\phi} = \frac{1}{(2\pi a^2)^{3/2}} \int d^3x e^{-\frac{\vec{x}^2}{2a^2}} \phi(\vec{x}, t) \tag{8.2}$$

The normalization factor is chosen so that the smeared object is identical to the original for constant fields. The square of this operator has a perfectly finite vacuum expectation value: in other words, $\bar{\phi}$ maps the vacuum state to a normalizable state

---

[1] The smeared field operators will still be unbounded operators, as are typically the position and momentum operators in ordinary quantum mechanics: hence, their domain will be a proper subset of the full Hilbert space, but at least it will not be empty! A systematic discussion of the use of smeared operators in field theory will be an indispensable part of our introduction to axiomatic quantum field theory in Chapter 9.

in the Hilbert space (in mathematical lingo, the vacuum state lies in the domain of the operator $\bar{\phi}$). A short calculation shows that

$$\langle 0|\bar{\phi}^2|0\rangle = \frac{1}{4\pi^2} \int_0^\infty \frac{k^2}{\sqrt{k^2 + m^2}} e^{-a^2 k^2} dk \sim \frac{1}{8\pi^2 a^2}, \quad a << \frac{1}{m} \tag{8.3}$$

which shows clearly the re-emergence of the quadratic divergence in the vacuum expectation value in the local limit $a \to 0$. We see that it is not even possible to define a dispersion $\Delta\phi(x)$ for the local field $\phi(x)$, while for the smeared field the dispersion is perfectly finite: $\Delta\bar{\phi} \equiv \sqrt{\langle 0|\bar{\phi}^2|0\rangle - \langle 0|\bar{\phi}|0\rangle^2} = \sqrt{\langle 0|\bar{\phi}^2|0\rangle}$. Physically, the restriction to smeared fields is also perfectly reasonable, as we can hardly expect any conceivable measurement apparatus to have infinitely fine resolution, either in space or in time.

The preceding discussion focussed on a spin-0 (hence bosonic) field. What about spin-$\frac{1}{2}$ fields, which, by the Spin-Statistics theorem, necessarily describe fermionic particles? We recall from the discussion of blackbody radiation in Chapter 1 that the classical "Rayleigh–Jeans" regime corresponds to the limit in which the denominator of the Planck distribution (1.22) vanishes, corresponding to a large occupation number in modes of any given frequency. We shall see in the next section that the classical limit for a quantum field necessarily requires such large occupation numbers. For a fermionic field, the corresponding particle quanta are restricted by the exclusion principle to an occupancy of either zero or one for each distinct quantum mode, *so a classical limit for such fermionic fields is simply impossible.* Of course, there are many particle states of such fields which exhibit classical behavior—the $\sim 10^{26}$ electrons, protons and neutrons bound together into a spherical billiard ball certainly behaves perfectly classically in many contexts—but the classical physics appropriate in this circumstance is particle mechanics, rather than classical field theory. In this chapter we are concerned with the approach to the latter, so fermionic fields will no longer be considered.

Returning, then, to bosonic fields, the only classical fields of importance are the electromagnetic and gravitational fields. As conventional quantum field theories of gravity are at best effective theories valid only at low energies, we shall concentrate on the electromagnetic field, which in any case deserves special consideration given its unique role in the gestation and birth of modern quantum mechanics and field theory. This leads us to the consideration of complementarity issues for the field $\vec{A}(x)$ (7.174) introduced in Section 7.5 for the description of massless spin-1 particles. We recall that this field undergoes a non-trivial modification under gauge transformations, leaving the physics invariant (classically, it is the spatial part of the four-vector potential). The associated physical (hence gauge-invariant) fields are the electric field and magnetic fields: in the radiation (or Coulomb) gauge in which we constructed $\vec{A}(x)$, these take the form $\vec{E} = \frac{\partial \vec{A}}{\partial t}$ and $\vec{B} = \vec{\nabla} \times \vec{A}$ (or $B_i = \epsilon_{ijk} \partial_j A_k$, $i, j, k = 1, 2, 3$). We can think of the electric field, involving a time-derivative, as the conjugate momentum field to the magnetic field, which involves only space derivatives, and hence only combinations of field values on a given time-slice. The commutation relation analogous to the result (8.1) for a canonical scalar field is (see Problem 1)

$$[E_i(\vec{x}, t), B_j(\vec{y}, t)] = i\epsilon_{ijk} \partial_k \delta^3(\vec{x} - \vec{y}) \tag{8.4}$$

The reader may easily verify that the corresponding equal-time commutators among electric and magnetic fields vanish separately.

Equal time commutation relations of the type (8.1) and (8.4) will later (cf. Chapter 12) play a central role in the development of a Lagrangian formalism for field theory, of enormous utility in the incorporation of spacetime and local gauge symmetries, which necessarily take a complicated and obscure form in the Hamiltonian approach we have followed so far. Here we note that the complementary role played by electric and magnetic fields leads to physical consequences of enormous importance in modern condensed matter physics and field theory. In the BCS theory of super-conductivity, for example, the appearance of a condensate of charged Cooper pairs in the superconducting state implies an essentially infinite dispersion in the local electric fields, requiring the dispersion of the local magnetic field to vanish in the interior of the superconductor- the famous Meissner effect. A *dual* Meissner effect, in which the chromoelectric field is forced to zero (in this case as a result of large chromomagnetic fluctuations) outside of thin tubes carrying the conserved flux required by Gauss's Law, is believed to lie at the core of color confinement in quantum chromodynamics (cf. Chapter 19). The non-zero zero-point energy of the electromagnetic field ($\frac{1}{2}\hbar\omega$ for each quantized mode of the field) can be viewed as a direct consequence of the non-existence of states in which both electric $\vec{E}$ and magnetic $\vec{B}$ fields have vanishing expectation values and dispersion, allowing zero energy ($\propto \vec{E}^2 + \vec{B}^2$). And so on . . .

In the next section we shall examine closely the conditions for "classical" behavior of a quantum field. This will turn out to be more easily accomplished in momentum space—i.e., by examining the multi-particle states for individual quantized modes of the field, rather than in terms of the spatiotemporally defined fields appearing in the commutators above. Before doing that, a brief historical digression is in order. As we recall from the historical account in Chapters 1 and 2, the birth of quantum field theory was accompanied by much uncertainty about the validity of a straightforward extension of quantum-mechanical principles to systems with infinitely many degrees of freedom. An early critique which caused a great deal of unease appeared in a paper of Landau and Peierls, appearing in 1930 and published the following year (see (Landau and Peierls, 1983)), in which the arbitrarily precise measurability of even *individual* components of the quantized electric or magnetic fields employing a charged point particle was denied (for fields smeared over a temporal extent Δt, Landau and Peierls claimed an intrinsic dispersion $\Delta E \simeq \Delta B \simeq \frac{1}{(\Delta t)^2}$). The basic difficulty was that the acceleration of the test particle in the presence of the field to be measured would lead to an uncontrollable radiation emission and concomitant energy-momentum loss. In a famous (and famously unread[2]) paper, Bohr and Rosenfeld (Bohr and Rosenfeld, 1983) subjected the measurability of the field components of the free electromagnetic field to a typically exhaustive examination, and concluded that the uncertainty relations holding among appropriately smeared averages of various components of the electric and magnetic field operators are precisely consistent with

---

[2] As Pais relates in his biography of Bohr (Pais, 1991): "It [the Bohr–Rosenfeld paper] has been read by very very few of the *aficionados*. . . As a friend of Bohr's and mine once said to me: 'It is a very good paper that one does not have to read. You just have to know that it exists.' "

the $\Delta p \Delta x \geq \frac{\hbar}{2}$ constraint holding for the dispersion of the momentum and position of an *extended* test body, sufficiently massive to allow the response of the test body to the field to lead to negligible accelerations, thereby minimizing the uncontrollable radiation emissions that had bothered Landau and Peierls.

The Bohr–Rosenfeld paper played an important historical role in reassuring the quantum community that the quantization of systems with infinitely many degrees of freedom did not lead to conceptual inconsistencies or phenomenologically pathological results, but that quantum field theories should indeed be viewed simply as quantum systems of a new type, where the constraints on measurability bear just the same relation to the underlying formal operator framework as in non-relativistic point-particle quantum mechanics. From a practical point of view, the enormous body of phenomenological information—the entire field of quantum optics—gathered over the three-quarters of a century that have elapsed since the Bohr–Rosenfeld paper has indeed allowed measurements of the quantum properties of light with unparalleled precision. However, these measurements typically address the behavior of states in which a limited number of quantized modes of the electromagnetic field are excited: in other words, we are concerned with the field in momentum rather than coordinate space. In the next section we shall see that the transition to classical behavior (as well as the non-classical deviations therefrom which are the bread and butter of quantum optics) is most easily studied in the occupation number space of these modes.

8.2   When is a quantum field "classical"?

A spatiotemporally-defined classical field is mathematically a c-number function (or collection of functions, if we are dealing with the components of vector or tensor fields) with sufficient mathematical regularity to allow the evaluation of a Fourier transform to wavevector-frequency $(\vec{k}, \omega)$ space. In other words, such a field can be viewed as a linear combination of modes of well-defined wavevector and frequency. For the electromagnetic field in a cavity, the field is more conveniently analysed in terms of standing wave modes, which involve linear combinations of oppositely traveling waves. In either case, the essential feature of a classical field is our ability to simultaneously specify the amplitude and phase of each mode of the field. The quantization of the field introduces an inescapable complementarity between these two properties. This is best illustrated first at the level of an individual mode: later (in Section 8.4) we shall see how to construct states of the field with arbitrary spatial dependence. It is important to realize at the outset that the restriction to a single mode is by no means physically unreasonable. Very early in the development of laser technology, ingenious mode-selection methods were devised (Smith, 1972) to ensure that lasing in a cavity produced multiple occupation of a *single quantum mode*. For example, introduction of a Fabry–Pérot interferometer with a suitable geometry into the cavity can ensure that only a single frequency mode fits into the positive gain portion of the laser gain profile. In such cases we may assume that the state of the electromagnetic field consists of zero photon number for all but a single mode. As indicated previously, such a mode will typically be a standing wave mode, but the essential features are the same if we simply consider the occupied mode to have well-defined photon momentum $\vec{k}$, polarization λ, and energy $E(k) = \omega$. In a cubical box of volume V, the integral over continuous wavevectors (i.e., momenta) becomes a discrete sum in the usual way, so that we can speak sensibly of a specific individual mode. Instead of imposing the physical constraints of an actual rectangular laser cavity (electric field vanishing at the plane boundaries, for example), we shall simply assume our cubical box of volume V to be a three-torus topologically, so that the fields satisfy simple periodic boundary conditions in each of the three Cartesian coordinates. The transition from continous to discrete momenta, and from creation–annihilation operators with δ distribution commutators to ones with Kronecker δs, is then made via

$$\int d^3k \rightarrow \frac{(2\pi)^3}{V} \sum_{\vec{k}} \,, \quad (k_x, k_y, k_z) = \frac{2\pi}{L}(n_x, n_y, n_z)$$

$$a(\vec{k}, \lambda) \rightarrow \sqrt{\frac{(2\pi)^3}{V}} a_{\vec{k},\lambda}$$

where we use subscripts (rather than functional dependence) to indicate discretely defined objects—for example, the creation–annihilation operators, which now satisfy

$$[a_{\vec{k},\lambda}, a^\dagger_{\vec{k}',\lambda'}] = \delta_{\lambda\lambda'}\delta_{\vec{k}\vec{k}'} \tag{8.5}$$

With these translations, and the assumption that only one mode is occupied, we can effectively truncate the quantized spin-1 field $\vec{A}$ given in (7.174) to

$$\vec{A}(x) = \sqrt{\frac{1}{2E(k)V}} \vec{\epsilon}_{\vec{k},\lambda}(a_{\vec{k},\lambda}e^{-ik\cdot x} + a^\dagger_{\vec{k},\lambda}e^{ik\cdot x}) \tag{8.6}$$

Note that we have also assumed a plane-polarized mode, so that $\vec{\epsilon}_{\vec{k},\lambda} = \vec{\epsilon}^{\,*}_{\vec{k},\lambda} = \hat{x}$, for example. The corresponding electric field operator is

$$\vec{E}(x) = i\sqrt{\frac{E(k)}{2V}} \vec{\epsilon}_{\vec{k},\lambda}(a_{\vec{k},\lambda}e^{-ik\cdot x} - a^\dagger_{\vec{k},\lambda}e^{ik\cdot x}) \tag{8.7}$$

To simplify the notation, we shall henceforth drop the subscripts $\vec{k}, \lambda$ indicating the specific mode under consideration, and write simply

$$\vec{E}(\vec{x}, t) = iC(ae^{i\vec{k}\cdot\vec{x}-i\omega t} - a^\dagger e^{-i\vec{k}\cdot\vec{x}+i\omega t})\vec{\epsilon} \tag{8.8}$$

Classically, the operators a (resp. $a^\dagger$) correspond to complex numbers $\alpha = Ae^{i\theta}$ (resp. α^*), giving a monochromatic wave of well-defined amplitude and phase:

$$\vec{E}(\vec{x}, t) = -2CA\sin(\vec{k}\cdot\vec{x} - \omega t + \theta)\vec{\epsilon} \tag{8.9}$$

If our state space were finite-dimensional, we could apply the polar decomposition theorem for finite-dimensional operators, which asserts that we can find a semipositive-definite hermitian operator $\mathcal{N}$ and unitary operator E, with

$$a = E\sqrt{\mathcal{N}}, \quad a^\dagger = \sqrt{\mathcal{N}}E^\dagger \tag{8.10}$$

$$\mathcal{N} = a^\dagger a \tag{8.11}$$

and (using $[a, a^\dagger] = 1$)

$$\mathcal{E}\mathcal{N} - \mathcal{N}E = E \tag{8.12}$$

If we further identify the unitary operator as the complex exponential of an hermitian phase operator Φ, $E = e^{i\Phi}$, then we see that the classical amplitude and phase concepts A, θ correspond to eigenvalues of the *non-commuting* operators $\sqrt{\mathcal{N}}, \Phi$ respectively. There is evidently a complementarity between the concepts of amplitude and phase at the quantum mechanical level. If one *assumes* a commutation relation

$$[\mathcal{N}, \Phi] = i \tag{8.13}$$

then the commutation relation (8.12), with $E = e^{i\Phi}$, indeed follows directly using the usual multiple-commutator formulas. In analogy to $[q, p] = i\hbar$ leading to the uncertainty relation $\Delta q \Delta p \geq \frac{\hbar}{2}$, one then concludes that $\Delta\mathcal{N}\Delta\Phi \geq \frac{1}{2}$, and it is immediately apparent that a classical limit, with both amplitude and phase defined to high relative precision, necessarily requires the number operator $\mathcal{N}$ to take on large values, with $\Delta\mathcal{N} << \bar{\mathcal{N}}$ (but $\Delta\mathcal{N} >> 1$, thus still allowing $\Delta\Phi << 2\pi$).

While this conclusion is basically correct, the argument leading to it is completely fallacious. For example, if we take the expectation value of the purported commutation relation (8.13) in the vacuum state $|0\rangle$ for our mode, we obtain

$$\langle 0|\mathcal{N}\Phi - \Phi\mathcal{N}|0\rangle = 0 = i \tag{8.14}$$

There are two sources of trouble here. First, the polar decomposition theorem does *not* extend in general to the existence of polar pairs $\mathcal{N}, E$ in infinite-dimensional spaces. Secondly, the transition from a unitary E to a phase operator Φ clearly cannot lead to the desired commutation relation (8.13), as the latter is clearly inconsistent. In fact, the definition and construction of a well-defined phase operator in quantum mechanics is a notoriously slippery subject—which is hardly surprising, given that even classically the phase variable is not uniquely defined, but given only modulo 2π.

We shall circumvent the first problem by dealing with a truncated system, where only states with a maximum occupation number N are considered. It will become apparent in the next section that the coherent states that most mimic classical behavior have very rapidly decreasing components for large occupation numbers N, with the probability of detecting more than N quanta falling more rapidly than $\left(\frac{e\bar{n}}{N}\right)^N$ for $N > \bar{n} \equiv \langle \mathcal{N} \rangle$ once we consider states with occupancy N exceeding the expectation value $\bar{n}$ of the number operator. Such states also have exponentially small ($\sim e^{-\bar{n}^2}$) probability of occupancy of the vacuum state $|0\rangle$. Thus, we expect to make only very small errors by simply truncating the state space at some very high, but finite, occupation number N. The second problem, defining a phase operator, can be avoided simply by observing that the specification of the phase θ modulo 2π amounts to specifying the real numbers $\cos\theta$ and $\sin\theta$, or equivalently, the complex number $e^{i\theta}$, corresponding to our unitary operator E above. There will be no difficulty obtaining a well-defined unitary E (or its hermitian "real" and "imaginary" parts $C \equiv \frac{1}{2}(E + E^\dagger)$ and $S \equiv \frac{1}{2i}(E - E^\dagger)$) once we have truncated the space.

In keeping with our demand that only a maximum number N of photons can occupy any given mode of the electromagnetic field, we define new creation operators by demanding that

$$a^\dagger|n\rangle = \sqrt{(n+1)}|n+1\rangle, \ n < N, \quad a^\dagger|N\rangle = 0 \tag{8.15}$$

The destruction operator a is then given by the adjoint of $a^\dagger$. The algebra is realized by $(N+1) \times (N+1)$ matrices, which are easily seen to give the modified commutator

$$[a, a^\dagger]_{nm} = \delta_{nm} - (N+1)\delta_{nN}\delta_{mN} \tag{8.16}$$

or, in operator language,

$$[a, a^\dagger] = 1 - (N+1)|N\rangle\langle N| \tag{8.17}$$

The restriction to states with extremely (factorially) small components in the $|N\rangle$ (or higher) mode means that the effect of the last term on the right, with a projection operator onto the highest mode, is effectively negligible. It is, of course, required, because the trace of the commutator of two finite-dimensional matrices must vanish. Note that we still have the usual number operator $\mathcal{N} = a^\dagger a$ with matrix elements $\mathcal{N}_{nm} = n\delta_{nm}, 0 \le n \le N$.

In this finite-dimensional space the application of the polar decomposition theorem is unproblematic, and we find that

$$a = E\sqrt{\mathcal{N}}, \quad a^\dagger = \sqrt{\mathcal{N}}E^\dagger \tag{8.18}$$

where E is the unitary operator corresponding to anticyclic permutation of the states $|0\rangle, |1\rangle, |2\rangle, ...|N\rangle$:

$$E = \sum_{n=0}^{N-1} |n\rangle\langle n+1| + |N\rangle\langle 0| \tag{8.19}$$

E is basically the lowering operator a without the square-root factors, except for the action on the vacuum, which, as mentioned above, has exponentially small occupancy in the quasi-classical states (see Section 8.3 below) in which we shall be interested. E is the quantum analog of the complex classical phase $e^{i\theta}$. It is a normal operator (i.e., being unitary, it commutes with its adjoint), so its hermitian and antihermitian parts $C \equiv \frac{1}{2}(E + E^\dagger)$ and $S \equiv \frac{1}{2i}(E - E^\dagger)$ commute

$$[C, S] = 0 \tag{8.20}$$

The commutator of the phase operator E with the number operator $\mathcal{N}$ is

$$[E, \mathcal{N}] = E - (N+1)|N\rangle\langle 0| \approx E \tag{8.21}$$

where once again we have neglected operators such as $|N\rangle\langle 0|$, which simply exchange the very small amplitudes for our state to have either 0 or N photons. Together with $[E^\dagger, \mathcal{N}] \approx -E^\dagger$, this then yields

$$[C, \mathcal{N}] = iS, \quad [S, \mathcal{N}] = -iC \tag{8.22}$$

For arbitrary hermitian operators $\mathbf{X}, \mathbf{Y}, \mathbf{Z}^3$ satisfying $[\mathbf{X}, \mathbf{Y}] = i\mathbf{Z}$, and defining $\langle \mathbf{X} \rangle = \langle \psi | \mathbf{X} | \psi \rangle$ for a unit-normalized state $|\psi\rangle$ (and likewise for $\mathbf{Y}$ and $\mathbf{Z}$), the fluctuation operators $\Delta \mathbf{X} \equiv \mathbf{X} - \langle \mathbf{X} \rangle$, $\Delta \mathbf{Y} \equiv \mathbf{Y} - \langle \mathbf{Y} \rangle$ satisfy $[\Delta \mathbf{X}, \Delta \mathbf{Y}] = i\mathbf{Z}$, whence, for arbitrary real λ,

$$0 \leq ((\Delta \mathbf{X} + i\lambda \Delta \mathbf{Y})|\psi\rangle, (\Delta \mathbf{X} + i\lambda \Delta \mathbf{Y})|\psi\rangle) \tag{8.23}$$

$$\Rightarrow 0 \leq \langle \psi | (\Delta \mathbf{X})^2 | \psi \rangle + i\lambda \langle \psi | [\Delta \mathbf{X}, \Delta \mathbf{Y}] | \psi \rangle + \lambda^2 \langle \psi | (\Delta \mathbf{Y})^2 | \psi \rangle$$

$$\Rightarrow 0 \leq \langle (\Delta \mathbf{Y})^2 \rangle \lambda^2 - \langle \mathbf{Z} \rangle \lambda + \langle (\Delta \mathbf{X})^2 \rangle, \quad \forall \lambda \tag{8.24}$$

The final inequality implies that the discriminant of the quadratic equation for λ must be negative or zero, i.e.,

$$\langle \mathbf{Z} \rangle^2 - 4 \langle (\Delta \mathbf{X})^2 \rangle \langle (\Delta \mathbf{Y})^2 \rangle \leq 0 \tag{8.25}$$

Writing $\sqrt{\langle (\Delta \mathbf{X})^2 \rangle} \to \Delta X$ for simplicity, (and likewise for $\mathbf{Y}$) we arrive at the uncertainty relation

$$\Delta X \cdot \Delta Y \geq \frac{1}{2} |\langle \mathbf{Z} \rangle| \tag{8.26}$$

Applying this general result to the commutation relations (8.22), we find the desired number-phase uncertainty relations

$$\Delta \mathcal{N} \cdot \Delta C \geq \frac{1}{2} |\langle S \rangle| \tag{8.27}$$

$$\Delta \mathcal{N} \cdot \Delta S \geq \frac{1}{2} |\langle C \rangle| \tag{8.28}$$

For a state of well-defined photon occupation number $|n\rangle$,

$$\langle n | C | n \rangle = \frac{1}{2} \langle n | E | n \rangle + \langle n | E^\dagger | n \rangle = 0 \tag{8.29}$$

$$\langle n | C^2 | n \rangle = \frac{1}{4} \langle n | E^2 + 2 + (E^\dagger)^2 | n \rangle = \frac{1}{2} \tag{8.30}$$

with a similar result for S. Evidently, the phase uncertainties are $\Delta C = \Delta S = \frac{1}{\sqrt{2}}$, corresponding to a phase spread uniformly over the unit circle $0 \leq \theta < 2\pi$. Note that in this case both $\Delta \mathcal{N}$ and $\langle C \rangle, \langle S \rangle$ are zero, so the uncertainty inequality is saturated trivially. Such states are "maximally quantal", having the least possible precise specification of phase. The states of interest in the classical limit—the so-called "coherent states" studied in the following section—provide a more interesting saturation of the uncertainty relations (8.27,8.28), in which both the number and phase operators have non-vanishing dispersion, with the product taking its minimum possible value, but with nevertheless an arbitrarily small *fractional* uncertainty in the number (i.e., amplitude) and phase expectation values.

---

[3] In the following, in an attempt to circumvent a confusing notational overload, we have introduced a bold-face notation for the operators to distinguish them from their c-number expectation values.

The answer to the question posed by the heading of this Section—namely, when is a quantum field classical?—is therefore quite straightforward. The behavior of such a field is classical only in the context of states with high occupation number for individual modes of the field. It is therefore necessary that the field (a) be bosonic in character, (b) be associated with stable particles, in order that a classical configuration, once established, be maintained over macroscopic time-scales, and (c) have a sensible non-interacting limit, so that the previous discussion, which assumed many-particle, but non-interacting, particle states is indeed applicable.

In the Standard Model, the only bosonic fields associated with *stable* elementary particles[4] are the gauge-vector bosons mediating electromagnetic (the photon) and strong (the gluons of quantum chromodynamics) interactions. We shall later see that the non-abelian gauge dynamics of the gluons of the strong interactions results in a qualitative alteration of the state space once interactions are turned on: namely, the phenomenon of color confinement (cf. Sections 19.3 and 19.4) in which the physical particle states of the theory are related to the underlying local (quark and gluon) fields in an extremely complicated way, so condition (c) is violated in this case. We are left (apart from gravity) with the quantized electromagnetic field as the single phenomenologically relevant example of a relativistic quantum field with a sensible classical limit. It is therefore hardly surprising, given the primary role played by correspondence principle arguments in the historical evolution of quantum mechanics and quantum field theory, that the unravelling of the quantum properties of light were at the core of the most critical developments in this history, as we have already seen in Chapters 1 and 2.

8.3   Coherent states of a quantum field

We have seen in the preceding section that the states in which a (bosonic) quantum field has approximately classical behavior, in the sense of having expectations of the amplitude (or number) and phase operators with small *fractional* quantum dispersion, necessarily involve linear combinations of high occupancy multi-particle states. Among such states there is a special subclass—called *coherent states*—which exhibit, in a certain sense, *ultraclassical* behavior. Such states will be defined by the requirement that the number-phase uncertainty relations (8.27,8.28) are saturated: i.e., the inequality becomes an equality, so that the mutual uncertainty is as small as it can possibly be. Such states are of more than theoretical interest, as they are in fact the states produced in tuned laser cavities for modes selected for high gain.

The properties of these distinguished coherent states are best obtained by starting with the inequality (8.23), but with the operators $\mathbf{X}, \mathbf{Y}$ chosen as the hermitian and antihermitian parts of the destruction operator for the mode in question:

$$\mathbf{X} = \frac{1}{2}(a + a^\dagger) \tag{8.31}$$

$$\mathbf{Y} = \frac{1}{2i}(a - a^\dagger) \tag{8.32}$$

---

[4] The Standard Model spin-0 Higgs is, of course, elementary but unstable: see Chapter 7, Problem 12.

with the commutator $\mathbf{Z} = -i[\mathbf{X}, \mathbf{Y}] = \frac{1}{2}$. We shall soon see that the states which saturate the mutual uncertainty relation for the hermitian operators $\mathbf{X}, \mathbf{Y}$ (called "quadrature" operators in the quantum optics literature) do the same for the number-phase uncertainty relations. The inequality (8.24) becomes an equality when the discriminant vanishes and $\lambda = \frac{\langle Z \rangle}{2(\Delta Y)^2}$. If we further require that the dispersions in X and Y are equal,[5] then $(\Delta X)^2 = (\Delta Y)^2 = \frac{1}{2}\langle \mathbf{Z} \rangle = \frac{1}{4}$, so $\lambda = 1$, and the state $|\psi\rangle$ must satisfy

$$0 = (\Delta\mathbf{X} + i\Delta\mathbf{Y})|\psi\rangle \Rightarrow (\mathbf{X} + i\mathbf{Y})|\psi\rangle = (\langle\mathbf{X}\rangle + i\langle\mathbf{Y}\rangle)|\psi\rangle \equiv \alpha|\psi\rangle \qquad (8.33)$$

where the eigenvalue α is in general complex. Of course, $\mathbf{X} + i\mathbf{Y} = a$, so the coherent states are simply the eigenstates of the destruction operator a. Expressed as linear combinations of multi-particle Fock states $|n\rangle$

$$|\psi\rangle = \sum_n c_n |n\rangle \qquad (8.34)$$

the eigenvalue equation $a|\psi\rangle = \alpha|\psi\rangle$ implies the recursion relation

$$c_{n+1} = \frac{\alpha}{\sqrt{n+1}} c_n \qquad (8.35)$$

with the unique solution $(c_0 \equiv C)$

$$c_n = C \frac{\alpha^n}{\sqrt{n!}} \qquad (8.36)$$

The requirement that $|\psi\rangle$ be unit normalized then implies $C = e^{-\frac{1}{2}|\alpha|^2}$, so

$$|\psi\rangle = e^{-\frac{1}{2}|\alpha|^2} \sum_{n=0}^{\infty} \frac{\alpha^n}{\sqrt{n!}} |n\rangle \qquad (8.37)$$

Henceforth, as is usual in quantum theory, we shall relabel the coherent state $|\psi\rangle$ by its eigenvalue with respect to the destruction operator a, thus $|\psi\rangle \rightarrow |\alpha\rangle$.

The interpretation of the complex number α becomes apparent if we return to the expression for the quantized single-mode electric field (8.8). The expectation values of the destruction a and creation $a^\dagger$ operators in the coherent state $|\alpha\rangle$ are α and α^* respectively, so writing $\alpha = Ae^{i\theta}$, the expectation value of the electric field in the coherent state is

$$\langle \alpha | \vec{E}(\vec{x}, t) | \alpha \rangle = -2CA \sin(\vec{k} \cdot \vec{x} - \omega t + \theta)\vec{\epsilon} \qquad (8.38)$$

i.e., exactly the classical monochromatic wave form (8.9). Thus the norm A and phase θ of the complex eigenvalue α encode the amplitude and phase of the corresponding classical field. We shall shortly see how to extend this result to construct coherent

---

[5] Minimum uncertainty states in which $\Delta X \neq \Delta Y$ are also of considerable interest in quantum optics: they are the so-called "squeezed" states.

states with an *arbitrary preassigned spatiotemporal behavior* for the field expectation value, even in a fully interacting theory!

Note that we have so far not imposed a mode cutoff at some large occupation number N as in the previous section. Our next task, then, is to verify the assertions made concerning the negligible effects of such a cutoff in the classical regime. First, note that the expectation value of the number operator $\mathcal{N}$ in the state $|\alpha\rangle$ is

$$\bar{\mathcal{N}} = \sum_n nc_n^2 = e^{-|\alpha|^2} \sum_n n\frac{|\alpha|^{2n}}{n!} = e^{-|\alpha|^2}|\alpha|^2 \frac{\partial}{\partial|\alpha|^2}e^{|\alpha|^2} = |\alpha|^2 \tag{8.39}$$

Thus, the probability $P(n)$ of finding exactly n photons in our coherent state characterized by an average photon number $\bar{\mathcal{N}}$ is precisely the Poisson distribution $P(n) = e^{-\bar{\mathcal{N}}}\frac{(\bar{\mathcal{N}})^n}{n!}$. Moreover, the classical limit requires that we choose $|\alpha|^2 = \bar{\mathcal{N}} \gg 1$ in order to achieve simultaneously a small fractional dispersion in the classical amplitude (related to $\bar{\mathcal{N}}$) and in the phase (which requires $\Delta\mathcal{N} \gg 1$). In this limit the Poisson distribution becomes approximately Gaussian,

$$P(n) \sim \frac{1}{\sqrt{2\pi\Delta\mathcal{N}}}e^{-\frac{1}{2}\frac{(n-\bar{\mathcal{N}})^2}{(\Delta\mathcal{N})^2}} \tag{8.40}$$

with $\Delta\mathcal{N} = (\bar{\mathcal{N}})^{1/2} \ll \bar{\mathcal{N}}$.

Introducing a mode cutoff at $N \gg \bar{\mathcal{N}} \gg 1$, we find that the components of our coherent state both at the bottom of the spectrum

$$\langle 0|\alpha\rangle = c_0 = e^{-|\alpha|^2/2} = e^{-\bar{\mathcal{N}}/2} \ll 1 \tag{8.41}$$

and at the top, using Stirling's formula for the factorial,

$$|\langle N|\alpha\rangle| \sim \frac{1}{(2\pi N)^{1/4}}e^{-\bar{\mathcal{N}}/2}\left(\frac{e\bar{\mathcal{N}}}{N}\right)^{N/2} \ll 1 \tag{8.42}$$

are negligibly small, allowing us to use freely the phase operators E, C, and S introduced in the previous section without worrying about the mutilation of the basic commutation relations (8.17) and (8.21) induced by our mode cutoff.

With the mode cutoff in place, the unitary operator E defined in (8.19) has eigenstates of the form

$$|\phi\rangle = \frac{1}{\sqrt{N+1}}\sum_{n=0}^{N}e^{in\phi}|n\rangle \tag{8.43}$$

with $E|\phi\rangle = e^{i\phi}|\phi\rangle$, provided the eigenphases ϕ take the discrete values $\phi = \frac{2\pi m}{N+1}, m = 0, 1, 2, \ldots N$. The corresponding eigenvalues of the hermitian (resp. antihermitian) parts C (resp. S) of E are, of course, $\cos\phi$ (resp. $\sin\phi$). The amplitude for measuring a phase ϕ in a coherent state $|\alpha\rangle$ is therefore

$$\langle\phi|\alpha\rangle = \frac{e^{-|\alpha|^2/2}}{N+1}\sum_{n=0}^{N}e^{-in\phi}\frac{\alpha^n}{\sqrt{n!}} = \frac{e^{-|\alpha|^2/2}}{N+1}\sum_{n=0}^{N}e^{-in(\phi-\theta)}\frac{|\alpha|^n}{\sqrt{n!}} \tag{8.44}$$

where we recall that θ is defined to be the phase of α: $\alpha = |\alpha|e^{i\theta}$. In the classical regime of interest, the factor $e^{-|\alpha|^2/2}\frac{|\alpha|^n}{\sqrt{n!}} = \sqrt{P(n)}$ becomes approximately Gaussian:

$$e^{-|\alpha|^2/2}\frac{|\alpha|^n}{\sqrt{n!}} \sim \frac{1}{(2\pi)^{1/4}\sqrt{\Delta\mathcal{N}}}e^{-\frac{1}{4}\frac{(n-\mathcal{N})^2}{(\Delta\mathcal{N})^2}} \tag{8.45}$$

The phase amplitude (8.44) can therefore be evaluated in the limit of large N as a Fourier sum approximated by a Gaussian integral

$$\langle\phi|\alpha\rangle \propto \int e^{-in(\phi-\theta)}e^{-\frac{1}{4}\frac{(n-\mathcal{N})^2}{(\Delta\mathcal{N})^2}}\,dn \propto e^{-i\mathcal{N}(\phi-\theta)}e^{-(\Delta\mathcal{N})^2(\phi-\theta)^2} \tag{8.46}$$

and the distribution in phase $P(\phi)$ of our coherent state is Gaussian

$$P(\phi) \equiv |\langle\phi|\alpha\rangle|^2 \propto e^{-2(\Delta\mathcal{N})^2(\phi-\theta)^2} \tag{8.47}$$

corresponding to a half-width in phase of $\Delta\phi = \frac{1}{2\Delta\mathcal{N}}$. The corresponding dispersion in $C = \cos\phi$ is evidently $\Delta C = |\sin\phi|\Delta\phi = \frac{|S|}{2\Delta\mathcal{N}}$ whence

$$\Delta\mathcal{N}\cdot\Delta C = \frac{1}{2}|S| \tag{8.48}$$

and we see that the number-phase uncertainty relation (8.27) is indeed saturated for the coherent state $|\alpha\rangle$, as expected.

The coherent state formalism is even more powerful then the preceding considerations may lead us to expect, enabling the explicit construction of states in fully interacting theories with an *exactly prescribed* spatial dependence for the field observables of the theory. To illustrate this point we shall consider a self-interacting scalar field ϕ, with dynamics specified by the ϕ^4-type interaction discussed earlier (cf. Chapter 6, (6.90)): namely, with a field energy density given by the normal-ordered operator

$$\mathcal{H}(x) =: \frac{1}{2}\dot\phi(x)^2 + \frac{1}{2}|\vec\nabla\phi(x)|^2 + \frac{1}{2}m^2\phi(x)^2 + \frac{\lambda}{4!}\phi(x)^4 : \tag{8.49}$$

and self-conjugate local scalar field

$$\phi(x) = \frac{1}{(2\pi)^{3/2}}\int\frac{d^3k}{\sqrt{2E(k)}}(a(\vec k)e^{-ik\cdot x} + a^\dagger(\vec k)e^{ik\cdot x}) \equiv \phi^{(+)}(x) + \phi^{(-)}(x) \tag{8.50}$$

We shall soon be focussing on the full Hamiltonian H, which is conserved, so we will evaluate all operators with the time taken to be zero in (8.49): thus $\phi(x) = \phi(\vec x, 0)$. We would like to construct coherent states in which the field operator $\phi(\vec x, 0)$ has a prescribed static spatial dependence $f(\vec x)$ for its expectation, with functionals of ϕ, such as the energy density above, having expectation values given as the corresponding functional of the prescribed c-number function $f(\vec x)$. The single-mode coherent states for photons discussed previously give us a clue for how to achieve this: they are eigenstates of the annihilation operator a for the given mode. Similarly, we expect general coherent states of a scalar quantum field ϕ to be eigenstates of the positive frequency part $\phi^{(+)}$ of the field, built entirely from destruction operators $a(q)$. To construct such states, introduce the operator

$$S = \exp(\frac{1}{2} \int d^3q (2\pi)^{3/2} \sqrt{2E(q)} \tilde{f}(q) a^\dagger(q)) \qquad (8.51)$$

where $\tilde{f}(q)$ is for the present an unspecified c-number function of spatial momentum q (as usual, we simplify the notation by omitting arrows for spatial vectors, e.g., $\vec{q}$, where the context is clear). A straightforward calculation using the Baker–Campbell–Hausdorff[6] expansion formula gives

$$S^{-1} a(k) S = a(k) + \frac{1}{2} (2\pi)^{3/2} \sqrt{2E(k)} \tilde{f}(k) \qquad (8.52)$$

and hence

$$S^{-1} \phi^{(+)}(\vec{x}, 0) S = \phi^{(+)}(\vec{x}, 0) + \frac{1}{2} f(\vec{x}) \qquad (8.53)$$

where $f(\vec{x})$ is the Fourier transform of $\tilde{f}$ in (8.51). Thus, if we define the coherent state

$$|f\rangle \equiv S|0\rangle \qquad (8.54)$$

we see (since $\phi^{(+)}|0\rangle = 0$) that our coherent state is indeed an eigenstate of the positive frequency part of the field

$$\phi^{(+)}(\vec{x}, 0)|f\rangle = \frac{1}{2} f(\vec{x})|f\rangle$$

$$\langle f|\phi^{(-)}(\vec{x}, 0) = \langle f|\frac{1}{2} f(\vec{x}) \qquad (8.55)$$

where we have taken f to be real. Thus, as promised, we have constructed a state $|f>$ with a field expectation value exactly equal to the preassigned function $f(\vec{x})$:

$$\frac{\langle f|\phi(\vec{x}, 0)|f\rangle}{\langle f|f\rangle} = f(\vec{x}) \qquad (8.56)$$

From the definition of the normal product it now follows that

$$\frac{\langle f| : \phi^2(\vec{x}, 0) : |f\rangle}{\langle f|f\rangle} = f(\vec{x})^2$$

$$\frac{\langle f| : \phi^4(\vec{x}, 0) : |f\rangle}{\langle f|f\rangle} = f(\vec{x})^4$$

$$\frac{\langle f| : |\vec{\nabla}\phi(\vec{x}, 0)|^2 : |f\rangle}{\langle f|f\rangle} = |\vec{\nabla}f(\vec{x})|^2 \qquad (8.57)$$

---

[6] Namely, $e^{-A} B e^A = B - [A, B] + \frac{1}{2!}[A, [A, B]] - \frac{1}{3!}[A, [A, [A, B]]] + \dots$

The time-derivative term in the free Hamiltonian is a little trickier, although, as we are dealing with coherent states with a static expectation value, the final result is, unsurprisingly, zero. By computations analogous to (8.52, 8.53) one finds

$$S^{-1}\pi^{(+)}(\vec{x},0)S = \pi^{(+)}(\vec{x},0) - \frac{i}{2}\dot{f}_E(\vec{x}) \tag{8.58}$$

where $\pi^{(+)}(\vec{x},0)$ is the positive frequency part of $\dot{\phi}(\vec{x},0)$, and

$$\dot{f}_E(\vec{x}) \equiv \int d^3k E(k)\tilde{f}(k)e^{i\vec{k}\cdot\vec{x}} \tag{8.59}$$

is real if $f(\vec{x})$ is. Then

$$\pi^{(+)}(\vec{x},0)|f\rangle = -\frac{i}{2}\dot{f}_E(\vec{x})|f\rangle, \quad \langle f|\pi^{(-)}(\vec{x},0) = \frac{i}{2}\langle f|\dot{f}_E(\vec{x}) \tag{8.60}$$

from which it follows that

$$\langle f|:\dot{\phi}^2:|f\rangle = \langle f|(\pi^{(+)2} + 2\pi^{(-)}\pi^{(+)} + \pi^{(-)2})|f\rangle = 0 \tag{8.61}$$

Thus the expectation value of the full Hamiltonian density in the state $|f\rangle$ is

$$\frac{\langle f|\mathcal{H}(\vec{x},0)|f\rangle}{\langle f|f\rangle} = \frac{1}{2}|\vec{\nabla}f(\vec{x})|^2 + \frac{1}{2}m^2 f(\vec{x})^2 + \frac{\lambda}{4!}f(\vec{x})^4 \tag{8.62}$$

As the expectation value of the full Hamiltonian, obtained as the spatial integral of (8.62), is time-independent, it is given, at any time, by the exact functional

$$\frac{\langle f|H|f\rangle}{\langle f|f\rangle} = \int d^3x(\frac{1}{2}|\vec{\nabla}f(\vec{x})|^2 + \frac{1}{2}m^2 f(\vec{x})^2 + \frac{\lambda}{4!}f(\vec{x})^4) \tag{8.63}$$

It is rather remarkable that the coherent-state formalism allows us to construct a state in a fully interacting quantum field theory with an exactly prescribable expectation value for the energy density. Strictly speaking, (8.63) contains disguised divergences (the famous ultraviolet divergences of local field theory), due to the fact that the coefficients m^2, λ (the so-called bare mass and coupling) of the quadratic and quartic terms in the expression (8.49) for the full Hamiltonian are only well-defined if the theory is cutoff at short distance (or large momenta). The need for such cutoffs in interacting field theories will be exhaustively discussed later in the book, beginning in Chapter 10. A well defined version of the above arguments would therefore require a regularization analogous to that performed in our previous examination of the single-mode case: namely, a cutoff not just of mode occupation number but also of the allowed momenta k, with $|\vec{k}| < \Lambda$, where Λ is some very-high-energy/momentum scale, far beyond those of accessible phenomena. Likewise, assuming $\tilde{f}(k)$ of compact support in momentum space (say, $\tilde{f}(k) = 0, |\vec{k}| > \Lambda$), the preceding arguments carry through with no essential modifications.

8.4 Signs, stability, symmetry-breaking

One further important physical constraint on a local quantum field theory, not mentioned previously, is readily addressed with the tools developed in this chapter: the necessity that the spectrum of the theory be bounded below, or, in other words, the existence of a state (or states) of minimum energy. Otherwise, the system is unstable, with any finite-energy initial state decaying endlessly to lower-energy states with the emission of infinitely many particles. Such a requirement imposes constraints on the signs of couplings appearing in the Hamiltonian, as we can easily see when employing the coherent state formalism of the preceding section. For example, stability requires that the coupling λ in (8.63) be *positive*. If $\lambda < 0$, then by choosing f spatially constant in some region of fixed volume and arbitrarily large in magnitude we can produce a state of arbitrarily negative energy, so the spectrum is unbounded below.

It may be thought that the positive sign of the coefficient of f^2 in (8.63) (i.e., of ϕ^2 in (8.49)) is also sacred: after all, what would a negative squared mass (or imaginary mass) mean? In fact, the theory with Hamiltonian

$$H = \int d^3x : \frac{1}{2}\dot{\phi}^2 + \frac{1}{2}|\vec{\nabla}\phi|^2 - \frac{1}{2}m^2\phi^2 + \frac{\lambda}{4!}\phi^4 : \tag{8.64}$$

is perfectly sensible, although the physical interpretation will involve a new concept: spontaneous symmetry-breaking. If we plot the expectation value of the Hamiltonian energy density $\langle H \rangle/V$ for a system quantized in a box of finite volume V, for the coherent state $|f\rangle$ (with f spatially constant over the box) we find a double well shape, symmetric around the zero-field point (see Figure 8.1).

It is apparent that there are two distinct states of minimum energy, with

$$\langle \phi \rangle = f_0 = \pm\sqrt{\frac{6}{\lambda}}m \tag{8.65}$$

On the other hand, the state with vanishing f (i.e., the conventional Fock-space vacuum $|0\rangle$, cf. (8.54)) clearly has *larger* energy than either of these: it is an excited state! In such a system, a low-energy state—in other words, a state with a small number

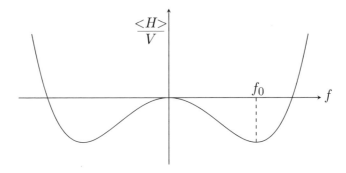

Fig. 8.1 Energy density for various coherent states with spontaneous symmetry-breaking.

of particles with finite energy—may be regarded as a perturbation of the situation in which the field takes the constant value f_0, or the constant value $-f_0$, over all space, with the symmetry of the Hamiltonian (8.64) guaranteeing equivalent physics in either case. In order to write the field operator in terms of creation and annihilation operators with respect to a new, truly minimum-energy, vacuum, we must first subtract off the vacuum-expectation-value (VEV) of the original field variable, (8.65). (Clearly, the canonical field (8.50) automatically has zero VEV with respect to the Fock vacuum $|0\rangle$ defined by $a(k)|0\rangle = 0, \forall k$). Thus, we introduce a "shifted" field $\hat{\phi}$:

$$\phi(x) = \hat{\phi}(x) + f_0 = \hat{\phi}(x) + \sqrt{\frac{6}{\lambda}}m \tag{8.66}$$

Rewriting the Hamiltonian (8.64) in terms of the shifted field, one finds

$$H = \int d^3x : \frac{1}{2}\dot{\hat{\phi}}^2 + \frac{1}{2}|\vec{\nabla}\hat{\phi}|^2 + m^2\hat{\phi}^2 + m\sqrt{\frac{\lambda}{6}}\hat{\phi}^3 + \frac{\lambda}{4!}\hat{\phi}^4 : \tag{8.67}$$

(An additive constant in the energy density, irrelevant for our present discussion, but of profound importance in modern theories of cosmological inflation, has been discarded).

Note that a physically sensible *positive* sign has reappeared in front of the quadratic mass term. Now, however, the theory contains a cubic as well as quartic interaction term. The original $\phi \to -\phi$ symmetry of (8.64), which would have guaranteed conservation of evenness or oddness of the number of particles in any scattering process, has been broken. The breaking is really due to an asymmetrical choice of an off-centered minimum of a *symmetric* potential energy curve. In other words, the symmetry is broken by the ground state, not by the underlying dynamics. If we had chosen *random* values for the coefficients of the $\hat{\phi}^2, \hat{\phi}^3, \hat{\phi}^4$ terms in (8.67), the symmetry-breaking would have been explicit in the dynamics: indeed, in this case, there would be a unique lowest-energy state (vacuum), and no non-trivial symmetry operation (such as $\phi \to -\phi$) connecting distinct degenerate vacuum states. Only for the special case where these couplings and masses are related as indicated in (8.67) can the theory be regarded as possessing an underlying symmetry broken only by the choice of a single asymmetric vacuum—from a set of degenerate ones connected by the symmetry—dictated by historical circumstance, just as the direction of spontaneous magnetization in a ferromagnet lowered below its Curie temperature will depend on the presence of small external magnetic fields to resolve the rotational ambiguity. Symmetry-breaking of this type is referred to as "spontaneous symmetry-breaking".

The symmetry of the theory defined by the Hamiltonian (8.64) is a discrete one: $\phi \to -\phi$. If the symmetry undergoing spontaneous breakdown is *continuous*, as in the case of rotational symmetry mentioned above in the context of ferromagnetism, there is a remarkable new feature, first noticed by Goldstone (Goldstone, 1961): the appearance of an exactly massless scalar particle, or "Goldstone boson". Imagine a theory with three scalar particles of identical mass m. (For example, the neutral and charged pions $\pi^0, \pi^\pm$ form such a triplet, if we ignore the slight differences in mass of the neutral and charged types due to the weak and electromagnetic interactions.) The free Hamiltonian for such a system may be written thus:

$$H_0 = \int d^3x : \frac{1}{2}\dot{\vec{\phi}}^2 + \frac{1}{2}|\vec{\nabla}\vec{\phi}|^2 + \frac{1}{2}m^2\vec{\phi}^2 : \qquad (8.68)$$

Here the three scalar fields have been written as the three "components" ϕ_1, ϕ_2, ϕ_3 of a three-dimensional field vector $\vec{\phi}$. This Hamiltonian possesses a continuous symmetry of rotations in field space, whereby $\phi_i \to R_{ij}\phi_j$, with R_{ij} a 3×3 orthogonal matrix, i.e., an element of the fundamental representation of the group $O(3)$, the rotation group in three dimensions. Note that the symmetry is a "global" one: exactly the same rotation is applied to the field vector at all spacetime points (otherwise, the spatial gradient or time-derivative terms in the Hamiltonian would not be left invariant). Such a symmetry is commonly referred to as an "isospin" symmetry. One may write down interactions which also respect the isospin symmetry of the free Hamiltonian, e.g., $\mathcal{H}_{int} = \frac{\lambda}{4!}(\vec{\phi}^2)^2$, which is clearly also invariant under rotations in field space. Now suppose that the full Hamiltonian is actually

$$H = \int d^3x : \frac{1}{2}\dot{\vec{\phi}}^2 + \frac{1}{2}|\vec{\nabla}\vec{\phi}|^2 - \frac{1}{2}m^2\vec{\phi}^2 + \frac{\lambda}{4!}(\vec{\phi}^2)^2 :$$

$$\equiv \int d^3x : \frac{1}{2}\dot{\vec{\phi}}^2 + \frac{1}{2}|\vec{\nabla}\vec{\phi}|^2 + P(\vec{\phi}) : \qquad (8.69)$$

Once again, the state of minimum energy does *not* correspond to zero vacuum expectation value of the fields $\vec{\phi}$. Rather, there are a whole family of minimum-energy states corresponding to the various field orientations with fixed magnitude

$$|\vec{\phi}| = \sqrt{\frac{6}{\lambda}}m \qquad (8.70)$$

A physical Fock space is constructed by imagining (and here the analogy to ferromagnetism becomes essentially complete) that some fluctuation has "tickled" the system into choosing a definite direction in field space for $\langle 0|\vec{\phi}|0\rangle$. By the rotational invariance of (8.69), we may as well call this direction the "3" direction, and rewrite the theory in terms of a shifted field:

$$\phi_i(x) = \sqrt{\frac{6}{\lambda}}m\delta_{i3} + \hat{\phi}_i(x) \qquad (8.71)$$

As in the theory (8.64), the physical vacuum state will be one in which the vacuum expectation value of the shifted field $\hat{\phi}$ vanishes. With this substitution we find that the "potential energy" part $P(\vec{\phi})$ of H becomes

$$P \to m^2\hat{\phi}_3^2 + \sqrt{\frac{\lambda}{6}}m\hat{\phi}_3\hat{\phi}_i\hat{\phi}_i + \frac{\lambda}{4!}(\hat{\phi}_i\hat{\phi}_i)^2 \qquad (8.72)$$

Note the appearance once again of a cubic term which violates the rotational symmetry (because the third direction in field space is singled out via $\hat{\phi}_3$). But in contrast to the discrete case, the theory now contains only a single massive particle, corresponding to $\hat{\phi}_3$: there is no mass term for $\hat{\phi}_1, \hat{\phi}_2$! These two directions correspond to the "flat" directions in field space near the point $\vec{\phi}_i = \sqrt{\frac{6}{\lambda}} m \delta_{i3}$. Note that the Hamiltonian after the shift is still invariant under rotations *around* the 3-axis (which leave $\hat{\phi}_3$ fixed), so the symmetry-breaking has reduced the symmetry group of the theory from O(3) (rotations around the 1, 2, or 3 axes in field space) to O(2) (rotations only around the 3 axis, represented by 2x2 matrices mixing the 1 and 2 coordinates). The two generators corresponding to the lost symmetries (rotations around the 1 and 2 directions) may be associated with the two massless Goldstone bosons appearing in (8.72).

The appearance of a massless scalar for every broken generator of a continuous global symmetry is a very general property of local quantum field theories. Note that our arguments in this section are based on an identification of the physical masses of particles with the coefficients of quadratic terms in the Hamiltonian, which is only, strictly speaking, valid to lowest order in perturbation theory in the interactions: we shall see later that the interactions will in general *renormalize* the physical masses. Nevertheless, the masslessness of the modes induced by spontaneous breaking of global continuous symmetries turns out to be exact to all orders of perturbation theory, as we shall show when we return to the subject in Section 14.3.[7] Some more examples of spontaneous symmetry-breaking in theories with continuous global symmetries are given in Problems 4, 5, and 6 at the end of this chapter.

One may wonder why the Goldstone theorem has assumed such central importance in modern particle physics: after all, there do not appear to be any massless spinless particles in Nature! There are two reasons:

(1) There are spontaneously broken continuous global symmetries which *would be exact* except for the presence of small explicit terms breaking the symmetry in the Hamiltonian. In such a case (see Problem 4) one gets scalar particles of small mass, called "pseudo-Goldstone bosons". This is the case for pions in the strong interactions, which are the pseudo-Goldstone bosons of broken approximate chiral symmetry.

(2) The spontaneous symmetry-breaking may occur in a theory where the continuous symmetry is a local *gauge* symmetry, and in this case each Goldstone boson appearing due to a broken generator of the symmetry group is in a sense "absorbed" into the gauge field of the corresponding gauge particle, making the latter massive (the so-called "Higgs" mechanism, discussed in detail in Section 15.6, which explains the mass splitting between the photon and the massive W and Z bosons of modern electroweak theory).

We shall return later, in the "Symmetries" section of the book (cf. Chapter 14), to a much more detailed examination of the physics of degenerate vacua and spontaneous symmetry-breaking, both for global and local symmetries of the underlying theory. For the time being, the preceding discussion provides a suggestive example of the

[7] The first general proof of this was given by Goldstone, Salam, and Weinberg, *Phys. Rev.* 127 (1962), 965.

power and utility of the coherent state formalism in connecting certain macroscopic properties of a quantum field theory with the microscopic dynamics.

8.5 Problems

1. Starting with the vector field $\vec{A}$ for a massless spin-1 particle given in (7.174), with the electric field defined as $\vec{E} = \frac{\partial \vec{A}}{\partial t}$ and the magnetic field $\vec{B} = \vec{\nabla} \times \vec{A}$ (or $B_i = \epsilon_{ijk}\partial_j A_k$, $i,j,k = 1,2,3$), derive the commutation relation:

$$[E_i(\vec{x},t), B_j(\vec{y},t)] = i\epsilon_{ijk}\partial_k \delta^3(\vec{x}-\vec{y})$$

2. Show that in the state (8.37), the probability of detecting more than N quanta falls more rapidly than $(\frac{e\bar{n}}{N})^N$ for $N > \bar{n} \equiv \langle \mathcal{N} \rangle$ once we consider states with occupancy N exceeding the expectation value $\bar{n}$ of the number operator.

3. Show that the states (8.43) are eigenstates of the operator (8.19), with eigenvalues $e^{i\phi}$, $\phi = \frac{2\pi m}{N+1}$, $m = 0,1,2,...N$.

4. Let σ and $\vec{\pi}$ (the vector symbol over π refers to an internal "flavor" index, *not* to ordinary space!) be a set of four self-conjugate scalar fields, interacting via the Hamiltonian

$$H = \int d^3x \{: \frac{1}{2}(\frac{\partial\sigma}{\partial t})^2 + \frac{1}{2}|\vec{\nabla}\sigma|^2 + \frac{1}{2}\sum_{i=1}^{3}((\frac{\partial\pi_i}{\partial t})^2 + |\vec{\nabla}\pi_i|^2)$$

$$- \frac{1}{2}\mu^2(\sigma^2 + \vec{\pi}^2) + \lambda(\sigma^2 + \vec{\pi}^2)^2 :\}$$

(a) Show that the ground state of this system has non-vanishing expectation value for one of the fields. (It is conventional to call this the σ field—why is this purely a matter of convention?). Show that this corresponds to a breaking of the global symmetry group (what is it?) of H. Show that there are massless particles (how many?) in the theory.

(b) What happens to the spectrum if a term $k\sigma$ (k a real constant) is added to H?

(c) If μ=80 MeV, λ=0.04, k=0, calculate the difference in energy density in J/m^3 between the false vacuum with $\langle\sigma\rangle = \langle\vec{\pi}\rangle = 0$ and one of the coherent states minimizing the energy.

5. In the theory of the preceding Problem, show that the σ particle is unstable and calculate its lifetime (inverse of the total decay rate) to lowest order in λ.

6. Let $\vec{\chi}_1$ and $\vec{\chi}_2$ be two three-vectors of self-conjugate scalar fields interacting via a Hamiltonian density with polynomial part

$$P(\vec{\chi}_1, \vec{\chi}_2) = -\frac{1}{2}\mu^2(\vec{\chi}_1^2 + \vec{\chi}_2^2) + \lambda_1(\vec{\chi}_1^2 + \vec{\chi}_2^2)^2 + \lambda_2\vec{\chi}_1 \cdot \vec{\chi}_2(\vec{\chi}_1^2 + \vec{\chi}_2^2)$$

where $\lambda_1, \lambda_2 > 0$, and $\lambda_1 > \lambda_2/2$.

(a) What is the global symmetry group of the Hamiltonian?

(b) Find the configuration that minimizes the polynomial P and show that it breaks the global symmetry (i.e., find $\langle \vec{\chi}_1 \rangle$ and $\langle \vec{\chi}_2 \rangle$). What is the residual global symmetry in this case after spontaneous symmetry-breaking? How many generators of the original symmetry group were broken?

(c) Identify (as linear combinations of the original fields) the massless and the massive fields after spontaneous symmetry-breaking, and find the masses of those with non-zero mass.

9
Dynamics VII: Interacting fields: general aspects

Our attempts to build a framework incorporating the essential features of quantum theory, special relativity, and the cluster principle have been motivated by an examination of the properties of the resultant scattering matrix, as given by the formal perturbative expansion (5.73). In particular, the entire discussion has taken place in the context of the interaction-picture representation of the dynamics, in which a computationally convenient division of the full Hamiltonian is made, and the dynamics of the field operators of the theory assigned to one portion (the "free" Hamiltonian), with the complicated dynamics induced by interactions assigned entirely to the states.

All of the conclusions reached heretofore must therefore be regarded as limited to those theories in which a perturbative expansion is both mathematically sensible and physically appropriate. On the other hand, one is faced with the unpleasant fact that in the case of the strong interactions (to give a particularly glaring example), for all but a very special class of processes (in certain asymptotic kinematic regimes), the perturbative expansion of the theory is useless—not simply for reasons of quantitative inadequacy, but because the qualitative implications of perturbation theory completely mislead us as to the physical structure of the theory.

We shall therefore turn our attention to a more general formulation of local quantum field theory, in which the results are divorced to the maximum degree possible from the special features and assumptions of perturbation theory. If we insist on a unitary dynamics, in which no attempt is made to divide the Hamiltonian H into "free" and "interaction" parts, we then have, as discussed in Chapter 4, basically two choices for the description of the dynamics of the theory: the *Heisenberg* picture, in which the entire time-dependence induced by H is transferred to the field operators of the theory, with the state of the system fixed once and for all by appropriate boundary conditions; or, the *Schrödinger* picture, in which the states carry the full time-dependence of H, while the field operators are time-independent. As we saw in Section 4.3, the discussion of scattering theory, even for ordinary non-relativistic potential scattering theory, is particularly transparent in the former framework, with the Heisenberg representation $|\alpha\rangle_{\text{in}}, |\beta\rangle_{\text{out}}$ states incorporating naturally the boundary conditions of a typical scattering process. In the case of relativistic quantum field theory, the Heisenberg representation is also preferred for reasons of manifest Lorentz covariance: we shall see that the Heisenberg field operators, while carrying the time development specified in a particular inertial frame, nevertheless remain local covariant fields (while the states are fixed), in contradistinction to the Schrödinger picture, where a definite choice of inertial frame is necessary in order to specify the relevant time variable for the state dynamics, and the field operators depend only on the spatial variables, to the detriment of manifest Lorentz covariance of the theory. Our next task will therefore be to reformulate local quantum field theory in the Heisenberg picture, avoiding as far as possible any reference to perturbative expansions of the quantities discussed.

This is a long chapter, and we beg the reader's patience insofar as the treatment requires perhaps a somewhat higher level of mathematical sophistication than previously necessary. The arguments have been spelled out in great detail to avoid confusions, and the required additional mathematics (mainly distribution theory) has been explained just enough to make the proofs comprehensible. However, the effort is justified by the central importance of the topics discussed, which encompass the *essential conceptual content* of interacting local quantum field theories, expressed, in the absence of an explicit "from the ground up" mathematical construction of four-dimensional field theories, in as precise a form as mathematical physicists have been able to achieve to the present date. In particular, a clear understanding of the variety of connections between particles and fields, discussed at length in Section 9.6, is really not possible without the insight into the nature of the interpolating field given by the asymptotic formalism developed in Sections 9.3 and 9.4.

9.1 Field theory in Heisenberg representation: heuristics

The transformation of a general quantum-mechanical operator from interaction picture to Heisenberg picture is given *formally* by (4.18) from Chapter 4:

$$O_H(t) = U^\dagger(t,0)O_{\text{ip}}(t)U(t,0) \tag{9.1}$$

with $U(t,0) = e^{iH_0 t}e^{-iHt}$. In the case of a quantum-mechanical system with a finite number of degrees of freedom, there are no hidden subtleties here: the operator $U(t,0)$ is properly unitary, acting within a single Hilbert space spanned by either the complete set of eigenstates of the free Hamiltonian H_0 or the full Hamiltonian H. We shall see later (cf. the discussion of Haag's theorem, Section 10.5) that none of these nice properties obtain for a quantum field theory with infinitely many degrees of freedom. Thus, strictly speaking, the manipulations that follow in this section are only valid if the field theory is *fully regularized*—by which we mean that both infrared (finite-volume V) and ultraviolet (short-distance a, or high-momentum Λ) cutoffs are imposed, reducing the number of allowed momentum modes to a finite number ($\sim \frac{V}{a^3}$). In fact, we have already noted (cf. Section 8.3, final paragraph) that such cutoffs are necessary to have a well-defined expression for the Hamiltonian in terms of the fields.

The attentive reader will, of course, complain that such cutoffs, while perhaps restoring quantum-mechanical sanity, will clearly do violence to the two other sacred ingredients which we have taken such pains to implement: Lorentz-invariance and locality. And it is certainly far from obvious that the cutoffs can be removed at the end of our calculations in a way that fully restores these desiderata. These are all important issues, to which we shall return on several occasions later in the book. But for the time being we shall throw caution to the winds and proceed as though the conventional manipulations of non-relativistic quantum theory make sense in the field theory case as well. A rigorous scattering theory, constructed without recourse to a

$$\left[\rho_{\mu}(\Lambda^{-1}x) \, c\rho_{\mu}(\Lambda^{-1}y) \right] = U \left[\rho_{\mu}^{\dagger}(x) \, \rho_{\mu}(y) \right] U$$

242   *Dynamics VII: Interacting fields: general aspects*   =

mathematically dubious interaction picture, will be the subject of Sections 9.3 and 9.4. In the meantime we shall assume that the interaction picture makes sense. Also, we shall restrict the discussion to the case of a single, self-conjugate massive scalar field with polynomial self-interactions, as the complications entailed by non-zero spin are completely irrelevant to the physical issues of importance here.

We therefore define a *Heisenberg field operator* $\phi_H(\vec{x}, t)$ in terms of the corresponding interaction-picture field via (9.1) as

$$\phi_H(\vec{x}, t) \equiv U^{\dagger}(t, 0)\phi(\vec{x}, t)U(t, 0) \tag{9.2}$$

As $U(t, 0)$ is by assumption unitary, the locality (by construction) of the free interaction-picture field implies the vanishing of the equal-time commutators of ϕ_H at distinct spatial points:

$$[\phi_H(\vec{x}, t), \phi_H(\vec{y}, t)] = U^{\dagger}(t, 0)[\phi(\vec{x}, t), \phi(\vec{y}, t)]U(t, 0) = 0, \quad \vec{x} \neq \vec{y} \tag{9.3}$$

The extension of this equal-time result to full microcausality (i.e., space-like commutativity) requires examination of the Lorentz transformation properties of ϕ_H, to which we now turn.

We recall that the construction of the interaction part V of the Hamiltonian as the spatial integral of a density implies the momentum conservation property (5.90), $[\vec{P}^{(0)}, V] = [\vec{P}^{(0)}, H_0] = 0$, where $\vec{P}^{(0)}$ is the spatial momentum operator on the Fock space of eigenstates of H_0 (the so-called "bare" states of field theory). As the interaction does not alter spatial momentum, we can drop the (0) subscript, as the same operator will measure the momentum of all states after interactions are switched on. In particular, the commutator of this spatial momentum operator with the Heisenberg field defined in (9.2) is

$$i[P_i, \phi_H(\vec{x}, t)] = U^{\dagger}(t, 0)i[P_i, \phi(\vec{x}, t)]U(t, 0) = U^{\dagger}(t, 0)\frac{\partial}{\partial x^i}\phi(\vec{x}, t)U(t, 0) = \frac{\partial}{\partial x^i}\phi_H(\vec{x}, t) \tag{9.4}$$

as the spatial momentum operator P_i commutes with H, H_0, hence with $U(t, 0)$. Of course, the energy component of the four-vector energy-momentum operator which generates the time-evolution of the Heisenberg field is the full Hamiltonian H:

$$i[P_0, \phi_H(\vec{x}, t)] = i[H, \phi_H(\vec{x}, t)] = \frac{\partial}{\partial t}\phi_H(\vec{x}, t) \tag{9.5}$$

The finite field translation property (5.93) for the interaction-picture field ϕ therefore takes the obvious analogous form for the Heisenberg picture field:

$$e^{iP_{\mu}a^{\mu}}\phi_H(x)e^{-iP_{\mu}a^{\mu}} = \phi_H(x + a) \tag{9.6}$$

with $P_{\mu} = (H, P_i) = (H, P_i^{(0)})$.

Our previous discussion of scattering theory in Heisenberg representation (cf. Section 4.3) was based on the specification of the Heisenberg state of a scattering system in terms of the behavior of the system in the far past (the "in-states" $|\alpha\rangle_{\text{in}}$) or the far future (the "out-states" $|\beta\rangle_{\text{out}}$). The former correspond to a set of far separated incoming particles such as those prepared in the beams of a high-energy collider, the latter to the set of outgoing detected particles, in interesting cases, following a collision. In the case of non-relativistic potential scattering theory, and assuming the absence of bound states of the incoming or outgoing particles, either the in- or the out-states can be shown to provide a complete set of eigenstates of the full Hamiltonian H. In other words, the Hilbert space of the theory $\mathcal{H}$ can be identified with either the space $\mathcal{H}_{\text{in}}$ spanned by the $|\alpha\rangle_{\text{in}}$ *or* the space $\mathcal{H}_{\text{out}}$ spanned by the $|\beta\rangle_{\text{out}}$ states. We have no option but to assume that the corresponding property remains valid in relativistic field theory, where a basis for the in (resp. out) states is provided by the (continuum normalized) multi-particle states $|k_1, k_2, ...k_N\rangle_{\text{in}}$ (resp. $|k'_1, k'_2, ...k'_M\rangle_{\text{out}}$) with

$$P^{\mu}|k_1, k_2, ...k_N\rangle_{\text{in}} = \sum_{n=1}^{N} k_n^{\mu}|k_1, k_2, ...k_N\rangle_{\text{in}} \tag{9.7}$$

$$P^{\mu}|k'_1, k'_2, ...k'_M\rangle_{\text{out}} = \sum_{n=1}^{M} k_n'^{\mu}|k'_1, k'_2, ...k'_M\rangle_{\text{out}} \tag{9.8}$$

Here again we emphasize that we are taking for simplicity the theory of a single stable spinless particle (with no bound states), so the states are fully specified by listing the momenta of the particles.

The *principle of asymptotic completeness* asserts that the Hilbert spaces $\mathcal{H}_{\text{in}}$, $\mathcal{H}_{\text{out}}$ are *identical*, and can be identified with the full Hilbert space $\mathcal{H}$ of the theory. The physical reasoning behind this assumption is in fact very straightforward. Imagine an arbitrary finite-energy state of the system at, say, time $t = 0$, corresponding to the moment of collision of some arbitrary set of formerly separated particles. The state of the system near $t = 0$ is, of course, extremely complicated: all we know (in a massive theory with short range interactions) is that the energy and momentum of the field(s) is concentrated in a small spatial region around the interaction vertex. It is physically clear that any such state will eventually evolve (absent bound states) into a linear combination of Fock states consisting of sets of a finite number ($\leq E/m$, where E is the total energy and m the mass of the field quantum) of outgoing stable particles receding to infinity, with, of course, the total energy and momentum of the system conserved at every stage. In other words, since the entire history of a system is encapsulated (in Heisenberg representation) by its specification at some arbitrary time (in this case, the far future), an arbitrary Heisenberg state of the system must be resolvable into a linear combination of multi-particle out-states: the $|\beta\rangle_{\text{out}}$ span the physical Hilbert space. Note that in a theory with unstable particles, these *do not* form part of the asymptotic Hilbert space: only the stable particles persisting at late time (including stable bound states, if such exist) are to be included in the list of $|\beta\rangle_{\text{out}}$. We shall return later in this chapter to a more detailed discussion of the nature of the Hilbert space in relativistic field theory. For the time being we note that the completeness of the in-states is assured given that of the out-states by the TCP theorem of local field theory (cf. Section 13.4): the joint operation of charge conjugation (particle–antiparticle interchange), parity and time reversal must leave the physics invariant, whence the completeness of the in-states defined in the far past follows unproblematically from the corresponding property of the out-states, argued heuristically above. Although the property of asymptotic completeness seems physically almost unassailable, the ability to explicitly and rigorously establish this property for even the most technically controllable interacting field theories is quite another matter, as we shall see later.

The states of our system carry, of course, a representation of the HLG (homogeneous Lorentz group), implemented by unitary operators $U_H(\Lambda)$ defined to have the action (on covariantly normalized states)

$$U_H(\Lambda)|k_1, k_2, ...k_N\rangle_{\text{in}} \equiv |\Lambda k_1, \Lambda k_2, ..., \Lambda k_N\rangle_{\text{in}} \qquad (9.9)$$

whence, from (9.7), we must have

$$U_H^\dagger(\Lambda)P^\mu U_H(\Lambda) = \Lambda^\mu{}_\nu P^\nu \qquad (9.10)$$

Choosing a countable basis[1] $|\alpha\rangle_{\text{in}}$ for $\mathcal{H}_{\text{in}}$, the discrete matrix $S_{\beta\alpha}$ is unitary (cf. Section 4.3.1), and Lorentz-invariance of the theory, $_{\text{out}}\langle\Lambda\beta|\Lambda\alpha\rangle_{\text{in}} = {}_{\text{out}}\langle\beta|\alpha\rangle_{\text{in}}$, then implies (see Problem 1) the corresponding transformation property on the (once again, continuum-normalized) out-states:

$$U_H(\Lambda)|k_1, k_2, ...k_N\rangle_{\text{out}} = |\Lambda k_1, \Lambda k_2, ..., \Lambda k_N\rangle_{\text{out}} \qquad (9.11)$$

Note the subscript H on $U_H(\Lambda)$: these are *not* the same unitary operators as the $U(\Lambda)$ introduced earlier (cf. (5.23)) implementing Lorentz transformations on the bare multi-particle states $|\alpha\rangle$ which form a basis of eigenstates of the free Hamiltonian H_0, and which satisfy

$$U^\dagger(\Lambda)P^{(0)\mu}U(\Lambda) = \Lambda^\mu{}_\nu P^{(0)\nu} \qquad (9.12)$$

As discussed above, since for the scalar theory under discussion, $P_i^{(0)} = P_i$, while $P_0^{(0)} = H_0 \neq P_0 = H$, it is apparent that the $U(\Lambda)$ and $U_H(\Lambda)$ operators are different.

We now return to the question of the Lorentz covariance properties of the Heisenberg field ϕ_H defined by unitary tranformation of the interaction-picture field ϕ in (9.2). The latter is by construction a local, Lorentz scalar field, but the Heisenberg field is obtained by a unitary transformation involving a specific choice of time variable and inertial frame, so the issues of space-like commutativity and scalar transformation property of ϕ_H under the HLG are not immediately obvious. To proceed further, we shall again rely on formal arguments assuming the existence of the interaction picture: in other words, we shall proceed as though our field theory is fully regularized (i.e., is cutoff in the infrared *and* the ultraviolet, leaving only a finite number of quantum mechanical degrees of freedom), so that the interaction-picture time-development operators operate in the same space as the Heisenberg states of

---

[1] The Fock space of field theory, as a countable direct sum of finite tensor products of one-particle spaces, is, somewhat surprisingly, a separable Hilbert space; see (Streater and Wightman, 1978).

the theory. After obtaining the relevant result linking matrix elements of interaction and Heisenberg picture operators, we shall return to the issue of removal of the cutoff (essential, of course, for restoring the locality and Lorentz transformation properties of the theory).

The formula we need, of fundamental importance in both relativistic field theory and many-body theory, is due to Gell–Mann and Low (Gell-Mann and Low, 1951). We begin with the unitary Møller wave operators $\Omega^{\mp} = U(0, \mp\infty)$, connecting bare states $|\alpha\rangle$ (eigenstates of H_0) to the corresponding in- and out-states (cf. Section 4.3.2):

$$|\alpha\rangle_{\text{in}} = U(0, -\infty)|\alpha\rangle, \quad |\alpha\rangle_{\text{out}} = U(0, +\infty)|\alpha\rangle \tag{9.13}$$

Now consider a general matrix element between arbitrary in- and out-states of a time-ordered product of m Heisenberg fields

$$_{\text{out}}\langle\beta|T\{\phi_H(x_1)\phi_H(x_2)..\phi_H(x_m)\}|\alpha\rangle_{\text{in}} =_{\text{out}} \langle\beta|\phi_H(y_1)\phi_H(y_2)..\phi_H(y_m)|\alpha\rangle_{\text{in}} \tag{9.14}$$

where the set $(y_1, y_2, ...y_m)$ of spacetime coordinates are obtained by subjecting the original set $(x_1, x_2, ...x_m)$ to a permutation in order to effect the desired time-ordering $t_1 \equiv y_1^0 > t_2 \equiv y_2^0 > ... > t_m \equiv y_m^0$. Using (9.2, 9.13) and the semigroup property of the $U(t, t_0)$ (cf. Problem 1 in Chapter 4), we find

$$_{\text{out}}\langle\beta|\phi_H(y_1)...\phi_H(y_m)|\alpha\rangle_{\text{in}} = \langle\beta|U(+\infty, t_1)\phi(y_1)U(t_1, t_2)\phi(y_2)...\phi(y_m)U(t_m, -\infty)|\alpha\rangle \tag{9.15}$$

Note that the left-hand side of (9.15) involves purely Heisenberg states and operators, while the right-hand side contains only interaction-picture states and operators. Further progress requires that we limit ourselves to the formal perturbative expansion of the theory (the reasons for which will become apparent below):

$$\tau(y_1, ...y_m) \equiv \sum_{n=0}^{\infty} \frac{(-i)^n}{n!} \cdot$$

$$\int \langle\beta|T\{\phi(y_1)\phi(y_2)..\phi(y_m)\mathcal{H}_{\text{int}}(z_1)\mathcal{H}_{\text{int}}(z_2)..\mathcal{H}_{\text{int}}(z_n)\}|\alpha\rangle d^4z_1 d^4z_2..d^4z_n \tag{9.16}$$

As the integrations over the spacetime coordinates $z_1, z_2, ...z_n$ in (9.16) are performed, the time-ordering symbol will redistribute the n $\mathcal{H}_{\text{int}}$ operators among the already time-ordered field operators $\phi(y_1)$, $\phi(y_2)$, etc. The full integration can evidently be subdivided into subregions in which n_0 of the $\mathcal{H}_{\text{int}}$ operators occur at times later than $t_1 \equiv y_1^0$, n_1 occur between times t_1 and t_2, and so on, with n_m interactions prior to the earliest field time t_m, and $n = n_0 + n_1 + ...n_m$. There are $\frac{n!}{n_0!n_1!...n_m!}$ equivalent ways of selecting the particular $\mathcal{H}_{\text{int}}$ operators to be placed in these temporal intervals. Accordingly, our m-point function $\tau(y_1, ...y_m)$ (commonly referred to as a "Feynman m-point function") in (9.16) may be re-expressed (relabeling the $z_1, .., z_n$ coordinates)

$$\tau(y_1, ...y_m) = \sum_{n=0}^{\infty} \frac{(-i)^n}{n!} \sum_{n_0,n_1,..n_m;n_0+n_1+\cdots n_m=n} \frac{n!}{n_0!n_1!...n_m!}$$

$$\cdot \int \langle\beta|T\{\mathcal{H}_{\mathrm{int}}(z_{0,1})..\mathcal{H}_{\mathrm{int}}(z_{0,n_0})\}\phi(y_1)T\{\mathcal{H}_{\mathrm{int}}(z_{1,1})..\mathcal{H}_{\mathrm{int}}(z_{1,n_1})\}\phi(y_2)\cdots$$

$$\cdots \phi(y_m)T\{\mathcal{H}_{\mathrm{int}}(z_{m,1})..\mathcal{H}_{\mathrm{int}}(z_{m,n_m})\}|\alpha\rangle$$

$$\cdot \,\theta(z^0_{0,j} - t_1)\theta(t_1 - z^0_{1,j})...\theta(t_m - z^0_{m,j})\prod_{i,j} d^4 z_{i,j} \quad (9.17)$$

The sums over the individual n_0, n_1, etc. T-products of interaction operators can now be reassembled into the corresponding interaction-picture time-evolution operators. For example.

$$\sum_{n_0} \frac{(-i)^{n_0}}{n_0!} \int \theta(z^0_{0,i} - t_1)T\{\mathcal{H}_{\mathrm{int}}(z_{0,1})..\mathcal{H}_{\mathrm{int}}(z_{0,n_0})\}\prod_j d^4 z_{0,j} = U(+\infty, t_1) \quad (9.18)$$

and similarly for the sums involving time-ordered operators sandwiched in the other temporal regions between the $\phi(y_i)$ operators. In the end we recover exactly

$$\tau(y_1, y_2, .., y_m) = \langle\beta|U(+\infty, t_1)\phi(y_1)U(t_1, t_2)\phi(y_2)...\phi(y_m)U(t_m, -\infty)|\alpha\rangle \quad (9.19)$$

i.e., the right-hand side of (9.15). In other words, with the provisos given earlier *vis-à-vis* regularization of the theory, we have the *Gell–Mann–Low formula*

$$_{\mathrm{out}}\langle\beta|\phi_H(y_1)\phi_H(y_2) \cdots \phi_H(y_m)|\alpha\rangle_{\mathrm{in}} = \sum_{n=0}^{\infty} \frac{(-i)^n}{n!} \int \langle\beta|T\{\phi(y_1)\phi(y_2)..\phi(y_m)$$

$$\cdot \,\mathcal{H}_{\mathrm{int}}(z_1)\mathcal{H}_{\mathrm{int}}(z_2)..\mathcal{H}_{\mathrm{int}}(z_n)\}|\alpha\rangle d^4 z_1 d^4 z_2..d^4 z_n \quad (9.20)$$

or, restoring the original non-time-ordered variables $x_1, x_2, ...x_m$:

$$_{\mathrm{out}}\langle\beta|T\{\phi_H(x_1)\phi_H(x_2)..\phi_H(x_m)\}|\alpha\rangle_{\mathrm{in}} = \sum_{n=0}^{\infty} \frac{(-i)^n}{n!} \int \langle\beta|T\{\phi(x_1)\phi(x_2)..\phi(x_m)$$

$$\cdot\mathcal{H}_{\mathrm{int}}(z_1)\mathcal{H}_{\mathrm{int}}(z_2)..\mathcal{H}_{\mathrm{int}}(z_n)\}|\alpha\rangle d^4 z_1 d^4 z_2..d^4 z_n \quad (9.21)$$

Note that as a special case, taking $m = 0$ (no ϕ_H fields), we recover our previous perturbative expression (5.73) for the S-matrix $S_{\beta\alpha} \equiv {}_{\mathrm{out}}\langle\beta|\alpha\rangle_{\mathrm{in}}$.

With this result in hand we can return to the question of the Lorentz transformation properties of the Heisenberg field $\phi_H(x)$. Of course, the presence of regularizing cutoffs reducing our field theory to a finite number of degrees of freedom explicitly breaks the invariance under the continuous HLG (for example, if we formulate the theory on a discrete spacetime lattice, the usual prelude to attempts at a rigorous construction of the continuum limit of a relativistic field theory). We shall see in Part 4 of this book, with our treatment of covariant renormalized perturbation theory, that the restoration of Lorentz-invariance can indeed be proven rigorously in a number of four-dimensional field theories (the so-called "perturbatively renormalizable" theories), *but only in the context of the formal perturbative expansion* of the theory in a suitably chosen cutoff-independent coupling parameter(s). The rigorous demonstration of the existence of Heisenberg fields in continuum spacetime with the desired locality properties and behavior under the Poincaré group has only been possible in a limited class of theories in two or three spacetime dimensions ((Simon, 1974),(Glimm and Jaffe, 1987)).

We shall return later to the question of what is rigorously known about the existence in the continuum of four-dimensional interacting field theories: here, we stay within the confines of perturbation theory, and assume the validity of the perturbative Gell–Mann–Low formula (9.21) absent ultraviolet and infrared cutoffs, with the full HLG implemented both at the interaction-picture level (via $U(\Lambda)$) and for the Heisenberg fields and states (via $U_H(\Lambda)$). For the special case $m = 1$, (9.21) reads

$$_{\text{out}}\langle\beta|\phi_H(x)|\alpha\rangle_{\text{in}} = \sum_{n=0}^{\infty} \frac{(-i)^n}{n!}$$

$$\cdot \int \langle\beta|T\{\phi(x)\mathcal{H}_{\text{int}}(z_1)\mathcal{H}_{\text{int}}(z_2)..\mathcal{H}_{\text{int}}(z_n)\}|\alpha\rangle d^4z_1 d^4z_2..d^4z_n \qquad (9.22)$$

We recall that the interaction-picture fields satisfy (cf. (5.76))

$$U^\dagger(\Lambda)\mathcal{H}_{\text{int}}(x)U(\Lambda) = \mathcal{H}_{\text{int}}(\Lambda^{-1}x), \quad U^\dagger(\Lambda)\phi(x)U(\Lambda) = \phi(\Lambda^{-1}x) \qquad (9.23)$$

We wish to establish the corresponding Lorentz scalar field transformation property for the fully interacting Heisenberg field $\phi_H(x)$, viz.

$$U_H^\dagger(\Lambda)\phi_H(x)U_H(\Lambda) = \phi_H(\Lambda^{-1}x) \qquad (9.24)$$

Let us begin with (9.22), for arbitrary in $|\alpha\rangle_{\text{in}}$ and out $|\beta\rangle_{\text{out}}$ states (recall that these *separately* form complete sets, by the principle of asymptotic completeness). Introducing an arbitrary fixed Lorentz transformation Λ

$$_{\text{out}}\langle\beta| U_H^\dagger(\Lambda)\phi_H(x)U_H(\Lambda)|\alpha\rangle_{\text{in}} = {}_{\text{out}}\langle\Lambda\beta|\phi_H(x)|\Lambda\alpha\rangle_{\text{in}}$$

$$= \sum_{n=0}^{\infty} \frac{(-i)^n}{n!} \int \langle\Lambda\beta|T\{\phi(x)\mathcal{H}_{\text{int}}(z_1)..\mathcal{H}_{\text{int}}(z_n)\}|\Lambda\alpha\rangle d^4z_1..d^4z_n$$

$$= \sum_{n=0}^{\infty} \frac{(-i)^n}{n!} \int \langle\beta|U^\dagger(\Lambda)T\{\phi(x)\mathcal{H}_{\text{int}}(z_1)..\mathcal{H}_{\text{int}}(z_n)\}U(\Lambda)|\alpha\rangle d^4z_1..d^4z_n$$

$$\qquad (9.25)$$

By the arguments of Section 5.5, the T-product of ultralocal scalar fields (and we are here assuming that $\mathcal{H}_{\text{int}}(x)$ is ultralocal) is Lorentz-covariant: i.e., the similarity transformation by $U(\Lambda)$ may be taken inside the T-product,[2] giving

---

[2] The reader will recall that the basic reason for this is that the only cases in which Λ can interchange the time-ordering of two fields involve space-like separations where the fields already commute, by locality.

$\{\varphi_\mu(x,t)\,\varphi_\mu(y,t)\} = 0 \qquad U^\dagger(\Lambda)[\varphi_\mu(x,t)\,\varphi_\mu(y,t)]U(\Lambda)$.

$x \neq y$

$[\varphi_\mu(\Lambda'x)\,\varphi_\mu(\Lambda'y)] = 0$

$[\varphi_\mu(x')\,\varphi_\mu(y')] = 0$

Q

$$_{\text{out}}\langle\beta|\,U_H^\dagger(\Lambda)\phi_H(x)U_H(\Lambda)|\alpha\rangle_{\text{in}}$$

$$= \sum_{n=0}^\infty \frac{(-i)^n}{n!} \int \langle\beta|T\{U^\dagger(\Lambda)\phi(x)\mathcal{H}_{\text{int}}(z_1)..\mathcal{H}_{\text{int}}(z_n)U(\Lambda)\}|\alpha\rangle d^4z_1..d^4z_n$$

$$= \sum_{n=0}^\infty \frac{(-i)^n}{n!} \int \langle\beta|T\{\phi(\Lambda^{-1}x)\mathcal{H}_{\text{int}}(\Lambda^{-1}z_1)..\mathcal{H}_{\text{int}}(\Lambda^{-1}z_n)\}|\alpha\rangle d^4z_1..d^4z_n \quad (9.26)$$

Changing integration variables from z_i to $w_i \equiv \Lambda^{-1}z_i$, and recalling that the Λ are unimodular $(\det(\Lambda) = 1)$,

$$_{\text{out}}\langle\beta|\,U_H^\dagger(\Lambda)\phi_H(x)U_H(\Lambda)|\alpha\rangle_{\text{in}}$$

$$= \sum_{n=0}^\infty \frac{(-i)^n}{n!} \int \langle\beta|T\{\phi(\Lambda^{-1}x)\mathcal{H}_{\text{int}}(w_1)..\mathcal{H}_{\text{int}}(w_n)\}|\alpha\rangle d^4w_1..d^4w_n$$

$$= {}_{\text{out}}\langle\beta|\phi_H(\Lambda^{-1}x)|\alpha\rangle_{\text{in}}, \quad \forall\alpha, \beta, \Lambda \tag{9.27}$$

As α, β run over arbitrary members of complete sets, this establishes (at least in the context of perturbation theory!) the desired operator scalar field property (9.24) for the fully interacting Heisenberg field ϕ_H. The extension of the equal-time commutativity property (9.3) to full space-like commutativity, $[\phi_H(x), \phi_H(y)] = 0$, $(x - y)^2 < 0$ is now a straightforward exercise (see Problem 2).

The Heisenberg field $\phi_H(x)$ plays a fundamental role in field theory for a very simple reason: knowledge of its matrix elements is tantamount to a complete description of the dynamics of the theory. In particular—and this is the primary topic of this chapter—it contains all the information needed to reconstruct both the Heisenberg in-states $|\alpha\rangle_{\text{in}}$ and the out-states $|\beta\rangle_{\text{out}}$, and therefore, the exact scattering matrix $S_{\beta\alpha} = {}_{\text{out}}\langle\beta|\alpha\rangle_{\text{in}}$ of the theory. In order to see how this comes about, it is convenient to introduce some auxiliary fields which incorporate the kinematical structure of the asymptotic states. We shall do this explicitly for the in-states, with the understanding that the entire procedure can be carried through, *mutatis mutandis*, for the out-states by the simple device of replacing "in" with "out" everywhere. We first note that creation and annihilation operators can be defined on the multi-particle Fock space $\mathcal{H}_{\text{in}}$ exactly as for the bare states $|k_1..k_N\rangle$ in (6.29, 6.30)

$$a_{\text{in}}(k)|k_1, k_2, ..k_N\rangle_{\text{in}} \equiv \sum_{r=1}^N (\pm)^{r-1}\delta^3(k - k_r)|k_1, ..k_{r-1}, k_{r+1}, ..k_N\rangle_{\text{in}} \tag{9.28}$$

$$a_{\text{in}}^\dagger(k)|k_1, k_2, ..k_N\rangle_{\text{in}} \equiv |k, k_1, k_2, ..k_N\rangle_{\text{in}} \tag{9.29}$$

We shall be concerned for the time being with the simplest case, a bosonic spinless particle, so only positive signs need be taken in (9.28). A *free field* $\phi_{\text{in}}(x)$ can then be defined in complete analogy to the interaction-picture field $\phi(x)$ in (6.73) (with $a = a^c$, as we are also restricting ourselves to a self-conjugate scalar field) by setting

$$\phi_{\text{in}}(x) = \frac{1}{(2\pi)^{3/2}} \int \frac{d^3 k}{\sqrt{2E(k)}} (a_{\text{in}}(k)e^{-ik\cdot x} + a_{\text{in}}^\dagger(k)e^{ik\cdot x}) \tag{9.30}$$

We shall see later that the mass m appearing in the interaction-picture field ϕ (and in the coefficient of ϕ^2 in the interaction Hamiltonian density) and denoting the energy of single-particle eigenstates of H_0 at zero momentum is by no means guaranteed to agree with the actual physical mass m_{ph}: i.e., the energy eigenvalue of the *full* Hamiltonian H on a single particle in (or out) state with zero spatial momentum. On the other hand, the energy function $E(k)$ appearing in (9.30) is given by $\sqrt{k^2 + m_{\text{ph}}^2}$, and the four-momenta appearing in the complex exponentials in the integral are on-mass-shell for the physical mass, $k \cdot k = m_{\text{ph}}^2$. This implies that the in-field ϕ_{in} satisfies the free *Klein–Gordon equation*, but with the physical mass:

$$(\Box + m_{\text{ph}}^2)\phi_{\text{in}}(x) = 0 \tag{9.31}$$

The reader may easily verify that the spacetime translation of ϕ_{in} is implemented by the full energy-momentum vector P_μ, of which the in-states in (9.28, 9.29) are eigenstates:

$$e^{iP_\mu a^\mu} \phi_{\text{in}}(x) e^{-iP_\mu a^\mu} = \phi_{\text{in}}(x + a) \tag{9.32}$$

So the in-field ϕ_{in}, despite satisfying a free field (i.e., covariant linear) equation, is most definitely a Heisenberg picture operator. As indicated earlier, a Heisenberg out-field $\phi_{\text{out}}(x)$ may be introduced in an exactly analogous fashion, starting from creation and destruction operators for the out-states: it will likewise satisfy the free Klein–Gordon equation (9.31) (replacing, of course, "in" with "out"). The Heisenberg field ϕ_H, defined by transformation from the free interaction-picture field ϕ, on the other hand, certainly does not satisfy a free Klein–Gordon equation, as is apparent from (9.22) giving its matrix elements: only the first, $n = 0$ term in the perturbative expansion of these matrix elements, satisfies the Klein–Gordon equation (albeit with the mass m associated with the free Hamitonian H_0), while higher terms involving insertions of the interaction Hamiltonian have a much more complicated behavior.

It is apparent that knowledge of the ϕ_{in} *and* ϕ_{out} fields allows us to reconstruct the complete set of in- and out-states, and hence their overlap, the S-matrix. This follows from the formula (6.132) (Problem 8 in Chapter 6), appropriately generalized for the in- (or out-)field case:

$$a_{\text{in}}(k) = i \int d^3 x \left\{ g_k^*(\vec{x}, t) \overset{\leftrightarrow}{\frac{\partial}{\partial t}} \phi_{\text{in}}(\vec{x}, t) \right\} \tag{9.33}$$

$$a_{\text{in}}^\dagger(k) = -i \int d^3 x \left\{ g_k(\vec{x}, t) \overset{\leftrightarrow}{\frac{\partial}{\partial t}} \phi_{\text{in}}(\vec{x}, t) \right\} \tag{9.34}$$

with $g_k(x) = \dfrac{1}{\sqrt{(2\pi)^3 2E(k)}} e^{-ik\cdot x}$ (k an on-mass-shell four-vector, with $k_0 = \sqrt{\vec{k}^2 + m_{\text{ph}}^2}$), and with obviously similar formulas for $a_{\text{out}}, a_{\text{out}}^\dagger$ (by replacing "in" with "out" everywhere). The double derivative $\overset{\leftrightarrow}{\dfrac{\partial}{\partial t}}$ is defined as in (6.95).

The central importance of the single interacting Heisenberg field ϕ_H lies in its significance as an *interpolating operator for the basic quanta of the field*: i.e., the stable particles whose multi-particle states form the complete asymptotic Hilbert space (either "in" or "out") of the theory. By this we mean that ϕ_H in some sense "behaves like" ϕ_{in} in the far past, for $t \to -\infty$, and like ϕ_{out} in the far future, for $t \to +\infty$. If the behavior is sufficiently "nice", the discussion of the previous paragraph makes it completely plausible that a complete construction of the in- and out-states of the theory, and hence of their overlap, the all-important S-matrix, should be possible *from a knowledge of the Heisenberg field ϕ_H alone*. We conclude this section by giving a *purely formal*[3] argument for this interpolating property of ϕ_H: in the forthcoming sections the whole asymptotic formalism will be rederived in a mathematically rigorous framework, with absolutely no reference to perturbation theory or the interaction picture. For now, we shall proceed as though the interaction picture, and the associated interaction-picture operators, all make perfect sense.[4] For the free interaction-picture field ϕ, the analog of (9.34) is

$$a^\dagger(k) = -i \int d^3x \left\{ g_k(\vec{x},t) \overset{\leftrightarrow}{\dfrac{\partial}{\partial t}} \phi(\vec{x},t) \right\} \tag{9.35}$$

First, observe that the creation operator on the left-hand side is time-independent: the reader should verify, by a short calculation involving an integration by parts, that this is a direct consequence of the fact that $\phi(\vec{x},t)$ satisfies the Klein–Gordon equation. The operator defined from the Heisenberg field ϕ_H in complete analogy to (9.35),

$$a_H^\dagger(k;t) \equiv -i \int d^3x \left\{ g_k(\vec{x},t) \overset{\leftrightarrow}{\dfrac{\partial}{\partial t}} \phi_H(\vec{x},t) \right\} \tag{9.36}$$

will, by contrast, be time-dependent (and in a very complicated way!). Next, note that the relationship between the creation–destruction operators and the respective fields is linear. This means that we can imagine smearing $a^\dagger(k)$ (by smearing the pure exponential g_k) with a normalizable momentum-space wavefunction peaked around the central momentum k, so that $a^\dagger(k)$ corresponds to a well-defined operator in the Hilbert space (in particular, acting on normalizable n-particle states, it produces a normalizable $n+1$-particle state). We shall assume without further comment that we are dealing with such smeared wave-packet type states in this section; in the following

---

[3] This expression is typically used in theoretical physics, as indeed here, as a euphemism for "mathematically incorrect, but nevertheless suggestive".

[4] Although the steps that follow are mathematically unjustified, the result is in fact correct within the context of renormalized perturbation theory, *if we choose a split between the free and interacting Hamiltonians which eliminates, order by order in perturbation theory, persistent interactions modifying the mass and normalization of the single-particle states.* In particular, we shall assume that the mass term in H_0 is taken to be the physical mass m_{ph}.

$z = 0, \vec{z}$ $|z^\circ, z^i)$

$z' = \Lambda z$ $z^2 = -|\vec{z}|^2 < o$

$(\Lambda z)^2$

sections the need for smearing (both fields and states) in any careful mathematical treatment of field theory will be met head on and explicitly.

Observe that the unitary rotation effecting the transformation from interaction to the Heisenberg picture

$$\phi_H(\vec{x}, t) = U^\dagger(t, 0)\phi(\vec{x}, t)U(t, 0) \tag{9.37}$$

applies as well to the time-derivative of ϕ_H, by using the equations of motion for the time-development operator $U(t, 0)$, namely $\frac{\partial}{\partial t}U(t, 0) = -iV_{\text{ip}}(t)U(t, 0)$, $\frac{\partial}{\partial t}U^\dagger(t, 0) = +iU^\dagger(t, 0)V_{\text{ip}}(t)$,

$$\frac{\partial}{\partial t}\phi_H(\vec{x}, t) = U^\dagger(t, 0)(\frac{\partial\phi(\vec{x}, t)}{\partial t} + i[V_{\text{ip}}(t), \phi(\vec{x}, t)])U(t, 0) \tag{9.38}$$

The commutator term vanishes if $V_{\text{ip}}(t) = \int d^3y \mathcal{H}_{\text{int}}(\vec{y}, t)$, provided the interaction Hamiltonian density is an ultralocal scalar field, as discussed in Chapter 5:

$$[\mathcal{H}_{\text{int}}(\vec{y}, t), \phi(\vec{x}, t)] = 0 \Rightarrow [V_{\text{ip}}(t), \phi(\vec{x}, t)] = 0 \tag{9.39}$$

so we have, defining (cf. Section 8.1) $\pi_H(\vec{x}, t) \equiv \dot{\phi}_H(\vec{x}, t)$, $\pi(\vec{x}, t) \equiv \dot{\phi}(\vec{x}, t)$,

$$\pi_H(\vec{x}, t) = U^\dagger(t, 0)\pi(\vec{x}, t)U(t, 0) \tag{9.40}$$

Thus, both the Heisenberg field and its time-derivative are obtained as unitary transforms of the corresponding interaction-picture free fields. This implies a corresponding unitary relation between the operators $a_H^\dagger(k; t)$ and $a^\dagger(k)$ built from these fields via (9.35) and (9.36).

In our earlier discussion of scattering theory in Chapter 4 (cf. (4.155, 4.156)), the in-state $|\alpha\rangle_{\text{in}}$ arises as a strong limit from the corresponding bare state $|\alpha\rangle$ via

$$|\alpha\rangle_{\text{in}} = \lim_{t \to -\infty} U^\dagger(t, 0)|\alpha\rangle \tag{9.41}$$

which suggests that the following limits obtain for the action of the interacting operator $a_H^\dagger(k; t)$ on an arbitrary normalizable in-state $|\alpha\rangle_{\text{in}}$[5]:

$$\lim_{t \to -\infty} a_H^\dagger(k; t)|\alpha\rangle_{\text{in}} = \lim_{t \to -\infty} U^\dagger(t, 0)a^\dagger(k)U(t, 0)|\alpha\rangle_{\text{in}} \tag{9.42}$$

$$= \lim_{t \to -\infty} U^\dagger(t, 0)|k, \alpha\rangle \tag{9.43}$$

$$= |k, \alpha\rangle_{\text{in}} \tag{9.44}$$

$$= a_{\text{in}}^\dagger(k)|\alpha\rangle_{\text{in}} \tag{9.45}$$

with a similar result for the action of the operators $a_{\text{in}}(k), a_H(k; t)$. As the Heisenberg field ϕ_H and the in-field ϕ_{in} are constructed in a precisely analogous fashion from their corresponding creation and destruction parts, this line of argument suggests that the

---

[5] Were $a(k)$ a bounded operator, this, together with the unitarity (and hence automatic boundedness) of U and $U^\dagger$ would justify going from (9.42) to (9.43). As we shall see below, it is not, so again this argument is suggestive, not rigorous. A mathematically correct theory of scattering will be given in Section 9.3.

Heisenberg field *approximates* the free in-field ϕ_{in} in the far past. A completely analogous argument leads to the corresponding result for the far future: for $t \to +\infty$, the action of ϕ_H is equivalent to that of the free-field ϕ_{out}. In other words, the Heisenberg field ϕ_H *interpolates* between the free fields $\phi_{\text{in}}, \phi_{\text{out}}$ describing far-separated free particles in the distant past and similar sets of far separated free particles in the far future: for this reason, the Heisenberg field is sometimes referred to as an *interpolating field* for the particle in question. Roughly speaking, we may write

$$\phi_H(\vec{x}, t) \to \phi_{\text{in}}(\vec{x}, t), \quad t \to -\infty$$

$$\phi_H(\vec{x}, t) \to \phi_{\text{out}}(\vec{x}, t), \quad t \to +\infty \tag{9.46}$$

although we must caution the reader that we are here glossing over important mathematical subtleties, which will be fully dealt with in our treatment of Haag–Ruelle scattering theory in Section 9.3.

We conclude this section by noting an important connection between the S-matrix, defined previously as the set of overlap amplitudes between in- and out-states, and the unitary operator implementing the mapping between in- and out-creation and annihilation operators. Defining an unitary operator S by

$$S|\alpha\rangle_{\text{out}} = |\alpha\rangle_{\text{in}}, \quad {}_{\text{out}}\langle\beta| = {}_{\text{in}}\langle\beta|S \tag{9.47}$$

we see that our scattering amplitudes $S_{\beta\alpha} \equiv {}_{\text{out}}\langle\beta|\alpha\rangle_{\text{in}}$ are just the matrix elements of S in the in-basis:

$$_{\text{out}}\langle\beta|\alpha\rangle_{\text{in}} = S_{\beta\alpha} = {}_{\text{in}}\langle\beta|S|\alpha\rangle_{\text{in}} \tag{9.48}$$

With $|\alpha\rangle_{\text{out}}$ an arbitrary out-state (these, by asymptotic completeness, form a complete set), we have

$$Sa^\dagger_{\text{out}}(k)|\alpha\rangle_{\text{out}} = S|k, \alpha\rangle_{\text{out}} = |k, \alpha\rangle_{\text{in}} = a^\dagger_{\text{in}}(k)|\alpha\rangle_{\text{in}} = a^\dagger_{\text{in}}(k)S|\alpha\rangle_{\text{out}} \tag{9.49}$$

which implies the intertwining identity

$$Sa^\dagger_{\text{out}}(k) = a^\dagger_{\text{in}}(k)S \tag{9.50}$$

or, using the unitary property of S,

$$a^\dagger_{\text{in}}(k) = Sa^\dagger_{\text{out}}(k)S^\dagger$$

$$a_{\text{in}}(k) = Sa_{\text{out}}(k)S^\dagger \tag{9.51}$$

This implies that the in- and out-fields (cf. (9.30), together with the analogous expression for the out-field) are similarly related by a unitary rotation implemented by the S-operator:

$$\phi_{\text{in}}(x) = S\phi_{\text{out}}(x)S^\dagger \tag{9.52}$$

It should be emphasized that the results (9.47–9.52) rely only on asymptotic completeness, and make absolutely no reference to either the interaction picture or the perturbative expansion thereof.

9.2 Field theory in Heisenberg representation: axiomatics

In the preceding section we have examined some features of the description of quantum field theory in the Heisenberg representation of the dynamics (in which the field operators carry the full dynamical evolution) from an heuristic point of view. Given the fact that our arguments rested on a formal transition from the interaction picture to the Heisenberg picture, and the unpleasant fact (now generally referred to as "Haag's theorem"; cf. Section 10.5) that the two pictures are unitarily inequivalent (i.e., cannot be related by a well-defined unitary transformation) in any continuum field theory with non-trivial perturbations,[6] the reader may well wonder whether any of our conclusions can be considered valid for the four-dimensional field theories of phenomenological interest. In the forthcoming sections of this chapter, and until further notice, we shall abandon all references to perturbation theory or the interaction picture: in particular, there will be no need to adopt an artificial split of the Hamiltonian of the theory into "free" (H_0) and "interaction" (V) parts. Instead, we shall adopt, following Wightman, a minimal set of assumptions (or "axioms") which incorporate the essential features of a relativistic quantum field theory.

These axioms may conveniently be divided into three categories: (i) axioms specifying the structure of the underlying (Hilbert) space of states of the system, (ii) axioms specifying the primitive properties of the local field(s) of the theory, and (iii) axioms which establish a physical particle–field duality: i.e., a connection between the dynamical (Heisenberg) field(s) of the theory and the asymptotic particle states which are the systems prepared and detected in actual high-energy experiments. The task of establishing the *consistency* of these assumptions, in effect by a mathematically rigorous construction of well-defined operators with the desired properties in an explicit Hilbert space, is the task of *constructive quantum field theory*, and will not be addressed in this book—partly because this project has only succeeded fully in a small class of field theories defined in less than four spacetime dimensions. However, we shall see that with a very natural set of assumptions delineating the properties of the interacting fields of our theory, a complete scattering theory can be erected which completely avoids the mathematical irregularities and pitfalls associated with perturbation theory or the interaction picture. This rigorous scattering theory is primarily the work of Haag and Ruelle, and we shall present an overview of their results in the following section.

As we are no longer concerned with a free Hamiltonian H_0, or interaction term $V \equiv H - H_0$, it will be convenient to drop the "H" subscript (for "Heisenberg representation") in what follows, as we shall be referring throughout to Heisenberg fields (and states). Thus the Heisenberg field will be simply $\phi(x)$ (which, in our previous notation referred to the interaction-picture field). Also, the only particle mass which appears will be the actual physical mass m_{ph} of the given (stable) particle, which we shall therefore denote simply as m. Finally, to strip away all inessential complications, we shall initially consider a theory containing a single, stable, massive, spinless, and self-interacting particle, with no bound states. A more general discussion, taking into

---

[6] Here, "non-trivial perturbation" can be as innocent as a shift in the mass of a free field, as we shall see in Section 10.5.

account essentially all possibilities (in particular, the situations encountered in the Standard Model of modern particle physics), will follow in Section 9.6. We shall state the axioms of our theory, divided into the three groups outlined above, with each axiom followed by (sometimes extensive!) explanatory comments.

We begin with the *State axioms*, specifying features of the state space of the theory:

1. **Axiom Ia**: The state space $\mathcal{H}$ of the system is a separable Hilbert space. It carries a unitary representation $U(\Lambda, a)$ (Λ an element of the HLG, a a coordinate four-vector) of the proper inhomogeneous Lorentz group (i.e., the Poincaré group). Thus, for all $|\alpha\rangle \in \mathcal{H}$, $|\alpha\rangle \to U(\Lambda, a)|\alpha\rangle \in \mathcal{H}$, with the $U(\Lambda, a)$ satisfying the Poincaré algebra $U(\Lambda_1, a_1)U(\Lambda_2, a_2) = U(\Lambda_1\Lambda_2, a_1 + \Lambda_1 a_2)$ (cf. 5.14).

   Comments: Our Hilbert space $\mathcal{H}$ is a countable direct sum of multi-particle spaces corresponding to a definite number of particles. The multi-particle space corresponding to a fixed finite number of particles is a finite tensor product of separable L^2 spaces, each with a countable basis, and is therefore itself separable. The separability of $\mathcal{H}$ follows trivially. The reader is free to visualize $\mathcal{H}$ as the space of in-states $\mathcal{H}_{in}$ (or out-states, $\mathcal{H}_{out}$) described in the preceding section, with the action of the $U(\Lambda, a)$ given by $e^{iP \cdot a} U_H(\Lambda)$ with $P^\mu, U_H(\Lambda)$ as defined in (9.7, 9.9).

2. **Axiom Ib**: The infinitesimal generators P_μ of the translation subgroup $T(a) = U(1, a)$ of the Poincaré group have a spectrum p_μ restricted to the forward light-cone, $p_0 \geq 0, p^2 \geq 0$.

   Comments: In accordance with our intuition of asymptotic completeness— that all Heisenberg states of the system correspond to field disturbances which eventually resolve into a finite number of well-separated stable particles of finite energy, and with individual four-momenta on or within the forward light-cone— the total energy-momentum p_μ of any state of the system must be resolvable into a sum of four-vectors, each of which is within the forward light-cone, and must therefore itself lie in this region.

3. **Axiom Ic**: There is a unique state $|0\rangle$ (up to a unimodular phase, of course), called the "vacuum", with the isolated eigenvalue $p_\mu = 0$ of P_μ. It is unit normalized, $\langle 0|0\rangle = 1$.

   Comments: In particular, we assume the absence of spontaneous symmetry-breaking, as discussed in Section 8.4. The need to accommodate simultaneously a discretely normalized vacuum and continuously normalized multi-particle states in an infinite-volume interacting theory will be seen later to be intimately connected with the difficulties associated with the interaction picture and the famous "Haag's theorem" (Section 10.5).

4. **Axiom Id**: The theory has a mass gap: the squared-mass operator $P^2 = P_\mu P^\mu$ has an isolated eigenvalue $m^2 > 0$, and the spectrum of P^2 is *empty* between 0 and m^2. The subspace $\mathcal{H}_1$ of $\mathcal{H}$ corresponding to the eigenvalue m^2 carries an irreducible spin-0 representation of the HLG. These are the single-particle states of the theory. The remaining spectrum of P^2 is continuous, and begins at $(2m)^2$.

   Comments: This axiom specifically excludes quantum electrodynamics with a massless photon (the only known massless particle): indeed, the structure of the state space and the asymptotic dynamics, and in particular the definition of

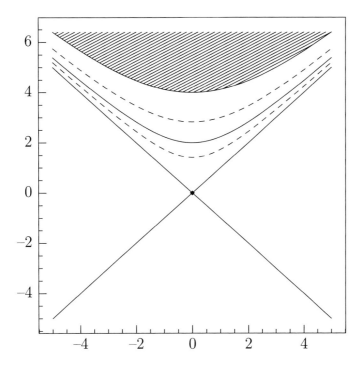

Fig. 9.1 The spectrum of the energy-momentum operator P^μ (P^0 vertical axis, arbitrary spatial component P^i horizontal axis, $m = 2$). The dashed lines enclose the support region for the function $\tilde{f}^{(1)}(p)$ introduced in Section 9.3, with $a = \frac{1}{2}$, $b = 2$.

a sensible S-matrix, is extremely subtle in the presence of an exactly massless particle (cf. Chapter 19). Fortunately, it is possible to give the photon a tiny mass without altering any critical features of the theory: the massless limit can then be performed smoothly *after* all calculation of rates, cross-sections, etc., are performed with the appropriate instrumental constraints. Thus, the assumption of a mass gap is not unduly constraining. In the two-particle subspace (say, for $|p_1, p_2\rangle_{\text{in}}$), the squared mass operator gives $(p_1 + p_2)^2 = 2m^2 + 2p_1 \cdot p_2$, with $p_1 \cdot p_2 \geq m^2$, so the spectrum of P^2 in this subspace is $[(2m)^2, \infty)$. Overall, the spectrum of P^2 is therefore $\{0, m^2, [(2m)^2, \infty)\}$. The spectrum of the four-vector operator P^μ is indicated in Fig. 9.1: it consists of the origin (the vacuum state, indicated by the central point), the one-particle mass hyperboloid $p^2 = m^2$, and the continuum, the shaded region contained within and above the hyperboloid $p^2 = 4m^2$. We assume no bound states, e.g., one-particle mass hyperboloids at $p^2 = 4m^2 - \mathcal{E}$.

We turn next to the assumptions concerning the existence and properties of a Heisenberg field incorporating the complete interacting dynamics of our theory: these will be termed the *Field axioms* of the theory.

1. **Axiom IIa:** An operator-valued (tempered) distribution $\phi(x)$ exists such that for any Schwartz test function $f(x)$,[7] the smeared field

$$\phi_f \equiv \int f(x)\phi(x)d^4x \tag{9.53}$$

is an unbounded operator defined on a dense subset $D \subset \mathcal{H}$. Moreover, $\phi_f D \subset D$, allowing the definition of arbitrary (finite) products of smeared fields.

Comments: The assumed properties are motivated by the equivalent statements for a free scalar field, for which (for real $f(x)$)

$$\phi_f = \int d^3p(\tilde{f}^*(\vec{p})a(\vec{p}) + \tilde{f}(\vec{p})a^\dagger(\vec{p})) \tag{9.54}$$

where

$$\tilde{f}(\vec{p}) \equiv \frac{1}{(2\pi)^{3/2}\sqrt{2E(p)}} \int f(x)e^{iE(p)x^0 - i\vec{p}\cdot\vec{x}}d^4x \tag{9.55}$$

Standard theorems ensure that the Fourier transform function $\tilde{f}$ is also a function (of the spatial momentum vector $\vec{p}$) of Schwarz type. A dense subset of $\mathcal{H}$ can be obtained by considering all normalizable n-particle states of the form

$$|g_1, g_2, ..., g_n\rangle = \int d^3q_1 d^3q_2..d^3q_n g_1(q_1)..g_n(q_n)|q_1, q_2, ..q_n\rangle \tag{9.56}$$

where $\int d^3q|g_i(q)|^2 < \infty, i = 1, 2, ..n$ (for example, the g_i may be chosen from a complete basis of L^2). The reader may easily verify (see Problem 4) that the state $\phi_f|g_1, g_2, ..g_n\rangle$ has finite norm. It is also easy to see that a further application of a second smeared field, ϕ_{f_1} say, still produces a finite norm state (thus, $\phi_f D \subset D$). That the smeared fields are unbounded operators is also unsurprising if we recall the discussion of Section 8.3, where normalized coherent states were constructed with arbitrary (and therefore, arbitrarily large) expectation values of the field. Note that smearing of the original local field $\phi(x)$ is unnecessary if we are only concerned with matrix elements $\langle\beta|\phi(x)|\alpha\rangle$ with $|\alpha\rangle, |\beta\rangle \in D$, which are perfectly well-defined functions of x. However, we must frequently consider the states (in the Hilbert space) obtained by sequential application of field operators, in which case the smeared (or, in the language of Haag, "almost local") operators ϕ_f are unavoidable. Finally, we note that in keeping with our restriction to a system of a single self-conjugate particle, the hermiticity of $\phi(x)$ translates to the following property for the smeared fields (for $f(x)$ in general complex):

$$(|\beta\rangle, \phi_f|\alpha\rangle) = (\phi_{f^*}|\beta\rangle, |\alpha\rangle), \quad \forall|\alpha\rangle, |\beta\rangle \in D \tag{9.57}$$

Axiom IIa implies the existence of the vacuum-expectation-value (VEV) of products of smeared fields, which can be written as overlaps of the famous

---

[7] The space of Schwartz test functions consists of C^∞ (i.e., infinitely times continuously differentiable) functions of fast decrease (i.e., falling faster than any power as the spacetime coordinates go to ∞).

Wightman distributions (Wightman, 1956) (somewhat loosely, the terminology "Wightman functions" is, in fact, more commonly used) $W(x_1, x_2, ..., x_n)$ with Schwarz test functions:

$$\langle 0|\phi_{f_1}...\phi_{f_n}|0\rangle = \int f_1(x_1)..f_n(x_n)\langle 0|\phi(x_1)..\phi(x_n)|0\rangle d^4x_1..d^4x_n$$

$$= \int f_1(x_1)..f_n(x_n)W(x_1, ..., x_n)d^4x_1..d^4x_n \qquad (9.58)$$

Again, the situation with free fields motivates the assumption that the distributions $W(x_1, x_2, ...x_n)$ are *tempered*: namely, continuous linear functionals on the space of fast-falling Schwarz test functions in the $4n$-dimensional space of the combined spacetime coordinates $x_1, x_2, ...x_n$. We remind the reader[8] that a tempered distribution $W(z)$ (here z is a vector in the combined coordinate spacetime of all the test functions in (9.58)) can always be written as a finite derivative of a polynomially bounded continuous function:

$$W(z) = D^m F(z) \qquad (9.59)$$

$$|F(z)| < C(1 + |z|_E^2)^p \qquad (9.60)$$

Here D^m is a generic notation for m derivatives with respect to spacetime coordinates (components of z), the norm in (9.60) is *Euclidean*, $|z|_E^2 \equiv \sum_{i,\mu}(x_i^\mu)^2$, and the constant C and power p depend, of course, on the particular Wightman distribution $W(z)$ under consideration. A simple example of a tempered distribution is the Dirac δ distribution, which can be written $\delta(x) = \frac{1}{2}\frac{d^2}{dx^2}|x|$, clearly satisfying (9.59, 9.60).

2. **Axiom IIb**: Under the unitary representation of the Poincaré group $U(\Lambda, a)$ introduced in Axiom Ia, the smeared fields transform as

$$U(\Lambda, a)\phi_f U^\dagger(\Lambda, a) = \phi_{f_{\Lambda,a}}, \quad f_{\Lambda,a}(x) \equiv f(\Lambda^{-1}(x - a)) \qquad (9.61)$$

Comments: Our standard transformation law on unsmeared fields (cf. (5.94))

$$U(\Lambda, a)\phi(x)U^\dagger(\Lambda, a) = \phi(\Lambda x + a) \qquad (9.62)$$

immediately leads to (9.61) after multiplying both sides by $f(x)$, integrating over the four-coordinate x, and appropriate changes of variable in the integration. The reader may also verify that $U(\Lambda, a)\phi_f = \phi_{f_{\Lambda,a}}U(\Lambda, a)$, following directly from (9.61), implies that the domain D is Poincaré-invariant, namely $U(\Lambda, a)D = D$, as $f_{\Lambda,a}$ is of Schwarz type if f is. The infinitesimal generators of translations are just the energy-momentum operators whose spectral properties are delineated in the first set of axioms above: thus, $U(1, a) = e^{iP \cdot a}$, while the general Poincaré group element can be expressed $U(\Lambda, a) = U(1, a)U(\Lambda, 0) = U(1, a)U(\Lambda)$, with

---

[8] For a review of the essential facts concerning tempered distributions, see (Streater and Wightman, 1978).

$U(\Lambda)$ the generators of the HLG (just the $U_H(\Lambda)$ of the preceding Section). The transformation law for a general (unsmeared) covariant field, given as a set of component fields $\phi_n(x)$ transforming according to a finite-dimensional representation $M(\Lambda)$ of the HLG, is a straightforward generalization of (9.62) (cf. (7.5)):

$$U(\Lambda, a)\phi_n(x)U^\dagger(\Lambda, a) = M_{nm}(\Lambda^{-1})\phi_m(\Lambda x + a) \tag{9.63}$$

For Lorentz tranformations infinitesimally close to the identity, $\Lambda^\mu{}_\nu = g^\mu{}_\nu + \omega^\mu{}_\nu$, we recall that $M_{nm}(\Lambda) = \delta_{nm} + \frac{i}{2}\omega_{\mu\nu}J^{\mu\nu} + O(\omega^2)$, corresponding to a unitary realization $U(\Lambda)$ likewise infinitesimally close to the identity operator in the Hilbert space of the theory:

$$U(\Lambda) = 1 + \frac{i}{2}\omega_{\mu\nu}\mathcal{M}^{\mu\nu} + O(\omega^2) \tag{9.64}$$

where the $\mathcal{M}^{\mu\nu}$ are a set of six independent self-adjoint operators which are the infinitesimal generators of HLG on the state space, in effect the representatives of the finite-dimensional $J^{\mu\nu}$ matrices, and hence satisfying the same commutator algebra (7.21)

$$[\mathcal{M}^{\mu\nu}, \mathcal{M}^{\rho\sigma}] = i(g^{\mu\sigma}\mathcal{M}^{\rho\nu} + g^{\nu\sigma}\mathcal{M}^{\mu\rho} - g^{\rho\mu}\mathcal{M}^{\sigma\nu} - g^{\rho\nu}\mathcal{M}^{\mu\sigma}) \tag{9.65}$$

The Poincaré group composition rule implies $U(\Lambda)U(1,a)U^\dagger(\Lambda) = U(1,\Lambda a)$, whence, using $U(1,a) = e^{iP\cdot a}$ and taking a infinitesimal, we obtain

$$U(\Lambda)P^\rho U^\dagger(\Lambda) = \Lambda_\sigma{}^\rho P^\sigma \tag{9.66}$$

Taking $\Lambda_\sigma{}^\rho = g_\sigma{}^\rho + \omega_\sigma{}^\rho$ in (9.66), a short calculation gives the commutation relations between the generators of translations (the energy-momentum four-vector) and those of the HLG:

$$[P^\rho, \mathcal{M}^{\mu\nu}] = i(g^{\rho\nu}P^\mu - g^{\rho\mu}P^\nu) \tag{9.67}$$

The full Lie algebra of the Poincaré group is completed by observing that the subgroup of translations is abelian, so the generators P^μ commute with each other:

$$[P^\rho, P^\sigma] = 0 \tag{9.68}$$

3. **Axiom IIc**: Let f_1, f_2 be Schwarz functions of *compact support*: thus, if f_1 vanishes outside a compact region v_1 of spacetime, and f_2 vanishes outside of the compact region v_2, and if $x_1 - x_2$ is space-like for all $x_1 \in v_1, x_2 \in v_2$, then

$$[\phi_{f_1}, \phi_{f_2}] = 0 \tag{9.69}$$

Comments: If the fields in question are fermionic, then, of course, the commutator in (9.69) should be replaced by an anticommutator. This requirement is an obvious consequence of our previous assumption of space-like commutativity at the level of unsmeared fields, e.g., (7.9). The restriction to compact supports for the smearing fields is not strictly necessary: it suffices that the supports of f_1 and f_2 are non-overlapping and mutually space-like.

4. **Axiom IId**: The set of states obtained by applying arbitrary polynomials in the smeared fields ϕ_f (with all possible Schwarz functions f) to the vacuum state $|0\rangle$ (cf. Axiom Ic) is dense in the Hilbert space $\mathcal{H}$. This axiom is sometimes expressed as the "cyclicity of the vacuum".

   Comments: In other words, an arbitrary physical state can be approximated arbitrarily well by linear combinations of states of the form $\phi_{f_1}\phi_{f_2}....\phi_{f_n}|0\rangle$ (including, of course, the case $n = 0$, i.e., the vacuum itself). This so-called *Cyclicity axiom* incorporates our desire that the complete dynamical information of the theory is contained in the single field ϕ. Of course, in theories containing several different types of particles (which are not bound states of each other, for example), all such independent fields must be allowed to act on the vacuum to produce the full Hilbert space of the theory. For the time being though, as emphasized above, we wish to focus on the simplest possible case, that of a single spinless particle associated with a single scalar field. The reader may easily verify that for free fields, a dense set of normed states of the form (9.56) can indeed be obtained by applying smeared (free) fields to the vacuum. Axiom IId is sometimes formulated in a different but equivalent way: as the statement that the set of (appropriately smeared) fields ϕ_f are *irreducible*—or in other words, that the only bounded operator that commutes with all the ϕ_f is a multiple of the identity.

Before stating the axioms of the third category—those necessary to complete the connection between particles and fields—we shall discuss two fundamental results which already follow from the axioms previously stated: specifically, from a combination of the spectral and locality properties of the theory. The first result concerns the analyticity properties of the Wightman distributions: indeed, we have already seen in Section 6.6 that a close connection exists between locality and analyticity of amplitudes in field theory. The second result is the Ruelle Clustering theorem, which is the precise correlate, in field-theoretic terms, of the clustering property of the S-matrix which was one of our primary motivations in developing the formalism of local field theory (cf. Chapter 6). The exact connection will become apparent after our discussion of Haag–Ruelle scattering theory in the next Section.

The analyticity domain of the Wightman functions follows directly from the translation property IIb, together with the spectral properties of the energy-momentum operator P expressed in Axioms Ib and Ic. Thus, writing $\phi(x_i) = e^{iP \cdot x_i} \phi(0) e^{-iP \cdot x_i}$, we may write (defining $\xi_i \equiv x_i - x_{i+1}$)

$$W(x_1, x_2, ...x_n) = \langle 0|\phi(0)e^{-iP \cdot (x_1-x_2)}\phi(0)e^{-iP \cdot (x_2-x_3)}....e^{-iP \cdot (x_{n-1}-x_n)}\phi(0)|0\rangle$$

$$= \langle 0|\phi(0)e^{-iP \cdot \xi_1}\phi(0)e^{-iP \cdot \xi_2}\phi(0)...e^{-iP \cdot \xi_{n-1}}\phi(0)|0\rangle \equiv W(\xi_1, ..\xi_{n-1})$$

$$(9.70)$$

By assumption, $W(\xi_1, ..\xi_{n-1})$ is a tempered distribution, with a well-defined Fourier transform $\tilde{W}(k_1, ..., k_{n-1})$:

$$W(\xi_1, ..\xi_{n-1}) = \int \tilde{W}(k_1, ..., k_{n-1}) e^{-i \sum_{i=1}^{n-1} k_i \cdot \xi_i} \prod_{i=1}^{n-1} \frac{d^4 k_i}{(2\pi)^4} \qquad (9.71)$$

Inserting (9.70) in the inverse Fourier formula for $\tilde{W}$ and performing the ξ_i integrals, we find

$$\tilde{W}(k_1, .., k_{n-1}) = (2\pi)^{4(n-1)} \langle 0|\phi(0)\delta^4(P - k_1)\phi(0)\delta^4(P - k_2)....\delta^4(P - k_{n-1})\phi(0)|0\rangle \qquad (9.72)$$

The spectrum of the energy-momentum operator P is restricted to the forward light-cone $k_i^2 \geq 0, k_i^0 > 0$, so $\tilde{W}(k_1, .., k_{n-1})$ vanishes unless all k_i are in the forward cone. The analytical continuation of (9.70) to complex values of the ξ_i

$$\zeta_i = \xi_i - i\eta_i \qquad (9.73)$$

with all η_i positive time-like ($\eta_i^2 > 0$) (but the ξ_i arbitrary real) therefore leads to the already well-defined integral (9.71) acquiring an additional real exponential damping factor (as $k_i \cdot \eta_i > 0$ for all i). As long as we stay away from coincident point singularities therefore, we see that the Wightman functions are analytic in this multi-dimensional complex domain, called the *forward tube* $\mathcal{T}_{n-1}$. An extremely important theorem of axiomatic field theory, due to Hall and Wightman (Hall and Wightman, 1957), shows that a further analytic continuation is possible, by the simple device of extending the Lorentz-invariance property for real coordinates and real Lorentz transformations Λ

$$W(\xi_1, ..., \xi_{n-1}) = W(\Lambda\xi_1, ..., \Lambda\xi_{n-1}) \qquad (9.74)$$

to *complex* Lorentz transformations (i.e., complex 4x4 matrices Λ with $\det(\Lambda) = 1$ and $\Lambda g \Lambda^T = g$), and defining the analytic extension to the *extended tube* $\mathcal{T}'_{n-1}$ consisting of points of the form $(\Lambda\zeta_1, ..., \Lambda\zeta_{n-1}), \zeta_i \in \mathcal{T}_{n-1}$ by

$$W(\Lambda\zeta_1, ..., \Lambda\zeta_{n-1}) = W(\zeta_1, ..., \zeta_{n-1}), \quad (\Lambda\zeta_1, ..., \Lambda\zeta_{n-1}) \in \mathcal{T}'_{n-1} \qquad (9.75)$$

The Hall–Wightman theorem gives an explicit characterization of the Wightman functions in this extended domain: they are simply analytic functions of the complex scalar dot-products $\zeta_i \cdot \zeta_j$ over the complex domain spanned by these dot-products. The analyticity of the Wightman functions in the extended tube is an extremely powerful constraint : for example, we can conclude immediately that for all $(\zeta_i) \in \mathcal{T}'_{n-1}$,

$$W(\zeta_1, ..., \zeta_{n-1}) = W(-\zeta_1, ..., -\zeta_{n-1}) \qquad (9.76)$$

This is obvious given that, by the Hall–Wightman theorem, the Wightman functions depend only on the dot-products $\zeta_i \cdot \zeta_j$, but can also be seen by the fact that we can connect the unit Lorentz transformation $\Lambda = 1$ to the spacetime reflection $\Lambda = -1$ (with $\det(\Lambda) = +1!$) by a continuous analytic path $\Lambda(\theta), \theta : 0 \to \pi$ in the complex Lorentz group, by taking

$$\Lambda^{\mu}{}_{\nu}(\theta) = \begin{pmatrix} \cos(\theta) & 0 & 0 & -i\sin(\theta) \\ 0 & \cos(\theta) & \sin(\theta) & 0 \\ 0 & -\sin(\theta) & \cos(\theta) & 0 \\ -i\sin(\theta) & 0 & 0 & \cos(\theta) \end{pmatrix} = e^{i\theta(J_3 - iK_3)} = e^{2i\theta A_3} \quad (9.77)$$

The reader should check that this is indeed a Lorentz transformation (i.e., satisfies $\det(\Lambda) = +1, \Lambda^T g\Lambda = g$)! It is complex insofar as the boost angle (coefficient of iK_3 in the exponent) is imaginary. The result (9.76), involving a reversal of both the temporal (T) and spatial (P) coordinates, will turn out to be a crucial part of the demonstration of the TCP theorem in the very general framework of axiomatic field theory given in Section 13.4. An obvious generalization of (9.76) to the Wightman function for a product $\phi^{A_1 B_1}(x_1)\phi^{A_2 B_2}(x_2)\cdots\phi^{A_n B_n}(x_n)$ of fields in arbitrary representations of the HLG follows directly from (9.77):

$$W^{A_1 B_1, A_2 B_2, ..., A_n B_n}(\zeta_1, ..., \zeta_{n-1}) = (-1)^{\sum_i 2A_i} W^{A_1 B_1, A_2 B_2, ..., A_n B_n}(-\zeta_1, ..., -\zeta_{n-1}) \quad (9.78)$$

by analytic continuation from $\theta = 0$ to $\theta = \pi$ in (9.77), and recalling that the 3-component of an angular momentum A_{i3} differs from the A_i value by an integer.[9]

Using local commutativity (Axiom IIc), it can be also shown that the analyticity domains in the difference variables $\xi_1, .., \xi_{n-1}$ for different permutations of the original arguments $x_1, \ldots x_n$ can be connected to obtain analyticity in a *permuted extended tube* (see (Streater and Wightman, 1978) for further details). A particular case of great importance is the *imaginary time continuation* of the Wightman functions, yielding the *Euclidean Schwinger functions* of the theory

$$S(x_1, x_2, ...x_n) \equiv W((\vec{x}_1, -ix_1^4), (\vec{x}_2, -ix_2^4), ..., (\vec{x}_n, -ix_n^4)) \quad (9.79)$$

where the $x_i^{\alpha}, \alpha = 1, 2, 3, 4$ are real. Unlike the Wightman functions, the Schwinger functions are *permutation symmetric* in their arguments (the Wightman functions involve field operators which do not commute at time-like separation). This can be shown directly from the axioms, but will become essentially trivial in the path-integral formalism, so we shall postpone further discussion of the Schwinger functions to Section 10.3, where we shall explore the properties of functional integrals in field theory.

We turn now to the issue of clustering—in particular, the result originally obtained by Ruelle (Ruelle, 1962) on the large distance asymptotics of the Wightman functions. In order to state this result, we need to introduce the concept of a smeared field localized around a point x, $\phi_f(x)$. Our original definition, $\phi_f \equiv \int f(x)\phi(x)d^4x$, with $f(x)$ a function falling faster than any power as x moves away from the origin $x = 0$, should be regarded as producing a smeared field localized in a very small, but finite, region near $x = 0$. A corresponding field (or product of fields; see below) localized

---

[9] A different choice of complex continuation from $\Lambda = 1$ to $\Lambda = -1$ could be made, resulting in the factor $(-1)^{\sum_i 2B_i}$. This is, however, identical to $(-1)^{\sum_i 2A_i}$ as the spin j_i associated with each field differs from $A_i + B_i$ by an integer, and the vacuum expectation value of a product of fields can only differ from zero if the spins they carry can couple to zero.

around an arbitrary coordinate x (in Haag's language (Haag, 1992), an *almost local field*) can be obtained by the standard process of translation:

$$\phi_f(x) \equiv e^{iP\cdot x}\phi_f e^{-iP\cdot x} = \int f(y-x)\phi(y)d^4y \tag{9.80}$$

In fact, we can define more general types of smeared fields as smeared *polynomials* in the basic field $\phi(x)$. For example, taking $f(x_1, x_2)$ to be a Schwarz function of rapid decrease in the pair of coordinates (x_1, x_2), we can define a smeared bilocal operator, localized again in the neighborhood of the origin, by

$$\phi_f \equiv \int f(x_1, x_2)\phi(x_1)\phi(x_2)d^4x_1 d^4x_2 \tag{9.81}$$

with a corresponding almost local field $\phi_f(x)$ localized in the neighborhood of an arbitrary coordinate point x by translation, as in (9.80). *Everything that follows applies with equal validity to almost local operators of this more general type.*

It is somewhat unintuitive, but nevertheless true, that the Fourier transform of a smeared field, $\tilde{\phi}_f(p)$

$$\tilde{\phi}_f(p) \equiv \int \phi_f(x)e^{-ip\cdot x}d^4x \tag{9.82}$$

carries a precise four-momentum: it alters the momentum of any state on which it acts by precisely p. Indeed, if $|\alpha\rangle, |\beta\rangle$ are states of well-defined four-momentum P_α, P_β, then evidently

$$\langle\beta|\int \phi_f(x)e^{-ip\cdot x}d^4x|\alpha\rangle = \int e^{-ip\cdot x}\langle\beta|e^{iP\cdot x}\phi_f e^{-iP\cdot x}|\alpha\rangle$$

$$= \int e^{+i(P_\beta - P_\alpha - p)\cdot x}\langle\beta|\phi_f|\alpha\rangle d^4x \propto \delta^4(P_\beta - (P_\alpha + p)) \tag{9.83}$$

Thus, $\langle\beta|\tilde{\phi}_f(p)|\alpha\rangle$ is non-zero only if $P_\beta = P_\alpha + p$.

As a preliminary to the statement and (partial) proof of the Ruelle Clustering theorem, we first show that the locality postulate (Axiom IIc) implies that the commutator of two smeared fields falls faster than any power as the localization points are separated in a space-like direction. We shall need this result only for the vacuum-expectation-value of the commutator. Thus, let $\vec{a}$ be a spatial three-vector, with the corresponding four-vector $a \equiv (0, \vec{a})$. With f_1, f_2 fast-decreasing Schwarz functions as usual, we shall show that for any power N, there exists a constant C such that

$$|\langle 0|[\phi_{f_1}(-a), \phi_{f_2}(+a)]|0\rangle| < C|\vec{a}|^{-N}, \quad |\vec{a}| \to \infty \tag{9.84}$$

This result depends on the fast decrease property of the smearing functions f_1, f_2 *as well as* the tempered distribution character of the Wightman distributions. We shall explain the proof in somewhat greater detail than usual in order to give the reader some idea of the flavor of the reasoning in axiomatic quantum field theory. First, note that

$$\langle 0|[\phi_{f_1}(-a), \phi_{f_2}(+a)]|0\rangle = \int f_1(x_1) f_2(x_2) \langle 0|[\phi(x_1 - a), \phi(x_2 + a)]|0\rangle d^4x_1 d^4x_2$$

$$(9.85)$$

The estimates we shall need rely on introducing a *completely Euclidean* metric on the multiple coordinate space of all the fields under consideration. Thus, we shall define eight-vectors $z = (x_1, x_2)$, $\alpha = (a, -a)$, with the Euclidean norm $|z|_E^2 = \sum_\mu ((x_1^\mu)^2 + (x_2^\mu)^2)$, $|\alpha|_E^2 = 2|\vec{a}|^2$. The product of smearing functions f_1, f_2 is likewise a fast falling function of z: $F(z) \equiv f_1(x_1) f_2(x_2)$. The VEV of the commutator is the difference of two Wightman distributions, $W(x_1 - a, x_2 + a) - W(x_2 + a, x_1 - a)$, which we shall write as a single tempered distribution of the eight-dimensional variable $z - \alpha$:

$$\langle 0|[\phi_{f_1}(-a), \phi_{f_2}(+a)]|0\rangle = \int F(z) W(z - \alpha) d^8z \qquad (9.86)$$

The tempered nature of $W(z)$ implies (cf. (9.59, 9.60)

$$W(z) = D^m \mathcal{P}(z - \alpha), \quad \mathcal{P}(z - \alpha) < C_1 (1 + |z - \alpha|_E^2)^p < C_1 (1 + |z|_E^2)^p (1 + |\alpha|_E^2)^p \qquad (9.87)$$

for any power p and some constant C_1, with D^m consisting of m coordinate derivatives. Substituting (9.87) into (9.86), and performing m integrations by parts,

$$|\langle 0|[\phi_{f_1}(-a), \phi_{f_2}(+a)]|0\rangle| = |\int \mathcal{P}(z - \alpha) D^m F(z) d^8z| \qquad (9.88)$$

A little geometric reasoning shows that if $|z|_E < |\vec{a}|$, the spacetime points $x_1 - a$ and $x_2 + a$ for the local fields appearing in the commutator on the right-hand side of (9.85) must be space-like, so that the integrand vanishes unless $|z|_E \geq |\vec{a}|$: accordingly, the polynomially-bounded function $\mathcal{P}$, and hence the integrand as a whole, must vanish except for $|z|_E > |\vec{a}|$:

$$|\langle 0|[\phi_{f_1}(-a), \phi_{f_2}(+a)]|0\rangle| = |\int_{|z|_E > |\vec{a}|} \mathcal{P}(z - \alpha) D^m F(z) d^8z| \qquad (9.89)$$

The fast decrease property of the smearing functions implies that for any power N there exists a constant C_2 such that

$$|D^m F(z)| < C_2 |z|_E^{-(N+2p+9)} (1 + |z|_E^2)^{-p} \qquad (9.90)$$

where p is the power already appearing in (9.87). Inserting (9.90) in (9.89), and using $d^8z = S_8 |z|_E^7 d|z|_E$, with S_8 the surface area of the eight-dimensional unit sphere,

$$|\langle 0|[\phi_{f_1}(-a), \phi_{f_2}(+a)]|0\rangle| < C_1 C_2 S_8 (1 + |\alpha|_E^2)^p |\int_{|z|_E > |\vec{a}|} |z|_E^{-N-2p-2} d|z|_E$$

$$< \frac{C_1 C_2 S_8}{N + 2p + 1} (1 + 2|\vec{a}|^2)^p |\vec{a}|^{-N-2p-1} \qquad (9.91)$$

which establishes the desired behavior for large $|\vec{a}|$ as promised in (9.84).

We now turn to a discussion of the clustering properties of our field theory. In analogy to the discussion of connected parts of the S-matrix in Sections 6.1 and 6.2, a set of connected[10] Wightman distributions $W^c(x_1, x_2, \ldots x_n)$ can be defined inductively, starting from the full distributions $W(x_1, \ldots x_n) = \langle 0|\phi(x_1)\phi(x_2) \ldots \phi(x_n)|0\rangle$, as follows:

$$W^c(x_1) = W(x_1) \tag{9.92}$$

$$W^c(x_1, x_2) = W(x_1, x_2) - W^c(x_1)W^c(x_2) \tag{9.93}$$

$$W^c(x_1, x_2, x_3) = W(x_1, x_2, x_3) - (W^c(x_1)W^c(x_2, x_3) + \text{perms})$$
$$- W^c(x_1)W^c(x_2)W^c(x_3) \tag{9.94}$$

and so on. The inversion of this set of equations,

$$W(x_1) = W^c(x_1)$$
$$W(x_1, x_2) = W^c(x_1)W^c(x_2) + W^c(x_1, x_2), \text{ etc.} \tag{9.95}$$

shows us that the full amplitude corresponding to the vacuum expectation value of a product of fields has a *cluster decomposition* into a sum of terms in which the fields are partitioned in all possible ways into clusters, with the fields in each cluster appearing inside a connected matrix element. Connected versions of the vacuum-expectation-values of the corresponding smeared ("almost local") operators will be defined similarly, thus

$$\langle 0|\phi_f(x_1)\phi_f(x_2)|0\rangle_c \equiv \langle 0|\phi_f(x_1)\phi_f(x_2)|0\rangle$$
$$- \langle 0|\phi_f(x_1)|0\rangle\langle 0|\phi_f(x_2)|0\rangle \tag{9.96}$$

and so on. The *Ruelle Clustering theorem* amounts to the intuitively plausible assertion that the connected expectation values so defined decrease rapidly when the points of localization of the fields are taken far apart spatially:

Theorem 9.1 *For a set of n spacetime coordinates $x_1, x_2, \ldots x_n$ at equal time, $x_i = (t, \vec{x}_i)$, with spatial diameter $d \equiv \max_{i,j}|\vec{x}_i - \vec{x}_j|$, for any positive N there exists a constant C_N such that for large d*

$$\langle 0|\phi_{f_1}(x_1)\phi_{f_2}(x_2)..\phi_{f_n}(x_n)|0\rangle_c < C_N d^{-N} \tag{9.97}$$

It will be an immediate consequence of the beautiful scattering theory due to Haag and Ruelle—to be discussed in the next section—that the clustering properties posited as a phenomenological necessity for the S-matrix in Section 6.1 *follow transparently* from the aforestated Clustering theorem for the fields. We shall not prove the Ruelle

---

[10] In much of the axiomatic quantum field theory literature, the term "truncated" is used instead of "connected" for the distributions defined in (9.92–9.94). In the interests of honesty, we should admit here that the clustering expansion of (9.92, etc.) is, strictly speaking, well-defined only for n-point functions that are permutation-symmetric on their arguments. We shall shortly be restricting ourselves to the special case where the spacetime coordinates $x_1, \ldots, x_n$ are far space-like separated, in which limit the necessary symmetry of the Wightman functions is restored.

Clustering theorem in its full generality here,[11] but indicate (following the extremely streamlined discussion of Haag, (Haag, 1992)) how the argument goes for the case $n = 2$.

A crucial ingredient in the argument leading to clustering is the assumption made in Axiom Id: the theory has a *mass gap*, with the minimum energy of states orthogonal to the vacuum equal to $m > 0$. Choose positive numbers a, b such that $0 < a < b < m$. Define the function $F_+(p_0)$ on the energy variable p_0 (for a four-vector p) as

$$F_+(p_0) \equiv \frac{e^{-\frac{K}{(p_0-a)^2}}}{e^{-\frac{K}{(p_0-a)^2}} + e^{-\frac{K}{(p_0-b)^2}}}, \quad a < p_0 < b \tag{9.98}$$

$$\equiv 0, \quad p_0 \le a \tag{9.99}$$

$$\equiv 1, \quad p_0 \ge b \tag{9.100}$$

with K an arbitrary real positive constant. Note that by definition $0 \le F_+ \le 1$, that $F_+(p_0) = 0$ for $p_0 < a$ and $F_+(p_0) = 1$ for $p_0 > b$, and that (see Problem 5) F_+ is infinitely many times continuously differentiable with respect to the energy variable p_0 for all p_0. Next, define

$$F_-(p_0) \equiv F_+(-p_0) \tag{9.101}$$

which is also C^∞, but vanishes for $p_0 > -a$. Finally, we define

$$F_0(p_0) \equiv 1 - F_+(p_0) - F_-(p_0) \tag{9.102}$$

which clearly vanishes for $|p_0| > b$. The three functions F_+, F_0 and F_- form a partition of unity, and are illustrated in Fig. 9.2 for a specific choice of parameters. It is easily

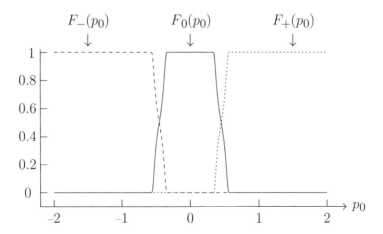

Fig. 9.2 C^∞ partition of unity for $a = 0.3$, $b = 0.6$, $m = 1.0$ ($K = 0.01$).

---

[11] For an accessible, but careful and rigorous proof, see Jost (Jost, 1965), Chapter VI, especially pp. 126–130.

seen that, like F_+, the functions F_0 and F_- are also C^∞ (infinitely many times continuously differentiable).

We now consider the vacuum-expectation-value of a pair $\phi_{f_1}(x_1), \phi_{f_2}(x_2)$ of almost local operators localized around spacetime points x_1, x_2 at equal time (the $n = 2$ case of the Clustering theorem). Any such operator can be split into three parts as follows, utilizing the energy functions $F_{\pm,0}$ introduced above:

$$\phi_f(x) = \phi_f^{(+)}(x) + \phi_f^{(0)}(x) + \phi_f^{(-)}(x)$$

$$\phi_f^{(+)}(x) \equiv \int F_+(p_0)\tilde\phi_f(p)e^{ip\cdot x}\frac{d^4p}{(2\pi)^4} = \int F_+(p_0)\tilde f(p)\phi(y)e^{ip\cdot(x-y)}d^4y\frac{d^4p}{(2\pi)^4}$$

$$\phi_f^{(0)}(x) \equiv \int F_0(p_0)\tilde\phi_f(p)e^{ip\cdot x}\frac{d^4p}{(2\pi)^4} = \int F_0(p_0)\tilde f(p)\phi(y)e^{ip\cdot(x-y)}d^4y\frac{d^4p}{(2\pi)^4}$$

$$\phi_f^{(-)}(x) \equiv \int F_-(p_0)\tilde\phi_f(p)e^{ip\cdot x}\frac{d^4p}{(2\pi)^4} = \int F_-(p_0)\tilde f(p)\phi(y)e^{ip\cdot(x-y)}d^4y\frac{d^4p}{(2\pi)^4}$$

$$(9.103)$$

with $\tilde f(p)$ the Fourier transform of the Schwarz smearing function $f(x)$. Note that $F_{\pm,0}(p_0)\tilde f(p)$ is C^∞ if $\tilde f(p)$ is, so the smearing of these three operators in coordinate space is with fast-decreasing functions: i.e., they are almost local if $\phi_f(x)$ is. By the momentum shift argument of (9.83), and the support in energy of the functions $F_{\pm,0}$, we know the following:

1. The operators of form $\phi_f^{(+)}(x)$ increase the energy of any state they act on by at least $a > 0$.
2. The operators of form $\phi_f^{(0)}(x)$ change the energy of any state they act on by at most $b < m$.
3. The operators of form $\phi_f^{(-)}(x)$ decrease the energy of any state they act on by at least $a > 0$.

The State axioms (specifically, the spectral assumptions Ib,Ic, and Id) then imply

$$\langle 0|\phi_f^{(+)}(x)|\alpha\rangle = 0, \ \forall|\alpha\rangle \ \Rightarrow \ \langle 0|\phi_f^{(+)}(x) = 0 \qquad (9.104)$$

$$\phi_f^{(0)}(x)|0\rangle = C|0\rangle, \ \ C = \langle 0|\phi_f^{(0)}(x)|0\rangle \qquad (9.105)$$

$$\phi_f^{(-)}(x)|0\rangle = 0 \qquad (9.106)$$

Accordingly,

$$\langle 0|\phi_{f_1}(x_1)\phi_{f_2}(x_2)|0\rangle = \langle 0|(\phi_{f_1}^{(0)}(x_1) + \phi_{f_1}^{(-)}(x_1))(\phi_{f_2}^{(+)}(x_2) + \phi_{f_2}^{(0)}(x_2))|0\rangle$$

$$= \langle 0|\phi_{f_1}^{(0)}(x_1)\phi_{f_2}^{(0)}(x_2)|0\rangle + \langle 0|[\phi_{f_1}^{(-)}(x_1), \phi_{f_2}^{(+)}(x_2)]|0\rangle$$

$$+ \langle 0|[\phi_{f_1}^{(0)}(x_1), \phi_{f_2}^{(+)}(x_2)]|0\rangle + \langle 0|[\phi_{f_1}^{(-)}(x_1), \phi_{f_2}^{(0)}(x_2)]|0\rangle \qquad (9.107)$$

By (9.84), the three commutators on the right-hand side of (9.107) fall faster than any power of $d \equiv |\vec{x}_1 - \vec{x}_2|$, while, by (9.105),

$$\langle 0|\phi_{f_1}^{(0)}(x_1)\phi_{f_2}^{(0)}(x_2)|0\rangle = \langle 0|\phi_{f_1}^{(0)}(x_1)|0\rangle\langle 0|\phi_{f_2}^{(0)}(x_2)|0\rangle$$
$$= \langle 0|\phi_{f_1}(x_1)|0\rangle\langle 0|\phi_{f_2}(x_2)|0\rangle \tag{9.108}$$

Thus

$$\langle 0|\phi_{f_1}(x_1)\phi_{f_2}(x_2)|0\rangle_c \equiv \langle 0|\phi_{f_1}(x_1)\phi_{f_2}(x_2)|0\rangle$$
$$- \langle 0|\phi_{f_1}(x_1)|0\rangle\langle 0|\phi_{f_2}(x_2)|0\rangle$$
$$< C_N d^{-N}, \quad d \to \infty \tag{9.109}$$

which is just the $n = 2$ case of Theorem 9.1. The extension to higher values of n (for which we refer the reader to the above-cited book of Jost (Jost, 1965)) requires an inductive argument relying on the fact that as the spatial diameter d of the set of n points becomes large, we may always find two subsets of points separated by an arbitrarily large spatial distance. It is also important to realize that the above argument is equally valid for the more general class of composite almost local operators: e.g., of the form (9.81). Indeed, we shall employ the Clustering theorem in just such a circumstance in the next section, when it is combined with the Haag–Ruelle scattering theory to establish the clustering properties of the S-matrix which served as a crucial motivation in our first stabs at constructing local field theory.

We can now state our final set of axioms, those that connect the particle(s) and field(s) of the theory. We may call them the *particle–field duality axioms*. They are essential in the development of any comprehensible theory of particle scattering in the context of field theory.

1. **Axiom IIIa**: For some one-particle state $|\alpha\rangle = \int g(\vec{k})|\vec{k}\rangle d^3k \ (g(\vec{k}) \in L^2)$ with discrete eigenvalue m^2 of the squared-mass operator (cf. Axiom Id), the smeared field $\phi_f(x)$ has a non-vanishing matrix element from this single-particle state to the vacuum, $\langle 0|\phi_f(x)|\alpha\rangle \neq 0$.
   Comments: If this situation holds, we call $\phi_f(x)$ an *interpolating Heisenberg field* for the given particle.

2. **Axiom IIIb**: (Asymptotic completeness.) The Hilbert space $\mathcal{H}_{\text{in}}$ (resp. $\mathcal{H}_{\text{out}}$) corresponding to multi-particle states of far-separated, freely moving stable particles in the far past (resp. far future) are unitarily equivalent, and may be identified with the full Hilbert space $\mathcal{H}$ of the system (which, from the cyclicity axiom IId, can be regarded as the space generated by application of the smeared fields to the vacuum). Thus, this axiom again connects particle concepts (the asymptotic in- and out-states) with a space $\mathcal{H}$ defined in terms of the action of the basic field(s) of the theory.
   Comments: As discussed in the preceding section, this assumption is almost unavoidable physically, as it incorporates a vast amount of phenomenological experience of particle interactions. On the other hand, from a brutally utilitarian point of view, it may be thought to be partially unnecessary: if there are physical states which do not correspond to the states produced and detected in high-

energy accelerators, who needs to know, given that essentially all of our insights into the nature of fundamental microphysical interactions are obtained from such accelerator experiments? The assumed unitarity of the S-matrix would only then require $\mathcal{H}_{\text{in}} = \mathcal{H}_{\text{out}}$, with both of these asymptotically defined spaces being (perhaps) proper subsets of the full Hilbert space $\mathcal{H}$. Indeed, the Haag–Ruelle scattering theory of the next section can only establish the existence of the asymptotic spaces as such subsets. Moreover, even in the few cases where we have maximum mathematical control: e.g., the explicitly constructed field theories corresponding to polynomially self-coupled scalar fields in two spacetime dimensions, where the validity and consistency of the axioms of type I and II can be explicitly checked by construction of the Hilbert space and operators of the theory, the validity of Axiom IIIb remains, in the words of Jaffe and Glimm ((Glimm and Jaffe, 1987), p. 275), "a very deep (and open) mathematical question". Our attitude for the remainder of this book, in the absence of conclusive evidence to the contrary, will simply be to assume the validity of asymptotic completeness, and to treat the asymptotic spaces $\mathcal{H}_{\text{in,out}}$ as equivalent to the full physical Hilbert space of the field theory.

The axiomatic framework outlined in this section is the basis for famous proofs (Streater and Wightman, 1978) of the Spin-Statistics and TCP theorems which establish rigorously these fundamental properties of local field theory (already discussed qualitatively in Chapter 3) in a very general context. We have already discussed (in Section 7.3) the Spin-Statistics theorem, albeit in the context of free interaction-picture fields. We shall return to both the Spin-Statistics and TCP theorems later, in the "Symmetries" section of the book (cf. Section 13.4). Another deep and beautiful result of the axiomatic approach—the *Wightman reconstruction theorem* ((Streater and Wightman, 1978), Section 3-4)—allows us to recover all essential features of the particle-oriented Fock-space formulation of field theory starting only with a set of Wightman functions satisfying the above axioms. Here, the cyclicity of the vacuum embodied in Axiom IID is critical, ensuring that arbitrary states of the Hilbert space of the system can be approximated to arbitrary accuracy by applying polynomials of the smeared field to the vacuum. An arbitrary matrix element of the smeared field between physical states can consequently be approximated by linear combinations of Wightman functions (integrated with smearing functions) to arbitrary accuracy. The appropriate transformation properties of the states and field operator under the Poincaré group are also explicitly demonstrated in the process of the reconstruction. We shall be performing a similar reconstruction shortly using the scattering theory of Haag and Ruelle. In this approach an explicit formula for the asymptotic in- and out-states of the theory is given in terms of limits of appropriately smeared Heisenberg fields acting on the vacuum. To the extent that we accept asymptotic completeness (Axiom IIIB), the construction of the asymptotic in- or out-spaces is tantamount to recovering the full physical Hilbert space of the theory.

9.3   Asymptotic formalism I: the Haag–Ruelle scattering theory

In this section we shall establish, in a rigorous way, the promised connection between the underlying Heisenberg field $\phi(x)$ and the asymptotic states of the theory: in particular, we shall show, following the seminal work of Haag and Ruelle, how to *construct* the asymptotic Hilbert spaces $\mathcal{H}_{\text{in,out}}$ as appropriate limits involving products of the Heisenberg field applied to the vacuum. The only conceptual input will be the axioms of the preceding Section. The treatment is restricted for simplicity to a theory of massive, self-interacting spinless particles, with no bound states in the theory. Our main result will be the fundamental *Asymptotic theorem of Haag*.

We begin by showing that a special type of smeared field can be constructed with the properties that (a) acting on the vacuum it produces single-particle states, and single-particle states *only*, and (b) these states are time-independent. The smearing is carried out in two stages. First, starting with the original local field $\phi(x)$, assumed to be an interpolating field for the stable particle of the theory, we construct a field $\phi_1(x)$ with a special type of smearing, with the property that it produces only one-particle states when acting on the vacuum. Next, this field is further smeared in such a way that the resultant field $\phi_{1,g}$ produces a time-independent (one-particle) state with a definite momentum-space wavefunction when acting on the vacuum. The first field, $\phi_1(x)$ is defined exactly as in (9.80), but with a function $f^{(1)}(x)$ chosen as the Fourier transform of a function $\tilde{f}^{(1)}(p)$ with support in the region $am^2 < p^2 < bm^2$ with $0 < a < 1$, $1 < b < 4$ (the numbers "1" and "4" appearing here are purely convenient choices, as will be apparent shortly). The function $\tilde{f}^{(1)}(p)$ will be as usual C^∞ and of fast decrease for large spatial momentum $\vec{p}$, although we shall not allow it to vanish anywhere on the one-particle mass shell $p_0 = \sqrt{\vec{p}^2 + m^2}$, as we shall want to construct wave-packets for particles centered around arbitrary momenta. An example of a possible region of support for $\tilde{f}^{(1)}(p)$ is shown in Fig. 9.1 as the region between the dotted lines. Note that this region includes the one-particle mass hyperboloid, but not the vacuum or any multi-particle states.[12] Consequently, the field $\phi_1(x)$ can only produce—and by Axiom IIIa, will produce!—one-particle states when acting on the vacuum, by the argument leading to (9.83). The field $\phi_1(x)$ is therefore an almost local field in the sense described in the preceding section, with the usual translation properties, $e^{iP\cdot a}\phi_1(x)e^{-iP\cdot a} = \phi_1(x+a)$ (with its infinitesimal version $i[P_\mu, \phi_1(x)] = \frac{\partial \phi_1}{\partial x^\mu}$). Acting on the vacuum, it can at best produce a one-particle state, and by Axiom IIIa we will henceforth assume that it is an interpolating field for the spinless particle of the theory, with a non-vanishing vacuum to single-particle matrix element. The latter is in fact determined up to a single overall normalization constant of the field, by Lorentz-invariance. Using covariantly normalized one-particle states, $_{\text{cov}}\langle \vec{k}'|\vec{k}\rangle_{\text{cov}} = 2E(k)\delta^3(k'-k)$,

$$_{\text{cov}}\langle\vec{k}|\phi_1(x)|0\rangle = \int f^{(1)}(y-x)\,_{\text{cov}}\langle\vec{k}|\phi(y)|0\rangle d^4y$$

$$= \int f^{(1)}(y-x)\,_{\text{cov}}\langle\vec{k}|e^{iP\cdot y}\phi(0)e^{-iP\cdot y}|0\rangle d^4y$$

$$= \int f^{(1)}(y-x)e^{ik\cdot y}\,_{\text{cov}}\langle\vec{k}|\phi(0)|0\rangle d^4y$$

$$= \,_{\text{cov}}\langle\vec{k}|\phi(0)|0\rangle \tilde{f}^{(1)}(\vec{k})e^{ik\cdot x} \tag{9.110}$$

---

[12] Also, bound states of two particles, with $p^2 = 4m^2 - \mathcal{E}$, are specifically excluded.

The Fourier transform $\tilde{f}^{(1)}(\vec{k})$ is written as a function of the spatial momentum $\vec{k}$ only, as it is the on-mass-shell value (at $k_0 = E(k) = \sqrt{\vec{k}^2 + m^2}$) which is relevant here. By Lorentz-invariance,

$$_{\text{cov}}\langle \vec{k}|\phi(0)|0\rangle = \ _{\text{cov}}\langle \vec{k}|U^\dagger(\Lambda)\phi(0)U(\Lambda)|0\rangle = \ _{\text{cov}}\langle \Lambda\vec{k}|\phi(0)|0\rangle \tag{9.111}$$

Choosing Λ to be the boost which takes momentum $\vec{k}$ to zero, we obtain the matrix element $_{\text{cov}}\langle \vec{0}|\phi(0)|0\rangle$. This matrix element (again, assumed non-zero by Axiom IIIa) is thus a constant dependent on the normalization of the basic Heisenberg field $\phi(x)$. For convenience, we may choose the normalization here to agree with that of a free field, for which $_{\text{cov}}\langle \vec{k}|\phi(0)|0\rangle = \frac{1}{(2\pi)^{3/2}}$, but it must be remembered that the normalization of the field is often conventionally fixed by other requirements—such as commutation relations—which will lead to a different normalization (more on this at the end of this section, when we derive the asymptotic condition). Thus, we may simply take, switching back to non-covariantly normalized states (recall $|\vec{k}\rangle_{\text{cov}} = \sqrt{2E(k)}|\vec{k}\rangle$, cf (5.15)),

$$\langle \vec{k}|\phi_1(x)|0\rangle = \frac{1}{(2\pi)^{3/2}\sqrt{2E(k)}}\tilde{f}^{(1)}(\vec{k})e^{ik\cdot x} \tag{9.112}$$

We also note at this point that $\frac{\partial}{\partial t}\phi_1(x)$ is an almost local field if $\phi_1(x)$ is, as the time-derivative of the fast-decreasing C^∞ smearing function $f^{(1)}$ is still fast decreasing.

Next, let $g(\vec{x}, t)$ be a positive-energy solution of the Klein–Gordon equation:

$$g(\vec{x}, t) = \int \tilde{g}(\vec{p})e^{i(\vec{p}\cdot\vec{x} - E(p)t)}\frac{d^3p}{2E(p)}, \quad E(p) \equiv \sqrt{\vec{p}^2 + m^2} \tag{9.113}$$

The momentum wavefunction $\tilde{g}(\vec{p})$ will be chosen to be C^∞ and rapidly decreasing (faster than any power) for large $\vec{p}$. Also, the time-derivative of $g(\vec{x}, t)$ has the same properties as $E(p)\tilde{g}(\vec{p})$ is also C^∞ and rapidly decreasing for large $\vec{p}$. Our final smeared field $\phi_{1,g}(t)$ is now defined as

$$\phi_{1,g}(t) \equiv -i \int d^3x \ \{g(\vec{x}, t) \overset{\leftrightarrow}{\frac{\partial}{\partial t}} \phi_1(\vec{x}, t)\} \tag{9.114}$$

Both terms in (9.114) therefore correspond to the spatial smearing of an almost local field with a single particle wavefunction solution of the Klein–Gordon equation. The admittedly somewhat clumsy subscript "1" is maintained as a reminder that this field has been engineered to produce only one-particle states when it acts on the vacuum: we shall later define a similar field without this restriction, and will need to be able to distinguish the two. Note the similarity of (9.114) to the creation operator defined in (9.36) of Section 9.1. It is easy to see that the state obtained by applying this field to the vacuum is time-independent:

$$\frac{\partial}{\partial t}\phi_{1,g}(t)|0\rangle = -i\int d^3x\,\{g(\vec{x},t)\frac{\partial^2}{\partial t^2}\phi_1(\vec{x},t) - \phi_1(\vec{x},t)\frac{\partial^2}{\partial t^2}g(\vec{x},t)\}|0\rangle \qquad (9.115)$$

$$= -i\int d^3x\,\{g(\vec{x},t)\frac{\partial^2}{\partial t^2}\phi_1(\vec{x},t) - \phi_1(\vec{x},t)(\vec{\nabla}^2 - m^2)g(\vec{x},t)\}|0\rangle \quad (9.116)$$

In going from (9.115) to (9.116) we have used the fact that the single particle wavefunction $g(\vec{x},t)$ satisfies the Klein–Gordon equation $(\frac{\partial^2}{\partial t^2} - \vec{\nabla}^2 + m^2)g = 0$. Transferring the spatial gradients by an integration by parts (using the fast decrease of g in $\vec{x}$-space), we find

$$\frac{\partial}{\partial t}\phi_{1,g}(t) = -i\int d^3x\,g(\vec{x},t)(\Box + m^2)\phi_1(\vec{x},t)|0\rangle \qquad (9.117)$$

As the energy-momentum operator P_μ annihilates the vacuum

$$P_\mu P^\mu \phi_1(\vec{x},t)|0\rangle = -\Box\phi_1(\vec{x},t)|0\rangle \qquad (9.118)$$

so we see that (9.117) involves the operator $P_\mu P^\mu - m^2$ acting on the one-particle state (of mass m) $\phi_1|0\rangle$. The result is clearly zero, establishing the time-independence of $\phi_{1,g}(t)|0\rangle$:

$$\frac{\partial}{\partial t}\phi_{1,g}(t)|0\rangle = 0 \qquad (9.119)$$

A little thought reveals that this property is no longer maintained for *multiple* applications of the field $\phi_{1,g}$ (at the same time t) to the vacuum. However, the resultant time-dependent multi-particle states will be shown below to have a well-defined (strong) limit for $t \to \pm\infty$. This is, in fact, the central result of the Haag–Ruelle approach to scattering.

The momentum wavefunction of the one-particle state $\phi_{1,g}(t)|0\rangle = \phi_{1,g}(0)|0\rangle$, defined as the overlap of this state with the non-covariantly (continuum) normalized state $|\vec{k}\rangle$, follows straightforwardly from (9.112, 9.114):

$$\psi_{1,g}(\vec{k}) \equiv \langle\vec{k}|\phi_{1,g}(t)|0\rangle = -i\int \tilde{g}(\vec{p})(e^{-ip\cdot x}\overset{\leftrightarrow}{\frac{\partial}{\partial t}}e^{ik\cdot x})\tilde{f}^{(1)}(\vec{k})d^3x\frac{d^3p}{2E(p)}$$

$$= (2\pi)^{3/2}\frac{\tilde{g}(\vec{k})\tilde{f}^{(1)}(\vec{k})}{\sqrt{2E(k)}} \qquad (9.120)$$

Any desired (fast-decreasing) momentum-space wavefunction of our single-particle state can evidently be obtained as a product of appropriately chosen factors $\tilde{g}(\vec{k})$ and $\tilde{f}^{(1)}(\vec{k})$.

The scattering theory we shall develop will involve the study of the limits of states constructed from application to the vacuum of fields of the type $\phi_{1,g}(t)$ in the limits $t \to \pm\infty$. This will require an understanding of the large time asymptotic behavior of the single-particle wavefunctions $g(\vec{x},t)$. Define a velocity vector $\vec{v}$ in the obvious way, $\vec{v} \equiv \frac{\vec{x}}{t}$, so that

$$g(\vec{x}, t) = \int \tilde{g}(\vec{p}) e^{it(\vec{p} \cdot \vec{v} - E(p))} \frac{d^3 p}{2E(p)} \tag{9.121}$$

For fixed $\vec{v}$, large t, this integral is dominated[13] by a stationary phase point at

$$\frac{\partial}{\partial \vec{p}}(\vec{p} \cdot \vec{v} - E(p)) = 0 \;\Rightarrow\; \vec{v} = \frac{\vec{p}}{E(p)} \;\Rightarrow\; \vec{p} = m\gamma\vec{v}, \;\; \gamma \equiv \frac{1}{\sqrt{1 - v^2}} \tag{9.122}$$

Expanding the integrand around the stationary phase point through quadratic terms one finds

$$\vec{p} \cdot \vec{v} - E(p) \sim -m/\gamma - \frac{1}{2} \sum_{i,j=1}^{3} (p_i - m\gamma v_i)\mathcal{M}_{ij}(p_j - m\gamma v_j) \tag{9.123}$$

with $\mathcal{M}_{ij} = \frac{1}{m\gamma}(\delta_{ij} - v_i v_j)$ a symmetric 3x3 matrix with eigenvalues $\frac{1}{m\gamma}$, $\frac{1}{m\gamma}$ and $\frac{1}{m\gamma^3}$, and hence determinant $\frac{1}{m^3\gamma^5}$. The Gaussian integration around the stationary phase point gives us the desired asymptotic behavior at large t:

$$g(\vec{x} = \vec{v}t, t) \sim C|t|^{-3/2} e^{-imt/\gamma}(\gamma^{3/2}\tilde{g}(m\gamma\vec{v}) + O(1/t)) \tag{9.124}$$

with C an irrelevant constant containing the mass m, π, etc. What if the momentum-space wavefunction $\tilde{g}(\vec{p})$ vanishes at (and in some neighborhood of) the stationary phase point $\vec{p} = m\gamma\vec{v}$? For example, $\tilde{g}$ may have compact support in momentum space, and simply vanish in some neighborhood of $m\gamma\vec{v}$, in which case not only the leading, but also all higher-order terms in the stationary phase expansion, will vanish. In this case it can be shown that $g(\vec{v}t, t)$ vanishes for large t faster than any power of t.[14] We shall, however, only need the weaker result encapsulated in (9.124). Finally, we note that for $\vec{x}$ (or $\vec{v}$) pointed along the direction of the particle's motion, the $t^{-3/2}$ falloff is just the expected spreading of the wave-packet due to the non-zero spread in momentum space, which leads to the particle being delocalized over a region of linear dimension $\propto t$, with $|g|^2 t^3 \sim$ constant at large times.

Before statement and proof of the Haag Asymptotic theorem, we shall need two important preliminary results. The first is a fairly direct consequence of the Ruelle Clustering theorem discussed in the preceding section. The essential physical content of the description of a physical system in terms of in/out-states is the intuition that for large times, past or future, the particles become physically isolated and cease to interact significantly with one another. In the field context, this turns out to be exactly equivalent to the fast decrease of connected VEVs of almost local fields at large spatial separation: the large-distance falloff of field expectation values is converted to a large time falloff via the asymptotic kinematics of the single particle wavefunctions given in (9.124).

First, let us temporarily use the notation $\phi_{1,g}(t)$ to generically denote any field obtained by spatially smearing an almost local field $\phi(\vec{x}, t)$ with a single particle

---

[13] The observation that such wave-packets obey the correct relativistic kinematics was the primary motivation for de Broglie's introduction of the wave hypothesis for particles, and as such the seminal development initiating the path to Schrödinger's wave mechanics.

[14] For a tight proof of this result, originally due to Ruelle, see the above-cited book of Jost (Jost, 1965), Chapter 6.

wavefunction $g(\vec{x}, t)$:

$$\phi_{1,g}(t) = \int \phi(\vec{x}, t) g(\vec{x}, t) d^3 x \qquad (9.125)$$

We note that our previously defined field $\phi_{1,g}(t)$ in (9.114) is just the sum of two such fields, with the almost local field being either ϕ_1 or $\frac{\partial}{\partial t}\phi_1$, and the single-particle wavefunction either $g(\vec{x}, t)$ (defined in (9.113)) or its time-derivative $\frac{\partial g(\vec{x}, t)}{\partial t}$. Our first preliminary result states

Lemma 9.2 *For large times, $t \to \pm\infty$,*

$$\mathcal{M}_{m,n}(t) \equiv \langle 0 | \phi^{\dagger}_{1,g'_1}(t)..\phi^{\dagger}_{1,g'_m}(t)\phi_{1,g_1}(t)...\phi_{1,g_n}(t) | 0 \rangle$$

$$\longrightarrow \begin{cases} O(|t|^{-3/2}), & m \neq n \\ \sum_{\text{pairs } P} \prod_{p \in P} \langle 0 | \phi^{\dagger}_{1,g'_{i_p}}(t)\phi_{1,g_{j_p}}(t) | 0 \rangle + O(t^{-3}), & m = n \end{cases} \qquad (9.126)$$

To establish this result, we begin by noting that the amplitude $\mathcal{M}_{m,n}(t)$ has a cluster expansion as a sum of terms in which the $m + n$ fields are distributed into N_c separate clusters, with the m_r $\phi^{\dagger}_{1,g'}(t)$ and n_r $\phi_{1,g}(t)$ fields in the rth cluster inside a separate *connected* VEV of the form (9.97). The rth cluster will give a contribution of the form

$$\int \langle 0 | \phi_1(\vec{x}_1, t).....\phi_1(\vec{x}_{m_r+n_r}, t) | 0 \rangle_c G_1(\vec{x}_1, t)...G_{m_r+n_r}(\vec{x}_{m_r+n_r}, t) d^3 x_1 ... d^3 x_{m_r+n_r}$$

$$= \int \langle 0 | \phi_1(\vec{x}_1, t)\phi_1(\vec{x}_1 + \vec{\xi}_2)...\phi_1(\vec{x}_1 + \vec{\xi}_{m_r+n_r}) | 0 \rangle_c$$

$$\cdot \; G_1(\vec{x}_1, t)...G_{m_r+n_r}(\vec{x}_1 + \vec{\xi}_{m_r+n_r}, t) d^3 x_1 d^3 \xi_2 .. d^3 \xi_{m_r+n_r}$$

$$= \int d^3 x_1 G_1(\vec{x}_1, t) \int \langle 0 | \phi_1(0, t)\phi_1(\vec{\xi}_2, t)..\phi_1(\vec{\xi}_{m_r+n_r}, t) | 0 \rangle_c$$

$$\cdot \; G_2(\vec{x}_1 + \vec{\xi}_2, t)..G_{m_r+n_r}(\vec{x}_1 + \vec{\xi}_{m_r+n_r}, t) d^3 \xi_2 .. d^3 \xi_{m_r+n_r} \qquad (9.127)$$

Here we use the generic notation $G_i(\vec{x}, t)$ to denote either a positive-energy wavefunction $g(\vec{x}, t)$ (appearing in the $\phi_{1,g}$ fields) or a negative-energy wavefunction $g^*(\vec{x}, t)$ (appearing in the $\phi^{\dagger}_{1,g}$ fields). Note that the asymptotic behavior at large time of the g^* wavefunctions is given directly by complex conjugating (9.124), and involves the same $|t|^{-3/2}$ falloff as the g functions. In passing from the first to the second line we have shifted integration variables by defining $\vec{x}_n \equiv \vec{x}_1 + \vec{\xi}_n, n = 2, \ldots m_r + n_r$. The last line is obtained using the translation property of the ϕ_1 fields.

At this point we invoke the Ruelle Clustering theorem, which asserts that the connected vacuum expectation value appearing in the penultimate line of (9.127) is a *fast decreasing* function of the $\vec{\xi}$ variables. In the asymptotic limit of large t then, each such variable can be regarded as restricted to a finite range. If we change to velocity space for the $\vec{x}_1 \equiv \vec{v}_1 t$ variable, we note that the G_n functions become asymptotically, using (9.124),

$$|G_n(t\vec{v}_1 + \vec{\xi}_n, t)| \sim |t|^{-3/2}\gamma_1^{3/2}|\tilde{G}_n(m\gamma_1\vec{v}_1)|, \quad t \to \pm\infty \qquad (9.128)$$

where $\gamma_1 \equiv (1 - v_1^2)^{-1/2}$. After changing variables $\int d^3x_1 \to t^3 \int_{|\vec{v}_1|<1} d^3v_1$, we obtain, after the $\vec{\xi}$ integrals are performed, a factor of t^3 from the $\vec{x}_1$ integral, and a factor of $t^{-3/2}$ from each of the $m_r + n_r$ wavefunctions. Note that the integral over velocity space $\vec{v}_1$ is over the unit ball, with the momentum-space wavefunctions $\tilde{G}(m\gamma_1\vec{v}_1)$ decreasing rapidly (in particular, faster than any power of γ_1) as the boundary is reached. All the integrals are therefore well-defined, allowing us to replace the wavefunctions by their asymptotic values and giving the overall asymptotic behavior for the rth cluster

$$\left| \int \langle 0|\phi_1(\vec{x}_1,t)....\phi_1(\vec{x}_{m_r+n_r},t)|0\rangle_c \prod_{i=1}^{m_r+n_r} G_i(\vec{x}_i,t)d^3x_1...d^3x_{m_r+n_r} \right|$$

$$\sim |t|^{3-\frac{3}{2}(m_r+n_r)} \tag{9.129}$$

Multiplying this behavior for all N_c clusters, with $\sum_{r=1}^{N_c} m_r \equiv m$ and $\sum_{r=1}^{N_c} n_r \equiv n$, we find that the contribution to $\mathcal{M}_{m,n}(t)$ from N_c clusters with $N = m+n$ total fields behaves asymptotically like $t^{3N_c - \frac{3}{2}N}$. As $\langle 0|\phi_1|0\rangle = 0$ ($\phi_1|0\rangle$ is a single-particle state, hence orthogonal to the vacuum), all clusters must have at least two fields. If any cluster has three fields, the total number of fields must satisfy $N \geq 2(N_c - 1) + 3$, which implies $3N_c - \frac{3}{2}N \leq -\frac{3}{2}$: i.e., a vanishing $t^{-3/2}$ behavior as $t \to \pm\infty$. Evidently, the only way to obtain a non-vanishing result at large time is to have all clusters contain exactly two fields, in which case $N = 2N_c$ and the power falloff is eliminated. Moreover, the only pairings that survive at large time involve a $\phi_{1,g}^\dagger$ field paired with a $\phi_{1,g}$ field. If we take instead two $\phi_{1,g}$ fields:

$$\int \langle 0|\phi_1(\vec{x}_1,t)\phi_1(\vec{x}_2,t)|0\rangle_c \, g_1(\vec{x}_1,t)g_2(\vec{x}_2,t)d^3x_1 d^3x_2$$

$$= \int \Delta(\vec{x}_1 - \vec{x}_2)\tilde{g}_1(\vec{p}_1)\tilde{g}_2(\vec{p}_2)e^{i(\vec{p}_1\cdot\vec{x}_1 - E(p_1)t + \vec{p}_2\cdot\vec{x}_2 - E(p_2)t)}\frac{d^3p_1}{2E(p_1)}\frac{d^3p_2}{2E(p_2)}d^3x_1 d^3x_2$$

$$\tag{9.130}$$

Here $\Delta(\vec{x}_1 - \vec{x}_2) \equiv \langle 0|\phi_1(\vec{x}_1,t)\phi_1(\vec{x}_2,t)|0\rangle_c$ is fast decreasing for large $\vec{x}_1 - \vec{x}_2$ by the Ruelle Clustering theorem, so that its Fourier transform $\tilde{\Delta}(\vec{p})$ is smooth (C^∞). Changing to center-of-mass variables $\vec{X} \equiv \frac{\vec{x}_1 + \vec{x}_2}{2}$, $\vec{x} \equiv \vec{x}_1 - \vec{x}_2$, and performing the spatial integrals, one finds

$$\langle 0| \phi_{1,g_1}(t)\phi_{1,g_2}(t)|0\rangle_c = \int \langle 0|\phi_1(\vec{x}_1,t)\phi_1(\vec{x}_2,t)|0\rangle_c \, g_1(\vec{x}_1,t)g_2(\vec{x}_2,t)d^3x_1 d^3x_2$$

$$= (2\pi)^3 \int \tilde{g}_1(\vec{p}_1)\tilde{g}_2(\vec{p}_2)e^{-i(E(p_1)+E(p_2))t}\tilde{\Delta}(\frac{\vec{p}_1 - \vec{p}_2}{2})\delta^3(\vec{p}_1 + \vec{p}_2)\frac{d^3p_1}{2E(p_1)}\frac{d^3p_2}{2E(p_2)}$$

$$= (2\pi)^3 \int \tilde{g}_1(\vec{p}_1)\tilde{g}_2(-\vec{p}_1)\tilde{\Delta}(\vec{p}_1)e^{-2iE(\vec{p}_1)t}\frac{d^3p_1}{4E(p_1)^2} \tag{9.131}$$

The smooth momentum dependence of all factors in the integral then implies the fast decrease (faster than any power) of (9.131) as $t \to \pm\infty$. A similar result obtains for a cluster consisting of two $\phi_{1,g}^{\dagger}$ fields. On the other hand, if we take a pairing of a $\phi_{1,g}^{\dagger}$ with a $\phi_{1,g}$ field, the complex time exponentials cancel, and the result is non-vanishing, and time-independent. Finally, we note that if $m = n$, allowing a complete pairing, as in the second line of Lemma 9.2, the remainder term must involve at least two clusters with three (or more) fields, and hence a falloff of t^{-3} at large time. This concludes the demonstration of Lemma 9.2.

Our second preliminary result concerns the symmetry under permutation of the states obtained by applying the $\phi_{1,g}$ fields to the vacuum. We state it as the following Lemma.

Lemma 9.3 *Define the time-dependent state $|\Psi, t\rangle$ as follows:*

$$|\Psi, t\rangle \equiv \phi_{1,g_1}(t)\phi_{1,g_2}(t)...\phi_{1,g_m}(t)|0\rangle \tag{9.132}$$

With P an arbitrary permutation of the sequence $1, 2, ...m$, define

$$|\Psi', t\rangle \equiv \phi_{1,g_{P(1)}}(t)\phi_{1,g_{P(2)}}(t)...\phi_{1,g_{P(m)}}(t)|0\rangle \tag{9.133}$$

Then for large time t, the distance between these two state vectors has the asymptotic behavior

$$\sqrt{(|\Psi, t\rangle - |\Psi', t\rangle, |\Psi, t\rangle - |\Psi', t\rangle)} \sim \frac{1}{|t|^{3/2}} \tag{9.134}$$

This result follows as an immediate consequence of Lemma 9.2, as the squared distance between the states is

$$(|\Psi, t\rangle - |\Psi', t\rangle, |\Psi, t\rangle - |\Psi', t\rangle) = \langle \Psi, t|\Psi, t\rangle + \langle \Psi', t|\Psi', t\rangle$$
$$- \langle \Psi, t|\Psi', t\rangle - \langle \Psi', t|\Psi, t\rangle \tag{9.135}$$

Each of the inner products appearing on the right-hand side of (9.135) is an amplitude of the form $\mathcal{M}_{m,m}(t)$, so by Lemma 9.2 approaches asymptotically a sum of m cluster pairs which is symmetric under permutation of any two fields. Thus the leading terms cancel, leaving a remainder of order t^{-3}, whence Lemma 9.3.

The proof of the *Haag Asymptotic theorem* follows very quickly from these results. First, the theorem itself.

Theorem 9.4 *The time-dependent state vector*

$$|\Psi, t\rangle \equiv \phi_{1,g_1}(t)\phi_{1,g_2}(t)...\phi_{1,g_n}(t)|0\rangle \tag{9.136}$$

converges strongly in the limit $t \to -\infty$ to the n-particle in-state

$$|\Psi\rangle_{\text{in}} = |g_1, g_2, .., g_n\rangle_{\text{in}} \equiv \int \psi_{1,g_1}(\vec{k}_1)...\psi_{1,g_n}(\vec{k}_n)|\vec{k}_1, .., \vec{k}_n\rangle_{\text{in}} d^3 k_1...d^3 k_n \tag{9.137}$$

with momentum wavefunctions $\psi_{1,g_1}(\vec{k}), ..\psi_{1,g_n}(\vec{k})$ (defined in (9.120)); thus, the states in (9.137) have the inner product structure corresponding to the contin-uum non-covariant normalization of (5.21) (with plus signs everywhere as we are *considering only bosons here). All of the above holds with the replacements $t \to +\infty$ and "in" $\to$ "out" everywhere.*

The strong convergence is easily established by taking the time-derivative of $|\Psi, t\rangle$:

$$\frac{\partial}{\partial t}|\Psi, t\rangle = (\frac{\partial}{\partial t}\phi_{1,g_1}(t))\phi_{1,g_2}(t)..\phi_{1,g_n}(t)|0\rangle + \phi_{1,g_1}(t)(\frac{\partial}{\partial t}\phi_{1,g_2}(t))..\phi_{1,g_n}(t)|0\rangle$$

$$+ \ldots + \phi_{1,g_1}(t)\phi_{1,g_2}(t)..(\frac{\partial}{\partial t}\phi_{1,g_n}(t))|0\rangle \tag{9.138}$$

The final term on the right-hand side of (9.138) vanishes by (9.119) as the time-derivative of ϕ_{1,g_n} acts directly on the vacuum state. However, all the other $(n-1)$ terms correspond to permutations of a similar term in which the field with the time-derivative is moved to the extreme right, and therefore, by Lemma 9.3, have norm of order $|t|^{-3/2}$. However, $||\frac{\partial}{\partial t}||\Psi, t\rangle|| < \frac{C}{|t|^{3/2}} \Rightarrow |||\Psi, t\rangle - ||\Psi, t'\rangle|| < \frac{2C}{|T|^{1/2}}, \ t, t' > T$. Thus the sequence of states $|\Psi, t\rangle$ for large t is a Cauchy sequence in the Hilbert space and must converge in norm to a limit vector $|\Psi\rangle_{\text{in}}$.

The second part of the theorem, establishing that the in- (or out-)states defined as such limits have the appropriate inner-product structure to define a Fock space of independent many-particle states, follows directly from Lemma 9.3, as the inner product $_{\text{in}}\langle g_1' g_2'...g_n'|g_1 g_2...g_n\rangle_{\text{in}}$ is simply the limit for $t \to -\infty$ of the amplitude $\mathcal{M}_{n,n}(t)$ defined in Lemma 9.2, the result of which is a symmetric sum of overlaps of single-particle states (see Problem 6).

The physical interpretation of the states resulting from the limiting processes of Theorem 9.4 is fairly clear. The smeared fields correspond to wave-packets which overlap less and less as the state is run either forwards or backwards in time. The asymptotic convergence can be greatly improved,[15] from the $t^{-3/2}$ behavior used above, to faster than any power of the time, if the momentum wavefunctions of the particles have non-overlapping support in momentum space: thus $\tilde{g}_i(\vec{p}) \neq 0$ iff $\tilde{g}_j(\vec{p}) = 0$ for $i \neq j$. In this situation, the particle velocities are "pointed" in different directions and the separation at large time is ensured, without recourse to wave-packet spreading. In particular, in this case, the reader may easily verify (Problem 7) that the coordinate space overlap of the single-particle wavefunctions of different particles remains exactly zero at all times. Thus the states constructed by the Haag procedure satisfy our intuitive picture of widely separated free particles in either the far past or the far future. We also note here without proof that the in- and out-states as defined in Theorem 9.4 can be shown to have the correct transformation properties under the Poincaré operators $U(\Lambda, a)$.

It is *extremely* important to realize that the construction of asymptotic multi-particle states by the limiting procedure of Theorem 9.4 remains perfectly valid if the underlying field $\phi(x)$ is itself a more general type of almost local field, such as the bilocal operator of (9.81), provided only that Axiom IIIa holds: namely, that the one particle state of the stable particle whose in- and out-states we wish to

---

[15] For further details, the reader is encouraged to consult the technical literature: e.g., the above-cited book of Jost (Jost, 1965).

construct has a non-vanishing vacuum to single-particle matrix element of this field, $\langle \vec{k}|\phi(x)|0\rangle \neq 0$. For example, the field $\phi(x)$ may have to be constructed from products of the "elementary" fields appearing in the Hamiltonian (or, more commonly, the Lagrangian, cf. Chapter 12) specifying the dynamics of the theory, if it corresponds to a stable bound state of the theory. We shall return to a detailed discussion of these issues in Section 9.6.

The Haag–Ruelle approach to scattering is ideally suited for understanding the emergence of the clustering property of the S-matrix from the underlying field-theoretic behavior. Consider the process depicted in Fig. 6.4, in which $n_1 + n_2$ particles scatter, and for simplicity of notation we have assumed that the scattering processes are number-conserving. Moreover, the wave-packets are constructed in such a way that n_1 of the particles (with initial and final wavefunctions $g_i, g_i', i = 1, ..n_1$) are localized, both before and after their interaction, around the spacetime point $(0, -\vec{\Delta})$, and the other n_2 particles (with wavefunctions $h_i, h_i', i = 1, 2, ..n_2$) around the point $(0, +\vec{\Delta})$. The desired clustering property of the S-matrix amounts to the statement that when $|\vec{\Delta}|$ is much greater than the size of the spacetime regions over which the two sets of particles interact, the full S-matrix amplitude should factorize into a product of independent scattering amplitudes for the "g" and "h" particles. Now, by the Haag Asymptotic theorem, the amplitude for the full process (a $n_1 + n_2 \to n_1 + n_2$ scattering amplitude) is given by

$$S_{n_1+n_2 \to n_1+n_2} = \lim_{T \to \infty} \langle 0|\phi^\dagger_{1,g_1'}(+T)..\phi^\dagger_{1,g_{n_1}'}(+T)\phi^\dagger_{1,h_1'}(+T)..\phi^\dagger_{1,h_{n_1}'}(+T)$$

$$\cdot \; \phi_{1,g_1}(-T)..\phi_{1,g_{n_1}}(-T)\phi_{1,h_1}(-T)..\phi_{1,h_{n_1}}(-T)|0\rangle \quad (9.139)$$

The limit $T \to \infty$ is, of course, a mathematical formality: in a typical high-energy scattering experiment the particles interact only in a spacetime region of microscopic dimensions. So the limit is very rapidly attained already when T is some very small value (e.g., 10^{-23} seconds for a typical strong interaction scattering event). We shall therefore fix T at some very small but finite value, at which point the S-matrix amplitude (9.139) has achieved its limit value to any preassigned level of precision, and enquire about the behavior of the combined scattering amplitude when the two groups of particles are separated by distance $2|\vec{\Delta}| >> T$. In this limit the commutator of any almost local field of "g" type appearing in (9.139) with a field of "h" type falls faster than any inverse power of $|\vec{\Delta}|$, so we may rearrange (9.139) (for fixed T) as follows:

$$S_{n_1+n_2 \to n_1+n_2} = \langle 0|\phi^\dagger_{1,g_1'}(+T)..\phi^\dagger_{1,g_{n_1}'}(+T)\phi_{1,g_1}(-T)..\phi_{1,g_{n_1}}(-T)$$

$$\cdot \; \phi^\dagger_{1,h_1'}(+T)..\phi^\dagger_{1,h_{n_1}'}(+T)\phi_{1,h_1}(-T)..\phi_{1,h_{n_1}}(-T)|0\rangle + o(|\vec{\Delta}|^{-N})$$

$$\equiv \langle 0|\Phi_g(0, -\vec{\Delta})\Phi_h(0, +\vec{\Delta})|0\rangle + o(|\vec{\Delta}|^{-N}) \quad (9.140)$$

Note that the two groups of fields, those involving "g" wavefunctions and those involving "h" wavefunctions, can be combined into the single operators $\Phi_g(0, -\vec{\Delta})$ and $\Phi_h(0, +\vec{\Delta})$ which are almost local operators localized around the indicated spacetime points. For example, the product field $\Phi_g(0, \vec{\Delta})$ can be viewed as the smearing of the multi-local product $\phi(x_1)\phi(x_2)...\phi(x_{2n_1})$ with a smearing function of Schwarz type falling fast for $|\vec{x}_i - \vec{\Delta}|$ much larger than all other distance scales in the problem. Now, by definition,

$$\langle 0|\Phi_g(0,-\vec{\Delta})\Phi_h(0,+\vec{\Delta})|0\rangle = \langle 0|\Phi_g(0,-\vec{\Delta})|0\rangle\langle 0|\Phi_h(0,+\vec{\Delta})|0\rangle$$
$$+ \langle 0|\Phi_g(0,-\vec{\Delta})\Phi_h(0,+\vec{\Delta})|0\rangle_c \qquad (9.141)$$

The connected term on the right-hand side of (9.141), by the Ruelle Clustering theorem 9.1, falls faster than any inverse power of the cluster separation $|\vec{\Delta}|$, whence the desired factorization of the S-matrix amplitude for the combined process into S-matrix amplitudes representing the separate scattering of "g"-type and "h"-type particles.

We conclude this section by employing the Haag–Ruelle theory to derive the long promised direct connection between the interpolating Heisenberg field $\phi(x)$ and the free in (resp. out) fields $\phi_{\text{in}}(x)$ (resp. $\phi_{\text{out}}(x)$) defined in Section 9.1. This connection, usually referred to as the *Asymptotic Condition*, was already "derived" heuristically by manipulations involving interaction-picture operators (see (9.45)). Here we shall see that the precise result we need, which serves as the starting point for the extremely important scattering theory formalism of Lehmann, Symanzik, and Zimmermann to which we turn in the next section, follows from the Haag–Ruelle theory (and hence, from the axioms of Section 9.2) without any reference to an interaction picture or perturbation theory. We begin by defining a smeared field $\phi_g(t)$, analogous to our $\phi_{1,g}(t)$ fields, except that the initial smearing function $f^{(1)(x)}$ is now taken to be a general Schwarz function $f(x)$ of fast decrease, with four-dimensional Fourier transform $\tilde{f}(k)$ *which is not restricted to a region of support sandwiching the one-particle mass hyperboloid as previously*. Eventually, in fact, we may even allow $f(x)$ to approach a δ-function (i.e., take $\tilde{f}(k)$ constant). For the time being though, our new field $\phi_g(t)$ will be obtained by smearing the almost local field $\phi_f(x)$, defined exactly as in (9.80), with a positive-energy single particle wavefunction $g(\vec{x},t)$ as in (9.113). We note that $\phi_g(t)|0\rangle$ is *not* any more a single particle state, nor is it time-independent. However, as far as the preconditions for Lemma 9.2 are concerned, $\phi_g(t)$ is just as good as our previous $\phi_{1,g}(t)$ field. We may therefore conclude that, picking for definiteness the limit for large negative time $t \to -\infty$,

$$\langle 0|\phi_{1,g_1'}^\dagger(t)...\phi_{1,g_m'}^\dagger(t)\phi_g(t)\phi_{1,g_1}(t)...\phi_{1,g_n}(t)|0\rangle$$

$$\to \sum_{i=1}^{m}\langle 0|\phi_{1,g_i'}^\dagger(t)\phi_g(t)|0\rangle_c\langle 0|\phi_{1,g_1'}^\dagger(t)..\widehat{\phi_{1,g_i'}^\dagger}(t)...\phi_{1,g_m'}^\dagger(t)\phi_{1,g_1}(t)...\phi_{1,g_n}(t)|0\rangle$$

$$+ \sum_{j=1}^{n}\langle 0|\phi_g(t)\phi_{1,g_j}(t)|0\rangle_c\langle 0|\phi_{1,g_1'}^\dagger(t)...\phi_{1,g_m'}^\dagger(t)\phi_{1,g_1}(t)..\widehat{\phi_{1,g_j}}(t)...\phi_{1,g_n}(t)|0\rangle,$$

$$(t \to -\infty) \qquad (9.142)$$

with remainder terms of relative order $|t|^{-3/2}$. Fields omitted from the second vacuum expectation value on each line (and coupled to the far past field $\phi_g(t)$) are indicated by the hat notation. The vanishing of $\langle 0|\phi_{1,g}(t)|0\rangle$ was previously assured by the fact that $\phi_{1,g}(t)|0\rangle$ is a one-particle state. This is no longer true for the new field ϕ_g: here we must explicitly assume the vanishing of the VEV of $\phi(x)$. For example, we may assume that our basic interpolating field transforms non-trivially under some symmetry unbroken by the vacuum (e.g., there is no spontaneous symmetry-breaking along the lines discussed in Section 8.4: if there is such a symmetry-breaking, we must shift the field as described there to remove its vacuum expectation value). Thus, the dominant terms at large time involve only clusters with pairings of the $\phi_g(t)$ field with either a $\phi^\dagger_{1,g'_i}(t)$ field or a $\phi_{1,g_j}(t)$ field, which are then omitted from the rest of the amplitude, as indicated by the notation $[\phi^\dagger_{1,g'_i}(t)]$ or $[\phi_{1,g_j}(t)]$. We have reassembled the clusters not containing the special field $\phi_g(t)$ into full amplitudes (i.e., without the "c" subscript). Furthermore, we may also eliminate the connected requirement on the two-field amplitudes appearing in (9.142), as $\langle 0|\phi^\dagger_{1,g'_i}(t)|0\rangle = 0$, $\langle 0|\phi_{1,g_j}(t)|0\rangle = 0$,

$$\langle 0|\phi^\dagger_{1,g'_i}(t)\phi_g(t)|0\rangle_c = \langle 0|\phi^\dagger_{1,g'_i}(t)\phi_g(t)|0\rangle \tag{9.143}$$

$$\langle 0|\phi_g(t)\phi_{1,g_j}(t)|0\rangle_c = \langle 0|\phi_g(t)\phi_{1,g_j}(t)|0\rangle \tag{9.144}$$

However, recalling that $\phi_{1,g}$ fields connect only to one-particle states, $\phi_{1,g_j}(t)|0\rangle$ is simply the time-independent state $|g_j\rangle_{\text{in}} = \int \psi_{1,g_j}(\vec{k})|\vec{k}\rangle_{\text{in}}d^3k$. Likewise, $\langle 0|\phi^\dagger_{1,g'_i}(t) = \int \psi^*_{g'_i}(\vec{k})\ _{\text{in}}\langle\vec{k}|d^3k$. On the other hand, the vacuum to one-particle matrix elements of $\phi_g(t)$ are determined up to normalization by Lorentz-invariance, as ϕ_g is obtained by smearing a local scalar field $\phi(x)$, assumed to be an interpolating field for the particle in question, so that, exactly as in (9.112), but replacing $f^{(1)} \to f$,

$$_{\text{in}}\langle\vec{k}|\phi_f(x)|0\rangle = \frac{Z^{1/2}}{(2\pi)^{3/2}\sqrt{2E(k)}}\tilde{f}(k)e^{ik\cdot x} \tag{9.145}$$

where we have allowed an arbitrary unfixed normalization factor, conventionally called $Z^{1/2}$, in case the normalization of the basic local field $\phi(x)$ is fixed by some independent (non-linear) constraint. The four-momentum k appearing in (9.145) is on mass shell for the particle in $|\vec{k}\rangle_{\text{in}}$, i.e., $k_0 = E(k) = \sqrt{\vec{k}^2 + m^2}$, $e^{ik\cdot x} = e^{i(E(k)t - \vec{k}\cdot\vec{x})}$. The subsequent smearing of ϕ_f with the single-particle positive-energy solution g to yield $\phi_g(t)$ implies

$$\langle 0|\phi^\dagger_{1,g'_i}(t)\phi_g(t)|0\rangle = \int \psi^*_{1,g'_i}(\vec{k})\ _{\text{in}}\langle\vec{k}|\phi_g(t)|0\rangle d^3k$$

$$= -i\int \psi^*_{1,g'_i}(\vec{k})\{g(\vec{x},t)\overset{\leftrightarrow}{\frac{\partial}{\partial t}}\ _{\text{in}}\langle\vec{k}|\phi_f(\vec{x},t)|0\rangle\}d^3x d^3k$$

$$= -i\int \psi^*_{1,g'_i}(\vec{k})\tilde{g}(\vec{p})\{e^{i(\vec{p}\cdot\vec{x}-E(p)t)}\overset{\leftrightarrow}{\frac{\partial}{\partial t}}\ _{\text{in}}\langle\vec{k}|\phi_f(\vec{x},t)|0\rangle\}d^3x d^3k\frac{d^3p}{2E(p)} \tag{9.146}$$

Next we need the orthogonality properties of the solutions of the Klein–Gordon equation (both positive and negative energy) which follow from the integrals

$$\int e^{i(\vec{p}\cdot\vec{x}-E(p)t)}\,\frac{\overleftrightarrow{\partial}}{\partial t}\,e^{i(E(k)t-\vec{k}\cdot\vec{x})}d^3x = 2i(2\pi)^3 E(k)\delta^3(\vec{p}-\vec{k}) \qquad (9.147)$$

$$\int e^{i(\vec{p}\cdot\vec{x}-E(p)t)}\,\frac{\overleftrightarrow{\partial}}{\partial t}\,e^{-i(E(k)t-\vec{k}\cdot\vec{x})}d^3x = 0 \qquad (9.148)$$

The second integral vanishes, as it is proportional to $(E(k)-E(p))\delta^3(\vec{p}+\vec{k})$. Inserting (9.147) in (9.146) we obtain

$$\langle 0|\phi^\dagger_{1,g'_i}(t)\phi_g(t)|0\rangle = Z^{1/2}\int \psi^*_{1,g'_i}(\vec{k})\psi_g(\vec{k})d^3k \qquad (9.149)$$

with $\psi_g(\vec{k})$ defined analogously to $\psi_{1,g}(\vec{k})$ as in (9.120), but with $\tilde{f}(k)$ replacing $\tilde{f}^{(1)}(\vec{k})$,

$$\psi_g(\vec{k}) = (2\pi)^{3/2}\frac{\tilde{g}(\vec{k})\tilde{f}(k)}{\sqrt{2E(k)}} \qquad (9.150)$$

On the other hand, $\langle 0|\phi_g(t)\phi_{1,g_j}(t)|0\rangle = \langle 0|\phi_g(t)|g_j\rangle_{\text{in}}$ involves the integral in (9.148) and vanishes identically.

Returning once again to (9.142), we see that the second line vanishes identically, so by applying Theorem 9.4 to the left-hand side, and to the multi-particle amplitudes multiplying $< 0|\phi^\dagger_{1,g'_i}(t)\phi_g(t)|0 >$, we obtain

$$_{\text{in}}\langle g'_1,..,g'_m|\phi_g(t)|g_1,..,g_n\rangle_{\text{in}} \rightarrow Z^{1/2}\sum_{i=1}^{m}\int \psi^*_{1,g'_i}(\vec{k})\psi_g(\vec{k})d^3k$$

$$\cdot \;_{\text{in}}\langle g'_1,..[g'_i]..,g'_m|g_1,g_2,..g_n\rangle_{\text{in}}, \quad t \rightarrow -\infty \quad (9.151)$$

We now consider a smeared field $\phi_{\text{in},g}(t)$ defined in complete analogy to $\phi_g(t)$ (with the same smearing functions), but starting from the free local field $\phi_{\text{in}}(x)$ defined in (9.30) rather than the interacting field $\phi(x)$. Recall that both $\phi_{\text{in}}(x)$ and $\phi(x)$ are Heisenberg fields, evolving with the dynamics specified by the full Hamiltonian. The contribution to $\phi_{\text{in},g}(t)$ from the destruction operator in ϕ_{in} is found to vanish using (9.148), and the creation term becomes time-independent:

$$\phi_{\text{in},g}(t) = \int \psi_g(\vec{k})a^\dagger_{\text{in}}(k)d^3k \qquad (9.152)$$

Sandwiching this result between the bra- and ket-states of (9.151), and recalling that creation operators acting to the left *destroy* particles,

$$_{\text{in}}\langle g'_1,..,g'_m\,|\phi_{\text{in},g}(t)|g_1,..,g_n\rangle_{\text{in}} = \int \psi^*_{1,g'_1}(\vec{k'_1})\psi^*_{1,g'_2}(\vec{k'_2})\cdots\psi^*_{1,g'_m}(\vec{k'_m})\psi_g(\vec{k})$$

$$\cdot \;_{\text{in}}\langle k'_1,k'_2,..k'_m|a^\dagger_{\text{in}}(k)|g_1,...,g_n\rangle_{\text{in}}d^3k'_1..d^3k'_m d^3k$$

$$= \sum_{i=1}^{m} \int \psi_{1,g_1'}^*(\vec{k}_1') \cdots \psi_{1,g_m'}^*(\vec{k}_m')\psi_g(\vec{k})\delta^3(\vec{k} - \vec{k}_i')$$

$$\cdot \ _{\text{in}}\langle k_1'..[k_i']..k_m'|g_1, ...g_n\rangle_{\text{in}}\, d^3k_1'...d^3k_m'\, d^3k$$

$$= \sum_{i=1}^{m} \int \psi_{1,g_i'}^*(\vec{k})\psi_g(\vec{k})d^3k \cdot \ _{\text{in}}\langle g_1', ..[g_i']..,g_m'|g_1, g_2, ..g_n\rangle_{\text{in}} \qquad (9.153)$$

which is precisely the same as the limiting behavior on the right-hand side of (9.151), up to the normalization factor of $Z^{1/2}$. The bra and ket in-states in (9.151) and (9.153) run over a dense subset of the Hilbert space $\mathcal{H}_{\text{in}}$ (indeed, choosing the $\psi_{1,g}(\vec{k})$ from a countable basis of $L^2(R^3)$, they run over a countable basis of $\mathcal{H}_{\text{in}}$), so, provided Axiom IIIb (asymptotic completeness) holds, and we are allowed to identify $\mathcal{H}_{\text{in}}$ with the full Hilbert space $\mathcal{H}$ of the theory, the stated equality in the limit amounts to *weak convergence* (i.e., matrix element by matrix element) of the smeared interpolating field $\phi_g(t)$ to the smeared (and time-independent) free in-field $\phi_{\text{in},g}$ as $t \to -\infty$. All of the above holds, of course, in the far future limit $t \to +\infty$, with "in" replaced by "out" everywhere. We also note that as we shall be employing the asymptotic limit of the Heisenberg field $\phi(x)$ only in matrix elements between (normalizable) states, the initial smearing of $\phi(x)$ is unnecessary (see the comments following Axiom IIa in the preceding section): we may set $\tilde{f}(k) = 1, \phi_f(x) = \phi(x)$ in (9.145).

The weak equivalence of $\phi_g(t)$ and $\phi_{\text{in},g}(t)$ (resp. $\phi_{\text{out},g}(t)$) at large negative (resp. positive) times, in other words, the limiting behavior just established, with complete mathematical rigor

$$_{\text{in}}\langle\beta|\phi_g(t)|\alpha\rangle_{\text{in}} \to Z^{1/2}\ _{\text{in}}\langle\beta|\phi_{\text{in},g}(t)|\alpha\rangle_{\text{in}}, \ t \to -\infty \qquad (9.154)$$

with the corresponding result for the far future limit, is commonly referred to as the *Asymptotic Condition*, and will be the critical starting point for our treatment of the scattering theory of Lehmann, Symanzik, and Zimmermann (LSZ) in the following section. It replaces our previous heuristic result (9.46). The Asymptotic Condition assures us that *all* of the information contained in the in- and out-states of the theory, and in particular in their overlap, the S-matrix, is already implicit in the behavior of the interpolating Heisenberg field(s) of the theory. The LSZ theory, and in particular the explicit link it provides between the S-matrix and vacuum expectation values of the associated interpolating fields, is of absolutely central importance in modern field theory. Indeed, the formula it gives us for the S-matrix in terms of expectation values of *time-ordered* Heisenberg fields will be of much greater practical utility, both within the confines of perturbation theory and beyond, than expressions of the type (9.139) obtained directly from the Haag–Ruelle approach.

9.4 Asymptotic formalism II: the Lehmann–Symanzik–Zimmermann (LSZ) theory

Our discussion of the Haag–Ruelle theory has allowed us to establish, on the basis of the very general axioms listed in Section 9.2, a precise connection between a local Heisenberg field $\phi(x)$ and the asymptotic (in- and out-)multi-particle states of the theory, *provided* that the given Heisenberg field has a non-vanishing matrix element from the vacuum to the single-particle state of the particle in question:

$$_{\text{in}}\langle \vec{k}|\phi(x)|0\rangle = \frac{Z^{1/2}}{(2\pi)^{3/2}\sqrt{2E(k)}}e^{ik\cdot x}, \quad Z\neq 0 \tag{9.155}$$

We have been able to verify (cf. 9.154), without any recourse to the interaction picture or perturbation theory, that for arbitrary normalizable in-states $|\alpha>_{\text{in}}, |\beta>_{\text{in}}$, with $g(\vec{x},t)$ a positive-energy solution of the Klein–Gordon equation, as in (9.113),

$$_{\text{in}}\langle\beta|-i\int d^3x\,\{g(\vec{x},t)\overset{\leftrightarrow}{\frac{\partial}{\partial t}}\phi(\vec{x},t)\}|\alpha\rangle_{\text{in}}$$

$$\rightarrow Z^{1/2}{}_{\text{in}}\langle\beta|-i\int d^3x\{g(\vec{x},t)\overset{\leftrightarrow}{\frac{\partial}{\partial t}}\phi_{\text{in}}(\vec{x},t)\}|\alpha\rangle_{\text{in}}, \quad t\rightarrow-\infty \tag{9.156}$$

The basic reason for this limiting behavior is that the smeared Heisenberg field on the left, sandwiched between in-states which correspond physically in the far past to states with widely separated free particles, samples a localized region of spacetime which is effectively the vacuum, and when appropriately folded with a positive-energy solution of the Klein–Gordon equation, acts like a free field in creating an additional free particle in that region. Although the Haag Asymptotic theorem provides an explicit formula for the S-matrix in terms of large time limits of appropriately smeared Wightman distributions, it turns out that the matrix elements specified by the theorem are only computable in a rather cumbersome way in perturbation theory, so while this result is of great conceptual value, it is of rather limited practical utility.[16] In this section we shall derive alternative expressions for the S-matrix which are particularly suitable for perturbative evaluation, while still allowing the application of non-perturbative methods in those situations where perturbation theory is invalid.

Comparing (9.34) and (9.113,9.120), we see that in the limit of plane wave solutions of well-defined momentum $\vec{k}$, $\tilde{g}(\vec{p}) = \frac{\sqrt{2E(p)}}{(2\pi)^{3/2}}\delta^3(\vec{p}-\vec{k})$, $\psi_g(\vec{p}) = \delta^3(\vec{p}-\vec{k})$ and the smeared in-field operator on the right-hand side of (9.156) becomes simply the creation operator $a^\dagger_{\text{in}}(\vec{k})$ for a particle of well-defined momentum. With a realistic particle wavefunction with some dispersion in momentum, and momentum-space wavefunction $\psi_g(\vec{p})$, we may denote the corresponding creation operator $a^\dagger_{\text{in},g} = \int \psi_g(\vec{p})a^\dagger_{\text{in}}(\vec{p})d^3p$, so

$$-i\int d^3x\, g(\vec{x},t)\overset{\leftrightarrow}{\frac{\partial}{\partial t}}{}_{\text{in}}\langle\beta|\phi(\vec{x},t)|\alpha\rangle_{\text{in}} \rightarrow Z^{1/2}{}_{\text{in}}\langle\beta|a^\dagger_{\text{in},g}|\alpha\rangle_{\text{in}}, \quad t\rightarrow-\infty \tag{9.157}$$

---

[16] Specifically, the matrix elements in an expression like (9.139) involve *non-time-ordered* fields, due to the $f^{(1)}(x)$ smearing of the original local fields. The graphical techniques of perturbation theory are, on the other hand, tailor-made for time-ordered products. This is clear both in the functional framework (cf. Section 4.2), in which functional integrals naturally yield such time-ordered operator matrix elements, or from the Gell-Mann–Low theorem proved in Section 9.1.

As $|\alpha\rangle_{\text{in}}, |\beta\rangle_{\text{in}}$ are arbitrary, we may take the conjugate of (9.157) to obtain

$$i \int d^3x \, g^*(\vec{x}, t) \frac{\overleftrightarrow{\partial}}{\partial t} \, {}_{\text{in}}\langle\beta|\phi(\vec{x}, t)|\alpha\rangle_{\text{in}} \rightarrow Z^{1/2} \, {}_{\text{in}}\langle\beta|a_{\text{in},g}|\alpha\rangle_{\text{in}}, \ t \rightarrow -\infty \quad (9.158)$$

We now observe that, *assuming asymptotic completeness (Axiom IIIb)*, the Hilbert spaces $\mathcal{H}_{\text{in}}$ and $\mathcal{H}_{\text{out}}$ coincide, with each other (and, although we do not need it here, with the full physical Hilbert space $\mathcal{H}$ of the theory). In other words, any (normalizable) $|\beta\rangle_{\text{in}}$ state is also an element of $\mathcal{H}_{\text{out}}$. Accordingly, the ${}_{\text{in}}\langle\beta|$ bra states in (9.157,9.158) may be replaced by arbitrary out-states:

$$-i \int d^3x \, g(\vec{x}, t) \frac{\overleftrightarrow{\partial}}{\partial t} \, {}_{\text{out}}\langle\beta|\phi(\vec{x}, t)|\alpha\rangle_{\text{in}} \rightarrow Z^{1/2} \, {}_{\text{out}}\langle\beta|a^{\dagger}_{\text{in},g}|\alpha\rangle_{\text{in}}, \ t \rightarrow -\infty \quad (9.159)$$

$$i \int d^3x \, g^*(\vec{x}, t) \frac{\overleftrightarrow{\partial}}{\partial t} \, {}_{\text{out}}\langle\beta|\phi(\vec{x}, t)|\alpha\rangle_{\text{in}} \rightarrow Z^{1/2} \, {}_{\text{out}}\langle\beta|a_{\text{in},g}|\alpha\rangle_{\text{in}}, \ t \rightarrow -\infty \quad (9.160)$$

Precisely analogous arguments imply, in the far future limit $t \rightarrow +\infty$,

$$-i \int d^3x \, g(\vec{x}, t) \frac{\overleftrightarrow{\partial}}{\partial t} \, {}_{\text{out}}\langle\beta|\phi(\vec{x}, t)|\alpha\rangle_{\text{in}} \rightarrow Z^{1/2} \, {}_{\text{out}}\langle\beta|a^{\dagger}_{\text{out},g}|\alpha\rangle_{\text{in}}, t \rightarrow \infty \quad (9.161)$$

$$i \int d^3x \, g^*(\vec{x}, t) \frac{\overleftrightarrow{\partial}}{\partial t} \, {}_{\text{out}}\langle\beta|\phi(\vec{x}, t)|\alpha\rangle_{\text{in}} \rightarrow Z^{1/2} \, {}_{\text{out}}\langle\beta|a_{\text{out},g}|\alpha\rangle_{\text{in}}, t \rightarrow \infty \quad (9.162)$$

with $a_{\text{out},g}, a^{\dagger}_{\text{out},g}$ defined in complete analogy to the corresponding in operators, but starting from the free field $\phi_{\text{out}}(x)$. The asymptotic conditions (9.159–9.162) will be the starting points for our derivation of the famous (and indispensable) LSZ reduction formulas for the S-matrix.

We begin with the S-matrix element for the scattering of n incoming scalar particles, described by momentum-space wavefunctions $\psi_{g_1}(\vec{k}), ..\psi_{g_n}(\vec{k})$, into m outgoing particles, with wavefunctions $\psi_{g'_1}(\vec{k}), ..\psi_{g'_m}(\vec{k})$. These wavefunctions are assumed to have disjoint support in momentum space: in particular, no incoming particle wavefunction has non-vanishing overlap with an outgoing particle wavefunction, as we wish to exclude uninteresting disconnected contributions to the S-matrix in which a particle passes through without interaction. After deriving the reduction formula, we shall take the limit in which the particle wave-packets approach plane waves (i.e., the $\psi(\vec{k})$ approach δ-functions), to make contact with the LSZ formulas as usually stated in field-theory textbooks. Thus

$$S_{g'_1 .. g'_m, g_1 .. g_n} = {}_{\text{out}}\langle g'_1, ..., g'_m | g_1, ..., g_n \rangle_{\text{in}} = {}_{\text{out}}\langle g'_1, ..., g'_m | a^{\dagger}_{\text{in},g_1} | g_2, ..., g_n \rangle_{\text{in}} \quad (9.163)$$

Next, we note that by our assumption of non-overlapping wavefunctions,

$$a_{\text{out},g_1} | g'_1, ..., g'_m \rangle_{\text{out}} = 0 \quad (9.164)$$

as the application of the destruction operator leads to a sum of terms involving overlap integrals $\int g_1^*(\vec{k})g_i'(\vec{k})d^3k$, which all vanish by assumption. We therefore have

$$S_{g_1'\cdots g_m',g_1\cdots g_n} = {}_{\text{out}}\langle g_1', ..., g_m'|g_1, ..., g_n\rangle_{\text{in}}$$

$$= {}_{\text{out}}\langle g_1', ..., g_m'|(a_{\text{in},g_1}^\dagger - a_{\text{out},g_1}^\dagger)|g_2, ..., g_n\rangle_{\text{in}} \qquad (9.165)$$

as the $a_{\text{out},g_1}^\dagger$ operator acting to the left as a destruction operator gives zero by the preceding argument. Matrix elements of $a_{\text{in},g_1}^\dagger$, $a_{\text{out},g_1}^\dagger$ are given as the asymptotic limits (9.159) and (9.161), so

$$S_{g_1'\cdots g_m',g_1\cdots g_n} = iZ^{-1/2}\left(\lim_{t\to+\infty} - \lim_{t\to-\infty}\right)\int d^3x\, g_1(\vec{x},t)\overset{\leftrightarrow}{\frac{\partial}{\partial t}} {}_{\text{out}}\langle g_1', ..., g_m'|\phi(\vec{x},t)|g_2, ..., g_n\rangle_{\text{in}}$$

$$= iZ^{-1/2}\int d^3x \int_{-\infty}^{+\infty} dt\, \frac{\partial}{\partial t}\{g_1(\vec{x},t)\overset{\leftrightarrow}{\frac{\partial}{\partial t}} {}_{\text{out}}\langle g_1', ..., g_m'|\phi(\vec{x},t)|g_2, ..., g_n\rangle_{\text{in}}\}$$

$$(9.166)$$

The time-derivative inside the integrand (9.166) can be rewritten recalling that the wavefunction $g_1(\vec{x},t)$ is a solution of the Klein–Gordon equation (cf. (9.113)):

$$\frac{\partial}{\partial t}(g_1(\vec{x},t)\overset{\leftrightarrow}{\frac{\partial}{\partial t}}\langle ....\rangle) = g_1(\vec{x},t)\frac{\partial^2}{\partial t^2}\langle ....\rangle - \frac{\partial^2 g_1(\vec{x},t)}{\partial t^2}\langle ....\rangle$$

$$= g_1(\vec{x},t)\frac{\partial^2}{\partial t^2}\langle ....\rangle - (\vec{\nabla}^2 - m^2)g_1(\vec{x},t)\langle ....\rangle \qquad (9.167)$$

Inserting (9.167) in (9.166), and integrating by parts to transfer the spatial gradients from the wavefunction g_1 (the fast spatial decrease of which ensures the absence of surface terms[17]) to the matrix element, we obtain

$$S_{g_1'\cdots g_m',g_1\cdots g_n} = iZ^{-1/2}\int d^4x\, g_1(\vec{x},t)(\Box_x + m^2)\, {}_{\text{out}}\langle g_1', ..., g_m'|\phi(x)|g_2, ..., g_n\rangle_{\text{in}}$$

$$(9.168)$$

where $\Box_x \equiv \frac{\partial}{\partial x^\mu}\frac{\partial}{\partial x_\mu}$. We note that in (9.168), the number of particles in the incoming state has been *reduced* by one, and been replaced by an appropriately smeared Heisenberg field operator sandwiched between the (remaining) incoming and outgoing states. A result of this type is called a "LSZ reduction formula". The notion that a smeared Heisenberg field can be used to create (or destroy) in- or outgoing particles should hardly be surprising, given the Haag Asymptotic theorem of the preceding section, but we note the important difference here that the S-matrix element is given in terms of an integral of a matrix element of such a field over all spacetime, and in particular over all time, rather than as a limit for large time.

---

[17] Recall that the matrix elements of $\phi(x)$ are tempered distributions—i.e., finite derivatives of a polynomially bounded continuous function—while the C^∞ function $g_1(\vec{x},t)$ decreases faster than any power of $|\vec{x}|$ at any given t.

The process of "reducing" particles from the incoming or outgoing state can be continued, as follows. We focus our attention next on an outgoing particle—say, the one with wavefunction $\psi_{g_1'}$. Begin with the matrix element under the integral in (9.168):

$$_{\text{out}}\langle g_1', ..., g_m'|\phi(x)|g_2, ..., g_n\rangle_{\text{in}} = {}_{\text{out}}\langle g_2', ..., g_m'|a_{\text{out},g_1'}\phi(x)|g_2, ..., g_n\rangle_{\text{in}}$$

$$= iZ^{-1/2} \lim_{t' \to +\infty} \int d^3x' g_1'^*(\vec{x}', t') \overset{\leftrightarrow}{\frac{\partial}{\partial t'}} {}_{\text{out}}\langle g_2', ..., g_m'|\phi(x')\phi(x)|g_2, .., g_n\rangle_{\text{in}} \quad (9.169)$$

As the spacetime point x is fixed in (9.169), the product of Heisenberg fields appearing in the matrix element is automatically time-ordered in the stated limit, so we may write

$$_{\text{out}}\langle g_1', ..., g_m'|\phi(x)|g_2, ..., g_n\rangle_{\text{in}}$$

$$= iZ^{-1/2} \lim_{t' \to +\infty} \int d^3x' g_1'^*(\vec{x}', t') \overset{\leftrightarrow}{\frac{\partial}{\partial t'}} {}_{\text{out}}\langle g_2', ..., g_m'|T(\phi(x')\phi(x))|g_2, .., g_n\rangle_{\text{in}} \quad (9.170)$$

If the far-future time limit in (9.170) is replaced by one in the far past, so that $t' \to -\infty$, we note that the time-ordering would imply

$$iZ^{-1/2} \lim_{t' \to -\infty} \int d^3x' g_1'^*(\vec{x}', t') \overset{\leftrightarrow}{\frac{\partial}{\partial t'}} {}_{\text{out}}\langle g_2', ..., g_m'|T(\phi(x')\phi(x))|g_2, .., g_n\rangle_{\text{in}}$$

$$= iZ^{-1/2} \lim_{t' \to -\infty} \int d^3x' g_1'^*(\vec{x}', t') \overset{\leftrightarrow}{\frac{\partial}{\partial t'}} {}_{\text{out}}\langle g_2', ..., g_m'|\phi(x)\phi(x')|g_2, .., g_n\rangle_{\text{in}}$$

$$= {}_{\text{out}}\langle g_2', ..., g_m'|\phi(x)a_{\text{in},g_1'}|g_2, .., g_n\rangle_{\text{in}} = 0 \quad (9.171)$$

using the asymptotic condition (9.160), and the fact that the in-state particle wave-functions are non-overlapping with $\psi_{g_1'}$. The expression in (9.170) may therefore be replaced by one in which the limits at $t' \to +\infty$ and $t' \to -\infty$ are subtracted, leading to an integral over t' of the time-derivative, just as in (9.166):

$$_{\text{out}}\langle g_1', ..., g_m'|\phi(x)|g_2, ..., g_n\rangle_{\text{in}}$$

$$= iZ^{-1/2} \int d^3x' \int_{-\infty}^{+\infty} dt' \frac{\partial}{\partial t'} \{g_1'^*(\vec{x}', t') \overset{\leftrightarrow}{\frac{\partial}{\partial t'}} {}_{\text{out}}\langle g_2', ..., g_m'|T(\phi(x')\phi(x))|g_2, .., g_n\rangle_{\text{in}}\} \quad (9.172)$$

Once again, using the fact that $g_1'(x')$ is a solution of the Klein–Gordon equation, and integrating by parts, one may convert this to the form

$$_{\text{out}}\langle g_1', ..., g_m'|\phi(x)|g_2, ..., g_n\rangle_{\text{in}}$$

$$= iZ^{-1/2} \int d^4x' g_1'^*(\vec{x}', t')(\Box_{x'} + m^2) {}_{\text{out}}\langle g_2', ..., g_m'|T(\phi(x')\phi(x))|g_2, .., g_n\rangle_{\text{in}}\} \quad (9.173)$$

Inserting (9.173) into (9.168), we obtain a result in which two particles—one incoming, the other outgoing—have been "reduced out" of the original $n \to m$ amplitude:

$$S_{g'_1 \cdots g'_m, g_1 \cdots g_n} = (iZ^{-1/2})^2 \int d^4x \, d^4x' \, g_1(\vec{x},t) g_1'^*(\vec{x}',t') (\Box_x + m^2)(\Box_{x'} + m^2)$$

$$\cdot \,_{\text{out}}\langle g'_2, .., g'_m | T(\phi(x')\phi(x)) | g_2, .., g_n \rangle_{\text{in}} \tag{9.174}$$

This process may evidently be continued (and we encourage the reader to carry it at least one step further; see Problem 9), removing *all* the incoming and outgoing particles from the initial and final states, and leading to the final *LSZ reduction formula*, giving the multi-particle S-matrix element in terms of an integral involving the vacuum-expectation-value of the time-ordered-product of $n + m$ Heisenberg interpolating fields (the $n + m$ point *Feynman amplitude*) for the particle undergoing scattering:

$$S_{g'_1 \cdots g'_m, g_1 \cdots g_n} = (iZ^{-1/2})^{m+n} \int \prod_{i=1}^{n} \prod_{j=1}^{m} g_i(x_i) g_j'^*(x_j') (\Box_{x_i} + m^2)(\Box_{x_j'} + m^2)$$

$$\cdot \,_{\text{out}}\langle 0 | T(\phi(x_1')..\phi(x_m')\phi(x_1)..\phi(x_n)) | 0 \rangle_{\text{in}} \, d^4x_i \, d^4x_j' \tag{9.175}$$

It is conventional to go over to the limit in which our wave-packets approximate plane wave solutions of well defined momentum $k_1, ..k_n, k_1', ..k_m'$. Thus, the single particle wavefunctions appearing in (9.175) are replaced by pure exponentials $g_k(x) = \frac{1}{\sqrt{(2\pi)^3 2E(k)}} e^{-ik \cdot x}$, and the LSZ formula gives the S-matrix element as a Fourier transform of the distribution obtained by applying Klein–Gordon operators $\mathcal{K}_x \equiv \Box_x + m^2$ to the Feynman amplitude for $n + m$ fields:

$$S_{k_1' \cdots k_m', k_1 \cdots k_n} = (iZ^{-1/2})^{m+n} \int \prod_{i=1}^{n} \prod_{j=1}^{m} \frac{1}{(2\pi)^{3/2}\sqrt{2E(k_i)}} \frac{1}{(2\pi)^{3/2}\sqrt{2E(k_j')}} e^{+ik_j' \cdot x_j' - ik_i \cdot x_i}$$

$$\cdot \, \mathcal{K}_{x_i} \mathcal{K}_{x_j'} \,_{\text{out}}\langle 0 | T(\phi(x_1')..\phi(x_m')\phi(x_1)..\phi(x_n)) | 0 \rangle_{\text{in}} \, d^4x_i \, d^4x_j' \tag{9.176}$$

It will be convenient to define an intermediate quantity from which the S-matrix amplitude can be extracted via (9.176). Leaving out for the time being the normalization factors and Klein–Gordon operators, we define the Feynman Green functions in both coordinate and momentum space in the obvious way:

$$G(x_1', \ldots x_n) \equiv \,_{\text{out}}\langle 0 | T(\phi(x_1')..\phi(x_m')\phi(x_1)..\phi(x_n)) | 0 \rangle_{\text{in}} \tag{9.177}$$

$$\tilde{G}(k_1', \ldots k_n) \equiv \int e^{+i\sum k_j' \cdot x_j' - i\sum k_i \cdot x_i} G(x_1', \ldots x_n) d^4x_1' \ldots d^4x_n \tag{9.178}$$

Note that the momenta appearing in (9.178) may be arbitrary four-vectors, not necessarily satisfying the on-mass-shell condition $k_i \cdot k_i = k_j' \cdot k_j' = m^2$. In other words, the LSZ formula provides us with a natural *off-mass-shell extension* of S-matrix elements. If we integrate by parts over the spacetime coordinates x_i, x_j' in (9.176) we may write the S-matrix element as

$$S_{k'_1..k'_m,k_1..k_n} = \prod_{i=1}^{n}\prod_{j=1}^{m} \frac{-iZ^{-1/2}(k_i^2 - m^2)}{(2\pi)^{3/2}\sqrt{2E(k_i)}} \frac{-iZ^{-1/2}(k_j'^2 - m^2)}{(2\pi)^{3/2}\sqrt{2E(k'_j)}} \tilde{G}(k'_1, \ldots k_n) \quad (9.179)$$

We see that if the on-mass-shell S-matrix element is to be *finite and non-vanishing*, the momentum-space Green function $\tilde{G}(k'_1, \ldots k_n)$ must contain a simple pole in the off-shellness variable $k^2 - m^2$ for each incoming and outgoing particle. The residue of the term containing a single such pole in each external variable, modified appropriately by the normalization factors in (9.179), then gives the on-mass-shell physical S-matrix for the scattering process. The appearance of simple poles for each external particle will be clarified in the context of perturbation theory in Chapter 10, when we see that they are associated graphically with the external legs of the Feynman diagrams corresponding to the Green function $\tilde{G}(k'_1, \ldots k_n)$.

There are a number of important points to be made in connection with the interpretation and use of the LSZ reduction formalism.

1. The asymptotic conditions used to derive the formula hold, by the Haag–Ruelle theory, for any almost local field $\phi(x)$ with a non-vanishing vacuum to single particle matrix element (9.155). In particular, they hold for almost local composite fields (i.e., multi-local combinations of the local fields appearing in the Hamiltonian defining the dynamics of the theory, as in (9.81)) with such a non-vanishing matrix element. Such fields must be used, as we shall see in more detail in the next section, if the particle in question is a bound state. Even if the particle corresponds to an elementary local field in the theory, *there is no unique interpolating field giving the correct S-matrix for its scattering!* For example, if $\phi(x)$ is a local interpolating field for the particle in question, with $\langle \vec{k}|\phi(x)|0\rangle \neq 0$, then for general values of $a, b, c, ..$ we certainly would expect that $\phi'(x) = a\phi(x) + b\phi(x)^2 + c\phi(x)^3..$ would also have a non-vanishing vacuum to single particle matrix element, and the LSZ formula will hold equally well using this field instead of $\phi(x)$. Of course, the Green function $G(x'_1, \ldots x_n)$ (and the normalization constant Z) will clearly be different with different fields: only the multiple pole residue of the on-mass-shell limit of its Fourier transform is guaranteed to be independent of the choice of field, as it gives the presumably unique physical S-matrix amplitude for the scattering of a specific stable particle.

2. The existence of simple poles in the off-shellness variables $k^2 - m^2$ for all external (incoming and outgoing) particles is a rigorous consequence of the Haag–Ruelle/LSZ theory, and depends critically on the assumed mass gap in the theory. In a theory such as QED, with a strictly massless photon, this result no longer holds. In fact, the singularities of charged-particle Green functions in the on-shell limit are softer than simple poles, and connected S-matrix elements for specified finite numbers of such incoming and outgoing particles vanish, as we shall see in Section 19.1. The problem is that, with a strictly massless particle, it takes essentially no energy to produce any number of extra very-low-energy particles, so that the probability of finding a strictly finite number in any process where a physical interaction has occurred is zero. Of course, in actual experiments the

detector resolution is finite, and ultra-soft photons are undetectable. Giving the photon a very small mass (smaller than the detector resolution) restores sanity: a non-vanishing S-matrix, and sensible cross-sections, rates, etc. We shall return to this subject in Chapter 19.

3. If the fields $\phi(x)$ appearing in (9.176) are ultralocal (cf. Section 5.5), the T-product defines a Lorentz scalar Green function, and the Lorentz invariance of the resulting S-matrix is manifest (recall that the non-covariant energy square-root factors are associated with our choice of non-covariantly normalized states). However, as just discussed, it is perfectly possible to use fields which are almost local (e.g., composite fields) but not strictly local, in which case the off-shell Green functions, both in coordinate and momentum space, are not Lorentz-invariant. However, the Lorentz-invariance of the S-matrix, which follows rigorously from the Haag–Ruelle theory, assures us that this property still holds in the on-shell limit (i.e., for the residue of the multi-pole term). This situation, in which a symmetry of the theory is only recovered in the on-shell limit, is actually quite common in field theory, as we shall see later in Part 3 of the book when we study symmetries in field theory in detail.

4. The extension to particles and fields with non-vanishing spin is straightforward. One begins from the generalization of (9.33, 9.34), which reconstruct the destruction and creation operators from the relevant free covariant (in- or out-)field, and applies the asymptotic condition precisely as above. An example for spin-$\frac{1}{2}$ Dirac fields is given in Problem 10.

5. The crossing symmetry of S-matrix amplitudes discussed in a few simple examples in Section 7.6 is seen to be an almost trivial consequence of the basic LSZ formula (9.176). In the case of the self-conjugate scalar field for which this formula applies, particles and antiparticles are identical, so the statement that initial-state particles (resp. antiparticles) can be exchanged with final-state antiparticles (resp. particles) by the simple expedient of inserting a minus sign in the corresponding four-momentum follows from (a) the symmetry of the T-product under exchange of the spacetime coordinates of the fields, and (b) the form of the Fourier transform, which contains a factor $e^{+ik'_j \cdot x'_j}$ for final-state particles (or antiparticles) and a factor $e^{-ik_i \cdot x_i}$ for initial-state particles (or antiparticles). In the event that we are dealing with non-self-conjugate fields, with distinct particles and antiparticles, the need to interchange particles and antiparticles when we cross from initial to final states is a simple consequence of the fact that the in- and out-fields contain positive frequency parts corresponding to particle destruction operators and negative-frequency parts corresponding to antiparticle creation operators. If the reader retraces the derivation of the LSZ formula in such a case, starting with the obvious generalization of the basic asymptotic conditions (9.159–9.162) for the complex field case (replacing, for example, $a^\dagger_{\text{in,g}}$ in (9.159) with $a^{c\dagger}_{\text{in,g}}$), the general form of the crossing rule for the S-matrix will become immediately evident.

6. We note that the reduction formula (9.176), containing the Green function $_{\text{out}}\langle 0|T(\phi(x'_1)..\phi(x'_m)\phi(x_1)..\phi(x_n))|0\rangle_{\text{in}}$, is precisely in a form amenable to perturbative treatment via the Gell–Mann–Low formula (9.21), taking $|\alpha\rangle, |\beta\rangle$ to be the

vacuum state. This convenient form explains why the LSZ, rather than the Haag–Ruelle, approach has dominated the treatment of scattering processes in field theory. It will be the starting point for our treatment of covariant perturbation theory in Chapter 10.

7. Obviously, any calculation of the S-matrix amplitude using (9.176) must include a knowledge of the normalization constant Z, appearing in (9.155). This constant is conventionally, and somewhat misleadingly, referred to as the "wavefunction renormalization constant" for the particle, although it is clear from the way we have introduced it that it is more properly associated with the choice of interpolating field. We shall see in Section 9.5 how to extract it from the behavior of the two-point Feynman amplitude $G(x_1, x_2)$—commonly called the "full Feynman propagator" of the theory.

8. The translation property of the Green function

$$G(x'_1, ..., x_n) = G(x'_1 - a, ..., x_n - a) \qquad (9.180)$$

for any fixed four-vector displacement a, valid for both local and almost-local fields, implies (see Problem 9) energy-momentum conservation: namely

$$S_{k'_1..k'_m, k_1..k_n} \propto \delta^4(k'_1 + .. + k'_m - k_1 - ... - k_n) \qquad (9.181)$$

9.5   Spectral properties of field theory

We have already seen that an enormous amount can be learned by starting from some very general assumptions concerning the nature of the particle states and interacting fields of a local quantum field theory. In particular, one object of direct phenomenological interest, the scattering matrix, has been related directly to the Fourier transform of a Feynman Green function, defined as the vacuum expectation value of a time-ordered product of Heisenberg fields. Of course, further progress requires that we develop methods to calculate these Green functions. One obvious option is perturbation theory, which we study in detail in the next Chapter. But the usefulness of perturbation theory is contingent on the existence of a split of the Hamiltonian of the theory into a solvable "free" part which isolates the "large" parts of the time evolution, in such a way that the resultant asymptotic expansion in the "interaction" part of the Hamiltonian produces an (initially) rapidly convergent series of approximants to the desired quantity, such as the S-matrix. In quantum electrodynamics this program has been brilliantly successful, leading to some of the most accurate predictions of quantitative science.

However, in the strong interactions the appearance of an effectively large interaction at some point in any hadronic process usually means that perturbation theory is at best only qualitatively useful. In such cases, we either have to give up on the project of a complete analytic calculation of the Green functions of the theory and rely on general properties of the theory (e.g., the axioms of Section 9.2) to put *contraints* on their behavior which translate into phenomenologically testable properties of the theory; or we resort, as in the case of lattice field theory, to a direct, but necessarily approximate, numerical calculation of the Green functions, making no reference to perturbation theory. In this section we describe some exact *non-perturbative* results along the lines of the first approach mentioned above, adhering mainly to the simplest possible case: the two-point Wightman $W(x_1, x_2)$ or Feynman $G(x_1, x_2)$ functions for a Heisenberg field interpolating for a stable self-conjugate massive spinless particle (self-interacting, with no bound states). Here, the spectral axioms play a central role.

Our first task is to derive a general representation for the Fourier transform of the Wightman two-point function $W(x_1, x_2) = \langle 0|\phi(x_1)\phi(x_2)|0\rangle$. The existence of a well-defined Fourier transform follows from the fact that $W(x_1, x_2) = W(x, 0), x \equiv x_1 - x_2$ is a tempered distribution. The structure of this representation will be determined simply by Lorentz-invariance and unitarity. In this context, the latter property means the existence of a complete orthonormal basis $|\alpha\rangle$ in our positive norm Hilbert space, where the states are chosen to be eigenstates of the energy-momentum four-vector P, $P^\mu|\alpha\rangle = P_\alpha^\mu|\alpha\rangle$. Inserting such a complete set between the two fields, we obtain

$$W(x_1 - x_2) = \sum_\alpha \langle 0|\phi(x_1)|\alpha\rangle\langle\alpha|\phi(x_2)|0\rangle$$

$$= \sum_\alpha \langle 0|e^{iP\cdot x_1}\phi(0)e^{-iP\cdot x_1}|\alpha\rangle\langle\alpha|e^{iP\cdot x_2}\phi(0)e^{-iP\cdot x_2}|0\rangle$$

$$= \sum_\alpha e^{-iP_\alpha \cdot x}\langle 0|\phi(0)|\alpha\rangle\langle\alpha|\phi(0)|0\rangle$$

$$= \int \{\sum_\alpha |\langle 0|\phi(0)|\alpha\rangle|^2\delta^4(p - P_\alpha)\}e^{-ip\cdot x}d^4p \tag{9.182}$$

where we assume that the integral over four-momentum can be interchanged with the sum over states (see below). Next, note that the function in brackets in (9.182),

$$f(p) \equiv \sum_\alpha |\langle 0|\phi(0)|\alpha\rangle|^2\delta^4(p - P_\alpha) \tag{9.183}$$

has, by the spectral axioms of our theory (cf. Section 9.2, Axioms Ib, Ic, Id), support only for $p^2 \geq 0$, and if we assume $\langle 0|\phi(0)|0\rangle = 0$, only for $p^2 \geq m^2$, where m is the single-particle mass. In fact, the support of $f(p)$ is restricted to the one-particle mass hyperboloid $p^2 = m^2$ and the multi-particle continuum starting at $p^2 = (2m)^2$ (or, if there is a symmetry $\phi \to -\phi$, at $p^2 = (3m)^2$). Moreover,

$$f(p) = \sum_\alpha |\langle 0|U^\dagger(\Lambda)U(\Lambda)\phi(0)U^\dagger(\Lambda)U(\Lambda)|\alpha\rangle|^2\delta^4(p - P_\alpha)$$

$$= \sum_\alpha |\langle 0|\phi(0)|\Lambda\alpha\rangle|^2\delta^4(p - P_\alpha)$$

$$= \sum_\alpha |\langle 0|\phi(0)|\alpha\rangle|^2\delta^4(p - \Lambda^{-1}P_\alpha)$$

$$= \sum_\alpha |\langle 0|\phi(0)|\alpha\rangle|^2\delta^4(\Lambda p - P_\alpha) = f(\Lambda p) \tag{9.184}$$

Accordingly, $f(p)$ is a Lorentz-invariant function of p, which vanishes for $p_0 < 0$. We may therefore write

$$\sum_\alpha |\langle 0|\phi(0)|\alpha\rangle|^2 \delta^4(p - P_\alpha) = \frac{1}{(2\pi)^3}\theta(p_0)\rho(p^2) \tag{9.185}$$

where the *spectral function* $\rho(p^2)$ is positive (or zero) with support on the spectrum of the squared mass operator P^2 as indicated above, and the normalization factor $\frac{1}{(2\pi)^3}$ is chosen for later convenience. We finally obtain, writing $\rho(p^2) = \int \delta(p^2 - \mu^2)\rho(\mu^2)d\mu^2$,

$$W(x) = \int \theta(p_0)\rho(p^2)e^{-ip\cdot x}d^4p$$

$$= \int_0^\infty \rho(\mu^2)W_0(x;\mu)d\mu^2 \tag{9.186}$$

where

$$W_0(x;\mu) \equiv \int \theta(p_0)\delta(p^2 - \mu^2)e^{-ip\cdot x}\frac{d^4p}{(2\pi)^3} = \frac{1}{(2\pi)^3}\int \frac{d^3p}{2E(p)}e^{-ip\cdot x} = \Delta_+(x;\mu) \tag{9.187}$$

Here $\Delta_+(x;\mu)$ is the invariant function arising from the two-point function of a *free*, canonically normalized scalar field of mass μ (cf. Chapter 6, (6.63)). This remarkable result—that the Wightman two-point function of an arbitrary scalar interacting Heisenberg field can be written as the positively weighted average of the corresponding free field Wightman functions for fields of varying mass, with a positive weight-function containing all the non-trivial interaction physics of the theory—is called the *Källen–Lehmann* representation of the two-point function. We note that it implies the vanishing of the VEV of the space-like commutator, $\langle 0|[\phi(x_1),\phi(x_2)]|0\rangle = 0$, $(x_1 - x_2)^2 < 0$, as the invariant function $\Delta_+(x;\mu)$ is symmetric at space-like points, $\Delta_+(x;\mu) = \Delta_+(-x;\mu)$, $x^2 < 0$, even though we have not assumed locality of our field. Vanishing of the matrix element of the space-like commutator between arbitrary states would require locality (Axiom IIc, Section 9.2).

The spectral representation (9.186) implies that the Fourier transform $\tilde{W}(p)$ of $W(x)$ is basically the invariant spectral function $\theta(p_0)\rho(p^2)$: as $W(x)$ is a well-defined tempered distribution, by the basic axioms of Section 9.2, its Fourier transform is likewise well-defined, and the defining sum for the spectral function (9.184) must therefore be convergent. We may therefore expect that the interchange of integration and summation performed above is in this case quite legal. As we shall see below, this is not necessarily the case for the spectral representation of other two-point functions.

Although it would require an exact solution of the interacting field theory to calculate the full spectral function $\rho(p^2)$, the contribution from one-particle states is calculable up to a normalization constant from (9.155). Thus

$$\frac{1}{(2\pi)^3}\theta(p_0)\rho_{1\text{part}}(p^2) = \int d^3k |\langle 0|\phi(0)|k\rangle_{\text{in}}|^2\delta^4(p - k) = \frac{Z}{(2\pi)^3 2E(p)}\delta(p_0 - E(p))\theta(p_0)$$

$$\Rightarrow \rho_{1\text{part}}(p^2) = Z\delta((p_0 - E(p))(p_0 + E(p)) = Z\delta(p^2 - m^2) \tag{9.188}$$

A spectral representation for the Feynman two-point Green function of the theory $G(x_1, x_2) = \langle 0|T(\phi(x_1)\phi(x_2))|0\rangle$, sometimes called the *full propagator*, can be derived following the pattern for $W(x_1, x_2)$. In fact, as

$$G(x_1, x_2) = \theta(t_1 - t_2)W(x_1, x_2) + \theta(t_2 - t_1)W(x_2, x_1) \tag{9.189}$$

and the spectral representation for $W(x_1, x_2)$ gives a linear superposition of free field Wightman functions for mass μ weighted by $\rho(\mu^2)$, one finds the obvious result (again, defining $x = x_1 - x_2$):

$$G(x) = \int \rho(\mu^2) G_0(x; \mu) d\mu^2 \tag{9.190}$$

where the corresponding free-field time-ordered Green function $G_0(x; \mu)$ for a particle of mass μ is just i times the free Feynman propagator $\Delta_F(x; \mu)$ introduced in Section 7.6 (cf. (7.215)). The Fourier transform is accordingly a weighted average of the momentum-space free Feynman propagator:

$$-i\tilde{G}(p) \equiv \hat{\Delta}_F(p^2) = \int \rho(\mu^2) \frac{1}{p^2 - \mu^2 + i\epsilon} d\mu^2 \tag{9.191}$$

The one-particle contribution to the full propagator, which we now denote $\hat{\Delta}_F(p^2)$, can be isolated and displayed explicitly, using (9.188):

$$\hat{\Delta}_F(p^2) = \frac{Z}{p^2 - m^2 + i\epsilon} + \int_{M^2_{\text{multi}}}^{\infty} \rho(\mu^2) \frac{1}{p^2 - \mu^2 + i\epsilon} d\mu^2 \tag{9.192}$$

where M^2_{multi} is the lowest squared-mass threshold for multi-particle states (i.e., $4m^2$ if $\langle 0|\phi(0)|k_1, k_2\rangle_{\text{in}} \neq 0$, $9m^2$ if the first non-vanishing multi-particle matrix element occurs for three particle states, $\langle 0|\phi(0)|k_1, k_2, k_3\rangle_{\text{in}}$, and so on). This result gives us the promised interpretation of the normalization constant Z appearing in the LSZ reduction formulas: it is simply the residue of the momentum-space full propagator of the interpolating field at the single-particle pole. The same procedures, whether perturbative or non-perturbative, which are applied to the calculation of the $n + m$ point Green function $\tilde{G}(k'_1, .., k'_m, k_1, ..k_n)$ in the LSZ formula can be used to calculate the two-point function and extract the required constant Z.

An important difference in the spectral representations of the two-point Wightman and Feynman functions is immediately apparent in (9.191): the representation for the Fourier transform of the *time-ordered* two-point function contains an integral, the convergence of which evidently requires that

$$\int_{M^2_{\text{multi}}}^{\infty} \frac{\rho(\mu^2)}{\mu^2} d\mu^2 < \infty \tag{9.193}$$

Unfortunately, the axioms introduced in Section 9.2 are not adequate to ensure the existence of this integral in all cases. Basically, the problem is that although $\theta(x_1^0 - x_2^0)$ and $\langle 0|\phi(x_1)\phi(x_2)|0\rangle$ are separately fine distributions (with well-defined Fourier transforms), their product, occurring in the definition of the time-ordered Green function, is *not necessarily* a well-defined distribution, with an unambiguous

Fourier transform. A classic example is given by the product of the θ and δ-functions, $\theta(x)\delta(x) = ?$, which is well-defined (namely, zero) on the subset of test functions vanishing at $x = 0$, but when extended to a distribution on the full Schwarz space necessarily involves an undetermined constant, $\theta(x)\delta(x) = C\delta(x)$ (we can think of the constant C as our (arbitrary) choice for the "value" of the step function $\theta(x)$ at $x = 0$). We can also regard convergence problems in the final integral result (9.191) as due to an unjustified interchange of summation and integration in the process of the "derivation", as mentioned above.

What do we actually know about the asymptotic behavior of the spectral function $\rho(\mu^2)$ at large μ^2? This behavior will clearly hinge on (i) the specific field(s) appearing in the time-ordered product (e.g., elementary versus composite), and (ii) the details of the dynamics (i.e., interactions) in the theory, which determine the multi-particle matrix elements in the sum (9.185) defining the spectral function. For the specific example under consideration here, our two-point function involves an elementary hermitian spinless field, and the dynamics is assumed to be specified by an interaction Hamiltonian leading to a perturbatively renormalizable theory,[18] such as $\mathcal{H}_{\text{int}} = \frac{\lambda}{4!}\phi^4$ (or more generally, $\mathcal{H}_{\text{int}} = \frac{\lambda_3}{3!}\phi^3 + \frac{\lambda_4}{4!}\phi^4$). In such theories, as we shall see later in our study (cf. Chapter 18) of the scaling properties of local field theories, the use of renormalization group techniques allows us to derive the asymptotic behavior both of $\hat{\Delta}_F(p^2)$ and $\rho(\mu^2)$, to any finite order of perturbation theory.[19] The result is that in each order of perturbation theory, the spectral function falls like $\frac{1}{\mu^2} \times$ powers of $\ln\mu^2$, ensuring the convergence of the spectral representation. For the super-renormalizable ϕ^3 theory, the falloff is even faster (Barton, 1965): $\rho(\mu^2) \simeq \frac{1}{\mu^4}$ (times logarithms). Thus, in these cases, at least within the context of perturbation theory, the Lehmann representation (9.192) is on a quite firm footing. Note that this representation implies that the momentum-space full Feynman propagator $\hat{\Delta}_F(p^2)$ can be regarded as the limit of an analytic function $F(w)$ as $w \to p^2 + i\epsilon$: i.e., as the complex variable w approaches the real axis from above, with

$$F(w) = \frac{Z}{w - m^2} + \int_{M^2_{\text{multi}}}^{\infty} \rho(w')\frac{1}{w - w'}dw' \tag{9.194}$$

---

[18] The definition and study of renormalizable field theories will be one of our primary objects in Part 4 of this book. For the time being, the reader is invited to think of such theories as ones in which a well-defined continuum limit exists *at the perturbative level*: namely, there is a well-defined asymptotic expansion of Feynman functions of the theory in powers of suitably defined coupling parameter(s), with the contributions at each order specified in terms of a finite number of parameters.

[19] There is a large amount of circumstantial evidence—though as yet no complete proof—that renormalizable self-interacting scalar field theories in four spacetime dimensions do not possess a non-trivial—i.e., interacting—continuum limit, even though the perturbative expansion is order-by-order well-defined. In other words, there is no set of Wightman functions satisfying all axioms of Section 9.2 whose asymptotic expansion in a suitably defined coupling constant agrees with the renormalized perturbative expansion of a ϕ^4 theory. Such theories, as we shall explain in Part 4 of the book, still have a perfectly sensible interpretation as *effective field theories*. For super-renormalizable theories, such as ϕ^3 theory in four dimensions (alas, with a spectrum unbounded below; cf. Section 8.4), or ϕ^4 theory in two or three spacetime dimensions, the continuum limit exists. Asymptotically free theories such as QCD, based on a non-abelian gauge group, are also thought to have a well-defined continuum limit beyond perturbation theory.

With $\rho(w')$ falling at least as fast as $1/w'$, this representation implies that $F(w)$ is a real analytic function in the complex plane of w, with a simple pole at $w = m^2$, and cuts on the positive real axis beginning at $w = M_{\text{multi}}^2$. There are multiple cuts because the spectral function $\rho(w')$ is the sum of n-particle contributions $\rho_n(p^2)$ which switch on at progressively higher values of w': at $w' = (2m)^2$ for the two-particle states $|\alpha\rangle$ in (9.185), at $w' = (3m)^2$ for the three-particle states, and so on. Using the familiar identity

$$\frac{1}{w - w' \pm i\epsilon} = \mathcal{P}\frac{1}{w - w'} \mp i\pi\delta(w - w') \tag{9.195}$$

the discontinuity of $F(w)$ across the cut for positive real $w = p^2$ is given by

$$F(p^2 + i\epsilon) - F(p^2 - i\epsilon) = -2i\pi\rho(p^2) = -2i\pi \sum_{n=2}^{\infty} \rho_n(p^2) \tag{9.196}$$

clearly indicating the presence of distinct branch points at $p^2 = n^2 m^2$. As the discontinuity of F across the cut is (for a real analytic function) equal to $2i\text{Im}(F)$ (where by convention the imaginary part is taken on the upper lip of the cut), the representation (9.194) can be rewritten as a *dispersion relation*

$$F(w) = \frac{Z}{w - m^2} + \frac{1}{\pi} \int_{M_{\text{multi}}^2}^{\infty} \frac{\text{Im}(F(w'))}{w' - w} dw' \tag{9.197}$$

allowing the reconstruction of the full analytic function $F(w)$ anywhere in the complex w-plane from knowledge of its residue at the single-particle pole and its discontinuity along the cut(s) on the positive real axis.

One further consequence of the Lehmann representation (9.192) is worth commenting on at this point. The positivity of the spectral function (which goes back, of course, to the underlying positivity of the metric of our Hilbert space) clearly implies that the $1/p^2$ behavior of the free propagator at large p^2 *cannot be damped* (to a more rapid decrease) by interactions, as the contribution of the integral is non-negative. At one point, attempts were made to construct a renormalizable theory of quantum gravity by introducing higher derivative terms in the Lagrangian with the effect of damping the high-momentum behavior of the graviton propagator in order to eliminate the proliferation of ultraviolet divergences in higher orders of perturbation theory that plague the conventional Einstein–Hilbert theory. The Lehmann spectral representation shows that such a damping can be possible only in the presence of negative metric states in the theory.

The derivation of the spectral representation given above depended, apparently, only on a few very basic properties of the Heisenberg field $\phi(x)$ appearing in the time-ordered product: specifically, hermiticity and the appropriate transformation properties of the fields under the Poincaré group, together, of course, with the completeness sum appropriate to a positive metric Hilbert space. However, as we indicated previously, the derivation also involves interchanges of summation and integration which are potential sources of disaster. In this case, disaster means a non-convergent spectral representation. In such a circumstance, the resultant dispersion relation needs a *subtraction*, resulting in the appearance of one (or more) undetermined arbitrary constants not already fixed by the properties of the individual fields. The appearance of such additional parameters is a consequence of the potential ambiguities in the product of two distributions mentioned previously.

Before giving a specific example it will be useful to consider an heuristic argument leading to a specification of the form these ambiguities can be expected to take in the momentum-space Feynman functions under consideration. We are concerned with a sum (corresponding to the two time orderings) of products of θ-functions of the form $\theta(x_1^0 - x_2^0)$ with vacuum expectation values of the product of two local operators $\langle 0|O_1(x_1)O_2(x_2)|0\rangle$ (O_1 and O_2 may or may not be the same operator). The only ambiguities that can occur involve taking the times x_1^0 and x_2^0 coincident (otherwise, we are dealing with well-defined Wightman distributions of the fields), and also spatial coincidence of the fields $\vec{x_1} \to \vec{x_2}$, as otherwise (at equal time) a small boost could be used to separate the times while (by locality) leaving the VEV of the two fields essentially unchanged. Therefore (cf. our discussion in Section 5.5 of the need for ultralocal interaction Hamiltonians) we may expect the appearance in general of contact type singularities when the spacetime points x_1 and x_2 coincide. In coordinate space we may expect that these correspond to the δ-function $\delta(x_1 - x_2)$ or spacetime-derivatives thereof; in momentum space, such contact terms correspond to constants or polynomials in the four-momentum variable. If the Green function in question is a Lorentz scalar (such as our scalar propagator $\hat{\Delta}_F(p^2)$ above), we therefore expect ambiguities of the form $C_0 + C_1 p^2 + \dots$. A spectral representation of the form (9.192), with $\hat{\Delta}_F(p^2) \to 0, p^2 \to \infty$, is clearly not possible in this case. Instead, the dispersion relation must be written in a subtracted form which restores the necessary convergence.

A full discussion of the role of dispersion relations in quantum field theory would require more space than is available here: instead, we give a simple example to make the issues more concrete. The cross-section for annihilation of an electron and positron into arbitrary hadronic final states (the so-called inclusive $e^+ - e^-$ annihilation process) played a pivotal role in the history of perturbative quantum chromodynamics (QCD): it represents the archetypal process in which perturbative calculations can be shown to yield reliable results at high energy in a strongly interacting theory such as QCD which displays the property of "asymptotic freedom" (roughly speaking, a damping of interaction strength at short distances or large momenta/energy). This cross-section turns out to be related to the spectral density for the two-point function of the hadronic electromagnetic current $J_{\text{em,had}}^\mu(x)$, which in momentum space takes the form

$$ i \int \langle 0|T\{J_{\text{em,had}}^\mu(x)J_{\text{em,had}}^\nu(0)|0\rangle e^{iq\cdot x}d^4x = (g^{\mu\nu}q^2 - q^\mu q^\nu)\omega(q^2) \qquad (9.198) $$

We shall see later, in Chapter 15, that the current $J_{\text{em,had}}^\mu(x)$ is a composite field, involving terms quadratic in quark fields. The transverse tensor on the right-hand side simply expresses the conservation of the current $\partial_\mu J_{\text{em,had}}^\mu(x) = 0$, so the interesting physics is contained in the scalar function $\omega(q^2)$. A dispersion relation for $\omega(q^2)$ can be "derived" along the same lines as the Lehmann representation (9.191) for the two-point function of elementary scalar fields (insertion of a complete set of states, interchange of summation and integration, etc.), but in this case the result involves a divergent integral, as the spectral density (imaginary part of $w(q^2)$) is found not to vanish for large q^2, but rather to go to a constant (in higher orders of perturbation theory, times logarithms). On the other hand, if we subtract the value of $w(q^2)$ at zero momentum (say, the subtraction can be made at any fixed momentum point), the resultant spectral representation involves a more rapidly convergent integral, so that a convergent dispersion relation can be written for this subtracted quantity. Using the notation introduced in (9.197)—namely, $w(q^2) \to F(w)$—and dropping the single-particle pole term,[20] we have the *once-subtracted dispersion relation*

$$F(w) - F(0) = \frac{1}{\pi} \int_{M^2_{\text{multi}}}^{\infty} \text{Im}(F(w'))(\frac{1}{w'-w} - \frac{1}{w'})dw' \tag{9.199}$$

or, equivalently,

$$F(w) = F(0) + \frac{w}{\pi} \int_{M^2_{\text{multi}}}^{\infty} \frac{\text{Im}(F(w'))}{w'(w'-w)}dw' \tag{9.200}$$

The constant $F(0) = w(q^2 = 0)$ appearing on the right-hand side indicates the appearance of an ambiguity in the definition of the T-product of two currents, even though the currents themselves are individually perfectly well-defined. In this case the ambiguity involves just the constant C_0 discussed previously (i.e., $C_1, C_2, .. = 0$), and a single subtraction produces sufficient inverse powers of the integration variable w' to ensure convergence of the spectral integral.[21]

The appearance of ambiguities in the time-ordered products of composite operators like the electromagnetic current discussed above, with the concomitant need for subtractions in the associated dispersion relation, may well provoke feelings of unease in the attentive reader. Our development of the LSZ reduction formalism in Section 9.4 was motivated by the desire to achieve a computationally convenient representation of general S-matrix amplitudes: the completely rigorous (and well-defined!) representation (9.139) following from the Haag Asymptotic Theorem 9.4, involving the vacuum expectation value of ordinary (i.e., not time-ordered) products of smeared field operators, is extremely cumbersome to implement either in perturbation theory or with available non-perturbative techniques. On the other hand, the LSZ formula (9.175) can be developed straightforwardly in perturbation theory, as we shall explain in detail in the next chapter, using the Gell–Mann–Low formula (9.21). Moreover, ground-state (i.e., vacuum) expectation values of time-ordered products of Heisenberg operators may be readily transcribed into a path-integral formulation (cf.

[20] The current $J^{\mu}_{\text{em,had}}(x)$ is assumed to contain Heisenberg fields with only strong interactions—the electromagnetic interactions are switched off in these fields—so that there is no pole corresponding to a stable particle (e.g., the photon) for which the current interpolates. The rho resonance appears as a pole of w, but on the unphysical second sheet, as the rho is unstable.

[21] Again, as in the case of the scalar-field two-point function, renormalization group techniques allow us to derive the relevant asymptotic behavior, in the case of an asymptotically free theory like QCD, to all orders of perturbation theory.

Section 4.2), which we shall see is of enormous importance in both perturbative and non-perturbative approaches to field theory.

Unfortunately, we now realize that the time-ordered products appearing in LSZ-type formulas, once developed perturbatively *à la* Gell–Mann–Low (thus introducing composite interaction Hamiltonian operators into the time-ordered products), are likely to contain undefined ambiguities! The ambiguities, of course, arise from the multiplication of distributions and are manifested as short-distance singularities in the resultant products, or, in momentum space, as the familiar ultraviolet divergences appearing in perturbative loop integrals for the Feynman functions of the theory. If the field theory is regulated at short distance, say by replacing continuous spacetime by a discrete spacetime lattice (thereby effectively introducing a high-momentum *ultraviolet* cutoff in the theory), the ambiguities are eliminated. Of course, to recover the full (continuous) Poincaré invariance of the theory, the spacing of the lattice points must eventually be taken to zero, or equivalently, the ultraviolet cutoff taken to infinity, and the question then arises as to the existence of a well-defined and unambiguous limit for the S-matrix amplitudes when this is done. From a physical point of view, the sensitivity of a field theory to the insertion of an ultraviolet (UV) cutoff is equivalent to the question of the sensitivity of the low-energy (low means momenta much smaller than the UV cutoff) predictions of the theory to our inevitable ignorance of *new* physics at much higher momenta (i.e., much smaller distance scales). The study of the sensitivity of local field theories at low energies to alterations in their short-distance structure will be the primary focus of Chapters 16 and 17 of this book. For the present, the reader should be reassured that for the class of field theories called "perturbatively renormalizable theories", the ambiguities appearing in the continuum limit (UV cutoff going to infinity) in the Feynman Green functions appearing in the LSZ formula can be shown to be completely absorbable in a finite set of low-energy parameters (masses and couplings) which uniquely determine (order by order, for all orders of perturbation theory) the S-matrix amplitudes of the theory, up to terms which fall as a power (usually, at least quadratic) of the low-energy mass and momentum scales divided by the UV cutoff. The latter correction terms are not unique, but depend on the precise details of the regularization (i.e., how the UV cutoff is introduced), reflecting the aforesaid ambiguities present in the underlying time-ordered products.

9.6   General aspects of the particle–field connection

In most of our discussion of interacting fields in this chapter so far, we have chosen to illustrate the general features for a particular class of field theory, involving a single stable spinless particle, whose self-interactions do not induce the formation of bound states. In this section, all such restrictive assumptions will be relaxed, as we wish to present a rather general discussion of various aspects of the particle–field connection in local relativistic quantum field theories. Our discussion of the Haag–Ruelle and LSZ theories already gives us the essential conceptual basis for this discussion. Some of the statements made here—for example, concerning color confinement in non-abelian gauge theories—will have to be taken on trust for the time being: a fuller discussion awaits in Part 4 of the book. This "borrowing" is justified by the desire to make the discussion of particle–field duality as complete as possible, by indicating the wonderful variety of ways in which the concepts of particle and field are linked in the panoply of field theories which are known to be relevant in high-energy physics.

We shall start at the particle end, by noting two important classifications which can be applied to particles. By "particle", we mean simply a state which to some appropriately high degree of approximation can be regarded as an eigenstate of the squared-mass operator P^2 and the spin (i.e., $\vec{J}^2, J_z$ in the frame where $\vec{P} = 0$). Particles can therefore be associated with irreducible representations of the HLG, as discussed in Sections 5.2 and 5.3. Beginning with this rather vague specification, one finds that the zoo of particles encountered in the Particle Data Book may be broken down into subcategories on the basis of the following two fundamental classifications:

1. Stable particles versus unstable particles (or "resonances").
2. Elementary particles versus composite particles.

It should be said at the outset that neither of these distinctions is to be regarded as absolute, for reasons that will shortly become clear.

We begin with the issue of stability. The physical Hilbert space of the theory actually only contains the *stable* particles—those which survive to asymptotically large times in the future, or can survive after preparation in the far past to reach an interaction event at finite times. This is the content of the asymptotic completeness principle discussed at length in the preceding sections. Some examples may be useful here to indicate what this really implies, which can be quite surprising at first glance.[22] As a consequence of weak and electromagnetic interactions, the only stable hadron is the proton (and its antiparticle, the antiproton). Accordingly, the asymptotic in- and out-space for hadrons consists entirely of multi-particle proton–antiproton states!

What, then, is a neutral pion—which as we know is unstable to decay electromagnetically to two photons? It may be regarded as a resonant state in the two-particle scattering channel of photons. Likewise, the neutron (unstable to β-decay) is a resonance in the three-particle (proton, electron, anti-electron–neutrino) channel! We shall explain more precisely below in field-theoretic terms what such a resonance means, but here the point to emphasize is that an unstable particle in the theory should be thought of as a transitory physical event representable only in a rather complicated way in terms of physical state vectors which can only involve stable particles. If we were to switch off the weak and electromagnetic interactions, then the strong interaction physical Hilbert space would consist of the multi-particle Fock space of in- (or out-) states of all stable particles, which would now include (in addition to the proton) the neutron and pion triplet. The rho particle, on the other hand, would still be regarded as a messy sort of temporary two-pion state: a resonance in the two-pion scattering channel. But there would be no single- or multi-rho states in the asymptotic Hilbert (in- or out-)space of the theory. As it is, in the Standard Model, the peculiar fact is that the physical Fock space of the theory, as defined in Sections 9.2, 9.3, and 9.4, contains single- (and multi-)particle states of the hydrogen-atom ground state (mass 938783014(80) eV, spin 0), but no (single- or multi-)neutron states! From the point

---

[22] We shall remain within the context of Standard Model physics in our examples, to avoid too many ifs, ands, or buts! Thus, the possibility of "exotic" processes such as proton decay is ignored.

of view of the asymptotic formalism of field theory, a stable composite particle is on just the same footing, in being present in the in- and out-Fock spaces of the theory, as a stable elementary particle of the theory.

Because of the wide difference in scales of various interactions, it may, of course, be perfectly sensible to view an unstable particle, with an average lifetime much larger than the other time-scales of interest in the processes under study (the neutron mean-life is 15 minutes—essentially infinite on the time-scale of subatomic processes!) as stable, and include it in the asymptotic states of the theory. In order to preserve unitarity, and avoid double-counting, the most honest way to do this is as indicated above: one switches off the guilty, destabilizing interaction responsible for decay. In the resultant theory, asymptotic states of the newly stable particle appear perfectly legally in the physical Hilbert space, and can be treated *à la* LSZ, as described in Section 9.4. The stable/unstable dichotomy is therefore essentially an issue of *isolating the important time-scales in the process of interest.*

Secondly, there is the question of elementarity. A natural definition of an elementary, as opposed to composite, particle would be negative in nature, and imply the absence of a detectible substructure. The presence of the qualifier "detectible" suggests the obvious caveat that future probing at much smaller distance scales (or higher momenta/energies) than presently accessible might well reveal a substructure where none is presently evident. Thus, there was in the first decades of the twentieth century no reason to suppose that the proton should not be regarded as elementary, as the quark-gluon substructure did not become evident until many years later. And there remains the possibility that all of the presently identified, and apparently "point-like", elementary particles of the conventional Standard Model (quarks, leptons, gauge bosons and Higgs bosons) might display, at sufficiently small distance scales (on the order of the Planck length), a "stringy" substructure. Nevertheless, at the scales presently accessible, it makes perfectly good sense to draw a line between the particles with an empirically evident composite character and those which are effectively "point-like", and of course, once isolated from other particles, characterized by mass and spin alone. A little later, as we describe the particle–field connection in detail, a very simple and precise definition of an elementary particle will be given in terms of the nature of the dynamics obeyed by the associated interpolating field.

To summarize briefly the above discussion, once specific temporal and spatial scales are identified, we may imagine sorting all known particle states into four categories, by specifying stable/unstable and elementary/composite. In the Standard model, all of these categories contain exemplars. Now, although the discussion of Heisenberg field theory in the present chapter has so far assumed for simplicity that our Heisenberg field $\phi(x)$ was the interpolating field for a *stable, elementary* spinless particle, the association of local fields with particles in the other three categories (i.e., unstable elementary, stable composite, and unstable composite) is perfectly possible. For example, the scalar field associated with the Higgs particle in the Standard Model is a perfectly sensible local field appearing in the Standard Model Lagrangian which specifies the full dynamics of the theory, even though, by the discussion above, it does not strictly speaking act as interpolating field for an asymptotic (stable) particle state of the theory. The spectral representation for the two-point function of this field will have (among others) a cut starting at $4m_e^2$ (where m_e is the electron mass) as one of the possible decay products of the Higgs consists of an electron–positron pair, but there will be no single particle pole, as the Higgs itself is unstable. Instead, the Higgs resonance is revealed as a pole below the real axis on the second Riemann sheet at a much higher energy (recent results suggest a value near 125 GeV).[23] In the case of stable composite particles, e.g., the proton, interpolating local (or almost local, cf. Section 9.2) fields can be constructed which go right into the Haag–Ruelle or LSZ formulas to determine the S-matrix for proton interactions. The reader will recall that the Haag Asymptotic theorem in particular was perfectly valid if the underlying field involved an appropriately smeared product of local fields: in the case of the proton, taking quantum chromodynamics as the underlying dynamical theory, the appropriate interpolating field involves the product of three quark fields (two up and one down) coupled to zero color.

It is now possible to give a more precise definition of the concept of "elementary", for either fields or particles. We shall define as elementary any field appearing in the fundamental Hamiltonian (or Lagrangian) specifying the exact dynamics of the theory at the distance scales in question. It is assumed that such a Hamiltonian can be given in an *explicit analytic form*: it can be written down, without approximation, on a finite piece of paper! Any particle interpolated for by such a field can be rightly called "elementary": the "point-like" character of the particle is reflected in the fact that its exact interacting dynamics has a precise *finite* expression in terms of products of local fields at a single spacetime point. With this terminology, the quark, lepton, Higgs and gauge boson fields of the Standard Model are elementary. Correspondingly, the leptons, Higgs and electroweak gauge bosons of the Standard Model are elementary particles as well.

What about the quarks and gluons, which we have omitted from our elementary particle list? Quantum chromodynamics (QCD), which we shall discuss in detail later in the book, provides a particularly stark example of the perils of the naive dictum "for every field, a particle". In confining theories such as QCD, the dynamics is specified via a Lagrangian involving quark and gluon fields which do not interpolate for any of the physical particles ("hadrons") in the theory.[24] In fact, these fields strictly speaking are not defined on the physical Hilbert space at all! However, multi-local combinations of the quark and gluon fields, appropriately coupled to be invariant under the local gauge symmetry of the theory, form well-defined almost local operators which do interpolate for physical particles, and are well defined on the physical state space. Thus, the product of three quark fields provides us with an interpolating field for the proton, the product of a quark field and its conjugate a pion field, and so on.

Although we have decided, in agreement with the definition proposed at the beginning of this section, to withhold the appellation "particle" from quarks and

---

[23] For an excellent discussion of the role and properties of unstable particles in a field-theory context, see (Brown, 1992), Section 6.3. An elementary discussion in standard quantum theory, including the interpretation of resonances as poles on the second Riemann sheet, can be found in (Baym, 1990), Chapter 4.

[24] Remarkably, there are even difficulties in the description of charged particle states in quantum electrodynamics, due to the masslessness of the photon: thus, the electron does not possess a conventional set of asymptotic states *à la* Haag–Ruelle theory—in the language of Schroer, it is an "infraparticle". More on this in Sections 19.1 and 19.2.

gluons, and speak only of quark and gluon *fields*, it is certainly true that perturbative calculations in QCD can be performed treating the omnipresent quarks and gluons as particles in just the same way that electrons and photons are so treated in quantum electrodynamics. Again, this is a question of the relevant spatiotemporal scales of the process: in a sense which can be made mathematically precise, QCD possesses a property of *asymptotic freedom* which renders the interaction arbitrarily weak at progressively smaller spatiotemporal scales (or equivalently, higher-energy scales), so that perturbative calculations become correspondingly more accurate. How these calculations can be connected to an actual S-matrix amplitude, in which necessarily the field energy must be allowed to dissipate over a (relatively) large spatial and temporal region, resulting in a *hadronization* of the underlying quark and gluon degrees of freedom, requires an understanding of an important scale separation property of renormalizable field theories, called "factorization", which will be an important topic of investigation in Part 4 of this book.

In summary, we once again emphasize the absence of any sacred one-to-one connection between particles and fields. A given particle (stable or unstable, elementary or composite) may be "represented" by many different local or almost local fields: if the particle is stable, and therefore represented in the asymptotic in- and out-states of the theory, any field with a non-vanishing vacuum to single-particle matrix element serves as an appropriate interpolating field for it. On the other hand, it may be convenient, and in the case of gauge theories indispensable, to introduce fields which interpolate for none of the physical particles of the theory, but in which products of such fields do interpolate for these particles. In such *confining* theories (cf. Section 19.3) we may loosely speak of the physical particle states as (stable or unstable) bound states of the underlying quark and gluon "particles", even though there is no attainable physical circumstance in which these latter can be realized as isolated entities with the characteristics expected of a particle: well-defined mass, spin, energy-momentum, and (in the case of QCD) a definite color quantum number derived from the putative interpolating field.

We conclude this section with a discussion of a subject which naturally arises when one thinks carefully about the nature of the particle–field connection in quantum field theory: the uniqueness (or otherwise) of the dynamical evolution specified by the theory, given the enormous fluidity of field representations available for the same particle, which nevertheless yield precisely the same S-matrix for scattering when all is said and done. At the pure particle level, the time development of the theory is certainly uniquely specified, in an almost trivial sense, once we assume asymptotic completeness. The multi-particle in-states $|k_1, k_2, .., k_N\rangle_{\text{in}}$, for example, form by assumption a complete set, and the time-evolution of these states is trivially unique, as they are eigenstates of the full Hamiltonian H (with eigenvalue $E = \sum_{i=1}^{N} E(k_i)$). However, this obviously begs the question of what such states "look like" at any times subsequent to the far past. In particular, knowledge of the particular combination of outgoing sets of freely receding particles that a given in-state represents in the far future requires that we have at our disposal the complicated connection between the in- and out-fields (ϕ_{in} and ϕ_{out}) of the theory, which then allows construction of the S-matrix, yielding the desired scattering amplitudes. As we have seen, such a connection is automatically afforded by the Heisenberg interpolating field $\phi(x)$ of the theory, which converges (weakly) to ϕ_{in} (resp. ϕ_{out}) in the far past (resp. far future). But the interpolating field is subject to a considerable freedom of choice. There are clearly an infinity of possible fields which interpolate between any two specified in- and out-fields: we need only ensure that the given Heisenberg field has a non-vanishing vacuum to single particle matrix element for the particle in question. Still, there is an intuitive feeling that a proper theory should, at least in principle, uniquely specify the physical situation not just at asymptotic times, but at finite intermediate times as well: for example, at the time $t \simeq 0$ at which a scattering event takes place.

The conceptual difficulty here primarily springs from the need to specify more clearly what one means by the phrase "physical situation" above. In standard quantum mechanics, the specification of the physical state at some time would require a measurement of a complete set of compatible observables. In field theory this is enlarged to the notion of *local* measurements, exploring properties of the system in bounded (microscopic) regions of spacetime. In other words, we would need to be able to measure matrix elements of local (or almost local) operators in or between specified physical states. To make this more concrete, let us take a famous example, of enormous phenomenological importance: the hadronic electromagnetic current $J^{\mu}_{\text{em,had}}(x)$ discussed in Section 9.5. A measurement of a general matrix element (typically called a "form factor") $_{\text{out}}\langle \beta | J^{\mu}_{\text{em,had}}(x) | \alpha \rangle_{\text{in}}$ for general spacetime points x is actually possible given the fact that the current couples linearly and gauge-invariantly to the photon field (cf. Chapter 15), which in turn is coupled in an accurately computable (via perturbation theory) way to leptons. The momentum transferred from scattered leptons (electrons, say: see Fig. 9.3) varies over the entire space-like domain, giving directly the Fourier transform (to momentum variable k) with respect to x of this matrix element. Unlike the situation with the S-matrix, this corresponds

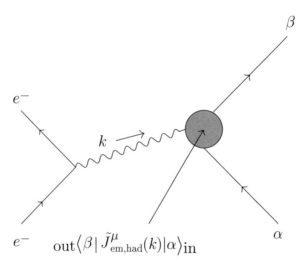

Fig. 9.3 Electron scattering off a hadronic state, allowing extraction of the space-like form factor.

to *off-mass-shell* (for the photon) information. It is precisely the availability of such off-mass-shell information (with momentum variable k not restricted to the light-cone $k^2 = 0$) that of course allows us to infer direct local behavior (in x space) of the matrix element. For example, if $|\alpha\rangle$ and $|\beta\rangle$ are single proton states (the in- and out-labels then become equivalent), the measurement of $J^0_{\text{em,had}}$ gives us a direct "peek" at the instantaneous charge density distribution inside a proton, at any given time. The LSZ formalism makes it clear that off-shell information of this sort is exquisitely sensitive to the precise choice of interpolating fields, so we are invited to enquire as to what extent we really "know" the local operator in question—in this case, the electromagnetic current for hadrons. Let us address this question in a simplified version of reality, in which only up and down quarks are present, which interact only with gluons and photons (i.e., weak interactions switched off). The asymptotic space of the theory contains, in addition to photons, protons, neutrons (which are now stable, absent weak decays) and a stable pion triplet. A unique hadronic electromagnetic charge operator can be written down trivially employing the asymptotic fields of the charged hadrons of the theory: namely, the proton in-field $p_{\text{in}}(x)$ and the charged pion field $\pi_{\text{in}}(x)$. These are free fields, so the desired charge operator on the asymptotic states (and therefore, by asymptotic completeness, on all states) is (cf. Problem 3 in Chapter 6, and Problem 4 in Chapter 7):

$$Q_{\text{had}} = \int d^3x (e : \bar{p}_{\text{in}}(x)\gamma^0 p_{\text{in}}(x) : +ie : \pi^\dagger_{\text{in}}(x) \overset{\leftrightarrow}{\frac{\partial}{\partial x_0}} \pi_{\text{in}}(x) :) \qquad (9.201)$$

which is, of course, time-independent, and must therefore equal $\int d^3x J^0_{\text{em,had}}(\vec{x}, t)$ for any t. However, we have the problem that in the presence of the extremely complicated interactions of QCD which end up binding the elementary quark and gluon degrees of freedom into the observed hadrons, any attempt to specify the full current operator $J^\mu_{\text{em,had}}(x)$ directly in terms of the free asymptotic fields of the theory must lead to an extremely complicated expression—certainly not one we are in any position to write down![25] Instead, we must take recourse in our previous definition of elementarity: the elementary fields of the theory are the *unique* (up to trivial redefinitions afforded by the symmetries of the theory) interpolating fields in terms of which an exact dynamics (e.g., Lagrangian or Hamiltonian) can be written down in some closed, finite expression. In other words, once we write down the full Lagrangian of QCD (as we will in Chapter 15) in terms of quark and gluon fields, the exact expression for the hadronic electromagnetic current follows immediately in terms of the interpolating up and down quark fields (which are the only elementary electrically charged hadronic objects in our truncated toy world)

$$J^\mu_{\text{em,had}}(x) = \frac{2}{3}e : \bar{u}(x)\gamma^\mu u(x) : -\frac{1}{3}e : \bar{d}(x)\gamma^\mu d(x) : \qquad (9.202)$$

and the physically directly measurable local matrix elements of this object are then determined (in principle!) uniquely by the fact that the dynamics of the quark

---

[25] Of course, the completeness of the asymptotic states assures that such an expression is in principle possible.

fields (and gluon fields through which they interact) is specified by a Hamiltonian/Lagrangian of definite form. Of course, the connection between the elementary quark and gluon fields in (9.202) and the asymptotic nucleon and pion fields appearing in (9.201) is exceedingly complicated, involving the intricacies of confinement in a strongly coupled theory. Nevertheless, great progress in direct calculation of form factors starting from the QCD Lagrangian has been possible using the techniques of lattice gauge theory at low energy, or perturbative QCD, at high energy (cf. Chapters 18 and 19).

The above example makes clear that local quantum field theory, although conceptually stimulated, as we saw in Chapters 5 and 6, by the requirements of S-matrix theory (in particular, the desire to construct Hamiltonians which lead to Lorentz-invariant and clustering S-matrices), specifies, in certain cases essentially uniquely, details of the dynamics which go beyond a pure S-matrix philosophy, which treats scattering amplitudes connecting the behavior of systems long before and long after interactions occur, but remains agnostic with regard to what happens "in between". Indeed, given a precise specification of the dynamics in terms of elementary fields, local observables can be constructed, and in some cases even measured, which give us a window into the behavior of the theory in finite regions of spacetime—behavior which should certainly be a central component of the conceptual content of any self-respecting field theory.[26]

9.7   Problems

1. Show, using the Lorentz-invariance of the S-matrix, that the transformation property (9.9) under the HLG of the in-states of the theory transfers to the corresponding out-state transformation property (9.11).

2. Show that the scalar field transformation property (9.24) of the Heisenberg field allows us to extend the equal-time commutativity property (9.3) to full space-like commutativity:

$$[\phi_H(x), \phi_H(y)] = 0, \quad (x - y)^2 < 0$$

3. Show that the full Heisenberg field can be reconstructed from the time-dependent creation and destruction operators $a_H(k; t)$, $a_H(k; t)$ defined by (9.36) via

$$\phi_H(x) = \int d^3k (g_k(x) a_H(k; t) + g_k^*(x) a_H^\dagger(k; t)) \tag{9.203}$$

where $g_k(x) \equiv \dfrac{1}{\sqrt{(2\pi)^3 2E(k)}} e^{-ik \cdot x}$. (A precisely analogous formula relates $\phi_{in}(x)$ (resp. $\phi_{out}(x)$) to the in (resp. out) creation and destruction operators $a_{in}^\dagger(k), a_{in}(k)$ (resp. $a_{out}^\dagger(k), a_{out}(k)$).)

4. With $|g_1, g_2, ..g_n\rangle$ defined as in (9.56), show that $\phi_f|g_1, g_2, ..g_n\rangle$ is a state of finite norm, with ϕ_f a free scalar field smeared with a Schwarz-type function.

---

[26] For a discussion of the role of almost local operators in devising thought experiments involving particle detectors which act as probes of particles in arbitrary localized regions of spacetime, see Section II.4.3 of (Haag, 1992).

5. Show that the function $F_+(p_0)$ (and hence, trivially, $F_0(p_0)$ and $F_-(p_0)$) defined in (9.98, 9.99, 9.100) is infinitely times continuously differentiable for arbitrary real p_0. (Note that the differentiability is trivial except at the two singular boundary points $p_0 = a, b$.)

6. Show that the sum of overlap integrals for the inner product of two in-state vectors obtained from Lemma 9.3 in the large time limit agrees with the inner product following from the Fock space metric (5.21) on continuum-normalized states, given the definition (9.137).

7. Show that if two momentum-space wavefunctions are non-overlapping, $g_i^*(\vec{p})g_j(\vec{p}) = 0$, the coordinate space overlap of the corresponding single-particle wavefunctions $g_{i,j}(\vec{x}, t)$ (cf. (9.113)) vanishes for all time.

8. The asymptotic conditions (9.157, 9.158) can be used to derive an important explicit relation (called the *Yang–Feldman* equation) between the interacting Heisenberg field $\phi_H(x)$ and the asymptotic free in-field $\phi_{in}(x)$.

   (a) First, show that with $a_H^\dagger(k; t)$ and $g_k(x)$ as in Problem 3, for arbitrary states $|\alpha\rangle_{in}, |\beta\rangle_{in}$,

   $$_{in}\langle\beta|a_H^\dagger(k; t)|\alpha\rangle_{in} = \sqrt{Z}\ _{in}\langle\beta|a_{in}^\dagger(k)|\alpha\rangle_{in}$$
   $$- i \int_{-\infty}^t dt' \int d^3x'\, g_k(x')(\Box_{x'} + m_{ph}^2)_{in}\langle\beta|\phi_H(x')|\alpha\rangle_{in}$$

   (Hint: as in the proof of the reduction formulas, a function at time t is expressed as the value at time $-\infty$ plus the integral of the time-derivative from $-\infty$ to t.)

   (b) Using the result of Problem 3 above, and the formula obtained in part (a) (with its obvious analog for matrix elements of $a_H(k; t)$), show that

   $$_{in}\langle\beta|\phi_H(x)|\alpha\rangle_{in} = \sqrt{Z}\ _{in}\langle\beta|\phi_{in}(x)|\alpha\rangle_{in}$$
   $$+ i \int d^4x'\, \Delta_R(x - x')(\Box_{x'} + m_{ph}^2)_{in}\langle\beta|\phi_H(x')|\alpha\rangle_{in}$$
   $$\Delta_R(x - x') \equiv \theta(t - t')(\Delta_+(x - x'; m) - \Delta_+(x' - x; m))$$

   with Δ_+ the fundamental invariant function defined in (6.63). As the states $|\alpha\rangle_{in}, |\beta\rangle_{in}$ form a complete set (asymptotic completeness), we have the operator identity (Yang–Feldman equation)

   $$\phi_H(x) = \sqrt{Z}\ \phi_{in}(x) + i \int d^4x'\, \Delta_R(x - x')(\Box_{x'} + m_{ph}^2)\phi_H(x') \quad (9.204)$$

9. Starting with the formula (9.174), in which one incoming particle and one outgoing particle have been reduced out of the asymptotic states, carry the LSZ process one step further by reducing out the incoming particle with wavefunction g_2 to obtain an expression with the time-ordered product of three Heisenberg fields.

10. Derive the following reduction formula for reducing out a single incoming Dirac particle:

$$S_{k'_1\sigma'_1..k'_m\sigma'_m,k_1\sigma_1..k_n\sigma_n}$$

$$= \frac{i}{Z^{1/2}} \int {}_{\text{out}}\langle k'_1\sigma'_1..k'_m\sigma'_m|\bar{\psi}(x_1)|k_2\sigma_2..k_n\sigma_n\rangle_{\text{in}}(i\overleftarrow{\partial}_{x_1} + m) \quad (9.205)$$

$$\cdot \frac{1}{(2\pi)^{3/2}} \sqrt{\frac{m}{E(k_1)}} u(k_1,\sigma_1)e^{-ik_1\cdot x_1} d^4x_1 \quad (9.206)$$

You should first derive the following Fourier transform formula for the creation operator for a Dirac particle in the in-state, in terms of the Dirac in-field defined with normalization as in (7.104):

$$b_{\text{in}}^\dagger(k,\sigma) = \int \psi_{\text{in}}^\dagger(x)\frac{1}{(2\pi)^{3/2}}\sqrt{\frac{m}{E(k)}} u(k,\sigma)e^{-ik\cdot x} d^3x$$

11. Derive the translation property

$$G(x'_1, ..., x_n) = G(x'_1 - a, ..., x_n - a)$$

for the n-point Green functions, and use it to establish the energy-momentum conservation property for the S-matrix (as given by LSZ (9.176)):

$$S_{k'_1..k'_m,k_1..k_n} \propto \delta^4(k'_1 + ..k'_m - k_1 - ... - k_n)$$

10
Dynamics VIII: Interacting fields: perturbative aspects

Local quantum field theory, as incorporated in the currently accepted framework of the Standard Model of elementary particle interactions, has provided a quantitative description of microphysical processes down to distance scales on the order of 10^{-18} meters, backed by an enormous quantity of empirical evidence supplied by experiments carried out over a huge range of energy scales. In some cases, the level of quantitative agreement is astonishing: the precision allowed in the calculation of certain quantum electrodynamic processes (such as the anomalous magnetic moment $g-2$ of the electron) using relativistic field theory is unmatched in any other area of physical science. Such precision is only possible for weakly coupled field theories such as quantum electrodynamics (QED), where a natural expansion parameter, the fine-structure constant $\alpha \simeq 1/137$, is sufficiently small that the asymptotic expansion in powers of α provided by perturbation theory gives a rapidly converging sequence of approximants to the desired physical quantity. Calculations of $g-2$ have been carried out through order α^4, leading to theoretical predictions valid to ten significant figures.

However, even in theories such as quantum chromodynamics (QCD), where the strength of the interaction would seem to vitiate the validity of a perturbation approach, the property of asymptotic freedom, which we shall discuss in detail in Part 4, allows in many cases the application of perturbative techniques in the derivation of the high-energy asymptotic behavior of many interesting strong interaction processes. Finally, even when not directly applicable, perturbation theory, as expressed in the very intuitive graphical formulation introduced by Feynman in the late 1940s, provides many important insights into the structure of local quantum field theory which turn out to be crucial in advancing the development of more general non-perturbative techniques. We shall therefore be devoting this chapter to an account of the perturbative aspects of field theory. The technical background, both operator and functional (path integral), necessary for the development of amplitudes in graphical expansions will be our main subject.

Perturbation theory for time-dependent processes (such as scattering) in non-relativistic quantum theory is usually formulated in terms of an interaction picture in which the Hamiltonian is split into free and interacting parts (see Section 4.1.1), and the generally complicated effects of the interaction restricted to the time development of the states, while the operators evolve with the free dynamics. Unfortunately, as we have mentioned on several occasions previously, and will discuss in more detail in the final Section of this chapter, in a quantum field theory with infinitely many degrees of freedom, the interaction picture typically *does not exist*—a result usually referred to as Haag's theorem. The problem can be removed by a full regularization of the theory in which both short- and long-distance cutoffs are introduced, leading to a theory with a finite (indeed, arbitrarily large!) number of independent quantum-mechanical degrees of freedom. Unfortunately, such regularizations inevitably result in a (one hopes, temporary) loss of the full Poincaré symmetry of the theory, and the task then remains to establish the return of this symmetry as the regularization is removed. In fact, any application of the usual formal "theorems" (such as the Gell–Mann–Low theorem of Section 9.1, (9.21)) of interaction-picture perturbative expansions in field theory necessarily require the insertion of a regularization in order to obtain unambiguous results, due to unavoidable ambiguities in time-ordered Feynman Green functions involving composite operators (such as interaction Hamiltonian densities), as discussed at the end of Section 9.5. On the other hand, the LSZ formula (9.176) derived in Section 9.4 is *completely independent* of any reasoning relying on interaction-picture arguments. Of course, the problem remains of actually calculating the Feynman n-point amplitudes contained in the LSZ formula (i.e., the VEV of the time-ordered products of n Heisenberg fields) in order to obtain the desired S-matrix elements. These Green functions can only be obtained analytically in a handful of toy field theories in 1+1 spacetime dimensions: in any realistic case, we necessarily must have recourse to approximative methods. These are basically of two kinds:

1. Perturbative evaluation of the desired amplitudes in an asymptotic expansion in a (finite number) of parameters defined in terms of *physically measurable amplitudes* of the theory. We imagine doing this in a fully regularized version of the theory so that well-defined results are obtained order by order in the perturbative expansion. The question of the sensitivity of the results so obtained to the presence of cutoffs (in particular, a short-distance or "UV" cutoff) will be the primary object of study in Part 4 of the book.

2. Non-perturbative approximative schemes may be employed. By far the most successful of these approaches—not necessarily with respect to the quantitative accuracy of the results obtained, but insofar as the approximations employed are *systematically improvable*—is lattice quantum field theory. The field theory is fully regularized on a (Euclidean) spacetime lattice, and the corresponding n-point functions (called *Schwinger* functions in the Euclidean case) evaluated numerically from a path-integral representation. In many cases the information so obtained in Euclidean space can be transferred directly to information about the actual Minkowskian physics of the theory (e.g., spectrum, matrix elements of various local operators, etc.).

In this chapter we concentrate on developing the basic techniques required to implement the first item above. The required perturbative expansions will be seen to have a natural interpretation in terms of graphical objects (Feynman graphs), with simple rules (Feynman rules) allowing the evaluation of the amplitudes in terms of elementary algebraic expressions associated with each graphical element (line, vertex, etc.). This will be done first using operatorial methods: the matrix elements of Heisenberg picture operators needed for the LSZ formula will be expanded using the

Gell–Mann–Low theorem, and the resultant expressions evaluated using a technical tool known universally as *Wick's theorem*. Later, we shall see how the resultant graphical objects also arise naturally in a path-integral formulation of the field theory. Finally, as promised, we discuss the significance of Haag's theorem in an attempt to assuage the reader's natural anxiety concerning the validity of results obtained in perturbation theory via interaction-picture methods.

10.1 Perturbation theory in interaction picture and Wick's theorem

Our primary object of interest in the following will be the S-matrix, which in a theory of self-interacting scalar fields is given by the LSZ formula derived in Section 9.4:

$$
S_{k'_1..k'_m,k_1..k_n} = (iZ^{-1/2})^{m+n} \int \prod_{i=1}^{n} \prod_{j=1}^{m} \frac{1}{(2\pi)^{3/2}\sqrt{2E(k_i)}} \frac{1}{(2\pi)^{3/2}\sqrt{2E(k'_j)}} e^{+ik'_j \cdot x'_j - ik_i \cdot x_i}
$$

$$
\cdot \, \mathcal{K}_{x_i} \mathcal{K}_{x'_j} \, _{\text{out}}\langle 0 | T(\phi_H(x'_1)..\phi_H(x'_m)\phi_H(x_1)..\phi_H(x_n)) | 0 \rangle_{\text{in}} d^4 x_i d^4 x'_j
$$

$$(10.1)$$

The VEV of the time-ordered product of $m + n$ fields appearing here (in common parlance, the $m + n$-point Green function of the theory) has a formal perturbative expansion via the Gell–Mann–Low theorem (9.21),

$$
_{\text{out}}\langle 0 | T\{\phi_H(y_1)\phi_H(y_2) .. \phi_H(y_m)\} | 0 \rangle_{\text{in}} = \sum_{p=0}^{\infty} \frac{(-i)^p}{p!} \int \langle 0 | T\{\phi(y_1)\phi(y_2)..\phi(y_m)
$$

$$
\cdot \, \mathcal{H}_{\text{int}}(z_1)\mathcal{H}_{\text{int}}(z_2)..\mathcal{H}_{\text{int}}(z_p)\} | 0 \rangle d^4 z_1 d^4 z_2 .. d^4 z_p \qquad (10.2)
$$

which provides the needed formal expansion of the Green function appearing in the LSZ formula in powers of the interaction. We once again emphasize that in a continuum field theory with no short-distance cutoff, the T-products appearing on the right-hand side of (10.2) contain ambiguities, so a suitable regularization (e.g., on a spacetime lattice) is implied to make the individual terms in the perturbative expansion meaningful. All the fields appearing on the right are, of course, free fields, as they are in the interaction picture.

Our task in this section is to derive an important technical result, called *Wick's theorem*, which will facilitate the computation of these T-products of free fields. Although, for the purposes of the LSZ formula, we clearly only need the VEV of the T-products, we shall derive the more general result contained in Wick's theorem, giving the T-products as a sum of terms involving products of normal-ordered products of the fields and c-number two-point Green functions (i.e., free Feynman propagators). As a normal-ordered product of fields vanishes when it encounters the vacuum either on the left (bra-state) or right (ket-state), only terms involving products of Feynman propagators (and no normal-ordered products) will actually survive in the VEV appearing on the right of (10.2). Nevertheless, the more general operator result derived below is important in other contexts, and worth the small additional effort required.

The reader should also note that we have reverted to the notation used prior to Section 9.2, wherein unsubscripted fields, such as $\phi(x)$, are free interaction-picture fields (as in Chapters 7 and 8), while Heisenberg fields are explicitly distinguished by a subscript "H", as in $\phi_H(x)$.

Wick's theorem is usually proved by an induction procedure, starting with the result for two fields—the simplest non-trivial case. A short calculation, using the fact that the positive (destruction) and negative (creation) frequency parts of the free field operator $\phi(x)$ have a c-number commutator, shows that the difference between the time-ordered and normal-ordered product of two fields is itself a c-number. The T-product of $\phi(x_1)$ and $\phi(x_2)$ is symmetric in its arguments, so with no loss of generality we may assume $x_1^0 > x_2^0$, whence

$$T(\phi(x_1)\phi(x_2)) - :\phi(x_1)\phi(x_2): $$
$$= \phi(x_1)\phi(x_2) - :(\phi^{(+)}(x_1) + \phi^{(-)}(x_1))(\phi^{(+)}(x_2) + \phi^{(-)}(x_2)): $$
$$= (\phi^{(+)}(x_1) + \phi^{(-)}(x_1))(\phi^{(+)}(x_2) + \phi^{(-)}(x_2)) $$
$$- (\phi^{(+)}(x_1)\phi^{(+)}(x_2) + \phi^{(-)}(x_1)\phi^{(+)}(x_2) + \phi^{(-)}(x_2)\phi^{(+)}(x_1) + \phi^{(-)}(x_1)\phi^{(-)}(x_2)) $$
$$= [\phi^{(+)}(x_1), \phi^{(-)}(x_2)] = \text{c} - \text{number} \tag{10.3}$$

On the other hand, a c-number is equal to its VEV, and by definition the normal-ordered product vanishes when sandwiched between vacuum states, so taking the VEV of the operator difference above we find

$$T(\phi(x_1)\phi(x_2)) - :\phi(x_1)\phi(x_2): \; = \; \langle 0|T(\phi(x_1)\phi(x_2))|0\rangle \tag{10.4}$$

or

$$T(\phi(x_1)\phi(x_2)) = \; :\phi(x_1)\phi(x_2): + \langle 0|T(\phi(x_1)\phi(x_2))|0\rangle$$
$$= \; :\phi(x_1)\phi(x_2): + i\Delta_F(x_1, x_2) \tag{10.5}$$

giving the desired resolution of the time-ordered product in terms of normal-ordered products and c-number functions—in this case, simply the Feynman propagator (times a factor if i; see (7.214)). We shall not pursue the inductive proof further here,[1] but rather provide an "all-at-once" proof of Wick's theorem for scalar fields using functional methods that closely parallel the way in which the theorem emerges in the path-integral formulation of field theory (discussed below), as a combinatoric property of the (functional) derivatives of a Gaussian functional. The proof shows that Wick's theorem is basically a manifestation of the Baker–Campbell–Hausdorff (BCH) formula, which is frequently found to be lurking in the background when operator reordering issues arise. We remind the reader that if two operators A and B have a commutator $[A, B]$ which itself commutes with both A and B, then

$$e^{A+B} = e^{-\frac{1}{2}[A,B]}e^A e^B \tag{10.6}$$

---

[1] See Bjorken and Drell (Bjorken and Drell, 1965), for the standard operatorial proof.

We now consider a very simple interaction Hamiltonian, linear in the canonical scalar field, so that

$$V_{\text{ip}}(t) = \int \mathcal{H}_{\text{int}}(\vec{x}, t) d^3x = \int j(\vec{x}, t)\phi(\vec{x}, t)d^3x \tag{10.7}$$

where $j(\vec{x}, t)$ is an unspecified real c-number source function, with respect to which we shall later wish to perform functional derivatives. This hermitian interaction Hamiltonian determines a unitary evolution operator $U(t, t_0)$ in the usual way (cf. (4.28)):

$$U(t, t_0) = T\{\exp\left(-i \int_{t_0}^{t} V_{\text{ip}}(\tau)d\tau\right)\} \tag{10.8}$$

with

$$\frac{\partial U(t, t_0)}{\partial t} = -i(\int j(\vec{x}, t)\phi(\vec{x}, t)d^3x)U(t, t_0) \tag{10.9}$$

The expansion (4.27) shows that time-ordered products of arbitrarily many $\phi(x)$ fields can be obtained from knowledge of $U(t, t_0)$ by differentiating it with respect to the source functions $j(x)$. Next, note that if we define a new (unitary) operator $E(t, t_0)$ by *omitting the time-ordering*

$$E(t, t_0) \equiv \exp\left(-i \int_{t_0}^{t} V_{\text{ip}}(\tau)d\tau\right) = e^{-i \int j(x)\phi(x)d^4x} = e^{-i \int j(x)(\phi^{(+)}(x)+\phi^{(-)}(x))d^4x} \tag{10.10}$$

the connection to a normal-ordered quantity is immediate using the BCH formula (10.6):

$$: E(t, t_0) : = : e^{-i \int j(x)(\phi^{(+)}(x)+\phi^{(-)}(x))d^4x} := e^{-i \int j(x)\phi^{(-)}(x)d^4x} e^{-i \int j(x)\phi^{(+)}(x)d^4x}$$
$$= e^{-\frac{1}{2} \int j(x_1)j(x_2)[\phi^{(-)}(x_1),\phi^{(+)}(x_2)]d^4x_1 d^4x_2} E(t, t_0) \tag{10.11}$$

where the time-integrals associated with the spacetime coordinates x_1, x_2 are implicitly assumed to go from t_0 to t. Our objective of finding a connection between the time-ordered and normal-ordered field products is therefore accomplished if we can find a simple relation between $U(t, t_0)$ and $E(t, t_0)$. We do this by studying the time-evolution equation satisfied by $E(t, t_0)$:

$$\frac{\partial E(t, t_0)}{\partial t} = \lim_{\Delta t \to 0} \frac{1}{\Delta t}(e^{-i\Delta t V_{\text{ip}}(t) - i \int_{t_0}^{t} V_{\text{ip}}(\tau)d\tau} - e^{-i \int_{t_0}^{t} V_{\text{ip}}(\tau)d\tau}) \tag{10.12}$$

Now, define $A \equiv -i\Delta t\, V_{\text{ip}}(t)$, $B \equiv -i \int_{t_0}^{t} V_{\text{ip}}(\tau)d\tau$, and apply (10.6), expanded to first order in Δt, so that

$$e^{A+B} = (1 - \frac{1}{2}[A, B] + A + O((\Delta t)^2))e^B \tag{10.13}$$

Inserting this in (10.12) gives the desired first time-derivative of $E(t, t_0)$:

$$\frac{\partial E(t, t_0)}{\partial t} = (-iV_{\text{ip}}(t) + \frac{1}{2}\int_{t_0}^{t}[V_{\text{ip}}(t), V_{\text{ip}}(\tau)]d\tau)E(t, t_0) \tag{10.14}$$

It now follows directly that the operator

$$E'(t, t_0) \equiv e^{-\frac{1}{2}\int_{t_0}^{t}d\tau_1\int_{t_0}^{\tau_1}d\tau_2[V_{\text{ip}}(\tau_1), V_{\text{ip}}(\tau_2)]}E(t, t_0) \tag{10.15}$$

satisfies (see Problem 1) the same first-order differential equation as $U(t, t_0)$

$$\frac{\partial E'(t, t_0)}{\partial t} = -iV_{\text{ip}}(t)E'(t, t_0) \tag{10.16}$$

and the same initial condition, $E(t_0, t_0) = U(t_0, t_0) = 1$, whence

$$U(t, t_0) = E'(t, t_0) = e^{-\frac{1}{2}\int_{t_0}^{t}d\tau_1\int_{t_0}^{\tau_1}d\tau_2[V_{\text{ip}}(\tau_1), V_{\text{ip}}(\tau_2)]}E(t, t_0)$$

$$= e^{-\frac{1}{2}\int_{t_0}^{t}d\tau_1\int_{t_0}^{t}d\tau_2\theta(\tau_1-\tau_2)[V_{\text{ip}}(\tau_1), V_{\text{ip}}(\tau_2)]}E(t, t_0) \tag{10.17}$$

Now let $t \to +\infty$ and $t_0 \to -\infty$, so that in effect we are studying the functionals $U(+\infty, -\infty) = S[j]$, $E(+\infty, -\infty) = E[j]$, with

$$S[j] = T\{\exp -i\int j(x)\phi(x)d^4x\} \tag{10.18}$$

$$E[j] = : \exp -i\int j(x)\phi(x)d^4x : \tag{10.19}$$

and the integrals extend over all spacetime. Of course, $S[j]$ is just the S-matrix (more precisely, the S-operator whose matrix elements constitute the S-matrix) for the system with interaction Hamiltonian (10.7). Using the normal-ordering result (10.11) relating E to $: E :$, we find

$$S[j] = e^{\frac{1}{2}\int d^4x_1 d^4x_2 j(x_1)j(x_2)\{[\phi^{(-)}(x_1),\phi^{(+)}(x_2)]-\theta(x_1^0-x_2^0)[\phi(x_1),\phi(x_2)]\}}E[j] \tag{10.20}$$

The exponent in (10.20) can be considerably simplified. As it is a c-number (involving only commutators of the free field $\phi(x)$), it is equal to its VEV:

$$\langle 0|\ [\phi^{(-)}(x_1),\phi^{(+)}(x_2)] - \theta(x_1^0 - x_2^0)[\phi(x_1), \phi(x_2)]|0\rangle$$

$$= \langle 0| - \phi^{(+)}(x_2)\phi^{(-)}(x_1) - \theta(x_1^0 - x_2^0)[\phi(x_1), \phi(x_2)]|0\rangle$$

$$= \langle 0| - \phi(x_2)\phi(x_1) - \theta(x_1^0 - x_2^0)[\phi(x_1), \phi(x_2)]|0\rangle$$

$$= \langle 0| - (\theta(x_1^0 - x_2^0) + \theta(x_2^0 - x_1^0))\phi(x_2)\phi(x_1)$$

$$- \theta(x_1^0 - x_2^0)(\phi(x_1)\phi(x_2) - \phi(x_2)\phi(x_1))|0\rangle$$

$$= -\langle 0|\theta(x_1^0 - x_2^0)\phi(x_1)\phi(x_2) + \theta(x_2^0 - x_1^0)\phi(x_2)\phi(x_1)|0\rangle$$

$$= -\langle 0|T(\phi(x_1)\phi(x_2))|0\rangle = -i\Delta_F(x_1, x_2) \tag{10.21}$$

Inserting this result in (10.20) we find the desired final result (Wick's theorem in functional notation)

$$T\{\exp -i \int j(x)\phi(x)d^4x\} =: \exp -i \int j(x)\phi(x)d^4x : e^{-\frac{1}{2}\int j(x_1)\phi\widehat{(x_1)\phi(x_2)}j(x_2)d^4x_1 d^4x_2}$$

(10.22)

where we have introduced the concept of the "contraction of the fields $\phi(x_1)$ and $\phi(x_2)$", written $\widehat{\phi(x_1)\phi(x_2)}$ and defined in this case simply as the Feynman two-point function of the fields in question:

$$\widehat{\phi(x_1)\phi(x_2)} \equiv \langle 0|T(\phi(x_1)\phi(x_2))|0\rangle = i\Delta_F(x_1, x_2)$$

(10.23)

The time-ordered and normal-ordered products of fields are recovered simply by taking the desired number of functional derivatives of (10.22) with respect to the c-number source function $j(x)$. Thus, the explicit expansion

$$T\{\exp -i \int j(x)\phi(x)d^4x\} = \sum_{n=0}^{\infty} \frac{(-i)^n}{n!} \int d^4x_1 d^4x_2 .. d^4x_n j(x_1)..j(x_n) T\{\phi(x_1)..\phi(x_n)\}$$

(10.24)

implies

$$T\{\phi(x_1)\phi(x_2)..\phi(x_n)\} = i^n \frac{\delta^n}{\delta j(x_1)\delta j(x_2)..\delta j(x_n)} T\{\exp -i \int j(x)\phi(x)d^4x\}|_{j=0}$$

(10.25)

and similarly

$$: \phi(x_1)\phi(x_2)..\phi(x_n) := i^n \frac{\delta^n}{\delta j(x_1)\delta j(x_2)..\delta j(x_n)} : \exp -i \int j(x)\phi(x)d^4x : |_{j=0} \quad (10.26)$$

The special case $n = 2$ derived above, (10.5), emerges immediately by taking the second functional derivative of (10.22) with respect to $j(x_1)$, $j(x_2)$ (and setting $j = 0$). In general, the application of functional derivatives to (10.22) clearly results in an expansion giving the T-product of n fields as a sum of terms where normal products of all possible subsets of the fields are multiplied by products of contractions (i.e., Feynman propagators) of the remaining fields. The reader is strongly encouraged to verify this explicitly for the case $n = 4$ (see Problem 2).

The preceding formulation of Wick's theorem connects extremely naturally with the functional (path-integral) formulation of field theory which we shall discuss in Section 10.3 (indeed, we have chosen to derive Wick's theorem in the functional language for precisely this reason), and can be extended to fermionic fields by using the Grassmann algebra technology to be discussed in Section 10.3.2 (Evans *et al.*, 1998). For fermions, the two-point Feynman Green functions are necessarily defined with a minus sign in the anti-time-ordered part (cf. Chapter 7, Problems 5 and 11); for example, for a Dirac field

$$S_F(x_1 - x_2)_{mn} = \langle 0|T(\psi_m(x_1)\bar\psi_n(x_2))|0\rangle \equiv \widehat{\psi_m(x_1)\bar\psi_n}(x_2)$$
$$\equiv \theta(x_1^0 - x_2^0)\langle 0|\psi_m(x_1)\bar\psi_n(x_2)|0\rangle - \theta(x_2^0 - x_1^0)\langle 0|\bar\psi_n(x_2)\psi_m(x_1)|0\rangle$$

while the corresponding two-point contractions for two ψ or two $\bar{\psi}$ fields vanish:

$$\psi_m \overbrace{(x_1)\psi_n}(x_2) = \bar{\psi}_m \overbrace{(x_1)\bar{\psi}_n}(x_2) = 0 \qquad (10.27)$$

The normal-ordered product of two fermionic fields likewise contains additional minus signs when creation or annihilation parts of the fields are interchanged to effect the normal-ordering. With these changes, and introducing[2] c-number sources $\eta(x), \bar{\eta}(x)$ (instead of the commuting $j(x)$ above) which *anticommute* for arbitrary spacetime points x (e.g., $\{\eta(x), \eta(y)\} = 0$, $\forall x, y$), one again recovers the basic result (10.22).[3] The result for time-ordered products of fermionic fields, i.e., once the anticommuting sources are removed by functional differentiation, is as above—namely, an expansion containing all possible subsets of fields under the normal product, multiplied by all possible contractions of the remaining fields—with the proviso that an extra minus sign must be included in each term where fermionic fields on the right-hand side (both in the normal-ordered parts as well as in the contractions) appear in an order which is an odd permutation of the order in which they appear in the time-ordered product on the left (as a result of the difference in sign obtained by reordering the attached sources on the left versus the right).

Another important generalization of Wick's theorem, more transparently obtained in an operatorial proof, states that in the expansion of time-ordered products involving (under the time-ordering symbol) *already normal-ordered* products of fields, contractions of the fields *within* each such normal-ordered product are omitted in the Wick expansion. The interested reader is referred to the standard texts, e.g., (Bjorken and Drell, 1965), for a complete operatorial proof of this extension.

10.2 Feynman graphs and Feynman rules

We return now to our original objective: the derivation of a perturbative expansion for the S-matrix elements as given by the LSZ formula (10.1), where the T-product of Heisenberg fields is expanded perturbatively via the formal expansion (10.2). At p'th order of this expansion, the relevant VEV of interaction-picture fields involves the time-ordered product

$$\langle 0|T\{\phi(x_1')..\phi(x_m')\phi(x_1)..\phi(x_n)\mathcal{H}_{\text{int}}(z_1)...\mathcal{H}_{\text{int}}(z_p)\}|0\rangle \qquad (10.28)$$

The calculation is conveniently divided into two stages: (a) a Wick expansion of the T-product appearing in (10.28), yielding the *coordinate space Feynman rules* of the theory, and (b) application of the Klein–Gordon operators and evaluation of the final (Fourier) integrals over x_i, x_j' in (10.1), yielding the *momentum-space Feynman rules* of the theory.

---

[2] A detailed description of the properties of such fermionic c-number functions is deferred to Section 10.3.2, when we consider fermionic functional integration.

[3] The desired generalization of Wick's theorem, where bosonic and/or fermionic fields appear under the time-ordered product, can also be accomplished by using an inductive operatorial proof. See, for example, Bjorken and Drell, *Relativistic Quantum Fields*, Section 17.4 (Bjorken and Drell, 1965).

For the moment, assume that we are dealing with a ϕ^4 theory (specifically, Theory A of Section 7.6, with $\mathcal{H}_{\text{int}}(x) = \frac{\lambda}{4!}\phi(x)^4$) of self-interacting massive scalar particles.[4] Thus the matrix element in (10.28) contains the T-product of $N = m + n + 4p$ free scalar fields, at spacetime points which we may relabel temporarily $y_1, y_2, ...y_N$ (with some of the y_i repeated four times). Wick's theorem then gives the VEV in (10.28) as a sum of terms of the form

$$\phi\overbrace{(y_1)\phi(y_2)}... \phi\overbrace{(y_{p+1})\phi(y_{p+2})}...\phi\overbrace{(y_{N-1})\phi(y_N)}, \quad N \text{ even}, \quad 0 \text{ otherwise} \qquad (10.29)$$

Here the scalar fields $\phi(y_1), ...\phi(y_N)$ are a permutation of the fields occurring in the particular term of interest extracted from the product

$$\phi(x'_1)..\phi(x'_m)\phi(x_1)..\phi(x_n)\mathcal{H}_{\text{int}}(z_1)..\mathcal{H}_{\text{int}}(z_p)$$

so the spacetime coordinates $y_1, y_2, ..y_N$ are selected from the $x_i, i = 1, 2, ..n, x'_j$, $j = 1, 2, ..m, z_k, k = 1, 2, ..p$. Recall from (10.1) that the spacetime coordinates x_i are associated with the n incoming particles, the coordinates x'_j with the m outgoing particles, and the z_k coordinates with the spacetime points at which interactions occur. Thus, the contractions occurring in (10.29) are of three kinds:

1. Contractions between an *external* spacetime point (i.e., one of the x_i or x'_j) and an interaction point z_k.
2. Contractions between two interaction points z_k.
3. Contractions between two external spacetime points.

We may dispose immediately of the last case, in which external particles are connected directly rather than through the intermediacy of an interaction, as it actually leads to a vanishing disconnected contribution.[5] Taking the external points to be x_1, x'_1 for simplicity, the integrals over x_1, x'_1 factorize from the rest of the expression in (10.1)

$$\int e^{ik'_1 \cdot x'_1 - ik_1 \cdot x_1} \mathcal{K}_{x_1}\mathcal{K}_{x'_1} \phi\overbrace{(x_1)\phi(x'_1)}d^4x_1 d^4x'_1$$

$$= \int e^{ik'_1 \cdot x'_1 - ik_1 \cdot x_1} \mathcal{K}_{x_1}\mathcal{K}_{x'_1} i\Delta_F(x_1 - x'_1)d^4x_1 d^4x'_1$$

$$= i\int e^{ik'_1 \cdot x'_1 - ik_1 \cdot x_1} \mathcal{K}_{x_1}\delta^4(x_1 - x'_1)d^4x_1 d^4x'_1$$

$$= i(k_1^2 - m_{\text{ph}}^2)\int e^{i(k_1-k'_1) \cdot x_1}d^4x_1 \rightarrow 0, \quad k_1^2 \rightarrow m_{\text{ph}}^2 \qquad (10.30)$$

---

[4] We shall see later that a sensible perturbation theory—one in which the amplitudes are expanded in terms of physically accessible low-energy parameters—requires a split of free and interacting Hamiltonians in which quadratic terms, called counterterms, are also transferred from the free to the interaction Hamiltonian. Our theory may also contain ϕ^3 terms, of course. The discussion given here is readily generalized to include the corresponding additional graphs in which two or three field lines connect at a spacetime point. Below, we consider the case of two distinct interacting scalar fields ϕ, ψ (Theory B from Section 7.6) in some detail.

[5] The reader will recall that although disconnected contributions in which a particle passes through without interacting with the others are certainly present in general S-matrix amplitudes, we explicitly removed them in the process of deriving the LSZ formula, so we should not expect to see them re-emerging here!

and vanish once the on-mass-shell limit is taken for the external momentum k_1 (note that we have integrated the Klein–Gordon operator by parts onto the exponential in the step leading to (10.30)).

The first two types of contraction described above form the building blocks of a graphical representation of the perturbative amplitudes of the theory first introduced by Feynman in his seminal work (Feynman, 1949*a*) on quantum electrodynamics in the late 1940s. Contractions between fields at an external point and an interaction point (or *vertex*) are referred to as the *external legs* of the graph. Contractions between two interaction vertices are the *internal lines* of the graph. For a simple illustration, consider Theory B in Section 7.6, with interaction Hamiltonian

$$\mathcal{H}_{\text{int}}(x) = \lambda\psi^\dagger(x)\psi(x)\phi(x) \tag{10.31}$$

with ϕ a self-conjugate, ψ a complex scalar field. In this theory, the only non-zero contractions are $\overbrace{\phi(x_1)\phi(x_2)} = i\Delta_F(x_1 - x_2; M)$ and $\overbrace{\psi(x_1)\psi^\dagger(x_2)} = i\Delta_F(x_1 - x_2; m)$. In Section 7.6.3 we computed the second-order ($n=2$) contribution to $\psi - \psi^c$ scattering in a theory with this interaction Hamiltonian. In this order the relevant LSZ formula contains the T-product $\langle 0|T\{\psi_H^\dagger(x_1')\psi_H(x_2')\psi_H(x_1)\psi_H^\dagger(x_2)\}|0\rangle$, the expansion of which to second order employing the Gell–Mann–Low formula (10.2) contains the following T-product of ten interaction-picture fields:

$$\langle 0|T\{\psi^\dagger(x_1')\psi(x_2')\psi(x_1)\psi^\dagger(x_2)\psi^\dagger(z_1)\psi(z_1)\phi(z_1)\psi^\dagger(z_2)\psi(z_2)\phi(z_2)|0\rangle \tag{10.32}$$

In the application of Wick's theorem to this expression, we recall that only connected contributions, in which any spacetime point can be connected with any other by a continuous sequence of contractions, are phenomenologically relevant. It will soon become clear that these are precisely the contributions which give rise (after the Fourier transformation effected by integrating over the external particle exponential factors) to a *single overall* δ-function of four-momentum conservation (cf. Chapter 6). In the particular case here, there are just four possible connected contractions, which correspond to two topologically distinct graphs, duplicated by the symmetry $z_1 \leftrightarrow z_2$, which simply produces a factor of 2 when the integrals over z_1, z_2 are performed. Thus, the relevant part of the Wick expansion of (10.32) is

$$\{\overbrace{\psi(x_1)\psi^\dagger}(z_1)\psi\overbrace{(z_1)\psi^\dagger}(x_2)\psi\overbrace{(x_2')\psi^\dagger}(z_2)\psi\overbrace{(z_2)\psi^\dagger}(x_1')$$

$$+ \overbrace{\psi(x_1)\psi^\dagger}(z_1)\psi\overbrace{(z_2)\psi^\dagger}(x_2)\psi\overbrace{(x_2')\psi^\dagger}(z_2)\psi\overbrace{(z_1)\psi^\dagger}(x_1')\}\overbrace{\phi(z_1)\phi}(z_2) + (z_1 \leftrightarrow z_2)$$

$$= \{i\Delta_F(x_1 - z_1; m) \cdot i\Delta_F(z_1 - x_2; m) \cdot i\Delta_F(x_2' - z_2; m) \cdot i\Delta_F(z_2 - x_1'; m)$$

$$+ i\Delta_F(x_1 - z_1; m) \cdot i\Delta_F(z_2 - x_2; m) \cdot i\Delta_F(x_2' - z_2; m) \cdot i\Delta_F(z_1 - x_1'; m)\}$$

$$\times i\Delta_F(z_1 - z_2; M) + (z_1 \leftrightarrow z_2)$$

$$\equiv G_{\text{scatt}}(x_1, x_2, x_1', x_2', z_1, z_2) \tag{10.33}$$

The contractions appearing here can be represented graphically as indicated in Fig. 10.1 (note the similarity to Fig. 7.2). At this point the reader may well be wondering about our frequently expressed cautions concerning the inevitability of ill-defined results in calculations performed *ab initio* in the continuum: our T-products of

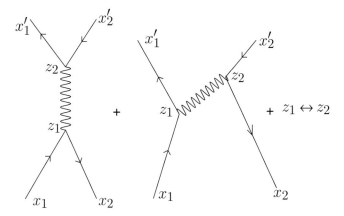

Fig. 10.1 Coordinate space contractions for $\psi - \psi^c$ scattering in second order.

fields appearing in the fundamental LSZ formula (10.1), which need only the further application of some spacetime-derivatives (in the Klein–Gordon operators) and Fourier integrals to yield the desired scattering amplitude, seem by Wick's theorem to dissolve into sums of at first sight perfectly well-defined products of free propagators (i.e., the Feynman functions $\Delta_F(x)$). These propagators are indeed—as distributions—well defined. For example, their Fourier transforms exist, also as well-defined distributions (namely, the familiar $1/(q^2 - m^2 + i\epsilon)$ factors). The problem arises from the fact that *products* of well-defined distributions are not *necessarily* well-defined, although in particular cases they may well be so. Problems with the multiplication of distributions typically arise when the distributions being multiplied have coincident singularities. In the situation here discussed, (10.33), no two Feynman propagators have the same coordinate space arguments, so this situation does not arise. The fourth-order graph shown in Fig. 10.2, on the other hand, evidently contains the square of the propagator

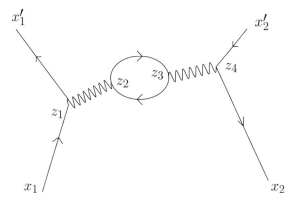

Fig. 10.2 Coordinate space contractions for $\psi - \psi^c$ scattering in fourth order, containing a closed loop.

$$\Delta_F(z_2 - z_3; m)\Delta_F(z_3 - z_2; m) = \Delta_F^2(z_2 - z_3; m) \qquad (10.34)$$

and this object is *not* a well-defined distribution! How do we know this? From the fundamental theorem guaranteeing the existence of the Fourier transform of any decent tempered distribution. If we attempt to compute this Fourier transform for the indicated squared propagator, we find (writing z for $z_2 - z_3$)

$$\Pi(q) \equiv \int \Delta_F^2(z; m)e^{iq \cdot z} d^4 z$$

$$= \int \frac{e^{-ik_1 \cdot z}}{k_1^2 - m^2 + i\epsilon} \frac{e^{-ik_2 \cdot z}}{k_2^2 - m^2 + i\epsilon} \frac{d^4 k_1}{(2\pi)^4} \frac{d^4 k_2}{(2\pi)^4} e^{iq \cdot z} d^4 z$$

$$= \int (2\pi)^4 \delta^4(q - k_1 - k_2) \frac{1}{k_1^2 - m^2 + i\epsilon} \frac{1}{k_2^2 - m^2 + i\epsilon} \frac{d^4 k_1}{(2\pi)^4} \frac{d^4 k_2}{(2\pi)^4}$$

$$= \int \frac{1}{k_1^2 - m^2 + i\epsilon} \frac{1}{(q - k_1)^2 - m^2 + i\epsilon} \frac{d^4 k_1}{(2\pi)^4} = \infty! \qquad (10.35)$$

The infinity arises when we integrate (as, in a continuum theory with no short-distance or high-momentum cutoff, we must) over all of four-dimensional momentum space: we then discover that the integrand behaves, for $k_1 \gg q$, like $\int d^4 k_1/k_1^4$, and is therefore logarithmically divergent. The (infinite) ambiguity is in this case of a very simple form: it is removed by a single subtraction at an arbitrary value of q, and therefore amounts to a single overall additive constant in the definition of the product distribution. For example, if we subtract at $q = 0$, we obtain

$$\Pi(q) - \Pi(0) = \int \frac{2q \cdot k_1 - q^2}{(k_1^2 - m^2 + i\epsilon)^2((q - k_1)^2 - m^2 + i\epsilon)} \frac{d^4 k_1}{(2\pi)^4} \qquad (10.36)$$

which is perfectly finite, as the integrand now behaves like $1/k_1^5$ at large values of k_1. We shall shortly see that the integrated momentum k_1 appears in graphs containing loops as in Fig. 10.2: accordingly, such momenta are usually called "loop momenta". Going back to coordinate space (by an inverse Fourier transform), we see that the (additive) ambiguity in the square of our Feynman propagator amounts to the Fourier transform of an undetermined constant, i.e., to a four-dimensional δ-function $\delta^4(z)$, corresponding as expected to the short-distance limit in which the spacetime vertices z_2 and z_3 coincide. On the other hand, if we work—as we shall henceforth in this chapter assume we are doing—in a regularized version of the theory[6] with a suitably chosen high-momentum cutoff, the divergence in (10.35) is removed and our amplitudes will be well-defined at all stages. Soon, we shall see that ultraviolet divergences arising from ill-defined products of coordinate-space distributions are associated with graphs containing *loops*, leading to unbounded integrations in momentum space.

Returning to our initial task, the evaluation of the nth order perturbative contributions to the S-matrix element in (10.1), we see that the final result is obtained

---

[6] A detailed account of suitable regularizations is postponed to Chapter 16, where we begin the discussion of the sensitivity of field-theory amplitudes to short-distance cutoffs in the theory.

by applying Klein–Gordon operators $\Box + m_{\text{ph}}^2$ (recall: m_{ph} is the *actual physical mass* of the particle) to the outer vertex on each of the external legs (for both incoming and outgoing particles). Equivalently, we may integrate the derivatives by parts onto the plane-wave exponentials, obtaining factors of $m_{\text{ph}}^2 - k_i^2$ (resp. $m_{\text{ph}}^2 - k_i'^2$) for each of the incoming (resp. outgoing) particles in the process. These, together with the indicated phase-space factors involving the square-root of the particle energies, then multiply the Fourier transform with respect to the external momenta of the coordinate space product of contractions (i.e., free propagators) arising from Wick's theorem as discussed above. In our explicit example (10.33), for example, the reader may easily verify (see Problem 3) that the Fourier transform of the indicated contractions yields (with a factor of two due to the $z_1 \leftrightarrow z_2$ symmetry), and integrating also over the interaction vertex points z_1, z_2, as required by (10.2),

$$\int e^{ik_1' \cdot x_1' + ik_2' \cdot x_2' - ik_1 \cdot x_1 - ik_2 \cdot x_2} G_{\text{scatt}}(x_1, x_2, x_1', x_2', z_1, z_2) d^4 x_i d^4 x_i' d^4 z_i$$

$$= 2 \times i\tilde{\Delta}_F(k_1'; m) \cdot i\tilde{\Delta}_F(k_2'; m) \cdot i\tilde{\Delta}_F(k_1; m) \cdot i\tilde{\Delta}_F(k_2; m)$$

$$\times \, (i\tilde{\Delta}_F(k_1 + k_2; M) + i\tilde{\Delta}_F(k_1 - k_1'; M)) \times (2\pi)^4 \delta^4(k_1 + k_2 - k_1' - k_2') \quad (10.37)$$

in terms of the momentum-space Feynman propagators $\tilde{\Delta}_F(k; m) = \frac{1}{k^2 - m^2 + i\epsilon}$. In the process of performing the integrations over the locations of the interactions vertices z_i, four-dimensional δ-functions implementing energy-momentum conservation at each vertex are generated. After integrating over the momenta carried by each propagator, we are left in this case (as the graph is connected: cf. Chapter 6) with a *single* overall δ-function enforcing energy-momentum conservation for the entire process. The products of Feynman propagators appearing in (10.37) have a clear graphical interpretation: the relevant graphs are in fact just those displayed in Fig. 7.2, in our original discussion of this scattering process in Chapter 7. We may summarize the ingredients of the above calculation in a general way valid for the perturbative calculation of arbitrary S-matrix amplitudes of our theory (with interaction Hamiltonian $\mathcal{H}_{\text{int}}(x) = \lambda \psi^\dagger(x)\psi(x)\phi(x)$), as a set of *Feynman rules* associating specific algebraic expressions with each of the graphical elements:

1. At each 4-vertex, a factor $-i\lambda$, and a four-momentum conservation factor which ensures that the sum of incoming momenta to the vertex equals the sum of outgoing momenta: namely, $(2\pi)^4 \delta^4(\Sigma)$ (where Σ is a shorthand notation for the difference in the total incoming and outgoing four-momenta at the vertex).
2. For each line carrying momentum q, a factor $i\tilde{\Delta}_F(q) = \frac{i}{q^2 - m^2 + i\epsilon}$.
3. Integrate over all internal momenta q_i: i.e., those associated with the internal lines of the diagram, connecting two interaction vertices. The external lines corresponding to propagators beginning at one of the external points x_i, x_i' are fixed at the corresponding external particle momenta. In general there will be more internal momenta present than δ-functions available to fix them, leaving some number L of remaining four-momentum integrals. If we think (as a consequence of rule 1) of the four-momentum as a conserved "fluid" flowing through the graph, it is immediately apparent that the number L of such remaining integrals must correspond to the number of independent closed loops around which an arbitrary

amount of four-momentum can flow without vitiating energy-momentum conservation at any interaction vertex on the loop. Evidently, ultraviolet divergences can only arise if there are one or more closed loops ($L \geq 1$) in the graph. Thus, the diagram in Fig. 10.2 has $L = 1$ and is referred to as a "one-loop diagram": as we saw earlier, it has a logarithmic ultraviolet divergence. We shall soon see that the organization of the Feynman graphs of the theory by loop number has a deep physical significance: it amounts to an ordering in increasing powers of Planck's constant $\hbar$. Graphs such as Fig. 10.1, with zero loops, are commonly called "tree diagrams" and correspond in some sense to the classical content of the theory. Later (Section 10.4) we shall see that these tree diagrams encode a formal perturbative solution of the *classical* field equations of the theory.

4. For each topologically distinct graph, apply a combinatoric factor taking into account the number of ways this contribution arises from the Wick contractions in (10.2). Although it is possible to write down an explicit formula for this factor in any specific field theory, in the author's experience it is generally easier to retreat to an undisclosed location and then examine the Wick contractions to identify the necessary factor.

5. For each external line corresponding to a particle of momentum k, a factor $\frac{-i}{Z^{1/2}}(k_i^2 - m_{\text{ph}}^2)\frac{1}{(2\pi)^{3/2}\sqrt{2E(k_i)}}$. When the momentum k_i is placed on-shell (i.e., k_i^2 is set equal to m_{ph}^2), the factor $(k_i^2 - m_{\text{ph}}^2)$ vanishes, so the remaining part of the amplitude must provide a simple pole (i.e., be proportional to $1/(k_i^2 - m_{\text{ph}}^2)$) in the on-shell limit for our perturbative expansion of the LSZ formula to yield sensible results. How this can be ensured will be addressed shortly. The normalization constant Z is extracted, as per our discussion of the Lehmann spectral representation in Section 9.5, from the (perturbatively computed) residue of the single-particle pole of the full Feynman propagator (i.e., the Fourier transform of the two-point Feynman Green function of the Heisenberg field).

The extension of these rules to other field theories, containing fields of non-zero spin (e.g., Dirac or vector fields), and other types of (polynomial) self-interactions, is straightforward. Internal lines for a Dirac particle connecting vertices at y_1 and y_2 become propagators $iS_F(y_1 - y_2)$, and so on. There is one additional rule, perhaps not immediately obvious from the preceding, concerning the contraction of fermionic fields in a closed loop: in such cases, an additional minus sign must be inserted in the amplitude, as a consequence of fermionic statistics. Consider, for example, Theory C of Section 7.6, with interaction Hamiltonian

$$\mathcal{H}_{\text{int}}^{(C)}(x) = \lambda \bar{\psi}_a(x)\psi_a(x)\phi(x) \tag{10.38}$$

where $\psi(x)$ is a massive Dirac field (with the Dirac index here denoted $a = 1,2,3,4$) and $\phi(x)$ a self-conjugate scalar field. Among the various Wick contractions of a second-order contribution to the S-matrix expansion (10.1) with this interaction will occur terms corresponding to the graph illustrated in Fig. 10.2, where now the bold lines represent fermions and the wiggly lines the scalar particle. In this case we have fermion fields at two separate vertices (at spacetime points z_2 and z_3) fully contracted so that the two Dirac propagators form a closed loop:

$$T(\bar{\psi}_{a_1}(z_2)\psi_{a_1}(z_2)\bar{\psi}_{a_2}(z_3)\psi_{a_2}(z_3).....) = -\psi_{a_1}\overbrace{(z_2)\bar{\psi}}_{a_2}(z_3)\psi_{a_2}\overbrace{(z_3)\bar{\psi}}_{a_1}(z_2) \quad (10.39)$$

$$= -\mathrm{Tr}\{iS_F(z_2 - z_3) \cdot iS_F(z_3 - z_2)\} \quad (10.40)$$

The minus sign arises because the reordering of the fields between the left and right-hand side of (10.39) involves an odd permutation, and hence, by the fermionic extension of Wick's theorem discussed in the previous Section, an additional minus sign. In graphical calculations this requirement is summarized in the simple rule: *closed fermion loops get an extra minus sign.*[7]

There are several non-trivial issues which arise in interpreting the results of a perturbative calculation of S-matrix amplitudes along the lines discussed above, some of which have already been touched on above. Basically, obstructions to obtaining well-defined results at each given order of perturbation theory may arise both in the ultraviolet (short distance) or infrared (long time) domains. Even then, when a well-defined regularized result has been obtained order by order, it must be kept in mind that the resultant (infinite) series is never a convergent Taylor expansion, but only, at best, an asymptotic expansion of the exact S-matrix amplitude. Let us examine these three sets of problems in more detail.

We have already touched on the difficulties at short distance—the famous "ultraviolet divergences" of perturbative quantum field theory—which go back to the ambiguities and/or divergences which arise when distributions with coincident spacetime singularities are multiplied. These divergences are most easily seen in momentum space, in the Fourier transforms of the products of coordinate space Green functions, as ultraviolet divergences of loop integrals, such as (10.35), due to insufficiently rapid falloff of the product of momentum-space propagators at large momentum. In order to give meaning to the amplitudes we must introduce some appropriate *regularization* of the loop integrals appearing in the amplitudes: namely, a large momentum (or short distance) cutoff rendering all amplitudes finite and well defined in all orders of perturbation theory. Here, "appropriate" means a regularization doing the least violence manageable to the important underlying symmetries (such as Poincaré invariance) of the theory, in such a way that eventually removal of the cutoff can be accomplished with the restoration of these symmetries. Equivalently, we need to show that the sensitivity of the Feynman amplitudes of the theory to the presence of the short-distance cutoff, which can be viewed as an expression of our ignorance of as yet unexplored new physics at very short distance scales, is for all practical purposes negligible. These issues of scale sensitivity of field theory—and the whole machinery of renormalization theory needed to address them adequately—will be the primary topic of discussion in Part 4 of this book. For the time being we will simply assume, when deriving or using perturbative results in the interaction picture, that the field

---

[7] Of course, exactly as discussed previously for the scalar theory of Fig. 10.2, the multiplication of two Dirac propagators with the same singularity (at $z_2 \to z_3$) results in an undefined result, visible as a divergence in the Fourier transform due to UV contributions to the corresponding loop integral. In this case the divergence is even more severe, resulting in the appearance of *two* arbitrary constants. More on all this in Chapters 16 and 17.

theory has been suitably regularized and that only well-defined expressions occur in the evaluation of the Gell–Mann–Low expansion (10.2).

Another set of problems for our perturbative analysis arise from the large-distance (or large-time) regime: specifically from the presence in a relativistic field theory of *persistent interactions* which continue to affect the propagation of particles even when (long before or long after a scattering interaction) they are well-separated from one another and moving freely. Firstly, persistent interactions are present even in the absence of scattering particles, i.e., in the evolution of the vacuum state. Such "disconnected vacuum bubbles" appear with equal probability at all spacetime points and lead to a spacetime volume dependent phase in the overlap of the in- and out-vacua of the theory. Secondly, persistent interactions can occur as "radiative corrections" on any of the incoming or outgoing particle lines in a scattering process. In Fig. 10.3 we display a Feynman diagram in $\lambda\phi^4$ theory in which both types of persistent interactions are present. We shall discuss the proper treatment of the external line radiative corrections below. The role of the vacuum fluctuations leads to one of the most interesting issues in modern cosmology: the cosmological constant, and its "abnormally low" value (the so-called "cosmological constant problem"). The fact that the energy of the discrete vacuum state is shifted by interactions is a perfectly normal situation in any quantum-mechanical theory. The actual level of the ground-state energy is not physically relevant in flat-space quantum field theory as it is an unobservable quantity, but in the presence of gravity the absolute level of energies (and energy-densities: specifically, the components of the energy-momentum tensor) is clearly relevant. Staying with the flat-space theories which are our focus in this book, it is easy to see that the net effect of the vacuum bubbles, defined as the set of graphs arising from the $m = 0$ special case of the Gell–Mann–Low theorem (10.2),

$$\text{out}\langle 0|0\rangle_{\text{in}} = \sum_{p=0}^{\infty} \frac{(-i)^p}{p!} \int \langle 0|T\{\mathcal{H}_{\text{int}}(z_1)\mathcal{H}_{\text{int}}(z_2)..\mathcal{H}_{\text{int}}(z_p)\}|0\rangle d^4z_1 d^4 z_2..d^4 z_p \quad (10.41)$$

is to introduce an overlap phase factor (see Problem 4 for an explicit example)

$$\text{out}\langle 0|0\rangle_{\text{in}} = \langle 0|U(+T/2, -T/2)|0\rangle = e^{-iTV\delta\mathcal{E}} \quad (10.42)$$

which is singular in the limit where spatial volume V and temporal extent T go to infinity. Here $\delta\mathcal{E}$ is the interaction-induced shift in the vacuum spatial energy-density

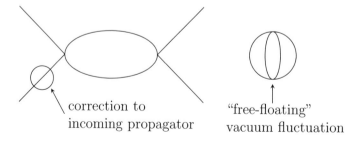

Fig. 10.3 A disconnected Feynman diagram in $\lambda\phi^4$ theory displaying persistent interactions.

(which in fact typically contains *ultraviolet* divergences). The presence of the spacetime volume factor VT is to be expected, as the disconnected vacuum bubbles, such as that shown in Fig. 10.3, are free to float over the entire spacetime volume, so the evaluation of the Feynman diagram necessarily produces a factor of VT. In fact, this divergence is the *only* singularity of an infrared nature in a theory of massive particles, and may be removed by the simple device of taking only connected contributions to S-matrix elements, which automatically dispenses with the noisome vacuum fluctuations. Effectively, this amounts to using the same vacuum state (either in or out) on both sides of (10.1), or equivalently, to dividing a general S-matrix element $_{\text{out}}\langle\beta|\alpha\rangle_{\text{in}}$ (as given by the LSZ formula, say) by the vacuum phase $_{\text{out}}\langle 0|0\rangle_{\text{in}}$.

Returning now to the second class of persistent interactions, those giving rise to radiative corrections on the external legs of the diagrams, we recall from the discussion of scattering theory in Section 9.4 that the generation of the appropriate in- and out-states whose overlap gives the desired S-matrix element is formally accomplished by taking the on-(physical)mass-shell limit for the four-momentum associated with each external particle, $k_i^2 \to m_{\text{ph}}^2$. In addition, the momentum-space $m + n$ point Green function (where m, n are the number of outgoing or incoming particles, respectively) appearing in the LSZ formula (cf. (10.1) must be multiplied by factors $k_i^2 - m_{\text{ph}}^2$ for each external particle, so that the on-mass-shell limit gives a well-defined finite result if and only if this Green function has *simple poles* in each of the off-shellness variables $k_i^2 - m_{\text{ph}}^2$, with the final S-matrix amplitude given as the residue of these poles. A careless execution of perturbation theory will result in amplitudes which, at any fixed order of perturbation theory, fail to have the required pole behavior! In particular, taking ϕ^4 scalar theory as an explicit example, if we begin with the Hamiltonian

$$H = \int d^3x : \frac{1}{2}\dot{\phi}^2 + \frac{1}{2}|\vec{\nabla}\phi|^2 + \frac{1}{2}m^2\phi^2 + \frac{\lambda}{4!}\phi^4 : \tag{10.43}$$

and define an interaction picture via the "obvious" separation $H = H_0 + V$ with

$$H_0 = \int d^3x : \frac{1}{2}\dot{\phi}^2 + \frac{1}{2}|\vec{\nabla}\phi|^2 + \frac{1}{2}m^2\phi^2 : \quad , \quad V = \int d^3x : \frac{\lambda}{4!}\phi^4 : \tag{10.44}$$

the free momentum-space Feynman propagator $\tilde{\Delta}_F(k)$ will clearly have a pole at $k^2 = m^2$, where m is the so-called "bare mass", corresponding to the coefficient of ϕ^2 in the Hamiltonian, but *not* to the actual physical mass m_{ph}, which differs from m as a consequence of the interaction term. This is hardly unexpected: the energy of states in quantum mechanics (in this case, the single particle at rest) is typically shifted from the unperturbed value once interactions are switched on. We can restore mass stability in our perturbation theory by splitting the full Hamiltonian in a different way, thereby *forcing* the free part H_0 to yield a single particle state with the correct physical mass, by defining

$$m^2 = m_{\text{ph}}^2 + \delta m^2 \tag{10.45}$$

and transferring the $\delta m^2 \phi^2$ "mass counterterm" into the interaction part of the Hamiltonian, so that we now have

$$H_0 = \int d^3x : \frac{1}{2}\dot{\phi}^2 + \frac{1}{2}|\vec{\nabla}\phi|^2 + \frac{1}{2}m_{\text{ph}}^2\phi^2 : \tag{10.46}$$

$$V = \int d^3x : \frac{1}{2}\delta m^2\phi^2 + \frac{\lambda}{4!}\phi^4 : \tag{10.47}$$

with the coefficient δm^2 adjusted order by order in the expansion in the "bare coupling" λ to ensure that the pole of the full Heisenberg propagator remains at the required physical value (namely, at $k^2 = m_{\text{ph}}^2$).[8] The resultant perturbative amplitudes are said to be "on-shell renormalized". It is also possible (and frequently convenient) to employ a more general class of perturbative splits, in which the poles of the free propagator are *not* at the physical mass, but at some "intermediate mass" (coinciding neither with the bare mass in the Hamiltonian nor the physical particle mass): this requires the reorganization of the full Green functions into a factorized product of full propagators on the external legs (with poles at the correct physical mass, by the Lehmann representation of Section 9.5) and "amputated" Green functions which can be computed perturbatively with intermediate mass renormalization. How this can be accomplished will be described below in Section 10.4, where the relevant topological concepts are introduced.

Finally, there remains the "inconvenient truth" that a perturbative expansion of field-theoretic amplitudes as a formal expansion in some suitably defined coupling constant(s) of the theory at best provides an asymptotic expansion to the exact amplitude, not a convergent (Taylor) expansion capable of yielding results of arbitrary accuracy (at least in principle) by pushing the calculation to sufficiently high orders. This should not be surprising if we recall that the same is true even in very simple (non-field-theoretic) models in non-relativistic quantum theory. The discrete energy eigenvalues $E_n(\lambda)$ of the one-dimensional anharmonic oscillator Hamiltonian,

$$H = \frac{p^2}{2m} + \frac{1}{2}m\omega^2 x^2 + \lambda x^4 \tag{10.48}$$

when developed in an expansion in powers of the anharmonicity λ, produce an asymptotic expansion $E_n(\lambda) \simeq \sum_p C_p^{(n)}\lambda^p$ in which the coefficients $C_p^{(n)}$ grow factorially with p, so that the series in fact has zero radius of convergence. The lack of analyticity in the λ variable is hardly surprising, if we consider the dramatically different behavior of the theory if λ is taken negative (real), however small: the Hamiltonian then lacks a ground state, as the λx^4 part of the potential energy eventually becomes arbitrarily negative for large enough x. The Rayleigh–Schrödinger perturbation expansion is still useful of course, treated as an asymptotic expansion, provided λ is sufficiently small, allowing an accurate estimate of the energy to be obtained by summing the first few terms of the series (before the summands begin to increase). In certain cases (the

---

[8] We remind the reader that we are assuming a fully regularized theory: in particular, all UV divergences have been taken care of, so the perturbative expansion of the Green functions yields finite well-defined results everywhere.

so-called "Borel summable" expansions; cf. Section 11.3) the information contained in the perturbative coefficients in fact suffices to determine the exact amplitudes, but this is not the case in most field theories of interest (in particular, in the gauge theories which underly the Standard Model of particle physics), and we must face the fact that such theories contain qualitatively important "non-perturbative physics", going beyond any information which can be gleaned from a purely perturbative approach. We will return to the issue of the non-convergence of perturbation theory, and the extent to which "truly non-perturbative" information can be accessed by alternative methods, in Chapter 11.

10.3   Path-integral formulation of field theory

Although it is perfectly possible to derive essentially all important field-theoretic results in an operator-based formalism, there are many instances in which a functional formalism yields equivalent results with a fraction of the effort required by an operatorial calculation. Moreover, a precise specification of the interacting dynamics of a quantum field theory, valid beyond the formal expansions of perturbation theory, is typically achieved only by formulating the theory on a finite spacetime lattice, at which point the functional integral formalism provides an unambiguous specification of the amplitudes of the theory, even in situations in which perturbation theory is useless. The only successful "first principles" calculations of the spectrum of a strongly interacting theory (in particular, of quantum chromodynamics (QCD), the field theory of strongly interacting particles) are those which have been done in this way, by stochastic estimates of the latticized path integral of the theory. In this section we shall build up the basic elements of the path-integral formalism for local quantum field theory, with our discussion of path integrals for quantum mechanical point particles in Section 4.2 serving as a natural starting point.

10.3.1   Path integrals for bosonic fields

In our review of basic quantum mechanical formalism in Chapter 4 (Section 4.2), we saw that ground-state expectation values of (time-ordered) Heisenberg representation coordinate operators describing a non-relativistic particle could be re-expressed in a functional, or "path-integral", framework. For example, we found (cf. (4.134, 4.135)) a functional integral expression for a generating functional $Z[j]$ whose functional derivatives yield the Feynman Green functions (i.e., ground-state expectation value of time-ordered Heisenberg operators) of the theory:

$$Z[j] \equiv \int \mathbf{D}p\mathbf{D}q \; e^{i \int_{-\infty}^{+\infty} (p(t)\dot{q}(t) - H(p(t),q(t)) - j(t)q(t))dt} \tag{10.49}$$

The analogous quantity for a free field theory, a generating functional for the vacuum expectation value of time-ordered free field operators, has already been introduced and discussed in Section 10.1 from an operatorial standpoint. Thus (introducing the subscript 0 to indicate a free field theory, and taking a free massive scalar field to be specific) we expect that the functional

$$Z_0[j] \equiv \langle 0|S[j]|0\rangle = \langle 0|T\{\exp -i \int j(x)\phi(x)d^4x\}|0\rangle \qquad (10.50)$$

plays a role analogous to the $Z[j]$ in (10.49). We emphasize at this point that an overall multiplicative constant in $Z_0[j]$ is irrelevant when the normalized n-point functions (compare (4.135)) are computed, as these involve derivatives of Z_0 *divided by* Z_0. We seek a functional integral representation for this object analogous to (10.49): such a representation clearly involves canonical "momenta" complementary to the coordinate degrees of freedom, which in this case are evidently the field operators at any spatial point (on a given time-slice). The appropriate complementary quantities were discussed in our treatment of the classical limit in Section 8.1: for our simple scalar theory, the field operator and its time-derivative form quantum-mechanically conjugate variables:

$$[\pi(\vec{x}, t), \phi(\vec{y}, t)] = -i\delta^3(\vec{x} - \vec{y}), \quad \pi(\vec{x}, t) \equiv \frac{\partial \phi(\vec{x}, t)}{\partial t} \qquad (10.51)$$

We shall proceed by formally imitating the representation (10.49), with the appropriate modifications for field theory, and then checking explicitly to ensure that the resultant functional indeed yields the desired Feynman functions of the free field theory. The generalization of the path-integral formula to include interactions will then be straightforward.

Essentially, the only modification needed to convert (10.49) into a field theory formula is to include a spatial integral, as our "coordinates" and "momenta" are now the values of the field $\phi(\vec{x}, t)$ and its time-derivative $\pi(\vec{x}, t)$ at all spatial points (at any given time). This immediately yields the path integral

$$Z_0[j] = \int \mathbf{D}\phi \mathbf{D}\pi e^{i \int (\pi(x)\dot{\phi}(x) - \mathcal{H}_0(\pi(x),\phi(x)) - j(x)\phi(x))d^4x} \qquad (10.52)$$

where, as expected, the time integrals of (10.49) have been augmented to spacetime integrals. As discussed in Section 4.2.4, expressions of this kind are purely formal in nature—they must be given a precise meaning by the following maneuvers:

1. An appropriate regularization of the spacetime continuum, so that we are dealing with a multiple integral over a well-defined discrete set of (many!) integration variables. An obvious way to do this is simply to imagine defining the theory on a finite discrete spacetime lattice: in other words, in addition to discretizing the time variable as was done in Section 4.2.1, we also replace the spatial integrals by sums over a finite spatial lattice.[9] Space and time-derivatives are, of course, replaced by appropriate difference quantities. The need for a short-distance cutoff to obtain well-defined results is, of course, familiar by now from our discussion of the difficulties that can arise from multiplying distributions in perturbation theory- here we imagine also a long-distance cutoff (our spacetime lattice is of finite extent) so that the functional integral becomes a multi- but finite-dimensional one. At an appropriate point, once the path integrals have been

---

[9] We have previously referred to a field theory regularized in this way as a *fully regularized* theory.

performed, the continuum (i.e., zero lattice spacing) limit can be taken to obtain the desired continuum results, and the spatial volume can then be allowed to go to infinity.

2. Even after the functional integral is converted by the aforesaid discretization into a conventional multi-dimensional integral, a further regularization, of a different kind, is required to obtain well-defined results. Again, as discussed in Chapter 4, the integral (10.52) as it stands is not absolutely convergent, as the integrand involves a complex undamped exponential. Just as in the quantum-mechanical case, we shall see that the inclusion of an appropriate $i\epsilon$ factor is needed to ensure absolute convergence of the integral in all directions in field space.

Does the formula (10.52) reproduce the correct Green functions, even for our very simple case of a free (massive) scalar field? The Hamiltonian density in this case is given by (6.89)[10]

$$\mathcal{H}_0 = \frac{1}{2}\pi(x)^2 + \frac{1}{2}|\vec{\nabla}\phi(x)|^2 + \frac{1}{2}m^2\phi(x)^2 \tag{10.53}$$

We shall assume that spacetime has been appropriately discretized and gradients and time-derivatives realized in such a way that the shift-invariance of the path integral is preserved. To avoid unnecessarily complicating the notation, however, we shall continue to use continuum notation, and to write the functional integral measure as $\mathbf{D}\phi$ (rather than $\prod_i d\phi(x_i)$, for example, where x_i indicates the discretized spacetime lattice points). By shift-invariance, we mean simply

$$\int \mathbf{D}\phi F[\phi + \chi] = \int \mathbf{D}\phi F[\phi] \tag{10.54}$$

$$\int \mathbf{D}\pi F[\pi + \chi] = \int \mathbf{D}\pi F[\pi] \tag{10.55}$$

for any fixed (c-number) function $\chi(x)$, and functional F, assuming that the integral exists (is absolutely convergent) in the first place. Inserting (10.53) into (10.52), we find that the dependence of the integrand on the momentum field $\pi(x)$ is Gaussian:

$$e^{i\int(\pi(x)\dot{\phi}(x)-\frac{1}{2}\pi(x)^2)d^4x} = e^{i\int\frac{1}{2}\dot{\phi}(x)^2d^4x} \cdot e^{-\frac{i}{2}\int(\pi(x)-\dot{\phi}(x))^2d^4x} \tag{10.56}$$

If we perform the functional integral over the $\pi(x)$ field first, holding the $\phi(x)$ field fixed, then the shift invariance property assumed for the $\mathbf{D}\pi$ integral implies (taking $\chi = -\dot{\phi}$ in (10.55)) that the latter decouples completely as a multiplicative factor $C \equiv \int \mathbf{D}\pi e^{-\frac{i}{2}\int\pi(x)^2d^4x}$, leaving only the integral over the "coordinate" quantities, i.e., the field $\phi(x)$:

---

[10] The normal-ordering, here omitted, amounts to a shift of the Hamiltonian by a fixed c-number: this affects the path integral by a multiplicative factor, which evidently cancels in formulas like (4.135) giving the n-point Green functions of the theory.

$$Z_0[j] = C \int \mathbf{D}\phi e^{i \int (\frac{1}{2}\dot\phi(x)^2 - \frac{1}{2}|\vec\nabla\phi(x)|^2 - \frac{1}{2}m^2\phi(x)^2 - j(x)\phi(x))d^4x} \tag{10.57}$$

$$\equiv C \int \mathbf{D}\phi e^{i \int (\mathcal{L}_0(\dot\phi, \vec\nabla\phi, \phi) - j\phi)d^4x}, \quad \mathcal{L}_0 = \frac{1}{2}\partial_\mu\phi\partial^\mu\phi - \frac{1}{2}m^2\phi^2 \tag{10.58}$$

The result of integrating out the field momentum variables is a path integral only over the field coordinate variables $\phi(x)$, with the integrand containing a function (the "free field Lagrangian") $\mathcal{L}_0(\dot\phi, \vec\nabla\phi, \phi)$ only of the field and its space and time-derivatives, in complete analogy to (4.105) where the mechanical Lagrangian of the quantized point particle makes a similar appearance. The appearance of the Lorentz-invariant scalar combination $\partial_\mu\phi(x)\partial^\mu\phi(x) = \dot\phi(x)^2 - |\vec\nabla\phi(x)|^2$ at this stage is certainly an encouraging development, and we shall see later in Chapter 12, when we develop the canonical formalism for field theory, that it is precisely the Poincaré invariant properties of the action (defined as the spacetime integral of the Lagrangian) that render the Lagrangian such a useful, even indispensable, tool in field theory.

Of course, we still have to face the fact that the integral (10.57) contains an oscillating integrand of absolute value unity, and is therefore certainly not absolutely convergent (much less uniquely specified). The integral can be given a definite meaning by introducing an appropriately signed small imaginary part in front of every term (quadratic in the field) in the Lagrangian which becomes unbounded in the course of the integration. Thus, we replace the real Lagrangian function appearing in (10.57) by

$$\mathcal{L}_{0\delta}(\dot\phi, \vec\nabla\phi, \phi) = \frac{1}{2}e^{i\delta}\dot\phi^2 - \frac{1}{2}e^{-i\delta}(|\vec\nabla\phi(x)|^2 + m^2\phi^2) \tag{10.59}$$

where δ is a small positive quantity to be taken to zero at the end of the calculations. This corresponds, as the reader may easily verify, to just the "rotation" of the time variable $t \to e^{-i\delta}t$ needed to obtain a well-defined real-time functional integral (cf. Section 4.2.1). The real part of $i\mathcal{L}_{0\delta}$ is readily seen to be negative definite, so the integrand is exponentially damped[11] in all regions where either the field or its spacetime-derivatives become large. We can now explicitly verify that the resultant path integral is well-defined by *evaluating it*, using once again the shift property (10.54). Define the differential operator

$$\mathcal{K} \equiv e^{i\delta}\frac{\partial^2}{\partial t^2} - e^{-i\delta}(\vec\nabla^2 - m^2) \tag{10.60}$$

The inverse of this operator (or kernel) is a Green function $G(x, y)$ defined by

$$\mathcal{K}G(x, y) = \delta^4(x - y) \tag{10.61}$$

---

[11] Of course, we should include a similar convergence factor in the integral over $\pi(x)$ performed previously, which should strictly speaking be taken to be the absolutely convergent integral $\int \mathbf{D}\pi e^{-\frac{i}{2}e^{-i\delta}\int \pi(x)^2 d^4x}$. See the discussion below for the interacting case, where all appropriate convergence factors are inserted *ab initio*.

where the operator $\mathcal{K}$ acts on the spacetime coordinate x. Writing $G(x, y)$ as a Fourier transform

$$G(x, y) = \int \tilde{G}(k) e^{-ik \cdot (x-y)} \frac{d^4 k}{(2\pi)^4} \tag{10.62}$$

and substituting this and (10.60) in (10.61), one readily finds

$$\tilde{G}(k) = -\frac{1}{e^{i\delta} k_0^2 - e^{-i\delta} (\vec{k}^2 + m^2)}$$

$$= -e^{-i\delta} \frac{1}{k_0^2 - e^{-2i\delta} (\vec{k}^2 + m^2)}$$

$$\rightarrow -\frac{1}{k^2 - m^2 + i\epsilon (\vec{k}^2 + m^2)}, \quad \epsilon \equiv 2\delta, \ \delta \rightarrow 0 \tag{10.63}$$

If the limit $\epsilon \rightarrow 0$ is taken after all integrals are performed, the (positive!) non-covariant factor $\vec{k}^2 + m^2$ is irrelevant and we obtain for $G(x, y)$ a well-defined covariant distribution.[12] The Green function $G(x, y)$, or equivalently, the *inverse* of the operator $\mathcal{K}$, is therefore, up to a sign, just our old friend the free Feynman propagator:

$$G(x, y) = -\int \frac{e^{-ik \cdot (x-y)}}{k^2 - m^2 + i\epsilon} \frac{d^4 k}{(2\pi)^4} = -\Delta_F (x - y) \tag{10.64}$$

Note that the presence of a slightly non-zero δ (hence ϵ) is essential so that the poles of the integrand at $k_0 = \pm\sqrt{\vec{k}^2 + m^2}$ be avoided (by displacement to $\pm\sqrt{\vec{k}^2 + m^2} \mp i\epsilon$), *once we remove the regulators and return to an infinite continuous spacetime*, as our use of continuum notation suggests we have done. Were δ zero, the energy integral (over k_0) in the continuum theory would run directly along the real axis through these poles, and the result of the integration would be ill-defined. We shall now see that the evaluation of the path integral (10.57) involves exactly the Green function $G(x, y)$, so that the original lack of definition in the functional integral can be traced precisely to the need to generate a well-defined distribution for the two-point function of the field. To complete the evaluation of (10.57), we observe that, using integration by parts,

$$\int (\mathcal{L}_\delta(\dot{\phi}, \vec{\nabla}\phi, \phi) - j\phi) d^4 x = -\int (\frac{1}{2}\phi(x)\mathcal{K}\phi(x) + j(x)\phi(x)) d^4 x \tag{10.65}$$

$$= -\int \{\frac{1}{2}(\phi(x) + \mathcal{K}^{-1}j(x))\mathcal{K}(\phi(x) + \mathcal{K}^{-1}j(x))$$

$$-\frac{1}{2}j(x)\mathcal{K}^{-1}j(x)\} d^4 x \tag{10.66}$$

---

[12] We shall see, however, in Chapter 17, that the momentum dependence of the ϵ term is crucial in guaranteeing the absolute convergence of Minkowski-space Feynman integrals, needed to establish rigorously the efficacy of the subtraction procedures used to renormalize the Minkowski-space amplitudes of a perturbatively renormalizable theory.

The integral over ϕ in (10.58) can now be shifted using (10.54) to eliminate the dependence on the source function j, giving a (convergent!) constant factor

$$C' = \int \mathbf{D}\phi e^{-i\int \phi(x)\mathcal{K}\phi(x)d^4x} \tag{10.67}$$

multiplying a term Gaussian in the source function $j(x)$:

$$Z_0[j] = CC'e^{\frac{i}{2}\int j(x)\mathcal{K}^{-1}j(x)d^4x} = CC'e^{-\frac{i}{2}\int j(x)\Delta_F(x-y)j(y)d^4x d^4y} \tag{10.68}$$

The reader will easily confirm that this result is precisely Wick's theorem for the Feynman n-point Green functions in (10.22, 10.23), previously derived by operatorial methods (note that the normal-ordered exponential on the right-hand side of (10.22) simply becomes unity once the vacuum expectation value is taken: once expanded, only the first term in the expansion survives). As indicated above, the overall multiplicative constant CC' is irrelevant once the functional derivatives needed to extract the n-point Green function are taken,

$$\langle 0|T\{\phi(x_1)\phi(x_2)..\phi(x_n)\}|0\rangle = i^n \frac{1}{Z_0[j]} \frac{\delta^n Z_0[j]}{\delta j(x_1)\delta j(x_2)..\delta j(x_n)}\bigg|_{j=0} \tag{10.69}$$

as the generating functional $Z_0[j]$ appears in both the numerator and the denominator. We may therefore take simply, for the generating functional of Green functions of our free scalar field theory,

$$Z_0[j] = e^{-\frac{i}{2}\int j(x)\Delta_F(x-y)j(y)d^4x d^4y} \tag{10.70}$$

We note that the removal of the implicit lattice-regularization of spacetime is completely unproblematic in this free field theory: we simply reinterpret the above integrals as fully continuous, confident that the resultant distribution $\Delta_F(x)$ (and its Fourier transform) are perfectly well-defined as a consequence of the "epsilonic" regularization of the path integral. The removal of the spacetime regularization of the theory will turn out to be a far more intricate matter once interactions are included—not surprisingly, given the discussion in Section 10.2 of the distributional singularities with which perturbative expansions in a non-Gaussian (higher than quadratic) interaction term are infected. A full resolution of this problem is the central subject of renormalization theory, and will be dealt with in Part 4 of the book.

At least formally, the inclusion of interactions in the path-integral formalism is perfectly straightforward: one simply replaces the free interaction Hamiltonian density $\mathcal{H}_0$ in (10.52) with the full Hamiltonian density $\mathcal{H}$. For example, in a $\lambda\phi^4$ theory we would have

$$Z[j] = \int \mathbf{D}\phi \mathbf{D}\pi e^{i\int (\pi(x)\dot{\phi}(x) - \mathcal{H}_\delta(\pi(x),\phi(x)) - j(x)\phi(x))d^4x} \tag{10.71}$$

$$\mathcal{H}_\delta = e^{-i\delta}\{\frac{1}{2}\pi(x)^2 + \frac{1}{2}|\vec{\nabla}\phi(x)|^2 + \frac{1}{2}m(a)^2\phi(x)^2 + \frac{1}{4!}\lambda(a)\phi(x)^4\} \tag{10.72}$$

with the full Feynman Green functions of the interacting theory given by functional derivatives of the functional $Z[j]$:

$$\langle 0|T\{\phi_H(x_1)\phi_H(x_2)..\phi_H(x_n)\}|0\rangle = i^n \frac{1}{Z[j]} \frac{\delta^n Z[j]}{\delta j(x_1)\delta j(x_2)..\delta j(x_n)}\bigg|_{j=0} \qquad (10.73)$$

The regularization needed to produce a well-defined path integral is indicated implicitly by the presence of the spacetime lattice spacing a in the dependence of the bare mass $m(a)$ and bare coupling $\lambda(a)$ on the short-distance cutoff: unlike the case for the free theory, the continuum limit of an interacting field theory requires that these coefficients be given the appropriate dependence on a in order that the low-energy amplitudes of the theory approach well-defined limits (cf. Chapter 16). How to do this will be the subject of Chapters 16 and 17: for the time being, we suppose that all functional integrals are defined on a bounded spacetime lattice (although for notational simplicity we continue to use a continuous notation $\int d^4x...$ for the implicitly discrete sums over a spacetime lattice), and therefore amount to finitely multi-dimensional normal (i.e., Stieltjes) integrals. The further regularization needed to produce an *absolutely convergent* integral is effected by the introduction of the overall complex factor $e^{-i\delta}$ multiplying the Hamiltonian density. The integral over the momentum field $\pi(x)$ can now be performed exactly as previously in the free case (it is absolutely convergent for δ small and positive), yielding

$$Z[j] = C \int \mathbf{D}\phi\, e^{i \int (\frac{1}{2}e^{i\delta}\dot\phi(x)^2 - \frac{1}{2}e^{-i\delta}|\vec\nabla\phi(x)|^2 - \frac{1}{2}e^{-i\delta}m(a)^2\phi(x)^2 - \frac{\lambda(a)}{4!}e^{-i\delta}\phi(x)^4 - j(x)\phi(x))d^4x}$$

$$\equiv C \int \mathbf{D}\phi\, e^{i \int (\mathcal{L}_\delta(\dot\phi,\vec\nabla\phi,\phi) - j\phi)d^4x} \qquad (10.74)$$

Note that the factors of $e^{+i\delta}$ and $e^{-i\delta}$ appear in precisely the right places to guarantee the absolute convergence of the remaining integral over $\phi(x)$: not surprisingly, as the original integral, before the $\pi(x)$ field was integrated out, was clearly absolutely convergent (with the absolute value of the integrand falling exponentially in the large field regime).

If we temporarily restore Planck's constant in the Lagrangian form of the path integral (it divides the action: namely, the spacetime integral of the Lagrangian),

$$Z[j] = C \int \mathbf{D}\phi\, e^{i \int (\frac{1}{\hbar}\mathcal{L}_\delta(\dot\phi,\vec\nabla\phi,\phi) - j\phi)d^4x} \qquad (10.75)$$

one immediately sees that $\frac{1}{\hbar}$ multiplies the Klein–Gordon operator $\mathcal{K}$ in the quadratic part of the action, and therefore the free propagator $\Delta_F(x)$, essentially the inverse of $\mathcal{K}$, should be regarded as proportional to a factor of $\hbar$. The interaction vertices of the theory, associated with the higher than quadratic part of $\mathcal{L}_\delta$, should each carry a factor of $1/\hbar$. These factors of $\hbar$ will be of importance in Section 10.4 in sorting out the "classical" (i.e., lowest order in $\hbar$) from "quantum" contributions to the amplitudes of the theory. The latter will turn out to be associated with the presence of closed loops in the graphs.

The reader may recall that in our treatment of path integrals for a quantum mechanical particle in Chapter 4, we described an alternative technique for eliminating the noisome oscillations in the Minkowski space (i.e., real-time) formulation. In this approach one analytically continues the amplitudes to imaginary time, basically by the "Wick rotation" $t \to -it$. Equivalently, the path integral is derived *ab initio* for matrix elements of the bounded hermitian operator e^{-Ht}, rather than the usual unitary real-time development operator e^{-iHt}. One then obtains, in the quantum-mechanics case, path integrals such as

$$K_E(q_f, t_f; q_i, t_i) = \int e^{-\int_{t_i}^{t_f}(\frac{1}{2}m\dot{q}^2(t) + \frac{1}{2}m\omega^2 q^2(t))dt}\mathcal{D}q(t) = \int e^{-S_E(\dot{q},q)}\mathcal{D}q(t) \quad (10.76)$$

which involves a purely real integrand which is exponentially damped for large $q(t)$ and $\dot{q}(t)$, and therefore yields a well-defined absolutely convergent (multi-dimensional) integral once the usual discretization of the time interval $t_i \leq t \leq t_f$ is carried out. In the field-theory case we observe that taking $\delta = \frac{\pi}{2}$ in (10.71,10.72) (which clearly leads to a convergent path integral) is equivalent to an analytical continuation to imaginary time $t \to -it$ of the Minkowski path integral: we simply reinterpret the $e^{-i\delta}$ factor in $\mathcal{H}_\delta$ as part of the time variable in $d^4x = d\vec{x}dt$. Note that the term $\pi(x)\dot{\phi}(x)d^4x = \pi(x)d\phi(x)d\vec{x}$ term is unchanged in this continuation. We then arrive at the Euclidean generating functional

$$Z_E[j] = \int \mathbf{D}\phi\mathbf{D}\pi e^{\int (i\pi(x)\dot{\phi}(x) + j(x)\phi(x))d^4x - H[\pi,\phi])d^4x} \quad (10.77)$$

$$H[\pi,\phi] \equiv \int \{\frac{1}{2}\pi(x)^2 + \frac{1}{2}|\vec{\nabla}\phi(x)|^2 + \frac{1}{2}m(a)^2\phi(x)^2 + \frac{1}{4!}\lambda(a)\phi(x)^4\}d^4x \quad (10.78)$$

Integrating out the π field in a (by now) familiar fashion, we obtain, up to an irrelevant multiplicative constant,

$$Z_E[j] = \int \mathbf{D}\phi e^{-S_E[\phi] + \int j(x)\phi(x)d^4x} \quad (10.79)$$

$$S_E[\phi] \equiv \int \{\frac{1}{2}\dot{\phi}(x)^2 + \frac{1}{2}|\vec{\nabla}\phi(x)|^2 + \frac{1}{2}m(a)^2\phi(x)^2 + \frac{1}{4!}\lambda(a)\phi(x)^4\}d^4x \quad (10.80)$$

which can be regarded as the "Euclidean Lagrangian" version of the functional integral for this field theory (cf. (4.99, 4.100)). The positive real *Euclidean Action* functional $S_E[\phi]$ appears here in analogy to the corresponding quantity $S_E(\dot{q}, q)$ for the quantum-mechanical particle appearing in (10.76). The functional derivatives of $Z_E[j]$ yield the *Euclidean Schwinger functions* $S(x_1, x_2, ..., x_n)$ of the theory, which we briefly discussed in Section 9.2:

$$S(x_1, x_2, ..., x_n) = \frac{1}{Z_E[j]} \frac{\delta^n Z_E[j]}{\delta j(x_1)\delta j(x_2)..\delta j(x_n)}\Big|_{j=0} \quad (10.81)$$

They are clearly permutation symmetric in the arguments $x_1, x_2, \ldots x_n$, as they correspond to the path integral

$$S(x_1, x_2, ..., x_n) = \frac{1}{Z_E[0]} \int \mathbf{D}\phi \ \phi(x_1)\phi(x_2)...\phi(x_n)e^{-S_E[\phi]} \equiv \langle \phi(x_1)\phi(x_2)...\phi(x_n) \rangle$$

(10.82)

involving the weighted average of the product of c-number functions $\phi(x_1)\phi(x_2)...\phi(x_n)$.

One sometimes encounters the assertion that the Minkowski formulation of the path integral, with a complex oscillating integrand, is intrinsically ill-defined, and that the Minkowski amplitudes which it generates should instead be regarded as obtained by analytic continuation from the Euclidean Green functions obtainable from the clearly well-defined (once spacetime is appropriately discretized) Euclidean version (10.79). This is simply incorrect: as we have emphasized above, once the regularizing $e^{\pm i\delta...}$ factors are included, the regularized Minkowski path integral yields perfectly definite and unambiguous results. However, it is certainly true that non-perturbative approaches to field theory involving a direct numerical attack on the evaluation of the path integral, as in lattice QCD for example, are necessarily restricted to the Euclidean version of the path integral. Such approaches typically proceed by a stochastic Monte Carlo estimation of the multi-dimensional lattice-regularized integral, using the real positive e^{-S_E} factor as a Boltzmann probability weight for generating field configurations. This method is not practicable in the Minkowski domain: even if we separate the negative-definite real part of the action (due to $i\delta$ terms) off as a Boltzmann weight, the residual wildly oscillatory part of the integrand leads to an intolerably small signal-to-noise ratio in the Monte Carlo estimates (the infamous "sign problem"). This is unfortunate, as the *approximate* evaluation of Euclidean amplitudes can be converted directly to Minkowski space information only in a very limited number of cases (e.g., in computations of the mass spectrum, and in evaluation of certain matrix elements of local operators). The problem, of course, goes back to the fundamental difficulty that a finite number of approximate numerical estimates of an analytic function is almost never sufficiently restrictive to allow a reasonably accurate analytic continuation over a substantial range in the complex plane, and in particular over the $\frac{\pi}{2}$ Wick rotation needed to recover Minkowski-space amplitudes from Euclidean results. There is a glimmer of hope in a stochastic approach to the direct simulation of the field-theoretic Minkowski path integral based on a complex extension of the Langevin equation, originally proposed by Parisi (Parisi, 1983) and Klauder (Klauder, 1984), but the application of this method to date has been more a matter of art than science, as the simulations not infrequently fail to converge to unambiguous results, and in some cases are known to converge, but to the wrong answer![13]

Returning to our original Minkowski formulation (10.74), an immediate advantage of the path-integral approach over operator-based formulations of field theory becomes apparent if we recall the Gell–Mann–Low theorem (9.21), which gives the perturbative interaction-picture expansion for the full Feynman Green functions of the theory. This

---

[13] For a careful recent study of the mathematical status of complex Langevin methods, see (Aarts *et al.*, 2010).

result is basically a triviality in the functional formalism. If we write the Lagrangian for a general self-coupled scalar theory as the difference of a free quadratic part $\mathcal{L}_0$[14] and an interaction polynomial $\mathcal{H}_{\text{int}}$, as in (10.74), the path integral for the m-point Green function of the theory takes the form

$$\int \mathbf{D}\phi \; \phi(x_1)..\phi(x_m) e^{i \int (\mathcal{L}_0(\phi,\partial_\mu \phi(z)) - \mathcal{H}_{\text{int}}(\phi(z)))d^4 z}$$

$$= \int \mathbf{D}\phi \; \phi(x_1)...\phi(x_m) \sum_{n=0}^{+\infty} \frac{(-i)^n}{n!} \int \mathcal{H}_{\text{int}}(z_1)..\mathcal{H}_{\text{int}}(z_n) d^4 z_1..d^4 z_n \; e^{i \int \mathcal{L}_0 d^4 z}$$

$$\simeq \sum_{n=0}^{+\infty} \frac{(-i)^n}{n!} \int d^4 z_1..d^4 z_n \int \mathbf{D}\phi \; \phi(x_1)...\phi(x_m) \mathcal{H}_{\text{int}}(z_1)..\mathcal{H}_{\text{int}}(z_n) \; e^{i \int \mathcal{L}_0 d^4 z}$$

$$(10.83)$$

If the interaction part $\mathcal{H}_{\text{int}}$ is associated with a perturbative parameter (such as the coupling constant λ_0 in the ϕ^4 theory considered previously), then the final line amounts to a perturbative expansion, with the nth order result giving a functional integral which clearly reproduces the interaction-picture *free field* Green function found on the right-hand side of the Gell–Mann–Low formula (9.21), where the initial/final states $|\alpha\rangle, |\beta\rangle$ are the vacuum. The second line in (10.83) follows rigorously from the first, as the exponential expansion is perfectly convergent *for fixed values of the field $\phi(x)$* (on the regularizing spacetime lattice). However, the unbounded character of the integration $\mathbf{D}\phi$ over the field values renders the subsequent interchange of summation and integration invalid, in the sense that the resultant perturbative series is no longer convergent, but can at best be regarded as an asymptotic expansion for the full amplitudes in powers of the coupling constant parameterizing the size of the interaction term. The non-convergence resulting from the exchange of an infinite sum with an unbounded integral is already familiar in very simple one-dimensional cases: the reader is encouraged to verify it explicitly in a simple non-Gaussian integral in Problem 6. We shall have much more to say about this issue in the following chapter, in the context of non-perturbative approaches to field theory.

The preceding discussion for bosonic scalar (spin-0) fields can be generalized in a straightforward fashion for bosonic fields of non-zero spin: specifically, spin-1 vector fields. Sensible interacting field theories of spin-1 particles typically possess important additional (beyond Poincaré) *local gauge symmetries* (resulting in subtleties in the path-integral formulation) which are much more conveniently discussed in the Lagrangian framework which we shall construct in Chapter 12, so we shall postpone further discussion of bosonic path integrals for higher spin fields until that point.

Before going on to the generalization of the path-integral method to fermionic theories, a few remaining points concerning the structure of Gaussian bosonic path integrals should be addressed, especially as the contrast they display to the corre-

---

[14] In order to avoid overburdening the notation, we henceforth omit $e^{i\delta}$ factors, and the δ subscript on the Lagrangian functional.

sponding behavior for the fermionic functional integrals discussed below is of great importance in numerical approaches to non-perturbative quantum field theory. Let us take another look at the basic (source-free) Gaussian path integral for a free self-conjugate scalar field, in the Euclidean formulation, as given by

$$Z_0[0] = \int \mathbf{D}\phi \; e^{-\frac{1}{2}\int \phi(x)\mathcal{K}_E\phi(x)d^4x} \tag{10.84}$$

where the Euclidean version of the Klein–Gordon operator is

$$\mathcal{K}_E = -\Box + m^2, \quad \Box \equiv \sum_{i=1}^{4} \frac{\partial^2}{\partial x_i^2} \tag{10.85}$$

The implicit regularization of this path integral on a finite Euclidean spacetime lattice (with N points) implies that the operator $\mathcal{K}_E$ is actually replaced by a real, symmetric, positive-definite NxN matrix $\mathcal{K}_{ij}$, so our integral actually reads

$$Z_0[0] = \int \prod_i d\phi_i e^{-\frac{1}{2}\sum_{i,j} \phi_i \mathcal{K}_{ij}\phi_j} \tag{10.86}$$

where the field variables ϕ_i representing the value of $\phi(x)$ at the lattice point x_i are real (and range from $-\infty$ to $+\infty$). Let O_{ij} be the orthogonal matrix (of unit determinant) which diagonalizes $\mathcal{K}_{ij}$:

$$\mathcal{K}_{ij} = \sum_k \lambda_k O_{ik} O_{jk} \Rightarrow \sum_{i,j} \phi_i \mathcal{K}_{ij}\phi_j = \sum_k \lambda_k \hat{\phi}_k^2 \tag{10.87}$$

where we have introduced new field variables $\hat{\phi}_k = \sum_i \phi_i O_{ik}$ related to the original ones by a unit Jacobian. The eigenvalues λ_k are all positive: otherwise our Euclidean path integral would be divergent! Changing integration variables to the $\hat{\phi}_k$ then,

$$Z_0[0] = \int d\hat{\phi}_k e^{-\frac{1}{2}\sum_{k=1}^{N} \lambda_k \hat{\phi}_k^2} = \frac{(2\pi)^{N/2}}{\prod_{k=1}^{N} \sqrt{\lambda_k}} = (2\pi)^{N/2} \det(\mathcal{K})^{-\frac{1}{2}} \tag{10.88}$$

Of course, this source-free integral is a field and source-independent constant—indeed, just the constant C' in (10.67)—which we have argued is irrelevant to the calculation of the Green functions of the theory via the functional formula (10.69), as it cancels in the numerator and the denominator. However, in the fermionic case we shall encounter cases of Gaussian functional integrals in which the corresponding operator (to $\mathcal{K}$) contains further bosonic fields, so the determinant is field-dependent and must be retained as a non-trivial component of the full path integral of the theory. The Gaussian integrals occurring in the fermionic case are typically over complex (i.e., non-self-conjugate) fields, so an even closer analogy is obtained by studying the Gaussian path integral for a complex scalar field $\psi(x)$, which can always be rewritten in terms of two (equal-mass) real (self-conjugate) scalar fields ϕ_1, ϕ_2 by

$$\psi(x) = \frac{1}{\sqrt{2}}(\phi_1(x) + i\phi_2(x)) \tag{10.89}$$

The Gaussian integral corresponding to (10.84) for the doublet of fields ϕ_1, ϕ_2 can be rewritten as an integral over the single complex field ψ (and its conjugate field ψ^*, treated as an independent variable)

$$\int \mathbf{D}\phi_1 \mathbf{D}\phi_2 e^{-\frac{1}{2}\int \sum_{i=1}^{2} \phi_i(x)\mathcal{K}_E\phi_i(x)d^4x} = \int \mathbf{D}\psi \mathbf{D}\psi^* e^{-\int \psi^*(x)\mathcal{K}_E\psi(x)d^4x}$$

$$\rightarrow (2\pi)^N \det(\mathcal{K})^{-1} \qquad (10.90)$$

where the final line gives the explicit evaluation of the regularized path integral, in this case yielding the inverse determinant of the Euclidean Klein–Gordon operator $\mathcal{K}_E$, without the square-root (due to the doubling of the degrees of freedom in the case of a complex scalar field). We shall shortly see that the entire content of the distinction between Bose–Einstein and Fermi–Dirac statistics for fields in the functional approach lies in the power of the determinant that appears in the basic Gaussian integral, which is negative for bosonic fields but positive for fermionic ones. This at first sight minor distinction leads, in fact, to an enormous increase in difficulty of numerical evaluation of the path integral for fermionic as opposed to bosonic fields.

10.3.2   Path integrals for fermionic fields

The equal-time anticommutation relation for Dirac fields (cf. Problem 5 in Chapter 7)

$$\{\psi_n(\vec{x}, t), i\psi_m^\dagger(\vec{y}, t)\} = i\delta^3(\vec{x} - \vec{y})\delta_{nm} \qquad (10.91)$$

suggests (by analogy to the corresponding result (8.1) for scalar fields) that the conjugate "momentum" field to $\psi(x)$ is $\pi(x) = i\psi^\dagger(x)$. If we formally imitate the Hamiltonian path integral for scalar fields (10.52), where now the free Hamiltonian density $\mathcal{H}_0$ is that appropriate for the Dirac field (Problem 3 in Chapter 7),

$$\mathcal{H}_0 = i\bar{\psi}\vec{\gamma} \cdot \vec{\nabla}\psi + m\bar{\psi}\psi \qquad (10.92)$$

we find, introducing c-number source functions $\eta, \bar{\eta}$ whose functional derivatives will (one hopes!) produce the desired time-ordered Feynman Green functions of products of $\bar{\psi}$ and ψ, the following expression for the generating functional $Z_0[\eta, \bar{\eta}]$ of the free Dirac theory:

$$Z_0[\eta, \bar{\eta}] = \int \mathbf{D}\psi \mathbf{D}\psi^* \, e^{i\int (i\psi^\dagger\dot{\psi} - i\bar{\psi}\vec{\gamma}\cdot\vec{\nabla}\psi - m\bar{\psi}\psi - \bar{\eta}\psi - \bar{\psi}\eta)d^4x} \qquad (10.93)$$

$$= \int \mathbf{D}\psi \mathbf{D}\bar{\psi} \, e^{i\int (\bar{\psi}(i\partial\!\!\!/ - m)\psi - \bar{\eta}\psi - \bar{\psi}\eta)d^4x} \qquad (10.94)$$

where operator adjoints $\psi^\dagger$ have been reinterpreted as c-number complex conjugates ψ^*. As usual, such an expression is assumed to be regularized on a finite spacetime lattice: the functional integrals over $\psi(x)$ and $\psi^\dagger(x)$, viewed as independent variables, are then finite in number, corresponding to the a discrete finite set of lattice spacetime coordinates $x_i, i = 1, 2, \ldots N$. Note that overall multiplicative factors, such as those induced by dropping the i in the definition $\pi(x) = i\psi^\dagger(x)$, are ignored just as in the bosonic case. In the second line the replacement of $\psi^\dagger$ with $\bar{\psi} = \psi^\dagger\gamma_0$ in the functional measure is completely innocent, as $\det(\gamma_0) = 1$. Note that the action (the non-source part of the exponent in (10.94)) has already assumed a Lorentz-invariant form: for fermionic theories the Lagrangian is obtained directly by including the $\pi\dot{\psi}$ term with the Hamiltonian, without the need for an integration over the conjugate momentum π field. Again, the reason for this will become clear in our discussion of the canonical formalism in Chapter 12.

Defining the differential operator

$$\mathcal{D} \equiv i\gamma^\mu \partial_\mu - m = i\slashed{\partial} - m \qquad (10.95)$$

in analogy to the Klein–Gordon operator $\mathcal{K}$ introduced previously for the scalar field, we find that the functional Z_0 can be formally evaluated by the same procedure of completion of the square used in the bosonic case:

$$Z_0[\eta, \bar{\eta}] = \int \mathbf{D}\psi \mathbf{D}\bar{\psi} \; e^{i \int \{(\bar{\psi} - \bar{\eta}\mathcal{D}^{-1})\mathcal{D}(\psi - \mathcal{D}^{-1}\eta) - \bar{\eta}\mathcal{D}^{-1}\eta\} d^4 x} \qquad (10.96)$$

$$= C e^{-i \int \bar{\eta}\mathcal{D}^{-1}\eta d^4 x} \qquad (10.97)$$

where we have assumed that the integrals over $\psi, \bar{\psi}$ are shift-invariant, as in the bosonic case. This already looks quite promising, as the Feynman–Dirac propagator defined by $iS_F(x - y) = \langle 0|T(\psi(x)\bar{\psi}(y))|0\rangle$ (cf. Problem 5 in Chapter 7)

$$S_F(x - y) = \int \frac{\slashed{p} + m}{p^2 - m^2 + i\epsilon} e^{-ip\cdot(x-y)} \frac{d^4 p}{(2\pi)^4} = \int \frac{e^{-ip\cdot(x-y)}}{\slashed{p} - m + i\epsilon} \frac{d^4 p}{(2\pi)^4} \qquad (10.98)$$

is, in fact, a Green function for the operator $\mathcal{D}$:

$$\mathcal{D}S_F(x - y) = (i\slashed{\partial} - m)S_F(x - y) = \delta^4(x - y) \qquad (10.99)$$

so that we may write (10.97), in analogy to (10.70), as

$$Z_0[\eta, \bar{\eta}] = C e^{-i \int \bar{\eta}(x)S_F(x-y)\eta(y) d^4 x d^4 y} \qquad (10.100)$$

with (α, β are Dirac indices, running 1,2,3,4)

$$\frac{i^2}{Z_0} \frac{\delta^2 Z_0}{\delta\eta_\beta(y)\delta\bar{\eta}_\alpha(x)}\bigg|_{\eta=\bar{\eta}=0} = i(S_F(x - y))_{\alpha\beta} = \langle 0|T(\psi_\alpha(x)\bar{\psi}_\beta(y))|0\rangle \qquad (10.101)$$

Unfortunately, plausible as it seems, the above argument conceals some deep flaws in the analogistic reasoning used to arrive at (10.100). In particular, the crucial minus sign embedded in the definition of the fermionic T-product

$$T(\psi_\alpha(x)\bar{\psi}_\beta(y)) \equiv \theta(x^0 - y^0)\psi_\alpha(x)\bar{\psi}_\beta(y) - \theta(y^0 - x^0)\bar{\psi}_\beta(y)\psi_\alpha(x) \qquad (10.102)$$

is *not* reproduced by the double functional derivative in (10.101), if the field sources $\eta, \bar{\eta}$ take values in a commutative field (such as the complex numbers). The change of sign when fermion fields are reordered implies a similar change of sign when the functional derivatives are correspondingly reordered: in other words, we must have

$$\frac{\delta^2}{\delta\eta_\beta(y)\delta\bar{\eta}_\alpha(x)} = -\frac{\delta^2}{\delta\bar{\eta}_\alpha(x)\delta\eta_\beta(y)} \tag{10.103}$$

This problem can (indeed, must) be fixed by insisting that the fermionic source functions $\eta, \bar{\eta}$ take values in a *Grassmann algebra*, any two elements of which anticommute with each other. This then implies that the c-number fermionic fields $\psi, \bar{\psi}$ appearing in (10.94) must also be anticommuting Grassmann numbers. Otherwise, when the source terms in the exponent are expanded, only linear terms would survive, and we would conclude that Green functions involving more than one ψ (or $\bar{\psi}$) field vanish! For example, if $\psi(x), \bar{\psi}(x)$ are conventional complex-valued c-number functions, commuting with the Grassmann source functions,

$$\left(\int \bar{\psi}(x)\eta(x)d^4x\right)^2 = \int \bar{\psi}(x)\bar{\psi}(y)\eta(x)\eta(y)d^4x d^4y = 0 \tag{10.104}$$

as the product $\eta(x)\eta(y)$ is antisymmetric under exchange of x and y, while the product $\bar{\psi}(x)\bar{\psi}(y)$ is symmetric. This is avoided if we also require $\bar{\psi}(x)$ (and likewise $\psi(x)$) to take values in a Grassmann algebra. As indicated above, we have implicitly defined the theory on a spacetime lattice, so that the fields (and sources) are defined on a finite discrete set of spacetime coordinates $x_i, i = 1, 2, ...N$. Our Grassmann algebra consists of the set of all multi-nomials (with complex coefficients) generated by the $4N$ Grassmann numbers $\psi(x_i), \bar{\psi}(x_i), \eta(x_i), \bar{\eta}(x_i)$. Denoting these generically by χ_i, the *only* properties attributed to these numbers are

$$\chi_i\chi_j = -\chi_j\chi_i, \quad \chi_i\chi_i = -\chi_i\chi_i = 0 \tag{10.105}$$

As the square of any given Grassmann number vanishes, the possible multi-nomials involve each of the $4N$ independent Grassmann numbers at most once. The linear space of allowed multi-nomials is therefore 2^{4N}-dimensional, consisting of the one-dimensional space with no Grassmanns (i.e., the complex numbers C), the $4N$ space spanned by the Grassmann generators appearing singly, the $\frac{4N(4N-1)}{2}$ space spanned by products of two distinct Grassmanns $\chi_i\chi_j, i < j$, and so on.

Of course, in order to pursue this strategy we must next ensure that we understand the meaning of functional derivatives (i.e., in our discretized system, partial differentiation with respect to a given Grassmann element χ_i) and functional (path) integrals (in the case of Z_0 above, multi-dimensional integration over the $\psi(x_i), \bar{\psi}(x_i)$ holding $\eta(x_i), \bar{\eta}(x_i)$ fixed). The definition of the derivative is obvious once we recall that functions $F(\chi_i)$ of the χ_i can at most depend linearly on any given χ_j. Any term containing χ_j can therefore be rearranged so that the single factor of χ_j is moved to the extreme left (with a concomitant factor of –1 if the rearrangement involves an odd permutation of Grassmann numbers), giving

$$F(\chi_i) = A(\chi_i, i \neq j) + \chi_j B(\chi_i, i \neq j) \tag{10.106}$$

The partial derivative with respect to χ_j is then defined in the natural way as

$$\frac{\partial F}{\partial \chi_j} \equiv B(\chi_i, i \neq j) \tag{10.107}$$

The reader is invited to check that, with this definition, the desired antisymmetry of the second derivatives expressed in (10.103) indeed obtains: namely

$$\frac{\partial^2 F}{\partial \chi_i \partial \chi_j} = -\frac{\partial^2 F}{\partial \chi_j \partial \chi_i} \tag{10.108}$$

While differentiation with respect to Grassmann variables is superficially very similar (modulo the occasional minus sign) to ordinary differentiation with respect to commuting variables, the meaning of integration over Grassmann variables turns out to be completely different from that of conventional Stieltjes (or Lebesgue) integration. One formal property of the integral over ψ and $\bar{\psi}$ used in the heuristic argument leading to (10.100) was the shift invariance requirement

$$\int \mathbf{D}\psi F[\psi - \chi] = \int \mathbf{D}\psi F[\psi] \tag{10.109}$$

$$\int \mathbf{D}\bar{\psi} F[\bar{\psi} - \bar{\chi}] = \int \mathbf{D}\bar{\psi} F[\bar{\psi}] \tag{10.110}$$

For a single Grassmann variable this implies

$$\int d\psi_i (\psi_i - \chi_i) = \int d\psi_i \psi_i \Rightarrow \left(\int d\psi_i \right) \chi_i = 0 \Rightarrow \int d\psi_i = 0 \tag{10.111}$$

where we have also assumed that a constant Grassmann number can be factored out of the integral. Thus any term *lacking* a particular Grassmann variable ψ_i vanishes when integrated over ψ_i. As any term containing the Grassmann variable ψ_i more than once vanishes, the only potentially non-vanishing integral is that of the variable ψ_i appearing linearly, which we can normalize to any value we please (as overall constants in the generating functional Z are irrelevant, as emphasized repeatedly above). It is conventional to define

$$\int d\chi_i \, \chi_i = 1 \tag{10.112}$$

for any Grassmann variable χ_i: i.e., for any of the field variables $\psi_i, \bar{\psi}_i$ appearing in the regularized path integral.

With this interpretation of the fermionic integrations in (10.94) we may justify the result for the free generating function (10.100), at least at the level of lattice-regularized fields. At this level, in contrast to the situation for bosonic fields, there are *no convergence problems due to oscillating integrands*: the expansion of the exponential terminates at a finite order, with terms containing each ψ_i and $\bar{\psi}_i$ variable once and only once, and the value of the path integral (as a function of the remaining source quantities $\eta_i, \bar{\eta}_i$) is just the coefficient of $\prod_{i=1}^{N} \bar{\psi}_i \psi_i$ in this expanded quantity.[15] So where does the requirement for an $i\epsilon$ in the denominator of the Fourier transformed Feynman–Dirac propagator come from? In the bosonic case, this small imaginary displacement arose naturally from the requirement of regularizing the oscillating Minkowski integrand, as the integration range for the field variables was infinite, leading to a failure of absolute convergence of the (finitely) multi-dimensional integral. Here the finite-dimensional fermionic integral gives a perfectly finite result provided that the 4Nx4N matrix (4 Dirac, N spacetime degrees of freedom) representing the Dirac operator $\mathcal{D}$ on our finite spacetime lattice is invertible. The discrete energies and momenta allowed on such a lattice mean that the singularity at $k_0 = \sqrt{\vec{k}^2 + m^2}$ encountered by a *continuous* energy integral over k_0 is "missed" when the integral is converted to a finite sum. In the fermionic case, the need for inclusion of an $i\epsilon$ prescription for avoiding this singularity only appears once the lattice spacing is taken to zero and the sum goes over to a continuous integration along the real energy axis. The need for including the $i\epsilon$ in the correct way at this point—i.e., as a negative imaginary part in the mass (see (10.98))—is dictated by our desire to recover the correct causal (i.e., time-ordered) propagator in agreement with the operator version of the theory.

For the interacting field theories of primary interest in modern particle physics— the gauge field theories of the Standard Model—the dependence of the Hamiltonian (or Lagrangian) on the Fermi fields is always quadratic, so we may in general write the fermionic part of the full functional integral (which may, of course, also contain further integrations over bosonic fields),

$$\int \mathbf{D}\bar{\psi}\mathbf{D}\psi e^{i\int (\bar{\psi}(x)\mathcal{D}(\phi)\psi(x) - \bar{\eta}(x)\psi(x) - \bar{\psi}(x)\eta(x))d^4x} \rightarrow \int \prod_{i=1}^{4N} d\bar{\psi}_i d\psi_i e^{i(\bar{\psi}_i \mathcal{D}(\phi)_{ij}\psi_j - \bar{\eta}_i \psi_i - \bar{\psi}_i \eta_i)}$$

$$(10.113)$$

where the notation $\mathcal{D}(\phi)$ indicates the possible dependence of the differential operator in the quadratic part on a generic set of bosonic fields ϕ, and the arrow a suitable discretization of the continuum theory in which the fermion fields are placed on a spacetime lattice. The indices i, j run from 1 to 4N, where N is the number of spacetime lattice points and the 4 comes from the discrete Dirac index. After completion of the square, exactly as previously for the free theory, this becomes

$$e^{-i\bar{\eta}_i \mathcal{D}(\phi)_{ij}^{-1} \eta_j} \int \prod_{i=1}^{4N} d\bar{\psi}_i d\psi_i e^{i\bar{\psi}_i \mathcal{D}(\phi)_{ij}\psi_j} = e^{-i\bar{\eta}_i \mathcal{D}(\phi)_{ij}^{-1} \eta_j} \det[\mathcal{D}(\phi)] \qquad (10.114)$$

The integral of the (exponential of the) source-free quadratic fermion action gives simply the determinant of the 4Nx4N matrix of the discretized operator $\mathcal{D}(\phi)$! The reason for this is quite simple: in order to obtain a non-vanishing contribution, the rules for Grassmann integration described above imply that only the term in which the expansion of the exponential contains each ψ_i and each $\bar{\psi}_i$ once and only once contributes. The coefficient of such terms involves the multiplication of 4N matrix

---

[15] Alternatively, we see that the value of the multi-dimensional Grassmann integral is also given by taking a derivative of the integrand with respect to each and every one of the ψ_i and $\bar{\psi}_i$: for Grassmann quantities, integration = differentiation!

elements $\mathcal{D}_{ij}$ with each row and column of the matrix appearing exactly once, and with a sign indicating the sign of the permutation needed to place the ψ_i in the same order as the $\bar{\psi}_i$. This is exactly the definition of the determinant of the matrix $\mathcal{D}_{ij}$. Note that the result is precisely the inverse of that obtained for the integration of Gaussian complex bosonic path integrals (cf. (10.90)). As indicated previously, the path integral for the interacting gauge field theories contained in the Standard Model are at most quadratic in the elementary spin-$\frac{1}{2}$ fields (leptons and quarks) of the theory, so that in principle the fermionic integrations can all be performed yielding determinantal functionals of the remaining bosonic fields, leaving only bosonic path integrals to be performed. In the case of lattice QCD, by far the greatest part of the numerical difficulties encountered with stochastic estimations of the resultant bosonic path integrals derives from the evaluation of the fermionic determinant associated with integrating out the quark fields.

There is one further important property of Grassmann integrals which we shall need to take into account in our discussion of anomalies in Chapter 15. The peculiar property of fermionic path integrals, whereby the fermion determinant appears to a positive power (as in (10.114)) when a Gaussian integral is performed, has a correlate in the behavior of the Grassmann integral under a change of fermionic variables. Suppose we wish to change variables from a set of Grassmann quantities $\psi_i, i = 1, 2, ..N$ to $\psi'_i = C_{ij}\psi_i$, where the C_{ij} are bosonic in character (i.e., complex scalars). The requirement that the new variables satisfy

$$1 = \int \prod_{i=1}^{N} d\psi'_i \prod_{i=1}^{N} \psi'_i = \int \prod_{i=1}^{N} d\psi'_i C_{1i_1}\psi_{i_1} C_{2i_2}\psi_{i_2} \cdots C_{Ni_N}\psi_{i_N} \tag{10.115}$$

implies that the Jacobian J of the change of variables, with $\int \prod_{i=1}^{N} d\psi'_i = J \int \prod_{i=1}^{N} d\psi_i$ is given by

$$J = \det(C)^{-1} \tag{10.116}$$

instead of by $\det(C)$, as would be the case for a multi-dimensional bosonic integral.

10.4 Graphical concepts: *N*-particle irreducibility

Our earlier discussion of the diagrammatic representation of field-theory amplitudes pointed out two potential sources[16] of singular behavior in the order-by-order implementation of perturbation theory. First, there is the infrared (more precisely, long time) problem of persistent interactions, which lead, in a careless application of perturbation theory, to an incorrect pole structure in external momentum of the *n*-point functions of the theory, making the extraction of sensible S-matrix elements via the LSZ formula impossible. Secondly, even when this first problem has been satisfactorily addressed, there remains the problem of ultraviolet divergences associated with (in coordinate space) the lack of definition of products of Feynman propagator distributions, or equivalently (in momentum space) loop integrals which diverge as the ultraviolet cutoff (associated, say, with the short-distance lattice-regularization of the theory) is

---

[16] The third difficulty, or "inconvenient truth", identified in Section 10.2, the intrinsic non-convergence of perturbative expansions, will be addressed in detail in the subsequent chapter.

removed. The resolution of these two issues is greatly facilitated by the introduction of graphical concepts which allow us to reorganize the Feynman diagrams of the theory in an intuitively powerful way. In the first case, the relevant concept is that of an amputated diagram, in the second, of proper, or *one-particle irreducible* vertices.

The graphical representation of the Green functions of an interacting field theory, as described in Section 10.2, suggests that important features of these amplitudes can be directly visualized and correlated with corresponding properties of the Feynman graphs which represent these amplitudes at any given order of perturbation theory. The most basic such property is that of *connectedness*, first discussed in the context of S-matrix elements in Section 6.1. This is a special case of the more general concept of "N-particle irreducibility". A graph (or set of graphs) contributing to a given n-point function, or (via the LSZ formula) S-matrix element, is said to be N-particle irreducible if it remains connected when any N internal lines of the graph are cut. Thus, the connected diagrams are 0-particle irreducible: they contain the subset of one-particle irreducible (or 1PI for short) diagrams which remain connected if only a single internal line is cut. The 1PI diagrams in turn contain as a subset the 2PI diagrams which remain connected even when two internal lines are cut, and so on. The physical significance of the connected contributions to S-matrix elements, and to the Green functions of the theory was already discussed at length in Sections 6.1 and 9.3. The physical interpretation of the 1PI and higher irreducible graphs will be discussed in due course below. Our task here is to connect these concepts to the functional approach to field theory introduced in the preceding section.

The functional derivatives of $Z[j]$ (divided by $Z[0]$) yield the full set of contributions to the Green functions of the theory—both connected and disconnected.[17] Exactly as for the generating functional of S-matrix elements, where the connection between full and connected S-matrix amplitudes is extremely simple when expressed in terms of generating functionals, as discussed in Section 6.1,

$$\mathcal{S}(j^*, j) \equiv \exp\left(\mathcal{S}^c(j^*, j)\right) \tag{10.117}$$

the generating functional $W[j]$ whose functional derivatives yield the *connected* Green functions of the theory is given by

$$Z[j] = \exp W[j], \quad W[j] \equiv \ln Z[j] \tag{10.118}$$

where, for our usual example of a self-interacting real scalar field, $Z[j]$ is given (cf. (10.74)) by the Minkowski functional integral

$$Z[j] = \int \mathbf{D}\phi\, e^{-i\int (\frac{1}{2}\phi(x)\mathcal{K}\phi(x) + P(\phi))d^4x + \int j(x)\phi(x)d^4x} \tag{10.119}$$

with $P(\phi)$ a polynomial (higher than quadratic) in the fields, the individual terms of which induce the three-point, four-point, etc. interaction vertices of the theory. As a concrete example we shall imagine in the following that both trilinear and quadrilinear interactions are present—we may take, for example, $P(\phi) = \frac{\lambda_3}{3!}\phi^3 + \frac{\lambda_4}{4!}\phi^4$. Thus, the

---

[17] However, as discussed previously, the division by $Z[0]$ eliminates the disconnected vacuum fluctuations accompanying any scattering process.

graphs of the theory contain elementary vertices from which either three or four lines emerge. The usual factors of $e^{i\delta}$ needed for convergence of the Minkowski path integral are suppressed here to avoid overburdening the notation.

The derivatives of (10.118), in analogy to the functional derivatives of $Z[j]$ for the full Green functions (see (10.73), lead directly to the cluster decomposition recursion formulas (cf. (9.92, 9.93, 9.94)) relating connected to full amplitudes

$$\frac{i\delta W}{\delta j(x_1)} = \frac{i}{Z} \frac{\delta Z}{\delta j(x_1)} \tag{10.120}$$

$$\frac{i^2 \delta^2 W}{\delta j(x_1)\delta j(x_2)} = \frac{i^2}{Z} \frac{\delta^2 Z}{\delta j(x_1)\delta j(x_2)} - \frac{i\delta W}{\delta j(x_1)} \frac{i\delta W}{\delta j(x_2)} \tag{10.121}$$

$$\frac{i^3 \delta^3 W}{\delta j(x_1)\delta j(x_2)\delta j(x_3)} = \frac{i^3}{Z} \frac{\delta^3 Z}{\delta j(x_1)\delta j(x_2)\delta j(x_3)} - \left(\frac{i\delta W}{\delta j(x_1)} \frac{i^2 \delta^2 W}{\delta j(x_2)\delta j(x_3)} + \text{perms} \right)$$
$$- \frac{i\delta W}{\delta j(x_1)} \frac{i\delta W}{\delta j(x_2)} \frac{i\delta W}{\delta j(x_3)} \tag{10.122}$$

and so on. After the functional derivatives are performed, the desired n-point functions are obtained by setting the source $j = 0$. These formulas display directly the removal of disconnected contributions from the full set of graphs generated by the generating functional $Z[j]$. We shall henceforth *only consider connected amplitudes*, relieving us of the obligation to further complicate the notation by a sub(or super)script "c", to distinguish the connected Green functions generated by $W[j]$ from those (containing all graphs) generated by $Z[j]$.

Before deriving the general form of the generating functional for the one-particle irreducible diagrams of the theory, we shall illustrate the basic idea with some examples, which will help to bring home the critical importance of these new graphical concepts in exposing exactly the pole structure in the external momenta needed to make the LSZ formula for S-matrix elements work. Recall from our discussion of persistent interactions in Section 10.2 that the appearance of the correct set of poles in the external particle momenta is by no means automatic in perturbation theory. The heuristic account that follows is intended to reveal in a visually intuitive way the appropriate reorganization of perturbation theory which will allow the direct and unambiguous application of the LSZ formula to the extraction of S-matrix elements.

Consider the set of all connected contributions to the four-point Green function in the theory with generating functional (10.119). All such graphs can be displayed in the generic form exhibited in Fig. 10.4: the blobs labeled $\hat{\Delta}_F$ on the four external lines represent all possible contributions to the full interacting two-point function (Feynman propagator): they contain all possible self-interactions of the external particles coming into and receding from the central collision process. The central blob, labeled $G_{\text{amp}}^{(4)}$, is the "amputated" connected four-point function of the theory, and represents the part of the process where the initial and final particles can no longer be considered to be moving freely and independently of one another. Thus, the spacetime points labeled z_1, z_2, z_1', z_2' represent interaction vertices, indeed the first interactions (as we move in towards the central collision process from the external points x_1, x_2, x_1', x_2') which

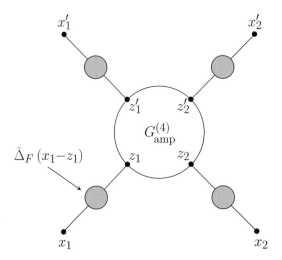

Fig. 10.4 General structure of connected four-point function $G^{(4)}(x_1, x_2, x_1', x_2')$.

can no longer be regarded as inducing persistent self-interactions of isolated, freely moving particles. We shall shortly see how to construct a generating functional which automatically yields amputated Green functions. But even before doing this, it is easy to see how the introduction of amputated graphs allows us to solve the problem of persistent interactions in the context of the LSZ formula mentioned above.

The decomposition of the four-point function illustrated in Fig. 10.4 (into external legs carrying self-interactions of the external particles and a central legless, or "amputated", interaction region) amounts to writing, in coordinate space

$$G^{(4)}(x_1', x_2', x_1, x_2) = \int \hat{\Delta}_F(x_1' - z_1')\hat{\Delta}_F(x_2' - z_2')\hat{\Delta}_F(x_1 - z_1)\hat{\Delta}_F(x_2 - z_2)$$
$$\cdot G^{(4)}_{\text{amp}}(z_1', z_2', z_1, z_2)d^4 z_1' d^4 z_2' d^4 z_1 d^4 z_2 \tag{10.123}$$

Fourier transforming the full propagators

$$\hat{\Delta}_F(x) = \int e^{iq\cdot x}\hat{\Delta}_F(q^2)\frac{d^4 q}{(2\pi)^4} \tag{10.124}$$

and the full Green function $G^{(4)}$ (cf. (9.178))

$$\tilde{G}^{(4)}(k_1', k_2', k_1, k_2) \equiv \int e^{+i\sum k_j'\cdot x_j' - i\sum k_i\cdot x_i}G^{(4)}(x_1', x_2', x_1, x_2)d^4 x_1' d^4 x_2' d^4 x_1 d^4 x_2 \tag{10.125}$$

we find, as expected, that the convolutions in coordinate space become algebraic products in momentum space

$$\tilde{G}^{(4)}(k_1', k_2', k_1, k_2) = \hat{\Delta}_F(k_1^2)\hat{\Delta}_F(k_2^2)\hat{\Delta}_F(k_1'^2)\hat{\Delta}_F(k_2'^2)\tilde{G}^{(4)}_{\text{amp}}(k_1', k_2', k_1, k_2) \tag{10.126}$$

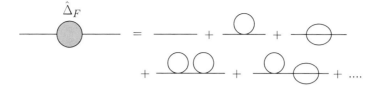

Fig. 10.5 Feynman Graphs contributing to the full Feynman propagator $\hat{\Delta}_F$ in ϕ^4 theory.

The 2-2 scattering amplitude in this theory is given by the LSZ formula (9.179): we must multiply the momentum-space four-point function $\tilde{G}^{(4)}$ by external leg factors of $\frac{-iZ^{-1/2}(k^2-m_{\mathrm{ph}}^2)}{(2\pi)^{3/2}\sqrt{2E(k)}}$ for every external momentum k, and then take the on-mass-shell limit $k^2 \to m_{\mathrm{ph}}^2$. By the Lehmann representation (9.192), each of the external leg full propagators in (10.126) produce in this limit a simple pole, with residue Z, at exactly the (squared) physical mass m_{ph}^2 (which may not be the same as the squared masses employed in the free propagators from which the diagrams are constructed!), so that the factors of $(k^2 - m_{\mathrm{ph}}^2)$ for each initial or final-state particle are cancelled, and we obtain for the 2-2 S-matrix element:

$$S_{k_1',k_2',k_1,k_2} = \prod_{i=1}^{2} \frac{-iZ^{1/2}}{(2\pi)^{3/2}\sqrt{2E(k_i)}} \prod_{j=1}^{2} \frac{-iZ^{1/2}}{(2\pi)^{3/2}\sqrt{2E(k_j')}} \tilde{G}_{\mathrm{amp}}^{(4)}(k_1',k_2',k_1,k_2) \quad (10.127)$$

with the amputated four-point function $\tilde{G}_{\mathrm{amp}}^{(4)}$ evaluated at on-mass-shell momenta (i.e., $k_i^2 = k_i'^2 = m_{\mathrm{ph}}^2$). The potentially singular behavior associated with the on-mass-shell limit (or equivalently, the infinite time propagation of self-interacting particles into and out of the process) has been taken care of in (10.127): we may continue on with the evaluation of $\tilde{G}_{\mathrm{amp}}^{(4)}(k_1',k_2',k_1,k_2)$ in perturbation theory, confident that the pole residue corresponding to the desired S-matrix element has been properly extracted. In particular, there is no need to use an on-mass-shell renormalization scheme of the sort described in Section 10.2, in which the pole of the free propagator is shifted to the physical value order by order in perturbation theory by the choice of suitable counterterms. The mass appearing in the free propagators constituting the perturbative expansion of the amputated Green function $\tilde{G}_{\mathrm{amp}}^{(4)}$ may be conveniently chosen at some "intermediate" value, depending on the particular renormalization scheme employed, as we shall see later in our detailed discussion of renormalization theory in Chapter 17.

Before going on to a discussion of proper, or "one-particle-irreducible" (1PI), Green functions, a short digression on some elementary graphical counting rules will equip us with some results which facilitate the reorganization of perturbation theory implied by the introduction of the 1PI condition. Recall that a *tree diagram* is defined as a connected graph which is rendered disconnected by cutting a single internal line. Let such a graph have I internal lines and V vertices. We imagine a "pruning" process whereby vertices are removed one by one by cutting a single internal line, starting at the outermost "branches" of the tree. Each such removal leaves the quantity $I - V + 1$

unchanged, as a single vertex and a single internal line have been removed. Eventually we arrive at the remnant core of the graph, with a single vertex, and no remaining internal lines, for which the quantity $I - V + 1$ is zero. We conclude that *for any tree graph, $I - V + 1 = 0$.*

More generally, a connected graph may contain L independent loops, resulting in momentum space in L independent four-momenta integrations corresponding to the free flow of momentum around each loop. We again consider the quantity $I - V + 1$, this time reducing the number of loops one at a time by cutting a single internal line in each loop, without altering the total number of vertices, and leaving the graph connected at each stage. In this case we reduce the number of loops L, and the quantity $I - V + 1$, by one at each cut, until we arrive at a connected graph with no loops- i.e., a tree graph, with $I - V + 1 = 0$. This establishes that the number of loops L in our initial graph is given precisely by the combination $I - V + 1$. The reader is invited to verify this rule by drawing a few simple graphs.

The study of the short-distance (or ultraviolet) singularity structure of field theory amplitudes is greatly facilitated by the introduction of the concept of one-particle-irreducible diagrams, and in particular, by the use of a functional, analogous to $W[j]$ for connected graphs, which automatically generates such 1PI diagrams. In addition, the graphs produced will be amputated: in other words, they are subsets of the graphs describing the connected amputated functions $G_{\text{amp}}^{(n)}$ discussed previously, where those diagrams which can be disconnected by cutting a single internal line are discarded. It is intuitively obvious that the full set of amputated connected diagrams (in coordinate space) can be reconstituted from the 1PI diagrams by convolving products of the latter with full Feynman propagators $\hat{\Delta}_F$ connecting the separate 1PI pieces.

For example, in Fig. 10.6, we see that a class of one-particle-reducible contributions to the connected four-point function have the structure of two three-point 1PI graphs $G_{\text{amp}}^{(3)}$ connected by a single full Feynman propagator. Note that $G_{\text{amp}}^{(3)}$ is automatically 1PI (why?). In momentum space the full graph decomposes algebraically into a product of the momentum-space amplitudes for the two $G_{\text{amp}}^{(3)}$ functions times the momentum-space Feynman propagator $\hat{\Delta}_F(p^2)$, where p is the definite four-momentum (fixed by

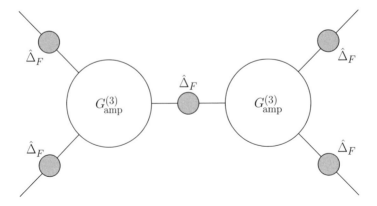

Fig. 10.6 A one-particle-reducible contribution to the connected four-point function.

energy-momentum conservation) passing between the two separate 1PI pieces. Thus the singularity structure of the full connected amplitudes—in momentum space, the dependence on the ultraviolet cutoff of the loop integrals in any given graph—is decomposable into independent pieces: namely, the proper (or 1PI) subgraphs into which it can be decomposed.[18] The study of the short-distance sensitivity of general amplitudes in field theory can therefore be reduced to a study of the cutoff dependence of the proper, or 1PI, Green functions of the theory.

In addition to simplifying the study of ultraviolet behavior, the introduction of the 1PI concept leads to two further important insights into the physics of local field theories. First, we shall see that the generating functional for 1PI graphs (sometimes referred to as the "effective action") also has a direct energetic interpretation which plays a critical role in the analysis of spontaneous symmetry-breaking, to which we shall return in Part 3. In fact, it plays a role in field theory analogous to that played by the free energy in thermodynamics. Secondly, if we reintroduce Planck's constant as an explicit signature of quantum effects, an expansion of the effective action in powers of $\hbar$ is found to (a) yield the classical action as the zeroth order term, and (b) correspond precisely to a reorganization of the perturbation theory according to the number of loops.

We turn now to the task of constructing a functional for the 1PI graphs of the theory. To avoid annoying factors of i we shall work in the Euclidean formulation. Furthermore, for the reasons adduced immediately above, Planck's constant will be reintroduced both in the path integral and in the definition of the generating functionals. Thus, the exponents appearing in the path integral acquire an explicit factor of $\frac{1}{\hbar}$, which can be traced back to the reintroduction of $\hbar$ in the Hamiltonian evolution, $e^{-iHt} \to e^{-iHt/\hbar}$, or, in the imaginary time formulation, $e^{-Ht} \to e^{-Ht/\hbar}$. Our Euclidean path integral now reads

$$Z[j] = \int \mathbf{D}\phi e^{-\frac{1}{\hbar}\int(\frac{1}{2}\phi(x)\mathcal{K}\phi(x)+P(\phi(x))-j(x)\phi(x))d^4x} = \int \mathbf{D}\phi e^{-\frac{1}{\hbar}(S[\phi]-\int j(x)\phi(x)d^4x)}$$

(10.128)

where now $\mathcal{K} = -\Box + m^2$ is the Euclidean Klein–Gordon operator, which is self-adjoint and positive-definite, and the coordinate integrations are over a Euclidean four-space. Of course, we expect to be able, at least in principle, to return to the physical Minkowski-space amplitudes by the process of Wick rotation. The generating functional for the connected diagrams is now defined as

$$W[j] = \hbar \ln Z[j] = \sum_n \frac{1}{n!} \int G^{(n)}(x_1, x_2, ..., x_n)j(x_1)j(x_2)...j(x_n)d^4x_1 d^4x_2...d^4x_n$$

(10.129)

where we shall for simplicity adopt the same notation $G^{(n)}$ for the Euclidean n-point Green functions (previously denoted $S(x_1, .., x_n)$, in (10.81)) of the theory as that

---

[18] In coordinate space it is clear that the ultraviolet singularities, which arise from multiplication of Feynman propagators (which are distributions) at coincident vertices, must be localized within the 1PI pieces, as the propagators connecting separate 1PI parts of the diagram appear independently.

used previously for Minkowski-space (Feynman) amplitudes, and remember that we are dealing throughout with *connected* amplitudes.

Note the additional overall factor of $\hbar$ included in the definition of $W[j]$: it is there to ensure the existence of a well-defined classical limit for $\hbar \to 0$. Indeed, we note that in any particular graphical contribution to $G^{(n)}$, each free propagator $\Delta_F = \mathcal{K}^{-1}$ is now accompanied by a factor $\hbar$, each vertex (term in $P(\phi)$) with a factor $\frac{1}{\hbar}$, and each external line with a factor $\frac{1}{\hbar}$ (as each source function expanded down in the path integral is accompanied by a factor $\frac{1}{\hbar}$). Thus, a particular graphical contribution to $G^{(n)}$ comes with $\hbar$ raised to a power equal to the total number of lines $= n + I$ (external plus internal), minus the number of vertices V, minus the number of source functions n, and plus one (from the overall factor of $\hbar$ in (10.129)). This gives exactly $I - V + 1$, which we saw previously is just the number of loops in the graph!

In particular, *the tree (loopless) graphs correspond to the leading contribution, of order $\hbar^0$, in the classical limit $\hbar \to 0$*. At least within the context of perturbation theory, quantum effects in a quantum field theory amplitude can be directly associated with the presence of loops in the corresponding graphs. From the path-integral expression (10.128) it is apparent that an expansion in $\hbar$ amounts to a saddle-point expansion of the integral, in which the leading term corresponds to finding a minimum of the exponent in field space. This minimum occurs at the extremal point of the classical action $S[\phi]$, at $\phi(x) = \phi_{\text{cl}}(x)$ where

$$\frac{\delta S}{\delta \phi(x)} = \frac{\delta}{\delta \phi(x)} \left(\frac{1}{2} \phi(x) \mathcal{K} \phi(x) + P(\phi(x)) - j(x)\phi(x) \right) \bigg|_{\phi = \phi_{\text{cl}}} = 0 \tag{10.130}$$

which amounts to

$$\mathcal{K} \phi_{\text{cl}}(x) + P'(\phi_{\text{cl}}(x)) = j(x) \tag{10.131}$$

In the source-free limit $(j = 0)$, this equation is simply the non-linear classical field equation corresponding to the classical least action principle (in Euclidean space, of course). In the presence of a source, it determines the classical field ϕ_{cl} implicitly as a functional of the external source j. The leading saddle-point approximation to $Z[j]$ is given by simply evaluating the exponential at its extremal point:

$$Z_{\text{cl}}[j] = e^{-\frac{1}{\hbar} \int (\frac{1}{2} \phi_{\text{cl}}(x) \mathcal{K} \phi_{\text{cl}}(x) + P(\phi_{\text{cl}}(x)) - j(x)\phi_{\text{cl}}(x)) d^4 x} \tag{10.132}$$

whence we obtain, by (10.129), the classical limit of the generating functional W for connected amplitudes

$$W_{\text{cl}}[j] = -\int (\frac{1}{2} \phi_{\text{cl}}(x) \mathcal{K} \phi_{\text{cl}}(x) + P(\phi_{\text{cl}}(x)) - j(x)\phi_{\text{cl}}(x)) d^4 x \tag{10.133}$$

where ϕ_{cl} is supposed to be determined as a functional of j via (10.131). Notice that $\hbar$ has disappeared at this point, as we are dealing with completely classical quantities. Note also that by using the (functional) chain rule for differentiation of a functional that depends on $j(x)$ both explicitly and implicitly through $\phi_{\text{cl}}(x)$,

$$\frac{\delta}{\delta j(x)} = \int d^4y \frac{\delta \phi_{\text{cl}}(y)}{\delta j(x)} \frac{\delta}{\delta \phi_{\text{cl}}(y)} + \left. \frac{\delta}{\delta j(x)} \right|_{\phi_{\text{cl}}} \tag{10.134}$$

we find that

$$\frac{\delta W_{\text{cl}}[j]}{\delta j(x)} = \int d^4y \frac{\delta \phi_{\text{cl}}(y)}{\delta j(x)} (-\mathcal{K}\phi_{\text{cl}}(y) - P'(\phi_{\text{cl}}(y)) + j(y)) + \phi_{\text{cl}}(x)$$

$$= \phi_{\text{cl}}(x) \tag{10.135}$$

where the term in brackets in the integral has vanished in virtue of the field equation (10.131).

The classical field equation (10.131) can be solved iteratively in increasing powers of the source function j. Let us temporarily restrict the polynomial $P(\phi)$ to correspond to ϕ^4 theory—we take $P(\phi) = \frac{\lambda}{4!}\phi^4$, $P'(\phi) = \frac{\lambda}{3!}\phi^3$. Next, we rewrite (10.131) as follows:

$$\phi_{\text{cl}}(x) = \mathcal{K}^{-1}j(x) - \frac{\lambda}{3!}\mathcal{K}^{-1}(\phi_{\text{cl}}^3)(x) \tag{10.136}$$

or more explicitly, using the Green function for $\mathcal{K}$, which is just the Euclidean propagator $\Delta_E(x)$ (with Fourier transform $\frac{1}{k^2+m^2}$, so that $\mathcal{K}\Delta_E(x) = \delta^4(x)$),

$$\phi_{\text{cl}}(x) = \int \Delta_E(x - x_1)j(x_1)d^4x_1 - \frac{\lambda}{3!}\int \Delta_E(x - z)\phi_{\text{cl}}(z)^3 d^4z \tag{10.137}$$

Reinserting the left-hand side result for ϕ_{cl} on the right, we find, through order j^3,

$$\phi_{\text{cl}}(x) = \int \Delta_E(x - x_1)j(x_1)d^4x_1 - \frac{\lambda}{3!}\int \Delta_E(x - z)$$

$$\cdot \Delta_E(z - x_1)j(x_1)\Delta_E(z - x_2)j(x_2)\Delta_E(z - x_3)j(x_3)d^4x_1 d^4x_2 d^4x_3 d^4z + O(j^5) \tag{10.138}$$

Ignoring signs, coupling constants, and combinatoric factors, these terms can be graphically represented as indicated in Fig. 10.7: it is apparent that one has generated a set of tree graphs with a preferred external point (the argument x of the classical field $\phi_{\text{cl}}(x)$), and source functions $j(x_i)$ attached to all other external points of the graph. The third term in Fig. 10.7 shows also one of the terms arising at order j^5 (of second order in the bare coupling λ, as there are two interaction vertices, at z_1 and z_2), which we obtain by iterating (10.138) one more time.

If we insert this iterative solution for ϕ_{cl} into (10.133), we obtain the connected generating functional in the classical limit as an explicit functional of the source function $j(x)$, expanded formally in increasing powers of $j(x)$:

$$W_{\text{cl}}[j] = \int \frac{1}{2}j(x_1)\Delta_E(x_1 - x_2)j(x_2)d^4x_1 d^4x_2$$

$$- \frac{\lambda}{4!}\int (\prod_{i=1}^{4} \Delta_E(z - x_i)j(x_i)d^4x_i)d^4z + O(j^6) \tag{10.139}$$

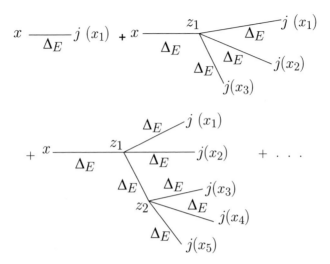

Fig. 10.7 Formal expansion of the classical field $\phi_{cl}(x)$ in powers of j.

with the graphical representation indicated in Fig. 10.8 (again ignoring combinatoric factors, signs, etc., and with one of the $O(j^6)$ terms not shown explicitly in (10.139) indicated graphically). As expected, only *connected tree graphs* are present, some of which are, however, clearly one-particle-reducible (i.e., become disconnected when a single internal line is cut). At the level of tree graphs, of course, the only 1PI graphs involve at most a single interaction vertex: as soon as more than one vertex is

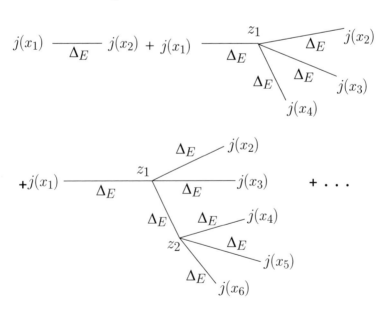

Fig. 10.8 Formal expansion of $W_{cl}[j]$ in powers of j.

present, there must be an internal line connecting two vertices (the graph as a whole is connected!), the removal of which will disconnect the diagram (as no loops are present). Thus any generating functional for 1PI diagrams can, at the classical (leading order in $\hbar$) level, only contain a finite set of terms corresponding to the interaction vertices of the theory (the monomial terms in $P(\phi)$).

The key to finding such a functional turns out to be the Legendre transform: we re-express the information contained in the functional $W[j]$ as a functional $\Gamma[\phi]$ of the derivatives $\frac{\delta W[j]}{\delta j(x)} \equiv \phi(x)$, in such a way that no information is lost in trading in the source j for the field ϕ, or *vice versa*. The field $\phi(x)$ obtained by functionally differentiating $W[j]$ with respect to the source $j(x)$ is sometimes (confusingly) called the "classical field", although properly speaking that term should be applied to its $\hbar \to 0$ limit $\phi_{\rm cl}(x) = \frac{\delta W_{\rm cl}[j]}{\delta j(x)}$ (cf. (10.135)). Referring to the path-integral representation for $W[j] = \hbar \ln Z[j]$, and recalling that derivatives of $Z[j]$ generate vacuum-expectation-values of the corresponding Heisenberg fields (here, continued to Euclidean space), we see that $\phi(x) = \frac{1}{Z[j]} \frac{\delta Z[j]}{\delta j(x)}$ is just the normalized VEV of the Heisenberg field $\phi_H(x)$ in the presence of the source term. It can therefore be viewed as a c-number source function, like $j(x)$, and in that sense is "classical".[19] As usual, the Legendre transform allowing us to incorporate the complete information in $W[j]$ in a recoverable way[20] as a functional of the derivative $\phi(x) = \frac{\delta W[j]}{\delta j(x)}$ is given by

$$\Gamma[\phi] \equiv -W[j] + \int j(x)\phi(x)d^4x \qquad (10.140)$$

where, of course, the dependence on $j(x)$ on the right-hand side must be eliminated in favor of $\phi(x)$ by inverting the equation $\phi(x) = \frac{\delta W[j]}{\delta j(x)}$. The Legendre transformation is defined in order to preserve the information encoded in $W[j]$. Using the chain rule (namely, (10.134), with the roles of j and ϕ interchanged), we find

$$\frac{\delta \Gamma[\phi]}{\delta \phi(x)} = j(x) + \int d^4y \frac{\delta j(y)}{\delta \phi(x)} \frac{\delta}{\delta j(y)} (-W[j] + \int j(z)\phi(z)d^4z)|_\phi$$

$$= j(x) + \int d^4y \frac{\delta j(y)}{\delta \phi(x)} (-\frac{\delta W[j]}{\delta j(y)} + \phi(y)) = j(x) \qquad (10.141)$$

so that $W[j] = -\Gamma[\phi] + \int j(x)\phi(x)d^4x$ allows reconstruction of $W[j]$ from $\Gamma[\phi]$ once ϕ is re-expressed in terms of j via (10.141). Moreover, we see that the two-point functions

$$\frac{\delta^2 W[j]}{\delta j(x)\delta j(y)} = \frac{\delta\phi(y)}{\delta j(x)} \qquad (10.142)$$

---

[19] We shall follow the usual confusing practice of using the same symbol $\phi(x)$ to refer to (a) the quantum (operator) free field, (b) the c-number field integrated over in the path-integral formalism, and (c) the independent field variable for the Legendre transform $\Gamma[\phi]$ of $W[j]$.

[20] See (Callen, 1960), Section 5.2, for a lucid geometrical introduction to Legendre transforms.

and

$$\frac{\delta^2 \Gamma[\phi]}{\delta\phi(x)\delta\phi(y)} = \frac{\delta j(x)}{\delta\phi(y)} \tag{10.143}$$

are functional inverses of one another. In particular, their discretized versions correspond to matrices which are inverses of one another.

It should be noted here that the assumption of invertibility (i.e., solvability of $\frac{\delta W[j]}{\delta j} = \phi$ for j in terms of ϕ, or of $\frac{\delta\Gamma[\phi]}{\delta\phi} = j$ for ϕ in terms of j) is not automatically assured for arbitrary functionals. In the case of the Legendre transformation connecting Lagrangians to Hamiltonians in mechanics (or field theory), for example, there are important cases (in the case of field theories, in situations involving local gauge symmetries, for example) where the form of the Lagrangian is such that it is not possible to solve uniquely for the velocities (or time-derivatives of the fields) in terms of the momenta defined as $p = \frac{\partial L}{\partial \dot{q}}$ (or, for field theory, $\pi \equiv \frac{\delta\mathcal{L}}{\delta\dot{\phi}}$), so the Hamiltonian is not defined uniquely by a Legendre transformation of the Lagrangian. The modifications needed in the canonical formalism in the presence of such eventualities will be described in Chapter 15. Even for a function $W(J)$ of a single variable J, it is apparent that the equation $W'(J) = \phi$ can be solved uniquely for J in terms of ϕ only if $W'(J)$ is a monotonic function of J: i.e., if the original function $W(J)$ is *convex*. The required convexity of $W[j]$ (and $\Gamma[\phi]$) in the field theory case can in fact be established quite generally (beyond perturbation theory), as we shall see when we once again take up the study of the effective action in the study of spontaneous symmetry-breaking in Chapter 14. For our perturbative purposes in this Chapter, it will suffice to show that the invertibility is valid order by order, to all orders, in a formal expansion in the source $j(x)$ or the classical field $\phi(x)$.

We return now to a study of the properties of the effective action $\Gamma[\phi]$. As a first step, we may enquire into the meaning of the classical limit, for $\hbar \to 0$, of this object. The classical limit is obtained by setting $W[j] = W_{cl}[j]$, so that, by (10.135), $\phi(x) = \phi_{cl}(x)$, and we find immediately from (10.133):

$$\Gamma_{cl}[\phi] = \int (\frac{1}{2}\phi(x)\mathcal{K}\phi(x) + P(\phi(x)))d^4x = S[\phi] \tag{10.144}$$

We see that in the classical limit, the generating functional $\Gamma[\phi]$ coincides with the *classical action* $S[\phi]$ of the theory. For this reason, the full (i.e., $\hbar \neq 0$) $\Gamma[\phi]$ is generally referred to as the "effective action" of the field theory, which in some sense generalizes the classical action to include quantum effects. If in the usual way we regard $\Gamma[\phi]$ as a generating functional of n-point vertex functions $\Gamma^{(n)}(x_1, x_2, ..., x_n)$ of the theory,

$$\Gamma^{(n)}(x_1, x_2, ..., x_n) = \frac{\delta^n \Gamma[\phi]}{\delta\phi(x_1)\delta\phi(x_2)..\delta\phi(x_n)}\bigg|_{\phi=0} \tag{10.145}$$

we see that in the classical limit (for the specific example $P(\phi) = \frac{\lambda}{4!}\phi^4$), the only surviving vertex functions are for $n = 2$ and $n = 4$:

$$\Gamma_{\text{cl}}^{(2)}(x_1, x_2) = \mathcal{K}_{x_1} \delta^4(x_1 - x_2) \qquad (10.146)$$

$$\Gamma_{\text{cl}}^{(4)}(x_1, x_2, x_3, x_4) = \lambda \delta^4(x_1 - x_2)\delta^4(x_2 - x_3)\delta^4(x_3 - x_4) \qquad (10.147)$$

Note that all the one-particle-reducible terms present in the connected tree graphs generated by $W_{\text{cl}}[j]$ have disappeared: instead we have only a two-point vertex given by the Klein–Gordon operator $\mathcal{K}$ (or equivalently, the inverse propagator Δ_E^{-1}), and a four-point fully amputated vertex corresponding to a single interaction point: in other words, the only amputated one-particle-irreducible graphs possible at the tree level. The appearance of the inverse propagator in $\Gamma_{\text{cl}}^{(2)}$ can be regarded as a result of the amputation process whereby a factor of the inverse propagator Δ_E^{-1} is applied for each external point, leading in the case of the two-point function to $\Delta_E^{-1} \cdot \Delta_E \cdot \Delta_E^{-1} = \Delta_E^{-1}$.

To go beyond the classical limit (i.e., the tree diagrams of the theory) we need to look a little more carefully at the structure of the effective action functional $\Gamma[\phi]$. A convenient way to do this is to start with the functional $W[j]$, as an expansion in powers of the source j, and carry out the Legendre transformation to $\Gamma[\phi]$ order by order in this formal expansion. We shall now revert to the usual procedure followed through the rest of the book, and reinstate natural units $\hbar = 1$, keeping in mind that an expansion in explicit powers of Planck's constant amounts to nothing more than a reorganization of the perturbation theory according to the number of loops. Moreover, to keep the resultant formulas from expanding to intolerable lengths on the printed page, we shall assume that spacetime has been discretized and work in a purely discrete framework, in which spacetime points x are replaced by indices i ($i = 1, 2, \ldots N$, where the spacetime lattice x_i has N points), and sources and fields are localized by attaching the relevant index- $J_i \equiv j(x_i)$ or $\phi_j \equiv \phi(x_j)$, for example. Integrals over spacetime will be replaced by summations over (typically) repeated indices, and operators $\mathcal{K}$ (resp. Green functions $\Delta_E(x, y)$, $G^{(3)}(x, y, z)$,etc.) by appropriately multi-indexed objects $\mathcal{K}_{ij}$ (resp. $\Delta_{ij}, G_{ijk}^{(3)}$,etc.). Connected contributions are readily identified as terms which cannot be algebraically factored into two or more parts involving non-overlapping sets of summed indices. We also allow for polynomial interactions $P(\phi)$ including both even and odd powers of the field ϕ, so that amplitudes $G^{(n)}$ are in general non-vanishing, even for odd n. By definition, $W[j]$ is the generating functional of the n-point connected amplitudes $G^{(n)}$, so in this discrete notation we have simply

$$W[J] = \frac{1}{2} J_i \hat{\Delta}_{ij} J_j + \frac{1}{3!} G_{ijk}^{(3)} J_i J_j J_k + \frac{1}{4!} G_{ijkl}^{(4)} J_i J_j J_k J_l + \ldots \qquad (10.148)$$

whence

$$\phi_i = \frac{\partial W}{\partial J_i} = \hat{\Delta}_{ij} J_j + \frac{1}{2} G_{ijk}^{(3)} J_j J_k + \frac{1}{3!} G_{ijkl}^{(4)} J_j J_k J_l + O(J^4) \qquad (10.149)$$

Note that here $\hat{\Delta}_{ij}$ is the discrete form of the *full Euclidean propagator* $\hat{\Delta}_E$, as $W[j]$ generates the complete interacting connected Green functions of the theory, and $\hat{\mathcal{K}}$ will now be defined as the (discrete) inverse of this full two-point function (which of course, in the free field limit, reduces to a discretized version of the Euclidean Klein–Gordon operator $\mathcal{K} = -\Box + m^2$).

The last relation can now be inverted, order by order in increasing powers of ϕ (which is of order J, by (10.149)), to yield J_i as a function of ϕ_i. Through terms of order ϕ^3 we find (see Problem 9), using the shorthand $(\hat{\mathcal{K}}\phi)_i \equiv \hat{\mathcal{K}}_{ij}\phi_j$,

$$J_i = (\hat{\mathcal{K}}\phi)_i - \frac{1}{2}\hat{\mathcal{K}}_{ij}G^{(3)}_{jkl}(\hat{\mathcal{K}}\phi)_k(\hat{\mathcal{K}}\phi)_l$$

$$+\frac{1}{2}\hat{\mathcal{K}}_{ij_1}G^{(3)}_{j_1k_1l_1}(\hat{\mathcal{K}}\phi)_{l_1}\hat{\mathcal{K}}_{k_1k_2}G^{(3)}_{k_2l_2j_2}(\hat{\mathcal{K}}\phi)_{l_2}(\hat{\mathcal{K}}\phi)_{j_2}$$

$$-\frac{1}{3!}\hat{\mathcal{K}}_{ij}G^{(4)}_{jklm}(\hat{\mathcal{K}}\phi)_k(\hat{\mathcal{K}}\phi)_l(\hat{\mathcal{K}}\phi)_m + O(\phi^4) \tag{10.150}$$

Reinserting this result in the expansion for $W[J]$, (10.148), we find after some straightforward algebra (Problem 10), the effective action through terms of order ϕ^4:

$$\Gamma[\phi] = -W[J(\phi)] + J_i\phi_i$$

$$= \frac{1}{2}\phi_i\hat{\mathcal{K}}_{ij}\phi_j - \frac{1}{3!}G^{(3)}_{ijk}(\hat{\mathcal{K}}\phi)_i(\hat{\mathcal{K}}\phi)_j(\hat{\mathcal{K}}\phi)_k$$

$$-\frac{1}{4!}\{G^{(4)}_{ijkl}(\hat{\mathcal{K}}\phi)_i(\hat{\mathcal{K}}\phi)_j(\hat{\mathcal{K}}\phi)_k(\hat{\mathcal{K}}\phi)_l$$

$$-3G^{(3)}_{i_1j_1k_1}(\hat{\mathcal{K}}\phi)_{j_1}(\hat{\mathcal{K}}\phi)_{k_1}\hat{\mathcal{K}}_{i_1i_2}G^{(3)}_{i_2j_2k_2}(\hat{\mathcal{K}}\phi)_{j_2}(\hat{\mathcal{K}}\phi)_{k_2}\} + O(\phi^5) \tag{10.151}$$

We observe that the individual terms appearing in $\Gamma[\phi]$ are again connected, consisting of the connected $G^{(n)}$ amplitudes singly, or tied together by $\hat{\mathcal{K}}$ operators, which remove a single internal (full) propagator when two separate $G^{(n)}$ are connected (so as to avoid doubling the connecting Feynman propagator, as the $G^{(n)}$ are not amputated). Moreover, the inverse (full) propagators $\hat{\mathcal{K}}$ attached to all external legs (i.e., to factors of the external field ϕ) ensure the removal of all external legs, so that $\Gamma[\phi]$ *generates amputated diagrams*—exactly the amplitudes leading to a smooth perturbative evaluation of the LSZ formula, as discussed previously. If we define a general n-point *proper vertex* $\Gamma^{(n)}$ as the n-th derivative at zero field of $\Gamma[\phi]$ (with a minus included to take care of the sign change in going from W to Γ):

$$\Gamma^{(n)}_{i_1i_2...i_n} \equiv -\frac{\partial^n\Gamma[\phi]}{\partial\phi_{i_1}\partial\phi_{i_2}..\partial\phi_{i_n}}\bigg|_{\phi=0} \tag{10.152}$$

we see that the two-point proper vertex is just $\Gamma^{(2)}_{ij} = -\hat{\mathcal{K}}_{ij}$, i.e., (minus) the inverse full propagator. Similarly, $\Gamma^{(3)}_{ijk}$, the three-point proper vertex, is just the amputated three-point connected Green function

$$\Gamma^{(3)}_{ijk} = \hat{\mathcal{K}}_{ii'}\hat{\mathcal{K}}_{jj'}\hat{\mathcal{K}}_{kk'}G^{(3)}_{i'j'k'} \tag{10.153}$$

which, for obvious reasons, is automatically 1PI. At order ϕ^4 a new feature appears, as the proper vertex is given by a combination of terms, specifically, the full connected four-point Green function $G^{(4)}$, with the four external legs amputated (by the $\hat{\mathcal{K}}$ factors), minus three other terms which are clearly one-particle-reducible in character:

$$\Gamma_{ijkl}^{(4)} = \hat{\mathcal{K}}_{ii'}\hat{\mathcal{K}}_{jj'}\hat{\mathcal{K}}_{kk'}\hat{\mathcal{K}}_{ll'}G_{i'j'k'l'}^{(4)} - \hat{\mathcal{K}}_{ii'}\hat{\mathcal{K}}_{kk'}G_{mi'k'}^{(3)}\hat{\mathcal{K}}_{mn}G_{nj'l'}^{(3)}\hat{\mathcal{K}}_{jj'}\hat{\mathcal{K}}_{ll'}$$

$$- \hat{\mathcal{K}}_{ii'}\hat{\mathcal{K}}_{jj'}G_{mi'j'}^{(3)}\hat{\mathcal{K}}_{mn}G_{nk'l'}^{(3)}\hat{\mathcal{K}}_{kk'}\hat{\mathcal{K}}_{ll'} - \hat{\mathcal{K}}_{ii'}\hat{\mathcal{K}}_{ll'}G_{mi'l'}^{(3)}\hat{\mathcal{K}}_{mn}G_{nj'k'}^{(3)}\hat{\mathcal{K}}_{jj'}\hat{\mathcal{K}}_{kk'}$$

$$(10.154)$$

A glance at Fig. 10.9, which shows a decomposition of the amputated $G^{(4)}$ amplitude into parts built from 1PI amplitudes (connected with full propagators), clarifies the meaning of the subtracted terms in (10.154): they serve to precisely remove from the full connected four-point amplitude its one-particle-reducible pieces. In other words, the first term on the right-hand side of Fig. 10.9, consisting of the 1PI four-point contributions to the full connected four-point amplitude, is identical to $\Gamma_{ijkl}^{(4)}$. So, at least through vertices with four external legs, the effective action constructed by Legendre transformation does indeed seem to generate the amputated one-particle-irreducible diagrams of the theory, and only such diagrams. Of course, it is hardly obvious from the above that this remains true for n-point amplitudes with arbitrarily large n.

A simple proof that the Legendre transform $\Gamma[\phi]$ indeed generates all the one-particle-irreducible, and only the one-particle-irreducible, graphs for arbitrary powers of ϕ can be given using a trick described by Zinn–Justin (Zinn–Justin, 1989). The individual contributions on the right-hand side of (10.154) are recognized as corresponding to connected graphs in the sense that any external index (like i) can be connected to any other (like l) by a continuous sequence of indices connected by the multi-index connected Green functions $\hat{\mathcal{K}}, G^{(3)}, G^{(4)}$, etc. For example, taking the second term on the right-hand side of (10.154), corresponding to the second graph on the right-hand side in Fig. 10.8, the index i can be connected to l by the sequence $i \to i' \to m \to n \to l' \to l$, passing sequentially through $\hat{\mathcal{K}}_{ii'}, G_{mi'k'}^{(3)}, \hat{\mathcal{K}}_{mn}, G_{nj'l'}^{(3)}$, and $\hat{\mathcal{K}}_{ll'}$. We can investigate the effect of cutting internal lines on the graphs generated by $\Gamma[\phi]$ by introducing a separable perturbation of the Klein–Gordon operator in the quadratic part of the action, as follows:

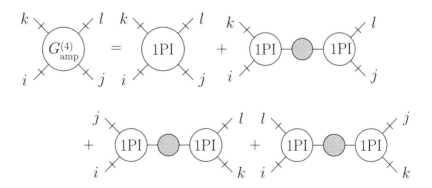

Fig. 10.9 Decomposition of $G_{\mathrm{amp}\ ijkl}^{(4)}$ in terms of 1PI vertices: crossbars indicate that the external legs are "amputated".

$$S_\epsilon[\phi] = S[\phi] + \frac{\epsilon}{2} \sum_{i,j} \phi_i \phi_j = S[\phi] + \frac{\epsilon}{2} \left(\sum_i \phi_i\right)^2 \tag{10.155}$$

The quadratic part of the discretized action is now

$$\frac{1}{2} \sum_{i,j} \phi_i (\mathcal{K}_{ij} + \epsilon M_{ij}) \phi_j, \quad M_{ij} = 1, \forall i,j \tag{10.156}$$

with a free propagator Δ_ϵ given by the inverse of $\mathcal{K} + \epsilon M$:

$$(\Delta_\epsilon)_{ij} = \Delta_{ij} - \epsilon v_i v_j + O(\epsilon^2), \quad v_i \equiv \sum_j \Delta_{ij} \tag{10.157}$$

The contributions of first order in the perturbing parameter ϵ evidently correspond to the replacement of a single free propagator Δ_{ij} by the separable piece $v_i v_j$ for every possible internal line,[21] and would therefore necessarily contain *disconnected graphs* unless the initial amplitude was itself 1PI. For non-zero ϵ, the discretized generating functional $Z[J]$ becomes, to first order in ϵ,

$$Z_\epsilon[J] = \int \prod_i d\phi_i \, e^{-S[\phi] - \frac{1}{2}\epsilon(\sum_i \phi_i)^2 + J_i \phi_i} = \left(1 - \frac{\epsilon}{2} \sum_{i,j} \frac{\partial^2}{\partial J_i \partial J_j}\right) Z[J] \tag{10.158}$$

and its logarithm $W_\epsilon[J]$ is given to the same order by

$$W_\epsilon[J] = W[J] - \frac{\epsilon}{2} \frac{1}{Z[J]} \sum_{i,j} \frac{\partial^2 Z[J]}{\partial J_i \partial J_j}$$

$$= W[J] - \frac{\epsilon}{2} \sum_{i,j} \left(\frac{\partial^2 W[J]}{\partial J_i \partial J_j} + \frac{\partial W[J]}{\partial J_i} \frac{\partial W[J]}{\partial J_j}\right) \tag{10.159}$$

Recall here that in going over from the external source J_i to the classical field ϕ_i, we have $\frac{\partial W[J]}{\partial J_i} = \phi_i$, while $\frac{\partial^2 W[J]}{\partial J_i \partial J_j}$ is the inverse of the matrix $\frac{\partial^2 \Gamma[\phi]}{\partial \phi_i \partial \phi_j}$ (cf. (10.142) and (10.143)). Now, the chain rule formula

$$\left.\frac{\partial}{\partial \epsilon}\right|_{\phi_i} = \left.\frac{\partial}{\partial \epsilon}\right|_{J_i} + \left.\frac{\partial J_i}{\partial \epsilon}\right|_{\phi_i} \frac{\partial}{\partial J_i} \tag{10.160}$$

applied to the Legendre relation for the perturbed effective action $\Gamma_\epsilon[\phi]$

$$\Gamma_\epsilon[\phi] + W_\epsilon[J] - J_j \phi_j = 0 \tag{10.161}$$

---

[21] As $\Gamma[\phi]$ generates amputated graphs, the only lines present are internal lines!

gives

$$
0 = \frac{\partial \Gamma_\epsilon[\phi]}{\partial \epsilon}\bigg|_{\phi_i} + \frac{\partial W_\epsilon[J]}{\partial \epsilon}\bigg|_{J_i} + \frac{\partial J_i}{\partial \epsilon}\bigg|_{\phi_i} \frac{\partial}{\partial J_i}(W_\epsilon[J] - J_j \phi_j)
$$

$$
= \frac{\partial \Gamma_\epsilon[\phi]}{\partial \epsilon}\bigg|_{\phi_i} + \frac{\partial W_\epsilon[J]}{\partial \epsilon}\bigg|_{J_i} + \frac{\partial J_i}{\partial \epsilon}\bigg|_{\phi_i}(\phi_i - \phi_i) = \frac{\partial \Gamma_\epsilon[\phi]}{\partial \epsilon}\bigg|_{\phi_i} + \frac{\partial W_\epsilon[J]}{\partial \epsilon}\bigg|_{J_i}
$$

so that the first-order shift in the effective action $\Gamma_\epsilon[\phi]$ is just minus that in $W_\epsilon[J]$, whence, from (10.159),

$$
\Gamma_\epsilon[\phi] = \Gamma[\phi] + \frac{\epsilon}{2}\left(\sum_i \phi_i\right)^2 + \frac{\epsilon}{2}\sum_{i,j}\left(\frac{\partial^2 \Gamma[\phi]}{\partial \phi_i \partial \phi_j}\right)^{-1} \tag{10.162}
$$

using the fact that the second derivative matrices of $\Gamma[\phi]$ and $W[J]$ are inverses. The second term on the right-hand side of (10.162) is just the perturbation originally introduced in the action (10.155), which must, of course, appear in the $O(\hbar^0)$ (classical) part of the full effective action $\Gamma_\epsilon[\phi]$. It generates the expected disconnected part of the two-point function (inverse full propagator). The third term contains the effect of disconnecting a single internal line in all the higher-order vertices, and a little thought shows that it consists entirely of connected diagrams. Indeed, as a function of the J_i sources, it expands into obviously connected graphs, as it is just $\sum_{i,j}\frac{\partial^2 W}{\partial J_i \partial J_j}$. But when each of the $J_i = \frac{\partial \Gamma[\phi]}{\partial \phi_i}$ is inserted as a function of the ϕ_i into the latter expression, it simply expands the previous connected graph (from $W[J]$) by a connected extension, so even when expanded as a function of the ϕ_i only connected diagrams are obtained. Note that the combinatoric factors relating different 1PI contributions are preserved in $\Gamma[\phi]$: the process of going from W to Γ involves (as we see from (10.154), and in Fig. 10.9) simply (a) amputating external lines, and (b) removing *en bloc* any one-particle-reducible graphs. This concludes the proof that the graphs generated by $\Gamma[\phi]$ remain connected even when any single internal line is cut, and hence must correspond exactly to the amputated 1PI graphs appearing in the connected n-point functions generated by $W[j]$.

Historically, the concept of one-particle-irreducibility was first introduced by Dyson (Dyson, 1949) in the context of the three-point function (electron–electron–photon vertex) in quantum electrodynamics, where the 1PI graphs contributing to this three-point function were referred to as the "proper vertex part". As we shall see later in Part 4, the systematic discussion of renormalizability pioneered by Dyson makes critical use of the fact that the ultraviolet sensitivity of the theory (in other words, the divergence structure of the loop integrals appearing in a general graph) can be fully analysed in terms of 1PI diagrams, as loop integrals in separate 1PI pieces of a larger graph are algebraically decoupled from one another.

A more systematic treatment of n-particle irreducibility was carried out by Symanzik a decade later (Symanzik, 1960), emphasizing the fact that the study of the singularity structure of amplitudes, as a function of the external momenta, was intimately related to the ability to decompose the contributing graphs by cutting one, two, or more internal lines. For example, the only amputated graphs capable of

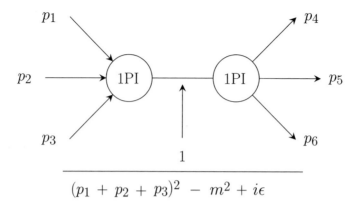

$$\frac{1}{(p_1 + p_2 + p_3)^2 - m^2 + i\epsilon}$$

Fig. 10.10 A one-particle-reducible contribution to the 6-point function $G^{(6)}(p_1, p_2, .., p_6)$.

producing single-particle pole singularities of the form $1/((\sum p_i)^2 - m^2)$, where the summed momenta p_i represent some non-trivial subset of the (appropriately signed) external momenta of the graph, involve one-particle-*reducible* diagrams such as the one shown in Fig. 10.10 in ϕ^4 theory (we are back in Minkowski space, hence the $i\epsilon$ in the propagator). The 1PI four-point vertices on either side of the central line cannot contain any single-particle poles, which arise only when single lines connect separate connected parts of the graph. Similarly, the ability to decompose a graph into two pieces by cutting two internal lines, as in Fig. 10.11, implies a branch-point structure in the variable $t \equiv (p_1 - p_2)^2$ when the two-particle threshhold is reached (at $t = 4m^2$).[22] Correspondingly, such two-particle threshholds are *absent* from 2PI (or two-particle-irreducible) amplitudes, which can only be disconnected by cutting at least *three* internal lines of the graphs contributing to such amplitudes, which are commonly referred to as "Bethe–Salpeter kernels". They play, as we shall see in the next chapter, an important role in understanding the physics of threshhold bound

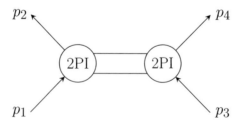

Fig. 10.11 A two-particle-reducible contribution to the four-point function $G^{(4)}(p_1, p_2, p_3, p_4)$.

---

[22] The reader may recall that exactly such branch-points appeared in our discussion of the Lehmann representation for the full propagator in momentum space in Section 9.5, with new cuts appearing precisely at squared-momentum values $w \equiv p^2$ at which new intermediate multi-particle states become kinematically possible.

states in quantum field theory. Finally, with the recognition of the importance of spontaneous symmetry breaking in quantum field theory in the 1960s, the central role of the effective action in understanding the energetics of broken symmetry became clear, starting with the seminal paper of Jona-Lasinio (Jona-Lasinio, 1964)—a topic to which we return in Chapter 14.

10.5   How to stop worrying about Haag's theorem

We have already indicated on numerous occasions, without providing specific justification, that there are difficulties in the implementation of an interaction picture in the case of continuum field theories, which can be circumvented by a temporary *full regularization* of the theory (i.e., by introduction of both large-distance (IR) and small-distance (UV) cutoffs) which reduces the number of independent dynamical variables to a finite number—for example, the fields, and their time-derivatives (which play the role of conjugate momenta)—on a finite number of spacetime points. The price one pays is, of course, the loss (one hopes, temporary) of the full continuous Poincaré symmetry of the theory. The unpleasant fact (now commonly referred to as "Haag's theorem" (Haag, 1955)) that the formulation of an interaction picture of time-development in an infinite-volume continuum field theory is mathematically untenable has confused many generations of students of quantum field theory, who look quite naturally to the spectacular successes of renormalized perturbation theory in quantum electrodynamics (all based on interaction-picture formulas) and ask, "Where's the problem?" In this section we shall attempt to allay the understandable fear that perturbative methods, and the Feynman graph approach described earlier in this chapter, are founded on mathematical quicksand, and therefore in some sense unreliable despite their obvious empirical success over the years.

We begin with a very simple example, found already in Haag's seminal paper (Haag, 1955), which nevertheless has all the essential features of the general case, but allows explicit calculation and a concrete display of the source of the difficulty. We shall start with a free scalar field of mass m_1 and introduce a perturbation of an extremely innocent kind—a shift to a free scalar field of mass m_2. Thus the Hamiltonian is (cf. (6.89)):

$$H = H_0 + V \tag{10.163}$$

$$H_0 = \int d^3x \mathcal{H}_0(\vec{x}, t) = \int d^3x : \{\frac{\dot{\phi}^2}{2} + \frac{1}{2}|\vec{\nabla}\phi|^2 + \frac{1}{2}m_1^2\phi^2\} : \tag{10.164}$$

$$V = \int d^3x \mathcal{H}_{\text{int}}(\vec{x}, t) = \int d^3x \frac{1}{2}\delta m^2 : \phi^2 : , \quad \delta m^2 \equiv m_2^2 - m_1^2 \tag{10.165}$$

Thus, our unperturbed field is a free field $\phi_1(x)$ of mass m_1, while our full "interacting" Heisenberg field is likewise a free scalar field $\phi_2(x)$, but of mass m_2. We should hardly expect to encounter difficulties with the interaction picture in so simple a case, but as we shall soon see, appearances in these matters are definitely deceiving!

On the basis of our treatment of the interaction picture in Section 4.3, we would expect that the ground states of the two Hamiltonians H_0 (i.e., the vacuum $|0\rangle_1$ for scalar particles of mass m_1) and H (the vacuum $|0\rangle_2$ for particles of mass m_2) would be straightforwardly related by a unitary operator. For example, choosing the in-vacuum as the ground state $|0\rangle_2$ for H, we have (cf. (4.156))

$$|0\rangle_2 = \Omega^-|0\rangle_1 \tag{10.166}$$

with Ω^- the unitary Møller wave operator defined in Section 4.3. Of course, $|0\rangle_1$ is the Fock vacuum with respect to the destruction operators obtained in the usual way from $\phi_1(x)$:

$$a_1(\vec{k})|0\rangle_1 = 0, \ \ \forall \vec{k} \tag{10.167}$$

$$a_1(\vec{k}) = \frac{i}{\sqrt{(2\pi)^3 2E_1(k)}} \int d^3x e^{ik \cdot x} \frac{\overleftrightarrow{\partial}}{\partial t} \phi_1(\vec{x}, t)$$

$$= \frac{i}{\sqrt{(2\pi)^3 2E_1(k)}} \int d^3x e^{-i\vec{k}\cdot\vec{x}} \{\dot{\phi}_1(\vec{x}, 0) - iE_1(k)\phi_1(\vec{x}, 0)\} \tag{10.168}$$

The formula for the destruction operator in terms of the field is time-independent, so we have chosen to set $t = 0$ in the last line. We recall from Chapter 4 that, by convention, the various pictures of time-development in quantum theory (interaction, Schrödinger, Heisenberg) are presumed to coincide at time $t = 0$. In particular, our interaction-picture field $\phi_1(\vec{x}, t)$ and "Heisenberg" field $\phi_2(\vec{x}, t)$ must coincide at $t = 0$, as must (cf. (9.40)) their respective time-derivatives:

$$\phi_1(\vec{x}, 0) = \phi_2(\vec{x}, 0) \tag{10.169}$$

$$\dot{\phi}_1(\vec{x}, 0) = \dot{\phi}_2(\vec{x}, 0) \tag{10.170}$$

where our "Heisenberg" field $\phi_2(x)$ is expressed in the usual way in terms of the destruction and creation operators $a_2(\vec{k}), a_2^\dagger(\vec{k})$ appropriate for particles of mass m_2:

$$\phi_2(x) = \int \frac{d^3q}{\sqrt{2E_2(q)(2\pi)^3}} (a_2(\vec{q})e^{-iq\cdot x} + a_2^\dagger(\vec{q})e^{iq\cdot x}) \tag{10.171}$$

Using the identity relations (10.169, 10.170), inserting (10.171) into (10.168), and carrying out the $\vec{x}$ and $\vec{q}$ integrations, we find the connection between the creation–destruction operators for the particles of mass m_1 and m_2:

$$a_1(\vec{k}) = \alpha(k)a_2(\vec{k}) + \beta(k)a_2^\dagger(-\vec{k}) \tag{10.172}$$

$$\alpha(k) = \frac{1}{2}\left(\sqrt{\frac{E_1(k)}{E_2(k)}} + \sqrt{\frac{E_2(k)}{E_1(k)}}\right) \tag{10.173}$$

$$\beta(k) = \frac{1}{2}\left(\sqrt{\frac{E_1(k)}{E_2(k)}} - \sqrt{\frac{E_2(k)}{E_1(k)}}\right) \tag{10.174}$$

Note that the commutation algebra of the creation–destruction operators is preserved by this transformation (commonly referred to as a *Bogoliubov transformation*), as a consequence of $\alpha(k)^2 - \beta(k)^2 = 1$; in particular,

$$[a_1(\vec{k}), a_1^\dagger(\vec{k}')] = [a_2(\vec{k}), a_2^\dagger(\vec{k}')] = \delta^3(\vec{k} - \vec{k}') \tag{10.175}$$

In a system with a finite number of degrees of freedom (for example, if the system were quantized on a discrete spacetime lattice with a finite number of points, with discrete values for the spatial momenta $\vec{k}$), standard results going back to von Neumann ensure the existence of a well-defined unitary transformation relating the $a_1(\vec{k})$ and $a_2(\vec{k})$. If the vacua $|0\rangle_1, |0\rangle_2$ (satisfying respectively $a_1(\vec{k})|0\rangle_1 = 0, a_2(\vec{k})|0\rangle_2 = 0, \forall\vec{k}$) are indeed related by a proper unitary transformation, as in (10.166), we should expect to be able to expand the $|0\rangle_1$ vacuum in terms of Fock states for particle 2, and obtain thereby a unit-normalized state:

$$|0\rangle_1 = g_0|0\rangle_2 + \int d^3k_1 g_1(\vec{k}_1)|\vec{k}_1\rangle_2 + \frac{1}{2}\int d^3k_1 d^3k_2 g_2(\vec{k}_1, \vec{k}_2)|\vec{k}_1, \vec{k}_2\rangle_2 + \dots \tag{10.176}$$

with $g_0 = {}_2\langle 0|0\rangle_1$ and

$${}_1\langle 0|0\rangle_1 = 1 = |g_0|^2 + \int d^3k_1 |g_1(\vec{k}_1)|^2 + \frac{1}{2}\int d^3k_1 d^3k_2 |g_2(\vec{k}_1, \vec{k}_2)|^2 + \dots \tag{10.177}$$

On the other hand, from (10.172), we know that

$$\alpha(k)a_2(\vec{k})|0\rangle_1 = -\beta(k)a_2^\dagger(-\vec{k})|0\rangle_1 \tag{10.178}$$

If we successively take the overlap of (10.178) with ${}_2\langle 0|, {}_2\langle\vec{k}_1|, {}_2\langle\vec{k}_1\vec{k}_2|, \dots$, we find easily, using the expansion (10.176),

$$g_1(\vec{k}) = 0 \tag{10.179}$$

$$g_2(\vec{k}_1, \vec{k}_2) = -\frac{\beta(k)}{\alpha(k)}\delta^3(\vec{k}_1 + \vec{k}_2) \, {}_2\langle 0|0\rangle_1 \tag{10.180}$$

$$g_3(\vec{k}_1, \vec{k}_2, \vec{k}_3) = 0, \text{ and so on} \tag{10.181}$$

Inserting the result (10.180) into (10.177) however, the integral diverges, as it involves the square of a δ-function. As $\beta(k) \neq 0$ if $m_1 \neq m_2$, the finite norm of $|0\rangle_1$ therefore requires the overlap ${}_2\langle 0|0\rangle_1 = g_0$ to vanish. The recursion implied by (10.178) then leads to the conclusion that $g_2 = g_4 = g_6 = g_{2n} = 0$. As the states with odd numbers of particles do not appear, we have arrived at an evident contradiction: ${}_1\langle 0|0\rangle_1 = 0$! In fact, the argument extends to arbitrary multi-particle states of particle 1, which all have vanishing overlap with those of particle 2. It is apparent that the Fock spaces of scalar particles of differing mass cannot be consistently incorporated within the same separable Hilbert space, allowing the desired unitary transformation between corresponding states. This result is evidently a direct consequence of the continuum normalization of the two-particle states, which in turn can be traced back to the fact that we have quantized our system in an infinite spatial box. Our problems in this particular example arise from the fact that we cannot obtain a properly normalized discrete vacuum state by mixing another such state with multi-particle states built from pairs of continuum-normalized opposite-momentum particles.

We may attempt to circumvent the above difficulties and re-establish a consistent treatment of both the "free" (particle 1) and "interacting" (particle 2) Fock spaces within a single Hilbert space by quantizing the system in a cubical box of finite spatial volume $V = L^3$ (with periodic boundary conditions), so that the allowed spatial momenta are now discrete, $k_i = \frac{2\pi n_i}{L}$, n_i integer, $i = 1,2,3$, and the corresponding multi-particle states discretely normalized, like the vacuum state from which they arise by application of creation operators. To distinguish discrete from continuous momenta, we shall indicate the former by subscripts (rather than arguments) in the following. Thus, the scalar field ϕ_1 now takes the form

$$\phi_1(x) = V^{-1/2} \sum_{\vec{k}} \frac{1}{\sqrt{2E_{\vec{k}}}} (a_{1\vec{k}} e^{-ik \cdot x} + a_{1\vec{k}}^\dagger e^{ik \cdot x}) \qquad (10.182)$$

with a similar equation for ϕ_2. The infinite volume limit is obtained from (10.182) with the usual transcription $\frac{1}{V} \sum_{\vec{k}} \to \int \frac{d^3k}{(2\pi)^3}$ and the normalization $a_{1\vec{k}} \to \frac{(2\pi)^{3/2}}{V^{1/2}} a_1(\vec{k})$ (so that $[a_{1\vec{k}}, a_{1\vec{k}'}^\dagger] = \delta_{\vec{k}\vec{k}'}$). The discrete analog of (10.172–10.174) is

$$a_{1\vec{k}} = \alpha_k a_{2\vec{k}} + \beta_k a_{2,-\vec{k}}^\dagger \qquad (10.183)$$

$$\alpha_k = \frac{1}{2}\left(\sqrt{\frac{E_{1k}}{E_{2k}}} + \sqrt{\frac{E_{2k}}{E_{1k}}}\right) \qquad (10.184)$$

$$\beta_k = \frac{1}{2}\left(\sqrt{\frac{E_{1k}}{E_{2k}}} - \sqrt{\frac{E_{2k}}{E_{1k}}}\right) \qquad (10.185)$$

with $E_{ik} = \sqrt{\vec{k}^2 + m_i^2}$ as before, but with $\vec{k}$ discrete. Recognizing that the expansion (10.176) involves only pairs of opposite momentum particles $\vec{k}, -\vec{k}$ (which can appear multiply for each discrete $\vec{k}$), we divide the set of all pairs into a set of independent pairs $\vec{k}, -\vec{k}$ with $\vec{k}$ in the right hemisphere $k_x > 0$ in order to avoid double-counting. The Bogoliubov transformation can be carried out separately on each such pair as the creation–annihilation operators for distinct $\vec{k}$ commute. Focussing our attention on a specific discrete $\vec{k}$, we can expand the $|0\rangle_1$ vacuum as follows:

$$|0\rangle_1 = \sum_N c_N |N\rangle_2, \quad |N\rangle_2 \equiv \frac{1}{N!}(a_{2\vec{k}}^\dagger a_{2,-\vec{k}}^\dagger)^N |0\rangle_2, \quad {}_2\langle N|N\rangle_2 = 1 \qquad (10.186)$$

The requirement $a_{1\vec{k}}|0\rangle_1 = 0$, using (10.183) then leads directly to the simple recursion relation (see Problem 11):

$$c_n = -\frac{\beta_k}{\alpha_k} c_{n-1} \Rightarrow c_N = (-\frac{\beta_k}{\alpha_k})^N c_0 \qquad (10.187)$$

The normalization condition ${}_1\langle 0|0\rangle_1 = 1 = \sum_N |c_N|^2 = |c_0|^2 \frac{1}{1-\beta_k^2/\alpha_k^2}$ thus gives

$$|c_0|^2 = \frac{\alpha_k^2 - \beta_k^2}{\alpha_k^2} = \frac{1}{\alpha_k^2} = \frac{4E_{1k}E_{2k}}{(E_{1k} + E_{2k})^2} \qquad (10.188)$$

The overlap between the two vacua $_2\langle 0|0\rangle_1$ is then given, up to an irrelevant phase, by the product[23] of c_0 factors of the form (10.188) for all independent pairs, labeled by $\vec{k}$ values in the right hemisphere:

$$_2\langle 0|0\rangle_1 = \prod_{\vec{k}} \frac{2\sqrt{E_{1k}E_{2k}}}{E_{1k} + E_{2k}} = \exp \sum_{\vec{k}} \ln \frac{2\sqrt{E_{1k}E_{2k}}}{E_{1k} + E_{2k}} \tag{10.189}$$

In the large volume limit, $\sum_{\vec{k}} \to \frac{V}{2(2\pi)^3} \int d^3k$ (the extra factor of $1/2$ arises from the restriction of $\vec{k}$ to the right hemisphere) and this becomes

$$_2\langle 0|0\rangle_1 \simeq \exp\{-\frac{V}{2(2\pi)^3} \int d^3k \ln\{\frac{(E_1(k) + E_2(k))}{2\sqrt{E_1(k)E_2(k)}}\}\} \tag{10.190}$$

For large k (i.e., $k >> m_1, m_2$) one easily verifies that

$$\ln\{\frac{(E_1(k) + E_2(k))}{2\sqrt{E_1(k)E_2(k)}}\} \sim \ln\{1 + \frac{(m_1^2 - m_2^2)^2}{32k^4} + O(1/k^6)\} \sim \frac{(m_1^2 - m_2^2)^2}{32k^4} \tag{10.191}$$

so that the momentum integral in (10.190) is ultraviolet convergent for spatial dimensions three or less (it is manifestly convergent for small k). However, the presence in the exponent of a volume factor multiplying this finite integral (in three dimensions) shows immediately the vanishing of the overlap of the two vacua in the infinite volume limit found earlier. In the usual terminology employed by axiomatic field theorists, the Fock spaces built on these two vacua are said to be "unitarily inequivalent" spaces. It should be noted that the ultraviolet dependence of the integral appearing in expressions for the vacuum overlap is extremely theory (and spacetime dimension) dependent: for example, an analogous computation for two free spin-$\frac{1}{2}$ fields of different mass gives a momentum integral divergent for spatial dimensions greater than one (rather than three, in the scalar case). The important point is that, irrespective of the ultraviolet behavior, the overlap between normalized vacua for free fields of differing mass *necessarily vanishes* in the infinite volume limit as a consequence of the geometrical/kinematical mismatch in Hilbert space between discretely normalized and continuum normalized states.

The above example has been generalized to a much broader statement concerning the non-existence of the interaction picture in essentially all cases in which field theory Hamiltonians H_0 ("free") and H ("interacting"), defined in an infinite-volume continuum spacetime supporting the full Poincaré group of spacetime invariances, differ non-trivially (i.e., by the integral of some local operator density) from each other. We shall state below (though not prove fully) the modern version of this "Haag's theorem": it is an inescapable consequence of the spectral and field axioms (Ia-d, IIa-d) of Section 9.2.

Before going on to discuss a more general proof, let us emphasize what Haag's theorem *does not say* about interacting field theories. In particular, there is no difficulty

---

[23] The vacuum of our theory may be regarded as the direct product of vacua for each momentum mode separately, so that overlaps of the form $_2\langle 0|0\rangle_1$ factorize as indicated.

whatsoever in establishing a well-defined unitary relation between the in- and out-states of an interacting field theory: the overlaps $_\text{out}\langle\beta|\alpha\rangle_\text{in} = S_{\beta\alpha}$ are taken between states living in spaces spanned by a complete basis of eigenstates of *the same Hamiltonian operator H*. Indeed, the Haag–Ruelle and LSZ scattering theories developed in Sections 9.3 and 9.4 lead to a perfectly well-defined, and *unitary*, S-matrix, on the basis of exactly the same axiomatic framework which can be used to establish the validity of Haag's theorem. The LSZ formula, for example, gives a rigorous connection between well-defined Green functions (time-ordered products of the full Heisenberg fields) and this unitary S-matrix. Direct non-perturbative evaluation of the Green functions of the theory (say, by lattice field theory methods) therefore completely circumvents any difficulty with the non-existence of interaction picture, as the latter is simply not employed at any point.

Of course, in many cases the only sensible approach to the evaluation of the Green functions is via perturbation theory, which is inescapably rooted in an interaction-picture formalism, as we discussed earlier in this chapter in Sections 10.1 and 10.2. If we now return to our initial example, with unperturbed and perturbation Hamiltonians given in (10.164, 10.165), we discover with some surprise that the formal execution of the perturbative evaluation *in interaction-picture* of the n-point Green functions of particle 2 (described by the "full" Hamiltonian H) leads to no particular difficulties. For example, using the Gell–Mann–Low theorem (9.21), we have the formal perturbative expansion for the two-point Feynman function $_2\langle 0|T(\phi_2(x)\phi_2(y))|0\rangle_2 = i\Delta_F^{(2)}(x-y)$ of the "interacting" field ϕ_2 in terms of the propagators of the "free" field ϕ_1:

$$
\begin{aligned}
_2\langle 0|T(\phi_2(x)\,\phi_2(y))|0\rangle_2 = {}& _1\langle 0|T(\phi_1(x)\phi_1(y))|0\rangle_1 \\
& -i\frac{\delta m^2}{2}\int {}_1\langle 0|T(\phi_1(x)\phi_1(y):\phi_1(z_1)^2:|0\rangle_1 d^4 z_1 \\
& +\frac{1}{2}(\frac{-i\delta m^2}{2})^2\int {}_1\langle 0|T(\phi_1(x)\phi_1(y):\phi_1(z_1)^2::\phi_1(z_2)^2:|0\rangle_1 d^4 z_1 d^4 z_2 \\
& +...
\end{aligned}
\tag{10.192}
$$

Expanding the vacuum-expectations inside the integrals using the Wick theorem (recall that fields inside a normal product symbol are not contracted), one finds

$$
\begin{aligned}
i\Delta_F^{(2)}(x-y) = {}& i\Delta_F^{(1)}(x-y) \\
& -i\frac{\delta m^2}{2}(2)\int i\Delta_F^{(1)}(x-z_1)i\Delta_F^{(1)}(z_1-y)d^4 z_1 \\
& +\frac{1}{2}(\frac{-i\delta m^2}{2})^2(8)\int i\Delta_F^{(1)}(x-z_1)i\Delta_F^{(1)}(z_1-z_2)i\Delta_F^{(1)}(z_2-y)d^4 z_1 d^4 z_2 \\
& +...
\end{aligned}
\tag{10.193}
$$

The factors (2) and (8) in the second and third lines are combinatoric factors reflecting the number of equivalent ways by which the fields can be contracted to give the displayed product of propagators. On transforming to momentum space, the convolutions over spacetime arguments become simple products, and the perturbative series reveals itself as a simple geometric expansion:

$$\Delta_F^{(2)}(k) = \Delta_F^{(1)}(k) + \delta m^2 (\Delta_F^{(1)}(k))^2 + (\delta m^2)^2 (\Delta_F^{(1)}(k))^3 + \dots$$

$$= \frac{1}{k^2 - m_1^2 + i\epsilon} + \delta m^2 (\frac{1}{k^2 - m_1^2 + i\epsilon})^2 + (\delta m^2)^2 (\frac{1}{k^2 - m_1^2 + i\epsilon})^3 + \dots$$

$$= \frac{1}{k^2 - m_1^2 - \delta m^2 + i\epsilon}$$

$$= \frac{1}{k^2 - m_2^2 + i\epsilon} \tag{10.194}$$

—exactly as expected. And this despite the use of an interaction-picture expansion which, as we have seen earlier, is mathematically inadmissible, given that we have used fields defined in infinite-volume space throughout!

The origin of our unexpected good fortune is not hard to uncover if we repeat the above calculation with our fields quantized in a box of finite volume V, as in (10.182). In this case the interaction picture is well-defined as the overlap between the "free" and "interacting" vacua is non-zero, and we may use the Gell–Mann–Low formula with no qualms. The Feynman propagators are now given, as a straightforward calculation along the lines of the argument leading from (7.217–7.218) and (7.214–7.215) reveals, by discrete sums (rather than integrals) over the spatial momenta. For example, for particle 1:

$$i\Delta_F^{(1)}(x_1 - x_2) = \frac{1}{V} \sum_{\vec{k}} \frac{1}{2E_1(k)} (\theta(t_1 - t_2)e^{ik\cdot(x_2 - x_1)} + \theta(t_2 - t_1)e^{-ik\cdot(x_2 - x_1)})$$

$$= i\frac{1}{V} \sum_{\vec{k}} \int \frac{dk_0}{2\pi} \frac{e^{-ik\cdot(x_1 - x_2)}}{k^2 - m_1^2 + i\epsilon} \tag{10.195}$$

Here the time component k_0 is still continuous (as the time dimension is still unbounded), but the spatial components of $\vec{k}$ are discrete as appropriate for periodic modes in our finite box. We may now repeat essentially *verbatim* the steps leading from (10.192) to (10.194) to obtain for the "full" propagator,

$$\Delta_F^{(2)}(x_1 - x_2) = \frac{1}{V} \sum_{\vec{k}} \int \frac{dk_0}{2\pi} \frac{e^{-ik\cdot(x_1 - x_2)}}{k^2 - m_2^2 + i\epsilon} \tag{10.196}$$

—precisely the expression for the free propagator of a particle of mass m_2, also quantized in a box of volume V. The desired infinite volume result for the "interacting" theory can then be obtained trivially by taking the $V \to \infty$ limit in (10.196), thereby returning to the conventional result (7.215). The recovery of the full Poincaré invariance of the infinite-volume continuum theory, with continuous and unbounded four-momenta, is perfectly straightforward in this case, as we are dealing with theories without non-trivial interactions: in particular, the execution of the perturbation expansion does not lead to ultraviolet-divergent loop integrals. Also, there are no divergent phases in the evolution of the vacuum states (either for particle 1 or particle 2) due to disconnected vacuum graphs. Nevertheless, we emphasize once again that, from the point of view of Haag's theorem, the interaction picture for our toy system in infinite volume is just as pathological as in cases in which non-trivial interactions (involving higher than quadratic terms in the fields) are present.

The situations in which the interaction picture is typically employed, to compute the scattering amplitudes of an interacting field theory say, involve perturbations of the free theory of a much more complicated nature than the innocent shift of mass discussed above, and we should hardly be surprised if the negative conclusions reached above concerning the existence of proper unitary transformations relating the Fock states of the free and interacting field remain in force in these more physically relevant circumstances. The proof of the generalized Haag theorem is usually accomplished in two steps. One first establishes that the unitary equivalence of two irreducible field operators (recall that from Axiom IId of Section 9.2, this implies that the only operator commuting with all fields is a multiple of the identity) implies the equality of their *equal-time* Wightman functions (VEVs of field products). The second, and more difficult, step uses analyticity properties of the Wightman functions following from the spectral and field axioms of Section 9.3 (and embodied in the *Hall–Wightman* theorem) to extend this equality to Wightman functions for arbitrary spacetime arguments. We shall outline the proof of the first part here, and refer the interested reader to the literature for the more technically challenging second step. An alternative proof, due to Jost and Schroer, and involving only the two-point function, will be relegated to an exercise for the reader (see Problem 13).

Let us suppose that we have two scalar[24] fields $\phi_1(\vec{x}, t)$ and $\phi_2(\vec{x}, t)$, with associated canonical momentum fields $\pi_1(\vec{x}, t) \equiv \dot{\phi}_1(\vec{x}, t), \pi_2(\vec{x}, t) \equiv \dot{\phi}_2(\vec{x}, t)$, and related at any given time t by a well-defined unitary operator $V(t)$ defined in a single Hilbert space accomodating both fields (and equal to the interaction-picture operator $U(t, 0)$ in our previous notation: the change is occasioned by the desire to avoid confusion with the $U(R, \vec{a})$ unitary representatives of the Euclidean group introduced below):

$$\phi_2(\vec{x}, t) = V^\dagger(t)\phi_1(\vec{x}, t)V(t) \tag{10.197}$$

$$\pi_2(\vec{x}, t) = V^\dagger(t)\pi_1(\vec{x}, t)V(t) \tag{10.198}$$

The Euclidean subgroup of the Poincaré group, consisting of spatial rotations R and translations $\vec{a}$ is realized on the Hilbert space by unitary operators $U_i(R, \vec{a}), i = 1, 2$ which have the usual action on our local fields (cf. (5.94)):

$$U_i(R, \vec{a})\phi_i(\vec{x}, t)U_i^\dagger(R, \vec{a}) = \phi_i(R\vec{x} + \vec{a}, t) \tag{10.199}$$

$$U_i(R, \vec{a})\pi_i(\vec{x}, t)U_i^\dagger(R, \vec{a}) = \pi_i(R\vec{x} + \vec{a}, t) \tag{10.200}$$

If we think of ϕ_1 (resp. ϕ_2) as our free interaction-picture field ϕ (resp. Heisenberg field ϕ_H), then U_1 (resp. U_2) are the inhomogeneous analogs of the unitary representatives

---

[24] The generalization to fields transforming under non-trivial representations of the Lorentz group is unproblematic.

$U(\Lambda)$ (resp. $U_H(\Lambda)$) introduced in Section 9.1. Note that we have at this stage already committed ourselves to continuum-normalized multi-particle states, by insisting on the invariance of the theory under the continuous Euclidean group. Finally, we shall assume that there is a unique invariant state (vacuum) for each set of Euclidean group representatives:

$$U_i(R,\vec{a})|0\rangle_i = |0\rangle_i, \quad {}_i\langle 0|0\rangle_i = 1, \quad i = 1,2 \tag{10.201}$$

From (10.197, 10.199) it follows that

$$
\begin{aligned}
U_1^\dagger(R,\vec{a})V(t)U_2(R,\vec{a})V^\dagger(t)\phi_1(\vec{x},t) &= U_1^\dagger(R,\vec{a})V(t)U_2(R,\vec{a})\phi_2(\vec{x},t)V^\dagger(t)\\
&= U_1^\dagger(R,\vec{a})V(t)\phi_2(R\vec{x}+\vec{a},t)U_2(R,\vec{a})V^\dagger(t)\\
&= U_1^\dagger(R,\vec{a})\phi_1(R\vec{x}+\vec{a},t)V(t)U_2(R,\vec{a})V^\dagger(t)\\
&= \phi_1(\vec{x},t)U_1^\dagger(R,\vec{a})V(t)U_2(R,\vec{a})V^\dagger(t) \tag{10.202}
\end{aligned}
$$

so that the operator $U_1^\dagger(R,\vec{a})V^\dagger(t)U_2(R,\vec{a})V(t)$ commutes with all fields $\phi_1(\vec{x},t)$ on timeslice t. An exactly similar sequence of manipulations (using (10.198,10.200)) establishes that this commutativity holds also with the $\pi_1(\vec{x},t)$ operators. The creation and annihilation operators appropriate for free field ϕ_1 can be reconstructed from the $\phi_1(\vec{x},t)$ and $\pi(\vec{x},t)$ so commutativity of $U_1^\dagger(R,\vec{a})V^\dagger(t)U_2(R,\vec{a})V(t)$ is thus established with all such operators, which implies that it must act as a multiple of the identity in the Fock space of field ϕ_1 (this is the irreducibility property, here invoked for fields and their conjugate momenta on a single time-slice). Thus

$$U_1^\dagger(R,\vec{a})V(t)U_2(R,\vec{a})V^\dagger(t) = c(R,\vec{a}) \tag{10.203}$$

with $c(R,\vec{a})$ a unimodular c-number. Hence

$$U_2(R,\vec{a}) = c(R,\vec{a})V^\dagger(t)U_1(R,\vec{a})V(t) \tag{10.204}$$

The fact that $U_{1,2}(R,\vec{a})$ form (infinite-dimensional) representations of the Euclidean group (cf. (5.14):

$$U_i(R_1,\vec{a}_1)U_i(R_2,\vec{a}_2) = U_i(R_1R_2,\vec{a}_1 + R_1\vec{a}_2) \tag{10.205}$$

implies that the $c(R,\vec{a})$ must likewise form a *one-dimensional* representation of the Euclidean group:

$$c(R_1,\vec{a}_1)c(R_2,\vec{a}_2) = c(R_1R_2,\vec{a}_1 + R_1\vec{a}_2) \tag{10.206}$$

Some simple group-theoretic reasoning (see Problem 12) leads to the conclusion that the only such representation is the trivial one, $c(R,\vec{a}) = 1$, whence

$$U_2(R,\vec{a}) = V^\dagger(t)U_1(R,\vec{a})V(t) \tag{10.207}$$

Finally, the uniqueness of the vacuum states implies, using (10.207),

$$U_2(R,\vec{a})V^\dagger(t)|0\rangle_1 = V^\dagger(t)U_1(R,\vec{a})|0\rangle_1 = V^\dagger(t)|0\rangle_1 \tag{10.208}$$

so that, up to an irrelevant unimodular phase, $V^\dagger(t)|0\rangle_1$ *is the vacuum for field* ϕ_2:

$$|0\rangle_2 = V^\dagger(t)|0\rangle_1 \tag{10.209}$$

The equality of equal-time vacuum-expectation-values for the two fields now follows trivially:

$$\begin{aligned}
&{}_1\langle 0|\phi_1(\vec{x}_1,t)\phi_2(\vec{x}_2,t)...\phi_1(\vec{x}_n,t)|0\rangle_1 \\
&= {}_1\langle 0|V(t)V^\dagger(t)\phi_1(\vec{x}_1,t)V(t)V^\dagger(t)\phi_1(\vec{x}_2,t)..V(t)V^\dagger(t)\phi_1(\vec{x}_n,t)V(t)V^\dagger(t)|0\rangle_1 \\
&= {}_2\langle 0|\phi_2(\vec{x}_1,t)\phi_2(\vec{x}_2,t)...\phi_2(\vec{x}_n,t)|0\rangle_2 \tag{10.210}
\end{aligned}$$

Recall that our fields ϕ_1, ϕ_2 are supposed to represent free and fully interacting fields, respectively, so this result is already astonishing, as we should certainly not expect, even at equal time, the free-field vacuum-expectation-values to coincide with the corresponding very complicated interacting ones. The final nail in the coffin of the interaction picture is inserted by the realization that the very strong analyticity constraints on the spacetime Wightman functions (cf. Section 9.2) allow the equality expressed in (10.210) to be extended to *arbitrary values of the spacetime coordinates of the fields.* Note that these analyticity properties follow from the full panoply of Wightman axioms (of type I and II) discussed in Section 9.2: in particular, locality, full Poincaré (not just Euclidean group) invariance, and the usual spectral properties. The insertion of θ-functions leads to a similar conclusion for the Feynman (time-ordered) Green functions of fields ϕ_1 and ϕ_2, from which we conclude (via LSZ) that the S-matrix of the interacting field ϕ_2 is equal to that of the free field ϕ_1: namely, unity. Thus, non-trivial interactions are excluded once we make the evidently overly strong assumption of well-defined ("proper") unitary equivalence of the representations for the two fields. The interested reader is encouraged to follow the more detailed accounts of this second step in the argument leading to Haag's theorem, involving an application of the fundamental Hall–Wightman theorem (Hall and Wightman, 1957)(cf. also Section 9.2) on analytic domains of Wightman functions, in (Barton, 1963) and (Greenberg, 1959).

A slightly different route (Streater and Wightman, 1978) to Haag's theorem utilizes the two-point function only, the analyticity properties of which are essentially trivial, as we have seen in Section 9.5. Taking $n = 2$ in (10.210), and with $\phi_1(x)$ a canonically normalized free scalar field of mass m, we have for the "interacting field" ϕ_2

$$_2\langle 0|\phi_2(\vec{x}_1,t)\phi_2(\vec{x}_2,t)|0\rangle_2 = \Delta_+(\vec{x}_1 - \vec{x}_2, 0; m) \tag{10.211}$$

where Δ_+ is the invariant function of (6.63). For x_1, x_2 any pair of space-like separated points, the corresponding times t_1, t_2 can be brought to equality by an appropriate Lorentz boost, so from the Lorentz-invariance properties of Δ_+ we conclude that the two-point Wightman function of the ϕ_2 field must coincide with that of the free field for space-like separations of $x_1 - x_2$. The equality can be analytically extended to the time-like domain: for example, we need only appeal to the Lehmann representation for the two-point function derived in Section 9.5, where the spectral function is already fully determined by knowledge of the two-point function in the space-like region.

Finally, one utilizes a theorem of Jost and Schroer (Jost, 1961) (see also Problem 13), wherein it is shown that any field whose two-point function coincides with that of a free field must itself be a free field (evidently, of the same mass).

The non-existence of the interaction picture for any Poincaré invariant local field theory with essentially any non-trivial split (other than a trivial c-number one) into free and interacting parts of the Hamiltonian is, of course, an unpleasant fact of life given the enormous utility of perturbative Feynman graph technology in modern particle physics. The attitude of the present author to this circumstance has already been outlined above, in the discussion of the perturbative expansion of the two-point function in the toy model of a scalar mass shift. The interaction-picture formalism can be reinstated with complete mathematical rigor by a full regularization of the field theory, in which both spatial infrared (i.e., finite volume) and ultraviolet (i.e., finite lattice spacing) cutoffs are introduced. The resultant theory, at the price of loss of Poincaré invariance, is now a quantum-mechanical system with a finite number of independent degrees of freedom, and the interaction picture makes perfect sense. The problem is now transferred to the issue of regaining sensible (in particular, Poincaré invariant!) results in the limit when these cutoffs are removed, *after the perturbative expansion of the n-point functions needed for evaluation of the S-matrix has been performed*. Note that the perturbative contributions obtained at each finite order of perturbation theory are completely well-defined in this cutoff theory (although, as emphasized previously, the expansion is only an asymptotic one, with the sum of perturbative contributions diverging because of factorial growth of the coefficients, as we shall see in the next chapter).

We consider first the behavior of the cutoff perturbative amplitudes as the spatial volume of the system is allowed to go to infinity. From the discussion in Section 10.2 of persistent interactions, we know that in a theory of massive particles, the only volume singularity of the n-point Feynman functions appears in the phase $_{\text{out}}\langle 0|0\rangle_{\text{in}}$ accumulated in the vacuum due to the vacuum energy density shift induced by interactions. This phase is removed by the simple expedient of considering only connected contributions to the S-matrix: it would in any event disappear subsequently once the S-matrix amplitudes are squared to determine the probability of scattering processes. The infinite volume limit is perfectly smooth in the remaining connected amplitudes, as the appearance of momentum integrals extending down to zero momentum is unproblematic in a theory with massive particles due to the absence of infrared divergences (on this matter, cf. Chapter 19).

It is important to realize that contrary to assertions one sometimes encounters in discussions of Haag's theorem, the vacuum fluctuations encountered in interaction picture, corresponding to the interaction-induced shift in the ground-state energy of the theory (and present even when the theory is fully regulated), are *not* the root cause of the non-existence of the interaction picture. Indeed, Haag's theorem applies in full force to supersymmetric field theories (see Section 12.6) in which the vacuum energy fluctuations cancel identically between bosonic and fermionic contributions. Nevertheless, the interacting n-point functions of these theories most certainly differ from their free limits, guaranteeing the non-existence of the interaction picture by the arguments given above. Indeed, there are typically mass shifts (equal for bosonic particles and their fermionic superpartners of course) induced by the interactions,[25] so we should certainly expect, in analogy to our toy model, a vanishing overlap between the free (or "bare") and interacting vacua of these theories.

The final step in restoring the Poincaré-invariant status of the theory (as well as the unitarity of the S-matrix, which is typically destroyed[26] by the omission of high-momentum states incurred by an ultraviolet cutoff) requires a removal of the short-distance cutoff, for example, in the case of a lattice regularization, by taking the lattice spacing a to zero. At this point the usual ultraviolet singularities of perturbative field theory, occasioned by high-momentum divergences in the loop integrals appearing in the Feynman graph expansion, come into play. It should be emphasized that quite apart from the difficulties engendered by Haag's theorem, the presence of an ultraviolet regularization is a *precondition for even defining a complete dynamical structure* for interacting local quantum field theories in 3+1-dimensional spacetime: the Hamiltonian density defining the dynamics of such theories in terms of local fields necessarily involves coefficients (the "bare" masses $m(a)$, couplings $\lambda(a)$, etc.) which typically are singular (or in some cases, as in asymptotically free theories, go to zero) in the continuum limit $a \to 0$. The dynamics of the interacting theory is reconstructed by a limiting process—renormalization—in which the sensitivity of the cutoff amplitudes to the short-distance physics at the scale of a lattice spacing is shown to be negligible (in the sense that the continuum amplitudes are typically corrected by terms of power order $(m_{\mathrm{ph}}a)^2, (pa)^2$, with p a typical momentum in the process), once the amplitudes are reparameterized in terms of couplings and masses determined by physical measurements at low energy (or distances much larger than the short-distance cutoff). The violations of Poincaré invariance and unitarity due to the presence of a short-distance cutoff are likewise negligible in such cases. Field theories for which power insensitivity to the ultraviolet cutoff can be established order by order in perturbation theory (and to all orders thereof) are termed "renormalizable": their study, and more generally the study of the sensitivity of field theory amplitudes to physics at widely different scales, will be the main object of our attention in Part 4 of this book. The field theories of the Standard Model of particle physics, and its supersymmetric extensions, are all of this type. In the opinion of this author, therefore, the proper response to Haag's theorem is simply a frank admission that the same regularizations needed to make proper mathematical sense of the dynamics of an interacting field theory at each stage of a perturbative calculation will do double duty in restoring the applicability of the interaction picture at intermediate stages of the calculation. The restoration of the complete panoply of desirable invariances and properties of a continuum (flat space) field theory must then be studied using the elegant technologies which have been devised for exploring the sensitivity of field theories to modifications of the spacetime structure either at short or large distances, and to which we turn in the final four chapters of the book.

---

[25] The non-renormalization theorems of the superpotential ensure the absence of renormalization in the mass terms in the Lagrangian, but there are typically shifts in the poles of the propagator due to non-trivial wavefunction renormalizations. See Section 12.6 for an explanation of these arcane terms.

[26] An important exception to this occurs with Hamiltonian lattice formulations of field theory, provided, of course, that the lattice Hamiltonian is constructed to be properly hermitian.

10.6 Problems

1. Verify that the operator $E'(t, t_0)$, defined in (10.15), satisfies the same first-order equation (10.16) and initial condition as $E(t, t_0)$, whence $E'(t, t_0) = E(t, t_0)$.

2. Determine the Wick expansion of $T(\phi(x_1)\phi(x_2)\phi(x_3)\phi(x_4))$ by taking the fourth functional derivative of (10.22) with respect to $j(x_1), j(x_2), j(x_3), j(x_4)$ and setting the sources to zero.

3. Perform the indicated spacetime integrations in (10.37) to obtain the momentum-space expression for 2-2 scattering in the theory with interaction (10.31).

4. The object of this exercise is to work out the lowest-order perturbative contributions to the vacuum energy density shift $\delta\mathcal{E}$ (see (10.42)) induced by a $\lambda\phi^4$ interaction.

   (a) First assume that the interaction is *not* normal-ordered, $\mathcal{H}_{\text{int}} = \frac{\lambda}{4!}\phi^4$. There is then a contribution to the vacuum-to-vacuum amplitude of first order in λ. Show that after an overall integral over spacetime (interpreted as $V \cdot T$, spatial volume times temporal extent) is extracted, the energy density shift is found to be

   $$\delta\mathcal{E} = \frac{\lambda}{8}(i\Delta_F(0))^2 = -\frac{\lambda}{8}\left(\int \frac{1}{k^2 - m^2 + i\epsilon} \frac{d^4k}{(2\pi)^4}\right)^2 \qquad (10.212)$$

   Show that by Wick rotating the momentum integral to Euclidean space a real result, albeit ultraviolet-divergent in the absence of a high-momentum cutoff, is obtained. Inserting a large-momentum cutoff Λ, show that $\delta\mathcal{E}$ is of order Λ^4 as $\Lambda \to \infty$.

   (b) Now suppose the interaction to be normal-ordered, $\mathcal{H}_{\text{int}} = \frac{\lambda}{4!} : \phi^4 :$. Show that the lowest-order contribution to $\delta\mathcal{E}$ is now of order λ^2, corresponding to the disconnected vacuum graph of Fig. 10.3. Find an expression for $\delta\mathcal{E}$ in terms of a three-loop Minkowski integral, and verify, by Wick rotation, that the (again, ultraviolet-divergent) result is real, and again of order Λ^4 when the loop integrals are cut off at a large momentum Λ.

5. The object of this exercise is to rederive the logarithmically divergent Coulomb self-energy of an electron originally found by Weisskopf in 1939 (cf. Section 2.3, Equations (2.53–2.62)). One first needs the charge density correlation function in a single electron state $\tilde{G}(\vec{\xi})$, from which the desired self-energy correction (for an electron at rest) follows immediately via (2.56). We will use non-covariantly (continuum) normalised electron states, with $\langle \vec{k}\sigma | \vec{k}'\sigma' \rangle = \delta_{\sigma\sigma'}\delta^3(\vec{k} - \vec{k}')$. Thus, we need to calculate (for a fixed spin choice σ)

   $$\int \langle \vec{k}\sigma | : e\psi^\dagger\psi(\vec{r} - \vec{\xi}/2, 0) : : e\psi^\dagger\psi(\vec{r} + \vec{\xi}/2, 0) : |\vec{0}\sigma\rangle d^3r = \tilde{G}(\vec{\xi})\delta^3(\vec{k}) \quad (10.213)$$

   where the fields are at time zero. The calculation is greatly simplified by shifting the first charge density (at $\vec{r} - \vec{\xi}/2$) to a slightly positive time ϵ, whereupon the product of charge densities can be taken to be time-ordered. The Wick expansion now yields a sum of two terms (of the form $iS_F : \psi\psi^\dagger :$, with S_F the Feynman propagator for a Dirac particle), which can be reduced (using the Fourier formula for S_F) to a three-dimensional momentum integral for $\tilde{G}(\vec{\xi})$ by explicitly performing the energy integral (after which the time shift ϵ can be set to zero). Show that one obtains the result (2.61) quoted previously,

$$\tilde{G}(\vec{\xi}) = e^2 \int \frac{d^3q}{(2\pi)^3} \frac{m}{E(q)} e^{i\vec{q}\cdot\vec{\xi}} \tag{10.214}$$

leading to a logarithmically divergent Coulomb self-energy.

6. (a) Consider the one-dimensional integral

$$Z(\lambda, m) \equiv \int_{-\infty}^{+\infty} e^{-\frac{1}{2}m^2 x^2 - \frac{\lambda}{4}x^4}\, dx \sim \sum_n c_n \lambda^n \tag{10.215}$$

where the sum in powers of λ is obtained by expanding the exponential *inside the integral*. Show that the coefficients c_n are given by

$$c_n = (-1)^n \sqrt{2} \frac{1}{(m^2)^{2n+1/2}} \frac{\Gamma(2n + \frac{1}{2})}{\Gamma(n+1)} \tag{10.216}$$

Evidently, the c_n increase factorially with n, so that the series in powers of λ has zero radius of convergence, and must be regarded as an asymptotic expansion only (cf. Section 11.1).

(b) Instead of an expansion in powers of λ, we may rescale the integration variable $y \equiv \lambda^{1/4}x$ and expand $Z(\lambda, m)$ in powers of $\frac{m^2}{\sqrt{\lambda}}$. Show that one then obtains a convergent Taylor series, with coefficients d_n decreasing factorially with n:

$$Z(\lambda, m) = \lambda^{-1/4} \sum_{n=0}^{\infty} d_n \frac{1}{\lambda^{n/2}}, \quad d_n = (-1)^n \frac{1}{\sqrt{2}} m^{2n} \frac{\Gamma(\frac{n}{2} + \frac{1}{4})}{\Gamma(n+1)} \tag{10.217}$$

Note that this result shows that $Z(\lambda, m)$ is a real analytic function of λ, with a cut of standard fractional power type along the negative real axis, and behaving for large $|\lambda|$ like $|\lambda|^{-1/4}$.

7. Let $\phi(x)$ be a free real scalar field with Feynman propagator $\Delta_F(x)$. Show that $\langle 0|T(e^{i\phi(x)}e^{-i\phi(0)})|0\rangle = e^{i(\Delta_F(x) - \Delta_F(0))}$ by

(a) Expanding out the operator exponentials and using Wick's theorem, and

(b) by using path-integral methods (i.e., evaluate the path integral for $Z_0(j)$ with $j(z) = \delta(z) - \delta(z - x)$).

8. Calculate the four-point function of four free Dirac fermion fields
$$\langle 0|T(\psi(x)\psi(y)\bar{\psi}(z)\bar{\psi}(w))|0\rangle$$
by taking four functional derivatives with respect to Grassmann sources $\eta(x), \bar{\eta}(x)$ of $Z_0[\eta, \bar{\eta}]$, the generating functional for the free fermion field theory. Check that the relative signs for the terms you obtain agree with Wick's theorem.

9. Verify the expansion for the sources J_i in terms of the ϕ_i in (10.150).

10. Verify the result (10.151) for the effective potential through terms of order ϕ^4.

11. Verify the recursion relation (10.187) giving the amplitudes for multiple pairs of the quanta of field ϕ_2 in the vacuum of field ϕ_1, where ϕ_1 (resp. ϕ_2) are free scalar fields of mass m_1 (resp. m_2).

12. Starting with the representation equation (10.206), show

    (a) That for $R = 1$, the only solutions are $c(1, \vec{a}) = e^{i\vec{p}\cdot\vec{a}}$, with $\vec{p}$ some fixed three-vector.

    (b) Using the fact that the one-dimensional (zero angular momentum) representation of the rotation group is trivial, $c(R, 0) = 1$, and the result of part (a), show that for any $R, \vec{a}$, $c(R, \vec{a}) = 1$.

13. Suppose that the two-point Wightman function of a Heisenberg field ϕ_H is known to coincide with that for a canonically normalized free field of mass m

$$_{\text{in}}\langle 0|\phi_H(x)\phi_H(y)|0\rangle_{\text{in}} = \Delta_+(x - y; m) \tag{10.218}$$

Show that the Heisenberg field ϕ_H must satisfy the Klein–Gordon equation

$$\mathcal{K}_x\phi_H(x) = (\Box_x + m^2)\phi_H(x) = 0 \tag{10.219}$$

(Hint: apply the Klein–Gordon operators $\mathcal{K}_x\mathcal{K}_y$ to (10.218).) Now, note that by the Yang–Feldman equation (9.204) derived in Chapter 9, Problem 8, using the asymptotic condition, the Heisenberg field must be identical up to normalization to the free in-field:

$$\phi_H(x) = \sqrt{Z}\,\phi_{\text{in}}(x) \tag{10.220}$$

delivering the result of Jost and Schroer once again, this time by appeal to the asymptotic theory developed in Sections 9.3 and 9.4.

11

Dynamics IX: Interacting fields: non-perturbative aspects

The extraordinary successes, beginning in the early 1950s, of perturbative quantum electrodynamics (QED) in predicting with astonishing accuracy the measured hydrogen fine structure and anomalous magnetic moment of the electron (and muon) played an important role in the acceptance of local quantum field theory as an appropriate framework for the description of subatomic phenomena. However, this success was almost immediately tempered by the realization that the meson field theories devised to provide a fundamental description of the dynamics of the strong interactions were completely unable to yield comparable quantitative precision, and the resultant frustration would lead to an abandonment, for more than a decade, of Lagrangian field theory in strong interaction physics, in favor of a purely S-matrix approach in which one attempted to constrain the form of strong interaction amplitudes on the basis of very general principles of unitarity, analyticity, and crossing symmetry, with the hope that a sufficiently clever exploitation of these properties might eventually allow a unique determination of the desired amplitudes, in terms of a finite number of measurable strong interaction parameters. This project would eventually collapse in the mid-1970s with the discovery of an appropriate field theory description of the strong interactions—quantum chromodynamics (QCD)—and the realization that the previous difficulties with field theory approaches to the strong interactions derived from (a) the use of an incorrect fundamental field theory, involving pions and nucleons, instead of the truly elementary quark and gluon underlying degrees of freedom, and (b) the lack of sufficiently powerful *non-perturbative* techniques needed to extract even the most basic qualitative features of a strongly coupled field theory.

The power and intuitive transparency of the perturbative, graph-based approach to quantum field theory explored in the preceding chapter is undeniable, and we shall see later that this framework allows us to extract quantitative results even in circumstances which at first sight would seem to involve strongly coupled theories where a perturbative expansion could hardly be valid. However, it is equally important to distinguish the cases in which perturbation theory is *intrinsically incapable* of capturing even the most elementary qualitative aspects of the physics. Broadly speaking, we can identify three categories of problems from the point of view of the role played by perturbative calculation:

1. Most obviously, we have the classic processes in which a small coupling (in the case of QED, or more generally, the standard model of electroweak interactions,

this would be the fine structure constant $\alpha \sim 1/137$) allows extremely accurate calculations of a given process simply by evaluating and summing the perturbative contributions up to a finite loop order. In the case of QED this has been done up to the four-loop order (α^4) for quantities such as the electron anomalous magnetic moment.

2. Next, we have situations in which a physical quantity, albeit in a weakly coupled theory, necessarily involves an infinite number of interactions between the constituent particles, and hence an infinite number of Feynman graphs. Bound states such as the hydrogen atom in a weakly coupled theory such as QED clearly fall into this category, as the permanent association of the proton and electron clearly requires that they exchange photons over an infinite time span, in contrast to the situation in unbound electron–proton scattering, where the scattering amplitudes can be perturbatively evaluated, with exchange of many photons suppressed by higher powers of α (uncompensated by kinematic enhancements due to the bound-state threshold, as we shall see below). Of course, the calculation of all Feynman diagrams contributing to a process is beyond our calculational powers, and it will soon become clear that even if that were possible, the resultant series is in fact a divergent asymptotic expansion, and cannot therefore be summed directly to yield a meaningful answer! Instead, we shall see that for a certain class of bound-state problems, the kinematic region important for the permanent binding of the constituent particles identifies a dominant component of the (infinite) set of perturbative amplitudes which can be convergently summed, and which represent the leading contributions to the bound-state properties (in an expansion in the available weak coupling). For lack of a better term, we may refer to such situations as "perturbatively non-perturbative" processes in field theory.

3. Finally, there are those physical processes in which the relevant coupling strength is large, so that an asymptotic expansion, even if formally available to high order, is simply useless in extracting quantitative (and in many cases, even qualitative) features of the physics. Quark confinement and chiral symmetry-breaking in QCD are archetypal examples of this type. We may (again, for lack of a better term) refer to these cases as the "essentially non-perturbative" processes in field theory. The Feynman graph approach is of little if any utility here: instead, numerical approaches in which the Euclidean functional integral of the discretized theory is evaluated directly by statistical Monte Carlo techniques (as in lattice gauge theory) provide the most fruitful line of attack. Indeed, the use of such methods has allowed us to obtain, starting from the QCD Lagrangian, and with accuracy now approaching in many cases the level of a few percent, many detailed predictions of hadron spectrum and structure.

In this chapter, after explaining the nature of the (unavoidable) divergence in perturbation theory, we shall give some examples of the various types of non-perturbative phenomena encountered in the field theories of importance in the Standard Model of particle physics. The physics of weakly coupled threshhold bound states will be explained, as it is crucial for understanding the classic successes of QED in the hydrogen atom spectrum, for example. The limitations of perturbation theory in strongly coupled theories, and the extent to which perturbation theory can even in principle be regarded as determining the exact amplitudes of the theory, will be explained, as well as the role played by Borel-summability (or its absence) of the perturbative series. Conventional wisdom holds that perturbative information by itself is virtually useless in non-Borel-summable theories, but we shall see that in at least one iconic case (the anharmonic "double-well" oscillator) purely perturbative information can be "massaged" to obtain a rigorously convergent sequence of approximants to the amplitudes of a non-Borel theory. We conclude the chapter with some brief remarks on numerical approaches to non-perturbative field theory.

11.1   On the (non-)convergence of perturbation theory

The fact that formal expansions of field theory amplitudes in powers of the coupling constant(s) of the theory are generally at best asymptotic, rather than convergent Taylor expansions, can be readily understood on the basis of a simple physical argument first given by Dyson (Dyson, 1952) in the context of quantum electrodynamics. If the amplitudes of the theory were in fact described by *analytic* functions $Z(\lambda)$ of the coupling constant (which we here generically denote λ), with $\lambda = 0$ a point of analyticity, thereby allowing a convergent Taylor expansion in powers of λ, with a finite radius of convergence, we should clearly expect the physics of the theory to change smoothly if the coupling λ is moved continuously from a small positive to a small negative value. In fact, such a change typically introduces dramatic instabilities which completely alter the spectral properties of the theory. For self-interacting ϕ^4 scalar theories, with a Hamiltonian given in d spacetime dimensions by

$$H = \int d^{d-1}x \{ \frac{1}{2}\dot\phi^2 + \frac{1}{2}|\vec\nabla|^2 + \frac{1}{2}m^2\phi^2 + \frac{\lambda}{4}\phi^4 \} \tag{11.1}$$

we have already seen in Section 8.4 that the spectrum of the theory becomes unbounded below if the sign of the coupling λ becomes negative. For this theory, the functional integral representation of the Euclidean generating functional (or vacuum to vacuum amplitude, setting the source function to zero)[1]

$$Z(\lambda, m) = \int \mathbf{D}\phi e^{-\int d^d x \{ \frac{1}{2}\vec\nabla\phi\cdot\vec\nabla\phi + \frac{1}{2}m^2\phi^2 + \frac{\lambda}{4}\phi^4 \}} \tag{11.2}$$

clearly diverges if we allow the real part of λ to become negative. Even for the $d = 0$-dimensional case, where the integral (11.2) degenerates to a one-dimensional integral (as the field is defined at a single point, where we may denote its value x, and there is no gradient term)

$$Z(\lambda, m) \equiv \int_{-\infty}^{+\infty} e^{-\frac{1}{2}m^2 x^2 - \frac{\lambda}{4}x^4} dx \tag{11.3}$$

---

[1] Here, the ∇ operator is the Euclidean d-gradient. As usual, we assume that the functional integral is made well-defined—regularized—by an appropriate discretization, both in the infrared and the ultraviolet: e.g., on a finite lattice.

simple arguments show (cf. Problem 6 in Chapter 10) that the function $Z(\lambda, m)$ is analytic in the complex plane of λ with a cut on the negative real axis extending up to the origin, so that a formal expansion around $\lambda = 0$ cannot converge. Instead, if we expand the "interaction" x^4 term inside the integral, we arrive at an asymptotic expansion[2]

$$Z(\lambda, m) \sim \sum_n c_n \lambda^n, \quad c_n = (-1)^n \sqrt{2} \frac{1}{(m^2)^{2n+1/2}} \frac{\Gamma(2n + \frac{1}{2})}{\Gamma(n+1)} \tag{11.4}$$

with oscillating coefficients c_n which increase factorially in magnitude at large orders n. Note that the divergent series encountered here has *absolutely nothing* to do with the ultraviolet divergences encountered in continuum field theories for $d \geq 2$, but occur even if the theory is fully regularized, as we assume throughout this section, in the ultraviolet (and, if necessary, in the infrared as well). The same cut-plane analyticity holds for the regularized functional integral (11.2) in any spacetime dimension, as we may analytically continue the integral to $\mathrm{Re}(\lambda) < 0$ by simultaneously rotating the phase of the coupling $\lambda \to |\lambda| e^{i\theta}$ with the global phase of the field integration variable $\phi(x) \to |\phi(x)| e^{-i\theta/4}$ so that the damping induced by the quartic term is maintained. The result is that the integral defines an analytic function of λ for $|\theta| < \pi$, with different (complex-conjugate) results obtained on the negative real axis cut depending on whether the continuation is performed clockwise or anticlockwise around the origin. It turns out, as we shall soon see, that the large-order divergence of perturbation theory is intimately related to the behavior of the (imaginary) discontinuity for negative coupling. Physically, this is due to the connection between the imaginary part developed by the energies of quantum mechanical states when the system becomes unstable and the lifetime of these states, so once again we see that there is a deep connection between the large-order behavior of perturbation theory and stability issues.

The divergence of the perturbation theory for quantum electrodynamic quantities expanded in powers of the fine-structure constant $\alpha = \frac{e^2}{\hbar c}$ was first shown by Dyson using exactly such a stability argument (Dyson, 1952). One may once again consider the zero-source generating functional $Z(\alpha)$ of the theory (i.e., the vacuum-to-vacuum amplitude absent external fields), the formal expansion of which involves the sum of vacuum diagrams, each of which must contain an even number of vertices (as all photon lines are internal), so that we have a formal expansion

$$Z(\alpha) \sim \sum_n c_n \alpha^n, \quad \alpha \equiv \frac{e^2}{\hbar c} \tag{11.5}$$

If we analytically continue the theory to negative α, the vacuum becomes unstable to the production of electron–positron pairs, by the following simple energetic argument. As a result of the negative sign of $\alpha \propto e^2$, like charges now attract and unlike charges repel, so that we can concoct a state of negative energy (into which the vacuum can

---

[2] The reader is reminded that the series $\sum_n c_n \lambda^n$ is said to be asymptotic to a function $f(\lambda)$ if $|f(\lambda) - \sum_{n=0}^N c_n \lambda^n| = O(\lambda^{N+1})$ as $\lambda \to 0$ for any fixed N.

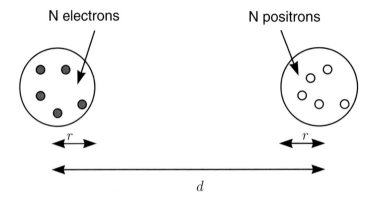

Fig. 11.1 Electron–positron pair assembly leading to energetic instability (for $\alpha < 0$).

therefore decay) by creating N electron–positron pairs and arranging them spatially (see Fig. 11.1) into two groups, with the electrons and positrons separated into two groups of spatial dimension r separated by a distance $d \gg r$. The energetic cost of this configuration, if the electrons and positrons are relativistic with momentum $p \sim \hbar/r$, is of order $+N\frac{\hbar}{r}c$ in kinetic energy, and of order $-N^2\frac{e^2}{r}$ for electrostatic potential energy (ignoring the repulsive contribution between the two groups if $d \gg r$), so we have instability once

$$N^2\frac{e^2}{r} > N\frac{\hbar}{r}c \Rightarrow N > \frac{\hbar c}{e^2} \sim \frac{1}{\alpha} \tag{11.6}$$

On the other hand, an asymptotic expansion of the form (11.5), with factorially growing coefficients $c_n \sim n! \sim n^n$, will consist of a series of terms which decrease until $\alpha n \sim O(1)$ (i.e., $n \sim N \sim 1/\alpha$) and then increase beyond that point.[3] This is exactly the behavior expected once graphs involving the generation of intermediate states with N electron–positron pairs (and therefore N powers of α) are included in the computed vacuum amplitude, inducing via (11.6) an energetic instability for negative α. Of course, the fact that the series gives increasingly accurate approximants until we reach on the order of 137 $(1/\alpha)$ loops means that the formal divergence of perturbation theory in pure quantum electrodynamics is of absolutely no practical consequence: we are dealing with the first type of theory discussed in the introduction, where an asymptotic expansion is perfectly adequate to the task of extracting results of an accuracy (in the case of QED) far beyond our capacity to empirically verify them.

    It turns out that for spacetime dimensions less than four, the large-order behavior of perturbation theory for scalar theories is essentially semiclassical in nature, and

---

[3] The heuristic argument given here, while correct for scalar electrodynamics—i.e., for theories with spinless charged "electrons" and "positrons"—ignores Fermi exchange effects which suppress configurations with fermions in highly overlapping states. More careful arguments, first given by (Parisi, 1977), lead to perturbative coefficients in QED rising like $\sqrt{n!}$. In the path-integral approach described below, the dominant behavior arises from saddle-points in a combined effective action arising from the free photon contribution and the determinant obtained by integrating out the electron field: see (Ioffe *et al.*, 2010), Section 5.8.

is determined by exactly the instability considerations indicated qualitatively above. In particular, as we shall see in our study of scale sensitivity in Part 4 of the book, scalar ϕ^4 theory in $d =$ two or three spacetime dimensions is *super-renormalizable*: the ultraviolet behavior becomes progressively softer (i.e., less divergent) as we go to higher orders of perturbation theory. As a result, the large-order behavior of the renormalized and cutoff theories is essentially the same. In four dimensions the renormalizability (rather than super-renormalizability) of the theory, with logarithmic ultraviolet dependence of the cutoff amplitudes, results in a more complicated situation once the theory is rewritten (renormalized) in terms of cutoff-independent physical quantities, as the strength of the ultraviolet sensitivity remains unchanged as we go to higher orders of perturbation theory. For $d \leq 3$, a seminal analysis initiated by Lipatov (Lipatov, 1977) shows that the leading behavior at large orders of perturbation theory is determined by classical extrema of the Euclidean action appearing in the functional integral (11.2). Considerations of space limit us to merely an outline of the argument here: for further details the reader is referred to the excellent discussion of Zinn–Justin ((Zinn–Justin, 1989), especially Chapters 33 and 37).

For simplicity we begin with the case $d = 1$ (with the single spacetime dimension treated as time): relabeling the field $\phi(x) \rightarrow q(t)$, (11.2) becomes (taking $m = 1$),

$$Z(\lambda) = \int \mathbf{D}q(t) e^{-\int dt \{ \frac{1}{2} \dot{q}(t)^2 + \frac{1}{2} q(t)^2 + \frac{\lambda}{4} q(t)^4 \}} \equiv \int \mathbf{D}q(t) e^{-S_E(q)} \tag{11.7}$$

which is exactly the Euclidean functional integral appropriate for an anharmonic oscillator defined by the Hamiltonian $H = \frac{p^2}{2} + \frac{1}{2}q^2 + \frac{1}{4}\lambda q^4$ (with $m = \omega = 1$ for simplicity: cf. Equations (4.100), (4.99, and (4.108–4.110)). If we take the functional integral to run over periodic functions satisfying $q(-\beta/2) = q(+\beta/2)$, this functional integral gives, as discussed in Section 4.2.1, the finite-temperature partition function

$$Z(\lambda, \beta) = \text{Tr} e^{-\beta H} = \sum_{n=0}^{\infty} e^{-\beta E_n(\lambda)} \sim e^{-\beta E_0(\lambda)}, \quad \beta \rightarrow \infty \tag{11.8}$$

so that for large β (low temperature) the ground-state energy dominates. The analog of the generating functional of connected graphs W becomes in this limit $W \equiv \ln Z \sim -\beta E_0(\lambda)$—in other words, just the ground-state energy (up to a factor of $-\beta$). The instability of the system as we analytically continue from positive real λ to negative λ is manifested, as usual in quantum mechanics, in the appearance of an imaginary part in the analytically continued energy eigenvalue $E_0(\lambda)$: indeed, just as in the zero-dimensional toy integral case discussed previously, the imaginary part is simply the signal of a cut appearing along the negative real axis in the complex plane of λ, with $Z(\lambda = -|\lambda| + i\epsilon) = Z^*(\lambda = -|\lambda| - i\epsilon)$ as $Z(\lambda)$ (dropping the at present uninteresting β dependence) is real-analytic (real for positive real λ). As for the one-dimensional integral, the value of Z on the top and bottom lips of the cut can be computed by rotating the phase of the "field" variable $q(t)$ in tandem with the phase of the coupling so as to preserve a negative real part (and hence convergence) in the exponent $-S_E$ appearing in (11.7). Thus, we let $\lambda \rightarrow e^{i\phi}|\lambda|$, $q(t) \rightarrow q(t)e^{-i\phi(\frac{1}{4} - \delta/\pi)}$ (with δ small positive) and arrive after rotating $\phi \rightarrow \pm\pi$ with the well-defined functional integrals

$$Z(\lambda = -|\lambda| \pm i\epsilon) = \int \mathbf{D}q(t) e^{-\int dt \{\frac{1}{2}\dot{q}_\theta(t)^2 + \frac{1}{2}q_\theta(t)^2 + \frac{\lambda}{4}q_\theta(t)^4\}} \qquad (11.9)$$

where $q_\theta(t) = e^{i\theta}q(t)$, $\theta = \mp(\frac{\pi}{4} - \delta)$. The discontinuity of Z across the cut on the negative axis is then given by subtracting the two integrals (11.9) for the two signs of θ. Lipatov realised that the resultant difference of path integrals can be deformed further to pass through saddle points (extrema of the Euclidean action $S_E(q)$) which dominate the result for small negative $\text{Re}(\lambda)$. The extremum of the action corresponds to trajectories of our particle $q_{\text{cl}}(t)$ satisfying the equation

$$\frac{\delta S_E(q)}{\delta q(t)}|_{q=q_{\text{cl}}} = -\ddot{q}_{\text{cl}}(t) + q_{\text{cl}}(t) + \lambda q_{\text{cl}}(t)^3 = 0 \qquad (11.10)$$

with the contribution of the saddle-point to Z proportional to $e^{-S_{\text{cl}}}$, $S_{\text{cl}} \equiv S_E(q_{\text{cl}})$. Physically, this is just Newton's Second Law for a unit-mass classical particle moving in the potential $V(q) = -\frac{1}{2}q^2 - \frac{\lambda}{4}q^4$, pictured in Fig. 11.2.

The equation (11.10) has an obvious time-independent solution (recall that this is for λ real negative), corresponding to the particle sitting motionless at a local minimum $q = \pm\sqrt{\frac{-1}{\lambda}}$:

$$q_{\text{cl}}(t) = \pm\sqrt{\frac{-1}{\lambda}}, \quad S_{\text{cl}} = \int_{-\beta/2}^{\beta/2} (\frac{1}{2}\dot{q}_{\text{cl}}(t)^2 + \frac{1}{2}q_{\text{cl}}(t)^2 + \frac{\lambda}{4}q_{\text{cl}}(t)^4)dt = -\beta/(4\lambda) \qquad (11.11)$$

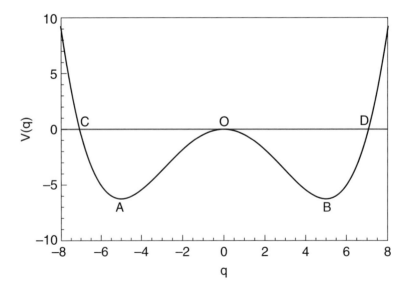

Fig. 11.2 $V(q) = -\frac{1}{2}q^2 - \frac{\lambda}{4}q^4$, $\lambda = -1.0$.

with a contribution $e^{-S_{cl}}$ to Z which (for negative real λ) vanishes exponentially in the large β limit. In order to obtain a finite contribution in the large β limit, we must find a saddle-point solution such that the integral in (11.11) remains finite for $\beta \to \infty$, which clearly requires that (a) $\dot{q}_{cl}(t) \to 0$ for large t, and (b) $\frac{1}{2}q_{cl}(t)^2 + \frac{\lambda}{4}q_{cl}(t)^4 \to 0 \Rightarrow q_{cl}(t) \to 0$ for large t. The Newtonian interpretation of (11.10) means that this solution corresponds to the particle with zero total energy beginning at rest at the origin at time $t = -\beta/2 \to -\infty$ and rolling down the potential well, reaching $q = \pm\sqrt{\frac{-2}{\lambda}}$ at some (arbitrary) finite time $t_0, -\beta/2 < t_0 < \beta/2$, then returning asymptotically, as $t \to +\infty$, to the origin. There are evidently two independent saddle-point solutions, depending on whether the particle rolls to the left (going from the origin O through the minimum at A to point C and then back) or to the right (O to D through the minimum at B and back) in Fig. 11.2. The reader may easily verify that the following explicit solution satisfies Newton's equation (11.10):

$$q_{cl}(t) = \pm\sqrt{\frac{-2}{\lambda}} \frac{1}{\cosh(t - t_0)} \tag{11.12}$$

with the finite (in the $\beta \to \infty$ limit) Euclidean action

$$S_{cl} = \int_{-\infty}^{+\infty} S_E(q_{cl}(t))dt = -\frac{4}{3\lambda} \tag{11.13}$$

The contribution of this saddle-point to Z is clearly proportional to $e^{-S_{cl}} = e^{\frac{4}{3\lambda}}$, which is exponentially small as $\lambda \to 0_-$. Finite action solutions of the Euclidean equations of motion such as (11.12) have been dubbed "pseudoparticles", or "instantons" in the literature. It is clear that $Z(\lambda)$ contains an *essential singularity* at the origin—a feature which persists quite generally in interacting field theories. A complete evaluation of the leading-order contribution to $\text{Im}(Z(\lambda)$ requires a careful evaluation of the Gaussian fluctuations around the saddle-point defined by (11.12): in addition to an obvious factor of β, due to the fact that the time t_0 at which the particle reaches its maximum displacement from the origin can be chosen anywhere between $-\beta/2$ and $+\beta/2$ (edge effects are unimportant for β large), the resultant Gaussian integral yields[4] a contribution of the form $Ke^{-\beta/2}\sqrt{\frac{-1}{\lambda}}$, with K an uninteresting numerical constant. To summarize, the leading contribution to the discontinuity of $Z(\lambda)$ for small negative λ and large β takes the form (with corrections of relative order λ and $e^{-\beta}$)

$$\text{Im}(Z(\lambda = -|\lambda| + i\epsilon)) \sim -K\beta e^{-\beta/2}\sqrt{\frac{-1}{\lambda}}e^{\frac{4}{3\lambda}} \tag{11.14}$$

On the other hand, from (11.8), the imaginary part of the partition function at large β and small coupling (where $\text{Re}(E_0(\lambda)) \to \frac{1}{2}$, the zero-point energy for the harmonic oscillator) must behave like

$$\text{Im}(Z) \sim -\beta \, \text{Im}(E_0(\lambda))e^{-\beta \text{Re}(E_0(\lambda))} \sim -\beta e^{-\beta/2} \, \text{Im}(E_0(\lambda)) \tag{11.15}$$

---

[4] See (Zinn–Justin, 1989), op. cit., for further details.

Comparing (11.15) and (11.14), we conclude that

$$\text{Im}(E_0(\lambda)) \sim K\sqrt{\frac{-1}{\lambda}}e^{\frac{4}{3\lambda}}(1 + O(\lambda)), \quad \lambda \to 0_- \tag{11.16}$$

The imaginary part of the energy displayed in (11.16) is directly connected to the tunneling rate for a zero-energy particle initially localized around the origin to escape through the barrier formed by the potential $\frac{1}{2}q^2 - \frac{|\lambda|}{4}q^4$ appearing in our original functional integral (11.7), as one may verify with a simple WKB calculation (see Problem 1). The connection of the instability generated by such tunneling to the large-order behavior of the coefficients appearing in the Rayleigh–Schrödinger perturbation theory for the ground-state energy $E_0(\lambda) \sim \sum_n c_n \lambda^n$ is established by use of analyticity, which allows us to connect the behavior of $E_0(\lambda)$ for positive real λ to the discontinuity across the cut on the negative real axis.

First, note that for large coupling λ, the energy $E_0(\lambda)$ has the asymptotic behavior $\lambda^{1/3}$, by a simple scaling argument (see Problem 2). This means that the function $f(\lambda) \equiv (E_0(\lambda) - \frac{1}{2})/\lambda$ is (a) analytic in the complex plane of λ, cut along the negative real axis, and (b) behaves for large $|\lambda|$ like $\lambda^{-2/3}$. Accordingly, $f(\lambda)$ satisfies the Cauchy formula

$$f(\lambda) = \frac{1}{2\pi i}\oint_C \frac{f(\lambda')}{\lambda' - \lambda}d\lambda' \tag{11.17}$$

where C is the contour indicated in Fig. 11.3. Replacing $f(\lambda)$ by $E_0(\lambda)$, and expanding the contour C to infinite size, whereupon the curved parts go to zero, while the two straight portions along the negative real axis combine to give the imaginary part of E_0, giving

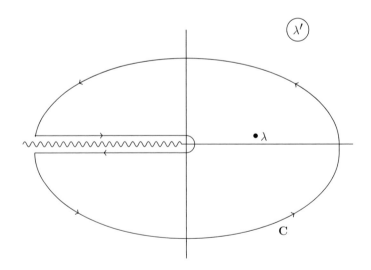

Fig. 11.3 Contour integral for dispersion relation in complex λ plane.

$$E_0(\lambda) = \frac{1}{2} + \frac{\lambda}{\pi} \int_{-\infty}^{0} \frac{\mathrm{Im} E_0(\lambda')}{(\lambda' - \lambda)\lambda'} d\lambda' \qquad (11.18)$$

Expanding the denominator factor $(\lambda' - \lambda)^{-1}$ in powers of λ inside the integral yields the desired asymptotic expansion and an explicit formula for the leading behavior at large order n of the coefficients c_n as an integral over the cut discontinuity of E_0:

$$c_n = \frac{1}{\pi} \int_{-\infty}^{0} \frac{\mathrm{Im} E_0(\lambda)}{\lambda^{n+1}} d\lambda \sim K(-1)^{n+1} (\frac{3}{4})^{n+\frac{1}{2}} \Gamma(n + \frac{1}{2})(1 + O(1/n)) \qquad (11.19)$$

where the corrections of order $1/n$ arise from the corrections of relative order λ to the leading behavior of the discontinuity for small negative λ. The important features to note here are first, that the coefficients rise factorially with order, as anticipated by our intuitive arguments, and second, *that they alternate in sign*. The latter property will move to center stage in Section 11.3, when we discuss the Borel summability (or absence thereof) of perturbative expansions in field theory.

For spacetime dimensions $d = 2$ or 3, we move into the realm of field theory proper, but the large-order analysis proceeds along much the same lines as for the anharmonic oscillator: the dominant contribution to the discontinuity in the partition function (11.2) when analytically continued to negative (small) λ is given by the saddle-point contribution to the functional integral arising from finite-action solutions (as before, called "instantons") of the classical Euclidean field equation describing the extrema of the action:

$$-\Delta \phi_{\mathrm{cl}}(x) + m^2 \phi_{\mathrm{cl}}(x) + \lambda \phi_{\mathrm{cl}}^3(x) = 0 \qquad (11.20)$$

where Δ is the Laplacian in d dimensions. The extremal value of the action S_{cl} at a solution of (11.20) can be simplified, using (11.20) to eliminate the derivative terms:

$$S_{\mathrm{cl}} = \int d^d x \{ -\frac{1}{2} \phi_{\mathrm{cl}}(x) \Delta \phi_{\mathrm{cl}}(x) + \frac{1}{2} m^2 \phi_{\mathrm{cl}}^2(x) + \frac{1}{4} \lambda \phi_{\mathrm{cl}}^4(x) \} = -\frac{\lambda}{4} \int d^d x \phi_{\mathrm{cl}}^4(x) > 0 \qquad (11.21)$$

The leading contribution is proportional to $e^{-S_{\mathrm{cl}}}$, so we are really looking for the finite-action solution with the minimum value for S_{cl}. From (11.21) it is clear that finite action requires that $\phi_{\mathrm{cl}}(x) \to 0, x \to \infty$. It can be further be shown (see (Zinn–Justin, 1989), op. cit) that the minimal action solutions correspond to spherical symmetry, $\phi_{\mathrm{cl}}(x - x_0) = \frac{m}{\sqrt{-\lambda}} u(r), r \equiv m|x - x_0|$, where the instanton solution is centered at an arbitrary (Euclidean) spacetime point x_0 (exactly analogous to the time t_0 appearing in the instanton solution (11.12) for $d = 1$). The rescaled dimensionless function $u(r)$ satisfies the ordinary non-linear differential equation

$$\frac{d^2 u(r)}{dr^2} = -\frac{d-1}{r} \frac{du(r)}{dr} - \frac{d}{dr}(-\frac{1}{2} u^2(r) + \frac{1}{4} u^4(r)) \qquad (11.22)$$

with the boundary condition $u(r) \to 0, r \to \infty$. In the previously discussed case of the anharmonic oscillator $(d = 1)$ the radial coordinate r corresponded to the time, and the equation (11.22) had a ready-made mechanical interpretation in terms of Newton's Law for a unit mass particle in a potential $V(u) = -\frac{1}{2} u^2 + \frac{1}{4} u^4$. This remains the case here if we interpret r as a time coordinate, for dimensions $d = 2, 3$, with the additional feature that a retarding frictional force term (proportional to velocity, and increasing with dimension) is present. The result is that the boundary condition $u(r) \to 0$ as "time" $r \to \infty$ is achieved (without an "overshoot" to negative r) only if we start our fictional particle at time $r = 0$ at $|u(0)| > \sqrt{2}$ (the value for the anharmonic oscillator case, $d = 1$, must be increased to overcome the additional frictional resistance: so we must "start" our particle to the right of point D, or the left of point C, in Fig. 11.2) at a discretely determined value—the minimal such value of $|u(0)|$ determining, as a detailed analysis confirms, the minimal action S_{cl}. A simple numerical ODE solver for (11.22) yields the desired results for $d = 1$, 2, or 3, displayed in Fig. 11.4 (only the $d = 1$ case allows an analytic solution: namely, $u(r) = \sqrt{2}/\cosh(r)$, cf. Eq. (11.12)). The corresponding classical action is given by inserting the result obtained from the solution of (11.22) into (11.21), which gives

$$S_{cl} = -\frac{\lambda}{4} \int d^d x \, \phi_{cl}^4(x) = -\frac{1}{\lambda} A_d, \quad A_d = \frac{\pi^{d/2}}{2\Gamma(d/2)} m^{4-d} \int_0^\infty r^{d-1} u^4(r) dr \qquad (11.23)$$

The physical interpretation of these classical solutions is a natural extension of that given earlier for the "rollover" solution (11.12) in the anharmonic oscillator case: $\phi_{cl}(x)$ represents a spatiotemporal picture of the dominant (i.e., most likely) tunneling behavior of a field configuration initially centered around the minimum-energy configuration for positive λ (i.e., zero field) through the energy barrier that appears once the coupling is continued to a negative real barrier.

Although we have focussed on the large-order behavior of the source-free generating function Z in (11.2)—in other words, as we saw in the preceding chapter, on the vacuum graphs of the theory—the saddle-point analysis above extends in a straight-forward way to the functional integral giving the Euclidean Schwinger functions of the

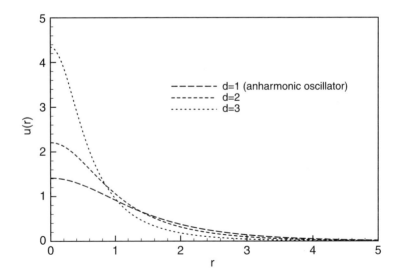

Fig. 11.4 Instanton solutions for ϕ^4-theory in dimensions $d = 1, 2, 3$.

theory (cf. (10.82)):

$$S(x_1, x_2, \ldots, x_n; \lambda) = \frac{1}{Z(\lambda, m)} \int \mathbf{D}\phi \; \phi(x_1)\phi(x_2)...\phi(x_n)e^{-\int d^d x \{\frac{1}{2}\nabla\phi\cdot\nabla\phi + \frac{1}{2}m^2\phi^2 + \frac{\lambda}{4}\phi^4\}}$$

$$(11.24)$$

The discontinuity across the cut for negative λ, which determines the large-order behavior by the dispersion analysis we have described, is again obtained by expanding around the instanton saddle-points corresponding to instantons $\phi_{cl}(x - x_0)$ centered around arbitrary spacetime points x_0, with a prefactor determined by the Gaussian integration around the saddle-point (see (Zinn–Justin, 1989) for details):

$$\mathrm{Im}S(x_1, x_2, ..., x_n; \lambda = -|\lambda| + i\epsilon) \sim K_d \frac{e^{\frac{A_d}{\lambda}}}{(-\lambda)^{d/2}} \int d^d x_0 \prod_{i=1}^{n} \phi_{cl}(x_i - x_0) \qquad (11.25)$$

Just as in the perturbative expansion (11.19) for the anharmonic ground-state energy, the presence of the essential singularity induced by $e^{\frac{A_d}{\lambda}}$ will lead to factorially growing contributions (with oscillating sign) at large order, so the perturbation theory (even when ultraviolet cutoffs are in place) is divergent.

With the technology we have described in this section in place, there is a great temptation to draw the conclusion that the dominant behavior at large orders of perturbation theory somehow ought to determine the dominant *physical* behavior of the corresponding amplitudes. This temptation must be strenuously resisted, for (at least) two reasons. First, one must bear in mind that even if a unique resummation procedure were available to convert the information contained in the perturbative coefficients (to all order) into well-defined convergent approximants to the exact field-theoretic amplitudes, there is absolutely no guarantee that the dominant portions of the large-order perturbative coefficients actually translate into the dominant parts of the amplitudes in the physical regime of interest. In fact, we shall see in the next section that bound states in field theory provide an immediate counterexample to any such claim.

Secondly, the analysis performed above was entirely carried out in the context of the Euclidean functional integral: actual physical amplitudes need to be obtained from those calculated in Euclidean space by an analytic continuation to Minkowski space. Unfortunately, it is well known that asymptotic estimates of analytic functions *cannot* in general be analytically continued: in other words, it is not true that the analytic continuation of an asymptotic series yields in general a correct asymptotic expansion for the analytically continued function. Of course, this would be possible were the expansions in question convergent Taylor series, but we have just seen that this is essentially never the case in a non-trivial interacting field theory. These obstacles have unfortunately severely limited the extraction of quantitatively useful physical information from a large and elegant body of work on instanton solutions, especially in quantum chromodynamics, where the tunneling processes described by instantons are almost certainly connected in a deep way to the chiral symmetries (and the breaking thereof) of the theory. A fortunate exception is in the theory of critical phenomena, in which the Borel resummation (cf. Section 11.3) of the large-order behavior of ϕ^4 theory in $d = 3$ dimensions has been used to extract highly accurate results (often to four or five significant figures) on critical exponents for second-order transitions in models in the same universality class as scalar field theory (LeGuillou and Zinn-Justin, 1980).

11.2 "Perturbatively non-perturbative" processes: threshhold bound states

The fact that the perturbative expansion for field-theory amplitudes is divergent means that the complete set of Feynman graphs contributing to a process—even if we could calculate them all exactly (an impossible task in practice)—do not define an exact amplitude for the process. In an asymptotic series $\sum c_k g^k$, the property that successive terms only decrease in magnitude for a finite number of orders (say, N) before starting to increase again, means that the series is only practically useful if the coupling constant g is "small", which here should be taken to mean that the smallest term in the series $c_N g^N$ is much smaller than the sum of the series $\sum_{k=0}^{N} c_k g^k$ up to that point.

On the other hand, certain physical processes *necessarily require* the inclusion of an infinite number of interactions. The formation of a stable bound state, for example, in which the constituent particles by definition continue to interact over an infinite time period, is clearly an example of such a situation. Clearly, we cannot simply imagine literally summing the infinite set of Feynman graphs contributing to the continuing interaction of the constituents: as we have seen in the preceding section, this would give an infinite result! Nevertheless, we shall see that for a certain class of bound states, defined by the appearance of non-relativistic threshold singularities, the dominant features of the bound-state formation, and the properties of the resultant bound state, are determined by a summable subset of graphs which predominate in the kinematical region appropriate for these systems. Moreover, in the event that the basic binding interaction is weak, the perturbative amplitudes can be reorganized into a series of terms, each containing a tower of infinitely many contributions, which determine to successively higher accuracy (i.e., higher order in the weak coupling) the properties of the resultant bound state. For lack of a better term, we may refer to such processes as "perturbatively non-perturbative".

Consider a theory with two fields $A(x), B(x)$ (for kinematic simplicity, we assume these fields to be equal mass in the following) interacting in such a way that a stable single particle state $|P\rangle$ (we omit spin labels, if present, for notational simplicity), of mass M_B, arises, with

$$\langle 0|T(A(x)B(y))|P\rangle \neq 0 \tag{11.26}$$

We then say that the state $|P\rangle$ is a *bound state* of the particles interpolated for by A and B. In the terminology of Section 9.2, the bilocal operator

$$C_x(X) = T(A(X + \frac{x}{2})B(X - \frac{x}{2})) \tag{11.27}$$

(suitably smeared over x) acts as an almost local field (centered at the point X) which interpolates for the bound-state particle. The matrix element

$$\Phi_P(x) \equiv (2\pi)^{3/2}\sqrt{2E(P)}\langle 0|T(A(\frac{x}{2})B(-\frac{x}{2})|P\rangle, \quad E(P) \equiv \sqrt{\vec{P}^2 + M_B^2} \quad (11.28)$$

is called the "Bethe–Salpeter wavefunction" of the bound state. The energy square-root factor for the bound state is included for convenience as our bound-state ket is non-covariantly normalized. As we shall see, it plays a role analogous to that of the Schrödinger wavefunction in non-relativistic quantum theory: in particular, the bound-state mass is determined by an eigenvalue equation involving this function.[5] The Källen–Lehmann representation (cf. Section 9.5) tells us that the existence of a single particle asymptotic state of mass M_B implies a pole of the form $1/(P^2 - M_B^2)$ in the Feynman two-point function of any Heisenberg field that interpolates for the bound-state particle (see (9.192)). The pole arises in the usual way from the contribution of single-particle intermediate states to the two-point function

$$G(X - Y) \equiv \langle 0|T(C_x(X)C_y^\dagger(Y))|0\rangle \quad (11.29)$$

where we have temporarily suppressed the relative coordinates x, y which are held finite and fixed while the pole in the Fourier transform of G

$$\tilde{G}(P) = \int \langle 0|T(C_x(X)C_y^\dagger(0))|0\rangle e^{iP \cdot X} d^d X \quad (11.30)$$

is evaluated. We have left the spacetime dimension d unspecified here, as we shall consider examples in various spacetime dimensions below.[6] The single particle contribution to the spectral function $\rho_{1\,\text{part}}$ (see (9.188)) is

$$\frac{1}{(2\pi)^3}\theta(P_0)\rho_{1\text{part}}(P^2) = \int d^{d-1}k\langle 0|C_x(0)|k\rangle\langle k|C_y^\dagger(0)|0\rangle\delta^4(P - k)$$

$$= \frac{1}{(2\pi)^3}\frac{1}{2E(P)}\phi_P(x)\phi_P^*(y)\delta(P_0 - E(P))\theta(P_0)$$

$$\Rightarrow \rho_{1\,\text{part}}(P^2) = \phi_P(x)\phi_P^*(y)\delta(P^2 - M_B^2) \quad (11.31)$$

which then gives rise (see (9.191, 9.192)) to the expected single-particle pole in the bound-state propagator

$$-i\tilde{G}(P) = -i\int e^{iP \cdot X}\langle 0|T\{A(X + x/2)B(X - x/2)A^\dagger(y/2)B^\dagger(-y/2)\}|0\rangle$$

$$\rightarrow \frac{\Phi_P(x)\Phi_P^*(y)}{P^2 - M_B^2 + i\epsilon}, \quad P^2 \rightarrow M_B^2 \quad (11.32)$$

---

[5] However, one must be careful, in the relativistic field theory case, not to attach the usual probabilistic interpretation to this function: recall the difficulties entailed in attempts to define a position operator in field theory, discussed in detail in Section 6.5.

[6] The mathematically fastidious reader may imagine smooth smearing functions of rapid decrease in x and y attached to the equations that follow, so that we are really dealing with almost local operators $C_f(X)$ in the sense of Section 9.2.

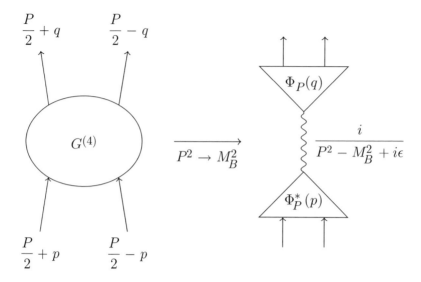

Fig. 11.5 Bound-state pole contribution to the scattering amplitude in a binding channel.

At this point it will be convenient to go over completely to momentum space, by Fourier transforming the remaining coordinate space relative variables x, y. We recall that the Feynman four-point function representing the scattering amplitude for elastic A-B scattering, with the external momenta chosen as in Fig. 11.5, is given by

$$\tilde{G}^{(4)}(\frac{P}{2}+q, \frac{P}{2}-q, \frac{P}{2}+p, \frac{P}{2}-p)$$

$$= \int e^{i(\frac{P}{2}+q)\cdot x_1 + i(\frac{P}{2}-q)\cdot x_2 - i(\frac{P}{2}+p)\cdot y_1} \langle 0|T\{A(x_1)B(x_2)A^\dagger(y_1)B^\dagger(0)\}|0\rangle d^d x_1 d^d x_2 d^d y_1$$

$$= \int e^{iP\cdot X + iq\cdot x - ip\cdot y} \langle 0|T\{A(X+x/2)B(X-x/2)A^\dagger(y/2)B^\dagger(-y/2)\}|0\rangle d^d X d^d x d^d y$$

$$\tag{11.33}$$

where we have omitted a spacetime integral over the fourth field (which deletes the uninteresting $(2\pi)^4 \delta^4(\sum P)$ energy-momentum conservation factor from the amplitude) in the second line, and made the change of variables $x = x_1 - x_2$, $X = \frac{x_1 + x_2 - y_1}{2}$, $y = y_1$ in the last line. Comparing (11.33) with (11.32), we see that the 2-2 scattering amplitude of our A and B particles will display a simple pole as the total incoming momentum P is taken onto the mass shell for the bound state $P^2 \to M_B^2$:

$$\tilde{G}^{(4)}(\frac{P}{2}+q, \frac{P}{2}-q, \frac{P}{2}+p, \frac{P}{2}-p) \to i\frac{\Phi_P(q)\Phi_P^*(p)}{P^2 - M_B^2 + i\epsilon}, \quad P^2 \to M_B^2 \tag{11.34}$$

This result may be given a convenient graphical expression, as in Fig. 11.5.

The existence of bound states which are potentially amenable to a systematically improvable analysis based on perturbation theory depends on the presence of a sufficiently weak coupling in which the bound-state properties can be expanded, yielding increasingly accurate results, at least up to some finite order of the expansion. On the other hand, it seems at first sight strange that a bound state would form at all in the presence of an arbitrarily weak coupling, as one would expect that the Feynman graphs representing increasing numbers of interactions between the binding constituents would be suppressed by higher powers of a small quantity. Indeed, a bound state can only form for arbitrarily weak coupling if the strength of each successive interaction is enhanced kinematically by a correspondingly large factor, so that amplitudes containing any number (indeed, an infinite number) of interactions contribute comparably to the maintenance of the identity of the bound state over an infinite time period. We shall see that the necessary enhancement arises from the presence of threshold singularities, arising from kinematic regions in which the constituent particles of the bound state are very close to being on mass shell. Moreover, the set of graphs which provide the strongest threshold enhancement at any given order of perturbation theory amount to a *summable* subset of the complete set of Feynman diagrams of the theory: indeed, they have the mathematical structure of a convergent geometric expansion, rather than a divergent asymptotic one. This fact is of tremendous historical importance, given the critical role played by precision calculations of bound state effects, such as the Lamb shift, in the development of quantum electrodynamics—or, for that matter, of quantum mechanics, as in the case of the hydrogen atom.

In Section 10.4 we saw that the singularities displayed by Feynman amplitudes in field theory are closely related to the connectedness structure of the amplitudes—in particular, to the way in which the amplitude is built up from p-irreducible amplitudes defined by connectivity after p internal lines are cut. Expressing the full amplitude in terms of p-irreducible vertices allows us to isolate the threshold singularities associated with intermediate states involving p (and exactly p) particles. For the present case, where the bound state arises from two constituents (the quanta of the A and B fields), the relevant vertex is the two-particle irreducible (2PI) scattering amplitude $K_P(q, p)$ (or *kernel*), the iteration of which, with two particle intermediate states connecting successive kernels, generates the full 2-2 amplitude, which we now slightly relabel and rescale to simplify the notation as

$$\tilde{G}_P(q, p) \equiv -\frac{1}{(2\pi)^d} \tilde{G}^{(4)}(\frac{P}{2} + q, \frac{P}{2} - q, \frac{P}{2} + p, \frac{P}{2} - p) \qquad (11.35)$$

The iteration is depicted graphically in Fig. 11.6, where the crossbars on the outgoing (top) propagators indicate that they have been amputated, with the result that the disconnected part simply becomes a δ-function equating the initial and final relative momenta p and q. Note that two-particle irreducibility in this context is defined as the property that the graph remains connected when a single A line and a single B line is cut: in other words, we imagine cutting the graphs in Fig. 11.6 *horizontally* (in the so-called "s-channel" for the scattering amplitude). Particle lines bearing an arrow in Fig. 11.6 should be regarded as full propagators (including self-energy corrections, etc.) for the A and B fields (see below).

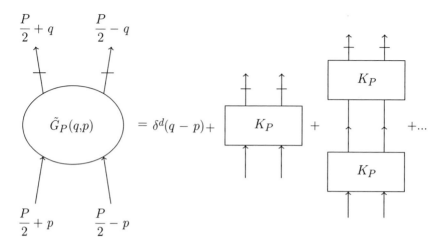

Fig. 11.6 2-2 scattering amplitude expressed as an iteration of two-particle irreducible segments.

The series of 2PI kernels in Fig. 11.6 can be re-expressed as an integral equation for $\tilde{G}_P^{(4)}$ with kernel K_P and inhomogeneous part $I(q, p) = \delta^d(q - p)$

$$\hat{\Delta}_F^{-1}\left(\frac{P}{2} + q\right)\hat{\Delta}_F^{-1}\left(\frac{P}{2} - q\right)\tilde{G}_P(q, p) = I(q, p) + \int K_P(q, k)\tilde{G}_P(k, p)\frac{d^d k}{(2\pi)^d} \qquad (11.36)$$

as depicted in Fig. 11.7. Here, $\hat{\Delta}_F(\frac{P}{2} + p)$ (resp. $\hat{\Delta}_F(\frac{P}{2} - p)$) represent *full* propagators for the A (resp. B) fields (i.e., the Feynman two-point functions for the fully interacting Heisenberg fields; cf. Section 9.5). To avoid overburdening the notation, we use the same symbol for both propagators, even though the fields A and B may be distinct: which propagator is meant will be clear from the context. The iteration of (11.36) clearly generates the succession of 2PI segments indicated in Fig. 11.6.

Note that the kernel $K_P(q, k)$ is defined to be amputated with respect to *both* incoming and outgoing legs (with momenta $\frac{P}{2} \pm k$ and $\frac{P}{2} \pm q$), in order to avoid doubling the internal propagators. The relevance of this representation of the 2-2 amplitude is that *the kernel $K_P(q, k)$ does not contain a single particle bound state pole*, inasmuch as two-particle intermediate states are absent by definition in the graphical expansion of K_P. Thus, if we take the on-mass-shell limit $P^2 \to M_B^2$ for the bound state in (11.36), only the $\tilde{G}_P$ factors contain the pole term arising from (11.34), so that, identifying the residues of this pole on both sides of (11.36), we find the famous *Bethe–Salpeter equation* (Salpeter and Bethe, 1951):

$$\hat{\Delta}_F^{-1}\left(\frac{P}{2} + q\right)\Phi_P(q)\hat{\Delta}_F^{-1}\left(\frac{P}{2} - q\right) = \int K_P(q, k)\Phi_P(k)\frac{d^d k}{(2\pi)^d} \qquad (11.37)$$

This result has the obvious graphical depiction indicated in Fig. 11.8. An alternative version, in which the Bethe–Salpeter wavefunction is itself amputated: i.e.,

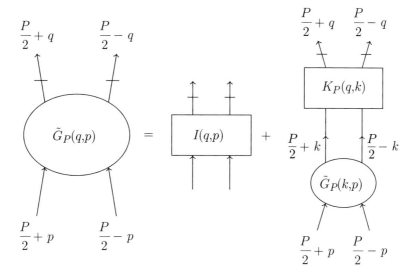

Fig. 11.7 Integral equation satisfied by the 2-2 scattering amplitude.

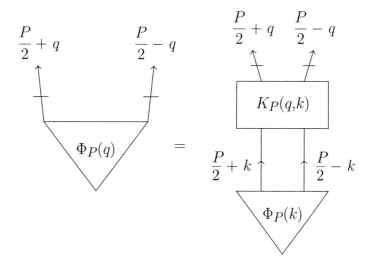

Fig. 11.8 Bethe–Salpeter equation determining the bound state wavefunction $\Phi_P(q)$.

$$\Psi_P(q) \equiv \hat{\Delta}_F^{-1}(\frac{P}{2} + q)\Phi_P(q)\hat{\Delta}_F^{-1}(\frac{P}{2} - q) \tag{11.38}$$

leads to a slightly different Bethe–Salpeter equation for $\Psi_P(q)$:

$$\Psi_P(q) = \int K_P(q, k)\hat{\Delta}_F(\frac{P}{2} + k)\Psi_P(k)\hat{\Delta}_F(\frac{P}{2} - k)\frac{d^d k}{(2\pi)^d} \tag{11.39}$$

The simplest example of a bound state arising in a weakly coupled system, and amenable to perturbative analysis, is found in our old friend, the scalar ϕ^4 theory, in two or three spacetime dimensions. Thus, we imagine a self-interacting non-self-conjugate massive scalar field ϕ with interaction Hamiltonian density

$$\mathcal{H}_{\text{int}}(z) = \frac{\lambda}{4} : (\phi^\dagger(z)\phi(z))^2 : \tag{11.40}$$

and free momentum-space propagator $\Delta_F(p) = \frac{1}{p^2 - m^2 + i\epsilon}$. The normal ordering means that graphs in which a scalar line leaves and returns to the same vertex are excluded (cf. Section 10.1), so any loop integral contains at least two scalar propagators and is ultraviolet convergent in $d = 2$ or 3 (as $\int d^d k / k^4 < \infty$ as far as the large momentum contribution is concerned, if $d < 4$). The theory is therefore ultraviolet finite *ab initio*, although, of course, there will be finite corrections which convert the bare mass m appearing in the free propagator to the physical mass m_{ph} of our scalar particle. These renormalization effects are not particularly relevant to the physics of the bound-state formation which is our primary interest here, and will not be emphasized in the following, although in a real calculation of bound-state properties one would need to re-express the final results in terms of the measurable physical mass. In any case, for the theory in question, we now take $A(x) = \phi(x)$, $B(x) = \phi^\dagger(x)$, so we are studying the possibility of particle–antiparticle binding in the $\phi - \phi^c$ channel.

The lowest-order contributions, through order λ^2, to the kernel K_P are indicated graphically in Fig. 11.9; analytically, one finds that the following terms correctly generate the 2-2 scattering amplitude $\tilde{G}_P$ through $O(\lambda^2)$ when (11.36) is iterated:

$$K_P(q, p) = i(\lambda - \lambda^2 \mathcal{F}((p - q)^2) - \frac{1}{2}\lambda^2 \mathcal{F}((p + q)^2)) + O(\lambda^3), \tag{11.41}$$

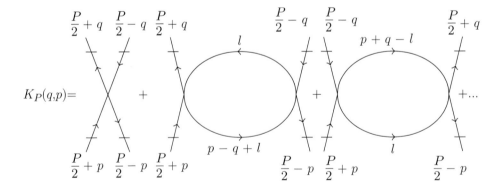

Fig. 11.9 Low-order contributions to the 2PI kernel $K_P(q, p)$ (arrows indicate charge flow: up for particles, down for antiparticles).

where we have defined

$$
\mathcal{F}(k^2) \equiv -i \int \Delta_F(l)\Delta_F(k-l)\frac{d^d l}{(2\pi)^d}
$$

$$
= -i \int \frac{1}{(l^2 - m^2 + i\epsilon)((k-l)^2 - m^2 + i\epsilon)} \frac{d^d l}{(2\pi)^d} \tag{11.42}
$$

The loop integral defining $\mathcal{F}(k^2)$ is perfectly finite in dimensions $d = 2$ or 3 (see Problem 3), and one obtains, in the region $0 < k^2 < 4m^2$,

$$
\mathcal{F}(k^2) = \frac{1}{\pi} \frac{1}{\sqrt{k^2(4m^2 - k^2)}} \arctan\sqrt{\frac{k^2}{4m^2 - k^2}}, \quad d = 2, \tag{11.43}
$$

$$
\mathcal{F}(k^2) = \frac{1}{4\pi} \frac{1}{\sqrt{k^2}} \ln\left(\frac{\sqrt{k^2} + 2m}{\sqrt{4m^2 - k^2}}\right), \quad d = 3 \tag{11.44}
$$

The value of $\mathcal{F}(k^2)$ for general k^2 can be obtained by analytic continuation of these formulas: in particular, one finds that $\mathcal{F}(k^2)$ is a real analytic function of k^2 with a cut on the positive real axis for $4m^2 \le k^2$ of square-root (resp. logarithmic) type in $d = 2$ (resp. 3), and no other singularities (the apparent square-root branch point at $k^2 = 0$ is spurious: only even powers of $\sqrt{k^2}$ appear in the Taylor expansion around $k = 0$).

We now search for the appropriate conditions for a bound-state pole to develop in the 2-2 scattering amplitude, where for convenience we work in the rest frame of the bound state and set $P^0 = \sqrt{P^2} = 2m - \kappa^2/m$ at the bound-state pole. Thus, the binding energy $\mathcal{E}_{\text{bind}} = \kappa^2/m$ of the bound state is parameterized in terms of the variable κ, which is a measure of the distance from the bound-state pole to the two-(free-)particle threshold at $P^0 = 2m$. The iteration of the leading-order kernel (just the constant $i\lambda$) clearly generates a tower of bubble graphs indicated in Fig. 11.10(a). These graphs form a geometric series; excluding the trivial disconnected contribution to $\tilde{G}_P(q,p)$, one has the sum (n is the number of loops)

$$
\tilde{G}_P^{(0)}(q,p) = \frac{i\lambda}{(2\pi)^d} \sum_{n=0}^{\infty} (-\lambda\mathcal{F}(P^2))^n \Delta_F\left(\frac{P}{2}+q\right)\Delta_F\left(\frac{P}{2}-q\right)\Delta_F\left(\frac{P}{2}+p\right)\Delta_F\left(\frac{P}{2}-p\right)
$$

$$
= \frac{i\lambda}{(2\pi)^d} \frac{1}{1+\lambda\mathcal{F}(P^2)} \Delta_F\left(\frac{P}{2}+q\right)\Delta_F\left(\frac{P}{2}-q\right)\Delta_F\left(\frac{P}{2}+p\right)\Delta_F\left(\frac{P}{2}-p\right)
$$

$$
\tag{11.45}
$$

The superscript (0) here indicates that this is the leading contribution to a reorganized set of perturbative contributions to $\tilde{G}_P$: we shall soon see that this tower of graphs determines the leading-order properties of the bound state for weak coupling. For a true bound state to be present at weak coupling (small λ) the value of the bubble integral $\mathcal{F}(P^2)$ must increase correspondingly at the bound-state pole to allow contributions of arbitrary order to remain comparable, thereby keeping the constituents bound for an infinite time. If the bound state is to be present at *arbitrarily weak coupling*, this means that $\mathcal{F}$ must become singular: this can only happen if the bound-

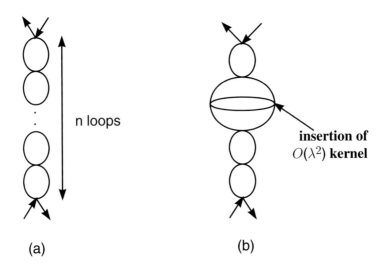

(a)   (b)

Fig. 11.10 (a) 2-2 amplitude from iteration of leading-order kernel; (b) subleading tower from a single insertion of a higher-order kernel.

state momentum $P = (2m - \kappa^2/m, \vec{0})$ approaches the two-particle threshold, i.e., as $\kappa/m \to 0$, when the bubble integral has the asymptotic behavior

$$\mathcal{F}(P^2) \sim \frac{1}{8m\kappa} + O(1), \quad \kappa/m \to 0, \quad d = 2, \tag{11.46}$$

$$\mathcal{F}(P^2) \sim \frac{1}{8m\pi} \ln\left(\frac{2m}{\kappa}\right) + O(\kappa^2 \ln\left(\frac{2m}{\kappa}\right)), \quad \kappa/m \to 0, \quad d = 3 \tag{11.47}$$

In other words, the bound state must become *non-relativistic*, with binding energy much smaller than the rest energy of the system. Since $\mathcal{F}$ is positive, we see also that the existence of a pole in (11.45) requires that the coupling λ be negative, corresponding to an attractive local point interaction between the constituent scalars. Of course, this would lead if taken literally to a theory with a spectrum unbounded below (cf. Section 8.4), but in $d = 2$ or 3 we are free to add a ϕ^6 interaction with an arbitrarily weak coupling to restore spectral sanity, without sensibly altering the bound-state properties (or the renormalizability of the theory), so we shall ignore this difficulty and proceed henceforth with negative λ.

Referring to (11.45) we see that the summed bubble graphs of Fig. 11.10(a) have a pole at $|\lambda|\mathcal{F}(P^2) = 1$, which for small λ, using the asymptotic forms (11.46, 11.47), occurs when

$$\kappa \sim \frac{|\lambda|}{8m}, \quad \mathcal{E}_{\text{bind}} \sim \frac{\lambda^2}{64m^3}, \quad d = 2 \tag{11.48}$$

$$\kappa \sim 2m \exp\left(-\frac{8\pi m}{|\lambda|}\right), \quad \mathcal{E}_{\text{bind}} \sim 4m \exp\left(-\frac{16\pi m}{|\lambda|}\right), \quad d = 3 \tag{11.49}$$

In one space and one time dimension ($d = 2$), the local $\frac{\lambda}{4}(\phi^\dagger\phi)^2$ interaction generates a δ-function potential $V(x) = g\delta(x)$ in the non-relativistic limit, with the dimensionless (in natural units) coupling given by $g = \frac{\lambda}{4m^2}$ (note that the coupling λ has dimension mass squared in $d = 2$). The reader may easily verify that such a potential in one spatial dimension does indeed lead (for $\lambda < 0$) in a system of reduced mass $m/2$ to a single bound state with the stated binding energy. The threshold singularity in two-space, one-time dimensions ($d = 3$) is much weaker—only logarithmic rather than linear—with the result that the binding energy vanishes exponentially for small coupling, with an essential singularity in the dependence of binding energy on coupling. This is actually the situation that arises for Cooper pairs in the BCS theory of super-conductivity, where an arbitrarily weak phonon-induced attractive coupling results in an exponentially small binding (and energy gap) in *three* spatial dimensions, but with the system effectively reduced to the two-dimensional Fermi surface of available electron states (see (Ziman, 1964), p. 330).

The asymptotic behavior indicated in (11.46, 11.47) is readily understood with a simple power-counting argument. The region of the loop integral responsible for the dominant contribution at weak coupling to the one-loop bubble integral $\mathcal{F}(P^2)$ (with $\mathcal{F}$ defined in (11.42)) corresponds to the non-relativistic scaling $\vec{l} \sim \kappa, l_0 \sim \kappa^2$ (the latter following if we perform the l_0 integration picking up the pole at $l_0 = \frac{\kappa^2}{2m} - m + \sqrt{\vec{l}^2 + m^2} - i\epsilon$, with $\vec{l} \sim \kappa << m$). Thus in $d = 2$ dimensions each of the denominators in the loop integral is of order κ^2, while the d^2l phase-space is of order κ^3, leading to the overall $1/\kappa$ threshold singularity for small κ. In $d = 3$ dimensions, the power-counting leads to κ^0, corresponding to a logarithmic dependence on κ when the spatial integral over $\vec{l}$ is performed. Each insertion of a higher-order piece of the 2PI kernel K_P in the sequence of bubble graphs, such as the diagram indicated in Fig. 11.10(b), reduces the strength of the threshold divergence at any given order in powers of λ. For example, in $d = 2$, the order λ^7 graph indicated in Fig. 11.10(b) produces only a $1/\kappa^5$ threshold singularity, one less power of $1/\kappa$ than the corresponding λ^7 graph in the leading tower of bubble graphs shown in Fig. 11.10(a). The reader may easily verify that adding terms of divergent structure $\lambda^{n+1}/\kappa^{n-1}$ to the geometric series in (11.45) (in contrast to the leading series, with terms of order λ^{n+1}/κ^n), corresponds to an order λ^2 contribution to the value of κ at the pole, and hence a higher order (by one power of λ) contribution to the binding energy of the bound state (i.e., in the $d = 2$ case, a contribution of order λ^3 to $\mathcal{E}_{\text{bind}}$ in (11.48)). In fact, the inclusion of successively more complicated kernel contributions is necessary to compute the bound-state properties to successively higher accuracy, with the result that $\mathcal{E}_{\text{bind}}$ (for example) becomes a divergent asymptotic expansion in λ.

The factorial divergence of perturbation theory in ϕ^4 theory discussed in the preceding section has not, of course, been eliminated: we must expect the total number of graphs contributing to the 2-2 amplitude $\tilde{G}_P$ to grow factorially with the power n of λ. But the remarkable simplification allowed by the existence of threshold singularities which ensure the persistence of the binding at arbitrarily weak coupling leads to the (highly fortunate) result that *to first approximation* the properties of a non-relativistic threshold bound state are determined by a tiny subset of Feynman graphs (the bubble diagrams of Fig. 11.10(a)) which form (in this scalar binding case) the simplest of all summable series: a geometric expansion! The factorial growth with order of the full set of graphs contributing to the scattering amplitude translates, when the graphs are reordered into towers on the basis of the strength of the threshold singularities they exhibit, into an infinite asymptotic expansion for the ground-state properties in powers of the weak coupling λ.

Threshold singularities of power strength, and therefore qualitatively similar to the $d = 2$ self-coupled scalar situation, reappear in massless gauge theories (both abelian and non-abelian varieties) in 3+1 spacetime dimensions. Some low-order contributions to the kernel in the case of an "onium" bound state in quantum electrodynamics (e.g., positronium, the bound state of an electron and positron) are shown in Fig. 11.11. In this case, our field $A(x)$ is the electron Dirac field $\psi(x)$ and $B(x)$ is $\bar{\psi}(x)$. As we have seen, threshold bound states are intrinsically non-relativistic in the weak coupling region where perturbative resummation is useful, and in gauge theories this singles out a particular gauge—Coulomb (or "radiation") gauge—as particularly useful in isolating the graphs with the strongest threshold singularities (Duncan, 1976). In this case the leading properties of onium bound states are determined by iteration of the 2PI kernel corresponding to exchange of a single Coulomb photon (Fig. 11.11(a)), with the two additional powers of κ per loop arising from the two extra space dimensions cancelled by the $1/\kappa^2$ behavior of the momentum-space Coulomb propagator $1/\vec{l}^2$ (for $l \sim \kappa$, where, of course, we need the exchanged photon to be *massless*). The iteration of this kernel leads to a series of ladder graphs (see Fig. 11.12)), with the property that each additional loop brings an extra factor of electron charge squared (and therefore α, the fine-structure constant), as well as an additional $1/\kappa$ power threshold singularity. Just as in ϕ^4 theory in $d = 2$, we consequently expect a pole to develop for $\kappa \sim \alpha m$, giving a binding energy of order $\alpha^2 m$.

We shall discuss the Feynman rules for gauge theories in Chapter 15: for present purposes, we need only know that in Coulomb gauge the A_0 propagator is instantaneous, corresponding to a factor $i/\vec{l}^2$ where $\vec{l}$ is the spatial momentum carried by the Coulomb line, and attaches with a factor $-ie\gamma_0$ at the charged fermion (i.e., electron or positron) line. The result is that the leading set of threshold singularities is generated by iteration of the (fully amputated) kernel indicated in Fig. 11.11(a): namely, $K_P(q, k) \sim \frac{ie^2}{|\vec{q}-\vec{k}|^2}(\gamma_0)(\gamma_0)$. Note that the kernel has four Dirac subscripts (not

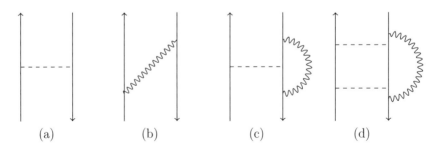

(a)   (b)   (c)   (d)

Fig. 11.11 (a) Coulomb exchange kernel; (b) transverse photon exchange kernel; (c, d) kernels contributing to Lamb shift.

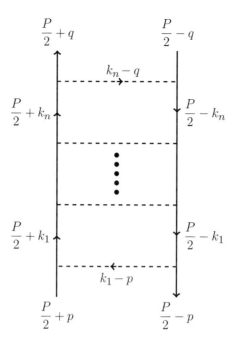

Fig. 11.12 Tower of ladder graphs generating the leading threshold singularities (of order α^{n+1}/κ^n) in a massless gauge theory. Arrows denote charge flow; the momentum flow is upwards on both fermion lines.

shown): the corresponding Dirac index dependence is given by the direct product of two γ_0 matrices. The full fermion propagator (in accordance with our notation from Chapter 7) is now written $\hat{S}_F(\frac{P}{2} + k)$ for the electron line and $\hat{S}_F(-\frac{P}{2} + k)$ for the corresponding positron line (which is an electron propagator pointing downward, hence with momentum reversed): in ladder approximation these full propagators become just the free ones. Moreover, the leading threshold singularities are generated in the non-relativistic kinematic domain where $k_0 \sim O(\kappa^2)$, $\vec{k} \sim O(\kappa)$, and we are as usual in the frame where $P = (2m - \kappa^2/m, \vec{0})$, so that we can replace

$$\hat{S}_F\left(\frac{P}{2} + k\right) \rightarrow \frac{\frac{\slashed{P}}{2} + \slashed{k} + m}{(\frac{P}{2} + k)^2 - m^2 + i\epsilon} \rightarrow P_+ \frac{1}{k_0 - \frac{\vec{k}^2 + \kappa^2}{2m} + i\epsilon}, \quad P_+ \equiv \frac{1 + \gamma_0}{2}$$

$$\hat{S}_F\left(-\frac{P}{2} + k\right) \rightarrow \frac{-\frac{\slashed{P}}{2} + \slashed{k} + m}{(-\frac{P}{2} + k)^2 - m^2 + i\epsilon} \rightarrow -P_- \frac{1}{k_0 + \frac{\vec{k}^2 + \kappa^2}{2m} - i\epsilon}, \quad P_- \equiv \frac{1 - \gamma_0}{2}$$

$$(11.50)$$

Note that the appearance of the P_+ (resp. P_-) projection operator on the electron (resp. positron) line explains why transverse photon exchange between the lines is absent in the leading non-relativistic approximation: a transverse gluon vertex comes with a spatial γ matrix $\vec{\gamma}$, and the projection operators on either side will then cause the amplitude to vanish, as $P_+\vec\gamma P_+ = P_-\vec\gamma P_- = 0$. In the extreme limit of infinitely massive fermions, $m \to \infty$, the propagators lose even their dependence on the spatial momentum $\vec k$, which corresponds in coordinate space to the static limit in which $S_F(x) \propto \delta^3(\vec x)$. This limit will be of interest to us in Chapter 19 in our discussion of quark confinement. The Bethe–Salpeter equation (in the version (11.39)) for this system evidently takes the form, making the substitutions relevant in the leading threshold approximation,

$$
\Psi_P(q) = \int K_P(q,k) S_F\!\left(\frac{P}{2}+k\right)\Psi_P(k) S_F\!\left(-\frac{P}{2}+k\right)\frac{d^4k}{(2\pi)^4}
$$

$$
= -i \int \frac{e^2}{|\vec q - \vec k|^2}\gamma_0 P_+ \Psi_P(k) P_- \gamma_0 \frac{1}{k_0 - \frac{\vec k^2+\kappa^2}{2m} + i\epsilon}\frac{1}{k_0 + \frac{\vec k^2+\kappa^2}{2m} - i\epsilon}\frac{d^4k}{(2\pi)^4}
$$

$$
= \int \frac{e^2}{|\vec q - \vec k|^2}\frac{m}{\vec k^2 + \kappa^2} P_+ \Psi_P(k) P_- \frac{d^3k}{(2\pi)^3} \tag{11.51}
$$

In passing from the second to the third line, we have used $\gamma_0 P_+ = P_+, P_-\gamma_0 = -P_-$ (the relative minus sign ensuring the attractive coupling between the electron and its antiparticle), and performed the integration over k_0 to pick up the pole at $k_0 = -\frac{\vec k^2+\kappa^2}{2m} + i\epsilon$. Note that the absence of q_0 in the leading order Coulomb kernel implies that $\Psi_P(q)$ (in the leading approximation) is really a function only of spatial momentum, $\Psi_P(\vec q)$, so that the only k_0 dependence in the integral is that displayed explicitly in the fermion propagators. From (11.51) we conclude immediately that (in the ladder approximation) (a) $\Psi_P(q)$ is in fact only a function of the spatial vector $\vec q$, and (b) that the Dirac structure of $\Psi_P(q)$ must be (in the representation (7.100)), as a consequence of the projection operators $P_\pm$ in (11.51),

$$
\Psi_P(q) \propto \begin{pmatrix} A(\vec q) & -A(\vec q) \\ A(\vec q) & -A(\vec q) \end{pmatrix} \tag{11.52}
$$

Writing $A(\vec q) = (\vec q^2 + \kappa^2)\phi(\vec q)$, we find that (11.51) reduces to

$$
\frac{\vec q^2 + \kappa^2}{m}\phi(\vec q) = \int \frac{e^2}{|\vec q - \vec k|^2}\phi(\vec k)\frac{d^3k}{(2\pi)^3} \tag{11.53}
$$

or in coordinate space (with $\phi(\vec r)$ the Fourier transform of $\phi(\vec q)$),

$$
-\frac{1}{m}\vec\nabla^2\phi(\vec r) - \frac{e^2}{4\pi r}\phi(\vec r) = -\frac{\kappa^2}{m}\phi(\vec r) \tag{11.54}
$$

—exactly the non-relativistic Schrödinger equation for two equal mass m particles binding via an attractive Coulomb potential $V(r) = -\frac{e^2}{4\pi r}$ to form a bound state with binding energy $-\frac{\kappa^2}{m}$.

Just as in the case of self-coupled scalar theories, the higher-order contribution to two-particle irreducible kernels (such as the graphs displayed in Fig. 11.11(b,c,d)) reduce the strength of the threshold singularities at any given order of perturbation theory, and consequently result in higher-order (in α) shifts in the binding energy (and other bound-state properties). For example, a single insertion of transverse photon exchange (graph in Fig. 11.11(b)) in a ladder of Coulomb exchanges will produce (for the same power of α) two extra factors of κ, as a spatial momentum term (of order κ) must be used (instead of the $P_\pm$ in (11.50)) in the numerator of a Dirac propagator adjacent to the transverse photon vertex (with a vertex factor $-ie\vec{\gamma}$) on both the electron and positron line, to avoid getting $P_\pm \vec{\gamma} P_\pm = 0$. As $\kappa \sim \alpha$ at the bound state pole, this results in a correction to the binding energy suppressed by α^2 relative to the leading term: i.e., a contribution of order α^4. This is in fact the magnetic hyperfine splitting.[7] For positronium, in which the electron and positron magnetic moments are comparable, the hyperfine splitting is of the same order as the relativistic $O(\alpha^4)$ corrections which arise from considering only Coulomb ladder graphs, but with fully relativistic propagators (instead of the non-relativistic limits (11.50)) on the electron and positron lines.[8] More complicated kernels, such as emission and absorption of a photon from an electron, with multiple Coulomb interactions in the interim (e.g., Fig. 11.11(c,d)), lead to $O(\alpha^5)$ shifts: in the case of the hydrogen atom, this is the famous *Lamb shift*. A systematic analysis of the reorganization of the scattering amplitude into towers of graphs of given threshold behavior for both abelian and non-abelian gauge theories, allowing the systematic extraction of the various contributions to the binding energy of onium states in both types of theory, has been given in (Duncan, 1976).

Of course, not all bound states in quantum theory appear as a consequence of the kinematic enhancement of an intrinsically weak coupling near a threshold, allowing the systematic "sorting" of perturbative contributions into summable towers of graphs as described above. In non-relativistic quantum mechanics, for example, an attractive Yukawa potential $V(r) = -g^2 e^{-mcr/\hbar}/r$ in three space dimensions will only produce a bound state once the coupling strength g exceeds a minimum critical value. The field-theoretic analog of this problem involves the exchange of a massive boson (mass m) between heavy fermions (e.g., pions exchanged between nucleons in the meson field theories of the 1950s). The fact that all particles are massive means that there are no infrared threshold singularities to effectively enhance the coupling by moving the bound-state pole close to the two-particle threshold, and the formation of the bound state is an essentially non-perturbative process: there is no systematic way to extract a dominant summable subset of graphs contributing to the bound-state properties. The situation is even more dramatically non-perturbative in the case of quantum chromodynamics, where the constituent binding particles (quarks and gluons) do not even exist as asymptotic states in the full theory (in contrast to perturbation theory), due to color confinement. The process by which quarks bind to form the low-mass hadrons necessarily involves a large effective coupling of order unity, and perturbative calculations give us no useful information in this regime.

---

[7] For the explicit calculation, see (Duncan, 1976), Appendix.

[8] Of course, in the case of the hydrogen atom, taking the proton mass to infinity, so that the nucleus simply serves as a source of electrostatic Coulomb energy, the summation of ladder diagrams with a fully relativistic electron propagator leads to a Bethe–Salpeter equation which is just the Dirac equation in the presence of a Coulomb potential, with bound-state energy given by (2.63). See (Weinberg, 1995a), Chapter 14.

11.3 "Essentially non-perturbative" processes: non-Borel-summability in field theory

The "inconvenient truth" that perturbative expansions in field theory (and in quantum mechanics, more generally) are at best asymptotic expansions rather than convergent Taylor expansions leads to an important question of principle: namely, to what extent does the information contained in a perturbative (i.e., graphical Feynman) expansion of the amplitudes of the theory fully specify, even in principle, the exact dynamics of the theory? Were the perturbation theory a convergent expansion, with a finite radius of convergence in the (complex) coupling constant plane, the answer to this question would be immediate and affirmative, as standard theorems of complex analysis would assure the possibility of reconstruction, at least in principle, of the full amplitudes for arbitrary values of the coupling by analytic continuation starting with the power series around the point of vanishing coupling. In turns out that in certain cases, the perturbative information (i.e., the coefficients of the divergent asymptotic expansion of an amplitude) does indeed suffice for a reconstruction of the full amplitude, so that in some sense the graphical expansion is "the whole story" for such theories. On the other hand, in most four-dimensional field theories of phenomenological interest, and in particular in the gauge theories which form the backbone of the Standard Model of elementary particle physics, this is unfortunately not the case. In addition to the question of principle raised here, there is the practical matter that in a strongly coupled theory, with a coupling constant of order unity, an asymptotic expansion, the partial summands of which begin to grow essentially immediately (at low orders of perturbation theory), is just unable to supply us with any quantitatively useful information with regard to the exact amplitudes of the theory.

As usual in discussions of (non-)convergence of perturbation theory, the toy integral (11.3) corresponding to 0-dimensional ϕ^4 theory provides a convenient starting point:

$$Z(\lambda, m) \equiv \int_{-\infty}^{+\infty} e^{-\frac{1}{2}m^2 x^2 - \frac{\lambda}{4}x^4} dx \qquad (11.55)$$

We can rewrite this quantity as an integral transform as follows

$$Z(\lambda, m) = \int_0^\infty t^{-1/2} e^{-t} B(\lambda t) dt, \qquad (11.56)$$

$$B(\lambda t) \equiv t^{1/2} \int_{-\infty}^{+\infty} \delta(\frac{1}{2}m^2 x^2 + \frac{\lambda}{4}x^4 - t) dx \qquad (11.57)$$

$$= \sqrt{\lambda t} \int_{-\infty}^{+\infty} \delta(\frac{m^2}{2}y^2 + \frac{1}{4}y^4 - \lambda t) dy \qquad (11.58)$$

Evaluating the last integral by locating the roots of the quadratic function in the δ-function, one finds (setting $z - \lambda t$),

$$B(z) = \frac{2\sqrt{z}}{\sqrt{(\sqrt{m^4 + 4z} - m^2)(m^4 + 4z)}} \qquad (11.59)$$

from which we conclude that the function B appearing in (11.56) is an analytic function of its argument at the origin, with a convergent Taylor series with radius of convergence $\frac{1}{4}m^4$ (due to the square-root branch point appearing at $z = -\frac{1}{4}m^4$). The integral transformation in (11.56) (called a *Borel transform*) has therefore accomplished the impressive feat of taming the divergence of the asymptotic expansion of $Z(\lambda, m)$ in powers of λ, essentially by dividing each coefficient of this asymptotic expansion by a factorial which reduces the rate of growth of the coefficients from factorial to power (and therefore to a series with a finite radius of convergence):

$$B(z) = \sum_{n=0}^{\infty} d_n z^n \Rightarrow Z(\lambda) \equiv \int_0^{\infty} t^{-1/2} e^{-t} B(\lambda t) dt \sim \sum_{n=0}^{\infty} c_n \lambda^n \qquad (11.60)$$

$$d_n = \frac{c_n}{\Gamma(n + \frac{1}{2})} \qquad (11.61)$$

If we substitute the explicit result (11.4) for the c_n into (11.61) we find the dominant asymptotic behavior $d_n \sim (-1)^n (\frac{4}{m^4})^n$, leading to the radius of convergence quoted above (using the ratio test). Note that the appearance of an oscillating sign factor in the coefficients ensures that the singularity of $B(z)$ occurs on the *negative* real axis for z: the Borel transform is well-defined (by analytic continuation of its power series around $z = 0$) and non-singular on the entire positive real axis where the integral (11.56) reconstructing $Z(\lambda, m)$ runs. In such cases one refers to the original divergent asymptotic expansion as *Borel summable*: the full partition function is recoverable in such cases from a knowledge of the perturbative expansion coefficients, which after division by a factorial, yield Taylor coefficients with power behavior at large order, and define a non-singular function $B(z)$ (for positive real z) which can at least in principle be used to reconstruct the desired $Z(\lambda, m)$ for arbitrary coupling λ.

The extension of the above ideas to dimensions $d \geq 1$ (i.e., the anharmonic oscillator for $d = 1$, fully regularized ϕ^4 theories in $d \geq 2$) with a positive sign mass term is quite straightforward. The Euclidean action of the discretized theory may be written

$$S(\phi_i) = \frac{1}{2} \sum_{i,j} \phi_i K_{ij} \phi_j + \frac{\lambda}{4} \sum_i \phi_i^4 \qquad (11.62)$$

where the quadratic form $\mathcal{K}$ is positive-definite. A Borel transform is then obtained by introducing a δ-function as previously. One finds (with N the number of spacetime points) after appropriately rescaling the fields,

$$B(z) = z^{-N/2} \int \prod_i d\psi_i \delta(\frac{1}{2} \psi_i K_{ij} \psi_j + \frac{1}{4} \sum_i \psi_i^4 - z) \qquad (11.63)$$

The appropriate analyticity of $B(z)$ and hence the property of Borel summability has been rigorously established for such theories, although for considerations of space we shall not attempt to provide a proof here.[9]

---

[9] For discussion and further references, see (Glimm and Jaffe, 1987), Section 23.2.

Unfortunately, the highly desirable feature of Borel-summability turns out to be rather fragile. For example, if we simply change the sign of the mass term, the Borel transform develops a singularity on the positive real axis, vitiating the reconstruction of the partition function via (11.56) (as the integral runs directly over a non-integrable singularity). We recall from Chapter 8 that a negative mass term in the Hamiltonian (or Euclidean action) of a ϕ^4 field theory simply amounts to a situation in which the system develops degenerate ground states (see Fig. 8.1). For $d = 1$ (the anharmonic oscillator case), the sign change leads to a double-well potential, with degenerate minima of the classical potential leading to the well-known tunneling phenomena and a unique symmetric ground-state wavefunction. Returning once again to our toy integral, if we start with an action function with a negative squared-mass term (and shifted by a constant to set the minimum action at zero)

$$Z(\lambda, m) = \int e^{-S(x)} dx, \quad S(x) = \frac{m^4}{4\lambda} - \frac{m^2}{2} x^2 + \frac{\lambda}{4} x^4 \qquad (11.64)$$

the Borel transform $B(z)$ becomes

$$B(z) = 2\sqrt{z} \int_0^\infty \delta(z - \tilde{S}(y)) dy, \quad \tilde{S}(y) \equiv \frac{1}{4}(m^2 - y^2)^2 \qquad (11.65)$$

and one finds that the contribution to the y integral from the root of the δ-function at $y = \sqrt{m^2 - 2\sqrt{z}}$ leads to a term in $B(z)$ with a simple pole on the positive real axis at $z = \frac{m^4}{4}$:

$$B(z) \sim \frac{1}{\sqrt{m^2 - 2\sqrt{z}}} = \frac{\sqrt{m^2 + 2\sqrt{z}}}{m^4 - 4z} \qquad (11.66)$$

The singularity at positive real values is a consequence of the fact that the coefficients in a power series expansion (in this case, in powers of $\sqrt{z}$), while only growing at a power rate (rather than factorially) now lack the $(-1)^n$ oscillating sign factor present in the Borel-summable case, which resulted in a singularity of $B(z)$ for *negative* real z, safely away from the contour of integration of the Borel transform (11.56). For $d \geq 1$ (quantum mechanics or field theory), the appearance of a singularity of the Borel transform on the positive real axis is associated with the presence of tunneling phenomena for physical (i.e., positive) values of the coupling λ: recall that, for the Borel summable cases discussed in Section 11.1, energetic instabilities exemplified by the instanton solutions responsible for the leading large-order behavior only occurred once we had analytically continued the coupling λ to negative real values. With a negative squared-mass term, tunneling between distinct local minima of the potential energy function already occurs for physical (i.e., positive) values of λ. The presence of instanton solutions (extrema of the Euclidean action) for physical values of the gauge coupling in non-abelian gauge theories like QCD mean that such theories must also necessarily develop Borel singularities on the positive axis, and are therefore not Borel-summable.

The discussion up to this point has implicitly assumed that the exact amplitudes of our theory are expressed in terms of a well-defined path-integral representation: i.e., the theory is fully regularized in both the ultraviolet and the infrared to reduce the number of degrees of freedom (effectively) to a finite level. In Part 4 of the book we shall examine the process of reorganizing weak-coupling perturbation theory in order to eliminate the dependence of the amplitudes on these cutoffs in favor of *renormalized amplitudes* defined in terms of physically accessible low-energy parameters of the theory. In particular, the formal expansion of the amplitudes of the theory in powers of the bare coupling parameter(s) appearing in the cutoff theory is replaced by an expansion in powers of cutoff-independent renormalized coupling(s). This reorganization of the perturbation series can result in the appearance of new singularities of the Borel transform on the positive real axis, called "UV renormalons", which once again vitiate the reconstruction of the non-perturbative amplitude from perturbative information. Even if the ultraviolet cutoff is maintained, in the infinite volume limit for massless field theories, *infrared* divergences can appear which similarly induce positive real singularities of the Borel transform (in this case, called "IR renormalons"), again destroying the Borel summability of the theory. The lesson from all of this is clear: the property of Borel summability is an extremely fragile one, and one which we can hardly ever expect to be present in interesting relativistic field theories.

The failure of the Borel resummation technique suggests that the question formulated earlier in this section—whether or not the information encoded in a formal perturbative expansion contains sufficient information to reconstruct the exact generating function(al) of the theory—should be answered in the negative in all such cases. This conclusion is unwarranted, as we shall now see. The Borel transform is only one of a variety of reconstruction techniques which attempt to connect perturbative computations with the exact amplitudes of theories defined by path integrals. In particular, the path integral for scalar field theories, with *either* sign of the mass term, may be reconstructed by use of methods which go under the generic name of "optimized perturbation theory". The particular form of optimized perturbation theory which we shall describe here is sometimes referred to as the "linear δ expansion". The basic idea is to construct a series of approximants to the path integral which only require perturbative calculations (defined here as path integrals with Gaussian exponents only: hence analytically computable), but nevertheless can be shown to converge rigorously to the exact answer. The basic idea is to interpolate between a "tunable" Gaussian approximation and the exact action by introducing an auxiliary interpolating variable $\delta, 0 \leq \delta \leq 1$, and a variational parameter μ—effectively a variable bare mass.

Thus, one writes, for a theory with Euclidean action $S(\phi, m, \lambda)$, an interpolating action

$$S_\delta = \delta S(\phi, m, \lambda) + (1 - \delta) S_0(\phi, m, \lambda; \mu) \tag{11.67}$$

where S_0 is quadratic in the field variable. The partition function can then be formally expanded in the δ variable

$$Z = \int \mathcal{D}\phi e^{-\delta S - (1-\delta)S_0} \sim \sum_{n=0}^{\infty} c_n(m, \lambda; \mu) \delta^n \tag{11.68}$$

where the evaluation of the c_n involve the usual perturbative manipulations: i.e., integrals with a Gaussian action S_0. The correct theory is, of course, recovered in the limit $\delta = 1$, and an Nth order approximant to the theory at $\delta = 1$ is clearly given by

$$Z_N \equiv \sum_{n=0}^{N} c_n(m, \lambda; \mu) \tag{11.69}$$

If we hold the parameter μ (the dependence on which, of course, disappears in Z at $\delta = 1$) fixed, the sequence of approximants is factorially divergent in the usual way. On the other hand, if we *change* μ at each increased order N by a *principle of minimal sensitivity (PMS)* (Stevenson, 1981)—namely, the requirement that the partial sum Z_N be extremal with respect to μ—then a miracle occurs: the sequence $Z_N(\mu_N)$, with the μ_N determined by the aforesaid PMS condition, can be rigorously shown to converge to the exact partition function $Z(\delta = 1)$! In particular, we can show that the *remainder* term at order N, defined by

$$R_N \equiv Z - Z_N(\mu_N) \tag{11.70}$$

goes exponentially to zero for large N. We shall provide the full proof here only for the case of the quartic toy integral: the full argument for $d = 1$ (the anharmonic oscillator, either single or double-well) can be found in (Duncan and Jones, 1993).

We begin with a useful identity which isolates the Nth approximant Z_N defined in (11.69) (see Problem 6)

$$Z_N = \int \mathcal{D}\phi \oint_C \frac{dz}{2\pi i} \frac{1}{z^{N+1}} \frac{1 - z^{N+1}}{1 - z} e^{-zS - (1-z)S_0} \tag{11.71}$$

where C is a small circular contour enclosing the origin (and excluding $z = 1$). The condition of minimal sensitivity requires that we extremize Z_N with respect to the variational parameter μ, the dependence on which is entirely contained in the "free" action S_0 (cf. (11.67)):

$$\frac{\partial Z_n}{\partial \mu} = 0 \Rightarrow \int \mathcal{D}\phi \oint_C \frac{dz}{2\pi i} \frac{\partial S_0}{\partial \mu} \frac{1}{z^{N+1}} e^{-zS - (1-z)S_0}$$

$$0 = \int \mathcal{D}\phi \frac{\partial S_0}{\partial \mu} (S - S_0)^N e^{-S_0} \tag{11.72}$$

We shall also need the following identity, which will facilitate the evaluation of the remainder term (see Problem 7):

$$e^{-f} - \sum_{n=0}^{N} \frac{(-f)^n}{n!} = e^{-f} \frac{1}{N!} \int_0^{|f|} e^{\xi \operatorname{sign}(f)} \xi^N d\xi \tag{11.73}$$

valid for *odd* N. The large order asymptotics discussed henceforth will implicitly assume that we are dealing with odd orders only. Using this identity, it follows immediately that the remainder term R_N at order N (odd) can be written as the sum of two terms:

$$R_N = A_N + B_N \tag{11.74}$$

$$A_N \equiv \frac{1}{N!} \int \mathcal{D}\phi\,\theta(S - S_0)e^{-S} \int_0^{S-S_0} e^\xi \xi^N\, d\xi \tag{11.75}$$

$$B_N \equiv \frac{1}{N!} \int \mathcal{D}\phi\,\theta(S_0 - S)e^{-S} \int_0^{S_0-S} e^{-\xi} \xi^N\, d\xi \tag{11.76}$$

For large (resp. small) fields ϕ, we have $S > S_0$ (resp. $S < S_0$), so we can refer to A_N (resp. B_N) as the strong (resp. weak) field contributions to the remainder term.

In the zero-dimensional case, there is a single spacetime point, the single field variable ϕ is called x, and our free and full actions become

$$S_0 = m^2 x^2 + \lambda\mu x^2, \quad S = m^2 x^2 + \lambda x^4 \tag{11.77}$$

where inessential factors of $\frac{1}{2}$ and $\frac{1}{4}$ in our original quartic toy integral (11.55) have been dropped. The PMS condition (11.72) determining μ_N becomes in this case

$$\int_0^\infty x^2 (x^4 - \mu_N x^2)^N e^{-(m^2 + \lambda\mu_N)x^2}\, dx = 0 \tag{11.78}$$

or, changing to the rescaled variable $u \equiv x^2/\mu_N$,

$$0 = \int_0^\infty u^{N+\frac{1}{2}}(u - 1)^N e^{-N\alpha u}\, du, \quad \alpha_N \equiv \frac{1}{N}(m^2\mu_N + \lambda\mu_N^2) \tag{11.79}$$

$$= \int_0^\infty \sqrt{u}\,\operatorname{sign}(u - 1)e^{-NS_N(u)}\, du, \quad S_N(u) \equiv \alpha_N u - \ln|u(u - 1)| \tag{11.80}$$

For large N, the integral in (11.80) is dominated by two saddle points at which $S'_N = 0$, one at $u = u_> > 1$ contributing with a positive sign, the other at $u = u_<, 0 < u_< < 1$ contributing with a negative sign. One readily finds

$$u_> = \frac{1}{2} + \frac{1}{\alpha_N}\left(1 + \sqrt{1 + \frac{\alpha_N^2}{4}}\right) \tag{11.81}$$

$$u_< = \frac{1}{2} + \frac{1}{\alpha_N}\left(1 - \sqrt{1 + \frac{\alpha_N^2}{4}}\right) \tag{11.82}$$

As the two contributions must cancel to satisfy the PMS condition, the values of the effective action function $S_N(u)$ at these two points must agree in the large N limit:

$$S_N(u_>) = S_N(u_<) \Rightarrow 2\sqrt{1 + \frac{\alpha_N^2}{4}} = \ln\frac{\sqrt{1 + \frac{\alpha_N^2}{4}} + 1}{\sqrt{1 + \frac{\alpha_N^2}{4}} - 1} \Rightarrow \alpha_N = 1.325487... \equiv \alpha_0$$

$$\tag{11.83}$$

so that (recalling that $\alpha_N \equiv \frac{1}{N}(m^2\mu_N + \lambda\mu_N^2)$) the desired PMS scaling of μ_N becomes asymptotically

$$\mu_N \sim \sqrt{\frac{\alpha_0}{\lambda}}N^{1/2} - \frac{m^2}{2\lambda} + O(N^{-1/2}) \tag{11.84}$$

We see already at this point—and this is crucial for the effectiveness of optimized perturbation theory in handling both the Borel ($m^2 > 0$) and non-Borel ($m^2 < 0$) cases with equal facility—that the dependence of the asymptotic behavior of the variational parameter μ_N on the sign of the mass term is *subdominant*: in either case, we must scale $\mu_N \propto N^{1/2}$ at large N. We shall now see that in either case, with this scaling, the remainder term R_N goes exponentially to zero, so the Z_N (for odd N) provide a rapidly convergent sequence of approximants to the exact integral Z.

We saw previously that the remainder term R_N can be written as the sum of strong-field and weak-field contributions A_N and B_N, defined in (11.75,11.76). For the quartic integral, $S - S_0 = \lambda(x^4 - \mu_N x^2) > 0$ when $x^2 > \mu_N$, and the expression for the strong-field contribution becomes

$$A_N = \frac{1}{N!}\int_{\sqrt{\mu_N}}^{\infty} dx\, e^{-(m^2x^2 + \lambda x^4)}\int_0^{\lambda(x^4 - \mu_N x^2)} e^{\xi}\xi^N d\xi \tag{11.85}$$

Defining $x^2 = \mu_N u$ and rescaling the ξ integral by

$$\xi = \lambda(x^4 - \mu_N x^2)\sigma = \lambda\mu_N^2 u(u-1)\sigma, \quad 0 < \sigma < 1 \tag{11.86}$$

we find

$$A_N = \frac{(\lambda\mu_N^2)^{N+1}}{2N!}\sqrt{\mu_N}\int_1^{\infty} du\, e^{-m^2\mu_N u - \lambda\mu_N^2 u^2}u^{N+\frac{1}{2}}(u-1)^{N+1}\int_0^1 \sigma^N e^{\lambda\mu_N^2 u(u-1)\sigma}d\sigma \tag{11.87}$$

Given that $u > 1$ in the integral above, the σ integral satisfies the obvious inequality

$$\int_0^1 \sigma^N e^{\lambda\mu_N^2 u(u-1)\sigma}d\sigma < e^{\lambda\mu_N^2 u(u-1)} \tag{11.88}$$

which, when inserted in (11.87), gives

$$A_N < \frac{(\lambda\mu_N^2)^{N+1}}{2N!}\sqrt{\mu_N}\int_1^{\infty} du\, u^{\frac{1}{2}}(u-1)e^{-(m^2\mu_N + \lambda\mu_N^2)u + N\ln(u(u-1))} \tag{11.89}$$

Inserting the PMS scaling behavior found earlier in (11.79, 11.80, 11.83), this becomes

$$A_N < \frac{(\lambda\mu_N^2)^{N+1}}{2N!}\sqrt{\mu_N}\int_1^{\infty} du\, u^{\frac{1}{2}}(u-1)e^{-NS_N(u)} \tag{11.90}$$

and we see that the u integral is dominated by exactly the same saddle-point at $u = u_>$ found earlier in implementing the PMS condition. In other words, at large (odd) N, the u integral gives a contribution proportional to $\frac{1}{\sqrt{N}}e^{-NS_N(u_>)}$, with $u_>$ given in

(11.81) (with $\alpha_N = \alpha_0 = 1.325..$). Inserting this result and using Stirling's formula for the factorial, a short calculation gives the desired asymptotic behavior

$$A_N \sim CN^{\frac{1}{4}} e^{N(1+\ln\alpha_0 - S_N(u_>))} = CN^{\frac{1}{4}} e^{-0.6627..N} \tag{11.91}$$

irrespective of the sign of m^2. A similar calculation for the weak-field contribution B_N gives exactly the same asymptotic behavior (up to a constant)—not surprisingly, given that we earlier saw that the PMS scaling is tantamount to equality of the effective actions at the corresponding saddle-points, $S_N(u_<) = S_N(u_>)$. The conclusion is that the sequence of approximants Z_N obtained by carrying out the δ expansion to order N (a process involving only Gaussian integrals and therefore graphically equivalent to conventional perturbation theory) and then evaluating the result at the PMS value for the variational parameter μ_N, converges to the exact answer with *exponential rapidity*, whether or not the conventional asymptotic expansion of the theory in powers of the coupling λ is Borel summable.

The optimized δ expansion can also be shown to lead to convergent results in dimensions $d = 1$ (anharmonic oscillator) and $d \geq 2$ (ϕ^4 field theory), again irrespective of the sign of the mass term (Duncan and Jones, 1993). For the anharmonic oscillator, the finite (Euclidean) time partition function is given by the anharmonic version of (4.110):

$$\text{Tr}(e^{-\beta H_{\text{anhar}}}) = \int dQ \langle Q | e^{-\beta H_{\text{anhar}}} | Q \rangle, \quad H_{\text{anhar}} = \frac{1}{2}(p^2 + m^2 q^2) + \lambda q^4 \tag{11.92}$$

$$= \int_{q(\beta)=q(0)} e^{-S(q)} \mathcal{D}q(t) \tag{11.93}$$

$$S(q) = \int_0^\beta \{\frac{1}{2}(\dot{q}^2 + m^2 q^2) + \lambda q^4\} dt \tag{11.94}$$

which can be subjected to an optimized δ expansion by choosing

$$S_0 = \int_0^\beta \{\frac{1}{2}\dot{q}^2 + \frac{1}{2}(m^2 + 2\mu\lambda)q^2\} dt \tag{11.95}$$

$$S = \int_0^\beta \{\frac{1}{2}\dot{q}^2 + \frac{1}{2}m^2 q^2 + \lambda q^4\} dt \tag{11.96}$$

In this case, the PMS scaling turns out to be $\mu_N \sim N^{2/3}$. The strong-field contribution to the remainder A_N dies exponentially as in the quartic integral case (i.e., like $e^{-\text{const}\cdot N}$), while the weak-field contribution goes like

$$B_N < \frac{\lambda\beta\mu_N^2}{4\sqrt{2\pi N}} e^{-\frac{1}{\lambda\beta}(N/mu_N)^2} \sim N^{5/6} e^{-CN^{2/3}/(\lambda\beta)} \tag{11.97}$$

In the case of the anharmonic oscillator, techniques are available for the calculation of the δ expansion perturbation theory coefficients to high order (e.g., $N \sim 75$), and these convergence results can therefore be checked explicitly. In the field-theory case, of course, higher loop calculations become simply impractical, so one has to hope for convergence at moderate values of N (less than 5, say).

Note that the convergence of the optimized approximants in (11.97) is lost at large Euclidean time extent β—a problem which is, of course, exacerbated in higher dimension where β becomes βV, with V the spatial volume. This means that with the particular interpolation chosen here (a variable bare mass), the optimized perturbation theory is not really useful in the field theory context, even though *in principle* it implies that exact results are reconstructible from "perturbative information" in the finite-volume theory for either sign of the mass term. One might hope to eliminate the volume dependence by attempting an optimized expansion for the connected amplitudes of the theory—by studying an optimized expansion of $\ln Z$ rather than Z, for example—but the convergence proof with the PMS optimized interpolation approach as described above breaks down in this case (Duncan and Jones, 1993). Of course, this procedure relied on a very specific choice for the interpolation between free and full actions (using a variational bare mass term, in particular), and it may very well be possible that a more ingenious interpolation scheme would allow a convergent reorganization of the perturbative expansion for connected amplitudes even in non-Borel field theories.

The just-described examples of the δ expansion indicate that the question posed at the beginning of this section—is the full content of field theory already present in perturbatively computable amplitudes (perhaps in a highly disguised form!)?—cannot be answered definitively in the negative, even for non-Borel-summable theories. However, as a practical matter, we must admit that for "essentially non-perturbative" processes in a strongly coupled non-Borel-summable field theory, those in which no summable subset of perturbation theoretic contributions can be shown to incorporate the dominant contribution to the desired amplitudes, a description in terms of Feynman diagrams yields at best a crude qualitative (and very possibly misleading) picture of the underlying physics. Very little can be learned, for example, about the physics of quark confinement by studying Feynman graphs of interacting quarks and gluons, however complicated.

In cases like these, where perturbation theory completely fails us, how can we hope to make progress in making reliable, *quantitatively accurate*, predictions in a relativistic field theory? If we give up on the most ambitious goal—explicitly calculating the full amplitudes of the theory from a finite (and small) set of experimentally determinable masses and couplings—much can be achieved simply by exploiting general structural features which we expect our strongly coupled field theory to possess. In the 1950s and 1960s, for example, the failure of local field theory models to provide a quantitative description of strong interaction processes (as they could hardly do prior to the discovery of quantum chromodynamics (QCD), and the development of appropriate non-perturbative techniques for dealing with QCD) led many theorists to adopt a highly positivistic approach, in which one attempted to constrain strong interaction scattering amplitudes (incorporated in the S-matrix of the theory), as the only directly measurable objects, on the basis of a set of "sacred" principles, primarily Lorentz-invariance, unitarity, crossing invariance, and a principle of *maximal analyticity*, which asserted the analyticity of scattering amplitudes as functions of the complexified kinematical variables *except* at points where singularities were necessitated by the appearance of thresholds. The clever application of dispersion relations which could be derived on the basis of these fundamental assumptions led to many important and experimentally verifiable predictions in strong interaction physics, despite the fact that the correct underlying local field theory had yet to be identified. Typically, in dispersion theory one derives relations between amplitudes: the complete calculation of a specific amplitude from first principles cannot, of course, be expected in the absence of a specific microtheory, although at the time there were hopes that a self-consistent "bootstrap" program for hadronic amplitudes would suffice to "almost uniquely" determine the S-matrix for hadronic scattering.

The development of current algebra in the late 1960s provides another example of the profitable exploitation of general symmetry assumptions to derive important relations between amplitudes, this time on the basis of an assumed commutator algebra of the currents of the theory associated with chiral symmetry. The results of current algebra follow purely from the assumed current commutators, and are compatible with a variety of underlying field-theoretic models (or "effective Lagrangians") which share the same current algebra (cf. Section 16.6), so the verification of current algebra predictions for low-energy multipion scattering (say) brings us no closer to a unique underlying dynamics than the results of the S-matrix approach. Consequently, if, as we now believe, the dynamics of the strong interactions is just as precisely defined by an underlying local quantum field theory (quantum chromodynamics) as the interactions of electrons and photons were found to be by quantum electrodynamics in the 1950s, a full test of such a theory must necessarily include a sufficiently accurate determination of enough of the phenomenological content of the theory to allow us to conclude that the specific quantum field theory chosen is indeed the correct one.

Fortunately, the last 30 years has seen the development of powerful new numerical techniques for reliably extracting much of the non-perturbative content of strongly coupled field theories. These techniques, which go under the general heading of "lattice field theory", mimic at a numerical level the rigorous construction of a continuum field theory (in those cases where a construction is possible; see (Glimm and Jaffe, 1987)), starting with a full regularization of the theory on a finite spacetime lattice (with lattice spacing a and spacetime volume V) and then taking the continuum limit $a \to 0$ and the infinite volume limit $V \to \infty$, sometimes referred to as the "thermodynamic limit" (in that order). In the case of four-dimensional massless Yang–Mills theories (coupled to N_f fermionic quarks, provided the number of quark types N_f does not exceed a critical number; cf. Chapter 15) this limit is believed to yield well-defined Green functions—in particular, a set of Wightman functions for the local operators of the theory with zero color which satisfy the Wightman axioms, and a theory with a non-zero mass gap in the spectrum. A rigorous proof of this assertion is likely to be extremely difficult: it is one of the seven Millenium Prize Problems announced by the Clay Mathematics Institute in May 2000!

Nevertheless, assuming the existence of the continuum and thermodynamic limits, a sequence of approximants to the exact Euclidean Schwinger functions of the theory can be obtained by evaluating the corresponding functional integrals numerically (typically, by Monte Carlo simulation methods) on a finite hypercubical $\mathcal{L} \times \mathcal{L} \times \mathcal{L} \times \mathcal{L}$ spacetime lattice (with $\mathcal{L} = La$, L integer, a the lattice spacing), and then increasing L as the lattice spacing is appropriately scaled towards zero. The statistical errors incurred in such a numerical approach can be determined by standard statistical techniques. The systematic errors are of two kinds: short-distance errors due to the finite lattice spacing a, and long-distance, due to the finite extent of the lattice both spatially and temporally. The former turn out to be simply of a power nature a^p, with the power p depending on the particular observable being measured. For a theory with a mass gap m, the finite volume corrections fall exponentially, at least as fast as e^{-mL}, as the physical size of the lattice is increased. In any event, the existence of a continuum limit ensures that the approximants to the desired Schwinger functions systematically approach the correct non-perturbative results, unlike the situation with partial summands of the formal perturbative expansion for these functions. In particular, using the methods of lattice gauge theory (described in greater detail in Sections 19.3 and 19.4), it has been possible to (a) verify the presence of a linearly rising potential at long distances (and Coulombic behavior at short distances) between static color sources (quark confinement), as illustrated in a typical quenched (i.e., pure gauge theory) calculation in Fig. 11.13 (from (Duncan *et al.*, 1995)), and (b) compute the spectrum of the low-lying hadrons from first principles (i.e., starting from the QCD Lagrangian) to within a few percent and verify agreement with the observed particle masses. (For a summary of some recent results, see (Kuramashi, 2008).)

Despite the enormous progress that has been made in obtaining quantitatively reliable non-perturbative information with the methods of lattice field theory (especially in the case of lattice QCD), the restrictions imposed in this approach to numerical estimates of the Euclidean path integral lead to some serious drawbacks. There are two main areas where lattice field theory leaves much to be desired:

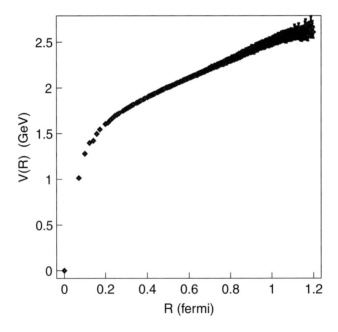

Fig. 11.13 Static quark–antiquark potential (pure gauge theory) from lattice simulation.

1. Unlike the situation in conventional multi-loop perturbation theory, where exact analytic results are frequently obtainable, and if not, numerical evaluation of the Feynman integrals is usually possible to essentially arbitrary accuracy, the use of Monte Carlo techniques means that one is necessarily dealing with statistical errors at each stage of a lattice calculation (i.e., with a given lattice and choice of bare masses and couplings), in addition to the inevitable systematic errors incurred by a finite lattice spacing and spatiotemporal volume. The statistical errors (usually containing subtle correlations which must be carefully understood) add an inescapable level of uncertainty to the final results. Moreover, they become uncontrollably large in several instances of great physical importance and interest, such as in the study of field theories at finite temperature and chemical potential, or in Minkowski space. This leads us to the second important deficiency of lattice field theory.

2. The behavior of the Euclidean Schwinger functions can be directly related to many quantities of central physical importance. For example, the lowest mass in each channel of well-defined conserved quantum numbers can be extracted from the Euclidean large-distance behavior of the appropriate two-point function via the Kållen–Lehmann representation (cf. Section 9.5). It is also possible to obtain certain matrix elements of local operators from the asymptotic behavior of the Euclidean Green functions of the theory. However, many features of high-energy scattering in QCD, for example, necessarily involve intrinsically Minkowski-space amplitudes which cannot be recovered from their only approximately known (by numerical estimation) Euclidean analytical continuations. On the other hand, although the functional integral can be formulated directly in Minkowski space as an absolutely convergent integral (cf. Section 10.3), the resultant multi-dimensional integrand undergoes essentially independent phase fluctuations at each lattice site, leading to an intractably low signal to noise ratio if one applies standard Monte Carlo sampling techniques: hence the blowup of statistical errors referred to above.[10] This is the infamous "sign problem", which is the bane of the numerical simulation approach in many important problems in condensed-matter physics as well as in relativistic field theory.

In summary, it is clear that we are still far from having a comprehensive and universally applicable strategy—a "magic bullet", as it were—for dealing with strongly-coupled field theories. For the time being we must instead make do with a patchwork of techniques which provide complementary (but far from complete) information about the physics of such theories.

11.4 Problems

1. The instability of the ground state for an anharmonic oscillator with negative λ (i.e. $V(q) = \frac{1}{2}q^2 - \frac{|\lambda|}{4}q^4$) can be studied by the standard WKB formula. The tunneling amplitude for a particle of zero energy to tunnel from the origin $q = 0$ to

---

[10] See, however, footnote 13 of Chapter 10 for a potentially useful Langevin simulation approach to complex actions.

the other side of the barrier (at $q_t = \pm\sqrt{2/|\lambda|}$) is proportional to $e^{-\int_0^{q_t} \sqrt{2V(q)}\,dq}$.
Evaluate the integral and compare with the exponential term in (11.16) (note:
the imaginary part of the energy is related to the decay rate, i.e. the square of
the tunneling amplitude).

2. In the Hamiltonian for the anharmonic oscillator $(m = \hbar = 1)$

$$H = -\frac{1}{2}\frac{d^2}{dq^2} + \frac{1}{2}q^2 + \frac{\lambda}{4}q^4 \tag{11.98}$$

show that a rescaling $x \equiv \lambda^{1/6}q$ leads to a new Hamiltonian

$$H' = \lambda^{-1/3}H = -\frac{1}{2}\frac{d^2}{dx^2} + \frac{1}{4}x^4 + \frac{1}{2}\lambda^{-2/3}x^2 \tag{11.99}$$

The last term is a Kato perturbation (see (Kato, 1995)) of the first two,
so the expansion of $\lambda^{-1/3}E(\lambda)$ in powers of $\lambda^{-2/3}$ is analytic: $E(\lambda) = \lambda^{1/3}\sum_{n=0}^{\infty} a_n\lambda^{-2n/3}$ is a convergent series.

3. Verify the one-loop results (11.43,11.44) for the scalar loop integrals in space-time
   dimensions 2 and 3, respectively (the identity (16.67) is useful).

4. Consider fermion–antifermion $(\psi - \psi^c)$ scattering via exchange of a massive
   spinless boson ϕ, as in Theory C of Section 7.6, in spacetime dimension $d = 4$.

   (a) Show that the one-loop graph (fourth order in the Yukawa coupling λ) arising
       from two successive exchanges of a ϕ (i.e., the graph displayed in Fig. 11.12
       with $n = 1$) is *infrared finite* in the on-threshold limit $p, q \to 0$. In this theory
       a bound state cannot form at weak coupling by infrared enhancement of the
       coupling strength: instead, the coupling itself *must* become large to encourage
       the persistent rescattering needed for bound-state formation.

   (b) Repeat the steps leading to (11.54) in this theory (i.e., study the Bethe–
       Salpeter equation in the ladder approximation, treating the fermion propa-
       gators non-relativistically) to show that the resultant Schrödinger equation
       contains a Yukawa potential with range $1/M$, where M is the mass of the
       ϕ. Of course, in this theory, even if a bound state exists, the ladder graphs
       do *not* play a preferred role in the formation of the bound state, unlike the
       situation for threshold bound states. The formation of the bound state in
       this case is an *essentially non-perturbative* phenomenon.

5. The effect of higher-order kernels in shifting the mass of a threshold bound state
   can be calculated perturbatively by the following procedure. We shall imagine
   inserting a single higher-order kernel (as, for example, in Fig. 11.10(b)) in the
   graphs for the 2-2 scattering amplitude.

   (a) Show that the first-order change in the amplitude $\tilde{G}_P(q,p)$ resulting from a
       shift $\Delta K_P(q,p)$ in the kernel $K_P(q,p)$ is given by

       $$\Delta\tilde{G}_P(q,p) = \int \tilde{G}_P(q,q')\Delta K_P(q',q'')\tilde{G}_P(q'',p)\frac{d^dq'\,d^dq''}{(2\pi)^d} \tag{11.100}$$

(b) By taking P to the bound-state pole and extracting the pole-term on both sides of (11.100) using (11.34), show that the first-order shift in the (squared) bound-state mass induced by $\Delta K_P(q, p)$ is

$$\Delta M_B^2 = -i \int \Phi_P^*(q') \Delta K_P(q', q'') \Phi_P(q'') \frac{d^d q' d^d q''}{(2\pi)^{2d}} \qquad (11.101)$$

This formula is the field-theory analog of the familiar expression for the energy shift in non-degenerate first-order perturbation theory in non-relativistic quantum mechanics.

6. Verify the contour-integral identity (11.71) for the partial summand of the asymptotic expansion of the partition function of a general scalar field theory.

7. Verify the identity (11.73) needed for the estimation of the remainder term in an asymptotic expansion.

12

Symmetries I: Continuous spacetime symmetry: why we need Lagrangians in field theory

For most beginning students of quantum field theory, an early surprise is in store when they encounter, for the first time since facing unpleasant problems in classical mechanics, typically involving absurdly complicated devices requiring the insertion of peculiar constraints, the notion of a Lagrangian as the fundamental object specifying the dynamical behavior of the theory. Certainly, such a creature plays little or no role in *non-relativistic* quantum theory, where the Hamiltonian, the explicit determinant of time evolution of the system, reigns supreme. Our first objective in this chapter is to understand the peculiar, and indispensable, utility of the Lagrangian approach to dynamics in relativistic quantum field theories. Our emphasis initially will be to underscore the facility with which a Lagrangian approach incorporates the desired—in fact, indispensable—spacetime symmetries of a relativistic field theory. In later chapters we shall see that the Lagrangian is an equally useful object in simplifying the treatment of local gauge symmetries, which are in some sense an amalgam of internal and spacetime symmetry.

12.1 The problem with derivatively coupled theories: seagulls, Schwinger terms, and T^* products

In our first attempts to construct a Lorentz-invariant theory of scattering in Section 5.5, we saw that the simple expedient of choosing the interaction Hamiltonian density $\mathcal{H}_{\text{int}}(x)$ of the theory to be a Lorentz scalar field seemed to lead us directly to a Lorentz-invariant S-matrix, provided $\mathcal{H}_{\text{int}}(x)$ was also local (i.e., commuting with itself at space-like separations). Unfortunately, as the discussion in Section 5.5 revealed, mere locality was not quite enough, as the behavior of the commutators of $\mathcal{H}_{\text{int}}(x)$ as the spacetime points of the commuting operators approach coincidence leads to delicate singularity issues which can destroy the desired Lorentz-invariance. Let us illustrate the difficulty with a simple example—a theory of a fermionic spin-$\frac{1}{2}$ field ψ interacting with a pseudoscalar[1] field ϕ via a derivative coupling:

$$\mathcal{H}_{\text{int}}(x) = g\bar{\psi}(x)\gamma^{\mu}\gamma_5\psi(x)\partial_{\mu}\phi(x) \tag{12.1}$$

---

[1] The assignment of parity quantum numbers to local fields will be explained in Section 13.1.

which is an interaction sometimes used to describe the effective low-energy (parity-conserving) strong-interaction coupling of nucleons to pions. Note that the interaction Hamiltonian density in (12.1) is clearly a Lorentz scalar field, as it is obtained by contracting an axial-vector field with the gradient of a pseudoscalar. From Wick's theorem, we see that the 2-2 scattering amplitude of the ψ particles contains the following contribution to second order in the coupling g (we suppress the spin indices for the fermions for simplicity):

$$\frac{(-ig)^2}{2} \int d^4 z_1 d^4 z_2 \langle p_1' p_2' | : \bar{\psi}\gamma^\mu \gamma_5 \psi(z_1) \bar{\psi}\gamma^\nu \gamma_5 \psi(z_2) : |p_1 p_2\rangle \cdot \langle 0|T(\partial_\mu \phi(z_1)\partial_\nu \phi(z_2))|0\rangle \tag{12.2}$$

So far, on the surface, we seem to be dealing with a perfectly covariant expression, with the Lorentz indices μ, ν properly contracted, but of course trouble lurks potentially in the time-ordered product, where an explicit choice of inertial frame is presupposed. Without the derivatives on the scalar field, this T-product is just the free scalar propagator $i\Delta_F(z_1 - z_2)$, which we saw in Chapter 7 is perfectly Lorentz-invariant (as a function of the spacetime separation $z_1 - z_2$). In this case, we find

$$\frac{\partial}{\partial z_2^\nu}\langle 0|T(\phi(z_1)\phi(z_2))|0\rangle = \langle 0|T(\phi(z_1)\partial_\nu \phi(z_2))|0\rangle + \delta_{\nu 0}\delta(z_1^0 - z_2^0)[\phi(z_2), \phi(z_1)] \tag{12.3}$$

$$= \langle 0|T(\phi(z_1)\partial_\nu \phi(z_2))|0\rangle \tag{12.4}$$

where the commutator appearing in (12.3) (in virtue of time-derivatives acting on the θ-functions defining the T-product) vanishes by locality of $\phi(z)$. Inserting the second derivative, however, we find

$$\langle 0|T^*(\partial_\mu \phi(z_1)\partial_\nu \phi(z_2))|0\rangle \equiv \frac{\partial}{\partial z_1^\mu}\frac{\partial}{\partial z_2^\nu}\langle 0|T(\phi(z_1)\phi(z_2))|0\rangle \tag{12.5}$$

$$= \langle 0|T(\partial_\mu \phi(z_1)\partial_\nu \phi(z_2))|0\rangle + \delta_{\mu 0}\delta(z_1^0 - z_2^0)[\phi(z_1), \partial_\nu \phi(z_2)] \tag{12.6}$$

$$= \langle 0|T(\partial_\mu \phi(z_1)\partial_\nu \phi(z_2))|0\rangle + i\delta_{\mu 0}\delta_{\nu 0}\delta^4(z_1 - z_2) \tag{12.7}$$

where in going from (12.6) to (12.7) we have used the equal-time commutator (8.1) of ϕ with $\dot{\phi}$. Now the T*-product defined in (12.5) is itself perfectly covariant, as it is simply the second spacetime-derivative of the Lorentz-invariant scalar propagator:

$$\langle 0|T^*(\partial_\mu \phi(z_1)\partial_\nu \phi(z_2))|0\rangle = i\int \frac{k_\mu k_\nu}{k^2 - m^2 + i\epsilon}e^{-ik\cdot(z_1-z_2)}\frac{d^4 k}{(2\pi)^4} \tag{12.8}$$

Accordingly, the T-product appearing in the second-order amplitude (12.2) contains a *non-covariant contact term* (sometimes called a *Schwinger term*)

$$\langle 0|T(\partial_\mu \phi(z_1)\partial_\nu \phi(z_2))|0\rangle = \langle 0|T^*(\partial_\mu \phi(z_1)\partial_\nu \phi(z_2))|0\rangle - i\delta_{\mu 0}\delta_{\nu 0}\delta^4(z_1 - z_2) \tag{12.9}$$

which in turn means that the 2-2 scattering amplitude (12.2) contains, in addition to a perfectly covariant contribution (from the T^*-product), the non-covariant piece

$$i\frac{g^2}{2}\int d^4z\langle p_1'p_2'| : \bar{\psi}\gamma_0\gamma_5\psi(z)\bar{\psi}\gamma_0\gamma_5\psi(z) : |p_1p_2\rangle \tag{12.10}$$

Referring back to Section 5.5, the reader may easily verify (see Problem 1) that the difficulty here was already identified in the general expression (5.81), where we showed that the appearance of spatial derivatives of a δ-function in the equal-time commutator of a non-ultralocal interaction Hamiltonian density spelled potential disaster for the Lorentz covariance of the theory. In this case, the cure is easy to find, as the non-covariant piece is itself local, and can be cancelled by augmenting the interaction Hamiltonian in (12.1) by the four-fermion operator in (12.10), with an opposite sign:

$$\mathcal{H}_{\text{int}}(x) = g\bar{\psi}\gamma^{\mu}\gamma_5\psi\partial_{\mu}\phi(x) + \frac{1}{2}g^2(\bar{\psi}\gamma_0\gamma_5\psi)^2(x) \tag{12.11}$$

The contribution to first order of the second term in (12.11) to the 2-2 scattering amplitude is easily seen to exactly cancel the undesired non-covariant piece (12.10). Moreover, this new term in the interaction Hamiltonian—dubbed the "seagull" vertex in the original literature—appears in one-to-one association with every internal scalar line in the Feynman graphs of the theory, serving to cancel the non-covariant Schwinger term in the scalar propagator wherever the latter chooses to pop up. The appearance of a non-covariant term in the Hamiltonian should not cause alarm: the energy density is itself not a covariant object (as in the free Hamiltonian density, cf. (6.89)).

In general, when non-covariant terms appear in a theory (via an interaction Hamiltonian failing to be an ultralocal scalar field), it is a non-trivial task to guess the appropriate seagull terms needed to restore Lorentz-invariance of the theory. In certain cases (e.g., gauge theories with certain choices of gauge) the required terms may even be non-local! We clearly need an effective means of assuring Lorentz-invariance of the theory *ab initio*—which, we shall soon see, is exactly what the canonical formalism, in its *Lagrangian* version, is guaranteed to supply.

12.2   Canonical formalism in quantum field theory

The dynamics of quantum theories generally (and quantum field theories in particular) is typically specified in terms of a Hamiltonian, the quantization of which requires that Planck's constant be inserted in the theory via a set of commutation relations between conjugate variables ("coordinates" and "momenta"). The specification of an explicit dynamics for a quantum field theory has been achieved so far in this book by constructing a free Hamiltonian for the desired particles, in terms of interaction-picture fields, and then attempting to construct a suitable interaction Hamiltonian (as the integral of a local interaction density) in terms of these free interaction-picture fields. For example, for the derivative coupled theory discussed above, and including the covariantizing seagull term, the total Hamiltonian density for the theory would be

$$\mathcal{H} = \mathcal{H}_0 + \mathcal{H}_{\text{int}}$$

$$= \frac{1}{2}\dot{\phi}^2 + \frac{1}{2}|\vec{\nabla}\phi|^2 + \frac{1}{2}m^2\phi^2 + \bar{\psi}(i\vec{\gamma}\cdot\vec{\nabla} + M)\psi$$

$$+ g\bar{\psi}\gamma^\mu\gamma_5\psi\partial_\mu\phi + \frac{1}{2}g^2(\bar{\psi}\gamma^0\gamma_5\psi)^2 \tag{12.12}$$

$$= \frac{1}{2}\dot{\phi}^2 + \frac{1}{2}|\vec{\nabla}\phi|^2 + \frac{1}{2}m^2\phi^2 + \bar{\psi}(i\vec{\gamma}\cdot\vec{\nabla} + M)\psi$$

$$+ g\bar{\psi}\gamma^0\gamma_5\psi\dot{\phi} - g\bar{\psi}\vec{\gamma}\gamma_5\psi\cdot\vec{\nabla}\phi + \frac{1}{2}g^2(\bar{\psi}\gamma^0\gamma_5\psi)^2 \tag{12.13}$$

In the interaction picture, the identification of conjugate field variables is trivial, as the equal-time commutation relations of the free field operators are exactly computable. For scalar fields, we have

$$[\phi(\vec{y},t), \pi^\phi(\vec{x},t)] = i\delta^3(\vec{x}-\vec{y}), \quad \pi^\phi(\vec{x},t) \equiv \dot{\phi}(\vec{x},t) \tag{12.14}$$

$$[\phi(\vec{y},t), \phi(\vec{x},t)] = [\pi^\phi(\vec{y},t), \pi^\phi(\vec{x},t)] = 0 \tag{12.15}$$

For Dirac fields, we have *anticommutators* (cf. Chapter 7, Problem 5):

$$\{\psi_n(\vec{y},t), \pi^\psi_m(\vec{x},t)\} = i\delta_{mn}\delta^3(\vec{x}-\vec{y}), \quad \pi^\psi(\vec{x},t) \equiv i\psi^\dagger(\vec{x},t) = i\bar{\psi}(\vec{x},t)\gamma^0 \tag{12.16}$$

$$\{\psi_n(\vec{y},t), \psi_m(\vec{x},t)\} = \{\pi^\psi_n(\vec{y},t), \pi^\psi_m(\vec{x},t)\} = 0 \tag{12.17}$$

The development of the canonical Hamiltonian formalism in field theory, as in classical mechanics, requires that we re-express the Hamiltonian of the theory in terms of the conjugate field pairs ϕ, π^ϕ and ψ, π^ψ. In the field-theory context, this replacement allows us to move trivially between different representations of the time-development of the theory: in particular, to obtain an expression for the Hamiltonian in the Heisenberg picture, which is the natural environment for discussing the exact dynamics of the theory, inasmuch as no artificial splits (motivated purely by calculational convenience) are made between "free" and "interaction" terms. Indeed, if we recall that the interaction-picture and Heisenberg fields are connected by the unitary operator $U(t,0) \equiv \Omega(t) = e^{iH_0t}e^{-iHt}$ (cf. (9.1)), we see that the Heisenberg fields and field momenta defined by

$$\phi_H(\vec{x},t) = \Omega^\dagger(t)\phi(\vec{x},t)\Omega(t), \quad \pi^\phi_H(\vec{x},t) = \Omega^\dagger(t)\pi^\phi(\vec{x},t)\Omega(t) \tag{12.18}$$

$$\psi_H(\vec{x},t) = \Omega^\dagger(t)\psi(\vec{x},t)\Omega(t), \quad \pi^\psi_H(\vec{x},t) = \Omega^\dagger(t)\pi^\psi(\vec{x},t)\Omega(t) \tag{12.19}$$

satisfy (in virtue of $\Omega^\dagger(t)\Omega(t) = 1$) *exactly* the same equal-time (anti)commutation relations as the original interaction-picture fields:

$$[\phi_H(\vec{y},t), \pi^\phi_H(\vec{x},t)] = i\delta^3(\vec{x}-\vec{y})$$

$$\{\psi_H(\vec{y},t), \pi^\psi_H(\vec{x},t)\} = i\delta^3(\vec{x}-\vec{y}) \tag{12.20}$$

where we have dropped the Dirac indices on the fermionic fields to avoid notational overload. This means that if we replace $\dot{\phi}$ by π^{ϕ} and $\bar{\psi}$ by $-i\pi^{\psi}\gamma^0$ in (12.13),

$$\mathcal{H} = \frac{1}{2}(\pi^{\phi})^2 + \frac{1}{2}\mid\vec{\nabla}\phi\mid^2 + \frac{1}{2}m^2\phi^2 - i\pi^{\psi}\gamma^0(i\vec{\gamma}\cdot\vec{\nabla} + M)\psi$$

$$- ig\pi^{\psi}\gamma_5\psi\pi^{\phi} + ig\pi^{\psi}\gamma^0\vec{\gamma}\gamma_5\psi\cdot\vec{\nabla}\phi - \frac{1}{2}g^2(\pi^{\psi}\gamma_5\psi)^2 \qquad (12.21)$$

and recall that the total Hamiltonian $H = \int \mathcal{H}(\vec{x}, t)d^3x$ is a constant of the dynamics, we can re-express the total Hamiltonian density immediately in terms of Heisenberg fields simply by subscripting all the fields with H:

$$\mathcal{H} = \frac{1}{2}(\pi_H^{\phi})^2 + \frac{1}{2}\mid\vec{\nabla}\phi_H\mid^2 + \frac{1}{2}m^2\phi_H^2 - i\pi_H^{\psi}\gamma^0(i\vec{\gamma}\cdot\vec{\nabla} + M)\psi_H$$

$$- ig\pi_H^{\psi}\gamma_5\psi_H\pi_H^{\phi} + ig\pi_H^{\psi}\gamma^0\vec{\gamma}\gamma_5\psi_H\cdot\vec{\nabla}\phi_H - \frac{1}{2}g^2(\pi_H^{\psi}\gamma_5\psi_H)^2 \qquad (12.22)$$

The fact that the full Hamiltonian of the theory has been expressed in terms of pairs of fields satisfying canonical commutation (or anticommutation relations) (12.20) means, as we shall soon see, that the dynamical equations of the theory can be rewritten as differential functional equations, the field equivalent of the first-order (in time) Hamiltonian equations of classical mechanics. In the next section we shall see that a simple Legendre transformation of these functional equations will lead us to a very simple criterion for ensuring the eventual Lorentz-invariance of our field theory. This is a particularly pressing objective, given that $\mathcal{H}$ in (12.22) displays no vestige whatsoever of the underlying Lorentz-invariance of the theory! Nor indeed should it: as indicated previously, $\mathcal{H}$ is a spatial energy density, clearly a frame-dependent, non-Lorentz-invariant object.

The first step in the derivation of the Hamiltonian field equations is a generalization of a familiar identity in ordinary quantum mechanics: the fact that the commutation of the momentum with a function of the coordinate operator is equivalent to a derivative of the latter. We shall begin with the case of bosonic fields, satisfying commutation, rather than anticommutation, relations. For example, from (12.20) one finds (for n integer)

$$[\phi_H^n(\vec{y}, t), \pi_H^{\phi}(\vec{x}, t)] = in\phi_H^{n-1}(\vec{y}, t)\delta^3(\vec{x} - \vec{y}) \qquad (12.23)$$

and

$$[\vec{\nabla}\phi_H(\vec{y}, t)\cdot\vec{\nabla}\phi_H(\vec{y}, t), \pi_H^{\phi}(\vec{x}, t)] = 2i\vec{\nabla}\phi_H(\vec{y}, t)\cdot\vec{\nabla}_y\delta^3(\vec{x} - \vec{y}) \qquad (12.24)$$

One easily generalizes these simple cases to establish that for any polynomial function $\mathcal{F}$ of $\phi_H, \pi_H^{\phi}, \vec{\nabla}\phi_H, \vec{\nabla}\pi_H^{\phi}$,

$$[\mathcal{F}(\phi_H(\vec{y}, t), \pi_H^{\phi}(\vec{y}, t), \vec{\nabla}\phi_H(\vec{y}, t), \vec{\nabla}\pi_H^{\phi}(\vec{y}, t)), \pi_H^{\phi}(\vec{x}, t)] = i(\frac{\partial\mathcal{F}}{\partial\phi_H} + \frac{\partial\mathcal{F}}{\partial\vec{\nabla}\phi_H}\cdot\vec{\nabla}_y)\delta^3(\vec{x} - \vec{y})$$

$$\qquad (12.25)$$

A similar argument, interchanging the roles of the field ϕ_H and its conjugate momentum π_H^ϕ, gives

$$[\mathcal{F}(\phi_H(\vec{y},t), \pi_H^\phi(\vec{y},t), \vec{\nabla}\phi_H(\vec{y},t), \vec{\nabla}\pi_H^\phi(\vec{y},t)), \phi_H(\vec{x},t)]$$

$$= -i\Big(\frac{\partial \mathcal{F}}{\partial \pi_H^\phi} + \frac{\partial \mathcal{F}}{\partial \vec{\nabla}\pi_H^\phi} \cdot \vec{\nabla}_y\Big)\delta^3(\vec{x}-\vec{y}) \tag{12.26}$$

In particular, taking for our function $\mathcal{F}$ the total Hamiltonian density $\mathcal{H}$ of a scalar theory, with the total Hamiltonian H given as a (time-independent) functional of the fields

$$H = \int \mathcal{H}(\phi_H(\vec{y},t), \pi_H^\phi(\vec{y},t), \vec{\nabla}\phi_H(\vec{y},t), \vec{\nabla}\pi_H^\phi(\vec{y},t))d^3y \tag{12.27}$$

the results (12.25, 12.26) show that commutation of a field with the full Hamiltonian can be re-expressed as a *functional* derivative[2] with respect to the canonically conjugate field:

$$[H, \pi_H^\phi(\vec{x},t)] = i\frac{\delta H}{\delta \phi_H(\vec{x},t)} = i\Big(\frac{\partial \mathcal{H}}{\partial \phi_H(\vec{x},t)} - \vec{\nabla}_x \cdot \frac{\partial \mathcal{H}}{\partial \vec{\nabla}\phi_H(\vec{x},t)}\Big) \tag{12.28}$$

$$[H, \phi_H(\vec{x},t)] = -i\frac{\delta H}{\delta \pi_H^\phi(\vec{x},t)} = -i\Big(\frac{\partial \mathcal{H}}{\partial \pi_H^\phi(\vec{x},t)} - \vec{\nabla}_x \cdot \frac{\partial \mathcal{H}}{\partial \vec{\nabla}\pi_H^\phi(\vec{x},t)}\Big) \tag{12.29}$$

On the other hand, we know that in Heisenberg representation, commutation with the full Hamiltonian simply acts as a time-derivative: $[H, A_H(\vec{x},t)] = -i\dot{A}_H(\vec{x},t)$ for any Heisenberg field A_H. Thus (12.28,12.29) amount to exact field analogs of the Hamiltonian equations $\frac{\partial H}{\partial q} = -\dot{p}, \frac{\partial H}{\partial p} = \dot{q}$ of classical mechanics:

$$\frac{\delta H}{\delta \phi_H(x)} = -\dot{\pi}_H^\phi(x), \quad \frac{\delta H}{\delta \pi_H^\phi(x)} = \dot{\phi}_H(x) \tag{12.30}$$

where we have returned to spacetime coordinate notation $((\vec{x},t) \to x)$.

Completely analogous arguments lead to similar conclusions for the fermionic fields of the theory. The only new features here are (a) that the functionals of fields considered contain an even number of fermionic fields in each term, and (b) that the functional derivatives with respect to the fermionic fields ψ_H and π_H^ψ must be defined[3] to insert a minus sign at any point during the execution of the product rule where the derivative operation is interchanged with a fermionic field, in order that the result agree with that obtained by operator commutation. Thus we again find

---

[2] In evaluating the functional derivatives in this formula, the full Hamiltonian H is assumed to be written as a spatial integral over the fields on time-slice t. As H is conserved in time, we are of course free to choose any time-slice on which to express the Hamiltonian as a spatial integral.

[3] Cf. the discussion of Grassmann functional derivatives in Section 10.3.2.

$$\frac{\delta H}{\delta \psi_H(x)} = -\dot{\pi}_H^\psi(x), \qquad \frac{\delta H}{\delta \pi_H^\psi(x)} = \dot{\psi}_H(x) \tag{12.31}$$

Our toy model with Hamiltonian density (12.22) furnishes an immediate and convenient example: for the scalar field, we find (suppressing the spacetime coordinate, and writing π_H^ψ in terms of the original $\bar{\psi}_H$)

$$\frac{\delta H}{\delta \phi_H(x)} = m^2 \phi_H - \vec{\nabla} \cdot (\vec{\nabla}\phi_H - g\bar{\psi}_H\vec{\gamma}\gamma_5\psi_H) = -\dot{\pi}_H^\phi \tag{12.32}$$

$$\frac{\delta H}{\delta \pi_H^\phi(x)} = \pi_H^\phi + g\bar{\psi}_H\gamma^0\gamma_5\psi_H = \dot{\phi}_H \tag{12.33}$$

Note the presence of a *fermionic* term (indicated here in bold face) in the *scalar* conjugate momentum field $\pi_H^\phi = \dot{\phi}_H - g\bar{\psi}_H\gamma^0\gamma_5\psi_H$: this will very soon play a crucial role in cancelling the non-covariant effects of the seagull term. Combining these equations (by eliminating π_H^ϕ), we arrive at the *Lorentz-covariant* Heisenberg field equation for the scalar field ϕ_H:

$$(\Box + m^2)\phi_H(x) = g\partial_\mu(\bar{\psi}_H\gamma^\mu\gamma_5\psi_H(x)) \tag{12.34}$$

Note that the quartic (in the ψ_H field) seagull term in the Hamiltonian plays no role so far: the scalar Heisenberg field equation would still be Lorentz-covariant without this term. The situation is quite different for the fermionic equation of motion. The second of Eqs. (12.31) applied to (12.22) gives directly[4]

$$-i\gamma^0(i\vec{\gamma} \cdot \vec{\nabla} + M)\psi_H - ig\gamma_5\psi_H\pi_H^\phi + ig\gamma^0\vec{\gamma}\gamma_5\psi_H \cdot \vec{\nabla}\phi_H - \boldsymbol{g^2(\pi_H^\psi\gamma_5\psi_H)\gamma_5\psi_H} = \dot{\psi}_H \tag{12.35}$$

where the term arising from the quartic seagull contribution is highlighted in bold-face type. Multiplying both sides by $i\gamma^0$ and inserting (12.33) to eliminate the scalar momentum field π_H^ϕ, we find

$$(i\vec{\gamma} \cdot \vec{\nabla} + M)\psi_H + g\gamma^0\gamma_5\psi_H\dot{\phi}_H - g\vec{\gamma}\gamma_5\psi_H \cdot \vec{\nabla}\phi_H = i\gamma^0\dot{\psi}_H$$

$$\Rightarrow (i\gamma^\mu\partial_\mu - M)\psi_H = g\gamma^\mu\gamma_5\psi_H\partial_\mu\phi_H \tag{12.36}$$

so that the Heisenberg field equation for the Dirac field ψ_H is also manifestly Lorentz-covariant. It is precisely at this point that we see that the contribution from the seagull term in the interaction Hamiltonian has cancelled the non-covariant term introduced by the fermionic component of the scalar momentum field π_H^ϕ. In other words, the strange necessity for a four-fermion seagull interaction in order to preserve Lorentz-invariant S-matrix amplitudes for fermion scattering, is directly correlated with the construction of a Hamiltonian leading to Lorentz-covariant field equations for the fermionic Heisenberg field of the theory. In this toy theory it was not too difficult to guess the type of extra non-covariant term needed in the interaction

---

[4] The reader may verify that the first Hamiltonian equation simply produces an equation for $\psi_H^\dagger$ equivalent to the adjoint of the equation obtained from the second equation; see Problem 3.

density to restore the Lorentz-invariance of the theory, but in more complicated models this becomes essentially impossible, and we need a fool-proof algorithm for assuring from the very beginning that our theory possesses the full Lorentz-invariance that we demand. The Lagrangian formalism is exactly the tool needed, as we shall now see.

12.3 General condition for Lorentz-invariant field theory

Any formalism in which the Lorentz-invariance of a field theory is to be manifest from the very outset must clearly be one in which time and space are treated in a symmetric fashion, consistent with the requirements of special relativity. This is clearly not the case in the Hamiltonian formalism, but as we shall now see, the symmetry between space and time can be restored by the simple expedient of a Legendre transformation, taking us from the Hamiltonian to a Lagrangian encoding of the dynamics of the theory. We shall discuss the scalar field case only—the argument goes through in similar fashion in the fermionic case. Also, we shall henceforth drop the H subscript, as it will be assumed for the remainder of the chapter that we are dealing entirely with fields in Heisenberg representation. To further streamline the algebra, let us introduce, motivated by (12.25, 12.26), the total Euler derivative of a function $\mathcal{F}(\phi, \vec{\nabla}\phi, ...)$ with respect to a field ϕ (the ellipsis denotes a possible dependence of $\mathcal{F}$ on other independent fields):

$$\frac{d\mathcal{F}}{d\phi} \equiv \frac{\partial\mathcal{F}}{\partial\phi} - \vec{\nabla} \cdot \frac{\partial\mathcal{F}}{\partial\vec{\nabla}\phi} \tag{12.37}$$

In other words, the Euler derivative acting on the Hamiltonian density is equivalent to the functional derivative acting on the spatially integrated Hamiltonian density (or Hamiltonian). The Hamiltonian equations (12.28, 12.29) can thus be written as a pair of partial differential equations for the Hamiltonian density $\mathcal{H} = \mathcal{H}(\pi^\phi, \vec{\nabla}\phi, \phi)$, which we assume here can be written in such a way that the momentum fields π^ϕ appear without spatial derivatives, as is generally the case for theories of interacting spin-0 and spin-$\frac{1}{2}$ particles.[5] Namely,

$$\left.\frac{d\mathcal{H}}{d\phi}\right|_{\pi^\phi} = -\dot{\pi}^\phi, \tag{12.38}$$

$$\left.\frac{d\mathcal{H}}{d\pi^\phi}\right|_{\phi} = \dot{\phi} \tag{12.39}$$

The restoration of spacetime symmetry clearly requires that we reintroduce the time-derivative $\dot{\phi}$ in favor of the canonical momentum π^ϕ, inasmuch as spatial gradients of the field ϕ are already in evidence. This can be done without loss of dynamical information by the use of a Legendre transformation: one introduces a *Lagrange density*

---

[5] The canonical treatment of theories of spin-1 fields involving a local gauge symmetry introduces further subtleties which we shall defer to Chapter 15.

$$\mathcal{L}(\dot{\phi}, \vec{\nabla}\phi, \phi) \equiv \pi^\phi \dot{\phi} - \mathcal{H}(\pi^\phi, \vec{\nabla}\pi^\phi, \phi, \vec{\nabla}\phi) \tag{12.40}$$

where π^ϕ is to be expressed in terms of $\dot{\phi}$ (and possibly ϕ and $\vec{\nabla}\phi$ as well) by solving the second Hamiltonian equation

$$\left. \frac{d\mathcal{H}}{d\pi^\phi} \right|_\phi = \dot{\phi} \tag{12.41}$$

Why the process of Legendre transformation should not only succeed in reintroducing time-derivatives, but do so in just such a way as to manifest directly the Lorentz-invariance of the theory is not obvious *a priori* (although we are about to demonstrate this explicitly starting with the Hamiltonian equations of motion): the underlying reason for the crucial role played by the Legendre transform in connecting a Lorentz-invariant formulation of the theory with the *energy* operator of the theory will become clear in the next section, when we discuss the Action Principle and Noether's theorem.

Returning to (12.40), we find for the Euler derivative of the Lagrangian with respect to ϕ (holding $\dot{\phi}$ fixed),

$$\frac{d\mathcal{L}}{d\phi} = \frac{d\pi^\phi}{d\phi}\dot{\phi} - \frac{d\mathcal{H}}{d\pi^\phi}\frac{d\pi^\phi}{d\phi} - \left. \frac{d\mathcal{H}}{d\phi} \right|_{\pi^\phi} \tag{12.42}$$

$$= - \left. \frac{d\mathcal{H}}{d\phi} \right|_{\pi^\phi} \tag{12.43}$$

using (12.41) to cancel the first two terms in (12.42).

On the other hand,

$$\left. \frac{\partial \mathcal{L}}{\partial \dot{\phi}} \right|_\phi = \pi^\phi + \dot{\phi}\frac{\partial \pi^\phi}{\partial \dot{\phi}} - \frac{d\mathcal{H}}{d\pi^\phi}\frac{\partial \pi^\phi}{\partial \dot{\phi}} = \pi^\phi \tag{12.44}$$

where we have used (12.41). Taking a time-derivative, we obtain,

$$\left. \frac{d}{dt}\frac{\partial \mathcal{L}}{\partial \dot{\phi}} \right|_\phi = \dot{\pi}^\phi = - \left. \frac{d\mathcal{H}}{d\phi} \right|_{\pi^\phi} \tag{12.45}$$

where in (12.45) we have used the first Hamiltonian equation (12.38). Comparing (12.43) and (12.45), we conclude that

$$\frac{d}{dt}\frac{\partial \mathcal{L}}{\partial \dot{\phi}} = \frac{d\mathcal{L}}{d\phi} = \frac{\partial \mathcal{L}}{\partial \phi} - \vec{\nabla} \cdot \frac{\partial \mathcal{L}}{\partial \vec{\nabla}\phi} \tag{12.46}$$

$$\Rightarrow \partial_\mu \frac{\partial \mathcal{L}}{\partial(\partial_\mu \phi)} = \frac{\partial \mathcal{L}}{\partial \phi} \tag{12.47}$$

The manifestly Lorentz-covariant equation (12.47) is the celebrated *Euler–Lagrange* equation of the theory. It is, of course, equivalent to the Heisenberg field equation for the field ϕ (i.e., for our toy theory, (12.34), as it encodes the full dynamical information contained in the Hamiltonian equations of motion. It is also immediately clear that these field equations will themselves be automatically Lorentz-covariant *provided the Lagrange density is chosen to be a Lorentz-scalar function of ϕ and $\partial_\mu \phi$.*

At this point we should take note of a potential difficulty in carrying out the process described here: the existence (and reversibility) of the Legendre transform depends on certain convexity properties of the function to be transformed—otherwise put, the derivatives expressing the momentum variable π^ϕ in terms of the time-derivative $\dot\phi$ must be a monotone function of the latter to ensure that we can solve uniquely for $\dot\phi$ in terms of π^ϕ (and *vice versa*). In contrast to the case with the Legendre transform defining the effective action $\Gamma(\phi)$ (cf. Section 10.4), where the required convexity is assured by fundamental positivity properties of field theory (cf. Section 14.3), there are many cases in which the transition from Lagrangian to Hamiltonian fails as a result of the fact that constraints, or local gauge symmetries, are present, giving rise to a singular transformation between field- (time-)derivatives and the field momenta. A general theory of constrained Hamiltonian systems has been developed (beginning with Dirac, (Dirac, 1964)) to handle such situations: we shall defer further discussion of these issues to Chapter 15, when we shall have to face head-on the peculiar features of the canonical formalism in theories with local gauge symmetry.

For the fermion fields, the Legendre transform takes an algebraically trivial form. Fermionic Lagrangians (or Hamiltonians) are linear in spacetime-derivatives of the Fermi field, with the derivatives appearing only in the free part

$$\mathcal{L}_0 = \bar\psi(i\gamma^0\partial_0 - i\vec\gamma \cdot \vec\nabla - M)\psi \tag{12.48}$$

Thus, we always have $\pi^\psi = i\bar\psi\gamma^0$, and the Legendre transform (starting with the Hamiltonian, say) simply amounts to inserting $\pi^\psi\dot\psi$ into the (negative) Hamiltonian density $-\bar\psi(i\vec\gamma \cdot \vec\nabla + M)\psi$ (thereby covariantizing it), and replacing $\pi^\psi \to i\bar\psi\gamma^0$ throughout (in both free and interaction terms).

Once again, our toy derivative-coupled theory, defined by the Hamiltonian (12.22), provides a convenient explicit example. The Lagrangian is easily constructed: we have (now including the fermionic contributions)

$$\mathcal{L} = \pi^\phi\dot\phi + \pi^\psi\dot\psi - \mathcal{H} \tag{12.49}$$

where we must eliminate the momentum fields by putting

$$\pi^\phi = \dot\phi - g\bar\psi\gamma^0\gamma_5\psi \tag{12.50}$$

$$\pi^\psi = i\psi^\dagger = i\bar\psi\gamma^0 \tag{12.51}$$

A little algebra then reveals the Lagrangian density as a manifestly Lorentz-scalar object

$$\mathcal{L} = \frac{1}{2}(\partial_\mu\phi\partial^\mu\phi - m^2\phi^2) + \bar\psi(i\slashed\partial - M)\psi - g\bar\psi\gamma^\mu\gamma_5\psi\partial_\mu\phi \tag{12.52}$$

$$= \mathcal{L}_0 + \mathcal{L}_{\text{int}} \tag{12.53}$$

$$\mathcal{L}_0 \equiv \frac{1}{2}(\partial_\mu \phi \partial^\mu \phi - m^2 \phi^2) + \bar{\psi}(i\not{\partial} - M)\psi \tag{12.54}$$

$$\mathcal{L}_{\text{int}} \equiv -g\bar{\psi}\gamma^\mu \gamma_5 \psi \partial_\mu \phi \tag{12.55}$$

with an interaction term $\mathcal{L}_{\text{int}}$ which is clearly a Lorentz scalar, and just the negative of the interaction Hamiltonian density (12.1) with which we began the chapter, in an attempt to construct a derivatively-coupled theory of pions and nucleons. The free Lagrangian $\mathcal{L}_0$ is a sum of scalar and Dirac field contributions which the reader may easily verify lead directly (by reversing the Legendre transformation, to return to a Hamiltonian) to the usual free scalar and Dirac Hamiltonians incorporated in (12.22).

We see now clearly that our original criterion for Lorentz-invariance, introduced in Section 5.5—namely, to construct interaction Hamiltonians as scalar densities built from products of the underlying covariant fields—was not too far from the mark, except that the correct prescription in general requires that we choose the *Lagrangian* to be a Lorentz scalar. In scalar theories without derivative coupling, one has $\mathcal{H}_{\text{int}} = -\mathcal{L}_{\text{int}}$, so the two prescriptions in fact coincide. In the present case, the presence of time-derivatives of the ϕ field in the interaction result in an extra non-scalar contribution—the quartic seagull term!—to the Hamiltonian interaction density (cf. (12.13)). Precisely such a term is needed, as we saw in Section 12.1, to restore the Lorentz-invariance of the amplitudes of the theory, by cancelling the non-covariant Schwinger terms which appear in propagators of the gradient field $\partial_\mu \phi$ appearing in the interaction. Of course, in practice it is far easier to start with a Lorentz scalar Lagrangian and generate the correct Hamiltonian (including any non-covariant seagull interactions, if necessary) by an algebraically trivial[6] Legendre transformation than to try to guess the form of the interaction Hamiltonian needed to absorb non-covariant terms in the propagators. In the following section we shall give a much more general discussion of the Lagrangian formalism, in which it will become clear that it provides the natural framework for incorporating and expressing the symmetries of the theory (including symmetries beyond those directly associated with Lorentz/Poincaré invariance).

The preceding discussion has focussed on the operator formulation of field theory (in particular, on the Heisenberg field equations of the theory). A completely parallel discussion of the relation between Hamiltonian and Lagrangian formulations can be given using the path-integral formulation. In this approach, the relevance of a Lorentz-invariant Lagrangian to the appearance of fully Lorentz-invariant amplitudes can be seen in a much more direct fashion, as it allows us to circumvent completely the appearance of non-covariant Schwinger terms and seagull vertices, and demonstrate directly a set of Feynman rules with no non-covariant elements. We shall see that under fairly general circumstances, the Legendre transformation connecting a field-theoretic Hamiltonian and Lagrangian is exactly equivalent to a *functional Fourier transform*.

We begin with the bosonic case. Suppose that our Hamiltonian density is a function of N bosonic fields $\phi_n, n = 1, 2, \ldots N$, where the ϕ_n may be individually scalar fields,

---

[6] The asserted triviality is, however, absent in the presence of local gauge symmetries, where the canonical procedure becomes quite delicate, as we shall see in Chapter 15.

or components of fields transforming under more complicated representations of the HLG (e.g., vector fields). We shall assume that the full Hamiltonian density is at most quadratic in the conjugate momentum fields $\pi_n, n = 1, 2, .., N$ (we drop the ϕ superscript on π, as only bosonic fields are present). Thus

$$\mathcal{H} = \frac{1}{2}\pi_n K_{nm}\pi_m + L_n\pi_n + V(\phi_n, \vec{\nabla}\phi_n) \tag{12.56}$$

Here, the quadratic form K_{nm} (and the L_n) may be integro-differential operators: our only condition is that K_{nm} be symmetric and invertible. It may also depend on fields in the theory other than the ϕ_n themselves (in which case we have no ordering problems to concern us), as may the L_n.[7] These conditions ensure that we can solve for π_n in terms of $\dot{\phi}_n$ using (12.41), a prerequisite for performing the Legendre transformation:

$$\dot{\phi}_n = K_{nm}\pi_m + L_n \Rightarrow \pi_n = K_{nm}^{-1}(\dot{\phi}_m - L_m) \tag{12.57}$$

The Lagrangian density is now obtained directly:

$$\mathcal{L} = \pi_n\dot{\phi}_n - \mathcal{H} = \frac{1}{2}(\dot{\phi}_n - L_n)K_{nm}^{-1}(\dot{\phi}_m - L_m) - V(\phi_n, \vec{\nabla}\phi_n) \tag{12.58}$$

If we now instead consider the Minkowski space path integral (10.52) (with the full Hamiltonian (12.56) rather than a free one), the Gaussian functional integral over the momentum fields π_n can be easily evaluated by the usual trick of completing the square, and we find

$$\int \mathbf{D}\pi_n e^{i\int(\pi_n(x)\dot{\phi}_n(x) - \mathcal{H})d^4x} = \mathrm{Const} \cdot e^{i\{\int(\frac{1}{2}(\dot{\phi}_n - L_n)K_{nm}^{-1}(\dot{\phi}_m - L_m) - V(\phi_n, \vec{\nabla}\phi_n))d^4x\}}$$

$$= \mathrm{Const} \cdot e^{i\int \mathcal{L}(\dot{\phi}_n, \phi_n, \vec{\nabla}\phi_n)d^4x} \tag{12.59}$$

so, as promised, the functional Fourier transform induced by the integration over momentum fields in the path integral has effected precisely the *algebraic* Legendre transformation from the Hamiltonian to the Lagrangian. The full generating functional of the theory is then obtained by a further functional integration over the ϕ_n, including, as usual, for convenience, source functions to allow us to generate the n-point functions of the theory by functional differentiation,

$$Z[j] = \int \mathbf{D}\phi_n e^{i\int(\mathcal{L}(\dot{\phi}_n, \phi_n, \vec{\nabla}\phi_n) - j_n(x)\phi_n(x))d^4x} \equiv \int \mathbf{D}\phi_n e^{\frac{i}{\hbar}\mathcal{I}[\phi_n, \partial_\mu\phi_n] - i\int j_n(x)\phi_n(x)d^4x} \tag{12.60}$$

where we have reinserted the usually invisible factor of Planck's constant in the final expression, for reasons shortly to become apparent. For theories where the Lagrange

---

[7] The form of Hamiltonian density assumed here takes care, for example, of the situations encountered in the quantization of massive or massless abelian gauge vector fields coupled to scalar or fermionic matter fields (see Problems 4 and 5)—in the massless abelian case, under the proviso that the Hamiltonian has been evaluated in a "physical" gauge in which *all* the gauge freedom has been removed. The more subtle aspects of the canonical quantization procedure, which emerge in theories with a local gauge symmetry, will be discussed in detail in Chapter 15.

density $\mathcal{L}$ ends up being a Lorentz scalar therefore, the Feynman rules (vertices and propagators) generated by this path integral will clearly lead to Green functions (and eventually, via LSZ, to S-matrix amplitudes) behaving appropriately under Lorentz transformation. In practice, as emphasized previously, we *begin* by specifying the dynamics of the theory in terms of a Lorentz-invariant action $\mathcal{I}$ (= spacetime integral of the Lagrangian density). The classical Principle of Least Action amounts, as is apparent from the functional integral (12.60) to a stationary phase approximation in which the integral is dominated, in the limit of very small $\hbar$, by fields $\phi_{n\,\mathrm{cl}}(x)$ which lead to extremal values of the action integral $\mathcal{I}$. We shall see in the next section that these fields are precisely those satisfying the Euler–Lagrange equations. From the point of view of quantum field theory, the discussion here shows that the computation of Green functions and scattering amplitudes can in fact proceed entirely at the Lagrangian level, using the representation (12.60), with no need to refer to the Hamiltonian of the theory (which is frequently a much more complicated object than the Lagrangian, especially, as we shall see later, in gauge theories).

In the fermionic case, the situation is even simpler. We saw earlier that the fermionic Lagrangian $\mathcal{L}(\psi, \partial_\mu \psi)$ is *algebraically identical* to $\pi^\psi \dot{\psi} - \mathcal{H}(\pi^\psi, \psi, \vec{\nabla}\psi)$ (taking just a single Fermi field for simplicity). Thus the transition between the two formulations in the path integral context does not even require us to perform a functional integral: we simply replace $\int \mathbf{D}\pi^\psi \to \int \mathbf{D}\bar{\psi}$ and $\pi^\psi \to i\bar{\psi}\gamma^0$ to convert the Hamiltonian functional integral into the Lagrangian one:

$$\int \mathbf{D}\pi^\psi \mathbf{D}\psi e^{i\int (\pi^\psi \dot{\psi} - \mathcal{H}(\pi^\psi,\psi,\vec{\nabla}\psi))d^4 x} \to \int \mathbf{D}\bar{\psi}\mathbf{D}\psi e^{i\int \mathcal{L}(\psi,\partial_\mu\psi)d^4 x} \qquad (12.61)$$

Again, we need only demand that $\mathcal{L}$ be constructed in a Lorentz-invariant way out of covariantly transforming Dirac fields to ensure the Lorentz-invariance of the amplitudes generated by this functional representation.

12.4 Noether's theorem, the stress-energy tensor, and all that stuff

The Euler–Lagrange equation (12.47) may be regarded as the differential expression of a global *Action Principle*. In the classical context it incorporates the Principle of Least Action familiar from classical mechanics: the classical field evolves dynamically in such a way as to extremize the associated (c-number) action. Let us begin with the assumption that our action can be written as the integral over a Lagrange density over fields ϕ_n and their (at most) first spacetime-derivatives $\partial_\mu\phi_n$. As the action involves a spacetime integration, and our fields may be assumed to vanish at infinity (or alternatively, to satisfy periodic boundary conditions imposed at very far distances), we may freely integrate by parts (neglecting boundary terms) to redistribute derivatives, so that for example, we may replace $\phi_n(x)\Box\phi_n(x) \to -\partial_\mu\phi_n(x)\partial^\mu\phi_n(x)$. An assumption of this kind is in fact physically required for sensible four-dimensional field theories. Higher derivative terms (such as $(\Box\phi)^2$) in the kinetic (quadratic in fields) part of the Lagrangian can be shown to imply a failure of spectral positivity of the field theory: clearly, the inverse Feynman propagator $\Delta_F^{-1}(p)$ of such a field will contain a term proportional to p^4, so that the Feynman propagator of the theory falls faster than

$1/p^2$ at large p, which we saw in Section 9.5 is incompatible with the Källen–Lehmann spectral representation of the two-point function of a local field theory formulated on a positive-definite Hilbert space.[8] On the other hand, any interaction (higher than quadratic) terms in the Lagrangian involving more than single derivatives of the fields turn out (in four spacetime dimensions) to violate perturbative renormalizability, as we shall see in Part 4 of the book. Indeed, the renormalizable gauge field theories of the Standard Model all satisfy our basic assumption and involve Lagrangians which can be written in the form $\mathcal{L}(\phi_n, \partial_\mu \phi_n)$, where i is an index labeling the independent fields of the theory. The extremal condition for an action obtained from such a Lagrangian is therefore (again relying on the freedom to integrate by parts)

$$0 = \delta \mathcal{I} = \int \delta \mathcal{L} d^4 x$$

$$= \int \left(\frac{\partial \mathcal{L}}{\partial \phi_n(x)} \delta \phi_n(x) + \frac{\partial \mathcal{L}}{\partial(\partial_\mu \phi_n(x))} \partial_\mu \delta \phi_n(x) \right) d^4 x$$

$$= \int \left(\frac{\partial \mathcal{L}}{\partial \phi_n(x)} - \partial_\mu \frac{\partial \mathcal{L}}{\partial(\partial_\mu \phi_n(x))} \right) \delta \phi_n(x) d^4 x, \quad \forall \delta \phi_n(x) \qquad (12.62)$$

As the first-order variation of the action must vanish for arbitrary local variations $\delta \phi_n(x)$ of the independent fields of the theory, the Euler–Lagrange equations follow directly:

$$\partial_\mu \frac{\partial \mathcal{L}}{\partial(\partial_\mu \phi_n(x))} = \frac{\partial \mathcal{L}}{\partial \phi_n(x)} \qquad (12.63)$$

We have already seen that the covariant form of this expression ensures that the dynamical equations of the theory, as encapsulated in these Euler–Lagrange equations, will be rendered compatible with the demands of special relativity by the simple device of choosing the Lagrange density to be a Lorentz scalar constructed from the underlying fields ϕ_n.

Our task for the remainder of this section is to display the ubiquitous role played by the action formulation in the study of symmetries, employing as our basic tool the beautiful result of Emmy Noether, dating from 1918 (translation of original paper in (Noether, 1971)), that connects the symmetries of a theory defined in terms of an action functional with conserved currents expressing the exact conservation laws implied by the dynamics of the theory. The Noether theorem allows for the discussion of the symmetries of the theory—whether spacetime related or "internal"—in a completely unified way, and therefore simplifies enormously the task of reading off from a given action the symmetries of the theory, or conversely, the construction of actions representing theories with desired conservation laws.

Noether's theorem predates by several years the introduction of quantized fields (indeed, of quantum mechanics itself, in its post-Heisenberg–Schrödinger form), and concerns the symmetry and conservation properties of classical field theories, such

---

[8] For an explicit demonstration of the failure of positivity in the context of canonical quantization of higher-derivative theories, see (Bernard and Duncan, 1975).

as Maxwellian electromagnetism or general relativity (the properties of the latter with regard to energy-momentum conservation having been the immediate motivation for Noether's studies). Indeed, the Noether theorem as such is essentially a classical statement: while the conservation laws implied by the theorem are rigorously valid at the classical level, there are several instances in which the corresponding result fails for the quantized field theory, in which case one says that the symmetry in question displays a "quantum anomaly". We shall return to the question of anomalies in Chapter 15: for the present, it suffices to observe that the manipulations that follow in our classical derivation of Noether currents pay no attention whatsoever to the subtleties that necessarily arise when one deals with non-linear functionals of quantum fields, involving potentially ill-defined products of the fields at identical spacetime points. As a matter of fact, the examples given later in this section of Noether symmetries are (with the sole exception of dilatation symmetry) anomaly-free: the conservation laws derived here classically can be rigorously shown to survive (in the form of the so-called "Ward identities" of the field theory) at the quantum level, with conserved currents defined in terms of appropriately renormalized operator products. The demonstration of this statement, however, requires the more detailed familiarity with renormalization theory which will be developed in Part 4 of the book.

Returning to classical field theory, we start by considering an action functional $\mathcal{I}_\Omega$ based on a specified Lagrangian $\mathcal{L}$, depending on a set of fields ϕ_i, integrated over an arbitrary domain Ω of spacetime:

$$\mathcal{I}_\Omega \equiv \int_\Omega \mathcal{L}(\phi_n, \partial_\mu \phi_n) d^4 x \tag{12.64}$$

Suppose that $\mathcal{I}_\Omega$ is invariant under a particular simultaneous infinitesimal transformation of the spacetime coordinates and the fields,

$$x^\mu \rightarrow x'^\mu = x^\mu + \delta x^\mu \tag{12.65}$$

$$\phi_n(x) \rightarrow \phi'_n(x') = \phi_n(x) + \delta\phi_n(x) \tag{12.66}$$

As a concrete example, we recall that under an infinitesimal Lorentz transformation $\Lambda^\mu_\nu = g^\mu_\nu + \omega^\mu_\nu$, $\omega^{\mu\nu} = -\omega^{\nu\mu}$, and for a covariant field $\phi_n(x)$ in a definite matrix representation M_{nm} of the HLG,

$$x^\mu \rightarrow x'^\mu = x^\mu + \omega^\mu_\nu x^\nu \tag{12.67}$$

$$\phi_n(x) \rightarrow \phi'_n(x') = M_{nm}(\Lambda)\phi_m(x) \tag{12.68}$$

For example, if $\phi(x)$ is a scalar field, transforming like $\phi(x) \rightarrow \phi'(x') = \phi(x)$, its four-gradient vector field transforms like

$$\partial_\mu \phi(x) \rightarrow \frac{\partial}{\partial x'^\mu} \phi'(x') = \frac{\partial x^\nu}{\partial x'^\mu} \frac{\partial}{\partial x^\nu} \phi(x) = \Lambda_\mu^{\ \nu} \partial_\nu \phi(x) \tag{12.69}$$

in accordance with (12.68). Of course, in making the Lorentz transformation in (12.64), we must also transform the domain of integration Ω to Ω', where $x' \in \Omega'$ if and only if $x \in \Omega$, thereby ensuring the invariance of $\mathcal{I}_\Omega$ under Lorentz transformations *provided $\mathcal{L}$ is constructed as a Lorentz scalar composite of the component fields ϕ_n.*

Leaving aside temporarily the special case of Lorentz transformations, we see that invariance of the action under (12.65, 12.66) amounts to

$$\delta\mathcal{I}_\Omega = \int_{\Omega'} \mathcal{L}(\phi'_n(x'), \partial_\mu \phi'_n(x')) d^4x' - \int_\Omega \mathcal{L}(\phi_n(x), \partial_\mu \phi_n(x)) d^4x = 0 \qquad (12.70)$$

The Jacobian of the infinitesimal coordinate transformation is (using $\det(1+M) = e^{\mathrm{TrLn}(1+M)} = 1 + \mathrm{Tr}(M) + O(M^2)$ for M an infinitesimal matrix)

$$\det(\frac{\partial x'^\mu}{\partial x^\nu}) = 1 + \frac{\partial}{\partial x^\mu} \delta x^\mu \qquad (12.71)$$

so, changing variables from x' back to x in the first integral, and neglecting second-order infinitesimals,

$$\delta\mathcal{I}_\Omega = \int_\Omega [\mathcal{L}(\phi_n + \delta\phi_n, \partial_\mu \phi_n + \delta\partial_\mu \phi_n)(1 + \frac{\partial}{\partial x^\mu} \delta x^\mu) - \mathcal{L}(\phi_n, \partial_\mu \phi_n)] d^4x$$

$$= \int_\Omega [\frac{\partial\mathcal{L}}{\partial\phi_n}\delta\phi_n + \frac{\partial\mathcal{L}}{\partial(\partial_\mu \phi_n)}\delta(\partial_\mu \phi_n) + \mathcal{L}\frac{\partial}{\partial x^\mu}(\delta x^\mu)] d^4x, \quad \forall\Omega$$

$$\Rightarrow \frac{\partial\mathcal{L}}{\partial\phi_n}\delta\phi_n + \frac{\partial\mathcal{L}}{\partial(\partial_\mu \phi_n)}\delta(\partial_\mu \phi_n) + \mathcal{L}\frac{\partial}{\partial x^\mu}(\delta x^\mu) = 0 \qquad (12.72)$$

There is a subtlety in the middle term of the final expression (12.72), arising from the fact that the variation δ and spacetime-derivative ∂_μ do not in general commute:

$$\partial_\mu \phi'_n(x') = \frac{\partial}{\partial x'^\mu} \phi'_n(x')$$

$$= \frac{\partial}{\partial x'^\mu}(\phi_n(x) + \delta\phi_n(x))$$

$$= \frac{\partial x^\nu}{\partial x'^\mu} \frac{\partial}{\partial x^\nu}(\phi_n(x) + \delta\phi_n(x))$$

so we find

$$\partial_\mu \phi'_n(x') = (g^\nu_\mu - \frac{\partial\delta x^\nu}{\partial x^\mu})(\partial_\nu \phi_n(x) + \partial_\nu \delta\phi_n(x))$$

$$= \partial_\mu \phi_n(x) + \partial_\mu \delta\phi_n(x) - \frac{\partial\delta x^\nu}{\partial x^\mu}\partial_\nu \phi_n(x)$$

$$\Rightarrow \delta(\partial_\mu \phi_n(x)) = \partial_\mu \delta\phi_n(x) - \frac{\partial\delta x^\nu}{\partial x^\mu}\partial_\nu \phi_n(x) \qquad (12.73)$$

Inserting $\delta(\partial_\mu \phi_n)$ from (12.73) in the invariance condition (12.72) we obtain

$$\frac{\partial\mathcal{L}}{\partial\phi_n}\delta\phi_n + \frac{\partial\mathcal{L}}{\partial(\partial_\mu \phi_n)}(\partial_\mu \delta\phi_n - \partial_\nu \phi_n \frac{\partial\delta x^\nu}{\partial x^\mu}) + \mathcal{L}\frac{\partial}{\partial x^\mu}(\delta x^\mu) = 0 \qquad (12.74)$$

One final rearrangement of this identity leads to a convenient form for the statement of Noether's theorem. Define the *intrinsic change* in ϕ_n as $\delta^*\phi_n \equiv \delta\phi_n - \delta x^\mu \frac{\partial\phi_n}{\partial x^\mu}$: this is the change in the field *other* than that due to the shift in coordinates. Multiplying out the factors and cancelling terms, a little algebra shows

$$
\delta^* \phi_n \{ \frac{\partial \mathcal{L}}{\partial \phi_n} - \partial_\nu \frac{\partial \mathcal{L}}{\partial(\partial_\nu \phi_n)} \} + \partial_\mu \{ \frac{\partial \mathcal{L}}{\partial(\partial_\mu \phi_n)} \delta \phi_n + [g^\mu_\nu \mathcal{L} - \frac{\partial \mathcal{L}}{\partial(\partial_\mu \phi_n)} \partial_\nu \phi_n] \delta x^\nu \}
$$

$$
= \frac{\partial \mathcal{L}}{\partial \phi_n} \delta \phi_n + \frac{\partial \mathcal{L}}{\partial(\partial_\mu \phi_n)} (\partial_\mu \delta \phi_n - \partial_\nu \phi_n \frac{\partial \delta x^\nu}{\partial x^\mu}) + \mathcal{L} \frac{\partial}{\partial x^\mu} (\delta x^\mu) = 0 \qquad (12.75)
$$

For fields satisfying the Euler–Lagrange equations of motion, the term proportional to $\delta^* \phi_n$ vanishes, and we are left with the desired Noether theorem:

$$
\partial_\mu J^\mu(x) = 0, \quad J^\mu \equiv [\frac{\partial \mathcal{L}}{\partial(\partial_\mu \phi_n)} \partial_\nu \phi_n - g^\mu_\nu \mathcal{L}] \delta x^\nu - \frac{\partial \mathcal{L}}{\partial(\partial_\mu \phi_n)} \delta \phi_n \qquad (12.76)
$$

In other words, the four fields constituting J^μ form a divergenceless vector, or *conserved Noether current*, the zeroth component of which gives the density for a conserved charge, by the familiar maneuver:

$$
\partial_\mu J^\mu = 0 \Rightarrow \frac{\partial}{\partial t} J^0 = \vec{\nabla} \cdot \vec{J} \Rightarrow \frac{\partial}{\partial t} \int J^0(\vec{x}, t) d^3 x = 0 \qquad (12.77)
$$

$$
\text{i.e. } \dot{Q}(t) = 0, \quad Q(t) \equiv \int J^0(\vec{x}, t) d^3 x \qquad (12.78)
$$

To summarize: invariances of the action under infinitesimal variations taking the form (12.65, 12.66) stand in one-to-one correspondence with conserved currents J^μ, each of which in turn gives rise to a conserved charge Q, preserved under the dynamics entailed by the Euler–Lagrange equations of the theory. After quantization, the latter equations simply embody the dynamics of the Heisenberg fields of the theory, so we may at least hope that quantized currents formed by simply replacing the classical fields from which the Noether currents are built with the corresponding Heisenberg fields will also provide conserved objects at the quantum level. This is by no means guaranteed *a priori*: as we mentioned earlier, the necessity for careful definition of the composite operators appearing in the currents, due to ordering difficulties and/or short-distance singularities, may interfere with the implicit smoothness properties assumed in the purely classical derivation of the Noether theorem, leading to a "quantum anomaly", or violation of a classically conserved quantity at the quantum level. The examples given below (apart from dilatation symmetry) will not, however, be infected with the anomaly disease, to which we return in Chapter 15.

We shall shortly see that the combination of fields multiplying δx^ν in (12.76) plays a fundamental role for relativistically invariant theories. The free μ, ν indices suggest that we define a second-rank "energy-momentum" tensor T^μ_ν, the physical significance of which will shortly emerge, as

$$
T^\mu_\nu \equiv \frac{\partial \mathcal{L}}{\partial(\partial_\mu \phi_n)} \partial_\nu \phi_n - g^\mu_\nu \mathcal{L} \qquad (12.79)
$$

In terms of $T^\mu_{\ \nu}$, the Noether current J^μ takes the form

$$J^\mu = T^\mu_{\ \nu}\delta x^\nu - \frac{\partial \mathcal{L}}{\partial(\partial_\mu\phi_n)}\delta\phi_n \qquad (12.80)$$

There is an important generalization of the version of Noether's theorem given above which will be important in our discussion of supersymmetry in section 12.6. Let us suppose that we are dealing with a global symmetry with $\delta x^\nu = 0$, but that the Lagrangian density is only invariant up to a spacetime divergence under an infinitesimal symmetry transformation, $\delta\mathcal{L} = \partial_\mu K^\mu$, so that the action integral $\mathcal{I}_\Omega$ is invariant if the domain of integration Ω is all of spacetime, but not in general for arbitrary finite domains Ω, due to boundary terms arising from the four-dimensional version of Gauss's theorem. The variation of the Lagrangian density under an infinitesimal symmetry transformation $\phi_n \to \phi_n + \delta\phi_n$ is

$$\begin{aligned}
\delta\mathcal{L} &= \frac{\partial \mathcal{L}}{\partial\phi_n}\delta\phi_n + \frac{\partial \mathcal{L}}{\partial(\partial_\mu\phi_n)}\delta(\partial_\mu\phi_n) \\
&= \partial_\mu\left(\frac{\partial \mathcal{L}}{\partial(\partial_\mu\phi_n)}\right)\delta\phi_n + \frac{\partial \mathcal{L}}{\partial(\partial_\mu\phi_n)}\partial_\mu\delta\phi_n \\
&= -\partial_\mu J^\mu_{\text{Noeth}} \qquad (12.81)
\end{aligned}$$

where we have used the Euler–Lagrange equations of motion in the second line, and introduced the notation J^μ_{Noeth} to indicate the conventional Noether current of (12.80) (with $\delta x^\nu = 0$). But by assumption, the variation of the Lagrangian can be written as a divergence, $\delta\mathcal{L} = \partial_\mu K^\mu$, so the current that is actually conserved in this case is

$$J^\mu = J^\mu_{\text{Noeth}} + K^\mu \qquad (12.82)$$

12.5   Applications of Noether's theorem

Now for some examples of Noether symmetries of central importance in relativistic field theory. We treat first the purely spacetime symmetries—those associated with the invariance of the theory under the Poincaré group composed of spacetime translations and homogeneous Lorentz transformations. Then we examine the case of *internal* symmetries, in which $\delta x^\nu = 0$, and only the fields undergo a transformation. The symmetries considered in this section all involve a Noether current of the standard form (12.80): we shall see an example of the generalized version (12.82) in the next section, when we consider global supersymmetry.

Example 1: Spacetime translation invariance.

As long as the Lagrangian density of the theory derives its dependence on the spacetime coordinate x entirely through the fields, with no other explicit x-dependence (e.g., in the coefficients multiplying the fields), the action functional $\mathcal{I}_\Omega$ of (12.64) is clearly invariant under fixed translations (infinitesimal or finite) of x:

$$x^\mu \to x'^\mu = x^\mu + \epsilon g^\mu_{\ \sigma}, \quad \sigma = 0, 1, 2, 3 \qquad (12.83)$$

$$\delta\phi_n(x) = 0 \qquad (12.84)$$

Indeed, this is simply the statement of invariance of the spacetime integral under the change of variables embodied in (12.83). The quantity ϵ is an arbitrary positive infinitesimal constant, which will be divided out of the definition of the current (at the end. There are evidently four independent symmetries, corresponding to time ($\sigma = 0$) and space ($\sigma = 1, 2, 3$) translations, corresponding to the currents (cf. (12.80)

$$J^\mu_{\ \sigma} = T^\mu_{\ \sigma}, \quad \sigma = 0, 1, 2, 3 \tag{12.85}$$

The reason for attaching the name "energy-momentum tensor" to $T^\mu_{\ \nu}$ is now apparent. Relabeling the conserved charge associated with $J^\mu_{\ \sigma}$ as P_σ (rather than Q_σ, say), we have from (12.79),

$$P_\sigma = \int d^3x \{\frac{\partial\mathcal{L}}{\partial\dot\phi_n}\partial_\sigma\phi_n - g^0_{\ \sigma}\mathcal{L}\} \tag{12.86}$$

and in particular, for the time-component (recall that the conjugate momentum fields $\pi_n \equiv \frac{\partial\mathcal{L}}{\partial\dot\phi_n}$)

$$P_0 = \int d^3x \{\pi_n\dot\phi_n - \mathcal{L}\} = \int d^3x\, \mathcal{H}(\vec{x}, t) = H \tag{12.87}$$

the corresponding charge is simply the full Hamiltonian, *given as the Legendre transform of the Lagrangian.* The fundamental role played by the Legendre transform in connecting the Lagrangian and Hamiltonian forms is therefore seen to emerge inescapably from the Noether treatment of the time-translational symmetry of the theory. For the spatial components of the four-vector P_σ we find

$$\vec{P} = \int d^3x\, \pi_n(\vec{x}, t)\vec\nabla\phi_n(\vec{x}, t) \tag{12.88}$$

The interpretation of this spatial vector as the spatial momentum can be seen if we take the case of purely bosonic fields, and impose the standard equal-time commutator relations

$$[\pi_n(\vec{x}, t), \phi_m(\vec{y}, t)] = -i\delta^3(\vec{x} - \vec{y}) \tag{12.89}$$

whereupon one finds

$$[\vec{P}, \phi_m(\vec{y}, t)] = \int d^3x[\pi_n(\vec{x}, t), \phi_m(\vec{y}, t)]\,\vec\nabla\phi_n(\vec{x}, t) = -i\vec\nabla\phi_m(\vec{y}, t) \tag{12.90}$$

so that $\vec{P}$ generates spatial translations of the fields of the theory, as expected of the spatial momentum operator. The Noether charges P_σ given in (12.86) are therefore nothing but our old friend the energy-momentum four-vector of the theory, with the "columns" of the energy-momentum tensor $T^\mu_{\ \nu}$ giving the associated conserved currents (thus, $\partial_\mu T^\mu_{\ \nu} = 0$).

The physical interpretation of the energy-momentum tensor defined in (12.79) is actually quite subtle. From the point of view of "flat space" field theory (formulated in Minkowski space), the only relevant property of the tensor density $T^{\mu\nu}(x)$ is that the spatial integrals $\int T^0{}_\sigma d^3x$ reproduce the conserved energy-momentum four-vector components P_σ which implement spacetime translations on the Heisenberg fields of the theory. The whole axiomatic formulation of interacting field theory *à la* Wightman, for example, only relies on the existence of the conserved generators of the Poincaré group, not on the presence of a set of "charge" densities which can be integrated to give these operators. Thus, nothing is altered from the point of view of Minkowski field theory if we "redistribute" the energy and momentum density on any time-slice as long as the spatial integral preserves the total energy and momentum of the field as given by P_σ. For example, we can certainly alter the "canonical" energy-momentum tensor (12.79) (to which we now add a "c" subscript to indicate its special origin in the canonical Noether procedure) by adding a "superpotential" term:

$$T_c^{\mu\nu} \equiv \frac{\partial \mathcal{L}}{\partial(\partial_\mu \phi_n)} \partial^\nu \phi_n - g^{\mu\nu} \mathcal{L} \to T^{\mu\nu} = T_c^{\mu\nu} + \partial_\lambda S^{\lambda\mu\nu}, \quad S^{\lambda\mu\nu} = -S^{\mu\lambda\nu} \qquad (12.91)$$

as the divergence of the added term (on the μ index) is automatically zero due to the antisymmetry property of the superpotential, so that the modified tensor leads to exactly the same energy-momentum vector as the canonical version.

The actual local distribution of energy and momentum only acquires physical significance if there are fields in the theory which couple directly, and locally, to the energy-momentum tensor. In fact, once we include gravitational effects along the lines of general relativity, such a field appears immediately in the form of the now dynamical spacetime metric $g_{\mu\nu}(x)$. Once a generally covariant action functional is constructed for the particular matter fields in the background metric $g_{\mu\nu}(x)$, the variation of the action with respect to the metric is precisely the energy-momentum tensor $T^{\mu\nu}(x)$ of these fields. In particular, in the weak field limit for the gravitational field, where we expand $g_{\mu\nu}(x) = \eta_{\mu\nu} + h_{\mu\nu}(x)$, and now $\eta_{\mu\nu} = \text{diag}(1, -1, -1, -1)$ is the fixed Minkowski metric (which we have heretofore simply called $g_{\mu\nu}$!), $h_{\mu\nu}(x)$ acts as an interpolating field for gravitons, and the term $h^{\mu\nu}(x)T_{\mu\nu}(x)$ of first-order in the metric deviation is the appropriate interaction Lagrangian density if we wish to compute S-matrix amplitudes for processes involving one graviton and multiple matter particles. The specific choice of the spatial dependence of $T_{\mu\nu}(x)$ clearly becomes physically significant in this case. In particular, the $T_{\mu\nu}$ tensor obtained by metric variation of the generally covariant matter action is clearly a symmetric tensor, $T_{\mu\nu} = T_{\nu\mu}$, which is clearly not guaranteed by the Noether expression (12.79), and indeed there are cases where the need to construct a generally covariant action *necessitates* the addition of a superpotential term to the canonical tensor $T_c^{\mu\nu}$ in order to obtain a properly symmetric tensor. For scalar field theory (e.g., $\lambda\phi^4$ theory), the canonical tensor $T_c^{\mu\nu}$ is already symmetric,

$$T_c^{\mu\nu} = \partial^\mu \phi \partial^\nu \phi - g^{\mu\nu} \mathcal{L} \qquad (12.92)$$

but leads to ultraviolet divergences when used as the current coupled to gravitons in single graviton-multi-scalar scattering amplitudes, as shown by Callan, Coleman, and Jackiw (Callan *et al.*, 1970). The situation is remedied by adding a term $-\frac{1}{12}R(x)\phi^2(x)$ to the generally covariant Lagrangian density, where $R(x)$ is the curvature scalar (so that the added term vanishes in flat space), thereby modifying the energy-momentum tensor (which is the metric variation of the Lagrange density) by a superpotential term, and leading to a *"new, improved"* energy-momentum tensor:[9]

$$T^{\mu\nu}_{\text{impr}}(x) = T^{\mu\nu}_c(x) - \frac{1}{6}(\partial^\mu \partial^\nu - g^{\mu\nu}\Box)(\phi^2(x)) \tag{12.93}$$

We shall see shortly that this improved tensor also leads to simplified forms for the currents associated with dilatation and conformal symmetry.

Example 2: Invariance under the homogeneous Lorentz group.

In this case the invariance obtains once we have constructed an action as the spacetime integral of a scalar Lagrangian density, built out of properly contracted covariant fields. The infinitesimal transformations are those given in (12.94, 12.95), repeated here for convenience:

$$x^\mu \to x'^\mu = x^\mu + \omega^\mu_{\ \nu}x^\nu \tag{12.94}$$

$$\phi_n(x) \to \phi'_n(x') = M_{nm}(\Lambda)\phi_m(x) \tag{12.95}$$

By definition (cf. Section 7.2) the representatives $M_{nm}(\Lambda)$ of a Lorentz transformation Λ in a particular (finite-dimensional) representation (which may be a direct sum of irreducible representations) are given in terms of the generators $(J^{\mu\nu})_{nm}$ and the infinitesimal rotation angles and boost rapidities $\omega_{\mu\nu}$ by

$$M_{nm}(\Lambda) = \delta_{nm} + \frac{i}{2}\omega_{\mu\nu}(J^{\mu\nu})_{nm} + O(\omega^2), \quad \omega_{\mu\nu} = -\omega_{\nu\mu} \tag{12.96}$$

so

$$\delta\phi_n(x) \equiv \phi'_n(x') - \phi_n(x) = \frac{i}{2}\omega_{\mu\nu}(J^{\mu\nu})_{nm}\phi_m(x) \tag{12.97}$$

There are six independent choices for the $\omega_{\mu\nu}$ corresponding to rotations around or boosts along the three spatial axes. Let us pick a specific one by choosing a pair κ, λ with $0 \le \kappa < \lambda \le 3$, and setting

$$\omega_{\mu\nu} = g^\kappa_{\ \mu}g^\lambda_{\ \nu} - g^\kappa_{\ \nu}g^\lambda_{\ \mu}, \quad \delta x^\nu = g^{\kappa\nu}x^\lambda - g^{\lambda\nu}x^\kappa \tag{12.98}$$

and the corresponding conserved currents are

$$\mathcal{M}^{\mu\kappa\lambda} = (g^{\kappa\nu}x^\lambda - g^{\lambda\nu}x^\kappa)T^\mu_{\ \nu} - i\frac{\partial\mathcal{L}}{\partial(\partial_\mu\phi_n)}(J^{\kappa\lambda})_{nm}\phi_m$$

$$= x^\lambda T^{\mu\kappa} - x^\kappa T^{\mu\lambda} - i\frac{\partial\mathcal{L}}{\partial(\partial_\mu\phi_n)}(J^{\kappa\lambda})_{nm}\phi_m \tag{12.99}$$

Each conserved current in turn leads to a conserved charge, obtained by spatially integrating its time component: to avoid unnecessary proliferation of notation, we

---

[9] As shown by (Freedman and Weinberg, 1974), the coefficient $\frac{1}{6}$ is further modified at the two-loop level, and beyond, by renormalization effects.

shall use the same letter $\mathcal{M}$ to denote both the current and its charge:

$$\mathcal{M}^{\kappa\lambda} \equiv \int d^3x \, \mathcal{M}^{0\kappa\lambda}, \quad \dot{\mathcal{M}}^{\kappa\lambda} = 0 \qquad (12.100)$$

The interpretation of the various terms appearing in (12.99) becomes clearer if we examine the special case of spatial rotations, by picking $(\kappa, \lambda) = (i, j)$, $1 \leq i < j \leq 3$. The associated charge is then seen to be

$$\mathcal{M}^{ij} = x^j P^i - x^i P^j - i\pi_n (J^{ij})_{nm} \phi_m \qquad (12.101)$$

the first two terms of which clearly give the orbital angular momentum of the system, whereas the last term, present only for fields transforming non-trivially under the Lorentz group, must correspond to spin angular momentum. That the total charge $\mathcal{M}^{ij}$ indeed corresponds to the total angular momentum operator follows from the fact that it generates, by commutation with the field, and employing the usual equal-time commutation relations, the correct infinitesimal variation:

$$[\mathcal{M}^{ij}, \phi_n(\vec{x}, t)] = i(x^i \partial^j - x^j \partial^i)\phi_n(\vec{x}, t) - (J^{ij})_{nm}\phi_m(\vec{x}, t) \qquad (12.102)$$

A similar result can be obtained for the commutation with the boost operator $\mathcal{M}^{0i}$— an exercise we leave for the reader (see Problem 6). The result (12.102) indicates that our Noether charges are indeed the correct generators of the HLG in the state space of the quantum field theory. Indeed, for a general Lorentz transformation Λ the covariant field transformation law is (cf.(7.5))

$$U(\Lambda)\phi_n(x)U^\dagger(\Lambda) = M_{nm}(\Lambda^{-1})\phi_m(\Lambda x) \qquad (12.103)$$

Note that we are working entirely in Heisenberg representation here, so that these $U(\Lambda)$s are really the $U_H(\Lambda)$ of Section 9.1, acting on the fully interacting Heisenberg fields of the theory, and in the state space spanned by eigenstates of the full Hamiltonian: we are omitting the H subscript for simplicity of notation. If we take Λ infinitesimally close to the identity, $\Lambda^\mu{}_\nu = g^\mu{}_\nu + \omega^\mu{}_\nu$, the corresponding unitary operators are expressed in terms of the Hilbert space generators $\mathcal{M}^{\mu\nu}$ of infinitesimal Lorentz transformations (see (9.64), for which we use the same notation as the Noether charges found above, as they will shortly be seen to be identical:

$$U(\Lambda) = 1 + \frac{i}{2}\omega_{\mu\nu}\mathcal{M}^{\mu\nu} + O(\omega^2) \qquad (12.104)$$

Inserting (12.104) into (12.103), we find, on expanding every term to first order in ω, the commutation relation for the generators $\mathcal{M}^{\mu\nu}$ with our field ϕ_n:

$$[\mathcal{M}^{\mu\nu}, \phi_n(x)] = i(x^\mu \partial^\nu - x^\nu \partial^\mu)\phi_n(x) - (J^{\mu\nu})_{nm}\phi_m(x) \qquad (12.105)$$

which agrees with (12.102) if we set $(\mu, \nu) \rightarrow (i, j)$. One may also compute the commutators of the various Noether charges with each other, using again the equal-time commutators of the theory. In this way, the verification of the full Poincaré algebra, as given in (9.65, 9.67, 9.68), can be carried out explicitly starting from the expressions for the Noether charges given above.

Example 3: Dilatation and conformal symmetry.

The Poincaré group consisting of the homogeneous Lorentz transformations (with six independent real parameters, associated with the generators $\mathcal{M}^{\mu\nu}$), together with the (four-parameter) group of spacetime translations (with generators P^{μ}), can actually be enlarged to a fifteen-parameter group, the *conformal group*, by including a dilatation symmetry (with generator $\mathcal{D}$) and a four-parameter set of conformal transformations (with generators $\mathcal{K}^{\mu}$). The effect of finite dilatation and conformal transformations on the spacetime coordinates is[10]

$$x^{\mu} \to x'^{\mu} \equiv e^{-\rho} x^{\mu} \quad \text{dilatation} \tag{12.106}$$

$$x^{\mu} \to x'^{\mu} \equiv \frac{x^{\mu} - c^{\mu} x^2}{1 - 2c \cdot x + c^2 x^2} \quad \text{conformal} \tag{12.107}$$

which, for infinitesimal transformations (ρ and c^{μ} infinitesimal), become

$$x^{\mu} \to x^{\mu} - \rho x^{\mu} + O(\rho^2) \tag{12.108}$$

$$x^{\mu} \to x^{\mu} + c_{\sigma}(2x^{\sigma} x^{\mu} - g^{\sigma\mu} x^2) + O(c^2) \tag{12.109}$$

The four-momentum generators P^{μ} must transform like $-i\frac{\partial}{\partial x_{\mu}}$ under dilatation generators, i.e., with a factor $e^{+\rho}$, so under a finite dilatation generated by $e^{i\rho\mathcal{D}}$

$$e^{i\rho\mathcal{D}} P^{\mu} e^{-i\rho\mathcal{D}} = e^{\rho} P^{\mu} \Rightarrow e^{i\rho\mathcal{D}} P^2 e^{-i\rho\mathcal{D}} = e^{2\rho} P^2 \tag{12.110}$$

so that in any theory with exact dilatation symmetry (so that $e^{-i\rho\mathcal{D}}|\psi\rangle$ is a physical state if $|\psi\rangle$ is), the mass spectrum of particles must either be exactly zero or continuous: clearly not the world we live in! Even for massless theories, we shall see that the classical Noether dilatation (and conformal) currents and charges are in general broken by quantum effects (anomalies) once interactions are present, so the formal existence of the conformal extension of the Poincaré group may seem at first sight to be a matter of purely formal interest.[11] Nevertheless, the nature of the breaking of the conformal group in interacting field theories is now completely understood, and has deep and important connections to the renormalization group properties of such theories which we shall study in detail in Part 4 of the book. Accordingly, we shall give a brief description of the dilatation current and its connection to the trace of the energy-momentum tensor, starting again from the general Noether prescription for construction of a conserved current for a classical symmetry of the action.

First, we need to establish appropriate transformation rules for the fields of the theory, which we shall take for simplicity to be self-conjugate spin-zero scalars. The

---

[10] The rather strange—and highly non-linear!—expression (12.107) for the conformal transformation can be understood once we re-express it as a sequence: coordinate inversion $\mathcal{I}$- translation $\mathcal{T}$- coordinate inversion $\mathcal{I}$, where $\mathcal{I}x^{\mu} \equiv x^{\mu}/x^2$, $\mathcal{T}x^{\mu} = x^{\mu} - c^{\mu}$. It follows immediately that the conformal transformations form an abelian subgroup of the full conformal group.

[11] An important exception arises in the case of two-dimensional conformal field theories: it turns out that there is a rich plethora of such *non-trivial* field theories displaying exact conformal invariance, which have been the subject of intensive study in the last thirty years. The situation with exactly conformally invariant theories in four dimensions is murkier; cf. Section 15.5.

classical Noether action for a massless ϕ^4 theory,

$$\mathcal{I} = \int (\frac{1}{2}\partial_\mu \phi(x)\partial^\mu \phi(x) - \frac{\lambda}{4!}\phi(x)^4)d^4x \tag{12.111}$$

is easily seen to be invariant under the dilatation transformation of the fields

$$\phi'(x') = e^{d\rho}\phi(x), \quad x' = e^{-\rho}x \tag{12.112}$$

provided we choose the real number d (called the "scale dimension" of the field ϕ) equal to unity. For example,

$$\int \phi'(x')^4 d^4x' = \int e^{4d\rho}\phi(x)^4 e^{-4\rho}d^4x = \int \phi(x)^4 d^4x \quad \text{if} \ \ d=1 \tag{12.113}$$

On the other hand, this invariance is destroyed the moment we include terms in the Lagrangian with dimensionful coefficients, such as a mass term $\frac{1}{2}m^2\phi^2$ or interaction terms other than ϕ^4, i.e., $\lambda^{(n)}\phi^n$, $n \neq 4$.

Referring back to the specification of a Noether symmetry in terms of the infinitesimal transformations of coordinates (12.65) and fields (12.66), our general expression for the associated Noether current (12.80) gives, for the present case of dilatation symmetry, (taking ρ infinitesimal, and dividing by $-\rho$ to obtain the conventional normalization)

$$J^\mu_{\text{dil,c}} = T_c^{\mu\nu}x_\nu + d\frac{\partial\mathcal{L}}{\partial(\partial_\mu\phi)}\phi = T_c^{\mu\nu}x_\nu + (\partial^\mu\phi)\phi \tag{12.114}$$

The subscript "c" in the dilatation current and the energy-momentum tensor indicate that we are using the canonical energy-momentum tensor, as described previously, without the "improvement" necessary once we couple quantized matter fields to gravitation. For the theory defined by action (12.111), the canonical energy-momentum tensor is just

$$T_c^{\mu\nu} = \partial^\mu\phi\partial^\nu\phi - \frac{1}{2}g^{\mu\nu}\partial_\mu\phi\partial^\mu\phi + \frac{\lambda}{4!}g^{\mu\nu}\phi^4 \tag{12.115}$$

and the conservation of the dilatation current *at the classical level* follows immediately (using $\partial_\mu T_c^{\mu\nu} = 0$, and the classical field equation $\Box\phi + \frac{\lambda}{3!}\phi^3 = 0$))

$$\partial_\mu J^\mu_{\text{dil,c}} = T^\mu_{c\ \mu} + \partial_\mu(\phi\partial^\mu\phi) = -\partial_\mu\phi\partial^\mu\phi + 4\frac{\lambda}{4!}\phi^4 + \partial_\mu\phi\partial^\mu\phi + \phi\Box\phi = 0 \tag{12.116}$$

As usual, Noether currrents may be modified by the addition of superpotential terms which do not alter their conservation property (as the superpotential terms are by definition automatically divergence-free), and in the case of the dilatation current, unlike the energy-momentum tensor which couples to gravitons, there are no physical fields coupled to the current $J^\mu_{\text{dil,c}}$, so we are free to construct an "improved" dilatation current by defining

$$J^{\mu}_{\text{dil,impr}} \equiv J^{\mu}_{\text{dil,c}} + \frac{1}{6}\partial_{\sigma}(x^{\mu}\partial^{\sigma} - x^{\sigma}\partial^{\mu})(\phi^2)$$

$$= T^{\mu\nu}_c x_{\nu} + (\partial^{\mu}\phi)\phi + \frac{1}{6}\partial_{\sigma}(x^{\mu}\partial^{\sigma} - x^{\sigma}\partial^{\mu})(\phi^2)$$

$$= T^{\mu\nu}_{\text{impr}} x_{\nu} \tag{12.117}$$

where $T^{\mu\nu}_{\text{impr}}$ is the improved energy-momentum tensor (12.93), and conservation of the dilatation current now reduces simply to the tracelessness of this tensor

$$\partial_{\mu}J^{\mu}_{\text{dil,impr}} = T^{\mu}_{\text{impr }\mu} = 0 \tag{12.118}$$

The reader may easily verify that the inclusion of a mass term $\frac{1}{2}m^2\phi^2$ in the Lagrangian results in a non-vanishing divergence of the current, and trace of the energy-momentum tensor:

$$\partial_{\mu}J^{\mu}_{\text{dil,impr}} = T^{\mu}_{\text{impr }\mu} = 2m^2\phi^2 \tag{12.119}$$

These results, while classically valid, turn out to be incorrect once we quantize our field theory: there are additional terms, proportional to Planck's constant, which appear on the right-hand side of both (12.118) and (12.119). They provide our first example of the famous quantum anomalies (in the present case, the "trace anomaly") of interacting quantum field theory, which we shall discuss in detail in Chapter 15. The important lesson which we need to take away from the present discussion is that the extension of the Poincaré group to the larger conformal group cannot be carried through in interacting quantum field theories (in four dimensions[12]). This is in contrast to the supersymmetric extension of the Poincaré group which we shall discuss below, where it turns out to be perfectly possible to construct a wide class of interacting field theories with an exact global supersymmetry which extends the conventional Poincaré symmetry of relativistic field theory.

Example 4: Abelian internal symmetries—phase transformations and charge conservation.

Our remaining examples involve invariances of the action under transformations which leave the spacetime coordinates invariant, $\delta x^{\mu} = 0$, but alter the fields, typically by a spacetime-independent linear rearrangement: these are the so-called *global internal symmetries* of the theory. Suppose that our (for the time being, classical) Lagrangian $\mathcal{L}$ is invariant under simultaneous changes of phase of a set of complex fields $\phi_n(x)$ according to

$$\phi_n(x) \rightarrow e^{i\omega q_n}\phi_n(x), \quad \phi_n^*(x) \rightarrow e^{-i\omega q_n}\phi_n^*(x) \tag{12.120}$$

---

[12] It turns out that in two dimensions, interacting conformal field theories can be constructed. Also, there appears to be a very special class of supersymmetric field theories in four dimensions that possess exact conformal invariance, even though they are interacting.

The action functional $\mathcal{I}_\Omega$ in (12.64) will then be invariant (for ω infinitesimal) under the variations,

$$\delta\phi_n(x) = i\omega q_n \phi_n(x), \quad \delta\phi_n^*(x) = -i\omega q_n \phi_n^*(x), \quad \delta x^\mu = 0 \tag{12.121}$$

The set of transformations of the type (12.120) clearly form a commutative (abelian) group. Now $\mathcal{L}$, being real (classically- or hermitian, once quantized), must contain both ϕ_n and ϕ_n^*, so the associated Noether current (12.76) can be written

$$J^\mu = \sum_n \{ \frac{\partial\mathcal{L}}{\partial(\partial_\mu \phi_n)}(-iq_n\phi_n) + \frac{\partial\mathcal{L}}{\partial(\partial_\mu \phi_n^*)}(iq_n\phi_n^*)\} \tag{12.122}$$

corresponding to a conserved charge Q given by

$$Q = -i \int d^3x \sum_n q_n \{ \pi_n(\vec{x},t)\phi_n(\vec{x},t) - \pi_n^*(\vec{x},t)\phi_n^*(\vec{x},t)\} \tag{12.123}$$

The charge density J^0 has a simple equal-time commutation relation with the fields of the theory:

$$[J^0(\vec{y},t), \phi_n(\vec{x},t)] = -q_n\delta^3(\vec{x}-\vec{y})\phi_n(\vec{y},t) \tag{12.124}$$

The physical interpretation of this conserved quantity for the quantized theory is very simple: the field ϕ_n may be considered as interpolating for a particle of "charge" q_n. Each term in the Lagrangian must contain a product of fields for which the phase factor $e^{i\omega \sum(\pm q_n)} = 1$, so that the interaction terms in the Lagrangian lead to graphs where the charge inserted by the incoming lines exactly balances that removed by the outgoing lines. Depending on the particular Lagrangian under consideration, the charge Q may represent electric charge, baryon number, lepton number, strangeness, or indeed any globally conserved quantum number, depending on the particular set of phase transformations chosen. Of course, as in the case of the spacetime symmetries discussed previously, the Noether charge Q also serves as the infinitesimal generator of the transformation (12.121), as we discover by integrating (12.124) over $\vec{y}$:

$$[Q, \phi_n(x)] = -q_n\phi_n(x) \tag{12.125}$$

in accordance with the interpretation that ϕ_n contains a destruction operator for a particle with charge q_n (and a creation operator for the antiparticle with charge $-q_n$).

Example 5: Non-abelian internal symmetries.

A further generalization of the type of internal symmetries discussed in the preceding example arises when the (still spacetime-*independent*!) transformation of the fields involves mixing of different field components by a matrix transformation. Typically, the matrices involved must be unitary (for complex fields) or orthogonal (for real fields), to maintain the invariance of the kinetic part of the Lagrangian (which, for example, for scalar fields takes the form $\sum_n \partial^\mu \phi_n^* \partial_\mu \phi_n$, in the complex case). Thus consider a theory with an action invariant under $\delta x^\mu = 0$ for the coordinates and a symmetry transformation

$$\phi_n(x) \to \phi_n'(x) = (e^{i w_\alpha t_\alpha})_{nm} \phi_m(x) \equiv M_{nm}(w_\alpha) \phi_m(x) \tag{12.126}$$

where the t_α are a set of infinitesimal generators for some group of linear transformations. For real fields we shall assume that the t_α are hermitian pure imaginary, so that the resultant matrix transformation $e^{i w_\alpha t_\alpha}$ of the fields is a real orthogonal one, while if the fields are complex, the generators are hermitian and the matrix unitary. In the example to be considered shortly, the only complex fields are Dirac fields, and the Lagrangian only contains derivatives of the ϕ_ns, not of the ϕ_n^*s (i.e., the $\bar{\psi}$s), so the Noether current J_α^μ associated with the αth generator takes the same form for both real and complex fields

$$J_\alpha^\mu(x) = -i \sum_n \frac{\partial \mathcal{L}}{\partial(\partial_\mu \phi_n(x))} (t_\alpha)_{nm} \phi_m(x) \tag{12.127}$$

In analogy with (12.124) we have the equal-time commutation of the charge density with the fields

$$[J_\alpha^0(\vec{y}, t), \phi_n(\vec{x}, t)] = -\delta^3(\vec{x} - \vec{y})(t_\alpha)_{nm} \phi_m(\vec{y}, t) \tag{12.128}$$

The isospin-invariant effective meson field theory of the 1950s provides a suitable example: we assume that pions and nucleons interact via a (in this case, non-derivative coupled) Yukawa interaction, with basic fields of the theory taken as a nucleon doublet $N(x) = (p(x), n(x))$ (where $p(x)$ and $n(x)$ are Dirac fields for the proton and neutron, assumed to have identical mass M) and a triplet of real pion fields $\pi_\alpha(x), \alpha = 1, 2, 3$ (with π_3 interpolating for the neutral pion, and $\frac{1}{\sqrt{2}}(\pi_1 + i\pi_2)$ for the positively charged pion, all assumed to have identical mass m_π). The Lagrangian for the full system (with obvious implicit summations over internal indices) is then taken to be

$$\mathcal{L} = \bar{N}(i\partial\!\!\!/ - M)N + \frac{1}{2}(\partial_\mu \vec{\pi} \cdot \partial^\mu \vec{\pi} - m_\pi^2 \vec{\pi} \cdot \vec{\pi}) - ig\bar{N}\gamma_5 \vec{\tau} N \cdot \vec{\phi} \tag{12.129}$$

where the i in the interaction term is there for hermiticity (for g real). The 2x2 matrices $\vec{\tau}$ are one-half the usual Pauli matrices, $\tau_\alpha = \frac{1}{2}\sigma_\alpha, \alpha = 1, 2, 3$, so the matrix group in question is just SU(2). The reader may easily verify, using the Lie algebra of SU(2), $[\tau_\alpha, \tau_\beta] = i\epsilon_{\alpha\beta\gamma}\tau_\gamma$, that the Lagrangian (12.129) is invariant under the following set of global infinitesimal transformations ($\vec{w}$ are *spacetime-independent*)

$$N(x) \to (1 + i\vec{w} \cdot \vec{\tau})N(x) \tag{12.130}$$

$$\bar{N}(x) \to \bar{N}(x)(1 - i\vec{w} \cdot \vec{\tau}) \tag{12.131}$$

$$\vec{\pi}(x) \to \vec{\pi}(x) + \vec{\pi}(x) \times \vec{w} \tag{12.132}$$

from which one may read off directly the vector of conserved Noether currents (12.76) for this theory

$$\vec{J}^\mu = \bar{N}\gamma^\mu \vec{\tau} N + \vec{\pi} \times \partial^\mu \vec{\pi} \tag{12.133}$$

and a corresponding set of conserved "isospin charges" $\vec{I}$

$$\vec{I} = \int d^3x \{N^\dagger \vec{\tau} N(\vec{x}, t) + \vec{\pi} \times \dot{\vec{\pi}}(\vec{x}, t)\} \tag{12.134}$$

The (approximate) conservation of the isospin quantum numbers $\vec{I}^2, I_3$ in strong interaction processes leads to many powerful and valuable selection rules in strong interaction physics: the reader is referred to (Sakurai, 1964) for an exhaustive discussion of these important results.

As in the case of the spacetime Poincaré symmetries, the commutator relations of the Noether charges for internal symmetries replicate the Lie algebra of the underlying symmetry group. For example, using the equal-time commutation relations for the pion fields and the corresponding equal-time anticommutation relations for the nucleon fields, one may easily demonstrate that the charges in (12.134) satisfy

$$[I_\alpha, I_\beta] = i\epsilon_{\alpha\beta\gamma} I_\gamma \tag{12.135}$$

exactly the Lie algebra of the rotation group, allowing us to take over the entire machinery of angular momentum in the discussion of isospin symmetry and conservation. The proof of this result for a general global internal symmetry (for bosonic fields) is deferred to the exercises at the end of this chapter (see Problem 7).

Our discussion of Noether's theorem so far has been carried out for the most part in a classical context: issues of operator ordering, regularization of operator products, and so on, have been resolutely ignored. At first sight, it may seem possible to circumvent these issues by resorting to a path-integral approach, and indeed, it is both important and enlightening to understand the precise way in which invariance and conservation intertwine in the context of the functional integral quantization of field theory. Inasmuch as the functional integral involves an action built from c-number fields, we might at first expect that our discussion to this point will carry over fairly directly to quantum field theory realized via path-integral concepts. In fact, we shall later see in our discussion of anomalous currents in Chapter 15 that subtleties arising from operator regularization cannot simply be dodged in a functional formalism: rather, they reappear in an unexpected location (specifically, in the case of Noether's theorem, in the definition of the functional measure).

We shall conclude this section by giving a brief description of the functional version of Noether's theorem for the case of non-anomalous symmetries, where the aforesaid subtleties do not enter. The end result (the functional analog of (12.76)) will be a set of identities—the so-called *Ward–Takahashi* identities—satisfied by the Feynman Green functions of the theory. Let us start with a theory of N fields $\phi_n(x), n = 1, 2, ..., N$ with a global internal symmetry (12.126), with Lagrangian $\mathcal{L}(\phi_n, \partial_\mu \phi_n)$. The fields may in fact be bosonic or fermionic, but here we shall assume for simplicity only bosonic fields, to avoid having to keep careful track of minus signs arising from interchange of Grassmann fields or sources. The symmetry parameters ω_α in (12.126)) are spacetime constants, as we are dealing with a global symmetry of the theory, but if the Lagrangian can be written (as the notation $\mathcal{L}(\phi_n, \partial_\mu \phi_n)$ implicitly suggests) so that the fields appear with at most a single spacetime-derivative, it is easily seen to be invariant if the w_α are allowed to be spacetime functions, in the following sense:

$$\mathcal{L}(M_{nm}(w_\alpha(x))\phi_m, M_{nm}(w_\alpha(x))\partial_\mu\phi_m) = \mathcal{L}(\phi_n, \partial_\mu\phi_n), \quad \forall w_\alpha(x) \qquad (12.136)$$

Note that the (now spacetime-dependent) gauge parameters $w_\alpha(x)$ are not differentiated on the left-hand side of (12.136), so that the invariance of the Lagrangian density under global transformations will ensure the stronger invariance property given by this equality. The generating functional of Feynman Green functions for this theory is

$$Z[j_n] = \int \mathbf{D}\phi_n e^{i \int \{\mathcal{L}(\phi_n, \partial_\mu\phi_n) - j_n(x)\phi_n(x)\} d^4x} \qquad (12.137)$$

We now wish to examine the result of a change of the functional variables of integration—i.e., the fields $\phi_n(x)$—given precisely by the internal symmetry (12.126), but where the gauge parameters $w_\alpha(x)$ are now allowed to be spacetime-dependent functions. If the matrix transformation $M_{nm}(w_\alpha)$ is a unitary one (or orthogonal, for real fields), then we should expect the Jacobian of the functional change of variables to be a product of unity over all spacetime points, and hence itself equal to unity. We should warn the reader at this point that it is precisely this—on the surface, quite innocent—assumption which fails in the case of the anomalous currents which we shall meet in Section 15.5. However, proceeding on the basis of a unit functional Jacobian, we conclude that $Z[j]$ must also be equal to

$$Z[j_n] = \int \mathbf{D}\phi_n e^{i \int \{\mathcal{L}(M_{nm}(w_\alpha(x))\phi_m, \partial_\mu M_{nm}(w_\alpha(x))\phi_m) - j_n(x)M_{nm}(w_\alpha(x))\phi_m(x)\} d^4x}$$
$$(12.138)$$

for *arbitrary* $w_\alpha(x)$, and in particular for infinitesimal w_α, where we may replace $M_{nm} = \delta_{nm} + iw_\alpha(x)(t_\alpha)_{nm} + O(w^2)$. For infinitesimal w_α, as a consequence of (12.136), we may write the Lagrangian density appearing in the exponent of (12.138) to first order in w, with $\phi_n'(x) = \phi_n(x) + iw_\alpha(x)(t_\alpha)_{nm}\phi_m(x)$, as

$$\mathcal{L}(\phi_n', \partial_\mu\phi_n') = \mathcal{L}(\phi_n, \partial_\mu\phi_n) + i\partial_\mu w_\alpha(x)(t_\alpha)_{nm}\phi_m(x)\frac{\partial\mathcal{L}}{\partial(\partial_\mu\phi_n)} + O(w^2)$$

$$= \mathcal{L}(\phi_n, \partial_\mu\phi_n) - \partial_\mu w_\alpha(x)J_\alpha^\mu(x) + O(w^2) \qquad (12.139)$$

where we see that the Noether current J_α^μ of the global symmetry, (12.127), has re-emerged as the coefficient of the spacetime variation of the gauge parameters. Of course, the fact that the term involving the Noether current vanishes if the gauge parameters are constant is simply a restatement of the assumed exact global internal symmetry of our Lagrangian. Subtracting the two equivalent expressions (12.137) and (12.138) for $Z[j_n]$, we find, to first order in $w_\alpha(x)$,

$$\int \mathbf{D}\phi_n \{-i\int (\partial_\mu w_\alpha)J_\alpha^\mu(x)d^4x + \int w_\alpha(x)j_n(x)(t_\alpha)_{nm}\phi_m(x)\}e^{i\int\{\mathcal{L} - j_n\phi_n\}d^4x} = 0$$
$$(12.140)$$

As this identity must hold for arbitrary $w_\alpha(x)$, we may functionally differentiate with respect to w_α and obtain

$$\int \mathbf{D}\phi_n \{i\partial_\mu J_\alpha^\mu(x) + j_n(x)(t_\alpha)_{nm}\phi_m(x)\} e^{i\int \{\mathcal{L} - j_n\phi_n\} d^4 x} = 0 \qquad (12.141)$$

This result is effectively *Noether's theorem in functional form*. The Green functions of the theory are obtained by functionally differentiating with respect to the sources $j_n(x)$ (cf. (10.73)): if we apply $i^p \frac{\delta^p}{\delta j_{n_1}(y_1)\cdots\delta j_{n_p}(y_p)}$ to (12.141) and then set the sources j_n to zero, we obtain the *Ward–Takahashi identities* (in coordinate space) associated with the internal symmetry (12.126):

$$\frac{\partial}{\partial x^\mu} \langle 0|T(J_\alpha^\mu(x)\phi_{n_1}(y_1)....\phi_{n_p}(y_p))|0\rangle$$

$$= -\sum_{r=1}^{p} \delta^4(x - y_r)(t_\alpha)_{n_r m}\langle 0|T(\phi_{n_1}(y_1)..\phi_{n_{r-1}}(y_{r-1})\phi_m(x)\phi_{n_{r+1}}(y_{r+1})..\phi_{n_p}(y_p))|0\rangle$$

$$(12.142)$$

The equivalence of this expression to the operator statement of current conservation, $\partial_\mu J_\alpha^\mu = 0$, may not be immediately obvious to the reader. One sees, however, that there is no term on the right-hand side containing the four-divergence of the current *inside* the time-ordered product: just a sequence of "contact" terms in which the spacetime argument of the current is set equal in turn to the locations of each of the field operators.

In fact, the Ward–Takahashi identity (12.142) may be rederived in operator language via a short calculation (see Problem 8) using current conservation and the equal-time commutation relations (12.128) of the charge densities J_α^0 with the ϕ_n fields. One may also regard (12.142) as the *off-shell expression* of current conservation. The operator statement $\partial_\mu J_\alpha^\mu = 0$ is, of course, equivalent to the condition $\langle \beta|\partial_\mu J_\alpha^\mu|\alpha\rangle = 0$ for arbitrary multi-particle states $|\alpha\rangle, |\beta\rangle$. Such states may be generated starting from the T-product on the left-hand side of (12.142) via the LSZ formula: by reducing out each of the p fields to produce the desired initial and final-state particles. One does this (cf. Section 9.4) by Fourier transforming the matrix element (applying the factor $e^{\sum_r \pm i p_r \cdot y_r}$ and integrating over y_r), multiplying by factors $p_r^2 - m^2$ for each external-state particle, and then taking the on-mass-shell limit $p_r^2 \to m^2$. When this procedure is followed, we find that in each term on the right-hand side, the δ-function eliminates the necessary external propagator of the $\phi_{n_r}(y_r)$ field, needed to remove the vanishing factor $p_r^2 - m^2$ in the on-mass-shell limit. Accordingly, the left-hand side (giving the desired matrix element $\langle \beta|\partial_\mu J_\alpha^\mu(x)|\alpha\rangle$) is also zero after the LSZ on-shell projection is performed.

12.6 Beyond Poincaré: supersymmetry and superfields

The immediacy and clarity of the enormous phenomenological support for the symmetries of the Poincaré group, consisting of Lorentz transformations (specifically, the proper, orthochronous subgroup corresponding to physically realizable transformations) and spacetime translations, force us to incorporate these symmetries at the very foundations of the quantum field theories with which we attempt to describe the physics of the microworld. In particular, it was a basic requirement that every satisfactory field theory contain a set of conserved generators (self-adjoint operators in the Hilbert space of the theory) $\mathcal{M}^{\mu\nu}, P^\rho$ satisfying the algebra of the Poincaré group (cf. (9.65–9.67)):

$$[\mathcal{M}^{\mu\nu}, \mathcal{M}^{\rho\sigma}] = i(g^{\mu\sigma}\mathcal{M}^{\rho\nu} + g^{\nu\sigma}\mathcal{M}^{\mu\rho} - g^{\rho\mu}\mathcal{M}^{\sigma\nu} - g^{\rho\nu}\mathcal{M}^{\mu\sigma}) \qquad (12.143)$$

$$[P^\rho, \mathcal{M}^{\mu\nu}] = i(g^{\rho\nu}P^\mu - g^{\rho\mu}P^\nu) \qquad (12.144)$$

$$[P^\rho, P^\sigma] = 0 \qquad (12.145)$$

For the first fifty years of quantum field theory, until the mid-1970s, the Poincaré symmetries were thought to represent the maximal set of spacetime symmetries: in other words, it was implicitly assumed that one could not expand the algebra (12.143–12.145) consistently by introducing further generators with non-trivial commutation relations with the $\mathcal{M}^{\mu\nu}, P^\rho$. This prejudice was reinforced by the famous "no-go" theorem of Coleman and Mandula (Coleman and Mandula, 1967), which showed that the Poincaré algebra was indeed maximal in this sense, on the basis of assumptions which seemed unexceptionable at the time. An important implicit assumption was that the generators of the algebra were bosonic in character: acting on bosonic states, they produced bosonic states, and on fermionic ones, fermionic states. The relaxation of this last assumption is the critical step in allowing the existence of supersymmetry (SUSY) algebras which expand the original Poincaré algebra stated above. The simplest possible extension of the Poincaré algebra turns out[13] to involve the introduction of a Majorana 4-spinor set of generators Q_α ($\alpha = 1, 2, 3, 4$)

$$\begin{pmatrix} C_s Q^* \\ Q \end{pmatrix} \qquad (12.146)$$

where $Q_a, a = 1, 2$ is a $(0, \frac{1}{2})$ 2-spinor and C_s the 2-spinor conjugation matrix (7.40) (see Section 7.4.3, especially footnote 5). The (anti)commutation relations of the new spinorial generators are

$$\{Q_\alpha, \bar{Q}_\beta\} = 2(\gamma_\mu P^\mu)_{\alpha\beta} \qquad (12.147)$$

$$[P^\mu, Q_\alpha] = [P^\mu, \bar{Q}_\alpha] = 0 \qquad (12.148)$$

In effect, the Q_α generators can be thought of as providing us with a "square root" of the energy-momentum four-vector! In addition, there are the commutation relations of the Q_α with the $\mathcal{M}^{\mu\nu}$, which simply express the fact that the Majorana 4-spinor transforms appropriately under the HLG: we shall not use these further, and do not give them explicitly here. Altogether, the algebra (12.147, 12.148) taken together with the Poincaré algebra expressed by (12.143, 12.144, 12.145) (and the commutator of the Q_α with the $\mathcal{M}^{\mu\nu}$) constitute a graded Lie algebra (in this case, the *super-Poincaré* algebra)—one involving both generators of bosonic and fermionic type, and in which commutators involving one (resp. two) bosonic generators lead to a linear combination

---

[13] The most general form of the supersymmetry algebra was first derived by by Haag, Lopuszanski, and Sohnius, (Haag *et al.*, 1975): see below.

of generators of fermionic (resp. bosonic) type, while the anticommutators of the fermionic generators are expressible as a linear combination of bosonic generators.

Now, it is hardly obvious that the assortment of commutation and anticommutation rules obeyed by the generators $\mathcal{M}^{\mu\nu}, P^\rho, Q_\alpha$, as given above, are even mathematically consistent. Their consistency can be established by giving an explicit realization, and we shall do so by considering the simplest possible field theory exhibiting the supersymmetry algebra given here. Consider a theory consisting of a free massless complex scalar field ϕ together with a free massless Majorana field ψ (corresponding to a self-conjugate massless fermion). There are two massless bosonic degrees of freedom (we can write the complex field $\phi = \frac{1}{\sqrt{2}}(A + iB)$, where A and B are independent real scalar fields), and likewise two massless fermionic degrees of freedom (for the two spin states of the massless fermion). The Lagrangian is (see (7.127) for the origin of the one-half in the fermionic part)[14]

$$\mathcal{L} = \partial_\mu \phi^* \partial^\mu \phi + \frac{i}{2}\bar\psi \slashed\partial \psi \tag{12.149}$$

with the standard Noether expressions for the energy-momentum operators

$$P^0 = \int d^3x \{\dot\phi^* \dot\phi + \vec\nabla\phi^* \cdot \vec\nabla\phi + \frac{i}{2}\bar\psi\vec\gamma \cdot \vec\nabla\psi)$$

$$\vec P = \int d^3x \{\dot\phi^* \vec\nabla\phi + \dot\phi\vec\nabla\phi^* + \frac{i}{2}\psi^\dagger \vec\nabla\psi\} \tag{12.150}$$

The fields satisfy the equal-time (anti)commutation relations:

$$[\dot\phi(\vec x, t), \phi^*(\vec y, t)] = [\dot\phi^*(\vec x, t), \phi(\vec y, t)] = -i\delta^3(\vec x - \vec y) \tag{12.151}$$

$$\{\psi_\alpha(\vec x, t), \bar\psi_\beta(\vec y, t)\} = (\gamma^0)_{\alpha\beta}\delta^3(\vec x - \vec y) \tag{12.152}$$

$$\{\psi_\alpha(\vec x, t), \psi_\beta(\vec y, t)\} = (i\gamma^2)_{\alpha\beta}\delta^3(\vec x - \vec y) \tag{12.153}$$

The unusual form of the anticommutation relation (12.153) arises because of the Majorana property of ψ (namely, $\psi^* = i\gamma^2\psi$, implying the equivalence of (12.152) and (12.153)). The equations of motion $\Box\phi = 0, \slashed\partial\psi = 0$ imply conservation (i.e., zero divergence) of the *fermionic* current

$$J^\mu = \sqrt{2}\{(\slashed\partial\phi)\gamma^\mu\psi_R + (\slashed\partial\phi^*)\gamma^\mu\psi_L\}, \quad \partial_\mu J^\mu = 0 \tag{12.154}$$

where $\psi_L = P_L\psi = \frac{1+\gamma_5}{2}\psi$, $\psi_R = P_R\psi = \frac{1-\gamma_5}{2}\psi$ are the upper and lower 2-spinor components of ψ respectively. The current J^μ gives rise in the usual way to an associated conserved charge Q_α

$$Q_\alpha = \sqrt{2}\int d^3x(\slashed\partial\phi(\vec x, t)\gamma^0 P_R + \slashed\partial\phi^*(\vec x, t)\gamma^0 P_L)_{\alpha\beta}\psi_\beta(\vec x, t) \tag{12.155}$$

---

[14] Notational alert: In SUSY, it is conventional to use the $*$ symbol for both complex conjugation (of complex and Grassmann numbers), as well as hermitian conjugation (of operators). We shall adhere to this policy throughout this section.

These charges can be regarded as infinitesimal generators of a symmetry of the theory, as follows. In order to obtain the variation in the fields from a commutator (in both the bosonic and fermionic case) we introduce an infinitesimal Majorana 4-spinor ξ_α whose components are *constant Grassmann numbers*, commuting with the scalar field ϕ but anticommuting with the fermionic field ψ. As ξ is infinitesimal, we work only to first order in it, and define the variation of the fields under an infinitesimal SUSY transformation as[15]

$$\delta_\xi\phi(x) \equiv \frac{1}{\sqrt{2}}[\bar\xi Q, \phi(x)], \quad \delta_\xi\psi \equiv \frac{1}{\sqrt{2}}[\bar\xi Q, \psi(x)] \tag{12.156}$$

Using the (anti)commutation relations given above, one finds explicitly

$$\delta_\xi\phi(x) = -i\bar\xi P_L\psi(x)$$

$$\delta_\xi\phi^*(x) = -i\bar\xi P_R\psi(x)$$

$$\delta_\xi\psi(x) = -(\not\partial\phi(x)P_R + \not\partial\phi^*(x)P_L)\xi$$

$$\delta_\xi\bar\psi(x) = \bar\xi(\not\partial\phi(x)P_L + \not\partial\phi^*(x)P_R) \tag{12.157}$$

Notice that the supersymmetry transformation has interchanged bosonic and fermionic fields, so that we must expect a very specific balance between the bosonic and fermionic fields appearing in the Lagrangian if the theory is actually to be invariant under such a transformation. In particular, the first-order (in ξ) variation of the Lagrangian density (12.149) under these variations of the fields is found (see Problem 9) to be a pure divergence

$$\delta_\xi\mathcal{L} = -\frac{i}{2}\partial_\mu\{\bar\xi\gamma^\mu(\not\partial\phi(x)P_R + \not\partial\phi^*(x)P_L)\psi(x)\} \tag{12.158}$$

The *total* action of the theory, defined as the integral over all spacetime of $\mathcal{L}$ (with all fields assumed to vanish at infinity), is therefore invariant under the infinitesimal SUSY transformations generated by the Q_α. However, as discussed in the preceding Section, the construction of the Noether current in this case requires an additional term (cf. (12.82), also Problem 10). A further straightforward calculation shows (see Problem 11) that the generators Q_α given in (12.155) satisfy precisely the remarkable anticommutation relations (12.147). The fact that Q_α commutes with P^μ follows from the conservation of Q_α (for $\mu = 0$) and the fact that the Q_α is the spatial integral of a density (for μ spatial).

The preceding discussion would amount to little more than a mathematical curiosity were it not possible to realize the super-Poincaré algebra in an interacting theory. In fact, the generalization to non-zero mass and non-trivial interactions turns out to be quite straightforward, once the appropriate technical machinery is in place. Apart from some algebraic wizardry involving Grassmann Majorana spinors (see Appendix C), the important simplifying device turns out to be the introduction of superspace, an abstract extension of the four spacetime "bosonic" dimensions by a further four dimensions with Grassmannian (anticommuting) coordinates. The use of superspace

---

[15] We remove the annoying $\sqrt{2}$ here, inserted for later convenience in normalizing the charges.

allows us to interpret the supersymmetry transformation of the fields introduced above in a completely *ad hoc* fashion in an intuitively natural and geometrically motivated way—to the extent that this is possible working in a space with anticommuting coordinates! The remaining subsections of this Chapter constitute an all too brief introduction to the essential ideas of supersymmetry. For further details, the reader is referred to any of the many excellent texts on supersymmetry (for example, (Weinberg, 1995*b*)).

12.6.1 The homogeneous Lorentz group and SL(2,C)

The homogeneous Lorentz group (HLG), which we first introduced as the group with fundamental representation consisting of 4x4 matrices satisfying

$$\Lambda^\mu_{\;\nu}\Lambda^\rho_{\;\mu} = \delta^\rho_\nu \tag{12.159}$$

is locally isomorphic to the group SL(2,C) of unimodular (i.e., unit determinant) complex 2x2 matrices. Of course, SU(2) is contained in SL(2,C), so we know SL(2,C) is at least large enough to contain the spatial rotations. A general element of SL(2,C) can be written

$$\lambda \equiv \begin{pmatrix} \alpha & \beta \\ \gamma & \delta \end{pmatrix}$$

where $\alpha, \beta, \gamma, \delta$ are complex numbers satisfying $\alpha\delta - \beta\gamma = 1$. Let $\sigma_\mu, \mu = 0, 1, 2, 3$ be the extended set of 2x2 σ-matrices, where $\sigma_0 \equiv 1$ and $\sigma_i, i = 1, 2, 3$ are the usual Pauli matrices. From any four-vector p^μ we can then construct an hermitian 2x2 matrix

$$P \equiv p^\mu\sigma_\mu, \quad \det(P) = p^\mu p_\mu = p^2 \tag{12.160}$$

Under the similarity transformation

$$P \to P' \equiv \lambda P \lambda^\dagger \tag{12.161}$$

and clearly $\det(P') = \det(P)$ so the four-vector p' corresponding to P' is a Lorentz transform of p, with

$$\lambda(p^\mu\sigma_\mu)\lambda^\dagger = \Lambda^\nu_{\;\mu}\sigma_\nu p^\mu \tag{12.162}$$

$$= p'^\nu\sigma_\nu \tag{12.163}$$

The group SL(2,C) is six-dimensional: four complex numbers contain eight (real) degrees of freedom, but the two constraints setting the real part of $\alpha\delta - \beta\gamma$ to 1 and the imaginary part to zero reduce the dimensionality to 6, the correct number for the HLG. Infinitesimally, we can write

$$\lambda = 1 + (\frac{i}{4}\epsilon_{ijk}\omega^{ij} + \frac{1}{2}\omega^{0k})\sigma_k \tag{12.164}$$

corresponding to

$$\Lambda^\mu_{\;\nu} = g^\mu_{\;\nu} + \omega^\mu_{\;\nu} \tag{12.165}$$

in the fundamental representation of HLG.

The fundamental representation of SL(2,C) is a complex 2-spinor $Q_a, a = 1, 2$:

$$Q = \begin{pmatrix} Q_1 \\ Q_2 \end{pmatrix}$$

which transforms under SL(2,C) as

$$Q \to \lambda Q \tag{12.166}$$

Suppose we manage to find a pair of operators Q_a on the state space, with the transformation under HLG

$$U^\dagger(\Lambda)Q_a U(\Lambda) = \lambda_{ab} Q_b \tag{12.167}$$

Taking Λ, λ infinitesimal, this becomes

$$-\frac{i}{2}\omega_{\mu\nu}[J^{\mu\nu}, Q_a] = \frac{i}{4}\epsilon_{ijk}\omega^{ij}(\sigma_k)_{ab}Q_b + \frac{1}{2}\omega^{0k}(\sigma_k)_{ab}Q_b \tag{12.168}$$

implying the commutation relations (rotations)

$$[J_{ij}, Q_a] = -\frac{1}{2}\epsilon_{ijk}(\sigma_k)_{ab}Q_b \tag{12.169}$$

and (boosts)

$$[J_{k0}, Q_a] = -\frac{i}{2}(\sigma_k)_{ab}Q_b \tag{12.170}$$

In the $\vec{J}, \vec{K}$ notation introduced in Section 7.2, with $J_1 = J_{32}, K_1 = J_{10}$, etc., things are even simpler:

$$[J_i, Q_a] = \frac{1}{2}(\sigma_i)_{ab}Q_b \tag{12.171}$$

$$[K_i, Q_a] = -\frac{i}{2}(\sigma_i)_{ab}Q_b \tag{12.172}$$

Finally, recalling the (A,B) notation (cf. Section 7.2) where we define generators $\vec{A} \equiv \frac{1}{2}(\vec{J} - i\vec{K})$, $\vec{B} \equiv \frac{1}{2}(\vec{J} + i\vec{K})$, these become

$$[A_i, Q_a] = 0 \tag{12.173}$$

$$[B_i, Q_a] = \frac{1}{2}(\sigma_i)_{ab}Q_b \tag{12.174}$$

so the 2-spinor Q_a corresponds to what we previously (in Chapter 7) called the $(0, \frac{1}{2})$ representation of the HLG. The reader will recall that such a spinor is conventionally written as the *lower* half of a Dirac 4-spinor.

An important notational convention: In SUSY, it is conventional to use the $*$ *symbol for both complex conjugation (of complex and Grassmann numbers), as well as* *hermitian conjugation (of operators)—the dagger symbol is reserved for column vectors and matrices of operators. Moreover, the* ∗ *symbol applied to products of Grassmann objects is defined to reverse the order, in analogy to the property of hermitian adjoints of operators.*

The conjugation matrix $C_s \equiv i\sigma_2$ has the property

$$C_s \vec{\sigma}^* = -\vec{\sigma} C_s \tag{12.175}$$

Accordingly,

$$[J_i, (C_s)_{ab} Q_b^*] = -(C_s)_{ab}[J_i, Q_b]^* \tag{12.176}$$

$$= -\frac{1}{2}(C_s)_{ab}(\sigma_i^*)_{bc} Q_c^* \tag{12.177}$$

$$= \frac{1}{2}(\sigma_i)_{ab}(C_s)_{bc} Q_c^* \tag{12.178}$$

and

$$[K_i, (C_s)_{ab} Q_b^*] = -(C_s)_{ab}[K_i, Q_b]^* \tag{12.179}$$

$$= -\frac{i}{2}(C_s)_{ab}(\sigma_i^*)_{bc} Q_c^* \tag{12.180}$$

$$= \frac{i}{2}(\sigma_i)_{ab}(C_s)_{bc} Q_c^* \tag{12.181}$$

so $(C_s)_{ab} Q_b^*$ commutes with B_i, i.e., is in the $(\frac{1}{2},0)$ representation of the HLG. This means that we can construct a 4-spinor $Q_\alpha, \alpha = 1, 2, 3, 4$ (for which we confusingly use the same letter Q, as the 2-spinors will shortly disappear) with the $(\frac{1}{2}, 0) \oplus (0, \frac{1}{2})$ transformation properties of the Dirac 4-spinor by taking $Q_a, a = 1, 2$ as the lower two components of the 4-spinor and $(C_s)_{ab} Q_b^*, a = 1, 2$ as the upper two components. If the components of such a spinor are anticommuting c-numbers, or fermionic fields, we call such an object a *Grassmann Majorana spinor*.

Note that the unimodularity of SL(2,C) implies that $\lambda^T C_s \lambda = \lambda$ whence

$$(\lambda Q)_a (C_s)_{ab} (\lambda Q)_b = Q_a (C_s)_{ab} Q_b \tag{12.182}$$

so $Q C_s Q$ is a scalar: the C_s-matrix can be used to couple two $(\frac{1}{2},0)$ reps (or two $(0,\frac{1}{2})$ reps) to a Lorentz scalar.

In the forthcoming sections we shall be needing a number of simple algebraic properties of Grassmann Majorana spinors, which are defined and studied in Appendix C. The relevant results are gathered there, and we strongly recommend that the reader spend a few minutes at this point in gaining some familiarity with the essential properties of these objects, the basic ingredients from which we construct supersymmetric theories.

12.6.2   The supersymmetry algebra

The translation and Lorentz generators which together constitute the Lie algebra of the Poincaré group satisfy the commutator algebra stated in (9.65-9.67). Under the

HLG, the energy-momentum operator P^μ transforms as a four-vector, i.e., as a $(\frac{1}{2}, \frac{1}{2})$ representation, while the antisymmetric second rank tensor $\mathcal{M}^{\mu\nu}$ is a member of the reducible representation $(1,0)\oplus(0,1)$. Note that the simplest nontrivial representations of the HLG, the spinorial $(\frac{1}{2},0)$ and $(0,\frac{1}{2})$, do not appear in the Poincaré algebra—a deficiency (if we wish to regard it as such) which was eliminated by the introduction of supersymmetry.

In 1975, Haag, Lopuszanski, and Sohnius (Haag *et al.*, 1975) derived the most general extension of the Poincaré algebra, in which Grassmannian spinorial generators, which can be thought of as roughly as "square roots of P_μ", are introduced. If $Q_a \in (0,\frac{1}{2})$, then we saw that $C_s Q^* \in (\frac{1}{2},0)$. Recalling that the four-vector momentum P_μ transforms in the $(\frac{1}{2}, \frac{1}{2})$ representation of the HLG, it is at least a possibility that the anticommutator of Q and Q^* can be proportional to P_μ:

$$\{Q_a, Q_b^*\} = 2\sigma_{ab}^\mu P_\mu \qquad (12.183)$$

$$[P_\mu, Q_a] = [P_\mu, Q_a^*] = 0 \qquad (12.184)$$

$$\{Q_a, Q_b\} = \{Q_a^*, Q_b^*\} = 0 \qquad (12.185)$$

In the above, the indices a, b run over the values $(1,2)$: Q_a, Q_a^* are *2-spinors*. In fact, the Haag–Lopuszanski–Sohnius theorem allows for an even more general algebra with N independent Grassmannian generators, $Q_{ar}, a = 1, 2, r = 1, 2, \dots N$, with an algebra

$$\{Q_{ar}, Q_{bs}^*\} = 2\delta_{rs}\sigma_{ab}^\mu P_\mu \qquad (12.186)$$

$$\{Q_{ar}, Q_{bs}\} = (C_s)_{ab}Z_{rs} \qquad (12.187)$$

where the Z_{rs} commute with everything and are called "central charges". This algebra is called "N-extended" supersymmetry and leads to consistent field theories for $1 \le N \le 8$. The case of $N = 1$ ("simple supersymmetry") is of the greatest phenomenological importance, and is the only one we shall consider in our brief introduction to SUSY.[16]

The N=1 SUSY algebra, Eqs. (12.183, 12.184, 12.185), is more frequently written in a four-component Dirac notation: as described previously, we group the generators Q_a, Q_a^* into a single Majorana 4-spinor $Q_\alpha, \alpha = 1, 2, 3, 4$ (see (12.146)) and

$$\bar{Q}_\beta \to (Q^{*T}, -Q^T C_s) \qquad (12.188)$$

The basic SUSY anticommutation relations, Eqs. (12.183, 12.185), can now be expressed as a single equation:

$$\{Q_\alpha, \bar{Q}_\beta\} = \begin{pmatrix} 0 & -C_s\{Q^*, Q\}C_s \\ \{Q, Q^*\} & 0 \end{pmatrix}_{\alpha\beta}$$

$$= \begin{pmatrix} 0 & -2C_s(\sigma_\mu P^\mu)^T C_s \\ 2\sigma_\mu P^\mu & 0 \end{pmatrix}_{\alpha\beta} \qquad (12.189)$$

---

[16] The derivation of Eqs. (12.186, 12.187) is given in full in Weinberg (Weinberg, 1995*b*), Chapter 2.

Recalling $C_s = i\sigma_2$ and $C_s\vec{\sigma}^T C_s = \vec{\sigma}$, $C_s\sigma_0 C_s = -\sigma_0$,

$$\{Q_\alpha, \bar{Q}_\beta\} = 2P^0 \begin{pmatrix} 0 & 1 \\ 1 & 0 \end{pmatrix}_{\alpha\beta} + 2P^i \begin{pmatrix} 0 & -\sigma_i \\ \sigma_i & 0 \end{pmatrix}_{\alpha\beta} \tag{12.190}$$

giving finally

$$\{Q_\alpha, \bar{Q}_\beta\} = 2(\gamma_\mu P^\mu)_{\alpha\beta} \tag{12.191}$$

together with

$$[P^\mu, Q_\alpha] = [P^\mu, \bar{Q}_\alpha] = 0 \tag{12.192}$$

so we have recovered the supersymmetry algebra of the fermionic generators Q_α introduced in an *ad hoc* fashion at the beginning of the section.

The Majorana bispinor $Q_\alpha \in (\frac{1}{2}, 0) \oplus (0, \frac{1}{2})$ under HLG, so acting on any one-particle state of spin j and momentum p_μ we must get

$$Q_\alpha |\vec{p}, j> \to |\vec{p}, j \pm \frac{1}{2}> \tag{12.193}$$

In particular, if $\vec{p} = 0$ (particle of mass m at rest, $H|\vec{p} = 0, j >= m|\vec{p} = 0, j >$), Eq. (12.192) implies

$$HQ_\alpha|\vec{p} = 0, j >= Q_\alpha H|\vec{p} = 0, j >= mQ_\alpha|\vec{p} = 0, j > \tag{12.194}$$

implying the existence of mass degenerate pairs of particles ("superpartners") with spins differing by $\frac{1}{2}$, and hence of opposite statistics (fermions mass degenerate with bosons), assuming the validity of the Spin-Statistics theorem. The toy theory studied earlier of a free massless (complex) scalar boson and a free massless spin-$\frac{1}{2}$ Majorana particle is merely the simplest possible example of such a situation. We must now face the task of devising an efficient machinery for constructing theories with interacting particles and non-zero mass and with an action invariant under the super-Poincaré group generated by the Q_α together with the usual Poincaré generators P^μ and $M^{\mu\nu}$.

12.6.3   Superfields

The simplest way to construct field theories with supersymmetric invariant Lagrangians is to introduce a Grassmannian extension of ordinary spacetime:

$$(x^\mu) \to (x^\mu, \theta_\alpha) \quad \text{(superspace)} \tag{12.195}$$

and to think of the SUSY generators Q_α as the Grassmann analogs of P_μ, generating translations in "θ-space", as P_μ does in x^μ space. Fields are now viewed as functions both of x^μ and θ_α, and a general field can be expanded as a *polynomial* in the θ_α of degree four or less. Schematically:

$$S(x, \theta) = \sum S_n(x)\theta^n, \quad n \le 4 \tag{12.196}$$

If the superfield S has overall bosonic character, then the coefficient fields S_n will be bosonic for n even and fermionic for n odd. If (as we shall take here) the leading term (no θ's) is a scalar bosonic field, we call $S(x, \theta)$ a "scalar superfield". So a superfield is a handy way of grouping together bosonic and fermionic fields in a "supermultiplet". We need the analog of the bosonic spacetime (infinitesimal) translation property

$$[\epsilon^\mu P_\mu, \phi(x)] = -i\epsilon^\mu \partial_\mu \phi(x) = -i(\phi(x + \epsilon) - \phi(x)) \tag{12.197}$$

This property connects the Hilbert space four-momentum operators P_μ to a differential operator $-i\partial_\mu$ acting on the fields. In the same way, we seek differential operators (now involving Grassmann derivatives in superspace) which will be equivalent to commutation with the Hilbert space operators Q_α. We must ensure, of course, that the differential operators we construct satisfy the SUSY algebra (12.191–12.192). From the Majorana identity (C.5) (Appendix C), we know that $\theta_\gamma = -\bar{\theta}_\alpha(\gamma_5 \epsilon)_{\alpha\gamma}$, with ϵ the 4x4 matrix $\text{diag}(C_s, C_s)$, so the following operator (of overall fermionic character)

$$K_\alpha \equiv (\gamma_5 \epsilon)_{\alpha\gamma} \frac{\partial}{\partial \theta_\gamma} - i\gamma_{\alpha\gamma}^\mu \theta_\gamma \partial_\mu \tag{12.198}$$

can equivalently be written

$$K_\alpha = -\frac{\partial}{\partial \bar{\theta}_\alpha} - i(\gamma^\mu \theta)_\alpha \partial_\mu \tag{12.199}$$

The Dirac adjoint is

$$\bar{K}_\beta = K_\gamma (\gamma_5 \epsilon)_{\gamma\beta} \tag{12.200}$$

$$= (\gamma_5 \epsilon)_{\gamma\delta} \frac{\partial}{\partial \theta_\delta} (\gamma_5 \epsilon)_{\gamma\beta} - i\gamma_{\gamma\delta}^\mu \theta_\delta (\gamma_5 \epsilon)_{\gamma\beta} \partial_\mu \tag{12.201}$$

$$= \frac{\partial}{\partial \theta_\beta} + i(\gamma_5 \epsilon \gamma^\mu)_{\beta\gamma} \theta_\gamma \partial_\mu \tag{12.202}$$

The anticommutator algebra of the K and $\bar{K}$ is now easily computed:

$$\{K_\alpha, \bar{K}_\beta\} = \{(\gamma_5 \epsilon)_{\alpha\gamma} \frac{\partial}{\partial \theta_\gamma}, i(\gamma_5 \epsilon \gamma^\mu)_{\beta\delta} \theta_\delta \partial_\mu\} + \{-i\gamma_{\alpha\gamma}^\mu \theta_\gamma \partial_\mu, \frac{\partial}{\partial \theta_\beta}\} \tag{12.203}$$

$$= -i(\gamma_5 \epsilon \gamma^\mu \gamma_5 \epsilon)_{\beta\alpha} \partial_\mu - i\gamma_{\alpha\beta}^\mu \partial_\mu \tag{12.204}$$

$$= -2i\gamma_{\alpha\beta}^\mu \partial_\mu \tag{12.205}$$

This establishes that the K_α have the same algebra as the Q_α, in the sense that, given a superfield with the property

$$[Q_\alpha, S(x, \theta)] = -iK_\alpha S(x, \theta) \tag{12.206}$$

(from which follows $[\bar{Q}_\beta, S(x,\theta)] = [Q_\gamma(\gamma_5\epsilon)_{\gamma\beta}, S] = -i\bar{K}_\beta S)$, then the SUSY algebra (12.191) is properly realized via

$$
\begin{aligned}
[\{Q_\alpha, \bar{Q}_\beta\}, S(x,\theta)] &= Q_\alpha[\bar{Q}_\beta, S] + [Q_\alpha, S]\bar{Q}_\beta + \bar{Q}_\beta[Q_\alpha, S] + [\bar{Q}_\beta, S]Q_\alpha \\
&= -iQ_\alpha(\bar{K}_\beta S) - i(K_\alpha S)\bar{Q}_\beta - i\bar{Q}_\beta(K_\alpha S) - i(\bar{K}_\beta S)Q_\alpha \\
&= i\bar{K}_\beta[Q_\alpha, S] + iK_\alpha[\bar{Q}_\beta, S] \\
&= \{\bar{K}_\beta, K_\alpha\}S \\
&= -2i\gamma^\mu_{\alpha\beta}\partial_\mu S \\
&= 2\gamma^\mu[P_\mu, S] \qquad\qquad\qquad\qquad\qquad\qquad (12.207)
\end{aligned}
$$

A glance at (12.198) and (12.202) shows that $\{K_\alpha, K_\beta\} = 0 = \{\bar{K}_\alpha, \bar{K}_\beta\}$. Moreover, a *covariant (superspace) derivative* can be defined as follows, by a simple change of sign from (12.198):

$$
\mathcal{D}_\alpha \equiv (\gamma_5\epsilon)_{\alpha\gamma}\frac{\partial}{\partial\theta_\gamma} + i\gamma^\mu_{\alpha\gamma}\theta_\gamma\partial_\mu \qquad\qquad (12.208)
$$

The change of sign relative to the definition of the K_α implies the anticommutation relations

$$
\{K_\alpha, \mathcal{D}_\beta\} = \{K_\alpha, \bar{\mathcal{D}}_\beta\} = 0 \qquad\qquad (12.209)
$$

$$
\{\mathcal{D}_\alpha, \bar{\mathcal{D}}_\beta\} = +2i\gamma^\mu_{\alpha\beta}\partial_\mu \qquad\qquad (12.210)
$$

Covariant superspace derivatives are useful, as they allow us to construct new superfields by applying SUSY-invariant constraints to the basic scalar superfield. For example, if we impose the constraint

$$
\mathcal{D}_\alpha S(x,\theta) = 0 \qquad\qquad (12.211)
$$

on a superfield S, then this constraint will be preserved under an infinitesimal SUSY transformation generated by (12.206), as a consequence of (12.209). This will be important later in the construction of the so-called "chiral superfields".

12.6.4 Transformation properties of components of scalar superfields

The rather schematic expansion of a general scalar superfield given in (12.196) can be made more explicit. Expanding $S(x,\theta)$ to fourth (the maximum) order in the θ_α,

$$
\begin{aligned}
S(x,\theta) = {}& C(x) - i\bar{\theta}\gamma_5\omega(x) - \frac{i}{2}(\bar{\theta}\gamma_5\theta)M(x) \\
& - \frac{1}{2}(\bar{\theta}\theta)N(x) - \frac{1}{2}\bar{\theta}\gamma_5\gamma_\mu\theta V^\mu(x) \\
& - i(\bar{\theta}\gamma_5\theta)\,\bar{\theta}(\lambda(x) - \frac{i}{2}\not{\partial}\omega(x)) - \frac{1}{4}(\bar{\theta}\gamma_5\theta)^2(D(x) - \frac{1}{2}\Box C(x)) \quad (12.212)
\end{aligned}
$$

where $C(x), D(x), M(x), N(x)$ and $V^\mu(x)$ are bosonic fields while the fields $\lambda(x)$ and $\omega(x)$ are fermionic (Majorana) 4-spinors. The peculiar combinations used to define the $D(x)$ and $\lambda(x)$ terms are chosen to simplify the transformation laws for these fields, as we shall soon see. In analogy to (12.197), an infinitesimal SUSY transformation of S is generated by an infinitesimal Grassmann "translation" in superspace ξ_α as follows:

$$\delta S = \bar{\xi}_\alpha K_\alpha S = \bar{\xi}(-\frac{\partial}{\partial\bar{\theta}} - i\gamma^\mu\theta\partial_\mu)S(x,\theta)$$

$$= \bar{\xi}\{i\gamma_5\omega + i\gamma_5\theta M + \theta N + \gamma_5\gamma_\mu\theta V^\mu + 2i(\gamma_5\theta)\bar{\theta}(\lambda - \frac{i}{2}\slashed{\partial}\omega)$$

$$+i(\bar{\theta}\gamma_5\theta)(\lambda - \frac{i}{2}\slashed{\partial}\omega) + \gamma_5\theta(\bar{\theta}\gamma_5\theta)(D - \frac{1}{2}\Box C)\}$$

$$- i(\bar{\xi}\gamma^\mu\theta)\{\partial_\mu C - i\bar{\theta}\gamma_5\partial_\mu\omega - \frac{i}{2}(\bar{\theta}\gamma_5\theta)\partial_\mu M$$

$$-\frac{1}{2}\bar{\theta}\theta\partial_\mu N - \frac{1}{2}\bar{\theta}\gamma_5\gamma_\nu\theta\partial_\mu V^\nu - i(\bar{\theta}\gamma_5\theta)\bar{\theta}(\partial_\mu\lambda - \frac{i}{2}\partial_\mu\slashed{\partial}\omega)\} \qquad (12.213)$$

The term with no θs corresponds to the change in the $C(x)$ field, so we immediately can read off

$$\delta C(x) = i\bar{\xi}\gamma_5\omega(x) \qquad (12.214)$$

Notice that the SUSY transformation has turned a bosonic scalar field C into a fermionic spinor field ω. Next, terms with a single θ:

$$-i\bar{\theta}\gamma_5\delta\omega(x) = \bar{\xi}(-i\slashed{\partial}C(x) + i\gamma_5 M(x) + N(x) + \gamma_5 V(x))\theta$$

$$= \bar{\theta}(i\slashed{\partial}C(x) + i\gamma_5 M(x) + N(x) + \gamma_5 V(x))\xi \qquad (12.215)$$

whence

$$\delta\omega(x) = (-\gamma_5\slashed{\partial}C(x) - M(x) + i\gamma_5 N(x) + iV(x))\xi \qquad (12.216)$$

To deal with the terms with two θs we will need a Fierz rearrangement theorem—(C.26) from Appendix C:

$$\theta_\alpha\bar{\theta}_\beta = -\frac{1}{4}\delta_{\alpha\beta}\bar{\theta}\theta + \frac{1}{4}(\gamma_5\gamma_\mu)_{\alpha\beta}\bar{\theta}\gamma_5\gamma^\mu\theta - \frac{1}{4}(\gamma_5)_{\alpha\beta}\bar{\theta}\gamma_5\theta \qquad (12.217)$$

We can now rearrange the terms with two θs in (12.213) as follows:

$$2i(\bar{\xi}\gamma_5\theta)\bar{\theta}(\lambda - \frac{i}{2}\slashed{\partial}\omega) + i(\bar{\theta}\gamma_5\theta)\bar{\xi}(\lambda - \frac{i}{2}\slashed{\partial}\omega) - (\bar{\xi}\gamma^\mu\theta)\bar{\theta}\gamma_5\partial_\mu\omega$$

$$= 2i(-\frac{1}{4})\bar{\xi}\gamma_5(\lambda - \frac{i}{2}\slashed{\partial}\omega)\bar{\theta}\theta + 2i(\frac{1}{4})\bar{\xi}\gamma_5\gamma_5\gamma_\mu(\lambda - \frac{i}{2}\slashed{\partial}\omega)\bar{\theta}\gamma_5\gamma^\mu\theta$$

$$+ 2i(-\frac{1}{4})\bar{\xi}\gamma_5\gamma_5(\lambda - \frac{i}{2}\slashed{\partial}\omega)\bar{\theta}\gamma_5\theta + i\bar{\xi}(\lambda - \frac{i}{2}\slashed{\partial}\omega)\bar{\theta}\gamma_5\theta$$

$$+ \frac{1}{4}\bar{\xi}\gamma^\mu\gamma_5\partial_\mu\omega\bar{\theta}\theta - \frac{1}{4}\bar{\xi}\gamma^\mu\gamma_5\gamma_\nu\gamma_5\partial_\mu\omega(\bar{\theta}\gamma_5\gamma^\nu\theta) + \frac{1}{4}\bar{\xi}\gamma^\mu\gamma_5\gamma_5\partial_\mu\omega\bar{\theta}\gamma_5\theta$$

$$= -\frac{i}{2}\bar{\theta}\theta\,\bar{\xi}\gamma_5(\lambda - i\partial\!\!\!/\omega)$$

$$+ i\bar{\theta}\gamma_5\theta\{-\frac{1}{2}\bar{\xi}(\lambda - \frac{i}{2}\partial\!\!\!/\omega) + \bar{\xi}(\lambda - \frac{i}{2}\partial\!\!\!/\omega) - \frac{i}{4}\bar{\xi}\partial\!\!\!/\omega\}$$

$$+ i\bar{\theta}\gamma_5\gamma^\mu\theta\{\frac{1}{2}\bar{\xi}\gamma_\mu\lambda - \frac{i}{4}\bar{\xi}\gamma_\mu\partial\!\!\!/\omega + \frac{i}{4}\bar{\xi}\gamma^\nu\gamma_5\gamma_\mu\gamma_5\partial_\nu\omega\}$$

$$= -\frac{i}{2}\bar{\theta}\theta\,\bar{\xi}\gamma_5(\lambda - i\partial\!\!\!/\omega) \quad (\rightarrow -\frac{1}{2}\bar{\theta}\theta\delta N) \tag{12.218}$$

$$+ \frac{i}{2}\bar{\theta}\gamma_5\theta\bar{\xi}(\lambda - i\partial\!\!\!/\omega) \quad (\rightarrow -\frac{i}{2}\bar{\theta}\gamma_5\theta\delta M) \tag{12.219}$$

$$+ \frac{i}{2}\bar{\theta}\gamma_5\gamma^\mu\theta\{\bar{\xi}\gamma_\mu\lambda - \frac{i}{2}\bar{\xi}(\gamma^\nu\gamma_\mu + \gamma_\mu\gamma^\nu)\partial_\nu\omega\} \quad (\rightarrow -\frac{1}{2}\bar{\theta}\gamma_5\gamma_\mu\theta\delta V^\mu) \tag{12.220}$$

(12.218–12.220) lead immediately to the desired transformation rules

$$\delta M(x) = -\bar{\xi}(\lambda(x) - i\partial\!\!\!/\omega(x)) \tag{12.221}$$

$$\delta N(x) = i\bar{\xi}\gamma_5(\lambda(x) - i\partial\!\!\!/\omega(x)) \tag{12.222}$$

$$\delta V^\mu(x) = -i\bar{\xi}\gamma_\mu\lambda(x) - \bar{\xi}\partial_\mu\omega(x) \tag{12.223}$$

The terms with three θs in (12.213) give $\delta(\lambda(x) - \frac{i}{2}\partial\!\!\!/\omega(x))$. Using (C.12, C.13), and identities (1) and (2) from Appendix C, we have

$$(\bar{\xi}\gamma_5\theta)(\bar{\theta}\gamma_5\theta) = (\bar{\theta}\gamma_5\theta)(\bar{\theta}\gamma_5\xi) \tag{12.224}$$

$$(\bar{\xi}\gamma^\mu\theta)(\bar{\theta}\gamma_5\theta) = -(\bar{\theta}\gamma_5\theta)(\bar{\theta}\gamma^\mu\xi) \tag{12.225}$$

$$(\bar{\xi}\gamma^\mu\theta)(\bar{\theta}\theta) = -(\bar{\xi}\gamma^\mu\gamma_5\theta)(\bar{\theta}\gamma_5\theta) = (\bar{\theta}\gamma_5\theta)(\bar{\theta}\gamma_5\gamma^\mu\xi) \tag{12.226}$$

$$(\bar{\xi}\gamma^\mu\theta)(\bar{\theta}\gamma_5\gamma_\nu\theta) = -(\bar{\xi}\gamma^\mu\gamma_\nu\theta)(\bar{\theta}\gamma_5\theta) = -(\bar{\theta}\gamma_5\theta)(\bar{\theta}\gamma_\nu\gamma^\mu\xi) \tag{12.227}$$

Using these identities, we see that the terms with three θs in (12.213) can be rewritten

$$(\bar{\theta}\gamma_5\theta)\bar{\theta}\{(D - \frac{1}{2}\Box C)\gamma_5\xi + \frac{1}{2}\gamma^\mu\xi\partial_\mu M + \frac{i}{2}\gamma_5\gamma^\mu\xi\partial_\mu N - \frac{i}{2}\partial_\mu V\!\!\!/\gamma^\mu\xi\} \tag{12.228}$$

so that we can read off

$$\delta(\lambda - \frac{i}{2}\partial\!\!\!/\omega) = \{\frac{i}{2}\partial\!\!\!/M - \frac{1}{2}\gamma_5\partial\!\!\!/N + \frac{1}{2}\partial_\mu V\!\!\!/\gamma^\mu + i(D - \frac{1}{2}\Box C)\gamma_5\}\xi \tag{12.229}$$

which, together with the transformation law (12.216) for ω gives the desired SUSY transformation of the λ field

$$\delta\lambda(x) = (\frac{1}{2}[\partial_\mu V\!\!\!/(x), \gamma^\mu] + i\gamma_5 D(x))\xi \tag{12.230}$$

Finally, the (very important!) transformation law for the D field is obtained by looking at the term in δS with four θs (see (C.20) from Appendix C):

$$\delta S \sim -(\bar{\xi}\gamma^\mu\theta)\,\bar{\theta}\gamma_5\theta\,\bar{\theta}(\partial_\mu\lambda - \frac{i}{2}\partial_\mu\partial\!\!\!/\omega) \tag{12.231}$$

$$= -(\bar{\xi}\gamma^\mu)_\alpha(\bar{\theta}\gamma_5\theta)\theta_\alpha\bar{\theta}_\beta(\partial_\mu\lambda_\beta - \frac{i}{2}\partial_\mu(\partial\!\!\!/\omega)_\beta) \tag{12.232}$$

$$= \frac{1}{4}(\bar{\theta}\gamma_5\theta)^2\bar{\xi}\gamma^\mu\gamma_5(\partial_\mu\lambda - \frac{i}{2}\partial_\mu\partial\!\!\!/\omega) \tag{12.233}$$

$$= -\frac{1}{4}(\bar{\theta}\gamma_5\theta)^2\delta(D - \frac{1}{2}\Box C) \tag{12.234}$$

Combining this with the transformation rule (12.214) for the $C(x)$ field, we obtain

$$\delta D(x) = \bar{\xi}\gamma_5\partial\!\!\!/\lambda(x) \tag{12.235}$$

In other words, the D term of any scalar superfield *transforms as a total spacetime-derivative* under an infinitesimal SUSY transformation, so if K is any such field, or product of such fields, a SUSY invariant action can be obtained simply by taking the D part of K as the Lagrangian density:

$$I = \int d^4x[K]_D \tag{12.236}$$

A simpler way to see this is to realize that the above action is really the integral of K over all of superspace

$$I = \int d^4x d\theta_\alpha K(x,\theta) \tag{12.237}$$

and K transforms into a mixture of spacetime and θ derivatives under an infinitesimal SUSY transformation (see the second line of (12.213)). We will see soon that the kinetic term of the simplest SUSY models arises from just this type of SUSY invariant.

12.6.5   Chiral superfields

The real scalar superfield S described above is *not* an irreducible representation of the extended Poincaré–SUSY algebra. It can basically be split into two chiral halves which are irreducible (a rough analogy is the second-rank tensor $F^{\mu\nu}$ which is really $(1,0)+(0,1)$ under the HLG). Recall from (12.208–12.211) that the covariant derivative $\mathcal{D}_\alpha$ transforms superfields into superfields. If we consider separately the upper two ("left-handed") and lower two ("right-handed") components of $\mathcal{D}_\alpha$:

$$\mathcal{D}_{L\alpha} \equiv (\frac{1+\gamma_5}{2})_{\alpha\beta}\mathcal{D}_\beta \tag{12.238}$$

$$\mathcal{D}_{R\alpha} \equiv (\frac{1-\gamma_5}{2})_{\alpha\beta}\mathcal{D}_\beta \tag{12.239}$$

Then the constraint

$$\mathcal{D}_{R\alpha}\Phi(x,\theta) = 0 \tag{12.240}$$

defines a left-chiral superfield Φ which remains left-chiral under a SUSY transformation (12.213) (as a consequence of (12.209)). Likewise

$$\mathcal{D}_{L\alpha}\tilde{\Phi}(x,\theta) = 0 \tag{12.241}$$

defines a right-chiral field $\tilde{\Phi}$. From the definition

$$\mathcal{D}_\alpha \equiv (\gamma_5\epsilon)_{\alpha\gamma}\frac{\partial}{\partial\theta_\gamma} + i\gamma^\mu_{\alpha\gamma}\theta_\gamma\partial_\mu \tag{12.242}$$

we can easily extract the left- and right-handed parts:

$$\mathcal{D}_{L\alpha} = \epsilon_{\alpha\gamma}\frac{\partial}{\partial\theta_{L\gamma}} + i(\gamma^\mu\theta_R)_\alpha\partial_\mu \tag{12.243}$$

$$\mathcal{D}_{R\alpha} = -\epsilon_{\alpha\gamma}\frac{\partial}{\partial\theta_{R\gamma}} + i(\gamma^\mu\theta_L)_\alpha\partial_\mu \tag{12.244}$$

The easiest way to solve the constraints (12.240, 12.241) is to define chiral coordinates in superspace

$$x^\mu_\pm \equiv x^\mu \mp i\theta^T_R\epsilon\gamma^\mu\theta_L \tag{12.245}$$

These are cooked up so that x_+ vanishes under a right derivative, x_- under a left derivative:

$$\mathcal{D}_{R\alpha}x^\mu_+ = (-\epsilon_{\alpha\beta}\frac{\partial}{\partial\theta_{R\beta}} + i(\gamma^\nu\theta_L)_\alpha\partial_\nu)(x^\mu - i\theta_{R\gamma}(\epsilon\gamma^\mu)_{\gamma\delta}\theta_{L\delta})$$

$$= i(\epsilon^2\gamma^\mu)_{\alpha\delta}\theta_{L\delta} + i(\gamma^\mu\theta_L)_\alpha = 0 \tag{12.246}$$

and

$$\mathcal{D}_{L\alpha}x^\mu_- = (\epsilon_{\alpha\beta}\frac{\partial}{\partial\theta_{L\beta}} + i(\gamma^\nu\theta_R)_\alpha\partial_\nu)(x^\mu + i\theta_{R\gamma}(\epsilon\gamma^\mu)_{\gamma\delta}\theta_{L\delta})$$

$$= -i\epsilon_{\alpha\beta}\theta_{R\gamma}(\epsilon\gamma^\mu)_{\gamma\beta} + i(\gamma^\mu\theta_R)_\alpha$$

$$= i\theta_{R\gamma}(\epsilon\gamma^\mu\epsilon)_{\gamma\alpha} + i(\gamma^\mu\theta_R)_\alpha$$

$$= -i(\gamma^\mu\theta_R)_\alpha + i(\gamma^\mu\theta_R)_\alpha = 0 \tag{12.247}$$

where in the last step we have used the transposition property $\epsilon\gamma^\mu\epsilon = -\gamma^{\mu T}$.

The condition $\mathcal{D}_R\Phi = 0$ means (since $\mathcal{D}_R$ does not contain $\frac{\partial}{\partial\theta_L}$) that the left-chiral field Φ can be written as a function of x^μ_+ and θ_L:

$$\Phi(x,\theta) = \phi(x_+) - \sqrt{2}\theta^T_L\epsilon\psi_L(x_+) + \mathcal{F}(x_+)\theta^T_L\epsilon\theta_L \tag{12.248}$$

where the expansion must terminate at the term quadratic in θ_L (which has only two independent components !). Note that $\phi, \mathcal{F}$ are complex scalar fields (two real degrees of freedom each) and ψ_L is a complex spinor doublet, so Φ contains in all eight real degrees of freedom. The signs, square-roots of 2, etc., are all conventional normalizations chosen to make the final Lagrangian look decent. Likewise, the most general right-chiral field can be expanded

$$\tilde{\Phi}(x,\theta) = \tilde{\phi}(x_-) + \sqrt{2}\theta_R^T \epsilon \psi_R(x_-) - \tilde{\mathcal{F}}(x_-)\theta_R^T \epsilon \theta_R \tag{12.249}$$

If we expand the left-chiral field Φ around the bosonic x^μ coordinate, we find that it can be expressed as follows in terms of conventional spacetime fields:

$$\Phi(x,\theta) = \phi(x) - i\theta_R^T \epsilon \gamma^\mu \theta_L \partial_\mu \phi(x) + \frac{1}{2}(-i)^2 \theta_R^T \epsilon \gamma^\mu \theta_L \theta_R^T \epsilon \gamma^\nu \theta_L \partial_\mu \partial_\nu \phi(x)$$

$$- \sqrt{2}\theta_L^T \epsilon \psi_L(x) + i\sqrt{2}\theta_L^T \epsilon \partial_\mu \psi_L(x)\theta_R^T \epsilon \gamma^\mu \theta_L + \mathcal{F}(x)\theta_L^T \epsilon \theta_L \tag{12.250}$$

In order to compare the fields in (12.250) more easily with our original component fields for the full scalar superfield S, we will need the following identities:

$$\theta_R^T \epsilon \gamma^\mu \theta_L = \bar{\theta}\gamma_5 \gamma^\mu \frac{1+\gamma_5}{2}\theta = \frac{1}{2}\bar{\theta}\gamma_5 \gamma^\mu \theta \tag{12.251}$$

$$\theta_R^T \epsilon \gamma^\mu \theta_L \theta_R^T \epsilon \gamma^\nu \theta_L = -\frac{1}{4}g^{\mu\nu}(\bar{\theta}\gamma_5 \theta)^2 \tag{12.252}$$

$$\theta_L^T \epsilon \psi_L = \bar{\theta}\gamma_5 \psi_L = \bar{\theta}\psi_L \tag{12.253}$$

$$\theta_R^T \epsilon \gamma^\mu \theta_L \theta_L^T \epsilon \partial_\mu \psi_L = -\frac{1}{2}\bar{\theta}\gamma_5 \theta\bar{\theta}\gamma_5 \partial\!\!\!/\psi_L = \frac{1}{2}\bar{\theta}\gamma_5 \theta\bar{\theta}\partial\!\!\!/\psi_L \tag{12.254}$$

$$\theta_L^T \epsilon \theta = \bar{\theta}\gamma_5 \frac{1+\gamma_5}{2}\theta = \bar{\theta}\frac{1+\gamma_5}{2}\theta \tag{12.255}$$

Inserting these results in (12.250) we obtain

$$\Phi(x,\theta) = \phi(x) - \frac{i}{2}\bar{\theta}\gamma_5 \gamma^\mu \theta\partial_\mu \phi(x) - \sqrt{2}\bar{\theta}\psi_L(x) + \bar{\theta}\frac{1+\gamma_5}{2}\theta\mathcal{F}(x)$$

$$+ \frac{i}{\sqrt{2}}\bar{\theta}\gamma_5 \theta\bar{\theta}\partial\!\!\!/\psi_L(x) + \frac{1}{8}(\bar{\theta}\gamma_5 \theta)^2\Box\phi(x) \tag{12.256}$$

Similarly, for the right-chiral field

$$\tilde{\Phi}(x,\theta) = \tilde{\phi}(x) + \frac{i}{2}\bar{\theta}\gamma_5 \gamma^\mu \theta\partial_\mu \tilde{\phi}(x) - \sqrt{2}\bar{\theta}\psi_R(x) + \bar{\theta}\frac{1-\gamma_5}{2}\theta\tilde{\mathcal{F}}(x)$$

$$- \frac{i}{\sqrt{2}}\bar{\theta}\gamma_5 \theta\bar{\theta}\partial\!\!\!/\psi_R(x) + \frac{1}{8}(\bar{\theta}\gamma_5 \theta)^2\Box\tilde{\phi}(x) \tag{12.257}$$

An extremely important special case occurs when Φ and $\tilde{\Phi}$ are conjugates of each other. This will be the case if the bosonic fields are related in the obvious way, $\tilde{\phi} = \phi^*, \tilde{\mathcal{F}} = \mathcal{F}^*$, and the upper components of ψ_L are related to the lower components of ψ_R in the usual charge-conjugation way familiar from the 4-spinor version of a Majorana field:

$$\psi_R = \begin{pmatrix} 0 \\ 0 \\ \psi_1 \\ \psi_2 \end{pmatrix}$$

and

$$\psi_L = \begin{pmatrix} \psi_2^* \\ -\psi_1^* \\ 0 \\ 0 \end{pmatrix}$$

Powers of a left-chiral field Φ are clearly left-chiral (i.e., satisfy (12.240)); likewise, powers of right-chiral fields are right-chiral (satisfy (12.241)). However a product like $\Phi\tilde{\Phi}$ is not chiral, although it is still a scalar superfield (of the S-type). If $\tilde{\Phi} = \Phi^*$ as discussed above, it is also hermitian. An hermitian Lagrangian can also be obtained by taking the "real part" of a chiral field (or power of chiral fields), so consider

$$S_c(x, \theta) \equiv \frac{1}{\sqrt{2}}(\Phi + \Phi^*) \tag{12.258}$$

Decomposing the complex bosonic fields $\phi, \mathcal{F}$ in the usual way

$$\phi = \frac{A + iB}{\sqrt{2}} \tag{12.259}$$

$$\mathcal{F} = \frac{F - iG}{\sqrt{2}} \tag{12.260}$$

and with ψ_L, ψ_R the upper and lower components of a single Majorana fermion field ψ as indicated above, we see that taking the real part of the chiral field Φ yields a constrained real superfield with components

$$S_c = A(x) - \bar{\theta}\psi(x) + \frac{1}{2}\bar{\theta}\gamma_5\gamma^\mu\theta\partial_\mu B(x) + \frac{1}{2}\bar{\theta}\theta F(x) - \frac{i}{2}\bar{\theta}\gamma_5\theta G(x)$$

$$- \frac{i}{2}\bar{\theta}\gamma_5\theta\bar{\theta}\gamma_5\partial\psi(x) + \frac{1}{8}(\bar{\theta}\gamma_5\theta)^2\Box A(x) \tag{12.261}$$

Comparing with the component expression for the general scalar superfield (12.212)

$$S(x, \theta) = C(x) - i\bar{\theta}\gamma_5\omega(x) - \frac{i}{2}(\bar{\theta}\gamma_5\theta)M(x)$$

$$- \frac{1}{2}(\bar{\theta}\theta)N(x) - \frac{1}{2}\bar{\theta}\gamma_5\gamma_\mu\theta V^\mu(x)$$

$$- i(\bar{\theta}\gamma_5\theta)\,\bar{\theta}(\lambda(x) - \frac{i}{2}\partial\omega(x)) - \frac{1}{4}(\bar{\theta}\gamma_5\theta)^2(D(x) - \frac{1}{2}\Box C(x)) \tag{12.262}$$

So we see that the real part of our chiral field is just a real scalar superfield with the identifications

$$\lambda = D = 0 \tag{12.263}$$

$$C \to A \tag{12.264}$$

$$M \to G \tag{12.265}$$

$$N \to -F \tag{12.266}$$

$$V_\mu \to -\partial_\mu B \tag{12.267}$$

$$\omega \to -i\gamma_5 \psi \tag{12.268}$$

so that the $\mathcal{F}$ term corresponds to $-\frac{N+iM}{\sqrt{2}}$ in the old notation. Recalling the SUSY transformation rules

$$\delta M(x) = -\bar{\xi}(\lambda(x) - i\partial\!\!\!/\omega(x)) \tag{12.269}$$

$$\delta N(x) = i\bar{\xi}\gamma_5(\lambda(x) - i\partial\!\!\!/\omega(x)) \tag{12.270}$$

we see that once $\lambda = 0$, the variation of the $\mathcal{F}$ term is a total spacetime-derivative, so the integral over spacetime of such a term yields a SUSY-invariant action:

$$I_f = \int d^4x [f(\Phi) + f(\Phi^*)]_{\mathcal{F}} \tag{12.271}$$

We will see shortly that this is the term responsible for non-trivial interactions in the simplest SUSY models.

12.6.6 SUSY Lagrangians

So far we have seen that SUSY-invariant actions are obtainable either as the D-term of a general scalar ("S-type") superfield or as the F-term of a chiral superfield. Thus, we can write

$$I = \frac{1}{2} \int d^4x [K]_D + \int d^4x ([f(\Phi)]_{\mathcal{F}} + [f(\Phi^*)]_{\mathcal{F}}) \tag{12.272}$$

where K is a full (S-type) superfield, Φ a left-chiral superfield. Φ begins with a scalar field ϕ with dimension of mass. In powers of mass, the relation $x_+^\mu = x^\mu - i\theta_R^T \epsilon \gamma^\mu \theta_L$ implies that θ has dimension $-\frac{1}{2}$. Anticipating here our discussion of renormalizability in Chapter 17, we note that perturbatively renormalizable theories (in four spacetime dimensions) necessarily contain only interactions of mass dimension less than or equal to 4. However, the dimension of the D-term is $2+\dim(K)$ which implies that K is at most quadratic in Φ for renormalizability. Hence the K-term corresponds to the kinetic part of the Lagrangian. The dimension of the F-term of $f(\Phi)$ is $1+\dim(f(\Phi))$, implying that F must be a polynomial of degree 3 or less. In general we wish to allow for a multiplet of superfields Φ_n under some internal symmetry group, and the K-term can be some general (hermitian) quadratic form

$$K(\Phi_n, \Phi_n^*) = \sum_{n,m} \Phi_n^* g_{nm} \Phi_m, \quad g_{nm} = g_{mn}^* \tag{12.273}$$

For simplicity, let us consider the construction of a SUSY-invariant Lagrangian for just a single chiral field Φ. The kinetic term is given by extracting the D (four θ) component of $\frac{1}{2}K = \frac{1}{2}\Phi^*\Phi = C(x) + \ldots\ldots - \frac{1}{4}(\bar{\theta}\gamma_5\theta)^2(D(x) - \frac{1}{2}\Box C(x))$. Obviously, $C(x) = \frac{1}{2}\phi^*\phi$, and to get D we need the term in $\Phi^*\Phi$ quartic in θs. Here Φ^* is just the expression (12.257) with tildes replaced by conjugates ($\tilde{\phi} \to \phi^*$, etc.). Assembling the terms in $\Phi^*\Phi$ quartic in θs, we find

$$[\Phi^*\Phi]_{\theta^4} = \frac{1}{8}(\bar{\theta}\gamma_5\theta)^2(\phi^*\Box\phi + \phi\Box\phi^*) + \frac{1}{4}\bar{\theta}\gamma_5\gamma^\mu\theta\bar{\theta}\gamma_5\gamma^\nu\theta\partial_\mu\phi^*\partial_\nu\phi$$

$$- i\bar{\theta}\psi_R\bar{\theta}\gamma_5\theta\bar{\theta}\partial\!\!\!/\psi_L + i\bar{\theta}\psi_L\bar{\theta}\gamma_5\theta\bar{\theta}\partial\!\!\!/\psi_R + \bar{\theta}\frac{1-\gamma_5}{2}\theta\bar{\theta}\frac{1+\gamma_5}{2}\theta\mathcal{F}^*\mathcal{F} \tag{12.274}$$

The following identities, easily verifiable with the machinery described in Appendix C, come in handy at this point:

$$\bar{\theta}\gamma_5\theta\bar{\theta}\psi_R\bar{\theta}\partial\!\!\!/\psi_L = \frac{1}{4}(\bar{\theta}\gamma_5\theta)^2\bar{\psi}_R\partial\!\!\!/\psi_L \tag{12.275}$$

$$\bar{\theta}\gamma_5\theta\bar{\theta}\psi_L\bar{\theta}\partial\!\!\!/\psi_R = -\frac{1}{4}(\bar{\theta}\gamma_5\theta)^2\bar{\psi}_L\partial\!\!\!/\psi_R \tag{12.276}$$

$$\bar{\theta}(1-\gamma_5)\theta\bar{\theta}(1+\gamma_5)\theta = -2(\bar{\theta}\gamma_5\theta)^2 \tag{12.277}$$

$$\bar{\theta}\gamma_5\gamma^\mu\theta\bar{\theta}\gamma_5\gamma^\nu\theta = -g^{\mu\nu}(\bar{\theta}\gamma_5\theta)^2 \tag{12.278}$$

Using these in (12.274) we find

$$\frac{1}{2}[\Phi^*\Phi]_{\theta^4} = \frac{1}{8}(\bar{\theta}\gamma_5\theta)^2\{\frac{1}{2}(\phi^*\Box\phi + \phi\Box\phi^*) - \partial^\mu\phi^*\partial_\mu\phi - 2\mathcal{F}^*\mathcal{F} - i(\bar{\psi}_R\partial\!\!\!/\psi_L + \bar{\psi}_L\partial\!\!\!/\psi_R)\}$$

$$= \frac{1}{8}(\bar{\theta}\gamma_5\theta)^2\{\frac{1}{2}(\phi^*\Box\phi + \phi\Box\phi^*) - \partial^\mu\phi^*\partial_\mu\phi - 2\mathcal{F}^*\mathcal{F} - i\bar{\psi}\partial\!\!\!/\psi\} \tag{12.279}$$

which should be compared with $-\frac{1}{4}(\bar{\theta}\gamma_5\theta)^2(D(x) - \frac{1}{2}\Box C(x))$, with $C(x) = \frac{1}{2}\phi^*\phi$. This gives the desired D-term:

$$D = \partial^\mu\phi^*\partial_\mu\phi + \mathcal{F}^*\mathcal{F} + \frac{i}{2}\bar{\psi}\partial\!\!\!/\psi \tag{12.280}$$

which is exactly the free Lagrangian (12.149) studied earlier for a massless complex spin-0 scalar ϕ and a massless spin-$\frac{1}{2}$ Majorana field ψ, together with an auxiliary complex scalar $\mathcal{F}$ with (at this stage) no interesting dynamics.

The other term in the general action (12.272) comes from the F-term in a polynomial f, typically called the *superpotential* (and at most cubic for renormalizability) in Φ. Recall that the component field decomposition for a general chiral field reads $\Phi = \phi(x_+) - \sqrt{2}\theta_L^T \epsilon\psi_L(x_+) + \mathcal{F}(x)\theta_L^T \epsilon\theta_L$, so the F-term corresponds to the term quadratic in θ_L:

$$[f(\Phi)]_{\theta_L^2} = \frac{\partial^2 f(\phi(x))}{\partial\phi^2}\theta_L^T\epsilon\psi_L(x)\theta_L^T\epsilon\psi_L(x) + \frac{\partial f(\phi)}{\partial\phi}\mathcal{F}(x)\theta_L^T\epsilon\theta_L \tag{12.281}$$

Recall the form of a Majorana spinor $\theta^T = (\theta_2^*, -\theta_1^*, \theta_1, \theta_2)$, from which we obtain

$$\theta_L^T\epsilon\theta_L = 2\theta_1^*\theta_2^*$$
$$\theta_L^T\epsilon\psi_L = \theta_1^*\psi_2^* - \theta_2^*\psi_1^*$$
$$(\theta_L^T\epsilon\theta_L)^2 = 2\theta_1^*\theta_2^*\psi_2^*\psi_1^* = -\frac{1}{2}\theta_L^T\epsilon\theta_L(\bar{\psi})_L\psi_L$$

allowing us to read off the coefficient of $\theta_L^T\epsilon\theta_L$ in (12.281):

$$[f(\Phi)]_{\mathcal{F}} = \mathcal{F}(x)\frac{\partial f}{\partial\phi}(\phi(x)) - \frac{1}{2}\frac{\partial^2 f}{\partial\phi^2}(\phi(x))\bar{\psi}_L\psi_L \tag{12.282}$$

Adding the hermitian adjoint, the total Lagrange density corresponding to the action (12.272) thus becomes

$$\mathcal{L} = \partial_\mu\phi^*\partial^\mu\phi + \mathcal{F}^*\mathcal{F} + \frac{i}{2}\bar{\psi}\not\partial\psi + \mathcal{F}\frac{\partial f}{\partial\phi} + \mathcal{F}^*(\frac{\partial f}{\partial\phi})^*$$

$$- \frac{1}{2}\frac{\partial^2 f}{\partial\phi^2}\bar{\psi}_L\psi_L - \frac{1}{2}(\frac{\partial^2 f}{\partial\phi^2})^*(\bar{\psi}_L\psi_L)^* \tag{12.283}$$

The Euler–Lagrange equation allows us to eliminate the non-dynamical field $\mathcal{F}$

$$\frac{\partial\mathcal{L}}{\partial\mathcal{F}} = 0 \Rightarrow \mathcal{F}(x) = -(\frac{\partial f(\phi)}{\partial\phi})^* \tag{12.284}$$

whereupon the Lagrangian becomes

$$\mathcal{L} = \partial_\mu\phi^*\partial^\mu\phi + \frac{i}{2}\bar{\psi}\not\partial\psi - P(\phi) - \frac{1}{2}\frac{\partial^2 f}{\partial\phi^2}\bar{\psi}_L\psi_L - \frac{1}{2}(\frac{\partial^2 f}{\partial\phi^2})^*(\bar{\psi}_L\psi_L)^* \tag{12.285}$$

with $P(\phi) \equiv |\frac{\partial f(\phi)}{\partial\phi}|^2$.

In general, there will be a minimum of $f(\phi)$ where $\frac{\partial f}{\partial\phi} = 0$, putting the polynomial $P(\phi)$ at its absolute minimum. In this case, SUSY is an unbroken global symmetry:

$$< 0|[\bar{\xi}_L Q, \psi_L]|0 > \propto < 0|\mathcal{F}|0 > \propto < 0|\frac{\partial f}{\partial\phi}|0 >= 0 \tag{12.286}$$

implying that the generators Q annihilate the vacuum. There are ways to evade this however (e.g., see (Weinberg, 1995b), Section 26.5, for a discussion of O'Raifertaigh breaking). One can show that if SUSY is unbroken at the lowest order, it will remain unbroken to all orders of perturbation theory. Obviously, given the notable absence of superpartners of equal mass to the known elementary particles, broken supersymmetry is clearly the norm, if indeed supersymmetry is present in Nature at all.

Finally, note that since the fundamental SUSY algebra (12.191) implies that the Hamiltonian P_0 can be constructed as a product of Q and $\bar{Q}$ operators, the exact vacuum energy must vanish if SUSY is unbroken (i.e., the disconnected vacuum energy graphs which determine the shift in vacuum energy due to interactions must cancel identically between bosonic and fermionic loop contributions to all orders of perturbation theory).

We conclude our abbreviated survey of supersymmetry by giving a simple explicit example: historically, the first four-dimensional field theory in which supersymmetry was demonstrated, the Wess–Zumino model (1974), obtained by taking for the superpotential

$$f(\phi) = \frac{1}{2}m\phi^2 + \frac{\sqrt{2}}{3}\lambda\phi^3 \tag{12.287}$$

$$\frac{\partial f}{\partial \phi} = m\phi + \sqrt{2}\lambda\phi^2 \tag{12.288}$$

$$\frac{\partial^2 f}{\partial \phi^2} = m + 2\sqrt{2}\lambda\phi \tag{12.289}$$

We can re-express the complex scalar $\phi = \frac{1}{\sqrt{2}}(A + iB)$, where A, B are real (i.e., hermitian) scalar fields. Then

$$P(\phi) = |\frac{\partial f}{\partial \phi}|^2 = \frac{1}{2}m^2(A^2 + B^2) + m\lambda A(A^2 + B^2) + \frac{\lambda^2}{2}(A^2 + B^2)^2 \tag{12.290}$$

while the Yukawa interaction terms become

$$-\frac{1}{2}\frac{\partial^2 f}{\partial \phi^2}\bar{\psi}_L\psi_L - \frac{1}{2}\Big(\frac{\partial^2 f}{\partial \phi^2}\Big)^*(\bar{\psi}_L\psi_L)^* = -\frac{1}{2}m(\bar{\psi}_L\psi_L + \bar{\psi}_R\psi_R) - \sqrt{2}\lambda\frac{A + iB}{\sqrt{2}}\bar{\psi}_L\psi_L$$

$$- \sqrt{2}\lambda\frac{A - iB}{\sqrt{2}}\bar{\psi}_R\psi_R$$

$$= -\frac{1}{2}m\bar{\psi}\psi - \lambda A\bar{\psi}\psi - i\lambda B\bar{\psi}\gamma_5\psi \tag{12.291}$$

The complete Lagrangian for this theory is thus

$$\mathcal{L} = \frac{1}{2}(\partial_\mu A)^2 + \frac{1}{2}(\partial_\mu B)^2 - \frac{1}{2}m^2(A^2 + B^2) + \frac{i}{2}\bar{\psi}\partial\!\!\!/\psi - \frac{m}{2}\bar{\psi}\psi$$

$$- \lambda A\bar{\psi}\psi - i\lambda B\bar{\psi}\gamma_5\psi - m\lambda A(A^2 + B^2) - \frac{\lambda^2}{2}(A^2 + B^2)^2 \tag{12.292}$$

The overall $\frac{1}{2}$ in the fermion kinetic term is the appropriate normalization for a self-conjugate Majorana field (recall the similar factor of $\frac{1}{2}$ when we go from complex to real scalar fields). Given that, we see that the scalar and spin-$\frac{1}{2}$ fields correspond to particles of equal bare mass m.[17] Of course, if we set $m = \lambda = 0$, and return to the complex field $\phi = \frac{1}{\sqrt{2}}(A + iB)$, we recover the simple Lagrangian (12.149) for free massless scalars and a Majorana fermion which served as our first explicit example of a theory exhibiting the super-Poincaré algebra.

12.7 Problems

1. Show that the interaction Hamiltonian density (12.1) is non-ultralocal: i.e., its equal-time commutator $[\mathcal{H}_{\text{int}}(\vec{x}, t), \mathcal{H}_{\text{int}}(\vec{y}, t)]$ contains a gradient of a δ-function (cf. (5.81)).

2. In the derivatively-coupled theory defined by interaction Hamiltonian density (12.11), the interaction part V of the full Hamiltonian is given by

$$V = \int \{ g\bar{\psi}\gamma^\mu\gamma_5\psi\partial_\mu\phi(\vec{y}, 0) + \frac{1}{2}g^2(\bar{\psi}\gamma^0\gamma_5\psi(\vec{y}, 0))^2 \} d^3y \tag{12.293}$$

with all the (unsubscripted) fields in interaction picture and taken at time $t = 0$.

(a) Show that

$$\frac{\partial}{\partial t}\phi_H(\vec{x}, t) = e^{iHt}e^{-iH_0 t}\dot{\phi}(\vec{x}, t)e^{iH_0 t}e^{-iHt} + e^{iHt}i[V, \phi(\vec{x}, 0)]e^{-iHt}$$

$$= \pi^\phi_H(\vec{x}, t) + e^{iHt}i[V, \phi(\vec{x}, 0)]e^{-iHt} \tag{12.294}$$

(b) Use the equal-time commutation relations (12.14) to evaluate the commutator in part (a), thereby recovering (12.33):

$$\frac{\partial}{\partial t}\phi_H(\vec{x}, t) = \pi^\phi_H(\vec{x}, t) + g\bar{\psi}_H\gamma^0\gamma_5\psi_H(\vec{x}, t) \tag{12.295}$$

3. Show that the field equation for $\psi^\dagger_H$ (or $\bar{\psi}_H$) obtained by applying the first of the Hamiltonian equations (12.31) to the Hamiltonian (12.22) is equivalent, after taking its adjoint, to the Dirac Heisenberg field equation (12.36) for ψ_H.

4. The Hamiltonian density for a massive neutral vector field, described by a Lorentz vector field Z_μ, coupled to a four-vector source field J^μ, which involves fields other than Z_μ (for example, we might have $J^\mu = e\bar{\psi}\gamma^\mu\psi$, with ψ a Dirac field) is given by

$$\mathcal{H} = \mathcal{H}_0 + \mathcal{H}_{\text{int}} \tag{12.296}$$

$$\mathcal{H}_0 = \frac{1}{2}\{\vec{\pi}^2 + |\vec{\nabla} \times \vec{Z}|^2 + m^2\vec{Z}^2 + \frac{1}{m^2}(\vec{\nabla} \cdot \vec{\pi})^2\} \tag{12.297}$$

$$\mathcal{H}_{\text{int}} = -\frac{1}{m^2}J^0\vec{\nabla} \cdot \vec{\pi} - \vec{J} \cdot \vec{Z} + \frac{1}{2m^2}(J^0)^2 \tag{12.298}$$

---

[17] Again anticipating later discussions of renormalizability, it can be shown that the underlying SUSY symmetry ensures that counterterms are induced by radiative corrections only in the D-terms, *not* in the F-terms. Accordingly, there are *no mass or coupling renormalizations* in this theory: the bare m, λ can be chosen at their fixed physical values! However, as the kinetic part of the Lagrangian derives from the D-term of (12.272), there is a non-trivial wavefunction (field) renormalization: in other words, $Z \neq 1$.

where π_i is the momentum field conjugate to Z_i $(i = 1, 2, 3)$. From Hamilton's equation, we easily find $\pi_i = K_{ij}^{-1}(\dot{Z}_j - \frac{1}{m^2}\partial_j J^0)$, where $K_{ij} \equiv \delta_{ij} - \frac{1}{m^2}\partial_i\partial_j$.

(a) Perform the Legendre transform to arrive at a Lagrangian density given as a function of $\vec{Z}, \dot{\vec{Z}}, J^0$, and $\vec{J}$. The result is non-local, and horribly non-covariant in appearance (sanity will be restored in part (d))! Note that the Lagrange density may always be thought of as defining an action via a spacetime integral (e.g., in the Lagrangian form of the functional integral (12.60)), so you are always permitted to rearrange derivatives in each term by free use of integration by parts.

(b) Next, show that the result obtained in (a) is completely equivalent to that obtained starting with the manifestly local and Lorentz-invariant Lagrangian

$$\mathcal{L} = -\frac{1}{4}F_{\mu\nu}F^{\mu\nu} + \frac{m^2}{2}Z_\mu Z^\mu - J^\mu Z_\mu, \quad F_{\mu\nu} \equiv \partial_\mu Z_\nu - \partial_\nu Z_\mu \quad (12.299)$$

Proceed as follows. First, show that Z^0 is a *dependent field*: namely one that can be expressed, via the Euler–Lagrange equations of motion, uniquely and entirely in terms of the other canonical fields and their conjugate momenta at the same time. Do this by writing down the equation of motion for Z^0 and showing that it reduces to

$$Z^0 = \frac{1}{m^2}(J^0 - \vec{\nabla}\cdot\vec{\pi}), \quad \pi_i = \frac{\partial\mathcal{L}}{\partial\dot{Z}_i} = \dot{Z}_i - \partial_i Z^0 \quad (12.300)$$

Show that (12.300) implies the formula $\pi_i = K_{ij}^{-1}(\dot{Z}_j - \frac{1}{m^2}\partial_j J^0)$ obtained previously in the Hamiltonian framework. Finally, eliminate Z^0 completely from the Lagrangian (12.299), and show that the resultant expression agrees with that found in part (a).

(c) Now carry out the canonical procedure in the usual direction, by starting with the Lagrangian, and eliminating $\dot{\vec{Z}}$ in favor of $\vec{\pi}$ in $\mathcal{H} = \vec{\pi}\cdot\dot{\vec{Z}} - \mathcal{L}$, to check that the original Hamiltonian (12.296–12.298) is recovered.

(d) In the functional integral approach, the fact that Z^0 is a dependent field manifests itself in the Gaussian dependence of the Lagrangian density on Z^0 (and its *spatial* derivatives). Show that if we explicitly integrate out Z^0 in the path integral

$$\int \mathbf{D}Z^\mu e^{i\int(-\frac{1}{4}F_{\mu\nu}F^{\mu\nu} + \frac{1}{2}m^2 Z_\mu Z^\mu - J_\mu Z^\mu)d^4x} \rightarrow \int \mathbf{D}\vec{Z}e^{i\int \mathcal{L}'(\vec{Z},\dot{\vec{Z}},J^0,\vec{J})d^4x}$$

$$(12.301)$$

the resultant Lagrangian $\mathcal{L}'$ is exactly the expression found in part (a). (Hint: note the identity $K_{ij}^{-1} = \delta_{ij} + \frac{1}{-\Delta+m^2}\partial_i\partial_j$) Its disgustingly non-covariant (and non-local) appearance is seen to arise from the fact that the Legendre transform from the Hamiltonian side naturally produces a Lagrangian density with dependent fields eliminated—which we see clearly in this instance serves to disguise the underlying Lorentz-invariance of the theory. The same situation arises when *redundant fields*, associated with local gauge symmetries, are present, as we shall see in Chapter 15. Of course, the sensible way to ensure Lorentz-invariance is to start with a Lorentz-invariant Lagrangian, from which we go (if needed) to the Hamiltonian.

5. The Hamiltonian density for a massless gauge vector field A_μ, coupled to a conserved current J^μ (which, as in Problem 4, we assume to depend on a separate set of fields), is given in the axial gauge $A_3 = 0$ by

$$\mathcal{H} = \mathcal{H}_0 + \mathcal{H}_{\text{int}} \tag{12.302}$$

$$\mathcal{H}_0 = \frac{1}{2}\pi_i(\delta_{ij} + \frac{\partial_i \partial_j}{\partial_3^2})\pi_j + \frac{1}{2}|\vec{\nabla} \times \vec{A}|^2 \tag{12.303}$$

$$\mathcal{H}_{\text{int}} = \partial_i \pi_i \frac{1}{\partial_3^2}J^0 - \frac{1}{2}J^0\frac{1}{\partial_3^2}J^0 - J_i A_i \tag{12.304}$$

where the spatial indices i, j run over the values 1,2. Perform the Legendre transformation to obtain the Lagrange density

$$\mathcal{L} = \frac{1}{2}\dot{A}_i^2 + \frac{1}{2}(\partial_i \dot{A}_i - J^0)\frac{1}{\Delta}(\partial_j \dot{A}_j - J^0) - \frac{1}{2}|\vec{\nabla} \times \vec{A}|^2 + J_i A_i \tag{12.305}$$

(b) Starting instead with the Lagrangian

$$\mathcal{L}_{\text{QED}} = -\frac{1}{4}F_{\mu\nu}F^{\mu\nu} - J^\mu A_\mu, \quad F_{\mu\nu} \equiv \partial_\mu A_\nu - \partial_\nu A_\mu \tag{12.306}$$

show that the equation of motion for A_0 implies that it is a dependent field:

$$\Delta A_0 = \partial_i \dot{A}_i - J^0 \Rightarrow A_0 = \frac{1}{\Delta}(\partial_i \dot{A}_i - J^0) = \frac{1}{\partial_3^2}(\partial_i \pi_i - J^0) \tag{12.307}$$

Setting $A_3 = 0$ (as we are allowed to do via a suitable gauge-transformation $A_\mu \to A_\mu - \partial_\mu \Lambda$: we are anticipating here the detailed discussion of gauge symmetry and gauge transformations that will follow in Chapter 15), and then eliminating A_0 via (12.307), show that one recovers the result (12.305) found in part (a).

6. Show that the boost generators $M^{0i}, i = 1, 2, 3$, as given by the Noether formula, generate the correct transformation of a set of fields ϕ_n lying in a definite representation of the HLG (possibly reducible) by computing the commutator $[M^{0i}, \phi_N(\vec{x}, t)]$. You will find (12.25) and (12.26) useful: also, to avoid subtleties associated with dependent fields or constraints (cf. Problems 4 and 5), assume that the Hamiltonian density $\mathcal{H}(\pi_n, \phi_n, \vec{\nabla}\phi_n)$ does not depend on gradients of the conjugate momentum field (i.e., on the $\vec{\nabla}\pi_n$).

7. The charges of an internal symmetry duplicate the Lie algebra of the original matrix transformation group on the fields. Let t_α be the matrix generators of some Lie group (in some representation), with structure constants $f_{\alpha\beta\gamma}$:

$$[t_\alpha, t_\beta] = if_{\alpha\beta\gamma}t_\gamma \tag{12.308}$$

The corresponding Noether charges are $Q_\alpha = \int d^3x J_\alpha^0(\vec{x}, t) = -i\int d^3x \pi_n(t_\alpha\phi)_n$ (cf. (12.127)). Verify, using the canonical equal-time commutation relations of the fields ϕ_n and their conjugate momenta π_n, the commutation relations for the charge densities

$$[J_\alpha^0(\vec{x},t), J_\beta^0(\vec{y},t)] = if_{\alpha\beta\gamma}\delta^3(\vec{x}-\vec{y})J_\gamma^0(\vec{x},t) \tag{12.309}$$

from which it follows that the Q_α satisfy the Lie algebra

$$[Q_\alpha, Q_\beta] = if_{\alpha\beta\gamma}Q_\gamma \tag{12.310}$$

8. Rederive the Ward–Takahashi identity (12.142) using operator methods. One may assume without loss of generality that the $y_1, y_2, ..y_p$ are already time-ordered. One can then write the T-product as a sum of terms with explicit θ-functions enforcing the time-ordering of the current relative to the ϕ_n fields, thereby facilitating the application of the spacetime-derivative. The contact terms arise from the $\mu = 0$ derivative, which result in a series of equal-time commutators, at which point (12.128) may be employed.

9. Verify that the variation in the Lagrangian density (12.149) induced by the infinitesimal SUSY transformations listed in (12.157) is a space-time divergence, as given in (12.158). The Grassmann identity (C.12) from Appendix C will be useful here.

10. Here, we shall check that the SUSY current (12.154) is indeed the appropriate conserved current of the form (12.82), in a situation in which there is a non-trivial variation in the Lagrangian density (12.149) (given by a spacetime divergence). From the SUSY variations (again ignoring the noisome $\sqrt{2}$ normalization) of the fields given in (12.157), and $\frac{\partial\mathcal{L}}{\partial(\partial_\mu\phi)} = \partial^\mu\phi^*$, $\frac{\partial\mathcal{L}}{\partial(\partial_\mu\phi^*)} = \partial^\mu\phi$, and $\frac{\partial\mathcal{L}}{\partial(\partial_\mu\psi)} = \frac{i}{2}\bar\psi\gamma^\mu$, show that the conventional Noether current takes the form (including the Grassmann infinitesimal ξ)

$$J^\mu_{\text{Noeth}} = i(\partial^\mu\phi^*)\bar\xi P_L\psi + i(\partial^\mu\phi)\bar\xi P_R\psi + \frac{i}{2}\bar\psi\gamma^\mu(\partial\!\!\!/\phi)P_R\xi + \frac{i}{2}\bar\psi\gamma^\mu(\partial\!\!\!/\phi^*)P_L\xi \tag{12.311}$$

Show that the correct conserved current J^μ in (12.154) is obtained from this by adding in the K^μ correction term arising from (12.158). Some of the Grassmann identities from Appendix C (specifically, (C.12) and (C.13)) will be useful in interchanging the order $\bar\psi \cdot\cdot\xi \to \bar\xi \cdot\cdot\psi$.

11. In this exercise we shall verify explicitly the anticommutation algebra of the supersymmetry generators for the theory given by Lagrangian (12.149), as given in (12.147). The fundamental equal-time (anti)commutation relations of the theory (12.151,12.152,12.153) will be used: note that if ϕ_a, ϕ_b (resp. $\psi, \bar\psi$) are bosonic (resp. fermionic) fields, then the anticommutator $\{\phi_a\psi, \bar\psi\phi_b\}$ can be rearranged into $\phi_a\{\psi, \bar\psi\}\phi_b - \bar\psi[\phi_a, \phi_b]\psi$. A short calculation then shows that the anticommutator $\{Q_\alpha, \bar{Q}_\beta\}$ can be written as the sum of a bosonic and fermionic part, where for example

$$\{Q_\alpha, \bar{Q}_\beta\}_{\text{bos}} = 2\int d^3x (\partial\!\!\!/\phi\gamma^0 P_R\partial\!\!\!/\phi^* + \partial\!\!\!/\phi^*\gamma^0 P_L\partial\!\!\!/\phi)_{\alpha\beta} \tag{12.312}$$

The energy-momentum four-vector P^μ is given, from (12.147) by

$$P^\mu = \frac{1}{8}\gamma^\mu_{\beta\alpha}\{Q_\alpha, \bar{Q}_\beta\}$$

(12.313)

Using (12.313), show that the Hamiltonian P^0 and spatial momentum $\vec{P}$ obtained from the SUSY anticommutator agrees precisely with the expected Noether expressions (12.150). (Again, see Appendix C for useful Majorana identities).

13
Symmetries II: Discrete spacetime symmetries

The invariance of a quantum field theory under the physically realizable transformations embodied in the Poincaré group—the proper orthochronous elements of the homogeneous Lorentz group corresponding to a realizable change of inertial frame, together with spacetime translations—has been built into the foundations of the theory from the very beginning, and many decades of experimental investigation at the subatomic level (where effects of gravitation can safely be neglected) have confirmed that this symmetry, if broken, can only fail at an extremely subtle and quantitatively minute level. Formally, the Lorentz group admits an obvious extension if we allow the discrete operations of space and time reflection, corresponding to the improper (because $\det(\Lambda) = -1$) Lorentz transformations[1] $\Lambda_P = \mathrm{diag}(1, -1, -1, -1)$ and $\Lambda_T = \mathrm{diag}(-1, 1, 1, 1)$, respectively. These operations generate potential symmetries of a quantum field theory, but we must emphasize the term "potential" here, for the simple reason that only some of the interactions in Nature (specifically, the strong and electromagnetic) appear to possess an exact invariance under spatial (parity) and temporal reflection (time reversal). The failure of parity invariance in the weak interactions, discovered by Lee and Yang in the mid-1950s, came as a great shock at the time, but we have since come to realize that Nature does not share the human prejudice that the laws of Nature should take exactly the same form whether expressed in a left- or right-handed coordinate system.

   Although this chapter is entitled "discrete spacetime symmetries", it turns out that it is essential to include the discrete symmetry of charge conjugation—the interchange of particles and antiparticles—in almost the same breath when discussing parity and time reversal, even though the connection to "spacetime" is less than obvious for a symmetry involving the swapping of particles for antiparticles. The reason for this was already indicated in an heuristic fashion in Chapter 3, where we saw that the assumptions of relativistic invariance, quantum mechanics, and locality of interactions implied that exchange of a virtual particle is physically indistinguishable from the spatiotemporally reflected exchange of the corresponding antiparticle. Our main objective in this chapter will be to show that the *combination* of these three discrete operations—parity, time reversal, and charge conjugation—is *necessarily* an exact symmetry of any local relativistic quantum field theory, even when (as is the case in the Standard Model of elementary particles as it presently stands) none of

---

[1] Note that Λ_P and Λ_T still satisfy $\Lambda^T g \Lambda = g$.

the individual operations represents an exact symmetry of the theory. This result—commonly referred to as the *TCP theorem*—will be obtained first for the case of theories with a dynamics specified by a Lagrangian (or Hamiltonian) density, and secondly, in a more general framework, by appealing to the irreducible fundamental properties of local field theory as incorporated in the Wightman axioms of Section 9.2. As a natural concomitant of the axiomatic proof of the TCP theorem, we shall also indicate how the Spin-Statistics connection (previously discussed in Section 7.3, in the context of Hamiltonian densities built from polynomials in local fields) also arises rigorously from the spectral and locality axioms of the Wightman formulation of field theory, with no commitment needed to a specific Lagrangian/Hamiltonian dynamics.

13.1   Parity properties of a general local covariant field

Consider an improper (determinant –1) Lorentz transformation Λ_P defined by the transformation $\vec{x} \to -\vec{x}, x^0 \to x^0$. The reversal of spatial directions implies that both $\vec{x}$ and $\vec{p}$ change sign under parity, while the orbital angular momentum vector $\vec{x} \times \vec{p}$ is invariant, a property which we shall assume applies also to spin angular momentum. We shall start by defining the parity operation for the "bare" states $|\vec{k}, \sigma\rangle$ created by the interaction picture free fields. Define a unitary parity operator $\mathcal{P} \equiv U(\Lambda_P)$ as follows: on the vacuum, we have simply

$$\mathcal{P}|0\rangle = |0\rangle$$

while on a single particle state

$$\mathcal{P}|\vec{k}, \sigma\rangle = \eta^*|-\vec{k}, \sigma\rangle, \quad |\eta| = 1 \tag{13.1}$$

where the complex unimodular number η is called the "intrinsic parity" of the given particle. Likewise, for a two-particle state

$$\mathcal{P}|\vec{k}_1, \sigma_1; \vec{k}_2, \sigma_2\rangle = \eta_1^*\eta_2^*|-\vec{k}_1, \sigma_1; -\vec{k}_2, \sigma_2\rangle \tag{13.2}$$

$$\Rightarrow \mathcal{P}a^\dagger(\vec{k}_1, \sigma_1)|\vec{k}_2, \sigma_2\rangle = \eta_1^*a^\dagger(-\vec{k}_1, \sigma_1)\eta_2^*|-\vec{k}_2, \sigma_2\rangle \tag{13.3}$$

$$= \eta_1^*a^\dagger(-\vec{k}_1, \sigma_1)\mathcal{P}|\vec{k}_2, \sigma_2\rangle \tag{13.4}$$

and in general, given the obvious generalization of (13.2) to general multi-particle states, it is obvious that we will have

$$\mathcal{P}a^\dagger(\vec{k}, \sigma) = \eta^*a^\dagger(-\vec{k}, \sigma)\mathcal{P} \tag{13.5}$$

leading to the following pair of transformation equations for the creation and annihilation operators (note: $\mathcal{P}$ is unitary)

$$\mathcal{P}a^\dagger(\vec{k}, \sigma)\mathcal{P}^{-1} = \eta^*a^\dagger(-\vec{k}, \sigma) \tag{13.6}$$

$$\mathcal{P}a(\vec{k}, \sigma)\mathcal{P}^{-1} = \eta a(-\vec{k}, \sigma) \tag{13.7}$$

Note that we must in general allow for the charge conjugate antiparticle to have a different intrinsic parity (more on this later):

$$\mathcal{P}a^c(\vec{k},\sigma)\mathcal{P}^{-1} = \eta^c a(-\vec{k},\sigma) \tag{13.8}$$

Of course, an exactly similar procedure can be used to define parity operators $\mathcal{P}_{\text{in}}$ (resp. $\mathcal{P}_{\text{out}}$) acting on the in- (resp. out-)states of the theory, simply by appending a subscript "in" or "out" to the state vectors and creation–annihilation operators in the preceding. The presence of an exact parity symmetry ("parity conservation") in the theory then amounts to the statement that a *single* parity operator $\mathcal{P}_{\text{in}} = \mathcal{P}_{\text{out}}$ effects the parity transformation for both sets of asymptotic states. This occurs in the event that the *full* Hamiltonian of the theory commutes with the parity operator $\mathcal{P}$ defined on the interaction-picture states as above. Namely, if we can choose the intrinsic parity quantum numbers η for the participating particles such that $[\mathcal{P},H_0] = [\mathcal{P},V] = 0$, it follows in the usual way that $[\mathcal{P},V_{\text{int}}] = 0$, and hence $[\mathcal{P},S] = 0$. This in turn implies

$$S_{\beta\alpha} = \langle\beta|S|\alpha\rangle = \langle\beta|\mathcal{P}^{-1}S\mathcal{P}|\alpha\rangle = \eta_\beta\eta_\alpha^*\langle\beta'|S|\alpha'\rangle = \eta_\beta\eta_\alpha^* S_{\beta'\alpha'} \tag{13.9}$$

where the prime indicates states with reversed momenta but unchanged spins (see Fig. 13.1 for an example). Exactly the same result obtains by applying a similarity operation to the sequence of out annihilation and in creation operators whose vacuum expectation value expresses the above S-matrix element, and using the identity of $\mathcal{P}_{\text{in}}$ and $\mathcal{P}_{\text{out}}$. As a consequence of the unimodularity of the intrinsic parities, $|S_{\beta\alpha}| = |S_{\beta'\alpha'}|$, which generally prevents the appearance in the scattering amplitudes of mixtures of scalar and pseudoscalar functions of spin and momentum (for many explicit examples, see Chapter 3 of the excellent book of Sakurai (Sakurai, 1964)).

As usual, we build symmetries into the Hamiltonian (or Lagrangian) density $\mathcal{H}$ (resp. $\mathcal{L}$) by constructing fields which have simple transformation properties under the symmetry group—in this case, the discrete parity transformation. As above, we begin by considering the effect of the parity operation as defined on a free interaction-picture (or in- or out-)field. We determined in Section 7.3 the general form of a free local covariant field $\phi^{AB}(x)$ transforming according to an irreducible (AB) representation of the Lorentz group, and restate the final result (7.84) here:

$$\phi_{ab}^{AB}(x) = \int \frac{d^3k}{(2\pi)^{3/2}\sqrt{2E(k)}} \sum_\sigma (u_{ab}^{AB}(\vec{k},\sigma)e^{-ik\cdot x}a(\vec{k},\sigma)$$

$$+(-)^{2B}(-)^{j-\sigma}u_{ab}^{AB}(\vec{k},-\sigma)e^{ik\cdot x}a^{c\dagger}(\vec{k},\sigma)) \tag{13.10}$$

From (13.6–13.8) we find immediately for the transformation of this field under $\mathcal{P}$

$$\mathcal{P}\phi_{ab}^{AB}(\vec{x},t)\mathcal{P}^{-1} = \int \frac{d^3k}{(2\pi)^{3/2}\sqrt{2E}} \sum_\sigma (u_{ab}^{AB}(\vec{k},\sigma)e^{-ik\cdot x}\eta a(-\vec{k},\sigma)$$

$$+(-)^{2B+j-\sigma}u_{ab}^{AB}(\vec{k},-\sigma)e^{ik\cdot x}\eta^{c*}a^{c\dagger}(-\vec{k},\sigma)) \tag{13.11}$$

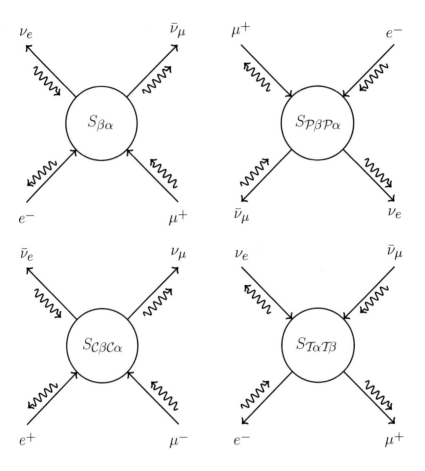

Fig. 13.1 A weak interaction process $(e^- + \mu^+ \to \nu_e + \bar{\nu}_\mu)$ and its $\mathcal{P}, \mathcal{C},$ and $\mathcal{T}$ transforms. The momentum (straight arrow) and spin (squiggly arrow) vectors are to be taken literally: we have chosen an example with incoming and outgoing particles of definite helicity.

Recall the explicit formula (7.69) derived earlier for the coefficient functions u_{ab}^{AB}:

$$u_{ab}(\vec{k}, \sigma) = \sum_{a'b'} (e^{-\theta \hat{k} \cdot \vec{A}})_{aa'} (e^{+\theta \hat{k} \cdot \vec{B}})_{bb'} \langle A B \, a' \, b' | j \, \sigma \rangle \tag{13.12}$$

We can reverse the sign of the momentum by inverting the boost rapidity angle θ, and also use the Clebsch–Gordon identity $\langle A B a' b' | j \sigma \rangle = (-)^{A+B-j} \langle B A b' a' | j \sigma \rangle$, obtaining

$$u_{ab}^{AB}(-\vec{k}, \sigma) = \sum_{a'b'} (e^{\theta \hat{k} \cdot \vec{A}})_{aa'} (e^{-\theta \hat{k} \cdot \vec{B}})_{bb'} (-)^{A+B-j} \langle B A b' a' | j \sigma \rangle = (-)^{A+B-j} u_{ba}^{BA}(\vec{k}, \sigma) \tag{13.13}$$

Changing integration variable $\vec{k} \to -\vec{k}$ in (13.11), and defining the parity inverted spacetime point $x' = (x^0, -\vec{x})$:

$$\mathcal{P}\phi_{ab}^{AB}(\vec{x},t)\mathcal{P}^{-1} = (-)^{A+B-j} \int \frac{d^3k}{(2\pi)^{3/2}\sqrt{2E}} \sum_\sigma (u_{ba}^{BA}(\vec{k},\sigma)e^{-ik\cdot x'}\eta a(\vec{k},\sigma)$$

$$+ (-)^{2B-2A}\eta^{c*}(-)^{2A+j-\sigma}u_{ba}^{BA}(\vec{k},-\sigma)e^{ik\cdot x'}a^{c\dagger}(\vec{k},\sigma)) \qquad (13.14)$$

The field appearing in (13.14) must be one of the covariant fields listed in Section 7.3 (cf. (7.84)), if a Hamiltonian constructed from such fields is to stand a chance of commuting with the parity operator $\mathcal{P}$. Indeed, it is almost the field ϕ_{ba}^{BA}! For this to work, however, we must choose

$$\eta^{c*}(-)^{2B-2A} = \eta \;\Rightarrow\; \eta^c = (-)^{2A-2B}\eta^* = (-)^{2j}\eta^* \qquad (13.15)$$

This establishes the well-known theorem that the intrinsic parity $\eta\eta^c$ of a particle–antiparticle system is $(-)^{2j}$ (hence negative for fermions)—a result critical for the understanding of positronium decay, for example.

Incorporating the constraint (13.15), we find the desired transformation law of an arbitrary covariant field under the parity operation:

$$\mathcal{P}\phi_{ab}^{AB}(\vec{x},t)\mathcal{P}^{-1} = \eta(-)^{A+B-j}\phi_{ba}^{BA}(-\vec{x},t) \qquad (13.16)$$

As a specific example, for a Dirac field, with $A + B = j = \frac{1}{2}$, the effect of the parity transformation evidently reverses the upper and lower components of the Dirac 4-spinor (as well as the spatial coordinate $\vec{x}$, of course), so (cf. (7.107)) we have simply

$$\mathcal{P}\psi(\vec{x},t)\mathcal{P}^{-1} = \eta\gamma_0\psi(-\vec{x},t) \qquad (13.17)$$

The transformation rule (13.16) for the free interaction-picture fields (or, with appropriate subscripts, in- or out-fields) applies whether or not the interacting dynamics of the theory is invariant under parity. Only if the interaction part of the Hamiltonian commutes with $\mathcal{P}$ however, can we extend (13.16) to the full Heisenberg fields $\phi_H^{AB}(\vec{x},t)$, as in this case $\mathcal{P}$ commutes with the transformation operator $U(t,0) = e^{iH_0t}e^{-iHt}$ effecting the transformation from interaction to Heisenberg picture (cf. (9.1)).[2] In this case, parity symmetry is equivalent to the statement that the full Heisenberg fields satisfy

$$\mathcal{P}\phi_{Hab}^{AB}(\vec{x},t)\mathcal{P}^{-1} = \eta(-)^{A+B-j}\phi_{Hba}^{BA}(-\vec{x},t) \qquad (13.18)$$

with $\mathcal{P}|0\rangle = |0\rangle$ where $|0\rangle = |0\rangle_{in} = |0\rangle_{out}$ is the vacuum of the full Hamiltonian of the theory. The transformation property (13.18) transfers immediately to a corresponding transformation rule for the Wightman functions of the theory, and thence, by the

---

[2] Of course, we must acknowledge here the usual embarrassment occasioned by Haag's theorem: in our appeal to the interaction picture, we presume some regularization which restores the existence of the interaction picture, which either preserves parity or a sufficiently close simulacrum thereto that the passage to a continuum limit is smooth *vis-à-vis* the parity properties of the theory.

Haag–Ruelle theory, to the exact S-matrix elements, and we recover the phenomeno-logical constraints (13.9). In a theory like the weak interactions, on the other hand, where parity is broken, there does not—indeed cannot—exist a single unitary operator $P(= P_{\text{in}} = P_{\text{out}})$ effecting the transformation (13.18) on the Heisenberg fields of the theory. Exactly the same situation holds with respect to the discrete symmetries of charge conjugation and time reversal which we are about to consider: the reader should keep in mind that although these operations are conveniently *defined* in terms of free fields and states, their extension to the fully interacting Heisenberg fields of the theory presupposes that the symmetry is in fact unbroken. As we shall see, the only operation for which this is in fact the case in the real world (assuming the general validity of the Wightman axioms) is the combined application of time-reversal, charge-conjugation, and parity (in any order), the $\mathcal{TCP}$ operator.

13.2 Charge-conjugation properties of a general local covariant field

Many interactions of interest in physics (in particular, the strong, electromagnetic and gravitational—though, as with parity, not the weak) are invariant under a symmetry operation which interchanges particles with antiparticles. This operation is called *charge conjugation*—somewhat misleadingly, as it can apply in situations where the particles have zero electrical charge (but some other global quantum number, e.g., strangeness, is non-zero, and therefore distinguishes the particle from the antiparticle). Define a unitary operator on the space of free multi-particle states as follows (ζ, ζ_c phases, or "charge conjugation quantum numbers"):

$$\mathcal{C}a(\vec{k}, \sigma)\mathcal{C}^{-1} = \zeta a^c(\vec{k}, \sigma)$$

$$\mathcal{C}a^\dagger(\vec{k}, \sigma)\mathcal{C}^{-1} = \zeta^* a^{c\dagger}(\vec{k}, \sigma)$$

$$\mathcal{C}a^c(\vec{k}, \sigma)\mathcal{C}^{-1} = \zeta_c a(\vec{k}, \sigma)$$

$$\mathcal{C}a^{c\dagger}(\vec{k}, \sigma)\mathcal{C}^{-1} = \zeta_c^* a^\dagger(\vec{k}, \sigma) \tag{13.19}$$

For example, a non-self-conjugate spinless field would transform under charge conjugation as follows:

$$\mathcal{C}\phi(x)\mathcal{C}^{-1} = \int \frac{d^3k}{(2\pi)^{3/2}\sqrt{2E}} (\zeta e^{-ik\cdot x} a^c(\vec{k}) + \zeta_c^* e^{ik\cdot x} a^\dagger(\vec{k})) \tag{13.20}$$

Provided we choose $\zeta = \zeta_c^*$, the resultant field is simply related to the hermitian conjugate of ϕ (allowing us to construct hermitian, local, and charge conjugation invariant interactions, by balancing $\phi(x)$ and $\phi^\dagger(x)$ factors in the interaction density):

$$\mathcal{C}\phi(x)\mathcal{C}^{-1} = \zeta\phi^\dagger(x) \tag{13.21}$$

The general analysis for non-zero spin is a bit messy, though straightforward. First, we shall need the conjugation property for a general rotation matrix D^j (see, for example, (Messiah, 1966), Appendix C, Eqs. (62, 64)):

$$Y^{(j)} D^j(R) Y^{(j)\dagger} = D^{j*}(R) \tag{13.22}$$

where $Y^{(j)}_{mm'} \equiv (-1)^{j+m} \delta_{m,-m'}$, and the complex conjugation on the right-hand side is applied to each matrix element (no transpose!). Taking the rotation R to be infinitesimal, $D^j = 1 - i\vec{\epsilon} \cdot \vec{J}$, this implies for the generators

$$Y^{(j)} \vec{J} Y^{(j)\dagger} = -\vec{J}^* \tag{13.23}$$

so that we have for the boost type operators

$$(e^{\theta \hat{k} \cdot \vec{J}})^* = Y^{(j)} e^{-\theta \hat{k} \cdot \vec{J}} Y^{(j)\dagger} \tag{13.24}$$

All of the above, of course, applies to any set of operators satisfying the angular momentum algebra, and in particular to our A- and B-type operators forming the HLG Lie algebra. Consequently:

$$(e^{-\theta \hat{k} \cdot \vec{A}})^*_{aa'} = (Y^{(A)} e^{\theta \hat{k} \cdot \vec{A}} Y^{(A)\dagger})_{aa'} = (-1)^{2A+a+a'} (e^{\theta \hat{k} \cdot \vec{A}})_{-a,-a'} \tag{13.25}$$

providing the information needed to derive the relevant complex conjugation property:

$$
\begin{aligned}
u^{AB}_{ab}(\vec{k}, \sigma)^* &= \sum_{a'b'} (-)^{2A+a+a'} (e^{\theta \hat{k} \cdot \vec{A}})_{-a,-a'} (-)^{2B+b+b'} (e^{-\theta \hat{k} \cdot \vec{B}})_{-b,-b'} \langle ABa'b'|j\sigma\rangle \\
&= (-)^{A+B+\sigma} (-)^{A+a} (-)^{B+b} \sum_{a'b'} (e^{-\theta \hat{k} \cdot \vec{B}})_{-b,b'} (e^{\theta \hat{k} \cdot \vec{A}})_{-a,a'} \langle AB - a' - b'|j\sigma\rangle \\
&= (-)^{A+B+\sigma} (-)^{A+a} (-)^{B+b} \sum_{a'b'} (e^{-\theta \hat{k} \cdot \vec{B}})_{-b,b'} (e^{\theta \hat{k} \cdot \vec{A}})_{-a,a'} \langle BAb'a'|j-\sigma\rangle \\
&= (-)^{A+B-j} (-)^{j+\sigma} (-)^{A+a} (-)^{B+b} u^{BA}_{-b,-a}(\vec{k}, -\sigma) \\
&= (-)^{A+B-j} Y^{(j)}_{\sigma\sigma'} Y^{(A)}_{aa'} Y^{(B)}_{bb'} u^{BA}_{b'a'}(\vec{k}, \sigma')
\end{aligned} \tag{13.26}
$$

The transformation property of the general covariant field (13.10) under the C-operation now follows straightforwardly:

$$
\begin{aligned}
\mathcal{C}\phi^{AB}_{ab}(x)\mathcal{C}^{-1} &= \int \frac{d^3k}{(2\pi)^{3/2}\sqrt{2E}} \sum_\sigma (u^{AB}_{ab}(\vec{k}, \sigma)\zeta a^c(\vec{k}, \sigma) e^{-ik\cdot x} \\
&\qquad + (-)^{2B}(-)^{j-\sigma} u^{AB}_{ab}(\vec{k}, -\sigma)\zeta^*_c a^\dagger(\vec{k}, \sigma) e^{ik\cdot x}) \\
&= (-)^{A+B-j} \sum_{\sigma\sigma'b'a'} \int \frac{d^3k}{(2\pi)^{3/2}\sqrt{2E}} Y^{(A)}_{aa'} Y^{(B)}_{bb'} \{Y^{(j)}_{\sigma\sigma'} u^{BA}_{b'a'}(\vec{k}, \sigma')\zeta^* a^{c\dagger}(\vec{k}, \sigma) e^{ik\cdot x} \\
&\qquad + (-)^{2B}(-)^{j-\sigma} Y^{(j)}_{-\sigma,-\sigma'} u^{BA}_{b'a'}(\vec{k}, -\sigma')\zeta_c a(\vec{k}, \sigma) e^{-ik\cdot x}\}^\dagger \\
&= (-)^{A+B-j} \sum_{b'a'} Y^{(A)}_{aa'} Y^{(B)}_{bb'} \int \frac{d^3k}{(2\pi)^{3/2}\sqrt{2E}} \sum_\sigma \{(-)^{2B}\zeta_c u^{BA}_{b'a'}(\vec{k}, \sigma) a(\vec{k}, \sigma) e^{-ik\cdot x} \\
&\qquad + \zeta^*(-)^{j+\sigma} u^{BA}_{b'a'}(\vec{k}, -\sigma) a^{c\dagger}(\vec{k}, \sigma) e^{ik\cdot x}\}^\dagger
\end{aligned} \tag{13.27}
$$

Using the relation $(-)^{j+\sigma} = (-)^{j-\sigma}(-)^{2j} = (-)^{j-\sigma}(-)^{2A+2B}$ (as $2j - 2\sigma$ is even, and A and B must be able to couple to j), this becomes

$$\mathcal{C}\phi_{ab}^{AB}(x)\mathcal{C}^{-1}$$

$$= (-)^{A+B-j} \sum_{b'a'} Y_{aa'}^{(A)} Y_{bb'}^{(B)} \int \frac{d^3k}{(2\pi)^{3/2}\sqrt{2E}} \sum_{\sigma} \{\zeta_c(-)^{2B} u_{b'a'}^{BA}(\vec{k},\sigma) a(\vec{k},\sigma) e^{-ik\cdot x}$$

$$+ \zeta^*(-)^{2B}(-)^{2A}(-)^{j-\sigma} u_{b'a'}^{BA}(\vec{k},-\sigma) a^{c\dagger}(\vec{k},\sigma) e^{ik\cdot x}\}^\dagger \tag{13.28}$$

Comparing the field in the last expression with the general result (7.84), we see that we recover the (hermitian adjoint) of a covariant field if and only if

$$\zeta_c = \zeta^* \tag{13.29}$$

in which case the field satisfies the transformation rule

$$\mathcal{C}\phi_{ab}^{AB}(x)\mathcal{C}^{-1} = \zeta(-)^{A-B-j} \sum_{a'b'} Y_{aa'}^{(A)} Y_{bb'}^{(B)} \phi_{b'a'}^{BA\dagger}(x) \tag{13.30}$$

As in the case of parity, if $A \neq B$ we must include *both* ϕ^{AB} and ϕ^{BA} fields in the Hamiltonian if the theory is to be invariant under C-conjugation. However, the combined operation of charge conjugation and parity, CP, does not mix these two types, so CP-invariant theories are possible in the presence of asymmetric (chiral) representations:

$$\mathcal{CP}\phi_{ab}^{AB}(c)(\mathcal{CP})^{-1} = \eta\zeta(-)^{2(B-j)} \sum_{a'b'} Y_{aa'}^{(A)} Y_{bb'}^{(B)} \phi_{a'b'}^{AB\dagger}(-\vec{x},t) \tag{13.31}$$

Again, in analogy to (13.17) for the parity operator, the general result (13.30) reduces in the Dirac field case $((\frac{1}{2},0) \oplus (0,\frac{1}{2})$ representation) to

$$\mathcal{C}\psi(x)\mathcal{C}^{-1} = \zeta C\psi^\dagger(x), \quad C \equiv i\gamma^0\gamma^2 \tag{13.32}$$

Another example: choosing ζ for a j=1, $(\frac{1}{2},\frac{1}{2})$ field. The connection between the four-vectorial notation ϕ^μ and the ϕ_{ab}^{AB} notation is as follows

$$\phi_{\frac{1}{2},\frac{1}{2}}^{\frac{1}{2}\frac{1}{2}} = \frac{1}{\sqrt{2}}(\phi^1 - i\phi^2)$$

$$\phi_{\frac{1}{2},-\frac{1}{2}}^{\frac{1}{2}\frac{1}{2}} = \frac{1}{\sqrt{2}}(\phi^0 - \phi^3)$$

$$\phi_{-\frac{1}{2},\frac{1}{2}}^{\frac{1}{2}\frac{1}{2}} = -\frac{1}{\sqrt{2}}(\phi^0 + \phi^3)$$

$$\phi_{-\frac{1}{2},-\frac{1}{2}}^{\frac{1}{2}\frac{1}{2}} = -\frac{1}{\sqrt{2}}(\phi^1 + i\phi^2) \tag{13.33}$$

If we now choose $\zeta = \pm 1$, for example, then, using (13.30)

$$\mathcal{C}\phi_{\frac{1}{2},\frac{1}{2}}^{\frac{1}{2}\frac{1}{2}}\mathcal{C}^{-1} = \mp\phi_{-\frac{1}{2},-\frac{1}{2}}^{\dagger} = \pm\frac{1}{\sqrt{2}}(\phi^{1\dagger} - i\phi^{2\dagger}) \tag{13.34}$$

and similarly for the other three components. In terms of the vectorial components, one easily finds

$$\mathcal{C}\phi^{\mu}(x)\mathcal{C}^{-1} = \pm\phi^{\mu\dagger}(x) \tag{13.35}$$

In the Hamiltonian of quantum electrodynamics, a four-vector field ($\phi^{\mu} = A^{\mu} =$ photon field, in this case self-conjugate) appears multiplying a four-vector current built from charged particle fields which changes sign under charge conjugation, so to ensure that the charge-conjugation operator $\mathcal{C}$ commutes with the Hamiltonian, we also assign $\zeta = -1$ to the photon field.

A comment on the underlying logic of all of this. We are free to *define* the $\mathcal{C}$ and $\mathcal{P}$ operators with any phase constants ζ, η we like (for any given field), but there are two possible outcomes: either

1. there exists a choice for ζ, η for each participating field such that some interaction (say the electromagnetic part of the Hamiltonian) then commutes with $\mathcal{C}$ (or $\mathcal{P}$), in which case we assign those values to the fields as the intrinsic parity and charge-conjugation quantum number of the associated particles; or
2. there is *no* way to pick ζ, η for all the fields so that $\mathcal{C}$ (or $\mathcal{P}$) commute with the Hamiltonian for the given interaction. In that case we say that the interaction *breaks* C (or P, or both). As pointed out above, this implies that the simple transformation properties (under C or P) of the interaction-picture fields *cannot* be extended to the full Heisenberg fields of the theory.

The point is that by assigning quantum numbers as in (1) above (if possible) we gain a new operator commuting with the Hamiltonian, and the more of these there are, the easier it is to solve any quantum theory.

One last caution: it is common to use the letter C to refer to the charge-conjugation quantum number (ζ above), as well as the unitary Hilbert space operator $\mathcal{C}$ implementing charge-conjugation on the states. I have avoided this for obvious reasons, while using the letter C as a general denotation of the charge-conjugation operation, but the reader should be aware of this when consulting the standard texts.

13.3 Time-reversal properties of a general local covariant field

The third, and final, discrete spacetime symmetry which remains to be discussed is that of invariance of the dynamical evolution under reversal of the direction of time flow. The basic properties of time reversal in quantum mechanics were discussed already in Section 4.1.3, which the reader may find useful to review at this point. We saw there that the operation of time reversal in quantum mechanics is implemented by an antiunitary operator $\mathcal{T}$ which reverses the direction of spatial momenta. Thus for a particle of spin j, the one-particle state transforms under the $\mathcal{T}$ operation as follows (cf. (4.60))

$$T|\vec{k}, \sigma\rangle = \tau^* Y^{(j)}_{\sigma\sigma'}|-\vec{k}, \sigma'\rangle, \quad |\tau| = 1 \tag{13.36}$$

As in the case of parity and charge conjugation, the antiunitary character of the time reversal operator allows for an arbitrary phase factor τ (resp. τ^c)—or intrinsic "time-parity"—for the particle (resp. antiparticle) in question. The transformation property (13.36) transfers in the usual way to a transformation property of creation and destruction operators under a similarity transformation with T:

$$T a^\dagger(\vec{k}, \sigma) T^{-1} = \tau^* Y^{(j)}_{\sigma\sigma'} a^\dagger(-\vec{k}, \sigma') \tag{13.37}$$

$$T a(\vec{k}, \sigma) T^{-1} = \tau Y^{(j)}_{\sigma\sigma'} a(-\vec{k}, \sigma') \tag{13.38}$$

$$T a^{c\dagger}(\vec{k}, \sigma) T^{-1} = \tau^{c*} Y^{(j)}_{\sigma\sigma'} a^{c\dagger}(-\vec{k}, \sigma') \tag{13.39}$$

$$T a^c(\vec{k}, \sigma) T^{-1} = \tau^c Y^{(j)}_{\sigma\sigma'} a^c(-\vec{k}, \sigma') \tag{13.40}$$

Recalling that T contains the charge-conjugation operator K (cf. (4.60)), one finds for the time-reversal transformation of the covariant field (13.10), using (13.38, 13.39),

$$T \phi^{AB}_{ab}(x) T^{-1} = \int \frac{d^3 k}{(2\pi)^{3/2} \sqrt{2E(k)}} \sum_\sigma (u^{AB*}_{ab}(\vec{k}, \sigma) e^{ik\cdot x} \tau Y^{(j)}_{\sigma\sigma'} a(-\vec{k}, \sigma')$$

$$+ (-)^{2B}(-)^{j-\sigma} u^{AB*}_{ab}(\vec{k}, -\sigma) e^{-ik\cdot x} \tau^{c*} Y^{(j)}_{\sigma\sigma'} a^{c\dagger}(-\vec{k}, \sigma')) \tag{13.41}$$

Changing integration variable $\vec{k} \to -\vec{k}$, defining $x_t = (\vec{x}, -t)$, and using the reflection and conjugation properties (13.13, 13.26) of the $u^{AB}_{ab}(\vec{k}, \sigma)$, a few lines of algebra (see Problem 3) leave us with

$$T \phi^{AB}_{ab}(x) T^{-1} = \tau Y^{(A)}_{aa'} Y^{(B)}_{bb'} \int \frac{d^3 k}{(2\pi)^{3/2} \sqrt{2E(k)}} \sum_\sigma (u^{AB}_{a'b'}(\vec{k}, \sigma) e^{-ik\cdot x_t} a(\vec{k}, \sigma)$$

$$+ \frac{\tau^{c*}}{\tau}(-)^{2B}(-)^{j-\sigma} u^{AB}_{a'b'}(\vec{k}, -\sigma) e^{ik\cdot x_t} a^{c\dagger}(\vec{k}, \sigma)) \tag{13.42}$$

The right-hand side of this expression is clearly a covariant field of type $\phi^{AB}_{a'b'}$, *provided we make the by now usual identification*

$$\tau^c = \tau^* \tag{13.43}$$

whereupon the desired transformation law for $\phi^{AB}_{ab}(x)$ emerges

$$T \phi^{AB}_{ab}(\vec{x}, t) T^{-1} = \tau Y^{(A)}_{aa'} Y^{(B)}_{bb'} \phi^{AB}_{a'b'}(\vec{x}, -t) = \tau(-)^{A+B+a+b} \phi^{AB}_{-a,-b}(\vec{x}, -t) \tag{13.44}$$

13.4 The TCP and Spin-Statistics theorems

Invariance under the combined operation of parity inversion, charge-conjugation, and time reversal enjoys a special status in local relativistic quantum field theories (see Fig. 13.2). Independently of the detailed dynamics of such theories, it follows from only the basic underlying ingredients outlined in Chapter 3: quantum theory, Lorentz-invariance, and locality (specifically, *weak local commutativity*, which asserts the

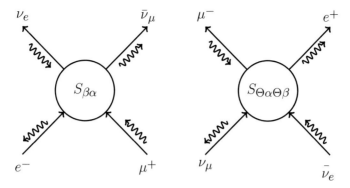

Fig. 13.2 The weak interaction process $(e^- + \mu^+ \to \nu_e + \bar{\nu}_\mu)$ of Fig. 13.1 and its $\Theta = \mathcal{TCP}$ transform. The TCP theorem ensures equality (up to phase) of the amplitudes for the process and its TCP transform in any (weakly) local relativistic quantum field theory describing the interactions responsible for the process.

vanishing of the vacuum expectation value of space-like commutators). We shall first demonstrate this result—the famous "TCP theorem"—in the context of a Lagrangian field theory, in which the dynamics is specified in terms of fields transforming under T, C, and P as indicated in the preceding sections, coupled together to form a Lorentz-scalar Lagrangian density. The more general argument, based on the axioms of axiomatic field theory discussed in Section 9.2, will then be outlined.

First note that for a complex scalar field ϕ (i.e., a field in the $A = 0, B = 0$ representation of the HLG), undergoing the P, C, and T transformations

$$\mathcal{P}\phi(\vec{x}, t)\mathcal{P}^{-1} = \eta\phi(-\vec{x}, t) \tag{13.45}$$

$$\mathcal{C}\phi(\vec{x}, t)\mathcal{C}^{-1} = \zeta\phi^\dagger(\vec{x}, t) \tag{13.46}$$

$$\mathcal{T}\phi(\vec{x}, t)\mathcal{T}^{-1} = \tau\phi(\vec{x}, -t) \tag{13.47}$$

the transformation property under the combined effect of $\Theta \equiv \mathcal{TCP}$ becomes remarkably simple:

$$\Theta\phi(\vec{x}, t)\Theta^{-1} = (\eta\zeta\tau)^*\phi^\dagger(-\vec{x}, -t) \tag{13.48}$$

We are at liberty to choose the phases η, ζ, τ defining the corresponding discrete symmetries as we please, and we shall henceforth choose

$$\eta\zeta\tau = 1 \tag{13.49}$$

whence (returning to spacetime coordinate notation)

$$\Theta\phi(x)\Theta^{-1} = \phi^\dagger(-x), \quad \Theta\partial_\mu\phi(x)\Theta^{-1} = -\partial_\mu\phi^\dagger(-x) \tag{13.50}$$

It follows that the action obtained as the spacetime integral of any *hermitian, Lorentz-scalar* Lagrangian density $\mathcal{L}(\phi, \partial_\mu\phi, \phi^\dagger, \partial_\mu\phi^\dagger)$ is invariant under the combined TCP operation, which simply interchanges ϕ and $\phi^\dagger$ (which must appear symmetrically if the Lagrangian is to be hermitian, and the dynamics unitary), and switches $x \to -x$ (the so-called "strong reflection" operation), which leaves the spacetime integral unaltered. Spacetime gradients must appear paired for the Lagrangian to be a scalar, so the minus signs in field derivatives must also cancel. The hermitian stress-energy tensor $T_{\mu\nu}(x)$ for such a theory (cf. Section 12.4),

$$T_{\mu\nu}(x) = \frac{\partial \mathcal{L}}{\partial(\partial_\mu \phi)} \partial_\nu \phi(x) + \frac{\partial \mathcal{L}}{\partial(\partial_\mu \phi^\dagger)} \partial_\nu \phi^\dagger(x) - g_{\mu\nu} \mathcal{L} \tag{13.51}$$

clearly satisfies the strong reflection property

$$\Theta\, T_{\mu\nu}(x)\, \Theta^{-1} = T_{\mu\nu}(-x) \tag{13.52}$$

Recalling that the Hamiltonian H of the theory is simply the spatial integral of the density $T_{00}(\vec{x}, 0)$,

$$H = \int d^3 x\, T_{00}(\vec{x}, 0) \Rightarrow \Theta\, H\, \Theta^{-1} = H \Rightarrow [\Theta, H] = 0 \tag{13.53}$$

so the Hamiltonian of our theory is also invariant under the TCP operation.

The above argument is easily generalized to fields $\phi_{ab}^{AB}(x)$ in arbitrary representations of the HLG: remarkably, we find that any such fields have (up to a sign) *exactly the same transformation property* as the scalar field above. Combining (13.16, 13.30, 13.44) (or more directly, (13.31) and (13.44)), a few lines of algebra lead to

$$\Theta\, \phi_{ab}^{AB}(x)\, \Theta^{-1} = (\eta\zeta\tau)^*(-)^{2A} \phi_{ab}^{AB\dagger}(-x) = (-)^{2A} \phi_{ab}^{AB\dagger}(-x) \tag{13.54}$$

where we have inserted the phase choice (13.49). The argument for invariance of a Lagrangian field theory constructed from such fields now follows directly. First, we note that the four-vector spacetime gradient ∂_μ, which transforms according to the $(\frac{1}{2}, \frac{1}{2})$ representation of the HLG, when applied to a ϕ^{AB}-type field multiplet, generates, by the usual Clebsch–Gordon machinery of angular momentum addition, a combination of irreducible fields of type $\phi^{A'B'}$ (with $A' = |A \pm \frac{1}{2}|$, $B' = |B \pm \frac{1}{2}|$), which individually transform under T, C, and P just as we have indicated above. Thus, we may assume henceforth that the (scalar!) Lagrangian density of our theory is simply a sum of terms involving products of ϕ^{AB} fields in which the independent A and B representation labels are separately coupled to zero spin. In any such product, the sign factors $(-)^{2A}$ must also multiply to give $+1$, and we therefore have the strong reflection property for the scalar Lagrangian density

$$\Theta\, \mathcal{L}(x)\, \Theta^{-1} = \mathcal{L}^\dagger(-x) = \mathcal{L}(-x) \tag{13.55}$$

where we have again assumed an hermitian Lagrangian in the last step. The invariance of the action of the theory, given as the spacetime integral of the Lagrangian density, now follows as before. A similar argument[3] ensures the strong reflection property for the Hamiltonian density $T_{00}(x)$, and hence the commutativity of the Hamiltonian with the Θ operator.

The preceding discussion of TCP symmetry is based on a specification of the dynamics in terms of a Lagrangian (or associated Hamiltonian) built from local fields. A much more general understanding of the origins of TCP symmetry can be given using the Wightman axiomatic formalism discussed in Section 9.2, with no commitment whatsoever to a detailed dynamics of the fields. The basic ingredients which, once present, *ensure* the exact TCP invariance of a field theory are (a) the spectral and Lorentz transformation properties needed to establish the analytic properties of the Wightman functions embodied in the Hall–Wightman theorem (specifically, the property (9.76) following directly from that theorem), and (b) a weaker version of space-like commutativity of the fields (called *weak local commutativity*, or WLC for short), in which the vanishing of the relevant commutators (or anticommutators) is only required in the vacuum expectation value. Without giving detailed proofs, we shall outline the argument here.

We see comparing (13.50) and (13.54) that the behavior of a general covariant field under the TCP operation is effectively identical to that of a scalar field, so to strip out inessential details and present the essence of the argument most clearly, we shall assume that we are dealing with a theory of a single real self-interacting scalar field $\phi(x)$. The invariance of the theory (and its vacuum) under TCP implies the existence of an antiunitary operator Θ satisfying (13.50): this then means that the n-point Wightman function defined by[4]

$$W(x_1, x_2, \ldots, x_n) \equiv \langle 0|\phi(x_1)\phi(x_2)\ldots\phi(x_n)|0\rangle \tag{13.56}$$

satisfies the constraint

$$W(x_1, x_2, \ldots, x_n) = W(-x_n, -x_{n-1}, \ldots, -x_1) \tag{13.57}$$

which follows immediately from the fact that $\langle 0|\Theta O \Theta^{-1}|0\rangle = \langle 0|O^\dagger|0\rangle$ for any operator O (by antiunitarity of Θ), if we substitute for O the product of field operators appearing in (13.56). By translation invariance, in terms of the displacements $\xi_i \equiv x_i - x_{i+1}$, this can then be written

$$W(\xi_1, \xi_2, \ldots, \xi_{n-1}) = W(\xi_{n-1}, \xi_{n-2}, \ldots, \xi_1) \tag{13.58}$$

We recall from Section 9.3 that the Haag–Ruelle scattering theory assures us that the entire phenomenological content of the theory contained in the S-matrix is uniquely recoverable from knowledge of the Wightman functions, so the essential physical content of TCP invariance (discussed in greater detail below) can be considered to be fully incorporated in the condition (13.58) on the Wightman functions of the theory.

---

[3] In this case, the stress-energy tensor density transforms like a symmetric rank 2 tensor—i.e., a combination of (0,0) and (1,1) fields. In either case, the sum of A quantum numbers in each term must be integer, so the product of $(-)^{2A}$ factors again gives unity.

[4] We shall be dealing with the full Wightman functions of the theory for the remainder of this chapter, and omit the "H" subscript indicating the omnipresent Heisenberg fields.

Next, we need to review briefly, and then extend somewhat, the analytic properties of Wightman functions discussed previously in Section 9.2. A set of $n-1$ complex 4-coordinates $(\zeta_1, \ldots, \zeta_{n-1})$ is said to lie in the *extended tube* $\mathcal{T}'_{n-1}$ if $\zeta_i = \Lambda z_i$ where Λ is an arbitrary complex Lorentz transformation and the $z_i, i = 1, 2, .., n-1$ lie in the *forward tube* (i.e., $z_i = \xi_i - i\eta_i$, $\eta_i^2 > 0, \eta_i^0 > 0$). The Hall–Wightman theorem establishes the analyticity of the Wightman functions continued into the whole of the extended tube, from which the reflection property (9.76) is easily deduced:

$$W(\zeta_1, \zeta_2, \ldots, \zeta_{n-1}) = W(-\zeta_1, -\zeta_2, \ldots, -\zeta_{n-1}) \tag{13.59}$$

Note that this result hangs only on the spectral and Lorentz transformation axioms of Section 9.2 (in particular Axioms Ia-d, IIa-b): in particular, there has as yet been no appeal to the locality axiom IIc (space-like commutativity or anticommutativity of the fields). A weakened version of the locality property is all that is needed to complete the proof of the TCP theorem, in the form (13.58). Let $(\xi_1, \xi_2, \ldots, \xi_{n-1})$ be a set of real space-like four-vectors with space-like convex hull, i.e.,

$$\left(\sum_{i=1}^{n-1} \lambda_i \xi_i \right)^2 < 0 \; \forall \lambda_i, \quad \sum_i \lambda_i = 1, \quad 0 \le \lambda_i \le 1 \tag{13.60}$$

Note that this implies that for any subset S of $(1, 2, \ldots, n-1)$, $\sum_{i \in S} \xi_i$ is also space-like, and hence, if $\xi_i = x_i - x_{i+1}$, then for $i < j$ we have $x_i - x_j = (x_i - x_{i+1}) + (x_{i+1} - x_{i+2}) + \ldots (x_{j-1} - x_j) = \xi_i + \ldots \xi_{j-1}$ also space-like, and hence $x_i - x_j$ is space-like for all $i \ne j$: the set ξ_i is *totally space-like*. Locality of the theory then implies that we can rearrange the operators $\phi(x_1)\phi(x_2) \ldots \phi(x_n)$ into the reversed order $\phi(x_n)\phi(x_{n-1}) \ldots \phi(x_1)$, and therefore, for the vacuum expectation value obtain

$$W(\xi_1, \ldots, \xi_{n-1}) = \langle 0 | \phi(x_1)\phi(x_2) \ldots \phi(x_n) | 0 \rangle$$
$$= \langle 0 | \phi(x_n)\phi(x_{n-1}) \ldots \phi(x_1) | 0 \rangle = W(-\xi_{n-1}, \ldots, -\xi_1) \tag{13.61}$$

The condition (13.61) on the Wightman functions is *strictly weaker* than the full requirement of space-like commutativity at the operator level, as we are only requiring the commutativity to be effective at the level of the vacuum expectation value: it is usually given the name "weak local commutativity" in the literature. Real points in $4(n-1)$-dimensional space satisfying (13.60) are called *Jost points*, and have a remarkable characteristic: they can all be shown to lie in the extended tube $\mathcal{T}'_{n-1}$! In other words, given any set $(\xi_1, \xi_2, \ldots, \xi_{n-1})$ satisfying (13.60), we can find a complex Lorentz transformation Λ such that $\xi_i = \Lambda z_i$, with z_i in the interior of the (complex) forward tube. Some explicit examples are given in Problem 4.[5] Now, the reader will doubtless recall that two analytic functions of a single complex variable that agree on any finite neighborhood of the real axis must continue to agree when analytically continued to the rest of the complex plane. An exactly similar phenomenon in the case of analytic functions of several complex variables ensures that the equality $W(\xi_1, \ldots, \xi_{n-1}) = W(-\xi_{n-1}, \ldots, -\xi_1)$ valid for real Jost points (and following from

---

[5] For the full proof of this assertion, see (Streater and Wightman, 1978), theorem 2-12.

weak local commutativity) can be extended to points in the entire extended tube $\mathcal{T}'_{n-1}$:

$$W(\zeta_1, \zeta_2, \ldots, \zeta_{n-1}) = W(-\zeta_{n-1}, -\zeta_{n-2}, \ldots, -\zeta_1)$$

$$= W(\zeta_{n-1}, \zeta_{n-2}, \ldots, \zeta_1), \quad \forall \zeta_i \in \mathcal{T}'_{n-1} \qquad (13.62)$$

where the final equality obtains in consequence of (13.59). The extended tube, of course, contains the forward tube, consisting of points $\zeta_i = \xi_i - i\eta_i$ with the η_i future time-like, and the ξ_i *arbitrary* spacetime points (in particular, not necessarily space-like!). Taking the boundary limit $\eta_i \to 0$ in (13.62) we replace the ζ_i with real points ξ_i everywhere and obtain

$$W(\xi_1, \xi_2, \ldots, \xi_{n-1}) = W(\xi_{n-1}, \xi_{n-2}, \ldots, \xi_1) \qquad (13.63)$$

which is precisely the desired TCP theorem, in the form (13.58). The extension of this result to fields in arbitrary representations of the HLG, and in the fermionic case, satisfying space-like anticommutativity (in the weak form) can be found in (Streater and Wightman, 1978) (theorem 4-7).

The immediate phenomenological consequences of TCP symmetry are among the most precisely tested predictions of quantum field theory. In particular, the fact that the exact Hamiltonian H of the world commutes with the $\mathcal{TCP}$ operator implies an exact relation between the static properties, such as mass and magnetic moment (if any), of a particle and its antiparticle. If $|\vec{k} = 0, \sigma\rangle$ is the one particle state of a stable particle at rest, with $H|\vec{k} = 0, \sigma\rangle = m_{\mathrm{ph}}|\vec{k} = 0, \sigma\rangle$ (m_{ph} the physical mass), then TCP invariance requires that the state $\mathcal{TCP}|\vec{k} = 0, \sigma\rangle$ be an eigenstate of H with exactly the same eigenvalue (i.e., m_{ph}). But this state is just the one particle state of the antiparticle at rest (with spin reversed, which by rotational invariance cannot alter the energy). Thus the particle and antiparticle masses must be exactly equal, any deviation implying a violation of TCP invariance. This equality has been tested in the case of the proton/antiproton to better than one part in 10^8. Even more precise tests of TCP invariance are available in the neutral kaon system, where the K^0 and $\bar{K}^0$ masses[6] are found to agree to within one part in 10^{18}!

In similar fashion, TCP symmetry implies that the magnetic moments of particle and antiparticle are equal in magnitude and opposite in sign. This follows from the fact that the single-particle state in the presence of a static magnetic field $\vec{B}$ must have the same energy as its TCP conjugate, which is an antiparticle of reversed spin. On the other hand, the magnetic field $\vec{B}$ is unchanged under TCP ($\vec{B}$ changes sign under C and T, as is clear by considering the behavior of steady currents giving rise to $\vec{B}$, but is unchanged under P, as it is an axial vector). Thus the magnetic moments, like the spins, must be equal and opposite, although the phenomenological checks in this case are much coarser than is the case for particle masses.

Consequences of TCP for typical scattering processes such as the weak process depicted in Fig. 13.2, in which none of the discrete symmetries T, C, or P are separately conserved, are typically much more difficult to check with comparable precision to the static properties described above: one typically finds that the TCP transform of a

---

[6] See (Bloch, 2006) for a recent review of the implications of TCP violation for neutral kaon physics.

phenomenologically accessible process is difficult to observe, and in any event, cross-sections of scattering processes are generally only measurable at a precision level far inferior to that of static properties such as mass or magnetic moment.

We conclude this chapter by revisiting the Spin-Statistics theorem, discussed already in Section 7.3 from the point of view of Hamiltonians expressed in terms of free, interaction-picture fields. As in the case of the TCP theorem, the Spin-Statistics connection can be derived rigorously from general underlying principles of relativistic quantum field theory, as incorporated in the Wightman axioms—with no commitment to a detailed dynamics as specified by a Hamiltonian (or Lagrangian). The strategy is almost identical to the proof of TCP invariance, employing the generalization of (13.59) to the Wightman function of a product of fields $\phi^{A_1 B_1}\phi^{A_2 B_2}\cdots\phi^{A_n B_n}$ in arbitrary representations of the HLG (see Section 9.2, (9.78)):

$$W^{A_1 B_1, A_2 B_2,..,A_n B_n}(\zeta_1,\ldots,\zeta_{n-1}) = (-1)^{\sum_i 2A_i} W^{A_1 B_1, A_2 B_2,..,A_n B_n}(-\zeta_1,\ldots,-\zeta_{n-1})$$

(13.64)

for all ζ_i in the extended tube. We shall now assume that the fields $\phi^{A_i B_i}, i = 1, 2, .., n$ appearing in the Wightman function above are either (a) Bose fields, defining as those with vanishing space-like commutator, or (b) Fermi fields, defined as those with vanishing space-like anticommutator. Bose fields are, of course, assumed to commute at space-like separation with any Fermi field. If we now specialize to totally space-like real Jost points, $\zeta_i = \xi_i$ (which the reader will recall are in the extended tube), the rearrangement of the fields will give a factor $(-1)^P$ where P is the number of transpositions of Fermi fields needed to effect the rearrangement. Thus, at the Jost points

$$W^{A_1 B_1, A_2 B_2,..,A_n B_n}(\xi_1,\ldots,\xi_{n-1})$$

$$= (-1)^{P+2} \sum_i A_i W^{A_n B_n, A_{n-1} B_{n-1},..,A_1 B_1}(\xi_{n-1},\ldots,\xi_1)$$

(13.65)

By analytic continuation to the full extended tube, followed by the return to the real boundary of the forward tube (exactly as above for TCP), we may therefore conclude that for arbitrary real four-coordinates $x_i, i = 1, 2, \ldots, n$ the Wightman functions (now written as a function of the full set of n coordinates, rather than the $n-1$ coordinate differences) satisfy

$$W^{A_1 B_1,...,A_n B_n}(x_1,\ldots,x_n) = (-1)^{P+2} \sum_i A_i W^{A_n B_n,...,A_1 B_1}(-x_n,\ldots,-x_1) \quad (13.66)$$

For $n = 2$, and taking the field $\phi^{A_2 B_2} = (\phi^{A_1 B_1})^\dagger$ (so $A_2 = B_1, B_2 = A_1$, cf. Section 13.2) this implies, in operator language,

$$\langle 0|\phi^{A_1 B_1}(x_1)(\phi^{A_1 B_1})^\dagger(x_2)|0\rangle = (-1)^{P+2(A_1+B_1)} \langle 0|(\phi^{A_1 B_1})^\dagger(-x_2)\phi^{A_1 B_1}(-x_1)|0\rangle$$

(13.67)

Multiplying both sides of (13.66) by test functions $f(x_1)f^*(x_2)$ and integrating over x_1, x_2, and noting that $(-1)^{2(A_1+B_1)} = (-1)^{2j}$, where j is the spin carried by the field $\phi^{A_1 B_1}$,

$$\int f(x_1)\langle 0|\phi^{A_1 B_1}(x_1)(\phi^{A_1 B_1})^\dagger(x_2)|0\rangle f^*(x_2)d^4x_1 d^4x_2$$

$$= (-1)^{P+2j}\int f^*(x_2)\langle 0|(\phi^{A_1 B_1})^\dagger(-x_2)\phi^{A_1 B_1}(-x_1)|0\rangle f(x_1)d^4x_2 d^4x_1 \quad (13.68)$$

which amounts to

$$||\int f^*(x)(\phi^{A_1 B_1}(x))^\dagger|0\rangle d^4x||^2 = (-1)^{P+2j}||\int f(-x)\phi^{A_1 B_1}(x)|0\rangle||^2 \quad (13.69)$$

As the squared norms in (13.69) cannot vanish (if the smeared fields give zero acting on the vacuum, the Wightman functions are zero and we have no theory!), we must have $P + 2j$ an even integer: in other words, Fermi fields (where $P = 1$) correspond to half-integer spin, while Bose fields ($P = 0$) correspond to integer spin.

13.5 Problems

1. Let $\mathcal{P}$ be the unitary operator

$$\mathcal{P} = e^{-i\theta \int d^3k\, a^\dagger(\vec{k})(a(\vec{k})+\lambda a(-\vec{k}))} \quad (13.70)$$

where $a(\vec{k})$ is the destruction operator for a spinless, self-conjugate boson. Find θ and λ (in terms of η) so that $\mathcal{P}$ will be the parity operator for a particle of intrinsic parity η:

$$\mathcal{P}a(\vec{k})\mathcal{P}^{-1} = \eta a(-\vec{k}) \quad (13.71)$$

(Hint: expand the left-hand side in a series of multiple commutators.)

(a) Use the result (13.30) for the charge-conjugation property of a general covariant field to derive the corresponding result for a Dirac 4-spinor field ψ, as constructed in Section 7.4:

$$\mathcal{C}\psi_n\mathcal{C}^{-1} = \zeta C_{nm}\bar{\psi}_m \quad (13.72)$$

where C_{nm} is a numerical 4x4 matrix. Write out this matrix explicitly.

(b) Show that the matrix C of part (a) has the following properties

$$C = -C^{-1} \quad (13.73)$$

$$C\gamma_\mu C = \gamma_\mu^T \quad (13.74)$$

(c) A four-vector current can be defined as follows

$$j_\mu(x) \equiv \frac{1}{2}[\bar{\psi}(x), \gamma_\mu\psi(x)] = \frac{1}{2}\sum_{nm}(\gamma_\mu)_{nm}(\bar{\psi}_n\psi_m - \psi_m\bar{\psi}_n) \quad (13.75)$$

Use the results of parts (a,b) above to establish that this current is odd under charge conjugation:

$$\mathcal{C}j_\mu(x)\mathcal{C}^{-1} = -j_\mu(x) \quad (13.76)$$

(d) Show that the expression for $j_\mu(x)$ given in part (c) is equivalent to normal-ordering, namely $j_\mu(x) =: \bar{\psi}(x)\gamma_\mu\psi(x):$ (where we are now dealing with interaction-picture free fields, of course).

2. Verify, using the reflection and positivity results (13.13,13.26), the steps leading from (13.41) to (13.42).

3. (a) Show that for a single real space-like vector ρ, there exists a complex Lorentz transformation Λ such that $\rho = \Lambda\zeta$, where ζ is in the forward tube (i.e., $\zeta = \xi - i\eta$ where η is future time-like, $\eta^2 > 0, \eta^0 > 0$. (Hint: choose a coordinate system in which $\rho^1 = \rho^2 = 0, |\rho^0| < \rho^3$. Then consider the transformation (9.77), for suitable choice of θ).

   (b) Now suppose the two real space-like vectors ρ_1, ρ_2 have space-like convex hull. Show that they form a point in the extended tube $\mathcal{T}_2'$: i.e., there exists a complex Λ such that $\rho_1 = \Lambda\zeta_1, \rho_2 = \Lambda\zeta_2$, with ζ_1, ζ_2 in the forward tube.

4. Suppose a set of n real points $(\rho_1, \rho_2, .., \rho_n)$ is in the extended tube $\mathcal{T}_n'$. Show that their convex hull (i.e., the set of spacetime points of form $\sum_i \lambda_i \rho_i, \sum_i \lambda_i = 1, 0 \leq \lambda_i \leq 1$) consists entirely of space-like points.

14

Symmetries III: Global symmetries in field theory

The symmetries discussed in the preceding two chapters have been primarily those involving the transformation properties of relativistic field theories under the continuous and discrete parts of the homogeneous Lorentz group. In this chapter our focus will be on internal *global* symmetries: those symmetries involving spacetime-independent transformations of the fields which leave the dynamics of the theory invariant (or almost invariant, in the case of weakly broken global symmetries). We have already seen some examples of such symmetries in the context of Noether's theorem in Section 12.4, where we considered symmetries under global phase transformations (cf. (12.120)) forming a commutative (abelian) group, or more generally, symmetries in which the transformation of a multiplet of fields under a linear (non-abelian) matrix group, as in (12.126), leaves the Lagrangian invariant.

The term "global" here refers to the fact that the same phase or matrix transformation is applied to the fields of the theory at all spacetime points: the far richer physics that emerges when the symmetry is *local*, allowing arbitrary dependence of the transformation on spacetime location, will be the subject of the next chapter. The *formal* application of Noether's theorem in the presence of a global symmetry leads, as we saw in Section 12.4, to the existence of a conserved current $J_\alpha^\mu(x)$ for each generator t_α of the global symmetry group, with the conserved charges associated with each current,

$$Q_\alpha \equiv \int J_\alpha^0(x) d^3x \qquad (14.1)$$

mimicking the commutation relations of the Lie algebra of the symmetry group:

$$[t_\alpha, t_\beta] = if_{\alpha\beta\gamma}t_\gamma \Rightarrow [Q_\alpha, Q_\beta] = if_{\alpha\beta\gamma}Q_\gamma \qquad (14.2)$$

We emphasize that the application of Noether's theorem, strictly speaking a classical result relying on adequate smoothness properties of the (classical) action and fields, to quantum field theories should be taken with a grain of salt: it is entirely possible for quantum effects to destroy the classical conservation of the current(s) $J_\alpha^\mu(x)$, leading to a non-zero divergence of order $\hbar$, $\partial_\mu J_\alpha^\mu(x) = O(\hbar)$, a symmetry violation, or *anomaly*, entirely invisible at the classical level. Typically, anomalies arise because of delicacies in the construction of well-defined composite operators (implementing the desired symmetry transformations) for the currents $J_\alpha^\mu(x)$ once the theory is quantized. From the functional integral point of view, anomalies arise when the functional change of variables needed in the demonstration of Noether's theorem in the path-integral formalism generates a non-trivial Jacobian once the functional integral is properly regularized. We shall postpone a detailed discussion of anomalies to Chapter 15, as the important cases typically involve interacting *local* gauge fields, which we introduce and study there, even though the anomalous symmetry may itself be only a global one.

Putting aside for the time being the possibility of quantum-breaking of a global symmetry via anomalies, an exact global symmetry (at the Lagrangian level) may be realized in two distinct ways, depending on the transformation properties of the ground state (in field theory, the vacuum) of the theory under the symmetry. Let us suppose that the dynamics is exactly invariant under a symmetry group G, with a Lie algebra spanned by the set of generators $t_\alpha, \alpha = 1, 2, ...n$, with corresponding conserved charges Q_α. Next, let us imagine that the generators t_α are chosen as appropriate linear combinations so that some subset of the corresponding charges, $Q_\alpha, \alpha = 1, 2, ..m \leq n$ annihilates the physical vacuum

$$Q_\alpha |0\rangle = 0 \tag{14.3}$$

In fact, the set of such generators must itself span the Lie algebra of a subgroup $H \subset G$, as $Q_\alpha |0\rangle = Q_\beta |0\rangle = 0 \Rightarrow [Q_\alpha, Q_\beta]|0\rangle = 0$. In this case, we say that the symmetry group H is realized in the *Wigner–Weyl* mode, while the generators $Q_\alpha, \alpha = m + 1, .., n$ under which the vacuum is *not* invariant correspond to the *Nambu–Goldstone* mode of symmetry realization. Since, by assumption, the full symmetry group G is preserved by the dynamics, the broken generators still commute with the Hamiltonian, and therefore lead (when exponentiated and applied to any minimum-energy vacuum state) to new states which are also minimum-energy states: i.e., the vacuum is degenerate. This is precisely the situation discussed previously in Chapter 8 under the heading "spontaneous symmetry-breaking".

By contrast, the surviving Wigner–Weyl symmetry group implies the existence of degenerate multiplets of particle states, which must span finite-dimensional representations of H: in this case the existence of the underlying symmetry is manifestly visible in the spectrum of the theory and in symmetry constraints on the transition amplitudes of the theory (as in the case of isospin symmetry, for example).[1] For the broken generators, the non-invariance of the ground state of the theory transfers to complicated transformation properties of the multi-particle states built on it, and the existence of an underlying exact dynamical symmetry can be far from obvious phenomenologically. However, the spontaneous breaking of an *exact* global symmetry leaves a remarkable phenomenological residue which can scarcely be missed: the appearance of an exactly massless particle, called a "Goldstone boson", for each broken generator of the original exact global group G. We shall prove this result—the *Goldstone–Salam–Weinberg* theorem—in Section 14.2.

In fact, for reasons to be discussed in Section 14.1, exact global symmetries (which are not associated with a local gauge symmetry) are very rare in Nature: indeed, it

---

[1] Of course, we may have $m = 0$, in which case H is null, the entire group G is spontaneously broken, and we have only the Nambu–Goldstone mode of realization, or $m = n$, in which case $H = G$, and the entire symmetry is realized in Wigner–Weyl mode.

is possible that there are none at all! Instead, we find many examples of *approximate* global symmetries, in which the breaking of the symmetry is in some (appropriate) sense small, allowing us to exploit the symmetry by taking it to be exact at zeroth order, and treating the effects of the symmetry-breaking perturbatively. The absence of exactly zero-mass Goldstone bosons in Nature also suggests that there are no dynamically exact (with exactly conserved charges) but spontaneously broken global symmetries. However, the chiral symmetry of quantum chromodynamics provides a clear example of a global symmetry which is (a) weakly broken by explicit non-symmetric terms in the Lagrangian, and (b) spontaneously broken by the vacuum. In this situation, one finds in place of the exactly zero-mass Goldstone bosons which would necessarily appear if the small symmetry-breaking terms in the Lagrangian were turned off, "light" spinless particles—*pseudo-Goldstone bosons*—associated with each spontaneously broken generator of the chiral group.[2] These light particles are just the pions, whose squared masses are just 2% of the squared mass of the proton, for example. The diagnostic techniques available for determining the presence or absence of spontaneous breaking of a global symmetry will be the subject of Section 14.3: it turns out that the concept of the effective action introduced as a graph theoretic concept in Chapter 10 plays an essential role here.

14.1 Exact global symmetries are rare!

Until the advent of Grand Unified Theories (GUTs) in the mid-1970s, the exact conservation of global quantum numbers such as baryon number (B) and lepton number (L) was considered an obvious feature of elementary particle interactions. In the Standard Model of electroweak interactions, for example, the particle content and gauge dynamics of the theory guarantees invariance of the Lagrangian with respect to phase transformations assigning quantum numbers $B = \frac{1}{3}$ to quarks, $B = 0$ to gluons and leptons, $L = 1$ to leptons (electrons, muons, and taus, and their associated neutrinos), and $L = 0$ to quarks and gluons. The classical Noether analysis of Section 12.4 then assures existence of divergenceless currents J_B^μ and J_L^μ, with associated exactly conserved charges (B and L respectively).

Even within the Standard Model, however, quantum effects produce an anomaly resulting in explicit violation of the current $J_B^\mu + J_L^\mu$ corresponding to the quantum number $B + L$. We defer discussion and derivation of these anomalies to Section 15.6: for our present purposes, it suffices to point out that the structure of the anomalous divergence of the $B + L$ current results in extremely small transition amplitudes violating $B + L$ conservation, of *non-perturbative order* $O(e^{-C/\alpha_{\text{wk}}})$, where α_{wk} is the weak fine-structure constant (of similar magnitude to $\alpha_{\text{em}} = 1/137$), as originally pointed out by 't Hooft. Such violations are so small as to be negligible, for all practical purposes.

However, there is considerable circumstantial evidence to suggest that much larger—indeed, potentially observable—violations of both baryon and lepton number conservation do indeed occur in Nature. Perhaps the most convincing argument for such violations come from modern cosmology. The observed asymmetry in the density

---

[2] For an explicit example of a pseudo-Goldstone boson, see Problem 4(b) in Chapter 8.

of matter with respect to antimatter in the visible Universe requires either that we introduce the asymmetry in an *ad hoc* fashion as an initial condition at some early point in the evolution of the Universe following the Big Bang, while retaining exact conservation (ignoring the effects of exponentially small anomalous violations), or, as assumed by essentially all present practitioners of early Universe cosmology, we must introduce a small breaking of baryon number (say) to account for the fact that the ratio of baryon to photon number densities $\frac{n_B}{n_\gamma}$ at the epoch of nucleosynthesis is on the order of 10^{-8}.

In fact, the structure of local quantum field theories, and in particular theories based on local gauge interactions, already implies that exact conservation of global symmetries is extremely fragile, and easily subject to violation on several grounds. This is in contradistinction to the presumed exact character of the *local* gauge symmetries described in the next chapter, both in the Standard Model as well as in Grand Unified (or even, superstring) extensions thereof. These symmetries may undergo spontaneous breaking, but are as far as we know exact symmetries at the dynamical level.

There are several reasons for the apparent fragility of global symmetries in local quantum field theories. Here is a (partial) list:

1. The dynamics of local field theories, specified in terms of a Lagrangian functional, for example, is dependent in form and structure on the distance (or energy) scales over which we wish to provide an accurate description of the transition amplitudes of the theory. A detailed understanding of the nature of this dependence will require the concepts embodied in the renormalization group which will be the central topic in Part 4 of this book. It will be shown there that Lagrangian field theories, treated so far as fundamental descriptions of the exact microphysics of some set of interacting particles, should rather be regarded as provisional *effective* descriptions of relativistic quantum phenomena, valid (to within a specifiable level of accuracy) in some range of energy scales. As we raise the energy horizon to which we expect our theory to hold, new microphysics typically emerges in the form of new degrees of freedom (fields/particles) with a new effective Lagrangian expressing the local interactions (valid down to the new lower-distance scale) of these new fields. The "ultimate" theory, if any, which emerges when this procedure is continued, say to or beyond the Planck energy scale characteristic of quantum gravity, may not even be a local quantum field theory at all: in the view of many particle physicists, it is a theory in which the basic entities may be one or higher-dimensional objects (strings, membranes, etc.).

    To the extent that the B and L (baryon and lepton) global symmetries of the Standard Model (SM) are "accidental", arising out of the paucity of renormalizable terms which can be included in the Standard Model Lagrangian given the assumed particle/field content of the theory, extensions of the Standard Model (BSM models, for "beyond the Standard Model"), constrained only by the requirement that they yield SM physics at sub-TeV energy scales will typically include Lagrangian interactions which violate the global symmetries of the Standard Model. The reason for this is simply that these symmetries, unlike the underlying local gauge symmetry which incorporate the essential physics of the Standard Model, are not imposed from the outset, or required by deep physical principles, but simply an accidental consequence of the limited number of low-mass fields (and gauge-symmetric interactions of such fields) needed to embody Standard Model physics at the sub-TeV scale. A concrete example is provided by Grand Unified Theories such as the SU(5) extension of the Standard Model, where explicit B and L violating terms automatically appear due to the enlarged particle content (although the combination $B - L$ is still preserved). In supersymmetric extensions of the Standard Model, the proliferation of particle species makes it even more likely that global symmetries of the SM (even the resistant $B - L$ symmetry) are explicitly violated by new terms in the supersymmetric Lagrangian of the extended theory. As we shall see in Part 4, the size of the symmetry violating contributions to transition amplitudes at a low-energy scale E is typically suppressed by a factor $(\frac{E}{M})^n$, where M is the characteristic mass/energy scale of the extended Lagrangian and the global symmetry violating terms correspond to operators in the low-energy effective Lagrangian of mass dimension $4 + n$.

2. A global symmetry of an effective Lagrangian at a low-energy scale may be promoted to a local symmetry of an extended Lagrangian describing the physics at a higher-energy scale, and then undergo spontaneous breaking, so that the symmetry is realized in the Nambu–Goldstone mode. This is the case, for example, with the global $B - L$ symmetry of the Standard Model and its SU(5) Grand Unified extension, when embedded in a SO(10) Grand Unified gauge theory. The global $B - L$ charge in the latter theory is (with the standard choice of gauge and fermion representations) then associated with a local gauge symmetry and a massive gauge boson (via the Higgs effect; cf. Chapter 15). The phenomenological consequences are similar to those arising in the case of explicit breaking: the transition amplitudes of the theory display $B - L$ violating processes at a level suppressed by inverse powers of the characteristic mass/energy of the new physics.

3. As discussed previously, depending on the precise set of representations (particle content) of the theory, quantum anomalies can arise which result in violations of global Noether symmetries in theories with gauge (or gravitational) interactions. Although the non-conservation is present and manifest in perturbation theory, the manifestation of the symmetry violation in actual S-matrix elements of the theory may involve exponentially suppressed (in the inverse gauge coupling) non-perturbative tunneling processes.

4. Quantum gravity effects seem to lead to a universal mechanism for the breaking of global symmetries, at a level involving inverse powers of the Planck mass, the characteristic scale of quantum gravity. Classical "no-hair" theorems assert the impossibility of associating global quantum numbers with a black hole. On the other hand, once quantum effects are turned on, black holes become unstable to decay via Hawking radiation. One can therefore imagine dumping baryons into a small black hole and having the baryon energy dissipated in Hawking radiation of photons or leptons. From an effective field theory point of view, the corresponding process can be represented by non-renormalizable effective operators with strength proportional to inverse powers of the Planck mass. Certain field theories coupled to gravity (Rey, 1989) give rise to wormhole instanton solutions where, as with black holes, global charge can also disappear

through the wormhole. The conventional wisdom (see (Peccei, 1988), Section 2.7) suggests that gravitational effects inevitably destroy global charge conservation. The present author is an agnostic on this issue, and would prefer to wait for a manifestly consistent quantum gravitational description of black holes and/or wormholes before pronouncing a definitive verdict.

14.2 Spontaneous breaking of global symmetries: the Goldstone theorem

In Section 8.4 we studied the effect of the spontaneous symmetry-breaking in a scalar field theory with a triplet $\vec{\phi}$ of equal mass scalar fields, in which the global symmetry group $G = O(3)$ of the Hamiltonian (or Lagrangian) is broken by the vacuum of the theory to a subgroup $H = O(2)$, leading to the appearance of two massless scalar particles corresponding to the two broken generators of G. The appearance of massless particles as the consequence of vacuum-breaking of an exact *continuous* global symmetry of the theory is a very general consequence of basic covariance and spectral properties of field theory: the theorem establishing this result goes back to a seminal paper of Goldstone, Salam, and Weinberg (Goldstone *et al.*, 1962). Later work by Kastler, Robinson, and Swieca established rigorously, using methods of axiomatic field theory (cf. Section 9.2), that in a theory with a mass gap (in which the lowest-mass single-particle state corresponds to non-zero mass), satisfying the usual spectral and Poincaré invariance axioms, spontaneous symmetry-breaking is impossible: an exact global symmetry of the Lagrangian must also be preserved by the vacuum. The Ruelle Clustering theorem of Section 9.2, in the stronger form available for theories with a mass gap, plays a critical role in establishing the spatial asymptotics needed in the proof of Kastler *et al.* (Kastler *et al.*, 1966). In this section we will describe the Goldstone theorem and outline (in a non-rigorous fashion) its proof from various angles, referring the reader to the aforementioned paper for a completely rigorous treatment.

Suppose that we have a (set of) four-vector conserved currents $J_\alpha^\mu(x), \partial_\mu J_\alpha^\mu = 0$ in a theory with a mass gap. Assuming the fields of the theory to transform covariantly under the HLG, the current(s) must satisfy

$$U^\dagger(\Lambda)J_\alpha^\mu(x)U(\Lambda) = \Lambda^\mu{}_\nu J_\alpha^\nu(\Lambda^{-1}x) \tag{14.4}$$

where Λ is an arbitrary Lorentz transformation. The vanishing of the four-divergence of J_α^μ ensures in the usual way that the associated charge

$$Q_\alpha(t) \equiv \int J_\alpha^0(\vec{x}, t)d^3x \tag{14.5}$$

is time-independent (and in fact, clearly commutes with the full energy-momentum four-vector P^μ of the theory). What is less obvious is that it also guarantees that $Q_\alpha(t) = Q_\alpha$ is invariant under the HLG: $U^\dagger(\Lambda)Q_\alpha U(\Lambda) = Q_\alpha$. This is best seen by first assuming Λ to be an infinitesimal Lorentz transformation: the general result for finite Λs then follows by exponentiation of the infinitesimal case in the usual way. Thus, let $\Lambda^\mu{}_\nu = g^\mu{}_\nu + \omega^\mu{}_\nu$ with $\omega^{\mu\nu} = -\omega^{\nu\mu}$ and ω infinitesimal. To first order in ω, we easily find, using (14.4) and (14.5),

$$U^\dagger(\Lambda)Q_\alpha U(\Lambda) = Q_\alpha + \int (\omega^0{}_\nu J^\nu_\alpha(x) - \omega^\mu{}_\rho x^\rho \partial_\mu J^0_\alpha(x))d^3x + O(\omega^2) \tag{14.6}$$

The $\rho = 0$ contribution to the second term in the integral is proportional to a total spatial derivative and hence vanishes: $x^0 \int \partial_i J^0_\alpha(x)d^3x = 0$, as does the contribution from spatial indices $\rho = i, \mu = j$, proportional to $\omega^j{}_i \int x^i \partial_j J^0_\alpha(x)d^3x \propto \delta_{ij}\omega^{ij} = 0$, by integration by parts. Finally, the contribution from $\rho = i, \mu = 0$ becomes, using current conservation,

$$\int -\omega^0{}_i x^i \partial_0 J^0_\alpha(x)d^3x = \int \omega^0{}_i x^i \partial_j J^j_\alpha(x)d^3x$$

$$= -\int \omega^0{}_i J^i_\alpha(x)d^3x$$

$$= -\int \omega^0{}_\nu J^\nu_\alpha(x)d^3x \tag{14.7}$$

cancelling the first term in the integral in (14.6) and giving the desired result

$$U^\dagger(\Lambda)Q_\alpha U(\Lambda) = Q_\alpha \Rightarrow [U(\Lambda), Q_\alpha] = 0 \tag{14.8}$$

If we now define a (set of) states $|\alpha\rangle \equiv Q_\alpha|0\rangle$, obtained by applying the charges Q_α to the Lorentz-invariant physical vacuum $|0\rangle$, $U(\Lambda)|0\rangle = |0\rangle$ of the theory, we conclude that since $U(\Lambda)|\alpha\rangle = |\alpha\rangle$

$$\langle\alpha|J^\mu_\alpha(x)|0\rangle = \langle\alpha|U^\dagger(\Lambda)J^\mu_\alpha(x)U(\Lambda)|0\rangle = \Lambda^\mu{}_\nu\langle\alpha|J^\nu_\alpha(\Lambda^{-1}x)|0\rangle \tag{14.9}$$

But the matrix elements in (14.9) can clearly be translated to $x = 0$, as the states $|\alpha\rangle$ have zero-energy-momentum (as mentioned above, $[P^\mu, Q_\alpha] = 0$), allowing us to conclude

$$\langle\alpha|J^\mu_\alpha(0)|0\rangle = \langle\alpha|J^\mu_\alpha(x)|0\rangle = 0 \tag{14.10}$$

Integrating (14.10) over all space then yields immediately

$$\int \langle\alpha|J^0_\alpha(x)|0\rangle d^3x = \langle\alpha|\alpha\rangle = 0 \tag{14.11}$$

which, by positive-definiteness of the Hilbert space, implies $|\alpha\rangle = Q_\alpha|0\rangle = 0$, so that the symmetry is in fact preserved by the vacuum state. The necessity for a mass gap in the preceding argument (first given in (Goldstone *et al.*, 1962)) is not immediately clear: it turns out that suitable asymptotic spatial convergence of the integrals defining the $|\alpha\rangle$ states in (14.11) is absent in theories with zero-mass particles. The arguments of Kastler *et al.* (Kastler *et al.*, 1966) fill in the required rigorous asymptotic bounds needed for a proper proof. We should also emphasize here that the argument fails in any theory with *local gauge interactions*:[3] as we shall see in the next chapter, in such

---

[3] Note that, as we shall see in our discussion of the Higgs phenomenon, it is perfectly possible for a theory with exact dynamical local gauge symmetry to have no massless particles—one simply arranges for the symmetry group G to be completely broken by the vacuum state, leading to a theory with a non-zero-

theories, the theory must be quantized either in a non-covariant "physical" gauge, with a positive-definite Hilbert space, or in a covariant ("Gupta–Bleuler") gauge in which the Hilbert space is *not* positive-definite (even though the negative or zero metric unphysical modes can be shown to decouple from physical transition amplitudes). As both explicit covariance of the fields and positive-definiteness of the Hilbert space are ingredients of the reasoning leading to (14.11), we must conclude that spontaneous breaking in theories with exact *local* gauge symmetries has distinctive features not present in the global case (in particular, we cannot expect a massless Goldstone boson to appear). In fact, we shall see that the peculiarities (generally dubbed the "Higgs mechanism") of theories with both local gauge symmetry and spontaneous symmetry-breaking, to be discussed in Section 15.7, play a central role in the electroweak sector of the Standard Model of elementary particle interactions.

The physics underlying the appearance of massless particles when a continuous global symmetry is broken by the vacuum becomes clearer if we approach the problem from another angle. Let us assume that our theory admits an exact continuous global symmetry G (which could be abelian or non-abelian) yielding Noether charges Q_α which generate the appropriate infinitesimal transformations on a finite set of local (or almost local) fields $\phi_n(x)$ spanning a representation of G with matrix generators t_α (cf. Section 12.4, Example 5):

$$[Q_\alpha, \phi_n(x)] = -(t_\alpha)_{nm}\phi_m(x) \tag{14.12}$$

The fields $\phi_n(x)$ here may be elementary or composite: we require only that they fill out a finite-dimensional representation of the global symmetry group G. If the vacuum is invariant under G, i.e., $Q_\alpha|0\rangle = 0$, $\forall\alpha$, we must have

$$\langle 0|[Q_\alpha, \phi_n(x)]|0\rangle = 0 \Rightarrow (t_\alpha)_{nm}\langle 0|\phi_m(x)|0\rangle = (t_\alpha)_{nm}\langle 0|\phi_m(0)|0\rangle = 0, \ \forall\alpha \tag{14.13}$$

On the other hand, if there are generator(s) t_α which do not annihilate the vector of vacuum expectation values,

$$N_{\alpha n} \equiv (t_\alpha)_{nm}\langle 0|\phi_m(0)|0\rangle \neq 0 \tag{14.14}$$

the symmetry generated by Q_α is spontaneously broken, and there must be a zero-mass particle in the theory.

In fact, a simple spectral argument of Källen–Lehmann type (cf. Section 9.5; see also (Itzhykson and Zuber, 1980), p. 520), shows that the conserved current J^μ_α itself acts as an interpolating field for this "Goldstone particle". Inserting a complete set of states $|n\rangle$, $P^\mu|n\rangle = P^\mu_n|n\rangle$ in the vacuum expectation value of the commutator $\langle 0|[Q_\alpha(t), \phi_n(0)]|0\rangle = N_{\alpha n}$, with the (conserved!) charge chosen at a definite time t, $Q_\alpha(t) = \int J^0_\alpha(\vec{x}, t)d^3x$, we find,

---

mass gap. In such cases, the possibility of evasion of the Goldstone theorem arises from the indicated mutual compatibility of manifest covariance and positive-definiteness in the local gauge case.

$$N_{\alpha n} = \sum_n \int \{\langle 0|J_\alpha^0(0)|n\rangle\langle n|\phi_n(0)|0\rangle e^{-iP_n \cdot x} - \langle 0|\phi_n(0)|n\rangle\langle n|J_\alpha^0(0)|0\rangle e^{iP_n \cdot x}\} d^3x$$

$$= (2\pi)^3 \sum_{n \neq |0>} \delta^3(\vec{P}_n)\{\langle 0|J_\alpha^0(0)|n\rangle\langle n|\phi_n(0)|0\rangle e^{-iE_n t} - \langle 0|\phi_n(0)|n\rangle\langle n|J_\alpha^0(0)|0\rangle e^{iE_n t}\}$$

$$\neq 0, \ \forall t \tag{14.15}$$

Note that the physical vacuum does not appear among the sum over states $|n\rangle$ in (14.15), as the two terms in brackets cancel in this case. In fact, the charge $Q_\alpha(t)$ is time-independent by assumption, so differentiating with respect to time, we must have, for all t,

$$\sum_{n \neq |0>} \delta^3(\vec{P}_n)E_n\{\langle 0|J_\alpha^0(0)|n\rangle\langle n|\phi_n(0)|0\rangle e^{-iE_n t} + \langle 0|\phi_n(0)|n\rangle\langle n|J_\alpha^0(0)|0\rangle e^{iE_n t}\} = 0$$

$$\tag{14.16}$$

The conditions (14.15, 14.16) together imply that there must be non-vacuum states $|n\rangle$ for which $\langle 0|J_\alpha^0(0)|n\rangle \neq 0$, $\langle n|\phi_n(0)|0\rangle \neq 0$ requires the vanishing of the energy E_n whenever the spatial momentum $\vec{P}_n$ of the state is zero. Such states can only be single-particle zero-mass states, and the requirement $\langle 0|J_\alpha^0(0)|n\rangle \neq 0$ is simply the statement that the current J_α^μ is an interpolating field for the corresponding Goldstone particle. If the components of J_α^μ are bosonic fields, the particle is a *Goldstone boson*. However, note that in supersymmetric theories, we can have global Noether currents of fermionic type (recall (12.154)), and the corresponding Goldstone mode will be a fermion, if the symmetry is spontaneously broken.

14.3 Spontaneous breaking of global symmetries: dynamical aspects

The explicit examples of spontaneous symmetry-breaking (SSB) in self-interacting scalar field theories described in Section 8.4 make the connection between the energetics of the theory—specifically, the properties of the lowest-energy state—and the presence or absence of SSB clear. In this section we shall discuss some useful diagnostic tools for determining whether SSB is present in the context of specific Lagrangian field theories. In contrast to the approach followed in Section 8.4, we shall primarily be using the (Euclidean) functional integral version of the quantized field theory. The physics of spontaneous symmetry-breaking, as we shall see, is essentially long-distance physics: the essential phenomena appear in the limit where the spatial volume of the system tends to infinity. We shall implicitly assume that a short-distance regularization of the theory is in place (e.g., a spacetime lattice) throughout the discussion, with the functional integral defined initially in a spacetime box of volume $\Omega = VT$, where V is the spatial volume and T the (imaginary) time extent of the system. The energetics of the system relevant to deciding on the existence of SSB then turns out to be conveniently encoded in the effective action functional $\Gamma[\phi]$ introduced in Section 10.4, or rather, on its specialization to constant fields $V(\phi) \equiv \Gamma[\phi(x) = \phi]$ called the *effective potential*. For simplicity, we shall work with a theory of a single real scalar field, with Euclidean action $S[\phi] = \int(\frac{1}{2}(\partial_\mu\phi)^2 - P(\phi))d^4x$, where $P(\phi)$ is a polynomial in ϕ.

The effective action $\Gamma[\phi]$ was defined in Section 10.4 as the functional Legendre transform of the generating functional $W[j]$ of connected Green functions:

$$Z[j] = \int \mathbf{D}\phi\, e^{-(S[\phi] - \int j(x)\phi(x)d^4x)} \tag{14.17}$$

$$W[j] = \ln Z[j]/Z[0] \tag{14.18}$$

$$\Gamma[\phi] = \int j(x)\phi(x)d^4x - W[j], \quad \phi(x) = \frac{\delta W[j]}{\delta j(x)} \tag{14.19}$$

The existence of the Legendre transform as defined in (14.19) presupposes that the functional relation $\phi(x) = \frac{\delta W[j]}{\delta j(x)}$ can be uniquely inverted to allow us to eliminate the source function $j(x)$ in favor of the classical field $\phi(x)$ in $\Gamma[\phi]$. Precisely this supposition becomes problematical in the situations in which spontaneous symmetry-breaking occurs, as we shall now see: instead, we shall introduce a more general definition of the Legendre transform, which (a) agrees with (14.19) in the absence of SSB, (b) is mathematically well-defined even in theories exhibiting SSB, and (c) is physically more directly related to the underlying energetic effects responsible for inducing SSB.

The basic idea is best exhibited in a simple toy model which we have already exploited in our discussion of Borel summability in Chapter 11: the zero-dimensional "field theory" obtained by setting all non-zero momentum modes of the field to zero, so that the path integral in (14.17) becomes a one-dimensional integral over the constant value of the field ϕ, in the presence of a constant source $j(x) = j$:

$$Z(j) = \int e^{\Omega(j\phi - P(\phi))}d\phi, \quad \exp\left(\Omega W(j)\right) = Z(j)/Z(0) \tag{14.20}$$

In the limit of infinite spacetime volume Ω, the integral is dominated by the point or points at which the maximal value of the exponent is achieved. In the examples to be considered below there will be a unique such point, and we have simply (for $\Omega \to \infty$)

$$W[j] = \sup_\phi(j\phi - P(\phi)) - \sup_\phi(-P(\phi)) = \sup_\phi(j\phi - P(\phi)) + \inf_\phi P(\phi) \tag{14.21}$$

For convenience, we shall choose the field polynomial $P(\phi)$ so that the second term on the right vanishes, and we have simply

$$W[j] = \sup_\phi(j\phi - P(\phi)) \tag{14.22}$$

For example, we may take the scalar theory discussed in Section 8.4, with either

$$P_+(\phi) = \frac{1}{2}m^2\phi^2 + \frac{\lambda}{4!}\phi^4 \tag{14.23}$$

with no SSB, or the theory with a negative sign in the mass term,

$$P_-(\phi) = -\frac{1}{2}m^2\phi^2 + \frac{\lambda}{4!}\phi^4 + \frac{3}{2\lambda}m^4 = \frac{\lambda}{4!}(\phi^2 - v^2)^2, \quad v \equiv \sqrt{\frac{6}{\lambda}}m \tag{14.24}$$

Taking first the positive-sign case, one easily sees that for any real j there is a unique maximum of $j\phi - P(\phi)$ when $\phi = vx, v \equiv \sqrt{\frac{6}{\lambda}}m$, $x^3 + x - \frac{6j}{\lambda v^3} = 0$, with the cubic equation having only one real root. The location of the root can be found readily for small j, and we find that the solution for ϕ depends analytically on j, with

$$\phi = \frac{6}{\lambda v^2} j - \frac{216}{\lambda^3 v^8} j^3 + O(j^5) \tag{14.25}$$

and

$$W_+(j) = \frac{3}{\lambda v^2} j^2 - \frac{54}{\lambda^3 v^8} j^4 + O(j^6) \tag{14.26}$$

Evidently, in this case, $W_+(j)$ is everywhere differentiable, and the relation $\phi = W'_+(j)$ reduces to the cubic equation above, with a unique 1-1 mapping between ϕ and j. The Legendre transform $\Gamma_+(\phi)$ of $W_+(j)$ can therefore be constructed in the usual fashion (as in (14.19)), and we find

$$\Gamma_+(\phi) = j\phi - W_+(j) = \frac{1}{2} m^2 \phi^2 + \frac{\lambda}{4!} \phi^4 = P_+(\phi) \tag{14.27}$$

which is just the classical action in this constant-field model ($S(\phi) \to P_+(\phi)$). There are no loop graphs in this model, as there are no non-zero momenta to integrate over, so this agrees with our discovery in Section 10.4 that the effective action reduces to the classical Lagrangian at tree level.

The situation is quite different in the case where the mass term has a negative sign, as the maximum of $j\phi - P_-(\phi)$ occurs at a solution of the cubic equation $\phi(\phi^2 - v^2) = \frac{6}{\lambda} j$, which has three real solutions for $|j| < \frac{\lambda}{9\sqrt{3}} v$ (for example, at $j = 0$ we find solutions at $\phi = 0, \pm v$). For small positive j, the solution giving the absolute maximum of $j\phi - P_-(\phi)$ is $\phi = v + \frac{3}{\lambda v^2} j + O(j^2)$, while for small negative j we must choose the solution at $\phi = -v + \frac{3}{\lambda v^2} j + O(j^2)$. There is clearly a discontinuity in ϕ as a function of j as we pass through $j = 0$, which asserts itself as a cusp (discontinuity in the first derivative) of $W_-(j)$:

$$W_-(j) = v|j| + \frac{3}{2\lambda v^2} j^2 + .. \tag{14.28}$$

The dramatic alteration in shape of $W(j)$ as a result of the change of sign in the mass term is clearly visible in Fig. 14.1. The failure of the derivative of $W_-(j)$ to be well defined at $j = 0$ invalidates the usual procedure for constructing the Legendre transform $\Gamma_-(\phi)$ of $W_-(j)$. Note that the cusp in $W_-(j)$ is a direct consequence of having taken the infinite volume limit $\Omega \to \infty$ in the defining integral (14.20): for finite Ω the integral defines an infinitely differentiable function of j, and there are absolutely no difficulties in defining a Legendre transform in the usual fashion. We can obtain a unique and well-defined *infinite-volume* $\Gamma_-(\phi)$, which, moreover, coincides with the usual definition in those cases where there are no difficulties with differentiability of $W(j)$, by defining the Legendre transform in general[4] as follows,

---

[4] For a beautiful introduction to the applications of convexity theory in classical thermodynamics, leading to the "sup" definition of the Legendre transform given here, see the Introduction to (Israel, 1978) by A. S. Wightman.

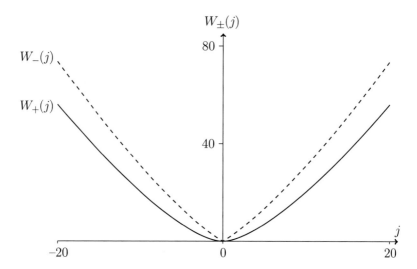

Fig. 14.1 The infinite volume connected generating functions $W_\pm(j)$ for the case of zero-dimensional scalar theories $P_\pm(\phi)$ and with positive or negative squared mass terms (parameters $m = 1, \lambda = 1.5, v = 2$). Note the cusp of $W_-(j)$ at $j = 0$.

$$\Gamma(\phi) = \sup_j (j\phi - W(j)) \tag{14.29}$$

A finite maximum obtains for arbitrarily large $|j|$ provided $W(j)$ rises at least linearly with j for large $|j|$, i.e., $W(j)$ is *convex* for large j. Recall that a real function $W(j)$ is convex if

$$W(\alpha j_1 + (1 - \alpha)j_2)) \leq \alpha W(j_1) + (1 - \alpha)W(j_2), \quad \forall \alpha, \ 0 < \alpha < 1 \tag{14.30}$$

An exactly analogous definition holds for convex functionals $W[j]$ on a function space of real functions $j(x)$. In fact, returning temporarily to the full field theory case, the convexity of the generating functional $W[j]$ (not just at large j) follows directly from its functional integral representation

$$\exp W[j] = \int \exp\left(\int j(x)\phi(x)d^4x\right)d\mu, \quad d\mu \equiv \exp(-S[\phi])\mathbf{D}\phi / \int \exp(-S[\phi])\mathbf{D}\phi \tag{14.31}$$

as an integral over a normalized positive measure $d\mu$, $\int d\mu = 1$. Integrals over such measures satisfy a *Hölder inequality* (see (Rudin, 1966), p. 62)

$$\int f^\alpha g^{1-\alpha} d\mu \leq \left(\int f d\mu\right)^\alpha \left(\int g d\mu\right)^{1-\alpha}, \quad 0 < \alpha < 1 \tag{14.32}$$

Setting $f = \exp\left(\int j_1(x)\phi(x)d^4x\right), g = \exp\left(\int j_2(x)\phi(x)d^4x\right)$, we see that (14.32) immediately implies, taking the logarithm,

$$W[\alpha j_1 + (1 - \alpha)j_2] \leq \alpha W[j_1] + (1 - \alpha)W[j_2], \quad 0 < \alpha < 1 \tag{14.33}$$

The convexity of $W(j)$ in our toy model (with or without SSB) is apparent from a glance at Fig. 14.1. It is easy to verify that the property of convexity of $W(j)$ carries over to $\Gamma(\phi)$, *provided that we use the "sup" definition of the latter.* Indeed,

$$
\begin{aligned}
\Gamma(\alpha\phi_1 + (1-\alpha)\phi_2) &\equiv \sup_j\{(j(\alpha\phi_1 + (1-\alpha)\phi_2) - W(j)\} \\
&= \sup_j\{(\alpha(j\phi_1 - W(j)) + (1-\alpha)(j\phi_2 - W(j))\} \\
&\leq \alpha\sup_{j_1}\{(j_1\phi_1 - W(j_1))\} + (1-\alpha)\sup_{j_2}\{(j_2\phi_2 - W(j_2))\} \\
&\leq \alpha\Gamma(\phi_1) + (1-\alpha)\Gamma(\phi_2)
\end{aligned}
\tag{14.34}
$$

The conventional definition (14.19) of the Legendre transform, when it is applicable, is *involutive*: the Legendre transform of the Legendre transform simply reproduces the original function. A glance at (14.22) shows that the generating function $W(j)$ in our toy model is in fact just the Legendre transform, using the "sup" definition, of the classical action of the model $P(\phi)$, so this involutive property would imply that $\Gamma(\phi)$ should be just our original action function $P(\phi)$. However, the above convexity argument shows that, in the case of a negative mass term, the Legendre transform of $\Gamma_-(\phi)$ cannot reproduce the classical action $P_-(\phi)$, as the latter is clearly not convex! One can show that $|\frac{dW_-(j)}{dj}| \geq v$ (with the equality holding only at $j = 0$: see (14.28) and Fig. 14.1), so that for $|\phi| < v$ the maximum in the definition (14.29) is attained at $j = 0$, and we find $\Gamma_-(\phi) = 0, -v \leq \phi \leq +v$. For $|\phi| > v$, the relation between j and ϕ is invertible and the "sup" definition of $\Gamma_-(\phi)$ reproduces $P_-(\phi)$ precisely (see Fig. 14.2). Note that the double-well structure of the potential $P_-(\phi)$ in the broken symmetry case has been eliminated in $\Gamma_-(\phi)$, which is in fact the *convex hull* (i.e., the boundary of the minimal convex set containing) of the set bounded below by the graph of $P_-(\phi)$, with a flat section connecting the points on the graph at $\phi = \pm v$, where $\Gamma_-(\phi)$ is evidently not differentiable. On the other hand, if we compute the effective potential at finite values of Ω (see Fig. 14.2), the resultant convex function is smooth and everywhere differentiable (inheriting these properties from the finite volume $W(j)$-see above). We shall shortly see that there is a direct physical interpretation of the flat (convexity restoring) section of $\Gamma_-(\phi)$ in terms of the energetics of a system with spontaneously broken symmetry and a degenerate vacuum in the infinite volume limit.

We now return to field theory proper, but ease the transition by considering first the case of one-time, zero-space dimensions, in which case our "field theory" amounts to the quantum mechanics of a one-dimensional anharmonic oscillator, and the path integral (14.17) defines the Euclidean kernel (cf. Section 4.2.1) over a finite time extent T, $\mathrm{Tr}\exp(-H_{\bar{j}}T)$ for the Hamiltonian

$$
H_{\bar{j}} \equiv \frac{p^2}{2} + \frac{\lambda}{24}(x^2 - v^2)^2 - \bar{j}x = -\frac{1}{2}\frac{d^2}{dx^2} + P(x) - \bar{j}x \equiv H_0 - \bar{j}x
\tag{14.35}
$$

Here the "field" ϕ has been replaced by the quantum coordinate x, and we have restricted the source function $j(t)$ to be independent of time, $j(t) = \bar{j} = $ constant. Evidently we are dealing with a "double-well" anharmonic oscillator, where we now also assume that we are in a regime where λ is taken large for v fixed, so that the two potential basins are separated by a large energy barrier (see Fig. 14.3). In this situation, the Gaussian wavefunctions $\psi_\pm(x) = \langle x|\pm\rangle$ centered on $x = \pm v$,

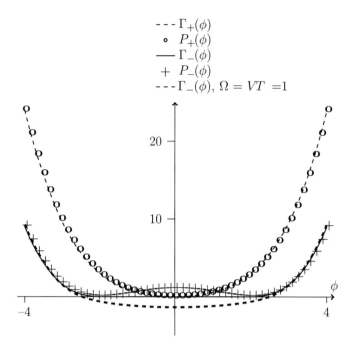

Fig. 14.2 The infinite volume effective potential $\Gamma_\pm(\phi)$ for theories defined by action $P_\pm(\phi)$ (same parameters as in Fig. 14.1). The effective potential $\Gamma_-(\phi)$ at finite volume, $\Omega = 1$, for the symmetry-broken case is also shown.

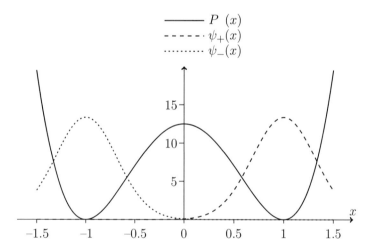

Fig. 14.3 Potential energy function $P(x)$ and approximate ground states $\psi_\pm(x)$ for a one-dimensional anharmonic oscillator ($\lambda{=}300$, $m = v = 1$).

$$\psi_\pm(x) \equiv Ce^{-\frac{\omega}{2}(x\mp v)^2}, \quad C = (\omega/\pi)^{1/4}, \quad \omega \equiv \sqrt{\frac{\lambda v^2}{3}} \tag{14.36}$$

are approximate degenerate ground-state eigenfunctions of H_0, with approximate eigenvalue $\omega/2$, due to the exponentially suppressed tunneling amplitude for the particle to transition between the two potential basins. A short calculation shows that $\gamma \equiv \langle +|H_0|-\rangle \sim O(\omega^2 e^{-\omega v^2})$ for λ (hence ω) large, and fixed v. We may therefore represent approximately the low-energy sector of this theory by truncating the Hilbert space to the two-dimensional subspace spanned by the states $|+\rangle$, $|-\rangle$, in which space the Hamiltonian $H_{\bar{j}}$ takes the matrix form

$$H_{\bar{j}} = \begin{pmatrix} \frac{\omega}{2} - v\bar{j} & \gamma \\ \gamma & \frac{\omega}{2} + v\bar{j} \end{pmatrix} \tag{14.37}$$

using the fact that the expectation value of the coordinate x in the highly localized states represented by wavefunctions $\psi_+(x), \psi_-(x)$ are $+v, -v$ approximately. The source-free Hamiltonian H_0 has, as is well known, a unique non-degenerate ground state, with (in the approximation (14.37)) the symmetric eigenfunction $\frac{1}{\sqrt{2}}(\psi_+(x) + \psi_-(x))$, and energy $\frac{\omega}{2} - \gamma$. The antisymmetric state with wavefunction $\frac{1}{\sqrt{2}}(\psi_+(x) - \psi_-(x))$ lies at an energy 2γ above this ground state, and will be suppressed in the partition function $Z(\bar{j})$ (for arbitrary $\bar{j}$) if we choose T large enough that $e^{-2\gamma T} \ll 1$, as we shall henceforth do. Switching on the source $\bar{j}$, the energy of the new ground state $|0, \bar{j}\rangle$ becomes

$$E_{0,\bar{j}} = \frac{\omega}{2} - \sqrt{\gamma^2 + v^2\bar{j}^2} \tag{14.38}$$

with $Z(\bar{j}) = e^{-TE_{0,\bar{j}}} + O(e^{-2T\gamma})$. The generating function $W(\bar{j})$ becomes, up to exponentially small corrections,

$$W(\bar{j}) = \frac{1}{T}\ln Z(\bar{j}) = -E_{0,\bar{j}} \tag{14.39}$$

The expectation value of the coordinate x (recall that this is the analog of the field operator ϕ in our zero-spatial-dimensional model) in the ground state of $H_{\bar{j}}$ can be calculated directly (by diagonalizing $H_{\bar{j}}$) or by taking the derivative of $W(\bar{j})$: in either case one finds

$$\bar{x} = \frac{dW(\bar{j})}{d\bar{j}} = \frac{v^2\bar{j}}{\sqrt{\gamma^2 + v^2\bar{j}^2}} \tag{14.40}$$

We pointed out previously that there are no difficulties in defining the Legendre transform in the usual way at finite spatiotemporal volume, which is certainly the case in our model (for finite T). And indeed, for small $\bar{j}$, when our approximations are valid, we see that there is no problem with differentiability, or with inverting the relation between $\bar{j}$ and $\bar{x}$, so we may define the effective potential $\Gamma(\bar{x})$ with the conventional Legendre transform,

$$\Gamma(\bar{x}) = \bar{j}\bar{x} - W(\bar{j}) = E_{0,\bar{j}} + \bar{j}\bar{x} = \langle 0, \bar{j}|H_0|0, \bar{j}\rangle \tag{14.41}$$

This is the promised energetic interpretation of the effective potential $\Gamma(\bar{x})$: *it is the expectation of the source-free Hamiltonian H_0 in the ground state of the sourced Hamiltonian $H_{\bar{j}}$ where the source $\bar{j}$ is chosen to give the value $\bar{x}$ for the expectation value of position in that ground state.* In our crude approximation, one finds (see Problem 3) that for $|\bar{x}| < v$, $\Gamma(\bar{x})$ is a convex function, with $\Gamma(\bar{x}) = \frac{\omega}{2} - \gamma\sqrt{(1 - \bar{x}^2/v^2)}$. The overlap matrix element γ is exponentially small in our model by our choice of a large quartic coupling λ, but it is *automatically* small in a true field theory, as the overlap of the states $|v\rangle$ (respectively, $|-v\rangle$) characterized by having field expectation values $\langle\phi(x)\rangle = +v$ (respectively $-v$) is exponentially suppressed in the spatial volume V: $\langle +v| - v\rangle \sim e^{-KV}$, essentially because this overlap involves the multiplication of order V tunneling amplitudes connecting the field variable at each spatial point[5] between the two vacuum values $+v$ and $-v$ (see Problem 4 for an explicit example). We see that in the infinite-volume limit (where $\gamma \to 0$) in the field theory cases, the effective potential develops a flat section, as in the zero-dimensional model (see Fig. 14.2), connecting the two classical minima of the field potential. Of course, in the quantum case there is an additional zero-point energy (which could be removed by normal-ordering): just the ubiquitous $\frac{\omega}{2}$ appearing in the preceding formulas.

Finally, we consider field theory proper, specifically the double-well scalar theory of Section 8.4, with Euclidean generating functional

$$Z[j] = \int \mathbf{D}\phi\, e^{-(S[\phi] - \int j(x)\phi(x)d^4x)}, \quad S[\phi] = \frac{1}{2}(\partial_\mu\phi)^2 - P_-(\phi) \qquad (14.42)$$

with $P_-(\phi)$ given in (14.24). In the discussion that follows we shall be discussing features of spontaneous symmetry-breaking at the lowest order of the loop expansion: the reader will recall from the discussion in Section 10.4 that the effective action at the leading order of a formal expansion in Planck's constant $\hbar$—which, as we showed there, was equivalent to the perturbative loop expansion—reproduces the classical action. We now know that in the spontaneously broken case (14.42), the properly defined (with either the "sup", or equivalently, the minimum-energy definition as in (14.41)) effective potential is convex, and at the leading order of $\hbar$ amounts to the convex hull of the double-well field potential $P_-(\phi)$. There will be higher-order loop corrections to the quantities discussed below, but they generally do not affect the qualitative features of global spontaneous symmetry breakdown.[6]

As previously, we begin by working at finite spacetime volume $\Omega = VT$, and restrict the source function $j(x)$ to be a spacetime constant, $j(x) = \bar{j}$. A connected finite-volume generating function $W(\bar{j})$ is then defined in the usual way by taking the infinite Euclidean time limit to project out the lowest-energy state $|0, \bar{j}\rangle$ in the presence of the source $\bar{j}$, as follows:

---

[5] We imagine throughout that our theory is regularized at short distance, say on a spatial lattice, so that V can be regarded simply as enumerating the finite number of spatial lattice points.

[6] An interesting exception is provided by the Coleman–Weinberg phenomenon (Coleman and Weinberg, 1973), in which a theory *without* spontaneous breaking at the classical level develops a non-vanishing field expectation value at the one-loop level due to radiative corrections. The features of vacuum structure, clustering, etc., discussed below, apply in full force to such theories, once these loop effects are included.

$$W_V(\bar{j}) \equiv \frac{1}{V} \lim_{T \to \infty} \frac{1}{T} \ln \frac{Z(\bar{j})}{Z(0)} \tag{14.43}$$

Note that the derivative $\frac{dW(\bar{j})}{d\bar{j}} \equiv \bar{\phi} = \langle 0, \bar{j} | \phi(x) | 0, \bar{j} \rangle = \langle 0, \bar{j} | \phi(0) | 0, \bar{j} \rangle$, as the expectation value of the field in the ground state of the translationally invariant sourced Hamiltonian is necessarily itself translation-invariant. The Legendre transform giving the effective potential is defined as above for the anharmonic oscillator, and as previously, is just the energy of the source-free Hamiltonian H_0 in the state $|0, \bar{j}\rangle$ in which the field has expectation $\bar{\phi}$

$$\Gamma_V(\bar{\phi}) = \bar{j}\bar{\phi} - W_V(\bar{j}) = \langle 0, \bar{j} | H_0 | 0, \bar{j} \rangle \tag{14.44}$$

For finite spatial volume V, $\Gamma_V(\bar{\phi})$ is convex and smooth, but develops cusps (discontinuities in the first derivative) and a flat section for $-v < \bar{\phi} < +v$ in the infinite volume limit. In this limit, the theory has two exactly degenerate minimum-energy states (in the absence of a source), which we may denote, following our previous notation, $|+v\rangle$ and $|-v\rangle$, with $\langle \pm v | \phi(x) | \pm v \rangle = \pm v$. As pointed out above, these states are strictly orthogonal in the infinite volume limit: indeed, the matrix element of any local operator taken between $|+v\rangle$ and $|-v\rangle$ also vanishes, as the local field cannot "twist" the scalar field expectation value from $-v$ to $+v$ over all spacetime points (see Problem 4). As

$$\langle -v | H_0 | - v \rangle = \langle +v | H_0 | + v \rangle \equiv E_0 \tag{14.45}$$

and

$$\langle -v | H_0 | + v \rangle = 0 \tag{14.46}$$

it follows that for $-v < \bar{\phi} < +v$, taking

$$|0, \bar{j}\rangle = \alpha | - v \rangle + \beta | + v \rangle, \quad |\alpha|^2 + |\beta|^2 = 1 \tag{14.47}$$

with

$$\bar{\phi} = \langle 0, \bar{j} | \phi(x) | 0, \bar{j} \rangle = |\alpha|^2(-v) + |\beta|^2(+v) \tag{14.48}$$

the infinite volume effective potential is immediately seen to be constant in the range $-v < \bar{\phi} < +v$:

$$\Gamma_\infty(\bar{\phi}) = \langle 0, \bar{j} | H_0 | 0, \bar{j} \rangle = |\alpha|^2 E_0 + |\beta|^2 E_0 = E_0 \tag{14.49}$$

This phenomenon is a very familiar one in the classical thermodynamics of coexisting phases of simple fluids (see Wightman's Introduction in (Israel, 1978)), where we also find flat sections in the boundaries representing the graphs of various thermodynamic functions (such as the internal energy U) when given as functions of the other extensive variables of the system (such as entropy S and volume V). It is apparent from this discussion that the coherent states $|f\rangle$ constructed in Section 8.3, with $\langle f | \phi(\vec{x}, 0) | f \rangle = f(\vec{x})$ and with energy reproducing the double-well structure of $P_-(\phi)$ (cf. (8.63)), do not in fact correspond to true ground states once $\bar{\phi}$ is constrained to take a value in the range $-v$ to $+v$: instead the system prefers the appropriate linear combination (14.47) of the two degenerate vacua of the system.

The mixed states[7] $\alpha| - v\rangle + \beta| + v\rangle \equiv |\bar{\phi}\rangle$ defined in (14.47) are perfectly well-defined normalized states in the Hilbert space of the theory, but, with the exception of the two extreme cases $|\pm v\rangle$ ($|\alpha| = 0$ or 1), they are not physically acceptable vacua. Indeed, a Fock space built on such states will necessarily result in a dramatic failure of the Ruelle clustering property discussed in Section 9.2, and hence in the basic property of cluster decomposition which constitutes one of the pillars on which we constructed the entire framework of local quantum field theory. To see this in a simple example, consider the connected part of the Wightman two-point function defined with respect to a mixed vacuum $|\bar{\phi}\rangle$, $-v < \bar{\phi} < v$:

$$\langle\bar{\phi}|\phi(x_1)\phi(x_2)|\bar{\phi}\rangle_c \equiv \langle\bar{\phi}|\phi(x_1)\phi(x_2)|\bar{\phi}\rangle - \langle\bar{\phi}|\phi(x_1)|\bar{\phi}\rangle\langle\bar{\phi}|\phi(x_2)|\bar{\phi}\rangle \tag{14.50}$$

In the first term on the right-hand side, we may insert a complete set of states

$$\langle\bar{\phi}|\phi(x_1)\phi(x_2)|\bar{\phi}\rangle = \langle\bar{\phi}|\phi(x_1)| - v\rangle\langle-v|\phi(x_2)|\bar{\phi}\rangle + \langle\bar{\phi}|\phi(x_1)| + v\rangle\langle+v|\phi(x_2)|\bar{\phi}\rangle$$

$$+ \sum_n{}' \langle\bar{\phi}|\phi(x_1)|n\rangle\langle n|\phi(x_2)|\bar{\phi}\rangle \tag{14.51}$$

where the primed sum runs over non-vacuum states, beginning with single-particle states separated (in this theory with a broken *discrete* symmetry) by a non-zero mass gap m from the degenerate vacuum states $|\pm v\rangle$. If x_1, x_2 are spacetime coordinates separated by a large space-like separation R, i.e., $(x_1 - x_2)^2 = -R^2$, then the primed sum has a Källen–Lehmann representation (cf. Section 9.5) as a spectral integral over free two-point functions with mass $\mu \geq m$, which fall at least as fast as e^{-mR} at large R. Up to these exponential falling terms, therefore, and using the vacuum orthogonality property $\langle\pm v|\phi(x)| \mp v\rangle = 0$, we find that in an infinite volume theory, for R large

$$\langle\bar{\phi}|\phi(x_1)\phi(x_2)|\bar{\phi}\rangle = |\alpha|^2\langle-v|\phi(x_1)| - v\rangle\langle-v|\phi(x_2)| - v\rangle$$

$$+ |\beta|^2\langle+v|\phi(x_1)| + v\rangle\langle+v|\phi(x_2)| + v\rangle + O(e^{-mR})$$

$$= |\alpha|^2\langle-v|\phi(0)| - v\rangle\langle-v|\phi(0)| - v\rangle$$

$$+ |\beta|^2\langle+v|\phi(0)| + v\rangle\langle+v|\phi(0)| + v\rangle + O(e^{-mR})$$

$$= |\alpha|^2(-v)^2 + |\beta|^2 v^2 = v^2 \tag{14.52}$$

whereas the second term in (14.50) becomes

$$\langle\bar{\phi}|\phi(x_1)|\bar{\phi}\rangle\langle\bar{\phi}|\phi(x_2)|\bar{\phi}\rangle = \bar{\phi}^2 \tag{14.53}$$

---

[7] The terminology "mixed" here being applied to non-extremal vacuum states should not be confused with the sense of "mixed" as distinguished from "pure" states in statistical physics: the states $|\bar{\phi}\rangle$ are pure states for all α, in the statistical sense, corresponding to definite rays in the Hilbert space. The thermodynamic analogy is with systems with coexisting phases: see Wightman, footnote 4, *op. cit.*

Combining (14.52) and (14.53), we find asymptotically,

$$\langle\bar{\phi}|\phi(x_1)\phi(x_2)|\bar{\phi}\rangle_c \to v^2 - \bar{\phi}^2 + O(e^{-mR}), \quad R \to \infty \tag{14.54}$$

As $v^2 - \bar{\phi}^2 \neq 0$ unless we choose $\bar{\phi} = \pm v$ (i.e., $|\alpha| = 1$ or 0), we see that clustering fails except for the two extreme points representing states where the field is globally oriented with expectation value $\bar{\phi}$ either at $+v$ or at $-v$. The preference for such states does not seem to be energetically based, as the states $|\bar{\phi}\rangle$ are degenerate in energy for all values of $\bar{\phi}$ with $-v \leq \bar{\phi} \leq +v$.

Indeed, the fact that we live in a Universe consisting of particle states built on a clustering vacuum has to be understood in "historical" terms—in other words, as a consequence of the cosmological evolution of the Universe. To take a familiar analogy from condensed matter physics, the choice of a direction of magnetization of an initially demagnetized region of ferromagnetic material as it falls below its Curie temperature T_c will typically depend on the presence of small random external fields, whose direction then gets "frozen in" as the material settles into a state with a non-zero expectation value for total electronic magnetic moment as the material cools below T_c. In our simple model with a discrete symmetry $\phi \to -\phi$, the choice of either the $|+v\rangle$ or the $|-v\rangle$ state as the physical vacuum would likewise depend on an accident of external fields which would "tickle" the field into one or the other state—at this early stage, for energetic reasons—from which the system cannot escape once the volume of the Universe grows and the temperature cools.

Of course, we can only expect the same orientation of the symmetry-breaking to obtain over regions which are causally connected when the freeze-out occurs. The fact that the entire present observable Universe appears to have the same direction of spontaneous symmetry-breaking for the putative Grand Unified Theory overlying the Standard Model of gauge interactions is one aspect of the deep and famous "horizon" problem of Big-Bang cosmology, in principle solved by the inflationary cosmology developed by Alan Guth and others. The history dependence of the low-energy dynamics of a system with spontaneous symmetry-breaking has a precise mathematical correlate in the non-uniformity of the zero-source and infinite-volume limits for $W_V[j]$: if the infinite-volume limit is taken after the sources are sent to zero, expectation values of observables necessarily respect the global symmetry (for example, $\langle\phi\rangle = 0$, corresponding to the symmetric, but non-clustering, ground state $\frac{1}{\sqrt{2}}(|-v\rangle + |+v\rangle)$), while if the infinite-volume limit is taken first, and then the source sent to zero from the positive (resp. negative) directions, the expectation values found from the functional integral refer to the "pre-magnetized" clustering states $|+v\rangle$ (resp. $|-v\rangle$).

The preceding discussion was restricted to a theory displaying a simple discrete global reflection symmetry, $\phi \to -\phi$, undergoing spontaneous breaking, which is signalled (in the infinite volume limit) by the appearance of a two-dimensional vacuum sector with two degenerate minimum-energy states $|+v\rangle, |-v\rangle$. The generalization of the discussion to a spontaneously broken continuous group is straightforward. In the second theory discussed in Section 8.4, for example, with a triplet of scalar field $\vec{\phi}$ subject to Hamiltonian (8.69),

$$H = \int d^3x : \frac{1}{2}\dot{\vec{\phi}}^2 + \frac{1}{2}|\vec{\nabla}\vec{\phi}|^2 + P(\vec{\phi}) :, \quad P(\vec{\phi}) = \frac{\lambda}{24}(\vec{\phi}^2 - v^2)^2, \quad v = \sqrt{\frac{6}{\lambda}m} \quad (14.55)$$

the global O(3) symmetry of the Hamiltonian, under the transformations $\vec{\phi} \to R\vec{\phi}$, with R an orthogonal rotation, is spontaneously broken, and the vacuum sector can be parameterized by the points on the surface of a sphere $|\langle\vec{\phi}\rangle| \le v$ in field space. There is thus a continuous infinity of orthogonal minimum-energy states in the infinite-volume limit, and the effective potential $\Gamma(\vec{\phi})$, which is again necessarily convex, is constant within and on the boundary of this sphere. A physically realistic theory, satisfying the constraints of clustering, again requires that we choose vacua corresponding to an extreme point, on the surface of the sphere, with $|\langle\vec{\phi}\rangle| = v$. Of course, Goldstone's theorem (our symmetry is now continuous!) asserts that the field theory constructed on such a vacuum state necessarily has zero-mass particle states. Denoting a particular clustering vacuum state by $|\vec{v}\rangle$, where $\vec{v}$ is a three-vector with magnitude $v = \sqrt{6/\lambda}m$, the proof of the Goldstone theorem outlined in the preceding section indicates that the state $\vec{J}^0(x)|\vec{v}\rangle$ contains single-particle Goldstone particle states, so that the spatial integral, giving the effect of the charges $\vec{Q}$ on $|\vec{v}\rangle$, produces a state with zero spatial momentum (and therefore energy) Goldstone particles. On the other hand, the effect of the O(3) charges is simply to perform an infinitesimal rotation in field space, so at least formally, the vacua $|\vec{v}\rangle$ with $|\langle\vec{\phi}\rangle| = v$ can be constructed from each other by application of the finite rotation $e^{i\vec{\omega}\cdot\vec{Q}}$—in other words, by constructing coherent states containing infinitely many zero-energy Goldstone modes. At infinite volume these states become orthogonal, and the associated formally unitary operations become improper, much as the interaction-picture operators in Haag's theorem (cf. Section 10.5).

Although the definition of an effective action (or the associated effective potential, for constant fields) in terms of a Legendre transform is very convenient for perturbative calculations (one simply sums the 1PI graphs), it is not particularly useful in situations where spontaneous symmetry-breaking occurs at a non-perturbative level, such as in strongly coupled scalar theories (in calculating upper bounds on the Higgs mass, for example) or in quantitative studies of chiral symmetry-breaking in quantum chromodynamics. A more convenient object in these cases, where we must resort to explicit numerical simulation of the lattice-regularized field theory, is provided by the *constraint effective potential* introduced in (O'Raifertaigh *et al.*, 1986). For the scalar theory (14.42), for example, define a functional $U_\Omega[\phi]$, which we shall call the *constraint effective action*, by the functional integral

$$e^{-U_\Omega[\phi]} \equiv \int \mathbf{D}\hat{\phi}\delta(\phi(x) - \hat{\phi}(x))e^{-S[\hat{\phi}]} \qquad (14.56)$$

where the field theory is defined at finite spacetime volume Ω. Note that the connected generating functional $W_\Omega[j]$ is completely reconstructible from the knowledge of $U_\Omega[\phi]$, by a functional Laplace transformation

$$W_\Omega[j] = \int \mathbf{D}\phi e^{-(U_\Omega[\phi] - \int j(x)\phi(x)d^4x)} \qquad (14.57)$$

Restricting ourselves to spacetime constant fields, we can similarly define the constraint effective potential $U_\Omega(\phi)$

$$e^{-U_\Omega(\phi)} \equiv \int \mathbf{D}\hat{\phi}\,\delta\!\left(\phi - \frac{1}{\Omega}\int \hat{\phi}(x)\,d^4x\right)e^{-S[\hat{\phi}]} \tag{14.58}$$

It is trivial to impose the δ-function constraint in (14.58) in a numerical (e.g., Monte Carlo) simulation of the lattice-regularized theory: for example, on a spacetime lattice of $\Omega = N$ points, we can just set $\hat{\phi}_N = N\phi - \sum_{i=1}^{N-1} \hat{\phi}_i$, and then simulate the remaining system of $N - 1$ field variables by standard statistical sampling methods. In the case of our zero-dimensional toy theory (14.20), one sees immediately that $U(\bar{\phi}) = P(\bar{\phi})$: the contraint effective potential therefore has the same double-well structure as the classical field potential in the symmetry-breaking case, and is evidently *not* convex, unlike the conventional effective potential $\Gamma(\bar{\phi})$. In proper field theory (models with kinetic terms), one finds (O'Raifertaigh *et al.*, 1986) that in the infinite volume limit, $\Omega \to \infty$, $U_\Omega(\bar{\phi})$ approaches the previously discussed convex function $\Gamma_\infty(\bar{\phi})$: in particular, the flat regions describing mixed vacua are recovered in this limit. For an application of the constraint effective potential approach to the problem of the Higgs boson mass limit in electroweak theory, see (Kuti and Shen, 1988).

14.4 Problems

1. Show that in the toy model defined by the integral (14.20), in the symmetry-breaking case with potential $P_-(\phi)$, the infinite volume limit of the effective potential (with the "sup" definition) $\Gamma_-(\phi)$ equals $P_-(\phi)$ for $|\phi| > v$, while $\Gamma_-(\phi)$ is constant for $-v \le \phi \le +v$.

2. Calculate the Hamiltonian matrix element $\gamma \equiv \langle +|H_0|-\rangle$ for the Hamiltonian H_0 in (14.35) between the Gaussian approximate ground states $\psi_\pm(x)$ given in (14.36).

3. Verify the result $\Gamma(\bar{x}) = \frac{\omega}{2} - \gamma\sqrt{(1 - \bar{x}^2/v^2)}$ for the effective potential for $|\bar{x}| < v$ in the anharmonic oscillator, using the two-dimensional truncation to the subspace spanned by $|\pm\rangle$.

4. The coherent states of a scalar field of mass m with different expectation values for the field become orthogonal in the infinite-volume limit. To see this, we begin by considering the field quantized at finite volume (at time zero):

$$\phi(\vec{x}, 0) = \frac{1}{\sqrt{V}} \sum_{\vec{k}} \frac{1}{\sqrt{2E_{\vec{k}}}} (a_{\vec{k}} e^{i\vec{k}\cdot\vec{x}} + a_{\vec{k}}^\dagger e^{-i\vec{k}\cdot\vec{x}}), \quad [a_{\vec{k}}, a_{\vec{k}'}^\dagger] = \delta_{\vec{k},\vec{k}'} \tag{14.59}$$

A coherent translationally-invariant (zero-momentum) state $|v\rangle$ with non-vanishing expectation value $\langle v|\phi|v\rangle = v$ can be constructed by applying an exponential of the creation operator $a_0^\dagger$ for zero-momentum modes of the field ϕ:

$$|v\rangle = Ke^{Cva_0^\dagger}|0\rangle \tag{14.60}$$

with $|0\rangle$ the vacuum with respect to the destruction modes of ϕ, $a_{\vec{k}}|0\rangle = 0, \forall \vec{k}$. K is a normalization constant to assure that $|v\rangle$ is unit normalized.

(a) Determine the constants C, K in terms of m, V, v.

(b) Show that $\langle -v| + v\rangle$ vanishes exponentially in the infinite volume limit $V \to \infty$.
(c) Show that $\langle -v|\phi(\vec{x}, 0)^2| + v\rangle$ vanishes exponentially in the infinite-volume limit $V \to \infty$. Note that this matrix element actually vanishes identically even at finite volume if we normal order the squared field: why? (See discussion of coherent states at end of Section 8.3.)

15

Symmetries IV: Local symmetries in field theory

The preceding three chapters have been devoted to examining the consequences of two main types of symmetry in quantum field theory: those in which the dynamics of the theory is invariant under certain transformations on the kinematical spacetime scaffolding of the theory (we have called these "spacetime symmetries"), and those in which the symmetry transformations operate globally (i.e., identically at all points in spacetime) and linearly on the set of independent fields present in the theory (calling these "internal global symmetries"). If only fields of spin zero and $\frac{1}{2}$ are present, these are in fact the only types of symmetry that are relevant in relativistic field theory. Once spin-1 fields are present, however, the situation changes radically. The formulation of renormalizable interacting field theories for spin-1 particles turns out to lead us inexorably to the introduction of a new type of symmetry—*local gauge symmetry*—which represents, in some sense, an amalgam of spacetime and internal symmetry.

Anticipating the discussion of scale dependence of Lagrangian field theories in Part 4 of the book, we shall see that the survival to low energies of non-trivial interactions of spin-1 particles guarantees the presence of local gauge invariance in the dynamics of the theory describing these low-energy processes. Of course, local gauge invariance was already fully present in the classical electrodynamics perfected by the great work of Maxwell in the 1860s, and the incorporation of a local gauge principle in relativistic quantum field theory was implicit in the very earliest works on quantum electrodynamics.[1] However, a full appreciation of the extraordinarily deep implications of local gauge symmetry for *all* the fundamental interactions in Nature had to await the development of the concepts and techniques of modern quantum field theory. In this chapter we begin our study of these implications.

15.1   Gauge symmetry: an example in particle mechanics

The basic idea of local gauge symmetry can be illustrated in a purely classical context with a simple example from point-particle mechanics. We imagine a particle, for convenience of unit mass, moving in one dimension subject to a potential: for reasons that will shortly become clear, we will denote the coordinate of motion r (rather than x, say), and suppose that the motion is restricted to the positive half-line $r > 0$ (the potential may be chosen to go to positive infinity as $r \to 0$, for example). The

---

[1] The modern "gauge" terminology, however, goes back to the work of Weyl, beginning in 1919.

Lagrangian for this system can be written:

$$\mathcal{L} = \frac{1}{2}\dot{r}^2 - V(r^2) \tag{15.1}$$

Now suppose that the motion of our particle is observed in a frame of reference attached to a turntable (situated just below the half-line along which the particle is moving) on which are inscribed perpendicular axes measuring two coordinates q_1 and q_2. The turntable is allowed to execute capricious rotations in the course of time, but at any time we have $r = \sqrt{q_1^2 + q_2^2}$. If we substitute this relation into the Lagrangian, we find a new Lagrangian in terms of the q_1, q_2 degrees of freedom which describe the dynamics of the system as observed in the frame of reference affixed to the turntable:

$$\mathcal{L} = \frac{1}{2}\frac{(q_1\dot{q}_1 + q_2\dot{q}_2)^2}{q_1^2 + q_2^2} - V(q_1^2 + q_2^2) \tag{15.2}$$

If we now wish to study the quantum mechanics of such a system, we must construct a Hamiltonian via a Legendre transformation, and impose canonical commutation relations on the conjugate momentum-coordinate pairs p_1, q_1 and p_2, q_2, where

$$p_1 = \frac{\partial \mathcal{L}}{\partial \dot{q}_1} = \frac{q_1(q_1\dot{q}_1 + q_2\dot{q}_2)}{q_1^2 + q_2^2} \tag{15.3}$$

$$p_2 = \frac{\partial \mathcal{L}}{\partial \dot{q}_2} = \frac{q_2(q_1\dot{q}_1 + q_2\dot{q}_2)}{q_1^2 + q_2^2} \tag{15.4}$$

It is immediately clear that the Legendre transform does not exist in this case, for the simple reason that it is impossible to solve uniquely for the velocities $\dot{q}_1, \dot{q}_2$ in terms of the conjugate momenta p_1, p_2, as the pair of equations (15.3, 15.4) are degenerate. In fact, we have the identity (or "primary constraint"—one following directly from the structure of the Lagrangian),

$$\chi(q_1, q_2, p_1, p_2) \equiv \chi(\vec{q}, \vec{p}) = q_1 p_2 - q_2 p_1 = 0 \tag{15.5}$$

Recalling our discussion of Legendre transforms in Section 14.3, we recognize a recurrence of the disease already encountered in systems with spontaneous symmetry-breaking, a lack of strict convexity in the quantity (in this case the Lagrangian) undergoing the Legendre transform. Unfortunately, the energetically motivated alternative "sup" definition of the Legendre transform introduced there to circumvent the difficulty is of no use here: we *must* have well-defined expressions for the velocities in terms of the momenta if we wish to identify conjugate canonical variables as a prelude to the quantization of the system and calculate a unique Hamiltonian dynamics at the quantum level. The "flat" regions of the Lagrangian function giving rise to the breakdown of the standard Legendre transform are easily identified: the Lagrangian in (15.2) is invariant under the time-dependent "gauge transformations"

$$q_1(t) \rightarrow q_1(t) \cos{(\theta(t))} + q_2(t) \sin{(\theta(t))}$$
$$q_2(t) \rightarrow -q_1(t) \sin{(\theta(t))} + q_2(t) \cos{(\theta(t))} \tag{15.6}$$

where $\theta(t)$ is an arbitrary differentiable function of time: in the turntable picture above, it corresponds to twisting the turntable in an arbitrary direction, given by the angle $\theta(t)$, at any given time t. Note that the constraint (15.5) is just the component of angular momentum in the "3" direction perpendicular to the q_1, q_2 axes: it is, in fact, the generator of the infinitesimal version of the gauge transformations (15.6). The "gauge-invariance" of the Lagrangian (15.2) amounts simply to the statement that the true physical coordinate $r = \sqrt{q_1^2 + q_2^2}$ is independent of $\theta(t)$, and (15.2) is simply (15.1) in disguised form.

The invariance property (15.6) means that we are free to rotate the coordinate pair $(q_1(t), q_2(t))$ into the "gauge" $q_2 = 0$ (say) by choosing $\theta(t) = \arctan\left(\frac{q_2(t)}{q_1(t)}\right)$, as the dynamics is independent of $\theta(t)$: in other words, by choosing the orientation of the turntable at any time so that the particle is situated on the q_1 axis. In the gauge $q_2(t) = 0$, the Lagrangian (15.2) reduces simply to

$$\mathcal{L} = \frac{1}{2}\dot{q}_1^2 - V(q_1^2) \tag{15.7}$$

which is precisely our original Lagrangian (15.1), with the trivial change of notation $r \rightarrow q_1$. For this Lagrangian, of course, there is absolutely no problem with quantization: we simply set $p_1 = \dot{q}_1$ and $H = \frac{1}{2}p_1^2 + V(q_1^2)$, with $[p_1, q_1] = -i\hbar$.

The Lagrangian (15.2) has an obvious generalization to three dimensions: with $\vec{q}$ a three-vector, we choose

$$\mathcal{L} = \frac{1}{2}\frac{(\vec{q} \cdot \dot{\vec{q}})^2}{\vec{q}^2} - V(\vec{q}^2) \tag{15.8}$$

In this case there are three primary constraints following immediately from $\vec{p} = \vec{q}\frac{\vec{q}\cdot\dot{\vec{q}}}{\vec{q}^2}$, $\vec{q} \times \vec{p} = 0$:

$$\chi_i(\vec{q}, \vec{p}) = \epsilon_{ijk} q_j p_k = 0 \tag{15.9}$$

which are just the angular momentum components L_i, generating rotations around the three spatial axes. In this case the set of gauge transformations

$$q_i(t) \rightarrow R_{ij}(t)q_j(t) \tag{15.10}$$

with $R(t)$ an orthogonal O(3) rotation clearly form a non-abelian group. Note that the commutators (or at the classical level, the Poisson brackets) of the primary constraints in this case form a closed algebra—indeed, just the Lie algebra of the rotation group. Constraints which close in this way are referred to as "first-class" constraints, and are always associated with the presence of superfluous "gauge" degrees of freedom which can be eliminated by an appropriate gauge-fixing procedure. In order to understand how to do this in a more general way, we must now turn to a brief discussion of the theory of constrained Hamiltonian systems.

We referred above to the constraint (15.5) as a "primary" constraint—one following directly from the definition of the momenta and the structure of the Lagrangian. The requirement that the constraints of the theory survive the time development—in other words, are consistent with the Euler–Lagrange equations of motion of the theory—may lead to further constraints, which are then termed "secondary". The distinction is *not* of fundamental importance, as primary and secondary constraints are to some degree interchangeable. Consider the Lagrangian, depending on three coordinates q_0, q_1, q_2,

$$\mathcal{L} = \frac{1}{2}(\dot{q}_1^2 + \dot{q}_2^2) + q_0(q_1\dot{q}_2 - q_2\dot{q}_1) - V(q_1^2 + q_2^2) \tag{15.11}$$

The absence of any dependence on $\dot{q}_0$ immediately implies the primary constraint

$$p_0 = \frac{\partial \mathcal{L}}{\partial \dot{q}_0} = 0 \tag{15.12}$$

However, the requirement that this constraint be preserved in the time evolution,

$$\frac{\partial p_0}{\partial t} = \frac{\partial}{\partial t}\frac{\partial \mathcal{L}}{\partial \dot{q}_0} = \frac{\partial \mathcal{L}}{\partial q_0} = 0 \tag{15.13}$$

where we have employed the Euler–Lagrange equation for the coordinate q_0, amounts to the further *secondary* constraint (setting the non-dynamical $q_0 = 0$)

$$\frac{\partial \mathcal{L}}{\partial q_0} = q_1\dot{q}_2 - q_2\dot{q}_1 = q_1 p_2 - q_2 p_1 = 0 \tag{15.14}$$

which is just the *primary* constraint (15.5) arising from the Lagrangian (15.2). In the next section it will become apparent that both Lagrangians have identical physical content as constrained Hamiltonian systems. The important distinction, as we shall see, is between those constraints whose Poisson brackets (or, in the quantum case, commutators) with each other vanish once the constraints themselves are imposed (so-called *first-class constraints*) and those with non-vanishing Poisson brackets on the constraint surface (*second-class constraints*, which we do not consider further here, as our primary interest lies in the Hamiltonian interpretation of local gauge symmetries).

15.2   Constrained Hamiltonian systems

The problems encountered in attempting to construct a meaningful Hamiltonian from singular Lagrangians such as (15.2, 15.8) suggest that a direct canonical interpretation of such theories in classical phase-space is simply impossible. Dirac was the first to show, in a masterful analysis presented in his *Lectures on Quantum Mechanics* (Dirac, 1964), that this conclusion is unwarranted, and that a well-defined Hamiltonian formalism can be constructed in the presence of constraints such as (15.5). A full introduction to the theory of constrained Hamiltonian systems would require far more space[2] than we can devote to it here, so we shall restrict ourselves to the elements of the theory directly relevant to the canonical treatment, and quantization, of theories with local gauge symmetries.

Let us return to the simple example described in the preceding section, with Lagrangian (15.2). Ignoring temporarily the inconvenient absence of a unique relation between velocities and momenta, we see, using (15.3, 15.4), that we can re-express the Hamiltonian function for this theory, initially given as

$$H = p_1\dot{q}_1 + p_2\dot{q}_2 - \mathcal{L} = \frac{1}{2}\frac{(q_1\dot{q}_1 + q_2\dot{q}_2)^2}{q_1^2 + q_2^2} + V(q_1^2 + q_2^2) \qquad (15.15)$$

in a number of equivalent ways: for example,

$$H = \frac{1}{2}p_1^2(1 + \frac{q_2^2}{q_1^2}) + V(q_1^2 + q_2^2) \qquad (15.16)$$

$$= \frac{1}{2}p_2^2(1 + \frac{q_1^2}{q_2^2}) + V(q_1^2 + q_2^2) \qquad (15.17)$$

$$= \frac{1}{2}(p_1^2 + p_2^2) + V(q_1^2 + q_2^2), \dots . \qquad (15.18)$$

The lack of a unique inversion for the velocities in terms of the momenta manifests itself in the multiplicity of equivalent expressions for the Hamiltonian in the three lines above, which are clearly equal once we take the primary constraint $\chi(q_1, q_2, p_1, p_2) = q_1p_2 - q_2p_1 = 0$ into account. In fact, the set of Hamiltonians given by (the "T" subscript denotes "total Hamiltonian", including constraints, in Dirac's language)

$$H_T = \frac{1}{2}(p_1^2 + p_2^2) + V(q_1^2 + q_2^2) - \lambda(t)\chi(q_1, q_2, p_1, p_2) \qquad (15.19)$$

with $\lambda(t)$ an *arbitrary* function of time (either explicitly, and/or through an arbitrary function of the coordinates $\vec{q}, \vec{p}$), are all equivalent in this sense. If we derive Hamiltonian equations of motion $\dot{q} = \frac{\partial H}{\partial p}$, $\dot{p} = -\frac{\partial H}{\partial q}$ in the usual way from (15.19), treating the variables q_1, q_2, p_1, p_2 as normal unconstrained variables, we obtain

$$\dot{q}_1 = p_1 + \lambda q_2, \quad \dot{p}_1 = -\frac{\partial V}{\partial q_1} + \lambda p_2$$

$$\dot{q}_2 = p_2 - \lambda q_1, \quad \dot{p}_2 = -\frac{\partial V}{\partial q_2} - \lambda p_1 \qquad (15.20)$$

The interpretation of the arbitrary function λ, which we shall see also plays the role of a Lagrange multiplier enforcing the constraint, becomes clear if we introduce new primed coordinates and momenta

$$\vec{q}(t) = R(t)\vec{q}\,'(t), \quad \vec{p}(t) = R(t)\vec{p}\,'(t) \qquad (15.21)$$

---

[2] For a careful and very thorough treatment of the full theory of constrained systems, with emphasis on gauge theories, see (Henneaux and Teitelboim, 1992).

with $R(t)$ the time-dependent rotation matrix

$$R(t) = \begin{pmatrix} \cos\left(\theta(t)\right) & \sin\left(\theta(t)\right) \\ -\sin\left(\theta(t)\right) & \cos\left(\theta(t)\right) \end{pmatrix} \tag{15.22}$$

transforming us from a stationary frame to the wobbly "turntable" frame of the previous section, with the angular velocity of the turntable $\dot{\theta}(t) = \lambda(t)$. One then finds that the primed coordinates and momenta satisfy the Hamiltonian equations *without* the additional λ-term in (15.19):

$$\dot{q}_1' = p_1', \quad \dot{p}_1' = -\frac{\partial V}{\partial q_1'} \tag{15.23}$$

$$\dot{q}_2' = p_2', \quad \dot{p}_2' = -\frac{\partial V}{\partial q_2'} \tag{15.24}$$

In other words, the arbitrariness of the constraint term in (15.19) precisely incorporates the gauge freedom in the solutions of the underlying one-dimensional problem when viewed in the floating turntable frame, if we interpret the Lagrange multiplier function $\lambda(t)$ as the angular velocity of the turntable $\dot{\theta}(t)$ at any given time.

At this point, it is useful to recall that the classical Hamiltonian equations of the theory, (15.20), can be expressed in a form which is particularly suggestive when one wishes to make the transition to quantum theory, in terms of the *Poisson bracket* $\{F, G\}$ defined on arbitrary functions $F(q_i, p_i), G(q_i, p_i)$ on the (unconstrained) phase-space as follows

$$\{F, G\} \equiv \frac{\partial F}{\partial q_i}\frac{\partial G}{\partial p_i} - \frac{\partial F}{\partial p_i}\frac{\partial G}{\partial q_i} \tag{15.25}$$

Using the Poisson brackets, the dynamical evolution on phase-space (i.e., Eqs. (15.20)) amounts to

$$\dot{q}_i = \{q_i, H_T\}, \quad \dot{p}_i = \{p_i, H_T\} \tag{15.26}$$

Equivalently, we can say that the total Hamiltonian acts as the generator of infinitesimal time translations: for example, $q_i(t + \delta t) = q_i(t) + \{q_i, \delta t \cdot H_T\}$, etc. The primary constraint $\chi(q_1, q_2, p_1, p_2) = q_1 p_2 - q_2 p_1 = 0$ is itself left invariant under Hamiltonian evolution, $\{\chi, H_T\} = 0$: this is physically obvious in our toy model, as the constraint is just the angular momentum which is preserved under the two-dimensional motion of our particle in the central potential $V(q_1^2 + q_2^2)$. Thus the constraint, once applied as an initial condition at time $t = 0$, will automatically be satisfied at any later time on trajectories following the Hamiltonian evolution (15.26). Or, in yet other words, the three-dimensional *constraint surface* obtained by restricting the four-dimensional phase-space (q_1, q_2, p_1, p_2) to points satisfying $\chi(q_1, q_2, p_1, p_2) = 0$ is invariant under Hamiltonian evolution. However, this three-dimensional space, as it is odd-dimensional, cannot act as a proper dynamical phase-space (with an equal number of "p's" and "q's"). In fact, it is clearly still too large, as it contains distinct points representing physically equivalent states of the system: those related by a gauge transformation

$$(\vec{q}, \vec{p}) \rightarrow (\vec{q}' = R\vec{q}, \vec{p}' = R\vec{p}) \tag{15.27}$$

where R is a 2x2 rotation matrix (see (15.22)). As the rotation R varies over all possible rotation angles $0 \le \theta < 2\pi$, the points $\vec{q}', \vec{p}'$ trace out a one-dimensional "gauge orbit" of physically equivalent points in phase-space. In the preceding section we saw that the gauge ambiguity of the system defined by Lagrangian (15.2) could be eliminated by imposing a "gauge condition" (such as $\psi(\vec{q}, \vec{p}) = q_2 = 0$), at which point we recover a non-singular Lagrangian (15.7) with perfectly regular canonical properties. An appropriately chosen gauge condition $\psi(\vec{q}, \vec{p})$ defines a surface in the original four-dimensional phase-space of our unconstrained system which intersects the gauge orbit passing through any given point in phase-space exactly once. The imposition of such a condition means that the gauge freedom of the unconstrained system has been completely eliminated: in the turntable model of the previous section, it means that we have specified unambiguously the orientation of the turntable at every moment in time.

Note that the constraint function $\chi(\vec{q}, \vec{p})$ acts as the infinitesimal generator of gauge transformations (i.e., O(2) rotations), as

$$\{q_1, \delta\theta \cdot \chi\} = -\delta\theta \, q_2, \quad \{q_2, \delta\theta \cdot \chi\} = +\delta\theta \, q_1 \tag{15.28}$$

$$\{p_1, \delta\theta \cdot \chi\} = -\delta\theta \, p_2, \quad \{p_2, \delta\theta \cdot \chi\} = +\delta\theta \, p_1 \tag{15.29}$$

Note also that a necessary condition for the gauge freedom to be completely eliminated is that the Poisson bracket $\{\psi, \chi\}$ of the gauge-fixing function and the constraint be non-zero: otherwise put, once on the gauge-fixed surface, any gauge transformation, and in particular any infinitesimal gauge transformation, must move us off that surface. Simple axial gauges such as $\psi = q_2$ clearly satisfy this requirement, as we see from (15.28).

The three-dimensional version of our toy model, (15.8), has three primary first-class constraints (15.9) whose Poisson brackets are just the Lie algebra of the gauge group O(3):

$$\{\chi_i, \chi_j\} = \epsilon_{ijk}\chi_k \tag{15.30}$$

The reader will recall that a set of constraints is said to be first-class if their Poisson brackets vanish once the constraints themselves are imposed, which is certainly the case if they form a closed Lie algebra as here. The gauge orbits in this model correspond to spheres of fixed radius $|\vec{q}|, |\vec{p}|$ for the coordinate and momentum vectors. Again, a complete gauge-fixing- amounting to selecting a single representative point on each gauge orbit- is easily achieved by the axial gauge corresponding to imposing, say, $\psi_1 = q_1 = 0, \psi_2 = q_2 = 0$, which at the Lagrangian level amounts to rotating the $\vec{q}$ vector at each time into the z-direction. Note that on the gauge-fixed surface, the constraint $\chi_3 = q_1 p_2 - q_2 p_1$ is *automatically* satisfied: there are only two independent first-class constraints, χ_1 and χ_2 which act non-trivially on this surface, and indeed the non-degeneracy of the determinant

$$\det\{\psi_m, \chi_n\} = q_3^2 \neq 0, \quad 1 \leq m, n \leq 2 \tag{15.31}$$

assures us that no non-zero linear combination of the gauge transformations implemented by χ_1 and χ_2 can leave us on the gauge surface defined by the gauge conditions $\psi_i = 0$. It is clear that the imposition of the two gauge conditions and two independent first-class constraints should reduce our originally six-dimensional phase-space to the two-dimensional phase-space appropriate for describing the underlying "true" one-dimensional physics of the model. We will now see, following the seminal discussion of Faddeev (Faddeev, 1969), how this can be accomplished in a very general way at the Hamiltonian level using the technique of canonical transformations. Our end result will be the famous *Dirac–Faddeev* formula giving a well-defined functional integral quantization of a Hamiltonian system with first-class (gauge) constraints.

Let us assume that our constrained Hamiltonian system is initially defined on a $2f$-dimensional phase-space with phase-space coordinates $(q_1, .., q_f, p_1, .., p_f)$, with (cf. (15.19) as an example)

$$H_T = h(q_1, .., q_f, p_1, .., p_f) + \sum_{m=1}^{r} \lambda_m \chi_m \tag{15.32}$$

and that the set of first-class constraints $\chi_m(q_i, p_i) = \chi_m(\vec{q}, \vec{p}), m = 1, \ldots, r$ generate gauge transformations whose orbits intersect uniquely the submanifold defined by a set of r gauge conditions $\psi_m(q_i, p_i) = 0, m = 1, \ldots, r$. As we saw earlier, this implies that the determinant of the Poisson bracket matrix $\{\psi_m, \chi_n\}$ be non-vanishing on the constraint surface. We shall assume that the gauge conditions are chosen to have vanishing Poisson brackets with each other:

$$\{\psi_m, \psi_n\} = 0 \tag{15.33}$$

The commutativity of the gauge-fixing conditions implies that we can find a canonical transformation to a new set of $2f$ coordinates and momenta, which we shall label $(Q_1^*, .., Q_{f-r}^*, Q_1, .., Q_r, P_1^*, .., P_{f-r}^*, P_1, .., P_r)$, where the ψ_m play the role of the last r momenta (which necessarily commute)

$$P_m \equiv \psi_m(\vec{q}, \vec{p}), \quad m = 1, .., r \tag{15.34}$$

We recall that a canonical transformation on phase-space is a change of coordinates which leaves the Poisson bracket invariant: in particular we must have $\{Q_i^*, P_j^*\} = \delta_{ij}$ for the first $f - r$ conjugate pairs and $\{Q_m, P_n\} = \delta_{mn}$ for the final r pairs of the gauge contraints with their conjugate coordinates.

For example, if our system is the $f = 2$-dimensional toy model defined by (15.19), and we wish to impose the (single) axial gauge condition $\psi = \alpha q_1 + \beta q_2 = 0, \alpha^2 + \beta^2 = 1$, a suitable set of new coordinates would be $Q_1^* = \beta q_1 - \alpha q_2, P_1^* = \beta p_1 - \alpha p_2, Q_1 = -\alpha p_1 - \beta p_2, P_1 = \psi = \alpha q_1 + \beta q_2$. Note that the determinant of the Poisson bracket matrix of χs and ψs is just the Jacobian of the change of variables from χ_m to Q_n:

$$\det\{\chi_m, P_n\} = \det(\frac{\partial \chi_m}{\partial Q_n}) \neq 0 \tag{15.35}$$

which ensures that our first-class constraints, re-expressed in the new variables,

$$\chi_m(Q_i^*, P_i^*, Q_m, P_m) = 0 \qquad (15.36)$$

can be solved uniquely for the r new coordinates $Q_m, m = 1, .., r$ as functions of the constrained starred variables only (once the $P_m = 0$ gauge conditions are applied):

$$Q_m = Q_m(Q_i^*, P_i^*, P_m = 0) \equiv f_m(Q_i^*, P_i^*) \qquad (15.37)$$

In the example given above, the single first-class constraint $q_1 p_2 - q_2 p_1 = 0$ becomes $Q_1 Q_1^* + P_1 P_1^* = 0$ which allows us to eliminate the coordinate $Q_1 = -\frac{P_1 P_1^*}{Q_1^*}$, which, of course, just amounts to $Q_1 = 0$ on the gauge-fixed surface $P_1 = \psi = 0$.

Returning to the general situation as given by (15.32), the usual lore on canonical transformations tells us that the physics of the system is uniquely captured by employing just the $2(f - r)$ set of constrained variables $(Q_1^*, .., Q_{f-r}^*, P_1^*, .., P_{f-r}^*)$ with a constrained Hamiltonian H^*, obtained by expressing the original Hamiltonian H_T in terms of the new variables and then implementing the gauge conditions $P_m = 0$ and using the constraints χ_m to eliminate the Q_m coordinates as in (15.37):

$$H^*(Q_i^*, P_i^*) = h(q_1, .., q_f, p_1, .., p_f)|_{P_m=0, Q_m=f_m(Q_i^*, P_i^*)} \qquad (15.38)$$

Once again, resorting to our simple toy model as a concrete example, we find that our original Hamiltonian (15.19) becomes, in terms of the constrained starred variables Q_1^*, P_1^*, in the gauge $\psi = \alpha q_1 + \beta q_2 = 0$,

$$H^*(Q_1^*, P_1^*) = \frac{1}{2}(P_1^*)^2 + V((Q_1^*)^2) \qquad (15.39)$$

This Hamiltonian has exactly the form we would obtain by the conventional canonical procedure beginning from the non-singular Lagrangian (15.7), which the reader will recall was obtained by exploiting the gauge symmetry of the singular Lagrangian (15.2) to eliminate the gauge freedom in the system *ab initio* (by setting $q_2 = 0$). The reader is strongly encouraged to carry through the gauge-fixing procedure in the three-dimensional version of the model, with Lagrangian (15.8) and an O(3) non-abelian gauge symmetry: the end result will be exactly the same Hamiltonian, representing a theory with a single physical degree of freedom.

The fully constrained Hamiltonians in (15.38, 15.39) can be subjected to quantization in the normal way, by imposing the canonical commutator condition[3]

$$[Q_i^*, P_j^*] = i\hbar \delta_{ij} \qquad (15.40)$$

Equivalently, we may formulate the theory in the path-integral framework by writing the functional integral for the propagation kernel (cf. (4.120))

$$K(t_f, t_i) = \int \prod_{i=1}^{f-r} \mathbf{D}Q_i^* \mathbf{D}P_i^* e^{\frac{i}{\hbar} \int_{t_i}^{t_f} (P_i^*(t)\dot{Q}_i^*(t) - H^*(Q_i^*(t), P_i^*(t)))dt} \qquad (15.41)$$

---

[3] As pointed out originally by Dirac, the classical to quantum transition in this context amounts simply to the replacement $\{\ldots, \ldots\} \rightarrow \frac{-i}{\hbar}[\ldots, \ldots]$.

We are now going to do something which at first sight seems very strange indeed: we wish to write an equivalent functional-integral representation for the kernel K, but in terms of the *full set of unconstrained variables* $q_1, \ldots, q_f, p_1, \ldots, p_f$ from which we started. In other words, we wish to restore the physically superfluous gauge degrees of freedom which we have just expended so much effort to eliminate! For the simple mechanical examples considered so far, such a maneuver would be completely unnecessary and, indeed, pointless, but for the gauge field theories which we are about to explore it is precisely the *unconstrained* version of the theory which manifests directly the critical (from a field-theoretic point of view) locality and Poincaré invariance properties which we are enjoined to preserve at all costs.

First, we restore the original set of $2f$ coordinates and momenta at any given time in the functional integral measure by inserting δ-functions which incorporate the procedures by which we originally eliminated the $2r$ coordinates and momenta Q_m, P_m to obtain the constrained Hamiltonian H^*:

$$\prod_{i=1}^{f-r} \mathbf{D}Q_i^* \mathbf{D}P_i^* \rightarrow \prod_{i=1}^{f-r} \mathbf{D}Q_i^* \mathbf{D}P_i^* \prod_{m=1}^{r} \mathbf{D}Q_m \mathbf{D}P_m \delta(P_m) \delta(Q_m - f_m(Q_i^*, P_i^*))$$

$$= \prod_{i=1}^{f} \mathbf{D}q_i \mathbf{D}p_i \prod_{m=1}^{r} \delta(\psi_m) \delta(Q_m - f_m(Q_i^*, P_i^*)) \tag{15.42}$$

where we have used the fact that the canonical measure $\prod_{i=1}^{f} \mathbf{D}q_i \mathbf{D}p_i$ is invariant under the canonical transformation to the $(Q_1^*, .., Q_{f-r}^*, Q_1, .., Q_r, P_1^*, .., P_{f-r}^*, P_1, .., P_r)$ variables. The δ-functions of the Q_m coordinates in (15.42) can be traded in for δ-functions of the first-class constraints χ_m at the cost of the Jacobian (15.35):

$$\prod_{m=1}^{r} \delta(Q_m - f_m(Q_i^*, P_i^*)) = \prod_{m=1}^{r} \delta(\chi_m) \frac{\partial(\chi_1, \chi_2, .., \chi_r)}{\partial(Q_1, Q_2, .., Q_r)} = \prod_{m=1}^{r} \delta(\chi_m) \cdot \det\{\chi_m, \psi_n\}$$

$$\tag{15.43}$$

The reader will recall (see, for example, (Goldstein, 2002), Section 9.1) that in Hamiltonian systems the combination $p_i \dot{q}_i - H$ is unchanged under a canonical transformation up to an additive total time-derivative dF (where F is the generating function of the canonical transformation), which in the exponent of the path integral will lead to an overall phase factor $e^{\frac{i}{\hbar}(F_f - F_i)}$, where F_i (resp. F_f) are the initial (resp. final) values of the generating function over the time evolution from t_i to t_f. In the field theory case, we shall be letting the initial and final times go to $-\infty$ and $+\infty$, where the fields can be safely switched off, so we may ignore this factor here. Recalling that the constrained Hamiltonian H^* in (15.41) is precisely obtained by subjecting the unconstrained Hamiltonian $h(q_1, .., q_f, p_1, .., p_f)$ to the δ-function constraints in (15.42), we see that our expression for the propagation kernel in terms of a path integral over constrained variables can be written

$$K(t_f, t_i) = \int \prod_{i=1}^{f} \mathbf{D}q_i \mathbf{D}p_i \prod_{m=1}^{r} \delta(\chi_m) \delta(\psi_m) \det\{\chi_m, \psi_n\} \, e^{\frac{i}{\hbar} \int_{t_i}^{t_f} (p_i \dot{q}_i - h(q_i, p_i)) dt}$$

$$= \int \prod_{i=1}^{f} \mathbf{D}q_i \mathbf{D}p_i \prod_{m=1}^{r} \mathbf{D}\lambda_m \delta(\psi_m) \det\{\chi_m, \psi_n\} \ e^{\frac{i}{\hbar} \int_{t_i}^{t_f} (p_i \dot{q}_i - h(q_i, p_i) - \lambda_m \chi_m) dt}$$

$$= \int \prod_{i=1}^{f} \mathbf{D}q_i \mathbf{D}p_i \prod_{m=1}^{r} \mathbf{D}\lambda_m \delta(\psi_m) \det\{\chi_m, \psi_n\} \ e^{\frac{i}{\hbar} \int_{t_i}^{t_f} (p_i \dot{q}_i - H_T(q_i, p_i)) dt} \quad (15.44)$$

In the second line we have implemented the first-class constraints $\chi_m = 0$ by introducing a set of auxiliary variables $\lambda_m, m = 1, .., r$, the functional integral over which reproduces the desired δ-functions $\delta(\chi_m)$. These variables play exactly the role of the Lagrange multipliers we introduced earlier in defining the total Hamiltonian H_T in (15.32), as we see in the final form (15.44), the previously announced *Dirac–Faddeev* formula. The Jacobian determinant $\det\{\chi_m, \psi_n\}$ appearing in this formula has become known in the field theory literature as the "deWitt–Faddeev–Popov" determinant, which we shall henceforth refer to as the "DFP" determinant. We shall now see that it plays an extremely important role in the functional integral quantization of gauge field theory.

15.3 Abelian gauge theory as a constrained Hamiltonian system

Maxwellian electrodynamics provides the classic example of a field theory with a local gauge symmetry, exemplifying just the features discussed in the previous two sections. The classical Lagrangian of this field theory

$$\mathcal{L}_{\mathrm{EM}} = -\frac{1}{4} F_{\mu\nu} F^{\mu\nu} - J^\mu A_\mu, \quad F_{\mu\nu} \equiv \partial_\mu A_\nu - \partial_\nu A_\mu \quad (15.45)$$

where the external current J^μ is conserved, $\partial_\mu J^\mu = 0$, has the Euler–Lagrange equation for the field $A_\nu, \nu = 0, 1, 2, 3$

$$\partial_\mu \frac{\partial \mathcal{L}_{\mathrm{EM}}}{\partial(\partial_\mu A_\nu)} = \partial_\mu(-F^{\mu\nu}) \quad (15.46)$$

$$= \frac{\partial \mathcal{L}_{\mathrm{EM}}}{\partial A_\nu} = -J^\nu \quad (15.47)$$

$$\Rightarrow \partial_\mu F^{\mu\nu} = J^\nu \quad (15.48)$$

whereupon, in (15.48), we recognize Maxwell's equations in covariant notation. The action defined by this Lagrangian,

$$\mathcal{I}_{\mathrm{EM}} = \int \mathcal{L}_{\mathrm{EM}} d^4 x \quad (15.49)$$

is invariant under the *local* (i.e., spacetime-dependent) gauge transformation

$$A_\mu(x) \to A_\mu^\Lambda(x) \equiv A_\mu(x) + \partial_\mu \Lambda(x) \quad (15.50)$$

where $\Lambda(x)$ is an arbitrary twice-differentiable function (as we wish the field strengths $F_{\mu\nu}$ to remain well-defined after the gauge transformation). The invariance of the term in the action involving J_μ is apparent after we use integration by parts to transfer the spacetime-derivative in the variation due to the gauge transformation to the conserved current J_μ, ignoring, as usual, boundary terms at spatiotemporal infinity where the fields may be switched off (or periodic boundary conditions imposed).

The set of transformations (15.50) evidently form an abelian group, as a gauge transformation induced by gauge-function $\Lambda_1(x)$ followed by another induced by $\Lambda_2(x)$ produces the same result if performed in the opposite order (Λ_2 before Λ_1). The transformations (15.50) are the field-theoretic analog of the time-dependent transformations (15.6) which preserved the Lagrangian of our mechanical toy example (15.2). Indeed, we shall shortly see that the gauge freedom implied by the insensitivity of the dynamics to the transformations (15.50) corresponds exactly to the variation induced by a set of first-class constraints, which in this case satisfy an abelian algebra: their Poisson brackets vanish identically, whether on or off the constraint surface. The analogy is almost, but not quite, exact: in the case of electrodynamics, first-class constraints arise as *secondary* constraints, as the condition that the primary constraints of the theory be preserved by the dynamics. In fact, the third mechanical example given in Section 15.1, with Lagrangian (15.11), and secondary first-class constraints (but otherwise equivalent in physical content to Lagrangian system (15.2)), is essentially identical in its constraint structure to Maxwell electrodynamics. To see this, we construct the conjugate momentum fields to the basic four-vector field A_μ in the usual way (cf. (12.44)):

$$\frac{\partial \mathcal{L}_{\text{EM}}}{\partial \dot{A}_\mu} = -F^{0\mu} \tag{15.51}$$

As $F^{00} = 0$, we have the primary constraint

$$\Pi_0 = -F^{00} = 0 \tag{15.52}$$

while for the spatial components of the field A_μ, the conjugate momentum fields are recognized as the electric field components of Maxwellian electrodynamics:

$$\Pi^i = -F^{0i} = F_{0i} = \partial_0 A_i - \partial_i A_0 \equiv E^i, \quad i = 1, 2, 3 \tag{15.53}$$

The vanishing of the momentum Π_0 field conjugate to A_0 is the automatic consequence of the absence of time-derivatives of A_0 in the Lagrangian: consequently, the Euler–Lagrange equation for A_0 (which is essentially the equation asserting that the primary constraint $\Pi_0 = 0$ is maintained in time) amounts to a *secondary* constraint relating fields (and/or their conjugate momenta) on a given time-slice:

$$0 = \dot{\Pi}_0 = \frac{\partial}{\partial t} \frac{\partial \mathcal{L}_{\text{EM}}}{\partial \dot{A}_0} = -\partial_i \frac{\partial \mathcal{L}_{\text{EM}}}{\partial(\partial_i A_0)} + \frac{\partial \mathcal{L}_{\text{EM}}}{\partial A_0} = \partial_i E^i - J^0 \tag{15.54}$$

which we recognize as Gauss's Law in classical electromagnetism. Note that this secondary constraint only depends on, and thereby restricts, the conjugate momentum fields $\Pi^i = E^i$ of the spatial vector potential A_i, in this case by setting the divergence of the electric field equal to the charge density (which we here take to be an externally prescribed classical function) at the same time. We now introduce unconstrained Poisson brackets on the classical phase-space defined by the conjugate pair $(A_i(\vec{x}), \Pi^i(\vec{x}))$

(at a given time)

$$\{F[A_i, \Pi^i], G[A_i, \Pi^i]\} \equiv \int (\frac{\delta F}{\delta A_i(\vec{x})} \frac{\delta G}{\delta \Pi^i(\vec{x})} - (F \leftrightarrow G)) d^3 x \qquad (15.55)$$

for arbitrary functionals F, G: in particular, the Poisson brackets for the spatial vector potential and its conjugate (electric) field are simply

$$\{A_i(\vec{x}), \Pi^j(\vec{y})\} = \delta_i^{\ j} \delta^3(\vec{x} - \vec{y}) \qquad (15.56)$$

The secondary constraints (plural, as we consider this equation as defining an independent constraint at each spatial point separately)

$$\chi(\vec{x}) \equiv J^0 - \partial_i E^i = J^0 - \partial_i \Pi^i = 0 \qquad (15.57)$$

are in fact *first-class*, as their Poisson brackets vanish identically:[4]

$$\{\chi(\vec{x}), \chi(\vec{y})\} = 0 \qquad (15.58)$$

In analogy to (15.28), an infinitesimal linear combination $\int \lambda(\vec{y}) \chi(\vec{y}) d^3 y$ of the first-class constraints acts as the infinitesimal generator of the associated *gauge transformations* of the theory,

$$\{A_i(\vec{x}), \int \lambda(\vec{y}) \chi(\vec{y}) d^3 y\} = -\int \lambda(\vec{y}) \partial_i \delta^3(\vec{x} - \vec{y}) d^3 y = \partial_i \lambda(\vec{x}) \qquad (15.59)$$

while the effect of a gauge-transformation on the momentum field variables (i.e., the electric field) is null:

$$\{\Pi^i(\vec{x}), \int \lambda(\vec{y}) \chi(\vec{y}) d^3 y\} = 0 \qquad (15.60)$$

—i.e., the electric field, unlike the vector potential, is *gauge-invariant*, and therefore possesses a direct physical meaning. Just as in the mechanical examples of the preceding sections, we are free to make such a transformation of the canonical fields independently at different times, so we recognize the gauge invariance generated by the first-class constraints of this theory as just the local gauge invariance (15.50)— restricted to the dynamical canonical fields A_i, of course.

We can now proceed to the construction of an unconstrained total Hamiltonian *à la* Dirac, imitating the procedure followed in the preceding section, where our starting

---

[4] The current J^μ should be regarded here as either a fixed external field, or built out of matter fields which have vanishing Poisson brackets with the A_μ fields.

Lagrangian is now (15.45):

$$
\begin{aligned}
H &= \int (\Pi^\mu \dot{A}_\mu - \mathcal{L})d^3x \\
&= \int (\Pi^i \dot{A}_i - \frac{1}{2}F^2_{0i} + \frac{1}{4}F^2_{jk} + J^0 A_0 + J^i A_i)d^3x \\
&= \int (\Pi^i(\Pi^i + \partial_i A_0) - \frac{1}{2}(\Pi^i)^2 + \frac{1}{4}F^2_{jk} + J^0 A_0 + J^i A_i)d^3x \\
&= \int (\frac{1}{2}(\vec{E}^2 + \vec{B}^2) - \vec{J} \cdot \vec{A} + A_0(J^0 - \vec{\nabla} \cdot \vec{E}))d^3x \qquad (15.61)
\end{aligned}
$$

We have used electric/magnetic field notation $E^i = \Pi^i, B^i = \frac{1}{2}\epsilon^{ijk}F_{jk}$ in the final line. Note that the time component of the four-vector potential A_0 now plays the role of the Lagrange multiplier term in the Dirac total Hamiltonian, as it multiplies the Gauss's Law first-class constraint:

$$
H = H_T = \int (\mathcal{H}_{\mathrm{EM}} + A_0(\vec{x})\chi(\vec{x}))d^3x \qquad (15.62)
$$

$$
\mathcal{H}_{\mathrm{EM}} = \frac{1}{2}(\vec{E}^2 + \vec{B}^2) - \vec{J} \cdot \vec{A} \qquad (15.63)
$$

The analogy of canonical electrodynamics to the mechanical example embodied in the Lagrangian (15.11) should now be clear. The derivation of a fully constrained Hamiltonian version of this theory in which all superfluous gauge degrees of freedom have been removed proceeds along the same lines as for the mechanical examples treated earlier: we must choose an appropriate set of gauge-fixing conditions which select a unique representative from each set of gauge equivalent field configurations. The canonical momenta Π^i are gauge-invariant, so we are here concerned only with the gauge freedom in the A_i fields, where (on a given time-slice) $A'_i(\vec{x})$ is gauge equivalent to $A_i(\vec{x})$ if there exists a function $\Lambda(\vec{x})$ such that $A'_i(\vec{x}) = A_i(\vec{x}) + \partial_i\Lambda(\vec{x})$. A common gauge condition is that leading to *Coulomb gauge*, where the field is rendered transverse (spatially divergence-free) at all times:

$$
\psi \equiv \partial^i A_i(\vec{x}, t) = 0 = \vec{\nabla} \cdot \vec{A}(\vec{x}, t) \qquad (15.64)
$$

Once appropriate boundary conditions are imposed (periodic boundary conditions with the system confined to a large spatial box, say), this condition evidently selects a unique representative on each gauge orbit, as

$$
\vec{\nabla} \cdot (\vec{A} + \vec{\nabla}\Lambda) = \vec{\nabla} \cdot \vec{A} = 0 \Rightarrow \vec{\nabla}^2\Lambda(\vec{x}) = 0 \Rightarrow \Lambda(\vec{x}) = \text{constant} \qquad (15.65)
$$

Alternatively, one could choose an *axial gauge*, where the gauge transformation is chosen to remove (say) the third component of the field,

$$\psi = A_3(\vec{x}, t) = 0 \tag{15.66}$$

which bears an obvious similarity to the gauge choices we made earlier in our mechanical examples (cf. (15.7)). Both of these conditions destroy the manifest Lorentz-invariance of the theory, of course. The Coulomb gauge is at least marginally superior in that it preserves at least the rotational symmetry of the theory (the O(3) subgroup of the HLG), and we shall adopt it for the time being in our construction of a fully constrained Hamiltonian theory.

For Coulomb gauge, we may now write down, using (15.44), the functional integral representation of the generating functional Z giving the vacuum persistence amplitude in the presence of the external current J^μ (we henceforth set $\hbar = 1$ as we return to field theory proper):

$$Z_{\rm EM} = \int \mathbf{D} E^i \mathbf{D} A_i \mathbf{D} A_0 \delta(\psi) \det\{\chi, \psi\} e^{i \int (E^i \dot{A}_i - \mathcal{H}_{\rm EM} - A_0 \chi) d^4 x} \tag{15.67}$$

with the Hamiltonian energy density given in (15.63). The DFP determinant appearing in this expression, involves the Poisson bracket of the Gauss's Law constraint (15.57) with the Coulomb gauge constraint (15.64) (at equal times, so we suppress time)

$$\{J^0(\vec{x}) - \frac{\partial}{\partial x^i} E^i(\vec{x}), \frac{\partial}{\partial y^j} A^j(\vec{y})\} = \frac{\partial}{\partial x^i} \frac{\partial}{\partial y^j} \{A^j(\vec{y}), E^i(\vec{x})\} = \vec{\nabla}^2 \delta^3(\vec{x} - \vec{y}) \tag{15.68}$$

so the corresponding determinant[5]

$$\mathcal{D}_{\rm FP} \equiv \det\{\chi(\vec{x}), \psi(\vec{y})\} = \det(\vec{\nabla}^2 \delta^3(\vec{x} - \vec{y})) \tag{15.69}$$

is a field-independent constant, which can be given a well-defined value by a full regularization of the theory (for example, by introducing IR and UV cutoffs via a lattice), but is of no physical significance, as we recall from Chapter 10 that overall multiplicative factors in the functional integral for the vacuum amplitude of a field theory disappear once we compute the connected Green functions of the theory. Discarding the determinant factor in (15.67), we see that the dependence of the exponent on the momentum (electric) fields is quadratic, so these may be integrated out in the usual fashion by completing the square (and again dropping irrelevant overall factors), leaving a functional integral over the original vector potential field variables $(A_0, A_i) = A_\mu$:

$$Z_{\rm EM} = \int \mathbf{D} A_\mu \mathbf{D} E^i \delta(\vec{\nabla} \cdot \vec{A}) e^{i \int (E^i (\partial_0 A_i - \partial_i A_0) - \frac{1}{2}(\vec{E}^2 + \vec{B}^2) - J^\mu A_\mu) d^4 x} \tag{15.70}$$

$$= \int \mathbf{D} A_\mu \delta(\vec{\nabla} \cdot \vec{A}) e^{i \int (\frac{1}{2} F_{0i}^2 - \frac{1}{4} F_{jk}^2 - J^\mu A_\mu) d^4 x} \tag{15.71}$$

---

[5] The determinant factor $\det\{\chi, \psi\}$ appearing in the functional integral (15.67) over *spacetime* fields is a product of the factor $\mathcal{D}_{\rm FP}$ given here at each discrete time, once the theory is regularized, for example, on a spacetime lattice: thus, for the present case of Coulomb gauge, $\det\{\chi, \psi\} = \mathcal{D}_{\rm FP}^{N_t}$, where N_t is the number of points in the time direction. Similarly, the δ-function gauge-fixing constraint $\delta(\psi)$ implicitly involves a product of δ-functions enforcing the constraint at each spacetime point: $\delta(\psi) = \prod_{\vec{x}, t} \delta(\vec{\nabla} \cdot \vec{A}(\vec{x}, t))$.

$$= \int \mathbf{D}A_\mu \delta(\vec{\nabla} \cdot \vec{A}) e^{i \int (-\frac{1}{4} F_{\mu\nu} F^{\mu\nu} - J^\mu A_\mu) d^4 x} \tag{15.72}$$

$$= \int \mathbf{D}A_\mu \delta(\vec{\nabla} \cdot \vec{A}) e^{i \int \mathcal{L}_{\mathrm{EM}} d^4 x} \tag{15.73}$$

Note that the original, manifestly local, gauge- and Poincaré-invariant action for our abelian electrodynamics has re-emerged: the only fly in the ointment is the δ-function enforcing the (non-Lorentz-invariant) Coulomb gauge restriction independently on each time-slice. The loss of Lorentz symmetry is only apparent: indeed, we may already anticipate that the theory must, despite appearances, preserve Lorentz symmetry, as the Euler–Lagrange equations (15.48) of our starting Lagrangian are perfectly covariant. We shall now demonstrate this highly desirable feature of the constrained formalism by showing that, despite appearances, the functional integral (15.67) is *independent* of the choice of gauge-fixing function ψ, leaving us free to replace the non-covariant Coulomb (or axial) gauge choices by a perfectly Lorentz-covariant choice—for example, "Landau (or Lorentz) gauge" $\partial_\mu A^\mu = 0$.

Recalling (cf. 15.50)) the notation $A_\mu^\Lambda \equiv A_\mu + \partial_\mu \Lambda$ for the effect of a finite gauge transformation on the gauge field A_μ, consider the functional $\Delta_{\mathrm{coul}}[A]$ of A_μ defined implicitly by

$$\Delta_{\mathrm{coul}}[A] \cdot \int \mathbf{D}\Lambda \delta(\vec{\nabla} \cdot \vec{A}^\Lambda(\vec{x}, t)) = 1 \tag{15.74}$$

We again remind the reader (see footnote 5) that the δ-function constraint in the functional integral implies a product of δ-functions at each and every spacetime point—a statement which can be given a precise meaning by regularizing the theory on a spacetime lattice. To the extent that the gauge fixing imposed by $\psi = 0$ picks a unique gauge field with $\psi(A^\Lambda) = 0$ on the orbit passing through an arbitrary field A, the functional integral over Λ in (15.74) receives its entire contribution from exactly one gauge function. In particular, for fields A already on the gauge surface $\vec{\nabla} \cdot \vec{A} = 0$, this must occur at $\Lambda = 0$, and we have

$$\Delta_{\mathrm{coul}}[A]^{-1} = \int \mathbf{D}\Lambda \prod_t \delta(\vec{\nabla}^2 \Lambda(\vec{x}, t))$$

$$= \int \prod_t \mathbf{D}\Lambda(\vec{x}, t) \delta(\int \mathcal{M}(\vec{x}, \vec{y}) \Lambda(\vec{y}, t) d^3 y) \tag{15.75}$$

$$\mathcal{M}(\vec{x}, \vec{y}) \equiv \vec{\nabla}^2 \delta^3(\vec{x} - \vec{y}) \tag{15.76}$$

where we have indicated explicitly the time-discretization in (15.75). The functional integral yields immediately (at each discrete time) the inverse determinant[6] of the operator $\mathcal{M}$ which we previously called $\mathcal{D}_{\mathrm{FP}}$ (cf. (15.69)), so we have that our functional

---

[6] This is clear if we discretize also the spatial dependence of the fields, and recall that for any matrix $\mathcal{M}_{ij}$, $\int d\Lambda_i \delta(\mathcal{M}_{ij}\Lambda_j) = (\det \mathcal{M})^{-1}$.

$$\Delta_{\text{coul}}[A] = \mathcal{D}_{\text{FP}}^{N_t} = \det\{\chi, \psi\} \tag{15.77}$$

is just the DFP determinant appearing in the unconstrained functional integral (15.67), *for fields already in Coulomb gauge* (which are, of course, just the fields appearing in said functional integral). Note that this functional is itself gauge-invariant, as, for any gauge function $\Lambda'(x)$,

$$\Delta_{\text{coul}}[A^{\Lambda'}]^{-1} = \int \mathbf{D}\Lambda\delta(\vec{\nabla}\cdot\vec{A}^{\Lambda+\Lambda'}(\vec{x},t)) = \int \mathbf{D}\Lambda\delta(\vec{\nabla}\cdot\vec{A}^{\Lambda}(\vec{x},t)) = \Delta_{\text{coul}}[A]^{-1}$$

where we have used the shift invariance of the functional integral over Λ under $\Lambda \to \Lambda - \Lambda'$ (for $\Lambda'(x)$ any fixed function on spacetime). The gauge-invariance is, of course, a triviality in this particular case (abelian gauge theory in Coulomb gauge), as we previously saw that the determinant is a field-independent constant. But this will not be the case once we repeat the procedure for non-abelian theories, so we shall proceed as though the DFP determinants we encounter are non-trivial functionals of the gauge field. One may similarly define a covariant DFP functional associated with the gauge choice $\psi = \partial^\mu A_\mu(x) - f(x) = 0$, where $f(x)$ is for the time being a perfectly arbitrary, but fixed, function of spacetime (thus, Landau gauge corresponds to taking $f(x) = 0$ identically):

$$\Delta_{\text{cov}}^f[A] \cdot \int \mathbf{D}\Lambda\delta(\partial^\mu A_\mu^\Lambda - f) = 1 \tag{15.78}$$

with, as usual, $A_\mu^\Lambda \equiv A_\mu + \partial_\mu\Lambda$. By exactly the same argument used above, involving the shift invariance of the functional integral, we see that $\Delta_{\text{cov}}^f[A]$ is a gauge-invariant functional of A, for any choice of the function f. A less obvious property (but one which will shortly become important) is that *for gauge fields on the gauge surface $\partial^\mu A_\mu(x) = f(x)$*, $\Delta_{\text{cov}}^f[A]^{-1} = \int \mathbf{D}\Lambda\delta(\Box\Lambda)$; i.e., the DFP determinant loses its dependence on the arbitrary function f (as well as on the gauge fields themselves).

We are now in a position to demonstrate the claimed independence of the path integral (15.67) to the specific gauge choice employed. We begin with the Coulomb gauge version of the path integral, replacing $\det\{\chi, \psi\}$ by the gauge-invariant functional $\Delta_{\text{coul}}[A]$, and then introduce a factor of unity via (15.78). We shall also slightly generalize our earlier discussion by including in the action external sources $K_i(x)$ coupled to an arbitrary set of gauge-invariant operators $O_i(x)$. The latter might, in the case of QED, for example, be the gauge-invariant fields $F_{\mu\nu}(x)$ which possess non-vanishing vacuum to single-photon matrix elements and whose T-products can therefore be related to S-matrix elements for multi-photon scattering via the LSZ formalism of Chapter 9. Our previous result (15.73) becomes

$$Z_{\text{EM}}[K_i] = \int \mathbf{D}A_\mu\delta(\vec{\nabla}\cdot\vec{A})\Delta_{\text{coul}}[A]e^{i\int(\mathcal{L}_{\text{EM}}+K_iO_i)d^4x} \tag{15.79}$$

$$= \int \mathbf{D}\Lambda\mathbf{D}A_\mu\Delta_{\text{cov}}^f[A]\delta(\partial^\mu A_\mu^\Lambda - f)\delta(\vec{\nabla}\cdot\vec{A})\Delta_{\text{coul}}[A]e^{i\int(\mathcal{L}_{\text{EM}}(A)+K_iO_i(A))d^4x}$$

$$= \int \mathbf{D}\Lambda \mathbf{D}A_\mu \Delta^f_{\text{cov}}[A^{-\Lambda}]\delta(\partial^\mu A_\mu - f)\delta(\vec{\nabla} \cdot \vec{A}^{-\Lambda})\Delta_{\text{coul}}[A^{-\Lambda}]$$

$$\cdot e^{i\int(\mathcal{L}_{\text{EM}}(A^{-\Lambda})+K_iO_i(A^{-\Lambda}))d^4x}$$

$$= \int \mathbf{D}\Lambda \mathbf{D}A_\mu \Delta^f_{\text{cov}}[A]\delta(\partial^\mu A_\mu - f)\delta(\vec{\nabla} \cdot \vec{A}^{-\Lambda})\Delta_{\text{coul}}[A]e^{i\int(\mathcal{L}_{\text{EM}}(A)+K_iO_i(A))d^4x}$$

$$= \int \mathbf{D}A_\mu \Delta^f_{\text{cov}}[A]\delta(\partial^\mu A_\mu - f)e^{i\int(\mathcal{L}_{\text{EM}}(A)+K_iO_i(A))d^4x} \tag{15.80}$$

In going from the second to the third equations we have made the functional change of variable $A_\mu \to A^\Lambda$ (there is no Jacobian as this amounts to an additive shift of the integration variables); in the fourth equation we have used the gauge-invariance of the functionals $\Delta_{\text{coul}}, \Delta^f_{\text{cov}}$ and of the Lagrangian $\mathcal{L}_{\text{EM}}$ and sourced fields O_i; and in the final line we have employed the definition (15.74) to remove all evidence of the original Coulomb gauge-fixing. Our final result (15.80) is manifestly Lorentz-covariant, as all spacetime indices are properly contracted. If we set $f = 0$ we recover the functional integral for abelian gauge theory in Landau gauge. Note that, as we pointed out earlier, despite appearances, $\Delta^f_{\text{cov}}[A]$ in (15.80) is in fact independent of f, and we may henceforth write it simply as $\Delta_{\text{cov}}[A]$. As our starting point (15.79) for Z_{EM} is clearly independent of the choice of the arbitrary function f, we may multiply it by an irrelevant constant factor obtained by a functional integral over all functions f with a damped Gaussian factor (ξ a positive real number)

$$C \equiv \int \mathbf{D}f e^{-\frac{i}{2\xi}\int f(x)^2 d^4x} \tag{15.81}$$

obtaining

$$Z_{\text{EM}}[K_i] = \int \mathbf{D}f\mathbf{D}A_\mu \Delta_{\text{cov}}[A]\delta(\partial^\mu A_\mu - f)e^{-\frac{i}{2\xi}\int f(x)^2 d^4x + i\int(\mathcal{L}_{\text{EM}}+K_iO_i)d^4x}$$

$$= \int \mathbf{D}A_\mu \Delta_{\text{cov}}[A]e^{i\int(\mathcal{L}_{\text{EM}}(A)-\frac{1}{2\xi}(\partial^\mu A_\mu)^2+K_iO_i)d^4x} \tag{15.82}$$

The functional integral (15.82) defines the partition function (or vacuum persistence amplitude) for abelian gauge theory in the so-called *covariant ξ-gauges* (first introduced for non-abelian theories by 't Hooft (T'Hooft, 1971)).

We reiterate that in the present case of abelian gauge theory, the DFP determinant factor $\Delta_{\text{cov}}[A]$ is in fact a field-independent constant, and may be omitted completely: we retain it here in anticipation of the fact that for non-abelian gauge theories, it develops a non-trivial structure and must be kept in order to arrive at unitary amplitudes. These gauges are extremely useful in performing perturbative calculations in gauge theories (abelian or non-abelian): the disappearance of the arbitrary constant ξ from all expressions for gauge-invariant quantities at the end of the calculation provides a very useful check on the intermediate manipulations. That the Green functions of the theory (vacuum expectation values of time-ordered products of the O_i fields, obtained by functionally differentiating $Z_{\text{EM}}[K_i]$ with respect to the source functions $K_i(x)$) are ξ-independent is clear from the fact that our original expression

(15.79) for $Z_{\rm EM}[K_i]$ did not contain the parameter ξ at all. It is a straightforward matter to extract the perturbative Feynman rules in the ξ-gauge for abelian quantum electrodynamics from (15.82), but we shall defer this task to the more interesting case of non-abelian gauge theory, where the full power of the deWitt–Fadeev–Popov approach becomes manifest. The Feynman rules for abelian gauge theory are in any event obtained trivially from the non-abelian ones by a simple reduction, so we lose no information by proceeding directly, as we shall shortly do, to the case of non-abelian gauge field theory.

We may promote our abelian gauge theory to a full-fledged quantum electrodynamics (QED), in which the external source is now the quantized four-vector current arising from Dirac fermions (e.g, the electron) of charge e and mass m,

$$J^\mu(x) = e\bar{\psi}(x)\gamma^\mu\psi(x) \tag{15.83}$$

We must also include the usual kinetic term in the Lagrangian for the fermions, thereby arriving at the Lagrangian for QED:

$$\mathcal{L}_{\rm QED} = -\frac{1}{4}F_{\mu\nu}F^{\mu\nu} + \bar{\psi}(i\slashed{D} - m)\psi, \quad D_\mu \equiv \partial_\mu + ieA_\mu \tag{15.84}$$

This Lagrangian is invariant under the local gauge transformations consisting of the following joint transformations on the A_μ and ψ fields:

$$A_\mu(x) \to A_\mu(x) + \partial_\mu\Lambda(x) \tag{15.85}$$

$$\psi(x) \to e^{-ie\Lambda(x)}\psi(x) \tag{15.86}$$

$$\bar{\psi}(x) \to e^{+ie\Lambda(x)}\bar{\psi}(x) \tag{15.87}$$

The invariance of the $F_{\mu\nu}F^{\mu\nu}$ and mass $m\bar{\psi}\psi$ terms under these transformations is obvious. The *covariant derivative* D_μ preserves the form of the gauge transformation of the charged field, as under (15.85–15.87),

$$D_\mu\psi(x) \to (\partial_\mu + ieA_\mu(x) + ie\partial_\mu\Lambda(x))e^{-ie\Lambda(x)}\psi(x)$$
$$= e^{-ie\Lambda(x)}(\partial_\mu + ieA_\mu(x))\psi(x) = e^{-ie\Lambda(x)}D_\mu\psi(x) \tag{15.88}$$

from which the invariance of the kinetic fermion term $\bar{\psi}\gamma^\mu D_\mu\psi$ follows immediately. Note that if we take the commutator of two covariant derivatives acting on the fermion field ψ, the derivative terms on ψ cancel, and we are left with the field tensor $F_{\mu\nu}$:

$$[D_\mu, D_\nu]\psi(x) = ie(\partial_\mu A_\nu(x) - \partial_\nu A_\mu(x))\psi(x) = ieF_{\mu\nu}(x)\psi(x) \tag{15.89}$$

As both $\psi(x)$ and $[D_\mu, D_\nu]\psi(x)$ must transform identically under (15.86), we see that the commutator must be gauge-invariant—an observation that will simplify our search below for a non-abelian generalization of the gauge-field kinetic term in (15.84).

The transformations embodied in (15.85–15.87) form an abelian (commutative) group, as successive transformations with gauge functions $\Lambda_1(x), \Lambda_2(x)$, etc., performed in any order lead to the same final result. The global gauge symmetry obtained by restricting the gauge functions $\Lambda(x)$ to spacetime constants clearly amounts to a U(1) phase transformation on the charged fermion field ψ, so we refer to a theory characterized by invariance under the transformations (15.85–15.87) as a "U(1) gauge theory".

The reader may be puzzled by the fact that in the construction of S-matrix elements involving (via the LSZ formula) T-products of the gauge and fermion fields associated with the desired external photons and electrons/positrons, we must introduce sources for these manifestly non-gauge-invariant fields into the path integral. For example, in Coulomb gauge, we clearly need to compute the functional

$$Z_{\text{QED}}[\vec{j}, \eta, \bar{\eta}] = \int \mathbf{D}A_\mu \mathbf{D}\psi \mathbf{D}\bar{\psi} \delta(\vec{\nabla} \cdot \vec{A}) e^{i \int (\mathcal{L}_{\text{QED}} + \vec{j}\cdot\vec{A} + \bar{\eta}\psi + \bar{\psi}\eta) d^4 x} \qquad (15.90)$$

Note that in Coulomb gauge we need only include a source $\vec{j}$ for the spatial part $\vec{A}$ of the four-vector potential, as this is the part of the gauge field that interpolates for the asymptotic (transverse) photon states (cf. the discussion at the end of Section 7.5). The source terms in (15.90) are clearly not locally gauge-invariant, so the reader may well wonder how we can manage the conversion of this functional into a manifestly Lorentz-covariant form along the lines of the maneuvers leading to the covariant functional (15.82) above, which required that the exponent in the path integral be exactly invariant under an arbitrary local gauge transformation of the fields. In fact, the Feynman Green functions (T-products of gauge and Dirac fields) are *not* gauge-invariant as such, nor do they need to be. Instead, we shall see that the physical information they contain, in the form of on-mass-shell S-matrix elements, is preserved under local gauge transformations. For example, we recall that the LSZ formula requires that we subject the generating functional $Z_{\text{QED}}[\vec{j}, \eta, \bar{\eta}]$ to the operation

$$\int d^4 x e^{ik\cdot x} \Box_x \vec{\epsilon}^{\,*}(\vec{k}, \lambda) \cdot \frac{\delta}{\delta \vec{j}(x)} Z_{\text{QED}} \qquad (15.91)$$

where k is an on-mass-shell four-vector for a photon, in order to extract the S-matrix amplitude for a scattering in which a photon with momentum k and polarization λ appears in the final state. In going from the third to the fourth line of (15.80), the gauge transformation induces a change in the source terms in the exponent,

$$\vec{j} \cdot \vec{A} + \bar{\eta}\psi + \bar{\psi}\eta \rightarrow \vec{j} \cdot (\vec{A} - \vec{\nabla}\Lambda) + e^{ie\Lambda}\bar{\eta}\psi + e^{-ie\Lambda}\bar{\psi}\eta \qquad (15.92)$$

and the shift in (15.91) due to Λ is seen to be proportional to

$$\int d^4 x e^{ik\cdot x} \Box_x \vec{\epsilon}^{\,*}(\vec{k}, \lambda) \cdot \vec{\nabla}\Lambda(x) \cdots \propto \vec{k} \cdot \vec{\epsilon}^{\,*}(\vec{k}, \lambda) = 0 \qquad (15.93)$$

where we have integrated by parts to transfer the spatial gradient to the complex exponential, and used the fact that the photon polarization vectors are transverse, $\vec{k} \cdot \vec{\epsilon}(\vec{k}, \lambda) = 0$. One can similarly establish that the Λ-dependence of the fermionic source terms visible in (15.92) does not affect the result once the on-mass-shell projection is made for initial- or final-state fermions as required in the LSZ formula (cf. (9.206); also Problem 1).

15.4 Non-abelian gauge theory: construction and functional integral formulation

The generalization of the idea of local gauge symmetry to a set of non-commuting transformations, in which the phase transformations on the fermionic fields of the theory form a non-abelian group, goes back to a seminal paper of Yang and Mills (Yang and Mills, 1954), in which the attempt is made to convert the global isotopic spin symmetry of the meson field theories used to describe the strong interactions in the mid-1950s (cf. Example 5 in Section 12.4) to a local gauge symmetry. The Lie group considered by Yang and Mills was just SU(2), but here we shall consider a completely general gauge group of dimension n_g, specified by a set of generators $t_\alpha, \alpha = 1, 2, \ldots, n_g$, and Lie algebra

$$[t_\alpha, t_\beta] = if_{\alpha\beta\gamma}t_\gamma \tag{15.94}$$

The matrix identity $[t_\alpha, [t_\beta, t_\gamma]] + [t_\beta, [t_\gamma, t_\alpha]] + [t_\gamma, [t_\alpha, t_\beta]] = 0$ then implies the Jacobi identity constraint on the structure constants $f_{\alpha\beta\gamma}$ of the group

$$f_{\alpha\delta\epsilon}f_{\beta\gamma\delta} + f_{\gamma\delta\epsilon}f_{\alpha\beta\delta} + f_{\beta\delta\epsilon}f_{\gamma\alpha\delta} = 0 \tag{15.95}$$

By choosing appropriate linear combinations of the generators, we can also arrange for the structure constants to be totally antisymmetric: i.e., to change sign under interchange of any two indices (the antisymmetry under exchange of the first two indices is, of course, guaranteed from (15.94)). We shall take the generator matrices to act in the fundamental representation of the group, of dimension N, and assume that the fermions $\psi_n, n = 1, 2, \ldots N$ of the theory also occupy this representation. Thus, we are considering a theory in which a global internal symmetry of the type (12.126) of Section 12.4 acting on a multiplet of matter fields $\psi_n(x)$ is extended to a local symmetry, with the parameters ω_α of the group element replaced by arbitrary functions $\Lambda_\alpha(x)$ of spacetime:

$$\psi_n(x) \to \psi'_n(x) \equiv U_{nm}(x)\psi_m(x), \quad U \equiv \exp\left(-ig\Lambda_a(x)t_\alpha\right) \tag{15.96}$$

For the gauge theories of the standard model, we are dealing with unitary groups: the generator matrices t_α are hermitian, the gauge functions Λ_α are real, and the finite group transformation matrices U are therefore unitary. In analogy to the abelian gauge transformation (15.86), it is conventional to include a factor of the gauge coupling constant g (analogous to the electric charge e in the QED case) in the gauge parameters defining U. The dynamical (as opposed to "gauge-kinematical") role of the gauge coupling constant will become clear shortly when we construct the full Lagrangian of the theory. The non-abelian groups associated with the strong and weak interactions are in addition "special" unitary, satisfying the additional constraint $\det(U) = 1$ (corresponding to the generators t_α being traceless, as we see immediately using the identity $\ln \det(U) = \mathrm{Tr}\ln(U)$). For the purposes of the present discussion, we may as well restrict ourselves to the special unitary groups SU(N) in which the matter fields $\psi_n(x)$ fill out the fundamental representation of the group. The dimension of SU(N) (i.e., the number of linearly independent traceless hermitian $N \times N$ generator matrices t_α) is $n_g = N^2 - 1$. Thus, for the gauge group SU(3), the gauge index α runs over the values 1,2,...,8. In addition to the fundamental representation of dimension N, the *adjoint representation*, of dimension n_g, will play a central role in the following. We remind the reader that a multiplet of real fields $V_\alpha(x), \alpha = 1, 2, .., n_g$ transforms according to the adjoint representation of SU(N) if, under the gauge transformation $U_{nm}(x)$ in (15.96),

$$t_\alpha V_\alpha(x) \to U(x) t_\alpha V_\alpha(x) U^\dagger(x) \equiv t_\alpha V'_\alpha(x) \tag{15.97}$$

Note that the similarity transformation in (15.97) preserves the traceless, hermitian character of the $t_\alpha V_\alpha$ matrix (if V_α are real fields, as we assume), so the fact that the generators t_α of SU(N) are a complete basis for all $N \times N$ traceless hermitian matrices ensures that the V'_α are well-defined by the above procedure.

The transformations (15.96) are the field-theoretic analogs of the non-abelian gauge transformations (15.10) leaving invariant the Lagrangian (15.8) in the mechanical example of Section 15.1. In that mechanical model, the coordinate vector $\vec{q}(t)$ of the point particle moves along a trajectory in three-dimensional space, but only the radial coordinate $r(t) \equiv \sqrt{\vec{q}(t) \cdot \vec{q}(t)}$ possesses physical significance: the individual Cartesian coordinates can "wobble" furiously, as though the entire system is being viewed from the standpoint of an inebriated experimentalist shaking (rotationally) the coordinate axes in a random fashion. Exactly the same arbitrariness attaches to the physical interpretation of the internal symmetry axes in the case of a gauge field theory.

To take a concrete example, consider the role of the color quantum number in quantum chromodynamics (QCD)—the local field theory describing the strong interaction sector of the Standard Model. QCD is a gauge theory exhibiting an exact invariance under local gauge transformations of the form (15.96), where independent unitary SU(3) rotations of the three quark fields $\psi_n(x)$ at arbitrary spacetime points leave the physics unchanged. If we label the three quarks (fancifully and, of course, arbitrarily) as "red", "green", and "blue", we see that the attachment of any particular color label to any particular quark at any given time is a completely arbitrary choice: the color "axes" may be unitarily rotated in a completely random way during the dynamical evolution of the system without altering any physical observable. Indeed, the physical observables—in a gauge field theory, those associated with local or almost local operators (in the language of Chapter 9)—are precisely those which (in analogy to mechanical quantities which depend only on the radial coordinate in the "turntable model" of Section 15.1) are *gauge-invariant*: that is, they are unchanged under the transformations (15.96). For the gauge group SU(N), such gauge-invariant observables, built from fermionic fields in the fundamental representation (the quark fields of QCD, for example) include composite operators such as

$$S(x) \equiv \bar{\psi}_n(x) \psi_n(x) \tag{15.98}$$

$$J^\mu(x) \equiv \bar{\psi}_n(x) \gamma^\mu \psi_n(x) \tag{15.99}$$

$$N(x) \equiv \epsilon_{n_1 n_2 ... n_N} \psi_{n_1}(x) \psi_{n_2}(x) \cdots \psi_{n_N}(x) \tag{15.100}$$

to present just a few examples. These constructs are easily seen to be invariant under $\psi(x) \to U(x) \psi(x)$, $\bar{\psi}(x) \to \bar{\psi}(x) U^\dagger(x)$ for $U \in$ SU(N): we need only take into account the fact that the action of the gauge matrices $U(x)$ leaves the implicit Dirac indices in the ψ_n fields unaltered (or in other words, the U matrices commute with the γ matrices implementing the Dirac algebra for our spin-$\frac{1}{2}$ fields). Local fields such as $S(x), J^\mu(x), N(x)$,, and so on, are said to be "colorless" or "color neutral", and represent the *only* local observables with unambiguous physical content in a local gauge field theory. From the axiomatic Wightman point of view discussed in Section 9.2, the Wightman functions (vacuum expectation values of products) of such gauge-invariant local fields contain the *entire physical content* of the theory: the vacuum is cyclic with respect to the algebra generated by all local gauge-invariant fields (cf. Section 9.2, Axiom IId). In particular, the complete S-matrix of a gauge theory like QCD with an exact (unbroken) non-abelian gauge symmetry is determined in principle from a knowledge of such functions, as the asymptotic Fock space of physical states consists entirely—as we shall see in Chapter 19—of colorless multi-particle states.

Just as in the abelian case, the construction of a gauge-covariant derivative for fermionic matter fields transforming non-trivially under the gauge group requires the existence of vector fields $A_{\alpha\mu}$ (one for each independent gauge transformation). Such fields are needed to absorb the term involving a spacetime-derivative of the local gauge functions $\Lambda_\alpha(x)$ in the kinetic part of the matter Lagrangian. It is convenient to "pack" these fields into the adjoint matrix (dimension $N \times N$)

$$A_\mu \equiv t_\alpha A_{\alpha\mu} \tag{15.101}$$

We then require, from (15.97), that under global (i.e., spacetime-independent) gauge transformations,

$$A_\mu \to U A_\mu U^\dagger, \quad U \in SU(N) \tag{15.102}$$

and can now build a covariant derivative in analogy to (15.84)

$$D_\mu = \partial_\mu + ig t_\alpha A_{\alpha\mu} = \partial_\mu + ig A_\mu \tag{15.103}$$

and demand that $D_\mu\psi$ transform identically to ψ for any set of matter fields in the fundamental representation:

$$\psi(x) \to U(x)\psi(x) \Rightarrow D_\mu\psi(x) \to U(x)D_\mu\psi(x) \tag{15.104}$$

The inclusion of an inhomogeneous term in the transformation rule for the gauge field A_μ under local gauge transformations,

$$A_\mu(x) \to A_\mu^U(x) = U(x)A_\mu(x)U^\dagger(x) + \frac{i}{g}(\partial_\mu U(x))U^\dagger(x) \tag{15.105}$$

is easily seen to do the trick:

$$\begin{aligned} D_\mu\psi(x) &\to (\partial_\mu + igU(x)A_\mu(x)U^\dagger(x) - (\partial_\mu U(x))U^\dagger(x))U(x)\psi(x) \\ &= (\partial_\mu U(x))\psi(x) + U(x)\partial_\mu\psi(x) + igU(x)A_\mu(x)\psi(x) - (\partial_\mu U(x))\psi(x) \\ &= U(x)(\partial_\mu + igA_\mu(x))\psi(x) = U(x)D_\mu\psi(x) \end{aligned} \tag{15.106}$$

thereby guaranteeing the local gauge-invariance of the fermionic part of the Lagrangian (as $\bar{\psi} \to \bar{\psi}(x)U^\dagger(x)$ under a local gauge transformation)

$$\mathcal{L}_{\text{ferm}} = \bar{\psi}(i\slashed{D} - m)\psi, \quad D_\mu \equiv \partial_\mu + igA_\mu \tag{15.107}$$

Note that the gauge invariance requires all members of the fermion multiplet to have the same mass: if m is a mass *matrix*, then $U^\dagger m U = m$ for all $U \in SU(N)$ implies m a multiple of the identity (by Schur's lemma).

A gauge-covariantly transforming field tensor suitable for constructing a kinetic Lagrangian for the adjoint gauge fields $A_{\alpha\mu}$ is constructed along exactly the same lines as (15.89) for the abelian theory. We note that derivatives of the matter field cancel in the commutator $[D_\mu, D_\nu]$:

$$[D_\mu, D_\nu]\psi = ig(\partial_\mu A_\nu - \partial_\nu A_\mu + ig[A_\mu, A_\nu])\psi \equiv igF_{\mu\nu}\psi \tag{15.108}$$

As the covariant derivatives (or products thereof) preserve the transformation property (15.96) of $\psi(x)$, under $\psi(x) \to U(x)\psi(x)$,

$$F_{\mu\nu}(x)\psi(x) \to U(x)F_{\mu\nu}(x)\psi(x) \Rightarrow F_{\mu\nu}(x) \to U(x)F_{\mu\nu}(x)U^\dagger(x) \tag{15.109}$$

so the matrix $F_{\mu\nu}(x) = t_\alpha F_{\alpha\mu\nu}$ transforms exactly as required for the adjoint representation (15.97), and is built from $n_g = N^2 - 1$ antisymmetric tensor fields $F_{\alpha\mu\nu}$ related to the underlying vector fields $A_{\alpha\mu}$ by (15.108):

$$F_{\mu\nu} = \partial_\mu t_\alpha A_{\alpha\nu} - \partial_\nu t_\alpha A_{\alpha\mu} + ig[t_\beta A_{\beta\mu}, t_\gamma A_{\gamma\nu}]$$
$$= t_\alpha(\partial_\mu A_{\alpha\nu} - \partial_\nu A_{\alpha\mu} - gf_{\alpha\beta\gamma}A_{\beta\mu}A_{\gamma\nu})$$
$$\Rightarrow F_{\alpha\mu\nu} = \partial_\mu A_{\alpha\nu} - \partial_\nu A_{\alpha\mu} - gf_{\alpha\beta\gamma}A_{\beta\mu}A_{\gamma\nu} \tag{15.110}$$

It is now a trivial matter to construct the non-abelian version of the kinetic gauge Lagrangian $-\frac{1}{4}F_{\mu\nu}F^{\mu\nu}$ in the abelian case: we simply take

$$\mathcal{L}_{\text{gauge}} = -\frac{1}{4}\text{Tr}(F_{\mu\nu}F^{\mu\nu}) = -\frac{1}{4}F_{\alpha\mu\nu}F_\alpha^{\mu\nu} \tag{15.111}$$

where it is conventional to normalize the group generators by $\text{Tr}(t_\alpha t_\beta) = \delta_{\alpha\beta}$. The invariance of $\mathcal{L}_{\text{gauge}}$ under local gauge transformations is now obvious from the transformation property (15.109) of the non-abelian field tensor. Including the matter fields, we have arrived at the *Yang–Mills Lagrangian*

$$\mathcal{L}_{\text{YM}} = \mathcal{L}_{\text{gauge}} + \mathcal{L}_{\text{ferm}} = -\frac{1}{4}\text{Tr}(F_{\mu\nu}F^{\mu\nu}) + \sum_a \bar{\psi}_a(i\slashed{D} - m_a)\psi_a \tag{15.112}$$

where we have made the obvious generalization of allowing for several fermionic multiplets ψ_a, of different mass, all transforming according to the fundamental representation of the gauge group $SU(N)$.

We note at this point the characteristic (and remarkable) feature of non-abelian gauge theories, which dramatically sets them apart from their abelian cousins such as quantum electrodynamics: *even in the absence of matter (scalar or Dirac fields), the gauge fields themselves form a highly non-trivial interacting field theory*, with the gauge Lagrangian $\mathcal{L}_{\text{gauge}}$ containing, in addition to quadratic kinetic terms, interaction terms that are cubic and quartic in the gauge fields. It is now generally accepted that the entire vast range of strong interaction phenomenology is a manifestation of an underlying local physics described by precisely the Lagrangian (15.112), where the gauge (or "color") group is SU(3), and the observed hadrons (strongly interacting particles) are colorless bound states of the eight gauge "gluon fields" $A_{\alpha\mu}, \alpha = 1, 2, .., 8$ and fermionic "quark field" multiplets $\psi_a, a = 1, 2, \ldots 6$ (corresponding to the up, down, strange, charmed, bottom, and top quarks). This Lagrangian defines the theory we now call quantum chromodynamics (QCD), in analogy to quantum electrodynamics (QED). Note that the Lagrangian (15.84) is formally identical to (15.112): however, the gauge group U(1) is abelian, with vanishing structure constants $f_{\alpha\beta\gamma}$, and the group multiplets are simply one-dimensional, with a single gauge field (the photon) and a single Dirac field associated with every charged spin-$\frac{1}{2}$ particle (electron, muon, quark, etc.).

We have given the explicit form of the local non-abelian gauge transformations for finite group elements: in other words, the gauge functions $\Lambda_\alpha(x)$ in (15.96) are finite spacetime functions. For transformations near to the identity ($\Lambda_\alpha \to \lambda_\alpha$ infinitesimal), simple algebra (Problem 2) gives the infinitesimal version of (15.96, 15.105, 15.109):

$$\psi(x) \to (1 - igt_\alpha\lambda_\alpha(x))\psi(x), \quad \bar{\psi}(x) \to \bar{\psi}(x)(1 + igt_\alpha\lambda_\alpha(x)) \quad (15.113)$$

$$A_{\alpha\mu}(x) \to A_{\alpha\mu}(x) + \partial_\mu\lambda_\alpha(x) + gf_{\alpha\beta\gamma}\lambda_\beta(x)A_{\gamma\mu}(x) \quad (15.114)$$

$$F_{\alpha\mu\nu}(x) \to F_{\alpha\mu\nu}(x) + gf_{\alpha\beta\gamma}\lambda_\beta(x)F_{\gamma\mu\nu}(x) \quad (15.115)$$

The analysis of the dynamics of the classical system defined by the Lagrangian (15.112) as a constrained Hamiltonian system proceeds in exact analogy to the abelian case: as there, we find that the fields $A_{\alpha 0}$ have vanishing conjugate momenta, since

$$\frac{\partial \mathcal{L}_{\text{YM}}}{\partial \dot{A}_{\alpha\mu}} = -F_\alpha^{0\mu} \quad (15.116)$$

As $F_\alpha^{00} = 0$, we have the primary constraints (one for each generator of the group)

$$\Pi_{\alpha 0} = -F_\alpha^{00} = 0 \quad (15.117)$$

while the equations of motion for the $A_{\alpha 0}$ amount to secondary constraints (which guarantee the preservation in time of the primary constraints) which are just the non-abelian version of Gauss's Law (see Problem 3):

$$0 = \dot{\Pi}_{\alpha 0} = \frac{\partial}{\partial t}\frac{\partial \mathcal{L}_{\text{YM}}}{\partial \dot{A}_{\alpha 0}} = -\partial_i\frac{\partial \mathcal{L}_{\text{YM}}}{\partial(\partial_i A_{\alpha 0})} + \frac{\partial \mathcal{L}_{\text{YM}}}{\partial A_{\alpha 0}} = \mathcal{D}_i E_\alpha^i - J_\alpha^0, \quad J_\alpha^\mu = g\sum_a \bar{\psi}_a t_\alpha \gamma^\mu \psi_a$$
$$(15.118)$$

The non-abelian "electric fields" E_α^i are in the adjoint representation with the adjoint covariant derivative $\mathcal{D}$ appearing in (15.118) defined as

$$\mathcal{D}_i E_\alpha^i \equiv \partial_i E_\alpha^i - gf_{\alpha\beta\gamma}A_{\beta i}E_\gamma^i \quad (15.119)$$

As in the abelian case, the E^i_α are the conjugate momentum fields for the $A_{\alpha i}$ fields

$$\Pi^i_\alpha = -F^{0i}_\alpha = \partial_0 A_{\alpha i} - \partial_i A_{\alpha 0} - g f_{\alpha\beta\gamma} A_{\beta 0} A_{\gamma i} \equiv E^i_\alpha, \quad i = 1, 2, 3 \tag{15.120}$$

satisfying the classical Poisson bracket relations (on a given time-slice)

$$\{A_{\alpha i}(\vec{x}), \Pi^j_\beta(\vec{y})\} = \delta_{\alpha\beta}\delta_i^{\ j}\delta^3(\vec{x} - \vec{y}) \tag{15.121}$$

The secondary constraints (non-abelian Gauss's Law) of the theory are

$$\chi_\alpha(\vec{x}) \equiv J^0_\alpha - \mathcal{D}_i E^i_\alpha = J^0_\alpha - \mathcal{D}_i \Pi^i_\alpha = 0 \tag{15.122}$$

Note that these constraints generate via Poisson brackets, as in the abelian case, and as in the mechanical examples of Sections 15.1 and 15.2, the infinitesimal local gauge transformations (15.113–15.115): for example, for the spatial gauge field (taking the Poisson bracket at equal times, and suppressing the time coordinate)

$$\{A^i_\alpha(\vec{y}), \int d^3x \lambda_\beta(\vec{x})\chi_\beta(\vec{x})\} = -\int d^3x \lambda_\beta(\vec{x})\{A^i_\alpha(\vec{y}), \partial_j E^j_\beta(\vec{x}) - g f_{\beta\gamma\delta} A_{\gamma j}(\vec{x}) E^j_\delta(\vec{x})\}$$

$$= \partial^i \lambda_\alpha(\vec{y}) + g f_{\alpha\beta\gamma}\lambda_\beta(\vec{y})A^i_\gamma(\vec{y}) \tag{15.123}$$

in agreement with (15.115). It follows that the Poisson bracket algebra of the constraints among themselves must imitate the Lie algebra of the underlying local gauge group,[7]

$$\{\chi_\alpha(\vec{x}), \chi_\beta(\vec{y})\} = i g f_{\alpha\beta\gamma}\delta^3(\vec{x} - \vec{y})\chi_\gamma(\vec{x}) \tag{15.124}$$

and therefore that the set of constraints (15.122) are indeed first-class as their Poisson brackets form a closed algebra. The reader will note the analogy to the constraint algebra in the mechanical non-abelian example (with gauge group O(3)) studied earlier, (15.30). Now that the primary and secondary constraints have been identified, we can proceed to the construction of the total Hamiltonian, following steps analogous to those leading to (15.61, 15.63). One obtains (see Problem 4), ignoring for the time being the free fermion kinetic parts (thus, we consider only the parts of the action involving the gauge field),

$$H = H_T = \int (\mathcal{H}_{YM} + A_{\alpha 0}(\vec{x})\chi_\alpha(\vec{x}))d^3x \tag{15.125}$$

$$\mathcal{H}_{YM} = \frac{1}{2}(\vec{E}^2_\alpha + \vec{B}^2_\alpha) - \vec{J}_\alpha \cdot \vec{A}_\alpha, \quad B^i_\alpha = \frac{1}{2}\epsilon^{ijk}F_{\alpha\,jk} \tag{15.126}$$

In order to proceed to a canonical quantization of this theory, we have, as usual, two choices: either (a) the gauge freedom is eliminated *ab initio* by imposing a physical gauge choice—one which reduces the number of degrees of freedom in the theory in accordance with the freedom implicit in the gauge symmetry to the point where

---

[7] Classical Poisson brackets can be defined also for the fermionic fields, with appropriate attention to signs: the charge densities J^0_α have vanishing Poisson brackets with the gauge fields and momenta, and the result is that they satisfy separately the algebra (15.124): see (12.309) for the corresponding quantum commutator result.

the remaining coordinates have well-defined conjugate momenta, and a Hamiltonian can be obtained by the standard Legendre transform procedure—or (b) we treat the system *à la* Dirac and Fadeev, with a total Hamiltonian written as in (15.126) in terms of unconstrained coordinates (i.e., fields) and the constraints and gauge conditions inserted via Lagrange multipliers, with the appropriate DFP determinant in the functional integral. We shall follow the latter approach.

In either case, we must make a gauge choice. In principle, one could impose a gauge condition either on the matter fields or the gauge fields of the theory. For example, if the gauge group is SU(2) and a doublet of complex scalar fields $\phi_n(x), n = 1, 2$ is present transforming under the fundamental representation of SU(2), one might fix the gauge by setting $\phi_1(x) = \phi_R(x), \phi_2(x) = 0$, with $\phi_R(x)$ real, which uniquely determines a local SU(2) transformation which rotates the field doublet into the gauge-fixed form. Such a "unitary" gauge, which is the natural generalization to field theory of the gauges considered earlier in our mechanical examples, is actually quite useful in exposing the physical degrees of freedom of the theory, especially when the gauge symmetry is spontaneously broken, as we shall see in Section 15.6. However, unitary gauges turn out to be highly inconvenient for perturbative calculations, as the Green functions of the theory in such a gauge are in general (even after renormalization of the Lagrangian parameters) ultraviolet-divergent, with only the on-shell limit (i.e., the S-matrix) possessing a sensible limit when the UV cutoffs in the theory are removed.

These problems are avoided by choosing a gauge in terms of a condition on the gauge vector fields of the theory. For example, as for QED, one may employ the gauge freedom to restrict the spatial components of the gauge fields $A_{\alpha i}$ by the axial gauge condition

$$\hat{n} \cdot \vec{A}_\alpha = 0, \quad \hat{n} \cdot \hat{n} = -1 \tag{15.127}$$

where it is conventional to simply choose the space-like unit vector $\hat{n}_i = \delta_{i3}$, so the gauge freedom is employed to move an arbitrary gauge field along a gauge orbit to the *unique* point where $A_{\alpha 3}(x) = 0$ (for all x) (see Problem 5). In this case, it is easy to see that the DFP determinant $\Delta_{\text{axial}}[A]$ is in fact a field-independent constant (as in the abelian case), and may therefore be omitted from the functional integral. Despite this simplifying feature, axial gauge is not a very popular choice, as it clearly destroys manifest Lorentz-invariance: not just boosts, but also (except for rotations around the $\hat{n}$-axis) rotational symmetry.

A less objectionable choice of physical gauge is supplied by Coulomb gauge, which at least retains manifest rotational symmetry, and is also physically desirable in situations where a non-relativistic limit plays an important physical role (as in bound-state problems with heavy quarks, for example):

$$\psi_\alpha \equiv \vec{\nabla} \cdot \vec{A}_\alpha(x) = 0 \tag{15.128}$$

with apologies for the need to temporarily appropriate the ψ symbol from the fermionic fields of the theory, in order to maintain conformity with our previous notation for the gauge fixing condition. In fact, the gauge choice (15.128) is faulty in one important respect: it fails to satisfy the important requirement that the gauge orbit, through an arbitrary configuration, intersects the gauge-fixed surface once, and only once. Instead, as pointed out by Gribov (Gribov, 1978), gauge orbits passing through "large fields" can intersect the gauge-fixed surface multiple times (the famous "Gribov copies"). In particular (cf. the discussion following (15.28, 15.29)), infinitesimal gauge transformations should definitely move us from a point on the gauge-fixed surface to a point off the surface. Instead, we find that for fields $\vec{A}_\alpha$ "sufficiently large" (in a sense soon to become clear), there exist infinitesimal transformations λ_α which preserve the Coulomb gauge condition (15.128). Referring to (15.114), we see that this is the case if (on a given time-slice, suppressing the time variable)

$$(\Delta\delta_{\alpha\beta} + gf_{\alpha\beta\gamma}\vec{A}_\gamma \cdot \vec{\nabla})\lambda_\beta(\vec{x}) = 0 \qquad (15.129)$$

for some non-trivial $\lambda_\beta(\vec{x})$. It is easy to exhibit examples of this phenomenon. For example, taking the gauge group to be SU(2), we have $f_{\alpha\beta\gamma} = \epsilon_{\alpha\beta\gamma}$, and making the Ansatz (all indices $\alpha, \beta, .., i, j, ..$ now run over the values 1,2,3):

$$A_\alpha^i(\vec{x}) = \epsilon_{\alpha ij}x_j V(r) \qquad (15.130)$$

$$\lambda_\alpha(\vec{x}) = x_\alpha R(r), \quad r \equiv \sqrt{\vec{x} \cdot \vec{x}} \qquad (15.131)$$

we find that the condition (15.129) amounts to

$$-\frac{1}{2}\Delta(x_\alpha R(r)) + gV(r)(x_\alpha R(r)) = E(x_\alpha R(r)), \quad E = 0 \qquad (15.132)$$

which is just the three-dimensional Schrödinger equation for a zero-energy $l = 1$ bound state of a particle of unit mass in the potential $gV(r)$. Making the potential $V(r)$ attractive and sufficiently deep (corresponding to a "strong" gauge field $\vec{A}_\alpha$)[8], there will exist normalizable *zero-energy* solutions for $R(r)$ (i.e., the gauge transformation function, which we require to vanish at infinity in accordance with the usual boundary condition of all vanishing fields there). A simple example is an attractive spherical well with angular momentum $l \geq 1$ (which includes our case of $l = 1$), where a normalizable zero-energy solution exists with $R(r) \sim r^{-l-1}$ at large r (Daboul and Nieto, 1994). The same phenomenon can be shown to occur in the covariant Landau gauge:

$$\partial_\mu A_\alpha^\mu = 0 \qquad (15.133)$$

Nevertheless, these gauges are perfectly appropriate if our interests are purely perturbative: in particular, if we are content with finding a consistent set of Feynman rules for generating the formal asymptotic expansion of the Green functions of the theory in powers of the coupling constant g. The reason for this is simply that the perturbative calculation of Green functions from the functional integral amounts to a saddle-point expansion around the Gaussian functional integrand represented by the free part of the action, and the results obtained to any finite order of a saddle-point expansion only depend on the structure of the integrand in the infinitesimal neighborhood of the saddle point. The preceding discussion makes it clear that in this neighborhood (of infinitesimally small fields) the troublesome Gribov copies are, in fact, absent.

Restricting our attention to perturbation theory therefore, we may construct an unconstrained functional integral in Coulomb gauge proceeding in analogy to the

---

[8] We recall that arbitrarily weak potentials do not bind in three dimensions, absent long-range Coulomb-like behavior.

abelian case (cf. (15.69)). In this case, however, the DFP determinant constructed from the Poisson bracket of the constraints with the Coulomb gauge conditions,

$$\{\chi_\alpha(\vec{x}), \psi_\beta(\vec{y})\} = \{-\mathcal{D}_i E_\alpha^i(\vec{x}), \partial_j A^j(\vec{y})\}$$
$$= (\Delta \delta_{\alpha\beta} + g f_{\alpha\beta\gamma} \vec{A}_\gamma \cdot \vec{\nabla}) \delta^3(\vec{x} - \vec{y}) \tag{15.134}$$

corresponds to a non-trivial functional of the gauge field

$$\Delta_{\text{coul}}[\vec{A}_\alpha] = \det(\Delta \delta_{\alpha\beta} + g f_{\alpha\beta\gamma} \vec{A}_\gamma \cdot \vec{\nabla}) \tag{15.135}$$

The reader will recognize here the reappearance of the same operator whose zero-mode eigenfunctions signaled the appearance of the Gribov ambiguity in (15.129). The non-abelian Dirac–Fadeev functional integral analogous to (15.67) (over gauge-field degrees of freedom only: the fermions will be inserted later) (cf.(15.73)), using the expression (15.125) for the total Hamiltonian density,[9] now becomes (cf.(15.73)):

$$Z_{\text{YM}} = \int \mathbf{D}A_{\alpha\mu} \mathbf{D}E_\alpha^i \Delta_{\text{coul}}[\vec{A}_\alpha] \delta(\vec{\nabla} \cdot \vec{A}_\alpha)$$
$$\cdot e^{i \int (E_\alpha^i (\partial_0 A_{\alpha i} - \partial_i A_{\alpha 0} - g f_{\alpha\beta\gamma} A_{\beta 0} A_{\gamma i}) - \frac{1}{2}(\vec{E}_\alpha^2 + \vec{B}_\alpha^2) - J_\alpha^\mu A_{\alpha\mu}) d^4 x} \tag{15.136}$$

$$= \int \mathbf{D}A_\mu \delta(\vec{\nabla} \cdot \vec{A}_\alpha) \Delta_{\text{coul}}[\vec{A}_\alpha] e^{i \int (\frac{1}{2} F_{\alpha 0i}^2 - \frac{1}{4} F_{\alpha jk}^2 - J_\alpha^\mu A_{\alpha\mu}) d^4 x} \tag{15.137}$$

$$= \int \mathbf{D}A_\mu \delta(\vec{\nabla} \cdot \vec{A}_\alpha) \Delta_{\text{coul}}[\vec{A}_\alpha] e^{i \int (-\frac{1}{4} F_{\alpha\mu\nu} F_\alpha^{\mu\nu} - J_\alpha^\mu A_{\alpha\mu}) d^4 x} \tag{15.138}$$

$$= \int \mathbf{D}A_\mu \delta(\vec{\nabla} \cdot \vec{A}_\alpha) \Delta_{\text{coul}}[\vec{A}_\alpha] e^{i \int \mathcal{L}_{\text{YM}} d^4 x} \tag{15.139}$$

where $\mathcal{L}_{\text{YM}}$ is the local and Lorentz scalar Yang–Mills Lagrangian (15.112), minus the fermion kinetic piece. Of course, manifest Lorentz-invariance is still broken by the δ-function enforcing the non-covariant Coulomb gauge condition and by the DFP functional Δ_{coul} which only depends on the spatial components of the gauge-field.

As for abelian gauge theories, a choice of gauge which preserves the manifest Lorentz symmetry and locality properties of the underlying dynamics is usually preferable to the non-covariant choices which lead to a straightforward canonical treatment. The conversion can be made (again, as in QED) by recognizing that the DFP determinant (15.135) can be written as the inverse of a functional integral which averages the gauge-fixing δ-function over all gauge transformations (cf. (15.74)). This "averaging over a group" requires the notion of Hurwitz measure: the integration over all elements U of a continuous Lie group G is uniquely defined by the two conditions

$$\int dU f(UV) = \int dU f(VU) = \int dU f(U), \quad V \in G \text{ (shift invariance)} \tag{15.140}$$

---

[9] Recall that the $A_{\alpha 0}$ fields are Lagrange multiplier fields enforcing the χ_α constraints: the second and third terms in the expression multiplying E_α^i in the exponent in (15.136) arise from the $\mathcal{D}_i E_\alpha^i$ part of χ_α, with an integration by parts on the ∂_i part of $\mathcal{D}_i$.

and

$$\int dU = 1 \quad \text{(normalization)} \tag{15.141}$$

For example, the Lie group $SU(2)$ is defined in the fundamental representation by unit determinant 2x2 unitary matrices $U = i\vec{\sigma} \cdot \vec{u} + u_4 \cdot 1$ with $\vec{u}^2 + u_4^2 = 1$: in other words, it is topologically the four-dimensional unit sphere. The Hurwitz measure turns out in this case to be the obvious choice: $\int dU = \frac{1}{2\pi^2} \int d\Omega$ where $d\Omega$ is the solid angle in four dimensions. It is easy to verify the shift invariance condition (see Problem 6) for this definition. For a local gauge symmetry we have the obvious generalization of the single Hurwitz integral to a *functional integral* $\int \mathbf{D}U(x)$ over independent gauge elements $U(x)$ at each spacetime point.

Now consider the functional $F[\vec{A}]$ defined by the Hurwitz functional integral

$$F[\vec{A}] \equiv \int \mathbf{D}U \delta(\vec{\nabla} \cdot \vec{A}_\alpha^U) \tag{15.142}$$

where $\vec{A}^U$ is the spatial gauge field after being subjected to the finite local gauge transformation $U(x)$ (as in (15.105)). We shall now evaluate this functional integral for gauge fields $\vec{A}_\alpha$ which are (a) sufficiently weak that no Gribov copies exist, and (b) in Coulomb gauge—i.e., satisfy $\vec{\nabla} \cdot \vec{A}_\alpha = 0$. It is clear that the δ-function in the integral is supported exactly at the identity value for the local gauge function $U(x) = e^{igt_\alpha \Lambda_\alpha(x)}$ (as we are already in Coulomb gauge). In the neighborhood of the identity, we may replace the finite group parameters Λ_α by their infinitesimal limits $\lambda_\alpha(x)$, and obtain (up to an irrelevant constant factor), using (15.114), and the identity from footnote 6,

$$F[\vec{A}] = \int \mathbf{D}\lambda_\alpha \delta(\vec{\nabla} \cdot (\vec{A}_\alpha + \vec{\nabla}\lambda_\alpha + g f_{\alpha\beta\gamma}\lambda_\beta \vec{A}_\gamma))$$

$$= \int \mathbf{D}\lambda_\alpha \delta((\Delta\delta_{\alpha\beta} + g f_{\alpha\beta\gamma}\vec{A}_\gamma)\lambda_\beta)$$

$$= \det{}^{-1}(\Delta\delta_{\alpha\beta} + g f_{\alpha\beta\gamma}\vec{A}_\gamma \cdot \vec{\nabla}) = \Delta_{\text{coul}}[\vec{A}_\alpha]^{-1} \tag{15.143}$$

Thus, we find, in analogy to (15.74) in the abelian case, that the DFP determinant provides a partition of unity over all gauge transformations

$$\Delta_{\text{coul}}[\vec{A}_\alpha] \int \mathbf{D}U \delta(\vec{\nabla} \cdot \vec{A}_\alpha^U) = 1 \tag{15.144}$$

and, by the shift-invariance property of the Hurwitz measure, is trivially gauge-invariant:

$$\Delta_{\text{coul}}[\vec{A}_\alpha^V]^{-1} = \int \mathbf{D}U \delta(\vec{\nabla} \cdot \vec{A}_\alpha^{VU}) = \int \mathbf{D}U \delta(\vec{\nabla} \cdot \vec{A}_\alpha^U) = \Delta_{\text{coul}}[\vec{A}_\alpha]^{-1} \tag{15.145}$$

We are now in a position to repeat the steps analogous to those leading from (15.79) to (15.80) in the abelian case, thereby making the transition from the non-covariant Coulomb gauge to a covariant gauge specified by the generalized Landau gauge condition

$$\partial^\mu A_{\alpha\mu}(x) = f_\alpha(x) \tag{15.146}$$

where the $f_\alpha(x)$ are, for the time being, arbitrary but fixed c-number functions. We shall include the fermion dynamics completely at this point, by including the usual integrals over Grassmannian integration variables, and begin with the functional

$$Z_{\text{YM}}[K_i] = \int \mathbf{D}A_{\alpha\mu}\mathbf{D}\psi\mathbf{D}\bar{\psi}\delta(\vec{\nabla}\cdot\vec{A}_\alpha)\Delta_{\text{coul}}[\vec{A}_\alpha]e^{i\int(\mathcal{L}_{\text{YM}}+K_i(x)O_i(x))d^4x} \tag{15.147}$$

where now $\mathcal{L}_{\text{YM}}$ is the full non-abelian Lagrangian (15.112), including fermion kinetic terms, and the $O_i(x)$ are an arbitrary set of gauge-invariant operators whose Green functions we wish to compute, by taking functional derivatives with respect to the associated sources $K_i(x)$. Defining, in analogy to (15.78), a covariant DFP functional by

$$1 = \Delta^f_{\text{cov}}[A] \cdot \int \mathbf{D}U\delta(\partial^\mu A^U_{\alpha\mu} - f_\alpha) \tag{15.148}$$

we can easily show (see Problem 7), just as in the abelian case, that for fields in the generalized Landau gauge (15.146), (a) $\Delta^f_{\text{cov}} \equiv \Delta_{\text{cov}}$ is in fact *independent* of the arbitrary functions $f_\alpha(x)$, (b) is a gauge-invariant functional, $\Delta_{\text{cov}}[A^V] = \Delta_{\text{cov}}[A]$ for any local gauge transformation $V(x)$, and (c) is the determinant of the covariant analog of the Coulomb operator (15.135),

$$\Delta_{\text{cov}}[A] = \det(\Box\delta_{\alpha\beta} + gf_{\alpha\beta\gamma}\partial^\mu A_{\gamma\mu}) \tag{15.149}$$

The steps leading from (15.79) to (15.80) can be repeated more or less *verbatim*, by inserting the partition of unity (15.148) (with the functional integral over abelian gauge transformations $\int \mathbf{D}\Lambda\dots$ now replaced by the corresponding non-abelian Hurwitz functional integral $\int \mathbf{D}U$), and using a shift of the integration variables which corresponds to a local (finite) gauge transformations on all the fields (leaving the Lagrangian $\mathcal{L}_{YM}$ and the gauge-invariant operators O_i unchanged)

$$A_{\alpha\mu}(x) \to A^{U^{-1}}_{\alpha\mu}(x), \quad \psi \to U^{-1}\psi(x) \tag{15.150}$$

Note that the shift of variables (15.150) has *unit Jacobian*. This is obvious for the fermionic fields, which undergo unitary rotation at each spacetime point by the matrix $U^{-1}(x)$ which has unit determinant. The gauge fields in the adjoint representation transform as follows (cf. (15.105):

$$t_\alpha A^{U^{-1}}_{\alpha\mu} = U^{-1}t_\alpha A_{\alpha\mu}U - \frac{i}{g}U^{-1}\partial_\mu U \tag{15.151}$$

As the second term on the right is an additive shift, and the first corresponds to an orthogonal rotation of the $A_{\alpha\mu}$ in the α indices, the Jacobian of the transformation on the gauge fields is also unity. The upshot is that we obtain (after removing the Coulomb DFP determinant in the form of the partition of unity (15.144)) a manifestly covariant expression for the *same* functional Z_{YM}:

$$Z_{\text{YM}}[K_i] = \int \mathbf{D}A_{\alpha\mu}\mathbf{D}\psi\mathbf{D}\bar{\psi}\delta(\partial^\mu A_{\alpha\mu} - f_\alpha)\Delta_{\text{cov}}[A]e^{i\int(\mathcal{L}_{\text{YM}}+K_i(x)O_i(x))d^4x} \quad (15.152)$$

As our starting point did not contain the functions f_α, the functional Z_{YM} must also be independent of the f_α (despite their appearance in the δ-function), and we may therefore multiply (15.152) by a physically irrelevant constant factor

$$C \equiv \int \mathbf{D}f_\alpha e^{-\frac{i}{2\xi}\int f_\alpha(x)^2 d^4x} \quad (15.153)$$

and interchange the functional integrals over the gauge and fermion fields with those over the f_α to obtain the non-abelian analog of (15.82):

$$Z_{\text{YM}} = \int \mathbf{D}A_{\alpha\mu}\mathbf{D}\psi\mathbf{D}\bar{\psi}\Delta_{\text{cov}}[A]e^{i\int(\mathcal{L}_{\text{YM}}-\frac{1}{2\xi}(\partial^\mu A_{\alpha\mu})^2+K_i O_i)d^4x} \quad (15.154)$$

The presence of the non-trivial (and unknown!) functional $\Delta_{\text{cov}}[A]$ in this formula makes it unsuitable for practical perturbative calculations. Instead, it is convenient to re-express this functional in terms of a functional integral representation, where the determinant in (15.149) is generated by integrating over complex Grassmannian "ghost fields" $\omega_\alpha(x), \bar{\omega}_\alpha(x)$, using the Gaussian fermionic integral (10.114),

$$\Delta_{\text{cov}}[A] = \det(\Box\delta_{\alpha\beta} + gf_{\alpha\beta\gamma}\partial^\mu A_{\gamma\mu}) = \int \mathbf{D}\omega_\alpha \mathbf{D}\bar{\omega}_\alpha e^{i\int \bar{\omega}_\alpha(\Box\delta_{\alpha\beta}+gf_{\alpha\beta\gamma}\partial^\mu A_{\gamma\mu})\omega_\beta d^4x} \quad (15.155)$$

We emphasize that the ghost fields are introduced here as a purely technical device: they evidently do not correspond to physical fields or particles in the theory, and in particular there are no asymptotic states associated with them. As the theory is based on an underlying hermitian Hamiltonian, we expect the perturbative S-matrix constructed from a Fock space of gauge and fermion particles (and *no* "ghost particles") to be unitary on its own.[10] Inserting (15.155) in (15.154), we obtain our final result for the generating functional of a theory of coupled Yang–Mills and fermionic matter fields, in a covariant ξ-gauge:

$$Z_{\text{YM}}[K_i] = \int \mathbf{D}A_{\alpha\mu}\mathbf{D}\psi\mathbf{D}\bar{\psi}\mathbf{D}\omega_\alpha\mathbf{D}\bar{\omega}_\alpha e^{i\int(\mathcal{L}_{\text{YM}}+\mathcal{L}_{\text{gh}}-\frac{1}{2\xi}(\partial^\mu A_{\alpha\mu})^2+K_i O_i)d^4x} \quad (15.156)$$

$$\mathcal{L}_{\text{YM}} = -\frac{1}{4}F_{\alpha\mu\nu}F_\alpha^{\mu\nu} + \sum_a \bar{\psi}_a(i\not{D} - m_a)\psi_a \quad (15.157)$$

$$\mathcal{L}_{\text{gh}} = \bar{\omega}_\alpha(\Box\delta_{\alpha\beta} + gf_{\alpha\beta\gamma}\partial^\mu A_{\gamma\mu})\omega_\beta \quad (15.158)$$

The advantage of the ghost field representation of the DFP determinant functional is apparent in (15.156): the path integral takes the standard form, as an integral

---

[10] By arguments analogous to those given in Section 15.3 for abelian gauge theory, the on-shell S-matrix can be shown to be the same in a physical gauge (such as Coulomb or axial gauge), where only manifestly physical gauge degrees of freedom are present, as in the covariant gauges under discussion. It is therefore clear that perturbative unitarity must therefore hold on a Fock space constructed from the interpolating fields associated with the gauge and matter fields of the theory, *sans* ghosts.

over an exponential of a Lagrangian with a polynomial structure in fields and their derivatives, from which a set of Feynman rules can easily be extracted.

A free Lagrangian is identified by setting the gauge coupling constant to zero, and the free propagators of the various fields (gauge, ghost, and fermion) can then be extracted, as in Section 10.3, as the Green functions associated with the differential operators in the corresponding quadratic parts of the Lagrangian. The ξ-dependent term in (15.156) will clearly enter in the determination of the gauge field propagator: a straightforward calculation (see Problem 8) shows that the two-point function for the free gauge fields is

$$\langle 0|T\{A_{\alpha\mu}(x)A_{\beta\nu}(y)\}|0\rangle = -i\delta_{\alpha\beta}\int \frac{(g_{\mu\nu} - (1-\xi)\frac{k_\mu k_\nu}{k^2})e^{-ik\cdot(x-y)}}{k^2 + i\epsilon} \frac{d^4k}{(2\pi)^4} \quad (15.159)$$

The choices $\xi = 0$ (resp. 1) are referred to as Landau (resp. Feynman) gauge, but as mentioned previously, it is often convenient to leave the gauge parameter ξ unfixed at intermediate stages, as its disappearance at the end in physically meaningful quantities is a powerful check (guaranteed by gauge invariance) on the correctness of the calculation.

The interaction vertices of the theory are associated with the cubic and quartic terms in the total Lagrangian. Thus, denoting the gauge, ghost and Fermi fields by A, ω, ψ generically, there are vertices in the graphs of the theory corresponding (schematically) to $A^3, A^4, \bar{\psi}\psi A$, and $\bar{\omega}\omega A$ field products (see Fig. 15.1 for the Feynman rules for the bosonic vertices of the theory; also, Problem 9). In addition, we must remember that the Grassmann nature of the ghost fields inserts minus signs (analogous to those for the physical fermion fields of the theory) when ghost fields are exchanged in Wick products, as well as in closed loops of ghost propagators (cf. (10.40)). As emphasized above, the ghost propagators *only appear as internal lines*, as the ghost fields do not correspond to physical asymptotic particle states.

A deeper understanding of the role of ghost fields in the quantization of local gauge theories has been provided by the beautiful theory developed by Becchi, Rouet, and Stora (Becchi *et al.*, 1976), and Tyutin (Iofa and Tyutin, 1976), where the first-class constraints appearing in a theory with local gauge symmetry are reinterpreted in terms of an exact *global* supersymmetry of the full Lagrangian density (e.g., the exponent in the path-integral expression (15.156), including ghost and gauge-fixing terms). The BRST theory (like its historical antecedent, the Gupta–Bleuler quantization method) is necessarily formulated on a state space with indefinite metric, but the existence of a global supersymmetry turns out to be exactly what is needed for the ghost states, together with longitudinal polarizations of massless gauge mesons, to decouple from the positive-definite subspace of physical states. The derivation of the Ward–Takahashi identities which summarize the content of the local symmetry at the level of the Green functions of the theory is also considerably simplified in the BRST approach. We shall not describe this approach further here, but refer the reader to the original papers and accounts in textbooks devoted specifically to quantization of gauge field theories.[11]

---

[11] For a readable account, see (Taylor, 1976), Chapter 12. The general graded cohomology BRST theory of Hamiltonian systems with first-class constraints can be found in (Henneaux and Teitelboim, 1992).

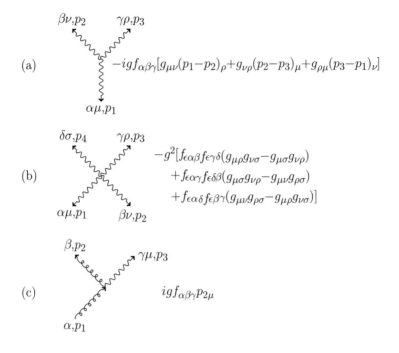

(a) $-igf_{\alpha\beta\gamma}[g_{\mu\nu}(p_1-p_2)_\rho+g_{\nu\rho}(p_2-p_3)_\mu+g_{\rho\mu}(p_3-p_1)_\nu]$

(b) $-g^2[f_{\epsilon\alpha\beta}f_{\epsilon\gamma\delta}(g_{\mu\rho}g_{\nu\sigma}-g_{\mu\sigma}g_{\nu\rho})$
$+f_{\epsilon\alpha\gamma}f_{\epsilon\delta\beta}(g_{\mu\sigma}g_{\nu\rho}-g_{\mu\nu}g_{\rho\sigma})$
$+f_{\epsilon\alpha\delta}f_{\epsilon\beta\gamma}(g_{\mu\nu}g_{\rho\sigma}-g_{\mu\rho}g_{\nu\sigma})]$

(c) $igf_{\alpha\beta\gamma}p_{2\mu}$

Fig. 15.1 Feynman rules for the (a) triple gluon, (b) quadruple gluon, and (c) ghost-ghost-gluon vertices in QCD. The arrows indicate direction of momentum flow.

As in the case of abelian gauge theories, we are frequently interested in calculating the perturbative S-matrix elements for scattering of the elementary quanta of the theory (e.g., quarks and gluons in QCD). This requires that sources be introduced for the non-gauge-invariant fields (the ψ and $A_{\alpha\mu}$) which interpolate for these objects in perturbation theory. The argument that the S-matrix is gauge-independent, and that we may therefore, for example, use the Feynman rules in an arbitrary ξ-gauge to compute the relevant scattering amplitudes, proceeds along similar lines to the discussion for abelian gauge theory at the end of the previous section.

Our functional integral quantization of non-abelian gauge theory has so far been carried out in Minkowski space. Many aspects of the Yang–Mills theory are more conveniently studied in the Euclidean space formulation of the theory. For example, the important role played by vacuum tunneling processes ("instantons") can only be exposed clearly in Euclidean space, and non-perturbative simulations of the lattice-regularized theory are all performed in the Euclidean version. The transition from Minkowski to Euclidean space is accomplished by the usual Wick analytical continuation:

$$x^0 \to -ix_4, \quad x^i \to x_i, \; i=1,2,3, \quad \partial_0 \to i\partial_4, \quad \partial_i \to \partial_i$$

$$A_{\alpha 0} \to \frac{i}{g}A_{\alpha 4}, \quad A_{\alpha i} \to \frac{1}{g}A_{\alpha i}, \; i=1,2,3,$$

$$F_{\alpha 0i} \to \frac{i}{g} F_{\alpha 4i}, \ i = 1, 2, 3, \quad F_{\alpha ij} \to \frac{1}{g} F_{\alpha ij}, \ i, j = 1, 2, 3$$

$$\gamma^0 \to \hat{\gamma}_4, \quad \gamma^i \to i\hat{\gamma}_i \ (i = 1, 2, 3), \quad \{\hat{\gamma}_\mu, \hat{\gamma}_\nu\} = 2\delta_{\mu\nu} \tag{15.160}$$

where we have also rescaled the gauge field by a factor of the inverse coupling constant. On performing these replacements in the Minkowski functional integral (15.156) we obtain (ignoring source terms, and taking for simplicity just a single Dirac field in the fundamental representation) the Euclidean functional integral:

$$Z_{\mathrm{YM,E}} = \int \mathbf{D} A_{\alpha\mu} \mathbf{D}\psi \mathbf{D}\bar{\psi} \mathbf{D}\omega_\alpha \mathbf{D}\bar{\omega}_\alpha e^{-\int (\mathcal{L}_{\mathrm{YM,E}} + \mathcal{L}_{\mathrm{gh}} + \frac{1}{2\xi g^2}(\partial_\mu A_{\alpha\mu})^2) d^4 x} \tag{15.161}$$

$$\mathcal{L}_{\mathrm{YM,E}} = \frac{1}{4g^2}(F_{\alpha\mu\nu})^2 - \bar{\psi}(i\hat{\slashed{D}} - m)\psi \tag{15.162}$$

$$F_{\alpha\mu\nu} = \partial_\mu A_{\alpha\nu} - \partial_\nu A_{\alpha\mu} - f_{\alpha\beta\gamma} A_{\beta\mu} A_{\gamma\nu}, \quad \hat{\slashed{D}} = \hat{\gamma}_\mu(-i\partial_\mu + t_\alpha A_{\alpha\mu}) \tag{15.163}$$

$$\mathcal{L}_{\mathrm{gh,E}} = \bar{\omega}_\alpha(\delta_{\alpha\beta}\partial_\mu\partial_\mu + f_{\alpha\beta\gamma}\partial_\mu A_{\gamma\mu})\omega_\beta \tag{15.164}$$

We note the following important features of this Euclidean functional representation:

1. The gauge field component of the Euclidean action,

$$S_{\mathrm{gauge,E}} \equiv \frac{1}{4g^2}\mathcal{F}_\xi[A_{\alpha\mu}], \quad \mathcal{F}_\xi[A_{\alpha\mu}] \equiv \int \{(F_{\alpha\mu\nu})^2 + \frac{2}{\xi}(\partial_\mu A_{\alpha\mu})^2\} d^4 x \tag{15.165}$$

is evidently positive-definite and provides the usual damping factor for large fields which ensures the existence of the functional integral (provided, as usual, the theory is fully regularized, say on a spacetime lattice). Perturbation theory corresponds to performing Gaussian functional integrals over the gauge fields keeping only the parts of $F_{\alpha\mu\nu}$ linear in $A_{\alpha\mu}$ in $\mathcal{F}_\xi$: i.e., functional integrals of polynomials of the fields with the Gaussian measure $e^{-\frac{1}{4g^2}\mathcal{F}_\xi^0}$, with

$$\mathcal{F}_\xi^0[A] = \int \{(\partial_\mu A_{\alpha\nu} - \partial_\nu A_{\alpha\mu})^2 + \frac{2}{\xi}(\partial_\mu A_{\alpha\mu})^2\} d^4 x \tag{15.166}$$

"Large" gauge fields A of the type responsible for Gribov copies (recall the example (15.130)), where the "potential" $V(r)$ must be of order unity) have a Gaussian action $\mathcal{F}_\xi^0[A] \geq C$, where C is a constant of order unity in the weak coupling limit $g \to 0$, and their contribution to the functional integral is therefore suppressed by the non-analytic dependence $e^{-\frac{C}{4g^2}}$, which has a *vanishing formal expansion* in powers of the coupling, to all orders of perturbation theory. This argument vindicates our earlier assertion that the Gribov ambiguity is irrelevant at the level of perturbative expansions.

2. The Dirac operator $\hat{\slashed{D}}$ appearing in the quadratic fermionic action is formally self-adjoint: the Euclidean γ matrices $\hat{\gamma}_\mu$ are hermitian, as are the gauge group generators t_α, the gauge fields $A_{\alpha\mu}$ are real, and the differential operators $-i\partial_\mu$ are self-adjoint when acting on functions in a suitable function space. In fact, if we quantize the theory in a box of finite spacetime volume (imposing, say, periodic

boundary conditions), this operator will have a purely discrete spectrum, with the usual orthogonality properties holding for its eigenfunctions:

$$\hat{\slashed{D}}\phi_i(x) = \lambda_i\phi_i(x), \; \lambda_i \text{ real}, \quad \int \phi_i^\dagger(x)\phi_j(x)d^4x = \delta_{ij} \tag{15.167}$$

where the eigenfunctions $\phi_i(x)$ carry implicitly discrete Dirac and fundamental representation gauge group indices (whence the † appearing in the orthonormality relation). These properties turn out to be crucial in the analysis of the chiral properties of gauge theory, as we shall see shortly in our discussion of the axial anomaly. The self-adjointness of $\hat{\slashed{D}}$ is also critical in non-perturbative evaluations of the Euclidean Green functions in QCD by Monte Carlo simulation of the lattice-regularized functional integral, as the fermion fields when integrated out yield the determinant of the Dirac operator $\mathcal{Q} \equiv i\hat{\slashed{D}} - m$ which can be shown to be real, as a consequence of

$$\gamma_5 \mathcal{Q}\gamma_5 = \mathcal{Q}^\dagger \Rightarrow \det(\mathcal{Q}) = \det(\gamma_5\mathcal{Q}^\dagger\gamma_5) = \det(\mathcal{Q}^\dagger) = \det(\mathcal{Q})^* \tag{15.168}$$

and even positive, and therefore can be treated as a probability measure in a stochastic evaluation of the path integral over the gauge fields by importance sampling methods (see Chapter 7, (Montvay and Münster, 1994)).

15.5 Explicit quantum-breaking of global symmetries: anomalies

The existence of global symmetries which are exactly conserved at the classical level but broken at the quantum level, corresponding to Noether currents with a divergence proportional to Planck's constant, was established in the late 1960s with the discovery of the famous *axial* (or *chiral*) *anomaly* by Adler (Adler, 1969), and by Bell and Jackiw (Bell and Jackiw, 1969). In fact, the existence of the chiral anomaly was foreshadowed in the much earlier work of Schwinger (Schwinger, 1951), where the decay rate of a pseudoscalar meson to two photons was computed perturbatively and shown to yield an amplitude which (much later) was realized (by Bell and Jackiw) to be exactly the source of the anomalously high decay rate of the neutral pion to two photons.

The origin of the failure of the classical Noether theorem due to quantum anomalies was clarified within the framework of the functional integral formulation by Fujikawa (Fujikawa, 1980, 1981). We shall discuss Fujikawa's treatment of the chiral anomaly in some detail, as it reveals the essential features of the problem, common to the entire class of quantum-induced symmetry-breaking anomalies in field theory. The reader will recall from the discussion in Section 7.4.5 that a massless Dirac field ψ decomposes naturally into the Weyl field components ψ_L (resp. ψ_R) which transform according to $(\frac{1}{2}, 0)$ (resp. $(0, \frac{1}{2})$) representations of the HLG. The kinetic part of the Lagrangian for such fields, $\bar{\psi}\gamma^\mu D_\mu \psi$ (with a covariant derivative D_μ appearing if gauge fields are present coupled to the fermions), lacks terms mixing the left and right parts of the Dirac field. Consequently, the fermionic Lagrangian possesses a global U(1)×U(1) symmetry under independent phase rotations of the left- and right-handed parts of the Dirac bispinor:

$$\psi_L(x) \to e^{i\omega_L}\psi_L(x)$$

$$\psi_R(x) \to e^{i\omega_R}\psi_R(x)$$

The "diagonal" subgroup corresponding to $\omega_L = \omega_R$ is just a global phase symmetry of the type discussed in the fourth example of Section 12.5: the associated Noether current $J^\mu = \frac{\partial \mathcal{L}}{\partial(\partial_\mu \psi)}\psi = \bar{\psi}\gamma^\mu\psi$ is not anomalous (as we shall see below), and is exactly conserved (it is usually referred to as a "vector" symmetry, as the associated current transforms as a spacetime vector, rather than as an axial vector). The *chiral* subgroup consisting of the global transformations

$$\psi_L(x) \to e^{i\omega}\psi_L(x)$$

$$\psi_R(x) \to e^{-i\omega}\psi_R(x)$$

$$\psi(x) \to e^{i\omega\gamma_5}\psi(x) \tag{15.169}$$

also leads to a Noether symmetry and, at the classical level, a conserved axial Noether current:

$$J_5^\mu = \bar{\psi}\gamma_5\gamma^\mu\psi \tag{15.170}$$

Of course, if a mass term $m\bar{\psi}\psi = m\bar{\psi}_L\psi_R + m\bar{\psi}_R\psi_L$ is added to the Lagrangian, the chiral symmetry is lost, and the current (15.170) will develop a non-zero divergence. The remarkable discovery of Adler, Bell, and Jackiw was that *even in the absence of a mass term* quantum effects inevitably produce a non-vanishing divergence of J_5^μ if the fermions are coupled to vector gauge fields via an exact local gauge symmetry. We shall (following Fujikawa) exhibit this result using the functional version of Noether's theorem: i.e., by deriving the Ward–Takahashi identities for the chiral current of fundamental representation fermions coupled to gauge fields (either abelian or non-abelian). We shall see that an anomalous divergence, proportional to Planck's constant $\hbar$—and therefore, explicitly a quantum effect—emerges once the Jacobian arising from a functional change of variable corresponding to a local chiral transformation is carefully evaluated. As a preparation for the discussion here, the reader may find it convenient to briefly review the derivation of the Ward–Takahashi identities for a non-anomalous current that concludes Section 12.5.

The chiral transformations (15.169) act only on the fermion fields of the theory, so we need only consider the fermionic part of the gauge theory path integral, with the gauge fields "frozen" at some fixed, but unspecified, values (which may later be integrated over after the pure gauge parts of the action are included). We shall work in Euclidean space, so our starting point is the fermionic functional integral (cf. (15.162))

$$Z_{\text{ferm}}[\eta, \bar{\eta}] = \int \mathbf{D}\psi \mathbf{D}\bar{\psi} \, e^{\frac{1}{\hbar} \int \mathcal{L}_{\text{ferm}} d^4 x + \int (\bar{\eta}(x)\psi(x) + \bar{\psi}(x)\eta(x))d^4 x} \tag{15.171}$$

$$\mathcal{L}_{\text{ferm}} = \bar{\psi}(x)(i\hat{\slashed{D}} - m)\psi(x) \tag{15.172}$$

We have explicitly indicated the factor of Planck's constant (normally set to unity) in the exponent of the functional integrand in order to clarify the quantum origins of the anomaly. We now retrace the procedure followed in Section 12.5 in deriving the functional form of the Noether theorem for a vector symmetry, by considering the effect of a functional change of variables induced by an infinitesimal *local* chiral transformation:

$$\psi(x) \to \psi'(x) \equiv e^{i\omega(x)\gamma_5}\psi(x) \sim (1 + i\omega(x)\gamma_5)\psi(x) + O(\omega^2)$$
$$\bar{\psi}(x) \to \bar{\psi}'(x) \equiv \bar{\psi}(x)(1 + i\omega(x)\gamma_5) + O(\omega^2) \tag{15.173}$$

Note that the Wick rotation to Euclidean space converts our previous Minkowski space definition of γ_5 to its Euclidean version $\gamma_5 = \hat{\gamma}_1\hat{\gamma}_2\hat{\gamma}_3\hat{\gamma}_4$, which turns out to be exactly the same matrix—i.e., (7.107). We can expand the general Grassmann fields $\psi(x), \bar{\psi}(x)$ in a complete set of normalized eigenfunctions of the self-adjoint Dirac operator $\hat{\not{D}}$ (cf. (15.167)), thereby giving a precise meaning to the functional integral in terms of a discrete multi-dimensional Grassmann integral

$$\psi(x) = \sum_i \phi_i(x)\psi_i \quad, \quad \bar{\psi}(x) = \sum_i \phi_i^\dagger(x)\bar{\psi}_i \tag{15.174}$$

$$\int \mathbf{D}\psi\mathbf{D}\bar{\psi} \equiv \int \prod_i d\psi_i d\bar{\psi}_i \tag{15.175}$$

Note that the eigenfunctions $\phi_i(x)$ contain a hidden (four-dimensional) Dirac index and a gauge group index corresponding to the fundamental representation of the gauge group (one-dimensional for the abelian case, such as QED, or N-dimensional for a SU(N) non-abelian gauge group). They depend on the c-number gauge fields buried in the covariant derivative $\hat{D}$. Under (15.173), the fermionic Lagrangian transforms to

$$\mathcal{L}_{\text{ferm}} \to \mathcal{L}_{\text{ferm}} - 2im\omega(x)\bar{\psi}\gamma_5\psi(x) - \partial_\mu\omega(x)\bar{\psi}\hat{\gamma}^\mu\gamma_5\psi(x) \tag{15.176}$$

while the source terms in (15.171) transform to

$$\int(\bar{\eta}(x)\psi(x) + \bar{\psi}(x)\eta(x))d^4x \to \int(\bar{\eta}(x)\psi(x) + \bar{\psi}(x)\eta(x))d^4x$$

$$+ i\int \omega(x)(\bar{\eta}(x)\gamma_5\psi(x) + \bar{\psi}(x)\gamma_5\eta(x))d^4x \tag{15.177}$$

The fermionic integration variables meanwhile undergo the following transformation (to first order in ω):

$$\psi'(x) = \sum_i \phi_i(x)\psi_i' = (1 + i\omega(x)\gamma_5)\sum_j \phi_j(x)\psi_j$$

$$\Rightarrow \psi_i' = \sum_j (\int \phi_i^\dagger(x)(1 + i\omega(x)\gamma_5)\phi_j(x)d^4x)\psi_j = \sum_j C_{ij}\psi_j$$

$$C_{ij} \equiv \int \phi_i^\dagger(x)(1 + i\omega(x)\gamma_5)\phi_j(x)d^4x = \delta_{ij} + i\int \phi_i^\dagger(x)\omega(x)\gamma_5\phi_j(x)d^4x$$

The Dirac conjugate fields $\bar{\psi}$ produce a similar result, as the sign of the ω term in (15.173) is the same as for ψ (due to the fact that $\bar{\psi} = \psi^\dagger \gamma_4$, giving an extra minus sign when the γ_5 is commuted through $\hat{\gamma}_4$):

$$\bar{\psi}'(x) = \sum_j \bar{\psi}_j C_{ji} \tag{15.178}$$

The effect of the change of variable (15.173) is therefore to introduce a Jacobian factor (cf. (10.116))

$$J = \det(C)^{-2} = e^{-2\text{Tr}\ln(C)}$$

$$= e^{-2i\sum_i \int \omega(x)\phi_i^\dagger(x)\gamma_5\phi_i(x)d^4x} \equiv e^{-2i\int \omega(x)\mathcal{A}(x)d^4x} \tag{15.179}$$

where we again remind the reader that we are working to first order in the gauge parameter $\omega(x)$. Note that if we had been working with the vector symmetry induced by the phase transformation $\psi \to e^{i\omega}\psi$ (with no γ_5), there would be no mass term in the variation of the Lagrangian, as in (15.176), and the functional Jacobian would be a product of the determinants of matrices C and $\bar{C}$ given by

$$C_{ij} = \delta_{ij} + i\int \phi_i^\dagger(x)\omega(x)\phi_j(x)d^4x, \quad \bar{C}_{ij} = \delta_{ij} - i\int \phi_i^\dagger(x)\omega(x)\phi_j(x)d^4x \tag{15.180}$$

which is simply (to order ω) unity, and we would recover the standard (non-anomalous) Ward–Takahashi identities (as in Section 12.5). For the chiral current, however, the functional Jacobian (15.179) is a non-trivial, though *gauge-invariant* functional of the gauge fields (as $\psi(x) \to U(x)\psi(x)$ induces the change $\phi_i(x) \to U(x)\phi_i(x)$, which leaves C_{ij} unchanged), which we shall shortly evaluate explicitly. Before doing that, let us assemble the various pieces needed to obtain the chiral Ward–Takahashi identity, which is simply the statement that the functional Z_{ferm} is unchanged by the functional change of variables (15.173)—provided, of course, that we take into account properly any non-trivial Jacobians induced by the change. After subjecting Z_{ferm} to the change of variable, we find that the first-order (in ω) change in the integral takes the form

$$0 = \delta Z_{\text{ferm}}[\eta, \bar{\eta}] = \int \omega(x)\mathcal{W}[\eta, \bar{\eta}; x] + O(\omega^2)$$

$$\Rightarrow 0 = \mathcal{W}[\eta, \bar{\eta}; x] = \int \mathbf{D}\psi \mathbf{D}\bar{\psi}\{i(\bar{\eta}(x)\gamma_5\psi(x) + \bar{\psi}(x)\gamma_5\eta(x)) - 2i\mathcal{A}(x)$$

$$+ \frac{1}{\hbar}(\partial_\mu(\bar{\psi}\hat{\gamma}_\mu\gamma_5\psi(x)) - 2im\bar{\psi}\gamma_5\psi(x))\}$$

$$\cdot e^{\frac{1}{\hbar}\int \mathcal{L}_{\text{ferm}}d^4x + \int (\bar{\eta}(x)\psi(x) + \bar{\psi}(x)\eta(x))d^4x} \tag{15.181}$$

which is the analog of the Noether functional theorem (12.141) for our anomalous symmetry. The Ward–Takahashi (WT) identities (analogous to (12.142)) are obtained by differentiating $\mathcal{W}[\eta, \bar{\eta}; x]$ (=0) with respect to the fermionic sources $\eta(x), \bar{\eta}(x)$ and then setting the sources to zero, whereupon we obtain a set of relations among the Euclidean n-point (Schwinger) functions of the theory, which are the analytic continuations of the multi-fermion Minkowski Green functions (T-products) of the theory. For example, applying $\frac{\delta^2}{\delta\bar{\eta}_\alpha(y)\delta\eta_\beta(z)}$, we obtain (cf. (12.142)),

$$\langle -i(\gamma_5\psi(x))_\alpha\bar{\psi}_\beta(z)\delta^4(x-y) + i(\bar{\psi}(x)\gamma_5)_\beta\psi_\alpha(y)\delta^4(x-z) + 2i\mathcal{A}(x)\psi_\alpha(y)\bar{\psi}_\beta(z)\rangle$$

$$- \frac{1}{\hbar}\langle(\partial_\mu(\bar{\psi}\hat{\gamma}_\mu\gamma_5\psi(x))) - 2im\bar{\psi}\gamma_5\psi(x))\psi_\alpha(y)\bar{\psi}_\beta(z)\rangle = 0 \tag{15.182}$$

If we analytically continue back to Minkowski space, the Euclidean correlators become T-products in the usual way, and we find the Minkowski space WT identity[12]

$$\langle 0|T\{-(\gamma_5\psi(x))_\alpha\bar{\psi}_\beta(z)\delta^4(x-y) + (\bar{\psi}(x)\gamma_5)_\beta\psi_\alpha(y)\delta^4(x-z)\}|0\rangle$$

$$- \frac{1}{\hbar}\langle 0|\frac{\partial}{\partial x^\mu}T\{\bar{\psi}\gamma^\mu\gamma_5\psi(x)\psi_\alpha(y)\bar{\psi}_\beta(z)\} - 2imT\{\bar{\psi}\gamma_5\psi(x)\psi_\alpha(y)\bar{\psi}_\beta(z)\}|0\rangle$$

$$+ 2i\mathcal{A}_M(x)\langle 0|T\{\psi_\alpha(y)\bar{\psi}_\beta(z)\}|0\rangle = 0 \tag{15.183}$$

where $\mathcal{A}_M$ is the Minkowski continuation of the functional anomaly (to be determined below). This result generalizes in the obvious way to arbitrary numbers of ψ and $\bar{\psi}$ fields, by taking the appropriate higher functional derivatives with respect to $\eta, \bar{\eta}$. As in the case of the non-anomalous WT identity (12.142), if we reduce out the fermions by taking the $\psi_\alpha(y)\bar{\psi}_\beta(z)$ fields on-shell and applying the LSZ formula, the contact terms disappear and we find that arbitrary matrix elements of the operator combination $\partial_\mu J_5^\mu(x) - 2im\bar{\psi}\gamma_5\psi(x) - 2i\hbar\mathcal{A}_M(x)$ vanish, and we therefore conclude that

$$\partial_\mu J_5^\mu(x) = 2im\bar{\psi}\gamma_5\psi(x) + 2i\hbar\mathcal{A}_M(x) \tag{15.184}$$

We see that even in the massless limit, when the mass term on the right-hand side vanishes, and the chiral symmetry (15.169) becomes an exact Noether symmetry at the classical level, there is a remaining non-zero contribution coming from the functional anomaly $\mathcal{A}_M$, the quantum origins of which are apparent in the explicit prefactor of Planck's constant multiplying the anomaly.

In order to compute the explicit form of the functional anomaly $\mathcal{A}$ (back in Euclidean space), we must recall that although we have already discretized the spectrum of $\hat{D}$ by placing the system in a finite spacetime volume, there is as yet no short-distance (or high-momentum) cutoff, and the determinant therefore involves an ill-defined product of infinitely many eigenvalues. We may regularize it in a gauge-invariant way by observing that the eigenvalues λ_i are gauge-invariant (see Problem 10), so that the inclusion of a factor $e^{-\lambda_i^2/\Lambda^2}$ in the trace in (15.179), where Λ is an ultraviolet cutoff, amounts to a smooth gauge-invariant tempering of the short-distance modes of the theory, which should be removed after evaluation of the determinant by letting the cutoff $\Lambda \to \infty$. Thus, we define the *regularized functional*

---

[12] The contact terms lose a factor of i as a consequence of the Wick rotation of the four-dimensional δ-functions, where one coordinate—the time—is rotated by a factor of i.

anomaly $\mathcal{A}_\Lambda$ as

$$\mathcal{A}_\Lambda \equiv \sum_i \phi_i^\dagger(x)\gamma_5 e^{-\lambda_i^2/\Lambda^2}\phi_i(x) = \sum_i \phi_i^\dagger(x)\gamma_5 e^{-\hat{\slashed{D}}^2/\Lambda^2}\phi_i(x) \qquad (15.185)$$

By the usual spectral analysis of self-adjoint operators, the discrete completeness sum is equivalent to one over plane wavefunctions, as for any operator $\mathcal{O}$, writing in Dirac notation $\phi_i(x) = \langle x|i\rangle$

$$\sum_i \phi_i^\dagger(x)\mathcal{O}\phi_i(x) = \sum_i \langle i|x\rangle\langle x|\mathcal{O}|i\rangle$$

$$= \langle x|\mathrm{Tr}(\mathcal{O})|x\rangle$$

$$= \int \phi_k(x)^\dagger \mathrm{Tr}(\mathcal{O})\phi_k(x)\frac{d^4k}{(2\pi)^4}, \qquad \phi_k(x) = e^{ik\cdot x} \qquad (15.186)$$

where the trace operation Tr extends over the discrete γ-matrix and internal gauge indices only. With $\mathcal{O} = e^{-\hat{\slashed{D}}^2/\Lambda^2}$ we therefore have

$$\mathcal{A}_\Lambda = \int \mathrm{Tr}(\gamma_5 e^{-ik\cdot x}e^{-\hat{\slashed{D}}^2/\Lambda^2}e^{ik\cdot x})\frac{d^4k}{(2\pi)^4} \qquad (15.187)$$

We may write the discrete trace $\mathrm{Tr} = \mathrm{tr_D}\mathrm{tr_G}$ where we explicitly separate the traces over Dirac ($\mathrm{tr_D}$) and fundamental representation gauge ($\mathrm{tr_G}$) indices. Now

$$\hat{\gamma}_\mu\hat{\gamma}_\nu = \frac{1}{2}\{\hat{\gamma}_\mu,\hat{\gamma}_\nu\} + \frac{1}{2}[\hat{\gamma}_\mu,\hat{\gamma}_\nu] = \delta_{\mu\nu} + \frac{1}{2}[\hat{\gamma}_\mu,\hat{\gamma}_\nu] \qquad (15.188)$$

whence (recall that in Euclidean space $\hat{\slashed{D}} = \hat{\gamma}_\mu D_\mu$ where $D_\mu = -i\partial_\mu + t_\alpha A_{\alpha\mu} = -i\partial_\mu + A_\mu$)

$$\hat{\slashed{D}}^2 = D_\mu D_\mu + \frac{1}{4}[\hat{\gamma}_\mu,\hat{\gamma}_\nu][D_\mu,D_\nu] = D_\mu D_\mu - \frac{i}{4}[\hat{\gamma}_\mu,\hat{\gamma}_\nu]F_{\mu\nu} \qquad (15.189)$$

The factors of $e^{-ik\cdot x}\ldots e^{ik\cdot x}$ in (15.187) merely serve to translate the covariant derivative D_μ by the four-vector k_μ,

$$e^{-ik\cdot x}D_\mu e^{ik\cdot x} = D_\mu + k_\mu \qquad (15.190)$$

from which we then see, using (15.189), that

$$e^{-ik\cdot x}e^{-\hat{\slashed{D}}^2/\Lambda^2}e^{ik\cdot x} = e^{-e^{-ik\cdot x}\hat{\slashed{D}}^2 e^{ik\cdot x}/\Lambda^2} = e^{-\frac{1}{\Lambda^2}((k_\mu+D_\mu)^2 - \frac{i}{4}[\hat{\gamma}_\mu,\hat{\gamma}_\nu]F_{\mu\nu})}$$

$$= e^{-k_\mu k_\mu/\Lambda^2}\cdot e^{-\frac{1}{\Lambda^2}(2k_\mu D_\mu + D_\mu D_\mu - \frac{i}{4}[\hat{\gamma}_\mu,\hat{\gamma}_\nu]F_{\mu\nu})} \qquad (15.191)$$

At this point we shall need some simple Dirac trace identities, which we leave to the reader to check (remembering that our Euclidean γ matrices are now all hermitian and satisfy $\mathrm{tr}_D(\hat{\gamma}_\mu \hat{\gamma}_\nu) = 4\delta_{\mu\nu})$. Specifically,

$$\mathrm{tr}_D(\gamma_5) = 0 \tag{15.192}$$

$$\mathrm{tr}_D(\gamma_5 \hat{\gamma}_\mu \hat{\gamma}_\nu) = 0 \tag{15.193}$$

$$\mathrm{tr}_D(\gamma_5 \hat{\gamma}_\mu \hat{\gamma}_\nu \hat{\gamma}_\rho \hat{\gamma}_\sigma) = 4\epsilon_{\mu\nu\rho\sigma} \tag{15.194}$$

where $\epsilon_{\mu\nu\rho\sigma}$ is the completely antisymmetric Euclidean four-tensor with $\epsilon_{1234} = 1$. It follows that when the expression (15.191) is traced with γ_5, the only terms which survive involve expanding out at least two factors of the term with $\frac{1}{\Lambda^2}[\hat{\gamma}_\mu, \hat{\gamma}_\nu]F_{\mu\nu}$ in the second exponential, in order to obtain the requisite minimum of four γ matrices needed to provide a non-vanishing trace. Pulling down additional factors, such as $(k_\mu D_\mu)^2/\Lambda^4$ or D_μ^2/Λ^2, lead to integrals over k which (together with the accompanying inverse powers of Λ) vanish in the infinite cutoff limit, for example:

$$\frac{1}{\Lambda^4} \int \frac{k^2}{\Lambda^4} e^{-k^2/\Lambda^2} \frac{d^4k}{(2\pi)^4} = \frac{1}{8\pi^2} \frac{1}{\Lambda^2} \to 0, \quad \Lambda \to \infty \tag{15.195}$$

$$\frac{1}{\Lambda^4} \int \frac{1}{\Lambda^2} e^{-k^2/\Lambda^2} \frac{d^4k}{(2\pi)^4} = \frac{1}{16\pi^2} \frac{1}{\Lambda^2} \to 0, \quad \Lambda \to \infty \tag{15.196}$$

Therefore, keeping only the term quadratic in $\frac{1}{\Lambda^2}[\hat{\gamma}_\mu, \hat{\gamma}_\nu]F_{\mu\nu}$, we find that for large Λ,

$$\mathcal{A}_\Lambda \to \mathrm{Tr}(\gamma_5 \frac{1}{2}(\frac{i}{4\Lambda^2})^2 [\hat{\gamma}_\mu, \hat{\gamma}_\nu][\hat{\gamma}_\rho, \hat{\gamma}_\sigma] F_{\mu\nu}F_{\rho\sigma}) \int e^{-k^2/\Lambda^2} \frac{d^4k}{(2\pi)^4}$$

$$= -\frac{1}{32\Lambda^4} \mathrm{tr}_D(\gamma_5 [\hat{\gamma}_\mu, \hat{\gamma}_\nu][\hat{\gamma}_\rho, \hat{\gamma}_\sigma]) \mathrm{tr}_G(F_{\mu\nu}F_{\rho\sigma}) \cdot \frac{\Lambda^4}{16\pi^2}$$

$$= -\frac{1}{16\pi^2} \mathrm{tr}_G(F_{\mu\nu}\tilde{F}_{\mu\nu}), \quad \tilde{F}_{\mu\nu} \equiv \frac{1}{2}\epsilon_{\mu\nu\rho\sigma}F_{\rho\sigma} \tag{15.197}$$

The result, gratifyingly, depends only on the field tensor $F_{\mu\nu}$, which transforms under local gauge transformations as $F_{\mu\nu} \to U(x)F_{\mu\nu}U^\dagger(x)$, so that the group trace in (15.197) is gauge-invariant, as desired. For the abelian case, we simply obtain (without the trace) a result proportional to $F_{\mu\nu}\tilde{F}_{\mu\nu}$ where $F_{\mu\nu}$ is the single field $\partial_\mu A_\nu - \partial_\nu A_\mu$. If we now rotate back to Minkowski space, a factor of $-i$ appears (in each term of $F_{\mu\nu}\tilde{F}_{\mu\nu}$ there are three space and one time indices), so the Minkowski anomaly is

$$\mathcal{A}_M = i\frac{1}{16\pi^2} \mathrm{tr}_G(F_{\mu\nu}\tilde{F}^{\mu\nu}) \tag{15.198}$$

and the axial current divergence (15.184) becomes

$$\partial_\mu J_5^\mu(x) = 2im\bar{\psi}\gamma_5\psi(x) - \hbar\frac{1}{8\pi^2} \mathrm{tr}_G(F_{\mu\nu}\tilde{F}^{\mu\nu}) \tag{15.199}$$

If we return to the canonical normalization of the gauge fields (by reversing the scaling $A_\mu \to \frac{1}{g}A_\mu$ in (15.160)), the anomaly term is seen to acquire an explicit factor of the squared coupling constant, and we obtain the usual form for the axial current divergence (setting, as usual in this book, $\hbar = 1$):

$$\partial_\mu J_5^\mu(x) = 2im\bar{\psi}\gamma_5\psi(x) - \frac{g^2}{8\pi^2}\text{tr}_G(F_{\mu\nu}\tilde{F}^{\mu\nu}) \tag{15.200}$$

We see that the coefficient of the anomaly is an exceedingly simple function of g, and can indeed be determined by a second-order calculation in perturbation theory, with no further modifications at higher order, as first demonstrated explicitly using graph-theoretic methods by Adler and Bardeen (Adler and Bardeen, 1969) (a result generally referred to as the "Adler-Bardeen non-renormalization theorem").

The chiral anomaly is of enormous importance in the physics of the Standard Model. It is at the heart of the current algebra derivation of the neutral pion decay rate to two photons (Bell and Jackiw, 1969), and of the resolution of the famous "U(1) problem", where the absence of a low-mass pseudoscalar isosinglet meson— despite the apparent prediction of such a particle via Goldstone's theorem applied to broken global chiral symmetry in QCD—is directly attributable to the anomalous breaking of chiral current conservation ('t Hooft, 1976). Later, in Part 4 of the book where we address issues of renormalizability, we shall see that quantum anomalies also place severe restrictions on the construction of theories with *local* gauge symmetry. Although our derivation of the chiral anomaly above was carried out in a gauge theory, the gauge particles were coupled to non-anomalous exactly conserved vector currents, and the anomalous axial current was associated with a global (non-gauged) symmetry of the theory. It turns out that interacting local gauge field theories require exact conservation of the associated currents (whose charges generate the symmetry transformations associated with the global restriction of the gauge group) in order to be consistent, although the local gauge symmetry may be spontaneously broken by the ground state of the theory (as we shall discuss in the next section). This requires that theories such as the electroweak component of the Standard Model, which are rife with axial currents coupled to the weak bosons of the theory, must satisfy an anomaly cancellation condition (see Problem 11) whereby the contributions to the anomaly in any gauged current from the various fermions coupled to that gauge boson are required to cancel in order to maintain consistency (and renormalizability) of the theory.

Another quantum anomaly of great importance in modern field theory is the *trace anomaly*, which arises in the divergence of the dilatation current, which we discussed briefly in the context of scalar field theories in Section 12.5. For a local gauge theory ((15.112) with a single fermion multiplet), the (appropriately improved (Freedman *et al.*, 1974)) energy-momentum tensor gives rise to a dilatation current, as in (12.117), with a non-vanishing trace even in the limit of massless fermions. In this case, in contrast to the axial anomaly, the trace anomaly receives contributions in all orders of perturbation, and a detailed graph theoretic analysis (using techniques of renormalization theory which we must defer to Part 4) gives ((Adler *et al.*, 1977), (Collins *et al.*, 1977)):

$$T^\mu{}_\mu = \frac{\beta(g)}{2g}\text{Tr}(F_{\mu\nu}F^{\mu\nu}) + (1 + \gamma(g))m\bar{\psi}\psi \tag{15.201}$$

Here the functions $\beta(g), \gamma(g)$ are well-defined functions of the renormalized coupling g which are related to the coupling and mass renormalizations of the theory and can be calculated explicitly order by order in perturbation theory. The $\beta(g)$ function in particular—the famous "β function" of the renormalization group—plays a critical role in understanding the scaling behavior of gauge theories, and will be discussed in detail later in the book. The intimate connection between the trace anomaly and the scaling behavior of interacting field theories should come as no surprise when we recall its origin in our attempt to formulate a Noether current for the classical dilatation symmetry of a Lagrangian with no dimensionful couplings (such as the Yang–Mills Lagrangian (15.112) with all fermion masses zero). The trace anomaly can also be understood at low orders (one loop) from a functional integral point of view (Fujikawa, 1981)—again, as for the chiral anomaly, the culprit is a non-trivial functional Jacobian—but it is difficult to obtain a rigorous all-orders result, as in (15.201), by this technique. Although the field theories of the Standard Model have non-vanishing $\beta(g)$ functions and are definitely not conformally (or dilatation) invariant, there are examples of supersymmetric field theories ($N{=}4$ supersymmetric Yang–Mills is the classic case) where the trace anomalies contributed by the various fields of the theory cancel and the β function *appears* to vanish to all orders of perturbation theory, suggesting an exactly conformally invariant (and even UV finite!) theory. The cautionary verb "*appears*" is used here because of the annoying fact that there is no known ultraviolet regulator which can be used to give a definite meaning to all the perturbative amplitudes of the theory while preserving both the local gauge invariance and the global supersymmetry which are essential ingredients in the formal arguments leading to the asserted conformal invariance.

As a final example of the important role played by quantum anomalies in modern particle theory, we may mention here the purely gravitational anomalies that arise in field theories in higher dimensions—in particular, in theories with local supersymmetry (*supergravity* theories). In this case, the anomalous current is the energy-momentum tensor itself! In any generally covariant theory of gravity, the graviton must couple to a covariantly conserved energy-momentum tensor arising from the matter fields, and it turns out (Alvarez-Gaumé and Witten, 1983) that the required cancellation of potential anomalies in the Ward identity expressing this conservation requires very careful choice of the fermionic representation content of the theory. The observation that the required gravitational anomaly cancellations corresponded to supergravity theories (in ten dimensions) which were the low-energy limits of a special class of superstring theories played a seminal role in the renaissance of string theory in the mid-1980s.

15.6 Spontaneous symmetry-breaking in theories with a local gauge symmetry

We saw in our study of spontaneous symmetry-breaking in Chapters 8 and 14 that while the dynamics (as specified, say, by a Lagrangian density) of a theory may possess an exact global symmetry, the energetics of the system may lead to a ground state which is not itself invariant under the symmetry transformation. If the symmetry is a continuous one, the Goldstone theorem then implies the appearance of exactly massless particles in the spectrum of the theory. In point of fact, spontaneously broken symmetries are much more prevalent than massless particles in Nature, so there must clearly exist a mechanism for avoiding the consequences of the Goldstone theorem in most cases. Sometimes, of course, the global symmetry is only approximate, so the associated Goldstone modes are merely "light" particles, rather than exactly massless ones. Such creatures are then referred to as "pseudo-Goldstone" particles.

But in the case of electroweak interactions in the Standard Model, we encounter a situation in which the spontaneous breakdown is associated with a local symmetry, corresponding to a Lagrangian which is *exactly locally gauge-invariant* but in which the vacuum (ground state) of the theory breaks the associated *global* charge. It should be emphasized that the underlying local gauge symmetry is always present, as it is simply the reflection of a redundancy in the field variables in the un-gauge-fixed Lagrangian: indeed, a famous theorem due to Elitzur (Elitzur, 1975) assures us that the vacuum-expectation-value of any non-gauge-invariant quantity always vanishes in a theory with an exact local symmetry, in the absence of gauge-fixing. Once a gauge is fixed, however, to remove the redundant degrees of freedom, the remaining (*discrete!*) global symmetry may undergo spontaneous symmetry-breaking exactly along the lines discussed in the previous chapter. The phrase "spontaneous breaking of local gauge symmetry" is therefore in some sense a misnomer, but a convenient one, if we think of it as a short circumlocution for "spontaneous breaking of remnant global symmetry after removal of redundant gauge degrees of freedom by appropriate gauge-fixing".

In the presence of local gauge symmetry, the conditions discussed in Section 14.2 for the applicability of the Goldstone theorem are not present, and instead of producing massless Goldstone particles we move in exactly the opposite direction, with the emergence of massive vector particles corresponding to gauge fields with no mass term in the Lagrangian! This remarkable phenomenon—discovered in the 1960s by Higgs (Higgs, 1964) (and independently, by several other workers), but already prefigured in the Ginzburg–Landau model of superconductivity (where the appearance of a photon "mass" underlies the exponential Meissner screening of the magnetic field in the superconducting medium)—is at the core of our present understanding of the electroweak sector of the Standard Model of elementary particles. The physical mechanism underlying the Higgs phenomenon can be completely understood in a simple abelian model (which Higgs himself used to illustrate the essential idea). We start with the Lagrangian for a complex scalar field $\phi(x)$ coupled gauge-invariantly to a vector field A_ν, with polynomial self-coupling $P(\phi^*\phi)$ for the scalar field:

$$\mathcal{L} = -\frac{1}{4}F_{\nu\rho}F^{\nu\rho} + (\partial_\nu - igA_\nu)\phi^* \cdot (\partial^\nu + igA^\nu)\phi - P(\phi^*\phi) \qquad (15.202)$$

which is clearly invariant under the local abelian transformations

$$\phi(x) \to e^{ig\Lambda(x)}\phi(x) \qquad (15.203)$$

$$\phi^*(x) \to e^{-ig\Lambda(x)}\phi^*(x) \qquad (15.204)$$

$$A_\nu(x) \to A_\nu(x) - \partial_\nu\Lambda(x) \qquad (15.205)$$

If the quadratic term in the scalar potential $P(\phi^*\phi)$ is positive,

$$P(\phi^*\phi) = +\mu^2\phi^*\phi + \lambda(\phi^*\phi)^2 \tag{15.206}$$

the vacuum state occurs for vanishing expectation value (VEV) of the scalar field, $\langle 0|\phi(x)|0\rangle = 0$, and the gauge symmetry is preserved by the vacuum. The theory then corresponds quite simply to the scalar quantum electrodynamics of a charged massive spinless particle coupled to a massless photon. If the quadratic coefficient is negative, on the other hand,

$$P(\phi^*\phi) = -\mu^2\phi^*\phi + \lambda(\phi^*\phi)^2 \tag{15.207}$$

the classical energy density is minimized for fields with magnitude $|\phi(x)| = \frac{\mu}{\sqrt{2\lambda}} \equiv v$, and we must expect the quantum scalar field to acquire a non-vanishing VEV as well, which to lowest order in the coupling is just the value v. In the absence of a coupling to the gauge field (i.e., setting $g = 0$) we would, of course, simply shift the scalar field by defining $\phi(x) = v + \hat{\phi}(x)$, and discover on rewriting the Lagrangian in terms of the shifted field that the real component of the field $\hat{\phi}_R$ possesses a sensible (positive) non-zero mass term, while the imaginary part $\hat{\phi}_I$ has no quadratic part and corresponds to the massless Goldstone mode.

For $g \neq 0$ the result is altogether different. The physical spectrum of the theory is most easily exposed in this case by employing the full—and *exact*—local gauge symmetry of the theory to rotate the complex field to a real value. Thus, writing $\phi(x) = \frac{1}{\sqrt{2}}(\phi_R(x) + i\phi_I(x))$, where ϕ_R, ϕ_I are self-conjugate (and with the canonical normalization of their kinetic terms), the gauge symmetry (15.203) can clearly be used to set $\phi_I(x) = 0$ identically. In this "unitary" gauge, the Lagrangian becomes

$$\mathcal{L} = -\frac{1}{4}F_{\nu\rho}F^{\nu\rho} + \frac{1}{2}(\partial_\nu - igA_\nu)\phi_R \cdot (\partial^\nu + igA^\nu)\phi_R - P(\frac{1}{2}\phi_R^2) \tag{15.208}$$

If we now shift the single remaining field ϕ_R by its VEV $v = \frac{\mu}{\sqrt{\lambda}}$ to reflect the appropriate VEV for the ground state

$$\phi_R(x) \equiv \frac{\mu}{\sqrt{\lambda}} + \psi_R(x) \tag{15.209}$$

the Lagrangian becomes

$$\mathcal{L} = -\frac{1}{4}F_{\nu\rho}F^{\nu\rho} + \frac{\mu^2 g^2}{2\lambda}A_\nu A^\nu + \frac{1}{2}(\partial_\nu - igA_\nu)\psi_R \cdot (\partial^\nu + igA^\nu)\psi_R - \mu^2\psi_R^2$$
$$+ \frac{\mu g^2}{\sqrt{\lambda}}A_\nu A^\nu \psi_R - \mu\sqrt{\lambda}\psi_R^3 - \frac{1}{4}\lambda\psi_R^4 \tag{15.210}$$

We recognize this as the theory of a *massive* Maxwell–Proca field A_ν, with mass (to lowest order) given by $\frac{\mu g}{\sqrt{\lambda}} = gv \equiv m_A$, and a *massive* real scalar field ψ_R, with mass (again to lowest order) given by $\sqrt{2}\mu$.[13] The remnant massive physical spin-zero particle associated with ψ_R has become known universally as the "Higgs particle". In addition to the usual scalar self-couplings, there are vertices in the theory corresponding to cubic $A_\nu A^\nu \psi_R$ and quartic $A_\nu A^\nu \psi_R^2$ interactions of the massive vector with the Higgs particle.

Note that if we count physical degrees of freedom, there is no discontinuity as we pass smoothly (by varying μ) from the symmetry-unbroken phase of theory (15.206) to the symmetry-broken phase (15.207). In the former theory we have a complex massive field with two independent modes (as the particle and antiparticle are distinct) and a massless spin-1 vector field, with two independent polarizations; in the latter, a massive spin-1 field corresponding to a particle with three independent polarizations and a self-conjugate spin-zero Higgs particle (one degree of freedom). One sometimes hears this transition described with the rather colorful language: "the massless gauge particle, by eating the would-be Goldstone boson induced by symmetry-breaking, converts itself into a massive vector particle!" This is really somewhat misleading: we are in either the symmetry unbroken phase, with a massless vector and massive complex scalar, or in the broken phase, where our initial choice of field variables in the Lagrangian (15.202) really corresponds to a misidentification of the correct, in this case completely massive, physical degrees of freedom of the theory.

Finally, we should note here, in anticipation of our discussions of renormalizability in Part 4, that the quantization of this abelian Higgs theory in unitary gauge, as given by the Lagrangian (15.210), gives rise to ultraviolet singularities in the (off-shell) Green functions of the theory which *cannot* be renormalized (i.e., absorbed into redefinitions of Lagrangian parameters). The potential for problems of this sort becomes apparent when we recall that the momentum-space propagator of our massive vector field contains a numerator factor $g^{\mu\nu} - k^\mu k^\nu / m_A^2$ (cf. Chapter 7, Problem 8) which leads (when divided by the $k^2 - m_A^2$ denominator factor) to a propagator of order unity at large momentum, with the consequent appearance of uncurable ultraviolet divergences in the perturbative expansion of the Green functions. We shall see below how this problem can be cured by a different choice of gauge, which of course does not alter the S-matrix, but provides a smooth (and ultraviolet convergent) off-shell extension of the Green functions of the theory.

Before proceeding to the general functional quantization (in renormalizable gauge) of a spontaneously broken local gauge theory, a less trivial example involving a non-abelian field may help to concretize some of the features which we should expect to appear in the general case. The theory in question is the Weinberg–Salam model of leptonic electroweak theory, which forms (with QCD) one of the two legs on which the modern Standard Model of particle physics stands. Let us take the local gauge group to be $\mathrm{SU}(2)\times\mathrm{U}(1)$, corresponding to a theory with four gauge mesons, which we will label $A_{\alpha\nu}, \alpha = 1, 2, 3$ or simply $\vec{A}_\nu$ (associated with the SU(2) group) and B_ν

---

[13] The quantization of this massive vector field can be carried out explicitly along the lines of Problem 4, Chapter 12. From the point of view of Dirac Hamiltonian theory, the primary constraint $\Pi_0 = 0$ gives rise to a secondary constraint (equation of motion for A_0) which contains the combination $m_A^2 A_0 + \vec{\nabla} \cdot \vec{\Pi}$, which has non-vanishing Poisson brackets with Π_0: i.e., we have a pair of *second-class constraints*. This is therefore a theory without the first-class constraints characteristic of a gauge theory—not surprisingly, as we have eliminated the gauge symmetry by a choice of gauge.

(associated with the abelian U(1) group). The $\vec{A}_\nu$ gauge fields are coupled with charge g under SU(2) to a single complex doublet of scalar fields $\phi_n, n = 1, 2$, which couples with U(1) charge g' to the B_ν field. Thus, the Lagrangian for the gauge and scalar fields takes the form

$$\mathcal{L} = -\frac{1}{4}\vec{F}_{\nu\rho} \cdot \vec{F}^{\nu\rho} - \frac{1}{4}G_{\nu\rho}G^{\nu\rho} - P(\phi)$$

$$+ [(\partial_\nu - i\frac{g'}{2}B_\nu - i\frac{g}{2}\vec{\tau} \cdot \vec{A}_\nu)\phi]^\dagger(\partial^\nu - i\frac{g'}{2}B^\nu - i\frac{g}{2}\vec{\tau} \cdot \vec{A}^\nu)\phi \qquad (15.211)$$

$$F_\alpha^{\nu\rho} = \partial^\nu A_\alpha^\rho - \partial^\rho A_\alpha^\nu + g\epsilon_{\alpha\beta\gamma}A_\beta^\nu A_\gamma^\rho, \quad G^{\nu\rho} = \partial^\nu B^\rho - \partial^\rho B^\nu \qquad (15.212)$$

$$P(\phi) = -\mu^2\phi^\dagger\phi + \lambda(\phi^\dagger\phi)^2 \qquad (15.213)$$

where $\vec{\tau}$ are the Pauli matrices, and we have adopted the conventional coupling sign and normalizations (involving a change of sign relative to (15.202) and a factor of $\frac{1}{2}$). Again, the physical spectrum is most readily revealed in unitary gauge, so we use the local SU(2) gauge freedom to rotate the scalar doublet field to eliminate the upper component and the imaginary part of the lower component, leaving only the real part of the lower component, which is then shifted to remove (at lowest order of perturbation theory) the VEV associated with the minimum of $P(\phi)$ at $|\phi| = \frac{\mu}{\sqrt{2\lambda}} \equiv \frac{1}{\sqrt{2}}v$:

$$\phi(x) = \begin{pmatrix} 0 \\ \frac{1}{\sqrt{2}}(v + H(x)) \end{pmatrix}$$

The single remaining self-conjugate scalar field $H(x)$ interpolates for the famous, but as yet undiscovered,[14] *Higgs particle* of the Standard Model. The vacuum expectation value of the scalar doublet

$$< \phi > = \begin{pmatrix} 0 \\ \frac{1}{\sqrt{2}}v \end{pmatrix}$$

generates in the scalar kinetic term in (15.211), as in the Higgs abelian model discussed previously, a mass term for the four vector bosons of the theory, in this case involving a squared mass matrix

$$\frac{1}{2}M_{\alpha\beta}^2 = < \phi^\dagger > T_\alpha T_\beta < \phi >, \quad \alpha, \beta = 1, 2, 3, Y \qquad (15.214)$$

where we have combined the four generators of SU(2)$\times$U(1) in a single notation, with the Y index referring to the *weak hypercharge* abelian $U(1)$ subgroup. Thus (using Y also to indicate the value of the "hypercharge" associated with the U(1) subgroup,

---

[14] As this book goes to press, there are intriguing indications at the Large Hadron Collider at CERN (Geneva, Switzerland) of a possible Higgs signal at a mass of approximately 125 GeV.

which must be assigned separately to the various field multiplets in the theory)

$$T_i = \frac{g}{2}\tau_i, i = 1, 2, 3, \quad T_Y = \frac{g'}{2}Y \left(= \frac{g'}{2} \text{ for } \phi\right), \quad [T_i, T_j] = ig\epsilon_{ijk}T_k, \quad [T_i, Y] = 0$$
$$(15.215)$$

The vector mass matrix separates into two uncoupled sectors, with the $(\alpha, \beta) = 1, 2$ sector giving

$$\frac{1}{2}M^2 = \frac{1}{8}g^2v^2 \begin{pmatrix} 1 & -i \\ i & 1 \end{pmatrix} \tag{15.216}$$

corresponding to a mass term

$$\frac{1}{4}g^2v^2 \cdot \frac{1}{\sqrt{2}}(A_{1\nu} - iA_{2\nu})^\dagger \cdot \frac{1}{\sqrt{2}}(A_1^\nu - iA_2^\nu) = M_W^2 W_\nu^\dagger W^\nu, \quad M_W^2 = \frac{1}{4}g^2v^2 \quad (15.217)$$

where we have defined a complex massive vector field $W_\nu = \frac{1}{\sqrt{2}}(A_{1\nu} - iA_{2\nu})$ with mass $gv/2$. In the 3-Y subspace we have the 2x2 squared mass matrix

$$\frac{1}{2}M^2 = \frac{1}{8}v^2 \begin{pmatrix} g^2 & -gg' \\ -gg' & g'^2 \end{pmatrix} \tag{15.218}$$

corresponding to a mass term

$$\frac{1}{8}v^2(g'B_\nu - gA_{3\nu})^2 + 0 \cdot (gB_\nu + g'A_{3\nu})^2 = \frac{1}{2}M_Z^2 Z_\nu Z^\nu \tag{15.219}$$

where the following field combinations (satisfying conventionally normalized canonical equal-time commutation relations if the A_3 and B fields do) have been defined

$$Z_\nu \equiv \frac{g'B_\nu - gA_{3\nu}}{\sqrt{g^2 + g'^2}} \tag{15.220}$$

$$A_\nu \equiv \frac{gB_\nu + g'A_{3\nu}}{\sqrt{g^2 + g'^2}} \tag{15.221}$$

where the self-conjugate field Z_ν has mass $m_Z = \frac{1}{2}v\sqrt{g^2 + g'^2}$, while the A_ν field is massless. The existence of a zero mode in the mass matrix (15.214) is clearly associated with the existence of a linear combination of generators $\frac{1}{2}(\tau_3 + Y) = \frac{1}{g}T_3 + \frac{1}{g'}T_Y$ which annihilates the VEV of the scalar doublet (which has $Y = 1$):

$$\frac{1}{2}(\tau_3 + 1) \begin{pmatrix} 0 \\ \frac{1}{\sqrt{2}}v \end{pmatrix} = 0$$

Thus there is an unbroken U(1) subgroup of the original SU(2)×U(1) local gauge symmetry, which must be associated with a massless gauge particle. This is, of course, the photon, in the modern electroweak theory. One may easily check that the $W, W^\dagger$ and Z fields transform under the generator $\frac{1}{g}T_3 + \frac{1}{g'}T_Y = \frac{1}{2}(\tau_3 + Y) \equiv Q$ as fields of electric charge -1, $+1$, and 0 respectively. The discovery in 1983 of a neutral massive vector Z boson in the weak interactions (in addition to the long-suspected charged weak carriers $W^{\pm}$) was a dramatic confirmation that the particular pattern of local symmetry-breaking described here indeed conforms to reality. Of course, the real value of such a model lies in its ability to accurately depict the weak interactions of the fundamental fermions of the theory: the leptons and quarks. We shall briefly describe the leptonic sector of the electroweak theory here, as proposed in Weinberg's seminal paper (Weinberg, 1967), before going on to discuss the functional quantization and derivation of Feynman rules for a general spontaneously broken local gauge theory.

The electroweak sector of the Standard Model is a *chiral gauge theory*: that is to say, left- and right-handed parts of the Dirac fermion fields of the theory (which the reader will recall from Chapter 7, fall into separate representations of the proper homogeneous Lorentz group) are placed in different representations of the gauge group. Thus, if ψ is a Dirac 4-spinor field, $\psi_L = P_L\psi = \frac{1+\gamma_5}{2}\psi$ and $\psi_R = P_R\psi = \frac{1-\gamma_5}{2}\psi$ are the left-handed and right-handed 2-spinor components of ψ respectively. This means that ψ_L and ψ_R may be in SU(2) multiplets of different dimensionality, and may be assigned different weak hypercharge quantum numbers Y_L and Y_R. One recovers the conventional $V - A$ structure of the charged weak currents by placing the left-handed part of the electron field e_L together with the purely left-handed Weyl electron neutrino field in a SU(2) doublet L_e (with weak hypercharge $Y_L = -1$),

$$
L_e(x) = \begin{pmatrix} \nu_e(x) \\ e_L(x) \end{pmatrix}
$$

and the right-handed part of the electron field e_R in a SU(2) singlet field R_e, with weak hypercharge $Y_R = -2$. We may also think of this chiral arrangement as corresponding to the inclusion of γ_5 factors (via chiral projection operators P_L, P_R) in the gauge group generators,

$$
T_i = \frac{g}{2}\tau_i P_L, \quad T_Y = \frac{g'}{2}(Y_L P_L + Y_R P_R) \tag{15.222}
$$

which, of course, satisfy the commutation relations (15.215). The charge operator then becomes $Q = \frac{1}{2}(\tau_3 - 1)P_L - P_R$, giving electric charge -1 to both components e_L and e_R of the electron field, and zero charge to the neutrino, as desired. The fermionic (leptonic) part of the Lagrangian, with these choices, becomes

$$
\mathcal{L}_{\text{lept}} = \bar{L}_e(i\partial\!\!\!/ - \frac{1}{2}g'\not\!B + \frac{1}{2}g\vec{\tau}\cdot\vec{A}\!\!\!/)L_e + \bar{R}_e(i\partial\!\!\!/ - g'\not\!B)R_e \tag{15.223}
$$

Note that there is so far no mass term for the electron field, as a direct coupling of the left- and right-handed parts of the electron field would violate the SU(2) symmetry. When the $\vec{A}_\nu$ and B_ν fields are rewritten in terms of the physical W_ν, Z_ν, A_ν fields, one recovers (see Problem 13), in addition to the long known $V - A$ structure for the charged weak currents (mediated by the W fields), a new set of neutral weak current interactions due to the massive Z boson, as well as, of course, conventional quantum electrodynamics for the interaction of the electron and photon fields. The muon and tau leptons (with their associated neutrinos) may be included by essentially "xeroxing" the structure above twice. Masses arise naturally in this model for the charged leptons once

Yukawa interactions, exactly invariant under the local gauge symmetry, are included between the leptons and the scalar field doublet:

$$\mathcal{L}_{\text{Yuk}} = -G_e\{\bar{L}_e \phi R_e + \bar{R}_e \phi^\dagger L_e\} \tag{15.224}$$

If we recall that the scalar field ϕ lies in a SU(2) doublet with weak hypercharge 1, with L_e and R_e having weak hypercharges –1 and –2 respectively, we see that the cubic Yukawa coupling here is invariant under both the SU(2) and U(1) parts of the gauge group. Moreover, once the field is shifted to extract the vacuum expectation value, a mass term $-G_e(\bar{e}_L \frac{v}{\sqrt{2}} e_R + \bar{e}_R \frac{v}{\sqrt{2}} e_L) = -m_e \bar{e}e, \ \ m_e = G_e \frac{v}{\sqrt{2}}$ emerges automatically for the electron. Muon and tau masses emerge similarly: they involve completely independent Yukawa couplings G_μ, G_τ, so we cannot expect any obvious connection between the charged lepton masses (on the basis of symmetry requirements), although, of course, the wide disparity (as yet, completely mysterious) of these masses is at the very least aesthetically disturbing.

The presence of γ_5 factors in the fermionic generators of our chiral SU(2)×U(1) gauge theory should alert us to the possibility of anomalies, and indeed the conservation of the Noether gauge currents of the purely leptonic electroweak theory described here is broken by quantum anomalies, which would render the theory non-renormalizable, and even prevent the execution of the functional quantization process to be described below (where we assume the absence of any non-trivial functional Jacobians in the functional integral). It is an extraordinary—and highly suggestive—feature of the electroweak theory that the quantum anomalies in the gauge currents are exactly cancelled once quark fields (in one-to-one correspondence with the lepton fields) are introduced with the appropriate quantum numbers (see Problem 11).

We now turn to the long-promised derivation of the Feynman rules for a spontaneously broken gauge theory. We shall emphasize the derivation of the propagators of the theory, as the possibility of obtaining a renormalizable theory hinges most directly on the high-momentum behavior of the propagators: in particular, we wish to show that the disastrous asymptotic behavior (of order $k^\mu k^\nu/k^2$) of the massive vector boson propagator in a unitary gauge can be removed by a choice of gauge which both (a) maintains manifest Lorentz covariance, and (b) damps the high-momentum behavior to the same level as that of a scalar propagator: i.e., $1/k^2$. Complete details of the derivation of the Feynman rules in broken gauge theories can be found in the classic articles by Weinberg (Weinberg, 1973) and Abers and Lee (Abers and Lee, 1973).

We begin by slightly altering the notation used in the examples discussed above: the generator matrices will now *not* contain factors of the coupling constant, and we return to our original sign conventions for the coupling(s), as incorporated in (15.113–15.115). We shall assume that our scalar field multiplets consist of purely real fields (we can, of course, always decompose a complex scalar field into two real fields by writing $\phi = \frac{1}{\sqrt{2}}(\phi_R + i\phi_I)$), with the generator matrices T_α real and antisymmetric, so that the covariant derivative on the scalar fields reads

$$\mathcal{D}^\mu \phi = (\partial^\mu - gT_\alpha A_\alpha^\mu)\phi, \quad [T_\alpha, T_\beta] = f_{\alpha\beta\gamma}T_\gamma \tag{15.225}$$

Note that, as in electroweak theory, the coupling g may vary from one simple subgroup of the full local gauge group G to another: to avoid overcomplicating the notation, we shall avoid indicating this explicitly below. The fermions fill, as usual, complex (but possibly chiral) representations of G, and we use, as previously, hermitian generators t_α in the fermionic representations, with covariant derivative

$$D^\mu \psi = \partial^\mu + i g t_\alpha A_\alpha^\mu \qquad (15.226)$$

Under the infinitesimal local gauge transformations

$$\phi(x) \to (1 + g T_\alpha \lambda_\alpha(x))\phi(x) \qquad (15.227)$$

$$\psi(x) \to (1 - i g t_\alpha \lambda_\alpha(x))\psi(x) \qquad (15.228)$$

$$A_\alpha^\mu(x) \to A_\alpha^\mu(x) + \partial^\mu \lambda_\alpha(x) + g f_{\alpha\beta\gamma} \lambda_\beta(x) A_\gamma^\mu(x) \qquad (15.229)$$

the Lagrangian

$$\mathcal{L} = -\frac{1}{4} F_{\alpha\mu\nu} F_\alpha^{\mu\nu} + \frac{1}{2} (\mathcal{D}_\mu \phi)^T \mathcal{D}^\mu \phi + \bar\psi (i\slashed{D} - m)\psi - \bar\psi \Gamma_i \psi \phi_i - P(\phi) \qquad (15.230)$$

is invariant, *provided* the fermion mass (matrix) m commutes with the generators, $[t_\alpha, m] = 0$, and the Yukawa couplings and the scalar polynomial are appropriately chosen: namely,

$$[t_\alpha, \Gamma_i] = -i(T_\alpha)_{ij} \Gamma_j \qquad (15.231)$$

$$\frac{\partial P}{\partial \phi_i} (T_\alpha)_{ij} \phi_j = 0 \qquad (15.232)$$

We shall now suppose that spontaneous breaking of the gauge symmetry occurs, induced by the presence of a non-trivial minimum of the scalar potential $P(\phi)$, and the appearance of a non-vanishing vacuum expectation value v_i for ϕ_i (at lowest order in the coupling)

$$\left. \frac{\partial P}{\partial \phi} \right|_{\phi_i = v_i \neq 0} = 0 \quad \Rightarrow \quad \langle 0|\phi_i(x)|0\rangle = v_i \qquad (15.233)$$

The vacuum expectation value will be removed in the usual fashion by defining a shifted field $\phi_i' \equiv \phi_i - v_i$, so that the action of an infinitesimal gauge transformation on the scalar and vector fields becomes

$$\phi(x) \to (1 + g T_\beta \lambda_\beta(x))\phi(x)$$

$$\Rightarrow \phi'(x) \to \phi'(x) + g T_\beta \lambda_\beta(x)(v + \phi'(x))$$

$$A_{\alpha\mu}(x) \to A_{\alpha\mu}(x) + \partial_\mu \lambda_\alpha(x) + g f_{\alpha\beta\gamma} \lambda_\beta(x) A_{\gamma\mu}(x) \qquad (15.234)$$

We shall now impose a gauge condition as a joint constraint on the gauge and scalar fields of the theory. The form of the constraint is at first sight rather peculiar, but will shortly be seen to give an algebraically convenient set of Feynman rules. We impose the local gauge condition

$$\partial^\mu A_{\alpha\mu}(x) - \xi g < v, T_\alpha \phi'(x) >= f_\alpha(x), \quad < v, T_\alpha \phi' >\equiv v_i (T_\alpha)_{ij} \phi_j' \qquad (15.235)$$

where ξ is an arbitrary positive real number, and we have introduced an obvious bracket notation to indicate dot-products with respect to the scalar field indices. The $f_\alpha(x)$ are arbitrary real functions, so in the absence of symmetry-breaking $(v = 0)$ our gauge choice reduces to the generalized Landau gauge used earlier (cf. (15.146)). Following steps analogous to those leading to (15.148, 15.149), we introduce the covariant DFP functional associated with this choice of gauge

$$1 = \Delta_{\text{cov}}[A, \phi'] \cdot \int \mathbf{D}U \delta(\partial^\mu A^U_{\alpha\mu} - \xi g < v, T_\alpha(\phi')^U > -f_\alpha)$$

$$\Delta_{\text{cov}}[A, \phi'] = \det(\Box \delta_{\alpha\beta} + g f_{\alpha\beta\gamma} \partial^\mu A_{\gamma\mu} - \xi g^2 < v, T_\alpha T_\beta v > -\xi g^2 < v, T_\alpha T_\beta \phi' >)$$

$$(15.236)$$

which, as previously (15.155, 15.156), can be represented more conveniently by introducing Grassmannian ghost fields ω_α with a ghost Lagrangian

$$\mathcal{L}_{\text{gh}} = \bar{\omega}_\alpha(\Box \delta_{\alpha\beta} + g f_{\alpha\beta\gamma} \partial^\mu A_{\gamma\mu} + \xi g^2 < T_\alpha v, T_\beta v > -\xi g^2 < v, T_\alpha T_\beta \phi' >)\omega_\beta$$

$$(15.237)$$

One notes here the appearance of (a) a ghost mass matrix $\xi g^2 < T_\alpha v, T_\beta v >$ and, (b) in addition to the ghost-vector vertex, a ghost-scalar coupling term. Precisely as in the unbroken case, one may establish that the generating functional of the theory is independent of the choice of the arbitrary functions f_α, which we may therefore integrate over, with a Gaussian modulating factor as in (15.153), to obtain the path integral (minus sources) for our spontaneously broken gauge theory:

$$Z_{SBGT} = \int \int \mathbf{D}A_{\alpha\mu} \mathbf{D}\psi \mathbf{D}\bar{\psi} \mathbf{D}\omega_\alpha \mathbf{D}\bar{\omega}_\alpha \mathbf{D}\phi' e^{i \int \mathcal{L}_{\text{tot}} d^4 x}$$

$$\mathcal{L}_{\text{tot}} = -\frac{1}{4} F_{\alpha\mu\nu} F^{\mu\nu}_\alpha - \frac{1}{2\xi}(\partial^\mu A_{\alpha\mu} - \xi g < v, T_\alpha \phi' >)^2$$

$$+ \frac{1}{2}(\mathcal{D}_\mu(v + \phi'))^2 - P(v + \phi') + \mathcal{L}_{\text{ferm}} + \mathcal{L}_{\text{gh}} \qquad (15.238)$$

The utility of the peculiar choice of gauge condition (15.235) now becomes apparent on examining the scalar field kinetic term,

$$\frac{1}{2}(\mathcal{D}_\mu(v + \phi'))^2 = \frac{1}{2}(\mathcal{D}_\mu \phi')^2 + \frac{g^2}{2} < T_\alpha v, T_\beta v > A_{\alpha\mu} A^\mu_\beta$$

$$- \frac{1}{2} < g T_\alpha v, (\partial_\mu - g T_\beta A_{\beta\mu})\phi' > A^\mu_\alpha + \frac{1}{2} < (\partial^\mu - \frac{1}{2} g T_\alpha A^\mu_\alpha)\phi', g T_\beta v > A_{\beta\mu}$$

$$= \frac{1}{2} < \mathcal{D}_\mu \phi', \mathcal{D}^\mu \phi' > + \frac{1}{2} M^2_{\alpha\beta} A_{\alpha\mu} A^\mu_\beta + g^2 A^\mu_\alpha A_{\beta\mu} < T_\alpha v, T_\beta \phi' > -g A^\mu_\alpha < T_\alpha v, \partial_\mu \phi' >$$

$$(15.239)$$

which evidently generates a non-trivial (but ξ-independent!) squared-mass matrix for the gauge vector bosons (cf. (15.214))

$$M^2_{\alpha\beta} \equiv g^2 < T_\alpha v, T_\beta v > \qquad (15.240)$$

The final term in (15.239), mixing the scalar and gauge fields, combines with the cross-term from the gauge-fixing part of the total Lagrangian to produce a total derivative, which then vanishes after integration over spacetime (recall that the T_α matrices are real antisymmetric):

$$-\frac{1}{2\xi}(-2\xi g \partial^\mu A_{\alpha\mu} <v, T_\alpha \phi'>) - gA_\alpha^\mu <T_\alpha v, \partial_\mu \phi'> = \partial^\mu(gA_{\alpha\mu}<v, T_\alpha \phi'>) \tag{15.241}$$

The propagators of the theory are associated with the parts of $\mathcal{L}_{\text{tot}}$ quadratic in the various fields, and now that unwanted mixing terms have been eliminated, these can easily be read off from the quadratic scalar, gauge, and fermion Lagrangians:

$$\mathcal{L}_{\text{scal}}^{\text{quad}} = \frac{1}{2}(\partial_\mu \phi_i')^2 - \frac{1}{2}\frac{\partial^2 P}{\partial \phi_i \partial \phi_j}(\phi = v)\phi_i' \phi_j' - \frac{\xi g^2}{2}(<v, T_\alpha \phi'>)^2 \tag{15.242}$$

$$\mathcal{L}_{\text{gauge}}^{\text{quad}} = -\frac{1}{4}(\partial_\mu A_{\alpha\nu} - \partial_\nu A_{\alpha\mu})^2 - \frac{1}{2\xi}(\partial_\mu A_\alpha^\mu)^2 + \frac{1}{2}M_{\alpha\beta}^2 A_{\alpha\mu}A_\alpha^\mu$$

$$\rightarrow \frac{1}{2}A_{\alpha\mu}((\Box\delta_{\alpha\beta} + M_{\alpha\beta}^2)g^{\mu\nu} + \delta_{\alpha\beta}(\frac{1}{\xi}-1)\partial^\mu \partial^\nu)A_{\beta\nu} \tag{15.243}$$

$$\mathcal{L}_{\text{ferm}}^{\text{quad}} = \bar{\psi}(i\not{D} - M_f)\psi, \quad M_f = m + \Gamma_i v_i \tag{15.244}$$

where the arrow in (15.243) refers to a rearrangement of the derivatives via an integration by parts. The squared-mass matrix for the scalar fields is easily read off from (15.242):

$$M_{ij}'^2 = \frac{\partial^2 P}{\partial \phi_i \partial \phi_j}(\phi = v) + \frac{1}{2}\xi g^2 (T_\alpha v)_i (T_\alpha v)_j \tag{15.245}$$

which is evidently dependent on the arbitrary gauge parameter ξ. The physical interpretation of this disconcerting feature is best seen if we choose a basis of the gauge group generators $T_\alpha, \alpha = 1, \ldots, N$ such that the first m generators span the unbroken subgroup, while $T_\alpha v \neq 0$ for $\alpha = m+1, .., N$ correspond to the broken directions (and, in the limit where the gauge coupling vanishes, to Goldstone bosons). We saw in the simple models examined earlier that the field redefinitions implicit in the unitary gauge effectively absorb the Goldstone scalar modes into the longitudinal parts of the massive gauge fields, so it should not be surprising to find that the "mass-matrix" for these modes is gauge-dependent, and linked to the behavior of the longitudinal part of the gauge vector propagator, which we shall shortly see is also ξ-dependent. The first term on the right-hand side of (15.245), on the other hand, is gauge-independent, and corresponds to the physical scalar modes. Indeed, this mass matrix contributes exactly zero in the Goldstone mode directions, as we see by differentiating the invariance condition (15.232) and setting $\phi(x) = v$,

$$\frac{\partial^2 P}{\partial \phi_k \partial \phi_i}(T_\alpha)_{ij}\phi_j + \frac{\partial P}{\partial \phi_i}(T_\alpha)_{ik} = 0 \Rightarrow \frac{\partial^2 P}{\partial \phi_k \partial \phi_i}(\phi = v)(T_\alpha v)_i = 0 \tag{15.246}$$

The gauge vector propagator can be read off easily from (15.243): we need the Green function for the operator $(\Box + M^2)g^{\mu\nu} + (\frac{1}{\xi} - 1)\partial^\mu\partial^\nu$, which the reader will easily verify corresponds to a Feynman propagator given by

$$\langle 0|T(A_{\alpha\mu}(x)A_{\beta\nu}(y)|0\rangle = -i \int \frac{d^4k}{(2\pi)^4} \left(\frac{g_{\mu\nu} - (1-\xi)\frac{k_\mu k_\nu}{k^2 - \xi M^2}}{k^2 - M^2 + i\epsilon}\right)_{\alpha\beta} e^{-ik\cdot(x-y)} \quad (15.247)$$

with the squared-mass matrix M^2 (carrying the α, β indices) given by (15.240). Notice that the ξ-dependence is entirely in the longitudinal part (proportional to $k_\mu k_\nu$) of the momentum-space propagator. The poles at $k^2 = \xi M^2$ cannot correspond to physical particle states as they depend on the arbitrary gauge parameter ξ: indeed, we know that the S-matrix is independent of ξ and therefore cannot have any such poles in single-particle cuts of amplitudes. However, these poles occur at exactly the same mass eigenvalues as those given by the Goldstone mode part of the scalar propagator, the second term on the right-hand side of (15.245). Indeed, suppose that ρ_i is an eigenvector of the latter matrix:

$$\xi g^2(T_\alpha v)_i (T_\alpha v)_j \rho_j = \lambda \rho_i \quad (15.248)$$

Multiplying both sides by $(T_\beta v)_i$, summing over i, and defining $\chi_\alpha \equiv (T_\alpha v)_j \rho_j$, we find

$$\xi M^2_{\beta\alpha} \chi_\alpha = \lambda \chi_\beta \quad (15.249)$$

This means that the unphysical poles in the longitudinal part of the gauge propagators can (indeed, by gauge invariance, *must*) be cancelled by poles at exactly the same locations in the scalar propagators. The Feynman $\xi = 1$ gauge choice gives a particularly simple momentum-space propagator, proportional to $g_{\mu\nu}$. Of course, as a check on more complicated higher-order perturbative calculations, it may be useful to retain the general form to ensure that all ξ-dependent terms cancel in the final physical result (see Problem 14).

The large momentum behavior of the gauge vector and scalar propagators is uniformly $1/k^2$ regardless of the value of the gauge parameter,[15] and whether or not we are in the symmetry broken ($M^2 \neq 0$) or unbroken ($M^2 = 0$) phase of the theory. This is in complete consonance with the intuition developed from our discussions of spontaneous symmetry-breaking in Chapter 14, as a phenomenon linked to the large-distance energetic properties of the theory, but essentially decoupled from the short-distance, or large momentum, properties of the amplitudes. The soft behavior of the vector propagators in these *renormalizable ξ-gauges* plays a crucial role in establishing the renormalizability of a spontaneously broken gauge theory with massive spin-1 particles, as we shall see in Part 4.

The discovery of the renormalizable SU(2)×U(1) gauge field theory for the weak and electromagnetic interactions in the early 1970s was followed in short order by

---

[15] The limit $\xi \to \infty$ is singular, and we see that we recover the unitary gauge momentum-space propagator in this limit, with the numerator factor $g_{\mu\nu} - k_\mu k_\nu/M^2$ characteristic of a massive Maxwell–Proca field, and with a propagator of order unity, rather than $1/k^2$ at large momentum.

the proposal of a similar gauge theory, *quantum chromodynamics* (QCD), as the underlying dynamical structure for the strong interactions. Earlier indications of a triplet structure for the fermionic "quark" constituents of hadrons were incorporated naturally by assuming that the gauge group in this case was SU(3), with the quark fields transforming in the fundamental three-dimensional representation. There were no available candidates for massive vector bosons transforming in an octet of the color group, so the gauge symmetry in this case was presumed to be exact, with the quark and gluon fields combining to give color-singlet interpolating fields for the physical hadronic particles. The absence of physical particles associated with the non-color-singlet fields of the theory was dubbed the "quark confinement hypothesis" (although a more accurate term would be "color confinement"). This remarkable phenomenon—in the author's view, matched only in twentieth-century physics by the exotic phenomena associated with superconductivity and superfluidity—is a consequence of the strongly coupled character of unbroken four-dimensional Yang–Mills theories in the infrared (at long distances), which we shall examine in detail in the final chapter of the book.

The "Standard Model" of modern particle physics is based on a Lagrangian field theory encompassing the interactions of leptons and quarks via gauge interactions associated with an exact local SU(3)×SU(2)×U(1) gauge theory, with a spontaneously broken SU(2) subgroup, in the simplest case by an elementary Higgs doublet field, as discussed above.[16] Remarkably, quark and lepton fields appear in the Standard Model (see Problem 11) with exactly the right quantum numbers to assure that none of the gauge currents receive anomalous contributions, via "miraculous" cancellations between the quark and lepton fields in each generation. We know, however, that this model cannot be complete, for (at least) two very convincing reasons: (i) the existence of non-zero neutrino masses, and (ii) the apparent existence of a massive *stable* weakly interacting dark-matter particle, which cannot be identified with any of the fields in the Standard Model. Needless to say, speculative extensions of the Standard Model abound. The possibility of unifying the three simple groups into a single larger group (SU(5), SO(10),..; see Problem 12), broken at a high-energy scale ($\sim 10^{15}$ GeV) to the gauge symmetries of the Standard Model, led early on to the development of Grand Unified Theories (GUTs), in which quarks and leptons appear in the same representation of the gauge group, thereby rendering less mysterious the cancellation of anomalies between seemingly unrelated quark and lepton representations. Another obvious extension of the Standard Model lies in the direction of supersymmetry: the minimal supersymmetric extension of the Standard Model (MSSM) is an attempt to build a supersymmetric theory with the minimal number of additional fields (super-partners to the conventional Standard Model fields). The breaking of supersymmetry is expected for reasons connected with the renormalization of the Higgs mass to appear at the now accessible TeV scale, and the search for supersymmetric partners of the Standard Model particles with masses in this range is a subject of intense interest at

---

[16] For an excellent and comprehensive introduction to the phenomenology of the Standard Model, see (Donoghue *et al.*, 1992).

the Large Hadron Collider (LHC), which should reach total energies of 10 or more TeV in the center-of-mass frame within the next few years.

15.7 Problems

1. The object of this exercise is to verify that the change in fermionic Green functions induced by a local gauge transformation in an abelian gauge theory does not affect the on-shell S-matrix amplitudes of the theory, as given by the LSZ formula. Suppose there is a single incoming (resp. outgoing) fermion carrying momentum p (resp. p'). Show that the *change* in the S-matrix amplitude induced by the gauge transformation $\Lambda(x)$ is proportional to

$$\int d^4x d^4x' \; e^{ip'\cdot x' - ip\cdot x} \big(e^{ie(\Lambda(x')-\Lambda(x))} - 1\big)$$

$$\cdot \; \bar{u}(p')(\not{p}'- m)\langle 0|T(\psi_H(x')\bar{\psi}_H(x)\ldots)|0\rangle(\not{p} - m)u(p) \quad (15.250)$$

where we have suppressed spin labels and the gauge transformation function $\Lambda(x)$ is assumed to go to zero rapidly at large x. The change (15.250) will vanish if the poles in the Fourier transform of the T-product of the form $1/(\not{p}'- m), 1/(\not{p} - m)$ are absent. Show that these poles are indeed absent by demonstrating that the Fourier transform $f(q', q)$ of $f(x, x') = e^{ie(\Lambda(x')-\Lambda(x))} - 1$ takes the form

$$\tilde{f}(q', q) = D(q')F(q) + \delta^4(q)F'(q') \quad (15.251)$$

where the functions $F(q)$ (resp. $F'(q')$) are smooth—and in particular, do not contain δ-functions $\delta^4(q)$ (resp. $\delta^4(q')$). Accordingly, when the Fourier transform of the T-product is convolved with $\tilde{f}$, the external propagator poles are smeared out and the external spinors $u(p), \bar{u}(p')$ are annihilated by the LSZ $(\not{p} - m), (\not{p}'- m)$ factors.

2. Verify the infinitesimal form of the non-abelian transformation rules (15.113–15.115), starting from (15.96, 15.105, 15.109).

3. Check that the Euler–Lagrange equation for $A_{\alpha 0}$ reduces to the non-abelian Gauss's Law (15.118), with the covariant derivative defined in (15.119).

4. Show that the non-abelian Hamiltonian density $\Pi^\mu_\alpha \dot{A}_{\alpha\mu} - \mathcal{L}_{YM}$ leads to the total Hamiltonian (keeping only the gauge fields) given in (15.125).

5. (a) Show that (for fields vanishing sufficiently fast at infinity) the axial gauge choice $A_{\alpha 3} = 0$ can be imposed with a *unique* choice of non-abelian gauge transformation.

   (b) Show that the Faddeev–Popov gauge functional $\Delta_{\mathrm{axial}}[A]$ for a non-abelian gauge theory is constant (i.e., field-independent) in axial gauge $A^3_\alpha = 0$.

6. Show that the Hurwitz integral for SU(2) (defined as the solid-angle integral over the 4-sphere) is shift-invariant:

$$\int dU f(UV) = \int dU f(U), \quad U, V \in SU(2) \quad (15.252)$$

7. Verify the expression (15.149) for the DFP functional in the covariant gauge $\partial^\mu A_{a\mu}(x) = f_a(x)$.

8. By examining the quadratic (in gauge fields) part of the action in the generating functional (15.156), show that the gauge field propagator in the covariant ξ-gauge is a Green function for the operator $\Box g^{\mu\nu} + (\frac{1}{\xi} - 1)\partial^\mu\partial^\nu$, and is given by (15.159).

9. By considering the lowest-order tree expressions for the three- and four-point functions for gauge vector scattering in momentum space, verify the Feynman vertex factors given in Fig. 15.1, parts (a) and (b).

10. Show that the eigenvalues λ_i of the Euclidean Dirac operator $\hat{D}[A] = \hat{\gamma}^\mu(-i\partial_\mu + A_\mu)$ are gauge-invariant: namely, show that if $\hat{D}[A]\phi_i(x) = \lambda_i\phi_i(x)$, then

$$\hat{D}[A^U]U(x)\phi_i(x) = \lambda_i U(x)\phi_i(x), \quad A_\mu^U = UA_\mu U^\dagger + i(\partial_\mu U)U^\dagger \qquad (15.253)$$

11. The appearance of an anomaly in the currents associated with a local gauge symmetry would destroy (at the quantum level) the local gauge symmetry of the theory, with dire consequences for the renormalizability of the theory, as we shall see in Part 4. In a chiral theory such as the electroweak sector of the Standard Model, the appearance of γ_5 factors in the generators of the local gauge group (due to the fact that left and right handed fermionic fields transform differently under the gauge group) signal the potential existence of such anomalies. Now suppose that the generators are decomposed into right and left handed parts (cf. (15.222)), $T_\alpha = t_\alpha^L P_L + t_\alpha^R P_R, P_L = \frac{1+\gamma_5}{2}, P_R = \frac{1-\gamma_5}{2}$, where now the t_α do not contain γ_5 factors. It can be shown (see (Donoghue *et al.*, 1992), for example) that the anomaly in the current $\bar{\psi}\gamma^\mu T_\alpha\psi$ is proportional to $\epsilon_{\mu\nu\rho\sigma}F_\beta^{\mu\nu}F_\gamma^{\rho\sigma}$ times a difference of traces over the left- and right-handed fields:

$$A_\alpha \propto \text{Tr}(t_\alpha^L\{t_\beta^L, t_\gamma^L\}) - \text{Tr}(t_\alpha^R\{t_\beta^R, t_\gamma^R\}) \qquad (15.254)$$

The cancellation of anomalies in the SU(2)×U(1) electroweak theory occurs via a magical cancellation between leptons and quarks in each generation. For the lowest generation, we have in addition to the electron and electron-neutrino fields discussed above, a left-handed quark doublet $(u_L(x), d_L(x))$ and right-handed singlets $u_R(x), d_R(x)$ under SU(2), with hypercharge assignments $\frac{1}{3}, \frac{1}{3}, \frac{4}{3}, -\frac{2}{3}$ for u_L, d_L, u_R, d_R respectively, in order to arrive at the desired electric charges $+\frac{2}{3}$ for the up quark and $-\frac{1}{3}$ for the down quark. The anticommutator of two Pauli matrices is a multiple of the identity and the trace of a single Pauli matrix vanishes, so we see immediately that the anomaly arising from three SU(2) indices $\alpha, \beta\gamma$ vanishes. The choice of one U(1) with two $SU(2)$ indices is easily seen to give an anomaly proportional to $\sum Y_L$, the sum of hypercharge quantum numbers for the left-handed-fields, while taking all three indices in the U(1) subgroup gives an anomaly proportional to $\sum Y_L^3 - \sum Y_R^3$. Show that, with the hypercharge assignments given above, both of these potential anomalies vanish. Remember that each quark field comes in 3 color versions, corresponding to the quantum numbers under the SU(3) color gauge group of the strong interactions, so the quark contributions need to be multiplied by a factor of 3.

12. Consider a broken gauge theory based on the gauge group G= SU(5). The symmetry-breaking is implemented by coupling the twenty-four-dimensional adjoint representation of gauge bosons to a real twenty-four-dimensional Higgs scalar representation, which can be conveniently represented as a traceless hermitian 5x5 matrix Φ, with $\Phi \rightarrow U^\dagger \Phi U$ giving the action of the group (where U is a 5x5 unitary matrix of determinant 1). The most general scalar polynomial of degree ≤ 4 symmetric under G is

$$P(\Phi) = a\text{Tr}(\Phi^2) + b(\text{Tr}(\Phi^2))^2 + c\text{Tr}(\Phi^4)$$

Show that for some range of parameters a, b, c a symmetry-breaking minimum of P is achieved where the ground-state vacuum expectation value of Φ is (for constant real α)

$$\langle 0|\Phi|0\rangle = \begin{pmatrix} 2\alpha & 0 & 0 & 0 & 0 \\ 0 & 2\alpha & 0 & 0 & 0 \\ 0 & 0 & 2\alpha & 0 & 0 \\ 0 & 0 & 0 & -3\alpha & 0 \\ 0 & 0 & 0 & 0 & -3\alpha \end{pmatrix}$$

What is the remaining unbroken symmetry group in this case? Find the spectrum of vector mesons in the symmetry-broken phase.

13. Work out the form of the neutral leptonic current sector (electron generation only) of the electroweak SU(2)×U(1) theory, by extracting the interaction terms in (15.223) containing the photon (A_μ) and Z-boson (Z_μ) fields. It is conventional to introduce the *Weinberg angle* θ_W, with $\frac{g'}{g} = \tan \theta_W$, so

$$A_{3\mu} = \sin (\theta_W) A_\mu - \cos (\theta_W) Z_\mu$$
$$B_\mu = \cos (\theta_W) A_\mu + \sin (\theta_W) Z_\mu \tag{15.255}$$

Show that the leptonic interactions involving these neutral fields take the form

$$\mathcal{L}_{\text{lept,neut}} = -e\,\bar{e}(x)\gamma^\mu e(x) A_\mu - e \tan (\theta_W)\bar{e}_R(x)\gamma^\mu e_R(x) Z_\mu$$
$$- \frac{e}{\sin (2\theta_W)}\bar{\nu}_e(x)\gamma^\mu \nu_e(x) Z_\mu + e \cot (2\theta_W)\bar{e}_L(x)\gamma^\mu e_L(x) Z_\mu$$
$$\tag{15.256}$$

14. The presence of neutral weak currents mediated by a massive Z meson necessitates the choice of SU(2)xU(1) as the electroweak gauge group. The basic QED annihilation process $e^+ + e^- \rightarrow \mu^+ + \mu^-$ now requires, in addition to the usual graph with an intermediate virtual photon, inclusion of a graph with an intermediate Z boson.

(a) Assuming the mass of the electron vanishes, show that the tree amplitude for this process is ξ-independent.

(b) Show that the presence of the Z graph leads to a forward-backward asymmetry in the process (i.e., terms linear in $\cos(\theta)$ in the center-of-mass differential cross-section).

(c) Show that the tree amplitude from the above two graphs is *not* ξ-independent if the electron and muon masses are not neglected. Explain what other graph or graphs have to be taken into account in this case to restore a gauge-invariant result.

16

Scales I: Scale sensitivity of field theory amplitudes and effective field theories

The history of the natural sciences since the late 1800s (if we temporarily set aside astronomy as primarily concerned with Nature "in the large") has to a great extent involved an attempt to decipher the behavior of matter at ever smaller distance scales. Qualitative descriptions of biological organisms at the macroscopic level have been supplanted by an astonishingly detailed understanding of the underlying biochemistry of life; the complex profusion of chemical phenomena revealed empirically by nineteenth-century and early-twentieth-century chemists are now understood to follow, in many cases with the detailed quantitative support of sophisticated algorithms of quantum chemistry, from a precise mathematical formulation based on Schrödinger's equation applied to atoms and molecules; the phenomenology of nuclei (fission, fusion, radioactivity, etc.) has been reduced to the behavior of more "elementary" constituents (quarks and gluons) obeying precise dynamical laws; and so on.

It is apparent from these examples that the *type* of theory or descriptive framework appropriate for the description of the *same* phenomena at different levels of "magnification" can vary enormously. Intricate details of the underlying microscopic dynamics of a physical process may simply be irrelevant to achieving an adequate "qualitative" understanding of the process as viewed at larger distance scales. Much of nuclear physics can be understood perfectly well by treating the proton and neutron as point-like fermions interacting non-relativistically via spin-dependent short-range potentials, with absolutely no understanding of the underlying non-abelian local gauge theory giving rise to these particles and their interactions.

A remarkable property of local quantum field theory, not shared by any of the larger-scale phenomenologies mentioned above (which we now, of course, believe to be consequences of the underlying field-theoretic phenomena), is that it is structurally amenable to a *precise mathematical description* of the way in which the form of the dynamical laws changes as the phenomena are examined at varying distance-scales. As we shall shortly see, the representation of the theory at a given distance- (or energy-)scale will turn out to be specified by an "effective Lagrangian" fixed in terms of an infinite vector of dimensionless couplings, and the variation of these couplings with the distance scale determined by a "renormalization group equation" which is in essence the infinitesimal Lie algebra corresponding to the "renormalization group" transformations associated with finite changes of the scale at which we examine the theory. In this chapter we shall begin the task of exploring these remarkable features of local quantum field theory.

16.1 Scale separation as a precondition for theoretical science

In our discussion of cluster decomposition in chapter 6, we emphasized the critical importance of the decoupling of amplitudes of far separated processes for the viability of experimental science. If this decoupling did not occur, the correct interpretation of the results of experiments carried out in some localized region of spacetime would be contingent on a specification of the state of the Universe far beyond the boundaries of the laboratory—a state of knowledge about our surroundings which we clearly can never possess. This (fortunate) insensitivity to our lack of knowledge of the state of the world at large distances from our experiments must be mirrored by a corresponding insensitivity of the phenomena to details of the dynamics at short distances which are, given the incomplete state of our theoretical knowledge, necessarily unknown at any given moment in the historical enterprise that constitutes the physical sciences. The source of this ignorance in elementary particle physics derives from the simple fact that short-distance details of the dynamics are associated (via the magic of Fourier transformation) with the behavior of the quantum amplitudes of the theory at high energy, and at any given point in history we have access to man-made accelerators with a limited energy range.[1] Thus, even the projected maximum center-of-mass energy (14 TeV) of the recently initiated LHC (Large Hadron Collider) in Geneva corresponds to an ability to probe the structure of elementary processes down to distances of about 10^{-5} fermis ($= 10^{-20}$ m). As small as this seems from a human perspective, it is still roughly 10^{15} times larger than the Planck scale of 10^{-20} fermis where we know that (at the very latest) the effects of quantum gravity may no longer be ignored and the nature of the quantum dynamics of elementary processes must undergo a dramatic, and as yet completely mysterious, alteration.

The last point raised above leads us to an interesting conclusion: the concept of an exact continuum limit for a local quantum field theory formulated on a flat Minkowski spacetime background (as we have done throughout our discussion of field theory to this point) has strictly speaking no correlate in physical reality. The Minkowski metric of special relativity—which, after all, underlies the most characteristic aspect of relativistic field theory, the space-like commutativity of local observables—is at best an emergent phenomenon, an approximate representation of an underlying theory in which kinematical and dynamical aspects are perhaps inextricably intertwined. Even the most ambitious attempts to come to grips with quantum gravity—as in M-theory, for example—seem to presuppose some sort of pre-existing spacetime background in which the elementary entities of the theory (be they strings, membranes, or what have you) move and interact.

---

[1] The existence of very-high-energy cosmic rays offers us a tantalizing, but unfortunately very narrow, window into physics at much higher energies than those accessible in accelerators, as do indirect cosmological arguments involving the very early Universe, but the vast majority of our detailed information about subatomic dynamics derives from the much more precise information gleaned from terrestrial accelerator experiments.

It is perfectly clear, however, that the unknown complexities of a final "Theory of Everything" (if such exists) cannot really be relevant in describing the physics of processes at energy scales much lower than those at which quantum gravity effects become important. This is not merely wishful thinking on our part: the extraordinary quantitative successes of quantum electrodynamics (QED), as in the anomalous magnetic moment of the electron for example (in complete quantitative agreement with experiment up to nine significant digits), assure us that the decoupling of the *effective low-energy physics* represented by a local abelian gauge theory of interacting electrons and photons from the unknown physics which must inevitably supersede QED at sufficiently high energy (or short distances) is not only present but extraordinarily efficacious. We might term this decoupling the "scale separation" property of local field theory: it lies at the heart of our ability to construct theoretical systems of finite complexity but adequate accuracy in the description of the "low"-energy physics which lies within our purview at any given moment of time (where "low" is, of course, a function of time). Our primary objective in this fourth and final part of the book will be to understand the origin, and consequences, of this fortunate, and remarkable, property of local quantum field theory. The adjective "remarkable" is hardly an exaggeration when we consider that scale separation breaks down completely in the arena of classical physics involving chaotic phenomena, where very small perturbations at small scales can rapidly propagate to much larger ones, essentially destroying our capacity to maintain accurate control over the temporal dynamics of many classical systems over long time-periods. The fact that quantum field theory is able to avoid this fate goes back ultimately, of course, to the linear character (and unitary temporal evolution) of the underlying quantum dynamical framework.

16.2   General structure of local effective Lagrangians

The demands of cluster decomposition—the requirement that quantum scattering amplitudes for far separated processes factorize in such a way as to ensure the statistical independence of spatially distant phenomena—were our first constraints in constructing the framework of local quantum field theory in Chapter 6. We saw there that the interaction Hamiltonian for any quantum system of self-interacting spinless particles satisfying the cluster decomposition principle necessarily must take the form (cf. (6.85))

$$\mathcal{H}_{int}(x) = \sum_{MM'} \frac{1}{M!M'!} \int \frac{d^3k_1'}{\sqrt{2E(k_1')}} .. \frac{d^3k_M}{\sqrt{2E(k_M)}}$$

$$\cdot f_{MM'}(k_1',..,k_M) e^{ik_1' \cdot x} a^\dagger(k_1')..e^{-ik_M \cdot x} a(k_M) \qquad (16.1)$$

where the functions $f_{MM'}(k_1',..,k_M)$ are smooth functions of momenta, expandable in joint Taylor expansions in the four-momenta $k_1',..k_{M'}', k_1,\ldots,k_M$. It will be convenient to rewrite this expression in terms of the positive- and negative-frequency components of the associated (for simplicity, self-conjugate) scalar field

$$\phi(x) = \frac{1}{(2\pi)^{3/2}} \int \frac{d^3k}{\sqrt{2E(k)}} (a(\vec{k})e^{-ik\cdot x} + a^\dagger(\vec{k})e^{ik\cdot x}) \equiv \phi^{(+)}(x) + \phi^{(-)}(x) \qquad (16.2)$$

as follows,

$$
\mathcal{H}_{\text{int}}(x) = \sum_{MM'} \frac{(2\pi)^{\frac{3}{2}(M+M')}}{M!M'!} f_{MM'}\left(-i\frac{\partial}{\partial x'_1}, \ldots, +i\frac{\partial}{\partial x_M}\right)\phi^{(-)}(x'_1)\cdots\phi^{(+)}(x_M)\Bigg|_{x_i=x'_i=x}
$$

(16.3)

where the spacetime-derivatives in the functions $f_{MM'}$ are converted to the appropriate momentum dependences if we insert the definition of the scalar field (16.2) in (16.1). We see that the interaction Hamiltonian density is necessarily an (infinite) expansion involving multi-nomials in the scalar field and all possible spacetime-derivatives thereof. At this stage, we have not yet inserted the demands of special relativity. The discussion in Chapter 12 reveals the appropriate further constraints which the as yet unspecified functions $f_{MM'}$ for this Hamiltonian must satisfy to yield relativistically invariant scatttering amplitudes: they must arise by the standard canonical procedure whereby the interaction Hamiltonian is derived via Legendre transformation of a *Lorentz scalar* Lagrangian density $\mathcal{L}(\phi, \partial_\nu\phi, \partial_\nu\partial_\rho\phi, \ldots)$, where now we must allow the Lagrangian to contain arbitrarily many derivatives and powers of the local scalar field $\phi(x) = \phi^{(+)}(x) + \phi^{(-)}(x)$ (with positive and negative frequency parts of the field paired throughout to satisfy the demands of locality) in order to obtain the general expansion for the (interaction) Hamiltonian density as indicated in (16.3). A Lagrangian of this type, containing effectively all possible terms consistent with cluster decomposition and Lorentz-invariance, is sometimes called a "Wilsonian effective Lagrangian",[2] in order to reflect the profound contributions made in the understanding of the scale sensitivity of local theories in the early 1970s by Ken Wilson (Wilson, 1971).

The reader with prior acquaintance with standard treatments of perturbative quantum field theory may object to the use of a scalar Lagrangian with "non-renormalizable terms" of higher than (mass) dimension 4 (in four spacetime dimensions) which are well known to lead to ultraviolet (large momentum) divergences in the loop integrals of perturbation theory which are not removable via the usual processes of mass, coupling, and wavefunction renormalization of the bare parameters of the theory. We shall return to the whole matter of perturbative renormalizability, and to its relation with the Wilsonian approach, in the next chapter. For now, this objection provides us with the opportunity to fully realize, and put into effect, the qualitative insights of the preceding section concerning the inescapable limitations on any local theory formulated on a flat Minkowski spacetime due to the unavoidable dissolution of this kinematic structure at very short distances (or large momenta). We therefore admit frankly that our Lagrangian field theory, with its associated path integral, must be

---

[2] At this point we should alert the reader to a dangerous source of terminological confusion. The use of the word "effective" in this chapter will be completely restricted to the sense indicated here, where we imagine writing an *exact* representation of *only part* of the physical content of the theory, basically by a change of variable in the functional integral defining the theory at short distances. The notion of an "effective action" Γ, as used in Chapter 10 in reference to the generating functional of the one-particle-irreducible n-point functions of the theory, plays no role here, and to avoid confusion with the aforesaid Γ we shall try to stick to the phrase "effective Lagrangian", while avoiding the perfectly natural term "effective action" for the spacetime integral thereof.

interpreted as a theory describing a field where the large momentum components of the field are *cut off*, either in some smooth, but largely arbitrary fashion (reflecting our ignorance of the ultimate underlying microphysics), or more simply, by a sharp cutoff which eliminates (sets to zero) Fourier modes of the field above some limiting value.

We shall want to ensure that the cutoff is imposed in such a way as to correspond to short distances both in the spatial and temporal directions of spacetime, and also to respect the Lorentz-invariance of the remaining "low-energy" (or "large-distance") theory. This is best done in Euclidean space, where the momentum modes of the theory can be divided cleanly into ultraviolet modes with $|k| = \sqrt{k \cdot k} > \Lambda$ (where k is a *Euclidean* four-vector), and infrared modes with $|k| < \Lambda$, with Λ a high-energy cutoff which the reader may assume for the time being to be several orders of magnitude below the Planck scale $\Lambda_{\text{Pl}} \sim 10^{19}$ GeV (to ensure that Minkowski space notions are still reasonable for the modes below this value) but many orders of magnitude higher than the energies presently accessible in accelerator experiments. Thus the scalar field $\phi(x)$ appearing in the infinite expansion specifying our Lagrangian should really be written with a subscript specifying the energy scale up to which it possesses Fourier modes:

$$\phi_\Lambda(x) = \int_{|k|<\Lambda} \tilde{\phi}(k) e^{-ik \cdot x} \frac{d^4 k}{(2\pi)^4} = \int \tilde{\phi}_\Lambda(k) e^{-ik \cdot x} \frac{d^4 k}{(2\pi)^4}, \quad \tilde{\phi}_\Lambda(k) \equiv \theta(\Lambda - |k|)\tilde{\phi}(k)$$

$$(16.4)$$

We emphasize that this cutoff *does not destroy Lorentz-invariance*, which in Euclidean space amounts to a four-dimensional O(4) rotation which clearly does not mix the ultraviolet and infrared modes. Thus we still expect that the cutoff theory should lead (after analytic continuation back from Euclidean space) to Lorentz-invariant low-energy amplitudes provided that the Euclidean effective Lagrangian is constructed to be invariant under Euclidean rotations (which implement the HLG for the Euclidean version of the theory).

The perturbation theory based on an effective Lagrangian built from the field $\phi_\Lambda(x)$ will lead to loop integrals in which the propagators $\Delta_\Lambda(k) = \langle \tilde{\phi}_\Lambda(k) \tilde{\phi}_\Lambda(-k) \rangle$ automatically vanish for $|k| > \Lambda$, so all such integrals (no matter how many the numerator factors of momentum arising from vertices with multiple derivatives on the fields) are by fiat ultraviolet finite. Of course, the amplitudes computed in this way will clearly depend on the scale Λ at which the effective Lagrangian is defined, and we must still hope that after further examination, this sensitivity to the short-distance structure of the cutoff theory will disappear when we restrict our attention to processes at momentum scales much smaller than Λ.

To summarize, our theory of interacting scalar particles is to be regarded as defined in the usual way (in Euclidean space) by a generating functional Z_Λ of Euclidean Schwinger functions given by the path integral

$$Z_\Lambda[j] = \int \mathbf{D}\phi_\Lambda e^{-S_E(\phi_\Lambda) + \int d^4 x j(x)\phi_\Lambda(x)}$$

$$(16.5)$$

where the Euclidean action S_E is given as an integral over a Lagrangian density containing all possible powers of the field and its derivatives. It is convenient (though not of great physical significance) to simplify the algebra by imposing a symmetry under $\phi \to -\phi$ restricting us to even powers of the field,

$$S_E(\phi_\Lambda) = \int d^4x \{ \frac{1}{2}(\partial_\nu \phi_\Lambda)^2 + \sum_{n \geq 0} a_n \phi_\Lambda^{2+n} + \sum_{n > 0} a'_n (\partial_\nu \phi_\Lambda)^2 \phi_\Lambda^n + \ldots \} \qquad (16.6)$$

$$\equiv \int d^4x \{ \frac{1}{2}(\partial_\nu \phi_\Lambda)^2 + \sum_{n \geq 0} a_n \mathcal{O}_n + \sum_{n > 0} a'_n \mathcal{O}'_n + \ldots \} = \int \mathcal{L}_\Lambda(\phi_\Lambda) d^4x$$

$$(16.7)$$

where the dots represent terms with a total of 4, 6, 8, etc., spacetime-derivatives acting on the fields (coupled, of course, to an overall Lorentz scalar).

We have used our freedom to rescale the field to set the coefficient of the free kinetic term $(\partial_\nu \phi_\Lambda)^2$ to be exactly $\frac{1}{2}$. The mass term is now concealed in the term $a_0 \phi_\Lambda^2$, while the coefficient a_2 corresponds to the usual dimensionless quartic coupling constant λ in our previous discussions of $\lambda \phi^4$ theory. Now, however, we have an infinite series of additional interaction terms (note: n is even), corresponding to new four-point vertices arising from the derivative coupling $(\partial_\nu \phi_\Lambda)^2 \phi_\Lambda^2$, six-point vertices from $\phi_\Lambda^6, (\partial_\nu \phi_\Lambda)^2 \phi_\Lambda^4$, and so on. Recalling that the action S_E in (16.5) must be dimensionless, implying mass dimension of 1 (from the kinetic term) for the field ϕ_Λ, we see that the coupling constants a_n (resp. a'_n) must have mass dimension $2 - n$ (resp. $-n$). It will be convenient to rescale these couplings in terms of dimensionless ones by extracting the appropriate powers of the cutoff (itself of mass dimension 1):

$$a_n \equiv g_n \Lambda^{2-n}, \quad n = 0, 2, 4, \ldots$$

$$a'_n \equiv g'_n \Lambda^{-n}, \quad n = 2, 4, 6, \ldots \qquad (16.8)$$

For the present, we shall be assuming that our theory is *weakly coupled*—in other words that the dimensionless couplings $g_n, g'_n, \ldots$ corresponding to interaction terms (i.e., those higher than quadratic in the field) are all of order unity, or perhaps somewhat smaller,[3] in which case a formal asymptotic expansion in these variables becomes quantitatively useful.

16.3 Scaling properties of effective Lagrangians: relevant, marginal, and irrelevant operators

We can begin to expose the physical content of the effective Lagrangian formulation of our theory by exploring the relative importance of the various terms in the action as a function of the energy scale of the phenomena under study. A convenient starting point is provided by the tree (or classical) approximation to the amplitudes of the theory. We recall from the discussion in Section 10.4 that the amplitudes of a local

---

[3] The concept of "order unity" possesses a somewhat elastic connotation in field theory, as it is not always obvious what the relevant expansion variable ought to be. The fine-structure constant $\alpha = e^2/4\pi = 1/137$ seems to be two orders of magnitude smaller than "order unity", but the electric charge $e \sim 0.3$ is clearly much closer to unity. Nevertheless, for many amplitudes in QED, an expansion in powers of α is appropriate, in the sense that the coefficients of powers of α, at low orders of perturbation theory, are fairly close to 1.

field theory can be expanded formally in powers of Planck's constant $\hbar$, with the lowest-order contributions corresponding to the tree (no-loop) graphs of the theory, the one-loop graphs contributing with one extra power of $\hbar$, the two-loop graphs with two extra powers, and so on. We can now ask about the perturbative contributions of the operators $\mathcal{O}_n, \mathcal{O}'_n$ appearing in the general action (16.7) to some n-point function of the theory, where we assume that the incoming and outgoing momenta of the process under consideration are all of order $E \ll \Lambda$. The only dimensionful scales present at the tree graph level are the energy scale of the process E (which permeates the internal propagators) and the UV cutoff Λ, with the dependence on the latter arising only from the explicit dependence of the couplings in (16.8) on Λ: in particular, at tree level there are no loop integrals extending up to and cut off at Λ to introduce further Λ-dependence. This means that the contribution of a particular operator at the energy scale E can be estimated by a trivial dimensional argument, essentially by just counting the mass dimension of the operator (integrated over spacetime), whence

$$\int d^4 x \mathcal{O}_n \equiv \int d^4 x \phi_\Lambda^{2+n} \sim E^{n-2} \tag{16.9}$$

$$\int d^4 x \mathcal{O}'_n \equiv \int d^4 x (\partial_\nu \phi_\Lambda)^2 \phi_\Lambda^n \sim E^n, \quad \text{etc.} \tag{16.10}$$

Including the coupling constants in (16.8) we see that these operators contribute to tree amplitudes at the relative order

$$\mathcal{O}_n \to g_n \Big(\frac{E}{\Lambda}\Big)^{n-2}, \quad n = 0, 2, 4, \ldots \tag{16.11}$$

$$\mathcal{O}'_n \to g'_n \Big(\frac{E}{\Lambda}\Big)^n, \quad n = 2, 4, 6, \ldots \tag{16.12}$$

and so on for the higher operators. This means that for the infrared physics with which we are concerned, where the energy E of the processes we are studying is much smaller than the ultraviolet cutoff Λ of the theory, the most important, or *relevant*, operator is $\mathcal{O}_0 = \phi^2$, the mass operator, whose effects grow quadratically as we lower the energy. This is hardly surprising if we consider the mass expansion of the free (Euclidean) propagator

$$\frac{1}{k^2 + m^2} \sim \frac{1}{k^2} - \frac{m^2}{k^4} + \frac{m^4}{k^6} + \ldots \tag{16.13}$$

where we see that increasing powers of the mass correspond to larger and larger contributions in the infrared region $k \ll m$. The quartic coupling operator $\mathcal{O}_2$, by contrast, contributes equally at all energy scales (in the tree amplitudes): it is therefore termed a "marginal" operator, which we should regard as a technical designation of its scaling behavior, and not (given its importance in generating non-trivial interactions) as a demeaning comment on its importance for the theory! Higher-dimension operators such as $\mathcal{O}_4 = \phi^6$ and $\mathcal{O}'_2 = (\partial_\nu \phi)^2 \phi^2$ (both of mass dimension 6 and contributing at order $(E/\Lambda)^2$ for $E \ll \Lambda$) contribute at a progressively smaller level to the low-energy

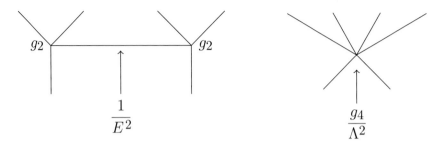

Fig. 16.1 Some tree graph contributions to the 2-4 scalar scattering amplitude.

physics, the higher their dimension, and are termed "irrelevant", from the point of view of tree amplitude scaling.

A simple example is given by the tree graphs displayed in Fig. 16.1, representing contributions to the 2-4 scattering amplitude—the first graph arising from the quartic coupling $g_2\phi^4$ in second order (and of order g_2^2/E^2, where the incoming and outgoing momenta are of order E) while the second graph, arising from the higher-dimension term $\frac{g_4}{\Lambda^2}\phi^6$ is very small, of order $\frac{E^2}{\Lambda^2}$ relative to the first (assuming all dimensionless couplings of order unity, or in any event much closer to unity than the ratio $\frac{E^2}{\Lambda^2}$), in agreement with the scaling deduced previously in (16.11).

The reader should once again guard against attaching the colloquial meaning of terms such as "irrelevant" to the physics generated by the corresponding operators: the dimension-six four-fermion operator of Fermi weak interaction theory, responsible for β-decay, for example, is "irrelevant" from this point of view, but the associated vast phenomenology of radioactivity is hardly so. A higher-dimension operator may initiate processes with very low amplitude (hence, rare processes), which may, however, be of a sufficiently different type from the processes induced by marginal or relevant operators as to stand out phenomenologically, and even to play an important role in uncovering details of the physics emerging at shorter distances (as in the case of the electroweak theory supplanting the Fermi theory of weak interactions). For reasons that will become clear in the next chapter, the classification into "relevant", "marginal", and "irrelevant" operators (of mass dimension <4, 4, and >4 respectively, in four spacetime dimensions) is mirrored in the terminology of renormalization theory by the terms "super-renormalizable", "strictly renormalizable", and "non-renormalizable", respectively.

When loop effects are included, the situation becomes more complicated, and much more interesting. If we take 2-2 scattering as a test case, the graphs in Fig. 16.2 illustrate some simple low-order contributions to the process: the lowest-order tree graph corresponding to the quartic coupling $g_2\phi^4$ (we will drop the Λ subscript on the fields here with the reminder that it simply instructs us to cut off the momenta on all internal propagators at $|k| = \Lambda$), the three one-loop graphs arising at second order in g_2, and the one-loop graph coming from the first-order contribution of the dimension 6 "irrelevant" operator $\frac{g_4}{\Lambda^2}\phi^6$. Setting $m^2 = 2a_0 = 2g_0\Lambda^2$, we shall assume that the momenta in the process and the unperturbed mass m are all much smaller

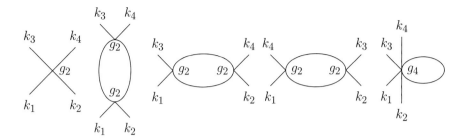

Fig. 16.2 Some tree and one-loop contributions to the 2-2 scalar scattering amplitude.

than the cutoff Λ. The final graph, arising from contracting two out of the six lines emerging from the six-point vertex associated with the higher-dimension ϕ^6 operator, contains the cutoff one-loop integral

$$\int \theta(\Lambda^2 - k^2) \frac{1}{k^2 + m^2} \frac{d^4k}{(2\pi)^4} = \frac{1}{8\pi^2} \int_0^\Lambda \frac{k^3}{k^2 + m^2} dk$$

$$= \frac{1}{16\pi^2} (\Lambda^2 - m^2 \ln \frac{\Lambda^2}{m^2}) + O(\frac{m^2}{\Lambda^2}) \qquad (16.14)$$

The loop integral (which in the absence of a cutoff would be quadratically divergent) therefore produces a large factor proportional to the cutoff squared, which will cancel the inverse factor of Λ^2 (in the coupling $\frac{g_2}{\Lambda^2}$) which we have previously used to argue for the "irrelevance" of the ϕ^6 operator at low energies. Of course, the result is just a momentum-independent constant contribution to the amplitude, of exactly the same form as the tree contribution proportional to g_2. In fact, a short calculation (see Problem 1) gives the following result for the truncated four-point function arising from the graphs in Fig. 16.2 (normalized to begin with g_2)

$$\Gamma^{(4)}(k_1, k_2, k_3, k_4) = g_2 - \frac{3}{4\pi^2} g_2^2 \{\mathcal{I}(s, m^2, \Lambda^2) + \mathcal{I}(t, m^2, \Lambda^2) + \mathcal{I}(u, m^2, \Lambda^2)\}$$

$$+ \frac{15}{16\pi^2} g_4 + O(\frac{m^2, k_i^2}{\Lambda^2}) \qquad (16.15)$$

$$\mathcal{I}(p^2, m^2, \Lambda^2) \equiv \int_0^1 (\ln(\frac{\Lambda^2}{x(1-x)p^2 + m^2}) - 1) dx \qquad (16.16)$$

$$s \equiv (k_1 + k_2)^2, \quad t \equiv (k_1 - k_3)^2, \quad u \equiv (k_1 - k_4)^2 \qquad (16.17)$$

We have already encountered the one-loop integral $\mathcal{I}$ in Chapter 10 (in Minkowski space, and with a slightly different notation, cf. (10.35)), where we pointed out that in a continuum theory it is necessarily ultraviolet-divergent, corresponding in coordinate space to an ill-defined multiplication of distributions at the same point: here, the divergence is explicit in the $\ln(\Lambda^2)$ factor, which blows up if we force $\Lambda \to \infty$. As our theory is cut off, momentum integrals terminate at Λ, and there are *no ultraviolet divergences at any point*. However, we see that the "large" integration range from the low-energy regime with momenta of order $k_i \sim E << \Lambda$ up to the UV cutoff Λ produces loop integrals which can promote the size of the contributions induced by the "irrelevant" operators (which only correct tree amplitudes by small amounts proportional to inverse powers of Λ) to values comparable to the marginal or relevant operators of dimension 4 or less. Indeed, the final term in (16.15) shows that, assuming that the dimensionless couplings $g_2, g_4, ..$ are of order unity[4], the momentum modes of the field between E and Λ, when integrated out in the path integral, produce order unity modifications (arising from higher-dimension operators) in the effective four-point coupling strength at the low-energy scale. This occurs simply because the explicit inverse powers of the cutoff in the coupling factors (16.8) can be cancelled by positive powers of the cutoff arising from loop-integrals containing vertices corresponding to these higher-dimension operators.

The "filtering down" effect from higher- to lower-dimension operators may seem fairly innocuous for the marginal couplings such as g_2, but it implies much more dramatic consequences for the coefficient of the *relevant* operators such as the mass term $g_0 \Lambda^2 \phi^2$. A classic example is given by the Higgs mass: if the Higgs turns out be described by an elementary scalar field, it must have a mass in the range of a few hundred GeV, to avoid violating the fairly precise agreement obtained between calculated electroweak radiative corrections and well-measured Standard Model weak processes. At first, this suggests that we assign a spectacularly small value to the dimensionless coupling g_0, of order $m_{\mathrm{Higgs}}^2 / \Lambda_{\mathrm{Pl}}^2 \sim 10^{-34}$ (using the Planck scale as our ultraviolet cutoff)! However, even if the dimensionless coupling is set to this value at some high-energy scale, the effect of integrating out field modes from this scale down to the energy scale of the electroweak theory will inevitably produce corrections of order unity, bringing the Higgs mass back up to the range of the UV cutoff, unless the value of the "bare" coupling g_0 at the UV scale is set with extraordinary precision. This is the famous fine-tuning issue associated with the "hierarchy" problem for scalar masses, which are not protected from large (i.e., power-like in the cutoff) radiative corrections, unlike, as we shall see later, fermion masses in gauge theories.

The instability of relevant operators to infection with large radiative corrections from large energy scales is, of course, worse the *lower* the dimension of the operator: consider, for example, the unit operator $\mathcal{O}_{-2} \equiv \phi^0 = 1$, of dimension zero, corresponding to an overall additive constant in the Lagrangian density (and to the zero-point energy in the Hamiltonian density), which has been ignored by our discussion so far, as it corresponds to a physically irrelevant additive shift in the energy in flat Minkowski space. If we imagine coupling our theory to gravity, such a term becomes physically relevant as a cosmological constant term in the Einstein field equations. On the other hand, the associated coefficient a_{-2} evidently receives contributions of order Λ^4 (due to the associated dimensionless coupling g_{-2} receiving contributions of order unity) from integrating out field modes from the UV scale Λ down to the much lower scale of astronomical phenomena, which, if we take the UV cutoff as the Planck scale, is 10^{120} times greater than the observed value (if we interpret the presently observed

---

[4] For the purposes of the present discussion, $1/137$ is a number of order unity, to be distinguished from the much tinier ratio of scales, $\sim 10^{-15}$, between, say, the LHC energy $E \sim 10^4$ GeV and the Planck energy $\Lambda \sim 10^{19}$ GeV.

dark-energy effects as arising from a cosmological constant). In comparison to this "cosmological constant problem", the fine-tuning required to achieve a suitable low-mass Higgs (say, of order 125 GeV) is hardly worth mentioning.

This mixing of higher- and lower-dimension operators would seem to complicate enormously our ability to give a direct quantitative interpretation to the terms in an effective Lagrangian such as (16.7). The problem would, of course, be greatly ameliorated if we could ensure that loop contributions to low-energy amplitudes could not contain integrations over a large momentum range capable of producing large factors (in the case under discussion, positive powers of $\frac{\Lambda}{E}$): in this case, the order of magnitude of the contributions of different operators in the effective Lagrangian could be deduced directly from the size of the coefficient couplings multiplying the respective operators.

We can see how to achieve this by taking note of a simple property of the Fourier transform change of functional field variables, which is a linear unitary one, allowing us to write the path integral (normally given in terms of the coordinate space fields) in terms of momentum-space modes of the field which can be divided in an obvious way into successive momentum "shells". Let us choose an energy scale $\mu << \Lambda$, much smaller than the UV cutoff of the effective theory but still above the energy scale at which we wish (or are able) to perform experiments, so $E < \mu$. In analogy to (16.4), we can define "sliced" (or even better, "peeled") fields:

$$\phi_{(\mu,\Lambda)}(x) \equiv \int_{\mu<|k|<\Lambda} \tilde{\phi}(k) e^{-ik\cdot x} \frac{d^4k}{(2\pi)^4} \tag{16.18}$$

$$\phi_\Lambda(x) = \phi_{(\mu,\Lambda)}(x) + \phi_\mu(x) \tag{16.19}$$

$$\int \mathbf{D}\phi_\Lambda = \int \prod_{|k|<\Lambda} \mathbf{D}\tilde{\phi}(k) = \int \prod_{|k|<\mu} \mathbf{D}\tilde{\phi}(k) \prod_{\mu<|k|<\Lambda} \mathbf{D}\tilde{\phi}(k)$$

$$= \int \mathbf{D}\phi_\mu \int \mathbf{D}\phi_{(\mu,\Lambda)} \tag{16.20}$$

Of course, in order to generate (by functional differentiation of the generating function Z_Λ) the desired n-point scattering amplitudes with momenta of magnitude up to, but not exceeding, μ, we must include a source function $j(x)$ which needs to contain momentum Fourier modes only up to, but not exceeding, the scale μ:

$$j_\mu(x) = \int_{|k|<\mu} \tilde{j}(k) e^{-ik\cdot x} \frac{d^4k}{(2\pi)^4} \tag{16.21}$$

As Fourier modes of different momentum are orthogonal (in coordinate space), the source term in the functional integral (16.5) depends only on the infrared field $\phi_\mu(x)$ (as the overlap of $\phi_{(\mu,\Lambda)}$ and j_μ vanishes):

$$\int d^4x \, j_\mu(x)\phi_\Lambda(x) = \int d^4x \, j_\mu(x)\phi_\mu(x) \tag{16.22}$$

If we factor the functional measure in the path integral (16.5) as indicated in (16.20), we see that the source term can be taken outside the integral over the momentum shell field $\phi_{(\mu,\Lambda)}$, and the resulting integral used to define a new effective Lagrangian $\mathcal{L}_\mu(\phi_\mu)$:

$$
\begin{aligned}
Z_\Lambda[j_\mu] &= \int \mathbf{D}\phi_\mu \int \mathbf{D}\phi_{(\mu,\Lambda)} e^{-\int d^4 x \mathcal{L}_\Lambda(\phi_\Lambda) + \int d^4 x j_\mu(x)\phi_\Lambda(x)} \\
&= \int \mathbf{D}\phi_\mu e^{\int d^4 x j_\mu(x)\phi_\mu(x)} \int \mathbf{D}\phi_{(\mu,\Lambda)} e^{-\int d^4 x \mathcal{L}_\Lambda(\phi_\mu + \phi_{(\mu,\Lambda)})} \\
&\equiv \int \mathbf{D}\phi_\mu e^{-\int d^4 x \mathcal{L}_\mu(\phi_\mu) + \int d^4 x j_\mu(x)\phi_\mu(x)}
\end{aligned}
\tag{16.23}
$$

where

$$
\mathcal{L}_\mu(\phi_\mu) \equiv -\ln\left(\int \mathbf{D}\phi_{(\mu,\Lambda)} e^{-\int d^4 x \mathcal{L}_\Lambda(\phi_\mu + \phi_{(\mu,\Lambda)})}\right)
\tag{16.24}
$$

The new effective Lagrangian, $\mathcal{L}_\mu$, can be expanded[5] in powers of the infrared field ϕ_μ and its derivatives, just as our original effective Lagrangian defining the theory at the high scale Λ, but of course, with coefficients which depend on the scale μ:

$$
\mathcal{L}_\mu = \frac{1}{2} a_0'(\mu)(\partial_\nu \phi_\mu)^2 + \sum_{n \geq 0} a_n(\mu)\mathcal{O}_n(\phi_\mu) + \sum_{n > 0} a_n'(\mu)\mathcal{O}_n'(\phi_\mu) + \dots
\tag{16.25}
$$

Note that the coefficient a_0' (often written Z, the wavefunction renormalization constant) of the kinetic term (which we were free to choose to be $\frac{1}{2}$ at the high scale, by rescaling the field) now becomes a function of μ as well. The *effective running couplings* $a_n(\mu), a_n'(\mu), \dots$ of course satisfy the boundary condition

$$
a_0'(\Lambda) = 1
$$
$$
a_n(\Lambda) = a_n = g_n \Lambda^{2-n}
$$
$$
a_n'(\Lambda) = a_n' = g_n' \Lambda^{-n}, \quad n \geq 2
\tag{16.26}
$$

Clearly, we can view a given effective Lagrangian as a point in an infinite-dimensional space of couplings $g_n, g_n', \dots$, and the process of integrating out (partially) the momentum modes of the field results in a well-defined flow through this space, where, given a definite starting point at the highest scale Λ, as indicated in (16.26), the structure of the effective Lagrangian is uniquely specified (corresponding to a unique point in coupling constant space) at any lower-energy scale μ. The size of the low-energy effective couplings is now directly associated with the importance of the corresponding interaction term for low-energy physics, as the large momentum range present in loop integrals (previously stretching all the way up to Λ) has been eliminated. In the next section we shall see how to write a precise set of coupled equations—the famous *renormalization group equations*—describing this flow.

---

[5] Strictly speaking, the locality of the effective Lagrangian defined by this procedure depends on certain smoothness properties which are not present with the sharp momentum cutoff envisaged here. In the next section we shall remedy this difficulty and derive an exact equation for the cutoff dependence of the local effective Lagrangian which arises once the momentum cutoff is appropriately chosen.

16.4   The renormalization group

The arguments of the preceding section suggest that the low-momentum, or large-distance physics, of our theory of a self-interacting scalar particle, described at a high scale by a specified effective Lagrangian, is determined at low energies by an effective Lagrangian of the same form but with modified couplings associated with each of the (infinitely many) operators appearing in the expansion of the Lagrangian. Given definite starting values for the couplings $g_n, g'_n, ..$ at the high scale Λ, the process of integrating out the intermediate modes between μ and Λ should therefore lead to a new effective Lagrangian at the lower scale with definite couplings $g_n(\mu), g'_n(\mu), ...$ We therefore expect that there should be an (infinite) coupled set of first-order differential equations describing the flow of the infinite set of couplings: first order, as we expect a unique solution simply by specifying the initial value of the couplings at the high scale Λ,

$$\mu \frac{\partial}{\partial \mu} g_n(\mu) = \beta_n(g_n(\mu), g'_n(\mu), ..)$$

$$\mu \frac{\partial}{\partial \mu} g'_n(\mu) = \beta'_n(g_n(\mu), g'_n(\mu), ..), \quad \text{etc.} \tag{16.27}$$

where the $\beta_n, \beta'_n, ..$ are dimensionless functions of the infinite set of *dimensionless* couplings $g_n, g'_n, ...$ Evidently, the process of successively integrating out momentum modes can be viewed in group-theoretical terms: the operation $\mathcal{R}(\mu_2, \mu_1)$ of integrating out modes between scales $\mu_2 < \mu_1$ followed by the subsequent process $\mathcal{R}(\mu_3, \mu_2)$ is evidently equivalent to the single operation $\mathcal{R}(\mu_3, \mu_1)$, and the set of all such processes is collectively termed the "renormalization group". The finite non-linear mappings in the infinite-dimensional coupling constant space of the theory represented by the abstract group elements $\mathcal{R}$ reduce infinitesimally to the differential form of the Lie generator of the renormalization group indicated in (16.27), which is called the "renormalization group equation" of the theory. As mentioned previously, it is the precise mathematical expression of the change in form of the physics as we examine the theory at different length (or momentum) scales.

The momentum shell approach described in the preceding section, while physically intuitive, turns out to be somewhat awkward from an analytical point of view in deriving the desired Lagrangian flow, as sharp cutoffs in momentum lead to singular terms when we execute the desired derivatives with respect to scale visible in (16.27). Instead, we shall (following Polchinski (1984)) use a continuous cutoff, by writing our general effective Lagrangian as a sum of free and interacting parts, as follows. In the free part, we introduce a cutoff via a damping function $\mathcal{D}$,

$$S_0[\phi, \Lambda] = \int d^4x \mathcal{L}_0(\phi, \lambda) = \int d^4x \frac{1}{2} \phi(x)(-\Box + m^2)\mathcal{D}(-\frac{\Box}{\Lambda})\phi(x)$$

$$= \int \frac{1}{2} \tilde{\phi}(k)(k^2 + m^2)\mathcal{D}(\frac{k^2}{\Lambda^2})\tilde{\phi}(-k)\frac{d^4k}{(2\pi)^4} \tag{16.28}$$

where the function $\mathcal{D}$ is essentially unity up to the scale $|k| = \Lambda$, then grows exponentially for $|k| > \Lambda$ so as to damp the propagator $\frac{1}{\mathcal{D}(k^2/\Lambda^2)(k^2+m^2)}$, effectively cutting off the loop integrals when any internal propagator exceeds momentum Λ. Note that the field ϕ in (16.28) is not cut off, but contains all momentum modes and is independent of the scale at which we are examining the theory, so derivatives with respect to scale do not affect the fields. (Alternatively, we may simply choose to pick once and for all a fixed "ultimate" UV cutoff for this field—the Planck scale Λ_{Pl}, say—reflecting our certain knowledge that a representation of the physics in terms of Minkowski space fields must fail at this point; see below.) The precise form of $\mathcal{D}$ is unimportant, but for definiteness we can take, for example,

$$\mathcal{D}(\rho) = 1 + \exp\left(\alpha(\rho - 1)\right) \tag{16.29}$$

with α a large positive constant. Thus, the inverse of the function $\mathcal{D}(\frac{k^2}{\Lambda^2})$ in (16.28) undergoes a rapid transition from unity for $|k|$ just below Λ to an exponentially small value for $|k|$ just above Λ. In particular, we can assume that the Λ derivative of $\mathcal{D}$ (or its inverse) is effectively zero for $|k| < \Lambda$. The interaction part of the effective Lagrangian at a given scale Λ, in analogy to (16.7), is given by the usual infinite expansion

$$\mathcal{L}_{\text{int}}(\phi, \Lambda) = \sum_{n \geq 0} (a_n(\Lambda)\mathcal{O}_n + a'_n(\Lambda)\mathcal{O}' + \ldots) \tag{16.30}$$

and contains *all possible local field dependent terms*: accordingly, the sums begin at $n = 0$, including the quadratic mass $\mathcal{O}_0 = \phi^2$ and kinetic $\mathcal{O}'_0 = (\partial_\nu \phi)^2$ terms, whose coefficients must be allowed to change as we change the scale. We shall assume, as previously, that we are only interested in the physics up to some scale μ much lower than a "top" UV scale Λ_{UV} at which the "bare" couplings g_n, g'_n are initially set, with $a_n = g_n \Lambda_{UV}^{2-n}, a'_n = g'_n \Lambda_{UV}^{-n}$ as before (cf. (16.26)). Accordingly, the external source $j(x)$ introduced to probe field modes associated with the desired scattering amplitudes need contain only momentum modes up to μ:

$$\tilde{\jmath}_\mu(k) = 0, \quad |k| > \mu \tag{16.31}$$

The generating functional describing the physics at any scale Λ is given by the path integral

$$Z_\Lambda[j_\mu] = \int \mathbf{D}\phi e^{-\int d^4 x (\mathcal{L}_0(\phi, \Lambda) + \mathcal{L}_{\text{int}}(\phi, \Lambda)) + \int d^4 x j_\mu(x)\phi(x)}$$

$$\equiv \int \mathbf{D}\phi e^{-S_0[\phi, \Lambda] - S_{\text{int}}[\phi, \Lambda] + \int \tilde{\jmath}_\mu(q)\tilde{\phi}(-q)\frac{d^4 q}{(2\pi)^4}} \tag{16.32}$$

We now claim that there exists a unique evolution with Λ of the interaction Lagrangian $\mathcal{L}_{\text{int}}(\phi, \Lambda)$ (from $\Lambda = \Lambda_{UV}$ down to $\Lambda = \mu$) such that the low-energy physics is *exactly preserved*—in other words, which leaves the generating functional $W[j_\mu] = \ln Z[j_\mu]$ of the connected low-momentum amplitudes of the theory invariant up to source-independent terms:

$$\frac{\partial}{\partial \Lambda} W_\Lambda[j_\mu] = \text{independent of } j_\mu \tag{16.33}$$

Thus, when we perform functional derivatives with respect to j_μ to extract the connected low-momentum n-point amplitudes of the theory, the Λ dependence disappears (for any Λ in the range $\mu < \Lambda < \Lambda_{UV}$), as long as we use the effective Lagrangian $\mathcal{L}_{\text{int}}(\phi, \Lambda)$ appropriate for that scale.

The derivation of the renormalization group flow equation for the Lagrangian $\mathcal{L}_{\text{int}}$ is facilitated by a functional integral identity, based on the observation that the functional integral of a total functional derivative vanishes provided the integrand has the usual falloff (in our case, exponential) for large values of the field. Namely, we have

$$\int \mathbf{D}\tilde{\phi} \, \frac{\delta}{\delta\tilde{\phi}(k)} \{ (\tilde{\phi}(k)\mathcal{D}(k^2/\Lambda^2) + \frac{1}{2}\frac{(2\pi)^4}{k^2+m^2}\frac{\delta}{\delta\tilde{\phi}(-k)}) e^{-S_0[\phi,\Lambda]-S_{\text{int}}[\phi,\Lambda]+\int \tilde{j}_\mu(q)\tilde{\phi}(-q)\frac{d^4q}{(2\pi)^4}} \}$$

$$= 0 \tag{16.34}$$

This functional identity holds for all values of the momentum k, but for reasons shortly to become apparent we shall apply it only in the regime $|k| > \mu$, where by assumption $\tilde{j}_\mu(k) = 0$. Thus, when working out the functional derivatives in (16.34), we can ignore any factors of $\tilde{j}_\mu(k)$ (or $\tilde{j}_\mu(-k)$) that appear. We shall also suppose that our fields are restricted to a large spacetime box of volume V, so that infrared singular functional derivatives such as

$$\frac{\delta}{\delta\tilde{\phi}(k)}\tilde{\phi}(k) = \delta^4(0) = \frac{1}{(2\pi)^4}\int d^4x \, e^{i0\cdot x} = \frac{V}{(2\pi)^4} \tag{16.35}$$

are given a definite meaning. (These terms will, in any case, later turn out to be irrelevant disconnected vacuum terms.) Carrying out the indicated functional derivatives in (16.34), and using

$$\frac{\delta S_0}{\delta\tilde{\phi}(-k)} = \frac{k^2+m^2}{(2\pi)^4}\mathcal{D}(k^2/\Lambda^2)\tilde{\phi}(k) \tag{16.36}$$

one finds after a short calculation (see Problem 2) that it can be rewritten as

$$\frac{1}{2}\mathcal{D}(\frac{k^2}{\Lambda^2})\delta^4(0)Z_\Lambda[j_\mu] = \int \mathbf{D}\tilde{\phi} \, \{ \frac{1}{2}\frac{k^2+m^2}{(2\pi)^4}\mathcal{D}(\frac{k^2}{\Lambda^2})^2\tilde{\phi}(k)\tilde{\phi}(-k)$$

$$+ \frac{1}{2}\frac{(2\pi)^4}{k^2+m^2}[\frac{\delta^2 S_{\text{int}}}{\delta\tilde{\phi}(k)\delta\tilde{\phi}(-k)} - \frac{\delta S_{\text{int}}}{\delta\tilde{\phi}(k)}\frac{\delta S_{\text{int}}}{\delta\tilde{\phi}(-k)}] \} e^{-S_0 - S_{\text{int}} + \int j_\mu \phi d^4 x}$$

$$\tag{16.37}$$

Note that the two terms in curly braces in (16.34) are chosen such that a contribution of the form $\tilde{\phi}(k)\mathcal{D}(k^2/\Lambda^2)\frac{\delta S_{\text{int}}}{\delta\tilde{\phi}(k)}$ cancels between them, leaving just the terms given here.

We can now return to our main focus: how to choose the effective Lagrangian $\mathcal{L}_{\text{int}}(\phi, \Lambda)$ at any given scale Λ to ensure that we obtain the *same* low-momentum amplitudes, by functionally differentiating the generating functional $Z_\Lambda[j_\mu]$. A glance at (16.32) shows that the differential variation of this functional with Λ arises from two sources: the Λ dependence of the propagator via the cutoff function $\mathcal{D}(k^2/\Lambda^2)$

embedded in the free Lagrangian $\mathcal{L}_0$, and the Λ-dependence of the "interaction" part $\mathcal{L}_{\text{int}}$ (through the Λ-dependent coupling parameters contained in the latter). In fact, we clearly have, differentiating (16.32),

$$\Lambda\frac{\partial Z_\Lambda[j_\mu]}{\partial\Lambda} = -\int \mathbf{D}\tilde{\phi}\{\int \frac{1}{2}(k^2+m^2)\tilde{\phi}(k)\tilde{\phi}(-k)\Lambda\frac{\partial\mathcal{D}(\frac{k^2}{\Lambda^2})}{\partial\Lambda}\frac{d^4k}{(2\pi)^4} + \Lambda\frac{\partial S_{\text{int}}}{\partial\Lambda}\}$$

$$\cdot e^{-S_0-S_{\text{int}}+\int j_\mu\phi d^4x} \tag{16.38}$$

Our choice of cutoff function $\mathcal{D}(k^2/\Lambda^2)$ implies (see (16.29)) that derivatives of $\mathcal{D}$ with respect to the scale Λ, if we keep Λ above the infrared scale μ, are exponentially negligible (of order $e^{-\alpha(1-k^2/\Lambda^2)}$) in the infrared region $|k| < \mu < \Lambda$. The support of both sides of the following identity is therefore precisely in the region of validity $|k| > \mu$ of (16.37):

$$\Lambda\frac{\partial\mathcal{D}(\frac{k^2}{\Lambda^2})}{\partial\Lambda} = -\mathcal{D}(\frac{k^2}{\Lambda^2})^2\Lambda\frac{\partial\mathcal{D}^{-1}}{\partial\Lambda} \tag{16.39}$$

Using (16.39), (16.38) can be re-expressed

$$\Lambda\frac{\partial Z_\Lambda[j_\mu]}{\partial\Lambda} = \int \mathbf{D}\tilde{\phi}\{\int \frac{1}{2}(k^2+m^2)\mathcal{D}(\frac{k^2}{\Lambda^2})^2\tilde{\phi}(k)\tilde{\phi}(-k)\Lambda\frac{\partial\mathcal{D}^{-1}}{\partial\Lambda}\frac{d^4k}{(2\pi)^4} - \Lambda\frac{\partial S_{\text{int}}}{\partial\Lambda}\}$$

$$\cdot e^{-S_0-S_{\text{int}}+\int j_\mu\phi d^4x} \tag{16.40}$$

Comparing (16.37) with (16.40), we see that by setting

$$\Lambda\frac{\partial S_{\text{int}}}{\partial\Lambda} = \frac{(2\pi)^4}{2}\int \Lambda\frac{\partial\mathcal{D}(k^2/\Lambda^2)^{-1}}{\partial\Lambda}\{\frac{\delta S_{\text{int}}}{\delta\tilde{\phi}(k)}\frac{\delta S_{\text{int}}}{\delta\tilde{\phi}(-k)} - \frac{\delta^2 S_{\text{int}}}{\delta\tilde{\phi}(k)\delta\tilde{\phi}(-k)}\}\frac{d^4k}{k^2+m^2} \tag{16.41}$$

and using the identity obtained by integrating both sides of (16.37) with the measure $\int d^4k\Lambda\frac{\partial\mathcal{D}^{-1}}{\partial\Lambda}..$, we obtain

$$\Lambda\frac{\partial Z_\Lambda[j_\mu]}{\partial\Lambda} = \frac{1}{2}\delta^4(0)\int \Lambda\frac{\partial\ln(\mathcal{D}(k^2/\Lambda^2))}{\partial\Lambda}d^4k\cdot Z_\Lambda[j_\mu] \tag{16.42}$$

or, equivalently, as desired (see (16.33)),

$$\Lambda\frac{\partial W_\Lambda[j_\mu]}{\partial\Lambda} = \frac{\Lambda}{Z_\Lambda}\frac{\partial Z_\Lambda[j_\mu]}{\partial\Lambda} = \frac{1}{2}\delta^4(0)\int \Lambda\frac{\partial\ln(\mathcal{D}(k^2/\Lambda^2))}{\partial\Lambda}d^4k = \text{independent of } j_\mu \tag{16.43}$$

The peculiar right-hand side appearing here, which must be associated with disconnected vacuum graphs[6] which do not contribute to the connected n-point functions obtained by differentiating W_Λ with respect to the low-momentum source j_μ, can easily

---

[6] The reader may find it convenient at this point to review the discussion of disconnected graphs and vacuum energy in Section 10.2.

be seen to arise from the cutoff-dependence of the zero-point energy associated with the free Lagrangian $\mathcal{L}_0$ (see Problem 3).

The equation (16.41) gives the desired variation in the form of the effective Lagrangian with the scale at which we probe the physics. Moreover, given that the starting effective Lagrangian (at the UV cutoff) yields a well-defined convergent functional integral representation for Z_Λ, this renormalization group equation is *non-perturbatively valid*, as it is based on exact manipulations of the functional integral. We note immediately that the space of *free Lagrangians* (i.e., those Lagrangians quadratic in the field, but with arbitrarily many spacetime-derivatives) is preserved under renormalization group transformations, as $\Lambda \frac{\partial S_{\text{int}}}{\partial \Lambda}$ is clearly quadratic in the fields (ignoring physically ignorable constant terms) if S_{int} is. However, if there are interactions present (with our $\phi \to -\phi$ symmetry, terms quartic or higher in the fields), the non-linear functional equation (16.41) produces an infinite-dimensional mixing of the operators in the general expansion (16.30). The reason for this is simply that this expansion, re-expressed in terms of momentum-space fields, can be written

$$
S_{\text{int}} = \sum_L \frac{1}{L!} \int h_L(k_1, k_2, \ldots, k_M; \Lambda)) \delta^4 \left(\sum_i k_i \right) \tilde{\phi}(k_1) \tilde{\phi}(k_2) \cdots \tilde{\phi}(k_L) \prod_i d^4 k_i
$$
(16.44)

where the functions h_M are scalar functions (under Euclidean rotations of their four-momentum arguments) expandable in powers of their momentum arguments. Inserting this form in (16.41) we find that the form of the effective Lagrangian is preserved under renormalization group transformation

$$
\Lambda \frac{\partial S_{\text{int}}}{\partial \Lambda} = \sum_L \frac{1}{L!} \int (\hat{h}_L^{(1)}(k_1, \ldots, k_L) - \hat{h}_L^{(2)}(k_1, \ldots, k_L)) \delta^4 \left(\sum_i k_i \right) \tilde{\phi}(k_1) \cdots \tilde{\phi}(k_L) \prod_i d^4 k_i
$$
(16.45)

where the functions

$$
\hat{h}_L^{(1)}(k_1, \ldots, k_L) = \sum_{M+N=L} \frac{L!}{M!N!} \int \mathcal{F}(k^2) h_{M+1}(k, k_1, \ldots, k_M)
$$

$$
\cdot h_{N+1}(-k, k_{M+1}, k_{M+2}, \ldots, k_L) d^4 k \quad (16.46)
$$

$$
\hat{h}_L^{(2)}(k_1, \ldots, k_L) = \int \mathcal{F}(k^2) h_{L+2}(k, -k, k_1, k_2, \ldots, k_L) d^4 k \quad (16.47)
$$

$$
\mathcal{F}(k^2) \equiv \frac{(2\pi)^4}{2} \frac{1}{k^2 + m^2} \Lambda \frac{\partial \mathcal{D}(k^2/\Lambda^2)^{-1}}{\partial \Lambda} \quad (16.48)
$$

are Taylor expandable in their momentum arguments, which holds in our case given our choice of a *smooth* cutoff function $\mathcal{D}(k^2/\Lambda^2)$ in (16.29). This analyticity property of the regularization used to smear the short-distance behavior theory is essential in maintaining the desired locality properties of our theory: i.e., in ensuring that the effective Lagrangian at any scale can be expressed as an infinite sum of local terms (products of the fields and their derivatives at a single spacetime point). The upshot of this whole discussion is that the cutoff variation of the effective Lagrangian amounts to an infinite set of non-linear first-order differential equations among the coefficients

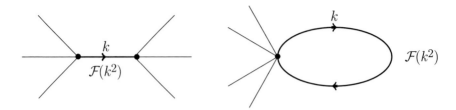

Fig. 16.3 Graphical representation of the renormalization group evolution of the effective Lagrangian.

of the local operator basis in terms of which we choose to express our cutoff theory, exactly as expressed in the renormalization group equations (16.27).

The physical interpretation of the two terms in (16.46, 16.47) is illustrated in Fig. 16.3. The graph on the left (corresponding to (16.46)) illustrates the differential change in a typical effective vertex (in this case, six-point) due to the differential variation $\mathcal{F}$ of the propagator, represented by the thick line, which in this case connects two vertices as an internal line in a tree graph. The graph on the right (corresponding to (16.47)) describes the variation in the effective vertex (in this case, a four-point vertex) due to the differential variation of the cutoff propagator in an internal line in a loop graph.

We have already seen an explicit example of the effect of the latter term in (16.15), where the irrelevant term $\frac{g_4}{\Lambda^2}\phi^6$ was shown to lead to an order unity modification of the marginal four-point vertex (unsuppressed by inverse powers of the high scale) as a consequence of the large loop integral in the final graph of Fig. 16.2. Let us see how to reproduce this result from the point of view of our new renormalization group equation. We shall assume that at the initial high scale Λ, all the dimensionless couplings except g_2 and g_4 are negligible, and we shall also ignore terms of order g_2^2, relative to g_2 and g_4. The evolution of the g_4 vertex is determined by terms of order g_2^2 (from (16.46)) or g_6 (from (16.47)) both of which we shall neglect: we thereby conclude that to the desired accuracy $\Lambda \frac{\partial}{\partial \Lambda} \frac{g_4}{\Lambda^2}$ is negligible, and we may replace $\frac{g_4(\Lambda)}{\Lambda^2}$ by $\frac{g_4(\mu)}{\mu^2}$ at any lower scale μ. The evolution of g_2 arising from (16.47) is determined by

$$
\begin{aligned}
\Lambda \frac{\partial g_2}{\partial \Lambda} &= -\frac{(2\pi)^4}{2} \int \frac{30}{(2\pi)^8} \frac{g_4(\Lambda)}{\Lambda^2} \Lambda \frac{\partial \mathcal{D}(k^2/\Lambda^2)^{-1}}{\partial \Lambda} \frac{d^4k}{k^2+m^2} \\
&\approx -\frac{(2\pi)^4}{2} \frac{g_4(\mu)}{\mu^2} \int \frac{30}{(2\pi)^8} \Lambda \frac{\partial \mathcal{D}(k^2/\Lambda^2)^{-1}}{\partial \Lambda} \frac{d^4k}{k^2+m^2}
\end{aligned} \tag{16.49}
$$

Integrating (16.49) from Λ down to the infrared scale μ,

$$
g_2(\mu) = g_2(\Lambda) + 15\frac{g_4(\mu)}{\mu^2} \int (\mathcal{D}(k^2/\Lambda^2)^{-1} - \mathcal{D}(k^2/\mu^2)^{-1}) \frac{1}{k^2+m^2} \frac{d^4k}{(2\pi)^4} \tag{16.50}
$$

We may approach the sharp momentum cutoff used in Section 16.3 by choosing a large value for the parameter α in (16.29), whereupon we may replace $\mathcal{D}(k^2/\Lambda^2)^{-1}$ by a step function $\theta(\Lambda - |k|)$. The integral in (16.50) is then restricted to the momentum shell $\mu < |k| < \Lambda$, and we find (replacing $\frac{g_4(\mu)}{\mu^2}$ by $\frac{g_4(\Lambda)}{\Lambda^2}$ as indicated above)

$$g_2(\mu) = g_2(\Lambda) + \frac{15}{16\pi^2} g_4(\Lambda) + O(\frac{m^2, \mu^2}{\Lambda^2}) \qquad (16.51)$$

which can be seen to agree with the order unity shift in g_2 induced by the six-point vertex obtained earlier in (16.15).

The renormalization group flow equation (16.45) gives an exact description of the appropriate form taken by the dynamics of the theory once phenomenologically inaccessible short-distance modes of the field are averaged out, but it nevertheless leaves us with a complicated, and not very practical, end result, as our effective Lagrangian contains an infinite number of terms. Although by lowering the cutoff from some very high (and experimentally unreachable) value Λ_{UV} to a value μ close to experimental energies we have ensured that large loop integrals involving powers of ratios of Λ_{UV} to the low scale μ have been eliminated, there remains the obvious difficulty that the calculation of a low-energy amplitude seems to require the inclusion of contributions from an infinite number of vertices in the low-energy Lagrangian.

It is a remarkable property of local quantum field theory that for a certain subset of theories, the sensitivity of the low-energy amplitudes to all but a finite number of coupling parameters—in particular, those associated with the marginal and relevant operators in the effective Lagrangian—is reduced to *inverse powers* of the high cutoff. From the renormalization group point of view, this occurs because the renormalization group flow has the property that the point describing the "location" of the Lagrangian in the infinite-dimension coupling space of the $g_n(\mu), g_n'(\mu), \ldots$ is attracted, for $\mu \ll \Lambda_{UV}$, onto a finite-dimensional submanifold (of dimension equal to the number of marginal and relevant operators), up to corrections of order $\frac{\mu}{\Lambda_{UV}}$ to some (typically even) power (modulo logarithms of $\frac{\mu}{\Lambda_{UV}}$). As a consequence, up to usually negligible corrections, we find that for these theories the low-energy amplitudes can be parameterized by just a finite set of parameters—namely, those needed to locate the theory on the finite-dimensional attractive submanifold, and which can in principle be determined by making an equal number of independent experimental measurements. The insensitivity asserted here is actually demonstrated in a perturbative setting: one shows that the formal expansion of an arbitrary scattering amplitude in powers of the marginal and relevant interaction couplings defined at low momentum, holding the irrelevant couplings fixed at the high cutoff scale Λ_{UV}, depend on the latter only by inverse powers of Λ_{UV}. We then say that the marginal and relevant operators of the effective theory form a "perturbatively renormalizable set". Exactly how this works will be the topic of Section 17.4 in the next Chapter.

Although the physical content of the renormalization group is most easily displayed using momentum cutoff regularization schemes of the type we have used up to this point, such schemes have distinct disadvantages from a calculational point of view once one goes beyond the lowest orders of perturbation theory. Moreover, in theories with local gauge symmetry, such cutoff schemes turn out to be incompatible with the local symmetry, with the unwanted result that the renormalization group evolution necessarily introduces non-gauge-invariant operators which greatly complicate the renormalization group analysis of the theory, and would in fact not be needed if the theory could be regularized in a manner compatible with local gauge symmetry. In the next section we shall discuss alternative approaches to the regularization of local field theory which allow a more efficient application of the insights of the renormalization group.

16.5   Regularization methods in field theory

We have so far been discussing the behavior of local field theories under change of scale in terms of fields and field products whose matrix elements are made well-defined by building in an explicit cutoff in the momentum-space Fourier transform modes of the fields. This sort of cutoff has a clear physical connection to our coordinate space intuition, whereby the ability to probe sensitivity to higher-momentum field modes is directly correlated to our ability to expose "finer" details of the interactions at ever shorter distances. Alternatively, we may use a spacetime lattice cutoff, whereby the continuum theory (should one exist) is approached by taking the spacing a between the points of our hypercubical lattice to zero, and where (on an infinitely extended lattice) the Fourier momenta assigned to the fields range continuously over the finite interval $-\frac{\pi}{a} < k_\mu < \frac{\pi}{a}$. Lattice cutoffs are especially valuable in non-perturbative formulations of field theory (via the Euclidean functional integral): indeed, rigorous constructive proofs of the existence of the continuum limit for super-renormalizable theory in less than four spacetime dimensions make extensive use of this method of provisionally defining the theory, as a prelude to the proof of existence of the continuum $(a \to 0)$ and "thermodynamic" $(V \to \infty$, where V is the spacetime volume) limits.[7]

A renormalization group approach based on lattice fields is also possible, in analogy to the "momentum shell" methods discussed above: one defines progressively "coarser" fields, with momentum components restricted at each discrete stage to one half the range of the previous fields, by forming "block averages" of the fields over hypercubical sublattices (consisting, in four dimensions, of the fields at the sixteen lattice points equidistant from the points of a dual lattice interspaced with the original one and with twice the lattice-spacing). This block renormalization technique was pioneered, and has been widely used, in the study of critical phenomena in spin systems, but as with the momentum-shell methods of the preceding sections, turns out to be somewhat clumsy, and beset with undesirable technical drawbacks when we come to the study of the renormalization group behavior of the four-dimensional field theories (especially gauge theories) of relevance to the Standard Model of elementary particle physics (and its potential extensions at higher energy).

In our discussion of effective Lagrangians up to this point, the theory is specified at any given cutoff scale Λ by a formal expansion containing an infinite number of scalar operators involving arbitrarily many powers of the field (and derivatives thereof), multiplied at the same spacetime point. From our discussion of the Wightman

---

[7] The use of the term "thermodynamic" here does not imply any connection to finite-temperature phenomena: it is a carry-over from the close formal analogy between the Euclidean quantum functional integral and the canonical partition sums of classical thermodynamics.

formalism, we are already familiar with the notion that local quantum fields should really be regarded as operator-valued distributions, and that the multiplication of such distributions can lead to ambiguities (or singularities) in the continuum theory. Thus, the matrix elements of the operators $\mathcal{O}_n, \mathcal{O}'_n$ etc., defined in Section 16.2 are actually infinite, if we insist in working in a continuum theory where the UV cutoff is infinite. Of course, the whole point of the philosophy espoused here is that a cutoff is not only technically but *physically* required, and with the momentum cutoff in place (say, by employing the modified propagator (16.29)), there are no ultraviolet divergences in any of the loop integrals we encounter, so the operators of the theory have perfectly well-defined matrix elements (and lead to well-defined perturbative corrections to n-point functions of the theory when inserted into the graphs for some process). The actual value of these matrix elements will, of course, depend on the cutoff, so we must keep in mind that operators such as $\mathcal{O}_0(x) \equiv \phi^2(x)$ or $\mathcal{O}'_0(x) \equiv (\partial_\nu \phi(x))^2$ have, strictly speaking, no meaning until we specify an ultraviolet regularization procedure, such as (in the momentum-shell framework) a value for the UV cutoff Λ, and the specific form of the cutoff (e.g., the smooth function (16.29)). The renormalization group flow equation (16.41) of the preceding section expresses the fact that the infinite set of operators so defined, at any given scale Λ, form a complete set, in the sense that we need only alter their coefficients in the effective Lagrangian in order to obtain an exactly equivalent description of the low-energy amplitudes of the theory at any other scale $\mu < \Lambda$. We once again remind the reader that this equation is an exact *non-perturbative* statement about the amplitudes of the effective field theory, assuming only that we start with a well-defined functional integral at the high scale: in the proof of (16.41), we have employed only exact functional integral identities, with no need to expand the exponent of the functional integral in a perturbative series.

Let us explore in a little more detail the freedom we have to choose different sets of operator products in an effective field theory without altering the physical content of the theory. At this point we shall resort to perturbation theory to gain some concrete intuition about the variability entailed by this freedom of choice. Staying for the time being with the momentum cutoff approach, let us consider the one particle to one particle matrix elements of $\mathcal{O}_0(x) \equiv \phi^2(x)$,

$$\langle k' | \mathcal{O}_0(x) | k \rangle = e^{iq \cdot x} \langle k' | \mathcal{O}_0(0) | k \rangle, \quad q \equiv k' - k \qquad (16.52)$$

Ignoring uninteresting initial and final-state factors, the matrix element $\langle k' | \mathcal{O}_0(0) | k \rangle$ receives, in addition to the tree-graph contribution (first graph in Fig. 16.4), a one-loop contribution of order g_2 from the graph on the right in Fig. 16.4:

$$\langle k' | \mathcal{O}_0(0) | k \rangle = 1 - 12g_2 \int \frac{\mathcal{D}(l^2/\Lambda^2)^{-1} \mathcal{D}((q-l)^2/\Lambda^2)^{-1}}{(l^2 + m^2)((q-l)^2 + m^2)} \frac{d^4 l}{(2\pi)^4} + \cdots \quad (16.53)$$

$$= 1 - 12g_2 \mathcal{I}(q^2; \Lambda^2, m^2) + \cdots \qquad (16.54)$$

where the dots represent other perturbative corrections which are not of interest to us presently. The one-loop integral $\mathcal{I}(q^2; \Lambda^2, m^2)$ can be expanded in powers of the momentum variable $q \ll \Lambda, m$ (see Problem 4):

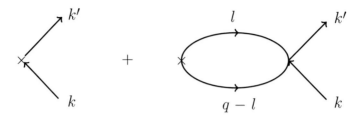

Fig. 16.4 Low-order contributions to a matrix element of $\phi^2(x)$.

$$\mathcal{I}(q^2; \Lambda^2, m^2) = \sum_n f_n(\frac{\Lambda^2}{m^2})(\frac{q^2}{m^2})^n \qquad (16.55)$$

With a little thought one establishes that the coefficient functions f_n, which contain the dependence on the UV cutoff Λ, and therefore incorporate the conventionality of our particular regularization of the operator $\mathcal{O}_0(x)$, contain at worst a logarithmic divergence $\ln \Lambda^2/m^2$ in the limit $\Lambda \to \infty$, plus vanishing corrections involving inverse powers of the cutoff. For example, taking the first term in the low-momentum expansion, and assuming the parameter α in (16.29) large, so that the cutoff is essentially a step function at Λ, we find

$$f_0(\frac{\Lambda^2}{m^2}) = \frac{1}{16\pi^2}(\ln (\Lambda^2/m^2) - 1 + O(\frac{m^2}{\Lambda^2})) \qquad (16.56)$$

with the $f_n(\frac{\Lambda^2}{m^2})$ for $n \geq 1$ given by dimensionless constants plus corrections of $O(\frac{m^2}{\Lambda^2})$. A similar calculation, again including just the two graphs appearing in Fig. 16.4, gives for the one-particle matrix element of $\mathcal{O}_0' = (\partial_\nu \phi)^2$ (the only difference being the appearance of a dot product of the four-momenta entering and leaving the two-point vertex of the $\mathcal{O}_0'$ operator),

$$\langle k'|\mathcal{O}_0'(0)|k\rangle = k \cdot k' - 12g_2 \int \frac{l \cdot (l-q)\mathcal{D}(l^2/\Lambda^2)^{-1}\mathcal{D}((q-l)^2/\Lambda^2)^{-1}}{(l^2 + m^2)((q-l)^2 + m^2)} \frac{d^4l}{(2\pi)^4} + \cdots$$
$$= k \cdot k' - 12g_2\mathcal{I}'(q^2; \Lambda^2, m^2) + \cdots \qquad (16.57)$$

where

$$\mathcal{I}'(q^2; \Lambda^2, m^2) = m^2 \sum_n f_n'(\frac{\Lambda^2}{m^2})(\frac{q^2}{m^2})^n \qquad (16.58)$$

In this case a quadratic dependence on the cutoff appears in the leading coefficient function f_0'. Again, taking α large so that $\mathcal{D}(l^2/\Lambda^2)^{-1}$ is approximately a step function $\theta(\Lambda - |l|)$, one finds

$$f_0'\left(\frac{\Lambda^2}{m^2}\right) = \frac{1}{16\pi^2}\left(\frac{\Lambda^2}{m^2} - 2\ln\left(\frac{\Lambda^2}{m^2}\right) + 1\right) \tag{16.59}$$

$$f_1'\left(\frac{\Lambda^2}{m^2}\right) = \frac{1}{16\pi^2}\left(-\frac{1}{2}\ln\left(\frac{\Lambda^2}{m^2}\right) + \frac{2}{3}\right) \tag{16.60}$$

while the f_n' are Λ independent constants for $n \geq 2$. Of course, for a more general choice of cutoff function (for example, keeping the parameter α in the cutoff function finite), the coefficients f_n, f_n' (and their generalizations to all orders of perturbation theory, as well as the corresponding coefficients for all possible local operators) will be different, although exactly the same low-energy physics can be reproduced by an appropriate (different) linear combination of the new set of regularized operators as defined by the altered cutoff method, as we have seen in the preceding section. The appearance of power-dependence (quadratic, in the case of (16.59)) on the ultraviolet cutoff Λ in loop integrals, as we have already seen in Section 16.2, is responsible for the mixing of operators of different mass dimension in the momentum cutoff approach to the renormalization group.

We shall now see that an alternative cutoff procedure can be used to give a precise meaning to the matrix elements of an arbitrary local operator at any order of perturbation theory, with the remarkable additional feature that all power-dependences on the cutoff are removed, leaving only terms with a logarithmic dependence on the cutoff scale (such as the $\ln\left(\frac{\Lambda^2}{m^2}\right)$ terms visible in (16.59,16.60)). First, note that the loop integral appearing in (16.53), with the cutoff functions $\mathcal{D}^{-1}$ omitted, would in fact be ultraviolet-convergent in any (integer) spacetime dimension less than four: it is only logarithmically divergent in four dimensions after all, right at the edge, as it were, of being a convergent integral at large momenta. This suggests a *dimensional regularization* approach whereby we temporarily imagine carrying out the integral in a general spacetime dimension $d < 4$, and then examine the behavior of the result as we analytically continue the resultant expression back to the physical spacetime dimension $d = 4$. To see how to do this, first note the expression for the radial phase-space in a general d-dimensional Euclidean integral

$$\int d^d l = \frac{2\pi^{d/2}}{\Gamma(d/2)}\int_0^\infty l^{d-1}dl \tag{16.61}$$

In order to preserve dimensional consistency, so that our expression for the regulated amplitude retains the same dimension in powers of mass regardless of the dimension d, we shall append the appropriate power of the ultraviolet scale Λ to each loop integral to maintain overall mass dimension 4: thus loop integrals will appear as $\Lambda^{4-d}\int d^d l \cdots$. Accordingly, we find that the one-loop integral in question, in d-dimensions, becomes

$$\mathcal{I}_d(q^2; \Lambda^2, m^2) \equiv \frac{\Lambda^{4-d}}{(2\pi)^d}\int \frac{d^d l}{(l^2 + m^2)((q-l)^2 + m^2)} \tag{16.62}$$

$$= \frac{\Lambda^{4-d}}{(2\pi)^d}\int_0^1 dx \int \frac{d^d l}{(l^2 - 2xq \cdot l + xq^2 + m^2)^2} \tag{16.63}$$

$$= \frac{\Lambda^{4-d}}{(2\pi)^d} \int_0^1 dx \int \frac{d^d l}{(l^2 + x(1-x)q^2 + m^2)^2} \qquad (16.64)$$

$$= \frac{\pi^{d/2} \Lambda^{4-d}}{(2\pi)^d \Gamma(d/2)} \int_0^1 dx \int_0^\infty \frac{2l^{d-1}}{(l^2 + \mathcal{M}^2)^2} dl, \qquad (16.65)$$

$$\mathcal{M}^2 \equiv x(1-x)q^2 + m^2 \qquad (16.66)$$

where we have used the familiar Feynman parameter identity

$$\frac{1}{AB} = \int_0^1 \frac{1}{((1-x)A + xB)^2} dx \qquad (16.67)$$

and performed a shift of integration variable $l \to l + xq$ to obtain (16.63). The remaining integral in (16.64) can be evaluated by making the change of variable $t = l^2$ and using the β function identity

$$\int_0^\infty \frac{t^{x-1}}{(t + \mathcal{M}^2)^{x+y}} dt = (\mathcal{M}^2)^{-y} B(x,y) = (\mathcal{M}^2)^{-y} \frac{\Gamma(x)\Gamma(y)}{\Gamma(x+y)} \qquad (16.68)$$

whence we find

$$\mathcal{I}_d(q^2; \Lambda^2, m^2) = \frac{\Gamma(2 - \frac{d}{2})}{(4\pi)^{d/2}} \int_0^1 \left(\frac{m^2 + x(1-x)q^2}{\Lambda^2}\right)^{\frac{d}{2}-2} dx \qquad (16.69)$$

We see that the right-hand side of (16.69) provides an analytic continuation of our originally four-dimensional loop integral to general complex spacetime dimensions d, with a finite result in the region $\mathrm{Re}(d) < 4$, and indeed analytic save at the poles of the Γ function at $d = 4, 6, 8, \ldots$. Of course, the poles of the Γ function at zero (and negative integer) values of its argument mean, not surprisingly, that the integral becomes divergent once we attempt to return to the physical spacetime dimension $d = 4$. At this point in the complex d-plane, our continued loop-integral has a Laurent expansion in the variable $\epsilon \equiv 4 - d$, the first few terms of which (using the Γ function property $\Gamma(z) \sim \frac{1}{z} - \gamma + O(z), z \to 0$, with γ the Euler–Mascheroni constant) are found to be

$$\mathcal{I}_d(q^2; \Lambda^2, m^2) \sim \frac{1}{16\pi^2}\left(\frac{2}{\epsilon} + \ln(4\pi) - \gamma + \ln(\Lambda^2/m^2) - \int_0^1 \ln\left(1 + x(1-x)\frac{q^2}{m^2}\right) dx + O(\epsilon)\right) \qquad (16.70)$$

We now define[8] the minimally subtracted dimensionally regularized matrix element of our $\mathcal{O}_0$ operator arising from the one-loop graph in Fig. 16.4 by simply omitting the pure pole term in (16.70), leaving the remaining "finite part" (henceforth indicated by the notation FP) in the $d \to 4$ limit:

---

[8] The ubiquitous appearance of the annoying factor of $\ln(4\pi) - \gamma$ accompanying the pole in ϵ has led to a modified minimal subtraction scheme, wherein the Laurent expansion is made in a shifted variable $\bar{\epsilon}$, with $\frac{2}{\bar{\epsilon}} \equiv \frac{2}{\epsilon} + \ln(4\pi) - \gamma$, and poles in $\bar{\epsilon}$ are then dropped. This is commonly referred to as the "MS-bar" scheme.

$$\text{FP } \mathcal{I}_d(q^2; \Lambda^2, m^2) \equiv \frac{1}{16\pi^2} (\ln (\Lambda^2/m^2) + \ln (4\pi) - \gamma - \int_0^1 \ln (1 + x(1-x)\frac{q^2}{m^2})dx)$$

$$(16.71)$$

The regularized amplitude (16.71) can be expanded in powers of the momentum as in (16.55): for example, referring to (16.56), we see that the leading coefficient $f_0(\frac{\Lambda^2}{m^2})$ has exactly the same logarithmic cutoff dependence in the momentum cutoff and dimensional regularization schemes, differing only by an overall additive constant, up to terms of $O(\frac{m^2}{\Lambda^2})$, suppressed by inverse powers of the cutoff. It can be easily shown (see Problem 5) that all higher coefficients $f_n, n \geq 1$ are in fact identical up to such terms in the two regularization schemes.

For the operator under discussion therefore, $\mathcal{O}_0 = \phi^2$, there would seem to be no important differences between the use of a momentum cutoff or the pole subtraction approach leading to (16.71). If we look instead at the operator $\mathcal{O}'_0 = (\partial_\nu \phi)^2$, with one-particle matrix elements given in (16.58), the situation is very different. Here, the one-loop integral, containing the extra factor of $l \cdot (l - q)$ in the numerator, has a quadratic dependence on the ultraviolet cutoff, resulting in the appearance of terms quadratic in Λ in the leading coefficient f'_0 (see (16.59)). On the other hand, the dependence of the dimensionally regularized amplitude on the cutoff Λ appears only through the prefactor $\Lambda^{4-d} = \Lambda^\epsilon$, and in developing the Laurent expansion of the regularized Feynman integral in powers of ϵ it is apparent that only powers of logarithms of the cutoff Λ can appear and not whole integer powers, via $\Lambda^\epsilon = 1 + \epsilon \ln \Lambda + \frac{1}{2}\epsilon^2(\ln \Lambda)^2 + \dots$.

A straightforward calculation, following exactly the steps used above to arrive at the minimally subtracted matrix element corresponding to (16.58), gives for the matrix element of $\mathcal{O}'_0$,

$$\mathcal{I}'_d(q^2; \Lambda^2, m^2) \equiv \frac{\Lambda^{4-d}}{(2\pi)^d} \int \frac{l \cdot (l-q)d^d l}{(l^2 + m^2)((q-l)^2 + m^2)} \qquad (16.72)$$

$$\text{FP } \mathcal{I}'_d(q^2; \Lambda^2, m^2) = \frac{1}{16\pi^2}\{Aq^2 + Bm^2 - (\frac{1}{2}q^2 + 2m^2)\ln (\Lambda^2/m^2)$$

$$+ \int_0^1 (3x(1-x)q^2 + 2m^2)\ln (1 + x(1-x)\frac{q^2}{m^2})dx\} \qquad (16.73)$$

$$A = \frac{1}{2}(\gamma - \ln (4\pi)) - \frac{1}{6}, \quad B = 2(\gamma - \ln (4\pi)) - 1 \qquad (16.74)$$

Comparing with the result (16.59) for the zero-momentum amplitude in the momentum cutoff scheme, we see that the term proportional to Λ^2 has, as expected, disappeared: only a logarithm of the cutoff appears, which is in fact restricted to the coefficients f'_0, f'_1, where it appears with the same coefficient in both the momentum and dimensional regularization schemes.

Our discussion of dimensional regularization has clearly been very restricted: we have considered only a few simple low-order perturbative contributions to a particular matrix element of the two simplest local operators of our theory. It would clearly be very desirable to derive a non-perturbative renormalization group equation for an effective Lagrangian defined in terms of such operators, along the lines of the derivation given in the preceding section for the momentum cutoff scheme. Unfortunately, the obviously very formal prescription given here for obtaining finite matrix elements of local operators, by simply eliminating the pure pole parts at $d = 4$ in Feynman loop integrals analytically continued to complex dimensionality, cannot be extended beyond the perturbatively expanded amplitudes of the field theory, as we simply have no way of giving a sensible *non-perturbatively valid* definition of a local quantum field theory in other than integer dimensions. For example, we do not know how to write down the analog of the functional integral (16.32) for the exact generating functional of a theory in *non-integer dimensions*, whose dynamics is specified in terms of an effective Lagrangian expanded in local operators, with perturbative matrix elements defined by dimensional pole subtraction.

Nevertheless, as we shall see in the next section, many important applications of effective field theories may be carried out completely in the context of perturbation theory, and in such cases the dimensional regularization approach is extraordinarily useful. We have already seen a glimmer of why this might be the case in the examples above: unlike the situation in a momentum cutoff scheme, integer powers of the renormalization scale which would otherwise result in the mixing of operators of different dimension as the scale is changed are simply absent in dimensional regularization—a fact which enormously simplifies the power-counting behavior of effective field theories. In particular, the contribution of higher-dimension "irrelevant" operators to low-energy amplitudes will remain "small" (in a precisely quantifiable sense) even when loop integrals are considered, provided we employ dimensional regularization methods to define these integrals.

Another very important advantage of dimensional regularization (over the momentum cutoff approach) is the ease with which it incorporates local vector gauge symmetries, which are formally preserved in this approach, as the form of the Lagrangian for such symmetries remains unaltered in (integer) dimensions other than the physical one. This turns out to have the very pleasant consequence that the Ward identities of the theory expressing the local gauge symmetry are preserved under dimensional regularization. We shall return to these issues in the subsequent chapters. In particular, a consistent definition of regularized local composite operators, extending the low-order examples given above, but valid to all orders of perturbation theory, requires graph-theoretical technology that we will develop in the next two chapters when we consider perturbative renormalization in more detail. The general procedure—the "normal product formalism" of Zimmermann—for obtaining well-defined local composite operators will be explained in detail in Section 18.1.

Before leaving the issue of regularization, we should comment on one potentially confusing issue which may already have crossed the reader's mind in connection with the absence of power-dependence on the cutoff scale in dimensionally regularized amplitudes. We previously emphasized the difficulty—due to just such power-dependences in a momentum cutoff scheme-with maintaining "small" values (i.e., much smaller than the cutoff scale of the theory) for the coefficients of the relevant operators in an effective Lagrangian defined by momentum cutoff. This "fine-tuning" difficulty is most dramatically manifested in the cosmological constant and hierarchy (Higgs mass) problems, briefly discussed earlier. These issues are *not* obviated by the existence of a regularization scheme (dimensional regularization), where power-dependence on the cutoff scale is automatically absent. In a sense, the "fine-tuning" at the UV scale necessary to remove, order by order in perturbation theory, large shifts in the coefficients of the relevant operators at low scales is just an automatic consequence of the structure of dimensionally regularized perturbative amplitudes: it is built *ab initio* into the definition of local operators (or rather, their *perturbative* matrix elements) in the dimensional scheme, and is therefore a matter of convention, not physics. In any event, the important applications of dimensional regularization in effective field theories, as we shall see in the next section, occur in situations (e.g., chiral Lagrangians in QCD) in which the UV scale is perhaps an order or two of magnitude above the scale of the interesting physics, so there is no issue of "fine-tuning" at the level of 10^{-15} or 10^{-120} as in the hierarchy or cosmological-constant problems.

16.6   Effective field theories: a compendium

Our study of effective field theory so far has concentrated on using the technology of effective Lagrangians to characterize in a precise mathematical language the change in the form of the local dynamics of an underlying microscopic theory when it is examined at progressively longer distance scales, "smearing out", as it were, the fine details of the interactions at shorter distances. We have chosen as our prime example a theory of a single self-interacting scalar field, and the aforesaid smearing process can be expressed in this case very simply in terms of the momentum-space Fourier modes $\tilde{\phi}(k)$ of the field, by progressively integrating out in the path-integral modes with momentum $|k| > \Lambda$, where Λ is a sliding ultraviolet cutoff.

The Wilsonian effective Lagrangian obtained by this procedure is just the simplest example of a much more general class of effective field theories, obtained in general by a combination of (a) change of functional variable of integration in the path integral defining the theory, and (b) a *partial* evaluation of the resultant path integral, whereby some, but not all, of the field variables are integrated out *in the absence of external sources*. The field modes removed by integration correspond to those which we are not interested (or unable) to probe, perhaps because they correspond to amplitudes which are inaccessible in presently available low-energy experiments. The dependence of the functional integrand obtained by this partial integration on the remaining field modes defines the *effective Lagrangian* which incorporates the accessible physics of the theory. It is essential to recognize that this effective field theory yields *exact* results for the n-point amplitudes of the remaining "smeared" fields.

Let us make the argument a little clearer by introducing some notation which captures the general context. Let $\phi_n, n = 1, 2, ..N$ represent the complete set of fields used to define the dynamics of the theory at some high-energy (short-distance) scale, and define a new set of fields $\Phi_m, m = 1, 2, ..M$ by

$$\Phi_m = f_m(\phi_n; \Lambda) \qquad (16.75)$$

Note that (a) the number M of smeared fields Φ_m may be different (typically, smaller) than the original number N of short-distance fields ϕ_n, and that (b) the smearing functions f_m may be linear or non-linear in character, may depend on a sliding energy scale Λ, and are not in general invertible—the smearing in this sense entailing a "loss of information" as regards the full local physics of the theory. If the theory is originally specified in terms of the original ϕ_n fields via a (Euclidean) Lagrangian $\mathcal{L}(\phi_n)$, then the *effective Lagrangian* $\mathcal{L}_{\text{eff}}(\Phi_m)$ *associated with the smeared fields* Φ_m is defined by

$$e^{-\int d^4x \mathcal{L}_{\text{eff}}(\Phi_m)} \equiv \int \prod_n \mathrm{D}\phi_n \delta(\Phi_m - f_m(\phi_n;\Lambda)) e^{-\int d^4x \mathcal{L}(\phi_n)} \tag{16.76}$$

Provided we are only interested in the n-point functions of the new fields Φ_m, we can discard completely the original microscopic Lagrangian $\mathcal{L}(\phi_n)$ in favor of the effective theory defined by $\mathcal{L}_{\text{eff}}(\Phi_m)$, as the generating functional for the Φ_m can be written entirely in terms of the latter,

$$Z[J_m] = \int \prod_n \mathrm{D}\phi_n e^{-\int d^4x \mathcal{L}(\phi_n) + \int d^4x J_m(x) f_m(\phi_n;\Lambda)} \tag{16.77}$$

$$= \int \prod_m \mathrm{D}\Phi_m e^{-\int d^4x \mathcal{L}_{\text{eff}}(\Phi_m) + \int d^4x J_m(x)\Phi_m(x)} \tag{16.78}$$

as we can see by introducing the definition (16.76) on the right-hand side of (16.78). We note here that the smearing functions are subject to certain smoothness requirements in order to ensure that the resultant effective Lagrangian (16.76) can be expanded in multi-nomials of local products of the smeared fields Φ_m (see the discussion following (16.47)).

The application of effective field theory methods has become a wide-spread industry in modern high-energy physics, and it would certainly require an entire additional volume to do justice to only the most widely used. We shall conclude our very brief introduction with a few examples that illustrate the main types, based on the nature of the smearing functions used to define the effective theory, and refer the reader to more detailed treatments available in the many excellent reviews of this subject for a more thorough discussion of the individual cases. Following the general philosophy exemplified by (16.75), we can choose functional change of variables which involve

1. a linear transformation of the modes of a single fundamental field of the theory,
2. a linear transformation involving several distinct fundamental fields, or
3. a non-linear change of field variables. In this latter case, one may be left with an effective field theory involving completely different fields than the underlying "microscopic" elementary fields which define the short-distance dynamics of the theory.

We have already encountered an example of the first type in our discussion of the renormalization group transformation of a scalar field theory with a momentum cutoff. Here the smearing function amounts to a cutoff of the Fourier modes $\tilde{\phi}(k)$ of a single scalar field. An extreme example of such a cutoff is the constraint effective potential discussed in Section 14.3 (cf. (14.58)), where the effective field Φ is just the zero-momentum mode $\tilde{\phi}(0)$ of the original field theory: all non-zero momentum modes are integrated out. As we saw in Chapter 14, the remaining (highly truncated!) effective theory is an important tool when examining the possibility of spontaneous symmetry-breaking of the underlying theory. A somewhat less trivial example is provided by *non-relativistic effective field theory* (NREFT),[9] where we are interested in the Fourier modes $\tilde{\phi}(k)$ of a massive field corresponding to non-relativistic quanta of the same: i.e., with $|\vec{k}| \sim \kappa << m, |k_0| \sim m + O(\kappa^2/m)$. We can expose these modes of the field by a simple linear transformation of the original field ϕ, which here we take to be a real scalar field with ϕ^4 interaction and short-distance (Minkowski) Lagrangian

$$\mathcal{L} = \frac{1}{2}\partial_\mu \phi \partial^\mu \phi - \frac{1}{2}m^2\phi^2 - \frac{\lambda}{4!}\phi^4 \tag{16.79}$$

Define a new field

$$\psi(x) \equiv \sqrt{(2m)}e^{imx^0}\int \theta(k_0)\tilde{\phi}(k)e^{-ik\cdot x}\frac{d^4k}{(2\pi)^4} \tag{16.80}$$

in terms of which the original field may be written

$$\phi(x) = \frac{1}{\sqrt{2m}}(e^{-imx^0}\psi(x) + e^{imx^0}\psi^\dagger(x)) \tag{16.81}$$

We shall assume that "relativistic" modes of the new field $\psi(x)$ have been integrated out,[10] and that the effective theory that remains contains only Fourier modes of $\tilde{\psi}(k)$ with $k_0 << m$. Accordingly, when (16.81) is substituted back into the Lagrangian (16.79), terms with unequal numbers of ψ and $\psi^\dagger$ fields are accompanied by time-dependent factors $e^{\pm 2inmx^0}$ with n a non-zero integer which must vanish when we integrate the Lagrange density over time to form the action, as they cannot be compensated by the remaining time-dependence of the ψ fields (which, given the assumed momentum scales $|k_0| \sim m + O(\kappa^2/m)$ of the original ϕ field, mean that the momentum modes relevant to $\tilde{\psi}(k)$ have $|\vec{k}| \sim \kappa << m, k_0 \sim O(\kappa^2/m) << m$). Substituting (16.81) in the Lagrangian (16.79), one finds for the Minkowski action of the resultant effective theory the leading terms

$$\mathcal{L} \to \int d^4x\{i\psi\dot{\psi}^\dagger + \frac{1}{2m}\psi^\dagger\vec{\nabla}^2\psi - \frac{\lambda}{16m^2}(\psi^\dagger\psi)^2\} + \cdots \tag{16.82}$$

where the dots refer to terms with unequal numbers of ψ and $\psi^\dagger$ (which will induce pair creation and annihilation processes which are unimportant in the non-relativistic regime), as well as terms like $\frac{1}{2m}\dot{\psi}^\dagger\dot{\psi}$ which scale like $\frac{k_0^2}{m}$ (subleading, as $k_0^2/m << k_0 \sim \vec{k}^2/m$, the scaling behavior of the free, quadratic part of the effective Lagrangian). The effective Lagrangian (16.82) can be used to establish the existence of bound states in $d = 2, 3$ dimensions (for negative λ) exactly as in Section 11.2, and the reader can verify that for weak coupling the bound-state energy is correctly determined to leading order

---

[9] The reader may find it convenient at this point to review our discussion of non-relativistic threshhold physics in Section 11.2.

[10] As in the renormalization group transformations of Section 16.3, this will result in a modification of the coefficients of the leading terms in the effective Lagrangian. In weakly coupled theories, as we imagine here, these modifications will be small and can be computed in perturbation theory. Practically, the determination of the coefficients in the effective Lagrangian for any given cutoff scheme is carried out by a "matching" process which we shall describe briefly at the end of this section.

in λ by the bubble diagrams generated by the non-relativistic Lagrangian (16.82) (see Problem 6).

The essential feature of the effective field theory chosen here is the careful choice of a change of field variable, followed by integrating out those Fourier modes which do *not* correspond to the dominant momentum regions involved in the physics of interest—in this case, the infrared regions generating the threshold singularities responsible for non-relativistic bound states in weakly coupled theories (cf. Section 11.2). The full development of the non-relativistic effective field theory for QCD (termed NRCQD) has led to important progress in understanding the physics of systems containing "heavy quarks" (i.e., charm, bottom, and top quarks) (Bodwin *et al.*, 1995). Other examples of effective field theories of this type are heavy quark effective theory (HQET) and soft collinear effective theory (SCET), which again involve changes of field variable and momentum mode restrictions appropriate for extracting the leading contributions to QCD processes involving, respectively, (HQET) processes with non-relativistic heavy quarks interacting with relativistic light quarks and gluons, and (SCET) processes in which highly energetic quarks interact with gluons or sets of gluons with small total squared four-momentum. Much more on the use of effective field theory methods in heavy quark physics generally can be found in the text of Manohar and Wise (Manohar and Wise, 2000).

The second type of effective field theory enumerated above corresponds to a situation in which the smearing function acts differently (though linearly) on different fields in the theory. The simplest case is one in which we simply integrate out completely the degrees of freedom corresponding to some subset of particles in the theory. Typically, this is useful when the particles can be divided into "light" and "heavy" subsets, with only the light particles accessible at available accelerator energies, so that the effects of the heavy particles occur only through their appearance in internal lines in the graphs of the theory. A simple toy model illustrating this situation consists of a light fermion field ψ (of mass m) coupled to a heavy scalar ϕ (of mass M) via a Yukawa interaction $g\bar{\psi}\psi\phi$. If we ignore self-interactions of the scalar field, the Euclidean-generating functional of the theory takes the form

$$ Z = \int \mathbf{D}\psi\mathbf{D}\phi\, e^{-\int d^4x(\mathcal{L}_\psi + \frac{1}{2}(\partial_\mu\phi)^2 + M^2\phi^2) + g\bar{\psi}\psi\phi)} \tag{16.83} $$

where $\mathcal{L}_\psi$ is the free Lagrangian for the light fermion (although we may also allow this field to have other interactions unconnected with the heavy scalar, e.g., gauge interactions, without altering what follows in any essential way). The exponent in this functional integral is at most quadratic in the scalar field, which we may therefore integrate out completely, obtaining

$$ Z = \int \mathbf{D}\psi\, e^{-\int d^4x\mathcal{L}_\psi - \frac{g^2}{2}\int d^4x\, d^4y\, S(x)\Delta_E(x-y)S(y)} \tag{16.84} $$

$$ S(x) \equiv \bar{\psi}(x)\psi(x) \tag{16.85} $$

$$ \Delta_E(z) = \int \frac{e^{-ik\cdot z}}{k^2 + M^2} \frac{d^4k}{(2\pi)^4} \tag{16.86} $$

If we formally expand the massive Euclidean scalar propagator Δ_E in inverse powers of the scalar mass,

$$\Delta_E(z) = \int e^{-ik \cdot z} \left(\frac{1}{M^2} - \frac{k^2}{M^4} + \frac{k^4}{M^6} \cdots \right) \frac{d^4 k}{(2\pi)^4}$$

$$= \frac{1}{M^2} \delta^4(z) + \frac{1}{M^4} \Box_z \delta^4(z) + \dots \tag{16.87}$$

we see that the net effect of the heavy scalar is to induce an effective Lagrangian in terms of the light fermion fields containing an infinite number of local terms of progressively higher dimension (from the spacetime-derivatives) and inverse powers of the large mass:

$$\mathcal{L}_{\text{ind}}(\psi) = G_0 S(x)^2 - G_2 (\partial_\mu S(x))^2 + \dots$$

$$= G_0 (\bar{\psi}(x)\psi(x))^2 - G_2 (\partial_\mu (\bar{\psi}(x)\psi(x)))^2 + \dots \tag{16.88}$$

$$G_0 \equiv \frac{g^2}{2M^2}, \quad G_2 \equiv \frac{g^2}{2M^4}, \dots \tag{16.89}$$

When we consider the low-energy amplitudes of the light fermion, higher terms will therefore be suppressed by (even) powers of the ratio of the low momenta (or masses) of the fermion amplitudes divided by the heavy mass M. Comparing the form of this induced Lagrangian with the general structure (16.7, 16.8), we see that in this case the heavy particle mass is playing the role of the ultraviolet cutoff Λ in the momentum cutoff theory. The neglect of heavy-particle self-interactions means that our toy model includes their effects only at tree level: the graphs generated by (16.88) correspond to the contraction of a single internal heavy-particle line to a four-fermion vertex, with the momentum dependence of the heavy-particle propagator appearing as an infinite series of higher derivative terms in coordinate space.

The Fermi weak interaction Lagrangian, quartic in fermion fields, arises in just this way, by integrating out the heavy W boson field in the electroweak model. In that case, of course, the quartic effective Lagrangian involves the square of vector and axial currents, as the W couples vectorially to the Fermi fields of the theory. Once loop effects are taken into account, things become more complicated (just as discussed in Section 16.3). The general situation was first delineated by Appelquist and Carrazzone (Appelquist and Carrazzone, 1975), in their *decoupling theorem*, which asserts that the only effect of integrating out a subset of heavy-particle fields in a local field theory, apart from terms suppressed by ratios of the momentum scales and masses of the light degrees of freedom to the heavy-particle masses, consists of a renormalization of the coefficients of the relevant and marginal operators in the remaining field theory, *provided that the latter form a perturbatively renormalizable set of operators*. The exact meaning of the latter condition will become clear in the following chapter, when we define and discuss in depth the concept of perturbative renormalization.

Our final category of effective field theories corresponds to situations in which the effective Lagrangian is specified in terms of fields which do not even appear in the microscopic Lagrangian specifying the short-distance dynamics: typically, these are non-linear functions of the original elementary fields of the theory. The classic example of this type is provided by the chiral Lagrangians describing the low-energy behavior of hadronic amplitudes. The underlying QCD Lagrangian containing quark and gluon fields in this case is replaced by an effective Lagrangian in terms of meson (and possibly baryon) fields. We shall illustrate the basic idea by taking a highly simplified version of the real world as our starting point: we assume there are only two quarks (the "up" and "down" quarks), and neglect, at least initially, their masses m_u and m_d (which are known to be much smaller than all other mass scales in QCD), so that our theory is described at the microscopic level (i.e., at distance scales much smaller than a fermi) by the Lagrangian (cf. (15.112))

$$\mathcal{L}_{\mathrm{QCD}} = -\frac{1}{4}\mathrm{Tr}(F_{\mu\nu}F^{\mu\nu}) + \sum_{a=1}^{2} \bar{q}_a i \not{D} q_a \tag{16.90}$$

The local gauge group will as usual be taken to be SU(3), although we shall see that specific details of the gauge group dynamics remain essentially hidden through the process of generating the effective field theory. We use the notation $q_a(x)$ for the quark fields, with $q_1(x) = u(x)$ the up-quark field and $q_2(x)$ the down-quark field. Defining left and right chiral parts of the quark fields in the usual way, $q_L(x) = \frac{1+\gamma_5}{2}q(x)$, $q_R(x) = \frac{1-\gamma_5}{2}q(x)$, the quark kinetic term in (16.90) can be rewritten as

$$\sum_a \bar{q}_a i \not{D} q_a = \sum_a (\bar{q}_{La} i \not{D} q_{La} + \bar{q}_{Ra} i \not{D} q_{Ra}) \tag{16.91}$$

as the $\gamma_0\gamma_\mu$ product separating $q^\dagger(x)$ from $q(x)$ in the quark kinetic bilinear commutes with γ_5. Formally, therefore, our fundamental Lagrangian is invariant under the eight-dimensional *chiral group* U(2)xU(2) (four generators from each U(2)) corresponding to the global linear field transformations

$$q_{La}(x) \rightarrow V_{Lab}\, q_{Lb}(x)$$
$$q_{Ra}(x) \rightarrow V_{Rab}\, q_{Rb}(x) \tag{16.92}$$

with V_L, V_R independent 2x2 unitary matrices. We may identify several important subgroups of this chiral group:

1. The abelian subgroup U(1)×U(1) corresponding to $V_L = e^{i\omega_L}$, $V_R = e^{i\omega_R}$ where $\omega_{L,R}$ are real phases. As we saw previously in Section 15.5 (cf. (15.169)), the diagonal subgroup with $\omega_L = \omega_R$ corresponds to an exact Noether symmetry— in our case, just "quark number conservation" (or equivalently, modulo a factor of 3, baryon number conservation), while the chiral subgroup with $\omega_L = \omega = -\omega_R$ is broken by the chiral anomaly at the quantum level, and is therefore *not* a symmetry of the full quantum field theory. Note that this latter symmetry would also be broken explicitly if a quark mass term (involving the bilinears $\bar{q}_L q_R, \bar{q}_R q_L$) were present.

2. The diagonal non-abelian SU(2) group defined by $V_L = V_R = V, V \in$ SU(2), also corresponds to an exact Noether symmetry of the theory: *isospin* symmetry (cf. Section 12.5).

3. The chiral non-abelian subgroup defined by $V_L = V_R^\dagger = V \in \mathrm{SU}(2)$ is (absent quark mass terms) an exact Noether symmetry of the theory: it is not anomalous (the anomaly (15.187) for the corresponding current would contain a single generator t_a of $\mathrm{SU}(2)$ in the trace, which would thereupon vanish). However, as we shall now discuss, it fails to be a symmetry of the vacuum: *the chiral symmetry of QCD is spontaneously broken.*

The final item above plays a central role in the low-energy dynamics of the theory, as we know from Goldstone's theorem that it immediately implies the presence of three massless spinless particles in the theory—one for each broken generator of the chiral $\mathrm{SU}(2)$ group. Of course, the fact that the up and down quark masses are not zero, but small, means that in the real world these particles are light, not massless, and since the late 1960s they have been identified with the pion iso-triplet. The hypothesis that the chiral symmetry is spontaneously broken via the appearance of a *quark condensate*, whereby the quark bilinear $\bar{q}_L q_R$ acquires a non-vanishing vacuum expectation value (VEV)

$$\langle 0|\bar{q}_{La} q_{Rb}|0\rangle = \mathcal{B}\delta_{ab} \tag{16.93}$$

where $\mathcal{B}$ is a constant of dimension mass^3 (from the dimensions of the quark fields) is no longer a matter for any serious debate: its validity has been more than adequately confirmed by extensive non-perturbative numerical computations using lattice QCD. Under a general chiral transformation (V_L, V_R),

$$\bar{q}_{La} q_{Rb} \rightarrow V_{La'a}^\dagger V_{Rbb'} \bar{q}_{La'} q_{Rb'} \tag{16.94}$$

from which we see that the VEV (16.93) leaves the diagonal isospin subgroup $V_L = V_R$ unbroken, but does indeed break the chiral $\mathrm{SU}(2)$ subgroup with $V_L = V_R^\dagger$. Of course, we are at liberty to redefine the quark fields by a (dynamically exact) chiral symmetry transformation $q_{La}(x) \rightarrow V_{ab}^* q_{Lb}(x)$, $q_{Ra}(x) \rightarrow V_{ab}^T q_{Rb}(x)$, for some $V \in \mathrm{SU}(2)$, whereupon the VEV becomes

$$\langle 0|\bar{q}_{La} q_{Rb}|0\rangle \rightarrow \mathcal{B}(V^2)_{ab} \equiv \langle V|\bar{q}_{La} q_{Rb}|V\rangle \tag{16.95}$$

where in the final equality we have chosen to parameterize the degenerate vacua of the theory by the chiral transformation V connecting the particular vacuum to the canonical one in (16.93) corresponding to $V = 1$. From our discussion of spontaneous symmetry-breaking in Section 14.3, we recall that well-defined amplitudes in an infinite-volume theory where an initially exact symmetry is spontaneously broken can be obtained only by introducing a small symmetry-breaking perturbation which "tickles" the system into a particular one of the infinitely many degenerate vacua, before the infinite volume limit is taken. As we shall see below, in real life this perturbation is provided by the quark masses which we have so far neglected.

In our proof of the Goldstone theorem in Section 14.2, we saw that the Noether current of a spontaneously broken symmetry serves as an interpolating field for the corresponding Goldstone boson (in other words, this operator possesses a non-vanishing vacuum to single-particle matrix element for the corresponding Goldstone boson, which by Haag–Ruelle theory means that it can be used to construct the exact

S-matrix scattering amplitudes of the Goldstone particle). In our case the (three) relevant currents are the axial vector currents $J_{5a\mu} \equiv \bar{q}\gamma_\mu\gamma_5\tau_a q$, $a = 1, 2, 3$ (where $\tau_{1,2,3}$ are the Pauli matrices: we avoid using the usual σ notation here as a field with this name will shortly make its appearance). We may just as well use the pseudoscalar operators $\bar{q}\gamma_5\tau_a q$, however: indeed, if a quark mass term is present (as it is, in the real world), these operators are proportional to the divergence of the axial vector currents $J_{5a\mu}$ (cf. (15.199), but with no anomaly term and an SU(2) generator matrix τ_a included both in the current on the left and the pseudoscalar divergence on the right). Consequently, they must have a non-vanishing vacuum to single pion matrix element if the $J_{5a\mu}$ do.

We shall therefore content ourself with writing a generating functional with sources for the $\bar{q}\gamma_5\tau_a q$ operators, as well as for the scalar density $\bar{q}q = \bar{q}_L q_R + $ h.c. whose VEV signals the spontaneous symmetry-breaking of the theory. Knowledge of this functional is tantamount (in virtue of the Haag–Ruelle or LSZ scattering theories discussed in Chapter 9) to knowledge of the full set of multi-pion scattering amplitudes in the theory. We therefore define (in Minkowski space, and glossing over the usual fine points *vis-à-vis* gauge fixing, DFP determinants, ghosts, etc., in our specification of the functional integral)

$$Z[s, \vec{p}] \equiv \int \mathbf{D}\bar{q}\mathbf{D}q\mathbf{D}A_\mu e^{i\int d^4x(\mathcal{L}_{\text{QCD}} - \bar{q}(x)(s(x) - i\gamma_5\vec{\tau}\cdot\vec{p}(x))q(x))} \qquad (16.96)$$

where the A_μ are the gauge vector gluon fields implementing the underlying color SU(3) local gauge symmetry, and $s(x), \vec{p}(x)$ are as usual c-number external sources coupled to the operators of interest in the theory. The expected spontaneous symmetry-breaking means that we must supplement this functional specification of the theory by a choice of vacuum, inserted via an infinitesimal "magnetizing" field—in this case, a small perturbing quark mass term $\epsilon\bar{q}(x)q(x)$ (ϵ small, real and positive), which can be implemented by taking the source field $s(x)$ to contain the spacetime constant term ϵ (plus, as usual, fields vanishing at infinity, or outside some compact region of support). The source term may be decomposed by chirally splitting the quark fields in the usual way:

$$\bar{q}(s - i\gamma_5\vec{\tau}\cdot\vec{p})q = \bar{q}_L(s + i\vec{\tau}\cdot\vec{p})q_R + \bar{q}_R(s - i\vec{\tau}\cdot\vec{p})q_L$$

$$= \bar{q}_L\chi q_R + \bar{q}_R\chi^\dagger q_L \qquad (16.97)$$

$$\chi(x) \equiv s(x) + i\vec{\tau}\cdot\vec{p}(x) \qquad (16.98)$$

From the form (16.97) it follows immediately that the source term (together with the Lagrangian $\mathcal{L}_{\text{QCD}}$, from our previous discussion) is invariant under the joint chiral transformation (16.92) (with V_L, V_R SU(2) matrices), together with the source field transformation

$$\chi(x) \rightarrow V_L\chi(x)V_R^\dagger \qquad (16.99)$$

If we consider the effect of a functional change of field variable in the path integral (16.96) consisting precisely of such a SU(2)×SU(2) chiral transformation (which, being anomaly-free, has unit functional Jacobian), we see that this invariance transfers directly to the functional $Z[s, \vec{p}]$ which we may just as well write as a functional $Z[\chi]$ of the 2x2 matrix source field $\chi(x)$:

$$Z[\chi] = Z[V_L \chi V_R^\dagger] \tag{16.100}$$

Next, let us define a new 2x2 matrix field $\Sigma(x) = \sigma(x) + i\vec{\tau} \cdot \vec{\pi}(x)$ which incorporates four fields—an isoscalar σ and an isovector $\vec{\pi}$—in terms of which we shall write our effective Lagrangian. The latter may be defined in terms of the functional Fourier transform of $Z[\chi]$, as follows:

$$\int d^4 x \mathcal{L}_{\text{eff}}[\Sigma] \equiv -i \ln \left(\int \mathbf{D}\chi e^{-\frac{i}{2} \int d^4 x \operatorname{Tr}(\chi^\dagger(x)\Sigma(x))} Z[\chi] \right) \tag{16.101}$$

where the factor of one-half arises as a consequence of the relation $\operatorname{Tr}(\chi^\dagger(x)\Sigma(x)) = \operatorname{Tr}(\chi(x)\Sigma^\dagger(x)) = 2(s(x)\sigma(x) + \vec{p}(x) \cdot \vec{\pi}(x))$. The inverse Fourier transform relation then becomes

$$Z[\chi] = \int \mathbf{D}\Sigma e^{i \int d^4 x (\mathcal{L}_{\text{eff}}[\Sigma] + \frac{1}{2} \operatorname{Tr}(\chi^\dagger(x)\Sigma(x)))} \tag{16.102}$$

Once again, the chiral SU(2)×SU(2) symmetry (16.100) transfers directly to our new effective Lagrangian $\mathcal{L}_{\text{eff}}[\Sigma]$ defined in (16.101),[11]

$$\mathcal{L}_{\text{eff}}[\Sigma] = \mathcal{L}_{\text{eff}}[V_L \Sigma V_R^\dagger] \tag{16.103}$$

The effective Lagrangian $\mathcal{L}_{\text{eff}}[\Sigma]$ is an *exact* transcription of the dynamics of QCD relevant for the determination of the full n-point Green functions of the quark bilinear fields $\bar{q}q$ and $\bar{q}\gamma_5\vec{\tau}q$ coupled to the sources χ: in particular, if we knew the exact form of this functional, we would be able to calculate arbitrary multi-pion scattering amplitudes exactly, and even determine the exact *nucleon* mass from the location of the nucleon–antinucleon threshold in $\pi^0 - \pi^0$ scattering, for example! Of course, all we know about this effective Lagrangian is that it is invariant under the chiral symmetry (16.103). However, by the same arguments of clustering and Lorentz-invariance which led to the general form (16.6) in Section 16.2, any (appropriately regularized) effective Lagrangian must be expandable in an infinite series of products of local operators and their spacetime derivatives. The chiral symmetry which we must impose on $\mathcal{L}_{\text{eff}}[\Sigma]$ implies that only certain combinations of the matrix field Σ can appear in this expansion: specifically, the Lagrangian must be given in terms of traces of products of Σ and $\Sigma^\dagger$ arranged to ensure the validity of (16.103). We thereby obtain the following expansion (rescaling the field Σ to fix the coefficient of the kinetic term at $\frac{1}{4}$, and including an infinitesimal mass term $\epsilon \operatorname{Tr}(\Sigma)$ from the source $s(x)$ as discussed above, with a coefficient rescaled to ϵ' after rescaling Σ)

$$\mathcal{L}_{\text{lin}}[\Sigma] = \frac{1}{4}\operatorname{Tr}(\partial_\mu \Sigma^\dagger \partial^\mu \Sigma) + \frac{\mu^2}{4}\operatorname{Tr}(\Sigma^\dagger \Sigma) - \frac{\lambda}{16}(\operatorname{Tr}(\Sigma^\dagger \Sigma))^2 + \epsilon' \operatorname{Tr}(\Sigma) + \dots \tag{16.104}$$

---

[11] The reader may easily check that the SU(2)xSU(2) chiral symmetry in (16.103) is equivalent to an O(4) rotation of the four fields $\sigma, \vec{\pi}$. One frequently finds discussions of chiral symmetry phrased in terms of this O(4) language.

where the dots refer to higher-dimension (and therefore, in the sense of Section 16.3, irrelevant) operators such as $(\mathrm{Tr}(\Sigma^\dagger\Sigma))^3$, $\mathrm{Tr}(\partial_\mu\Sigma^\dagger\partial_\nu\Sigma)\mathrm{Tr}(\partial^\mu\Sigma^\dagger\partial^\nu\Sigma)$, etc. We shall return to the role of higher-dimension operators in our theory below. Here, we note that the dimension 4 (or less) terms indicated in (16.104) constitute the so-called "linear σ model". As our field $\sigma = \frac{1}{2}\mathrm{Tr}(\Sigma)$ reproduces the matrix elements of $\bar{q}q$ we must ensure that it acquires a VEV in the vacuum, which implies the choice of sign of the second term in (16.104). The linear model leads in the usual way at tree level to a vacuum expectation value for the σ field at the unique minimum of the field polynomial (for infinitesimal positive ϵ')

$$P(\sigma, \vec{\pi}) = -\frac{\mu^2}{2}(\sigma^2 + \vec{\pi}^2) + \frac{\lambda}{4}(\sigma^2 + \vec{\pi}^2)^2 - 2\epsilon'\sigma \qquad (16.105)$$

which occurs at $\langle\sigma\rangle \equiv v = \mu/\sqrt{\lambda}$. Displacing the field $\sigma(x) = v + \hat{\sigma}(x)$ in the usual way, we find that the $\vec{\pi}$ fields lose their mass term and become massless Goldstone bosons as expected. This result, of course, continues to hold at the exact minimum of the non-derivative part of the (unknown!) full $\mathcal{L}_{\mathrm{lin}}[\Sigma]$, which by the chiral symmetry, is necessarily a function of $\sigma^2 + \vec{\pi}^2$: thus, at the minimum, we have $\vec{\pi} = 0$ and the flat directions are just those of the $\vec{\pi}$ fields. In addition to the massless Goldstone $\vec{\pi}$ fields, the model also contains the massive $\hat{\sigma}$ degree of freedom, with mass of order μ.[12]

As we explained in our general discussion at the beginning of this section, the derivation of an effective field theory usually entails, in addition to a functional change of variable, the partial elimination of degrees of freedom by integrating out field modes which are not important at the energy scales of interest. We now proceed to this second step, beginning with the (still, in principle exact) effective theory (16.104). We shall be interested in processes occurring at momentum scales much lower than the mass scale μ of the "heavy" degrees of freedom interpolated by the $\hat{\sigma}$ field. Precisely as in our discussion of heavy particle decoupling above, we can do this by integrating out the σ degree of freedom, leaving an effective Lagrangian depending only on the Goldstone fields $\vec{\pi}$. The most convenient way to do this involves a further change of variable, whereby we re-express the theory in terms of new fields $S(x), \vec{\Pi}$ via the non-linear transformation

$$\Sigma(x) = \sigma(x) + i\vec{\tau}\cdot\vec{\pi}(x) \equiv S(x)e^{i\vec{\tau}\cdot\vec{\Pi}/v}, \quad S(x) \equiv \sqrt{\sigma^2 + \vec{\pi}^2} = v + \hat{S}(x) \qquad (16.106)$$

The reader may easily verify (Problem 8) that this change of variable, when inserted in (16.104), leads, after expanding the exponential, to a Lagrangian with (canonically normalized) massive $\hat{S}$ field and massless $\vec{\Pi}$ fields. The chiral symmetry transfers directly to the unitary matrix field $U(x) \equiv e^{i\vec{\tau}\cdot\vec{\Pi}/v}$ (as $S(x) = \sqrt{\det(\Sigma)}$ is chirally invariant): the theory must be invariant under

---

[12] Note that while the theory contains physical (massless) pions, there is no *stable* particle associated with the σ field: the mass scale μ is naturally of the order of the other important physical hadronic scales, e.g., the rho resonance pole or nucleon mass, i.e., closer to 1 GeV, and quite a bit larger than the VEV v, which turns out to be just the pion decay constant $f_\pi \sim 100$ MeV (see Problem 7).

$$U(x) \rightarrow V_L U(x) V_R^\dagger \qquad (16.107)$$

The result of integrating out the massive $\hat{S}$ field must therefore be an effective Lagrangian $\mathcal{L}_{\text{nonlin}}[U]$, subject to the exact global symmetry (16.107), which our new effective theory inherits from the original symmetry (16.103) of the linear model. Following a by now familiar pattern, we therefore set about constructing the most general chirally invariant functional of the unitary matrix field U, as an expansion in terms involving traces of products of U, $U^\dagger$ and their spacetime-derivatives. As factors of U and $U^\dagger$ must appear adjacent in the traces to ensure invariance under (16.107), they must have derivatives to avoid evaporating via the unitarity constraint $U^\dagger U = U U^\dagger = 1$. The coefficient of the leading term in the expansion (two derivatives only) can be chosen to yield the canonical normalization of the kinetic term for the $\vec{\Pi}$ field, and we find the *non-linear σ model*

$$\mathcal{L}_{\text{nonlin}}[U] = \frac{v^2}{4} \text{Tr}(\partial_\mu U^\dagger \partial^\mu U)$$

$$+ L_1 \text{Tr}(\partial_\mu U^\dagger \partial^\mu U)^2 + L_2 \text{Tr}(\partial_\mu U^\dagger \partial_\nu U) \text{Tr}(\partial^\mu U^\dagger \partial^\nu U)$$

$$+ L_3 \text{Tr}(\partial^\mu U^\dagger \partial_\mu U \partial^\nu U^\dagger \partial_\nu U) + \dots \qquad (16.108)$$

where we have followed the notation of Gasser and Leutwyler (Gasser and Leutwyler, 1985) in notating the higher-order coefficients L_1, L_2, etc. If we expand the exponential $U = e^{i\vec{\tau} \cdot \vec{\Pi}/v}$, we find an effective Lagrangian with a massless kinetic term for our Goldstone pion fields $\vec{\Pi}$ and interaction terms which all contain two (or more) spacetime-derivatives. For example, the terms $\mathcal{L}_2$ with two derivatives only (from the first term on the right-hand side of (16.108)) become (see Problem 9)

$$\mathcal{L}_2 = \frac{1}{2}\partial_\mu \vec{\Pi} \cdot \partial^\mu \vec{\Pi} + \frac{1}{6v^2}(\vec{\Pi} \cdot \partial_\mu \vec{\Pi} \, \vec{\Pi} \cdot \partial^\mu \vec{\Pi} - \vec{\Pi}^2 \partial_\mu \vec{\Pi} \cdot \partial^\mu \vec{\Pi}) + \dots \qquad (16.109)$$

where the dots represent operators of dimension 8 or higher. The terms $\mathcal{L}_4, \mathcal{L}_6..$ with four, six,.. spacetime-derivatives may similarly be expanded in powers of the fields $\vec{\Pi}$ which interpolate for the massless pions of our toy theory. At least at tree level, we see that the interaction terms in this effective Lagrangian give rise to powers of the external momenta of the process corresponding to the spacetime-derivatives, and that at low momenta (much smaller than the scale v, say, which can be shown to correspond exactly to the pion decay constant f_π—see Problem 7) the multi-pion scattering amplitudes of our theory should be given to a good approximation by the graphs generated by $\mathcal{L}_2$, with progressively smaller contributions from $\mathcal{L}_4, \mathcal{L}_6 \dots$, etc.

But what about loops, which must certainly be included if we wish to calculate in a systematic way the complete amplitudes implied by our effective theory? And what about the infinite set of terms represented by the dots in (16.109), containing higher powers of the $\vec{\Pi}$ field, but still only two derivatives, which we may expect to contribute comparably to the indicated ones, by this argument?

In fact, as the discussion in Section 16.5 made clear, the formal expansion of an effective Lagrangian in local operators containing powers of the fields and their derivatives only acquires a precise meaning once a regularization procedure is provided which defines unambiguously the n-point functions of these operators. In the case of chiral Lagrangians, the dimensional regularization scheme described there is by far the best choice. The cutoff introduced in this scheme was called Λ previously, but in accordance with standard convention, we shall use μ henceforth, to be distinguished from the similarly notated scale of the "heavy" σ degrees of freedom of the linear σ model discussed above, although the scale of the dimensional regularization may perfectly well be chosen at such a value.

Note that the choice of μ is in a sense arbitrary, as it merely corresponds to a reshuffling of the definition of the (infinite set) of dimensionally regularized operators of the theory (as we shall see in Chapter 17), although it should certainly be chosen at a "sensible" value which minimizes the size of the logarithmic arguments in the amplitudes of interest (thereby improving the rapidity of convergence of the chiral perturbation theory). Indeed, as we saw in the preceding section, the only dependence of the regularized amplitudes of the theory on the regularization scale μ comes through logarithmic factors of the form $\ln\left(p^2/\mu^2\right)$, where p is a generic external momentum of the process, as the loop integrals contain only massless pion propagators carrying combinations of the external and loop momenta, and the latter are integrated out, with dependence on μ only arising from expanding the μ^{4-d} factors associated with each loop integral around the physical dimension $d = 4$.

Apart from the mass dimensions provided by the couplings associated with the various vertices in the $2(4,6,\dots)$ derivative Lagrangians $\mathcal{L}_2$ $(\mathcal{L}_4, \mathcal{L}_6,\dots)$, which (as U is dimensionless) must have mass dimension M^{4-n}, and the dimension M^{-1} associated with the $1/v$ factor accompanying each pion in an interaction vertex, the remaining mass dimension of a dimensionally regulated amplitude must arise solely from integer powers of the generic external momentum p (modulated by dimensionless logarithms involving the regularization scale μ). Consider a graph with E external pions and a total of N_π pion fields appearing at all the interaction vertices, of which there are N_2 from the $\mathcal{L}_2$ Lagrangian, N_4 from $\mathcal{L}_4$, etc. The total mass dimension of the (truncated) amplitude is $4 - E$, of which a mass power $\sum_n (4 - n) N_n - N_\pi$ is contributed by explicit factors of the dimensionful constants of the theory. The remaining D mass dimensions must therefore come from the powers of external momentum, and is

$$D = 4 - E - \left(\sum_n (4 - n) N_n - N_\pi\right) \tag{16.110}$$

The E external pion fields and N_π fields at the interaction vertices clearly produce graphs with the number of internal lines $I = \frac{N_\pi - E}{2}$ (as each internal line arises from the contraction of two pion fields). Moreover, we saw in Section 10.4 that a connected graph with L loops and $V = \sum_n N_n$ vertices has $I = L + V - 1$ internal lines. Eliminating $N_\pi = 2L + 2V + E - 2$ from (16.110), we find

$$D = 2 + \sum_n (n - 2) N_n + 2L \tag{16.111}$$

As the second and third terms on the right are zero or positive, we see that the leading behavior of an arbitrary multi-pion amplitude at low momentum (a) vanishes at least quadratically as the external momenta go to zero, and (b) that the leading behavior at low momentum ($D = 2$), is given completely by the lowest term $\mathcal{L}_2$ in our general effective Lagrangian, and indeed, by only the *tree graphs* ($L = 0$) obtained therefrom. Higher-order corrections at low momentum, of order p^4 say, require using the higher-order term $\mathcal{L}_4$ at tree level, or $\mathcal{L}_2$ to one-loop order; and so on. In any event, it is clear that only a finite and well-defined set of parameters (which must be determined by experimental fits, or calculated non-perturbatively, say by lattice QCD techniques) are relevant up to any given order of the low-momentum expansion of the amplitudes of the theory. We see again the quintessential advantage of an effective field theory: the sensitivity of the amplitudes of the theory in a restricted momentum regime is found to be restricted to a limited set of terms, allowing these amplitudes to be calculated in a systematic, though approximate, fashion in perturbation theory, starting from the leading terms in the effective Lagrangian for the theory.

The highly simplified toy model with which we have introduced the ideas of chiral effective Lagrangians must, of course, undergo substantial elaboration to provide an adequate description of low-energy hadronic physics in the real world. Our massless up and down quarks must be given (small) masses, and the somewhat higher strange quark mass also included if we wish to study the low-energy amplitudes of strange mesons. The methods described above can be generalized to deal with the inclusion of quark mass terms, treated perturbatively (so the amplitudes of the theory are developed in a double expansion in the small external momenta of the pseudo-Goldstone mesons and the light quark masses), and sources can be introduced for the vector and axial vector currents of the theory. Even the axial U(1) anomaly (of importance in the calculation of the neutral pion decay rate to two photons) can be included systematically in the resultant effective Lagrangian. The interested reader is encouraged to pursue these further developments, which are fully laid out in the seminal work of Gasser and Leutwyler (Gasser and Leutwyler, 1985) (in turn based on the pioneering contributions of Weinberg (Weinberg, 1968)).

Considerations of space require us to bring to a conclusion this all-too-brief survey of effective field theory: admittedly, we have barely scratched the surface of this rich, and, for modern elementary particle theory, profoundly important subject. A few more comments are in order, however. As a practical matter, most applications of effective field theory are carried out by a *matching procedure*, whereby the coefficients of the putative effective Lagrangian (itself determined by appropriate choice of a smearing of the underlying fields followed by integrating out unwanted modes, and fully exploiting any available symmetries to restrict the set of allowed operators) relevant to the physics regime of interest are determined by equating amplitudes computed by the use of the effective Lagrangian with the same amplitudes computed either perturbatively or non-perturbatively (e.g., by lattice methods) up to the appropriate level of sensitivity in an expansion in some available small momentum ratio. The technical details of this matching procedure will not be described further here: we refer the reader to any of a number of excellent reviews on effective field theory, for example, the review of Georgi (Georgi, 1993) and the TASI-2002 lectures of Rothstein (Haber and Nelson-eds, 2004).

16.7 Problems

1. Verify that the graphs in Fig. 16.2 give rise to the amplitude displayed in (16.15). Use the Feynman parameter identity (16.67) to combine the propagators in the loop integral, which you may evaluate with a sharp momentum cutoff (i.e., a factor of $\theta(\Lambda - |k|)$).

2. Carry out the steps leading from (16.34) to (16.37).

3. Show that the cutoff dependence of the connected functional W_0 arising from the functional integral of the (exponentiated) free cutoff Lagrangian $\mathcal{L}_0(\phi, \Lambda)$ in (16.28) corresponds to the source-independent term in (16.43).

4. Verify that the loop integral appearing in (16.53) can be expanded as indicated in (16.55).

5. Show that the expansion coefficients f_n of the one-loop integral (16.62) in powers of q^2/m^2 for $n > 0$ are identical up to powers of m^2/Λ^2 in dimensional regularization and sharp momentum cutoff schemes. Note that the integral subtracted at zero momentum (thereby leaving only the $f_n, n > 0$ terms) is convergent as $d \to 4$ in the dimensional scheme and has a finite limit (with $O(m^2/\Lambda^2)$ corrections) as $\Lambda \to \infty$ in the sharp momentum cutoff scheme.

6. Starting from the effective non-relativistic Lagrangian (16.82), show that the bubble diagrams contributing to the 2-2 amplitude (corresponding to the Fourier transform of the four-point function $\langle 0|T(\psi\psi\psi^\dagger\psi^\dagger)|0\rangle$) produce in $d = 2$ or $d = 3$ spacetime dimensions (provided $\lambda < 0$) a pole corresponding to a non-relativistic threshold bound state, of exactly the form found in the full relativistic theory in Section 11.2.

7. By identifying the Noether currents $J_{5a\mu} = \bar{q}\gamma_\mu\gamma_5\tau_a q$ (to lowest order) for chiral SU(2) transformations in the QCD Lagrangian with the corresponding current in the effective non-linear Lagrangian (16.108), we find that the σ model VEV v can be identified with the pion decay constant f_π. Here is the argument:

   (a) Show that the Noether current implementing infinitesimal chiral transformations (i.e., $V_L = V_R^\dagger = 1 + i\omega_a\tau_a$, ω_a infinitesimal) in the effective non-linear model is (to lowest order in the chiral expansion)

   $$J_{5a\mu}^{\text{eff}} = i\frac{v^2}{4}\text{Tr}(\tau_a(U^\dagger\partial_\mu U - U\partial_\mu U^\dagger)) \tag{16.112}$$

   (b) The pion decay constant is defined in terms of the vacuum to one-pion matrix element of the axial current $J_{5a\mu}$ (which appears in the hadronic part of the Fermi interaction Hamiltonian $\frac{G_F}{\sqrt{2}}\bar{u}\gamma^\mu(1 + \gamma_5)d\ \bar{\mu}\gamma_\mu(1 + \gamma_5)\nu_\mu$ responsible for charged pion decay to $\mu + \bar{\nu}_\mu$), as follows:

   $$\langle 0|J_{5a\mu}(0)|\pi_b(p)\rangle = if_\pi p_\mu \delta_{ab} \tag{16.113}$$

   Using (16.112) (with U suitably expanded in $\vec{\Pi}$ fields) in (16.113), show that we can identify f_π with v.

8. Carry out the change of field variables indicated in (16.106) in the terms given explicitly in (16.104), with the "tickling" term set to zero ($\epsilon' = 0$), to obtain

the form of the nonlinear Lagrangian (as a function of S and U fields) prior to integrating out the "heavy" S field.

9. Substituting the definition $U(x) \equiv e^{i\vec{\tau}\cdot\vec{\Pi}/v}$ in the first term in the non-linear chiral Lagrangian (16.108), and expanding out the exponential, show that one obtains the terms shown in (16.109). Use the indicated quartic interaction term to calculate, to lowest order at small pion momenta, the amplitude for $\pi^+ + \pi^- \to \pi^0 + \pi^0$ scattering.

17
Scales II: Perturbatively renormalizable field theories

In the previous chapter we emphasized the importance of the scale separation property of local quantum field theories, which expresses our ability to predict, at least to some reasonable level of accuracy, the features of particle interactions at long distances (or equivalently, at low energies) despite the fact that we are inevitably ignorant of the "ultimate" details of these interactions at arbitrarily small distance scales. Historically, this property was first realized in the context of the perturbative treatment of a specific quantum field theory, quantum electrodynamics, the local gauge theory describing the interactions of photons with electrons (and other charged leptons). As we saw in Chapter 2, the development of interacting quantum field theories in the two decades from the early 1930s to around 1950 was severely hampered by the ubiquitous infinities—more precisely, ultraviolet divergences in the integrals over the momenta of particles appearing in intermediate states in scattering amplitudes—which plagued all higher-order calculations in these theories.

In the late 1940s the development of graphical techniques for covariant perturbation theory proved to be the critical ingredient needed to surmount this impasse. The covariant graphical techniques introduced by Feynman revealed in a much more transparent way the structure of the scattering amplitudes of the theory, and in particular the "nested" character of the divergent contributions to these amplitudes, features which were then exploited by Dyson in his classic development of perturbative renormalization theory for quantum electrodynamics (Dyson, 1949). Dyson was able to show, order by order in perturbation theory, and *to all orders of perturbation theory*, that the distressing divergences disappeared provided the amplitudes of the theory were re-expressed in terms of a finite number of "renormalized" parameters corresponding to measurable (and therefore *ipso facto* finite) low-energy properties of the theory. Otherwise stated, if an ultraviolet cutoff Λ is introduced to regularize the theory (thereby making all loop integrals finite), the reparameterization of the amplitudes of the theory in terms of renormalized quantities softened the dependence on the cutoff Λ, yielding amplitudes which were finite in the limit $\Lambda \to \infty$, with a cutoff sensitivity at finite Λ corresponding to powers (typically quadratic) of the ratio of the masses and momenta in the scattering amplitude to the UV cutoff Λ.

The remarkable quantitative agreement of the quantum electrodynamic amplitudes (anomalous magnetic moment of the electron and muon, Lamb shift, etc.) computed using this procedure with the measured experimental values remain among the most impressive successes of physical science. Nevertheless, the overpowering impression persisted among many physicists that the procedures of Dyson amounted to an intellectually unsavory "fudge"—a mere sweeping under the rug of potential underlying inconsistencies in the theory of which the ultraviolet divergences were apparently the overt symptom. This unease began to dissipate in the early 1970s, with the development by Wilson of the effective field theory point of view which we discussed in Section 16.2. The inescapable presence of new physics (quantum gravity, string theory, or what have you) at short distances implies, as emphasized in the preceding chapter, that we must necessarily imagine a cutoff Λ of some kind at high energies: the question of infinities in the (unphysical!) limit where this cutoff is taken to infinity is then replaced by the issue of *sensitivity* of the measurable low-energy amplitudes of the theory to the value (and type) of this cutoff. Our object in this chapter is to review the techniques that have been developed to study this sensitivity in the context of the formal weak coupling perturbation expansions introduced in Chapter 10, and in particular to show that for a certain subset of field theories, the sensitivity is of the weak kind (inverse powers of the cutoff) first described by Dyson. We shall also see that these results fit naturally into the picture of the renormalization group flow of effective Lagrangians discussed in the previous chapter.

Before diving into the technical details, it may be helpful to the reader to give a simple example of the phenomenon outlined above, whereby the strong dependence of a regularized scattering amplitude on an ultraviolet cutoff can be dramatically weakened by a reparameterization of the theory in terms of "renormalized" parameter(s). Let us suppose that we are told that the 2-2 scattering amplitude $_{\text{out}}\langle k_3, k_4 | k_1, k_2 \rangle_{\text{in}}$ for some particle is described by the expression (given, as usual, up to terms involving inverse powers of the cutoff Λ)

$$\Gamma^{(4)}(k_1, k_2, k_3, k_4) = \frac{ig^2}{1 + Cg^2 \ln{(t/\Lambda^2)}} + \frac{ig^2}{1 + Cg^2 \ln{(u/\Lambda^2)}} + O(\frac{k_i^2}{\Lambda^2})$$

$$t \equiv -(k_3 - k_1)^2, \quad u \equiv -(k_4 - k_1)^2 \tag{17.1}$$

where g is the coupling constant for some interaction in the theory cut off at momentum Λ, and C is an uninteresting (real positive) numerical constant.[1] If we expand out the denominator factors in a perturbative series in powers of g^2, it is apparent that at any given finite order of perturbation theory, the amplitude displays logarithmic ultraviolet divergences, becoming infinite as a power of $\ln{(\Lambda^2)}$ as the cutoff Λ is taken to infinity. If we instead parameterize the amplitude in terms of a *renormalized coupling* g_R, defined simply in terms of the value of the 2-2 scattering amplitude at some experimentally accessible value of the momentum transfer variables $t = u = \mu^2 << \Lambda^2$, as follows:

---

[1] The curious reader may be interested in the origin of this simple expression, although it is not relevant to the present discussion. It represents the 2-2 amplitude for scattering in a theory of N massless Dirac fermions in two spacetime dimensions, with quartic interaction Lagrangian $\frac{1}{2}g^2(\sum_{i=1}^{N} \bar{\psi}_i \psi_i)^2$ (the so-called Gross–Neveu model (Gross and Neveu, 1974)), in the limit where $N \to \infty$ with $g^2 N$ fixed—the so-called "1/N" expansion. In this case the constant $C = N/(2\pi)$. It also gives, in a similar limit, the 2-2 scattering amplitude for a theory of N massless complex scalar fields in four spacetime dimensions with interaction Lagrangian $-\frac{1}{2}g^2(\sum_{i=1}^{N} \phi_i^\dagger \phi_i)^2$, but with the constant C now negative, $C = -N/(16\pi^2)$.

$$g_R^2 \equiv -\frac{i}{2}\Gamma^{(4)}(t = u = \mu^2) = \frac{g^2}{1 + Cg^2\ln{(\mu^2/\Lambda^2)}} \Rightarrow g^2 = \frac{g_R^2}{1 + Cg_R^2\ln{(\Lambda^2/\mu^2)}} \quad (17.2)$$

In the jargon of renormalization theory with which we must now begin to familiarize ourselves, such a definition is referred to as a *renormalization condition*: a complete set of such conditions (one for each free parameter needed to uniquely identify the low-energy theory) specifies a *renormalization scheme*. If we now re-express the 2-2 amplitude (17.1) in terms of the renormalized coupling g_R, we find

$$\Gamma^{(4)}(k_1, k_2, k_3, k_4) = \frac{ig_R^2}{1 + Cg_R^2\ln{(t/\mu^2)}} + \frac{ig_R^2}{1 + Cg_R^2\ln{(u/\mu^2)}} + O(\frac{k_i^2}{\Lambda^2}) \quad (17.3)$$

and we see that all dependence on the ultraviolet cutoff Λ has been removed to the level of the (in practice, for quantum electrodynamics) tiny inverse power terms which are, of course, harmless in the limit $(\Lambda \to \infty)$ in which the cutoff is removed entirely. In particular, the expansion of our 2-2 amplitude in powers of g_R gives a *renormalized perturbation expansion* in which each term is separately insensitive (again, up to harmless inverse power corrections) to the UV cutoff Λ. The essential components of the renormalization procedure in field theory can be seen already here, in this admittedly algebraically trivial example: first, the choice of a regularization method (effectively, a mathematical model of our ignorance of the theory at short distances), and second, the renormalization conditions identifying a specific reparameterization of the amplitudes of the theory (allowing us to conveniently inject accessible low-energy information about the theory).

A third feature of the renormalization program which becomes apparent once we evaluate amplitudes order by order in perturbation theory derives directly from the reparameterization step: the appearance of *subtracted amplitudes* at each order of perturbation theory. Let us illustrate this point at the first subleading order (g_R^4) in the expansion of our 2-2 amplitude (17.1) (corresponding to one-loop contributions in the underlying field theory). We may evidently define a coupling constant shift (or *counterterm*) δg_R^2 by the trivial identity

$$g^2 \equiv g_R^2 + \delta g_R^2 \quad (17.4)$$

where, by expanding (17.2), we have through order g_R^4,

$$\delta g_R^2 = -Cg_R^4\ln{(\Lambda^2/\mu^2)} + O(g_R^6) \quad (17.5)$$

Inserting (17.5) into the perturbative expansion of the amplitude (17.1) in powers of the bare coupling g, we find (neglecting terms suppressed by powers of the cutoff)

$$\begin{aligned}
\Gamma^{(4)}(k_1, k_2, k_3, k_4) &= i\{g^2 + Cg^4\ln{(\Lambda^2/t)} + \ldots\} + (t \to u) \\
&= i\{g_R^2 + Cg_R^4\ln{(\Lambda^2/t)} + \delta g_R^2 + O(g_R^6)\} + (t \to u) \\
&= i\{g_R^2 + Cg_R^4\ln{(\Lambda^2/t)} - Cg_R^4\ln{(\Lambda^2/\mu^2)} + O(g_R^6)\} + (t \to u) \\
&= i\{g_R^2 + Cg_R^4\ln{(\mu^2/t)} + O(g_R^6)\} + (t \to u) \quad (17.6)
\end{aligned}$$

We see in the penultimate line of (17.6) that the effect of the reparameterization in terms of the renormalized coupling g_R has been to introduce a *subtraction* of the $O(g_R^4)$

amplitude which precisely removes the logarithmic Λ-dependence of the latter in the original cutoff amplitude.

Understanding the structure of such subtractions will be crucial in developing a renormalization technology capable of exposing the cutoff sensitivity of field theory amplitudes at all orders of perturbation theory. In particular, let us note that although the reparameterization indicated in (17.2) has evidently succeeded in suppressing the cutoff dependence of the elastic scattering 2-2 amplitude (up to ignorable inverse power terms, as always), we have *not* demonstrated that the same reparameterization suffices to remove the cutoff sensitivity of *all* the amplitudes of the theory: 2-4, 2-6, etc., particle production amplitudes, for example. This much stronger requirement— that a reparameterization of a finite number of couplings (and masses) in terms of low-energy quantities can suppress the cutoff dependence of the *entire* set of S-matrix amplitudes of the theory, order by order, and to all orders of perturbation theory— is, amazingly, satisfied by a rich variety of local quantum field theories, which we collectively refer to as "perturbatively renormalizable theories": they are the subject of our enquiry in this chapter. Later, after developing the appropriate technology for perturbative renormalization theory, we shall see (in Section 17.4) that such theories appear naturally as low-energy limits of the effective Wilsonian field theories described in Chapter 16.

In the next section we shall examine in detail the structure of the cutoff dependence of general multi-loop Feynman integrals appearing in the perturbative expansion of amplitudes in a local quantum field theory. We shall see that the occurrence of divergent integrals (and subintegrals) in such loop amplitudes is associated with cutoff-dependent contributions which have a very simple (in fact, polynomial) momentum dependence. This latter fact will then be exploited in the subsequent section to demonstrate the equivalence of the set of subtractions needed to remove the leading cutoff dependence of an arbitrary Feynman amplitude (reducing it to the inverse power-dependence of the type seen above) to the result of a reparameterization of a set of coupling and mass parameters appearing in the Lagrangian of the theory. The intimate connection *reparameterization* $\Longleftrightarrow$ *subtractions*, of which we have just seen a particularly trivial example, is the essence of the proof of cutoff-insensitivity for perturbatively renormalizable theories.

17.1 Weinberg's power-counting theorem and the divergence structure of Feynman integrals

The sensitivity of field theory amplitudes to a short distance, or large momentum, cutoff arises in perturbation theory from the presence of loop integrals over the four-momenta carried by internal lines of the diagrams contributing to the amplitude in question. The level of the sensitivity is determined by the size of the contributions these integrals receive when some subset (or all) of the loop momenta are on the order of (or greater than) the imposed ultraviolet cutoff. Rigorous estimates of these contributions are possible if we work in Euclidean space, as first demonstrated by Weinberg (Weinberg, 1960). In fact, for the remainder of this chapter we shall be exclusively concerned with the cutoff sensitivity of the *Euclidean* amplitudes of field theory. Once well-defined Euclidean amplitudes are obtained (in the infinite cutoff limit) they can be shown (Zimmermann, 1969) to be analytically continuable to physically sensible Minkowski amplitudes, and hence to a well-defined S-matrix. Weinberg's theorem, which we shall state shortly, applies to a general multi-loop integral of the form

$$\mathcal{I}(k_1, k_2, .., k_E) = \int \frac{\mathcal{N}(k_1, k_2, .., k_E; l_1, l_2, .., l_L)}{\mathcal{D}(k_1, k_2, .., k_E; l_1, l_2, .., l_L)} \frac{d^d l_1 d^d l_2 \cdot \cdot d^d l_L}{(2\pi)^{dL}} \tag{17.7}$$

Here $k_1, k_2, .., k_E$ are the external momenta associated with the E external lines of the graph, and $l_1, l_2, .., l_L$ the L independent loop momenta appearing within the graph. The numerator factor $\mathcal{N}(k_1, k_2, .., k_E; l_1, l_2, .., l_L)$ is a multi-nomial of finite order in its four-momenta arguments, while the denominator factor is a product of the usual (Euclidean) Feynman propagator factors $p^2 + m_i^2$, which we shall here assume are all massive, to avoid concern over infrared divergences (which we shall be taking up specifically in Chapter 19), which are, of course, not relevant in a discussion of *short-distance* sensitivity of the theory. At present, we leave the spacetime dimension d free, as Weinberg's theorem applies for all (integral) dimensions.

A simple example is the one-loop graph of Fig. 17.1 (cf. also Fig. 16.2) contributing to 2-2 scattering in ϕ^4 theory, which we have already encountered on several occasions:

$$\mathcal{I}(k_1, .., k_4) = \int \frac{1}{(l^2 + m^2)((k_1 + k_2 - l)^2 + m^2)} \frac{d^d l}{(2\pi)^d} \tag{17.8}$$

For the purposes of the present discussion, we ignore overall numerical factors, powers of the coupling constant, etc. For large values of the loop momentum, $l \equiv |l| >> |k_i|, m$, the integral scales like

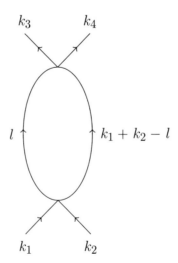

Fig. 17.1 A one-loop contribution to the 2-2 amplitude in ϕ^4-theory.

$$\int^\Lambda \frac{1}{l^4} l^{d-1} dl = \int^\Lambda l^{d-5} dl \qquad (17.9)$$

where Λ is a UV cutoff. Evidently, if $d < 4$ the integral is convergent at the upper end, and we may let $\Lambda \to \infty$ obtaining a finite result. If $d = 4$, the integral is logarithmically divergent, so the result at finite Λ contains a logarithmic sensitivity $\sim \ln \Lambda$ to the cutoff. For (integer) dimensions $d > 4$, the integral has a much larger, power growth dependence Λ^{d-4} on the cutoff. Evidently, the quantity $D \equiv d - 4$, which we shall term the *superficial degree of divergence of the loop integral*, indicates the demarcation point for ultraviolet convergence of the loop integral. Note that it is obtained very simply by counting the difference in powers of loop momenta in the numerator and denominator of the Feynman integrand, including a factor of l^d from the measure $d^d l$. For $D < 0$, we say that the integral is superficially convergent: in this case, the result at finite cutoff (much larger than the external momenta and masses) differs from the infinite cutoff limit by inverse powers of the cutoff (for d even, at least quadratically, perhaps with logarithmic modifications which are dominated by the power falloff), and we shall refer to such a loop integral as "UV-finite". For $D = 0$, we have a logarithmically divergent integral, with $\ln \Lambda$ sensitivity to the cutoff, while for $D > 0$ we have power-divergent integrals (linearly for $D = +1$, quadratically for $D = +2$, etc.) with much stronger dependence on the UV cutoff. Weinberg's theorem generalizes the particular case of a UV-finite integral to an arbitrary Euclidean Feynman integral of the form (17.7), as follows.

Theorem 17.1 (*Weinberg, 1960*) *The general Euclidean loop integral (17.7) is UV-finite provided the superficial degree of divergence associated with scaling any subset of loop momenta uniformly to infinity is negative. In other words, the overall integral is UV-finite if the superficial degree of divergence associated with integrating over any hyperplanar subspace of the full dL-dimensional Euclidean integration space is strictly negative.*

In the event that this condition is satisfied, we are assured that the sensitivity of the corresponding amplitude to a high-momentum cutoff Λ corresponds to the mild, and from the point of view of renormalization theory, ignorable inverse power-dependence on the cutoff that we have seen now on numerous occasions. We shall not attempt to reproduce the details of Weinberg's proof here, which depends on a technically sophisticated (but physically not particularly enlightening) application of real analysis, especially as the result appears completely natural, particularly once rephrased in the language of large momentum flows, as we shall see shortly.

Before going on to more general cases, let us just note that for the simple one-loop integral (17.8) of Fig. 17.1 one easily sees that in $d = 4$ dimensions, the degree of divergence associated with a hyperplane subspace of dimension $n \le 4$ is $D = n - 4$ so that the only divergence comes from the region corresponding to $n = 4$ in which all components of the loop four-momentum become large. This will typically be the case for more complicated multi-loop graphs: we will need only examine hyperplanes corresponding to all components of some subset (or possibly all) of the loop momenta becoming large. Physically this corresponds to regions of the integration in which large momentum "irrigates" various loops of the graph, individually or in some specified joint fashion. If this seems all a bit vague at the moment, we beg the reader's patience for a few moments longer: explicit examples will soon follow.

In our first foray into perturbative field theory in Chapter 10, we encountered the logarithmic divergence of (17.8) for infinite UV cutoff in four spacetime dimensions and recognized (cf. (10.35)) this infinity as the symptom in momentum space of the singular result of trying to multiply individually well-defined distributions at the same point in coordinate space (i.e., trying to obtain the squared coordinate space Feynman propagator $\Delta_F(z)^2$). We saw there (cf. the discussion following (10.36)) that the singularity amounts in coordinate space to an additive *local* δ-function term $\delta^4(z)$, or to a constant in the Fourier-transformed momentum space: indeed, subtracting the loop integral at any fixed value of the external momenta—for example, zero—gives a perfectly UV-finite result (cf. (10.36)). Let us revisit this result briefly from the point of view of Weinberg's theorem, as it serves as a useful prelude to the discussion in the more general multi-loop case. Taking $d = 4$ spacetime dimensions, and subtracting from the amplitude (17.8) its value with all external momenta set to zero, we obtain

$$\mathcal{I}_R(k_1, .., k_4) \equiv \mathcal{I}(k_1, .., k_4) - \mathcal{I}(0, .., 0) \tag{17.10}$$

$$= \int \{ \frac{1}{(l^2 + m^2)((k_1 + k_2 - l)^2 + m^2)} - \frac{1}{(l^2 + m^2)^2} \} \frac{d^4 l}{(2\pi)^4}$$

$$= \int \frac{2l \cdot (k_1 + k_2) - (k_1 + k_2)^2}{(l^2 + m^2)^2((k_1 + k_2 - l)^2 + m^2)} \frac{d^4 l}{(2\pi)^4} \tag{17.11}$$

and we see that the subtracted Feynman integral now has superficial degree of divergence $D = -1$ and is therefore convergent, by Weinberg's theorem (there is only a single loop here, so the only region of large momentum flow corresponds to l large, where the scaling of the subtracted integrand is as l^{-1}). Otherwise put, with a cutoff Λ in place, the subtracted integral depends on Λ at the level of power-suppressed terms of order $O(\frac{k_i^2, m^2}{\Lambda^2})$: in the conventional language of renormalization theory, it is *UV-finite.*[2] The origin of this convergence is simply that the leading behavior of the original ("unrenormalized") integrand (17.8) and the subtraction term at large l are identical, so the subtraction removes the leading asymptotic behavior of the integrand responsible for the logarithmic divergence (= logarithmic dependence of the cutoff integrals on Λ).

There is an alternative, and for our future purposes extremely important, way to understand the efficacy of this subtraction procedure in reducing the cutoff dependence of the loop amplitude. We shall introduce the notation

$$t^\Gamma \mathcal{I}(k_1, .., k_4) \equiv \mathcal{I}(0, .., 0) \tag{17.12}$$

for the subtraction term in (17.10). The operation t^Γ will be defined on all one-particle-irreducible (1PI) graphs Γ with superficial degree of divergence $D \geq 0$ as extracting the terms up to order D in the Taylor expansion in the external momenta k_i around

---

[2] The quadratic dependence arises from the fact that we can further improve the convergence of the integral by symmetric integration: i.e., taking the average of the integrand at l and $-l$ before integration.

zero momentum of the loop integral(s) $\mathcal{I}$ representing Γ. In the present case the loop integral has $D = 0$, and the operation t_Γ therefore simply takes the leading term of the Taylor expansion, i.e., the unsubtracted amplitude at zero momentum. Thus, our subtracted amplitude (17.11) amounts to

$$\mathcal{I}_R(k_1, .., k_4) = (1 - t^\Gamma)\mathcal{I}(k_1, .., k_4) \tag{17.13}$$

On the other hand, the right-hand side of (17.13) can be viewed as the sum of all the Taylor terms in the expansion of $\mathcal{I}(k_1, .., k_4)$ around zero momentum, linear or higher in the momenta. But as Γ is 1PI, all of its internal lines[3] are parts of loops, and therefore every differentiation of an internal propagator reduces the superficial degree of divergence of the integrand by 1. In our simple one-loop case, for example,

$$\frac{\partial}{\partial k_1^\mu} \frac{1}{(k_1 + k_2 - l)^2 + m^2)} = -2\frac{(k_1 + k_2 - l)_\mu}{((k_1 + k_2 - l)^2 + m^2)^2} \tag{17.14}$$

so the differentiation has reduced the scaling of the propagator at large loop momentum from $1/l^2$ to $1/l^3$. Accordingly, all the terms in the Taylor expansion of $\mathcal{I}_R$ in (17.13) have degree of divergence $D = -1$ or less, and are therefore UV-convergent.

This argument can clearly be generalized to 1PI graphs Γ with a superficial degree of divergence $D > 0$, by defining the Taylor operator t^Γ as the sum of the first $D + 1$ orders of the Taylor expansion in the external momenta (around zero): i.e., all terms up to homogeneous order k^D in the external momenta k_i of the graph. The Taylor operator is defined simply to be zero when applied to a superficially *convergent* graph or subgraph. As what remains after $t^\Gamma \mathcal{I}(k_i)$ is subtracted from $\mathcal{I}(k_i)$ are just the terms in the Taylor expansion with at least $D + 1$ derivatives with respect to the k_i, and each such derivative lowers the superficial degree of divergence of the Feynman integral $\mathcal{I}$ associated with the graph Γ (i.e., the scaling of the integrand when all loop momenta get uniformly large) by 1, we see that the subtracted amplitude $(1 - t^\Gamma)\mathcal{I}$ will again have $D < 0$, just as in our simple one-loop example.

Another critical point, the full implications of which will emerge in the next section, concerns the structure of the subtraction terms generated by the Taylor operator t^Γ: by definition, they are *polynomial* in the external momentum of the subtracted graph. This implies that they are equivalent to the terms that would be generated in perturbation theory by a *local* term in the Lagrangian with as many field operators as external lines of the graph in question (in this case, four), and with a finite number of spacetime-derivatives applied to the fields to generate the appropriate factors of momentum entering the graph. As we shall see in the following sections, it is exactly this property of the subtractions effected by the zero-momentum Taylor operators introduced here that allows us to connect the *subtractions* needed to remove the dominant cutoff dependence of the amplitudes of the theory with a precise set of *reparameterizations* of these amplitudes in terms of low-energy constants defined by appropriate renormalization conditions (as in the preceding section). It cannot be emphasized too forcefully that the entire essence of perturbative renormalization theory is implicit in the ideas introduced in this paragraph: the reader is strongly

---

[3] Recall from Chapter 10 that external legs are truncated by definition in 1PI amplitudes.

encouraged to engage in a thorough mental mastication of the arguments just given, before swallowing whole and proceeding to the next course.

Our discussion so far has been entirely in the context of the perturbative expansion of the Euclidean space n-point amplitudes (Schwinger functions) of the field theory, and indeed, the Weinberg power-counting theorem is explicitly formulated in this context. One sometimes encounters the assertion that a rigorous convergence analysis of Feynman amplitudes must *necessarily* be carried out in Euclidean space: only then, having obtained UV-finite, and suitably analytic, Euclidean space amplitudes, are we allowed to analytically continue back to the physically relevant Minkowski amplitudes. This is in fact, as first pointed out by Zimmermann (Zimmermann, 1968), incorrect. It is certainly true that Minkowski-based Feynman integrals based on the usually defined momentum-space free propagator $\Delta_F(k) = 1/(k^2 - m^2 + i\epsilon)$ are only conditionally convergent (as there is an infinite volume associated with the hyperboloid shells defined by finite ranges of the Minkowski squared momentum, say $a < k^2 < b$), and the power-counting theorems require absolute convergence. But if we return to the form (10.63) of the free propagator implied by the original definition of the Minkowski space functional integral in terms of an infinitesimally rotated time axis,

$$\Delta_F(k) = \frac{1}{(k^2 - m^2 + i\epsilon(\vec{k}^2 + m^2))} = \frac{1}{(k_0^2 - (\vec{k}^2 + m^2) + i\epsilon(\vec{k}^2 + m^2))} \quad (17.15)$$

absolute convergence of any Minkowski space Feynman integral (subtracted appropriately for UV-finiteness in Euclidean space) follows via the inequality (Problem 1)

$$\left| \frac{1}{(k_0^2 - (\vec{k}^2 + m^2) + i\epsilon(\vec{k}^2 + m^2))} \right| \leq \sqrt{1 + \frac{4}{\epsilon^2}} \frac{1}{k_0^2 + \vec{k}^2 + m^2} \quad (17.16)$$

In fact, for ϵ kept positive and non-zero, all of our Minkowski space Feynman integrands are majorized by the corresponding Euclidean ones, for which the arguments we have been presenting, using the power-counting prescriptions of the Weinberg theorem, are rigorously valid. The absolute convergence (and UV-finiteness) of the subtracted Euclidean amplitudes is therefore *a fortiori* correct for their Minkowski versions. Of course, proper mathematical rigor requires that we establish that the $\epsilon \to 0$ limit can indeed be carried out at the end, leading to well-defined *covariant* Minkowski space amplitudes. The covariance property in particular is not immediately obvious, given the presence of the non-covariant factor $\vec{k}^2 + m^2$ in the ϵ term in (17.15), but, once again, Zimmermann (Zimmermann, 1968) has done the hard work for us, and shown explicitly that the limit does indeed exist, with the resultant amplitudes well-behaved covariant tempered distributions. For the following, we shall return to Euclidean space, confident that the subtraction procedures devised there to remove the ultraviolet sensitivity of the amplitudes will be equally efficacious in Minkowski space.

The true power of Weinberg's theorem really emerges when we go to higher orders of perturbation theory, when we encounter multi-loop graphs with multiple independent regions of integration giving divergent contributions to the overall amplitude. Consider the two-loop diagram illustrated in Fig. 17.2 contributing to the 2-2 amplitude in ϕ^4 theory (again, in four spacetime dimensions). The corresponding loop integral is (again, ignoring overall numerical factors, couplings, etc.)

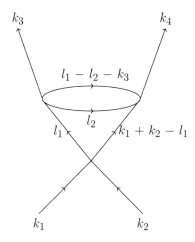

Fig. 17.2 A two-loop contribution to the 2-2 amplitude in ϕ^4-theory.

$$\mathcal{I}(k_i) = \int \frac{1}{(l_1^2 + m^2)((k_1 + k_2 - l_1)^2 + m^2)} \frac{d^4 l_1}{(2\pi)^4}$$

$$\cdot \int \frac{1}{(l_2^2 + m^2)((l_1 - l_2 - k_3)^2 + m^2)} \frac{d^4 l_2}{(2\pi)^4} \tag{17.17}$$

The superficial degree of divergence of the graph as a whole, obtained by taking both loop momenta l_1, l_2 large and counting total powers of these momenta in the numerator (namely the eight powers arising from the measure $d^4 l_1 d^4 l_2$) is evidently $D = 8 - 4 - 4 = 0$, so there is definitely an overall logarithmic divergence. But there is also a divergence coming from the region where the "outer" loop momentum l_1 is kept fixed and the "inner" loop momentum l_2 becomes large: along this hyperplane, the degree of divergence (counting just powers of l_2) is also $D = 0$. A divergence of this type, arising from large momentum flow through a part of, but not the whole, diagram, we shall term a "subdivergence". Note that there is also a region of the integration corresponding to a large momentum flow where l_1 and l_2 get large with $l_1 - l_2$ fixed, corresponding to large momentum flowing through the lines with momentum l_1, l_2 and $k_1 + k_2 - l_1$ in Fig. 17.2, but this region has degree of divergence $D = 4 - 2 - 2 - 2 = -2$ and therefore gives a convergent contribution to the full graph. In order to devise an appropriate set of subtractions capable of removing the leading logarithmic divergences in this situation, it will be convenient to introduce a notation for identifying various subgraphs $\gamma, \gamma_1, ..$ etc., of the full graph Γ associated with possible subdivergences of the full loop integral $\mathcal{I}$. We shall identify a graph or subgraph by indicating within square brackets the momenta of the lines contained therein, thus

$$\Gamma = [l_1, k_1 + k_2 - l_1, l_2, l_1 - l_2 - k_3]$$

$$\gamma = [l_2, l_1 - l_2 - k_3] \tag{17.18}$$

We have indicated here only the *renormalization parts* of the full graph Γ: namely, those connected 1PI subgraphs with superficial degree of divergence greater than or equal to zero (including possibly the entire graph, as here). Each possible subgraph γ can be assigned a degree of divergence $d(\gamma)$ corresponding to the power-counting associated with large momentum flow through all lines of that subgraph. In the present case, the only renormalization parts are the full graph Γ and the subgraph γ, with in both cases $d(\Gamma) = d(\gamma) = 0$. Taylor subtraction operators t^{Γ} and t^{γ} can be similarly associated with each renormalization part, in an obvious extension of the procedure followed in our previous one-loop example. For the subgraph γ, the external momenta now include, in addition to the k_i, the loop momentum l_1 associated with the lower pair of internal lines in Fig. 17.2 (more generally, the external momenta of a given renormalization part consists simply of those momenta which remain fixed when the large momentum flow giving rise to the degree of divergence of that subgraph is invoked). As in our present case both $d(\Gamma)$ and $d(\gamma)$ are zero, the Taylor operators amount to either setting $k_1 = k_2 = k_3 = k_4 = 0$ (t^{Γ}) or $k_1 = k_2 = k_3 = k_4 = l_1 = 0$ (t^{γ}). It seems entirely plausible, and the reader is encouraged to verify that the following subtracted integral, when examined in the context of Weinberg's theorem, is UV-finite:

$$\mathcal{I}_R(k_i) = (1 - t^{\Gamma})\bar{\mathcal{I}} = \bar{\mathcal{I}}(k_i) - \bar{\mathcal{I}}(0)$$

$$\bar{\mathcal{I}} \equiv \int \frac{1}{(l_1^2 + m^2)((k_1 + k_2 - l_1)^2 + m^2)} \frac{d^4 l_1}{(2\pi)^4}$$

$$\cdot (1 - t^{\gamma}) \int \frac{1}{(l_2^2 + m^2)((l_1 - l_2 - k_3)^2 + m^2)} \frac{d^4 l_2}{(2\pi)^4}$$

$$= \int \frac{2 l_2 \cdot (l_1 - k_3) - (l_1 - k_3)^2}{(l_1^2 + m^2)((k_1 + k_2 - l_1)^2 + m^2)(l_2^2 + m^2)^2((l_1 - l_2 - k_3)^2 + m^2)} \frac{d^4 l_1 d^4 l_2}{(2\pi)^8}$$

$$(17.19)$$

Indeed, the subtraction effected by t^{γ} ensures that the degree of divergence associated with large momentum flow through the subgraph γ is reduced to -1 (just as in our first one-loop example), while the overall subtraction effected by t^{Γ} reduces the degree of divergence of the graph as a whole (when both l_1 and l_2 become large) to -1. The particular combination of four loop integrals implied by the subtractions of (17.19) is therefore UV-finite by Weinberg's theorem, or equivalently, cutoff insensitive at the level of inverse powers of the cutoff. We have introduced a notation $\bar{\mathcal{I}}$ to indicate a subtracted amplitude containing Taylor subtraction operators for all proper subgraphs of the graph in question, but *not* the overall ("top level") subtraction needed if the full graph is superficially divergent.

At this point it will be convenient to introduce some graphical notations which will serve us in good stead as we attempt to generalize the insights gleaned from these first simple examples to arbitrary graphs. First note that the inner subtraction effected by the t^{γ} operator in (17.19) actually corresponds to replacing the integral over l_2 by a pure number (as t^{γ} sets l_1 and k_3 to zero), leaving a graph with only the lines $[l_1, k_1 + k_2 - l_1]$, which amounts to a graph in which the entire subgraph represented by γ

has been shrunk to a single vertex (in this case, without any momentum dependence, as the Taylor operator is of degree zero). We shall denote such a graph by Γ/γ, and more generally, graphs Γ in which some set of disjoint renormalization parts $\gamma_1, \ldots \gamma_n$ are shrunk to a point as a consequence of having been replaced by their Taylor operator evaluations as $\Gamma/\{\gamma_1, .., \gamma_n\}$. Also, the integrands associated with graphs (or subgraphs, or shrunk graphs) will be indicated by an obvious superscript notation. We may then abbreviate the expression (17.19) as follows:

$$\mathcal{I}_R^\Gamma(k_i) = (1 - t^\Gamma)\bar{\mathcal{I}}^\Gamma, \quad \bar{\mathcal{I}}^\Gamma = \mathcal{I}^\Gamma + \mathcal{I}^{\Gamma/\gamma}(-t^\gamma)\bar{\mathcal{I}}^\gamma \tag{17.20}$$

Note that the integrand $\bar{\mathcal{I}}^\gamma$ to which the inner subtraction t^γ is applied is just the original subgraph $\mathcal{I}^\gamma$ consisting of the lines $[l_2, l_1 - l_2 - k_3]$: the overbar in this case is superfluous, as this graph contains no proper superficially divergent subgraphs which need be subtracted. The formula (17.20) is our first example of a general recursion formula (originally due to Bogoliubov), the explicit solution of which (due to Zimmermann) will describe in a compact way the exact divergent structure of arbitrary Feynman integrals.

The two-loop example just discussed possesses a simplifying feature which may already have occurred to the reader: the divergent regions of the graph are *nested*— in other words, the inner subdivergence arising from the region of large momentum flow through the subgraph γ corresponds to a set of lines which are strictly contained in the set of lines that carry the large momentum leading to the overall divergence of the full graph Γ. This makes it easy to guess the proper set of sequential subtractions $(1 - t^\Gamma) \cdots (1 - t^\gamma) \cdots$ which render the graph UV-finite: one simply begins by subtracting off the innermost divergence and then proceeding outwards, at each stage performing the Taylor subtraction if a divergent subgraph is encountered. The generalization of this procedure to a large graph containing *non-intersecting* sets of nested divergences is also clear: the Taylor subtractions can be performed independently on the separate non-intersecting sets, leading to a subtracted integrand which possesses a finite $\Lambda \to \infty$ limit by Weinberg's theorem.

We shall use the terminology "non-overlapping" to describe the situations summarized here: two renormalization parts γ_1, γ_2 of a diagram (i.e., superficially divergent 1PI subgraphs) are said to be *non-overlapping* if either (a) one is entirely contained in the other (the "nested" case), or (b) they are completely non-intersecting (i.e., no lines in common). Our analysis of the divergence structure of Feynman graphs would be essentially concluded if we had only the non-overlapping case to consider.

There are, however, cases of divergent subgraphs that intersect partially, giving rise to the famous problem of *overlapping divergences*. Evidently, we must consider at least a two-loop diagram in order to find two distinct—but not simply nested— subdivergences. We shall illustrate the problem in a a self-interacting scalar theory which we have not heretofore studied- $\frac{\lambda}{3!}\phi^3$ theory in six spacetime dimensions—which has the advantage of being topologically similar to gauge theory, inasmuch as the basic interaction involves a trilinear coupling.[4] To fourth order in λ, one encounters the

---

[4] Note that the ϕ^3-theory has an unbounded spectrum below (cf. Section 8.4): all finite-energy states are unstable. Nevertheless, the renormalized perturbation theory of this model is perfectly sensible, to

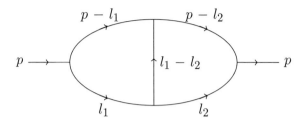

Fig. 17.3 A two-loop contribution to the scalar propagator in ϕ^3-theory.

self-energy contribution to the scalar propagator indicated by the graph in Fig. 17.3, corresponding to the Feynman integral (the external legs are truncated):

$$\mathcal{I}^\Gamma(p) = \int \frac{1}{(l_1^2 + m^2)((p - l_1)^2 + m^2)} \frac{d^6 l_1}{(2\pi)^6}$$

$$\cdot \int \frac{1}{(l_2^2 + m^2)((l_1 - l_2)^2 + m^2)((p - l_2)^2 + m^2)} \frac{d^6 l_2}{(2\pi)^6} \qquad (17.21)$$

The overall divergence degree of this graph is quadratic, $D = -2 - 2 + 6 - 2 - 2 - 2 + 6 = +2$, while the subdivergences corresponding to l_1 large (with l_2 fixed) or l_2 large (with l_1 fixed) have degree of divergence $D = -2 - 2 - 2 + 6 = 0$: i.e., logarithmic. Again, we introduce an abbreviation for the various renormalization parts of the diagram,

$$\Gamma = [l_1, p - l_1, l_2, l_1 - l_2, p - l_2],$$
$$\gamma_1 = [l_1, p - l_1, l_1 - l_2],$$
$$\gamma_2 = [l_2, l_1 - l_2, p - l_2] \qquad (17.22)$$

with $d(\Gamma) = 2, d(\gamma_1) = 0, d(\gamma_2) = 0$. It is apparent that in this case we are faced with two divergent subgraphs (γ_1 and γ_2) which have a non-trivial intersection (the line carrying momentum $l_1 - l_2$) but are not simply nested as in our previous two-loop example of Fig. 17.2. The correct subtraction procedure is hardly obvious in this case, especially as the subtraction terms obtained by applying the Taylor operators t^{γ_1} and t^{γ_2} depend on the order in which these operations are applied, as the reader may easily verify. It turns out that the correct way to "slice" the integrand in order to extract correctly the dominant contributions in all divergent subintegrations of the two-loop integral (17.21) involves subtractions only on non-overlapping renormalization parts. Specifically, we define the fully subtracted two-loop self-energy as

$$\mathcal{I}_R^\Gamma(p) = (1 - t^\Gamma)\bar{\mathcal{I}}^\Gamma(p) \qquad (17.23)$$
$$\bar{\mathcal{I}}^\Gamma(p) = \mathcal{I}^\Gamma + \mathcal{I}^{\Gamma/\gamma_1}(-t^{\gamma_1})\bar{\mathcal{I}}^{\gamma_1} + \mathcal{I}^{\Gamma/\gamma_2}(-t^{\gamma_2})\bar{\mathcal{I}}^{\gamma_2} \qquad (17.24)$$

any finite order, and the graph topology and divergence structure are similar to those appearing in gauge theories, making it a very useful laboratory for illustrating important issues of renormalization, unclouded by complications introduced by higher spin fields.

$$\mathcal{I}_R^\Gamma(p) = \mathcal{I}^\Gamma + \mathcal{I}^{\Gamma/\gamma_1}(-t^{\gamma_1})\bar{\mathcal{I}}^{\gamma_1} + \mathcal{I}^{\Gamma/\gamma_2}(-t^{\gamma_2})\bar{\mathcal{I}}^{\gamma_2} - t^\Gamma \mathcal{I}^\Gamma$$
$$- t^\Gamma(\mathcal{I}^{\Gamma/\gamma_1}(-t^{\gamma_1})\bar{\mathcal{I}}^{\gamma_1}) - t^\Gamma(\mathcal{I}^{\Gamma/\gamma_2}(-t^{\gamma_2})\bar{\mathcal{I}}^{\gamma_2})$$
$$\equiv \mathcal{I}_a(p) + \mathcal{I}_b(p) + \mathcal{I}_c(p) + \mathcal{I}_d(p) + \mathcal{I}_e(p) + \mathcal{I}_f(p) \tag{17.25}$$

Once again, as in (17.20), the overbars on $\bar{\mathcal{I}}^{\gamma_1}, \bar{\mathcal{I}}^{\gamma_2}$ are superfluous, as these subgraphs do not contain any further subdivergences to be subtracted. The required subtractions are indicated in Fig. 17.4, where we have introduced a simple notational device—a "o" attached to the external legs of a divergent 1PI graph or subgraph γ, to indicate the presence of the Taylor subtraction operator t^γ. A little thought shows that the subtractions indicated here are precisely those needed to render the fully subtracted amplitude $\mathcal{I}_R(p)$ UV-finite by Weinberg's theorem. One needs merely to check that the sum of terms indicated in (17.25) have negative degree of divergence in any of the possible subintegration hyperplanes of the integral in (17.21). For example, consider the region in which large momentum flows through γ_1: i.e., large l_1 with l_2 fixed. The subcombination $\mathcal{I}_a(p) + \mathcal{I}_b(p)$ has degree of divergence $D = -1$ as the logarithmic divergence for large l_1 (l_2 fixed) is subtracted in exactly the same way as in our previous examples, as does the combination $\mathcal{I}_d(p) + \mathcal{I}_e(p)$ (which is just minus the first three terms in the Taylor expansion of the already convergent $\mathcal{I}_a(p) + \mathcal{I}_b(p)$ around $p = 0$: recall that the full graph has degree of divergence $d(\Gamma) = 2$). The combination $\mathcal{I}_c(p) + \mathcal{I}_f(p)$ gives

$$\mathcal{I}_c(p) + \mathcal{I}_f(p) = -\int \frac{1}{(l_1^2 + m^2)} \Big\{ \frac{1}{((p - l_1)^2 + m^2)} - \frac{1}{(l_1^2 + m^2)} - \frac{2p \cdot l_1}{((l_1^2 + m^2)^2}$$
$$- \frac{4(p \cdot l_1)^2 - p^2}{((l_1^2 + m^2)^3} \Big\} \frac{d^6 l_1}{(2\pi)^6}$$
$$\cdot \int \frac{1}{(l_2^2 + m^2)^3} \frac{d^6 l_2}{(2\pi)^6} \tag{17.26}$$

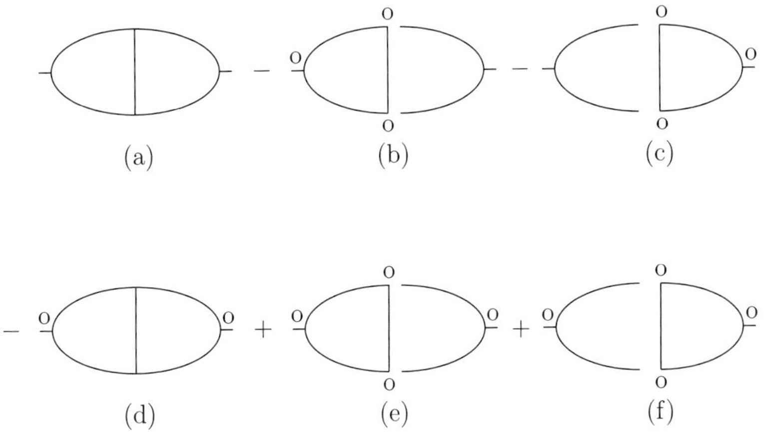

(a)   (b)   (c)

(d)   (e)   (f)

Fig. 17.4 Subtracted two-loop self-energy in ϕ^3 theory.

The subtracted propagator $1/((p - l_1)^2 + m^2) - 1/(l_1^2 + m^2) - \dots$ appearing between the curly braces is easily seen to have the asymptotic behavior $\sim 1/l_1^5$ for large l_1 (it consists of the terms in the Taylor expansion in p of the unsubtracted propagator with three or more derivatives), so the degree of divergence of the l_1 subintegration is reduced to $-2 - 5 + 6 = -1$, as desired. Note that we are specifically examining the region of l_1 large with l_2 *fixed*, so the logarithmic divergence in the final l_2 integral is of no consequence here. The efficacy of the subtractions in the subregion corresponding to large momentum flow through γ_2 is also obvious, by the symmetry of the graph. There remains the region of large momentum flow through the entire diagram (both l_1 and l_2 large). In this case it is easy to see that the dominant asymptotic contributions cancel pairwise between $\mathcal{I}_a$ and $\mathcal{I}_d$, $\mathcal{I}_b$ and $\mathcal{I}_e$, and $\mathcal{I}_c$ and $\mathcal{I}_f$.

The generalization of the subtraction procedure described here to arbitrary Feynman integrals was first obtained as a recursive formula by Bogoliubov and Parasiuk (Bogoliubov and Parasiuk, 1957)(with technical improvements by Hepp (Hepp, 1966)—hence the appellation "BPH scheme"). It is an obvious generalization of (17.24, 17.25):

$$\mathcal{I}_R^\Gamma = (1 - t^\Gamma)\bar{\mathcal{I}}^\Gamma \tag{17.27}$$

$$\bar{\mathcal{I}}^\Gamma = \mathcal{I}^\Gamma + \sum_{\gamma_1,\gamma_2,\dots\gamma_n;\gamma_i \cap \gamma_j = 0, i \neq j} \mathcal{I}^{\Gamma/\{\gamma_1,\dots,\gamma_n\}} \prod_{i=1}^{n} (-t^{\gamma_i})\bar{\mathcal{I}}^{\gamma_i} \tag{17.28}$$

where for conciseness we no longer indicate the dependence on external momenta. The sum in (17.28) is over all sets of disjoint renormalization parts γ_i strictly contained in Γ. It is a recursive formula inasmuch as the integrand $\bar{\mathcal{I}}^{\gamma_i}$ associated with each such renormalization part has all of its subdivergences already subtracted (by recursive use of the same formula, until renormalization parts are reached with no further subdivergences). If the graph as a whole is superficially convergent, the overall subtraction performed by the t^Γ operator in (17.27) is absent. If the full graph is divergent, then this t^Γ performs the subtraction needed to remove the divergent contribution from the asymptotic region in which large momentum irrigates the entire graph.

The proof of the Bogoliubov–Parasiuk formula is by induction on the number of loops. It is manifestly correct at one loop: either the graph is already convergent, or the $(1 - t^\Gamma)$ subtraction renders it finite; there are no subdivergences in a one-loop graph, so $\bar{\mathcal{I}}^\Gamma(p) = \mathcal{I}^\Gamma$ in (17.27). The subtractions given in (17.28) are then shown to be efficient in subtracting off the divergent contributions from subintegrations in which large momentum circulates between distinct renormalization parts γ_i, as we saw explicitly in our two-loop example of Figs. 17.3 and 17.4. A careful proof of (17.28) will not be given here (see, for example, (Hepp, 1966), (Van der Kolk and de Kerf, 1975)). From the point of view of physical intuition it is far more useful to examine its operation in explicit examples, where the subtractions are readily seen to induce cancellations between appropriate subclasses of diagrams for each distinct type of large momentum flow. We therefore encourage the reader to repeat the arguments given above for the integrals of (17.17, 17.21) for the additional examples given in the Problems at the end of this chapter.

The recursion formula (17.28) was explicitly solved by Zimmermann (Zimmermann, 1969), giving a transparent, and exceedingly simple, formula which generates automatically the complete set of subtractions needed to make an arbitrary Feynman diagram UV-finite. First, we define a *forest U of a graph* Γ as any set (including possibly the empty set) of *non-overlapping* renormalization parts of Γ. The set of all forests of Γ is denoted $\mathcal{F}(\Gamma)$. The origin of the name "forest" is apparent with a glance at Fig. 17.5. Here we show a forest consisting of N renormalization parts, $\gamma_1, \gamma_2, .., \gamma_N$ which are all proper subgraphs of some full graph Γ. Strictly speaking, forests containing only renormalization parts that are subgraphs but not (in the event that the full graph is superficially divergent) the full graph Γ are called "normal forests"— the set of which we shall denote $\mathcal{N}(\Gamma)$—while forests containing Γ (if divergent) are "full forests", comprising the set $\mathcal{F}(\Gamma)$. Fig. 17.5 displays the structure of a normal forest, with the nodes representing renormalization parts and the lines connecting two nodes indicating that the upper renormalization part is strictly contained in the lower one. Note that a normal forest has an *extremal set* of "biggest" renormalization parts (in Fig. 17.5, these are $\gamma_1, \gamma_2, .. \gamma_n$), none of which are contained in any other renormalization part. Typically, we denote forests of a graph by capital Roman letters, U, V, etc. Let $\mathcal{I}^{\Gamma}$ represent, as usual, the unsubtracted Feynman integrand for the graph Γ. Zimmermann's solution for the subtractions needed to generate directly $\bar{\mathcal{I}}^{\Gamma}$ in (17.28) is, for an arbitrary renormalization part γ (which may include Γ itself),

$$\bar{\mathcal{I}}^{\gamma} = \sum_{U \in \mathcal{N}(\gamma)} \prod_{\gamma_r \in U} (-t^{\gamma_r}) \mathcal{I}^{\gamma} \tag{17.29}$$

We shall show, by induction on the number of loops, that (17.29) solves (17.28). In the event that U is the empty forest, the product of (negative) Taylor operators is interpreted as unity: this term simply reproduces the original, unsubtracted integrand. All other terms involve at least one renormalization part and, therefore, a subtraction of the original integrand. The formula is evidently correct at one loop, as there are no possible subdivergences, the normal forests of γ are empty, and $\bar{\mathcal{I}}^{\gamma} = \mathcal{I}^{\gamma}$. We now proceed by induction, and assume that the Zimmermann formula has been established up to the (lower) number of loops contained in the subgraphs $\gamma_1, .., \gamma_n$ in (17.28). We may therefore insert (17.29) in (17.28) for the subgraphs $\Gamma_i, i = 1, 2, .., n$, obtaining

$$\bar{\mathcal{I}}^{\Gamma} = \mathcal{I}^{\Gamma} + \sum_{\gamma_1, \gamma_2, .. \gamma_n; \gamma_i \cap \gamma_j = 0, i \neq j} \mathcal{I}^{\Gamma/\{\gamma_1, .., \gamma_n\}} \prod_{i=1}^{n} (-t^{\gamma_i}) \sum_{U_i \in \mathcal{N}(\gamma_i)} \prod_{\gamma_r \in U_i} (-t^{\gamma_r}) \mathcal{I}^{\gamma_i} \tag{17.30}$$

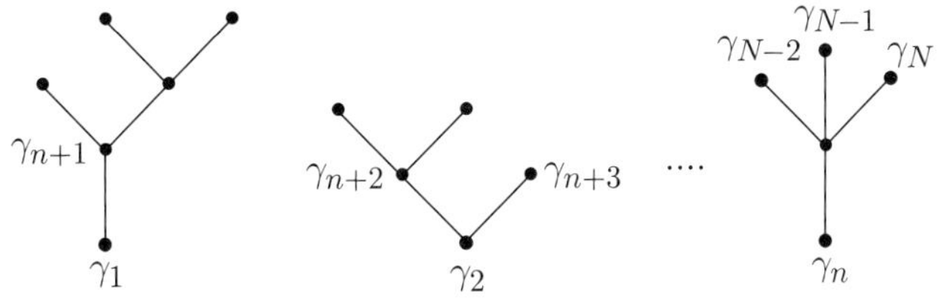

Fig. 17.5 Structure of renormalization parts in a normal Zimmermann "forest".

With a little thought, the reader will easily see that the sum in the second term on the right-hand side of (17.30) simply assembles all the non-empty normal forests of the graph Γ, sorted by their extremal elements $\gamma_1, \gamma_2, ..\gamma_n$, thereby reproducing the forest formula at the next higher level of induction. In the event that the full graph Γ is itself divergent, then the sum over all forests $\mathcal{F}(\Gamma)$ can be divided into pairs of normal forests U and full forests $[\Gamma, U]$ with an additional subtraction $-t^\Gamma$ for each full forest, yielding the *Zimmermann forest formula* for the fully subtracted amplitude given recursively in (17.27):

$$\mathcal{I}_R^\Gamma = \sum_{U \in \mathcal{F}(\gamma)} \prod_{\gamma_r \in U} (-t^{\gamma_r}) \mathcal{I}^\Gamma \tag{17.31}$$

Note that the nested nature of the subtractions in the forest formula (due to the absence of overlapping renormalization parts) allows us to write the complete Feynman integrand $\mathcal{I}^\Gamma$ (broken up into a product of reduced integrands $\mathcal{I}^{\Gamma/\{\gamma_1,...,\gamma_n\}}$ and individual renormalization parts $\mathcal{I}^{\gamma_i}$ in (17.30)) to the extreme right in (17.31): the Taylor operators in each of the "trees" in Fig. 17.5 then act sequentially, starting at the top (the smallest divergent subdiagram) and working downwards. The trickiest part of a proper proof[5] of the convergence of $\mathcal{I}_R^\Gamma$ in (17.31) lies in the demonstration that there exists an "admissible" routing of the loop momenta in the subdiagrams such that the result of the Taylor operations is unambiguous (as we necessarily encounter the situation in higher orders that the external momenta of one subgraph become the internal momenta of another).

As usual, nothing serves better to clarify how this works than an explicit example. In the case of the graph of Fig. 17.3, in addition to the empty forest, $\mathcal{F}(\Gamma)$ evidently contains the forests $[\gamma_1], [\gamma_2], [\Gamma], [\Gamma, \gamma_1], [\Gamma, \gamma_2]$—but *not* a forest containing both overlapping subdiagrams γ_1 and γ_2—giving rise to exactly the six terms indicated in (17.25) and displayed in Fig. 17.4, the convergence of which has already been explained. The reader is strongly encouraged to verify the efficacy of the Zimmermann forest formula with additional examples such as the graph of Figure 17.2, where one may also verify explicitly the equivalence of the Bogoliubov–Parasiuk recursion and Zimmermann formulas.

The zero-momentum subtraction method described so far in this section (referred to commonly as the "BPHZ" or Bogoliubov–Parasiuk–Hepp–Zimmermann scheme) is by no means the only way to arrive at a fully subtracted amplitude which is UV-finite according to the requirements of Weinberg's theorem. Clearly, the addition of finite (cutoff independent) constant terms to the amounts prescribed by our Taylor operators t^γ for each renormalization part would produce an equally UV-finite fully subtracted amplitude, differing from the BPHZ one by a finite amount. Indeed, soon after the

---

[5] The technical niceties involved in a proper choice of momentum routing are dealt with in full detail in (Zimmermann, 1969). Alternatively, one may employ an "α-parametric" representation, replacing Feynman propagators $1/(p^2 + m^2) \rightarrow \int_0^\infty e^{-\alpha(p^2+m^2)} d\alpha$, performing the loop momentum integrals explicitly, and formulating the subtractions directly for the resultant integrands in the multi-dimensional "α" space, in which case the ambiguities of momentum ordering can be avoided: see (Bèrgere and Lam, 1976).

introduction of the dimensional regularization scheme by 't Hooft and Veltman (cf. Section 16.5), it was demonstrated (Speer, 1973) that one may replace the t^γ Taylor subtraction operators, which act on the Feynman *integrands* of divergent subdiagrams by a dimensional pole part operation $t^\gamma_{\mathrm{DR}} \equiv \mathcal{PP}^\gamma$ (cf. Section 16.5) which removes the pure pole singularities in the Laurent expansion of the dimensionally continued Feynman integral of the subgraph γ in the variable $\epsilon = d_{\mathrm{ph}} - d$, where d is the dimensional continuation variable and d_{ph} the physical spacetime dimension. As the poles which would produce a divergent result when the limit $d \to d_{\mathrm{ph}}$ is performed at the end have been removed, this procedure automatically produces a result which contains no explicit UV cutoff Λ but, as we saw in Section 16.5, necessarily depends on a renormalization scale μ required to consistently define the dimensionally continued amplitudes.

In this dimensional subtraction approach, all the formulas we have written so far (in particular, the Bogoliubov–Parasiuk recursion formula and the Zimmermann forest formula) remain in force, with the simple replacement $t^\gamma \to \mathcal{PP}^\gamma$. As we shall see in the next section, the demonstration that the subtractions appearing in these formulas are precisely equivalent to a reparameterization of the amplitudes of the theory in terms of a well-defined finite set of low-energy parameters (themselves specific non-linear functions of the coupling and mass parameters appearing as coefficients in the Lagrangian of the theory) depends crucially on the fact that the counterterm $t^\gamma \bar{\mathcal{I}}^\gamma$ associated with a renormalization part γ is a polynomial of degree $d(\gamma)$ in the external momenta of the subgraph γ, where $d(\gamma)$ is the (by definition, non-negative) degree of divergence of γ. It takes exactly the same form, in other words, as the insertion of a local operator $\mathcal{O}(\gamma)$ at a single vertex replacing the entire subgraph γ, where $\mathcal{O}(\gamma)$ contains as many field operators as there are external lines of γ, with (a finite number of) spacetime-derivatives chosen to reproduce the powers of external momenta generated by the Taylor operator t^γ. While this is obvious in the BPHZ zero-momentum subtraction scheme, it is far from so when we employ pole subtractions to remove the large momentum divergences of a general Feynman integral, as the pole terms are extracted *after* the loop integrations are performed, and may contain *a priori* a complicated (and in particular, non-polynomial) dependence on the external momenta of the subgraph.

To see how the dimensional subtraction method avoids this potential trap, we first note that the subdivergences of γ are already subtracted by internal counterterms in $\bar{\mathcal{I}}^\gamma$. To show that the pole part $\mathcal{PP}^\gamma \bar{\mathcal{I}}^\gamma$ is indeed a polynomial in the external momenta of γ, and therefore equivalent to a local operator insertion, we need only show that differentiating $\bar{\mathcal{I}}^\gamma$ $d(\gamma) + 1$ times with respect to the external momenta of γ results in a UV-convergent integral, with vanishing pole part as $d \to d_{\mathrm{ph}}$—thereby establishing that the pole part is a polynomial of degree $d(\gamma)$ in these momenta, and reproducible by a local operator insertion. The arguments needed to establish this via Weinberg's theorem are not particularly difficult: the essential point is that the differentiation of $\bar{\mathcal{I}}^\gamma$, which the reader will recall already contains counterterms needed to remove all the proper subdivergences of γ, lowers the total degree of divergence of both the overall graph γ and its (already subtracted) subdivergences to the point where we have a completely UV-finite integral, with no pole in ϵ. As usual, the only situation where this is potentially a subtle issue occurs when we have overlapping divergences, so we encourage the reader to verify the asserted convergence for the third derivative (with respect to p) of the amplitude $\bar{\mathcal{I}}^\Gamma(p)$ in (17.28) for the superficially quadratically divergent graph of Fig. 17.3 (see Problem 2). This convergence implies that the final overall subtraction operator t^Γ of (17.23), when replaced by the dimensional pole part operation $\mathcal{PP}^\Gamma$, indeed produces a quadratic (even) polynomial $A + Bp^2$ in the momentum p, corresponding to the contribution which would be obtained by an insertion of the local operator $A\phi^2 + B(\partial_\mu \phi)^2$ at the vertex obtained by contracting the two-loop graph to a point.

Although the individual terms in the Zimmermann forest formula (17.31) correspond to UV-divergent integrals, and must therefore be regulated in some fashion if we wish to examine them individually, the combined sum of all the terms, by construction, produces an integrand which yields an absolutely convergent multi-loop integral, *even in the absence of a cutoff* (or any other type of regularization: e.g., dimensional). We shall shortly see that the "BPHZ-renormalized" amplitudes provided by the forest formula exactly correspond to the perturbative expansion of the amplitudes of a Lagrangian field theory in powers of a suitably reparameterized coupling constant, so the reader may be wondering why we need to introduce a regularization scheme in the first place.

The answer is twofold. First, as emphasized repeatedly in the preceding chapter, a physical regularization of the amplitudes at high momentum is present in any event, whether we wish it or not, due to the inevitable breakdown of flat-space Minkowski field theory once quantum gravity effects become appreciable, or simply because the field theory itself is only a low-energy effective theory to be supplanted at higher energies by a more comprehensive microscopic theory (for example, in the way in which the Standard Model may be replaced at short distance scales by a Grand Unified Theory with a larger local gauge group containing the gauge symmetries of the Standard Model). We must therefore always bear in mind that the finite results yielded by (17.31) are approximations to the actual physical amplitudes, with corrections of order inverse powers of the ratio of the masses and momenta of the particles described by the theory to the ultraviolet scale at which our low-energy theory, for whatever reason, begins to break down.

Secondly, for purely practical reasons, the integrands obtained by combining all the terms in the forest formula into a single integrand of the form (17.7), especially in higher orders (i.e., two or more loops) are usually extremely long and cumbersome expressions, making it impossible in practice to analytically perform the resultant integration. It is usually vastly simpler to introduce a regularization which renders the individual terms in the forest formula well-defined, perform the (much simpler) integrals corresponding to each term individually, and then combine the results, at which point one can verify that the singular dependence on the regularization variable (cutoff Λ in momentum cutoff schemes, or $1/\epsilon$ poles in dimensional regularization) cancels in the complete subtracted amplitude. In verifying the perturbative renormalizability of the theory, to which we now turn, it is very important, when choosing a regularization scheme, that the symmetries of the underlying Lagrangian are preserved at the intermediate stages of the calculation.

17.2 Counterterms, subtractions, and perturbative renormalizability

The considerations of the preceding section provide a complete characterization of the terms giving rise to cutoff-dependence in arbitrary multi-loop Feynman amplitudes, but as yet the meaning of these contributions in the context of the Lagrangian dynamics of a local quantum field theory is unclear. In fact, for a certain class of local field theories, the subtractions prescribed by the BPHZ procedure are exactly equivalent to a reparameterization of the perturbative Feynman amplitudes of the theory in terms of a finite set of low-energy parameters, which can be determined (order by order in perturbation theory) by an equivalent number of independent low-energy measurements. We shall see how this comes about first in the algebraically simplest case—that of self-coupled scalar particles. We recall that the general form taken by the Euclidean action giving a clustering relativistically invariant theory for such particles is (cf. (16.6)), in d spacetime dimensions,

$$S_E(\phi) = \int d^d x \{ \frac{1}{2} (\partial_\nu \phi)^2 + \sum_{n \geq 0} a_n \phi^{2+n} + \sum_{n > 0} a'_n (\partial_\nu \phi)^2 \phi^n + \ldots \}$$

$$\equiv \int d^d x \{ \frac{1}{2} (\partial_\nu \phi)^2 + \sum_{n \geq 0} a_n \mathcal{O}_n + \sum_{n > 0} a'_n \mathcal{O}'_n + \ldots \} = \int \mathcal{L}(\phi) d^4 x \quad (17.32)$$

where an ultraviolet momentum cutoff Λ (or alternatively some form of short-distance cutoff), is assumed implicitly present. The dots "..." represent operators with four or more spacetime-derivatives.

In what follows we shall frequently resort to dimensional analysis, so we remind the reader at this point that, using natural units ($\hbar = c = 1$), ensuring the dimensional consistency of our equations amounts to counting powers of mass, with momentum and energy having dimensions of mass, and space and time coordinates, inverse powers of mass. The Euclidean action S_E must be dimensionless (it appears in the exponent in the functional integral) so by examining the kinetic term in (17.32) (the first term on the right-hand side), we conclude that the scalar field ϕ must have, in d spacetime dimensions, dimension $m^{d/2-1}$. Having determined the dimension of the field, it is straightforward to examine the other terms in (17.32) to establish their mass dimension. We shall denote this "engineering dimension" of any quantity in powers of mass with the notation "dim", thus:

$$\dim(\phi) = \frac{d}{2} - 1$$

$$\dim(a_n) = 2 + n(1 - \frac{d}{2})$$

$$\dim(\mathcal{O}_n) = (n + 2)(\frac{d}{2} - 1)$$

$$\dim(a'_n) = n(\frac{d}{2} - 1) + d$$

$$\dim(\mathcal{O}'_n) = n(1 - \frac{d}{2}) + d \quad (17.33)$$

In perturbation theory, we as usual assign the quadratic field kinetic $(\frac{1}{2}(\partial_\nu\phi)^2)$ and mass $(a_0\phi^2)$ terms to the "free" part of the Lagrangian, and the remaining terms in (17.32) are placed in the "interaction", generating the vertices of the Feynman graphs of the theory.[6] In the present case, the interaction vertices are associated with the bare coupling constants $a_1, a_2, a_3, .., a'_1, a'_2, a'_3, ..$ etc. Let us now consider a 1PI graph with E external lines, with momentum-space amplitude $\Gamma^{(E)}(k_1, k_2, \ldots, k_E)$. We shall be concerned only with dimensional analysis here—counting powers of mass dimension— so overall numerical (dimensionless) factors may be ignored. Thus, using the symbol $\sim$ to indicate dimensional equivalence, we have

$$\delta^d(\sum_i k_i)\Gamma^{(E)}(k_1, k_2, \ldots, k_E) \sim \prod_{i=1}^{E}(k_i^2 + m^2) \int d^d x_1 \cdots d^d x_E \langle 0|\phi(x_1) \cdots \phi(x_E)|0\rangle$$

(17.34)

The product of $k^2 + m^2$ factors on the right-hand side serves to truncate the external propagators, and, of course, the 1PI character of $\Gamma^{(E)}$ means that only a subset of the terms contributing to the E-point function in the integral are included, but this does not alter the fact that dimensional consistency require the left- and right-hand sides of (17.34) to have the same engineering dimension. This implies

$$-d + \dim(\Gamma^{(E)}) = 2E - dE + E(\frac{d}{2} - 1) \Rightarrow \dim(\Gamma^{(E)}) - d + E(1 \quad \frac{d}{2})$$

(17.35)

Now let us consider a particular L-loop graph contributing to $\Gamma^{(E)}$ with $N_1, N_2, ..$ vertices of the interaction terms $\mathcal{O}_1, \mathcal{O}_2, ..$ etc., and similarly $N'_1, N'_2, ..$ vertices for the primed operators $\mathcal{O}'_1, \mathcal{O}'_2, ...$ With l representing a generic loop momentum, and with all loop momenta large compared to masses and external momenta k_i, we also have, dimensionally,

$$\Gamma^{(E)} \sim a_1^{N_1} a_2^{N_2} \cdots (a'_1)^{N'_1}(a'_2)^{N'_2} \cdots \int l^{D-dL} d^{dL}l$$

(17.36)

where D is the *superficial* degree of divergence of the graph in question, which we recall is computed precisely by counting powers of loop momentum in the multi-loop Feynman integrand, including, of course, the phase-space associated with L independent d-dimensional momentum integrations. Referring to (17.33), and comparing the dimensions of the left- and right-hand sides of (17.36), we find

$$\dim(\Gamma^{(E)}) = \sum_{n=1}(N_n(2 + n(1 - \frac{d}{2})) + N'_n n(1 - \frac{d}{2}) + ..) + D$$

(17.37)

Using our previous result (17.35) for the dimension of $\Gamma^{(E)}$, we therefore find for the superficial degree of divergence of this graph

---

[6] For the time being we shall ignore terms with more than two derivatives and quadratic in the field. Such "Pauli–Villars" terms can be included in the free propagator, and result in a damping at high momentum analogous to that employed in our treatment of the renormalization group flow of effective Lagrangians in Section 16.4. In other words, they can be viewed as a modification of the cutoff scheme employed to define individually divergent Feynman graphs.

$$D = d + E(1 - \frac{d}{2}) - \sum_{n=1}(N_n(2 + n(1 - \frac{d}{2})) + N'_n n(1 - \frac{d}{2}) + ..) \qquad (17.38)$$

In particular, in $d = 4$ spacetime dimensions, the result becomes

$$D = 4 - E + \sum_{n=1}((n - 2)N_n + nN'_n + ..) \qquad (17.39)$$

We note that the sum divides into (a) a single negative contribution ($n = 1$), corresponding to the operator ϕ^3, increasing insertions of which decrease the superficial degree of divergence of the graph, (b) a term with vanishing coefficient of N_2, which counts the number of appearances of the quartic vertex induced by the ϕ^4 term in the Lagrangian, increasing insertions of which therefore do not affect the superficial degree of divergence of the graph, and (c) terms with positive contributions to D, corresponding to operators $\mathcal{O}_3 = \phi^5, \mathcal{O}_4 = \phi^6, \ldots, \mathcal{O}'_1 = (\partial_\nu \phi)^2 \phi^2, \ldots$ with engineering dimension greater than 4, and for which increasing numbers of vertices result in increasing degree of divergence of the graph containing them. This division precisely corresponds to the classification of operators referred to in Section 16.3 as (a) relevant, (b) marginal, or (c) irrelevant, from the point of view of the renormalization group behavior of effective field theories described there. In the present context it will be more convenient to refer to the operators of type (a) as *super-renormalizable*, (b) as *renormalizable*, and (c) as *non-renormalizable*, for reasons that will shortly become clear. The essential point to be grasped at this juncture is that, while the number of super-renormalizable and renormalizable operators/vertices is finite, (indeed, there are only two possible interaction terms in 4-dimensions, corresponding to ϕ^3 and ϕ^4), the number of non-renormalizable vertices/operators is always infinite. We note here for future reference that in $d = 6$ spacetime dimensions, (17.39) becomes

$$D = 6 - 2E + \sum_{n=1}((2n - 2)N_n + 2nN'_n + ..) \qquad (17.40)$$

so the only renormalizable operator in this case is ϕ^3, with the quartic vertex from ϕ^4 already corresponding to a non-renormalizable term. For spacetime dimensions greater than 6, *all* interaction operators fall into the non-renormalizable category.

The result (17.39) allows us to identify immediately the renormalization parts for a scalar field theory in four dimensions. First, consider the case where only super-renormalizable and renormalizable terms are present in the Lagrangian. Thus, $N_n = 0, n > 2$, $N'_n = 0, n > 0$ etc., and we have simply

$$D = 4 - E \qquad (17.41)$$

We may identify this theory by the useful notation ϕ_4^4, where the subscript indicates the spacetime dimension, and the superscript indicates the highest dimension (renormalizable) interaction operator. Let us also assume the discrete symmetry $\phi \to -\phi$ which eliminates terms containing odd powers of the field. Then the only renormalization parts of the resultant ϕ^4 theory correspond to 1PI graphs with $E = 0, 2$, or 4 external lines. We have already seen in Chapter 10 that the vacuum graphs of the theory ($E = 0$) induce a physically irrelevant (in flat space) phase shift of the vacuum state, which cancels in the appropriately normalized amplitudes. Thus the renormalization parts of this theory correspond to quadratically divergent 1PI two-point subgraphs (termed "self-energy", or sometimes "vacuum polarization" graphs) and logarithmically divergent 1PI four-point subgraphs (which we shall call "vertex renormalization parts" for reasons shortly to become apparent). All graphs with more than four external lines are superficially convergent: they may, of course, contain internal subdivergences, but these must correspond to self-energy or vertex renormalization subgraphs. Similarly, in ϕ_6^3 theory, we conclude from (17.40) that with only renormalizable or super-renormalizable terms present, the renormalization parts correspond to 1PI graphs with 1, 2, or 3 external lines: termed "tadpole", "self-energy", and "vertex" renormalization parts respectively. Graphs with more than three external lines (e.g., 2-2 scattering graphs) are superficially convergent. Here we cannot impose the reflection symmetry $\phi \to -\phi$, as we wish to retain the one non-trivial renormalizable term, which is cubic in the field.

On the other hand, if non-renormalizable terms are present, there is an infinite number of renormalization parts in either case, as the superficial degree of divergence D of graphs with any number of external lines eventually becomes positive if enough vertices of non-renormalizable operators are inserted, due to the positive terms in the sums in (17.39, 17.40). We are about to see that the subtractions introduced in the preceding section to remove the dominant cutoff dependence of Feynman amplitudes are exactly equivalent to a precise set of reparameterizations of the parameters in the (cutoff) Lagrangian in terms of parameters defined at low energy. Evidently, the presence of non-renormalizable operators in the basic Lagrangian will require the reparameterization of an infinite number of Lagrangian parameters, if we wish to compute cutoff-insensitive amplitudes to arbitrary orders of perturbation theory. In such a case, we say that we are dealing with a "perturbatively non-renormalizable theory".

What if we exclude *ab initio* non-renormalizable terms from the Lagrangian? From the point of view of the Wilsonian effective Lagrangian theory discussed in the preceding chapter, this would seem to be a physically unreasonable procedure: after all, we saw there that the inevitable presence of new physics (such as quantum gravity) at very short distances necessarily induces an infinite set of higher-dimension operators in the effective Lagrangian for any theory cutoff at some high-momentum scale. We shall return to precisely this question in Section 17.4, where the implications of restricting ourselves to a Lagrangian with only super-renormalizable or renormalizable terms are examined from a general renormalization group point of view. For the time being, let us suppose (as we have frequently done in the book, making no excuses!) that we can describe the interactions of a real scalar field in four dimensions by a Lagrangian containing only three terms (and an implicit UV cutoff Λ),

$$\mathcal{L} = \frac{1}{2}(\partial_\mu \phi)^2 - \frac{1}{2}m^2\phi^2 - \frac{1}{4!}\lambda\phi^4 \qquad (17.42)$$

where we have returned to more conventional notation for the mass and coupling term: $a_0 = \frac{1}{2}m^2, a_2 = \frac{1}{4!}\lambda$. This Lagrangian, of course, represents *a class of physical theories*, with varying masses for the ϕ-particle, and for the strength of its self-coupling. The relevant physical theory must be fixed by performing experiments to measure the physical mass m_{ph} (related, but as we shall see, certainly not identical, to the parameter m^2 in the Lagrangian), and a physical coupling strength λ_{ph}. One might, for example, define the latter as the S-matrix element for elastic 2-2 scattering in the zero (spatial) momentum limit $p_0^\mu = (m_{\mathrm{ph}}, 0)$, *sans* uninteresting momentum-conservation and phase-space factors (cf. (7.196)):

$$S_{p_0 p_0, p_0 p_0} \equiv \lambda_{\mathrm{ph}} \cdot (2\pi)^4 \delta^4() \frac{1}{(2\pi)^6 (2m_{\mathrm{ph}})^2} \tag{17.43}$$

Evidently, for any given fixed value of the UV cutoff Λ, these definitions fix the bare parameters m^2, λ appearing in the Lagrangian as functions of measurable quantities

$$m^2 = m^2(\lambda_{\mathrm{ph}}, m_{\mathrm{ph}}, \Lambda), \quad \lambda = \lambda(\lambda_{\mathrm{ph}}, m_{\mathrm{ph}}, \Lambda) \tag{17.44}$$

Arbitrary Feynman amplitudes of the (cutoff) theory can then be computed as functions of m, λ, and Λ, and then *reparameterized* in terms of the physical mass and coupling $m_{\mathrm{ph}}, \lambda_{\mathrm{ph}}$. This can be done (indeed, in general, must be done) order by order in a formal expansion in the measured coupling λ_{ph}, as we typically are unable to solve the theory exactly. The choice of definition of λ_{ph} means that, to lowest order in perturbation theory (see (7.196)), the bare and physical coupling coincide, $\lambda_{\mathrm{ph}} = \lambda + O(\lambda^2)$ (equivalently, $\lambda = \lambda_{\mathrm{ph}} + O(\lambda_{\mathrm{ph}}^2)$), which also ensures that the bare and physical masses m and m_{ph} coincide to lowest order of perturbation theory (in either λ or λ_{ph}). In the first section of this chapter we saw a simple example in which such a reparameterization in fact completely removed the dominant cutoff-dependence of the amplitudes, leaving only a quantitatively ignorable dependence at the level of inverse powers of the cutoff. We are about to see that exactly this softening of ultraviolet sensitivity occurs, order by order in perturbation theory and (to all orders) in theories containing only renormalizable or super-renormalizable operators in their cutoff Lagrangian.

Before proceeding to the general argument, we need to point out an important feature of the Lagrangian (17.42), which we have heretofore glossed over. The Lagrangian contains three terms, and we should therefore in general expect it to represent a three-parameter family of theories, if we allow the coefficient of the kinetic term to vary, effectively by rescaling the field in (17.42), $\phi \equiv \sqrt{\hat{Z}} \phi_R$:

$$\mathcal{L} = \frac{1}{2} \hat{Z} (\partial_\mu \phi_R)^2 - \frac{1}{2} \hat{Z} m^2 \phi_R^2 - \frac{1}{4!} \hat{Z}^2 \lambda \phi_R^4 \tag{17.45}$$

We use the notation $\hat{Z}$, rather than the more common Z, to emphasize the fact that the "wavefunction renormalization" constant being introduced here may, but need not, coincide with the LSZ field normalization constant Z appearing in the vacuum to single particle matrix element of the field, as in (9.155). The new constant $\hat{Z}$ merely corresponds to our freedom to rescale the field by a constant, $\phi \to \kappa \phi$. The new rescaled field ϕ_R is generally termed the "renormalized field". From the point of view of S-matrix elements, such a rescaling is physically irrelevant: if two fields ϕ_A and ϕ_B differ by a simple scaling, $\phi_A = \kappa \phi_B$, then their LSZ normalization constants are related by $Z_A = \kappa^2 Z_B$, and the LSZ formula (9.176) gives identical results for the S-matrix using either field, as the change in $Z^{-1/2}$ factors exactly compensates for the rescaling of the fields in the Feynman Green function. Nevertheless, we shall want to remove cutoff sensitivity, if possible, from as many elements of the formalism as possible, and in particular, from the Feynman amplitudes (n-point functions) appearing in the LSZ formula, even before these are taken on-shell and appropriately scaled. This freedom will also be necessary to complete the connection of the reparameterization program described above with the BPHZ subtraction procedure of the preceding section.

The reparameterization of the theory suggested above in terms of a directly measurable physical mass and coupling is only one of an infinite variety of possible redefinitions of the parameters of the theory. In order to establish the connection of the reparameterized amplitudes with the subtracted ones discussed in the previous section, we shall make an alternative choice. Our "physical" (or "renormalized") coupling parameter, now and henceforth dubbed λ_R, will be defined as the zero four-momentum value of the 1PI Euclidean four-point function of the renormalized field ϕ_R, which we denote, for general external momenta, $\Gamma^{(4)}(k_1, k_2, k_3, k_4)$:

$$\lambda_R \equiv \Gamma^{(4)}(0, 0, 0, 0) \tag{17.46}$$

Note that we take $k_i^\mu = 0$ for all four components of the external momenta, which means that we are parameterizing the theory in terms of an off-shell value for a Green function. The overall numerical factor in $\Gamma^{(4)}$ is chosen so that its perturbative expansion in terms of the bare coupling λ begins with λ, with a coefficient of unity. The first few graphs contributing to $\Gamma^{(4)}$ in terms of the bare parameters of (17.45) are indicated in Fig. 17.6. Thus, we have a formal expansion:

$$\hat{Z}^2 \lambda \equiv \lambda_R + \delta\lambda, \quad \delta\lambda = c_2 \lambda_R^2 + c_3 \lambda_R^3 + \ldots \ldots \tag{17.47}$$

The shift $\delta\lambda$ between the bare and renormalized coupling is called a "coupling constant counterterm": the coefficients $c_2, c_3, \ldots$ appearing in the expansion of the counterterm in λ_R must be determined order by order in perturbation theory to ensure that the zero momentum value of $\Gamma^{(4)}$ remains pinned at λ_R, as it is defined to be. We shall see shortly how this is accomplished in practice. We likewise define counterterms for the other two (as yet) floating parameters in the Lagrangian:

$$\hat{Z} \equiv 1 + \delta\hat{Z} = 1 + a_1 \lambda_R + a_2 \lambda_R^2 + \ldots \tag{17.48}$$

$$\hat{Z}m^2 \equiv m_R^2 - \delta m^2 = m_R^2 - (b_1 \lambda_R + b_2 \lambda_R^2 + \ldots) \tag{17.49}$$

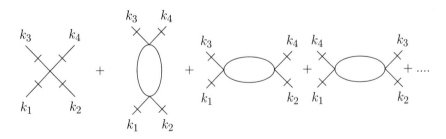

Fig. 17.6 Low-order 1PI contributions (through one loop) to $\Gamma^{(4)}(k_1, k_2, k_3, k_4)$ in ϕ^4 theory. The crossbars indicate that external legs are amputated.

The counterterms $\delta \hat{Z}, \delta m_R^2$, will be used to adjust the behavior of the self-energy (i.e., the 1PI two-point function of the renormalized field ϕ_R) at zero momentum, as follows. Note that the connected two-point function of the theory (in momentum space, the "full" Feynman propagator $\hat{\Delta}_F(p)$; cf. Section 10.4) can be graphically represented as a series of free propagators $\Delta_F(p)$, interspersed by 1PI self-energy corrections, as indicated in Fig. 10.5. Denoting the sum of all 1PI two-point self-energy graphs as $\Pi(p)$ (the graphs contributing to this Green function in ϕ^4 through two loops are displayed in Fig. 17.7), we have algebraically

$$\hat{\Delta}_F(p) = \Delta_F(p) + \Delta_F(p)\Pi(p)\Delta_F(p) + \Delta_F(p)\Pi(p)\Delta_F(p)\Pi(p)\Delta_F(p) + \ldots$$

$$= \frac{1}{\Delta_F^{-1}(p) - \Pi(p)} \tag{17.50}$$

In other words, the inverse (Euclidean) full propagator consists just of a tree contribution which is just the inverse free propagator, together with (minus) the 1PI loop graphs indicated in Fig. 17.7, just as we would expect, given that it coincides with the two-point function corresponding to the functional $\Gamma(\phi_R)$ generating the 1PI graphs of the theory (cf. Equations (10.142, 10.143)). We are at liberty to apportion the quadratic (mass term) part of the Lagrangian at liberty into a "free" and "interacting" part, and we shall *define* the coefficient of $\frac{1}{2}\phi_R^2$ in the free Lagrangian as m_R^2: thus $\Delta_F^{-1}(p) = p^2 + m_R^2$ in (17.50). Moreover, we shall choose $\delta \hat{Z}, \delta m_R^2$ order by order in perturbation theory to remove the first two terms in the Taylor expansion (in p^2) of $\Pi(p)$ (which is, of course, a scalar function of p^2, by Lorentz-invariance), as follows:

$$\Pi(0) = 0 \tag{17.51}$$

$$\left.\frac{\partial^2 \Pi(p)}{\partial p^2}\right|_{p=0} = 0 \tag{17.52}$$

Equivalently, these latter two conditions may be phrased, using (17.50), in terms of the full propagator,

$$\hat{\Delta}_F^{-1}(p=0) = m_R^2 \tag{17.53}$$

$$\left.\frac{1}{2}\frac{\partial^2 \hat{\Delta}_F^{-1}(p)}{\partial p^2}\right|_{p=0} = 1 \tag{17.54}$$

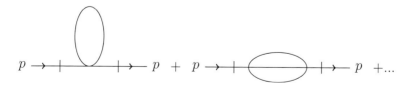

Fig. 17.7 Low-order 1PI contributions (through two loops) to $\Pi(p)$ in ϕ^4 theory. The crossbars indicate that external legs are amputated.

The three constraints (17.46, 17.51, 17.52) constitute the *renormalization conditions* for our theory, as described in the first section of this chapter: the associated amplitudes are said to be calculated in the BPHZ *renormalization scheme*.

Note that there is no reason to expect—and indeed it is not the case—that m_R corresponds to the physical mass m_{ph} of the particle: i.e., the location of the pole of the full propagator $\hat{\Delta}_F(p)$. However, as is apparent from (17.53), it is a perfectly sensible quantity (with dimensions of mass, of course) in terms of which to parameterize the amplitudes of the theory, and in terms of which a formula can be obtained (order by order in perturbation theory) for m_{ph}. The reparameterization procedure is most easily carried out by rewriting the Lagrangian (17.45) as a sum of three terms:

$$\mathcal{L} = \mathcal{L}_0 + \mathcal{L}_{\text{basic}} + \mathcal{L}_{\text{ct}} \tag{17.55}$$

$$\mathcal{L}_0 = \frac{1}{2}(\partial_\mu \phi)^2 - \frac{1}{2}m_R^2 \phi^2 \tag{17.56}$$

$$\mathcal{L}_{\text{basic}} = -\frac{1}{4!}\lambda_R \phi^4 \tag{17.57}$$

$$\mathcal{L}_{\text{ct}} = \frac{1}{2}\delta\hat{Z}(\partial_\mu \phi)^2 + \frac{1}{2}\delta m^2 \phi^2 - \frac{1}{4!}\delta\lambda \, \phi^4 \tag{17.58}$$

To avoid notational overload, we have dropped the "R" subscript on the field, with the understanding that here and henceforth we are concerned only with the Green functions of the rescaled field. The "free" Lagrangian $\mathcal{L}_0$, which determines the propagators appearing in our graphs, is now written in terms of this rescaled (or "renormalized") field (ϕ_R of (17.45)), and the renormalized mass parameter m_R (thus, the Euclidean propagator denominators are simply $k^2 + m_R^2$). Inasmuch as all the terms in the "basic vertex Lagrangian" $\mathcal{L}_{\text{basic}}$ and the "counterterm Lagrangian" $\mathcal{L}_{\text{ct}}$ contain at least one power of λ_R, the full interaction Lagrangian now contains vertices corresponding not just to the original four-point interaction vertex of (17.45), but additional two- and four-point vertices corresponding to the terms in the counterterm Lagrangian.

We now wish to establish the following remarkable property of the amplitudes (i.e., *n*-point Euclidean Green functions) of this theory: *once reparameterized in terms of the renormalized quantities m_R, λ_R as defined above, the amplitudes acquire precisely the zero momentum subtractions corresponding to the Zimmermann formula (17.31), thereby softening their UV cutoff dependence to the level of inverse powers, order by order in the perturbative expansion of the amplitudes in λ_R.* A theory of this kind, in which the redefinition of a finite number of Lagrangian mass and coupling parameters induces subtractions removing the UV cutoff dependence (up to inverse powers) of all the amplitudes of the theory, to all orders of perturbation theory, is called "perturbatively renormalizable".

Our demonstration will be inductive in the number of loops of the diagrams considered. Accordingly, our first task is to initiate the induction by demonstrating the validity of the italicized assertion above in the lowest non-trivial order, namely, for the one-loop amplitudes of the theory. In fact, we need only consider the 1PI amplitudes of the theory for our proof of renormalizability. Any (connected) amplitude which is not one-particle-irreducible can be divided into subgraphs that are, connected by single internal lines carrying fixed momenta (which are linear combinations of the external momenta of the process). Consequently, any potential loop divergences are isolated in the separate 1PI parts of the diagram: once these are appropriately subtracted, no further divergences can arise by connecting 1PI subdiagrams with internal lines whose momenta are fixed. First, let us consider the superficially divergent 1PI diagrams of the theory (i.e., the renormalization parts), and calculate these to one-loop order, imposing the BPHZ renormalization conditions (17.46, 17.51, 17.52). We first examine the four-point function $\Gamma^{(4)}(k_1, k_2, k_3, k_4)$, computed to order λ_R^2. In addition to the graphs of Fig. 17.6, involving only the basic vertex (17.57), there is a contribution at order λ_R^2 from the $c_2\lambda_R^2$ part of the $-\frac{1}{4!}\delta\lambda\,\phi^4$ piece of the counterterm Lagrangian. The parts of the counterterm Lagrangian quadratic in the fields result in the insertion of two-point vertices and generate one-particle reducible diagrams. Thus, to order λ_R^2 (cf. (16.15), with the change of notation $g_2 \to \lambda_R$), using a momentum cutoff Λ to regulate the individual diagrams, and neglecting terms suppressed by inverse powers of Λ,

$$\Gamma_1^{(4)}(k_1, k_2, k_3, k_4) = \lambda_R - \frac{1}{32\pi^2}\lambda_R^2\{\mathcal{I}(s, m_R^2, \Lambda^2) + \mathcal{I}(t, m_R^2, \Lambda^2) +$$

$$\mathcal{I}(u, m_R^2, \Lambda^2)\} - c_2\lambda_R^2 \tag{17.59}$$

$$\mathcal{I}(p^2, m^2, \Lambda^2) \equiv \int_0^1 (\ln\left(\frac{\Lambda^2}{x(1-x)p^2 + m_R^2}\right) - 1)dx \tag{17.60}$$

$$s \equiv (k_1 + k_2)^2, \quad t \equiv (k_1 - k_3)^2, \quad u \equiv (k_1 - k_4)^2 \tag{17.61}$$

where the subscript indicates explicitly that our four-point function is being calculated through one-loop order. The renormalization condition (17.46) implies that we must choose the coefficient c_2 in the vertex renormalization counterterm to precisely remove the contribution of the second term on the right-hand side of (17.60) at zero momentum, as the correct zero-momentum value for $\Gamma^{(4)}$ is already given by the lowest-order tree contribution:

$$c_2 = -\frac{1}{32\pi^2} \cdot 3\mathcal{I}(0, m_R^2, \Lambda^2) \tag{17.62}$$

Inserting (17.62) in (17.60), we find, through order λ_R^2 (i.e., to one loop),

$$\Gamma_1^{(4)}(k_1, k_2, k_3, k_4) = \lambda_R - \frac{1}{32\pi^2}\lambda_R^2\{\mathcal{I}_R(s, m_R^2) + \mathcal{I}_R(t, m_R^2) + \mathcal{I}_R(u, m_R^2)\} \tag{17.63}$$

with

$$\mathcal{I}_R(p^2, m_R^2) \equiv (1 - t^\gamma)\mathcal{I}(p^2, m_R^2, \Lambda^2) = \int_0^1 \ln\left(\frac{m_R^2}{x(1-x)p^2 + m_R^2}\right)dx \tag{17.64}$$

satisfying the required constraint $\mathcal{I}_R(0, m_R^2) = 0$. We remind the reader that the Taylor operator t^γ (with γ referring to the one-loop bubbles of Fig. 17.6) when applied to a one-loop integral with superficial degree of divergence zero, simply evaluates the graph at zero external momentum. In other words, the counterterm proportional to $c_2\lambda_R^2$ is exactly equivalent to the Taylor operation defined in the preceding section, as a consequence of the need to keep the four-point function pinned at the defined value λ_R at higher orders. As expected, the result is the removal from the amplitude of the logarithmic sensitivity to the cutoff Λ, leaving only terms of order $m_R^2/\Lambda^2, k_i^2/\Lambda^2$, which we have neglected above.

Next, we consider the one-loop subtractions induced by the field renormalization (17.48) and mass (17.49) counterterms, which require the insertion in our graphs, to first order, of interaction vertices generated by the associated one-loop counterterm Lagrangian

$$\mathcal{L}_{\text{ct,1 loop}} = \frac{1}{2}a_1\lambda_R(\partial_\mu\phi)^2 + \frac{1}{2}b_1\lambda_R\phi^2 \tag{17.65}$$

The perturbative expansion of the full propagator $\hat{\Delta}_F(p)$ through one loop now consists of the three graphs indicated in Fig. 17.8: in addition to the one-loop self-energy correction involving the basic vertex of $\mathcal{L}_{\text{basic}}$, there is a counterterm graph (Fig. 17.8(c)) in which the operator (17.65) induces a two-point vertex with the coefficient $a_1\lambda_R p^2 + b_1\lambda_R$. Thus, we have, through order λ_R,

$$\hat{\Delta}_{F,1}(p) = \Delta_F(p) + \Delta_F(p)\Pi_1(p)\Delta_F(p) + \Delta_F(p)(a_1\lambda_R p^2 + b_1\lambda_R)\Delta_F(p)$$

$$= \Delta_F(p) + \Delta_F(p)\Pi_{1,R}(p)\Delta_F(p) \tag{17.66}$$

$$\Pi_{1,R}(p) = \Pi_1(p) + a_1\lambda_R p^2 + b_1\lambda_R = -\frac{1}{2}\lambda_R\int\frac{1}{l^2 + m_R^2}\frac{d^4l}{(2\pi)^4} + a_1\lambda_R p^2 + b_1\lambda_R \tag{17.67}$$

Note that in this case the unsubtracted one-loop self-energy is given by a quadratically divergent integral which is, however, independent of the external momentum p. We must now choose the one-loop counterterm coefficents a_1, b_1 to impose the renormalization conditions (17.51,17.52) on the complete one-loop self-energy $\Pi_{1,R}(p)$, including counterterm contributions. As $\Pi_1(p)$ is independent of p in this case, we have simply

$$a_1 = 0 \tag{17.68}$$

$$b_1 = \frac{1}{2}\int\frac{1}{l^2 + m_R^2}\frac{d^4l}{(2\pi)^4} \tag{17.69}$$

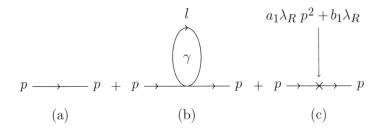

(a)   (b)   (c)

Fig. 17.8 Zeroth order (a), unsubtracted one-loop (b), and counterterm (c) contributions to the propagator in ϕ^4 theory, through one-loop order.

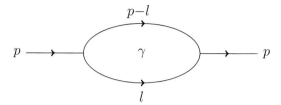

$p-l$

p → γ → p

l

Fig. 17.9 Unsubtracted one-loop propagator graph in ϕ_6^3 theory.

which (in this admittedly very special case) results in the complete cancellation of the self-energy to one-loop order: $\Pi_{1,R}(p) = 0$! More generally, the effect of the counterterm is clearly just to remove the first two terms in the Taylor expansion in p^2 of the unsubtracted self-energy (which for this special case degenerates to the first constant term)—namely, the quadratically divergent γ in Fig. 17.8(b):

$$\Pi_{1,R} = (1 - t^\gamma)\Pi_1 \tag{17.70}$$

In most theories, the one-loop self-energy $\Pi_1(p)$ will be a non-trivial function of momentum, and the Taylor operator t^γ will, as discussed in the preceding section, (a) remove the dominant UV-cutoff dependence of the graph, leaving only inverse power-dependence, and (b) leave a non-zero "renormalized" self-energy $\Pi_{1,R}(p)$, which contributes non-trivially to the full propagator. For example, in ϕ_6^3 theory,[7] the one-loop propagator graph in Fig. 17.9 corresponds to an unsubtracted self-energy

$$\Pi_1(p) = \frac{1}{2}\lambda_R^2 \int \frac{1}{(l^2 + m_R^2)((p-l)^2 + m_R^2)} \frac{d^6l}{(2\pi)^6} \equiv \lambda_R^2 \mathcal{I}(p) \tag{17.71}$$

and the counterterm Lagrangian $a_2\lambda_R^2(\partial_\mu\phi)^2 + b_2\lambda_R^2\phi^2$ induces a subtraction exactly equivalent to the Taylor operator t^γ, as we must choose, pursuant to (17.51, 17.52),

$$a_2 = -\frac{1}{2}\frac{\partial^2}{\partial p^2}\mathcal{I}(p)\Big|_{p=0} \tag{17.72}$$

$$b_2 = -\mathcal{I}(0) \tag{17.73}$$

whence

$$\Pi_{1,R}(p) = \lambda_R^2(\mathcal{I}(p) + a_2p^2 + b_2) = (1 - t^\gamma)\Pi_1(p) \tag{17.74}$$

as required. In this case, of course, $\Pi_{1,R}(p)$ is a non-trivial function of p.

To summarize, we have established that the reparameterization of the superficially divergent Euclidean Green functions of ϕ_4^4-theory—namely, the two point propagator and four point 2-2 amplitude—in terms of renormalized parameters consistent with the BPHZ renormalization conditions (17.46, 17.51, 17.52) has resulted in the appearance of exactly the Taylor subtractions required by the forest formula for these amplitudes,

---

[7] In ϕ_6^3-theory, as in gauge theories, each additional loop is accompanied by the square of the basic coupling λ_R, as the reader may easily verify by examining a few simple graphs.

when computed up to one-loop order. What about all the other one-loop amplitudes of the theory, with more than four external legs, which are, in virtue of (17.41), *superficially convergent* ? Such amplitudes may still contain ultraviolet divergences, through the presence of divergent one-loop 1PI subgraphs. However, the latter are exactly the two-point and four-point one-loop subgraphs whose UV divergence has just been shown to be subtracted by the one-loop counterterms generated by the reparameterization of the theory in terms m_R, λ_R (and rescaling of the field). By Weinberg's theorem, a superficially convergent graph is UV-finite if all its subdivergences (in this case, a single one-loop subdivergence) are subtracted, thereby appropriately lowering the degree of divergence of the subgraph to a negative value. The first step of our inductive argument is therefore complete.

The inductive step of the renormalization proof proceeds in a familiar fashion: we assume that it has been established that the subtractions implicit in the Bogoliubov–Parasiuk recursion formula (17.27, 17.28) (or its explicit solution (17.31)) are exactly those generated by the counterterms required to implement the renormalization conditions (17.46, 17.51, 17.52), *up to L loops*. The reader will recall (cf. Section 10.4) that reinserting $\hbar$ explicitly in Feynman amplitudes provides a convenient counting device for loops: an amplitude with L loops is proportional to $\hbar^L$. Consider a contribution of order $\hbar^{L+1}$ to a 1PI amplitude Γ on the left-hand side of (17.24). The renormalization parts $\gamma_1, ..\gamma_n$ in (17.27) are proper subgraphs of Γ, therefore of order $\hbar^L$ at most, and by the induction hypothesis their subtractions correspond to counterterms induced by the BPHZ renormalization conditions up to loop order L. If Γ is superficially convergent, we are done, as the t^Γ operation in (17.24) is defined to be zero, and the amplitude is already finite with no further subtractions needed. If Γ is a two- or four-point function (in ϕ_4^4 theory), then we must define the *as yet unfixed* counterterms $a_{L+1}, b_{L+1}, c_{L+2}$ to enforce the renormalization conditions, by subtracting off the constant and quadratic terms in the zero momentum expansion (in the case of the two-point function, with $d(\Gamma) = 2$), or just the zero momentum value for the four-point functions (with $d(\Gamma) = 0$). These counterterms, $a_{L+1}, b_{L+1}, c_{L+2}$, therefore correspond exactly to the new order $(L+1)$-loop subtraction $t^\Gamma \bar{\mathcal{I}}^\Gamma$ in (17.24). There is, as usual, no better way to convince oneself that nothing is being swept under the rug at this point than by examining an explicit example: the two-loop self-energy of Fig. 17.3 with overlapping divergences discussed earlier (see Problem 3). Although our argument has considered 1PI graphs, we may now extend it to general connected graphs by recalling, as discussed previously, that a general connected graph is simply the algebraic product of its 1PI components with connecting propagator factors at fixed momentum, so its UV-finiteness is assured once the component 1PI pieces are properly subtracted.

The UV-finiteness of amplitudes subtracted according to the recursive Bogoliubov–Parasiuk procedure (or the explicit forest formula) therefore implies that the amplitudes of ϕ_4^4, once reparameterized in terms of renormalized quantities, lose their dependence on an ultraviolet cutoff Λ, up to the usual inverse power terms of order $O(\frac{m^2, k_i^2}{\Lambda^2})$ (times possible powers of logarithms of $\frac{m}{\Lambda}, \frac{k_i}{\Lambda}$), which are considered negligible from the point of view of renormalization theory. Whether they are indeed so, from a quantitative point of view, depends, of course, on whether the range of energy scales over which our low-energy field theory remains valid is sufficiently large. Certainly, any perturbatively renormalizable local field theory which remains valid up to the scale of Grand Unified Theories (10^{15} GeV) or the Planck scale (10^{19} GeV) will contain contributions from the "new physics" at the high scale which are completely negligible, from the point of view of computing amplitudes *for processes allowed by the low-energy theory*, even at LHC energies of 10 TeV. Of course, these considerations are not relevant if we are dealing with processes which are simply forbidden by the low-energy theory (e.g., proton decay in Grand Unified Theories) and can only occur by virtue of the new physics appearing at high energy scales.

Once renormalized amplitudes $\Gamma_R^{(n)}(k_1, k_2, \ldots k_n; m_R, \lambda_R)$ are obtained in the above BPHZ zero momentum renormalization scheme, up to the desired loop order in perturbation theory, they may, of course, be re-expressed in terms of other low-energy parameterizations, which may bear a more direct relation to directly physically measurable quantities. For example, we may prefer to use an "on-shell" renormalization scheme in which amplitudes are parameterized in terms of the actual physical mass $m_{\rm ph}$ of our scalar particle (as given by the location in the pole of the full propagator; cf. Section 9.5) and the on-shell 2-2 scattering amplitude $\lambda_{\rm ph}$ at zero (or some other standard) spatial momentum. In addition, the field rescaling constant $\hat{Z}$ may be chosen to adjust the residue of the pole of the full propagator to be unity. These changes amount to non-linear reparameterizations of the BPHZ parameters $m_R, \lambda_R, \hat{Z}$ which, of course, do not alter the UV-finiteness of the amplitudes: more exactly, this means that we can re-expand the BPHZ parameters in power series in the new renormalized coupling $\lambda_{\rm ph}$ with coefficients which are finite (up to inverse power corrections, as usual) as the UV-cutoff Λ is taken much larger than all other scales in the theory. Alternatively, we may decide to compute our amplitudes *ab initio* in an on-shell scheme,[8] by an appropriate alteration of the renormalization conditions (17.46, 17.51, 17.52) used to determine the counterterms to each perturbation theory.

The zero-momentum renormalization scheme for ϕ_4^4-theory can be generalized in a fairly obvious way by fixing the renormalized coupling λ_R as the value of the four-point function $\Gamma^{(4)}(k_1, k_2, k_3, k_4)$ (all momenta outgoing) at a non-zero Euclidean momentum: the most convenient choice is the *Euclidean symmetric point*, defined by

$$k_i^2 = \mu^2, \quad k_i \cdot k_j = -\mu^2/3, \quad i \neq j \tag{17.75}$$

where the dot-products are determined by symmetry plus momentum conservation $\sum_{i=1}^4 k_i = 0$. Similarly, renormalization conditions for the self-energy (17.51,17.52) are imposed at $p^2 = \mu^2$ rather than at zero momentum. This scheme- really, a one parameter set of schemes, parameterized by the arbitrary scale μ- results in renormalized Green functions with an explicit dependence on the renormalization scale μ (exactly as we saw in the toy example at the beginning of this chapter). Nevertheless, physical S-matrix elements are independent of μ, once they are reparameterized in terms of measurable low-energy quantities (such as $\lambda_{\rm ph}, m_{\rm ph}$ of the on-shell scheme, for example). This means that in order to keep the physics invariant, we must change λ_R, m_R as a function of the scale μ: in this sense, the renormalization procedure induces a "running" coupling constant and mass.

---

[8] The reader should be warned that for massless theories there are difficulties, due to infrared divergences, in adopting such an on-shell renormalization procedure, as we shall see in Chapter 19.

In the case of gauge theories, the use of a dimensional renormalization scheme, as described at the end of the preceding section, is almost indispensable, due to the importance in maintaining the gauge symmetry in the regularized amplitudes, as we shall discuss in the next section. Here, as for the Euclidean point subtraction schemes above, the presence of the free dimensionful renormalization scale parameter μ (cf. Section 16.5) means that this procedure corresponds to a one-parameter class of renormalization schemes, where, for a given fixed physical theory, the renormalized parameters $m_R(\mu), \lambda_R(\mu)$ will depend on the choice of renormalization scale. Again, the dimensionally renormalized parameters of the theory m_R, λ_R (in the scalar case) may be re-expressed, if desired, in terms of "physically defined" mass and coupling parameters which do not depend at all on the choice of dimensional renormalization scale μ (any more than the physical scattering amplitudes of a theory computed via the BPHZ scheme depend on our arbitrary, if technically convenient, choice of a zero-momentum subtraction point).

The discussion of perturbative renormalizability for scalar theories given above generalizes without difficulty to other non-gauge theories. There is an important point which needs further amplification before we can move on to the case of local gauge theories, however. Perturbative renormalizability of an interacting field theory (in contrast to the classification of operators by their power dimension) is really a property of a *set of interaction operators*,[9] not of any given individual term that may appear in the interaction Lagrangian. A simple example will suffice to illustrate what is at issue here. Recall the theory of an isotriplet pion field interacting with a Dirac nucleon doublet via a Yukawa interaction term (previously discussed in the context of global isospin symmetry and Noether's theorem, cf. (12.129)), described by the bare (pre-reparameterization) Lagrangian

$$\mathcal{L} = \bar{N}(i\slashed{\partial} - M)N + \frac{1}{2}(\partial_\mu \vec{\pi} \cdot \partial^\mu \vec{\pi} - m_\pi^2 \vec{\pi} \cdot \vec{\pi}) - ig\bar{N}\gamma_5 \vec{\tau} N \cdot \vec{\phi} \qquad (17.76)$$

The engineering mass dimension of a Dirac fermion field in four spacetime dimensions is seen to be $3/2$, by examining the kinetic term in the Lagrangian, so the Yukawa term is of dimension 4, hence renormalizable, and we may therefore expect that the theory is perturbatively renormalizable, according to arguments which parallel those made above for pure scalar theories. One easily finds, for example (see Problem 4), that the degree of divergence of a graph with E_ψ external fermion lines and E_ϕ external scalar lines has superficial degree of divergence

$$D = 4 - \frac{3}{2}E_\psi - E_\phi \qquad (17.77)$$

so there are evidently only a finite number of types of superficially divergent graphs, as in ϕ^4-theory. In particular, once one begins to enumerate the possible renormalization parts in this theory, one encounters 1PI loop diagrams with four external scalar lines

---

[9] In Section 17.4 we shall see that this set corresponds to a finite-dimensional low-energy surface in the infinite-dimensional coupling constant space of Wilsonian effective theory, onto which the renormalization group flow necessarily contracts, at least in the neighborhood of zero coupling corresponding to formal perturbation theory.

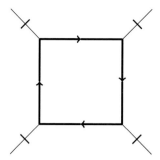

Fig. 17.10 Fermion loop contribution to $\vec{\pi} - \vec{\pi}$ scattering amplitude in Yukawa isospin theory (thick lines are Dirac propagators, thin lines are (amputated) scalar propagators).

with superficial degree of divergence zero, as in Fig. 17.10, where the four thick internal lines refer to Dirac nucleon propagators, each of order $1/(\text{loop momentum})$, hence giving rise to a logarithmically divergent loop integral $\sim \int d^4l/l^4$. The forest formula implies that counterterms corresponding to the operator $(\vec{\pi} \cdot \vec{\pi})^2$ must be present to generate the corresponding Taylor subtraction operator needed to remove the leading cutoff dependence of this diagram. Note that the *single* required operator satisfies the global O(3) isospin symmetry of the basic Lagrangian (17.76), *provided the regularization procedure employed to define the individual diagrams also does*. In the case of a global symmetry, such as the isospin symmetry present here, this is not difficult to manage: even a crude momentum cutoff will suffice, provided the mode cutoffs on the different components of the scalar and Dirac fields are done identically. As we shall see in the next section, cutoff procedures capable of maintaining a local gauge symmetry are much harder to come by: in this case, the spatiotemporal aspect of the symmetry requires a much more delicate treatment of the momentum modes of the fields.

Returning to (17.76), we see that the theory described by this Lagrangian is not perturbatively renormalizable as it stands, until we include in the interaction Lagrangian *all renormalizable operators (i.e., up to dimension 4) satisfying the global symmetries of the theory*. In the present case, this means that we must include a quartic scalar term *ab initio* in our bare Lagrangian:

$$\mathcal{L} = \bar{N}(i\slashed{\partial} - M)N + \frac{1}{2}(\partial_\mu\vec{\pi} \cdot \partial^\mu\vec{\pi} - m_\pi^2\vec{\pi} \cdot \vec{\pi}) - ig\bar{N}\gamma_5\vec{\tau}N \cdot \vec{\phi} - \frac{\lambda}{4!}(\vec{\pi} \cdot \vec{\pi})^2 \quad (17.78)$$

The reparameterization of the bare parameters and fields in the Lagrangian (17.78) now suffices to generate all the counterterms needed to implement the Taylor subtractions in the BPHZ renormalization scheme for this theory. In particular, the $\delta\lambda$ counterterm arising from the final interaction in (17.78) is needed to remove the logarithmic divergence of Fig. 17.10. We say that the original Yukawa term, together with the quartic scalar term, form a *perturbatively renormalizable set of operators*. An even simpler example of this phenomenon occurs with the ϕ_6^3 theory introduced earlier. Once non-renormalizable operators are excluded, we see from (17.40) that the renormalization parts of the theory consist of the 1PI diagrams with $E = 1, 2,$ or 3

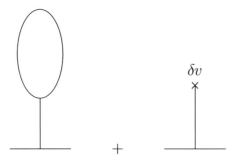

Fig. 17.11 Tadpole graph appearing in ϕ_6^3-theory, and its accompanying counterterm.

external lines (there is, of course, no $\phi \to -\phi$ symmetry to exclude odd powers of the field in this theory), so a naive choice of Lagrangian

$$\mathcal{L} = \frac{1}{2}(\partial_\mu \phi)^2 - \frac{1}{2}m^2\phi^2 - \frac{\lambda}{3!}\phi^3 \tag{17.79}$$

lacks the counterterm needed to remove UV divergences which appear in the "tadpole" graphs of the theory: i.e., the 1PI diagrams with $E = 1$ external scalar line, as in Fig. 17.11. Instead, we must write

$$\mathcal{L} = \frac{1}{2}(\partial_\mu \phi)^2 - \delta v\, \phi - \frac{1}{2}m^2\phi^2 - \frac{\lambda}{3!}\phi^3 \tag{17.80}$$

and choose the counterterm δv order by order in perturbation theory to remove the (momentum independent) tadpole terms as they appear at increasing loop order.[10] In this case, the generation of a complete perturbatively renormalizable set of operators requires the inclusion of the super-renormalizable operator ϕ. In the next chapter we shall see that the need for "completing" the set of operators included in the Lagrangian is intimately related with the "operator mixing" which is characteristic of the behavior of local operators in an interacting field theory.

The construction of a perturbatively renormalizable theory requires, as we have just seen, that "enough" operators be included to allow for all necessary counterterms, but it is also essential *not* to include even a single non-renormalizable operator in the basic Lagrangian if we wish to have a theory in which a finite number of reparameterizations suffice to remove the primary (divergent) cutoff dependence of the amplitudes. As we mentioned previously, if even a single such operator is included in the Lagrangian, the divergence counting formula (17.39) implies that renormalization parts appear with arbitrarily many external lines, simply by considering graphs with arbitrarily many insertions of the non-renormalizable vertex. Although the Zimmermann forest formula continues to yield formally UV-finite subtracted amplitudes to all orders of perturbation theory in such a case, the t^γ subtraction operators appearing therein now implicitly correspond to an infinite series of counterterms present in the reparameter-

---

[10] As the tadpole graphs lack momentum dependence, BPHZ subtraction removes them entirely, so we may simply drop any graph with a tadpole insertion in this scheme.

ized Lagrangian: we are now forced to expand the Lagrangian to include essentially *all* operators consistent with any global symmetries of the theory, and our amplitudes will depend on an infinite set of parameters, severely restricting the predictive content of the theory, at the very least. The restriction of the terms in the Lagrangian to only that finite set compatible with the given field content, with the imposed global and local symmetries, and with perturbative renormalizability, may have unintended consequences: in particular, the appearance of additional *accidental symmetries* which are only valid in virtue of the absence of non-renormalizable operators compatible with the initial imposed symmetries of the theory but not with renormalizability. Such accidental symmetries play an important role in the Standard Model (see Section 12.5, (Weinberg, 1995*a*)).

It may have occurred to the reader at this point that the insertion of an ultraviolet cutoff in a general clustering, Lorentz-invariant theory, as discussed in Chapter 16 effectively implies that our theory should indeed be described by an effective Wilsonian Lagrangian in which *all* operators—super-renormalizable, renormalizable, *and* non-renormalizable—appear. And the renormalization group discussion of the role of these three classes of operator given there suggested that in fact the non-renormalizable operators are in some sense the *least* important at low energy. The apparent inconsistency between these two points of view—the historically seminal restriction of acceptable field theories to the (finite) class of perturbatively renormalizable ones, and the much wider class of effective field theories encompassed in the Wilsonian approach—will be discussed in Section 17.4, where we shall see that perturbatively renormalizable theories emerge by evaluating the renormalization group flow of a general Wilsonian effective Lagrangian *with a particular choice of initial conditions at the ultraviolet end*, and utilizing the demonstrable insensitivity of the low-energy properties (at the level of perturbation theory) to this special choice.

17.3   Renormalization and symmetry

We have seen that the constraint of perturbative renormalizability selects Lagrangians with a finite but sufficient number of renormalizable and super-renormalizable terms to absorb the divergent parts of the graphs generated by these terms, while avoiding the appearance of an infinite set of independent renormalization parts whose divergent parts cannot be absorbed by the counterterms generated by reparameterizing the coefficients in the original Lagrangian. In some cases, the existence of a finite perturbatively renormalizable set of operators depends crucially on the existence of a symmetry restricting the type of operators which may appear in the Lagrangian. In the preceding section we have already seen an example—the isospin invariant Lagrangian (17.78), where the O(3) symmetry restricts the possible renormalizable interactions which can, and indeed *must*, be included in order to provide counterterms for all divergent subgraphs of the theory. However, in this case the symmetry is not itself essential for perturbative renormalizability: any theory with only scalar and Dirac fermion fields and with interactions of Yukawa type and scalar self-interactions quartic or less in the scalar fields will be renormalizable, provided all possible renormalizable (and super-renormalizable) terms are included. The presence of the global O(3) symmetry serves only to restrict the number of such terms which we must include in the Lagrangian, in the example cited above leaving us with only two non-trivial interaction terms. The situation is completely different once we include vector fields interpolating for spin-1 particles. In this case, perturbative renormalizability requires the presence of a *local* gauge symmetry, and, as we shall see now, the intertwining of symmetry and renormalization properties becomes much more intricate than in the global symmetry case.

The new features which appear in the treatment of renormalization for spin-1 fields can be traced back to the large momentum behavior of the propagator for a canonical (massive) field of spin-j, which contains a numerator factor of order k^{2j}. This leads to a scalar ($j = 0$) propagator with a $1/k^2$ falloff at large momentum k, a Dirac fermion propagator falloff $1/k$, but no falloff at all for a massive spin-1 (Euclidean) propagator $(g^{\mu\nu} - k^\mu k^\nu/m^2)/(k^2 + m^2) \sim k^0$, $k >> m$. The result is that, although the engineering dimension of a massive A_μ vector field is the *same* as that of a scalar field (namely, mass to the first power in four spacetime dimensions), internal vector lines contribute an extra factor of $+2$, compared to scalar lines, to the overall degree of divergence of any 1PI graph containing them. For a theory of fermions interacting with vectors by a trilinear coupling $g\bar{\psi}\gamma^\mu\psi A_\mu$ (with g dimensionless), this means that the power-counting formula (17.77) for Yukawa theories must be replaced by

$$D = 4 - \frac{3}{2}E_\psi - E_A + 2I_A \qquad (17.81)$$

for a graph with E_ψ external fermion lines, E_A (resp. I_A) external (resp. internal) vector lines. The positive term $+2I_A$ has the usual disastrous consequence: regardless of the number of external lines possessed by a subgraph, at high enough order of perturbation theory the subgraph will become divergent, and increasingly so as we go to even higher orders in the trilinear coupling, by inserting more and more internal gauge field lines, leading to a non-renormalizable theory with infinitely many independent renormalization parts.

Of course, we have already seen in Chapter 15 that in a theory with *massless* vector particle and an exact local gauge symmetry, there exist choices of gauge in which the vector propagator has the same falloff, $1/k^2$, as a scalar propagator, thereby removing the unfortunate $+2I_A$ contribution in (17.81). This is the case, for example, in the covariant ξ-gauges discussed in Section 15.4, where the (Minkowski) gauge field propagator takes the form (cf. (15.159)):

$$\langle 0|T\{A_{\alpha\mu}(x)A_{\beta\nu}(y)\}|0\rangle = -i\delta_{\alpha\beta} \int \frac{(g_{\mu\nu} - (1-\xi)\frac{p_\mu p_\nu}{p^2})e^{-ip\cdot(x-y)}}{p^2 + i\epsilon} \frac{d^4p}{(2\pi)^4} \qquad (17.82)$$

The equivalence of the theory formulated in Hamiltonian language (and therefore manifestly unitary) to the manifestly covariant one determined by the (Euclidean) functional integral (15.161) depends crucially, as we explained in Section 15.4, on the local gauge invariance of the Lagrangian. Just as in the case of global symmetries, where Noether's theorem in functional form leads to a set of Ward–Takahashi identities constraining the Green functions of the theory, the critical local gauge invariance here can be re-expressed in a set of Ward identities (in the non-abelian case, more frequently called *Slavnov–Taylor* identities), the preservation of which, both by our regularization and by our reparameterization procedures, will be crucial for ensuring the perturbative renormalizability of the gauge theory. We shall illustrate the basic point using an abelian gauge theory—for example, QED—to reduce the algebra to a manageable level. Thus, we begin with a Euclidean functional integral (recall that no ghosts are needed in the abelian case) for coupled photons and electrons, with e the electron charge,

$$Z[J] = e^{W[J]} = \int \mathbf{D}A_\mu \mathbf{D}\psi \mathbf{D}\bar{\psi} e^{-\int (\mathcal{L}_{\mathrm{E}} + \frac{1}{2\xi}(\partial_\mu A_\mu)^2 - J_\mu A_\mu) d^4 x} \tag{17.83}$$

$$\mathcal{L}_{\mathrm{E}} = \frac{1}{4}(F_{\mu\nu})^2 - \bar{\psi}(\partial\!\!\!/ - m)\psi - ie\bar{\psi}A\!\!\!/\psi \tag{17.84}$$

We have not included source terms for the fermion fields, as we shall be considering only Green functions with external gauge field lines in the following. The functional integral (17.83) is invariant under a change of variable of integration $A_\mu(x) \to A_\mu(x) + \partial_\mu \lambda(x)$, with $\lambda(x)$ infinitesimal; and as this also corresponds to a local gauge transformation leaving $\mathcal{L}_{\mathrm{E}}$ invariant, we have, to first order in λ:

$$0 = \int \mathbf{D}A_\mu \mathbf{D}\psi \mathbf{D}\bar{\psi} \int [\frac{1}{\xi}\partial_\mu A_\mu(x)\Box\lambda(x) - J_\mu \partial_\mu \lambda(x)] d^4 x \, e^{-\int (\mathcal{L}_{\mathrm{E}} + \frac{1}{2\xi}(\partial_\mu A_\mu)^2 - J_\mu A_\mu) d^4 x} \tag{17.85}$$

Defining $\omega(x) \equiv \Box\lambda(x)$, this becomes

$$0 = \int \mathbf{D}A_\mu \mathbf{D}\psi \mathbf{D}\bar{\psi} \int [\frac{1}{\xi}\partial_\mu A_\mu(x) + \frac{1}{\Box}\partial_\mu J_\mu(x)]\omega(x) d^4 x \, e^{-\int (\mathcal{L}_{\mathrm{E}} + \frac{1}{2\xi}(\partial_\mu A_\mu)^2 - J_\mu A_\mu) d^4 x} \tag{17.86}$$

whence, taking the functional derivative with respect to $\omega(x)$, we find

$$0 = \int \mathbf{D}A_\mu \mathbf{D}\psi \mathbf{D}\bar{\psi}[\frac{1}{\xi}\partial_\mu A_\mu(x) + \frac{1}{\Box}\partial_\mu J_\mu(x)] \, e^{-\int (\mathcal{L}_{\mathrm{E}} + \frac{1}{2\xi}(\partial_\mu A_\mu)^2 - J_\mu A_\mu) d^4 x} \tag{17.87}$$

Factors of $A_\mu(x)$ within the functional integral may be replaced by functional derivatives with respect to the source $J_\mu(x)$ acting on the generating functional $Z[J]$, so our result (17.87) can be rewritten as a functional differential equation for $Z[J]$,

$$\frac{1}{\xi}\partial_\mu \frac{\delta Z[J]}{\delta J_\mu(x)} + \frac{1}{\Box}\partial_\mu J_\mu(x) Z[J] = 0 \tag{17.88}$$

or better, for the generating functional of connected Green functions, $W[J]$,

$$\frac{1}{\xi}\partial_\mu \frac{\delta W[J]}{\delta J_\mu(x)} = -\frac{1}{\Box}\partial_\mu J_\mu(x) \tag{17.89}$$

As we have seen repeatedly in the preceding discussion of renormalization, the 1PI graphs of the theory play a primary role: they are, as it were, the "atomic" constituents in terms of which the divergence structure of the theory is most conveniently analyzed. Consequently, it is preferable to re-express the constraint (17.89) in terms of the generating functional $\Gamma[\mathcal{A}_\mu]$ which generates the 1PI graphs of the theory. In this case, it is defined by the Legendre transformation (cf. (10.140))

$$\Gamma[\mathcal{A}] = -W[J] + J_\mu \mathcal{A}_\mu \tag{17.90}$$

where the source variable J_μ and classical field variable $\mathcal{A}_\mu$ satisfy

$$J_\mu(x) = \frac{\delta \Gamma[\mathcal{A}]}{\delta \mathcal{A}_\mu} \tag{17.91}$$

$$\mathcal{A}_\mu = \frac{\delta W[J]}{\delta J_\mu} \tag{17.92}$$

Inserting these relations in (17.89) we obtain, without any further ado,

$$\frac{1}{\xi} \partial_\mu \mathcal{A}_\mu(x) = -\frac{1}{\Box} \partial_\mu \frac{\delta \Gamma[\mathcal{A}]}{\delta \mathcal{A}_\mu(x)} \tag{17.93}$$

This seemingly innocent equation contains a wealth of information about the n-point 1PI functions of the gauge field, as we recover statements about these simply by taking the appropriate number of functional derivatives of (17.93) with respect to $\mathcal{A}_\mu$. In particular, if we differentiate once with respect to $\mathcal{A}_\nu(y)$, we find

$$\frac{1}{\xi} \frac{\partial}{\partial x_\nu} \delta^4(x - y) = -\frac{1}{\Box_x} \frac{\partial}{\partial x_\mu} \frac{\delta^2 \Gamma[\mathcal{A}]}{\delta \mathcal{A}_\mu(x) \delta \mathcal{A}_\nu(y)} \tag{17.94}$$

Recalling that the second derivative of Γ yields the (full) *inverse* propagator of the theory, we may write down immediately the Fourier transform of (17.94), with the obvious translations $\partial_\mu \to i p_\mu$, $\Box \to -p^2$,

$$\frac{1}{\xi} p_\nu = \frac{1}{p^2} p_\mu \hat{\Delta}_F^{-1}{}_{\mu\nu}(p) \tag{17.95}$$

The full gauge field (i.e., photon) propagator can be obtained, just as in (17.50), as an iteration of free propagators interspersed with self-energy corrections. The only difference is that here the propagator and self-energy carry two vector indices, so the products in the iteration are matrix products. Consequently,

$$\hat{\Delta}_F^{-1}{}_{\mu\nu}(p) = \Delta_F^{-1}{}_{\mu\nu}(p) - \Pi_{\mu\nu}(p) = p^2 (\delta_{\mu\nu} + (\frac{1}{\xi} - 1) \frac{p_\mu p_\nu}{p^2}) - \Pi_{\mu\nu}(p) \tag{17.96}$$

where the first term on the right-hand side is the inverse of the free Euclidean propagator (cf. (17.82)) in the covariant ξ-gauge, and the self-energy $\Pi_{\mu\nu}(p)$ is given in the bare (i.e., pre-reparameterization) theory by graphs such as those indicated in Fig. 17.12. In particular, the one-loop graph of Fig. 17.12 is given by

$$\Pi_{1\,\mu\nu}(p) = e^2 \int \frac{\text{tr}[\hat{\gamma}_\mu((i\slashed{l} + m) \hat{\gamma}_\nu (i(\slashed{p} + \slashed{l}) + m)]}{(l^2 + m^2)((p + l)^2 + m^2)} \frac{d^4 l}{(2\pi)^4} \tag{17.97}$$

where the $\hat{\gamma}_\mu$ are the Euclidean γ-matrices defined in (15.160). Using the appropriate Euclidean trace identities (see Problem 5), this loop integral becomes

$$\Pi_{1\,\mu\nu}(p) = 4e^2 \int \frac{\delta_{\mu\nu}(m^2 + l \cdot (p + l)) - l_\mu(p + l)_\nu - l_\nu(p + l)_\mu}{(l^2 + m^2)((p + l)^2 + m^2} \frac{d^4 l}{(2\pi)^4} \tag{17.98}$$

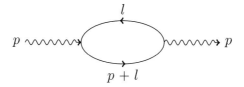

Fig. 17.12 One-loop photon self-energy graph.

which is at first sight quadratically divergent, in accordance with the power-counting rule (17.81). Now the Ward identity (17.95) evidently implies, on inserting (17.96),

$$p_\mu \Pi_{\mu\nu}(p) = 0 \qquad (17.99)$$

as the left-hand side of (17.95) is already given by just the free propagator part of (17.96). In other words, the local gauge symmetry embodied by the Ward identities of the theory requires the photon self-energy tensor to be *transverse*. This transversality must, of course, hold at each loop order in perturbation theory, as we may formally expand $\Pi_{\mu\nu}$ in powers of $\hbar$ and require that (17.99) hold at each order. Any regularization which violates this property will *ipso facto* do violence to the local gauge symmetry, and, as we shall now see, destroy the perturbative renormalizability of the theory. Let us examine this issue explicitly for the one-loop integral (17.98). Inserting a Feynman parameter in the usual way (cf. (16.67)), this becomes

$$\Pi_{1\,\mu\nu}(p) = 4e^2 \int_0^1 dx \int \frac{\delta_{\mu\nu}m^2 - 2l_\mu l_\nu + 2x(1-x)p_\mu p_\nu + \delta_{\mu\nu}(l^2 - x(1-x)p^2}{(l^2 + x(1-x)p^2 + m^2)^2} \frac{d^4l}{(2\pi)^4}$$
$$(17.100)$$

If we regularize the integral simply by imposing a momentum cutoff $|l| < \Lambda$, the integral has a leading contribution for $\Lambda \gg p, m$ given by

$$\Pi_{1\,\mu\nu}(p) \sim \frac{e^2}{8\pi^2}\Lambda^2\delta_{\mu\nu} \qquad (17.101)$$

which clearly does not satisfy (17.99). Such a divergence, were it truly present in the theory, could only be removed by a counterterm corresponding to an explicit photon mass term in the Lagrangian $\delta m^2 A_\mu A_\mu$, which of course also violates the local gauge invariance of the theory.

On the other hand, dimensional regularization of the integral maintains the desired transversality, as we might expect (or at least hope!), given that the formal structure (and hence, the propagators and vertices) of a locally gauge-invariant Lagrangian does not depend on the spacetime dimension in which it is formulated. Making the standard replacement

$$\frac{d^4l}{(2\pi)^4} \rightarrow \mu^{4-d}\frac{d^dl}{(2\pi)^d} \qquad (17.102)$$

in (17.100), with μ a regularization scale needed to maintain dimensional consistency of the expression while we are away from four dimensions, and following steps completely analogous to those leading from (16.62) to (16.65) (see Problem 6), we find

$$\Pi_{1\,\mu\nu}(p) = 8e^2 \frac{\pi^{d/2}}{(2\pi)^d}(p_\mu p_\nu - \delta_{\mu\nu}p^2)\Gamma(2 - \frac{d}{2})\int_0^1 x(1-x)(\frac{x(1-x)p^2 + m^2}{\mu^2})^{\frac{d}{2}-2}dx$$

$$(17.103)$$

which manifestly satisfies the transversality condition (17.99). The result is, of course, singular as we return to four dimensions, $d = 4 - \epsilon, \epsilon \to 0$, with the pole part given by

$$t_{DR}^\gamma \Pi_{1\,\mu\nu}(p) = 8e^2 \frac{\pi^2}{(2\pi)^4}(p_\mu p_\nu - \delta_{\mu\nu}p^2)\frac{2}{\epsilon}\int_0^1 x(1-x)dx = \frac{e^2}{6\pi^2}\frac{1}{\epsilon}(p_\mu p_\nu - \delta_{\mu\nu}p^2)$$

$$(17.104)$$

We have introduced the notation t_{DR}^γ for the dimensional subtraction operator on the renormalization part γ—in this case simply the one-loop graph of Fig. 17.12, which replaces the zero momentum Taylor operators of the BPHZ scheme when we follow a dimensional renormalization procedure as described in the preceding section. The critical point is that the subtraction operator amounts to exactly a counterterm $\delta Z \times \frac{1}{4}(F_{\mu\nu})^2$, which inserted to first order in the photon propagator clearly produces (Problem 7), in momentum space, just the transverse tensor seen in (17.104). But such a counterterm simply amounts to a rescaling of the photon field A_μ in our bare Lagrangian (17.84): in the more conventional (but somewhat misleading) language of renormalization theory, to a "wavefunction renormalization". Taking into account the subtraction effected by the t_{DR}^γ pole part operator, we are left with the renormalized one-loop self-energy (taking the spacetime dimension back to the physical value)

$$\Pi_{1,R\,\mu\nu}(p) = \lim_{\epsilon\to 0}(1 - t_{DR}^\gamma)\Pi_{1\,\mu\nu}(p) \qquad (17.105)$$

$$= -\frac{e^2}{2\pi^2}(p_\mu p_\nu - \delta_{\mu\nu}p^2)\int_0^1 (\gamma + \ln(\frac{x(1-x)p^2 + m^2}{4\pi\mu^2}))dx \quad (17.106)$$

$$= (p_\mu p_\nu - \delta_{\mu\nu}p^2)\Pi_{1,R}(p^2) \qquad (17.107)$$

$$\Pi_{1,R}(p^2) \equiv -\frac{e^2}{2\pi^2}\int_0^1 (\gamma + \ln(\frac{x(1-x)p^2 + m^2}{4\pi\mu^2}))dx \qquad (17.108)$$

Note that the dependence on the arbitrary renormalization scale μ (which plays the role of the UV cutoff in a momentum cutoff scheme) is logarithmic, not quadratic. In effect, the need to produce the transverse tensor $p_\mu p_\nu - \delta_{\mu\nu}p^2$ *outside* the loop integral has reduced the quadratic divergence of the loop integral (17.98) to a logarithmic one. Inserting the renormalized self-energy in the expression (17.96) for the inverse full photon propagator, and inverting once again, we find, for the renormalized Euclidean photon propagator (to one loop),

$$\hat{\Delta}_{F\mu\nu}(p) = \frac{\delta_{\mu\nu} - \frac{p_\mu p_\nu}{p^2}}{p^2(1 - \Pi_{1,R}(p^2))} + \xi\frac{p_\mu p_\nu}{p^4} \qquad (17.109)$$

The radiative corrections have induced a change only in the transverse ("physical") part of the propagator, but the propagator pole is still at $p^2 = 0$ (note that $\Pi_{1,R}(0)$ is finite and non-zero): our physical photon mass is still safely zero. The gauge-variant part of the propagator retains exactly its lowest-order, tree-level value—a property which can be shown to persist at higher orders of perturbation, thanks to the Ward identity. Of course, the residue at the pole—the LSZ "Z" constant—is not unity (it is, in this renormalization scheme, just $1/(1 - \Pi_{1,R}(0))$, and must therefore be included in the LSZ formula (9.179) when S-matrix elements are computed.

Similar improvements in the degree of divergence relative to the naive power-counting rule (17.81) occur in other dimensionally subtracted Green functions of the theory, all with the net result of limiting the renormalization parts of the theory to just those which correspond to the counterterms available by reparameterizing only the couplings and fields present in our initially locally gauge-invariant Lagrangian (17.84). For example, the analog of Fig. 17.10, the one-loop contribution to light by light (elastic 2-2 photon) scattering arising from an internal electron loop, would seem at first glance to be logarithmically divergent, as in the Yukawa case, and therefore to require the presence of a quartic counterterm $\delta\lambda(A_\mu A_\mu)^2$. Such a term, of course, destroys local gauge invariance, and once admitted, would then lead to the need for a photon mass counterterm as well, massive photon propagators, and the re-emergence of the fatal $+2I_A$ contribution in (17.81). In fact, the Ward identity obtained by differentiating (17.94) a further three times with respect to the classical photon field $\mathcal{A}$ amounts to the statement that the 1PI four-point photon Green function of the theory is divergence-less on any of its four spacetime indices, or in momentum space:

$$p_{1\mu}\Gamma^{(4)}_{\mu\nu\rho\sigma}(p_1, p_2, p_3, p_4) = p_{2\nu}\Gamma^{(4)}_{\mu\nu\rho\sigma}(p_1, p_2, p_3, p_4) = \ldots = 0 \qquad (17.110)$$

The transversality, just as in the case of the two-point function, is realized by the appearance of four transverse tensors on each of the external legs of the diagram, which then reduces the effective degree of divergence (in this case, by an astonishing four powers, although in the present circumstance one would suffice!) to a negative value. The diagram is therefore finite, and no counterterm is in fact needed.

A complete demonstration of perturbative renormalizability for a general non-abelian gauge theory,[11] quantized in a covariant ξ-gauge and defined by the functional integral (15.161), requires the systematic application of the appropriate generalizations of our simple Ward identity (17.94) for the general 1PI Green functions of the theory—involving gauge field, and fermion and ghost fields. These "Slavnov–Taylor" identities can then be shown to imply that the only renormalization parts arising to arbitrary orders of perturbation theory indeed correspond to counterterms associated with reparameterization of the original Lagrangian. The argument is somewhat lengthy

---

[11] It should be confessed at this point that the addition of a photon mass term in the abelian case does not in fact ruin perturbative renormalizability: the additional term induced in the Ward identity (basically, one displaces the gauge parameter $\frac{1}{\xi}$ by $\frac{m_A^2}{\Box}$) does not alter the transversality of the multi-photon 1PI amplitudes needed to exclude quartic non-gauge-invariant counterterms, although a mass counterterm is now necessary. In the non-abelian case, however, the non-linear gauge couplings result in the generation of an infinite number of non-gauge-invariant counterterms once an explicit mass term is included for the gauge fields.

and will not be reproduced here, as its conceptual essence is already visible in the abelian examples given above.[12]

We have seen that the explicit breaking of global symmetries of the Lagrangian does not in general alter the renormalizability status of the theory: typically, one simply has a larger number of renormalizable and super-renormalizable counterterms which must be included in the Lagrangian to absorb divergences in multi-loop diagrams. The same is true of spontaneously broken symmetries, which from a Lagrangian point of view can be thought of as explicitly broken theories with special relations between the Lagrangian parameters (cf. Section 8.4). Evidently, the renormalizability of gauge theories (especially in the non-abelian case) is much more intimately connected with the underlying local symmetry, and relies on the absence of any explicit symmetry-breaking terms, even if their engineering dimension places them in the category of (superficially) renormalizable or super-renormalizable operators.

Nevertheless, the perturbative renormalizability of a non-abelian gauge theory is unaffected by *spontaneous breaking* of the remnant discrete gauge symmetry (after gauge-fixing; cf. Section 15.6), which results in the appearance of physically massive gauge vector particles via the Higgs phenomenon. We saw in Chapter 14 that the spontaneous breakdown of a symmetry is quintessentially a long-distance (therefore, low-energy) phenomenon, so it should not be surprising that the short-distance scaling behavior of the theory is basically unaffected by the choice of vacuum state entailed by the field shifts used to implement the effects of the spontaneous breakdown. Once a perturbatively renormalizable theory is reparameterized in terms of a (finite!) set of well-defined low-energy parameters (and appropriately rescaled fields), the Lagrangian, and hence the associated Hamiltonian defines a dynamical evolution in the state space which is (order by order in perturbation theory, of course) insensitive at the power level to the UV cutoff in the theory. This remains true whether the Hamiltonian is applied to states obtained by applying field operators to the "false" non-ground-state "vacuum" of the unshifted, symmetric theory, or to the physically relevant states built on the true ground-state vacuum of the theory. Indeed, we have already seen in Section 15.6 that by exploiting the underlying exact local gauge symmetry of a spontaneously broken gauge theory (abelian *or* non-abelian) one may derive covariant Feynman rules in which the massive vector propagators have the soft $1/k^2$ falloff essential for the normal alignment of renormalizable (resp. non-renormalizable) operators with mass dimension 4 (resp. > 4). The non-abelian Slavnov–Taylor identities do the remaining job, just as in the unbroken case, of restricting the divergent counterterms to just those associated with reparameterization of the parameters of the original Lagrangian, which, we recall, is manifestly locally gauge-symmetric before this property is disguised by the field shifts employed to display the spontaneous breaking.

17.4 Renormalization group approach to renormalizability

The property of renormalizability—otherwise stated, the assertion that the complete dynamics of a local quantum field theory can be expressed in terms of a *finite* number

---

[12] The interested reader will find the complete proof in (for example) Section 12.4, (Itzhykson and Zuber, 1980).

of parameters defined in terms of the low-energy (or large-distance) properties of the theory—appears at first sight quite astonishing in the light of our discussion of Wilsonian effective Lagrangians in Sections 16.2 and 16.3. There, in view of the inevitable breakdown of Minkowski-based field theories at short distance due to (at the very least) quantum gravity effects, we insisted that any physically sensible theory should include a cutoff at some high-energy/momentum scale, and that the necessary result of such a cutoff was the appearance of an *infinite* number of operators, including the baleful non-renormalizable ones, in the Lagrangian density defining the dynamics of the cutoff theory. The cutoff μ used is a matter of convenience, as the resultant effective Lagrangian $\mathcal{L}_\mu$ is defined to yield exactly the same low-energy (i.e., sub-μ) physics irrespective of the choice of μ; but we saw in Section 16.3 that even if the coefficient of a particular operator is fixed to zero at some scale, it will no longer be so at lower scales, due to the renormalization group flow in the infinite-dimensional coupling constant space of a cutoff theory. Thus, the whole notion of working with a Lagrangian, with fields cutoff at some value Λ, but only a finite number of operators (the super-renormalizable and renormalizable ones) with non-zero couplings, and maintaining zero couplings for the (infinitely many) non-renormalizable operators as the cutoff Λ is progressively increased, seems at first sight incomprehensible from the Wilsonian point of view. Our object in this section is quite simply to explain the resolution of this apparent paradox, along the lines originally followed by Polchinski (Polchinski, 1984). Our discussion will make clear that the specific procedure of perturbative renormalization developed in the preceding sections of this chapter amounts really to just one special (though technically very convenient!) application of the much more general idea of a Wilsonian effective Lagrangian introduced in the preceding chapter.

We recall that a local scalar field theory with field modes above some momentum scale Λ integrated out is described by an effective Lagrangian (cf. (16.25)):

$$\mathcal{L}_\Lambda = \sum_n a_n(\Lambda)\mathcal{O}_n(\phi_\Lambda) \tag{17.111}$$

We have slightly altered notation here, by labeling all operators (irrespective of the number of spacetime-derivatives) by a single index n: thus, in addition to simple powers of the field, the kinetic operator is included in the list, together with all other operators with two, four, six, etc., spacetime-derivatives. The field $\phi_\Lambda(x)$ (cf. (16.4)) only contains Fourier modes of the field with Euclidean momentum $|k| < \Lambda$. Counting powers of mass dimension, we find that if the operator $\mathcal{O}_n$ has mass dimension $4 - d_n$, the associated coefficient a_n has mass dimension d_n, and we may define (cf. (16.8)) dimensionless couplings $g_n(\Lambda)$ by the simple scaling

$$g_n(\Lambda) \equiv a_n(\Lambda)\Lambda^{-d_n} \tag{17.112}$$

The scale Λ is, as previously stated, arbitrary, and we may imagine fixing the dimensionless couplings at some fixed very high ultraviolet scale Λ_{UV} (much higher than the physics we wish to explore, but safely smaller than the scale of quantum gravity effects, say),

$$\bar{g}_n \equiv g_n(\Lambda_{UV}) \tag{17.113}$$

and then using the non-linear first-order evolution equations (16.27) describing the renormalization group flow of the effective Lagrangian to determine the dimensionless couplings at any lower scale $\mu < \Lambda_{\mathrm{UV}}$, as a function of the dimensionless ratio $\mu/\Lambda_{\mathrm{UV}}$ and the initial high-energy parameters $\bar{g}_n$:

$$\mu\frac{\partial}{\partial\mu}g_n(\mu) = \beta_n(g_n(\mu)) \Rightarrow g_n(\mu) = g_n(\bar{g}_n; \mu/\Lambda_{\mathrm{UV}}) \qquad (17.114)$$

In Section 16.3 we divided the set of local operators $\mathcal{O}_n$ into the *relevant* operators corresponding to $d_n > 0$, the *marginal* operators with $d_n = 0$, and the *irrelevant* operators with $d_n < 0$ (corresponding, respectively, to the classification into super-renormalizable, renormalizable, and non-renormalizable operators in the language of perturbative renormalization). The relevant and marginal operators correspond to operators (in spacetime dimension 4) with mass dimension less than or equal to 4, and therefore constitute a finite set: let there be N of these (the actual number may depend on the type of fields and interactions, and imposed symmetries, of course). We shall distinguish the operators in this finite set by writing small Roman indices a, b, etc., and indicate the irrelevant operators of dimension greater than 4 (of which there are an infinite number) by Greek letters α, β, etc., reserving later Roman characters m, n, r, etc., for the general set of operators. In Section 16.3 we also saw in some simple examples that the dominant effect of integrating out non-renormalizable operators between a high UV cutoff scale and a low-energy scale (up to small corrections involving inverse powers of the large ultraviolet scale) was to produce modifications, potentially of order unity, in the couplings associated with marginal and relevant operators. The point of the following discussion is to reproduce this result in a much more general context. Our derivation will follow the streamlined approach described by Weinberg (Weinberg, 1995a), based on Polchinski's original arguments (Polchinski, 1984).

We shall demonstrate that in the regime of weakly coupled perturbation theory, the renormalization group flow implied by (17.114) maps an arbitrary initial surface $\bar{S}$ in the high-energy coupling constant space of the $\{\bar{g}_n\}$ to an N-dimensional surface S of the $\{g_n(\mu)\}$, a given point on which is uniquely determined by specifying N low-energy parameters, up to corrections which fall as inverse whole powers of the ratio of the UV cutoff Λ_{UV} to the low-energy scale μ. The demonstration relies on a linear stability analysis familiar in the treatment of non-linear dynamical systems. We first consider the effect of a small (infinitesimal) change δg_n in the parameters on the renormalization flow generated by (17.114). Defining

$$M_{nm}(g_n) \equiv \frac{\partial\beta_n}{\partial g_m} \qquad (17.115)$$

we have, to first order in the δg_n,

$$\mu\frac{\partial}{\partial\mu}\delta g_n(\mu) = M_{nm}\delta g_m(\mu) \qquad (17.116)$$

We may also define a matrix G_{nm} expressing the variation of the low-energy parameters under variation of the initial parameters $\bar{g}_n$ (see (17.114)):

$$G_{nm}(\mu) \equiv \frac{\partial g_n}{\partial \bar{g}_m} \tag{17.117}$$

Differentiating (17.114) with respect to the initial parameters, one finds also

$$\mu \frac{\partial}{\partial \mu} G_{nm} = M_{nr} G_{rm} \tag{17.118}$$

We shall assume that the finite $N \times N$ submatrix G_{ab} with rows and columns restricted to the marginal and relevant couplings is not singular, with well-defined inverse G_{ab}^{-1}, which is presumably the case with the exception of perhaps an isolated set of measure zero in the coupling constant space, which we assume our renormalization group flow avoids. By usual matrix algebra, one has

$$\mu \frac{\partial}{\partial \mu} G_{ab}^{-1} = -G_{ac}^{-1} (\mu \frac{\partial}{\partial \mu} G_{cd}) G_{db}^{-1} = -G_{ac}^{-1} M_{cn} G_{nd} G_{db}^{-1} \tag{17.119}$$

Note that the final pair of matrices $G_{nd} G_{db}^{-1}$ appearing here cannot (for general n) be collapsed to a Kronecker δ, as the summed index d only runs over a partial subset of the couplings. We now introduce a projected set of variations $\hat{\delta g}_\alpha$ in the irrelevant (non-renormalizable) couplings:

$$\hat{\delta g}_\alpha(\mu) \equiv \delta g_\alpha(\mu) - G_{\alpha a} G_{ab}^{-1} \delta g_b(\mu) \tag{17.120}$$

Effectively, as we shall soon see, $\hat{\delta g}_\alpha$ measures the extent to which variations in the (infinitely many) non-renormalizable couplings cannot be compensated for by variation in the N marginal/renormalizable couplings. The flow equation for the $\hat{\delta g}_\alpha$ follows directly from (17.116, 17.118, 17.119), and we find

$$\mu \frac{\partial}{\partial \mu} \hat{\delta g}_\alpha(\mu) = M_{\alpha n} \delta g_n - M_{\alpha n} G_{na} G_{ab}^{-1} \delta g_b$$

$$+ G_{\alpha a} G_{ac}^{-1} M_{cn} G_{nd} G_{db}^{-1} \delta g_b - G_{\alpha a} G_{ab}^{-1} M_{bn} \delta g_n$$

$$= M_{\alpha\beta} \delta g_\beta - M_{\alpha\beta} G_{\beta a} G_{ab}^{-1} \delta g_b$$

$$+ G_{\alpha a} G_{ac}^{-1} M_{c\beta} G_{\beta d} G_{db}^{-1} \delta g_b - G_{\alpha a} G_{ab}^{-1} M_{b\beta} \delta g_\beta$$

$$= (M_{\alpha\beta} - G_{\alpha a} G_{ab}^{-1} M_{b\beta})(\delta g_\beta - G_{\beta c} G_{cd}^{-1} \delta g_d) \tag{17.121}$$

$$\equiv \hat{M}_{\alpha\beta} \hat{\delta g}_\beta \tag{17.122}$$

The equivalence of the first and third lines follows from the fact that for values of the index n in the marginal/relevant subset, we have $G_{na} G_{ab}^{-1} = \delta_{nb}$, and the two terms on the first line then cancel: hence, the sum over n may be restricted to the non-renormalizable set labeled by β in the third line. A similar argument establishes the equivalence of the second and fourth lines, at which point (17.121) follows with some straightforward shuffling of indices. In the free field limit (all couplings corresponding to higher than quadratic operators set to zero) there are no loop integrals, and the entire scale dependence of the dimensionless parameters is due to the rescaling by engineering dimension in (17.112), and we therefore have

$$M_{nm} \sim \hat{M}_{nm} \sim -d_n \delta_{nm} \qquad (17.123)$$

If we then consider renormalization group flows in the infinitesimal neighborhood of the free field surface (and when we work in perturbation theory, effectively computing multiple derivatives of the amplitudes at zero coupling, this is exactly what we are doing), the projected variations $\hat{\delta g}_\alpha$ for the irrelevant couplings, for which $d_\alpha < 0$ must decay at low energy like inverse powers of the UV cutoff:

$$\hat{\delta g}_\alpha \sim (\frac{\mu}{\Lambda_{UV}})^{-d_\alpha} = (\frac{\mu}{\Lambda_{UV}})^{|d_\alpha|} \qquad (17.124)$$

or, equivalently,

$$\delta g_\alpha(\mu) \sim G_{\alpha a} G_{ab}^{-1} \delta g_b(\mu) + O((\frac{\mu}{\Lambda_{UV}})^{|d_\alpha|}), \quad \mu \ll \Lambda_{UV} \qquad (17.125)$$

The reason for the otherwise strange qualifier "irrelevant" applied to the non-renormalizable couplings and operators in the theory (indexed by α) should be apparent at this point: their effects at low energy may be entirely subsumed in variations of the marginal and relevant couplings (indexed by b).

The result (17.125) (see Fig. 17.13) expresses the desired result: arbitrary infinitesimal displacements of the initial (high-energy) point $\{\bar{g}_n\}$ (in other words, in the tangent space of any surface containing the initial point) amount to displacements of the low-energy parameters (at cutoff μ) in a finite, N-dimensional surface $\mathcal{S}$, as the displacements $\delta g_\alpha(\mu)$ are simply linear combinations of the N displacements $\delta g_b(\mu)$ of the marginal/relevant couplings of the theory. All low-energy amplitudes of the theory (i.e., with external Euclidean momenta less than μ) are given by the specification of the effective Lagrangian $\mathcal{L}_\mu$ determined by the $g_n(\mu)$, so fixing the low-energy physics uniquely amounts to specifying a finite number—in fact, exactly N—of independent low-energy amplitudes, which then locate a unique point on the attractive surface $\mathcal{S}$. This is exactly the procedure used in the preceding sections, where we have imposed N renormalization conditions as a prelude to the reparameterization of the theory in terms of the parameters defined by these conditions. It should be emphasized that the $g_\alpha(\mu)$ coefficients of the infinitely many non-renormalizable operators in $\mathcal{L}_\mu$ are *not zero*: indeed, they are needed to incorporate the effects of the marginal/relevant couplings once all the modes of the field between Λ_{UV} and μ are integrated out. In other words, using $\mathcal{L}_\mu$ to actually calculate field theory amplitudes would require including all the vertices for the operators $\mathcal{O}_n$, but restricting the loop integrals to internal propagator momenta $|k| < \mu$—clearly an impractical procedure, and not the way we actually proceed in renormalized perturbation theory. Instead, the process of renormalization as outlined in the preceding sections of this chapter implicitly amounts to computing field-theory amplitudes (order by order in perturbation theory) successively starting the renormalization group flow at the initial point

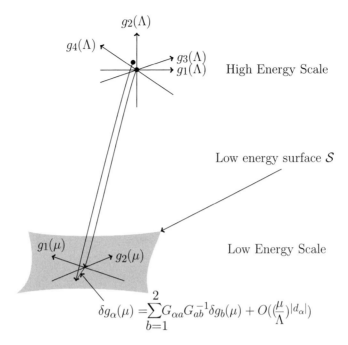

Fig. 17.13 Schematic illustration of an attractive renormalization group flow onto a finite-dimensional (N=2) low-energy surface.

$$g_\alpha(\Lambda_{\text{UV}}) = 0$$

$$g_a(\Lambda_{\text{UV}}) = \bar{g}_a(\Lambda_{\text{UV}}) \neq 0 \qquad (17.126)$$

with the "bare" couplings $\bar{g}_a(\Lambda_{\text{UV}})$ chosen to fix the physical theory at low energy at the desired end-point on the low-energy surface $\mathcal{S}$ (by imposing the renormalization conditions), and then taking the limit $\Lambda_{\text{UV}} \to \infty$, confident that the end-point of the renormalization group flow remains on the attractive surface $\mathcal{S}$ at the unique physical point identified by fixing the particular N low-energy amplitudes that define the renormalization scheme we have decided to employ. In the language of the renormalization group, we can say that the huge variety of theories defined by the infinitely many couplings $\bar{g}_n$ specified at the short-distance cutoff lie in a single *universality class* of theories, namely those which collapse to a finite-(N)-dimensional surface at low energies, as pictured in Fig. 17.13.

While the above argument asserts that the low-energy limit corresponds to some set of marginal/relevant operators, and hence to a theory that is perturbatively renormalizable by power-counting, it may be the case, depending on the field content and the nature of the interactions, that the resultant low-energy theory is in fact just a free theory. A classic example is the Fermi theory of weak interactions: if we integrate out all fields down to a Gev (say), and consider only the weak interactions of leptons and baryons (e.g., nuclear β decay), then we have only spin-$\frac{1}{2}$ fields to consider, and there are no renormalizable interactions in four dimensions involving only fermionic spin-$\frac{1}{2}$ fields: the four-fermion coupling term is dimension 6 and therefore non-renormalizable.[13] From the point of view of the argument just above, the low-energy theory is free, and we simply ignore the weak interactions, as a remnant of high-energy "new physics", contributing at the "negligible" $O((\frac{\mu}{\Lambda_{\mathrm{UV}}})^{|d_\alpha|})$ level of (17.125) (where here $\mu \sim$ MeVs, $\Lambda_{\mathrm{UV}} \sim$ the W mass, 80 GeV, and $d_\alpha = 2$). Of course, if we integrate only down to 1 TeV, we are left with all the fields of the Standard Model, including the W-, Z-, and (presumptive) Higgs bosons, which do form a perturbatively renormalizable theory—exactly the point of the great electroweak unification of the early 1970s.

We also note here the obvious point that we are not always so fortunate to have the couplings of the low-energy renormalizable theory—albeit a non-trivial one with interesting interactions—sufficiently small to render perturbation theory quantitatively useful. In the case of QCD, we do indeed end up with a renormalizable theory at low energy, but one with a gauge coupling constant of order unity, which means that a complete quantitative evaluation of the low-energy amplitudes of the theory necessarily requires explicitly non-perturbative methods, such as lattice gauge theory. In the next chapter we shall see that QCD however possesses the remarkable property that the running coupling *decreases* with increasing energy (the famous property of "asymptotic freedom"), so that the renormalization group actually provides an escape route allowing the perturbative calculation of certain high-energy amplitudes of the theory. Of course, in the case of QED we are very fortunate in this regard: the low-energy coupling of the electron to the photon provides an expansion coefficient of order 1/137 (the fine-structure constant), so perturbation theory is (initially) rapidly convergent, and the accuracy of the results obtained for the anomalous magnetic moment of the electron can be used to establish the absence of new physics (in the form of a dimension-5 operator $\bar{\psi}\sigma_{\mu\nu}\psi F^{\mu\nu}$) up to an energy scale of at least 10^7 GeV (Weinberg, 1995a).

It must once again be emphasized that the arguments given here, based as they are on a linear stability analysis, make sense only in the context of perturbation theory, where we are entitled to treat all couplings associated with interaction operators as infinitesimal. Infinitesimal couplings remain so under the renormalization group flow from Λ_{UV} down to the low scale μ, even though the flow is non-linear. No statement is made—nor can be made—concerning the actual existence of flows beginning at the specified UV starting point (17.126), with *finite* rather than infinitesimal couplings, and ending at some specified desired low-energy coupling strengths (say, for the renormalized coupling λ_R in ϕ^4-theory). This is a global issue which requires non-perturbative control over the theory, as one is in principle interested in situations where the renormalization group flow is diverted into regions where weak coupling perturbation theory is no longer valid.

---

[13] The same problem occurs in attempts to describe quantum gravity in terms of a local field theory: there are no renormalizable interactions of spin-2 gravitons. The leading low-energy residual of whatever microscopic theory adequately describes quantum gravity effects, the Einstein–Hilbert Lagrangian, leads to the appearance of infinitely many counterterms if we attempt a perturbative expansion. Indeed, the only coupling in the theory, Newton's constant G, has dimensions of mass^{-2}—the classic signature of a non-renormalizable interaction.

In fact, it may well be the case that the limit outlined above, in which Λ_{UV} is taken to infinity while holding a set of N low-energy amplitudes fixed leads to a high-energy theory which does not correspond to a physically sensible theory—for example, by having negative terms in the effective action which render the functional integral at the high-scale divergent. In such a case, the only scaling of the bare couplings at the ultraviolet end which leads to a sensible low-energy theory corresponds to sending all the interaction terms to zero—in other words, to the "trivial" result of a free field theory. There is considerable circumstantial evidence (we shall consider some in the next chapter) that both ϕ^4 theory and quantum electrodynamics in four spacetime dimensions, taken as self-standing field theories, are in fact trivial theories of this sort, even though they are, as we have seen earlier, formally perturbatively renormalizable to all orders of perturbation theory.[14]

There is no paradox here: recall (Section 11.1) that the formal perturbation series is always only a divergent, asymptotic one. A local field theory in which a well-defined continuum limit exists, where all ultraviolet cutoffs have been removed in a way consistent with full Poincaré invariance while retaining the hermiticity of the Lagrangian and the unitarity of the theory, is presumably one in which n-point functions exist satisfying the full panoply of Wightman axioms discussed in Chapter 9. However, it may well be the case that the perfectly well-defined all-orders expansions of the amplitudes of a field theory do not correspond to the asymptotic expansions of a set of Wightman functions satisfying the needed axioms. In this situation, the perturbation expansion may still be of enormous phenomenological utility (as in the case of QED): we must regard the relevant microscopic theory not as a continuum field theory, but as an effective Wilsonian theory valid up to a high-energy scale beyond which new physics comes into play, altering significantly the ultraviolet behavior. As long as the interaction couplings at the high scale are reasonably small, we may expect that the flow down to the low-energy scale where we are doing physical measurements produces an attraction onto a finite-dimensional surface on which we work in the setting of renormalization theory. In the case of the electroweak sector of the Standard Model, the measurements are at an energy scale on the order of hundreds of GeVs, and we are fortunate that the low-energy couplings are small here. In the case of QCD, the low-energy couplings attract to a finite-dimensional surface where the gauge coupling appropriate for hadronic phenomena in the sub-GeV regime is of order unity, and renormalized perturbation theory, though formally perfectly sensible, is not useful, and in fact, is qualitatively misleading with regard to the physics of the theory, as we shall see in Chapter 19.

The renormalization group approach to perturbative renormalizability has some quite striking advantages in comparison to the detailed analysis of divergence structure presented earlier in this chapter. The decoupling of ultraviolet sensitivity is seen to proceed by simple scaling arguments, with no reference to the complications of nested or overlapping subdivergences in the Feynman graphs associated with the perturbative amplitudes. There are, however, considerable disadvantages attached to this approach,

---

[14] For a detailed description of the mathematical issues involved in establishing triviality of field theories, see (Fernandez *et al.*, 1992).

at least from the point of view of high-energy theory (though less so in the many condensed-matter applications of the renormalization group). The main one arises from the need to impose momentum cutoffs, which we have seen do violence to the local gauge symmetry which plays a central role in all sectors of the Standard Model. This results in enormous technical complications when repeating the proof of renormalizability for gauge theories along the lines of Wilsonian effective theory, as given above for scalar field theories, although the method has been successfully applied (if painfully) even in this case (Kopper and Müller, 2009). Furthermore, the renormalization techniques developed earlier in the chapter provide the germs for further extensions which are indispensable in understanding the short-distance/high-energy behavior of field theories (where here we are talking about energy scales intermediate between the important dimensionful scales defining the theory at low energy, and the high-energy scale at which the theory becomes invalid). Here the notion of oversubtraction of amplitudes—a straightforward extension of the subtraction techniques of Section 17.2—becomes critical in understanding the factorization properties of field theory amplitudes in this regime. These ideas are difficult, if not impossible, to implement in the framework of renormalization group flow arguments of the type given in this section, although, as we are about to see, renormalization group ideas, appropriately reformulated for use within the framework of renormalized perturbation theory, play an indispensable role in extracting useful information about high-energy amplitudes once certain factorization properties of the latter have been established.

17.5   Problems

1. Verify the inequality, valid for $A, B \geq 0$, (implying (17.16), setting $A = k_0^2$, $B = \vec{k}^2 + m^2$):

$$|\frac{1}{(A - B + i\epsilon B)}| \leq \sqrt{1 + \frac{4}{\epsilon^2}\frac{1}{A + B}} \qquad (17.127)$$

2. Check that the result of differentiating the internally subtracted two loop graph in Figure 17.3 three times with respect to the external momentum p is UV-finite in dimensional regularization. One needs to show that the amplitudes resulting from the differentiation of the basic graph, together with the inner subtractions prescribed by the forest formula (i.e. (a), (b), and (c) in Figure 17.4), can be rearranged into a sum of individually UV-finite terms. For example, the terms in which the propagators carrying momentum $p - l_1$, $p - l_2$ both receive derivatives are manifestly UV-finite by Weinberg's Theorem. Thus, one must show that the terms in which all three derivatives are applied to a single propagator, together with associated subtraction terms, give a UV-finite result. The UV convergence of the third derivative implies that the pole part of the subtracted two loop diagram must be a polynomial, at most quadratic, in the external momentum p.

3. Determine, in terms of zero-momentum amplitudes, the self-energy and vertex counterterms responsible for renormalizing the two-loop self-energy of Fig. 17.3

in ϕ_6^3-theory, and show that these counterterms give rise to exactly the set of subtractions indicated in Fig. 17.4.

4. Imitating the arguments leading to (17.41), derive the formula (17.77) giving the superficial degree of divergence of a graph with E_ψ external fermion and E_ϕ external scalar lines in a theory of a Dirac fermion field Yukawa-coupled to a scalar field:

$$D = 4 - \frac{3}{2}E_\psi - E_\phi \qquad (17.128)$$

5. Show that the Euclidean γ matrices $\hat{\gamma}_\mu, \mu = 1, 2, 3, 4$ defined in (15.160) satisfy the trace identities:

$$\mathrm{Tr}(\hat{\gamma}_\mu \hat{\gamma}_\nu) = 4\delta_{\mu\nu} \qquad (17.129)$$

$$\mathrm{Tr}(\hat{\gamma}_\mu \hat{\gamma}_\nu \hat{\gamma}_\rho \hat{\gamma}_\sigma) = 4(\delta_{\mu\nu}\delta_{\rho\sigma} - \delta_{\mu\rho}\delta_{\nu\sigma} + \delta_{\mu\sigma}\delta_{\nu\rho}) \qquad (17.130)$$

6. Evaluate, after dimensional continuation via (17.102), the photon one-loop self-energy (17.100) integral, and show that one obtains the result displayed in (17.103).

7. Show that a counterterm of the form $\delta Z \times \frac{1}{4}(F_{\mu\nu})^2$, inserted to first order in the momentum-space photon propagator, produces exactly the transverse tensor in (17.104).

18
Scales III: Short-distance structure of quantum field theory

In the preceding chapter we saw that for a certain subclass of local quantum field theories, whose local dynamics is determined by a Lagrangian containing only a finite number N of operators, the low-energy amplitudes of the theory, order by order in perturbation theory, lose their leading sensitivity to ultraviolet modifications of the theory once reparameterized in terms of an equal number of independent low-energy quantities. We refer to such theories as "perturbatively renormalizable", and the requirement that the Lagrangian of such theories contain only operators of mass dimension less than (relevant/super-renormalizable) or equal to (marginal/renormalizable) that of the spacetime dimension is extremely restrictive, with the pleasant result that enormous phenomenological predictivity obtains with a minimum of input. More exactly, we have seen that in perturbatively renormalizable theories, a general Green function $\mathcal{M}$ of the theory, depending on generic momenta p, masses m, and bare Lagrangian couplings g, evaluated with ultraviolet cutoff Λ, becomes, once reparameterized in terms of renormalized masses m_R and couplings g_R (which may depend on a choice of renormalization scheme—for example, through a renormalization scale μ)

$$\mathcal{M}(p, m, g, \Lambda) \to \mathcal{M}_R(p, m_R, g_R, \mu) + O(\frac{p^2, m_R^2, \mu^2}{\Lambda^2}) \tag{18.1}$$

The preceding equation is to be interpreted as valid order by order in the formal asymptotic expansion of both sides in powers of the subset of the g_R corresponding to interaction (higher than quadratic) vertices of the theory. For simplicity, we have taken the couplings to be dimensionless. The power suppressed terms in (18.1) are to be thought of as incorporating "new physics" which may be interesting in its own right, but is not directly of interest in the calculation of the desired amplitude $\mathcal{M}$. In particular, we assume that these terms are quantitatively negligible: the masses and momenta of particles involved in the given amplitude are much less than the energy scale Λ at which new physics may emerge. The subtraction technology developed in Section 17.2 was precisely fitted to the task of extracting just the parts of the full amplitude which survive in this limit.

Our subject in this chapter will be to show that the subtraction procedure used to demonstrate (18.1) has a natural generalization to situations in which *three* distinct energy scales are present: the "low" masses m and momenta of some of the particles, a momentum scale $Q >> p, m$ much greater than the remaining momenta and the masses of the theory, and (as always) a high-energy frontier scale Λ reflecting our

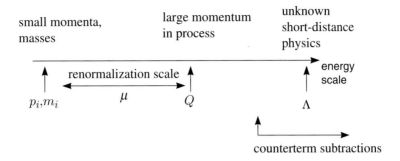

Fig. 18.1 Energy scales involved in UV subtraction of amplitudes (leading behavior for large Λ).

ignorance of the ultimate microphysics underlying our field theory (see Fig. 18.1). We shall see that it is possible in many cases to effectively repeat the procedures leading to (18.1) to extract the leading behavior of the renormalized amplitude (with Λ dependence now already discarded) and obtain a *factorized amplitude*

$$\mathcal{M}_R(p,Q,m_R,g_R,\mu) \to \sum_i \hat{\mathcal{M}}_{R,i}(p,m_R,g_R,\mu)C_i(Q,m_R,g_R,\mu) + O(\frac{p^2,m_R^2,\mu^2}{Q^2})$$

$$(18.2)$$

where now the "small" terms (usually called "higher twist" contributions) are of the order of inverse powers of the large momentum scale Q, which in some sense has taken over the role previously played by the "ultimate" cutoff Λ. The result (18.2), effectively decoupling the dependence of the full amplitude on the large and small momenta, is the expression in momentum space of a coordinate space property of local operators originally uncovered by Wilson (Wilson, 1969), and which has come to be known as the "Wilson operator-product expansion".

The proof of Wilson's hypothetical expansion, first given by Zimmermann, simply extends the subtraction procedure used to remove the leading Λ dependence of the amplitudes further down, to the "large" (but not too large!) scale Q, as schematically indicated in Fig. 18.2. The coefficient functions C_i are correspondingly termed "Wilson coefficients", while the set of amplitudes $\hat{\mathcal{M}}_{R,i}$ will turn out to involve insertions of appropriately defined local composite operators. We have seen on many occasions (cf. Section 16.5) that products of field operators contain ultraviolet divergences, and as (18.2) no longer contains any reference to a cutoff scale Λ, it is clear that the composite operators appearing here must come fully equipped with a prescription for subtracting off any additional Λ-dependence which their insertion in a graph might occasion. Our exploration of the factorization properties of amplitudes at high energy must therefore begin, naturally enough, with a more detailed treatment of the definition and properties of local composite operators. A full treatment of the factorization and renormalization group properties of high-energy amplitudes would easily require a separate (and sizeable!) book, so we must beg the reader's indulgence in providing merely an overview, with (in most cases) detailed proofs omitted.

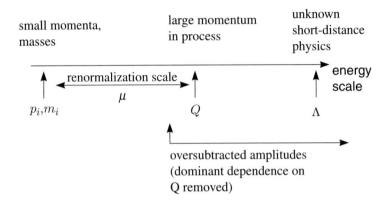

Fig. 18.2 Energy scales involved in oversubtraction of amplitudes (leading behavior for large Q).

18.1 Local composite operators in field theory

We have already alluded on many occasions to the ultraviolet divergences which appear in a continuum field theory when one attempts to multiply local field operators at the same spacetime point, in order to form composite operators, such as those needed in the formulation of either a renormalizable Lagrangian or the more general Wilsonian effective Lagrangians discussed in Chapter 16. In coordinate space the singular structure of these composite operators is directly associated with the singularities encountered when distributions are multiplied at the same argument (cf. Section 10.2). We shall concentrate on the structure of these divergences in momentum space, as the required technology is essentially already at our disposal given our previous study of the divergence structure of Feynman graphs in Sections 17.1 and 17.2. Our objective is to describe a systematic approach to the renormalization to all orders of composite operators, effectively generalizing the introductory discussion given in Section 16.5 in terms of some simple one-loop examples.

We shall work in ϕ_4^4-theory, which we assume to be renormalized by zero momentum BPHZ subtractions, implemented via the counterterms of (17.58). The renormalizability of the theory then implies the existence of the limit of the Euclidean Green functions of the theory $\langle \phi(x_1)\phi(x_2) \cdots \phi(x_n) \rangle$ (cf. (10.82)) or their momentum-space Fourier transforms as the UV cutoff Λ is taken to infinity. However, if two of the field operators in the above n-point function are taken at the same spacetime point (say, by taking $x_1 = x_2 = 0$), a singularity appears, which can be viewed as the re-emergence of a singular UV cutoff dependence for graphs involving an insertion of the composite operator $\phi^2(0)$ in an $(n-2)$-point amplitude. This singularity is inherited by the matrix elements of the composite operator (involving a total of $n-2$ incoming and outgoing particles), as such matrix elements are obtained (via LSZ) by taking the appropriately projected on-mass-shell limit of the momentum-space amplitude (continued back to Minkowski space, of course). To be specific, consider the momentum-space amplitude

$$\Gamma^{(2,1,1)}(k,k') \equiv \hat{\Delta}_F^{-1}(k)\hat{\Delta}_F^{-1}(k') \int d^4x\, d^4x'\, e^{ik\cdot x - ik'\cdot x'} \langle \phi^2(0)\phi(x)\phi(x') \rangle \qquad (18.3)$$

The superscript notation indicates that the Green function involves a composite (squared) operator and two separate field operators, and we use the Γ notation to make explicit the fact that the amplitude is taken to be 1PI, with the inverse propagator prefactors removing the external legs carrying momentum k in and k' out. Accordingly, the perturbative expansion of the amplitude begins with the constant unity, and the order λ_R contribution is given simply by the logarithmically divergent one-loop graph of Fig. 16.4. Note that there is no opportunity as yet for the order λ_R^2 counterterm contained in $\delta\lambda$ to appear to cancel the divergence, as we are working only to order λ_R (associated with the vertex on the right): the special vertex on the left associated with the insertion of the ϕ^2 operator (henceforth labeled V) does not, of course, carry a factor of the coupling constant. The presence of the composite operator $\phi^2(0)$ has evidently introduced *additional* UV divergences which are not taken care of by the normal counterterm subtractions. Nevertheless, a renormalized version of the amplitude (18.3) can be defined very simply, and to all loop orders, following the techniques of the BPHZ subtraction scheme. In order to do this we shall make a slight change in notation for the Taylor subtraction operator t^γ associated with a renormalization part γ of a graph (namely, a superficially divergent 1PI subgraph), writing

$$t^\gamma \to t^{D(\gamma)} \qquad (18.4)$$

where the degree function $D(\gamma)$ indicates the number of terms in the Taylor expansion around zero momentum to be included in the Taylor operator. For renormalization graphs *not containing* the composite vertex V, the subtraction degree is computed as usual

$$D(\gamma) = 4 - E_\gamma \qquad (18.5)$$

where E_γ is the number of external lines attached to γ, while for renormalization parts containing V (such as the one-loop graph in Fig. 16.4), we define

$$D(\gamma) = \delta - E_\gamma \qquad (18.6)$$

with δ an integer at least as large as the engineering dimension of the composite operator at the vertex V (in this case, 2). The Zimmermann forest formula (17.31) is now taken over exactly as in Chapter 17,

$$\mathcal{I}_{R,\delta}^\Gamma = \sum_{U \in \mathcal{F}(\gamma)} \prod_{\gamma_r \in U} (-t^{D(\gamma_r)}) \mathcal{I}^\Gamma \qquad (18.7)$$

to define a fully subtracted, and UV-finite, amplitude, with $\mathcal{I}^\Gamma$ the unrenormalized Feynman integrand associated with every graph Γ contributing to the amplitude in (18.3). For the one-loop graph in Fig. 16.4, taking $\delta = 2$, the logarithmically divergent one-loop subgraph contains the composite vertex, and therefore acquires a Taylor subtraction of degree $2 - 2 = 0$, exactly sufficient to remove the logarithmic divergence. With a little thought, one sees that the $D(\gamma)$ defined in this way exactly computes the superficial degree of divergence of all subgraphs, whether or not they contain the special vertex V. In fact, we can think of the vertex V as a normal four-point vertex, but with two of the external lines missing, whence the difference of 2 in the degree functions (18.5) and (18.6) (taking $\delta = 2$). However—and this freedom will become a crucial ingredient in the techniques to be developed in this chapter—we may also choose $\delta > 2$, thereby subtracting additional finite terms from the already adequately subtracted subintegrations associated with each γ, and obtaining an *oversubtracted* but nevertheless completely UV-finite amplitude. The sum of all graphs, renormalized according to the prescription (18.7), defines the insertion of a *renormalized composite ϕ^2 operator of degree δ*, henceforth denoted $N_\delta(\phi^2)$ (and frequently referred to as a "Zimmermann normal product operator")[1]:

$$\sum_\Gamma \mathcal{I}_{R,\delta}^\Gamma \equiv \hat{\Delta}_F^{-1}(k)\hat{\Delta}_F^{-1}(k') \int d^4x d^4x' e^{ik\cdot x - ik'\cdot x'} \langle N_\delta(\phi^2(0))\phi(x)\phi(x')\rangle \qquad (18.8)$$

In the event that we choose $\delta = 2$, the minimal value required to yield a UV-finite amplitude, the associated composite operator, $N_2(\phi^2)$ is called *minimally subtracted*. Composite operators, such as $N_4(\phi^2)$, containing more than the minimal number of subtractions required to remove the singular UV-dependence, are called *oversubtracted*.

All of the preceding may be carried out in a dimensional renormalization scheme simply by reinterpreting the Taylor operators $t^{D(\gamma_r)}$ in the forest formula as pole-part extraction operators, as described in the previous chapter. The renormalized composite operator $N_\mu(\phi^2)$ (for example) so defined implicitly depend on the renormalization scale μ used to define the dimensionally continued integrals, but we lose the ability to define oversubtracted operators in which additional momentum dependence is removed from the renormalization parts, with inconvenient consequences for the proof of the operator product expansion (for example). Nevertheless, as we shall see below, the dimensionally renormalized operators can be explicitly related to linear combinations of the more intuitive BPHZ normal product ones. As usual, one is dealing with the usual freedom available in choosing a particular "basis" of local operators from an infinite set of independent ones.

The physical interpretation of the subtractions implemented in our new forest formula (18.7) according to the degree function (18.5) is clear: these are just the subtractions generated by the appearance of counterterms in the Lagrangian once the theory is reparameterized in terms of a set of low-energy parameters identified through renormalization conditions (in the present scheme, at zero momentum), as we saw in the previous chapter. But the additional subtractions involving renormalization parts containing the vertex V associated with the insertion of the composite ϕ^2 operator, employing the degree function (18.6), clearly have nothing to do with these counterterms, and we may well be concerned that they involve an unacceptable mutilation of the composite operator, perhaps destroying important properties, such as locality (space-like commutativity), etc.

---

[1] The normal products defined here are to be distinguished, of course, from the "normal-ordered products" introduced in our discussion of Wick's theorem.

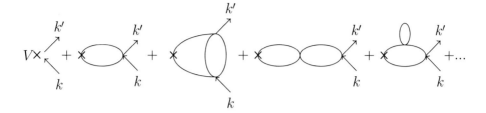

Fig. 18.3 Unsubtracted graphs contributing to $\Gamma^{(2,1,1)}$ up to two loops.

In fact, it is not hard to see that the extra subtractions introduced in (18.7) to render the amplitude $\Gamma^{(2,1,1)}$ UV-finite correspond simply to a *multiplicative renormalization* of the composite ϕ^2 operator, completely analogous to the previous rescaling of the basic field operator $\phi \to \sqrt{Z}\phi$ in (17.45), needed to absorb singular cutoff dependence in the two-point function of the theory. Rather than give a formal demonstration of this statement with the forest formula, we shall illustrate the basic point with an example.

In Fig. 18.3 we show the graphs contributing to $\Gamma^{(2,1,1)}$ in ϕ_4^4-theory through two loops. The corresponding subtractions induced by application of the forest formula are indicated in Fig. 18.4, where we remind the reader that the appearance of "o" symbols on the external legs of a subgraph indicate the application of the appropriate Taylor zero-momentum operation to that subgraph. In the present case, this effectively means just setting the momenta entering that subgraph to zero. The lowest-order graph (a) in Fig. 18.4 is by definition just unity. Also, the reader will recall from the

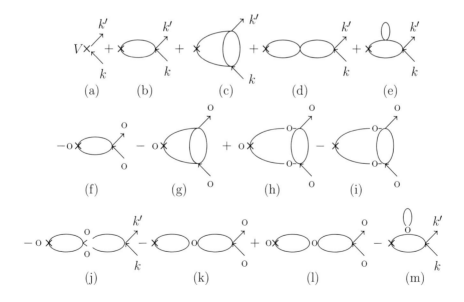

Fig. 18.4 Fully subtracted graphs contributing to $\Gamma^{(2,1,1)}$ up to two loops.

previous chapter that the one-loop propagator correction in graph (e) is momentum-independent in ϕ^4-theory, so graphs (e) and (m) in fact cancel identically. A brief inspection shows that the indicated subtractions indeed suffice to remove divergent UV contributions from all possible large momentum flows, as required by Weinberg's theorem. The graphs of Fig. 18.4 can be rearranged as indicated in Fig. 18.5. We see that the fully subtracted amplitude factorizes into a dimensionless number, which we shall call Z_{ϕ^2}, independent of the external momenta k, k', corresponding to the contents of the parenthesis on the top line, times the contributions to $\Gamma^{(2,1,1)}$ corresponding to the amplitude obtained by inserting the bare ϕ^2 operator into the 1-1 amplitude and carrying out all necessary counterterm subtractions (in this case, only the one-loop vertex subtractions) arising from the reparameterization of the theory. In other words, with a UV-cutoff Λ present to regularize the individual graphs in Figs. 18.3–18.5, the minimally subtracted ϕ^2 operator, giving UV-finite insertions into 1PI graphs (and hence, by LSZ, with finite matrix elements), is obtained by a cutoff-dependent rescaling of the bare operator:

$$N_2(\phi^2(0)) = Z_{\phi^2}(\lambda_R, \frac{\Lambda^2}{m_R^2})\phi^2(0) \tag{18.9}$$

This simple multiplicative relation between the renormalized normal product operator $N_2(\phi^2(0))$ and its bare counterpart $\phi^2(0)$ *suggests* that the renormalized operator will possess, in addition to UV-finite matrix elements, the desired Lorentz scalar and locality (space-like commutativity) properties, and indeed, these properties have been rigorously established (see (Zimmermann, 1970), and references cited therein). We should point out here that for certain particularly "nice" composite operators—the most important examples being those operators corresponding to conserved Noether currents associated with a Ward–Takahashi identity—the bare composite operator may already have finite matrix elements, allowing us to simply take the corresponding Z factor to be unity (see Problem 1).

More generally, defining renormalized composite operators may require a combination of bare operators, as a consequence of *operator mixing*. For example, in a theory with two independent scalar fields ϕ, χ, with basic interaction Lagrangian

$$\mathcal{L}_{\text{int}} = \lambda_1 \phi^4 + \lambda_2 \phi^2 \chi^2 + \lambda_3 \chi^4 \tag{18.10}$$

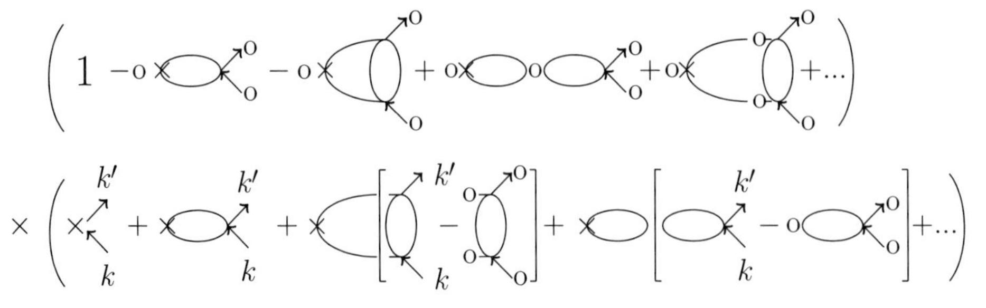

Fig. 18.5 Factorization of the fully subtracted graphs contributing to $\Gamma^{(2,1,1)}$ up to two loops.

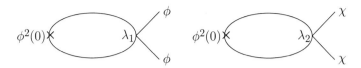

Fig. 18.6 One-loop renormalization parts appearing in the renormalization of $\phi^2(0)$ in the theory defined by (18.10).

the minimally subtracted operators $N_2(\phi^2)$, $N_2(\chi^2)$ are linear combinations of the bare ϕ^2 and χ^2 operators: thus, we have a 2×2 matrix of renormalization constants $(Z_{\phi\phi}, Z_{\phi\chi},$ etc.) connecting the bare operators with the renormalized ones. The need for including the χ^2 operator in the renormalization of ϕ^2 is apparent when one considers the graphs of Fig. 18.6, where we see that renormalization parts arising in the insertion of a ϕ^2 operator induce subtractions corresponding to a local χ^2 operator, as the zero momentum subtraction of the graph on the right amounts to a lowest-order insertion of a χ^2 operator.

We mentioned previously that the ability to define over-subtracted operators with more than the minimum number of subtractions needed to ensure the UV-finiteness of amplitudes containing these operators will be extremely important in understanding the underlying physics of operator product expansions. In fact, such operators are simply particular linear combinations of the minimally subtracted ones, as we shall now see, albeit combinations with particularly useful properties.

Consider, for example, the oversubtracted operator $N_4(\phi^2)$, where the Taylor operator acting on renormalization parts containing the special vertex V where the operator is inserted into the diagram contains two additional terms. Thus, the subtraction (f) of graph (b) in Fig. 18.4 contains, in addition to the constant term obtained by evaluating the one-loop integral at zero external momentum, the linear and quadratic terms in the Taylor expansion in external momenta of this graph. The graph evidently is a scalar function $\Pi(q^2)$ of the momentum $q = k' - k$ inserted by the composite operator, so the subtraction must take the form $a + bq^2 = a + b(k^2 + k'^2) - 2bk \cdot k'$, with a (cutoff-dependent) and b (finite) constants. The extra terms contained in the oversubtracted operator, proportional to b, are clearly just what we would get from a lowest-order insertion of the composite operator $\phi\Box\phi$ (giving the momentum dependence $-(k^2 + (k')^2)$ and $\partial_\mu\phi\partial_\mu\phi$ (giving the $2k \cdot k'$ term). A little time spent examining the effect of the oversubtraction at the next loop order shows that these new operators containing derivatives appear only minimally subtracted when their momentum dependence enters a renormalization part requiring subtraction. One also finds in higher order that the oversubtractions generate a term corresponding to the minimally subtracted $N_4(\phi^4)$ operator (see Problem 2). The result is the famous *Zimmermann identity*:

$$N_4(\phi^2) = N_2(\phi^2) + rN_4(\partial_\mu\phi\partial_\mu\phi) + sN_4(\phi\Box\phi) + tN_4(\phi^4) \qquad (18.11)$$

displaying, as predicted, the oversubtracted operator as a linear combination of minimally subtracted ones, with the constants $r, s,$ and t UV-finite functions of the renormalized parameters of the theory, as they must be, given that all operators appearing in the identity are fully renormalized. Alternatively, we may write

$$N_2(\phi^2) = N_4(\phi^2) - rN_4(\partial_\mu\phi\partial_\mu\phi) - sN_4(\phi\Box\phi) - tN_4(\phi^4) \qquad (18.12)$$

The general rule is very simple: we may write a minimally subtracted operator of degree D as a linear combination of all the independent operators of degree $D + \delta$, $\delta > 0$ with which it may mix (given symmetry constraints) under renormalization. A general proof of these Zimmermann identities involves straightforward, if lengthy, algebraic reshuffling of the forest formula (18.7), which we shall not give here. The interested reader is referred to the lectures of Lowenstein (Lowenstein, 1976) and Zimmermann (Zimmermann, 1970), in which all the details are given with proper mathematical rigor.

Higher than quadratic composite operators can be defined in a similar way to the above: the minimally subtracted $N_4(\phi^4)$ operator, for example, requires any renormalization part γ containing the four-point vertex corresponding to the operator insertion to be subtracted with the normal Taylor operator $t^{D(\gamma)}$, i.e., with $D(\gamma) = 4 - E_\gamma$. Moreover, these Zimmermann normal products satisfy some obvious (and convenient!) properties with respect to spacetime-derivatives—namely:

1. Derivatives may be passed through the normal product by the simple expedient of raising the degree of the subtractions by one for each derivative that enters the product: e.g.,

$$\partial_\nu N_\delta(\phi\partial_\mu\phi) = N_{\delta+1}(\partial_\nu(\phi\partial_\mu\phi)) \qquad (18.13)$$

The reason is simple. After Fourier-transforming to momentum space, we see that the interior ∂_ν derivative on the right-hand side of (18.13) corresponds to an extra factor of external momentum for any renormalization part containing the special vertex for the $N_{\delta+1}$ composite operator: thus, the Taylor subtraction operator on the right must perform an extra momentum differentiation to ensure that the left- and right-hand sides agree.

2. Usual Leibniz rules of differentiation obtain within normal products: e.g.,

$$N_\delta(\partial_\mu(\phi\partial_\mu\phi)) = N_\delta(\partial_\mu\phi\partial_\mu\phi) + N_\delta(\phi\Box\phi) \qquad (18.14)$$

Dimensionally subtracted composite operators satisfy similar properties, with the important change that there is no freedom to change the subtraction level, so the derivatives pass through the normal product with no change in its definition: e.g., $\partial_\nu N_\mu(\phi\partial_\mu\phi) = N_\mu(\partial_\nu(\phi\partial_\mu\phi))$.

Fully subtracted amplitudes involving multiple insertions of composite operators are constructed by an obvious generalization of the forest formula (18.7): one simply applies the appropriate subtraction degree formula for each renormalization part taking into account the sum of the degree increments (if some or all operators are oversubtracted) for all the special vertices appearing in that renormalization part. We thereby arrive at a very general and flexible formalism, with a remarkable formal benefit: it allows an extremely concise formulation of the renormalized Lagrangian dynamics of a perturbatively renormalizable theory, in terms of a *Zimmermann effec-*

tive Lagrangian.[2] The usual interaction-picture formulation of perturbation theory in terms of the bare (unrenormalized) fields and parameters is replaced by the (Euclidean) Lagrangian specification (subscript Z for "Zimmermann"), for ϕ_4^4-theory,

$$\mathcal{L}_Z = \mathcal{L}_{Z,0} + \mathcal{L}_{Z,\text{int}} \tag{18.15}$$

$$\mathcal{L}_{Z,0} = N_4(\frac{1}{2}(\partial_\mu \phi)^2 + \frac{1}{2}m_R^2\phi^2) \tag{18.16}$$

$$\mathcal{L}_{Z,\text{int}} = N_4(\frac{\lambda_R}{4!}\phi^4) \tag{18.17}$$

No counterterms appear here (the mass and coupling parameters are the BPHZ renormalized ones), but the perturbative expansion of a general BPHZ renormalized 1PI function in terms of correlation functions of free fields with dynamics specified by (18.16) is defined as

$$\Gamma_R^{(N)}(k_1, ..., k_N) = \sum_{r=0} \frac{(-1)^r \lambda_R^r}{r!(4!)^r} \int d^4z_1 d^4z_2 \cdots d^4z_r \langle N_4(\phi^4(z_1))N_4(\phi^4(z_2))\cdots N_4(\phi^4(z_r))$$
$$\cdot \tilde{\phi}(k_1)\cdots\tilde{\phi}(k_N)\rangle_{1\text{PI}} \tag{18.18}$$

where the graphs obtained by Wick expansion of the correlation function on the right are to be subjected to the forest formula subtraction formula corresponding to the indicated multiple insertion of the quartic interaction operator. The latter is minimally subtracted, and it is more or less obvious that this prescription precisely corresponds to the BPHZ renormalization scheme described in detail in the preceding chapter.

The reader may be somewhat puzzled by the fact that the mass operator in (18.16) appears in oversubtracted form, as a $N_4(\phi^2)$. The reason is easily seen if we examine the effect of a small change in the renormalized (squared) mass, or equivalently, compute the first derivative of a renormalized 1PI amplitude with respect to m_R^2. The effect is simply (with a change of sign) to double each internal propagator of the graph, as

$$\frac{\partial}{\partial m_R^2}\frac{1}{p^2 + m_R^2} = -\frac{1}{p^2 + m_R^2} \cdot \frac{1}{p^2 + m_R^2} \tag{18.19}$$

This doubling occurs, of course, not only in the basic unsubtracted graphs but also in each of the subtraction terms which pop up whenever there is a divergent subgraph. The point at which the propagator is doubled may be regarded as a new special vertex associated with the insertion of the ϕ^2 operator appearing in $\mathcal{L}_{Z,0}$. The result is as shown in Fig. 18.7 for a simple example: the 1PI four-point function at one loop, where "X" marks the point of the ϕ^2 insertion. Recall that only internal lines are present and differentiated, as we are dealing with a 1PI, and therefore automatically amputated, amplitude. It is clear that the mass derivative of this one-loop contribution to $\Gamma_R^{(4)}$ corresponds to an insertion of the *oversubtracted* $N_4(\phi^2)$ operator, as it is subtracted at zero momentum even though the overall degree of divergence of the

---

[2] We apologize once again to the reader for the lamentable overuse of the adjective "effective", which appears here now for the third time with a completely different connotation!

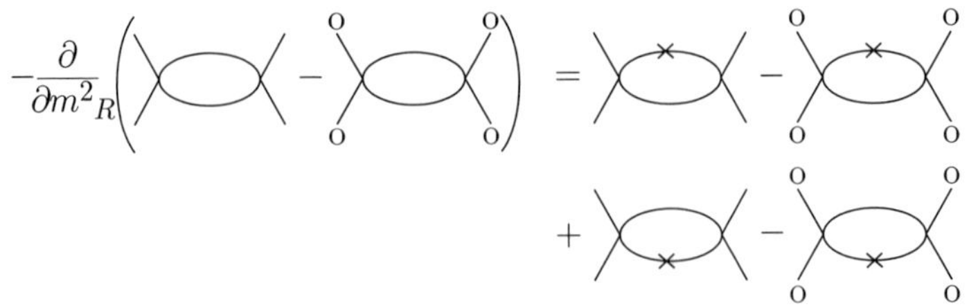

$$-\frac{\partial}{\partial m^2}_R \left(\bigotimes - \bigotimes \right) = \bigotimes - \bigotimes$$

$$+ \bigotimes - \bigotimes$$

Fig. 18.7 Mass derivative of the four-point one-loop renormalized amplitude $\Gamma_R^{(4)}$ in ϕ_4^4-theory.

one-loop graph with one of the propagators doubled is now –2 rather than zero, and a minimally subtracted $N_2(\phi^2)$ operator would by definition not require a subtraction of an already superficially convergent subgraph containing its vertex. In general, *all* the operators appearing in a Zimmermann effective Lagrangian of this type carry a degree subscript equal to the spacetime dimension, independent of their actual engineering dimension. This means that the operators corresponding to super-renormalizable terms are necessarily oversubtracted.

We must now return to the basic theme of this chapter—the use of renormalization techniques to study the short distance, or equivalently, large momentum behavior of amplitudes in a renormalizable local field theory. We indicated earlier that the concept of oversubtraction provides the key to unlocking this behavior. In particular, we wish to consider the situation in which there is a distinct large momentum scale Q present in the renormalized amplitudes being studied, with Q much larger than all other dimensionful quantities (masses, super-renormalizable couplings if any, and other momentum variables: but, of course, as the amplitudes have been renormalized, no UV cutoff Λ).

The simplest case concerns an amplitude in which *all* external momenta are of order Q. It was realized a long time ago by Symanzik (Symanzik, 1970) (and almost simultaneously, by Callan (Callan, 1970)) that in this regime the leading contribution to the amplitude at large Q (neglecting subdominant terms suppressed by inverse powers of Q (cf. (18.2)) satisfies an homogeneous partial differential equation, which in certain circumstances can be solved and used to extract the desired asymptotic behavior. The equation in question is now referred to universally as the Callan–Symanzik equation. We shall not follow the more involved methods used by either Symanzik or Callan to derive this equation here, as it is an almost immediate consequence of the Zimmermann identity discussed earlier, and the approach we use will generalize more easily to the case of factorized amplitudes to be treated in the following section. Also, we shall henceforth focus on the scalar ϕ_6^3-theory introduced in the previous chapter, for the same reasons indicated there: the topological structure of the diagrams is essentially identical to that of a four-dimensional gauge theory, and moreover, the structure and strength of the ultraviolet divergences are very similar to the gauge-theory case. Thus, instead of (18.15, 18.16, 18.17), we shall be dealing with an effective Zimmermann Euclidean Lagrangian (in six dimensions) given by the following free and interaction parts:

$$\mathcal{L}_{Z,0} = N_6(\frac{1}{2}(\partial_\mu\phi)^2 + \frac{1}{2}m_R^2\phi^2) \tag{18.20}$$

$$\mathcal{L}_{Z,\text{int}} = N_6(\frac{\lambda_R}{3!}\phi^3) \tag{18.21}$$

Note that the engineering dimension of the scalar field ϕ is 2 in six dimensions, so the kinetic and interaction terms are minimally subtracted and the mass operator oversubtracted, as usual. No linear term in the field is included, as the BPHZ zero-momentum subtractions automatically remove all tadpoles—a process equivalent to cancelling such graphs with a additive field shift order by order in perturbation theory (see Fig. 17.11). The corresponding graphs in a gauge theory such as QED, in which a photon line virtualizes into a electron–positron pair which subsequently disappears into the vacuum, are, of course, necessarily zero by (for example) angular momentum conservation.

The key to understanding the Callan–Symanzik equation lies in an important difference in the behavior of minimally subtracted (such as the $N_2(\phi^2)$ in ϕ_4^4-theory and $N_4(\phi^2)$ in ϕ_6^3-theory) and oversubtracted (e.g., $N_4(\phi^2)$ in ϕ_4^4-theory and $N_6(\phi^2)$ in ϕ_6^3-theory) operators when inserted into amplitudes at large (external) momentum $k_i = Q\hat{k}_i$, where Q is a large momentum scale and the $\hat{k}_i$ are Euclidean momenta of order unity. For example, the minimally subtracted ϕ^2 operators, as we have seen, simply introduce an additional internal propagator into the diagrams, without any additional subtractions (see Fig. 18.8(a)). The result is to lower the superficial degree

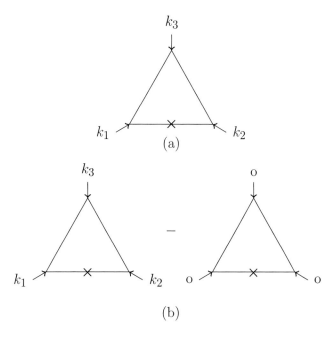

$$(a)$$

$$(b)$$

Fig. 18.8 (a) A one-loop graph corresponding to an insertion of $N_4(\phi^2)$ in $\Gamma_R^{(3)}$ in ϕ_6^3-theory. (b) Result of insertion of the oversubtracted $N_6(\phi^2)$ in the same graph.

of divergence of the overall (fully subtracted) graph by 2, which, by a corollary of Weinberg's theorem discussed in the previous chapter, lowers correspondingly the asymptotic dependence of the graph for large Q by (modulo logarithms) two powers of Q. This softening of the asymptotic behavior is guaranteed for all *non-exceptional momenta*: specifically, provided no partial subset of the momenta entering the graph vanishes (for the reason for this, and an explicit counterexample, see Problem 3).

On the other hand, the insertion of an oversubtracted ϕ^2 operator generates additional subtractions at zero momentum which do not fall off at large Q, as indicated in Fig. 18.8(b), where the insertion of the oversubtracted $N_6(\phi^2)$ operator into the one-loop 1PI three-point function $\Gamma_R^{(3)}(k_1, k_2, k_3)$, $k_i = Q\hat{k}_i$, in ϕ_6^3-theory produces the softened graph (falling like $1/Q^2$, modulo logs) corresponding to the insertion of the minimally subtracted operator, as in Fig. 18.8(a), but with an extra zero-momentum subtraction which is just a constant as $Q \to \infty$. For this reason, minimally subtracted operators are sometimes referred to as *soft* operators, while their oversubtracted counterparts are termed *hard* operators. Nevertheless, it is the oversubtracted operator, as we saw previously, that corresponds to mass derivatives of the renormalized amplitude. This observation, together with the Zimmermann identity connecting minimally and oversubtracted operators, will provide the key to our derivation of the Callan–Symanzik equation.

We begin with the formal expansion of the 1PI N-point function in ϕ_6^3-theory (analogous to (18.18) for ϕ_4^4-theory):

$$\Gamma_R^{(N)}(k_1, ..., k_N) = \langle \tilde{\phi}(k_1) \cdots \tilde{\phi}(k_N) \rangle_{1PI} \tag{18.22}$$

$$= \sum_{r=0} \frac{(-1)^r \lambda_R^r}{r!(3!)^r} \int d^4 z_1 d^4 z_2 \cdots d^4 z_r \langle N_6(\phi^3(z_1)) N_6(\phi^3(z_2)) \cdots N_6(\phi^3(z_r)) $$
$$\cdot \tilde{\phi}(k_1) \cdots \tilde{\phi}(k_N) \rangle_{1PI} \tag{18.23}$$

The fields on the first line are fully interacting (Heisenberg) fields (we omit the usual "H" subscript here to avoid overburdening the notation), whereas the second line corresponds to the interaction-picture expansion. Recall that the $\langle \cdots \rangle_{1PI}$ symbol in the second line is to be interpreted by first Wick-expanding the operators products inside the bracket to generate a set of bare (unsubtracted) 1PI irreducible graphs, each of which is then subjected to the forest formula to generate the appropriate subtractions. We now define a series of zero-momentum insertion operations on the general renormalized N-point function $\Gamma_R^{(N)}$ as follows:

$$\Delta_0 \Gamma_R^{(N)} \equiv \int d^6 z \langle N_4(\tfrac{1}{2}\phi^2(z)) \tilde{\phi}(k_1) \cdots \tilde{\phi}(k_N) \rangle_{1PI} \tag{18.24}$$

$$\Delta_1 \Gamma_R^{(N)} \equiv \int d^6 z \langle N_6(\tfrac{1}{2}\phi^2(z)) \tilde{\phi}(k_1) \cdots \tilde{\phi}(k_N) \rangle_{1PI} \tag{18.25}$$

$$\Delta_2 \Gamma_R^{(N)} \equiv \int d^6 z \langle N_6(\tfrac{1}{2}\partial_\mu\phi(z)\partial_\mu\phi(z)) \tilde{\phi}(k_1) \cdots \tilde{\phi}(k_N) \rangle_{1PI} \tag{18.26}$$

corresponding to insertions at zero-momentum (as a consequence of the $\int d^6 z$ integration) of the minimally subtracted mass operator, the oversubtracted mass operator, and the minimally subtracted kinetic term operator, respectively. Finally, there is the insertion operator for an additional minimally subtracted ϕ^3 interaction vertex (at zero momentum):

$$\Delta_3 \Gamma_R^{(N)} \equiv \sum_{r=0} \frac{(-1)^r \lambda_R^r}{r!(3!)^r} \int d^4 z \, d^4 z_1 \cdots d^4 z_r \langle N_6(\phi^3(z)) N_6(\phi^3(z_1)) \cdots N_6(\phi^3(z_r)) $$
$$\cdot \tilde\phi(k_1) \cdots \tilde\phi(k_N) \rangle_{1\mathrm{PI}} \tag{18.27}$$

First, note that insertions of $\int d^6 z \, N_6(\frac{1}{2}(\partial_\mu \phi(z)\partial_\mu \phi(z) + m_R^2 \phi^2(z)))$ introduce a factor of an inverse propagator at a two-point vertex (i.e., a factor of $p^2 + m_R^2$ for an internal line carrying momentum p) on each internal line of a given bare graph—in other words, just the factor unity for each internal line. This insertion, corresponding to the operation $m_R^2 \Delta_1 + \Delta_2$, therefore just multiplies each bare graph by the number of internal lines it contains, which by the usual graph topology arguments is just $\frac{1}{2}(3r - N)$ for a graph containing r basic 3-vertices. On the other hand, the power r associated with each term in (18.23) is evidently obtained by the differential operation $\lambda_R \frac{\partial}{\partial \lambda_R}$. Thus we obtain the *counting identity*

$$(m_R^2 \Delta_1 + \Delta_2)\Gamma_R^{(N)}(k_1, .., k_N) = -\frac{N}{2}\Gamma_R^{(N)}(k_1, .., k_N) + \frac{3}{2}\lambda_R \frac{\partial}{\partial \lambda_R}\Gamma_R^{(N)}(k_1, .., k_N) \tag{18.28}$$

On the other hand, from (18.23), we find

$$\frac{\partial}{\partial \lambda_R}\Gamma_R^{(N)}(k_1, .., k_N) = -\frac{1}{3!}\Delta_3 \Gamma_R^{(N)}(k_1, .., k_N) \tag{18.29}$$

so that (18.28) may be written

$$(m_R^2 \Delta_1 + \Delta_2)\Gamma_R^{(N)}(k_1, .., k_N) = -\frac{N}{2}\Gamma_R^{(N)}(k_1, .., k_N) - \frac{\lambda_R}{4}\Delta_3 \Gamma_R^{(N)}(k_1, .., k_N) \tag{18.30}$$

We pointed out earlier that an insertion of the oversubtracted ϕ^2 operator is equivalent to a mass derivative:

$$\Delta_1 \Gamma_R^{(N)}(k_1, .., k_N) = -\frac{\partial}{\partial m_R^2}\Gamma_R^{(N)}(k_1, .., k_N) \tag{18.31}$$

The final ingredient is the Zimmermann identity analogous to (18.12). It is convenient to use the operator $\partial_\mu(\phi \partial_\mu \phi) = \partial_\mu \phi \partial_\mu \phi + \phi \Box \phi$, instead of $\phi \Box \phi$, in our basis of operators, so with a slight change of notation, and with the subtraction degrees appropriate for ϕ_6^3-theory,

$$N_4(\phi^2(z)) = N_6(\phi^2(z)) + r(\lambda_R, m_R)N_6(\partial_\mu \phi(z)\partial_\mu \phi(z))$$
$$+ s(\lambda_R, m_R)N_6(\partial_\mu(\phi(z)\partial_\mu \phi(z))) + t(\lambda_R, m_R)N_6(\phi^3(z)) \tag{18.32}$$

Integrating over z, the pure derivative term proportional to $s(\lambda_R, m_R)$ vanishes, and we have, in terms of vertex insertion operators, the final relation

$$\Delta_0 = \Delta_1 + r(\lambda_R, m_R)\Delta_2 + \frac{1}{2}t(\lambda_R, m_R)\Delta_3 \qquad (18.33)$$

which is essentially the Callan–Symanzik equation in disguised form, as we shall now see. Note that the engineering dimension of the functions $r(\lambda_R, m_R), t(\lambda_R, m_R)$ must be –2 in powers of mass, so we have (as λ_R is dimensionless)

$$r(\lambda_R, m_R) = \frac{1}{m_R^2}f(\lambda_R), \quad t(\lambda_R, m_R) = \frac{1}{m_R^2}g(\lambda_R) \qquad (18.34)$$

where $f(\lambda_R)$ (resp. $g(\lambda_R)$) begin at order λ_R^2 (resp. λ_R^3) in perturbation theory. Combining (18.29, 18.30, 18.31, 18.33), we find the promised Callan–Symanzik equation

$$(m_R^2\frac{\partial}{\partial m_R^2} + \beta(\lambda_R)\frac{\partial}{\partial \lambda_R} - N\gamma(\lambda_R))\Gamma_R^{(N)}(k_i; \lambda_R, m_R) = \frac{m_R^2}{f(\lambda_R) - 1}\Delta_0\Gamma_R^{(N)}(k_i; \lambda_R, m_R)$$
$$(18.35)$$

where we have indicated explicitly the dependence of the N-point function on the renormalized coupling and mass, and defined the functions

$$\beta(\lambda_R) \equiv \frac{3}{2}\frac{\lambda_R f(\lambda_R) - 2g(\lambda_R)}{f(\lambda_R) - 1} \sim O(\lambda_R^3) \qquad (18.36)$$

$$\gamma(\lambda_R) \equiv \frac{1}{2}\frac{f(\lambda_R)}{f(\lambda_R) - 1} \sim O(\lambda_R^2) \qquad (18.37)$$

We have already seen that in the large-momentum (or short-distance) limit where $k_i = Q\hat{k}_i$ with Q large and the $\hat{k}_i$ non-exceptional and fixed, the insertion of the soft mass operator effected by the vertex operation Δ_0 on the right-hand side of (18.35) suppresses the asymptotic behavior of our N-point function by two powers of Q, so neglecting such contributions we find an *homogeneous equation* which must be obeyed by the leading high-momentum contributions to the amplitude (up to inverse powers of Q):

$$\mathcal{D}_{CZ}\Gamma_R^{(N)}(k_i; \lambda_R, m_R) \equiv (m_R^2\frac{\partial}{\partial m_R^2} + \beta(\lambda_R)\frac{\partial}{\partial \lambda_R} - N\gamma(\lambda_R))\Gamma_R^{(N)}(k_i; \lambda_R, m_R) \approx 0$$
$$(18.38)$$

In other words, the particular combination of mass and coupling derivatives contained in the *Callan–Symanzik operator* $\mathcal{D}_{CZ}$ is exactly equivalent, to all orders of perturbation theory, to an insertion of a soft mass operator, and must therefore suppress, by powers, the asymptotic behavior of any N-point (Euclidean)1PI amplitude provided all external momenta are taken large. Eq. (18.38) as it stands is not obviously useful, as we are hardly in a position to explore the response of physical amplitudes to a change in the mass of the particles being scattered. However, we can translate the dependence on mass into one on uniformly rescaled momenta for the process by simple dimensional analysis. Let d_N be the engineering dimension of $\Gamma_R^{(N)}$ in powers of mass (thus, $d_N = 4 - N$ for ϕ_4^4-theory, $6 - 2N$ for ϕ_6^3-theory). The total powers of mass and momentum in each term (the coupling λ_R is dimensionless) contributing to this 1PI amplitude must therefore be d_N, which we may express with the usual Euler derivative:

$$(\kappa \frac{\partial}{\partial \kappa} + m_R \frac{\partial}{\partial m_R})\Gamma_R^{(N)}(\kappa k_i; \lambda_R, m_R) = d_N \Gamma_R^{(N)}(\kappa k_i; \lambda_R, m_R) \tag{18.39}$$

We may therefore eliminate the mass derivative in (18.38) to obtain the asymptotic equation

$$(-\frac{1}{2}\kappa \frac{\partial}{\partial \kappa} + \beta(\lambda_R)\frac{\partial}{\partial \lambda_R} - N\gamma(\lambda_R) + \frac{1}{2}d_N)\Gamma_R^{(N)}(\kappa k_i; \lambda_R, m_R) = 0 \tag{18.40}$$

In Section 18.3 we shall return to (18.40), and show how to solve an homogeneous partial differential equation of this type to constrain the large-momentum asymptotic behavior of an arbitrary amplitude in a perturbatively renormalizable field theory. It will also be seen there that a very similar equation can be derived in a completely different way using renormalization group ideas, and the connection between the two (involving the concept of mass singularities) will be explained.

The derivation of the Callan–Symanzik equation for fully amputated 1PI amplitudes can be easily generalized to take care of the case when some or all of the external propagators are present. For example, if our amplitude contains all external legs, the number of lines in each basic graph is $\frac{1}{2}(3r + N)$ (instead of $\frac{1}{2}(3r - N)$ in the fully amputated case) for a graph containing r basic 3-vertices, and the end result is a Callan–Symanzik operator with a change of sign in the $N\gamma(\lambda_R)$ term in (18.38). Similarly, if the amplitude is "half-amputated", with only half the external legs amputated, the $\gamma(\lambda_R)$ term is absent.

The functions[3] $\beta(\lambda_R)$ (the famous "β function" of renormalization group lore) and $\gamma(\lambda_R)$ (which, for reasons to be seen later, is termed the "anomalous dimension" of the scalar field) could be computed order by order in perturbation theory by first determining the coefficient functions in the Zimmermann identity perturbatively (paying very careful attention to the forest formula!), but it is more convenient to extract them by simply applying the Callan–Symanzik equation to two independent 1PI N-point functions, in the asymptotic large momentum limit where the right-hand side may be neglected. For example, to one-loop order, we may use the two-point 1PI function (inverse propagator) and three-point function, for which a short calculation (Problem 4) reveals the form:

$$\Gamma_R^{(2)}(k) = k^2 + m_R^2 - \frac{\lambda_R^2}{128\pi^3} \int_0^1 \{(x(1-x)k^2 + m_R^2) \ln(1 + x(1-x)k^2/m_R^2)$$

$$- x(1-x)k^2\}dx + O(\lambda_R^4) \tag{18.41}$$

---

[3] Modulo a noisome factor of $\frac{1}{2}$; see below.

$$\Gamma_R^{(3)}(k_1, k_2, k_3) = \lambda_R + \lambda_R^3 \{ \int \frac{1}{l^2 + m_R^2} \frac{1}{(l + k_1)^2 + m_R^2} \frac{1}{(l + k_1 + k_2)^2 + m_R^2} \frac{d^6 l}{(2\pi)^6}$$

$$- \int \frac{1}{(l^2 + m_R^2)^3} \frac{d^6 l}{(2\pi)^6} \} + O(\lambda_R^5) \tag{18.42}$$

Note that the self-energy term in (18.41) (given by the integral over the Feynman parameter x) begins at order k^4 at small momentum, in keeping with the BPHZ subtraction of a quadratically divergent integral. The three-point function (triangle graph in Fig. 18.8(a), without the mass insertion) is logarithmically divergent and therefore receives a single subtraction at zero momentum, as indicated in (18.42). The Callan–Symanzik operator applied to $\Gamma_R^{(2)}(k)$ produces a dominant contribution proportional to k^2 at large k and order λ_R^2 with contributions from the mass derivative and anomalous dimension term:

$$m_R^2 \frac{\lambda_R^2}{128\pi^3} \frac{1}{6} \frac{k^2}{m_R^2} - 2\gamma(\lambda_R) k^2 = 0 \Rightarrow \gamma(\lambda_R) = \frac{\lambda_R^2}{1536\pi^3} + O(\lambda_R^4) \tag{18.43}$$

while the same considerations applied to the three-point function $\Gamma_R^{(3)}(k_1, k_2, k_3)$ (in this case the mass derivative of the first integral in (18.42) is asymptotically suppressed and can be thrown away), using

$$m_R^2 \frac{\partial}{\partial m_R^2} \int \frac{1}{(l^2 + m_R^2)^3} \frac{d^6 l}{(2\pi)^6} = -3m_R^2 \int \frac{1}{(l^2 + m_R^2)^4} \frac{d^6 l}{(2\pi)^6} = -\frac{1}{128\pi^3} \tag{18.44}$$

imply

$$\frac{1}{128\pi^3} \lambda_R^3 + \beta(\lambda_R) - 3\lambda_R \gamma(\lambda_R) = 0 \Rightarrow \beta(\lambda_R) = -\frac{3\lambda_R^3}{512\pi^3} + O(\lambda_R^5) \tag{18.45}$$

The significance of the (at first sight innocent) negative sign appearing in the β function can scarcely be overstated: it leads, as we shall see in Section 18.3, to the critical property of *asymptotic freedom*, implying that the theory becomes effectively weakly coupled at large momenta, restoring the quantitative usefulness of perturbation theory even in theories which are (at low momenta) strongly coupled. The unphysical ϕ_6^3-theory serving as our toy example here shares this remarkable property with non-abelian gauge theories generally, and with QCD in particular. The discovery and proper interpretation in 1973 of asymptotic freedom was rewarded in 2004 by the conferral of the Nobel Prize in Physics to Gross, Politzer, and Wilczek. But before going on to describe the special features of asymptotically free theories in more detail, we shall generalize our discussion of high-momentum behavior of amplitudes given so far to situations in which, as described in the introduction to this chapter, only a proper subset of the external momenta are large. The required generalization will lead us directly to the Wilson operator product expansion.

Considerations of space prevent us from describing many of the quite beautiful applications of the Zimmermann normal product formalism for local composite operators. Suffice it to say that the use of normal products allows one to write the Lagrangian (Heisenberg) field equations of the theory as rigorous relations, correct to all orders of renormalized perturbation theory, between well-defined renormalized composite operators. Likewise, the current conservation properties in theories with global or local symmetries of the theory can be expressed as precise operator statements, and, in the case in which anomalies appear in these currents, they can be seen to arise automatically as a natural consequence of a Zimmermann identity relating the minimally- and over-subtracted versions of the naive (classical) divergence of the current. For more on these fascinating and deep results, the reader is encouraged to consult the Erice lectures of Lowenstein (Lowenstein, 1976).

18.2 Factorizable structure of field theory amplitudes: the operator product expansion

At the beginning of this chapter we indicated that the methods of perturbative renormalization, originally formulated to isolate and extract the dominant dependence of field-theoretic amplitudes on a single large momentum scale Λ, beyond which we lose precise control of the dynamics of the theory, can be generalized to the problem of extracting the asymptotic dependence of already renormalized amplitudes on a physical large momentum scale Q, where Q is much larger than the masses, dimensionful couplings, and other momentum scales in the problem. The first indications that something of this kind might be possible go back to Wilson's hypothesis (Wilson, 1969) of a short-distance expansion for the product of local operators (in the original application, these were hadronic current operators) at nearby spacetime points:

$$A(x + \xi/2)A(x - \xi/2) \to \sum_i C_i(\xi)\mathcal{O}_i(x), \quad \xi \to 0, \xi^2 < 0 \qquad (18.46)$$

Here, x, ξ are Minkowski spacetime coordinates, and the limit $\xi \to 0$ is presumed to be taken from the space-like direction.[4]

The *Wilson coefficient functions* $C_i(\xi)$ (which may carry Lorentz indices) are UV-finite functions which are ordered in the sum to be decreasingly singular as $\xi \to 0$, while the local operators $\mathcal{O}_i$ are well-defined renormalized local composite operators of the kind discussed in the preceding section, ordered according to increasing engineering dimension. The expansion acquires a definite meaning in the weak convergence sense, once we sandwich both sides of (18.46) between definite initial and final states, at which point the remainder terms at any finite point in the expansion are asserted to vanish more rapidly as $\xi \to 0$ than the kept terms (see (Zimmermann, 1970)). The operators $\mathcal{O}_i$ appearing in this expansion are just those which are capable of mixing with the bilocal operator on the left under renormalization, subject to symmetries of the theory, in a sense to be made precise below. In order to extract useful information from an expansion of this sort, Wilson was obliged to assume (incorrectly) that the strong interactions behaved in a scale-invariant way, at which point the dependence of the coefficient function $C_i(\xi)$ on ξ becomes power-like, with the power related simply to the scale dimension of the associated composite operator $\mathcal{O}_i$.

---

[4] We shall see below that the structure of the expansion is considerably altered if the local limit is approached from the light-cone direction, with $\xi^2 = 0$.

We now know, from the rigorous work of Zimmermann, that the expansion is indeed correct in renormalized perturbation theory, but that the short-distance behavior of the coefficient functions is more complicated, involving, in general, logarithms as well as powers of ξ. Nevertheless, up to logarithms, the leading power behavior of the Wilson coefficient functions (in perturbation theory) is still associated in a simple way, as we shall see below, with the engineering dimension of the associated composite operator, in such a way that each additional power of mass dimension in the operator corresponds to a softening of the short-distance behavior of the associated Wilson coefficient by a power of ξ (modulo logarithms).

The connection of such an expansion—a sort of operator generalization of the Taylor expansion (though, as typical in field theory, at best an asymptotic and not a convergent one)—to the promised separation of large momentum behavior becomes clear once we consider a definite matrix element of (18.46) (with $x = 0$) and Fourier transform the ξ variable:

$$\mathcal{T}(q, k_i, k_i') \equiv \int d^4\xi\, e^{iq\cdot\xi} \langle k_i'|T\{A(\xi/2)A(-\xi/2)\}|k_i\rangle \qquad (18.47)$$

$$\mathcal{T}(q, k_i, k_i') \to \sum_i \tilde{C}_i(q)\langle k_i'|\mathcal{O}_i(0)|k_i\rangle, \quad q^2 \to -\infty \qquad (18.48)$$

The amplitude $\mathcal{T}(q, k_i, k_i')$ corresponds to a situation in which a large space-like momentum q is introduced and then removed (by the A fields) from a set of particles at fixed low momenta k_i, k_i', as indicated in Fig. 18.9.[5] Here, the $\tilde{C}_i(q)$ are the Fourier transforms of the coordinate space Wilson coefficient functions $C_i(\xi)$ appearing in (18.46): by the usual properties of Fourier transformation, successive coefficient functions $C_i(\xi)$ with (as discussed above) additional powers of ξ in their short-distance behavior correspond to additional *inverse* powers of $Q \equiv \sqrt{-q^2}$ in the asymptotics of

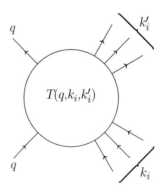

Fig. 18.9 Momentum space amplitude $\mathcal{T}(q, k_i, k_i')$ corresponding to the insertion of a bilocal operator.

[5] This is a Euclidean analog of the forward scattering amplitude $\mathcal{T}(k, q; k, q)$ discussed in Section 6.6: we shall see later how to extend the use of the OPE to Minkowskian situations of this sort.

$\tilde{C}_i(q)$ for large Q. Thus, the *leading* asymptotic behavior of $\mathcal{T}(q, k_i, k'_i)$ for large Q is determined by the leading term(s) in the expansion, in many cases, by a single operator of minimal engineering dimension, provided, of course, that such an operator possesses a non-vanishing matrix element between the initial and final states indicated in (18.48). A glance at this formula also shows that the general dependence of our amplitude on "large" q and "small" k_i, k'_i momenta has been *factorized* in the expansion. This factorization property of amplitudes is a deep consequence of the dynamics of local field theories, and the Wilson OPE is the most direct expression thereof.

Our discussion of the Wilson operator product expansion (henceforth, OPE) will take place entirely in the arena of momentum space: i.e., in the form given in (18.48). There are several reasons for this. First, we are primarily interested in the behavior of S-matrix amplitudes at high energy, which are naturally formulated directly as momentum-space objects. But more importantly, the physical intuition underlying the emergence of the factorization properties of amplitudes is far more easily acquired by an examination of the behavior of large momentum flows in graphical amplitudes than by direct consideration of the corresponding coordinate space amplitudes. Moreover, there are generalizations of the OPE expansion (the cut vertex formalism of Mueller (Mueller, 1981) is an example) in which non-local operators appear, and which do not even have a natural expression in terms of the coordinate space asymptotic behavior of amplitudes. For all these reasons, our discussion of the OPE, and more generally, the factorization property, will be given in terms of momentum-space amplitudes.

The basic strategy underlying Zimmermann's proof of the OPE can easily be illustrated with a simple example. As usual, the ϕ_6^3 theory provides a convenient stage for displaying the central idea. We consider the case where the local field A in (18.47) is just the canonical ϕ field, with a single incoming and outgoing particle carrying momentum k. In Euclidean space, the corresponding amplitude is given by the connected contributions to the correlation function

$$\mathcal{T}(q, k) = \int d^6\xi \, e^{iq\cdot\xi} \langle \phi(\xi/2)\phi(-\xi/2)\tilde{\phi}(k)\tilde{\phi}(-k)\rangle \tag{18.49}$$

with the lowest-order graph displayed in Fig. 18.10. The external propagators associated with the fields carrying momentum $\pm k$ are assumed to be truncated (as indicated

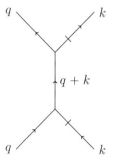

Fig. 18.10 Lowest-order contribution to $\mathcal{T}(q, k)$.

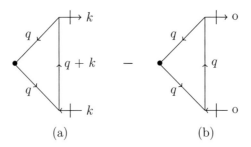

Fig. 18.11 (a) One-loop graph giving divergent contribution to $\phi^2(0)$. (b) Zero-momentum subtraction renormalizing $\phi^2(0)$.

by crossbars), but the propagators on the left, carrying momentum q, are not. The local limit $\xi \to 0$ of the operator product in (18.49) corresponds to integrating $\mathcal{T}(q, k)$ over q, thereby obtaining a δ-function setting ξ to zero. This corresponds, of course, to the one-loop graph indicated in Fig. 18.11(a), which is logarithmically divergent. This simply indicates that, as we have seen above, the composite operator $\phi^2(0)$ is ultraviolet-divergent and requires renormalization. The divergence is, of course, removed by the zero momentum subtraction indicated in Fig. 18.11(b), leading to a finite result which we interpret as arising from the minimally subtracted composite operator $N_4(\phi^2(0))$. This subtraction is effective in making the one-loop integral finite (and this is the crucial point!) precisely because it *removes the dominant dependence of the graph in Fig. 18.11(a) at large momentum q*. This suggests that we can introduce an oversubtracted bilocal operator $N_4(\phi(\xi/2)\phi(-\xi/2))$, with Fourier transform given to lowest order by the graphs indicated in Fig. 18.12, with a finite integral over q, and therefore with the finite local limit $N_4(\phi(\xi/2)\phi(-\xi/2)) \to N_4(\phi^2(0))$, $\xi \to 0$.

The term "oversubtraction" is appropriate here as the tree diagram giving the leading order contribution to the amplitude containing the bilocal operator is already finite and does not in that sense "need" a subtraction. However, the dominant asymptotic behavior for large q, k fixed, of $\mathcal{T}(q, k)$ is closely related to exactly the extra subtractions introduced to define this oversubtracted bilocal operator, which in the forest language correspond to counting as renormalization parts those 1PI subgraphs which *would become* divergent when the vertices $\pm\xi/2$ associated with the bilocal operator are pinched to a point (turning Fig. 18.12 into Fig. 18.11).

This insight, combined with clever use of forest formula techniques, allowed Zimmermann to provide a rigorous, all-orders proof of the OPE in a very general context. We shall sketch the proof here, but the basic steps will be translated from the language of Zimmermann forests into explicit graphical expressions where the structure of the subtractions, and their relation to the dominant large momentum flows in the diagrams, will be more physically intuitive. Also, we shall restrict ourself to the leading term in the expansion, as forest formula techniques are more or less indispensable in handling the combinatorics of the subleading terms in the expansion. We begin with the Euclidean case, before going on to the phenomenologically more important light-cone expansion.

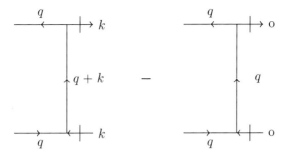

Fig. 18.12 Over-subtraction of the bilocal operator $\int d^6\xi e^{iq\cdot\xi} N_4(\phi(\xi/2)\phi(-\xi/2))$.

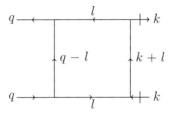

Fig. 18.13 One-loop two-particle-reducible contribution to $\mathcal{T}(q,k)$.

Our graphical demonstration of the OPE will depend on a crucial property of two-particle-irreducible diagrams (or "kernels": cf. Sections 10.4, 11.2). Consider first the one-loop "box" diagram contribution to $\mathcal{T}(q,k)$ indicated in Fig. 18.13. Recalling that external propagators are amputated on the right side only, this amplitude (ignoring combinatoric and coupling factors) takes the form

$$I_{1\,\text{loop}}(q,k) = \frac{1}{(q^2+m^2)^2} \int \frac{1}{(l^2+m^2)^2} \frac{1}{(q-l)^2+m^2} \frac{1}{(l+k)^2+m^2} \frac{d^6l}{(2\pi)^6} \quad (18.50)$$

For large $Q \equiv \sqrt{q\cdot q}$, the integral (ignoring the external propagator factors) is UV-finite and of order $1/Q^2$, as we can see by examining the contributions of the possible regions of large momentum (of order Q) flow through the diagram. In particular, we have:

1. The region where the loop momentum is large, i.e., $l_\mu \sim Q$, with phase-space volume Q^6 and integrand of order $1/Q^8$.
2. The region in which the large momentum q flows in and out of the diagram entirely through the propagator carrying momentum $q - l$, corresponding to loop momentum $l_\mu \sim k_\mu, m \ll Q$. This region also contributes asymptotic behavior $1/(q-l)^2 \sim 1/Q^2$.

The presence of the second region implies that a zero-momentum subtraction $I(q,k) - I(q,0)$ does not reduce the asymptotic behavior. On the other hand, in the corresponding over-subtracted graph, $I_{1\,\text{loop}}(q,k) - I_{1\,\text{loop}}(q,0)$, it is easy to see that the contribution of the first region, where large momentum permeates the entire graph, is suppressed to at least order $1/Q^3$, as the effect of the small incoming momentum k is subdominant once all the internal propagators of the loop are far off-shell (of order Q^2). The reader may easily verify this assertion explicitly by constructing the subtracted integrand and subjecting it to the simple power-counting analysis along the lines just followed above. The presence of a dominant contribution in regions where a subset of lines remain soft (low momentum) is clearly connected to the two-particle *reducibility* of our box diagram: the large momentum Q is afforded a rapid exit route from the diagram on the left, via the single left-most internal propagator. On the other hand, a two-particle-irreducible (2PI) diagram such as the one indicated in Fig. 18.14, while still of order $1/Q^2$ for large Q, receives its entire dominant contribution from the region of large momentum flow through the entire diagram: i.e., $l_\mu \sim Q$. The Feynman integral in this case is

$$K_{1\ \text{loop}}(q, k) = \frac{1}{(q^2 + m^2)^2} \int \frac{1}{l^2 + m^2} \frac{1}{(q - l)^2 + m^2}$$

$$\cdot \frac{1}{(l + k)^2 + m^2} \frac{1}{(k + l - q)^2 + m^2} \frac{d^6 l}{(2\pi)^6} \qquad (18.51)$$

and we see immediately that the contribution of the second region $l_\mu \sim k_\mu, m$ to the integral is of order $1/Q^4$. The reason is simply that in the 2PI case the large momentum q entering at the bottom left is forced to flow through at least *two* internal lines of the graph in order to exit the graph on the upper left-hand side.[6] This property generalizes to an arbitrary multi-loop 2PI contribution to $\mathcal{T}(q, k)$, so defining the sum of all such 2PI graphs as $K(q, k)$, we conclude that the zero-momentum subtraction $K(q, k) - K(q, 0)$ softens the asymptotic behavior by at least a power of Q, along the same lines as discussed above for the large-momentum region of the box diagram. This property is all that we shall need below to show that the oversubtractions introduced

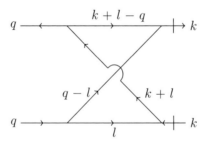

Fig. 18.14 One-loop two-particle irreducible contribution to $\mathcal{T}(q, k)$.

---

[6] One of the corollaries of Weinberg's power-counting theorem discussed in Section 17.1 provides a rigorous estimate for the asymptotic behavior of any convergent Feynman integral in terms of exactly the minimal routing argument given here, so the reader may be assured that the asserted behavior is on a very solid footing.

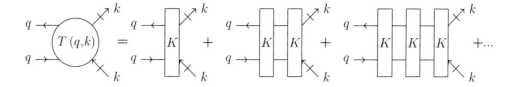

Fig. 18.15 Ladder expansion for $T(q,k)$ in terms of 2PI kernels K.

to define the bilocal operator $N_4(\phi(\xi/2)\phi(-\xi/2))$ are just those needed to obtain a factorized expression for the leading asymptotic behavior.

The introduction of two-particle irreducible kernels simplifies the description of the oversubtraction procedure needed to exhibit the emergence of an operator product expansion, by simplifying the graphical structure of the 1PI contributions to the amplitudes $T(q, k_i, k_i')$ of Fig. 18.9. In particular the 1PI contributions to these amplitudes[7] may be expressed as a sum of ladder graphs in which 2PI kernels are iterated, as shown in Fig. 18.15 (for the 2-2 case $T(q, k)$). Each 2PI kernel in Fig. 18.15 is itself a sum of infinitely many graphs, of which a few of the lowest-order ones are shown in Fig. 18.16.

It should be emphasized that Figs. 18.15 and 18.16 are *skeleton graphs*: each propagator line actually represents the full renormalized scalar propagator, including all possible self-energy corrections, with their associated BPHZ subtractions, and each vertex where three propagator lines meet at a point a full 1PI three-point vertex function including all BPHZ subtractions from any renormalization parts it may contain. In this way, we ensure that the set of ladder graphs in Fig. 18.15 indeed contains all the graphs making up the fully renormalized 1PI amplitude $T(q, k)$.

Another critical point here is one which we encountered earlier in our treatment of perturbative renormalization in Chapter 17: the subtractions induced by the counterterms of the theory produce UV-finite amplitudes with a dependence on the

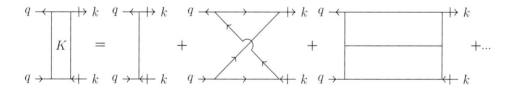

Fig. 18.16 Low-order skeleton graphs contributing to the 2PI kernel $K(q,k)$.

---

[7] We note here that in a ϕ_6^3 theory there are also one-particle reducible graphs contributing to $T(q, k_i, k_i')$, as the absence of a discrete $\phi \rightarrow -\phi$ symmetry allows the large momentum q to flow through a subgraph connected only by a zero-momentum propagator to the part containing the small momenta k_i, k_i'. This means that among the operators $\mathcal{O}_i$ appearing on the right-hand side of the OPE (18.48) in this theory is the scalar field ϕ itself. We shall ignore these graphs, which do not occur in the QCD/QED analogs of ϕ_6^3-theory, as an amplitude in which two photons carry large momentum q in and out of a graph cannot be connected by a single gluon (or photon) to the rest of the diagram, by Lorentz-invariance. Thus, we shall only consider the 1PI contributions to the $T(q, k_i, k_i')$ amplitudes in the following.

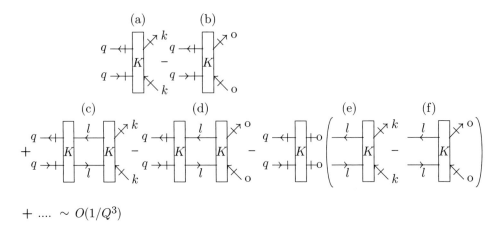

$$+ \dots \sim O(1/Q^3)$$

Fig. 18.17 Oversubtracted amplitude $T(q, k)$.

external momenta which is modified from that given by naive power-counting by at most powers of logarithms. As our discussion of factorization treats only the (integer) power of the large momentum Q as determinative of the leading asymptotic behavior, logarithmic factors are ignorable in isolating subdominant asymptotic terms. Thus, the replacement of free propagators and vertices by their full counterparts in Fig. 18.16 does not alter our previous conclusions *vis-à-vis* the flow of large momentum through 2PI kernels. In particular, a single subtraction of such a kernel suffices to lower the asymptotic behavior at large Q by at least a power of Q.

In Fig. 18.17 we show the first few diagrams (namely, the graphs (a) and (c)) in the skeleton expansion of $T(q, k)$, supplemented by a set of subtraction terms corresponding to the oversubtractions needed to define the $N_4(\phi(x)\phi(y))$ operator discussed above. The point of these subtraction terms is just to remove the leading asymptotic dependence of the basic terms (a) and (c) in the large Q limit. The uninteresting full propagators carrying momentum q in and out of the diagram on the left are, until further notice, truncated (as indicated by the crossbars), as they supply simply an overall Q dependence (of order $1/Q^4$) in all the diagrams. Accordingly, the asymptotic dependence of the basic graphs (a) and (c), prior to subtraction, is, modulo logarithms, $1/Q^2$.

We have previously explained why the subtraction effected by the graph (b) reduces the dependence of (a) by at least a single power of Q. An analysis of the possible large momentum flows through the graph (c) allows us to establish a similar suppression for the graphs on the second line of Fig. 18.17. For example, if the loop momentum $l_\mu \sim Q$, then the large momentum irrigates both 2PI kernels, and the subtraction is effective separately between graphs (c) and (d), and between graphs (e) and (f). On the other hand if $l_\mu \sim k_\mu \sim m$, the cancellation occurs between graphs (c) and (e), and between (d) and (f), as only the kernel on the left is irrigated by large momentum. The result is that the sum of diagrams in Fig. 18.17 is of order $1/Q^3$ at large Q, rather than $1/Q^2$. The extra subtractions appearing here are just those needed if we were to pinch the external vertices x and y in Fig. 18.15 together to construct an insertion of the composite minimally subtracted $N_4(\phi^2)$, which is finite precisely because the resultant loop integral over q has asymptotic behavior $\int (1/q^4 \cdot 1/q^3) d^6 q$ and is therefore UV-finite. The reader should verify this by explicitly constructing the forests appearing in the renormalized amplitudes for the $N_4(\phi^2)$ operator (see Problem 5).

If we examine the subtraction terms appearing in Fig. 18.17 closely, which we now realize incorporate exactly the leading asymptotic behavior of $T(q,k)$, a remarkable property emerges: they factorize algebraically into functions of the large momentum q and the remaining "small" momentum k. Indeed, transferring the subtraction terms to the right-hand side, we obtain the graphical equation indicated in Fig. 18.18, where the dependence on q is isolated in the set of graphs in the first parenthesis (which sum simply to $\mathcal{T}(q,0)$), while the second parenthesis contains exactly the graphs contributing to the insertion of the minimally subtracted composite $N_4(\phi^2)$ operator in the 1-1 matrix element. In terms of Euclidean correlation functions, reinserting the external legs carrying momentum q on the left, we have the asymptotic result

$$\langle \tilde{\phi}(q)\tilde{\phi}(-q)\tilde{\phi}(k)\tilde{\phi}(-k)\rangle_{1PI} \sim \tilde{C}_{\phi^2}(q)\langle N_4(\phi^2(0))\tilde{\phi}(k)\tilde{\phi}(-k)\rangle_{1PI} + O(1/Q^3) \quad (18.52)$$

where the Wilson coefficient function $\tilde{C}_{\phi^2}(q)$ is given in this case simply by setting the small momenta to zero in the full 1PI amplitude, and is clearly of order $1/Q^2$ for large Q. This result generalizes straightforwardly to amplitudes with more than two low-momentum fields,

$$\langle \tilde{\phi}(q)\tilde{\phi}(-q)\tilde{\phi}(k_1)\tilde{\phi}(k_3)\cdots \tilde{\phi}(k_n)\rangle_{1PI} \sim \tilde{C}_{\phi^2}(q)\langle N_4(\phi^2(0))\tilde{\phi}(k)\tilde{\phi}(k_2)\cdots \tilde{\phi}(k_n)\rangle_{1PI}$$
$$+ O(1/Q^3) \quad (18.53)$$

Indeed, in this case, the ladder expansion of the amplitude $\mathcal{T}(q, k_1, k_2, ..., k_n)$ terminates on the right with a 2PI kernel with two incoming lines on the left and $n > 2$ (small momentum) lines on the right. The reader may easily verify that such a kernel is automatically suppressed (to order $1/Q^4$) if large momentum flows through it. The subtraction terms needed for the oversubtraction are therefore just the ones discussed above for the 2PI 2-2 kernels to the left of this final 2-n kernel, and the reader may easily verify (Problem 6) that the factorization obtained is precisely as

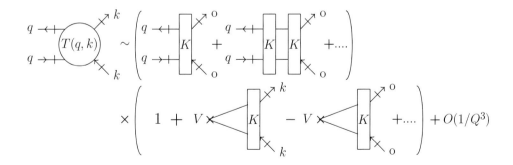

Fig. 18.18 Factorized structure of $\mathcal{T}(q,k)$ at large Q.

indicated in (18.53). The transition to an operator expansion can then be made in the usual fashion: the Euclidean correlation functions are analytically continued back to Minkowski space, and the resultant Minkowski amplitudes (vacuum-expectation-values of T-products) taken on-shell via the LSZ formula to yield the desired expansion (18.48) for matrix elements of the operator product.

The graphical arguments presented above are especially valuable in exposing the physical basis for factorization, and indeed are perfectly adequate in convincing oneself of the validity of the OPE at the leading order, especially in situations where, as in our case, only a single operator of lowest dimension contributes to the leading asymptotic behavior. To derive the general form of the expansion (18.48) however, including all the subdominant terms, the forest formalism of Zimmermann is indispensable. At this point, the power of the normal product formalism really manifests itself clearly: the ability to perform progressively higher degrees of oversubtraction by altering the subtraction degree of composite operators (as in (18.6)) turns out to be exactly the formal ingredient needed to organize an efficient proof of the general form of the expansion, in which local composite operators of increasing dimension are accompanied by coefficient functions of progressively more rapid falloff (by powers of the large momentum, modified by logarithms) at large momentum. Lack of space prevents us from going into the details here, but the interested reader may find the complete argument in the lectures of Zimmermann (Zimmermann, 1970).

Returning to our result (18.53), we see that the leading large momentum behavior of $\mathcal{T}(q, k)$ is contained in the coefficient function $\tilde{C}_{\phi^2}(q) = \mathcal{T}(q, 0)$, with the low-momentum k-dependence isolated in a correlation function of a renormalized composite operator. Note that the factorization has automatically produced UV-finite components: both the coefficient function and the composite operator contain all necessary subtractions to remove ultraviolet cutoff-dependence (which, after all, had already been removed in the original unfactorized amplitude). However, in a strongly coupled theory we are still no nearer to actually determining the precise form of the asymptotic behavior, as this would presumably require the computation of the full coefficient function, to all orders, and appropriately resummed to obtain a sensible finite result. Fortunately, just as in the case of the amplitudes discussed in the preceding section, in which all external momenta were large, a Callan–Symanzik equation controlling the large-q behavior of $\tilde{C}_{\phi^2}(q)$ can be derived. We shall see in the next section that the validity of such an equation in non-abelian gauge theories, together with the remarkable property of asymptotic freedom (namely, a negative β function at weak coupling), reduces the problem of large-momentum behavior to an entirely perturbative one.

As we shall now see, the desired Callan–Symanzik equation amounts to the statement of factorizability of the amplitude $\tilde{C}_{\phi^2}(q) = \mathcal{T}(q, 0)$ *after insertion of a soft mass operator*. When this holds, we say that the process (or amplitude) is "renormalization group controlled" (the connection to the renormalization group will also be explained more fully in the following section). Recall from our previous discussion that the insertion of the soft mass operator (in other words, of the minimally subtracted $N_4(\phi^2)$)in a 1PI amplitude with N external legs is equivalent, up to a multiplicative constant, to application of the Callan–Symanzik operator $\mathcal{D}_{CZ} \equiv m_R^2 \frac{\partial}{\partial m_R^2} + \beta(\lambda_R)\frac{\partial}{\partial \lambda_R} - N\gamma(\lambda_R)$

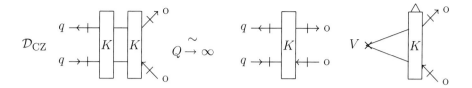

Fig. 18.19 Factorized structure of the two kernel contribution to $\mathcal{D}_{\mathrm{CZ}}T(q,0)$ at large Q.

to that amplitude. We note first that applying $\mathcal{D}_{\mathrm{CZ}}$ (with $N = 4$) to the first term in the skeleton expansion (see Fig. 18.15) of $T(q,0)$ (consisting of a single 2PI kernel) automatically produces an asymptotically suppressed amplitude: the insertion of a soft ϕ^2 vertex produces an extra internal propagator (with no additional subtractions) in an amplitude which receives its dominant asymptotic contribution from the regime in which all internal lines are far off-shell (of order Q^2). Moreover, in all higher terms in the skeleton expansion, the mass insertion must avoid the left-most kernel, as the large momentum cannot avoid flowing at least through this part of the graph.

Thus, in the term with two kernels, the application of $\mathcal{D}_{\mathrm{CZ}}$ leads to the factorization indicated in Fig. 18.19. We use the "hat" (or "caret") symbol to indicate the part of the graph containing the mass insertion, with the specification that the two propagators on the left of a kernel also receive mass insertions if the hat symbol is attached to that kernel.[8] Once a kernel receives the mass insertion, the momentum flowing in and out on the left is forced to be small, as momenta of order Q would again lead to an asymptotically suppressed contribution, by the previous argument. The standard OPE derived earlier may therefore be applied to the part of the graph to the left of the mass-inserted kernel (in Fig. 18.19, the 2PI kernel on the left), which thereupon loses its dependence on the small momentum connecting it to the right side of the graph. The result is that the mass-inserted kernel sees a momentum-independent amplitude to its left, leading to the appearance of a local vertex V, as indicated in Fig. 18.19. The same reasoning applied to the contribution with three 2PI kernels leads to the factorization indicated in Fig. 18.20, as the reader will confirm with a little thought. (Here, the OPE factorization of the two kernel subgraph on the left of the final graph on the top line is allowed by the fact that the loop momentum l connecting it to the mass-inserted kernel must be soft.)

Putting these results together, we arrive at the factorization indicated in Fig. 18.21 for the amplitude obtained by applying the Callan–Symanzik operator $\mathcal{D}_{\mathrm{CZ}} \equiv m_R^2 \frac{\partial}{\partial m_R^2} + \beta(\lambda_R)\frac{\partial}{\partial \lambda_R} - 4\gamma(\lambda_R)$ to the coefficient function $\tilde{C}_{\phi^2}(q)$. As usual, the vertex V indicates the point of insertion of a minimally subtracted $N_4(\phi^2)$ operator. We see that, up to terms suppressed by $1/Q^2$, the coefficient function satisfies an homogeneous Callan–Symanzik equation, with an additional term corresponding to $\tilde{C}_{\phi^2}(q)$ (the graphs on the top line) multiplied by the momentum-independent series of graphs on the second line. This term is (a) dimensionless and (b) only a function $\gamma_{\phi^2,CZ}$ of the

---

[8] For these terms, one takes $N = 0$ in $\mathcal{D}_{\mathrm{CZ}}$, as the graph is only "half amputated"—see the discussion following (18.38).

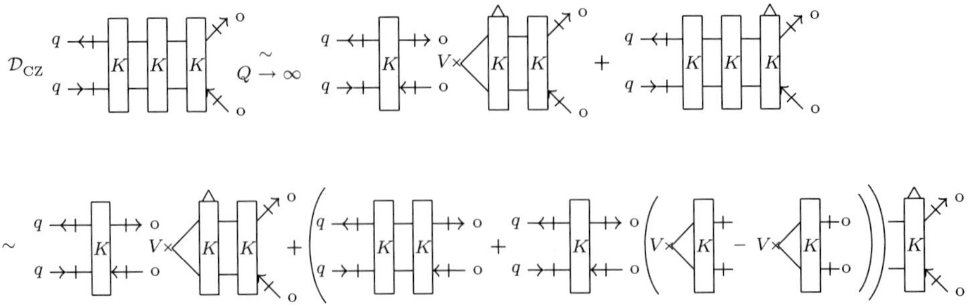

Fig. 18.20 Factorized structure of the three-kernel contribution to $\mathcal{D}_{CZ}T(q,0)$ at large Q.

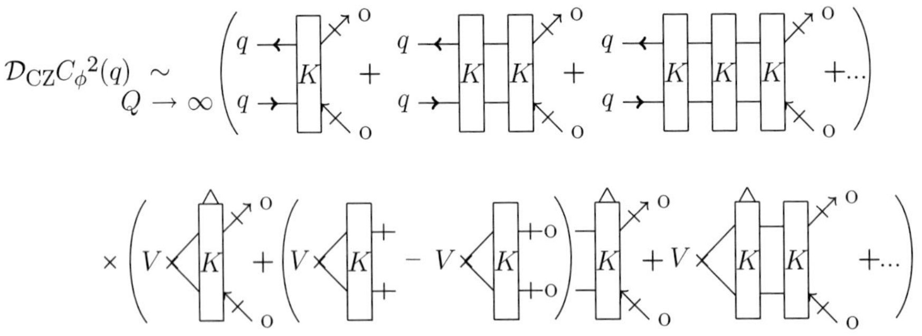

Fig. 18.21 Callan–Symanzik equation for coefficient function $\tilde{C}_{\phi^2}(q)$, in graphical form.

renormalized parameters m_R, λ_R, and therefore only of the dimensionless renormalized coupling λ_R. For reasons that will become apparent in the next section, it is called the *anomalous dimension of the composite operator* $N_4(\phi^2)$, and is evidently given by a single-particle matrix element (with a mass insertion) of this operator at zero momentum. In equation form, we have

$$\mathcal{D}_{CZ}C_{\phi^2}(q; m_R, \lambda_R) = (m_R^2 \frac{\partial}{\partial m_R^2} + \beta(\lambda_R)\frac{\partial}{\partial \lambda_R} - 4\gamma(\lambda_R))C_{\phi^2}(q; m_R, \lambda_R)$$

$$\approx \gamma_{\phi^2, CZ}(\lambda_R)C_{\phi^2}(q; m_R, \lambda_R) \tag{18.54}$$

where the approximation symbol ≈ 0 indicates the neglect of terms suppressed by powers of Q (see also Problem 7). In the next section we shall see how to solve this equation. At that point it will become apparent that in the important subclass of asymptotically free theories (including the present case of ϕ_6^3-theory, and, much more importantly, non-abelian gauge theories such as QCD), it determines the precise leading asymptotic behavior at large Q of the coefficient function (and therefore, of our factorized amplitude $T(q, k)$) on the basis of purely perturbative information.

Our discussion so far of the short-distance/large momentum factorization properties of amplitudes has focussed on the behavior of the Euclidean Green functions of the theory, allowing us to help ourselves plentifully to the physically transparent power-counting rules provided by Weinberg's theorem. Real physical processes, on the other hand, have a stubborn propensity to unfold in Minkowski spacetime, and we may therefore wonder whether any of the impressive formal results obtained so far have any meaningful phenomenological application. Fortunately, the basic principles of factorization are still found to be valid in the Minkowski regime for numerous high-energy processes. Although the topology of the (over)subtractions required to demonstrate this is essentially the same as in the Euclidean case, there are important differences in detail, as we shall now see.

We shall consider the archetypal Minkowski process exhibiting factorization, deep-inelastic scattering, as the argument in this case provides the basic template for establishing renormalization group control of many different types of high-energy amplitudes. The process is depicted in Fig. 18.22: a deeply space-like photon emitted by an incoming lepton (momentum p), with large momentum q, $q^2 \equiv -Q^2$, results in the fragmentation of an incoming hadron (e.g., a proton) of momentum k into an arbitrary hadronic final state, indicated n in the figure. We are interested in the *total inclusive cross-section* for this process, summing over all possible hadronic final states n. The factors contributed to the graph in Fig. 18.22 by the lepton lines and photon propagator are completely known, as the electromagnetic part of the process is treated to lowest-order perturbation theory. Apart from these boring kinematical factors, therefore, the amplitude in Fig. 18.22 is just given by a Fourier transform of the corresponding matrix element of the hadronic electromagnetic current $\langle n|J_{\text{em,had}}(x)|k\rangle$ (cf. (9.202)). The inclusive cross-section is obtained by squaring this amplitude and

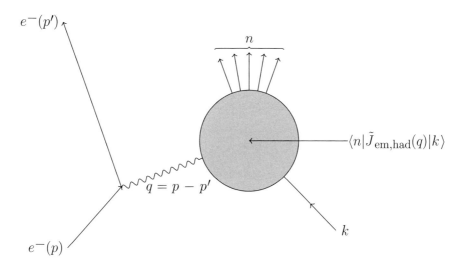

Fig. 18.22 Deep-inelastic electron–hadron scattering.

summing over all possible final states, subject to energy-momentum conservation, and is therefore proportional to the tensor[9]

$$\sum_n (2\pi)^4 \delta^4 (P_n - k - q) \langle k | J^\mu_{\text{em,had}}(0) | n \rangle \langle n | J^\nu_{\text{em,had}}(0) | k \rangle$$

$$= \text{Im}\{ i \int d^4 x \, e^{iq \cdot x} \langle k | T\{ J^\mu_{\text{em,had}}(x) J^\nu_{\text{em,had}}(0) \} | k \rangle \} \qquad (18.55)$$

where the second line follows by standard manipulations along the lines used to establish the Källen–Lehmann spectral representation in Section 9.5 (see Problem 8).

This result (basically the optical theorem of scattering theory, relating a total cross-section to the imaginary part of a forward scattering amplitude) allows us to concentrate our attention on the 2-2 amplitude on the second line, where a large space-like momentum q is inserted and then extracted on the left of the diagram, with the momentum k kept fixed (and eventually, sent on-mass-shell for the incoming and outgoing hadron). We saw in Section 6.6 that a forward scattering amplitude like (18.55) can also be written as the Fourier transform of a retarded commutator $\theta(x^0)[J^\mu_{\text{em,had}}(x), J^\nu_{\text{em,had}}(0)]$ (cf. (6.126–6.129)), so by locality, the integral over x is restricted to the forward light-cone. We shall now see that in a certain kinematic limit, the coordinate displacement x of the two current operators can be forced *onto* the light-cone, i.e., to the value $x^2 = 0$ (in Minkowski space), and that the product of operators can again be expanded, with a leading set of operators (with associated coefficient functions) providing the dominant asymptotic contribution.

Let us consider the *Bjorken limit*, in which $k \cdot q$ and q^2 are large (i.e., $>> k^2, m^2$) and comparable, with the ratio fixed:

$$\omega \equiv \frac{1}{x} \equiv -\frac{2k \cdot q}{q^2} \quad \text{fixed} \qquad (18.56)$$

We may automatically realize the Bjorken limit by choosing a convenient Lorentz frame. For a general Minkowski four-momentum p,[10] define light-cone coordinates $p_\pm \equiv \frac{1}{\sqrt{2}}(p_0 \pm p_3), \vec{p} = (p_1, p_2)$ in terms of which the invariant dot-product takes the form

$$p \cdot q = p_+ q_- + p_- q_+ - \vec{p} \cdot \vec{q} \qquad (18.57)$$

Note that the vector symbol here applies only to the two (or, in six spacetime dimensions, four) transverse dimensions orthogonal to the preferred spatial z-direction. We shall work in a frame in which $\vec{q} = 0$, $q_+ \sim Q^2/m_R$, and $q_- < 0 \sim m_R$. For $k_\mu \sim m_R$ fixed, in the Bjorken limit,

$$\omega = -\frac{2(k_- q_+ + k_+ q_-)}{2q_- q_+} \sim -\frac{k_-}{q_-} \qquad (18.58)$$

---

[9] For a fuller discussion, including the kinematic factors glossed over here, see Section 13.4, (Itzhykson and Zuber, 1980).

[10] In six dimensions we likewise take $p_\pm \equiv \frac{1}{\sqrt{2}}(p_0 \pm p_5)$, $\vec{p} = (p_1, p_2, p_3, p_4)$.

Now, by taking q_+ large, we force the Fourier transform to extract the dominant dependence of the retarded commutator of currents for x_- small (as the exponent is $q \cdot x = q_+ x_- + ..$). However, locality restricts us to the interior of the forward light-cone, $2x_+ x_- > \vec{x}^2$ so $x_- \to 0$ forces also $\vec{x} \to 0$, and the Bjorken limit naturally probes the region $x^2 \to 0$: i.e., the light-cone singularities of the operator product. Fortunately, as in the Euclidean case, where the structure of the amplitude simplifies (via factorization) in the Euclidean limit $x^2 \to 0$ ($\Rightarrow x \to 0$), the forward amplitude in (18.55) also displays a factorized structure in the Bjorken limit in Minkowski space. In this case, however, the leading contribution involves an infinite "tower" of operators and coefficient functions—not surprisingly, as the light-cone limit involves a surface, rather than a single point.

To expose the essential ideas, while avoiding the (not inconsiderable!) complications of spin and local gauge symmetry with which we would have to contend in QCD, we shall sketch the factorization procedure in our old standby, ϕ_6^3-theory. Thus, instead of the second line of (18.55), we consider exactly our previous amplitude $\mathcal{T}(q, k)$ (essentially, the Fourier transform of $\langle k|T\{\phi(x)\phi(0)\}|k\rangle$) of Fig. 18.15, but now *in Minkowski space, and in the Bjorken limit* (18.56). The factorization will be demonstrated for the full amplitude: the imaginary part can then be taken at the end, to obtain the desired inclusive cross-section.

We begin, as before, with the lowest-order tree diagram contributing to $\mathcal{T}(q, k)$, as indicated in Fig. 18.23(a). The large momentum q flows through a single propagator and the (fully amputated) graph is therefore (suppressing the ubiquitous $i\epsilon$ terms) proportional to

$$\frac{1}{(q+k)^2 - m_R^2} = \frac{1}{q^2 + 2k_- q_+ + 2k_+ q_- + k^2 - m_R^2} \tag{18.59}$$

In the Bjorken limit, the terms $2k_+ q_-$ and k^2 are of order m_R^2, suppressed by two powers of the large scale Q relative to the terms q^2 and $2k_- q_+$ which are both of order Q^2. This means that the leading asymptotic dependence of graph (a) on Q is unaltered if we set k_+ and $\vec{k}$ to zero (the latter meaning, in six spacetime dimensions, $k_1 = k_2 = k_3 = k_4 = 0$). We shall indicate that the $+$ and transverse vector components

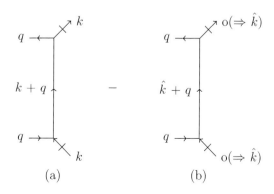

Fig. 18.23 Oversubtraction of lowest-order $\mathcal{T}(q, k)$ in the Bjorken limit.

of an external momentum entering a subgraph have been set to zero by once again appending the "o" symbol to the corresponding leg, and also define, for a general momentum $p_\mu = (p_+, p_-, \vec{p})$, the projected momentum $\hat{p}_\mu = (0, p_-, \vec{0})$. Accordingly, the subtraction effected by graph Fig. 18.23(b), with propagator

$$\frac{1}{(q + \hat{k})^2 - m_R^2} = \frac{1}{q^2 + 2k_- q_+ - m_R^2} \tag{18.60}$$

reduces the asymptotic behavior of the amplitude from $1/Q^2$ to $1/Q^4$.

At the one-loop level we encounter the box diagram indicated in Fig. 18.24(a). In analogy to the subtractions indicated in Fig. 18.17, we introduce the subtractions shown in graphs (b), (c), and (d). As previously, we only expect graph (b) to contain the leading contribution at large Q from the region in which all four internal lines are off-shell of order Q^2, with additional subtraction terms needed to take care of the case in which the loop momentum l is soft ($l^2 \ll Q^2$). The Feynman integral for all four terms combines to the expression

$$\mathcal{I}_{1\,\text{loop}}(q, k) = \int \frac{1}{(l^2 - m_R^2 + i\epsilon)^2} \Big\{ \frac{1}{(q - l)^2 - m_R^2 + i\epsilon} - \frac{1}{(q - \hat{l})^2 - m_R^2 + i\epsilon} \Big\}$$

$$\cdot \Big\{ \frac{1}{(k + l)^2 - m_R^2 + i\epsilon} - \frac{1}{(\hat{k} + l)^2 - m_R^2 + i\epsilon} \Big\} \frac{d^6 l}{(2\pi)^6}$$

$$= \int \frac{1}{(l^2 - m_R^2 + i\epsilon)^2} \frac{(l^2 - 2q_- l_+)}{((q - l)^2 - m_R^2 + i\epsilon)((q - \hat{l})^2 - m_R^2 + i\epsilon)}$$

$$\cdot \frac{(2k_+ l_- + k^2)}{((k + l)^2 - m_R^2 + i\epsilon)((\hat{k} + l)^2 - m_R^2 + i\epsilon)} \frac{d^6 l}{(2\pi)^6} \tag{18.61}$$

We now need to estimate the leading asymptotic behavior of this rather formidable expression at large Q. We no longer have the Weinberg theorem, and its corollaries, at our disposal, as we are in Minkowski space. In particular, denominators (such as $(q - \hat{l})^2 - m_R^2$) with only *linear* dependence on loop momentum components appear— a circumstance completely alien to the Euclidean space analysis. Nevertheless, the indicated subtractions do indeed do their job, and end up reducing the asymptotic dependence by two powers of Q, as desired. A "physicist's" proof of this assertion is

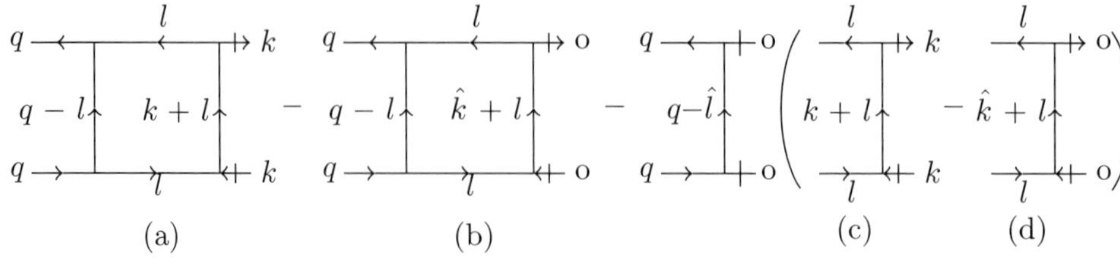

Fig. 18.24 Oversubtraction of a one-loop box graph contribution to $\mathcal{T}(q, k)$ in the Bjorken limit.

easily obtained by a straightforward scaling analysis, but the general result is confirmed by extensive computational experience in perturbative field theory, although a general power-counting theorem in Minkowski space of the scope and power of Weinberg's theorem for the Euclidean case has, to the author's knowledge, never been established.

Let us therefore proceed directly, by examining the contribution to (18.61) from a region of phase-space corresponding to arbitrary power scalings of the loop momentum components:

$$l_+ \sim Q^\alpha, \quad l_- \sim Q^\beta, \quad \vec{l} \sim Q^\gamma, \quad \alpha, \beta, \gamma > 0 \tag{18.62}$$

The volume of loop phase-space corresponding to this region evidently scales like $Q^{\alpha+\beta+4\gamma}$. By examining the scaling under (18.62) of each of the numerator and denominator terms in (18.61), we find that the subtracted amplitude receives a contribution of order $Q^{\mathcal{P}(\alpha,\beta,\gamma)}$, with the power given by

$$\mathcal{P}(\alpha, \beta, \gamma) = \alpha - \beta + 4\gamma - 2 - 3\max(\alpha + \beta, 2\gamma) - \max(2, \alpha + \beta, 2\gamma) \tag{18.63}$$

A short exercise (see Problem 9) shows that

$$\mathcal{P}(\alpha, \beta, \gamma) \leq -4 \tag{18.64}$$

so we may reasonably conclude that the subtractions in graphs (b), (c), and (d) have indeed succeeded in suppressing by two powers of Q the leading $1/Q^2$ dependence of the box diagram Fig. 18.24(a). The subtractions are effective because, in analogy to the Euclidean case, in the region where the loop momentum l is soft ($l_\mu \sim k_\mu, m_R$), corresponding to the large momentum flowing entirely through the left-most vertical line, the leading asymptotic dependence on Q cancels separately between graphs (a) and (c), and between (b) and (d); whereas, in the region of l "hard" (this means $l_+ \sim q_+ \sim Q^2, \vec{l} \sim Q, l_-$ fixed) where all lines are far off-shell, the cancellation occurs between graphs (a) and (b), and between (c) and (d), as the reader may easily check, using power-scaling arguments along the lines of (18.62–18.64).

The preceding examples suggest that we may proceed exactly as in the Euclidean case to introduce an oversubtracted skeleton expansion for the Minkowski amplitude $\mathcal{T}(q, k)$: the topological structure of these subtractions is exactly the same as in the Euclidean case, the only difference begin that the subtraction point is at a projected light-like momentum, rather than at zero momentum. The result is that the leading asymptotic behavior factorizes as indicated graphically in Fig. 18.25 (replacing Fig. 18.18). This result may be written explicitly as

$$\mathcal{T}(q, k) \sim \int \mathcal{T}(q, \hat{l})(\hat{\Delta}_F(l))^2 \mathcal{T}(l, k) \frac{d^6 l}{(2\pi)^6}, \quad Q \to \infty, \, \omega \text{ fixed} \tag{18.65}$$

The amplitude $\mathcal{T}(q, k)$ is a Lorentz scalar and therefore a function of $q^2, q \cdot k$, and k^2, or of Q^2, k^2 and the Bjorken variable $\omega = -\frac{k_-}{q_-}$. A slight change of notation makes this clear:

$$\mathcal{T}(q, k) \equiv \Gamma(\omega, Q^2, k^2) \Rightarrow \mathcal{T}(q, \hat{l}) = \Gamma(\hat{\omega}, Q^2, 0), \, \hat{\omega} = -\frac{l_-}{q_-} \tag{18.66}$$

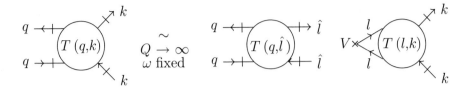

Fig. 18.25 Factorization of $\mathcal{T}(q,k)$ in the Bjorken limit.

One can show (see Problem 10) that $\Gamma(\omega, Q^2, k^2)$ is analytic in the cut plane of ω with cuts running from $-\infty$ to -1 and from 1 to $+\infty$. In the light-like case ($\hat{l}^2 = 0$), we can therefore expand

$$\Gamma(\hat{\omega}, Q^2, 0) = \sum_{n=0}^{\infty} \hat{\omega}^n C_n(Q^2) \tag{18.67}$$

Thus, the factorization (18.65) can be written, using $(\hat{\omega})^n = (\frac{l_-}{k_-})^n \cdot \omega^n$,

$$\Gamma(\omega, Q^2, k^2) \sim \sum_n C_n(Q^2) \int \mathcal{T}(l,k)(\hat{\Delta}_F(l))^2 (\hat{\omega})^n \frac{d^6 l}{(2\pi)^6} \tag{18.68}$$

$$= \sum_n \omega^n C_n(Q^2) \int \mathcal{T}(l,k)(\hat{\Delta}_F(l))^2 (\frac{l_-}{k_-})^n \frac{d^6 l}{(2\pi)^6} \tag{18.69}$$

$$= \sum_n \omega^n C_n(Q^2) \langle k|N_4(\phi(0)(\frac{i\overleftrightarrow{\partial}_-}{2k_-})^n \phi(0))|k\rangle \equiv \sum_n \omega^n v_n C_n(Q^2) \tag{18.70}$$

with $v_n \equiv \langle k|N_4(\phi(0)(\frac{i\overleftrightarrow{\partial}_-}{2k_-})^n \phi(0))|k\rangle$ the matrix elements of a tower of renormalized composite operators incorporating the low-energy physics of the process.[11]

Factors of loop momentum l_- appearing at the vertex V in Fig. 18.25 (arising from expanding the $\mathcal{T}(q,\hat{l})$ amplitude on the left) have been converted to the corresponding spatial derivatives appearing in minimally subtracted composite operators, of which there are clearly an infinite number contributing at leading order. Note that these operators receive only a single subtraction to remove a logarithmic divergence, irrespective of the power n of the loop component l_- present in the graph (see Problem 11). The result (18.70) is called the *light-cone expansion* of the amplitude $\mathcal{T}(q,k) \equiv \Gamma(\omega, Q^2)$. It shows that the leading asymptotic behavior of the amplitude in the Bjorken limit is given by an infinite set of factorized terms involving the product of coefficient functions depending only on the large scale Q, and matrix elements of renormalized composite operators. In the leading term, all operators of minimum *twist* (defined as

---

[11] Note that if the initial and final states are taken as light-like elementary scalars, with $k \to \hat{k}$, the matrix elements are unity, $v_n = 1$. This is just the renormalization condition for the composite operators equating the matrix element at the special subtraction point to its lowest-order value, as higher loop corrections vanish if taken at the subtraction point, where they are subtracted.

the engineering dimension of the operator minus its spin, the latter given in this case by the number of spacetime-derivatives ∂_-) appear. The subtraction degree of the operator is determined in a light-cone expansion not by the engineering dimension, as in the Euclidean case, but by the twist: thus, all the operators $N_4(\phi(0)(\frac{i\overset{\leftrightarrow}{\partial}_-}{2k_-})^n\phi(0))$ appearing in (18.70) have twist 4 (note: the factors of k_- are not included in the dimension) and appear at the same, leading twist, level in the OPE. Operators of higher twist will contribute to the amplitude at levels suppressed by powers of Q. In practice, one extracts individual terms in the infinite sum in (18.70) by taking moments with respect to the variable $x \equiv \frac{1}{\omega}$ of $\mathrm{Im}(\Gamma(\omega, Q^2))$, which is directly related to the inclusive cross-section for the deep inelastic scattering, as discussed above (see Problem 12).

The discussion of factorization for deep inelastic amplitudes in QCD follows completely analogous lines to the argument for ϕ_6^3-theory given here. In this case (see, for example, the review of Mueller (Mueller, 1981)), the leading contributions are given by a tower of operators of twist 2: namely, the quark composite operators

$$\mathcal{O}_n \equiv N_2(\bar{\psi}\gamma_-(\frac{i\overset{\leftrightarrow}{D}_-}{2k_-})^n\psi) \tag{18.71}$$

where D_μ is the gauge-covariant derivative (15.103) for the quark field ψ. In a general covariant gauge, these composite operators involve both quark and gauge fields, but the analysis simplifies, and becomes (modulo spin complications) extremely similar to the ϕ_6^3 case if we choose a light-cone gauge in which $A_-(x) = 0$, in which case the gauge-covariant derivatives may be replaced by ordinary ones, $D_- \to \partial_-$.

Just as in the Euclidean case, the asymptotic behavior of the coefficient functions $C_n(Q^2)$ is determined by a Callan–Symanzik equation. The derivation of this equation follows exactly the lines of the space-like factorization: one applies a soft mass insertion, via the Callan–Symanzik operator $\mathcal{D}_{CZ}$, to the large momentum amplitude $\mathcal{T}(q,\hat{l})$, which is then refactorized. One then obtains, in analogy to (18.54), for the asymptotic behavior of $C_n(Q^2)$ (as always, up to power suppressed terms),

$$\mathcal{D}_{CZ}C_n(Q^2) \sim \gamma_n C_n(Q^2) \tag{18.72}$$

where γ_n, in analogy to the Euclidean case, is given by a single (soft) mass insertion on the 1-1 matrix element of $N_4(\phi(0)(\frac{i\overset{\leftrightarrow}{\partial}_-}{2k_-})^n\phi(0))$ (see Problem 11), and is a dimensionless function of λ_R. In the case of asymptotically free theories such as QCD (or ϕ_6^3), (18.72) can be used to reduce the determination of the leading asymptotic behavior in Q to perturbative information, as we shall explain in the next section. In particular, the asymptotic behavior of individual moments of the inclusive amplitude can be explicitly computed and compared with experiment.[12]

---

[12] For more details on all of this, the reader is referred to the original literature: (Gross and Wilczek, 1974a), (Gross and Wilczek, 1974b). The general approach to factorization outlined here is covered in great detail in the review of Mueller (Mueller, 1981); see also (Buras, 1980).

18.3 Renormalization group equations for renormalized amplitudes

The problem of determining the asymptotic behavior of amplitudes when all or some of the external momenta are taken large can be approached from a quite different direction from the methods used above, which are deeply grounded in the subtraction technology of renormalization theory. Instead, we can rely on information provided by the renormalization group flow that underlies the whole dependence of field theory amplitudes on the energy scale at which these amplitudes are examined. This can be done both for the cases where all external momenta are taken large (a classic example of enormous phenomenological importance being inclusive electron–positron annihilation to hadrons, which amounts to calculating the imaginary part of the two-point Green function of the hadronic electromagnetic current, with both incoming and outgoing momentum q large), and when only some momenta are large (as in deep-inelastic scattering, discussed above). We shall see that results entirely equivalent to those discussed previously (in particular, the Callan–Symanzik equations regulating the large-momentum amplitudes) can be obtained, although the physical arguments involved are somewhat different.

In Section 16.4 we introduced the concept of the renormalization group in order to describe the evolution of a Wilsonian effective Lagrangian when we change the ultraviolet momentum scale up to which Fourier components of the field are included. The reader will recall (Section 16.2) that the physics of a clustering, Lorentz-invariant system of interacting particles under quite general assumptions can be described by a scalar Lagrangian function involving an infinite number of terms with arbitrary powers and spacetime-derivatives of the field(s). Such a Lagrangian, formulated on a flat Minkowski background spacetime, is assumed to be valid up to some ultraviolet momentum scale Λ, beyond which the physics of the model breaks down, for example, via quantum gravity effects. The UV scale Λ can be varied while leaving the low-energy/momentum predictions of the theory unchanged by appropriately varying the infinite set of (dimensionless) coefficients g_n of the terms in the effective Lagrangian, and the corresponding flow in the infinite parameter space of coupling coefficients is referred to as the renormalization group. The Lie algebra of this group, corresponding to infinitesimal group operations, is specified by an infinite set of coupled, non-linear first-order differential equations (16.27):[13]

$$\mu\frac{\partial}{\partial\mu}g_n(\mu) = \beta_n(g_n(\mu))\tag{18.73}$$

where here the floating cutoff of the theory is denoted, as common in treatments of the renormalization group, by the Greek letter μ.

In general, we can say little about the solution to these equations, but in the event that the renormalization group flow remains within the region of parameter space in which the couplings corresponding to interactions (i.e., terms higher than quadratic in the fields) are small enough to validate the use of perturbation theory, one is able to show, as we saw in Section 17.4, that the renormalization group

---

[13] As in Section 17.4, we have chosen to label all the coefficients with a single symbol here, not distinguishing between terms with different numbers of spacetime-derivatives.

flow, starting at a generic point in the infinite-dimensional coupling space at a high cutoff, attracts to a *finite(N)-dimensional* submanifold of the coupling space at low values of the cutoff. The dimensionality N of this submanifold—points on which can be regarded as associated in one-to-one fashion with physically distinct low-energy theories—corresponds simply to the number of renormalizable and super-renormalizable operators (or equivalently, relevant and marginal terms) in the full effective Lagrangian of the theory.

The fact that the "memory" of a high-energy cutoff is lost in the low-energy physics provided we parameterize the amplitudes of the theory in terms of renormalization conditions associated with just this finite subset of operators of the theory is referred to as "perturbative renormalizability": it is a property of a truncated Lagrangian containing solely renormalizable and super-renormalizable terms compatible with the symmetries and fields present in the theory. To the extent that we are concerned only with the low-energy properties of the renormalized amplitudes (Green functions) of such a perturbatively renormalizable theory, we may forget about its origins in a much more complicated Wilsonian action extending the physics of the theory up to much higher energies. The remnant information of the full renormalization group then amounts simply to the statement that the physics at low energy can be described in terms of any set of N independent parameters uniquely identifying the point on the N-dimensional submanifold onto which the flow converges at low energy. Although this freedom seems at first sight a bit trivial, it turns out that it allows us, in combination with information about the mass-dependence of the amplitudes of the theory, to derive highly non-trivial conclusions about the energy dependence of amplitudes, which (fortunately) turn out to be completely consistent with those implied by the Callan–Symanzik equations derived previously. The material in this section can therefore be regarded as an alternative approach to the study of energy/momentum dependence of amplitudes, one which provides additional physical insight and in some cases is technically more convenient than the operator inspired technology used in the preceding sections.

In order to examine the consequences of the reparameterization invariance of physical amplitudes on the low-energy attractor manifold we must, of course, allow ourselves the freedom of examining these amplitudes in a variety of renormalization schemes. Sticking to the zero momentum BPHZ scheme, for example, we learn nothing, as the parameterization is completely fixed by the unique choice of subtraction point. In principle, one could imagine very general reparameterizations which mix all N relevant and marginal couplings. Our chief aim will be to explore the connection of reparameterization invariance to the scaling of amplitudes as the overall energy scale is changed, so it suffices to examine a one-parameter subset of reparameterizations in which the particular renormalization scheme is identified by specifying a dimensionful parameter: for example, the momentum scale μ introduced in the renormalization scheme in which the propagator and interaction vertex are subtracted at a Euclidean momentum point, as in (17.75), or the scale μ introduced in the dimensional renormalization scheme, via (17.102). In either of these schemes, the Green functions necessarily acquire a dependence on the renormalization scale μ. Moreover, if we fix our attention on a definite physical theory, identified as a unique point on the low-energy attractor surface (corresponding to some definite endpoint of the renormalization group flow at some conventionally chosen low cutoff, and ignoring as usual variations in this surface of order inverse powers of the much higher UV cutoff), then the renormalized parameters m_R, λ_R (in ϕ_4^4-theory, say) must also vary with μ to keep the physics fixed.

We shall illustrate the derivation of the renormalization group equation for renormalized amplitudes taking self-coupled scalar field theory renormalized at Euclidean scale μ as our starting point (the reader may imagine our theory to be ϕ_4^4 although an essentially identical argument holds for ϕ_6^3 theory). The analysis of perturbative renormalization in Section 17.2 makes it clear that the renormalized 1PI Euclidean N-point function $\Gamma_R^{(N)}$ in such a theory is related to the corresponding 1PI function $\Gamma^{(N)}$ computed by using the bare parameters and field in (17.42) with a UV cutoff Λ by (a) rescaling the bare field $\phi \to \sqrt{\hat{Z}}\phi_R$, and (b) reparameterizing the resultant amplitude in terms of renormalized mass and coupling parameters defined at the Euclidean subtraction point. Thus, the upshot of our demonstration of perturbative renormalizability is the relation[14]

$$\Gamma_R^{(N)}(k_i; \lambda_R, m_R, \mu) = \hat{Z}(\lambda_R, m_R, \mu, \Lambda)^{-N/2}\Gamma^{(N)}(k_i; \lambda, m, \Lambda) + O(\frac{k_i^2, m_R^2, \lambda_R^2, \mu^2}{\Lambda^2})$$

$$(18.74)$$

In essence, a unique physical theory is specified by the boundary condition that we start with a cutoff theory at scale $\Lambda_{UV} = \Lambda$, in which the full Wilsonian Lagrangian is given by (17.42), and then reparameterize the amplitudes (and rescale the field) at low energy, by using the parameters m_R, λ_R and field rescaling $\hat{Z}$ defined by the renormalization conditions (17.46, 17.51, 17.52). The correction term on the right of (18.74) is just the usual, by assumption ignorable, sensitivity of the amplitudes of a perturbatively renormalizable theory to the UV cutoff of the theory: we shall henceforth drop it entirely (by taking $\Lambda \to \infty$ at the appropriate point, for example). Of course, the same physical theory can just as well be parameterized in terms of couplings defined at a different Euclidean scale $\tilde{\mu}$:

$$\Gamma_R^{(N)}(k_i; \tilde{\lambda}_R, \tilde{m}_R, \tilde{\mu}) = \hat{Z}(\tilde{\lambda}_R, \tilde{m}_R, \tilde{\mu}, \Lambda)^{-N/2}\Gamma^{(N)}(k_i; \lambda, m, \Lambda) \qquad (18.75)$$

Dividing (18.75) by (18.74), and defining

$$F(\tilde{\lambda}_R, \tilde{m}_R, \tilde{\mu}; \lambda_R, m_R, \mu) \equiv \frac{\hat{Z}(\tilde{\lambda}_R, \tilde{m}_R, \tilde{\mu}, \Lambda)}{\hat{Z}(\lambda_R, m_R, \mu, \Lambda)} \qquad (18.76)$$

we find

$$\Gamma_R^{(N)}(k_i; \tilde{\lambda}_R, \tilde{m}_R, \tilde{\mu}) = F(\tilde{\lambda}_R, \tilde{m}_R, \tilde{\mu}; \lambda_R, m_R, \mu)^{-N/2}\Gamma_R^{(N)}(k_i; \lambda_R, m_R, \mu) \qquad (18.77)$$

In other words, the 1PI Green functions of our theory transform covariantly (by multiplicative rescaling) under reparameterizations corresponding to an alteration of the subtraction scale μ. The absence of a dependence on the UV cutoff Λ in the

---

[14] The negative power of $\hat{Z}$ here arises as a consequence of the need to divide the basic N-point Green function $G^{(N)}$ by N full propagators in order to arrive at the fully amputated $\Gamma^{(N)}$: each such propagator gives a factor of $\hat{Z}$ on rescaling, converting the $\sqrt{\hat{Z}}$ associated with each of the N fields in $G^{(N)}$ to a $1/\sqrt{\hat{Z}}$.

renormalized amplitudes, of course, implies the same for the rescaling factor F in (18.76) and (18.77). The finite transformation expressed in (18.77) can clearly be viewed as an invertible element of a one-parameter continuous Lie group in which successive transformations of renormalization scale satisfy an obvious composition rule. This one parameter group is all that remains of the vastly more complicated renormalization group flow embodied in the equations (18.73), which themselves lead to the collapse to the low-energy attractor surface corresponding to perturbative renormalizability, as we saw in Section 17.4. Nevertheless, this equation—or rather, its infinitesimal Lie algebra version—once combined with information on the mass singularities of the amplitudes, will lead us back to the same powerful constraints on the Green functions of the theory derived previously in the form of the Callan–Symanzik equation.

The desired infinitesimal version of (18.77) is readily obtained: we simply keep $\tilde{\mu}, \tilde{\lambda}_R$ and $\tilde{m}_R$ fixed while applying $\mu \frac{\partial}{\partial \mu}$. As a result, λ_R and m_R must also be allowed to vary, and, with the understanding that everywhere

$$\mu \frac{\partial}{\partial \mu} \equiv \mu \frac{\partial}{\partial \mu}\bigg|_{\tilde{\lambda}_R \tilde{m}_R \tilde{\mu}} \tag{18.78}$$

we obtain

$$(\mu \frac{\partial}{\partial \mu} F^{-N/2}) \Gamma_R^{(N)} + F^{-N/2} (\mu \frac{\partial}{\partial \mu} + \mu \frac{\partial \lambda_R}{\partial \mu} + \mu \frac{\partial m_R}{\partial \mu}) \Gamma_R^{(N)} = 0 \tag{18.79}$$

Multiplying through by $F^{N/2}$, we find

$$(\mu \frac{\partial}{\partial \mu} + \mu \frac{\partial \lambda_R}{\partial \mu} + \mu \frac{\partial m_R}{\partial \mu}) \Gamma_R^{(N)} = \frac{1}{2} N (\mu \frac{\partial}{\partial \mu} \ln F) \Gamma_R^{(N)} \tag{18.80}$$

After taking the partial derivatives, we may set $\tilde{\mu} = \mu, \tilde{m}_R = m_R, \tilde{\lambda}_R = \lambda_R$ and define the dimensionless functions

$$\beta(\lambda_R, \frac{m_R}{\mu}) \equiv \mu \frac{\partial \lambda_R}{\partial \mu} \tag{18.81}$$

$$\gamma_m(\lambda_R, \frac{m_R}{\mu}) \equiv \frac{1}{m_R} \mu \frac{\partial m_R}{\partial \mu} \tag{18.82}$$

$$\gamma(\lambda_R, \frac{m_R}{\mu}) \equiv \mu \frac{\partial \ln F}{\partial \mu} \tag{18.83}$$

As the coupling λ_R is dimensionless, the dimensionless functions β, γ, γ_m can, of course, only depend on the ratio m_R/μ. Inserting these definitions in (18.80), we find the *renormalization group equation for 1PI amplitudes*:

$$\{\mu \frac{\partial}{\partial \mu} + \beta(\lambda_R, \frac{m_R}{\mu}) \frac{\partial}{\partial \lambda_R} + \gamma_m(\lambda_R, \frac{m_R}{\mu}) m_R \frac{\partial}{\partial m_R} - N \gamma(\lambda_R, \frac{m_R}{\mu})\}$$

$$\times \Gamma_R^{(N)}(k_i; \lambda_R, m_R, \mu) = 0 \tag{18.84}$$

We emphasize that this equation is *exact* (having taken the UV cutoff Λ of the theory to infinity, of course): we have so far not considered any simplifications arising in an asymptotic regime. An equation of exactly the same form holds if we use dimensional renormalization, where the parameter μ is introduced via (17.102), with the additional simplification that the renormalization group functions β, γ_m, γ lose their mass dependence (on m_R) and are therefore only functions of the dimensionless coupling λ_R. These functions are, moreover, to be regarded as *quantum effects*: the dependence on the subtraction scale μ appears only in loop diagrams requiring subtractions—in other words, in contributions to the amplitude containing non-zero powers of Planck's constant. In particular, the tree diagrams of the theory are independent of μ.

At this point, a superficial resemblance of (18.84) to the asymptotic version of the Callan–Symanzik equation (18.38) should already be apparent. Recall that the latter equation applies in the event that the external momentum set k_i is non-exceptional (no non-trivial subset of the Euclidean momenta k_i summing to zero), and that this is also the condition for the absence of mass singularities of the amplitude in the zero mass limit. In particular, all Lorentz-invariant dot-products $k_i \cdot k_j$ are non-zero (and uniformly large, say of order Q^2, with Q a large momentum scale, if we consider the asymptotic regime as previously for the Callan–Symanzik case), and Weinberg's theorem then assures us that the graphs contributing to Γ_R^N receive their dominant contribution from regions in which all internal propagators are far off-shell, with denominators of order Q^2, and therefore with a sensitivity to the mass of order m_R^2/Q^2. Neglecting the mass sensitivity, and setting m_R to zero, we arrive at the approximate asymptotic equation, valid to inverse powers of the large momentum scale,

$$\{\mu\frac{\partial}{\partial\mu} + \beta(\lambda_R)\frac{\partial}{\partial\lambda_R} - N\gamma(\lambda_R)\}\Gamma_R^{(N)}(k_i; \lambda_R, \mu) = 0 \qquad (18.85)$$

which is now formally identical to (18.38). (Note, however, that the renormalization scale μ in this zero mass theory now plays the role of the BPHZ renormalized mass m_R in the Callan–Symanzik equation.) The dependence on the renormalized mass (now set to zero) in $\Gamma_R^{(N)}(k_i; \lambda_R, \mu)$ and the renormalization group functions $\beta(\lambda_R)$ and $\gamma(\lambda_R)$ (which therefore also lose their dependence on μ) has been omitted, and we have an equation which can be made useful, in precise analogy to the steps leading from (18.38) to (18.40) by trading in the derivative with respect to renormalization scale μ (which describes a physically inaccessible dependence of the amplitudes) for one implementing a uniform rescaling of the external momenta, via the dimensional equation (d_N is the engineering dimension of $\Gamma_R^{(N)}$ in powers of mass)

$$(\kappa\frac{\partial}{\partial\kappa} + \mu\frac{\partial}{\partial\mu})\Gamma_R^{(N)}(\kappa k_i; \lambda_R, \mu) = d_N\Gamma_R^{(N)}(\kappa k_i; \lambda_R, \mu) \qquad (18.86)$$

Using (18.86) to eliminate the μ derivative in (18.85), we find

$$(-\kappa\frac{\partial}{\partial\kappa} + \beta(\lambda_R)\frac{\partial}{\partial\lambda_R} - N\gamma(\lambda_R) + d_N)\Gamma_R^{(N)}(\kappa k_i; \lambda_R, \mu) = 0 \qquad (18.87)$$

This equation is identical in form (modulo a redefinition of the functions $\beta(\lambda_R), \gamma(\lambda_R)$ by a factor of 2) with the Callan–Symanzik equation (18.40). At the tree level, as discussed above, the dependence on the subtraction scale disappears, as do the functions β, γ, and the resultant equation,

$$(-\kappa\frac{\partial}{\partial\kappa} + d_N)\Gamma^{(N)}_{\text{tree}}(\kappa k_i; \lambda_R) = 0 \Rightarrow \Gamma^{(N)}_{\text{tree}}(\kappa k_i; \lambda_R) = \kappa^{d_N}\Gamma^{(N)}_{\text{tree}}(k_i; \lambda_R) \qquad (18.88)$$

simply reduces to the statement that the tree amplitudes contributing to $\Gamma^{(N)}_{\text{tree}}$ have engineering dimension d_N, which therefore determines the scaling behavior with respect to momentum of the zero-mass theory. The modifications to this scaling induced by loop diagrams, with their concomitant subtractions, are the additional information inserted in (18.87) by the renormalization group functions β and γ.

As we are at liberty to set the renormalization scale μ to the value of the renormalized mass m_R in the BPHZ scheme appearing in (18.40), these renormalization group functions must in fact (apart from the trivial factor of 2) be identical, despite their very different definitions in the two approaches. We must therefore conclude that the scaling properties associated with the insertion of soft mass operators in BPHZ renormalization are in fact physically equivalent to the contraints imposed by the renormalization group, once supplemented with information about the mass singularities of the amplitudes under consideration.

The renormalization group equation (18.87) relates in a linear fashion the scaling behavior in momentum of the N-point amplitudes to the coupling constant dependence, with the derivative-free terms responsible for an overall scaling of the amplitude. To see this, define a *running (or effective) coupling* $\lambda_{\text{eff}}(\kappa)$ as the solution to the first-order ordinary differential equation and boundary condition

$$\kappa\frac{\partial}{\partial\kappa}\lambda_{\text{eff}}(\kappa) = \beta(\lambda_{\text{eff}}(\kappa)), \quad \lambda_{\text{eff}}(1) = \lambda_R \qquad (18.89)$$

Further, let

$$z(\kappa) \equiv \exp\left(\int_1^\kappa \gamma(\lambda_{\text{eff}}(\kappa'))\frac{d\kappa'}{\kappa'}\right) \qquad (18.90)$$

whence we find

$$\kappa\frac{\partial}{\partial\kappa}z^{-N}(\kappa) = -N\gamma(\lambda_{\text{eff}}(\kappa))z^{-N}(\kappa) \qquad (18.91)$$

With these definitions, it is easy to show (see Problem 13) that the general solution to (18.87) is

$$\Gamma^{(N)}_R(\kappa k_i; \lambda_R, \mu) = \kappa^{4-N}z^{-N}(\kappa)\Gamma^{(N)}_R(k_i; \lambda_{\text{eff}}(\kappa), \mu) \qquad (18.92)$$

The extraction of the large momentum behavior of the amplitudes (i.e., for κ large) is therefore transferred to a knowledge of the κ-dependence of the running coupling $\lambda_{\text{eff}}(\kappa)$, from which we may determine the scaling factor $z(\kappa)$, provided that we are able to determine the dependence of the full renormalized 1PI amplitude $\Gamma^{(N)}_R$ on the coupling (now replaced by its running counterpart). In a general theory where the coupling(s) may be large, rendering perturbation theory inapplicable, this solution is not particularly helpful. But in an important subclass of cases, the asymptotic behavior is determined by the weak-coupling regime of the theory, and we are able to make rigorous statements about the large momentum properties of the amplitudes.

Suppose that for some renormalized coupling $\bar{\lambda}$, and for all $0 < \lambda_R < \bar{\lambda}$, we have $\beta(\lambda_R) < 0$. Then it is apparent from (18.89) that with a physical renormalized coupling $\lambda_R < \bar{\lambda}$, the running coupling $\lambda_{\text{eff}}(\kappa)$ is monotone decreasing for $\kappa > 1$, and indeed, we have that $\lambda_{\text{eff}}(\kappa) \to 0$, $\kappa \to \infty$. Such a theory is said to be *asymptotically free*. The leading term in the perturbative expansion of the β function is necessarily negative in an asymptotically free theory, to enforce negativity of the β function at arbitrarily small couplings. We have already encountered an example in the ϕ_6^3-theory discussed previously, with a β function given to lowest order (including the factor of two in accordance with the new definition (18.81)) by

$$\beta(\lambda_R) = -\beta_0 \lambda_R^3 + O(\lambda_R^5), \quad \beta_0 = \frac{3}{256\pi^3} \tag{18.93}$$

Once κ is sufficiently large, the effective coupling will become sufficiently small that the β function is dominated by its leading term, so that the extreme large-momentum (or short-distance) asymptotics of the theory is determined by solving the defining equation (18.89), keeping only the leading term in (18.93). It is more convenient to solve for the squared effective coupling λ_{eff}^2, which satisfies, at this leading order

$$\kappa \frac{\partial}{\partial \kappa} \lambda_{\text{eff}}^2(\kappa) = -2\beta_0 \lambda_{\text{eff}}^4(\kappa) \tag{18.94}$$

the solution to which is

$$\lambda_{\text{eff}}^2(\kappa) \sim \frac{\lambda_R^2}{1 + 2\beta_0 \lambda_R^2 \ln(\kappa)}, \quad \kappa \to \infty \tag{18.95}$$

The running coupling in ϕ_6^3-theory therefore falls off logarithmically with the momentum rescaling variable: in effect, free field behavior is restored in the amplitude $\Gamma_R^{(N)}(\kappa k_i; \lambda_R, \mu)$ when all momenta are taken large, though admittedly very slowly. Exactly the same behavior obtains in QCD, with renormalized coupling g_R and lowest-order β function (for a SU(N) theory with N_q quark fields in the fundamental representation)

$$\beta(g_R) = -\beta_0 g_R^3 + O(g_R^5), \quad \beta_0 = \frac{1}{16\pi^2}(\frac{11}{3}N - \frac{2}{3}N_q) \tag{18.96}$$

The extraordinary progress made in the last three-and-a-half decades in bringing the high-energy behavior of many strong interaction amplitudes under analytic control is entirely dependent on this very fortunate property of the theory, first uncovered by Gross, Politzer, and Wilczek, and for which they received the 2004 Nobel Prize in Physics. Considerations of space preclude our delving further into this fascinating and hugely important area of modern particle physics.[15]

---

[15] Classic QCD applications of the renormalization group control of high-energy processes, such as the inclusive electron–positron annihilation cross-section to hadrons, and deep-inelastic scattering can be found in the reviews of Mueller and Buras cited earlier, as well as in any number of modern texts on Standard Model field theory.

Asymptotically free theories in four spacetime dimensions are rather rare: indeed, in the class of perturbatively renormalizable theories, the only field theories with this property are gauge theories with a non-abelian gauge group, and with not too many matter fields charged under the gauge group (e.g., in (18.96), we must have $N_q < \frac{11}{2} N$). Our other field-theoretic workhorse, ϕ_4^4-theory, variants of which are clearly present in the Standard Model in the event that the Higgs particle turns out to be an elementary scalar, is *not* asymptotically free. To lowest (one-loop) order, one finds (see Problem 14)

$$\beta(\lambda_R) = +\beta_0 \lambda_R^2 + O(\lambda_R^3), \quad \beta_0 = \frac{3}{16\pi^2} \tag{18.97}$$

and a running coupling

$$\lambda_{\text{eff}}(\kappa) \sim \frac{\lambda_R}{1 - \beta_0 \lambda_R \ln{(\kappa)}} \tag{18.98}$$

which evidently runs into a singularity (the famous "Landau pole"[16]) at a finite value of κ, beyond which the effective coupling changes sign, apparently destabilizing the theory. Of course, we are no longer entitled to rely on the perturbative one-loop form of the β function once the effective coupling becomes large, as it certainly does once the singularity is approached. Non-perturbative studies have provided considerable evidence for the hypothesis that this situation should nevertheless be interpreted as signaling the "triviality" of ϕ^4 theory (cf. the discussion of global aspects of the renormalization group flow in Section 17.4): the removal of the UV cutoff of the theory, corresponding to prescribing a well-defined Wilsonian effective Lagrangian at arbitrarily high-energy scales, necessarily implies the vanishing of the renormalized coupling λ_R defined at any fixed low-energy scale (see, for example, (Fernandez *et al.*, 1992), for a detailed treatment of the mathematical issues surrounding triviality). Nevertheless, given, as emphasized on many previous occasions, the unavoidable presence of a physical cutoff at sufficiently high energy, there is absolutely no reason to reject ϕ_4^4 theory as a perfectly adequate low-energy effective field theory whose amplitudes are accurately computable (if we are lucky enough to have a sufficiently small λ_R) by the standard technology of perturbative renormalization theory.

An alternative scenario to triviality for four-dimensional non-asymptotically-free theories has been the subject of much interest: the presence of a non-trivial ultraviolet fixed point at which the theory recovers an exact scale (even conformal) invariance. Consider a four-dimensional scalar theory where the β function is positive at small coupling, $\beta(\lambda) > 0$, $0 < \lambda < \lambda^*$, but with a zero at some positive coupling value $\beta(\lambda^*) = 0$, as indicated in Fig. 18.26. We also assume that the particular physical theory of interest has a renormalized coupling $\lambda_R < \lambda^*$, which serves as the starting point, $\lambda_{\text{eff}}(\kappa = 1) = \lambda_R$, for the renormalization group flow of the effective coupling defined in (18.89). As the β function is positive, the effective coupling increases monotonically with κ, with the rate of growth slowing as $\lambda_{\text{eff}}(\kappa)$ approaches the

---

[16] Quantum electrodynamics, and abelian gauge theories generally, exhibit a similar structure, with a positive β function at small coupling: the associated singular behavior, in the context of the photon propagator, was first pointed out in the 1950s by Landau.

value λ^*, which therefore acts as a *fixed point* of the flow: $\lambda_{\text{eff}}(\kappa) \to \lambda^*$, $\kappa \to \infty$. From (18.90), the scaling factor $z(\kappa)$ therefore behaves asymptotically as

$$z(\kappa) \sim \exp\left(A + \int^{\kappa} \gamma(\lambda^*)\frac{d\kappa'}{\kappa'}\right) \sim C\kappa^{\gamma(\lambda^*)}, \quad \kappa \to \infty \qquad (18.99)$$

which means that the scaling behavior κ^{4-N} of the tree amplitudes of the massless theory (cf. (18.88), which follows from the fact that the N scalar fields in the associated Green function each have engineering dimension 1 (in four dimensions), together with naive dimensional analysis, is altered by quantum loop effects to the scaling $\kappa^{4-N(1+\gamma(\lambda^*))}$. We interpret this by saying that the interactions have induced an anomalous scale dimension $\gamma(\lambda^*)$ for the underlying scalar field, but that scaling by fixed powers (rather than fractional powers of logarithms, as in the asymptotically free case) is restored at the fixed point. In fact, we saw earlier in our discussion of the trace anomaly (cf. (15.201)), the trace of the energy momentum tensor, which acts as the divergence of the would-be conserved current of scale transformations (cf. (12.118)), only vanishes in a massless interacting theory if the β function does so: i.e., at precisely the fixed point indicated in Fig. 18.26. The situation described here does not appear to arise in any of the field theories comprising the Standard Model of particle interactions in four dimensions, but conformally invariant interacting field theories in two dimensions have been the subject of intense scrutiny, partly as a consequence of their close connection to aspects of string theory. Moreover, the renormalization group treatment of critical phenomena in condensed-matter theory—specifically, the scaling behavior of thermodynamic quantities at a second-order transition—is based precisely on the existence of an infrared fixed point in scalar field theories which can be shown to describe the long-distance behavior of spin models near such transitions. A classic case is the β function of massless ϕ^4 theory in three dimensions, which has exactly the appearance of Fig. 18.26, but with an overall minus sign, indicating that power scaling behavior is obtained for the correlation functions in the infrared limit

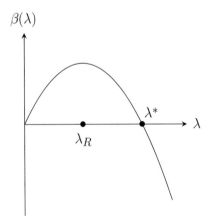

Fig. 18.26 β function for a theory with an ultraviolet fixed point.

(κ taken to zero).[17] The Wilson–Fisher analysis of critical exponents in second-order phase transitions relies on the existence of an infrared fixed point of just this kind.

Finally, we turn to the case in which the amplitude of interest has, in addition to a large-momentum scale Q, important low-momentum scales, and in which a factorization of the amplitude via an operator product expansion can be established. We examined in the previous section the case of an (Euclidean) amplitude with $N + 2$ scalar lines, two of which inject and remove a large momentum q, leading to the factorization indicated in (18.53). Moreover, the factorization of the mass-inserted amplitude led to an homogeneous Callan–Symanzik equation (18.54) for the coefficient function $\tilde{C}_{\phi^2}(q)$, which can evidently be solved in an asymptotically free theory by the techniques described above to produce an explicit form for the leading asymptotic behavior for large q. Here we wish to reproduce this result, but starting from a renormalization group analysis of the amplitude, in analogy to the arguments given previously for the case of uniformly large external momenta.

Let us suppose that we have already established the validity of an operator product expansion (OPE) like (18.53) for a scalar amplitude with $N + 2$ external legs, and suppose for simplicity that only a single operator $\mathcal{O}$ appears in the leading twist contribution (e.g., $\mathcal{O} = \phi^2$ in the case considered explicitly in Section 18.3). The right-hand side of the OPE contains a renormalized amplitude $\Gamma_R^{(N,\mathcal{O})}$ with N scalar legs and a single insertion of the renormalized composite operator, obtained by multiplicative renormalization (cf. (18.9) of the corresponding bare operator, $\mathcal{O}_R(x) = Z_{\mathcal{O}}(\lambda_R, m_R, \mu, \Lambda)\mathcal{O}(x)$. This amplitude, in analogy to (18.74), may be written in terms of the corresponding bare amplitude (neglecting terms vanishing like inverse powers of the UV cutoff),

$$\Gamma_R^{(N,\mathcal{O})}(k_i; \lambda_R, m_R, \mu) = Z_{\mathcal{O}}(\lambda_R, m_R, \mu, \Lambda)\hat{Z}(\lambda_R, m_R, \mu, \Lambda)^{-N/2}\Gamma^{(N,\mathcal{O})}(k_i; \lambda, m, \Lambda) \tag{18.100}$$

Repeating the procedure leading from (18.74) to (18.84), we find

$$\{\mu\frac{\partial}{\partial\mu} + \beta(\lambda_R, \frac{m_R}{\mu})\frac{\partial}{\partial\lambda_R} + \gamma_m(\lambda_R, \frac{m_R}{\mu})m_R\frac{\partial}{\partial m_R}$$

$$-N\gamma(\lambda_R, \frac{m_R}{\mu}) + \gamma_{\mathcal{O}}(\lambda_R, \frac{m_R}{\mu})\}\Gamma_R^{(N,\mathcal{O})}(k_i; \lambda_R, m_R, \mu) = 0 \tag{18.101}$$

where the new anomalous dimension arises from a logarithmic derivative of the renormalization constant $Z_{\mathcal{O}}(\lambda_R, m_R, \mu, \Lambda)$ associated with the composite operator $\mathcal{O}$. The OPE implies the asymptotic behavior

$$\Gamma_R^{(N+2)}(q, k_i; \lambda_R, m_R, \mu) \sim \tilde{C}_{\mathcal{O}}(q; \lambda_R, m_R, \mu)\Gamma_R^{(N,\mathcal{O})}(k_i; \lambda_R, m_R, \mu) \tag{18.102}$$

with $\Gamma_R^{(N+2)}(q, k_i; \lambda_R, m_R, \mu)$ satisfying the renormalization group equation (18.84) with $N \to N + 2$. Inserting the right-hand side of (18.102) into the latter, and using (18.101), one finds

---

[17] A comprehensive treatment of this important area can be found in the treatise of Zinn–Justin (Zinn–Justin, 1989).

$$\{\mu\frac{\partial}{\partial\mu} + \beta\frac{\partial}{\partial\lambda_R} + \gamma_m m_R\frac{\partial}{\partial m_R} - (2\gamma + \gamma_{\mathcal{O}})\}\tilde{C}_{\mathcal{O}}(q;\lambda_R, m_R, \mu) = 0 \qquad (18.103)$$

An equation of Callan–Symanzik type for the coefficient function is therefore obtained, *provided we can neglect the mass dependence in the coefficient function* and take $m_R = 0$—in other words, if the mass singularities of the full amplitude can be removed from the high-momentum (or "hard") end of the amplitude (the coefficient function) and completely incorporated in the low-momentum end (the matrix element(s) of the composite operator $\mathcal{O}$). We then obtain a renormalization group equation for the coefficient function,

$$\{\mu\frac{\partial}{\partial\mu} + \beta(\lambda_R)\frac{\partial}{\partial\lambda_R} - (2\gamma(\lambda_R) + \gamma_{\mathcal{O}}(\lambda_R))\}\tilde{C}_{\mathcal{O}}(q;\lambda_R, \mu) = 0 \qquad (18.104)$$

of the same form as the Callan–Symanzik equation (18.54) obtained previously (modulo differences in definition of anomalous dimensions in the two approaches). The suppression of soft mass insertions in hard amplitudes which lies at the core of the BPHZ approach to high-energy behavior is thus seen, in the renormalization group approach, to be tantamount to the property of factorization of mass singularities. In the case of QCD, the ability to isolate the extremely complicated mass singularity structure of hadronic amplitudes (containing for example, the intricate confinement physics inaccessible to perturbative treatment) is the basic precondition to extracting useful information (via asymptotic freedom) using renormalization group equations.

18.4   Problems

1. The conserved currents J_α^μ associated with a non-abelian symmetry satisfy the Ward–Takahashi identity (12.142), where all the operators are bare, unrenormalized ones (prior to rescaling). Rewriting the identity in terms of renormalized fields (with $N_3(J_\alpha^\mu) = Z_J J_\alpha^\mu$, $\phi_R = \hat{Z}^{-1/2}\phi$), show that the bare J_α^μ must already be ultraviolet finite, and that we may therefore simply set the associated renormalization factor $Z_J = 1$.

2. By examining the graph of Fig. 18.27(a), verify that the oversubtracted $N_4(\phi^2)$ operator in ϕ_4^4-theory contains a contribution from the (minimally subtracted) $N_4(\phi^4)$ operator, as implied by the Zimmermann identity (18.11).

3. The one-loop hexagon graph in ϕ_6^3 theory shown in Fig. 18.27(b) is taken at an exceptional momentum point as various subsets of the incoming external momenta combine to zero momentum. It is proportional to the (UV-finite) Feynman integral, with superficial degree of divergence –6,

$$\mathcal{I}(q_i, m) = \int \frac{1}{(l-q_1)^2 + m^2}\frac{1}{(l-q_2)^2 + m^2}\frac{1}{(l-q_3)^2 + m^2}\frac{1}{(l^2 + m^2)^3}\frac{d^6l}{(2\pi)^6} \qquad (18.105)$$

(a) By rescaling the loop integration variable by $Q \equiv \sqrt{q^2}$, show that

$$\mathcal{I}(q_i, m) = \frac{1}{Q^6}\mathcal{I}(\hat{q}_i, \frac{m}{Q}), \quad \hat{q}_i = q_i/Q, \quad \hat{q}_i^2 = 1, \quad i = 1, 2, 3 \qquad (18.106)$$

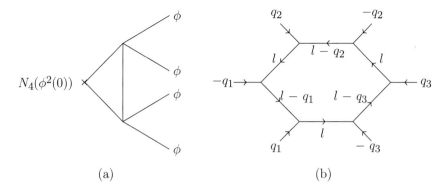

Fig. 18.27 (a) A one-loop graph in ϕ_4^4-theory needing an oversubtraction in the insertion of $N_4(\phi^2)$ in the 2-2 amplitude.(b) A one-loop graph in ϕ_6^3-theory at an exceptional external momentum point.

> Show that for large Q, the dependence of $\mathcal{I}(\hat{q}_i, \frac{m}{Q})$ on the vanishing rescaled mass $\frac{m}{Q}$ is logarithmic, due to a logarithmic infrared divergence when $l \to 0$ in the rescaled integral. This implies a logarithmic modification of the naive power scaling $1/Q^6$ when all q_I are taken large simultaneously (of order Q): $\mathcal{I}(q_i, m) \sim C \ln{(Q/m)}/Q^6$.
>
> (b) Now assume that a mass insertion is made on one of the lines carrying momentum l, thereby increasing the power of the propagator $\frac{1}{l^2+m^2}$ to four. Show that there is *no suppression* of the asymptotic dependence in this case, despite the reduction of the superficial degree of divergence of the graph to -8. Explain this result in terms of the enhanced mass singularity of the rescaled integral after the mass insertion.
>
> (c) Show that if the incoming external momenta are chosen non-exceptionally, $q_i, i = 1, 2, ...6$, $\sum_i q_i = 0$, all q_i of order Q, but no subset of the q_i summing to zero, a mass insertion on any line does reduce the asymptotic dependence for large Q by two powers of Q, in accordance with the superficial degree of divergence.

4. Verify the expressions (18.41,18.42) for the 1PI two- and three-point functions of ϕ_6^3-theory at one-loop order.

5. Construct the Zimmermann forests needed to subtract the 1-1 matrix element of $N_4(\phi^2)$ through 2-loop order, and show how they correspond to the subtractions introduced in Fig. 18.17 (here, the 2PI kernel factors can be replaced by single propagators, for simplicity).

6. By analysing the large momentum flows for the skeleton ladder expansion of the amplitude $\mathcal{T}(q, k_1, k_2, ..., k_n)$ (with $n > 2$), show that the required oversubtractions lead to the OPE factorization given in (18.53).

7. The point of this exercise is to check the validity of the Callan–Symanzik equation (18.54) for the coefficient function of the ϕ^2 operator in ϕ_6^3-theory. To do this, first calculate the coefficient function $\tilde{C}_{\phi^2}(q)$ through one loop, renormalized

by zero momentum BPHZ subtraction (use an interaction Lagrangian $\frac{\lambda_R}{3!}\phi^3$, and do not forget self-energy and vertex corrections). Then calculate $\mathcal{D}_{CZ}\tilde{C}_{\phi^2}(q)$ (using $\gamma(\lambda_R), \beta(\lambda_R)$ from (18.43,18.45)) and show that (18.54) holds, neglecting subdominant terms of order $1/q^4$, with $\gamma_{\phi^2,CZ} = -\frac{1}{128\pi^3}\lambda_R^2 + O(\lambda_R^4)$.

8. Verify the equality of the left- and right-hand sides of (18.55) (see also Section 6.6)).

9. Starting from (18.63), verify the inequality (18.64).

10. The analyticity of the amplitude $\Gamma(\omega, Q^2)$ (suppressing the k^2 dependence) in the complex ω-plane is limited by the presence of cuts due to physical thresholds in the sum over states n in (18.55). For a physical state with forward time-like four-momentum P^μ, we must have $P_\pm > 0$ (as $P_o > |\vec{P}|$). Show that this implies that the imaginary part of the amplitude vanishes unless $\omega > 1$. Moreover, by Bose symmetry (amplitude $\mathcal{T}(q, k)$ even under $q \to -q$), we must have $\Gamma(\omega, Q^2) = \Gamma(-\omega, Q^2)$. Consequently, $\Gamma(\omega, Q^2)$ is analytic in the complex ω-plane except for cuts running along the real axis from $-\infty$ to -1 and $+1$ to $+\infty$.

11. (a) Show that the one-loop term contributing to the right-most graph in Fig. 18.25, with the operator $N_4(\phi(0)(\frac{i\overleftrightarrow{\partial}_-}{2k_-})^n\phi(0))$ appearing at the vertex V, is logarithmically divergent regardless of the power n. The relevant Feynman integral is

$$\int \frac{1}{(l^2 - m^2 + i\epsilon)^2} \frac{1}{(k+l)^2 - m^2 + i\epsilon} (\frac{l_-}{k_-})^n \frac{d^6l}{(2\pi)^6} \tag{18.107}$$

Show that this integral is rendered finite when once subtracted at the light-like point $k = \hat{k}$ (i.e., setting $k_+ = \vec{k} = 0$). Hint: consider the integral with l_-^n replaced by the general tensor $l_{\mu_1}l_{\mu_2}...l_{\mu_n}$; after introducing Feynman parameters (cf. (16.67), easily generalized to higher powers of either propagator by differentiation with respect to A or B), and making the usual shift in integration variable, one sees that the only surviving term when all Lorentz indices are set equal to the $-$ light-cone coordinate is proportional to k_-^n (as $g_{--} = 0$), with a logarithmically divergent coefficient.

(b) Repeat for the one-loop graph studied in part (a), but with a single mass insertion (doubled propagator), which gives, up to uninteresting factors, the lowest-order contribution to the anomalous dimension γ_n in (18.72).

12. Use the analyticity of $\Gamma(\omega, Q^2)$ demonstrated in Problem 8 to derive the following Cauchy representation for the coefficients in (18.70):

$$v_n C_n(Q^2) = \frac{1}{2\pi i}\oint \frac{d\omega}{\omega^{n+1}}\Gamma(\omega, Q^2) \tag{18.108}$$

where the contour runs around a small circle enfolding the origin. By unfolding the contour of integration onto the cuts, and changing variables to $x \equiv 1/\omega$, show that

$$v_n C_n(Q^2) = \frac{1 + (-1)^n}{\pi} \int_0^1 x^{n-1}\text{Im}(\Gamma(x, q^2))dx \tag{18.109}$$

giving the coefficient functions in terms of moments of the inclusive deep-inelastic cross-section (cf. (18.55)).

13. Our objective in this problem is to verify the general solution (18.92) to (18.87). Starting from the definitions (18.89, 18.90), establish first the identities

$$\beta(\lambda_R)\frac{\partial\lambda_{\text{eff}}(\kappa)}{\partial\lambda_R} = \beta(\lambda_{\text{eff}}(\kappa)) \tag{18.110}$$

$$\frac{\partial z(\kappa)}{\partial\lambda_R} = \frac{1}{\beta(\lambda_R)}(\gamma(\lambda_{\text{eff}}(\kappa)) - \gamma(\lambda_R))z(\kappa) \tag{18.111}$$

where the derivatives with respect to λ_R are at fixed κ. Using these identities, together with (18.91), verify directly that the Ansatz (18.92) solves (18.87).

14. Calculate the Euclidean 1PI four-point vertex function $\Gamma_R^{(4)}(k_1, k_2, k_3, k_4)$ in zero mass ϕ_4^4-theory, with all momenta incoming, and subtracted at the symmetric Euclidean point (with $k_i^2 = \mu^2$, $k_i \cdot k_j = -\mu^2/3$, $i \neq j$), through one loop. Show that one obtains, to order λ_R^2,

$$\Gamma_R^{(4)}(k_i) = \lambda_R + \frac{\lambda_R^2}{32\pi^2}(\ln\left(\frac{3(k_1 + k_2)^2}{4\mu^2}\right) + \ln\left(\frac{3(k_1 + k_3)^2}{4\mu^2}\right)$$

$$+ \ln\left(\frac{3(k_1 + k_4)^2}{4\mu^2}\right)) + O(\lambda_R^3) \tag{18.112}$$

Use this result to verify the result (18.97) previously stated for the one-loop β function in ϕ_4^4-theory.

19
Scales IV: Long-distance structure of quantum field theory

In the final chapter of this book we wish to turn our attention to the behavior of field theory amplitudes in the long-distance regime, where "distance" is to be interpreted spatiotemporally: we are interested in those aspects of elementary processes described by a local relativistic field theory which involve large, even macroscopic, regions of spacetime. Evidently, this behavior is closely connected with the whole definition of the stable, asymptotic particle states of the theory, which are after all what survive when a long time has elapsed after an elementary interaction and the final-state particles are allowed to separate and become free from each other's influence.

In Section 9.3 we described in detail the construction of these asymptotic states and their connection to the local (or almost local) Heisenberg fields of the theory, along the lines of the Haag–Ruelle scattering theory. An essential input to the Haag–Ruelle formalism is the existence of a *mass gap*: the single particle state(s) of the theory correspond to isolated δ-function singularities $\delta(p^2 - m^2)$, $m \neq 0$ in the spectral density of the squared-mass operator $P_\mu P^\mu$, or equivalently, simple poles in the momentum-space full propagator defined by Fourier transforming the two-point Feynman function of any suitable interpolating field for the particle in question. In other words, the scattering theory unfolds in a straightforward and physically intuitive way provided we adhere to field theories of *purely massive particles*. Once exactly massless particles are present in the theory, the construction of a rigorous scattering theory—in particular, the construction of a separable asymptotic Fock space based on almost local operators, along the lines of Theorem 9.4—becomes considerably more complicated. Nevertheless, the desired extension of the Haag–Ruelle theory has been carried out by Buchholz for massless spin-0 bosons (Buchholz, 1977) and for massless spin-$\frac{1}{2}$ fermions (Buchholz, 1975), leading to a well-defined S-matrix and asymptotic in- and out-Fock spaces along more or less the usual lines, at least for theories in odd spatial dimensions where the Huyghens principle applies.

For massless spin-1 *gauge* bosons on the other hand, where the dynamics is subject to an exact local gauge symmetry, the situation is far more subtle. Strangely enough, the conceptual (if not calculational) difficulties are greater in the case of unbroken abelian theories such as QED than in non-abelian gauge theories such as QCD, for the simple reason that the latter theories do have a mass gap, despite the presence of massless fields in the Lagrangian, as a consequence of the non-perturbative confinement of non-gauge-invariant states, to be discussed below. Accordingly, the slippery issues which arise once massless gauge particles appear in the asymptotic spectrum are avoided in the case of unbroken non-abelian theories, as they also are in the case of spontaneously broken gauge theories (abelian or non-abelian) where the Higgs phenomenon results in the spin-1 gauge particles of the theory associated with the broken generators becoming massive. For an unbroken abelian gauge theory such as QED, on the other hand, with a massive charged particle (e.g., the electron) coupled to exactly massless photons, we shall see that the definition of a conventional Fock space and associated S-matrix fails in a fundamental way: strictly speaking, the S-matrix *vanishes identically* in such a theory. Indeed, the single-particle pole(s) in amplitudes associated with incoming or outgoing charged particle(s) are softened to branch points, making the LSZ formalism useless for obtaining finite scattering amplitudes.

19.1 The infrared catastrophe in unbroken abelian gauge theory

We shall calculate the amplitude for creation and absorption of an arbitrary number of real photons by an external c-number conserved current $J_\mu(x)$, coupled to a quantized photon field $A_\mu(x)$. Real photon emission and absorption from this classical external current is best studied in the transverse, or Coulomb, gauge, with the photon field satisfying

$$\vec{\nabla} \cdot \vec{A}(x) = 0 \tag{19.1}$$

corresponding to the Fourier expansion (cf. (7.174)), choosing a real basis for the polarization vectors,

$$\vec{A}(x) = \int \frac{d^3k}{(2\pi)^{3/2}\sqrt{2E(k)}} \sum_{\lambda=1}^{2} (\vec{\epsilon}(\vec{k},\lambda)a(\vec{k},\lambda)e^{-ik\cdot x} + \vec{\epsilon}(\vec{k},\lambda)a^\dagger(\vec{k},\lambda)e^{ik\cdot x}) \tag{19.2}$$

with $E(k) = |\vec{k}|$ for massless photons. The polarization vectors $\vec{\epsilon}(\vec{k},\lambda)$ are transverse, i.e., $\vec{k} \cdot \vec{\epsilon}(\vec{k},\lambda) = 0$, thereby ensuring the Coulomb gauge condition. It will be convenient to re-express the photon field in a four-dimensional Fourier representation, by reintroducing the energy integral, with the mass-shell condition inserted via the usual δ-function:

$$\vec{A}(x) = \int \frac{d^4k}{(2\pi)^4}((2\pi)^{5/2}\sqrt{2E(k)} \sum_{\lambda=1}^{2} (\theta(k_0)\delta(k^2)\vec{\epsilon}(\vec{k},\lambda)a(\vec{k},\lambda)e^{-ik\cdot x} + \text{h.c.}) \tag{19.3}$$

The Heisenberg field equation for the theory is just the quantized version of Maxwell's equation,

$$\Box \vec{A}_H(x) = \vec{J}_{\text{tr}}(x), \quad \vec{\nabla} \cdot \vec{J}_{\text{tr}}(x) = 0 \tag{19.4}$$

where $\vec{J}_{\text{tr}}(x)$ is the transverse part of the full external current J_μ (an explicit example will be considered shortly), the rest of J_μ (or "longitudinal" part) being associated with the Coulombic flux carried along with the incoming and outgoing charged particle(s) of the classical current, but not involved in the generation or removal of real photons present at outgoing or incoming null (i.e., light-like) infinity. The prescribed c-number current $\vec{J}_{\text{tr}}(x)$ is real and transverse, and therefore has the Fourier expansion

$$\vec{J}_{\mathrm{tr}}(x) = \int \frac{d^4k}{(2\pi)^{5/2}} \sum_{\lambda=1}^{2} \vec{\epsilon}(\vec{k}, \lambda)(\tilde{J}(\vec{k}, \lambda)e^{-ik\cdot x} + \tilde{J}^*(\vec{k}, \lambda)e^{ik\cdot x})) \tag{19.5}$$

with $\tilde{J}(\vec{k}, \lambda)$ a c-number function of momentum for each polarization. The strange power of 2π is chosen for later convenience.

The field equation (19.4) can be solved subject to the asymptotic conditions which relate the Heisenberg field to the in- and out-fields in the far past and future (cf. (9.46))[1]

$$\vec{A}_H(\vec{x}, t) \to \vec{A}_{\mathrm{in}}(\vec{x}, t), \quad t \to -\infty$$

$$\vec{A}_H(\vec{x}, t) \to \vec{A}_{\mathrm{out}}(\vec{x}, t), \quad t \to +\infty \tag{19.6}$$

by introducing retarded and advanced Green functions for the $\Box$ operator in (19.4):

$$\Delta_R(x) \equiv -\int \frac{e^{-ik\cdot x}}{k^2 + i\epsilon k_0} \frac{d^4k}{(2\pi)^4}, \quad \Delta_R(x) = 0, \ x^0 < 0 \tag{19.7}$$

$$\Delta_A(x) \equiv -\int \frac{e^{-ik\cdot x}}{k^2 - i\epsilon k_0} \frac{d^4k}{(2\pi)^4}, \quad \Delta_R(x) = 0, \ x^0 > 0 \tag{19.8}$$

$$\Box\Delta_R(x) = \Box\Delta_A(x) = \delta^4(x) \tag{19.9}$$

For example, the retarded Green function $\Delta_R(x)$ vanishes for negative time coordinates $x^0 < 0$ as the denominator has two simple poles $k_0 = \pm|\vec{k}| - i\epsilon$ which are both below the real axis, allowing us to close the k_0 integration contour in the upper-half-plane for $x^0 < 0$, avoiding both poles and giving zero by Cauchy's theorem. A similar argument gives the corresponding advanced property for $\Delta_A(x)$. The Green function property (19.9) is obvious, as the infinitesimal displacement factors $\pm i\epsilon k_0$ become irrelevant once the d'Alembertian operator $\Box$ is applied, generating a $-k^2$ factor in the integrand and cancelling the poles. These Green functions immediately provide the solution of (19.4) subject to the boundary conditions (19.6):

$$\vec{A}_H(x) = \vec{A}_{\mathrm{in}}(x) + \int d^4y \Delta_R(x - y)\vec{J}_{\mathrm{tr}}(y) \tag{19.10}$$

$$\vec{A}_H(x) = \vec{A}_{\mathrm{out}}(x) + \int d^4y \Delta_A(x - y)\vec{J}_{\mathrm{tr}}(y) \tag{19.11}$$

Subtracting these two equations, we find

$$\vec{A}_{\mathrm{out}}(x) = \vec{A}_{\mathrm{in}}(x) + \int d^4y \Delta(x - y)\vec{J}_{\mathrm{tr}}(y), \quad \Delta(x) \equiv \Delta_R(x) - \Delta_A(x) \tag{19.12}$$

---

[1] The field renormalization constant Z appearing in the asymptotic condition, non-trivial in theories with fully quantized local field interactions, is simply unity in our case, where the photon field is coupled to a c-number source. This follows from the fact that the Heisenberg field and the associated asymptotic in- and out-fields differ by c-number terms (see (19.10, 19.11)), and therefore satisfy identical equal-time commutators, which then forces $Z = 1$. We remind the reader that the limits indicated in (19.6) are to be interpreted as weak limits—for matrix elements of the indicated, suitably smeared, operators (see Section 9.3).

Employing the familiar identity $\frac{1}{k^2+i\epsilon k_0} = P(\frac{1}{k^2}) - i\pi\epsilon(k_0)\delta(k^2)$ (with $\epsilon(k_0)$ the sign function, $\epsilon(k_0) = \theta(k_0) - \theta(-k_0)$), the Green function $\Delta(x)$ is found to have the Fourier representation

$$\Delta(x) = i \int e^{-ik\cdot x} \epsilon(k_0)\delta(k^2) \frac{d^4 k}{(2\pi)^3} \tag{19.13}$$

Taking the Fourier transform of (19.13), and using the Fourier representations (19.3) and (19.5) for the field and current, we find

$$a_{\text{out}}(\vec{k}, \lambda) = a_{\text{in}}(\vec{k}, \lambda) + \frac{i}{\sqrt{2E(k)}} \tilde{J}(\vec{k}, \lambda) \tag{19.14}$$

This explicit relation between the out- and in-annihilation operators also determines (up to an overall phase) the S-matrix for the theory (cf. (9.51)),

$$a_{\text{out}}(\vec{k}, \lambda) = S^{\dagger} a_{\text{in}}(\vec{k}, \lambda) S \tag{19.15}$$

with a similar relation connecting the creation operators $a_{\text{out}}^{\dagger}, a_{\text{in}}^{\dagger}$. The desired S-matrix operator is easily found if we recall the Baker–Campbell–Hausdorff formula $e^B A e^{-B} = A + [B, A]$, valid if $[B, A]$ is a c-number. If we take S as the *formally* unitary operator (exponential of an anti-hermitian operator)

$$S = \exp\left(i \int \frac{d^3 k}{\sqrt{2E(k)}} \sum_{\lambda=1}^{2} (\tilde{J}(\vec{k}, \lambda) a_{\text{in}}^{\dagger}(\vec{k}, \lambda) + \tilde{J}^*(\vec{k}, \lambda) a_{\text{in}}(\vec{k}, \lambda))\right) \tag{19.16}$$

then (19.15) follows immediately using the creation–annihilation commutator algebra $[a_{\text{in}}(\vec{k}, \lambda), a_{\text{in}}^{\dagger}(\vec{k}', \lambda')] = \delta_{\lambda\lambda'}\delta(\vec{k} - \vec{k}')$.

The expression just obtained for the scattering operator is more easily interpreted by re-expressing it in Wick ordered form, with all creation operators placed to the left of all destruction operators. This is easily achieved using the standard identity,

$$e^{A+B} = e^{-\frac{1}{2}[A,B]} e^A e^B \tag{19.17}$$

valid if the commutator $[A, B]$ is a c-number. Choosing A and B as the first and second terms in the exponent in (19.16), we find

$$S = C \exp\left(i \int \frac{d^3 k}{\sqrt{2E(k)}} \sum_{\lambda=1}^{2} \tilde{J}(\vec{k}, \lambda) a_{\text{in}}^{\dagger}(\vec{k}, \lambda)\right) \cdot \exp\left(i \int \frac{d^3 k}{\sqrt{2E(k)}} \sum_{\lambda=1}^{2} \tilde{J}^*(\vec{k}, \lambda) a_{\text{in}}(\vec{k}, \lambda)\right) \tag{19.18}$$

with the commutator contribution

$$C \equiv \exp\left(-\frac{1}{2} \int \frac{d^3 k}{2E(k)} \sum_{\lambda=1}^{2} |\tilde{J}(\vec{k}, \lambda)|^2\right) \tag{19.19}$$

Recalling our original definition of S in Section 9.1 (9.47), we see that the photon state produced by our classical current, given a photon vacuum in the asymptotic past, takes the explicit form

$$S|0\rangle_{\text{in}} = C \exp\left(i \int \frac{d^3 k}{\sqrt{2E(k)}} \sum_{\lambda=1}^{2} \tilde{J}(\vec{k}, \lambda) a_{\text{in}}^{\dagger}(\vec{k}, \lambda)\right)|0\rangle_{\text{in}} \qquad (19.20)$$

as the exponential in (19.18) containing purely destruction operators simply reduces to unity when acting on the in-vacuum. This result shows that the effect of a classical current coupled to the quantized photon field is simply to generate a coherent state of the photon field $\vec{A}$ (cf. Section 8.3, especially (8.51))—a linear superposition of multi-photon states with fixed phase and amplitude relations between the components of the state with different numbers of photons (determined by the specific charge current inducing the radiation).

The results (19.16, 19.20) represent the complete solution of the problem of quantized electromagnetic radiation from a classical source current. The formula for the S-matrix (19.16) looks plausible and, at first sight, completely unproblematic: we have an apparently unitary S-operator given in a simple and explicit form in terms of the (Fourier-transformed) classical current. In fact, as we shall now see, for any process involving radiation from an accelerated classical charged particle, this operator is actually *zero*! The problem arises from the apparently innocent normalization prefactor C given by (19.19), and in particular from the behavior of the integral in the infrared regime of small k, where the behavior $E(k) = |\vec{k}|$ for massless photons will, as we shall see shortly, lead to an infrared divergence of the integral at low momentum.

The problem is easily exposed if we consider the simplest situation in which a (classical) massive (mass m) charged (electric charge e) particle generates the c-number current source $J_\mu(x)$. We shall assume that the particle enters and leaves a bounded spacetime region where it undergoes a temporary acceleration, for simplicity as a consequence of some non-electromagnetic interaction. Outside this region the particle moves freely, approaching the interaction zone with four-momentum p and leaving it with four-momentum p' (see Fig. 19.1). The spacetime trajectory of the particle, given as a function of the particle's proper time τ, therefore satisfies

$$x^\mu(\tau) = \frac{p}{m}\tau, \quad \tau < \tau_-$$

$$x^\mu(\tau) = \frac{p'}{m}\tau, \quad \tau > \tau_+ \qquad (19.21)$$

where we shall also assume the trajectory to be smooth (infinitely τ-differentiable) in the interaction zone. The four-current $J^\mu(x)$ is just the charge density $e\delta^4(x - x^\mu(\tau))$ times the four-velocity $\frac{dx^\mu}{d\tau}$, integrated over the proper time of the trajectory,

$$J^\mu(x) = e \int d\tau \frac{dx^\mu}{d\tau} \delta^4(x - x^\mu(\tau)) \qquad (19.22)$$

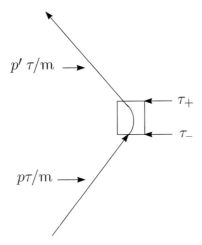

Fig. 19.1 spacetime trajectory of a classical charged particle undergoing a localized interaction.

with Fourier transform

$$\tilde{J}^\mu(k) = \frac{e}{(2\pi)^{3/2}} \int_{-\infty}^{+\infty} d\tau \frac{dx^\mu}{d\tau} e^{-ik\cdot x(\tau)} \tag{19.23}$$

where the unexpected power of 2π arises from our unconventional normalization of the Fourier transform in (19.5).

The low-momentum behavior of $\tilde{J}^\mu(k)$ will turn out to be singular, with the leading term determined by the asymptotic parts of the particle trajectory indicated in (19.21), as the interaction zone integral from τ_- to τ_+ is clearly perfectly finite as $k \to 0$. The contribution of the asymptotic portions to the integral are

$$\int_{-\infty}^{\tau_-} d\tau \frac{p^\mu}{m} e^{-ik\cdot p\tau/m} = \frac{p^\mu}{-ik\cdot p} e^{-ik\cdot p\tau_-/m} \to i\frac{p^\mu}{k\cdot p}, \text{ small } k \tag{19.24}$$

$$\int_{\tau_+}^{+\infty} d\tau \frac{p'^\mu}{m} e^{-ik\cdot p'\tau/m} = \frac{p'^\mu}{ik\cdot p'} e^{-ik\cdot p'\tau_+/m} \to -i\frac{p'^\mu}{k\cdot p'}, \text{ small } k \tag{19.25}$$

We conclude that the momentum-space current density in this situation must take the form

$$\tilde{J}^\mu(k) = \frac{ie}{(2\pi)^{3/2}} \left(\frac{p^\mu}{k\cdot p} - \frac{p'^\mu}{k\cdot p'} \right) \mathcal{M}(k) \tag{19.26}$$

where the residual amplitude $\mathcal{M}(k)$ satisfies (a) $\mathcal{M}(k) \to 1, k \to 0$, and (b) $\mathcal{M}(k)$ vanishes faster than any power of k for large k (as a consequence of the smoothness of the trajectory in the interaction zone). Note that the current conservation property $\partial_\mu J^\mu = 0 \Rightarrow k_\mu \tilde{J}^\mu(k) = 0$ is automatically satisfied by (19.26). The full momentum-space current in (19.26) can be decomposed into a longitudinal and transverse part

$$\tilde{J}^\mu(k) = k^\mu \tilde{J}_l(k) + \tilde{J}^\mu_{\text{tr}}(k) = k^\mu \tilde{J}_l(k) + \sum_{\lambda=1}^{2} \epsilon^\mu(\vec{k}, \lambda) \tilde{J}(\vec{k}, \lambda) \qquad (19.27)$$

where the Coulomb gauge polarization vectors satisfy $\epsilon^0(\vec{k}, \lambda) = 0$, $\vec{k} \cdot \vec{\epsilon}\,(\vec{k}, \lambda) = 0$, and therefore $k_\mu \epsilon^\mu(\vec{k}, \lambda) = 0$. The functions $\tilde{J}(\vec{k}, \lambda)$ are just (up to uninteresting overall constants) the objects introduced earlier in (19.5). The current conservation property $k_\mu \tilde{J}^\mu(k) = 0$ then follows directly as a consequence of the photon mass-shell condition $k^2 = 0$. We also have (again employing $k^2 = 0$, together with (19.26))

$$\tilde{J}^\mu(k) \tilde{J}^*_\mu(k) = -\sum_{\lambda=1}^{2} |\tilde{J}(\vec{k}, \lambda)|^2 = \frac{e^2}{(2\pi)^3}\left(\frac{p}{k \cdot p} - \frac{p'}{k \cdot p'}\right)^2 |\mathcal{M}(k)|^2 \qquad (19.28)$$

Note that the four-vector square here is *negative* because only the space-like transverse part survives. Returning to our expression for the S-operator, (19.18), we see that the exponent in the prefactor C displayed in (19.19) is given, for massless photons with $E(k) = |\vec{k}|$, and with the current satisfying (19.28), by an infrared (logarithmically) divergent integral, resulting in the vanishing of C, and thence, the S-operator itself (remember the implicit minus sign in the four-vector square!),

$$C = \exp\left(\frac{e^2}{2(2\pi)^3}\int \frac{d^3k}{2E(k)}\frac{1}{|\vec{k}|^2}\left(\frac{p}{E(p) - \hat{k}\cdot\vec{p}} - \frac{p'}{E(p') - \hat{k}\cdot\vec{p}'}\right)^2 |\mathcal{M}(k)|^2\right)$$

$$\sim \exp\left(-\infty\right) = 0 \qquad (19.29)$$

where for the photon $E(k) = |\vec{k}|$, while for the massive charged particle $E(p) = \sqrt{\vec{p}^2 + m^2}$, etc. Note that the integral is cut off at the upper end by the rapid falloff of the residual amplitude $\mathcal{M}(k)$: the problem is entirely at the infrared (low-momentum) end. Indeed, if we regulate the infrared divergence by temporarily inserting a photon mass, so that $E(k) = \sqrt{k^2 + m_\gamma^2}$, and cut the integral off at some momentum Λ (below which the amplitude $\mathcal{M}(k)$ may be approximated by unity), the integral in the exponent in (19.29) becomes

$$\langle\langle\left(\frac{p}{E(p) - \hat{n}\cdot\vec{p}} - \frac{p'}{E(p') - \hat{n}\cdot\vec{p}'}\right)^2\rangle\rangle_{\hat{n}} \int^{|\vec{k}|=\Lambda} \frac{d^3k}{2|\vec{k}|^2\sqrt{k^2 + m_\gamma^2}} \qquad (19.30)$$

where the quantity in angle brackets corresponds to an angular average over directions of the unit vector $\hat{n}$. The integral on the right is logarithmically divergent in the limit of vanishing photon mass:

$$\int^{|\vec{k}|=\Lambda} \frac{d^3k}{2|\vec{k}|^2\sqrt{k^2 + m_\gamma^2}} = 2\pi \int_0^\Lambda \frac{dk}{\sqrt{k^2 + m_\gamma^2}} \sim 2\pi \log\frac{\Lambda}{m_\gamma}, \quad m_\gamma \to 0 \qquad (19.31)$$

Note that while the infrared divergence manifests itself as a zero in the S-operator once we sum to all powers of the electric charge e, normal perturbation theory corresponds to an expansion in powers of e, in which case the divergence appears as logarithms of the photon mass—one for each power of the fine structure constant $\alpha \propto e^2$—and therefore an independent logarithmic divergence at each order of perturbation theory in any specific S-matrix element involving a definite number of incoming and outgoing photons. The only way to avoid this "infrared catastrophe" is, as we see clearly in (19.30), to take $p = p'$, and prevent our charged particle from receiving any transfer of momentum, however small!

The physical interpretation of these results is actually extremely simple. Unlike the situation for massless pions in a chirally symmetric theory (cf. Section 16.6), where emission of low-momentum pions is suppressed by powers of the low momentum, there is no penalty in quantum electrodynamics to the emission of low-momentum photons. Moreover, the emission of a massless photon with arbitrarily low momentum incurs an arbitrarily low-energy cost, and therefore we should hardly be surprised if the *slightest* momentum shift of a charged particle induces the emission of a very large number of extremely soft photons.

The result of this proliferation of emitted photons, as we have just seen, at least for the simplified case of a classical charged particle acting as the source, is that the coherent photon state thereby produced contains so many multi-photon states with arbitrarily many soft photons that the exclusive amplitude for our charged particle to emit *any definite finite number* of photons simply vanishes. In particular, the vacuum-persistence-amplitude $_{\text{out}}\langle 0|0\rangle_{\text{in}} = C$ itself vanishes. The situation is somewhat analogous to that discussed previously in the context of Haag's theorem (Section 10.5), in that we have unitarily inequivalent spaces, but in this case, not between the physical in- and out-asymptotic spaces and the computationally convenient (for perturbation theory) but physically dispensable interaction-picture space, but between the in- and out-spaces themselves, which by asymptotic completeness we have come to regard as identical to each other and to the basic physical Hilbert space of the theory in any "sensible" field theory.

The situation we are here encountering clearly suggests at the conceptual level a more serious disease than any we have previously uncovered in our studies of local field theory, and in fact it must be admitted that after much intense investigation there does not appear to be any way to resuscitate the concept of a normal separable Fock space with unitarily equivalent asymptotic spaces in an unbroken abelian gauge theory like QED, with massless photons coupled to charged particles. Nevertheless, we hasten to assure the depressed reader that a cure is at hand, even though it requires the abandonment of scattering amplitudes like the S-matrix as the fundamental phenomenological object of the theory, and the return to a direct evaluation of only carefully defined and directly *measurable* quantities. This "Bloch–Nordsieck resolution" of the infrared catastrophe of QED was already proposed in the 1930s, before the advent of modern covariant quantum electrodynamics a decade later, and will be the subject of the following section.

There may be some concern that the results we have obtained are subject to the restriction that while our photons are described in a fully quantum mechanical context, the charged source is classical and that perhaps this "hybrid" treatment is in some way introducing inconsistencies into the theory, resulting in the evaporation of our beloved S-matrix. In fact, essentially identical results are obtained in the fully quantized version of the theory. We shall briefly explain how this works, referring the reader to the beautiful article of Weinberg (Weinberg, 1965) for the combinatoric details needed to obtain the final result. Consider a process in QED in which a single incoming electron scatters off an arbitrary set of other particles, which for simplicity we take to be themselves uncharged (an example might be Compton scattering, where the electron scatters off a single hard photon). The basic process is indicated in Fig. 19.2, where we consider the effect of emission of a single soft photon, carrying four-momentum k (much smaller than all other momentum scales in the process), from either the incoming (Fig. 19.2(a)) or outgoing (Fig. 19.2(b)) electron. In the limit $k \to 0$, both of these amplitudes are found to diverge linearly. In the former case, for example, the amplitude is proportional to

$$\bar{u}(p', \sigma') \mathcal{M}(k) \frac{\not{p} - \not{k} + m}{(p-k)^2 - m^2} ie\gamma^\mu u(p, \sigma)$$

$$= ie\bar{u}(p', \sigma') \mathcal{M}(k) \frac{\gamma^\mu(-\not{p} + m) + 2p^\mu - \not{k}\gamma^\mu}{-2p \cdot k + k^2} u(p, \sigma)$$

$$\sim (-ie \frac{p^\mu}{p \cdot k}) \bar{u}(p', \sigma') \mathcal{M}(k) u(p, \sigma), \quad k \to 0 \tag{19.32}$$

Here the amplitude $\mathcal{M}(k)$ simply represents the "core" of the diagram, where the relevant momentum scales are much higher than k, so that for small k we may simply assume that it approaches some (for our purposes) uninteresting constant. The vector index μ of the emitted photon (associated with the vertex factor $ie\gamma^\mu$) is, of course, to be contracted with a polarization vector ϵ_μ in the event that the photon is a real one appearing in the final state, or with the corresponding vector index of an absorbed photon if it ends up being a virtual photon. The initial on-mass-shell spinor satisfies the Dirac equation $(\not{p} - m)u(p, \sigma) = 0$, and we have employed the Dirac algebra $\not{p}\gamma^\mu = 2p^\mu - \gamma^\mu\not{p}$, and neglected higher powers of k, in arriving at (19.32). The corresponding emission from an outgoing line in Fig. 19.2(b) produces similarly, in the small k limit,

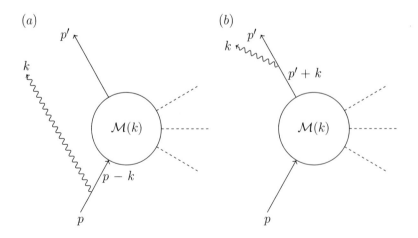

Fig. 19.2 Soft photon emission from a charged fermion.

a factorized amplitude with the dependence on the soft photon momentum isolated in a linearly divergent prefactor,

$$\bar{u}(p',\sigma')ie\gamma^\mu \frac{p\!\!\!/ + k\!\!\!/ + m}{(p'+k)^2 - m^2}\mathcal{M}(k)u(p,\sigma) \sim (+ie\frac{p'^\mu}{p'\cdot k})\bar{u}(p',\sigma')\mathcal{M}(k)u(p,\sigma) \quad (19.33)$$

Combining these results, we see that the emission of a single soft photon from the charged electron traversing the process produces a divergent prefactor which is identical (up to a sign) to our Fourier-transformed classical current (19.26):

$$\mathcal{M}_{\text{emit one photon}}(k) \sim -ie(\frac{p^\mu}{k\cdot p} - \frac{p'^\mu}{k\cdot p'})\bar{u}(p',\sigma')\mathcal{M}(k)u(p,\sigma) \quad (19.34)$$

The absorption of a photon on either the incoming or outgoing electron line leads to an exactly similar result, with a change of sign (as in this case we have $k \to -k$). A single virtual photon exchange requires both of these factors, together with the virtual photon propagator $-ig_{\mu\nu}/(k^2 + i\epsilon)$, and an integration over the photon four-momentum. Putting in the appropriate coupling and combinatoric factors, one finds that the low-momentum contribution arising from single virtual photon exchange results in the modification of the uncorrected core amplitude $\mathcal{M}$ by a multiplicative factor

$$J = i\frac{e^2}{2}\int \frac{d^4k}{(2\pi)^4}\frac{1}{k^2 + i\epsilon}(\frac{p}{k\cdot p} - \frac{p'}{k\cdot p'})^2 \quad (19.35)$$

while multiple virtual photon exchanges simply exponentiate this result leading to an overall amplitude in which soft photon effects appear as an exponential prefactor,

$$\mathcal{M}_{\text{virtual photon exchanges}} \sim \exp(J)\mathcal{M}(e=0) \quad (19.36)$$

The real suppression factor (19.29) previously obtained in our semiclassical analysis corresponds to the absolute magnitude $|\exp(J)| = \exp(\text{Re}(J))$.[2] One finds that the real part arises effectively from the mass-shell δ-function in the virtual photon propagator $1/(k^2 + i\epsilon) = \mathcal{P}(1/k^2) - i\pi\delta(k^2)$, leaving us with the three-dimensional d^3k integral visible in (19.29) (recall that in the classical problem, $\mathcal{M}(k) = 1$ in the infrared region where the factorization of the soft-photon effects is valid). The reader is referred to the previously mentioned article of Weinberg's for a full discussion of the combinatorics of the soft photon effects, and a careful analysis of the integral appearing in (19.35).

The annoying "evaporation" of S-matrix amplitudes as a consequence of the exponentiation of infrared divergences appearing in S-matrix amplitudes at finite orders of perturbation theory is a symptom of deep structural problems in the formulation of the physical state space of a theory like QED with massless gauge particles appearing in the asymptotic states of the theory. In fact, the phenomenon is even present in some low-dimensional non-gauge theories—the classic example being the model of a

---

[2] There is also a divergent phase factor, arising from the imaginary part of J. This is associated with divergences familiar from careful treatments of Coulomb scattering in non-relativistic quantum mechanics: see (Weinberg, 1965), Section V.

massless boson in two spacetime dimensions derivatively coupled to a massive Dirac fermion initially studied in this context by Schroer (Schroer, 1963). The Lagrangian is just that studied earlier in Chapter 12 (cf. (12.52)) in connection with the appearance of seagulls and Schwinger terms in a derivatively coupled theory, but in two dimensions and with the boson mass $m = 0$ (also, the γ_5 corresponding to a pseudoscalar boson is unimportant and can be dropped). The derivative of the massless boson field, $\partial_\mu \phi$, acts as a stand-in for the photon field in four dimensions, and one easily sees (see Problem 1) that virtual photon exchanges in processes analogous to Fig. 19.2 give rise to logarithmic infrared divergences arising from loop integrals behaving for small k like $\int d^2k/k^2$. In this model, Schroer was able to show that the fermion propagator in momentum space has a branch point rather than a simple pole in the on-mass-shell limit, $\tilde{S}_F(p) \sim (p^2 - M^2)^{-1 + g^2/2\pi}$, where g is the coupling constant in the derivative coupling term (see (12.52)). The softening of the pole structure means that any literal application of the LSZ formula will clearly result in a vanishing of S-matrix elements involving an on-mass-shell Dirac fermion in either the initial or final state.

But this embarassment is merely the most obvious manifestation of much deeper problems in the Fock-space formulation of field theory on which we have relied throughout this book. Indeed, the very general arguments of Section 9.5 (leading to the *Källen–Lehmann* representation) show that the existence of a simple pole in the two-point function of a massive field theory follows from very general principles (single particle representation of the Poincaré group, unitarity, etc.) which we would certainly be loath to abandon. Massive particles suffering the indignity of loss of their single particle pole in Green functions of the associated field have been dubbed *infraparticles* by Schroer, and it has been demonstrated rigorously (Buchholz, 1986) that the same situation obtains in any field theory with an exact local Gauss's Law, in particular, in four-dimensional quantum electrodynamics. In the latter case, the situation is made even messier by the gauge-variance of the electron propagator: in different gauges, the precise form of the "smearing out" of the simple pole singularity of the free electron field is dependent on the gauge used to quantize the theory.

A careful analysis of the mathematical structure of the state space in quantum electrodynamics leads to some dramatic—indeed, disturbing—conclusions,[3] to wit:

1. As a consequence of the above-cited Buchholz theorem (following from the existence of a local Gauss's Law), the mass-squared operator $P_\mu P^\mu$ lacks a discrete eigenvalue, with a corresponding normalizable one-electron state, at the value corresponding to the squared electron mass.

2. Single-particle electron states with differing momentum fall into unitarily inequivalent spaces: in particular, matrix elements of all local operators of the theory (and not just of the S-operator, as discussed above) vanish between such states. This unitary inequivalence is far more consequential than the example we have already encountered with Haag's theorem, where the unitarily inequivalent spaces concern the absence of well-defined unitary operators connecting the asymptotic physical in- (or out-)spaces of the theory to the computationally useful (in perturbation theory) but physically unmeaningful interaction-picture states. Here the

---

[3] For a recent review of the situation, see the discussion of Haag in (Haag, 1992), Sections VI.2, VI.3.

inequivalence is such that we are unable to even construct well-defined normalizable single-electron wave-packets as the electron states of different momentum lie in inequivalent sectors. Again, the unbroken abelian symmetry giving rise to an exact Gauss's Law is the underlying culprit: each charged particle state is associated with an asymptotic Coulomb flux,[4] with the form of the flux depending on the velocity of the charged particle. The local fields of the theory can be shown to leave the asymptotic flux unchanged, resulting in a non-denumerable infinity of inequivalent sectors of the full Hilbert space (which thereby becomes non-separable) characterized by different values of the charged particle velocity (and therefore, asymptotic flux). The presence of a non-denumerable infinity of charged particle sectors (distinguished by a superselection rule associated with the charged particle momentum) can also be regarded as a consequence of a spontaneous symmetry-breaking of the Lorentz boost symmetry.(Fröhlich *et al.*, 1979)

3. The usual Haag–Ruelle scattering theory (cf. Section 9.3) which depends heavily on the existence of a mass gap, and normalizable single-particle states, becomes inapplicable for infraparticle amplitudes. Despite many efforts to reinstate the concept of a well-defined S-matrix, effectively by considering charged particle states "dressed" with a coherent cloud of infinitely many photons, a physically transparent and computationally practical scheme for dealing with transition amplitudes of charged states in QED remains an unrealized goal. The construction of suitable asymptotic Maxwell fields, on the other hand, is not particularly problematic (Buchholz, 1982): indeed, one finds that there are no infrared divergences in the Feynman amplitudes associated with processes (such as light-by-light scattering) in which only electrically neutral photons, and no electrons, appear in the initial and final states. The S-matrix in the charge-free sector is therefore (subject to the usual need for ultraviolet renormalization) perfectly well defined.

The conceptual difficulties which infect the construction of a rigorous framework for charged-particle scattering in quantum electrodynamics might seem to run directly in the face of the stunningly precise phenomenological successes of this theory. Fortunately, we shall see that a careful analysis of the conditions under which actual measurements can be performed in quantum electrodynamic processes provides a secure pathway out of the quagmire. Just as the difficulties imposed by Haag's theorem can be evaded by avoidance of an ill-defined formal intermediary (the interaction picture for the unregulated theory), either by fully regulating the theory or by non-perturbative methods, the disconcerting evaporation of a well-defined S-matrix can also be circumvented once we formulate properly the inescapably *inclusive* transition probabilities which correspond to the actually measured quantities in any feasible experiment. Admittedly, the abandonment of the S-matrix as the central phenomenological object incorporating the sum total of available information concerning the scattering physics

---

[4] In our classical current model this flux has its origin in the longitudinal part of the current density (19.27). In a fully quantized theory the asymptotic flux can be shown to commute with all local operators and therefore to be a c-number: see (Buchholz, 1982).

of the theory seems at first a radical step. Nevertheless, the elimination of the infrared catastrophe provided by this "Bloch–Nordsieck" resolution, to which we now turn, provides the basis for the unambiguous calculation of the quantum electrodynamic component of essentially all high-energy processes, the vast majority of which involve charged particles in the initial or final state, in modern particle physics.

19.2 The Bloch–Nordsieck resolution

The physical origin of the infrared problems of a theory with exactly massless photons was clarified already in 1937, in the seminal paper of Bloch and Nordsieck (Bloch and Nordsieck, 1937), even prior to the development of the fully covariant perturbative formalism for quantum electrodynamics in the late 1940s. Massless photons of arbitrarily low momentum (and therefore energy) couple with equal strength to a charged source, unlike the situation with derivatively coupled massless pions in a chiral effective Lagrangian (cf. Section 16.6), where the emission or absorption amplitude for a pion vanishes linearly with the momentum. The result, as we have discussed in the previous section, is that the slightest acceleration of a charged source results in the radiation of an infinite number of very soft (low-energy) photons. This "cloud" of soft photons is the inevitable concomitant of any process in which an electrically charged particle undergoes any change of momentum, however small. On the other hand, actual measurements of quantum electrodynamic processes employ photodetectors with a finite resolution, which cannot register soft photons below a certain minimum energy Δ. Thus, measurements involving charged particles are inevitably *inclusive*: transition probabilities should be computed summing over all final states compatible with the trigger limitations of the apparatus used.

A simple example will suffice to illustrate the general situation. Let us imagine that we are interested in the probability of emission of a single "hard" photon, of momentum $\vec{q}$ and polarization λ, with $|\vec{q}| > \Delta$, from a charged particle undergoing a scattering (and therefore a change of momentum). We shall again resort to the analytically solvable semiclassical model of the preceding section, so that our charged particle is represented by a classical c-number current density, with transverse part as given in (19.5). The hard photon is registered by our photodetector irrespective of the presence of an arbitrary number n of additional soft photons with momenta $\vec{k}_1, \vec{k}_2, .., \vec{k}_n$ with $|\vec{k}_i| < \Delta$ which are emitted by the particle but remain unregistered by the detector. The total probability of a single trigger of the detector by the indicated hard photon is therefore, allowing for the emission of precisely n soft photons, so to speak, "flying below the radar" of our photodetector,

$$P_n(\vec{q}, \lambda; \Delta) = \frac{1}{n!} \int_{|\vec{k}_i| < \Delta} d^3k_1 \cdot \cdot d^3k_n \sum_{\lambda_i} |\text{out}\langle \vec{q}\lambda, \vec{k}_1\lambda_1, ..\vec{k}_n\lambda_n|0\rangle_{\text{in}}|^2 \qquad (19.37)$$

where the $1/n!$ factor takes into account multiple counting of identical soft photon states (by Bose symmetry). We may convert the out-state appearing in the matrix element to an in-state by introducing the scattering operator S, as in (9.47), where in our case S is given explicitly by (19.18):

$$P_n(\vec{q}, \lambda; \Delta) = |C|^2 \frac{1}{n!} \int_{|\vec{k}_i| < \Delta} d^3 k_1 \cdots d^3 k_n \sum_{\lambda_i}$$

$$\cdot \,_{\text{in}} \langle \vec{q} \lambda, \vec{k}_1 \lambda_1, .. \vec{k}_n \lambda_n | \frac{i^{n+1}}{n+1!} \{ \int \frac{d^3 k}{\sqrt{2E(k)}} \sum_\lambda \tilde{J}(\vec{k}, \lambda) a^\dagger_{\text{in}}(\vec{k}, \lambda) \}^{n+1} |0\rangle_{\text{in}} |^2 \quad (19.38)$$

The destruction operators in the exponential on the right in (19.18) act on the in-vacuum, and the exponential therefore reduces to unity, while the left exponential can be expanded as shown in (19.38), with only the term involving $n+1$ creation operators surviving. C is the vacuum persistence amplitude (amplitude for emission of no photons) given in (19.19). Any one of the $n+1$ creation operators can be used to the left to remove the single distinguished hard photon, giving an overall factor of $(n+1)\tilde{J}(\vec{q}, \lambda)/\sqrt{2E(q)}$ in the matrix element. Taking this outside the soft photon integrals, we find

$$P_n(\vec{q}, \lambda; \Delta) = |C|^2 \frac{|\tilde{J}(\vec{q}, \lambda)|^2}{2E(q)} \frac{1}{n!} \int_{|\vec{k}_i| < \Delta} d^3 k_1 \cdots d^3 k_n \sum_{\lambda_i} |_{\text{in}} \langle \vec{k}_1 \lambda_1, .. \vec{k}_n \lambda_n | \frac{(\mathcal{A}^\dagger_J(\Delta))^n}{n!} |0\rangle_{\text{in}} |^2$$

$$\mathcal{A}_J(\Delta) \equiv \int_{|\vec{k}| < \Delta} \frac{d^3 k}{\sqrt{2E(k)}} \sum_\lambda \tilde{J}^*(\vec{k}, \lambda) a_{\text{in}}(\vec{k}, \lambda) \quad (19.39)$$

where we are allowed to restrict the integral over $\vec{k}$ in the n-particle creation operator to the soft-momentum regime $|\vec{k}| < \Delta$, as the only photons present in the final state are now the soft ones. The restriction to soft momenta in the momentum integrals $d^3 k_1 \cdots d^3 k_n$ can now be relaxed, as the matrix element for any particle with momentum $|\vec{k}_i| > \Delta$ vanishes given the restriction to soft momenta in the creation operator $\mathcal{A}^\dagger_J(\Delta)$. Moreover, the sum over all n-particle states implied by these momentum integrals can be expanded to a complete set (by including states with $m \neq n$ photons) as the additional states manifestly have vanishing matrix elements to the vacuum of the indicated n-particle creation operator. Recalling the completeness relation (5.22) for a multi-particle bosonic Fock space, the sum over n-particle states can be augmented to a complete set of in-states:

$$P_n(\vec{q}, \lambda; \Delta) = |C|^2 \frac{|\tilde{J}(\vec{q}, \lambda)|^2}{2E(q)} \sum_\alpha |_{\text{in}} \langle \alpha | \frac{(\mathcal{A}^\dagger_J(\Delta))^n}{n!} |0\rangle_{\text{in}} |^2 \quad (19.40)$$

$$= |C|^2 \frac{|\tilde{J}(\vec{q}, \lambda)|^2}{2E(q)} \frac{1}{(n!)^2} \,_{\text{in}} \langle 0 | (\mathcal{A}_J(\Delta))^n (\mathcal{A}^\dagger_J(\Delta))^n |0\rangle_{\text{in}} \quad (19.41)$$

The matrix element in (19.41) is easily evaluated (see Problem 2),

$$_{\text{in}} \langle 0 | (\mathcal{A}_J(\Delta))^n (\mathcal{A}^\dagger_J(\Delta))^n |0\rangle_{\text{in}} = n! (\int_{|\vec{k}| < \Delta} \frac{d^3 k}{2E(k)} \sum_\lambda |\tilde{J}(\vec{k}, \lambda)|^2)^n \quad (19.42)$$

The reader will recall (cf. (19.19)) that the vacuum persistence (i.e., no-photon-emission) factor C is given by

$$C = \exp\left(-\frac{1}{2}\int \frac{d^3 k}{2E(k)} \sum_{\lambda} |\tilde{J}(\vec{k},\lambda)|^2\right) \tag{19.43}$$

with the integral in the exponent logarithmically divergent in the infrared (in the massless photon limit) for currents corresponding to a charged particle undergoing a change of momentum, thereby resulting in the vanishing of C. For any finite n, the probability $P_n(\vec{q},\lambda;\Delta)$ in (19.41) therefore also vanishes for massless photons, as the matrix element multiplying $|C|^2$ contains a finite power of this same infrared divergence, as we see from (19.42). On the other hand, if we calculate, as previously argued, the inclusive probability allowing for emission of arbitrarily many soft photons, we find that the divergence in the integral at small k is exactly cancelled, giving a finite probability for the detection of a single hard photon (momentum $\vec{q}$ and polarization λ) by a detector of finite resolution Δ:

$$P_{\text{tot}}(\vec{q},\lambda;\Delta) \equiv \sum_{n=0}^{\infty} P_n(\vec{q},\lambda;\Delta) = \frac{|\tilde{J}(\vec{q},\lambda)|^2}{2E(q)} \exp\left(-\int_{|\vec{k}|>\Delta} \frac{d^3 k}{2E(k)} \sum_{\lambda} |\tilde{J}(\vec{k},\lambda)|^2\right) \tag{19.44}$$

The infrared divergence in the integral in the exponent is now effectively cutoff by the detector resolution Δ: in fact, the exponent turns out to contain logarithms of the form $\ln(q^2/\Delta^2)$ with currents of the form (19.28) and $q = p - p'$. In other words, transition probabilities, and more generally all types of measurable cross-sections, for quantum electrodynamics processes can be expected to depend in an important, but fortunately calculable, way on the sensitivity of the measurement apparatus to the "haze" of low-energy photons which are inevitably present.

We have been illustrating the essential nature of the infrared problem in quantum electrodynamics with the aid of a semiclassical model in which the source current is treated classically (but with a quantized Maxwell field), and taking full advantage of a delicious property—complete analytic solvability—of this model. In particular, we have not needed to resort to perturbative approximations, as our results contain the exact emission probabilities to all orders in the particle charge e (which is hidden in the current $\tilde{J}(\vec{k},\lambda)$- cf. (19.26)). In the fully quantized version (QED) of quantum electrodynamics, in which the charged particle fields are also treated quantum mechanically, we must of course resort to perturbation theory.

Before describing the Bloch–Nordsieck resolution in QED proper, it is useful to take a look at the cancellation of infrared divergences visible in (19.40–19.44), from the point of view of a perturbative expansion in the squared charge, or fine-structure constant $\alpha = e^2/4\pi$, as the cancellations occurring in the fully quantized theory arise in a completely analogous way. We shall work to order α^2, or e^4, recalling that $\tilde{J} \sim O(e)$, and that the leading term in P_{tot} is of order α, as we insist on the emission of a single hard photon of momentum $\vec{q}$. It is clear from (19.39) that the n-soft-photon emission probability is of order α^{n+1}, so to order α^2 the total transition probability P_{tot}, which we already know to be infrared finite for finite detector resolution Δ, is given by just the contributions from $n = 0$ and $n = 1$:

$$P_{\text{tot}} = P_0 + P_1 + O(\alpha^3)$$

$$= \frac{|\tilde{J}(\vec{q}, \lambda)|^2}{2E(q)} \{1 - \int \frac{d^3k}{2E(k)} \sum_\lambda |\tilde{J}(\vec{k}, \lambda)|^2) + \int_{|\vec{k}|<\Delta} \frac{d^3k}{2E(k)} \sum_\lambda |\tilde{J}(\vec{k}, \lambda)|^2\} + O(\alpha^3)$$

$$(19.45)$$

The first two terms in the curly braces in (19.45) arise from the expansion (to order α) of the no-photon-emission probability $|C|^2$: in particular, the infrared divergent integral in the second term corresponds to the emission and reabsorption of a single virtual photon (of arbitrary momentum) on the charged-particle line, accompanying, of course, the hard photon emission described by the overall prefactor. The third term, also an infrared divergent integral (cut off on the ultraviolet end by the detector resolution Δ), corresponds to the total probability for the emission of a single undetected soft photon. We see that the cancellation in the infrared divergence between the two integrals appearing in (19.45) amounts to a cancellation in the total probability P_{tot} between infrared divergences arising from virtual photon contributions and real photon emission terms. If we introduce a photon mass m_γ to separately regularize each of the integrals in the infrared, the singular $\ln(m_\gamma)$ dependence in each integral evidently cancels exactly between virtual and real photon emission terms, at each order of the perturbative expansion in α, once the finite detector resolution is properly taken into account.

Returning now to the fully quantized version of quantum electrodynamics, one finds that precisely the same mechanism operates to produce well-defined transition probabilities (or cross-sections) once photon detector resolutions are taken into account, once again by cancellation between virtual and real photon contributions. The detailed analysis, which we must here omit for considerations of space, can be found in the seminal paper of Yennie, Frautschi, and Suura (Yennie *et al.*, 1961). However, the mechanism of the cancellation can be indicated with a simple example. Fig. 19.3 shows the low-order Feynman graphs contributing to the emission of a hard photon

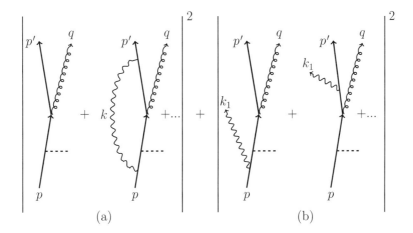

Fig. 19.3 Contributions to hard photon emission from an electron in QED (through $O(\alpha^2)$).

of momentum q (indicated by the spiral line) from a charged particle which has had its momentum altered[5] by an interaction indicated by the dashed line (which for simplicity we may take to be of non-electromagnetic character). To lowest order in the particle charge e (order $\alpha \sim e^2$ for the cross-section, or the squared amplitude), only the single hard photon need be taken into account, as in the left graph in Fig. 19.3(a), but to the next order we must take into account the possibility of a virtual photon exchange, as in the right graph in Fig. 19.3(a), or an additional emission of a real soft photon (momentum k_1) below the detector threshold, as in Fig. 19.3(b).

Once again one finds that the infrared divergence in the virtual photon diagram (actually, in the interference term obtained by squaring the amplitude indicated in Fig. 19.3(a)) cancels exactly with an infrared divergence in the real photon emission graphs of Fig. 19.3(b). The proof that this cancellation is effective for general processes, to all orders of perturbation theory, can be found in the aforecited paper of Yennie *et al.*

The essential point which we wish to emphasize here is that a careful specification of the limitations of any measurement process in a situation involving exactly massless abelian gauge particles automatically leads to well-defined *finite* transition probabilities and cross-sections once the measurement process is carefully specified, even if the intermediate quantities (S-matrix amplitudes) which we normally rely on in field theory have a singular structure in the zero mass limit. The formal difficulties (infrared divergences at finite orders of perturbation theory, vanishing of the S-matrix when the amplitudes are summed to all orders) appear because of mathematically convenient idealizations in the theoretical formulation which do not correspond to physical reality: specifically, the propagation of particles in a Minkowski space of infinite spatial volume, and the existence of detectors of infinitely precise resolution.

For example, in a finite spatial volume the momentum integrals become discretized sums, with a natural infrared cutoff of order the inverse spatial size of the "box". Thus, unlike the situation discussed above, where the average number of photons emitted (into infinite volume) by an accelerated charge is infinite, the average number of photons *per unit volume* in blackbody radiation is perfectly finite, as the reader may easily confirm by integrating $\rho(\nu, T)/h\nu$, with the energy density $\rho(\nu, T)$ given by the Planck formula (1.22), over all photon frequencies ν. Of course, if we consider an infinite volume box, the total number of photons is again infinite. Inasmuch as the formulation of a scattering theory typically presupposes the asymptotic propagation of incoming and outgoing particles through an infinite spatial volume, it is not surprising that we encounter formal difficulties due to the concomitant appearance of infinitely many very soft photons of arbitrarily long wavelength and correspondingly low energy.

19.3   Unbroken non-abelian gauge theory: confinement

Imagine a fictional physicist, fully informed of the basic relativistic and quantum theoretical frameworks underlying modern physics, but encountering for the first time the experimental discoveries of the past century. Two phenomena in particular stand out in their capacity to provoke astonishment—and even, at first sight, disbelief. The

---

[5] The classical term for this sort of process is *Bremsstrahlung*—German for "braking radiation".

first—already well established experimentally by the second decade of the twentieth century—was the phenomenon of superconductivity. The sudden collapse of electrical resistivity (to unmeasurably low values) in mercury cooled to 4.2 K must have seemed, frankly, miraculous at the time. As it turned out, a correct microscopic interpretation of superconductivity would have to wait until the mid 1950s, with the BCS theory developed by Bardeen, Cooper, and Schrieffer.

The second "miraculous" phenomenon was the discovery of quark confinement. In this case we are talking more about a gradual process of deepening understanding rather than a singular event of discovery. The notion that the observed hadronic particles and resonances could, at least from a group-theoretical point of view, be easily interpreted by regarding these objects as composites of fractionally charged "quarks", as proposed in the mid 1960s by Gell'Mann and Zweig, seemed at first a purely technical convenience, but the discovery of point-like substructures in deep-inelastic scattering from protons soon (by the early 1970s) gave rise to the quark-parton model, identifying the kinematical objects of Gell'Mann and Zweig with actual particle constituents of the observed hadrons, now thought of as bound states of the quarks/partons.

Unfortunately, there was soon overwhelming experimental evidence that the putative quark constituents of hadrons simply never appeared as isolated (fractionally charged!) objects in the final states of even the most energetic hadronic collisions. This in turn gave rise to the hypothesis of "quark confinement": basically, the appearance in asymptotic states of hadronic scattering processes of particles with the quantum numbers of quarks was simply excluded by fiat. Physically, this could be "explained" by the assumption that the separation of an isolated quark from other hadronic matter required an infinite (or at least, extremely large) amount of energy. Such a phenomenon certainly runs counter to the deep-seated intuition inherent in all local quantum field theories that far separated objects have negligible influence on each other: exponentially decreasing with distance for theories with only massive particles, or like an inverse power of the separation for massless theories. As we shall see, the two "miracles" of twentieth-century physics mentioned here—superconductivity and quark confinement—in fact share deep similarities in their underlying mechanisms.

From a formal point of view our apparent inability to observe isolated quarks can be re-expressed simply as the statement that the S-matrix vanishes for in- or out-states in which particles with quantum numbers of quarks appear. Indeed, we have just encountered a superficially similar circumstance, in the case of quantum electrodynamics in four spacetime dimensions, where the infrared divergences associated with the copious emission and absorption of low-energy photons sum to give a vanishing result for exclusive S-matrix amplitudes involving scattering of charged particles (which, for an abelian gauge theory, means the matter fields, e.g., in the case of QED, electrons or positrons). In the abelian case, however, we have seen that this vanishing does not imply the inability to produce asymptotic isolated electron states (for example), but only the impossibility that such charged particles can be treated in isolation from the inescapable low-energy "photon cloud" surrounding them.

It was realized very early—immediately following the surge in interest in non-abelian gauge theories following 't Hooft's proof of their renormalizability in 1971 and their subsequent employment (in the symmetry broken phase) in the development of the electroweak Standard Model—that the infrared divergences of unbroken (massless) non-abelian gauge theories are even more ferocious than those encountered in the abelian case. The reason is simple: in the non-abelian case, the massless gauge vector particles ("gluons") are themselves charged under the gauge symmetry group of the theory, so the emission of soft gluons occurs from gluons themselves, as well as from the fermionic matter fields (quarks). The self-interaction of gluons leads to a highly non-trivial interacting theory even if the matter fields are decoupled (say, by making them infinitely massive), unlike the situation in QED. In fact, the similarity between the long-distance structure of unbroken abelian and non-abelian gauge theories is only superficial: the underlying dynamics leads to dramatically different physics in the two cases, as we shall see.

The adoption in the mid 1970s of an unbroken non-abelian gauge theory as the most promising candidate for a field-theoretic description of the dynamics underlying hadronic processes came with the realization that the theory would have to provide a mechanism not just for eliminating the fermionic quark fields (transforming under the fundamental representation of the gauge group) from the asymptotic spectrum, but also the massless gauge fields (gluons, transforming under the adjoint representation), which were, of course, nowhere in evidence experimentally. So the obvious need for "quark confinement" was expanded into the more general requirement of "color confinement", which excludes all objects with non-vanishing charge under the non-abelian symmetry (referred to as "color" to avoid confusion with electric charge) from the asymptotic spectrum. From the point of view of a Wightman formulation of the theory, the physical states of the theory are obtained by application of gauge-invariant operators to the vacuum, which are the only local fields which can act as interpolating fields (i.e., possess non-vanishing vacuum to single particle matrix elements) for the stable particles appearing in the asymptotic spectrum. On the other hand, the underlying dynamics of the theory is specified in terms of a simple Lagrangian (15.112) in which only non-gauge-invariant quark and gluon fields appear.[6] The local or almost local fields describing observable stable hadronic particles (for pure QCD, based on a SU(3) gauge group, with the weak interactions switched off, these include the nucleon and pion fields) then correspond to composite, gauge-invariant fields built from the underlying quark and gluon fields.

The phenomenon of confinement of a charge associated with a local gauge symmetry is actually not limited to non-abelian theories, as we can see if we examine field theories in one or two spatial dimensions. The simplest case is provided by the analog of quantum electrodynamics in two spacetime dimensions, comprising the class of theories with action given by (in Minkowski space)

$$\mathcal{I} = \int d^2x \{ -\frac{1}{4} F_{\mu\nu}(x) F^{\mu\nu}(x) + \bar{\psi}(x)(i\slashed{D} - m)\psi(x) \}, \quad D_\mu \equiv \partial_\mu + ie A_\mu(x) \quad (19.46)$$

Two special cases in this class are of particular importance for the present discussion, corresponding to the fermionic field (which, in virtue of the similarities of the model

---

[6] A review of the discussion of the general relation between particles and fields provided in Section 9.6 may be useful at this point.

to four-dimensional QCD, we shall dub the "quark" field) being either (a) extremely massive, $m >> e$ (note that in two spacetime dimensions the gauge field is dimensionless and the charge coupling constant e has dimensions of mass, as required by a dimensionless action), or (b) massless $m = 0$—the famous "Schwinger model". First, note that in one spatial dimension, gauge-fixing to axial gauge A_1 (equivalent in this case to the transverse or Coulomb gauge $\partial_i A_i = \partial_1 A_1 = 0$) leaves only the auxiliary, non-dynamical A_0 field, responsible for the static Coulomb interaction, with the Green function (in one space dimension)

$$-\nabla^2 V(\vec{x}) = \delta(\vec{x}) \Rightarrow -\frac{\partial^2}{\partial x_1^2} V(x_1) = \delta(x_1) \Rightarrow V(x_1) = -\frac{1}{2}|x_1| \qquad (19.47)$$

leading to a full Hamiltonian (after elimination of A_0, see Problem 3) consisting of the usual free "quark" Hamiltonian plus a Coulomb energy contribution (cf. 2.53)[7]:

$$H_{\mathrm{coul}} = -\frac{1}{4} \int \rho(x_1, t)|x_1 - x_1'|\rho(x_1', t)dx_1 dx_1' \qquad (19.48)$$

with the charge density given by $\rho(x) = e\bar{\psi}(x)\gamma_0\psi(x)$. There are no transverse degrees of freedom in one space dimension, so real "photons" are absent in this model: the entire physics induced by the gauge field is incorporated in the Coulomb interaction (in Coulomb gauge). Even classically, this theory confines charge, as we see that the Coulomb potential grows linearly with the separation of charges: thus if we set $\rho(x_1) = +Q\delta(x_1) - Q\delta(x_1 - L)$, the static Coulomb energy of the opposite charged pair is $\frac{1}{2}Q^2L$, so an infinite amount of energy would be required to completely isolate either charge from the other. The result is easily understood from Gauss's Law: the electric flux leaves the $+Q$ charge with magnitude Q and energy density $Q^2/2$ and travels directly to the $-Q$ charge along the only available spatial axis, giving a total electrostatic energy $Q^2L/2$.

This is in contrast to the situation in three space dimensions, of course, where the static electric flux originating on a charge spreads out throughout the ambient three-dimensional volume, decreasing the energy density and giving rise to an electrostatic interaction energy of separated charges falling inversely with their separation. In the present case, the asymptotic spectrum cannot contain increasingly far-separated "quarks" (with no intervening charged particles) without incurring an arbitrarily large energy penalty.

On the other hand, we expect, in the limit of very heavy "quarks" ($m >> e$), to find non-relativistic bound quark–antiquark states (analogous to the "onium" mesons of QCD) of zero total charge. In fact, the only stable particles in the theory correspond to neutral bosons, which can undergo non-trivial scatterings. As we decrease the mass m relative to the coupling e, eventually reaching the "strong-coupling" regime of $e >> m$, the stable bosons of the theory become, somewhat paradoxically, *weakly* coupled (Coleman, 1976), and in the exactly massless limit for the quark (the original

---

[7] As pointed out by Coleman, the physics of the massive model is enriched in an interesting way by allowing for an external electric field, in which case additional terms appear in the Hamiltonian. Here we set this field to zero. See (Coleman, 1976) for a detailed discussion of the general case.

"Schwinger model") the spectrum of the theory collapses to that of a single *free*, neutral, massive boson (with mass $= e/\sqrt{\pi}$). In fact, the gauge-invariant operators of the theory can all be re-expressed in terms of a scalar field ϕ, in terms of which the Hamiltonian density reads $\mathcal{H} = \frac{1}{2}(\pi_\phi^2 + (\partial_1\phi)^2 + e^2\phi^2/\pi)$.

The linear form of the Coulomb potential in the Schwinger model is, of course, a kinematical consequence of the single spatial dimension available for the spread of electric flux. In two spatial dimensions the Coulomb potential (Green function of the Laplacian) grows logarithmically with distance, still providing charge confinement in the abelian case, although a "weaker" form than in one spatial dimension, while in three space dimensions we have the usual $1/r$ falloff, allowing us to isolate charged particles from one another, although not, as we have seen, from the ever present "cloud" of infrared low-energy photons. In non-abelian models, on the other hand, there are strong arguments to believe that *linear* confinement persists even in two or three space dimensions, as a consequence of the very non-trivial self-interacting dynamics of the gauge fields of the theory. In the remainder of this chapter we shall see how the methods of lattice gauge theory can be used to provide both analytic and numerical support for this hypothesis.

As in the case of spontaneous symmetry-breaking, the physics of confinement primarily concerns the long-distance properties of the theory, and we may therefore expect that the details of the theory at very short-distance scales are unimportant, as long as we take care to regularize the theory in a way that does not do violence to those features of the theory that are intimately connected with the long-distance phenomenon of interest. In our case, the features in question are those related directly to the local gauge symmetry of the Lagrangian, which we take to be unbroken in the Lagrangian, and with the associated remaining global symmetry (after gauge-fixing) preserved by the vacuum of the theory (in other words, we are not in a Higgs phase of the theory where the gauge fields of the theory are screened by a vacuum condensate of charged fields).

In addition, we shall work in Euclidean space, as it is easy to formulate a simple and direct criterion for confinement (or non-confinement) of matter fields in any particular representation of the gauge group in an imaginary-time formulation, as we shall soon see. The regularization of the theory at short distances will be performed by working on a four-dimensional hypercubic lattice with a large but finite number of points L in each (Euclidean) spacetime dimension, with a lattice spacing a separating nearest neighbors in each spacetime direction (see Fig. 19.4, where a small section of the lattice, in the μ, ν plane, is shown). We shall assume for definiteness that we are dealing with a single unbroken gauge group SU(N) (the abelian case U(1) can be treated in a completely analogous fashion), with the dynamics specified by a Euclidean functional integral, as in (15.161–15.164), for the continuum theory.

The matter fields of the theory will be identified with field variables localized on the sites of the lattice, which we will label with bold-faced Roman letters $\mathbf{n}, \mathbf{m}$, etc. Thus, a bosonic scalar (resp. Dirac) field in the fundamental representation will be specified at location $\mathbf{n}$ on the lattice as $\phi_\mathbf{n}$ (resp. $\psi_\mathbf{n}$), where the gauge group "color" index (and, in the fermion case, Dirac) indices are suppressed. The continuum vector gauge fields $A_{\alpha\mu}(x)$ of the theory lying in the adjoint representation (thus, the index $\alpha = 1, 2, ..., N^2 - 1$) are encoded in an $N \times N$ matrix field chosen to simplify the task

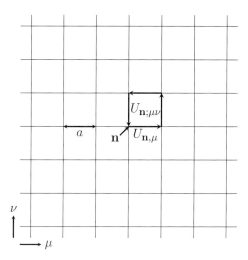

Fig. 19.4 A slice through a Euclidean hypercubic lattice supporting a lattice gauge field.

of constructing a gauge-invariant discretized action, as follows. Recall that the adjoint gauge fields can be conveniently "packed" into a single matrix field (cf. (15.101)):

$$A_\mu \equiv t_\alpha A_{\alpha\mu} \tag{19.49}$$

where the t_α are a set of hermitian generators of SU(N), normalized by $\mathrm{Tr}(t_\alpha t_\beta) = \delta_{\alpha\beta}$. We shall also slightly change the notation previously used for finite gauge transformations: thus, a finite local gauge transformation of a (e.g., scalar) matter field in the fundamental representation of SU(N) will be denoted (cf. (15.96))

$$\phi(x) \to \phi'(x) \equiv \Lambda(x)\phi(x), \quad \Lambda(x) \in \mathrm{SU}(N) \tag{19.50}$$

while the gauge field appearing in (19.49) transforms like (cf. (15.105))

$$A_\mu(x) \to A_\mu^\Lambda(x) = \Lambda(x)A_\mu(x)\Lambda^\dagger(x) + \frac{i}{g}(\partial_\mu\Lambda(x))\Lambda^\dagger(x) \tag{19.51}$$

The symbol U, formerly used for the local gauge transformations now denoted by Λ in (19.50), will instead be used to denote the *parallel gauge transporter*, which is defined, for the infinitesimal path $(x + dx, x)$ corresponding to the straight segment from x to $x + dx$, to be the transformation

$$U(x + dx, x) \equiv e^{-igA_\mu dx_\mu} = 1 - igA_\mu dx_\mu \tag{19.52}$$

One sees immediately, working to first order in dx_μ,

$$\Lambda(x + dx)U(x + dx, x)\Lambda^\dagger(x) = \Lambda(x + dx)(1 - igA_\mu dx_\mu)\Lambda^\dagger(x)$$
$$= (\Lambda(x) + \partial_\mu\Lambda(x)dx_\mu)(1 - igA_\mu dx_\mu)\Lambda^\dagger(x)$$
$$= 1 + dx_\mu((\partial_\mu\Lambda(x))\Lambda^\dagger(x) - ig\Lambda(x)A_\mu\Lambda^\dagger(x))$$

$$= 1 - ig(\Lambda(x)A_\mu\Lambda^\dagger(x) + \frac{i}{g}(\partial_\mu\Lambda(x))\Lambda^\dagger(x))dx_\mu$$

$$= 1 - igA_\mu^\Lambda(x)dx_\mu \equiv U^\Lambda(x+dx,x) \tag{19.53}$$

that the transporter $U(x+dx,x)$ serves to "shift" a matter field localized at point x to one transforming under the local gauge transformation appropriate for point $x+dx$:

$$U(x+dx,x)\phi(x) \to U^\Lambda(x+dx,x)\Lambda(x)\phi(x) = \Lambda(x+dx)U(x+dx,x)\phi(x) \tag{19.54}$$

The parallel transport property for the infinitesimal path $(x+dx,x)$ generalizes in an obvious way to finite paths specified by some continuous contour $\mathcal{C}_{a\to b}$ from spacetime point a to point b, as we may simply divide the path into infinitesimal segments and form the path-ordered product[8] of $U(x+dx,x)$ transporters to obtain a finite transporter

$$U(\mathcal{C}_{a\to b}) = \mathrm{P}\exp\left\{-ig\int_{\mathcal{C}_{a\to b}} A_\mu dx_\mu\right\} \tag{19.55}$$

transforming like

$$U(\mathcal{C}_{a\to b}) \to U^\Lambda(\mathcal{C}_{a\to b}) = \Lambda(b)U(\mathcal{C}_{a\to b})\Lambda^\dagger(a) \tag{19.56}$$

For a closed path, we end up with a transporter transforming covariantly under the adjoint representation of the gauge group,

$$U(\mathcal{C}_{a\to a}) \to \Lambda(a)U(\mathcal{C}_{a\to b})\Lambda^\dagger(a) \tag{19.57}$$

from which it follows (from $\Lambda^\dagger(a)\Lambda(a) = 1$) that the trace of any closed-path transporter is *gauge-invariant*:

$$\mathrm{Tr}(U(\mathcal{C}_{a\to a})) \to \mathrm{Tr}(\Lambda(a)U(\mathcal{C}_{a\to a})\Lambda^\dagger(a)) = \mathrm{Tr}(U(\mathcal{C}_{a\to a})) \tag{19.58}$$

For the special case of straight line contours, the parallel transporter satisfies a familiar type of first-order equation, analogous to the equation (4.20) satisfied by the time-ordered interaction-picture evolution operator (4.28). Suppose the contour is just a straight line path in the Euclidean "time" (fourth) direction, from the spacetime point $y \equiv (\vec{x}, y_4)$ to the point $x \equiv (\vec{x}, x_4)$, where $x_4 > y_4$. Then we have

$$\frac{\partial}{\partial x_4}U(\mathcal{C}_{y\to x}) = \lim_{\Delta x_4 \to 0}\frac{1}{\Delta x_4}(e^{-igA_4(\vec{x},x_4)\Delta x_4} - 1)U(\mathcal{C}_{y\to x})$$

$$= -igA_4(\vec{x},x_4)U(\mathcal{C}_{y\to x}) \tag{19.59}$$

Of particular interest in the discretized lattice version of the theory are the parallel transporters corresponding to links connecting nearest neighbor sites on the lattice. Thus, we define (see Fig. 19.4) $U_{\mathbf{n},\mu}$ as the parallel transporter for the path $(\mathbf{n} + a\hat{\mu}, \mathbf{n})$

---

[8] The finite path-ordered product is defined, analogously to the time-ordered product of (4.27), by expanding the exponential and ordering the gauge-field factors in each term so that "later" fields along the path are placed to the left.

extending from site **n** in the positive μ direction by one lattice spacing. As a consequence of (19.56), this object transforms under local SU(N) gauge transformations, specified by assigning an SU(N) element $\Lambda_{\mathbf{n}}$ to each lattice site **n**, as follows:

$$U_{\mathbf{n},\mu} \rightarrow \Lambda_{\mathbf{n}+a\hat{\mu}} U_{\mathbf{n},\mu} \Lambda_{\mathbf{n}}^{\dagger} \tag{19.60}$$

The smallest closed path on the lattice corresponds to a *plaquette*, or a square one lattice spacing on a side, corresponding to a path $\mathbf{n} \rightarrow \mathbf{n} + a\hat{\mu} \rightarrow \mathbf{n} + a\hat{\mu} + a\hat{\nu} \rightarrow \mathbf{n} + a\hat{\nu} \rightarrow \mathbf{n}$ (see Fig. 19.4). The corresponding transporter, $U_{\mathbf{n};\mu\nu}$, $\mu < \nu$, is given explicitly by

$$
\begin{aligned}
U_{\mathbf{n};\mu\nu} &= U_{\mathbf{n},\nu}^{\dagger} U_{\mathbf{n}+a\hat{\nu},\mu}^{\dagger} U_{\mathbf{n}+a\hat{\mu},\nu} U_{\mathbf{n},\mu} \\
&= e^{igaA_\nu(\mathbf{n})} e^{igaA_\mu(\mathbf{n}+a\hat{\nu})} e^{-igaA_\nu(\mathbf{n}+a\hat{\mu})} e^{-igaA_\mu(\mathbf{n})}
\end{aligned}
\tag{19.61}
$$

For the time being we shall assume that we are dealing with smooth classical fields, so that $aA_\mu(\mathbf{n} + a\hat{\nu})$ may be approximated in the exponent by $a(A_\mu(\mathbf{n}) + a\partial_\nu A_\mu(\mathbf{n}))$, neglecting terms of order a^3. The exponentials can be combined using a Baker–Campbell–Hausdorff formula

$$e^{aX} e^{aY} = e^{aX + aY + \frac{1}{2}a^2[X,Y] + O(a^3)} \tag{19.62}$$

and a short calculation (see Problem 4) then gives

$$U_{\mathbf{n};\mu\nu} = e^{-iga^2 F_{\mu\nu} + O(a^3)} \tag{19.63}$$

with $F_{\mu\nu}$ the $N \times N$ hermitian matrix field strength tensor (cf. (15.108))

$$F_{\mu\nu} = \partial_\mu A_\nu - \partial_\nu A_\mu + ig[A_\mu, A_\nu] = t_\alpha F_{\alpha\mu\nu} \tag{19.64}$$

As expected, the closed-path plaquette variable $U_{\mathbf{n};\mu\nu}$ inherits the covariant adjoint transformation property (19.57) from the field-strength tensor which has the same transformation behavior. The trace of this quantity is easily seen to be exactly invariant under the full set of local lattice gauge transformations specified by (19.60). Moreover, the real part of the trace of the plaquette transporter can be expanded for small lattice spacing, giving

$$\text{Re Tr}(U_{\mathbf{n};\mu\nu}) = \frac{1}{2}\text{Tr}(U_{\mathbf{n};\mu\nu} + U_{\mathbf{n};\mu\nu}^{\dagger}) = \text{Tr}(1) - \frac{1}{2}g^2 a^4 \text{Tr}(F_{\mu\nu}(\mathbf{n}) F_{\mu\nu}(\mathbf{n})) + O(a^5) \tag{19.65}$$

Note that the trace of the term linear in the exponent vanishes, as the exponent must be anti-hermitian ($U_{\mathbf{n};\mu\nu}$ is unitary). The second term on the right is nothing but the usual pure gauge Lagrangian density (15.111) (in Euclidean space, so there are no raised indices). The higher terms in (19.65), of order a^5 or higher, correspond by dimensional analysis to operators of dimension 5 or higher—exactly the ones which in a cutoff effective Lagrangian correspond to "irrelevant" operators, to which the low-energy physics should be insensitive, as we saw in Chapter 16. The continuum gauge action corresponding to (15.111)) (in Euclidean space) becomes, after a naive discretization on a hypercubic lattice,

$$S_{\text{gauge}} = \frac{1}{4} \int \text{Tr}(F_{\mu\nu} F_{\mu\nu}) d^4 x \to \sum_{\mathbf{n}} \frac{1}{4} a^4 \text{Tr}(F_{\mu\nu}(\mathbf{n}) F_{\mu\nu}(\mathbf{n})) \tag{19.66}$$

Comparing this with (19.65), we see that the usual continuum action, once regulated on a lattice, corresponds up to irrelevant (dimension 5 and higher) operators with the *Wilson lattice action* (Wilson, 1974)

$$S_{\text{Wils,latt}} = \beta \sum_{\mathbf{n},\mu<\nu} (1 - \frac{1}{N} \text{Re } \text{Tr}(U_{\mathbf{n};\mu\nu})), \quad \beta \equiv \frac{N}{g(a)^2} \tag{19.67}$$

The coupling constant $g(a)$ here refers to a dimensionless parameter appearing in an effective, *cutoff* action functional: the presence of a spacetime hypercubic lattice with spacing a between nearest neighbor sites implies that the Fourier momentum modes of the fields are cutoff at a high momentum $\Lambda \sim \frac{\pi}{a}$, so in the language of Sections 16.4 and 17.4, the value chosen for $g(a)$ (or equivalently, $g(\Lambda)$) must be chosen to "flow" as the lattice spacing a is taken to zero (to recover a continuum theory), or as the UV cutoff Λ is taken to infinity, while holding some suitable set of low-energy amplitudes fixed. We expect from the asymptotic freedom property (Section 18.3) of unbroken non-abelian gauge theories that this will require $g(a)$ to *vanish* logarithmically as the lattice spacing is sent to zero. In other words, the continuum limit of the theory is approached by taking β in (19.67) *large*, and then examining the correlation of observables over larger and larger separations in lattice units (corresponding to a fixed separation in physical units). We have a sensible continuum limit if a uniquely specified rescaling of lattice to physical units *simultaneously* results in well-defined and non-trivial (i.e., not simply free field value) continuum limits for all the distinct gauge-invariant observables of the theory.

We may remark here that abelian gauge theories based on a U(1) gauge group can be regularized gauge-invariantly on a lattice by a completely analogous procedure. In this case, the link variables are unimodular complex numbers, $U_{\mathbf{n},\mu} = \exp(i\theta_{\mathbf{n},\mu})$, $-\pi < \theta_{\mathbf{n},\mu} < +\pi$, with plaquette angles $\theta_{\mathbf{n};\mu\nu}$ constructed by summing the four angles around an elementary square. One may take, in analogy to the non-abelian Wilson action (19.67),

$$S_{\text{Wils,latt}} = \beta \sum_{\mathbf{n},\mu<\nu} (1 - \cos(\theta_{\mathbf{n};\mu\nu})) \tag{19.68}$$

The continuum limit again corresponds to taking the coupling β large, which forces the path integral to concentrate in the region of small $\theta_{\mathbf{n};\mu\nu}$. The specific choice of periodic function used here is to a large extent a matter of convenience: different functions with the same Gaussian behavior for small plaquette angles correspond to effective Lagrangians at the UV cutoff scale differing by higher-dimension operators (i.e., higher powers of $\theta_{\mathbf{n};\mu\nu} \sim F_{\mu\nu}$) which we expect to be in the same universality class (cf. Section 17.4) as the theory defined by the Wilson action, say. For example, a very useful choice for analytic computations is the *Villain U(1) action*:

$$S_{\text{Vill,latt}} = \sum_{\mathbf{n},\mu<\nu} S_{\text{Vill}}(\theta_{\mathbf{n};\mu\nu}) \tag{19.69}$$

$$S_{\text{Vill}}(\theta) \equiv -\log \sum_{m=-\infty}^{+\infty} \exp\{-\frac{\beta}{2}(\theta - 2m\pi)^2\} \tag{19.70}$$

The Wilson and Villain actions (for $\beta = 5$) are displayed in Fig. 19.5: it is apparent that the small θ behavior is identical; we are free to use either as our regularized version of the abelian gauge theory. Similarly, a non-abelian Villain lattice action (say, for the gauge group SU(2)) can be defined by choosing the single plaquette action:

$$S_{\text{Vill}}(U_{\mathbf{n};\mu\nu}) = -\log \sum_{m=-\infty}^{+\infty} \exp\{-\frac{\beta}{2}(\arccos(\frac{1}{2}\text{Tr}(U_{\mathbf{n};\mu\nu})) - 2m\pi)^2\} \tag{19.71}$$

The addition of charged (fundamental representation) scalar matter fields $\phi_{\mathbf{n}}$ local-ized on lattice sites is also straightforward: one may easily verify (Problem 5) that the discrete gauge-invariant actions $\sum_{\mathbf{n}\mu}(\phi_{\mathbf{n}+\hat{\mu}}^* U_{\mathbf{n},\mu}\phi_{\mathbf{n}} + \text{c.c.})$ and $\sum_{\mathbf{n}} P(\phi_{\mathbf{n}}^*\phi_{\mathbf{n}})$ (with P an at most quadratic polynomial) can be linearly combined to give a matter lattice action $S_{\text{matter,latt}}$ containing the most general set of gauge-invariant relevant and marginal continuum operators. The inclusion of charged fermionic fields is somewhat trickier, as a consequence of the infamous "doubling" problem, whereby naive discretizations of the usual Dirac action lead to a superfluity (by a power of two) of the fermionic degrees of freedom (see (Montvay and Münster, 1994), Section 4.2). We shall not treat this subject here, as we shall be concerned only with the confinement question in the static (infinite quark mass) limit, where the problem can be circumvented.

The Euclidean path integral for the lattice discretized gauge theory can now be written down directly. For a theory with scalar matter site fields $\phi_{\mathbf{n}}$ interacting with the gauge link fields $U_{\mathbf{n},\mu}$, the partition function of the theory is given by

$$Z_{\text{latt}} = \int \prod_{\mathbf{n}} d\phi_{\mathbf{n}} \prod_{\mathbf{n}\mu} dU_{\mathbf{n},\mu} e^{-(S_{\text{Wils,latt}}+S_{\text{matter,latt}})} \tag{19.72}$$

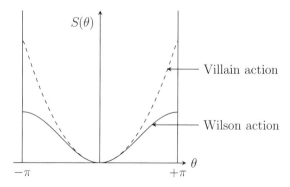

Fig. 19.5 Comparison of Wilson and Villain actions for a U(1) lattice gauge theory.

where the integrals over the link variables $U_{\mathbf{n},\mu} \in \mathrm{SU}(N)$ are the usual Hurwitz measure ones (cf. (15.140)). Note that the local gauge group is now compact, as it is simply the direct product of a finite number of independent $\mathrm{SU}(N)$ groups acting on each lattice site. The problem of an infinite gauge group volume which plagued the continuum formulation of the theory, and required the insertion of a gauge-fixing prescription to provide unambiguous finite results for correlation functions computed from the functional integral, has simply disappeared. We may therefore perform an unrestricted integration over the link variables $U_{\mathbf{n},\mu}$, provided the observables $\mathcal{O}[\phi_{\mathbf{n}}, U_{\mathbf{n},\mu}]$ being averaged in the functional integral are themselves gauge-invariant:

$$\langle \mathcal{O} \rangle = \int \prod_{\mathbf{n}} d\phi_{\mathbf{n}} \prod_{\mathbf{n}\mu} dU_{\mathbf{n},\mu} \mathcal{O}[\phi_{\mathbf{n}}, U_{\mathbf{n},\mu}] e^{-(S_{\mathrm{Wils,latt}} + S_{\mathrm{matter,latt}})}/Z_{\mathrm{latt}} \qquad (19.73)$$

with gauge-invariance requiring

$$\mathcal{O}[\phi_{\mathbf{n}}, U_{\mathbf{n},\mu}] = \mathcal{O}[\Lambda_{\mathbf{n}}\phi_{\mathbf{n}}, \Lambda_{\mathbf{n}+a\hat{\mu}}U_{\mathbf{n},\mu}\Lambda_{\mathbf{n}}^{\dagger}] \qquad (19.74)$$

With this non-gauge-fixed formulation, correlation functions of gauge-variant objects are not useful, and in fact, frequently vanish. The correlation function of two distinct link variables, for example, $\langle U_{\mathbf{n},\nu}U_{\mathbf{m},\mu} \rangle$, automatically vanishes as the integration over all link variables "includes" an averaging over local gauge transformations of the two link variables (during which the action exponent is constant), and the Hurwitz integral over gauge transformations Λ of a single link variable U shifted by Λ will vanish, as $\int d\Lambda(\Lambda U) = 0$. Thus, we are unable to compute a non-zero gauge field propagator in this approach. Likewise, the charged matter field Euclidean propagator, defined by the correlation function $\langle \phi_{\mathbf{n}}\phi_{\mathbf{m}}^{*} \rangle$, will automatically vanish for $\mathbf{n} \neq \mathbf{m}$.

The physical questions we ask of the theory must therefore be expressed in terms of explicitly gauge-invariant operators. For example, a quark–antiquark pair, obtained in standard perturbation theory by applying the bilocal operator $\bar{\psi}(x)\psi(y)$ to the vacuum, must be obtained here by the application to the vacuum of an almost local operator $\bar{\psi}(x)U(\mathcal{C}_{y \to x})\psi(y)$, in which the two quark fields are connected by a parallel transporter ("Wilson line") along some continuous path from y to x, and which is exactly gauge-invariant, using (19.56). The specific choice of path will not be important, as we will be interested below in an asymptotic argument in which the initial quark–antiquark state, however constituted at the outset, is allowed to rearrange itself into a compatible state (of equal conserved quantum numbers) with minimal energy by evolving the system over a long Euclidean time. In the Schwinger model discussed earlier, the physical significance of the Wilson line is clear: it represents the "string" of electric flux connecting the oppositely charged particles and giving rise to the linear dependence of energy with separation.

We wish to study the question of the static energy of a quark–antiquark pair in a non-abelian gauge theory, where the static condition is enforced by taking the quark mass M very large (effectively infinite). The pair is inserted into the vacuum at Euclidean time $x_4 = 0$ and removed at Euclidean time $x_4 = T$. Here, by "quark" we mean a spin-$\frac{1}{2}$ fermionic field transforming under the fundamental representation of the gauge group, which for definiteness we take to be $\mathrm{SU}(N)$. A suitable almost local gauge-invariant operator to perform the insertion is $\bar{\psi}(\vec{x}, 0)U(\mathcal{C}_{(\vec{y},0) \to (\vec{x},0)})\psi(\vec{y}, 0)$, while the removal is accomplished by $\bar{\psi}(\vec{y},T)U(\mathcal{C}_{(\vec{x},T)\to(\vec{y},T)})\psi(\vec{x},T)$ (see Fig. 19.6), where the contours are chosen for simplicity to be straight line spatial paths connecting the locations $\vec{x}$ and $\vec{y}$ at fixed time 0 or T.

We also assume, without explicitly indicating this in the notation, that the quark and antiquark (although of equal mass M) are of different "flavors"—i.e., one is not the antiquark of the other—to eliminate the possibility of mutual annihilation (into pure gauge energy). Instead, the two heavy objects are forced to propagate over a large Euclidean time T, after which they are removed from the system. The Euclidean amplitude for this process, written for the time being in the continuum theory, is represented schematically (ignoring overall normalization, gauge-fixing issues, etc.) by the functional integral

$$
\langle \bar{\psi}(c)U(\mathcal{C}_{b\to c})\psi(b)\bar{\psi}(a)U(\mathcal{C}_{d\to a})\psi(d)\rangle
$$

$$
= \int \mathbf{D}A_{\alpha\mu}\mathbf{D}\psi\mathbf{D}\bar{\psi}\; \bar{\psi}(b)U(\mathcal{C}_{b\to c})\psi(b)\bar{\psi}(a)U(\mathcal{C}_{d\to a})\psi(d)e^{-(S_{\text{gauge}}+S_{\text{matter}})}
$$

$$
= \int \mathbf{D}A_{\alpha\mu}\text{Tr}(S_E(d,c;A)U(\mathcal{C}_{b\to c})S_E(b,a;A)U(\mathcal{C}_{d\to a}))e^{-S_{\text{gauge}}} \tag{19.75}
$$

where a,b,c,d are the spacetime points $(\vec{x},0), (\vec{x},T), (\vec{y},T), (\vec{y},0)$, as indicated (super-imposed on a spacetime lattice) in Fig. 19.6. In the final line the integral over the fermionic quark fields has been performed for each fixed gauge field in the remaining $\mathbf{D}A_{\alpha\mu}$ functional integral. Thus, the function $S_E(b,a;A)$, for example, is the Euclidean Dirac propagator for the massive quark propagating in the background classical gauge field $A_{\alpha\mu}$. The trace appearing in (19.75) is over gauge group (i.e., fundamental representation) indices, and we have ignored an irrelevant overall minus sign arising from permuting the Grassmann quark fields.

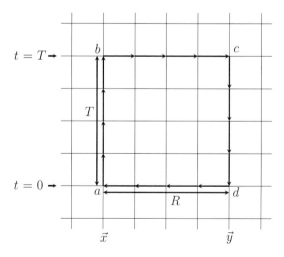

Fig. 19.6 Path corresponding to the Wilson loop observable $W(R,T)$.

We have already seen in Chapter 11 (see the discussion following (11.50)) that the exchange of transverse gauge particles between charged Dirac fermions is suppressed in the limit where the fermion mass(es) are taken large, and that moreover, in the extreme static limit (with the mass taken to infinity), the spatial momentum dependence of the propagator and the coupling to spatial gauge fields disappears entirely, plausibly enough, as an infinitely massive particle is insensitive to the transfer of any finite amount of momentum, and, in the absence of acceleration, cannot radiate or absorb real gauge quanta.

In the static limit therefore, the Euclidean propagator for our quarks is just the Green function for the Euclidean Dirac operator appearing in (15.162), but with the spatial derivatives and spatial components of the gauge field set to zero, leaving only the fourth (Euclidean "time") components

$$-i\not{D} + M = -i\hat{\gamma}_\mu(-i\partial_\mu + gt_\alpha A_{\alpha\mu}) + M \to -\hat{\gamma}_4 D_4 + M, \quad D_4 \equiv \partial_4 + igA_4 \quad (19.76)$$

where we have reverted in the final expression to the use of the matrix gauge field $A_4 = t_\alpha A_{\alpha 4}$, and removed a factor of the coupling constant from the field to maintain consistency with our notation throughout this section. Thus, our static propagator in a background continuum field satisfies

$$(-\hat{\gamma}_4 D_4 + M)S_E(x, y; A) = \delta^4(x - y) \quad (19.77)$$

Using (19.59), we may write down the solution to (19.77) (see Problem 6)

$$S_E(x, y; A) = e^{-M|x_4 - y_4|}\delta^3(\vec{x} - \vec{y})(P_+\theta(y_4 - x_4) + P_-\theta(x_4 - y_4))U(\mathcal{C}_{(\vec{y},y_4)\to(\vec{x},x_4)}) \quad (19.78)$$

with $P_\pm \equiv (1 \pm \hat{\gamma}_4)/2$ the projection operators appropriate for quark and antiquark propagation (in the static limit). Inserting (19.78) into (19.75) we see that the Euclidean propagation amplitude for our quark–antiquark pair is proportional (using cyclicity of the trace) to the *Wilson loop* variable $W(R, T) = \mathrm{Tr}(U_{\mathcal{C}_{a\to b\to c\to d\to a}}) \equiv \mathrm{Tr}(U_{\mathcal{C}(R,T)})$ corresponding to the gauge-invariant trace of the closed rectangular contour displayed in Fig. 19.6, averaged in the Euclidean functional integral over gauge fields distributed according to a Boltzmann weight $e^{-S_{\mathrm{gauge}}}$ determined by the pure gauge action. If our contour is very long in the Euclidean time direction $(T \gg R = |\vec{x} - \vec{y}|$ in Fig. 19.6), the Euclidean propagation amplitude must acquire, by reasoning familiar from Section 4.2, a factor $e^{-V(\vec{x}-\vec{y})T}$ where the *static potential energy* $V(\vec{x} - \vec{y})$ is defined as that of the minimum-energy state into which the quark–antiquark pair introduced at $t = 0$ (plus the gauge gluons with which they interact) can rearrange itself, or alternatively, the minimum-energy state with a non-vanishing matrix element of the bilocal operator $\bar{\psi}(\vec{x}, 0)U(\mathcal{C}_{(\vec{y},0)\to(\vec{x},0)})\psi(\vec{y}, 0)$ to the vacuum. For the lattice-regularized theory, this quantity, defined mathematically as

$$V(R) \equiv \lim_{T\to\infty} \left\{-\frac{1}{T}\log < W(R, T) > \right\} \quad (19.79)$$

can be numerically estimated by generating a large ensemble of statistically independent gauge field (i.e., link) configurations according to the Boltzmann weight arising from the Wilson action using Monte Carlo techniques, and then averaging the Wilson loop variable (for various choices of R and T) over this ensemble. This program, initiated in the late 1970s by Creutz (Creutz, 1980), has been pushed to quite large lattices and a high level of statistical precision, and there is by now overwhelming numerical evidence that $V(R)$, in addition to the expected Coulombic behavior at short distances (where we expect perturbative behavior due to the asymptotic freedom property at short distance described in the preceding chapter), possesses a linear dependence of $V(R)$ on R at large separations (see Fig. 11.13).

The appearance of a linearly rising static potential was already demonstrated by Wilson *in the strong coupling limit* where $\beta = N/g^2$ is taken small, in his seminal paper on lattice gauge theory (Wilson, 1974). We shall briefly explain the reason for this result here. For compact Lie groups G (such as the SU(N) groups considered here), any invariant function on the group $f(U)$, $f(V^{-1}UV) = f(U)\forall U, V \in G$ has a Fourier expansion in terms of the *character functions* $\chi_r(U)$ associated with a complete set of unitary representations of the group (labeled by the index r). The character function $\chi_r(U)$ is simply the trace of the unitary matrix representing the group element U in the rth representation, of dimension d_r,

$$\chi_r(U) \equiv \mathrm{Tr}(U), \quad \chi_r(1) = d_r \tag{19.80}$$

and any invariant function, and in particular the exponential function $e^{\beta(\frac{1}{N}\mathrm{ReTr}(U_P)-1)}$ appearing in the lattice functional integral from the Wilson gauge lattice action (where P denotes a particular plaquette) can be expanded

$$e^{\beta(\frac{1}{N}\mathrm{ReTr}(U_P)-1)} = \sum_r c_r(\beta)\chi_r(U_P) = c_0(\beta)(1 + \sum_{r\neq 0} \frac{c_r(\beta)}{c_0(\beta)}\chi_r(U_P)) \tag{19.81}$$

where we have separated out explicitly the contribution of the trivial representation ($r = 0$) with $\chi_0(U) = 1$. For concreteness, let us take the case of gauge group SU(2). The representations are labeled by the index j, which can be integer or half-integer, with the fundamental (spinor) representation corresponding to $j = \frac{1}{2}$. In this case the coefficient ratios appearing in the sum are

$$\frac{c_j(\beta)}{c_0(\beta)} = (2j+1)\frac{I_{2j+1}(\beta)}{I_1(\beta)} \sim \frac{\beta^{2j}}{(2j-1)!2^{2j}} + O(\beta^{2j+2}), \; \beta \to 0 \tag{19.82}$$

Note that for small β, the leading contributions arise from the use of representations with minimum dimensionality (which give a non-zero contribution to the desired amplitude). For SU(N), the lowest-dimensional non-trivial representation is the fundamental, which we shall denote with subscript "F" (thus for SU(2), $c_F = c_{j=1/2}$).

The *Schur orthogonality theorem* for the finite-dimensional irreducible unitary matrix representations of SU(N) (where the superscripts r, s identify the representation)

$$\int dU (U_{ij}^{(r)})^* U_{mn}^{(s)} = \frac{1}{d_r}\delta_{rs}\delta_{im}\delta_{jn} \tag{19.83}$$

implies a useful identity,

$$\int \chi_r(U^\dagger V)\chi_s(UW)dU = \frac{1}{d_r}\delta_{rs}\chi_r(VW) \tag{19.84}$$

which allows us to perform the gauge link integrations appearing in the lattice functional integral giving the expectation value of the Wilson loop observable $W(R,T)$ (= trace in the fundamental representation of the product of link variables around the rectangular loop of Fig. 19.6):

$$\langle W(R,T)\rangle = \frac{1}{Z}\int \prod_{\mathbf{n}\mu} dU_{\mathbf{n},\mu}\chi_F(U_{\mathcal{C}(R,T)})\prod_P e^{\beta(\frac{1}{N}\operatorname{ReTr}(U_P)-1)} \tag{19.85}$$

$$Z = \int \prod_{\mathbf{n}\mu} dU_{\mathbf{n},\mu}\prod_P e^{\beta(\frac{1}{N}\operatorname{ReTr}(U_P)-1)} \tag{19.86}$$

The character expansion (19.81) can be inserted in (19.85), whereupon we obtain

$$\langle W(R,T)\rangle = \frac{1}{Z}\int \prod_{\mathbf{n}\mu} dU_{\mathbf{n},\mu}\chi_F(U_{\mathcal{C}(R,T)})\sum_{r_P}\prod_P c_{r_P}\chi_{r_P}(U_P) \tag{19.87}$$

In the limit $\beta \to 0$, the leading contribution to $\langle W(R,T)\rangle$ comes from picking the minimum set of non-trivial representations for each plaquette appearing in the product over plaquettes in (19.87), compatible with obtaining a non-zero result for the integral over links. At a minimum, we must include a full set of RT plaquettes "tiling" the interior of the rectangular $R \times T$ contour given by the Wilson loop. Otherwise, there will be unmatched link variables (appearing only once) which integrate to zero. Moreover, this minimum set of plaquettes must all be associated with the fundamental representation in order to obtain a non-zero result, as the boundary links appearing in $\chi_F(U_{\mathcal{C}(R,T)})$ are in this representation, and integrals over products of characters in different representations vanish by Schur orthogonality, (19.84). The upshot of this reasoning (see (Montvay and Münster, 1994), Section 3.4, for the gory combinatoric details) is that the Wilson loop expectation value in the strong coupling limit satisfies[9]

$$\langle W(R,T)\rangle \propto (\frac{c_F(\beta)}{d_F c_0(\beta)})^{RT} \quad \text{(small } \beta) \tag{19.88}$$

For SU(2), for example, this becomes

$$\langle W(R,T)\rangle \propto u(\beta)^{RT}, \quad u(\beta) = \frac{I_2(\beta)}{I_1(\beta)} \sim \frac{\beta}{4} + O(\beta^3) \tag{19.89}$$

i.e. $\langle W(R,T)\rangle \sim e^{-KRT} \equiv e^{-TV(R)}, \quad V(R) = KR, \quad K \equiv -\log(u(\beta)) > 0$

$$\tag{19.90}$$

---

[9] The factors appearing here originate as follows. Each plaquette variable integration provides a $1/d_F$ factor, pursuant to (19.84), and a character coefficient $\frac{c_F(\beta)}{c_0(\beta)}$, as in (19.81), with the overall factors of $c_0(\beta)$ cancelling between the numerator and denominator (Z) in (19.85).

The falloff of the Wilson loop as a (negative) exponential of the area RT of the loop (the famous "area law") clearly indicates a linear rise in the static potential $V(R)$, as defined in (19.79). The coefficient K appearing in (19.90) is commonly referred to as the "string tension". It has dimensions of force, and turns out in QCD to take the interesting phenomenological value of approximately 15 metric tons: the gluon flux "string" extending from a single isolated quark could support a rather large truck (carrying the quantum numbers of a single anti-quark)!

The physical interpretation of the area-law dependence of the Wilson loop observable in the strong coupling limit is not hard to uncover, given our earlier discussion of confinement in the massive Schwinger model. From (19.65) we see that the lattice plaquette variable $\mathrm{Re}\,\mathrm{Tr}(U_{\mathbf{n};\mu\nu})$ corresponds to the square of the μ, ν component of the color field strength tensor (summed over colors to yield a gauge-invariant object). If we take our Wilson loop in Fig. 19.6 to be oriented in the Euclidean "x-t" plane, the plaquettes tiling the interior of the loop correspond to local values of the square of the color electric field F_{14} (Euclidean version of F_{10} in Minkowski space) in the x-direction. The necessity for including all these plaquettes in the expansion of the action (in order to obtain a non-vanishing contribution to the functional integral) amounts to the imposition of a non-abelian Gauss's Law whereby color flux originating on the quark must make its way to the anti-quark.

However, if we probe for the presence of color electric flux elsewhere in the volume of our lattice, by inserting, say, additional plaquette variables somewhere off the plane of the loop in the observable being measured in the path integral to check for the presence of electric flux elsewhere, we find that our result in (19.88) is suppressed by additional factors of β, which is, of course, small in the strong coupling limit under consideration. Similarly, subdominant contributions to the Wilson loop expectation are obtained from tilings in which the set of plaquettes bordered by the Wilson loop "bulge away" from the plane of the loop, corresponding to color electric flux "straying" away from the straight line connecting quark to anti-quark. In fact, the appearance of exactly linear confinement in 1+1-dimensional gauge theory is now seen to follow simply from the kinematic impossibility of such straying when only one spatial dimension is present: indeed, one can easily show that the result (19.88) is *exact* (for all β) in 1+1-dimensional gauge theory (see Problem 7).

A number of rigorous results have been established for the behavior of the Wilson loop observable in lattice gauge theories. Two results of particular interest can be stated here:

1. The strong coupling expansion (unlike the weak coupling expansions of perturbation theory, which as we saw in Chapter 11 are at best only asymptotic expansions) is a *Taylor expansion*: in other words, the lattice observables are analytic in β at $\beta = 0$. Moreover, for at least some finite range of β, the string tension $K(\beta)$ is non-vanishing (Osterwalder and Seiler, 1978).

2. The static potential $V(R)$ defined by the limit (19.79) cannot rise more rapidly than linearly with R at large R (Seiler, 1978): the area law is in this sense maximal. As we must expect at least constant terms in $V(R)$, Euclidean symmetry implies that a *perimeter* law $\langle W(R, T)\rangle < e^{-C(T+R)}$, for some constant C, is minimal.

As we have emphasized previously (see the discussion following (19.67)), the continuum limit of the lattice-regularized gauge theory is expected to appear in the limit in which the bare coupling $g(a)$ is taken to zero as the lattice spacing a is taken to zero: in other words, in the limit of $\beta = N/g(a)^2$ going to infinity. Thus the appearance of an area law at *small* β tells us absolutely nothing about the persistence of a linearly rising potential between static quarks once we insist on the validity of our stated action functional down to length scales much smaller than those implied by the low-energy parameters of the theory (mass of lightest hadron, string tension, etc.). Once β becomes large, there is a vast increase in the number of contributing terms in the strong coupling expansion of the Wilson loop observable, corresponding to "tilings" of the loops by essentially arbitrary surfaces of plaquettes bounded by the loop. In other words, field configurations in the path integral become important which correspond to the spreading out of the electric color flux lines from the "frozen" string going directly from the quark to the anti-quark (the so-called "roughening transition"), which clearly have the tendency to weaken the static potential felt by the quark pair.

Indeed, the strong coupling expansion of the loop observable in an *abelian* U(1) gauge theory also displays a linearly rising static potential (one simply has $u(\beta) = \frac{I_1(\beta)}{I_0(\beta)}$ in (19.89)), despite the fact that the continuum limit should yield a free pure gauge theory with the usual quadratic gauge action $F_{\mu\nu}^2$, and a static Coulomb potential between charges, $V(R) \propto 1/R$. In the abelian case there exists a rigorous proof (due to Guth (Guth, 1980)) that there is a phase transition at a finite critical coupling $\beta_c \sim 1$ such that the string tension K defined in (19.90) vanishes identically for $\beta > \beta_c$, leaving a perimeter law (as $e^{-(A+B/R)T} \to e^{-AT}$, $T >> R$) corresponding exactly to the desired Coulomb behavior at long distances, as long as we stay on the weak-coupling (large β) side of the transition.

For non-abelian gauge theories, we instead expect (or at least, fervently hope) that the string tension K remains non-zero for arbitrarily large β, and in fact scales appropriately with β so that the coefficient K of the linear potential, once converted to physical units (say, in units determined by the mass gap, or lightest hadron, of the theory) remains finite as we take the lattice spacing to zero. Clearly, this behavior will depend crucially on the highly non-trivial self-interacting dynamics of a non-abelian pure gauge theory, which must somehow act to "focus" the lines of electric color flux along the "string" connecting the static quark–antiquark pair. In the next section we shall show how a plausible mechanism for such focussing can be displayed in a simpler model: gauge theory in three spacetime (two space, one time) dimensions.

19.4   How confinement works: three-dimensional gauge theory

The lattice-regularization of gauge theory, together with a strong coupling expansion, as introduced by Wilson and described in the preceding section, gives a hint of the sort of mechanism which might be responsible for a linear rising potential between static quarks. We have seen that in the extreme strong coupling limit, the Euclidean path integral describing the propagation of a quark–antiquark pair over long Euclidean times is dominated by field configurations corresponding to "color" electric flux concentrated on the line between the quarks. The result is an area-law dependence of the Wilson loop observable.

An analytically precise, and physically intuitive, description of how this happens is available in at least one non-trivial case: compact abelian gauge theory in 2+1 spacetime dimensions. For this theory, a full semiclassical account of confinement has been developed, beginning with the work of Polyakov (Polyakov, 1977) and Banks *et al.* (Banks *et al.*, 1977), with important elaborations by Göpfert and Mack (Göpfert and Mack, 1982). As in four spacetime dimensions, there is no distinction in the confining behavior at strong coupling of abelian and non-abelian gauge theories in 2+1 dimensions: both display a linearly rising inter-quark potential for small β. However, we expect the continuum limit of the abelian theory in 2+1 dimensions to yield the usual perturbative Coulomb potential (Green function of the two-dimensional Laplacian operator), i.e., a potential which rises logarithmically, rather than linearly, at large distance. On the other hand, numerical simulations provide convincing evidence that the linear rise (=area law) persists for non-abelian gauge theories in 2+1 dimensions for lattice spacings much smaller than the scale of the theory, as set by the inverse squared coupling (in 2+1 dimensions, g has dimensions mass$^{1/2}$). In this section we shall present an account of confinement in 2+1-dimensional gauge theories, using an *interpolating model*, which contains both the abelian and non-abelian (with gauge group SU(2)) theories as special limiting cases.

We begin with a lattice action built from an SU(2) gauge field, realized on the lattice with link variables $U_{\mathbf{n},\mu} \in$ SU(2), and an isovector (i.e., in the three-vector representation of SU(2)) scalar field $\boldsymbol{\phi}_{\mathbf{n}}$ which is constrained to be unit length, $\boldsymbol{\phi}_{\mathbf{n}} \cdot \boldsymbol{\phi}_{\mathbf{n}} = 1$. We use the Villain form for the SU(2) gauge action, with coupling β_g, and introduce a second coupling constant β_h to indicate the strength of the scalar-gauge coupling:

$$
S_V = -\sum_P \log \sum_{m=-\infty}^{+\infty} \exp\left\{-\frac{\beta_g}{2}\left(\arccos\left(\frac{1}{2}\mathrm{Tr}(U_P)\right) - 2m\pi\right)^2\right\}
$$

$$
+ \frac{\beta_h}{2} \sum_{\mathbf{n}\mu} \mathrm{Tr}(\boldsymbol{\phi}_{\mathbf{n}} \cdot \boldsymbol{\sigma} U_{\mathbf{n},\mu} \boldsymbol{\phi}_{\mathbf{n}+\hat{\mu}} \cdot \boldsymbol{\sigma} U_{\mathbf{n},\mu}^\dagger) \tag{19.91}
$$

where we use the simplified index notation P for plaquettes (thus, P runs over $\{\mathbf{n}, \mu\nu\}$ with $\mu < \nu$).[10] In the limit of vanishing β_h, we are left with a pure non-abelian gauge theory (in this case, with gauge group SU(2)). On the other hand, when β_h is taken to infinity, the theory becomes a pure *abelian* U(1) gauge theory. We can see this by using the local SU(2) gauge symmetry on each lattice site to rotate the $\vec{\phi}$ field to the 3-direction, so that the second term in (19.91) becomes a sum over links of $\beta_h \mathrm{Tr}(\boldsymbol{\sigma}_3 U_{\mathbf{n},\mu} \boldsymbol{\sigma}_3 U_{\mathbf{n},\mu}^\dagger)$, which for $\beta_h \to \infty$ forces the non-abelian link variables to collapse onto the U(1) subgroup given by $U_{\mathbf{n},\mu} = \exp(i\boldsymbol{\sigma}_3 \theta_{\mathbf{n},\mu})$, at which point we have precisely the abelian Villain model specified earlier in (19.69, 19.70). The model therefore interpolates smoothly between lattice-regularized unbroken (compact)

---

[10] This is a lattice-regularized version of the so-called Georgi–Glashow model, with an adjoint scalar field coupled to non-abelian gauge vector fields, in the regime where the isovector "Higgs" field develops a non-vanishing vacuum-expectation value. The "frozen" magnitude of the scalar field can be viewed as arriving from a scalar potential $P(\vec{\phi}) = \lambda(\vec{\phi}^2 - 1)^2$ in the limit of large quartic coupling, $\lambda \to \infty$.

abelian and non-abelian gauge theories. From the discussion in the preceding section we know that linear confinement (an area law for the Wilson loop observable) obtains in both cases in the strong coupling expansion, and indeed, it is easy to verify, for all finite values of β_h, that we obtain a non-vanishing string tension for small β_g.

In this interpolating model there is substantial analytic and numerical evidence that linear confinement persists for all values of the lattice gauge parameter β_g. For example, one can determine numerically an "isotonic" line of constant string tension in the (β_g, β_h) plane (see Fig. 19.7) with just the qualitative features expected from the semiclassical model of monopole confinement which we shall shortly discuss (Duncan and Mawhinney, 1990). The essential point is that the physics of confinement evolves smoothly from the purely abelian case, where we have a mathematically rigorous and fairly complete physical picture of the confining mechanism, to the much more complicated non-abelian case, where a proper analytic treatment does not exist.

We begin by examining the physical mechanism responsible for linear confinement at the abelian end ($\beta_h \to \infty$), where the theory reduces to the Villain model. The path integral giving the desired Wilson loop observable is just (see Problem 8)

$$\langle W \rangle = \int \prod_{\mathbf{n}\mu} \frac{d\theta_{\mathbf{n},\mu}}{2\pi} \sum_{l_P} \prod_P \exp\left(-\frac{1}{2\beta_g} l_P^2 + i l_P \theta_P + i \sum_{\mathbf{n}\mu} J_{\mathbf{n},\mu} \theta_{\mathbf{n},\mu}\right) \qquad (19.92)$$

where the link angle variables $\theta_{\mathbf{n},\mu}$ are continuous, and the plaquette variables l_P are integer valued. The abelian Wilson loop is obtained by setting the lattice vector field $J_{\mathbf{n},\mu}$ equal to unity on all the ordered links comprising the perimeter of the Wilson loop (see Fig. 19.6), and zero on all other links. Also, we have used a shorthand index notation for plaquettes, where P indicates the plaquette in the μ, ν plane with the site

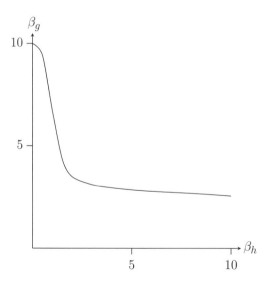

Fig. 19.7 A line of constant string tension in the $(\beta_g - \beta_h)$ plane.

n in the lower left-hand corner, $l_P = l_{\mathbf{n};\mu\nu}$. Performing a discrete integration by parts, we may write the second term in the exponent as

$$\sum_P l_P \theta_P = \sum_{\mathbf{n}\mu\nu} \theta_{\mathbf{n}\nu} \bar{\Delta}_\mu l_{\mathbf{n};\mu\nu} \tag{19.93}$$

Here, right (resp. left) discrete difference operators Δ_μ (resp. $\bar{\Delta}_\mu$), which interconvert under a discrete integration by parts, are defined as follows

$$\Delta_\mu \phi_{\mathbf{n}} \equiv \phi_{\mathbf{n}+\hat\mu} - \phi_{\mathbf{n}}, \quad \bar{\Delta}_\mu \phi_{\mathbf{n}} \equiv \phi_{\mathbf{n}} - \phi_{\mathbf{n}-\hat\mu} \tag{19.94}$$

The integrations over the link angles in (19.92) can now be performed, yielding a single (Kronecker) δ-function constraint for each link:

$$\int \frac{d\theta_{\mathbf{n},\mu}}{2\pi} e^{iJ_{\mathbf{n},\nu}\theta_{\mathbf{n},\nu} + i\theta_{\mathbf{n},\nu}\bar{\Delta}_\mu l_{\mathbf{n};\mu\nu}} = \delta(\bar{\Delta}_\mu l_{\mathbf{n};\mu\nu} + J_{\mathbf{n},\nu}) \tag{19.95}$$

Note that despite the continuum δ-function notation, the lattice fields appearing on the right all take integer values, so the constraint is of the Kronecker type. A general solution of the constraint for the plaquette variable $l_{\mathbf{n};\mu\nu}$ can be written as a sum of the general solution $\epsilon_{\mu\nu\lambda}\bar{\Delta}_\lambda\phi_{\mathbf{n}}$ (where $\phi_{\mathbf{n}}$ is an arbitrary integer-valued site field) of the homogeneous equation $\bar{\Delta}_\mu l_{\mathbf{n};\mu\nu} = 0$, and any particular solution of the inhomogeneous equation $\bar{\Delta}_\mu l_{\mathbf{n};\mu\nu} = -J_{\mathbf{n},\nu}$. To obtain a solution to the latter, we choose a unit vector ζ , pointing along either axis of the Wilson loop (for definiteness, say along the x-direction for a Wilson loop oriented in the x-y plane), and define an inverse operator as follows:

$$\frac{1}{\zeta \cdot \bar{\Delta}} s_{\mathbf{n}} \equiv \sum_{m_x=0}^{n_x} s_{m_x n_y n_z} \tag{19.96}$$

The right-hand side represents a well-defined and periodic integer-valued site field provided the site field $s_{\mathbf{n}}$ sums to zero when accumulated across the lattice in the ζ (i.e., x) direction. One then finds that a general solution of $\bar{\Delta}_\mu l_{\mathbf{n};\mu\nu} = -J_{\mathbf{n},\nu}$ may be written, using current conservation $\bar{\Delta}_\mu J_{\mathbf{n},\mu} = 0$,

$$l_{\mathbf{n};\mu\nu} = \zeta_\nu \frac{1}{\zeta \cdot \bar{\Delta}} J_{\mathbf{n},\mu} - \zeta_\mu \frac{1}{\zeta \cdot \bar{\Delta}} J_{\mathbf{n},\nu} + \epsilon_{\mu\nu\lambda}\bar{\Delta}_\lambda\phi_{\mathbf{n}} \tag{19.97}$$

where the first two terms on the right-hand side are the desired particular solution. The inverse operators in (19.97) are well-defined, as only the current in the plane of the Wilson loop orthogonal to ζ appears (say, the y-direction), by antisymmetry, and this component of the current clearly sums to zero across the lattice in the ζ direction (by combining current contributions from opposite sides of the loop). Using the δ constraint (19.95) in the Wilson loop average (19.92), and converting the sums over the integer valued site field $\phi_{\mathbf{n}}$ into integrals over a real-valued site field $\chi_{\mathbf{n}}$ via the Poisson identity,

$$\sum_{\phi=-\infty}^{+\infty} f(\phi) = \sum_{\rho=-\infty}^{+\infty} \int d\chi f(\chi) e^{2\pi i \rho \chi} \tag{19.98}$$

we find an equivalent expression for the Wilson loop average:

$$\langle W \rangle = \int \prod_{\{\rho_n\}} d\chi_n \sum_{\{\rho_n\}} \exp\left(-\frac{1}{4\beta_g} \sum_{n\mu\nu} (\zeta_\nu \frac{1}{\zeta \cdot \bar{\Delta}} J_{n,\mu} - \zeta_\mu \frac{1}{\zeta \cdot \bar{\Delta}} J_{n,\nu} + \epsilon_{\mu\nu\lambda} \bar{\Delta}_\lambda \chi_n)^2\right)$$

$$\cdot \exp\left(+2\pi i \sum_n \rho_n \chi_n\right) \tag{19.99}$$

The Gaussian integral over χ_n can be performed explicitly:

$$\int \prod_n d\chi_n e^{-\frac{1}{2\beta_g}\chi_n(-\Delta)\chi_n + \chi_n(2i\pi\rho_n - \frac{1}{\beta_g}\sigma_n)}$$

$$= e^{-2\pi^2\beta_g \sum_{nm}\rho_n(-\Delta^{-1})_{nm}\rho_m + 2\pi i \sum_n \rho_n \Delta^{-1}\sigma_n} \cdot e^{-\frac{1}{2\beta_g}\sum_n \sigma_n \Delta^{-1}\sigma_n} \tag{19.100}$$

where we have defined

$$\sigma_n = \epsilon_{\mu\nu\lambda}\zeta_\mu \Delta_\lambda \frac{1}{\zeta \cdot \bar{\Delta}} J_{n\nu} \tag{19.101}$$

Here Δ is the lattice Laplacian

$$\Delta = \bar{\Delta}_\mu \Delta_\mu \tag{19.102}$$

whose (negative) inverse gives the lattice Coulomb potential

$$\Delta_{nr} v_{rm}^{\text{coul}} = -\delta_{nm} \tag{19.103}$$

The site field ρ_n in fact describes a gas of *magnetic monopoles* interacting with the electric current loop carried by $J_{n,\mu}$. To see this, observe that an electric current J_μ produces a magnetic B-field via the curl operation

$$J_\mu = \epsilon_{\mu\nu\lambda}\bar{\Delta}_\nu B_\lambda \tag{19.104}$$

Setting

$$B_\lambda = \epsilon_{\lambda\nu\mu}\zeta_\mu \frac{1}{\zeta \cdot \bar{\Delta}} J_\nu \tag{19.105}$$

one easily finds that (19.104) is satisfied. We now see that the term coupling the ρ_n field to the Wilson loop current J_ν (the second term in the first exponential in (19.100)) is proportional to

$$\sum_n \rho_n \Delta^{-1}\sigma_n = \sum_{nm} \rho_n v_{nm}^{\text{coul}}(\Delta_\lambda B_\lambda)_m \tag{19.106}$$

In other words, the objects whose density is represented by the ρ_n field couple Coulombically to the divergence of the magnetic field generated by the current loop, and may therefore be properly regarded as magnetic monopoles. The first term in the exponent (quadratic in ρ_n) is just the Coulombic interaction energy of this gas of monopoles.

To summarize, we have shown that the Wilson loop average in (19.92) factorizes exactly into the product of two terms,

$$\langle W \rangle = \langle W \rangle_{\mathrm{mon}} \cdot \langle W \rangle_{\mathrm{SW}} \tag{19.107}$$

where the first term describes the interaction of a magnetic monopole gas (with monopole density $\rho_{\mathbf{n}}$) with an electric current $J_{\mathbf{n},\mu}$ running around the Wilson loop (which generates the field $\sigma_{\mathbf{n}}$),

$$\langle W \rangle_{\mathrm{mon}} = \sum_{\{\rho_{\mathbf{n}}\}} e^{-2\pi^2 \beta_g \sum_{\mathbf{nm}} \rho_{\mathbf{n}} (-\boldsymbol{\Delta}^{-1})_{\mathbf{nm}} \rho_{\mathbf{m}} + 2\pi i \sum_{\mathbf{n}} \rho_{\mathbf{n}} \boldsymbol{\Delta}^{-1} \sigma_{\mathbf{n}}} \tag{19.108}$$

and a second "spin-wave" term which is an explicit functional of the current $J_{\mathbf{n},\mu}$:

$$\langle W \rangle_{\mathrm{SW}} = e^{-\frac{1}{4\beta_g} \sum_{\mathbf{n}\mu\nu} (\zeta_\nu \frac{1}{\zeta \cdot \bar{\Delta}} J_{\mathbf{n},\mu} - \zeta_\mu \frac{1}{\zeta \cdot \bar{\Delta}} J_{\mathbf{n},\nu})^2 - \frac{1}{2\beta_g} \sum_{\mathbf{n}} \sigma_{\mathbf{n}} \boldsymbol{\Delta}^{-1} \sigma_{\mathbf{n}}} \tag{19.109}$$

This second, rather complicated-looking expression is just the usual (electrostatic) Coulomb term in disguise. This is most easily seen by choosing a specific orientation for the Wilson loop—say in the x-y plane—and setting $\zeta_\mu = \delta_{\mu 1}$. In this case, only $J_{\mathbf{n}1}$ and $J_{\mathbf{n}2}$ are non-zero. The exponent in (19.109) is then seen to become (apart from a prefactor $-\frac{1}{2\beta_g}$)

$$\sum_{\mathbf{n}} (\bar{\Delta}_1^{-1} J_{\mathbf{n}2} \bar{\Delta}_1^{-1} J_{\mathbf{n}2} + \Delta_3 \bar{\Delta}_1^{-1} J_{\mathbf{n}2} \boldsymbol{\Delta}^{-1} \Delta_3 \bar{\Delta}_1^{-1} J_{\mathbf{n}2}) \tag{19.110}$$

$$= \sum_{\mathbf{nm}} J_{\mathbf{n}2} (-\Delta_1^{-1} \bar{\Delta}_1^{-1} + \Delta_1^{-1} \bar{\Delta}_3 \boldsymbol{\Delta}^{-1} \Delta_3 \bar{\Delta}_1^{-1}) J_{\mathbf{m}2} \tag{19.111}$$

However, as the difference operators all commute, one may easily check that

$$-\Delta_1^{-1} \bar{\Delta}_1^{-1} + \Delta_1^{-1} \bar{\Delta}_3 \boldsymbol{\Delta}^{-1} \Delta_3 \bar{\Delta}_1^{-1} = -\boldsymbol{\Delta}^{-1} - \Delta_1^{-1} \Delta_2 \boldsymbol{\Delta}^{-1} \bar{\Delta}_2 \bar{\Delta}_1^{-1} \tag{19.112}$$

Finally, using current conservation,

$$\bar{\Delta}_2 \bar{\Delta}_1^{-1} J_2 = -J_1 \tag{19.113}$$

and we obtain

$$\langle W \rangle_{\mathrm{SW}} = e^{\frac{1}{2\beta_g} \sum_{\mathbf{nm}} J_{\mathbf{n},\mu} \boldsymbol{\Delta}_{\mathbf{nm}}^{-1} J_{\mathbf{m},\mu}} = e^{-\frac{1}{2\beta_g} \sum_{\mathbf{nm}} J_{\mathbf{n},\mu} v_{\mathbf{nm}}^{\mathrm{coul}} J_{\mathbf{m},\mu}} \tag{19.114}$$

This contribution to the Wilson loop is precisely (see Problem 9) the term that one expects to survive in the continuum limit of the pure abelian theory, and cannot therefore be responsible for an area-law behavior of the loop average in the limit of large loops (recalling that the Coulomb potential in two space dimensions only rises logarithmically, not linearly, at large distance). Returning to the monopole contribution (19.108), note that with the x-y orientation of the Wilson loop chosen above, the $\sigma_{\mathbf{n}}$ field (cf. (19.101)) can be written as the gradient in the z-direction orthogonal to the loop of a uniform density localized on the plane interior of the loop:

$$\sigma_{\mathbf{m}} = \Delta_3 \bar{\Delta}_1^{-1} J_{\mathbf{m},2} = \Delta_3 \sum_{n_x=0}^{m_x} J_{n_x m_y m_z, 2} \qquad (19.115)$$

In other words, the monopole density in (19.108) couples Coulombically to a magnetic potential corresponding to a dipole sheet localized on the plane of the Wilson loop, exactly as we would expect for an *electric* current loop traversing the perimeter. If we view the sum over monopole fields $\rho_{\mathbf{n}}$ in (19.108) as a finite temperature partition function for an electric loop interacting with a gas of magnetic monopoles, it is perfectly plausible that monopoles will tend to "condense" on the dipole sheet, producing a screening of the magnetic field produced by the current loop, and giving rise to a shift in energy of the system proportional to the area of the loop—in other words, to an area law for the Wilson loop observable. The string tension corresponds to the coefficient of the area, and will be suppressed exponentially at large β_g due to the fact that the individual monopoles come with a self-energy proportional to $\beta_g v_{\mathrm{nn}}^{\mathrm{coul}}$, so that in this limit the density of the monopole gas dies exponentially, and we are left with just the "spin-wave" (i.e., Coulombic) contribution discussed above. For further details on the abelian theory, the reader is encouraged to consult (Göpfert and Mack, 1982) for rigorous estimates, or (Duncan and Mawhinney, 1990) for a semiclassical evaluation of the monopole term employing the techniques of Debye–Hückel screening theory.

So far, we have been discussing the mechanism for inducing a linearly rising confining potential in the compact abelian U(1) gauge theory corresponding to the $\beta_h \to \infty$ limit of the interpolating model specified by the lattice action (19.91). For finite $\beta_h > 0$, we are dealing with a theory with an exact local SU(2) gauge symmetry, broken spontaneously to the U(1) subgroup under which the "Higgs" isovector field is invariant. In such theories, in four spacetime dimensions, it was shown a long time ago, by 't Hooft ('T Hooft, 1974), and independently by Polyakov (Polyakov, 1974), that there exist static solutions to the coupled scalar-gauge field equations corresponding to finite mass field configurations carrying non-zero magnetic charge and finite mass—the famous 't Hooft–Polyakov monopoles.

A time-independent (but spatially varying) solution of the 3+1-dimensional theory amounts, of course, to a solution of the field equations of the *Euclidean* version of the scalar-gauge theory in one less dimension (i.e., precisely the 2+1-dimensional theory we are considering here), where the finite mass of the monopole solutions in 3+1 dimensions translates to a configuration with finite (and, for a solution of the field equations, locally extremal) Euclidean action. These "instanton"-type Euclidean configurations are the non-abelian deformations of the abelian monopoles revealed by the transformations described above in the compact abelian limit. They can be exposed, and studied numerically, by a "cooling" procedure whereby scalar-gauge configurations on the lattice generated by Monte Carlo simulation techniques are subjected to a stochastic process which dampens all fluctuations on the scale of a lattice spacing, revealing the underlying "classical" configurations. The upshot is that the non-abelian monopoles are found to persist all the way down to $\beta_h = 0$—the limit in which we have a pure, unbroken non-abelian SU(2) gauge theory. The isotones— lines of constant string tension—of the interpolating theory are completely smooth curves (see Fig. 19.7) in the $\beta_g - \beta_h$ plane, as we indicated earlier, and the shape of the curves can even be understood qualitatively from known properties of 't Hooft–Polyakov monopoles (see (Duncan and Mawhinney, 1990)). The essential difference between the behavior of the theory at the abelian ($\beta_h \to \infty$) and non-abelian ($\beta_h \to 0$) ends lies simply in the fact that in the latter case the monopole action no longer diverges for large β_g: instead, the monopole density remains finite in the continuum limit, corresponding to the persistence of the area law, and linear confinement on distance scales much larger than the lattice cutoff. However, in the small β_h regime the monopole cores grow to the point where they overlap, and we no longer have a beautiful and analytically tractable transcription of the theory in terms of a dilute monopole gas as in the abelian case.

In four spacetime dimensions, an analogous treatment of the Villain version of compact U(1) gauge theory (see (Banks *et al.*, 1977)) shows that in the strong-coupling regime (i.e., for β_g smaller than the critical coupling shown by Guth (Guth, 1980) to mark the transition point to a non-confining phase) an area law is obtained as a consequence of the appearance of *magnetic vortices*—i.e., closed loops carrying magnetic current—which interlace with the electric current loop corresponding to the Wilson loop observable. These Euclidean configurations correspond in Minkowski space to virtual events in which magnetic monopole/anti-monopole pairs appear and subsequently annihilate. The concomitant large fluctuations in the local magnetic charge density lead to a suppression of the electric field in the bulk, and a "focussing" of the (by Gauss's Law, necessarily conserved) electric flux travelling between opposite electric charges onto a "flux tube" connecting the charges, with an energy cost proportional to the length of the tube. The situation is precisely the "dual" (in the sense of interchange of electric and magnetic fields) of the Meissner effect in superconductivity. There, large fluctuations in the local *electric* charge density in the superconducting state lead to a suppression of magnetic field in the bulk of the superconductor (recall from Section 8.1 that electric and magnetic fields are complementary quantities, with corresponding mutual uncertainty constraints). Indeed, if we had actual magnetic monopoles at our disposal, then inserting an oppositely (magnetically) charged pair into the bulk volume of a superconductor would lead precisely to the formation of a string of magnetic flux connecting the two, with an energy rising linearly with their separation. This is the "dual Meissner" interpretation of quark confinement in QCD, which remains, nearly forty years after its introduction, a perfectly reasonable *qualitative* picture of the underlying mechanism leading to a linearly rising potential between static quarks.

A glance at the recent literature surrounding confinement in four-dimensional Yang–Mills theories reveals a complicated nexus of competing, and at first sight incompatible, hypotheses advanced by theorists interested in the detailed physical mechanism leading to quark and/or color confinement (for an extensive review, see (Greensite, 2003)). The origins of this complexity lie in (at least) two directions. Firstly, the intrinsic fluidity of gauge theories, which allows physically equivalent, but sometimes superficially completely different, descriptions of the same physical phenomenon simply by changing the gauge, tends to induce a proliferation of hypothetical mechanisms even where the underlying relevant physics is the same. Secondly, the asymptotic freedom of the theory, so helpful in allowing the extraction of quantitative results from perturbation theory at high energies, proves a double-edged sword at low energy or long distances. Precisely in this regime, the field configurations responsible for the confinement phenomenon (as well as the dynamical chiral symmetry-breaking discussed in Section 16.5) necessarily correspond to strongly coupled modes of the theory. Specifically, this means, as indicated previously, that semiclassical methods, which depend on saddle-point expansions rendered sensible by the existence of some type of small expansion parameter, no longer prove useful except in the most crudely qualitative way. This means that we will probably never be able to arrive at an analytically tractable, as well as quantitatively accurate, description of four-dimensional non-abelian confinement along the lines discussed above for three-dimensional compact abelian gauge theory. It is indeed fortunate that the lattice formulation of four-dimensional Yang–Mills theory has at least given us the option of direct numerical evaluation, using Monte Carlo methods, of the (Euclidean) amplitudes of the theory, with results which leave us with no possible doubt that the overall picture of confined elementary quark and gluon constituents is indeed the correct framework for hadronic physics.

19.5   Problems

1. In the model used by Schroer to introduce the concept of infraparticles, a massless boson field is coupled to a massive fermion in 1+1 spacetime dimensions, via the Lagrangian

$$\mathcal{L} = \frac{1}{2}\partial_\mu\phi\partial^\mu\phi + \bar{\psi}(i\slashed{\partial} - M)\psi + ig\bar{\psi}\gamma^\mu\psi\partial_\mu\phi \qquad (19.116)$$

   Show that the one-loop contributions to the $\bar{\psi}\psi\phi$ vertex (with the external fermions on-mass-shell) contain logarithmic divergences from the infrared part of the loop integral.

2. Verify the result (19.42) for the matrix element appearing in (19.41).

3. Show that the action (19.46) for QED in 1+1 spacetime dimensions leads, after going to axial (=Coulomb) gauge $A_1 = 0$, to a Hamiltonian consisting of the usual massive free fermion piece, plus the Coulomb interaction term (19.48), after the dependent A_0 field is eliminated.

4. Using the identity (19.62), verify that the plaquette variable $U_{\mathbf{n};\mu\nu}$ defined in (19.61) reduces to (19.63).

5. Show that by choosing $C(a), D(a)$ as suitable functions of the lattice spacing a, the combination of lattice fields

$$C(a)\sum_{\mathbf{n}\mu}(\phi_{\mathbf{n}+\hat{\mu}}^* U_{\mathbf{n},\mu}\phi_{\mathbf{n}} + \text{c.c.}) + D(a)\sum_{\mathbf{n}} P(\phi_{\mathbf{n}}^*\phi_{\mathbf{n}}) \qquad (19.117)$$

   reproduces for $a \to 0$ the classical continuum scalar action for the most general renormalizable gauge-invariant theory with fundamental representation scalar fields coupled to the gauge vector fields. Here, P is a polynomial up to degree 2.

6. Show that the Green function (19.78) satisfies its defining equation (19.77).

7. Consider a SU(2) gauge theory defined on a two-dimensional $L \times L$ Euclidean lattice, with free boundary conditions for the links on the edge (thus, link variables at the boundary appear in only a single plaquette variable in the action). Show that the Wilson loop average $\langle W(R, T) \rangle$ is given *exactly* by the expression (19.88), at all values of β (and not just for small β, as in higher dimensions).

8. Using the Poisson identity (19.98), show that the abelian gauge action (19.92) is equivalent to our original version of the Villain action (19.70).

9. In this Problem we shall evaluate the Wilson loop average in a pure abelian gauge theory in three (Euclidean) dimensions in the continuum. As the Wilson loop observable is gauge-invariant, we may choose any convenient gauge: in this case, a Feynman gauge will be the preferred choice, corresponding to a Euclidean gauge action (cf. the discussion following (15.159))

$$S_{\text{gauge,E}} = \frac{1}{4g^2} \int (F_{\mu\nu}^2 + 2(\partial_\mu A_\mu)^2) d^3 x \qquad (19.118)$$

The Wilson loop variable in an abelian theory corresponds to the familiar Aharanov–Bohm phase factor $\exp\left(i \oint A_\mu dx_\mu\right) = \exp\left(i \int J_\mu A_\mu d^3 x\right)$, where $J_\mu(x)$ is a conserved current of unit strength localized on the Wilson loop. Show that

$$\int \mathbf{D} A_\mu e^{-S_{\text{gauge,E}} + i \int J_\mu A_\mu d^3 x} = C e^{+\frac{g^2}{2} \int J_\mu \Delta^{-1} J_\mu d^3 x} \qquad (19.119)$$

with $\Delta = \partial_\mu \partial_\mu$. This is exactly the continuum version of (19.114) (with the identification $\beta_g = \frac{1}{g^2}$).

Appendix A
The functional calculus

The concepts of functionals and functional derivatives play an indispensable role in many field-theoretic calculations, so here we shall collect the main results on which we rely throughout the book. A functional is simply a mapping from some space of functions to the real (or complex) numbers. Typically, we shall need to probe the variation of a functional in response to a small change in the function(s) on which it depends, in order to define, in analogy to derivatives of (real or complex) functions, a *functional derivative*. The limiting procedure needed to obtain a derivative requires that the function space be supplied with a norm, and the most general spaces of this type which are useful in physics are Banach spaces (of which the complex Hilbert spaces of quantum theory are a subclass). Some functionals commonly encountered in quantum field theory include *generating functionals*, such as

$$Z[j] = \sum_n \frac{1}{n!} \int G^{(n)}(x_1, x_2, \ldots, x_n) j(x_1) j(x_2) \ldots j(x_n) d^4 x_1 d^4 x_2 \ldots d^4 x_n \quad \text{(A.1)}$$

which encode knowledge of an infinite class of correlation functions $G^{(n)}(x_1, x_2, \ldots, x_n)$ (symmetric under permutation of their arguments) in a single functional $Z[j]$, from which we can recover, as we shall soon see, the correlation functions by functional differentiation. We shall also encounter *action functionals*, or spacetime integrals of Lagrangian densities, such as

$$\mathcal{I}[\phi] = \int \mathcal{L}(\phi, \partial_\mu \phi) d^4 x = \int (\frac{1}{2} \partial_\mu \phi(x) \partial^\mu \phi(x) - \frac{1}{2} m^2 \phi(x)^2 - \frac{\lambda}{4!} \phi(x)^4) d^4 x \quad \text{(A.2)}$$

The examples given here possess an obvious smoothness property with respect to small perturbations of the argument functions $j(x)$ or $\phi(x)$. Let $f(x)$ be a Schwarz test function (i.e., an infinitely differentiable function of compact support—properties which will facilitate the processes of integration by parts which are indispensable in allowing us to develop a useful functional calculus). For a general functional $Z[j]$ (with $j(x)$ a c-number function on spacetime, say) let us suppose that a distribution $\frac{\delta Z[j]}{\delta j(x)}$ exists such that for all such $f(x)$ the limit

$$\lim_{\epsilon \to 0} \frac{Z[j + \epsilon f] - Z[j]}{\epsilon} = \int \frac{\delta Z[j]}{\delta j(x)} f(x) d^4 x \quad \text{(A.3)}$$

exists. The existence of the limit implies that the *functional derivative* $\frac{\delta Z[j]}{\delta j(x)}$ (called a *Fréchet derivative* in the mathematical literature) is a uniquely defined distribution: in other words, a continuous linear functional on the space of test functions (with an appropriately defined norm, (Friedlander, 1982)). The definition (A.3) is the obvious expression of our desire to express the first variation of the functional in a form analogous to that familiar from ordinary multi-variable calculus:

$$\delta Z[j] \equiv Z[j + \delta j] - Z[j] = \int \frac{\delta Z[j]}{\delta j(x)} \delta j(x) d^4 x + O((\delta j)^2) \tag{A.4}$$

The reader may easily verify that the definition (A.3) applied to (A.1) gives, for the first functional derivative,

$$\frac{\delta Z[j]}{\delta j(y)} = \sum_n \frac{1}{n!} \int G^{(n)}(y, x_1, x_2, \dots, x_n) j(x_1) j(x_2) \dots j(x_n) d^4 x_1 d^4 x_2 \dots d^4 x_n \tag{A.5}$$

and that the correlation functions $G^{(n)}$ are recoverable from $Z[j]$ by taking the nth functional derivative and then setting the "source" functions j to zero:

$$G^{(n)}(y_1, y_2, \dots, y_n) = \left. \frac{\delta^n Z[j]}{\delta j(y_1) \delta j(y_2) \dots \delta j(y_n)} \right|_{j=0} \tag{A.6}$$

which is the functional analog of the usual formula for the Taylor coefficients of a Taylor-expandable function. For action functionals such as (A.2), the functional derivative gives the *total Euler derivative* of the Lagrange density

$$\mathcal{I}[\phi + \delta\phi] - \mathcal{I}[\phi] = \int (\mathcal{L}(\phi + \delta\phi, \partial_\mu \phi + \partial_\mu \delta\phi) - \mathcal{L}(\phi, \partial_\mu \phi)) d^4 x$$

$$= \int \left(\frac{\partial \mathcal{L}}{\partial \phi} - \partial_\mu \frac{\mathcal{L}}{\partial(\partial_\mu \phi)} \right) \delta\phi(x) d^4 x \tag{A.7}$$

$$\Rightarrow \frac{\delta \mathcal{I}[\phi]}{\delta \phi(x)} = \frac{\partial \mathcal{L}}{\partial \phi(x)} - \partial_\mu \frac{\mathcal{L}}{\partial(\partial_\mu \phi(x))} \tag{A.8}$$

where the integration by parts maneuvers required to reach the second line are validated by the smoothness and compact support of the test functions $\delta\phi(x)$. The actual mechanics of functional differentiation can be simplified by noting that

$$j(y) = \int \delta^4(y - x) j(x) d^4 x \Rightarrow \frac{\delta j(y)}{\delta j(x)} = \delta^4(y - x) \tag{A.9}$$

and applying the obvious generalization of the Leibniz rule to arbitrary products of the source function.

In Section 10.3 the concept of functional differentiation is extended to functionals of Grassmann functions, where both the argument functions and the functionals themselves take values in an anticommuting number field. The reader is referred to that section for an explanation of the basic properties of such functionals.

Appendix B
Rates and cross-sections

Prepare an incoming scattering state in the usual fashion:

$$|t\rangle = \int d\alpha' g(\alpha') e^{-iE_{\alpha'}t} |\alpha'\rangle_{in} \qquad (B.1)$$

where $g(\alpha')$ is sharply peaked around some state α with well-defined energy (E_α) and momentum. Recall that α is a shorthand notation for energy, momentum, *and* internal quantum numbers (if any) of all the incoming particles. The state (B.1) is unit normalized, $\langle t|t\rangle = 1$, provided

$$\int d\alpha' \mid g(\alpha') \mid^2 = 1 \qquad (B.2)$$

We are interested in the probability of appearance at late times $t \gg 0$ of a state $|\beta\rangle$ with no overlap with the original state $|\alpha\rangle$ (for example, we look for particles coming out at an angle to the beam). Accordingly, taking the overlap of the time-dependent state (B.1) with a free state $\langle\beta|$, and referring to (4.181, 4.183), the trivial $\delta_{\beta\alpha}$ part of the S-matrix in (4.182) does not contribute, and we have, after the collision:

$$< \beta \mid t > = \int d\alpha' g(\alpha') \frac{e^{-iE_{\alpha'}t}}{E_{\alpha'} - E_\beta + i\epsilon} T_{\beta\alpha'} \qquad (B.3)$$

The probability that we shall find a state in the range of final-state phase-space β to $\beta + d\beta$ is thus

$$dP(\beta, t) = |< \beta \mid t >|^2 \, d\beta$$

$$= \int d\alpha' d\alpha'' g(\alpha') g^*(\alpha'') \frac{e^{-i(E_{\alpha'} - E_{\alpha''})t} T_{\beta\alpha'} T^*_{\beta\alpha''}}{(E_{\alpha'} - E_\beta + i\epsilon)(E_{\alpha''} - E_\beta - i\epsilon)} d\beta \qquad (B.4)$$

The event rate is just the time-derivative of this: namely,

$$d\Gamma(\beta, t) = -i \int d\alpha' d\alpha'' g(\alpha') g^*(\alpha'') T_{\beta\alpha'} T^*_{\beta\alpha''}$$

$$\cdot \left(\frac{1}{E_{\alpha''} - E_\beta - i\epsilon} - \frac{1}{E_{\alpha'} - E_\beta + i\epsilon} \right) e^{-i(E_{\alpha'} - E_{\alpha''})t} d\beta \qquad (B.5)$$

Strictly speaking, this event rate vanishes as $t \to \infty$, as a result of the rapid oscillations of the exponential factor. A finite number of particles localized in wave-packets eventually separate, and we should not be surprised that the interaction rate then goes to zero. The situation in an accelerator is somewhat different: the machine provides a steady supply of incoming particles in a beam. This situation may be idealized as a monoenergetic incoming plane wave of infinite extent. In other words, to obtain a steady event rate, we may assume that the folding functions $g(\alpha'), g(\alpha'')$ are sharply focussed at energy E_α. The exponential time-dependent factor in (B.4) may then be dropped. The two energy denominators then become identical except for a change in the sign of ϵ: the difference is just a δ-function (see (4.188)). This leads to

$$d\Gamma(\beta) = 2\pi\delta(E_\alpha - E_\beta) \mid \int d\alpha' g(\alpha') T_{\beta\alpha'} \mid^2 d\beta \tag{B.6}$$

As a consequence of the clustering property of the S-matrix, as discussed in Chapter 6, the T-matrix element $T_{\beta\alpha'}$ is a smooth function of the momenta characterizing β and α', *with the exception of a single overall δ-function of momentum conservation* $\delta^3(P_\beta - P_{\alpha'})$. Thus we can write

$$T_{\beta\alpha'} \simeq \mathcal{T}_{\beta\alpha}\delta^3(P_\beta - P_{\alpha'}) \tag{B.7}$$

where in the smooth part $\mathcal{T}$ of T we can replace α' by α as the smearing function g is sharply peaked at α' near α. Since the smooth part no longer depends on the integration variable α', it can be pulled out of the integral in (B.6) to give

$$d\Gamma(\beta) \simeq 2\pi\delta(E_\alpha - E_\beta) \mid \mathcal{T}_{\beta\alpha} \mid^2 \mid \int d\alpha' g(\alpha')\delta^3(\vec{P}_\beta - \vec{P}_{\alpha'}) \mid^2 d\beta \tag{B.8}$$

Let α correspond to a state of N particles so $g(\alpha') = g(k'_1, ..., k'_N)$ is the momentum-space wavefunction of the initial packets. This is related to the coordinate space wavefunction in the usual way

$$g(k'_1, ...) = \int \frac{d^3x_1..d^3x_N}{(2\pi)^{3N/2}} \psi(x_1, .., x_N) e^{-i(k'_1 \cdot x_1 + ...)} \tag{B.9}$$

while the δ-function of momentum conservation may be written

$$\delta^3(P_{\alpha'} - P_\beta) = \frac{1}{(2\pi)^3} \int d^3x e^{i(k'_1 + ... + k'_N - P_\beta) \cdot x} \tag{B.10}$$

Combining (B.9) and (B.10), one finds

$$\int d^3k'_1..d^3k'_N g(k'_1, ..)\delta^3(k'_1 + ..k'_N - P_\beta)$$

$$= (2\pi)^{3N/2-3} \int d^3x e^{-iP_\beta \cdot x} \psi(x, x, .., x) \tag{B.11}$$

The initial wavefunction evaluated at coincident spatial points can be written

$$\psi(x, x, .., x) = e^{iP_\alpha \cdot x} f(x) \tag{B.12}$$

where $f(x)$ is a slowly varying envelope. Thus in the quantity

$$\left|\int d\alpha' g(\alpha')\delta^3(P_{\alpha'} - P_\beta)\right|^2 = (2\pi)^{3N-6}\int d^3x\,d^3y\,e^{i(P_\beta-P_\alpha)\cdot(y-x)}f^*(y)f(x)$$

we can replace $f^*(y)$ by $f^*(x)$ in the integral on the right and obtain

$$\left|\int d\alpha' g(\alpha')\delta^3(P_{\alpha'} - P_\beta)\right|^2 = (2\pi)^{3N-6}\int d^3x\,|f(x)|^2\int d^3y\,e^{i(P_\beta-P_\alpha)\cdot y}$$

$$= (2\pi)^{3N-3}\int d^3x\,|f(x)|^2\,\delta^3(P_\beta - P_\alpha)$$

$$= (2\pi)^{3N-3}\delta^3(P_\beta - P_\alpha)\int d^3x\,|\psi(x,x,..,x)|^2$$

$$= (2\pi)^{3N-3}\rho_{rel}\,\delta^3(P_\beta - P_\alpha) \tag{B.13}$$

where the relative density ρ_{rel} is defined as

$$\rho_{rel} \equiv \int d^3x\,|\psi(x,x,..x)|^2 \tag{B.14}$$

Our final result for the event rate is thus

$$d\Gamma(\beta) = (2\pi)^{3N-2}\rho_{rel}\delta^4(P_\beta - P_\alpha)\,|\,\mathcal{T}_{\beta\alpha}\,|^2\,d\beta \tag{B.15}$$

where we have combined the energy-conservation δ-function of (B.8) with the three-momentum δ-function of (B.13) to give a single four-dimensional δ-function of energy-momentum conservation.

There are two particularly important special cases of (B.15) which deserve special attention: N=1 (particle decay), and N=2 (two-particle scattering).

1. If there is only one particle in the initial state, we have immediately

$$\rho_{rel} = \int d^3x\,|\psi(x)|^2 = 1 \tag{B.16}$$

so the differential decay rate (into final-state phase-space between β and $\beta + d\beta$) is

$$d\Gamma(\beta) = 2\pi\delta^4(P_\beta - P_\alpha)\,|\,\mathcal{T}_{\beta\alpha}\,|^2\,d\beta \tag{B.17}$$

2. Two particle collisions correspond to N=2. Consider a volume V containing ρ_1 target particles at rest per unit volume, and ρ_2 projectile particles per unit volume travelling with speed v_2.(For the time being we shall assume that the target and projectile particles are distinguishable). The effective differential cross-section $d\sigma$ is defined on a classical analogy as the effective area presented to the projectile beam by each target particle leading to final states in the range β to $\beta + d\beta$. Since the flux of projectiles is $\rho_2 v_2$ and the total number of target particles is $\rho_1 V$, the event rate is

$$d\Gamma(\beta) = \rho_1 V d\sigma \cdot \rho_2 v_2 \qquad\qquad \text{(B.18)}$$

so

$$d\sigma = \frac{1}{\rho_1 \rho_2 v_2} \cdot \frac{d\Gamma(\beta)}{V} \qquad\qquad \text{(B.19)}$$

Quantum-mechanically we are dealing with a two-particle wavefunction $\psi(x_1, x_2)$ where the joint probability of finding both target and projectile in the same volume V is

$$\int d^3 x_1 \int_{V_1} d^3 x_2 \mid \psi(x_1, x_2) \mid^2 \qquad\qquad \text{(B.20)}$$

where V_1 is a box of volume V around x_1. This integral represents the probability of having a *single* projectile particle in volume V moving with speed v_2: i.e., to flux v_2/V, impinging on a *single* target particle. The corresponding rate, from definition (B.18), is

$$d\Gamma = d\sigma \cdot \frac{v_2}{V} \int d^3 x_1 \int_{V_1} d^3 x_2 \mid \psi(x_1, x_2) \mid^2 \qquad\qquad \text{(B.21)}$$

The projectile beam wavefunction usually varies slowly (in amplitude) over the interaction range with the target particle, so choosing V much larger than the interaction range but much smaller than the scale of variation of the projectile wave-packet envelope

$$\int d^3 x_1 \int_{V_1} d^3 x_2 \mid \psi(x_1, x_2) \mid^2 \simeq \int d^3 x_1 \int_{V_1} d^3 x_2 \mid \psi(x_1, x_1) \mid^2$$

$$= V \int d^3 x_1 \mid \psi(x_1, x_1) \mid^2$$

$$= V \rho_{rel} \qquad\qquad \text{(B.22)}$$

Inserting this result in (B.21), and using (B.15) for N=2, we obtain

$$d\sigma(\alpha \to \beta) = \frac{(2\pi)^4}{v_2} \delta^4(P_\beta - P_\alpha) \mid \mathcal{T}_{\beta\alpha} \mid^2 d\beta \qquad\qquad \text{(B.23)}$$

The above result was obtained in the frame in which one of the particles was at rest. Note that from (B.19) the differential cross-section $d\sigma$ was given as the quantity $d\Gamma(\beta)/V$, which is a Lorentz-invariant (# of events per unit time per unit spatial volume—i.e., # per unit spacetime volume), divided by the quantity $\rho_1 \rho_2 v_2$. It is customary to *define* the cross-section in a general frame to be exactly the same number as in (B.23), by generalizing the latter quantity in a Lorentz-invariant way. Note that in the rest frame of the target (particle 1) the densities of both particles are given by (c=1 everywhere!)

$$\rho_1 = \rho_1^{(0)}$$

$$\rho_2 = \frac{\rho_2^{(0)}}{\sqrt{1 - v_2^2}} \tag{B.24}$$

where $\rho_1^{(0)}, \rho_2^{(0)}$ are the particle densities in the rest frames of the particles themselves. Thus

$$\rho_1 \rho_2 v_2 = \rho_1^{(0)} \rho_2^{(0)} \frac{v_2}{\sqrt{1 - v_2^2}} = \rho_1^{(0)} \rho_2^{(0)} \frac{|\vec{p}_2|}{m_2} \tag{B.25}$$

The invariant quantity $\sqrt{(p_1 \cdot p_2)^2 - m_1^2 m_2^2}$ (p_1, p_2 four-vectors) becomes, in the frame where particle 1 is at rest, just $m_1 |\vec{p}_2|$, so the quantity equal to $\rho_1 \rho_2 v_2$ in the target rest frame may be written in any frame as

$$\rho_1^{(0)} \rho_2^{(0)} \frac{\sqrt{(p_1 \cdot p_2)^2 - m_1^2 m_2^2}}{m_1 m_2} = \rho_1 \sqrt{1 - v_1^2} \cdot \rho_2 \sqrt{1 - v_2^2} \cdot \frac{\sqrt{(p_1 \cdot p_2)^2 - m_1^2 m_2^2}}{m_1 m_2}$$

$$= \rho_1 \frac{m_1}{E_1} \rho_2 \frac{m_2}{E_2} \cdot \frac{\sqrt{(p_1 \cdot p_2)^2 - m_1^2 m_2^2}}{m_1 m_2}$$

$$= \rho_1 \rho_2 v_\alpha \tag{B.26}$$

with the relative velocity $v_\alpha \equiv \frac{\sqrt{(p_1 \cdot p_2)^2 - m_1^2 m_2^2}}{E_1 E_2}$ providing the appropriate generalization of v_2 in (B.23). Our final formula is thus

$$d\sigma = \frac{(2\pi)^4}{v_\alpha} \delta^4(P_\beta - P_\alpha) \, |\mathcal{T}_{\beta\alpha}|^2 \, d\beta \tag{B.27}$$

Although the above formulas were derived under the assumption of distinguishable particles in the initial state (in particular, we did not worry about niceties of symmetrizing or antisymmetrizing the initial-state wavefunction), it turns out that the final result is still valid in such cases. The derivation can be found in any of the standard texts on scattering theory (see (Newton, 1966), for example).

Appendix C
Majorana spinor algebra

Recall from the discussion in Section 7.4.3 that we may continue to employ the very convenient (and familiar) Dirac 4-spinor language even when describing the two-component Majorana fields that naturally interpolate for self-conjugate spin-$\frac{1}{2}$ particles. Here, we shall follow the conventions of that section and define a Majorana spinor as a Dirac spinor with the following relation between the upper and lower 2-spinors:

$$\phi^{(\frac{1}{2}0)} = C_s \phi^{(0\frac{1}{2})*} \tag{C.1}$$

where $C_s = i\sigma_2$ is the 2-spinor conjugation matrix introduced in Section 7.2. Thus, if Q_a is a $(0\frac{1}{2})$ spinor, in the fundamental representation of SL(2,C), we can form a Majorana spinor as follows:

$$\begin{pmatrix} C_s Q^* \\ Q \end{pmatrix} \tag{C.2}$$

Any Dirac spinor ψ can be decomposed $\psi = \frac{1}{\sqrt{2}}(\chi_1 + i\chi_2)$ where $\chi_{1,2}$ are Majorana:

$$\chi_1 \equiv \frac{1}{\sqrt{2}}(\psi - C\gamma_0\psi^*) \tag{C.3}$$

$$\chi_2 \equiv \frac{-i}{\sqrt{2}}(\psi + C\gamma_0\psi^*) \tag{C.4}$$

where

$$C \equiv i\gamma_2\gamma_0 = \begin{pmatrix} -C_s & 0 \\ 0 & C_s \end{pmatrix}$$

In general, a numerical Majorana spinor is any 4-spinor of the form

$$\begin{pmatrix} C_s \chi^* \\ \chi \end{pmatrix}$$

where χ is a complex 2-spinor.

Of particular interest in supersymmetry are Majorana spinors whose components are Grassmann in character. These components can either be c-number constants or anticommuting components of fermionic fields (with each other and with any c-number

Grassmann quantities present). Thus, consider

$$s = \begin{pmatrix} C_s \chi^* \\ \chi \end{pmatrix}$$

where

$$\chi = \begin{pmatrix} \chi_1 \\ \chi_2 \end{pmatrix}$$

and $\chi_1, \chi_2, \chi_1^*, \chi_2^*$ are independent Grassmann quantities (i.e., they square to zero and anticommute with each other).

The following 4x4 matrices will be very useful:

$$\epsilon \equiv \begin{pmatrix} C_s & 0 \\ 0 & C_s \end{pmatrix}$$

$$\gamma_5 \equiv \begin{pmatrix} 1 & 0 \\ 0 & -1 \end{pmatrix}$$

$$\beta = \gamma_0 \equiv \begin{pmatrix} 0 & 1 \\ 1 & 0 \end{pmatrix}$$

Using these, one finds that the adjoint of a Majorana spinor, $\bar{s} \equiv s^{*T}\beta$ can be written

$$\bar{s} = s^T \epsilon \gamma_5 \tag{C.5}$$

Let M be a general 4x4 numerical matrix (containing normal complex numbers), and s_1, s_2 two Grassmann Majorana spinors:

$$\bar{s}_1 M s_2 = s_{1\alpha}(\epsilon \gamma_5 M)_{\alpha\beta} s_{2\beta} \tag{C.6}$$

$$= -s_{2\beta}(M^T \gamma_5 \epsilon^T)_{\beta\alpha} s_{1\alpha} \tag{C.7}$$

$$= s_2^T M^T \gamma_5 \epsilon s_1 \tag{C.8}$$

$$= \bar{s}_2 (\epsilon \gamma_5)^{-1} M^T \epsilon \gamma_5 s_1 \tag{C.9}$$

But $\epsilon \gamma_5$ is basically the charge conjugation matrix C, and

$$C^{-1} M^T C = M, \quad M = 1, \gamma_5, \gamma_5 \gamma_\mu \tag{C.10}$$

$$= -M, \quad M = \gamma_\mu, [\gamma_\mu, \gamma_\nu] \tag{C.11}$$

which therefore gives us

$$\bar{s}_1 M s_2 = \bar{s}_2 M s_1, \quad M = 1, \gamma_5, \gamma_5 \gamma_\mu \tag{C.12}$$

$$= -\bar{s}_2 M s_1, \quad M = \gamma_\mu, [\gamma_\mu, \gamma_\nu] \tag{C.13}$$

For the special case where s_1, s_2 are the same spinor, we find

$$\bar{s}\gamma_\mu s = \bar{s}[\gamma_\mu, \gamma_\nu]s = 0 \tag{C.14}$$

It follows that the only non-vanishing bilinears that can be built from a single Grassmann Majorana s are $\bar{s}s, \bar{s}\gamma_5 s$, and $\bar{s}\gamma_5\gamma_\mu s$.

Inserting the explicit expression for s in terms of its Grassmann components, one finds

$$\bar{s}s = 2(\chi_1^*\chi_2^* - \chi_1\chi_2) \tag{C.15}$$
$$\bar{s}\gamma_5 s = 2(\chi_1^*\chi_2^* + \chi_1\chi_2) \tag{C.16}$$

There are only four independent objects cubic in $\chi_1, \chi_2, \chi_1^*, \chi_2^*$, so products of three ss can always be reduced to

$$\bar{s}\gamma_5 s \cdot s = \begin{pmatrix} 2\chi_1\chi_2\chi_2^* \\ -2\chi_1\chi_2\chi_1^* \\ 2\chi_1^*\chi_2^*\chi_1 \\ 2\chi_1^*\chi_2^*\chi_2 \end{pmatrix}$$

while the only surviving quantity involving four ss is

$$\chi_1^*\chi_2^*\chi_1\chi_2 = \frac{1}{8}(\bar{s}\gamma_5 s)^2 \tag{C.17}$$

We list here for convenience a number of useful identities:

$$(\gamma_5 s)_\alpha \bar{s}\gamma_5 s = -s_\alpha(\bar{s}s) \tag{C.18}$$
$$s_\alpha(\bar{s}\gamma_5\gamma_\nu s) = -(\gamma_\nu s)_\alpha \bar{s}\gamma_5 s \tag{C.19}$$
$$(\bar{s}\gamma_5 s)s_\alpha \bar{s}_\beta = -\frac{1}{4}(\gamma_5)_{\alpha\beta}(\bar{s}\gamma_5 s)^2 \tag{C.20}$$
$$(\bar{s}s)^2 = -(\bar{s}\gamma_5 s)^2 \tag{C.21}$$

An important consequence of these identities is the Fierz rearrangement property for bilinears of Grassmann spinors. Let θ_α be a Grassmann Majorana spinor as usual. It follows from the discussion above that the only non-vanishing objects bilinear in θ are $\bar{\theta}\theta, \bar{\theta}\gamma_5\theta$ and $\bar{\theta}\gamma_5\gamma^\mu\theta$. Hence, by Lorentz-invariance,

$$\theta_\alpha\bar{\theta}_\beta = A\delta_{\alpha\beta}\bar{\theta}\theta + B(\gamma_5\gamma_\mu)_{\alpha\beta}\bar{\theta}\gamma_5\gamma^\mu\theta + C(\gamma_5)_{\alpha\beta}\bar{\theta}\gamma_5\theta \tag{C.22}$$

Tracing with the identity (i.e., setting $\alpha = \beta$ and summing) gives

$$\mathrm{Tr}(1 \cdot \theta\bar{\theta}) = -\bar{\theta}\theta = 4A\bar{\theta}\theta \Rightarrow A = -\frac{1}{4} \tag{C.23}$$

Likewise

$$\text{Tr}(\gamma_5\gamma_\nu\theta\bar{\theta}) = -\bar{\theta}\gamma_5\gamma_\nu\theta = -4B\bar{\theta}\gamma_5\gamma_\nu\theta \Rightarrow B = \frac{1}{4} \tag{C.24}$$

$$\text{Tr}(\gamma_5\theta\bar{\theta}) = -\bar{\theta}\gamma_5\theta = 4C\bar{\theta}\gamma_5\theta \Rightarrow C = -\frac{1}{4} \tag{C.25}$$

To summarize, we have the following identity:

$$\theta_\alpha\bar{\theta}_\beta = -\frac{1}{4}\delta_{\alpha\beta}\bar{\theta}\theta + \frac{1}{4}(\gamma_5\gamma_\mu)_{\alpha\beta}\bar{\theta}\gamma_5\gamma^\mu\theta - \frac{1}{4}(\gamma_5)_{\alpha\beta}\bar{\theta}\gamma_5\theta \tag{C.26}$$

References

Aarts, G., Seiler, E., and Stamatescu, I. (2010). Complex Langevin method: When can it be trusted? *Physical Review D*, **81**, 054508.

Abers, E. S. and Lee, B. W. (1973). Gauge theories. *Physics Reports*, **9**, 1–141.

Adler, S. L. (1969). Axial vector vertex in spinor electrodynamics. *Physical Review*, **177**, 2426–2438.

Adler, S. L. and Bardeen, W. A. (1969). Absence of higher-order corrections in the anomalous axial-vector divergence equation. *Physical Review*, **182**, 1517–1536.

Adler, S. L, Collins, J. C., and Duncan, A. (1977). Energy-momentum-tensor trace anomaly in spin-1/2 quantum electrodynamics. *Physical Review D*, **15**, 1712–1721.

Alvarez-Gaumé, L. and Witten, E. (1983). Gravitational anomalies. *Nuclear Physics B*, **234**, 269–330.

Anderson, A. (1994). Canonical transformations in quantum mechanics. *Annals of Physics*, **232**, 292–331.

Appelquist, T. and Carrazzone, J. (1975). Infrared singularities and massive fields. *Physical Review D*, **11**, 2856–2861.

Banks, T., Myerson, R., and Kogut, J. (1977). Phase transitions in abelian lattice gauge theories. *Nuclear Physics B*, **129**, 493–510.

Barton, G. (1963). *Introduction to Advanced Field Theory* (1st edn). Interscience Publishers (John Wiley and Sons), New York.

Barton, G. (1965). *Introduction to Dispersion Techniques in Field Theory* (1st edn). W. A. Benjamin, New York.

Baym, G. (1990). *Lectures on Quantum Mechanics* (3rd edn). Westview Press, New York.

Becchi, C., Rouet, A., and Stora, R. (1976). Renormalization of gauge theories. *Annals of Physics*, **98**, 287–321.

Bell, J. S. and Jackiw, R. (1969). A PCAC puzzle: $\pi^0 \to \gamma\gamma$ in the σ-model. *Nuovo Cimento A*, **51**, 47–61.

Bèrgere, M. and Lam, Y. M. P. (1976). Bogoliubov-Parasiuk theorem in the α-parametric representation. *Journal of Mathematical Physics*, **17**, 1546–1557.

Bernard, C. and Duncan, A. (1975). Lorentz covariance and Matthew's theorem for derivative-coupled field theories. *Physical Review D*, **11**, 848–859.

Bjorken, J. D. and Drell, S. D. (1965). *Relativistic Quantum Fields* (1st edn). McGraw-Hill Book Company, New York.

Bloch, F. and Nordsieck, A. (1937). Note on the radiation field of the electron. *Physical Review*, **52**, 54–59.

Bloch, P. (2006). CPT invariance tests in neutral kaon physics. *Journal of Physics G: Nuclear and Particle Physics*, **33**, 666–667.

Bodwin, G., Braaten, E., and Lepage, G. P. (1995). Rigorous QCD analysis of inclusive annihilation and production of heavy quarkonium. *Physical Review D*, **51**, 1125–1171.

Bogoliubov, N. N. and Parasiuk, O. S. (1957). Über die Multiplikation der Kausalfunktionen in der Quantentheorie der Felder. *Acta Mathematica*, **97**, 227–266.

Bohr, N. and Rosenfeld, L. (1983). On the question of the measurability of electromagnetic field quantities. In *Quantum Theory and Measurement (eds. J. A. Wheeler and W. H. Zurek)*, Chapter IV.2, pp. 479–522. Princeton University Press, Princeton, New Jersey.

Boltzmann, L. (1898). Über vermeintlich irreversible Strahlungsvorgänge. *Berliner Berichte*, 182.

Born, M. and Fuchs, K. (1939a). On fluctuations in electromagnetic radiation. *Proceedings of the Royal Society of London, Series A*, **170**, 252–265.

Born, M. and Fuchs, K. (1939b). On fluctuations in electromagnetic radiation-correction. *Proceedings of the Royal Society of London, Series A*, **172**, 465–466.

Born, M., Heisenberg, W., and Jordan, P. (1926). Zur Quantenmechanik 2. *Zeitschrift für Physik*, **35**, 557–615.

Born, M. and Jordan, P. (1925). Zur Quantenmechanik. *Zeitschrift für Physik*, **34**, 858–888.

Brenig, W. and Haag, R. (1963). General quantum theory of collision processes. In *Quantum Scattering Theory (ed. M. Ross)*. Indiana University Press, Bloomington, Indiana.

Brown, L. S. (1992). *Quantum Field Theory* (1st edn). Cambridge University Press, New York.

Buchholz, D. (1975). Collision theory for massless fermions. *Communications in Mathematical Physics*, **42**, 269–279.

Buchholz, D. (1977). Collision theory for massless bosons. *Communications in Mathematical Physics*, **52**, 147–173.

Buchholz, D. (1982). The physical state space in quantum electrodynamics. *Communications in Mathematical Physics*, **85**, 49–71.

Buchholz, D. (1986). Gauss' law and the infraparticle problem. *Physics Letters B*, **174**, 331–334.

Buras, Andrzej J. (1980). Asymptotic freedom in deep inelastic processes in leading order and beyond. *Reviews of Modern Physics*, **52**, 199–276.

Callan, Curtis G. (1970). Broken scale invariance in scalar field theory. *Physical Review D*, **2**, 1541–1547.

Callan, C. G., Coleman, S., and Jackiw, R. (1970). A new, improved energy-momentum tensor. *Annals of Physics (N.Y.)*, **59**, 42–73.

Callen, H. B. (1960). *Thermodynamics* (1st edn). John Wiley and Sons, New York.

Coleman, S. (1976). More about the massive Schwinger model. *Annals of Physics*, **101**, 239–267.

Coleman, S. and Mandula, J. (1967). All possible symmetries of the S matrix. *Physical Review*, **159**, 1251–1256.

Coleman, S. and Weinberg, E. (1973). Radiative corrections as the origin of spontaneous symmetry breaking. *Physical Review D*, **7**, 1888–1910.

Collins, J. C., Duncan, A., and Joglekar, S. D. (1977). Trace and dilatation anomalies in gauge theories. *Physical Review D*, **16**, 438–449.

Condon, E. U. and Shortley, G. H. (1935). *The Theory of Atomic Spectra* (1st edn). Cambridge University Press, Bentley House, London.

Creutz, M. (1980). Monte carlo study of quantized su(2) gauge theory. *Physical Review D*, **21**, 2308–2315.

Daboul, J. and Nieto, M. M. (1994). Quantum bound states with zero binding energy. *Physics Letters A*, **190**, 357–362.

Dirac, P. A. M. (1927*a*). The physical interpretation of the quantum dynamics. *Proceedings of the Royal Society of London, Series A*, **113**, 621–641.

Dirac, P. A. M. (1927*b*). The quantum theory of the emission and absorption of radiation. *Proceedings of the Royal Society of London, Series A*, **114**, 243–265.

Dirac, P. A. M. (1928). The quantum theory of the electron, I. *Proceedings of the Royal Society (London) A*, **117**, 610–624.

Dirac, P. A. M. (1933). Théorie du positron. In *Septieme Conseil de Physique Solvay: Structure et propriétés des noyaux atomiques*, pp. 203–221. Gauthiers-Villars (Paris).

Dirac, P. A. M. (1945). On the analogy between classical and quantum mechanics. *Reviews of Modern Physics*, **17**, 195–199.

Dirac, P. A. M. (1964). *Lectures on Quantum mechanics*. Yeshiva University, New York.

Donoghue, J. F., Golowich, E., and Holstein, B. R. (1992). *Dynamics of the Standard Model* (1st edn). Cambridge University Press, Cambridge, UK.

Duncan, A. (1976). Fine structure in non-abelian gauge theories. *Physical Review D*, **13**, 2866–2880.

Duncan, A., Eichten, E., Flynn, J., Hill, B., Hockney, G., and Thacker, H. (1995). Properties of B mesons in lattice QCD. *Physical Review D*, **51**, 5101–5129.

Duncan, A. and Janssen, M. (2007*a*). Van Vleck and the correspondence principle (part one). *Archive for History of the Exact Sciences*, **61**, 553–624.

Duncan, A. and Janssen, M. (2007*b*). Van Vleck and the correspondence principle (part two). *Archive for History of the Exact Sciences*, **61**, 625–671.

Duncan, A. and Janssen, M. (2008). Pascual Jordan's resolution of the conundrum of the wave-particle duality of light. *Studies in History and Philosophy of Modern Physics*, **39**, 634–666.

Duncan, A. and Janssen, M. (2009). From canonical transformations to transformation theory, 1926–1927: The road to Jordan's Neue Begründung. *Studies in the History and Philosophy of Modern Physics*, **40**, 352–362.

Duncan, A. and Jones, H. F. (1993). Convergence proof for optimized δ expansion: Anharmonic oscillator. *Physical Review D*, **47**, 2560–2572.

Duncan, A. and Mawhinney, R. (1990). Semiclassical approach to confinement in three-dimensional gauge theories. *Physical Review D*, **43**, 554–565.

Dyson, F. J. (1949). The S-matrix in quantum electrodynamics. *Physical Review*, **75**, 1736–1755.

Dyson, F. J. (1952). Divergence of perturbation theory in quantum electrodynamics. *Physical Review*, **85**, 631–632.

Ehrenfest, P. (1911). Welche Züge der Lichtquantenhypothese spielen in der Theorie der Wärmestrahlung eine wesentliche Rolle? *Annalen der Physik*, **36**, 91–118.

Ehrenfest, P. (1925). Energieschwankungen im Strahlungsfeld oder Kristallgitter bei Superposition quantisierter Eigenschwingungen. *Zeitschrift für Physik*, **34**, 362–373.

Einstein, A. (1905*a*). Über einen die Erzeugung und Verwandlung des Lichtes betreffenden heuristischen Gesichtspunkt. *Annalen der Physik*, **17**, 132–148.

Einstein, A. (1905*b*). Zur Elektrodynamik bewegter Körper. *Annalen der Physik*, **17**, 891–921.

Einstein, A. (1909*a*). Über die Entwicklung unserer Anschauungen über das Wesen und die Konstitution der Strahlung. *Physikalische Zeitschrift*, **10**, 817–825.

Einstein, A. (1909*b*). Zum gegenwärtigen Stand des Strahlungproblems. *Physikalishe Zeitschrift*, **10**, 185–193.

Einstein, A. (1916). Zur Quantentheorie der Strahlung. *Mitteilungen der Physikalischen Gesellschaft, Zürich*, **18**, 47–62.

Einstein, A. (1917). Quantentheorie der Strahlung. *Physikalische Zeitschrift*, **18**, 121–128.

Elitzur, S. (1975). Impossibility of spontaneously breaking local symmetries. *Physical Review D*, **12**, 3978–3982.

Evans, T. S., Kibble, T. W. B., and Steer, D. A. (1998). Wick's theorem for nonsymmetric normal ordered products and contractions. *Journal of Mathematical Physics*, **39**, 5726–5738.

Faddeev, L. D. (1969). The Feynman integral for singular Lagrangians. *Theoretical and Mathematical Physics*, **1**, 1–13.

Fermi, E. (1929). Sopra l'elettrodinamica quantistica I. *Rendiconti d. R. Acc. dei Lincei*, **9**, 881–887.

Fermi, E. (1930). Sopra l'elettrodinamica quantistica II. *Rendiconti d. R. Acc. dei Lincei*, **12**, 431–435.

Fernandez, R., Fröhlich, J., and Sokal, A. D. (1992). *Random Walks, Critical Phenomena, and Triviality in Quantum Field Theory* (1st edn). Springer Press, New York.

Feynman, R. P. (1948). Space-time approach to non-relativistic quantum mechanics. *Reviews of Modern Physics*, **20**, 367–387.

Feynman, R. P. (1949*a*). Space-time approach to quantum electrodynamics. *Physical Review*, **76**, 769–78.

Feynman, R. P. (1949*b*). The theory of positrons. *Physical Review*, **76**, 749–759.

Fock, V. A. (1933). Zur theorie des positrons. *Doklady Akademii Nauk USSR*, **6**, 265–272.

Freedman, D. Z., Muzinich, I. J., and Weinberg, E. J. (1974). On the energy-momentum tensor in gauge field theories. *Annals of Physics*, **87**, 95–125.

Freedman, D. Z. and Weinberg, E. J. (1974). The energy-momentum tensor in scalar and gauge field theories. *Annals of Physics*, **87**, 354–374.

Friedlander, F. G. (1982). *Introduction to the theory of distributions* (1st edn). Cambridge University Press, Cambridge, UK.

Fröhlich, J., Morchio, G., and Strocchi, F. (1979). Charged sectors and scattering states in quantum electrodynamics. *Annals of Physics*, **119**, 241–284.

Fujikawa, K. (1980). Path integral for gauge theories with fermions. *Physical Review D*, **21**, 2848–2858.

Fujikawa, K. (1981). Energy-momentum tensor in quantum field theory. *Physical Review D*, **23**, 2262–2275.

Furry, W. H. and Oppenheimer, J. R. (1934). On the theory of the electron and positive. *Physical Review*, **45**, 245–262.

Gasser, J. and Leutwyler, H. (1985). Chiral perturbation theory: Expansions in the mass of the strange quark. *Nuclear Physics B*, **250**, 465–516.

Gell-Mann, M. and Low, F. (1951). Bound states in quantum field theory. *Physical Review*, **84**, 350–354.

Georgi, H. (1993). Effective field theory. *Annual Review of Nuclear and Particle Science*, **43**, 209–252.

Glimm, J. and Jaffe, A. (1987). *Quantum Physics: A Functional Integral Point of View* (2nd edn). Springer-Verlag, New York.

Goldstein, H. (2002). *Classical Mechanics* (3rd edn). Addison-Wesley, Reading, Massachusetts.

Goldstone, J. (1961). Field theories with superconductor solutions. *Nuovo Cimento*, **19**, 154–164.

Goldstone, J., Salam, A., and Weinberg, S. (1962). Broken symmetries. *Physical Review*, **127**, 965–970.

Göpfert, M. and Mack, G. (1982). Proof of confinement of static quarks in 3-dimensional u(1) lattice gauge theory for all values of the coupling constant. *Communications in Mathematical Physics*, **82**, 545–606.

Gordon, W. (1926). Der Comptoneffekt nach der Schrödingerschen Theorie. *Zeitschrift für Physik*, **40**, 117–133.

Greenberg, O. W. (1959). Haag's theorem and clothed operators. *Physical Review*, **115**, 706–710.

Greensite, J. (2003). The confinement problem in lattice gauge theory. *Progress in Particle and Nuclear Physics*, **51**, 1–83.

Gribov, V. N. (1978). Quantization of non-abelian gauge theories. *Nuclear Physics B*, **139**, 1–19.

Gross, D. and Wilczek, F. (1974a). Asymptotically free gauge theories i. *Physical Review D*, **8**, 3633–3652.

Gross, D. and Wilczek, F. (1974b). Asymptotically free gauge theories ii. *Physical Review D*, **9**, 980–993.

Gross, D. J. and Neveu, A. (1974). Dynamical symmetry breaking in asymptotically free field theories. *Physical Review D*, **10**, 3235–3253.

Guth, A. (1980). Existence proof of a nonconfining phase in four-dimensional u(1) lattice gauge theory. *Physical Review D*, **21**, 2291–2307.

Haag, R. (1955). On quantum field theories. *Kgl. Danske Videnskab. Selskab, Mat.-Fys. Medd.*, **29**, 1–37.

Haag, R. (1992). *Local Quantum Physics* (1st edn). Springer-Verlag, Berlin.

Haag, R., Lopuszanski, J. T., and Sohnius, M. (1975). All possible generators of supersymmetries of the S-matrix. *Nuclear Physics B*, **88**, 257–274.

Haber, H. and Nelson-eds, A. (2004). *Particle Physics and Cosmology (TASI 2002)*. World Scientific, Singapore.

Hall, D. W. and Wightman, A. S. (1957). A theorem on invariant analytic functions with applications to relativistic quantum field theory. *Kgl. Danske Videnskab. Selskab, Mat.-Fys. Medd.*, **31**, 1–41.

Heisenberg, W. (1925). Über eine quantentheoretische Umdeutung kinematischer und mechanischer Beziehungen. *Zeitschrift für Physik*, **33**, 879–893.

Heisenberg, W. (1931). Über Energieschwankungen in einem Strahlungsfeld. *Berichte über die Verhandlungen der Sächsischen Akademie der Wissenschaften zu Leipzig, mathematische-physikalische Klasse*, **83**, 3–9.

Heisenberg, W. (1938). Über die in der Theorie der Elementarteilchen auftretende universelle Länge. *Annalen der Physik*, **424**, 20–33.

Heisenberg, W. (1943a). Die beobachtbaren Grössen in der Theorie der Elementarteilchen I. *Zeitschrift für Physik*, **120**, 513–538.

Heisenberg, W. (1943b). Die beobachtbaren Grössen in der Theorie der Elementarteilchen II. *Zeitschrift für Physik*, **120**, 673–702.

Heisenberg, W. (1944). Die beobachtbaren Grössen in der Theorie der Elementarteilchen III. *Zeitschrift für Physik*, **123**, 93–112.

Heisenberg, W. and Pauli, W. (1929). Zur Quantendynamik der Wellenfelder I. *Zeitschrift für Physik*, **56**, 1–61.

Heisenberg, W. and Pauli, W. (1930). Zur Quantendynamik der Wellenfelder II. *Zeitschrift für Physik*, **59**, 168–190.

Henneaux, M. and Teitelboim, C. (1992). *Quantization of Gauge Systems* (1st edn). Princeton University Press, Princeton, N.J.

Hepp, K. (1966). Proof of the Bogoliubov-Parasiuk theorem on renormalization. *Communications in Mathematical Physics*, **2**, 301–326.

Higgs, P. W. (1964). Broken symmetries and the masses of gauge bosons. *Physical Review Letters*, **13**, 508–509.

Iofa, M. Z. and Tyutin, I. V. (1976). Gauge invariance of spontaneously broken non-Abelian theories in the Bogolyubov-Parasyuk-Hepp-Zimmermann method. *Theoretical and Mathematical Physics*, **27**, 316–322.

Ioffe, B. L., Fadin, V. S., and Lipatov, L. N. (2010). *Quantum Chromodynamics: Perturbative and Nonperturbative Aspects* (1st edn). Cambridge University Press, Cambridge, UK.

Israel, R. B. (1978). *Convexity in the Theory of Lattice Gases* (1st edn). Princeton University Press, Princeton, N.J.

Itzhykson, C. and Zuber, J-B. (1980). *Quantum Field Theory* (1st edn). McGraw-Hill, Inc, New York.

Jona-Lasinio, G. (1964). Relativistic field theories with symmetry-breaking solutions. *Nuovo Cimento*, **34**, 1790–1795.

Jordan, P. (1926). Über kanonischen Transformationen in der Quantenmechanik: II. *Zeitschrift für Physik*, **38**, 513–517.

Jordan, P. (1927a). Über eine Neue Begründung der Quantenmechanik. *Zeitschrift für Physik*, **40**, 809–838.

Jordan, P. (1927b). Über eine Neue Begründung der Quantenmechanik ii. *Zeitschrift für Physik*, **44**, 1–25.

Jordan, P. and Klein, O. (1927). Zum Mehrkörperproblem der Quantentheorie. *Zeitschrift für Physik*, **45**, 751–765.

Jordan, P. and Pauli, W. (1928). Zur Quantenelektrodynamik ladungsfreier Felder. *Zeitschrift für Physik*, **47**, 151–173.

Jordan, P. and Wigner, E. (1928). Über das Paulische äquivalenzverbot. *Zeitschrift für Physik*, **47**, 631–651.

Jost, R. (1961). Properties of Wightman functions. In *Lectures on Field Theory and the Many-Body Problem (ed. E. R. Caianiello)*, pp. 127–145. Academic Press, New York.

Jost, R. (1965). *The general theory of quantized fields* (1st edn). American Mathematical Society, Providence, Rhode Island.

Kaiser, D. (2005). *Drawing theories apart: the dispersion of Feynman diagrams in postwar physics* (1st edn). University of Chicago Press, Chicago, IL.

Kastler, D., Robinson, D. W., and Swieca, A. (1966). Conserved currents and associated symmetries; Goldstone's theorem. *Communications in Mathematical Physics*, **2**, 108–120.

Kato, T. (1995). *Perturbation Theory for Linear Operators* (2nd edn). Springer-Verlag, Berlin-Heidelberg-New York.

Kirchhoff, G. (1859). Über den Zusammenhang zwischen Emission und Absorption von Licht und Wärme. *Monatsberichte der Akademie der Wissenschaft zu Berlin*, **12**, 783–787.

Kirchhoff, G. (1860). On the relation between the radiating and absorbing powers of different bodies for light and heat. *Philosophical Magazine*, **20**, 1–21.

Kirschner, M. and Gerhart, J. C. (2005). *The Plausibility of Life*. Yale University Press, New Haven, Connecticut.

Klauder, J. R. (1984). Coherent-state Langevin equations for canonical quantum systems with applications to the quantized Hall effect. *Physical Review A*, **29**, 2036–2047.

Klein, M. (1970). *Paul Ehrenfest: The Making of a Theoretical Physicist* (1st edn). North-Holland Publishing Company, Amsterdam.

Klein, O. (1926). Quantentheorie und fünfdimensionale Relativitätstheorie. *Zeitschrift für Physik*, **37**, 895–906.

Kopper, Christoph and Müller, Volkhard F. (2009). Renormalization of spontaneously broken SU(2) Yang-Mills theory with flow equations. *Reviews of Mathematical Physics*, **21**, 781–820.

Kramers, H. A. (1927). La diffusion de la lumiere par les atomes. In *Atti Cong. Intern. Fisica (Transactions of the Volta Centenary Congress, Como)*, Volume 2, pp. 545–557.

Kuhn, T. (1978). *Black-body Theory and the Quantum Discontinuity 1894–1912* (1st edn). Oxford University Press, Oxford.

Kuramashi, Y. (2008). PACS-CS results for 2+1 flavor lattice QCD simulation on and off the physical point. In *Proceedings of Science: Lattice 2008*, pp. 18–31.

Kuti, J. and Shen, Y. (1988). Supercomputing the effective action. *Physical Review Letters*, **60**, 85–88.

Landau, L. and Peierls, R. (1983). Extension of the uncertainty principle to relativistic quantum theory. In *Quantum Theory and Measurement (eds. J. A. Wheeler and W. H. Zurek)*, Chapter IV.1, pp. 465–476. Princeton University Press, Princeton, New Jersey.

Leech, J. W. (1965). *Classical Mechanics* (2nd edn). Methuen and Co, London.

LeGuillou, J. C. and Zinn-Justin, J. (1980). Critical exponents from field theory. *Physical Review B*, **21**, 3976–3998.

Lehmann, H., Symanzik, K., and Zimmermann, W. (1955). Zur Formulierung quantisierter Feldtheorien. *Nuovo Cimento*, **1**, 205–225.

Lipatov, L. N. (1977). Divergence of the perturbation-theory series and pseudoparticles. *JETP Letters*, **25**, 104–107.

London, F. (1926). Winkelvariable und kanonische Transformationen in der Undulationsmechanik. *Zeitschrift für Physik*, **40**, 193–210.

Lorentz, H. A. (1916). *Les théories statistiques en thermodynamique: conferences faites au College de France, novembre 1912*. Teubner, Leipzig, Berlin.

Lowenstein, J. H. (1976). BPHZ renormalization. In *Erice Lectures 1975: Proceedings of the NATO Advanced Study Institute*, pp. 95–160. D. Reidel Publishing Company.

Majorana, E. (1937). Teoria simmetrica dell'elettrone e del positrone. *Nuovo Cimento*, **14**, 171–184.

Manohar, A. V. and Wise, M. B. (2000). *Heavy Quark Physics* (1st edn). Cambridge University Press, Cambridge, UK.

Maxwell, J. C. (1873). *A Treatise on Electricity and Magnetism* (1st edn). Clarendon Press, Oxford.

Maxwell, J. C. (1875). On the dynamical evidence of the molecular constitution of bodies. *Nature*, **11**, 357–374.

Messiah, A. (1966). *Quantum Mechanics: Volume 2* (1st edn). North Holland Publishing Company, Amsterdam.

Miller, Arthur I. (1994). *Early Quantum Electrodynamics- A Source Book* (1st edn). Cambridge University Press, Cambridge, U.K.

Montvay, I. and Münster, G. (1994). *Quantum Fields on a Lattice* (1st edn). Cambridge University Press, Cambridge, UK.

Mueller, A. H. (1981). Perturbative QCD at high energies. *Physics Reports*, **73**, 237–368.

Newton, R. G. (1966). *Scattering Theory of Waves and Particles* (1st edn). McGraw-Hill, New York.

Noether, E. (1971). Invariant variation problems (translation by m. a. tavel). *Transport Theory and Statistical Physics*, **1**, 183–207.

O'Raifertaigh, L., Wipf, A., and Yoneyama, H. (1986). The constraint effective potential. *Nuclear Physics B*, **271**, 653–680.

Osterwalder, K. and Seiler, E. (1978). Gauge field theories on a lattice. *Annals of Physics*, **110**, 440–471.

Pais, A. (1982). *Subtle is the Lord: The Science and Life of Albert Einstein* (1st edn). Oxford Univesity Press, Oxford.

Pais, A. (1991). *Niels Bohr's Times, in Physics, Philosophy and Polity* (1st edn). Oxford University Press, Oxford.

Parisi, G. (1977). Asymptotic estimates in perturbation theory with fermions. *Physics Letters B*, **66**, 382–385.

Parisi, G. (1983). On complex probabilities. *Physics Letters B*, **131**, 393–395.

Pauli, W. (1933). Die allgemeinen Prinzipien der Wellenmechanik. In *Handbuch der Physik* (2nd edn), Volume 24,1, pp. 83–272. Springer Verlag, Berlin.

Pauli, W. (1940). The connection between spin and statistics. *Physical Review*, **58**, 716–722.

Pauli, W. and Weisskopf, V. (1934). Über die Quantisierung der skalaren relativistischen Wellengleichung. *Helvetica Physica Acta*, **7**, 709–731.

Peccei, R. D. (1988). Discrete and global symmetries in particle physics. In *Broken Symmetries: Proceedings of the 37 International Universitäts Wochen für Kern- und Teilchenphysik*, pp. 1–50.

Peierls, R. E. (1934). The vacuum in Dirac's theory of the positive electron. *Proceedings of the Royal Society A, London*, **146**, 420–441.

Planck, M. (1899). Über irreversible Strahlungsvorgänge: Fünfte mitteilung (Schluss). *Berliner Berichte*, 440–480.

Planck, M. (1900*a*). Über irreversible Strahlungsvorgänge. *Annalen der Physik*, **1**, 69–122.

Planck, M. (1900*b*). Zur Theorie des Gesetzes der Energieverteilung im Normalspektrum. *Verhandlungen der Deutschen Physikalischen Gesellschaft*, **2**, 237–245.

Polchinski, J. (1984). Renormalization and effective Lagrangians. *Nuclear Physics B*, **231**, 269–295.

Polyakov, A. M. (1974). Particle spectrum in quantum field theory. *JETP Letters*, **20**, 194–195.

Polyakov, A. M. (1977). Quark confinement and topology of gauge theories. *Nuclear Physics B*, **120**, 429–458.

Proca, A. (1936). Sur la théorie ondulatoire des électrons positifs et négatifs. *Journal de Physique et le Radium*, **7**, 347–353.

Rayleigh, Lord (1900). Remarks upon the law of complete radiation. *Philosophical Magazine*, **49**, 539–540.

Reeh, H. and Schlieder, S. (1961). Bemerkungen zur Unitäräquivalenz von Lorentzinvarianten feldern. *Nuovo Cimento*, **22**, 1051–1068.

Rey, S-J. (1989). Axion dynamics in wormhole background. *Physical Review D*, **39**, 3185–3189.

Rudin, W. (1966). *Real and Complex Analysis* (1st edn). McGraw-Hill, Inc, New York.

Ruelle, D. (1962). On the asymptotic condition in quantum field theory. *Helvetica Physica Acta*, **35**, 147–163.

Sakurai, J. J. (1964). *Invariance Principles and Elementary Particles* (1st edn). Princeton University Press, Princeton, New Jersey.

Salpeter, E. E. and Bethe, H. A. (1951). A relativistic equation for bound-state problems. *Physical Review*, **84**, 1232–1242.

Schroer, B. (1963). Infrateilchen in der Quantenfeldtheorie. *Fortschritte der Physik*, **11**, 1–32.

Schweber, Silvan S. (1994). *QED and the men who made it: Dyson, Feynman, Schwinger, and Tomonaga* (1st edn). Princeton University Press, Princeton, New Jersey.

Schwinger, J. (1948*a*). On quantum electrodynamics and the magnetic moment of the electron. *Physical Review*, **73**, 416–417.

Schwinger, J. (1948*b*). Quantum electrodynamics. I. a covariant formulation. *Physical Review*, **74**, 1439–1461.

Schwinger, J. (1951). On gauge invariance and vacuum polarization. *Physical Review*, **82**, 664–679.

Seiler, E. (1978). Upper bound on the color-confining potential. *Physical Review D*, **18**, 482–483.

Simon, B. (1974). *The P(φ₂) Euclidean (Quantum) Field Theory* (1st edn). Princeton University Press, Princeton, New Jersey.

Smith, P. W. (1972). Mode selection in lasers. *Proceedings of the IEEE*, **60**, 422–440.

Speer, E. R. (1973). Renormalization and Ward identities using complex space-time dimension. *Journal of Mathematical Physics*, **15**, 1–6.

Stevenson, P. M. (1981). Optimized perturbation theory. *Physical Review D*, **23**, 2916–2944.

Streater, R. F. and Wightman, A. S. (1978). *PCT, Spin and Statistics, and All That* (2nd edn). W. A. Benjamin, New York.

Swanson, M. S. (1993). Phase-space anomalies and canonical transformations. *Physical Review A*, **47**, R2431–R2434.

Symanzik, K. (1960). On the many-particle structure of Green's functions in quantum field theory. *Journal of Mathematical Physics*, **1**, 249–273.

Symanzik, K. (1970). Small distance behaviour in field theory and power counting. *Communications in Mathematical Physics*, **18**, 227–246.

't Hooft, G. (1974). Magnetic monopoles in unified gauge theories. *Nuclear Physics B*, **79**, 276–284.

't Hooft, G. (1976). Symmetry breaking through Bell-Jackiw anomalies. *Physical Review Letters*, **37**, 8–11.

Taylor, J. C. (1976). *Gauge Theories of Weak Interactions* (1st edn). Cambridge University Press, Cambridge, UK.

Taylor, J. R. (1966). Cluster decomposition of S-matrix elements. *Physical Review*, **142**, 1236–1245.

t'Hooft, G. (1971). Renormalizable Lagrangians for massive Yang-Mills fields. *Nuclear Physics B*, **35**, 167–188.

Titchmarsh, E. C. (1948). *Introduction to the Theory of Fourier Integrals* (2nd edn). Clarendon Press, Oxford.

Toll, J. S. (1956). Causality and the dispersion relation: Logical foundations. *Physical Review*, **104**, 1760–1770.

Tomonaga, S. (1946). On a relativistically invariant formulation of the quantum theory of wave fields. *Progress of Theoretical Physics*, **1**, 1–13.

van der Kolk, C. M. and de Kerf, E. A. (1975). A simplified proof of the Bogoliubov-Parasiuk theorem. *Physica*, **80A**, 339–359.

von Neumann, J. (1996). *Mathematical Foundations of Quantum Mechanics* (2nd edn). Princeton University Press, Princeton, New Jersey.

Weinberg, S. (1960). High-energy behavior in quantum field theory. *Physical Review D*, **118**, 838–849.

Weinberg, S. (1964a). Feynman rules for any spin. *Physical Review*, **133**, 1318–1332.

Weinberg, S. (1964b). Systematic solution of multiparticle scattering problems. *Physical Review*, **133**, 232–256.

Weinberg, S. (1965). Infrared photons and gravitons. *Physical Review*, **140**, 516–524.

Weinberg, S. (1967). A model of leptons. *Physical Review Letters*, **19**, 1264–1266.

Weinberg, S. (1968). Nonlinear realizations of chiral symmetry. *Physical Review*, **166**, 1568–1577.

Weinberg, S. (1972). *Gravitation and Cosmology* (1st edn). John Wiley and Sons, New York. See Chapter 2, Section 13.

Weinberg, S. (1973). Perturbative calculations of symmetry breaking. *Physical Review D*, **7**, 2887–2910.

Weinberg, S. (1995*a*). *The Quantum Theory of Fields: Volume 1* (1st edn). Cambridge University Press, Cambridge, UK.

Weinberg, S. (1995*b*). *The Quantum Theory of Fields: Volume 3* (1st edn). Cambridge University Press, Cambridge, UK.

Weisskopf, V. (1939). On the self-energy and the electromagnetic field of the electron. *Physical Review*, **56**, 72–85.

Weisskopf, V. S. (1936). Über die Elektrodynamik des Vakuums auf Grund der Quantentheorie des Elektrons. *Kongelige Danske Videnskabernes Selskab, Mathematisk-fysiske Meddelelser*, **14**, 3–39.

Weyl, H. (1929). Elektron und Gravitation: I. *Zeitschrift für Physik*, **56**, 330–352.

Wichmann, E. H. and Crichton, J. H. (1963). Cluster decomposition properties of the S matrix. *Physical Review*, **132**, 2788–2799.

Wick, G. C., Wightman, A. S., and Wigner, E. P. (1952). The intrinsic parity of elementary particles. *Physical Review*, **88**, 101–105.

Wightman, A. (1956). Quantum field theory in terms of vacuum expectation values. *Physical Review*, **101**, 860–866.

Wigner, E. P. (1939). On unitary representations of the inhomogeneous Lorentz group. *Annals of Mathematics*, **40**, 149–204.

Wigner, E. P. (1959). *Group Theory and its Applications to the Quantum Mechanics of Atomic Spectra*. Academic Press.

Wigner, E. P. (1979*a*). Events, laws of nature and conservation laws. In *Symmetries and Reflections*. Ox Bow Press, Woodbridge, Connecticut.

Wigner, E. P. (1979*b*). Symmetry and conservation laws. In *Symmetries and Reflections*. Ox Bow Press, Woodbridge, Connecticut.

Wilson, K. G. (1969). Non-Lagrangian models of current algebra. *Physical Review*, **179**, 1499–1512.

Wilson, K. G. (1971). Renormalization group and critical phenomena. *Physical Review B*, **4**, 3174–3183.

Wilson, K. G. (1974). Confinement of quarks. *Physical Review D*, **10**, 2445–2459.

Wilson, K. G. and Kogut, J. (1974). The renormalization group and the epsilon expansion. *Physics Reports*, **12**, 75–199.

Wüthrich, A. (2010). *The Genesis of Feynman Diagrams* (1st edn). Springer (Archimedes series), Dordrecht-Heidelberg-London-New York.

Yang, C. N. and Mills, R. L. (1954). Conservation of isotopic spin and isotopic gauge invariance. *Physical Review*, **96**, 191–195.

Yennie, D. R., Frautschi, S. C., and Suura, H. (1961). The infrared divergence phenomena and high-energy processes. *Annals of Physics*, **13**, 379–452.

Ziman, J. M. (1964). *Principles of the Theory of Solids* (1st edn). Cambridge University Press, Cambridge, UK.

Zimmermann, Wolfhart (1968). The power counting theorem for Minkowski metric. *Communications in Mathematical Physics*, **11**, 1–8.

Zimmermann, W. (1969). Convergence of Bogoliubov's method of renormalization in momentum space. *Communications in Mathematical Physics*, **15**, 208–234.

Zimmermann, W. (1970). Local operator products and renormalization. In *Brandeis Lectures on Elementary Particles and Quantum Field Theory, Volume 1*, pp. 395–582. MIT Press.

Zinn–Justin, J. (1989). *Quantum Field Theory and Critical Phenomena* (1st edn). Oxford University Press, Oxford.

Index